# Experimenting on a Small Planet

William W. Hay

# Experimenting on a Small Planet

## A History of Scientific Discoveries, a Future of Climate Change and Global Warming

Second Edition

GREG AND BILL AT WORK ON THE NEW EDITION

William W. Hay
University of Colorado at Boulder
Estes Park, CO
USA

ISBN 978-3-319-27402-7        ISBN 978-3-319-27404-1    (eBook)
DOI 10.1007/978-3-319-27404-1

Library of Congress Control Number: 2015960780

This Springer imprint is published by SpringerNature
The registered company is Springer International Publishing AG Switzerland

*First, my mentors and colleagues at the University of Illinois, George W. White, Harold W. Scott, Ralph Langenheim, and George Swenson, who taught me the fine art of skepticism.*

*Then those at the University of Miami in Florida: Cesare Emiliani was a Renaissance Man—a true 'Polymath'—knowledgeable not only about his own area of science, but many other fields as well. Cesare was one of the first scientists to realize that planet Earth has a disease—uncontrolled growth of the human population. In addition to Cesare, there were many in the University's Rosenstiel School of Marine and Atmospheric Sciences who taught me things I didn't know: Chris Harrison spurred my interest in plate tectonics and Bob Ginsburg introduced me to modern sedimentology. It was my colleagues in Physical Oceanography and Atmospheric Science, particularly Claes Rooth and Eric Kraus, who had the audacity to imagine wholly different modes of atmospheric and oceanic circulation in the ancient past, and Eric Barron, the student who we sent to the National Center for Atmospheric Research to take on the job of seriously investigating our speculations.*

*It was Earl Kauffman at the University of Colorado who convinced me we didn't really understand the climate of the Cretaceous, forcing me into an unending and ongoing search for the answers.*

*I want to especially thank Jörn Thiede, founder of GEOMAR in Kiel, then Director of the Alfred Wegener Institute in Bremerhaven. Jörn gave me the opportunity to spend unfettered years pursuing my goal of understanding how the Earth's 'warm' climate system worked.*

*My colleagues at GEOMAR (now the Leibnitz Institut fur Meereswissenschaften): Roland von Huene, Erwin Suess, and Wolf Christian Dullo. Along with Michael Sarnthein at the University of Kiel's Department of Geological Sciences, they all furthered my ongoing education.*

*Then there are my personal heroes who worked on understanding the symptoms of our planet's disease: Roger Revelle, who eloquently described the 'Great Uncontrolled Experiment.' Wally Broecker, who warned us that if we didn't make rapid progress in understanding the ocean and its role in climate science, Mother Nature might get ahead of us. Bill Ruddiman, who realized that the human influence on the planet began not with the Industrial Revolution, but back at the beginnings of civilization.*

*And those who have meticulously documented the ongoing change of Earth's climate, but especially: Phil Jones of the Climate Research Unit of the University of East Anglia. Jim Hansen of NASA's Goddard Institute of Space Science and Columbia University in New York. Michael Mann of the Earth System Science Center of The Pennsylvania State University. All of them have been subjects of malicious vilification for simply reporting factual information, offering scientific interpretation, and suggesting possible means of amelioration to avoid the obvious disaster if trends continue.*

*Finally, Al Gore. The only US politician with enough background in science to fully understand the situation and the courage to bring it to public attention.*

# Foreword

I am again honored to be asked to write a Foreword to the second edition of Experimenting on a Small Planet. Of all the books on past and future climate change this one is unique in that it includes all of the background information on mathematics, physics, chemistry, biology and meteorology need for an in-depth understanding of what is happening to Spaceship Earth in a form easily understood by anyone with a high-school education.

Bill Hay was my graduate study advisor at the University of Illinois in Urbana in the late 1960s. In his Preface, he states that "This book is intended for the layperson in the United States, although I hope my scientific colleagues and those elsewhere might also find it interesting and amusing." Amusing, yes, because this highly readable book is laced throughout with quips, cartoons, and witticisms. He presents each topic in its historical context so that the reader can trace not only the key discoveries going back centuries but also be entertained by fascinating stories about the evolution of scientific thought on each topic. This volume provides the general public and specialists alike with far more new information than any one expert is likely to have at his fingertips.

Most of our knowledge of Earth's past climate history has been developed over the past 35 years. There are only a few books that discuss climate in the context of Earth's history and most of them are concerned with the ice ages of the immediate past. Bill's book discusses not only the Earth's recent cold past, but also the warmer Earth toward which we seem to be headed. It explores why the rate of climate change is accelerating beyond anything in Earth's past history. It fills a major and growing need not only for the public but teachers and researchers needing to understand the severity of the problem facing mankind.

This scholarly effort provides the academic world with a much needed textbook on past, present, and future climates and the nature of climate change. This new edition is in a larger, easier-to-read format, with many color illustrations and several new chapters that discuss the Coriolis Effect, and atmospheric and oceanic circulation in more detail.

The book also tells us much about the author himself via his own observations as well as vignettes about his personal life and career inserted as "intermezzi" between each chapter. These confirm what those of us who studied under him have long known, that he is a true Renaissance Man. Widely travelled and fluent in a number of European languages (French, German, including several otherwise unintelligible Swiss dialects, and Spanish), he is able to get by in a number of others (Dutch, Czech, and Russian). He is highly appreciative of modern and classical art, music, and literature, as well as good food and wine. This we were exposed to when he'd give a party for one of his master's or doctoral graduates or a visiting scientist at his second-story garage flat known as the "Hay Loft" in Champaign, Illinois. There he would skillfully cook the main course of a gourmet meal while apportioning to all of the guests the various tasks of preparing the accompanying bread, salad, vegetables, and dessert.

A hint of his good taste in wine and liquors is given in his Preface under "Pairings" where he suggests which of those would be good to sip while reading various sections of the book. In this second edition the last items in each chapter are recommendations for music to help one better contemplate the topic just read and a suggested "Libation."

The depth of his knowledge of wines, however, was best demonstrated to my wife, Cindy, and me, during my postdoctoral studies in Switzerland when we accompanied Bill to the University of Bern where he gave an invited lecture at their Geology Institute, following which his host, Prof. Franz Allemann, invited us all out to his place for a little "wine tasting." Fresh samples of six different Swiss wines in unmarked glasses were presented to him in turn, and he was asked to not only identify each but to give their geographic origins. To our collective amazement this he did without a miss, correctly naming the wines and the river valleys from which each came. He had acquired that knowledge while on a U.S. National Science Foundation (NSF) Postdoctoral Fellowship in Basel, Switzerland.

Bill's experiences abroad help explain his understanding of the need for background knowledge to understand science. They have contributed in no small way to the depth, breath, and intrigue of his current book. Even after his formal retirement from the University of Kiel in Germany in 2002, he continues to teach intensive short courses in paleoclimatology in The Netherlands, Italy, Austria, Germany, and, most recently, China. His insatiable scientific curiosity continues to this day as further investigations of past warm climates and through international collaboration.

Sherwood W. ("Woody") Wise Jr.
Department of Earth, Ocean, and Atmospheric Sciences
Florida State University, Tallahassee, FL, USA

Lyman D. Toulmin
Professor of Geology
Department of Earth, Ocean, and Atmospheric Sciences
Florida State University, Tallahassee, FL, USA

We're on our way

# Preface

## Preamble

More than half a century ago, Roger Revelle and Hans Suess, both scientists at the Scripps Institution of Oceanography in La Jolla, California, published what has become the classic paper on the possible future effects of human activities resulting in increasing the content of carbon dioxide ($CO_2$), a "greenhouse" gas, in the atmosphere. Their 1957 paper included the statement

> Within a few centuries we are returning to the atmosphere and oceans the concentrated organic carbon stored in sedimentary rocks over hundreds of millions of years. This experiment, if adequately documented, may yield a far-reaching insight into the processes determining weather and climate.

Roger and Hans thought we were going to make only a minor modification of Earth's climate with this addition of fossil carbon. However, the experiment was more complex than they realized. The addition of $CO_2$ to the atmosphere is only one of several changes humans are making to planet Earth. Over the past two and a half centuries since the beginning of the Industrial Revolution, human activities have become increasingly important in affecting conditions on the surface of the Earth. Five human-driven factors are forcing regional and global environmental change:

1. Increasing levels of atmospheric greenhouse gases;
2. Clearing of forests for agriculture and construction of buildings and roads on a scale such as to affect the amount of the Sun's energy reflected or absorbed by our planet;
3. Replacement of plants that freely transpire and return water to the atmosphere by water-conserving plants that grow faster and now form the basis for our food supply in many parts of the world;
4. Overuse of fertilizers and pesticides in agriculture; and
5. Mining of minerals and extraction of petroleum, natural gas, and coal with concomitant release of nutrients and poisonous materials into the environment.

All of these are closely related to the growth of the human population on the planet, the development of civilization, and the need for increasing food supplies. Of these five, only the first has been seriously addressed in attempts to assess its effect on future climate change. This is because greenhouse gas concentrations have a global effect. The other factors have regional effects but may, over the long run, have irreversible global implications.

To assess the global effects of the addition of greenhouse gases to the atmosphere, an Intergovernmental Panel on Climate Change (IPCC) was established in 1988. Its parents are the World Meteorological Organization (WMO) and the United Nations Environment Programme (UNEP), both agencies of the United Nations. Their thrust has been to document factors related to climate, and to make projections about what changes might be expected until the end of this century. The IPCC is an extremely conservative group to which climate scientists are advisory. It has issued four reports, the most recent of which projects a global temperature rise between 1.1 °C (2 °F) and 6.4 °C (11.5 °F) by the end of this century, accompanied by a sea-level rise between 18 and 59 cm (7–23¼ in.). These may not sound like large changes, but when the Earth was 6.4 °C cooler there was half a mile of ice over Chicago. Unfortunately, many climate scientists and most of the general public do not realize that even if there were concerted efforts undertaken to reduce greenhouse gas emissions and mitigate their effects, climate change would not end in 2100. Geochemists who have studied the $CO_2$ system are aware that, depending on the magnitude of the perturbation, the effects may continue for tens to hundreds of thousands of years.

## How to Read This Book

This book is intended for the layperson in the United States, although I hope that my scientific colleagues and those elsewhere might also find it interesting and amusing. It is organized to provide first an introduction to concepts that are important in thinking about changes on planet Earth. Then follows a general description of how the climate system works. Finally there is a series of chapters that discuss each aspect of the climate system, how they have varied over time, and whether or not human activity can affect them.

Depending on how much you already know, you can skim familiar topics to see if they contain something you didn't know; things you are thoroughly familiar with you can skip altogether; I was already working on my Ph.D. before someone explained to me that it is a waste of valuable time to read something you already know.

Most Americans, even those with an engineering background, have difficulty thinking in terms of the "metric system" used in the rest of the world. The Fahrenheit temperature scale is still used in only two countries: the US and Belize; the rest of the world uses the Celsius (= Centigrade) system. Accordingly, although the citations of data in the text will be in metric terms, where appropriate I include their US or Fahrenheit equivalents in parentheses.

When appropriate a timeline is given at the end of each chapter, followed by suggested reading "if you want to know more."

I also suggest music you may want to listen to while thinking about what you have read, and I suggest "libations" to aid your thinking.

I also found access to the Internet is more useful than pages of references to the technical literature. Where particularly appropriate I refer to some specific Internet sites, and at the end I list some books for those who wish to delve further into the science of climate change and other topics covered in this book. Also, the reader can find many useful illustrations and diagrams on the Internet by 'Googling' a topic. Googling for illustrations of "Siccar Point" will bring up a host of pictures of that (geologically) famous place; similarly, typing in the name of someone mentioned in this book, such as W. Broecker, and using *Google Scholar*, will bring up a list of publications, some of which may be downloaded for free. However, there is a caveat about using the Internet for investigating climate change. The number of websites that present disinformation or misinformation, particularly regarding carbon dioxide, greatly exceeds those presenting scientifically valid information. Someone wants to keep you burning as much fossil fuel as possible.

## Pairings

Strong coffee will be helpful throughout the book if you are a daytime reader. However, if you like to read in the evening, you may wish to skip my specific recommendations for a libation and settle for more general solutions. The earlier chapters pair well with a California Chardonnay or a French Chablis, the middle chapters with a substantial red, such as Merlot or Pinot Noir, and the later chapters with Cabernet Sauvignon or a fine Bordeaux. Toward the end you may require a Cognac, Armagnac, or a Marc. There will be individual sections where stronger drink, such as Scotch, Bourbon, or even a Martini is simply called for.

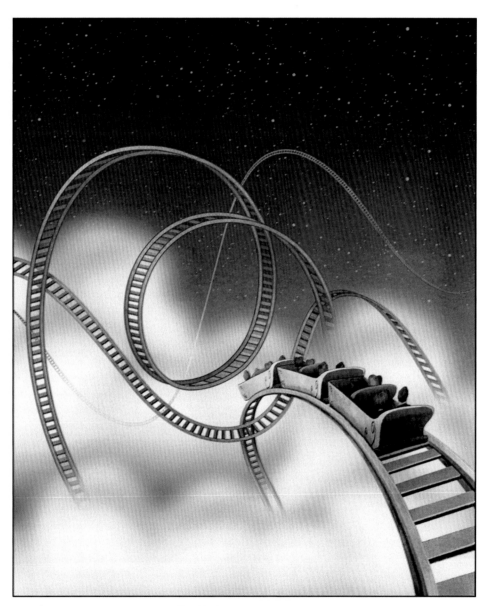

Looking at an uncertain future

# Prologue

## A Parable—Climate Change and Heresy

Heresy is defined as an opinion or doctrine that is at variance with an orthodox or accepted doctrine, or in other words, a belief or theory that is strongly at variance with established beliefs.

Perhaps the most famous scientific heretic was Galileo Galilei. In January, 1610, when he was 46-years old, Galileo used the newly invented telescope to discover that the planet Jupiter had four moons that rotated around it. In December he discovered that Venus goes through phases, like the Moon, and recognized that it must be orbiting the Sun. It was followed by observations of sunspots by Galileo and others that demonstrated that the Sun was not "perfect" as the Church believed, but that it rotated. He realized that these observations were strong support for Copernicus' theory of a century earlier, that the Earth orbited the Sun. Johannes Kepler, an astronomer in Germany, confirmed Galileo's observations, and was working out the mathematical laws of planetary motion. Jesuit astronomers in Rome also confirmed the observations and honored Galileo for his discoveries.

The established belief, based on literal interpretation of a few passages in the Bible was that the Earth was the center of the universe and everything in the skies revolved around it. It may be useful to remember that the Bible also indicates that the Earth is flat, rectangular, and rests on four pillars, although these ideas had quietly faded after Magellan's circumnavigation of the globe in the early sixteenth century. However, it was the time of the Reformation, and the idea that Earth was fixed and immobile was thought to be critical to the power of the Catholic Church. Although many important people, both within and outside the Church believed Galileo to be correct, "Holy scripture," was not to be questioned. Copernicus had made a tiny ripple in the pond of orthodoxy. Galileo's confirmation was a splash; it was an Inconvenient Truth that challenged the authority of the Church.

In December 1614 Tommaso Caccini, a Dominican friar, preached a sermon in Florence suggesting that Galileo and mathematicians who subscribed to Copernicus' view might be heretics. Although Caccini's superior apologized to Galileo a month later, Caccini went on to file a deposition with the Roman Inquisition. A committee of consultants to the Inquisition found that the proposition that the Sun is the center of the universe was absurd and formally heretical. Galileo was warned not to defend the Copernican view; he was also ordered not to discuss the topic either orally or in writing.

In 1624 Galileo went to Rome where Pope Urban VII told him he could write about the Copernican idea as long as he treated it as merely a mathematical hypothesis. Galileo rewrote some earlier unpublished manuscripts, and applied for a license to publish his book *Dialogo sopra I due massimi sistemi del mondo (Dialog between the two chief world systems)*. An anonymous complaint argued that Galileo's proposed book was heretical and threatened the foundations of the Church. In 1633 Galileo was summoned before the Inquisition in Rome, kept under arrest, and threatened with torture. Consultants examined the manuscript and concluded he made too strong an argument for the Copernican system. Galileo saved himself by recanting in a formal ceremony, but spent the rest of his life under house arrest. The *Dialog* was eventually published in Latin in Strasbourg in 1635, but banned by the Church until 1835.

Galileo was pardoned for his heretical views in 1992.

Today, in the United Sates we face a surprisingly similar situation. The majority belief in the United States seems to be that the Earth is immensely large and humans are very small. Nothing they do can affect natural processes. If there is climate change it is due to natural causes, unrelated to any of man's activities. Many Americans believe that the Earth is only 6000 years old, that the first humans were Adam and Eve, and that there is no such thing as evolution. Science that challenges these beliefs has come to be regarded with suspicion. Logic is rejected as a means of analyzing factual information. Some even believe that the climate is not really changing, but there is a secret conspiracy among thousands of scientists to fool the public; if it were true this would be the best kept secret in history.

Of course, short-term corporate profits are intimately involved in the discussion of climate change. There is powerful industry support for the idea that human activities have no impact on planet Earth. It is likely that more money is spent on denying climate change than is spent on climate research.

But today there are at least eight "heretical" views abroad:

1. That the introduction of large amounts of carbon dioxide ($CO_2$) into the atmosphere from the burning of fossil fuels, manufacture of cement, and deforestation will have an effect on climate. (Al Gore's 'Inconvenient Truth').
2. That changes in land use by agriculture and construction may be affecting climate.
3. That, in addition to $CO_2$, the increasing atmospheric concentrations of other greenhouse gases are due to man's activities.
4. That the climate system has a propensity for generating positive feedback to make a bad problem worse.
5. That global warming implies rising sea levels.
6. That the absorption of $CO_2$ by the ocean changes conditions for life in the sea.
7. That the geologic past has an interesting record of climatic catastrophes due to perturbations of the climate system that got out of hand.
8. That unlimited population growth may not be a good thing.

Galileo's heresy had no immediate impact on humankind. These modern "heresies" have enormous implications for humanity in the future. As a good friend, a geologist in the petroleum industry, has told me, "You can only poke a balloon with a pin so many times before something happens." That was years before the 2010 oil spill in the Gulf of Mexico. All of the above "heresies" are true and their implications are potentially far more dangerous than most of the public realizes.

I now believe that we passed a point of no-return not in 2007 as I stated in the first edition of this book, but back in 1982 when the melting of Arctic sea ice began to expose open water. Since then the summer and fall melting of the sea-ice cover of the Arctic has increased to a rate far beyond any predictions. In my view we passed over a major tipping point for the Earth's climate system, but this is certain to be regarded as an alarmist heresy by some. I believe that even if we could instantly remove the excess $CO_2$ we have added to the atmosphere, we can no longer return to the conditions that prevailed during most of the nineteenth and early twentieth centuries.

To understand why I believe all these "heresies" to be true, you need to understand how science works and what we know about climate today. That is what this book is about.

## Humanities Greatest Adventure

The 10,000-year-old cave paintings at Altamira (Spain) and Lascaux (France) show that humans had clearly differentiated themselves from the rest of the animal kingdom many thousands of years ago. But humanity really started on its great adventure about 8000 years ago, with the development of what we call "civilization." It was then, or perhaps even a bit earlier, that we began to organize ourselves into communities, dividing up the work, and gradually beginning to modify the landscape to suit our needs. Then in the late eighteenth century came the Industrial Revolution. It is as though we had stepped on the accelerator and pushed it to the floor. The human population increased, and along with it the use of natural resources. Alteration of planet Earth began on a grand scale. Scientists who study climate have known for more than forty years that we are perturbing the Earth out of its "natural state." But it is only within the last decade that we have discovered that the pace of change is far greater than anyone had anticipated. Forty years ago it was thought that the effects of climate change would first begin to be felt toward the end of the twenty-first century. But changes that were expected to happen then are already occurring. We are off on a wild ride into an uncertain future.

It is as though we accepted an invitation to be the first people to take a ride on a new experimental roller-coaster in an amusement park on the shore. It is said to be the biggest, wildest, most exciting roller coaster ever built. Some say it is going to be the most frightening roller coaster ever built, and there have been warnings that it might not be entirely safe. To make matters more interesting, by the time we arrive for the ride the fog has rolled in from the sea, and we can see only part of the first incline, the one we would be towed up by the cable under the car. Some way up the incline we come to the top of the fog and can see some really scary parts of the new coaster ride but we can't see how they are connected. Some of the people in the car say "We don't like this. I heard that guy down there say they aren't really sure the car will stay on the track. Maybe we should press the emergency button and stop the ride while we can still get off." While we are having this discussion and searching for the emergency button, we suddenly and unexpectedly go over the top of the incline, and start down for the thrilling ride. Unfortunately the fog is so thick below us that we cannot see what will come next. In 2007 humanity passed the point of no-return and started on just such a wild ride into the unknown.

GOOD LUCK—WE'LL NEED IT!

Perhaps more than we bargained for

Don

Woody & Cindy

Jim, Nina, Anthony, Peter & Bill

# Acknowledgments

First, I want to give special thanks to Don Marszalek, Pete the Rationalizer, Jim Stunkel, Nina Bailey, and Anthony Shaw who generously gave of their time to review and correct text for this second edition of the book. I also wish to thank the many professionals, too many to list, who offered valuable suggestions and encouragement.

I want to thank my old friend Greg Wray who produced the color "infotoons" that accompany each chapter. Greg has been a great source of support throughout this project. The infotoons are cartoons that relate to some significant idea or event in the chapter and are meant to convey impressions, both obvious and subliminal. These and other illustrations are available in poster size from Greg through his website Gregwray.com.

Another long-time acquaintance, Al Smuskievicz, acted as professional editor for several of the chapters.

A number of colleagues offered suggestions for improvements of the discussions, provided new material, and corrected some errors in the first edition: Wolf Berger, Sherwood and Cindy Wise, Marc Mangel, Erle McBride, Bilal Haq, Richard Hahn, Paul Berlanger, Harald Strauss, Colin Summerhayes, George deVries Klein, Tom Johnson, and Steven Sherwood.

Finally I would like to thank the dedicated staff at NASA, especially the Blue Marble Team and those at GISS, the Goddard Institute for Space Science, NOAA's Laboratories, especially the National Snow and Ice Data Center, the European Space Agency, the Hadley Centre in Britain, and other scientific information centers for the great work they do in making data, maps, and graphics available to the public.

# Contents

# About the Cover

In the eight or so millennia since humans have organized themselves into "civilization," they have made many changes in Earth's natural environment: burning off forests and grasslands as a means of hunting, clearing out the natural vegetation for agriculture, constructing villages, towns, and cities, mining ores and quarrying stone. But all of these changes have, until recently, had little impact on Earth's overall condition. Then in the late eighteenth century things changed; the Industrial Revolution brought prosperity and allowed rapid growth of the human population. The energy for the Industrial Revolution was fossil fuels, starting with coal, then including petroleum, and finally natural gas. These all produce a greenhouse gas, carbon dioxide, $CO_2$, when they are burned. The increased human population has an appetite for meat, and has increased the population of cattle, which produce methane that then is converted into carbon dioxide. Construction of cities and roads has required the use of concrete. It is made from the oxide of calcium produced by heating limestone, driving off carbon dioxide in the process. In contrast to man's earlier activities which only had regional impacts, the massive addition of a greenhouse gas to the atmosphere has a global effect. The insidious thing about carbon dioxide is that once in the atmosphere it remains there for hundreds of thousands of years.

The major modern sources of atmospheric $CO_2$ are shown as chemical experiments on the cover. Starting from the lower left and moving counterclockwise, they are: (1) burning of petroleum, exemplified by the least fuel efficient modern automobile, the Hummer; (2) the burning of coal, formerly for heating, but more recently to produce electricity; (3) the roasting of limestone to make lime for cement and concrete; and (4) the grazing of cattle, producing methane that ultimately, after a few years, becomes converted to $CO_2$. These are the major elements of the "grand uncontrolled geophysical experiment" being conducted on our small planet. In the upper right is a pink object in the sky, the Orion nebula. It is hydrogen gas re-radiating energy received from nearby stars. Hydrogen is the simplest atom, and the energy re-radiated from it is in the pink part of the visible spectrum. The process is the same as that of Earth's greenhouse gases that absorb infrared radiation from the planet's surface and re-radiate it, but in this case the gas molecules are much more complex. With greenhouse gases, like carbon dioxide and methane, both the incoming and outgoing radiation are in the infrared part of the spectrum and invisible to human eyes.

A TOAST TO OUR FUTURE

"IT'S GOING TO HAPPEN"

*Start by doing what's necessary; then do what's possible; and suddenly you are doing the impossible.*

Francis of Assisi

This book is a journey through what we have learned about climate and how it can change. We start with some personal reminiscences, a brief account of who I am, and a discussion of what science is and how it works.

## 1.1 Leningrad—1982

It was July 4, 1982. I was in Leningrad, walking back to my hotel on the other side of the Neva River after a fine dinner at the famous pre-revolutionary Astoria Hotel. After severe destruction in World War II, the city had been rebuilt much as it had been before, with only a few modern buildings interrupting the classical architecture of the city built by Peter the Great as Russia's window on the world. The 'white nights' at this time of year are truly wonderful. The Sun is just below the horizon, but the sky is bright, the diffuse light illuminating the city in a very special way. And the Russians are out in droves, enjoying their summer.

Shortly after taking office as President of the United States, Ronald Reagan decided to cut off scientific exchanges with Soviet colleagues, assuming that they were dangerous to our national security. He had ended cooperation in research programs that had developed over the preceding decade. However, there was one field in which his administration desperately wanted to find out what the Soviets knew—climate change. They suspected that the Russians were developing some sort of weapon that could change the weather and completely alter the way in which wars would be fought. My impression was that past attempts at short-term local weather modification, such as making rain, had failed, but I knew that our Soviet colleagues were very much interested in the possibility of long-term climate changes that would be bought on by human activities.

As a geologist, I was odd man out in that small group of (otherwise) distinguished climate scientists meeting in Leningrad with a small group of Soviet counterparts. We were supposed to somehow find out if they really had a weapon that could change the weather. The Soviet group was headed by their most distinguished climate scientist, Mikhail Budyko, director of the Division for Climate Change Research at the State Hydrological Institute in Leningrad. Budyko was not only well known as a climate scientist, he also played a major role in setting development policy in the Soviet Union.

Over the decades since World War II Soviet engineers had developed a grand plan for damming the Siberian Rivers that flow into the Arctic Ocean and diverting the water for irrigation in the south. Budyko had convinced the Soviet leadership that the flow of these rivers was critical to maintaining the sea ice cover on the Arctic Ocean. He had argued that the climatic consequences of losing the Arctic sea ice cover were unpredictable but certain to be very dramatic and quite possibly disastrous for the food supply of the Soviet Union. Mikhail Budyko was responsible for convincing the Soviets to abandon the grand diversion plan.

In 1979 Budyko and Alexander Ronov of the Vernadsky Institute of Geochemistry in Moscow had written a paper on the history of Earth's atmosphere, proposing that the major long term changes in the Earth's climate were ultimately the result of changes in the amount of the greenhouse gas carbon dioxide ($CO_2$) in the atmosphere. This idea had been briefly popular at the end of the 19th century, but largely forgotten since. They believed that the amount of carbonate rock (limestone) deposited at any given time was an index to the amount of carbon dioxide in the atmosphere (an idea we will discuss later). Ronov had spent more than 40 years compiling his inventory of the amounts of different kinds of sedimentary rock that had been deposited during each age of Earth's history. I had been chosen to be a member of the US delegation because I had written a paper pointing out an error in their argument.

Earlier in the day there had been presentations about the greenhouse effect, the role of $CO_2$, and the possible effect of human additions of that gas on the Earth's future climate. The general conclusion was that it could well have a

W.W. Hay, *Experimenting on a Small Planet*, DOI 10.1007/978-3-319-27404-1_1

significant effect but this would probably not begin to be felt until the middle or end of the 21st century. In 1982 that seemed a long way off. I had made my presentation discussing the error in their paper, but also pointing out that they had also included in their analysis other data which strongly supported their hypothesis.

At the dinner I found that Budyko had arranged for me to be seated next to him. He was what one would have expected to find as a Russian aristocrat of the old order, tall, thin, with fine features, a most distinguished air, and impeccably dressed in a blue suit. In stark contrast to the formality of our host, I was in a sport coat and wearing a Navajo bolo rather than a tie. Except when he was proposing the traditional toasts, we spent most of the evening in a private discussion about how to improve knowledge of the history of Earth's atmosphere. Our discussion then turned to the future. For me the future was a purely hypothetical topic. I asked "do you think it is really possible that the sea-ice cover on the Arctic Ocean might melt away if we don't do something to reduce $CO_2$ emissions?" Mikhail Ivanovitch took hold of my sleeve and looked me directly in the eyes —"Professor Hay, it is going to happen no matter what we do now. Stopping the plan for diversion of the rivers may have avoided sudden climate change for now, but even if we could start to get $CO_2$ emissions under control today, there will still be enough going into the atmosphere to melt the Arctic ice cover. All that we can hope for is that the climatic changes can be kept in bounds so that it doesn't become a complete disaster." An interesting way to end a fine dinner laced with vodka.

We were staying at the Leningrad (now the Saint Petersburg) Hotel on the Pirogovskaya Embankment on the north side of the Neva River, just opposite where the cruiser "Aurora" is moored. On October 25, 1917 the "Aurora" fired the shot that was the signal for the Communists to storm the Winter Palace and seize control of the government. Not far away on the other side of the hotel is the Finland Railway Station where Lenin arrived in his sealed train on his return from exile in Switzerland. I had been to the Soviet Union several times before and felt comfortable enough with my smattering of Russian that I enjoyed wandering around on my own. I decided that while the others took a cab, I would walk back to our hotel. The way back led through the Great Arch in the center of the wide yellow crescent of the old Imperial General Staff building into Palace Square, bounded by the glorious green and white facade of the Winter Palace on the far side. I walked around the west side of the Winter Palace to the Palace Embankment on the south side of the Neva. Across the river was the building housing Peter the Great's "curiosity cabinet," one of the world's first museums, and containing some of the most wonderful fossils ever found. I had been able to see them in detail on an earlier visit to Leningrad. Further down on the other side of the Neva was the Lieutenant Schmidt Embankment with its statue of Admiral Kruzenshtern. Catherine the Great had sent the Admiral out to circumnavigate the globe and measure the temperature of the ocean. This voyage, from 1803 to 1806 was the first scientific oceanographic expedition. Kruzenshtern made the discovery that the deep waters of the ocean were cold everywhere. St. Petersburg, Petrograd, Leningrad, now again St. Petersburg has long been a center for climate study.

The embankment along the Nevsky River is lined with huge blocks of very beautiful red granite with oval inclusions of contrasting black granite with white centers that look like giant owl's eyes—each one would be a museum specimen elsewhere. As a geologist I reveled in the sheer beauty of the sidewalk and balustrades made of this rare stone. Across the river from the Winter Palace the Petropavlovski Fortress with its tall golden spire gleamed in the soft light. I hurried past the (relatively) small Summer Palace, where Peter lived during construction of the city. I needed to get across the Troitskiy Bridge before it was raised at midnight. All the bridges over the Neva are raised from midnight to 6 A.M. to allow ships to go from the Gulf of Finland to Lake Ladoga. Figure 1.1 shows the bridges up.

From the middle of the bridge I looked west at the beautiful panorama of the city. This whole city had been reclaimed from the swamps at the east end of the Gulf of Finland at an enormous cost of both lives and money. I began to wonder what Peter's "Venice of the North" might look like in a 100 years. It had actually been modeled after Amsterdam, where Peter had studied shipbuilding. It was built in what was a swamp at great human and monetary expense. Almost the entire city is built on rock fill brought into raise the land surface to about 10 m (33 ft) above sea level. Amsterdam itself is much more vulnerable; it is only 2–3 m (7–10 ft) above sea level. However, melting the Arctic ice pack would not in itself cause a rise in sea level because the sea ice is floating.

Water has many strange properties; they will be discussed in detail in Chap. 17. Unlike almost everything else, when it freezes it expands. Because it expands it becomes less dense than water and floats. However, the water level doesn't change when ice freezes on the surface of a lake or the ocean. The surface of the ice will be higher than that of the water on which it floats, by about 1/9 of its thickness. The additional volume required when the water becomes ice doesn't alter the sea-level because the extra volume now displaces the air above sea-level; vice versa, melted sea ice doesn't increase sea-level because, although there is more liquid water in the ocean, the volume it occupies is smaller than when it was ice. You can check this out by looking at an ice cube floating in your iced drink. If you mark the water level on the side of the glass with the ice cube in it, and check it out after the ice has melted you will see that the water level stays the same.

**Fig. 1.1** Leningrad—The Petropavlovski (Peter and Paul) Fortress seen across the Neva after midnight during the 'White Nights.' Peter the Great is buried in the church with the tall spire

But would the ocean water itself expand as it warmed and cause sea level to rise, flooding this city? What about Greenland and the Antarctic? Would their ice sheets begin to melt, and if so how fast could that happen? What would the vegetation look like? My thoughts were interrupted by a policeman politely inviting me and the dozens of others to get off the bridge before it was raised.

I put my ideas aside to think about later. It is now more than a quarter century later and I have given the matter a lot of thought. The changes we then thought would not happen until 2050 or later are already well under way. In retrospect, as you will see later, 1982 was probably the year in which Earth went past a tipping point in climate change. It was the year in which the decline in sea-ice cover of the Arctic Ocean began as a response to the increasing levels of atmospheric carbon dioxide. From today's (2015) perspective I believe that it was in 1982 that planet Earth went past the point of no-return. Earth's future will be different from its past. Return to the conditions of the mid-20th century is no longer possible.

## 1.2  'Global Warming' or 'Global Weirding'

In 1975 Wally Broecker published a paper in the journal *Science* with the title *Climate Change; Are we on the Brink of a Pronounced Global Warming?* It had the following abstract:

If man-made dust is unimportant as a major cause of climatic change, then a strong case can be made that the present cooling trend will, within a decade or so, give way to a pronounced warming induced by carbon dioxide. By analogy with similar events in the past, the natural climatic cooling which, since 1940, has more than compensated for the carbon dioxide effect, will soon bottom out. Once this happens, the exponential rise in the atmospheric carbon dioxide content will tend to become a significant factor and by early in the next century will have driven the mean planetary temperature beyond the limits experienced during the last 1000 years.

It was a remarkably prophetic statement from 35 years ago as to what we now know is happening.

Many people labor under the misconception that "global warming" means a gradual rise in temperature by a few degrees. If that were the problem it would seem simple enough to take care of. Noel Coward noted in his song that "Only mad dogs and Englishmen go out in the noonday Sun." We could just stay indoors or in the shade. However, that is not all it means. At the 1982 annual meeting of the American Association for the Advancement of Science in Washington, D.C., German climatologist Herman Flohn predicted that "Long before global warming due to $CO_2$ can be proven, its effect will be noticed by extremes of weather, setting local records such as the highest and lowest temperatures, greatest and least rain and snowfall amounts, most powerful storms, unusually large numbers of tornados, etc. These indicate that the climate system is becoming unstable in response to the greenhouse effect."

By now it should be evident to everyone that our planet's climate is changing. 2015 was the warmest year on record and the extremes that Herman Flohn had predicted have become ever more common. Most significantly, the melting of Arctic sea ice increased to unprecedented, wholly unexpected rates in the fall of 2007, and although the ice-covered area has been somewhat larger since, the area with ice more than 1 year old is shrinking rapidly. The reduced ice cover has now begun to affect the weather over North America. Budyko's prediction is coming true, only it is a few decades sooner than expected.

In 1910 Tom Friedman, in a commentary in the New York Times, suggested that we replace the expression 'Global Warming' by 'Global Weirding' because the effect of planetary warming is not evenly distributed and gradual, but is reflected by the unexpected extreme weather conditions suggested by Herman Flohn.

Interestingly, a changing climate would not be a surprise to the early English colonists of North America. When they arrived here they found North America to be very different from their homeland. It was unbearably hot in summer, cold in winter, unpleasantly humid, and plagued by violent storms. Although inhabited by 'savages,' the colonists considered it questionable whether it was suitable for habitation by civilized human beings. Incidentally, until very recently Washington, D.C. was considered a 'hardship post' for British diplomats because of its climate. In the 17th and 18th centuries it was popularly believed among northern Europeans that real civilization required a northern European climate.

The colonists thought that the problem with North America was the forests. If they could clear enough land of trees, the climate would ameliorate and become more like that of Britain. They thought that the trees were responsible for the high humidity, and they attributed disease to the unsuitable climate. The idea that human activity could modify the climate was not a fantasy to them. It was their hope for creating reasonable living conditions. In 1785 Thomas Jefferson wrote that there had been a moderation of the climate in Virginia since the colonists had arrived; at least the winters were not so severe. He too thought that clearing the land of trees would improve the climate of North America. He believed that as the settlers moved west and cleared the area west of the Appalachians, the climate of the entire continent would improve. Noah Webster challenged Jefferson's ideas, saying that they were hearsay based on the recollections of old people, remembering the 'good old days.' Jefferson realized that memories and anecdotal evidence were not reliable and in 1824 he proposed that "Measurements of the American climate should begin immediately, before the climate has changed too drastically. These measurements should be repeated at regular intervals." (From a letter to Lewis Beck, cited by Fleming, 1998.)

Interestingly, the unsuitable climate problem does not seem to have arisen among the Spanish colonists, probably because the landscapes and climate of Mexico and western South America were much like those of the Iberian Peninsula.

The idea that human activity has no appreciable effect on climate is a fiction invented over the past few decades. It appears to have arisen from those interested in maintaining a clear path toward short term profits in the petroleum and coal industries, and it is largely a US phenomenon. Many people are aware that cities moderate the local climate. Both summer and winter temperatures are warmer than in the surrounding countryside. There is less intense sunshine. There may be more frequent fog, and even the rainfall may be different. The question is not whether humans are affecting the entire planet's climate, but how much and for how long?

Some science fiction movies are about 'terraforming' other planets, like Mars, so they will be more like Earth. As will become clear later in this book, we are 'anthropoforming' Earth. It is a truly grand experiment, and it is proceeding at a rate no one could have imagined 50 years ago. I only wish we could have done it to some other planet so we could find out how it works out before using our home planet as the guinea pig.

But before we go any further in our look at climate change, I should tell you a bit about myself, what science is about, how science operates, and a few basic ideas.

## 1.3  My Background

First of all, let me make clear that I never intended to become seriously involved with the problems of future climate change, but one thing leads to another, and another, and another (Fig 1.2).

My life has been governed by parents who had a great respect for education and by serendipity. Somehow, already in high school at St. Marks's in Dallas, Texas, I had figured out that I wanted to be a paleontologist and spend my life

**Fig. 1.2** Portrait of me taken by my old friend from the Munich days, Heinz Just, in his library in Lautrach, Germany, 2008

one language to another because their meanings carry not only literal, but emotional and connotational content specific to particular regions. Even worse, the meanings of words change with time. Today, in discussions of global warming, semantic manipulation has been raised to the level of an art form in presenting disinformation.

During my sophomore year I heard about the possibility of studying abroad. In 1953 I joined the small first group of American undergraduates to study at Ludwig-Maximilian's University in Munich, Germany, organized by Wayne State University in Detroit. Living in a city that had been thoroughly bombed out, was in the midst of rebuilding, and still had an intense cultural life, changed all of us as we recently discovered at our 50th reunion.

It was in Munich that I had my first courses in Geology, and it was where I decided to become a micropaleontologist. My first course in Earth history was taught by Richard Dehm, who also introduced me to the idea of continental drift. The idea that the continents had once been all together in one large block which had then separated into two, and subsequently into the seven we now recognize, had been proposed by the German meteorologist-climatologist Alfred Wegener in 1912. Dehm suggested that, whether right or wrong, it was a very useful device for geologists to be able to remember what kinds of rocks and fossils were where.

A paleontologist is a person who studies fossils, and a micropaleontologist is one who studies microscopic fossils. The practical value of these fossils is that they come up intact in the mud from drilling holes, such as oil wells. Although I never intended to go into the petroleum industry, I fell in love with these tiny beautiful objects when I saw them for the first time through a microscope in the Deutsches Museum, Munich's great museum of science and technology. I finished my Bachelor's degree at SMU and was able to study in Europe another year at the University of Zurich and Swiss Federal Institute of Technology courtesy of the "Swiss Friends of the USA" and my indulgent parents. In Zurich I had my introduction to alpine geology from Rudolf Trümpy, who was just starting his career. I found out that the Alps were the result of a collision between Africa and Europe and that the idea of mobile continents had its origins among alpine geologists before Alfred Wegener came on the scene. In 1958 I received my Master's from the University of Illinois at Urbana, where I learned that the idea of continental drift was nonsense. In 1960 I got my Ph.D. from Stanford, where geologists were strict believers in Uniformitarianism—the idea that geologic processes have always operated as slowly as they do today—and any idea of catastrophic events having affected the Earth was anathema. Strange in a place where earthquakes are not that infrequent. In 1967 we had one that felt like elephants were running through the building. My Ph.D. advisor at Stanford was a Swiss, Hans Thalmann, so discrete discussion of continental

studying fossils. I went to Southern Methodist University for my undergraduate degree; my family had a long history of association with SMU and it was a great place to launch my education. When I told my freshman advisor that I wanted to be a paleontologist, he told me that to have a career in such a special field I would need to eventually get a Ph.D. I already knew that. He advised me to do my undergraduate studies in fields that would provide me with the proper background, and to leave the direct study of paleontology to graduate school. So I majored in Biology and minored in Chemistry. I also took German to fulfill the foreign language requirement. I had already studied French and Latin in high school and thoroughly enjoyed learning a fourth language. My freshman year instructor in German was Alvin Jett. He was fascinated by semantics—the study of how meaning is conveyed by the usage and interrelationships of words, phrases, and sentences. I found this seemingly esoteric topic fascinating. I became aware that you do not understand the language you grow up with until you learn a second language. It is almost impossible to translate words, phrases and sentences from drift was allowed. Both my M.S. thesis and Ph.D. dissertation were on studies of microfossils. My Ph.D. was on the Cretaceous-Tertiary boundary in Mexico. We now believe that this was the time when an asteroid collided with the Earth, ending the reign of the dinosaurs and causing extinction of many organisms that lived in the open ocean. Luckily for me in my doctoral defense I just described it as an extraordinary time of extinction and origin of new species. If I had suggested a catastrophic event, I would have been thrown out of the room.

In 8 years as an undergraduate and graduate I had accumulated transcripts from six universities and encountered several different cultures and heard a lot of conflicting views on many topics. Exposure to conflicting views is a valuable learning experience.

After having completed work for my doctorate I had a year of postdoctoral study back in Europe, in Basel, Switzerland. The Geologic-Paleontologic Institute of the University of Basel had long been one of the great centers for micropaleontological research.

In 1960 I began my professional academic career at the University of Illinois in Urbana and soon found friends in the Departments of Chemistry, Art, Electrical Engineering, Microbiology, Astronomy, Economics, Physics, Architecture, and Agriculture. That is what an academic career is all about— receiving informal education in new fields from your peers.

As it turned out, microfossils have been one of the keys to understanding the history of the oceans. The deposits on the ocean floor are often made up largely of these tiny fossils, the remains of single-celled animals and plants that populate the surface layers. It also turned out that they not only tell us about the age of the sediments, they tell us about the climate of the past and, even stranger, we now know that they are major participants in Earth's carbon cycle, and have served as one of the chief mechanisms of regulating the Earth's climate.

Most of our knowledge about the oceans has come from recovering long sequences of cores of sediments from the deep sea floor. Many of them have been taken by the Deep Sea Drilling Project, which began operations in 1967, and by its continuation as the Ocean Drilling Program, and now the Integrated Drilling Ocean Program. By chance, I became one of the advisors to the Deep Sea Drilling Program in 1965, even before it started. I remained closely involved until I retired a few years ago. Again, just by chance, the particular group of microfossils I was studying at that time, the "calcareous nannoplankton," turned out to be critically important to the project. Calcareous nannoplankton fossils are extremely small (about 1/100 mm, which is a couple of ten thousandths of an inch!) but highly complex and beautiful platelets made of calcium carbonate (the same material as limestone); they are produced in vast numbers by microscopic plants floating in the ocean's surface waters. My students and I had a near monopoly on knowledge of this

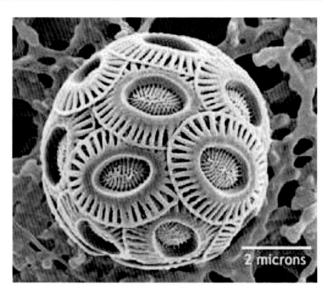

**Fig. 1.3** Scanning electron micrograph of a specimen of a calcareous nannoplankton species, *Emiliania huxleyi*. What you see is a 'coccosphere,' made up of elliptical plates called coccoliths. The organism, a single-celled plant, would be mostly inside the tiny coccosphere

particular group of fossils, and since they were the easiest tool for determining the age of the sediment, we were very popular. An example of these fossils, *Emiliania huxleyi*, is shown in Fig. 1.3.

*Emiliania huxleyi* is one of the most abundant organisms in today's oceans. The scientific name is in two parts. The second part, *huxleyi*, is the species, named by Erwin Kamptner of Vienna to honor Thomas Henry Huxley, one of Charles Darwin's greatest exponents and author of an essay "On a piece of chalk," often regarded as a masterpiece of scientific writing. Google that title and you should be able to find a copy of the essay you can download. The generic name, *Emiliania*, honors Cesare Emiliani, founder of the science of paleo-oceanography, the study of the oceans of the past. Swiss postdoctoral student Hans Mohler and I created this generic (or 'group') name in 1967 because this species was very unlike any other and seemed to have no close relatives alive today. Hans went on to larger things: he became Geologist to the Sultan of Oman. Cesare Emiliani later pointed out that there are only two species capable of changing the albedo (reflectivity) of planet Earth, *Emiliania huxleyi* and *Homo sapiens*. Cesare also calculated that the total number of human cells in the world and the number of cells of *E. huxleyi* are about equal, so humans have a single-celled competitor for domination of the planet.

When the Deep Sea Drilling Project started in 1967, most geologists thought that the ocean had the most boring history of any place on the planet, essentially an unchanging environment since shortly after the Earth had formed as a planet. A few months into the project it had become apparent that the oceans had a very interesting and complex history. Now, more than 40 years later, we are trying to understand how the climate of the Earth has evolved, and most importantly: why. My specific goal has been to try to understand the world of the Cretaceous Period, which lasted from 140 until 65 million years ago. It was a time when the Earth was warmer than it is today; it is a difficult problem and we are still putting the story together. Accordingly, in the 1970s my interest was concentrated on a time when Earth's sea level was much higher than today, when the Polar Regions were largely free of ice and snow, and when dinosaurs roamed the lands. The last 30 years of my life have been largely devoted to this topic and only now do I think we are beginning to understand what life on a "warm Earth" was like.

I spent my first sabbatical leave at the University of Miami's Rosenstiel School of Marine and Atmospheric Sciences and at the Hebrew University in Jerusalem. From 1968 until 1973 I was Professor at both Illinois and Miami, spending the fall semester in the north and the spring semester in the south. It was during this time that I changed my major interest from micropaleontology to trying to reconstruct the circulation of the oceans in the past. It was a short step to becoming involved in studies of Earth's ancient climates. In 1974 I moved full time to Miami as Chair of the Rosenstiel School's Division of Marine Geology and Geophysics, and then served as Dean from 1976 to 1980.

From 1980 to 1982 I worked in Washington, D.C. as head of a consortium of oceanographic institutions. The Deep Sea Drilling Program was about to end, and we were searching for a way to continue the program of scientific ocean drilling.

When the future of ocean drilling got settled as the 'Ocean Drilling Program,' I was anxious to return to the academic life I enjoyed so much. I moved to the University of Colorado at Boulder where I served as Director of the University Museum, Professor of Geology, and member of the Cooperative Institute for Research in Environmental Sciences. I also had a close working relationship with colleagues at the National Center for Atmospheric Research in Boulder, becoming ever more deeply involved in studies of Earth's paleoclimates.

I spent my second sabbatical leave, 1988–89, at Ludwig-Maximilian's University in Munich, returning to the city that had so influenced my early career. I had become interested in how investigators early in the 20th century had interpreted the glaciations of the Alps, and spent some time in the Bavarian and Austrian countryside, visiting the places they had studied.

Shortly after my return to the University of Colorado, I had an invitation to join the new marine geological institute, 'GEOMAR,' in Kiel, Germany. Invitations to other European Universities followed, and I took a series of leaves of absence to work at GEOMAR in Kiel, at the Universities of Utrecht in the Netherlands, Vienna in Austria, and at the Institute for Baltic Sea Research in Warnemünde and the University of Greifswald in the former East Germany. I was

about to return to the University of Colorado for good when I was made an offer I really couldn't refuse, again working toward figuring out how to continue scientific ocean drilling. So I retired from Colorado, and moved full time to GEOMAR in Kiel. During the 1990s I was exposed to many new ideas as our knowledge of the history of the oceans grew. Not many academics have the chance to meet, work with, and learn from the students and young faculty at so many different institutions. I finally retired from Kiel in 2002 and now live in the Colorado Rockies.

As a scientist one tends to take a detached view of the world, and it was not until a few years ago that I realized that future climate change is not an abstraction that might affect generations far off in the future. It is happening now, and, I believe, much faster than anyone had expected. Stranger still, we seem to be nudging the climate of the Earth back toward the conditions of the Cretaceous. As a geologist I would love to see what happens; but as a human I seriously wonder whether civilization will survive the next few hundred years.

We are all 'products of our environment' and that applies especially to the intellectual environment we have experienced. To relieve the tedium of learning, I have inserted 'Intermezzi' between the chapters of this book devoted to understanding the grand experiment humans are performing on planet Earth. The 'Intermezzi' recount some of the entertaining episodes from my life that have shaped my view of the world.

## 1.4 What Is Science?

Science is an unending voyage of discovery. Simply, it is a search for truth. Absolute truth seems to be restricted to mathematics. In all other fields we can only hope to make ever closer approximations to the truth. The search for truth involves making observations and performing experiments; in short, gathering factual information. The results lead to the development of one or more "hypotheses" to explain the observations or the outcomes of the experiments. Each hypothesis makes predictions about the outcome of further investigations or experiments. Hypotheses that have passed many tests achieve recognition as being close approximations to truth and are referred to as "theories." A theory which stands the test of time can become a "Law."

It is important to realize that the use of the word 'theory' in its scientific context is quite different from its use in everyday conversational English, where 'theory' is used to refer to a guess with little or no supporting evidence— essentially an hypothesis in its earliest stages of development and that has not been rigorously tested. 'Theoretical' implies great uncertainty. Fundamentalist Creationists take full advantage of the fact that the general public is not aware of the different usage when they say "evolution is, after all, just a theory." The fact that Darwinian evolution is considered a theory by scientists means that it has withstood many tests and been found over and over again to be the simplest explanation of observed facts.

Occam's razor is another important principle of guidance in science. It is attributed to a 14th-century English Franciscan friar, William of Ockham. Spelling was not very strict in the 14th century, and the name of the town from which he came, near Guildford southwest of London, was sometimes spelled Ockham, sometimes Occam. William of Ockham's idea was that the explanation of any phenomenon should make as few assumptions as possible, eliminating those that make no difference in the observable predictions of the explanation. In Latin this is known as the *lex parsimoniae* ("law of parsimony" or "law of succinctness"), or "*entia non sunt multiplicanda praeter necessitatem.*" This is often paraphrased as "All other things being equal, the simplest solution is the best."

Needless to say, this got him into serious trouble with the religious authorities of the time who much preferred the most complex, least understood explanation. Pope John XXII summoned him to Avignon in 1324 to stand trial for whatever crime this might have been. William of Ockham was never convicted and was able to flee to Munich, where he came under the protection of Ludwig IV of Bavaria. Ludwig IV was already considered a heretic by Pope John XXII, but Ludwig was not to be trifled with. He had himself elected Emperor of the Holy Roman Empire and appointed another Franciscan monk, Pietro Rainalducci, to be his own Antipope, Nicholas V, residing in Rome. It was a messy time in central European politics. Incidentally, if you have read Umberto Eco's fascinating medieval murder mystery *The Name of the Rose*, you will recall it takes place in these times, and the hero, the Franciscan Friar William of Baskerville, was said to be a good friend of Ockham.

In Munich, Occamstrasse is a short but very interesting street in the suburb of Schwabing, lined with *Kneipen* (he equivalent of a British pub) where you can drink beer and get into philosophical discussions with total strangers, cabarets where comedians make fun of politicians, and a movie theater that shows classic films, particularly *Fantasia* and *Gone with the Wind*. As a student, I lived three blocks away.

Supernatural explanations are not allowed in science because they can't be tested. They belong to the realm of religion. Furthermore, science is devoid of emotion; there is no good or evil, just an advance of knowledge in a search for the truth. However, Thomas Chrowder Chamberlin, a geologist who served as President of the University of Wisconsin from 1887 to 1892 and went on to start a great Department of Geology at the University of Chicago, recognized that some scientists become overly fond of their ideas. He wrote a paper, published in 1890, that has been reprinted many times: "The Method of Multiple Working Hypotheses," with the subtitle "With this method the dangers of parental affection for a favorite theory can be circumvented." He likened a scientist's hypothesis to a child. Seeing your idea die a sudden death as new information becomes available is much like losing a child, but it's not quite so bad if you have several of them and at least one survives.

The road to truth has many twists and turns, some of them quite unexpected. And there are roadblocks along the way. Scientific research is much like sailing a boat in the ocean. Yachtsmen describe sailing as a sport that consists of hours of tedious boredom interspersed with moments of stark terror. Gathering data can be one of the most unexciting activities imaginable. Many university researchers try to engage undergraduates by letting them participate in the data gathering exercise. For many it turns out to be such a mind-numbing experience that they never want to have anything more to do with science. I suspect that many of them go on to become politicians, or journalists. For the few who experience that moment when the data suddenly come together and reveal something wholly unexpected, the experience is unforgettable.

The number of observations or measurements one can make is infinite. The art of making progress in science is to ask the right questions. In particular it is important to devise a test that will demonstrate that one hypothesis is wrong and another is correct. Following this you can gradually work your way along the path to the truth. However, someone once pointed out to me that there are three ways to speed up the process and make rapid advances in science: (1) analyze samples using a technique or instrument that has never been used before; (2) analyze a sample from someplace no one has ever been before; (3) analyze the data in a way no one has ever done before. It's true; most leaps forward in science are the result of the opportunities created by one of these advances, but serendipity, chance meetings and opportunities also play a great role.

There is a governor on scientific progress, called the "peer review system." Contrary to what many believe, scientists are among the most conservative people on Earth. When you make a discovery, its validity must be proven to your colleagues (not all of them friendly) through publication. Everyone has a chance to dissect your ideas and tear them apart. Most scientific publication is through the "peer review process" whereby the manuscript documenting and interpreting your research is sent to several reviewers chosen by the editor or editorial board. As anyone who has published many papers will tell you, some of the reviews are helpful, but others are outrageous critiques by ignorant souls who haven't got the least understanding of what you are trying to say. The system works in such a way that really bad ideas almost never make it into print. Unfortunately, some really good ideas may get delayed for years by the process. Robert Muir Wood's 1985 book, with the ominous title "The Dark Side of the Earth," describes the development of modern ideas in geology, and notes that none of the seminal papers leading to the now universally accepted theory of plate tectonics managed to get through the peer review process.

The sciences can broadly be divided into two groups, those that depend on observations that lead to a hypothesis that is then confirmed by experiments (Physics, Chemistry, and Biology), and those that depend on observations leading to hypotheses that must be confirmed by further observations (Geology and Astronomy).

Perhaps it is useful here to remember that science is unique among professions in that it is dedicated to the search for truth. The quest for truth is by its nature antithetical to dogma or ideology. In the legal profession in the United States, the search for truth is the responsibility of a jury or a judge who will hear conflicting evidence from opposing lawyers. Expertise in the subject at hand is often cause for dismissal from a jury. The goal of lawyers is to win the case for their client; any relationship to the truth may be purely coincidental. The overwhelming majority of members of the US Congress are lawyers. It is helpful to recall the in his *Devil's Dictionary*, Ambrose Bierce appropriately defined the word: Lawyer, *n.* One skilled in circumvention of the law.

Scientists are by nature skeptics. As you will see, the scientific community is very conservative and often slow to accept new ideas. Scientific skepticism involves inquiry, critical investigation, and the use of reason in examining other scientists' results. The U.S. media often refer to those who assert that the current climate change is not caused by human intervention as "skeptics." They are not skeptics in the sense used by scientists. They are climate change "deniers." Denial is the rejection of ideas without objective consideration. Climate change denial is a business employing many lobbyists, or, in the case of politicians, a means of raising funds and appealing to their uninformed "base" (Table 1.1).

## 1.5 The Observational Sciences

Geology and astronomy are unusual sciences in that they depend almost entirely on observations. In most other fields of science, such as physics, chemistry, and biology, progress is made by performing experiments, but in geology and astronomy changes occur over such long periods of time that experiments are impossible. In astronomy the search for observational evidence involves the use of a variety of telescopes and satellites to collect data for analysis. In the early days of geology, observations were made with the eyes, a hammer to knock off a fresh chunk of rock, and sometimes a hand lens (a small magnifying glass) to take a close look at the rock. Since the middle of the last century increasingly more sophisticated methods have been developed; some of these are described in the later chapters of this book.

Geologists now perform laboratory experiments to study processes such as erosion and transport and deposition of

**Table 1.1** Words with special meanings to scientists

| Word | American public's use | Meaning in science |
|------|----------------------|--------------------|
| Accuracy | The state or quality of being accurate; the extent to which something is accurate; precision, exactness | The closeness of a measurement, calculation, or specification to the correct value |
| Anomaly | Abnormality | Difference from long-term average |
| Bias | An inclination, leaning, tendency, bent; a preponderating disposition or propensity; pre-dilection; prejudice | A systematic distortion of an expected statistical result due to a factor not allowed for in its derivation; also, a tendency to produce such distortion |
| Error | Something incorrectly done through ignorance or inadvertence; a mistake | The quantity by which a result obtained by observation or by approximate calculation differs from an accurate determination |
| Manipulation | The exercise of subtle, underhand, or devious influence or control. Tampering for nefarious purposes | Processing of data |
| Mean | Bad, excellent | Average |
| Positive feedback | Helpful suggestion, praise | Potentially dangerous self-reinforcing cycle |
| Precision | The fact, condition, or quality of being precise; exactness, accuracy | The degree of refinement in a measurement, calculation, or specification, esp. as represented by the number of digits given |
| Quite | Rather | Completely |
| Scheme | A self-seeking or an underhand project, a plot or a visionary or foolish project | A plan, design; a program of action; the designed scope and method of an undertaking |
| Sign | An indication, a token; something that represents something else; An action, mark, notice, etc., conveying information or instructions, and related senses | The property of a quantity of being positive or negative. A conventional symbol used to represent a particular instruction, operation, or concept, as in algebra, music, finance, etc |
| Theory | Speculation, conjecture | A hypothesis that has been confirmed or established by observation or experiment, and is propounded or accepted as accounting for the known facts |

(continued)

**Table 1.1** (continued)

| Word | American public's use | Meaning in science |
|------|----------------------|--------------------|
| Uncertainty | Ignorance | The amount of variation in a numerical result that is consistent with observation |
| Values | Worth or quality as measured by a standard of equivalence. The principles or moral standards held by a person or social group | Numerical measures of physical quantities |

Definitions based largely on the Oxford English Dictionary

sediment by rivers and along the shore. However, when these experiments are carried out, there is always a problem of determining whether the scale and rate of the processes can replicate larger features in the real world. Increasingly complex computer simulations are being developed to understand how the Earth changes over time, but it is work in progress.

## 1.6    The Complexity of Nature

As you read this book, my discussions may seem to wander off into unrelated topics. Later some of the connections will seem obvious, others obscure. John Muir (1838–1914) was great naturalist, and devoted his life to arguing for the preservation of wilderness. He is often called "The Father of the National Parks." As he said: "When we try to pick out anything by itself, we find it hitched to everything else in the universe."

## 1.7    Summary

Virtually all of our knowledge of the Earth's past history comes from observations, and interpretation of those observations changes as we develop better understanding of the processes that shape the planet.

A few decades ago the idea that humans could actually perform an experiment on a planetary scale was pure science-fiction. Then in the 1970s a few scientists realized that we were indeed performing an experiment on a planet, our own Earth. Over the past few decades more and more scientists and laymen have come to realize that humankind has unwittingly embarked on a grand unplanned experiment to change the climate and with it the environmental conditions on Earth's surface. Geology has now, unintentionally, become an experimental science.

However, there has been a strong conviction that humankind will reverse course in the nick of time, ending

the grand experiment. A few suggest that the climate of the past few centuries wasn't all that good anyway, and maybe a change is in order. Since we didn't expect the effects of climate change to become obvious until the middle or end of the 21st century, there seemed to be no need to take any corrective measures. At least those were the views in the United States, not shared by most of the rest of the world. Of course, there are a few—they call themselves 'conservatives' because they reject any idea of change on ideological grounds—who insist that nothing at all is happening.

A Timeline for this chapter:

| Year | Event |
|------|-------|
| 1324 | William of Ockham is put on trial for heresy for proposing his Law of Parsimony—a.k.a. Occam's Razor —that the simplest explanation is the most likely to be true |
| 1785 | Thomas Jefferson writes about the warming of the climate in Virginia |
| 1824 | Jefferson proposes that measurements of the American climate should begin immediately, before the climate has changed too drastically |
| 1890 | T. C. Chamberlin publishes *The Method of Multiple Working Hypotheses* |
| 1906 | Ambrose Bierce publishes *The Cynic's Word Book*, later editions to become *The Devil's Dictionary* |
| 1951 | I enter Southern Methodist University |
| 1953–54 | I attend Wayne State University's Junior Year in Munich, Germany |
| 1955 | I graduate from SMU with BS in Biology |
| 1955–56 | I attend the University of Zurich and the Swiss Federal Institute of Technology sponsored by the Swiss Friends of the USA |
| 1956–57 | I attend the University of Illinois at Urbana, MS in Geology 1958 |
| 1957–59 | I attend Stanford University, Ph.D. in Geology, 1960 |
| 1959–60 | I attend the University of Basel, Switzerland as an NSF Postdoctoral Fellow |
| 1960–73 | I am appointed Assistant Professor, then Associate Professor and then Professor, Department of Geology, University of Illinois, Urbana |
| 1967 | I spend my first Sabbatical Year at the University of Miami's Institute of Marine Sciences where I get to know Cesare Emiliani |
| 1968 | I spend the summer as Visiting Professor at the Hebrew University in Jerusalem |
| 1968–80 | I become Professor of Marine Geology and Geophysics, and then Dean at the Rosenstiel School of Marine and Atmospheric Sciences, University of Miami, Florida (formerly the Institute of Marine Sciences) |
| 1975 | Wally Broecker publishes *Climate Change; Are We on the Brink of a Pronounced Global Warming?* |
| 1979 | Mikhail Budyko and Alexander Ronov publish *Chemical evolution of the atmosphere in the Phanerozoic* arguing |

(continued)

(continued)

| Year | Event |
|------|-------|
|  | that the major changes in Earth's climate in the past were due to changing atmospheric $CO_2$ concentrations |
| 1980–82 | I serve as President of Joint Oceanographic Institutions in Washington, D.C. |
| 1982–87 | I serve as Director of the University of Colorado (Boulder) Museum |
| 1982 | Herman Flohn states "Long before global warming due to $CO_2$ can be proven, its effect will be noticed by extremes of weather, setting local records such as the highest and lowest temperatures, greatest and least rain and snowfall amounts, most powerful storms, unusually large numbers of tornados, etc. These indicate that the climate system is becoming unstable in response to the greenhouse effect." |
| 1982 | Meeting with Mikhail Budyko in Leningrad: 'It's going to happen.' |
| 1983–98 | I am appointed Professor of Geology and Fellow of CIRES at the University of Colorado |
| 1985 | Robert Muir Wood publishes *The Dark Side of the Earth* |
| 1988–89 | My Sabbatical Year spent at Ludwig-Maximilian's University in Munich |
| 1990–2010 | I am Guest Professor, then Professor at GEOMAR, University of Kiel, Germany, also Visiting Professor at the University of Utrecht, The Netherlands; the University of Vienna, Austria; and the University of Greifswald, Germany; and visitor at several other institutions. |
| 1998 | James Rodger Fleming publishes *Historical Perspectives on Climate Change* |
| 2007 | Massive melt-back of Arctic sea ice |
| 2010 | Tom Friedman coins the term 'Global Weirding' |

If you want to know more:

Broecker, W., 1975. Climate Change; Are We on the Brink of a Pronounced Global Warming? Science, v.189, pp.460–463.—The classic paper that sounded the early warning.

Budyko, M. I., Ronov, A. B., Yanshin, A. L., 1987. *History of the Atmosphere*. Springer, 139 pp.—A classic, written for scientists, emphasizing the role of atmospheric chemistry and particularly $CO_2$ in climates of the past.

Chamberlin, T. C., 1897. The Method of Multiple Working Hypotheses. The Journal of Geology, v. 5, p. 837–848.—The classic paper on hedging your bets. (You can download a free pdf from: http://www.mantleplumes.org/WebDocuments/Chamberlin1897.pdf

Fleming, J. R., 1998. *Historical Perspectives on Climate Change*. Oxford University Press, Oxford, U.K., 194 pp.—An illuminating history of the development of ideas about climate and climate change.

McCain, G., Segal, E. M., 1969. *The Game of Science*. Brooks/Cole Publishing Co., Belmont, CA, 178 pp.—A lighthearted layman's introduction to what science is all about.

An unabridged version of Ambrose Bierce's *Devil's Dictionary* can be downloaded here: http://www.dankunlimited.com/uno_pound_sharing/Know_files/-The-Unabridged-Devil-Dictionary.pdf

Music for contemplation of what you have read: Dmitri Shostakovich's Symphony No. 7 in C major, Op. 60 (*Leningrad*)—finished during the early part of the siege.

Libation: First, a bite of a nice side dish of marinated herring or other oily fish. Then accompanied by chilled Vodka (I like 'Russian Standard'—Русский Стандарт) from the freezer. Very un-Russian-like I prefer to sip and savor vodka very slowly. Alternating the fish and vodka will leave you able to walk afterward.

## Intermezzo I. A Slightly Unusual Family

I can't say that I was the product of a very ordinary family. The Hay family goes back a long way—into Scotland and possibly before that into Normandy. There are two versions of the early origins. One is that a Scottish peasant named Hay made himself particularly useful in defeating the Danes at the Battle of Luncarty in 990 AD. As reward he was made a noble and the extent of the lands given him was determined by the flight of a falcon released from the Hawk Stone in St. Madoes. The falcon lighted on a rock now known as the Falcon Stone just east of Rossie Priory, about 8 km west of Dundee on the Firth of Tay. The other, only slightly less romantic, version is that the family name was originally de la Haye and that they came to Britain with William the Conqueror in 1066. In any case, Sir William Hay became Earl of Erroll in 1453. In Sir James George Frazer's book *The Golden Bough: A Study in Magic and Religion*, published in 1834, it is reported that the well-being of the Hay clan is assured if, on Allhallows Eve (Halloween), a young son climbs high in the Oak of Erroll, and using a new dirk cuts off a piece of the mistletoe growing there. A dirk is a Scottish ritual dagger, absolutely essential for this sort of thing. Unfortunately the great oak tree of Erroll died many years ago, and falcons can no longer nest there. Presumably the mistletoe has spread to other trees and lost its magical powers. My brother got very interested in tracing back the family history, made the pilgrimage to the Falcon Stone, and snooped about in Edinburgh. Apparently our original Scottish ancestor, John Hay, listed in the archives in Edinburgh as 'the one who went to America' emigrated to Virginia in 1639.

My grandmother, Mary Norton Randle Hay, was born in 1864 in Washington-on-the Brazos, a small town, now gone, located a few miles south of modern

College Station, Texas. It was the site of the signing of the Texas Declaration of Independence on March 2, 1836. It was the first of five towns that served as the capital of the Republic of Texas in 1836. In 1837 Houston was designated to be the capital, and in 1840 it was moved to Austin. My grandmother's mother, Kathleen Emma Crawford Randle, had been born there too, during the time Texas was an independent republic. My great grandmother's parents had moved there before the Texas war of Independence, so that side of the family were real Texans. An interesting footnote to history is that many of the settlers from the United States who accompanied Stephen F. Austin to Mexican Texas in 1822 to establish the settlements along the Brazos River brought their slaves with them. Slavery was abolished in Mexico 1829, with the Texans being granted a 1 year extension to comply with the law. Many of the colonists simply converted their slaves into indentured servants. One of the lesser known reasons for the Texas War of Independence was that the colonists wanted to retain the right to hold slaves. My ancestors were no exception. However, there were those opposed to slavery, most notably Sam Houston.

My grandfather, Stephen John Hay, moved to the young town of Dallas from Georgia in 1887. He was Mayor of Dallas from 1907 to 1911. He was a political innovator, introducing the Commissioner form of government to Dallas, and saw to it that the involvement of national political parties in local affairs was minimal. He was also a champion of higher education and believed that Dallas needed a major institution of higher learning. He was intimately involved in the founding of Southern Methodist University in 1911. SMU was a joint venture between the Methodist Church and the city of Dallas. The city contributed land and money toward the first building, which was appropriately named Dallas Hall. It was designed to copy Thomas Jefferson's Rotunda at the University of Virginia and was finished in 1915. In spite of its name and support from the church, SMU was from the beginning non-sectarian and devoted to free inquiry and research. Today only about 25 % of the students identify themselves with the church.

When I was a student in the 1950s SMU had one requirement related to religion. You had to take a course in History of either the Bible or Philosophy. I really couldn't imagine anything much more boring than History of the Bible, except possibly Philosophy; so the Bible it was. It was one of the best series of lectures I ever had, going into where the texts came from, how they fitted into the context of the time in which they were written, and how the Council of Nicea in 325 decided what would be included in the Christian Bible, and what excluded. We even read the more interesting parts of the Apocrypha.

In 1916 my grandfather Hay died suddenly of meningitis, leaving behind his widow, Mary N. R. Hay to raise the family, one teenage son (my father) and three daughters, alone. Grandfather Hay had died leaving many debts, and my grandmother could have declared bankruptcy and written them off. But she insisted on paying them off to the last penny, and did so by becoming an insurance agent. The story goes that she would visit the businessmen of Dallas and simply tell them how much insurance they were going to buy. She was a determined and forceful woman; few turned her down. She became Dean of Women at SMU. It is probably to her that I owe my lifelong interests in both geology and classical music. She had a small rock and mineral collection and stories to go with each specimen. Her funeral was a truly memorable experience, held in the Highland Park Methodist Church on the SMU campus with an overflow crowd of former students. It was a concert devoted to the music of her favorite composer, Johann Sebastian Bach. The University named a women's dormitory in her honor.

One of the daughters, Elise, taught Music (voice) at SMU; another led a life as a housewife, and a third, Elizabeth was something very different. She was an airplane mechanic before the First World War (yes, you read that right). A Russian nobleman had come to the US to buy an airplane, they fell in love, and she moved to Russia, although they also spent time at their second home in Egypt. In Russia, they lived mostly in Moscow, where he had apartments in the Kremlin. Elizabeth witnessed the October revolution from that very special vantage point. Her husband was ultimately killed in the fighting, and she returned to the US. Back here she was employed as an undercover detective by Pinkerton's. She posed as the caretaker for the dog of a wealthy family, actually investigating losses of jewels and securities, thought to be an inside job. The family enjoyed traveling by rail; they had their own railway car. She told me that on these trips the dog had more luggage than she did. The robberies indeed turned out to be an inside job. My father sometimes referred to her as his 'crazy' sister.

My father worked in the East Texas oilfields in his youth, but probably learned the insurance business from his mother. He was in the first graduating class of SMU, and later became a Trustee of the University. Financed by friends, he started his own insurance company in the 1920s. It survived the depression and was finally sold to another company in the 1960s. He married my mother when they were students at SMU. He said it was love at first sight on a streetcar. My mother's parents had a manufacturing business in Dallas, and the six of us all lived together as a large family. I later realized that this made it possible to share expenses and live well. We were not wealthy, but I would say we were 'upper middle class.'

I was born in October, 1934. My brother was 4 years older than I and suffered from asthma. From the late 1930s through the 50s we lived on the north edge of Dallas in the suburb called Preston Hollow that is now home to George W. Bush. In 1938 we had recovered enough from the Great Depression so that my father bought a new car, a green Packard with running boards, that somehow seated six. The whole family took a grand tour to Yellowstone and the Colorado Rockies. The next summer we vacationed in Colorado, and my brother's health improved greatly. Then came the news of Pearl Harbor. Perhaps my oldest vivid memory was of that day when we sat by the radio in our home in Preston Hollow and listened to President Roosevelt's "Day of Infamy" speech. Because of my brother's asthma, my mother, my brother and I lived in Colorado through most of World War II when travel was almost impossible due to gasoline rationing. My father's business was in Dallas, and because of the travel restrictions we only saw him a few times each year. But it was in Colorado that my Grandmother Hay's rocks and minerals became related to mountains and landscapes, and I became interested in geology.

Although my father was a businessman and didn't have much time for concerts or theater after work, he saw to it that I was able to become involved in cultural affairs from an early age.

An art gallery opened in a shopping center a mile from my home, and I would drop in whenever they had new exhibitions. I learned that art is not just something to see in Museums, it is an important part of cultural life. The only thing I could afford to buy was the occasional book on art (they are still in my library), but when I was in high school the gallery owners began to invite me to openings. I met local artists and architects, some of whom have remained lifelong friends. The cultural interests developed then became important components of my life. I've often wondered what my life would have been like if that little gallery had not opened.

Another of my passions was Opera. How could a kid from Texas get interested in Grand Opera? On Saturday afternoons in the fall and winter, Texaco broadcast the Matinee of the Metropolitan Opera in New York. These were the days before television, and what better way to spend a dreary, cold, rainy afternoon than listening to great music. I guess half the kids I knew were addicted to it, especially because on Friday night everything would be explained by Reuben A. Bradford's program "*Opera Once Over Lightly.*" His program started out in Dallas, and was eventually broadcast over a network of 197 NBC stations. In northern Texas, where it originated, it achieved the highest ratings of any radio show. Bradford would discuss the plot in terms every Texan could relate to and play recordings of some of the arias. Some of his two line summaries are classic:

*La Forza del Destino*: *We're talking now of destiny's force, accompanied by singing and stabbing, of course*
*Cavalleria Rusticana*: *This is the picture now to be shown, "Love thy neighbor," but leave his wife alone.*
*Il Trovatore*: *Loud, long, and very gory, Murder and music; that's Il Trovatore*

In the end of the second act of Puccini's beautiful opera *Tosca*, Floria Tosca, a well known opera singer in Rome at the time of the Napoleonic Wars, makes a deal with the Chief of Police, Scarpia, to save her lover, Mario Cavaradossi, who has just been sent to the Castell San Angelo to await his execution in the morning. The deal involves what we would probably call consensual rape, a fake shooting of Mario, and a note allowing Floria and Mario to leave the city. Bradford describes what happens next:

> With the deal all wrapped up he makes a grab at the dame but he don't snatch fast enough. As she skids around a corner, Tosca grabs a paper-knife, spars for an opening then lets the cop have it where it will do the most good. After he hits the floor and stops wiggling she places two candlesticks at his head then quietly leaves the late Roman Police Chief staring at the scenery loft.

Of course Scarpia had already secretly reneged on the deal pre-mortem, so everybody dies in the end.

I had always thought Tosca was fantasy, but recently learned that it is based on real people and events, sanitized and cleaned up for the stage from the much bloodier reality of the Roman police state of 1811.

*Opera Once Over Lightly* was so popular because it was one of the funniest programs on the air. We were listening one evening when my father asked me "Would you like to meet Bradford?" It turned out he was an old family friend and had rented a room in my Grandmother Hay's house years before. He was simply

'Bradford' to everyone. He was as funny in person as he was on the air. Incidentally, Bradford hated Bach. In his book he blames Johann Sebastian Bach for inventing the fugue, which he describes as follows:

A fugue is a little curlique gimmick usually started by the violins. They give birth to the thing, nurse it, fool around with it a bit and toss it over to the bass section. This crowd looks it over, makes a few changes, turns it around, points it the other way then throws it back to the fiddles. The string section never seems pleased with the changes so they call in the horns and maybe a couple of flutes and oboes. They try to put the thing back something like it was at first but the brass boys are not having any of that. They blow harder, call in the drums and all come galloping over to rescue the orphan.

By now the poor little original tune is so chewed-up and bent out of shape that nobody wants it. There's not a whole note left, not even a half-note, only a bunch of 16th and 32nds flying around the hall. You pick up a pile of the pieces, put 'em in a hat and shake 'em, and brother, you've got a handful of Opus 49, Fuge 17, Box 214, Teaneck, N. J.

I often wondered what the discussions of music between Bradford and my grandmother must have been like.

What was amazing about the radio broadcasts was the audience. Truck drivers, factory workers, housewives and high-school students wrote to Bradford to say how much they liked it, and that they had learned to treasure Texaco's Saturday afternoon broadcasts from the Met. The scripts were collected into a book with the same name as the radio show, published in 1955. His silly program did a lot to bring music appreciation to the American public in the 1950s

This raises the question as to what grand opera is about anyway. It is certainly one of the strangest forms of entertainment. And if you think European/American opera is strange, you need to see Chinese Opera—that will open whole new vistas of strangeness. In the early days (1600s) European opera was a play with music and songs which became known as arias, rather like the modern 'musical.' Then by the end of the 1700s there were almost no spoken parts anymore. The music adds to the drama of the play and the music and plot become intertwined. Then came Richard Wagner, and his music carries with it ideas or memories that are not in any of the arias.

Then came the motion picture, using Wagner's idea of music in the background conveying ideas, memories and emotions. A whole new *genre*, 'film music' was created. Have you ever seen a film without music? Now we often have recorded music as a background while we work. How unnatural can you get?

As my old friend Greg Wray, whose illustrations you see in this book, says: "Humans are really strange critters."

THE DEVIL INVENTS MATHEMATICS

# The Language of Science

*If I were again beginning my studies, I would follow the advice of Plato and start with mathematics.*
Galileo Galilei

In this chapter we review the scientific terminology and the mathematics used in this book. Climate change is an incredibly complex topic involving many if not all different fields of science. This book discusses many aspects of science and their relation to climate change may not be immediately obvious. Many topics will creep back in later in an unexpected way. Science involves factual knowledge; concrete facts are often expressed and manipulated as numbers; the language of science is mathematics.

Although this book should be of interest to just about anyone with an interest in science, and the history of science, it is my assumption is that the reader will be an American. By "American" I mean a resident of the United States of America. There are lots of other Americans both to the north and south of the borders of the United States and some of them don't appreciate our usurpation of the term. As you will see, Americans are the only people on Earth stuck with using an archaic "English" system of weights and measures. We call it the 'English' or (rarely) 'American' system. It was also used throughout the British Commonwealth (formerly the British Empire) until the latter part of the 20th century; they called it the 'Imperial' system.

All other people—even the English—use the much simpler 'metric system.' As a result, many Americans treat science as some sort of bizarre field of inquiry based on a foreign language. The language of science is divorced from everyday use in the United States. That undoubtedly has a lot to do with the fact that U.S. students' science scores rank at or near the bottom of those of the industrialized countries. It also helps explain why more than half the members of the US Congress prefers to accept religious dogma over scientific facts when making decisions about climate change.

You were probably taught much of what is in this chapter in grade school or high school, but you may not have clearly understood some of it, and chances are that you have forgotten a lot of it. I suggest that you at least skim over it if you don't want to read it all now, and you can always come back to these topics later. However, I should point out that this chapter does include a lot of background information that you likely did not learn in school. It is especially useful for stimulating conversation at cocktail parties or getting to know people at the local bar.

A note about dates cited in this book: Here I use CE = Common Era (or Christian Era) rather than AD (Anno Domini) with its religious connotation and BCE = Before Common Era = BC of Christians.

Aristotle (384–322 BCE) defined mathematics as "the science of quantity," but it is much more than that; it is an adventure for the curious mind. It requires the rigorous application of logic to reach conclusions. It does not allow for "miracles," they are the property of religions.

In his 1992 book *Planet Earth* Cesare Emiliani noted that Americans have a singular aversion to mathematics. He suggested that this might be because many Americans consider themselves to be very religious Christians. There is only the simplest of arithmetic in the Bible and not much of even that. Since there is no math in the Bible, Cesare conjectured that God did not invent it. If God did not invent it, he concluded, it must be the work of the Devil!

## 2.1 Numbers and Symbols

As the universal language of science: mathematics involves the rigorous use of numerals, symbols, and concepts based on the application of logic to understand phenomena and make predictions. The simplest forms of math are called "arithmetic" including addition, subtraction, multiplication, and division. From these simple calculations developed the more complex equations we refer to as "algebra." Then followed the development of the more sophisticated technique of exploring changes in rates, termed "calculus."

At the beginning of the 20th century it was thought that the Greeks had "invented" mathematics in about 700 BCE.

© Springer International Publishing Switzerland 2016
W.W. Hay, *Experimenting on a Small Planet*, DOI 10.1007/978-3-319-27404-1_2

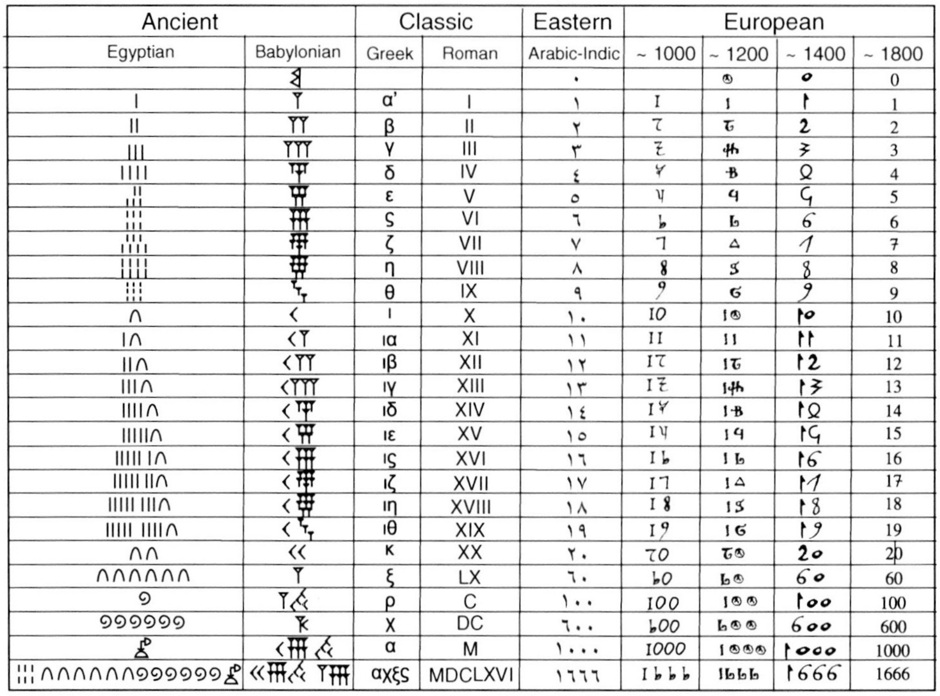

| Ancient | | Classic | | Eastern | European | | | |
|---|---|---|---|---|---|---|---|---|
| Egyptian | Babylonian | Greek | Roman | Arabic-Indic | ~ 1000 | ~ 1200 | ~ 1400 | ~ 1800 |
|  | 𒐕 |  |  | · |  | ◉ | o | 0 |
| I | 𒁹 | α' | I | ١ | I | ı | ı | 1 |
| II | 𒁹𒁹 | β | II | ٢ | 7 | ʊ | 2 | 2 |
| III | 𒁹𒁹𒁹 | γ | III | ٣ | 7̃ | ᵬ | ʒ | 3 |
| IIII | 𒐝 | δ | IV | ٤ | ४ | ᴃ | Ω | 4 |
| ıı ·ıı | 𒐝 | ε | V | o | ५ | ۹ | ɕ | 5 |
| ·ı· ı·ı | 𒐞 | ϛ | VI | ٦ | Ь | ᴦ | 6 | 6 |
| ·ıı· ·ıı | 𒐟 | ζ | VII | ٧ | ٦ | ⋏ | 1 | 7 |
| ıııı ıııı | 𒐠 | η | VIII | ٨ | 8 | ᴣ | 8 | 8 |
| ıı ııı | 𒐡 | θ | IX | ٩ | 9 | 6 | 9 | 9 |
| ∩ | 𒌋 | ι | X | ١٠ | 10 | 1◉ | ɩo | 10 |
| I∩ | 𒌋𒁹 | ια | XI | ١١ | 11 | 11 | ɩı | 11 |
| II∩ | 𒌋𒁹𒁹 | ιβ | XII | ١٢ | 1ʊ | 1ʊ | ɩ2 | 12 |
| III∩ | 𒌋𒁹𒁹𒁹 | ιγ | XIII | ١٣ | 1ʒ | 1ᴃ | ɩʒ | 13 |
| IIII∩ | 𒌋𒐝 | ιδ | XIV | ١٤ | 1४ | 1ᴃ | ɩΩ | 14 |
| IIIII∩ | 𒌋𒐝 | ιε | XV | ١٥ | 1५ | 1۹ | ɩɕ | 15 |
| IIIII I∩ | 𒌋𒐞 | ιϛ | XVI | ١٦ | 1Ь | 1ᴦ | ɩ6 | 16 |
| IIIII II∩ | 𒌋𒐟 | ιζ | XVII | ١٧ | 1٦ | 1⋏ | ɩ1 | 17 |
| IIIII III∩ | 𒌋𒐠 | ιη | XVIII | ١٨ | 18 | 1ᴣ | ɩ8 | 18 |
| IIIII IIII∩ | 𒌋𒐡 | ιθ | XIX | ١٩ | 19 | 16 | ɩ9 | 19 |
| ∩∩ | 𒌋𒌋 | κ | XX | ٢٠ | 70 | ʊ◉ | 2o | 20 |
| ∩∩∩∩∩∩ | 𒁹 | ξ | LX | ٦٠ | ᴦ0 | Ь◉ | 6o | 60 |
| ꝯ | 𒁹𒌍 | ρ | C | ١·· | 100 | 1◉◉ | ɩoo | 100 |
| ꝯꝯꝯꝯꝯꝯ | 𒌍 | χ | DC | ٦·· | Ь00 | Ь◉◉ | 6oo | 600 |
| ꝏ | 𒁹𒐠𒌍 | α | M | ١··· | 1000 | 1◉◉◉ | ɩooo | 1000 |
| ıı∩∩∩∩∩∩ꝯꝯꝯꝯꝯꝏ | 𒌋𒌋𒐠𒌍 𒁹𒐠 | αχξϛ | MDCLXVI | ١٦٦٦ | I Ь Ь Ь | 1Ьᴦᴦ | ɩ666 | 1666 |

**Fig. 2.1** Numerals used by different peoples at different times. The Egyptian system is derived from a more ancient system of simply scratching lines. Roman numerals are still used in special contexts. The Arabic-Indic numerals are used in a region extending from Egypt to India. The European numerals are commonly called "Arabic numerals" but they differ from those used by modern-day Arabs. They were derived from Hindu numerals via a circuitous route from Persia to Spain and finally into the rest of Europe at the beginning of the 12th century. Shapes of the numerals used in Europe did not become stabilized until the 18th century. The overall trend is for the symbols representing numbers to become simpler with time, so that it occupies less space and less time is spent writing the number

However, we now know that the Greeks' knowledge of math was derived from much older Egyptian and Mesopotamian sources.

The idea of using numbers as representations of objects seems to have originated at least 25,000 years ago. Two different systems evolved, both still in use today. One system, probably originating in tropical Africa and developed in Egypt was based on the number 10, probably because we have 10 fingers. This "decimal" system is the basis of modern arithmetic calculations. You might think that there is only one way of doing arithmetic. That is probably true for addition and subtraction as long as no fractions are involved. But multiplication and division are more complicated, and a number of methods were developed by different cultures. Luckily, Egyptian practical mathematics comes down to us in a couple of workbooks presenting problems and solutions for students, written in hieratic (the written form of hieroglyphics) on fragile papyrus dating from the Egyptian Middle Kingdom around 2000–1800 BCE. These are the most ancient textbooks of mathematics still in existence. The arithmetic operations of addition, subtraction and multiplication are easily handled. The problem comes with division. The Egyptian method seems incredibly complicated. Basically it involves starting with a guesstimate of the answer and then developing a series of approximations to narrow down the error. It consists of finding all the numbers that are not the right answer, until the one finally left is the correct answer or as close to it as one can get given the time allotted to solve the problem. A sample of Egyptian numbers written in hieroglyphics is shown in Fig. 2.1. Note that they wrote the larger numbers backwards from the way we do. If you enjoy mathematics you will find George Joseph's book on the non-European roots of mathematics, *The Crest of the Peacock*, an excellent read.

The other system, based on the number 60—the 'sexagesimal' system—originated with the Sumerians in Mesopotamia, the region along the Tigris and Euphrates Rivers some 6,000 years ago. It is best known from Babylonian mathematics. The Babylonians succeeded the Sumerians in that area about 3,800 years ago. We know a lot about Babylonian mathematics because they wrote in easily decipherable cuneiform numbers on clay tablets which don't readily deteriorate with age, as does papyrus. Many thousands of these clay tablets have been preserved. A sample of Babylonian numbers is shown in the second column of Fig. 2.1.

Although it has been suggested that the base 60 system might have originally had an astronomical origin related to an approximation of the number of days in a year it is more likely that it has to do with the problem of fractions in making calculations involving division. The base 10 is divisible by 1, 2, and 5, but the base 60 of the Babylonian system is divisible by 1, 2, 3, 4, 5, 6, 10, 12, 15, 20, and 30 avoiding many of the fractions that can make calculations of division troublesome. It is also easier to do many other mathematical operations in the "Babylonian" sexagesimal system. They were able to determine areas of land, calculate compound interest, and take square roots of numbers. Today we use the Babylonian system for time, geometry, and geography. A cuneiform tablet (YBC 7289) in the collections of Yale University, dated to between 1,900 and 1,700 BCE, gives the square root of 2 (in modern terminology) as 1.41421296 (Fig. 2.2); the modern value is 1.41421356…. Not bad.

Why would the Babylonians (or anyone else for that matter) want to know the square root of 2 or any other complicated number for that matter? It was important to them in measuring areas of land and determining the lengths of the sides of triangles in constructing buildings. Today not only land surveyors, but, engineers, architects, construction workers, carpenters, artists, and designers use square roots in any kind of work that involves triangles. In finance, the square root of time is used in determining the volatility of market prices. Your Hedge Fund manager should know all about this.

Figure 2.1 compares numerals used by different cultures. The Egyptian hieroglyphic numerals, similar to Hebrew numerals (not shown), are written with the smallest unit on the left and the largest on the right—opposite to the convention used in the other numeral systems shown. Note that the Egyptians, Greeks, and Romans and the Europeans during the Dark Ages did not have a numeral symbol for zero. Instead of a symbol, they simply inserted an empty space in the list of numerals making up the number. This could lead to problems if you didn't notice the space. The numeral zero was introduced to Europe in 1202 in the book *Liber Abaci* by Leonardo Bonacci, also known as Leonardo of Pisa or as Fibonacci (from fils Bonacci, meaning son of Bonacci). He had been born in North Africa and his numerals, derived from those used in India and Persia became known as "Arabic Numerals." The European versions of these Arabic numerals that are in use across the world today are a little more than 200 years old. Ironically, the numerals commonly used in most of the Arab world today are quite different from these "Arabic" symbols.

Although the Babylonians had a symbol for zero, they recognized that it was a very strange number. Add or subtract zero from another number and the answer is the other number. Multiply zero by a number and the answer is zero. Division by zero was long a mystery. Any number divided by zero has the same answer: infinity (beyond the finite numbers). It took centuries to figure this mystery out, eventually resulting in a special symbol for infinity: ∞.

One evening I was seated next to Buckminster Fuller, the famous architect, inventor of the geodesic dome, and all-round thinker/philosopher, at a dinner party. I was able to

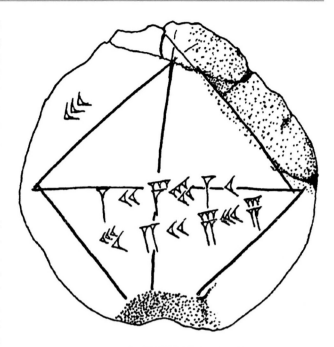

**Fig. 2.2** Babylonian Tablet YBC 7289 showing the number for the square root of 2 ($\sqrt{2}$) in the context of a square with two diagonals connecting opposite corners, forming two 'wedges'. More than a 1000 years later this would be transformed into a simple *triangle* as Pythagoras theorem that the sum of the squares of the two sides adjacent to the right angle is equal to the square of the hypotenuse. (Discussed in detail in the section of the next chapter: Trigonometry.)

ask him questions that had troubled me since I was a child. "Why is a right angle 90°, not 100°?" His answer was my introduction to Babylonian mathematics: "Because it's a quarter of the way around a circle. The circle has 360° because that was an old approximation for the number of days in a year. Incidentally, the Babylonians liked to use 60 as the base for a numbering system, so that's why there are 60 s in a minute, 60 min in a degree (or hour) etc. The 'second' is quite ancient, based on the human pulse rate (normal = 60 per minute). We carry a clock around internally. So the Babylonians favored multiples of 60 for different things." My next question to Bucky was even more profound. "Why do we build rooms as rectangles, with square corners?" Answer: "I think it's a carryover from making the angle between the wall and the floor 90°. People found out early on that if that angle is not 90° the wall falls over." My question and his answer were not really as stupid as they sound. There are some ancient round buildings shaped like a dome that avoided these rules, but Bucky Fuller was the first person to really throw these ancient ideas out the window and start out with new geometric rules for architecture. In his structures, such as the geodesic domes, the walls are not 90° from the horizontal, and they don't fall over. Bucky coined the term 'spaceship Earth' and he thought we ought to accept responsibility for keeping it in working order. If you want a mind-bending experience read one of his books entitled '*Synergetics*.'

## 2.2  Arithmetic, Algebra, and Calculus

The derivations and Oxford English Dictionary (OED) definitions of these terms are as follows:

Arithmetic (from Latin *arithmetica*): the science of numbers; the art of computation by figures. Americans learn this in grade school and most find it fun.

Algebra: [from the Arabic *al jabr*, meaning "reunion of broken parts". Introduced ca. 825 by Baghdad mathematician Abu Ja'far Muhammad ibn Musa al-Khwarizmi in the title of his book on equations *Al Kitab Fi Hisab Al Jabr Wa Al Muqabala* الكتاب الْمُختصر في حساب الْجبر وَ الْمُقابلة, (The Compendious Book on Calculation by Completion and Balancing—The Science of Transposition and Cancelation).] It was al-Khwarizmi who introduced "Arabic" numerals to the West. It was originally defined as the branch of mathematics in which letters are used to represent numbers in formulae and equations, but later defined as that field of mathematics in which symbols are used to represent quantities, relations, operations, and other concepts, and operations may be applied only a finite number of times. Taught in the United States in High School, algebra often leaves students apprehensive about mathematics.

Calculus: (from Latin calculus meaning "reckoning or account.") The methods were developed independently late in the 17th century by Gottfried Wilhelm Leibniz (1646–1716) and Isaac Newton (1642–1727) for very different purposes; Leibniz used calculus as a mathematical exercise used among other things to determine the ever-changing curvatures used in Baroque architecture; Newton used it to investigate physics. The OED defines it as a system or method of calculation, 'a certain way of performing mathematical investigations and resolutions;' a branch of mathematics involving or leading to calculations, as the differential or integral calculus, etc. The differential calculus is often spoken of as 'the calculus.'

Calculus is taught in the United Sates in a few High Schools but mostly to College and University students planning careers in Science or Engineering. Mention of the word strikes terror into the hearts of most Americans. Never mention this word at a cocktail party unless you want to drink alone. On the other hand, Russians find it fun (Fig. 2.3).

**Fig. 2.3** Francesco Borromini's baroque church, the Oratorio dei Filippini, built between 1637 and 1650 in Rome. The façade is intended to suggest a human with outstretched welcoming arms. An early example of baroque architecture with the kinds of curves that would later be refined and elaborated by architects using Leibniz's calculus

**Fig. 2.4** Interior of the pilgrimage church Vierzehnheiligen, built by architect Balthasar Neumann between 1743 and 1772. Calculus was used to compute the elaborate curves

You can immediately recognize calculus by the symbols it uses, originally suggested by Leibnitz:

In differential calculus:

$$\frac{dy}{dt} = z$$

Meaning the change in y ($dy$) over the change in time ($dt$) is $z$

In integral calculus:

$$\int_{0}^{\infty} z$$

Meaning that $z$ is the total amount of a continuous quantity taken between the limits $x$ (0) and $y$ ($\infty$). Leibniz's original differential and integral signs, M and $\int$ are a reflection of the beautiful curvatures of baroque design. One of the Baroque churches that required calculus for its design is the pilgrimage church of Vierzehnheiligen, near Bamberg, Germany (Fig. 2.4).

Leibniz remarked that "Music is a hidden arithmetic exercise of the soul, which does not know that it is counting." Much of the music of baroque composer Johann Sebastian Bach (1685–1750) is like the mathematical progressions of calculus expressed in sound.

## 2.3 Orders of Magnitude and Exponents

The numbers used in discussions about the Earth can get very large. So, rather than writing them out completely, scientists often denote them as small numbers multiplied by some power of ten. The power of ten is the number of zeros that come after the first numeral(s) in the number or before the decimal point in the number. For example, 10 can be written $1 \times 10^{1}$, but that actually takes up more space. This system becomes more useful when the number is very large. Written out, one trillion is 1,000,000,000,000. In scientific notation it is $1 \times 10^{12}$ (with 12 designating the number of zeros that come after the 1). Using exponents saves space and means the reader does not need to count the number of digits.

Incidentally, humans inherently have a hard time with numbers as large as ten. A good friend, Lev Ropes, who specializes in designing presentations at scientific meetings, explained to me that humans generally have difficulty with numbers past seven. We can remember seven items easily enough, but things get difficult after that. This is probably why there are 7 days in a week. Seven is also a special number in Judeo-Christian religion, with seven candles on the Menorah, seven deadly sins, seven seals and seven trumpets in the book of Revelation, and several other sevens. You may have also noticed that after the first 8 months of the year, January through August, the rest are named for numbers, albeit starting with seven (September). I was surprised to learn that we have a hard time keeping more than seven colors straight. A person looking at a diagram with ten colors will spend most of their time trying to figure out how many colors there are rather than understanding what the diagram is meant to show.

Scientists, particularly geologists and astronomers, deal with large numbers. Geologists' numbers refer to time; astronomers' numbers refer to distances—which, incidentally, are measured by the length of time light takes to cover a distance (light-years). Chemists and physicists deal with both very large and very small numbers.

Powers of ten can be expressed with words. $1 \times 10^1$ is ten or a "deca" (as in decade), $1 \times 10^2$ is a hundred or "hecto," $1 \times 10^3$ is a thousand or "kilo." Names for larger powers represent increases of three decimal places. Thus, $1 \times 10^6$ is a million or "mega," $1 \times 10^9$ is a billion or "giga," $1 \times 10^{12}$ is a trillion or "tera," $1 \times 10^{15}$ is a quadrillion or "peta," $1 \times 10^{18}$ is a quintillion or "exa," $1 \times 10^{21}$ is a sextillion or "zeta," $1 \times 10^{24}$ is a septillion or "yotta." The latter terms have yet to find much use. If you know a bit of Latin and understand the prefixes used you will realize that these 'illion' terms used in the United States make no sense. A quintillion simply has three more zeros than a quadrillion. The "zillion" you sometimes hear has no specific meaning. It is simply slang used to mean a large number. Table 2.1 shows the terms in common use.

On the other hand, numbers that are used to describe atoms can get very small. So, going in the other direction the base 10 scale, $10^0$ is 1, $10^{-3}$ is a "milli," $10^{-6}$ is a "micro," $10^{-9}$ is a "nano," and $10^{-12}$ is a "pico." One thousandth of a meter is a millimeter (mm). The limit of visual acuity of the unaided human eye is about 0.1 mm. One millionth of a

meter is a micrometer or micron ($\mu$). The limit of resolution of a light microscope is about 1/4 $\mu$m. One billionth of a meter is a nanometer (nm). The wavelengths of visible light are between 400 and 700 nm. (The wavelength ($\lambda$) is the distance from crest to crest or trough to trough of a series of waves). An Ångström is 0.1 nm ($10^{-10}$ m). Atoms are a few Ångströms in diameter.

Because we have a hard time visualizing large numbers, they may tend to blur, and a trillion may not sound much different from a billion. Politicians seem to suffer the most from this inability to distinguish between large numbers. When I was living in Washington, D.C., in the early 1980s, many politicians were fond of using the expression: "A billion here, a billion there, before you know it, it begins to add up to real money." Members of Congress seem to have mastered the transition from a billion to a trillion with no effort at all. Somehow the US national debt (as of February 3, 2015) of $18,091,422,494,095.17 seems less formidable when rounded off and written as $18.091 $\times 10^{12}$, but it is just as large. The daily increase in debt of $2,370,000,000 looks trivial when written as $2.4 $\times 10^9$.

The three most commonly used measurement terms in climate studies are "hectopascals" (often incorrectly spelled "hectapascals") to describe atmospheric pressure (replacing the older term millibars); gigaton, equal to 1 billion ($10^9$) metric tons, useful in discussing the amounts of material used by humans, and petawatt, equal to 1 quadrillion ($10^{15}$) watts, used in discussions of Earth's energy budget.

Countries other than the United States have different terms for numbers larger than a million. In Great Britain, a billion used to be the same as a U.S. trillion, but the British meaning has recently been changed to conform to the US meaning. Outside the United States the prefix "bi" means twice as many zeros as in a million; "tri" means three times as many zeros, and so on. A European trillion is an American quintillion. Sooner or later US politicians will discover they can make budget deficits sound much smaller by using the European system, just as UK politicians have realized they can make investment in public services sound bigger by using the American system!

Curiously, the otherwise logical scientific community uses the US system of nomenclature for large numbers.

**Table 2.1** Commonly expressions for orders of magnitude used in the Earth sciences

| Number | Scientific notation | Name | Prefix | Number | Scientific notation | Name | Prefix |
|---|---|---|---|---|---|---|---|
| 10 | $1 \times 10^1$ or $1 \times 10\wedge 1$ | Ten | deca- | 1 | $1 \times 10^0$ or $1 \times 10\wedge 0$ | One | – |
| 100 | $1 \times 10^2$ or $1 \times 10\wedge 2$ | Hundred | hecto- | 0.1 | $1 \times 10^{-1}$ or $1 \times 10\wedge{-1}$ | Tenth | – |
| 1,000 | $1 \times 10^3$ or $1 \times 10\wedge 3$ | Thousand | kilo- | 0.01 | $1 \times 10^{-2}$ or $1 \times 10\wedge{-2}$ | Hundredth | – |
| 1,000,000 | $1 \times 10^6$ or $1 \times 10\wedge 6$ | Million | mega- | 0.001 | $1 \times 10^{-3}$ or $1 \times 10\wedge{-3}$ | Thousandth | milli- |
| 1,000,000,000 | $1 \times 10^9$ or $1 \times 10\wedge 9$ | Billion | giga- | 0.000001 | $1 \times 10^{-6}$ or $1 \times 10\wedge{-6}$ | Millionth | micro- |
| 1,000,000,000,000 | $1 \times 10^{12}$ or $1 \times 10\wedge 12$ | Trillion | tera- | 0.000000001 | $1 \times 10^{-9}$ or $1 \times 10\wedge{-9}$ | Billionth | nano- |
| 1,000,000,000,000,000 | $1 \times 10^{15}$ or $1 \times 10\wedge 15$ | Quadrillion | peta- | 0.000000000001 | $1 \times 10^{-12}$ or $1 \times 10\wedge{-12}$ | Trillionth | pico- |

However, to avoid confusion between the two systems, the numbers are often expressed simply and clearly by denoting the power of ten.

Incidentally, as discussed below, the exponent on the 10 is also called the logarithm to the base 10 ($\log_{10}$). So, for example, the logarithm to base 10 for 1 billion is 9. To multiply numbers, you simply add the logarithms or exponents to the base 10. For one billion times one billion, the calculation would be $9 + 9 = 18$, giving you $10^{18}$.

To get a better grasp of large numbers you might want to consider planning to rob a bank. You plan to escape by driving immediately to the airport and hopping on a commercial flight. If you ask for $1,000,000 in $20 bills, how long would it take the cashier to count them out, and could you fit them in a bag you could carry? A good cashier could probably count one bill per second, so it would take 50,000 s. That is almost 14 h. You might miss your flight. The bills would take up a volume of a bit more than 27 ft$^3$. You might already anticipate that this is not going to fit into the overhead bin on the airplane nor under the seat in front of you. In fact an ordinary suitcase holds less than 3 ft$^3$ so you would need 10 suitcases, and the airline will only accept two.

Alternatively, you could ask for $100 bills, but have you ever tried to cash one? It would be more reasonable to ask for gold. At current prices, $1 \times 10^6$ is equal to 91 pounds of gold. This would fit neatly into a sturdy briefcase, but security might want to open it to make sure it's not uranium or plutonium for a bomb. If you try to check it through, the airline will want to charge for overweight luggage, and you can bet it will get lost. At this point you will probably have concluded that it would be easier to ask the bank to make a wire transfer of the money to your Swiss bank account.

One symbol you will see in discussions about the ocean is ‰. You already know that. % means percent, or parts in a hundred. The ‰ is read as per mille, meaning parts in a thousand. It is used in discussions of the salinity or saltiness of the ocean in which a kilogram (1,000 g = 2.20462262 lbs) of normal seawater contains about 34.5 g (=1.21695 oz) of salt. Seawater's average salinity is said to be 34.5 ‰. Note that you will get the same 34.5 ‰ if you convert ounces to pounds and divide that number by the weight of a kilogram expressed in pounds. The metric system seems much simpler, doesn't it?

By the way, if you are an aficionado of gin, like me, you are probably familiar with Hendrix Gin, which the makers claim to be preferred by 1 person in a thousand or 1‰ of gin drinkers. It's the cucumber that makes it special.

Here is a practical example of the importance of orders of magnitude: When I was a graduate student at Stanford University near San Francisco several of us had rented a house in Palo Alto, and we passed around cooking and refreshment duties. One member of our group was a chemistry graduate student, who decided it would be a great idea to make that forbidden drink, absinthe. Absinthe contains wormwood and a supposedly psychoactive agent. It was famously consumed by many avant-garde painters and writers in the late 19th and 20th centuries, and social conservatives of the time blamed it for the progressive ideas of their adversaries. The drink was, until quite recently, banned in the United States. Our chemist had found a recipe for absinthe somewhere, and all we needed were grain alcohol and the necessary herbs. San Francisco has a number of herbalist shops, so we headed up to the big city. We selected the herbalist who had cat skins on display in his window, thinking he would certainly have everything we might need. Rather secretively our 'chemist' read off his list of items and their required amounts, one by one. The herbalist weighed out each item; all of them dried herbs, and poured them into envelopes. Then we got to wormwood, the crucial ingredient. The herbalist looked at us sternly and said "that's a hundred times more than you need." He seemed to have divined what we were up to. Our chemist replied "Well that's what our recipe calls for." "OK," said the herbalist after a bit of thought. "I'll sell it to you, but it's way, way too much."

We took our ingredients back to our house, and we carefully mixed them all up with the alcohol, according to the recipe. I think we allowed it to age about a week. Then came the test. One drop on the tongue and you couldn't taste anything else for hours. Wormwood has one of the bitterest tastes imaginable. Apparently we had made a mistake of about two orders of magnitude. Even diluted, it was awful. Incidentally, wormwood is not only one of the ingredients of absinthe, it is the flavoring in vermouth, though in concentrations many orders of magnitude less than what was in our concoction.

Orders of magnitude become very important when we think about time, especially the vastness of geologic time. When trying to imagine geologic changes on the surface of Earth we are talking about long, long, long periods of time. Unimaginable lengths of time. Millions of years. Geologists have adopted several abbreviations to save words. A thousand years is designated by ky, kyr, Kyr, and ka for Kiloyears (1,000 years). A million years can be designated by any of the following abbreviations: my, myr, or ma. Ma stands for millions of years ago. A billion years is designated by Ga.

## 2.4 Logarithms

In the discussion above you saw how exponents of the number 10 can be used to indicate the number of times to move the decimal point to the left or right. They indicate the number of times you have to multiply a number by itself to get a given value. You read '$10^3$' as '10 to the third power.' They save a lot of space in writing out very large or very small numbers and make comparisons of numbers much easier.

However, there is another way of doing the same thing, through the use of 'logarithms.' Two scholars, Scottish Baron John Napier (1550–1617) and Swiss clockmaker Joost Bürgi (1552–1632), independently developed systems that embodied the logarithmic relations and, within years of each other, published tables for their use. But Napier published first, in 1614, and Bürgi published in 1620, so Napier gets the credit for inventing them. Also, Bürgi forgot to include instructions for the use of his tables of logarithms; they were published later.

A logarithm is the power to which a standard number, such as 10, must be raised to produce a given number. The standard number on which the logarithm is based is called, of all things, the base. Thus, in the simple equation $n^x = a$, the logarithm of a, having n as its base, is x. This could be written $\log_n a = x$. As an example from the preceding section about orders of magnitude, that equation might be $10^3 = 1,000$; written the other way it becomes $\log_{10} 1,000 = 3$. With logarithms to the base 10, simple orders of magnitude are very easy to figure out. What is not so easy to figure out is what the $\log_{10}$ of 42 would be. (You may recall from Douglas Adams' *The Hitchhiker's Guide to the Galaxy* that 42 is the answer to the question 'what is the meaning of life, the Universe, and everything?') To determine $\log_{10}$ 42 you need to do some real work. When I was a student, you could buy books of tables of logarithms to the base 10, and look it up. So there. Now most calculators and computer spreadsheet programs, such as Excel, will do this for you. For your edification $\log_{10}$ 42 is 1.623249. Then you will ask: why would anyone want to convert the numbers to logarithms anyway? A very simple reason: multiplying two numbers together is the same as adding their logarithms together. Better yet, dividing one number by another is the same as subtracting the logarithm of the second number from that of the first. Before the days of electronic calculators and computers, this was a very efficient way of multiplying and dividing large numbers, but most of us had to have a book of tables to do it.

Logarithms have crept into the way we describe a number of things. Logarithms to the base 10 are called 'common logarithms' and you already know how they are used. You may be aware that the acidity or alkalinity of a solution is described by its 'pH' which is a number ranging from 0 to 10. Now for the bad news: pH is the *negative* logarithm to the base 10 of the concentration of the hydrogen ion ($H^+$) in the solution. This means that a small number means there are lots of $H^+$ ions in the solution, and a large number means

that there are only a few. How's that for convoluted logic? Now for the good news, it should make sense later in the book when protons and electrons are discussed.

## 2.5  Logarithms and Scales with Bases Other than 10

Some other very useful logarithms don't use 10 as their base. Computers depend on electrical circuits that are either open or closed—two possibilities. So computers use a system with base 2, the 'binary logarithm.'

Then there is a magical number that is the base of a logarithm that will allow you to calculate compound interest, or how much of a radioactive element will be left after a given time or anything that grows or decays away with time. This base is an 'irrational number;' that is, it has an infinite number of decimal places. It starts out 2.7182818128459045... and goes on forever; it has a special name in mathematics, it is '$e$'. Its discovery is usually attributed to Swiss mathematician Leonhard Euler in the 18th century, but it also appears, figured out to several decimal places, on Babylonian clay tablets. Logarithms with $e$ as their base are called 'natural logarithms.' The uses of these magical logarithms will be discussed in later chapters. It comes back up again in the discussions of radioactive decay and the impact of the growth of the human population on our small planet. Incidentally, the repetition of 1828 at the beginning of the number has struck mathematicians as something very odd and raised the question whether other parts sequence might repeat at some point. But it has now been calculated to 869,894,101 decimal places and so far there have been no further significant repetitions.

## 2.6  Earthquake Scales

A good example is of the use of logarithms with a base other than 10 is when describing the intensity of earthquakes. This has recently caused considerable confusion in the press.

Earthquakes are events in which the ground moves. They were rather mysterious until the 1970s when it was realized that the surface of the Earth is constructed of large rigid plates that move with respect to each other. Most large earthquakes occur along the boundaries where these plates converge as one is forced down beneath the other, or in geological terminology, 'subducted.' Recent large earthquakes are shown in Fig. 2.5.

There are currently three different scales used to describe the intensity of earthquakes; two of these scales are logarithmic. In 2010 confusion about these scales was rampant as journalists tried to compare the January 10 earthquake in

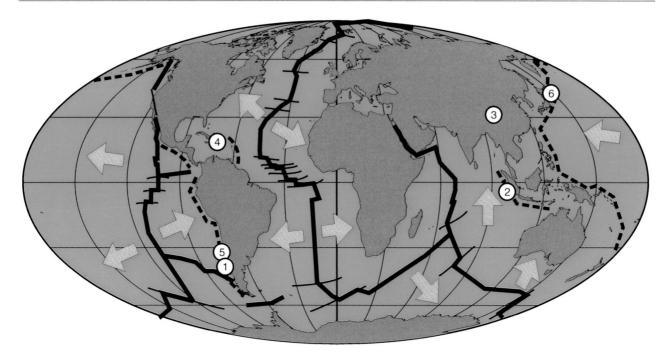

**Fig. 2.5** Major plates of Earth's crust. The plates grow along the mid-ocean ridge system (*heavy black lines*) which is sometimes offset by 'transform faults' (*thin black lines*) where the plates slide past each other. Most large earthquakes occur where the plates converge (*heavy dashed lines*) and one is forced beneath the other, a condition geologists call 'subduction.' A few earthquakes occur due to stresses within the plates ('intra-plate' earthquakes). The sites of some recent major earthquakes are shown ($M_W$ = Moment Magnitude Scale of earthquake intensity): *1* Valdivia, Chile, May 22, 1960, subduction, $M_W$ = 9.5, deaths = 5,700 (est.); *2* Sumatra, December 26, 2004, subduction, $M_W$ = 9.2, deaths = 227,898, mostly by drowning from the ensuing tsunami; *3* Eastern Sichuan, China, May 12, 2008, intra-plate, $M_W$ = 7.9, deaths = 87,587; *4* Haiti, January 12, 2010, $M_W$ = 7.0, transform fault, deaths = 222,570; *5* Maule, Chile, February 27, 2010, subduction, $M_W$ = 8.8, deaths = 577; *6* Tohoku, Japan, March 11, 2011, subduction, $M_W$ = 9.0, more than 13,000 deaths, mostly by drowning from the ensuing tsunami

Haiti (7.0 $M_W$ transform fault), which killed over 200,000 people, with the February 27 earthquake in Chile (8.8 $M_W$ subduction) which killed less than 600 people.

The classification of intensity of earthquakes began in the latter part of the 19th century with a scale developed by Michele Stefano Conte de Rossi (1834–1898) of Italy and François-Alphonse Forel (1841–1912) of Switzerland. The Rossi-Forel scale was purely descriptive ranging from I, which could be felt by only a few people, through V, which would be felt by everyone, and might cause bells to ring, to X, which would be total disaster. It was modified and refined by the Italian volcanologist Giuseppe Mercalli (1850–1914) toward the end of the century, and a revision ranging from I to XII became widely accepted in 1902, just in time for the San Francisco earthquake; the city had some areas near the top of the scale. It has been modified a number of times since to become the 'Modified Mercalli Scale' with intensities designated MM I to MM XII based on human perceived movement of the ground and damage to structures. On this scale the Haiti earthquake ranged from MM XII at the epicenter near the town of Léogâne and MM VII in Port-au-Prince about 16 miles (25 km) to the east. The MM scale is purely descriptive of the damage.

In 1935, the seismologist Charles Richter (1900–1985) developed a more 'scientific' scale based on the trace of the earthquake recorded in a particular kind of seismograph, the Wood-Anderson torsion seismometer. (A seismograph is an instrument that measures the motion of the ground.) Originally developed for use in southern California, Richter's scale has become known as the 'local magnitude scale' designated $M_L$. It is based on logarithms to the base 10, so that an earthquake with $M_L$ 7 is ten times as strong as one with an intensity of $M_L$ 6. Richter and his colleague Beno Gutenberg estimated that on this scale the San Francisco earthquake of 1906 had a magnitude of about $M_L$ 8. Unfortunately, the Wood-Anderson torsion seismometer doesn't work well for earthquakes of magnitude greater than $M_L$ 6.8.

In 1979, seismologists Thomas C. Hanks and Hiroo Kanamori published a new scale, based on the total amount of energy released by the earthquake. Development of this new scale was made possible by more refined instruments and an increasingly dense network of seismic recording stations. The new scale, in use today, is called the 'Moment Magnitude Scale' (MMS) and is designated $M_W$. It is based on determining the size of the surface that slipped during an earthquake, and the distance of the slip. It too is a logarithmic scale, but its base is 31.6. That is, an earthquake of $M_W$ 7 is 31.6 times as strong as one of $M_W$ 6.

On this scale the Haiti earthquake was $M_W$ 7.0, and the Chile earthquake $M_W$ 8.8. Most journalists had learned about the 'Richter scale' ($M_L$) when they were in school, and had never heard of the $M_W$ scale. The result was that news reports comparing the Chile and Haiti earthquakes cited very different numbers to compare their intensity, indicating that the Chile quake was from 50 to 1,000 times more powerful. The confusion arose because of the different scales ($M_W$ versus $M_L$) used by different sources to report the magnitude of the quake. In terms of the ground motion ($M_L$) the Chile earthquake was about 70 times stronger, but in terms of the actual energy released ($M_W$) it was 500 times as large as the Haiti earthquake.

Why were the death tolls so different? Because there had been no significant earthquakes in Haiti for over a century, no special precautions had been taken in constructing homes and larger buildings. Those built on solid rock received relatively little damage, but much of the city of Port-au-Prince had been built on unconsolidated sediments which amplified the shaking and caused many structures to collapse. The Haiti earthquake was the result of slow convergence of the North Atlantic and Caribbean tectonic plates.

Then, during the afternoon of March 11, 2011, Japan suffered an $M_W$ 9 earthquake with an epicenter in the sea east of Tohoku. The earthquake was followed by a catastrophic tsunami. The death toll exceeded 13,000. It was one of the five largest earthquakes since 1900.

In contrast, there had been 13 earthquakes of magnitude $M_W$ 7 or greater since 1973 along the southeastern Pacific margin. The 2010 epicenter was located about 230 km north of the magnitude $M_W$ 9.5 Valdivia earthquake that occurred on May 22, 1960 and 870 km south of that of a magnitude $M_W$ 8.5 earthquake that occurred on November 11, 1922. All of these and the many other smaller earthquakes in this region are a result of the convergence of the South American and Nazca tectonic plates.

## 2.7 The Beaufort Wind Force Scale

In 1805 Francis Beaufort (1774–1857), an Irish hydrographer and officer in the Royal Navy, devised a scale to relate wind conditions to their effect on the sails of a frigate, ranging from "just sufficient to give steerage" to "that which no canvas sails could withstand." The scale became a standard for the Royal Navy's ship's log entries in the late 1830s. Then, in 1846, British astronomer and physicist Thomas Romney Robinson (1792–1882) built a device to accurately measure the velocity of the wind, an anemometer.

It consisted of four hemispherical cups attached to a pole, which would spin in the wind, and a device for determining the speed of rotation. The conversion from speed of rotation of the anemometer to wind speed took some time. The four-cup anemometer had the problem that when the concave cup was facing the wind on one side, the convex side of the cup was facing the cup on the other side. Nevertheless, in the 1850s the Beaufort scale was calibrated to the four-cup anemometer rotations reflecting wind speeds.

In 1906 George Clarke Simpson (1878–1965) issued his *Report of the Director of the (British) Meteorological Office upon an Inquiry into the Relations between the Estimates of Wind Force According to Admiral Beaufort's Scale and the Velocities Recorded by Anemometers Belonging to the Office.* In this Report he shows that the measured wind speeds correspond closely to the Beaufort scale using the empirical relationship:

$$v = 1.87B^{3/2} \quad \text{or} \quad B = 0.659v^{2/3} \quad \text{or} \quad B = (2/3)v^{2/3}$$

where $v$ is the wind speed in miles per hour at 33 ft above the sea surface (the average height of the deck of a frigate above the sea) and $B$ is Beaufort scale number.

In metric terms this becomes

$$v = 0.836B^{3/2} \quad \text{or} \quad B = 1.127v^{2/3}$$

where $v$ is the wind speed in meters per second at 10 m above the sea surface and $B$ is Beaufort scale number.

In 1916, most sailing ships having been retired out of service, George Simpson, Director of the British Meteorological Office, made slight changes to the scale to better reflect how the sea behaved and he also extended it to land observations. The revision was standardized in 1923 and was subsequently improved to become more useful for meteorologists and communication with the general public.

In 1926 a Canadian physicist and meteorologist, John Patterson (1872–1956), devised a three-cup anemometer. Each cup produced maximum torque when it was at 45° to the wind flow. This new anemometer also had a more constant torque and responded more quickly to gusts than the four-cup anemometer.

With adoption by many countries of the metric system, with wind speeds expressed simply in m/s or km/h, the Beaufort scale has fallen out of use except for severe weather warnings to the public. In 1946 the Beaufort scale was extended, with forces 13–17 added for special cases, such as tropical cyclones. However, today the extended scale is used solely for typhoons affecting eastern Asia and its offshore islands (Table. 2.2).

The logarithmic base for these scales is independent of whether the units are imperial or metric and has the value 3/2 or 1.5. For example, $B = 4$ gives a wind speed of 6.7 m/s, B7

**Table 2.2** The Beaufort Wind Scale, used to relate wind and sea state conditions on the open sea

| BN | Wind conditions | Maximum wind speed | | | Maximum wave height | | Wave description |
|---|---|---|---|---|---|---|---|
| | | Knots | mph | m/s | Ft | m | |
| 0 | Calm | 0.6 | 0.7 | 0.3 | 0 | 0 | Flat |
| 1 | Light air | 3.0 | 3.4 | 1.5 | 1 | 0.2 | Ripples w/o crests |
| 2 | Light breeze | 6.4 | 7.4 | 3.3 | 2 | 0.5 | Small wavelets |
| 3 | Gentle breeze | 10.6 | 12.2 | 5.5 | 3.5 | 1 | Large wavelets, crests begin to break, few whitecaps |
| 4 | Moderate breeze | 15.5 | 17.9 | 8 | 6 | 2 | Small waves, breaking crests, frequent whitecaps |
| 5 | Fresh breeze | 21 | 24.1 | 10.8 | 9 | 3 | Moderate waves, many whitecaps, some spray |
| 6 | Strong breeze | 26.9 | 31 | 13.9 | 13 | 4 | Long waves, white foam crests, some airborne spray |
| 7 | Near gale | 33.4 | 38.4 | 17.2 | 19 | 5.5 | Sea heaps up, foam blown into streaks, airborne spray |
| 8 | Fresh gale | 40.3 | 46.3 | 20.7 | 25 | 7.5 | Moderately high waves forming spindrift, considerable airborne spray |
| 9 | Strong gale | 47.6 | 54.8 | 24.5 | 32 | 10 | High waves, crests sometimes roll over, dense foam, airborne spay reduces visibility |
| 10 | Storm | 55.3 | 63.6 | 28.4 | 41 | 12.5 | Very high waves with overhanging crests, foam on wave crests give sea a white appearance |
| 11 | Violent storm | 63.4 | 72.9 | 32.6 | 52 | 16 | Exceptionally high waves, very large patches of foam cover sea surface, airborne spray severely reduces visibility |
| 12 | Hurricane–typhoon | > | > | > | >52 | >16 | Huge waves, sea completely white with foam, air filled with driving spray greatly reducing visibility |
| 13 | Typhoon | 80 | 92 | 41.4 | – | – | – |
| 14 | Typhoon | 89 | 103 | 46.1 | – | – | – |
| 15 | Typhoon | 99 | 114 | 50.9 | – | – | – |
| 16 | Severe typhoon | 108 | 125 | 56.0 | – | – | – |
| 17 | Severe typhoon | 118 | 136 | 61.2 | – | – | – |

*BN* Beaufort scale number

is 15.5 m/s, B11 is 30.5 m/s—all squarely in the middle of the range for each class. Using this formula the highest wind thus far recorded in a tropical storm, a wind gust of 113.2 m/s (220 kt; 253 mph) on 10 April 1996 at 64 m above sea level on Barrow Island off Australia during Tropical Cyclone Olivia would be B26 in a projection of the scale.

## 2.8 Extending the Beaufort Scale to Cyclonic Storms

As will be discussed in detail in Chap. 24 on atmospheric circulation, cyclones are storms that develop from atmospheric low pressure systems. Responding to Earth's rotation, air at in the lower part of the atmosphere spirals in toward the center of low pressure. It forms a gyre rotating counterclockwise in the Northern Hemisphere, and clockwise in the Southern Hemisphere. Intense cyclonic storms form both in the tropics and at high latitudes.

In 1971 meteorologist Tetsuya Fujita of the University of Chicago and Allen Pearson, head of the U.S. National Severe Storms Forecast Center introduced a scale for rating the intensity of tornados in the United States. It is based primarily on the damage to different kinds of human-built structures and vegetation but includes estimates of the wind velocities. Therefore the Fujita (or Fujita-Pearson) scale is neither linear nor logarithmic with respect to the estimated wind speeds.

In 1973 wind engineer Herbert Saffir and meteorologist Robert Simpson, Director of the U.S. National Hurricane Center, published a scale (acronym: SSHS = Saffir-Simpson Hurricane Scale, or SSHWS Saffir-Simpson Hurricane Wind Scale) for alerting the public to the danger posed by a hurricane according to its intensity. They chose the maximum sustained surface wind speed [defined as the peak 1 min wind at the unobstructed standard meteorological observation height of 10 m (33 ft)] associated with the storm as the basis of the scale. The categories are based on the perceived

**Table 2.3** The TORRO tornado intensity scale

| TORRO scale (B scale equivalent) | Wind speed | | | Tornado description | Damage description |
|---|---|---|---|---|---|
| | Mph | km/h | m/s | | |
| FC | | | | Funnel cloud aloft (not a tornado) | No damage to structures, unless on tops of tallest towers, or to radiosondes, balloons, and aircraft. No damage in the country, except possibly agitation to highest tree-tops and effect on birds and smoke |
| T0 B8 | 39–54 | 61–86 | 17–24 | Light | Loose light litter raised from ground-level in spirals. Tents, marquees seriously disturbed; most exposed tiles, slates on roofs dislodged. Twigs snapped; trail visible through crops |
| T1 B10 | 55–72 | 87–115 | 25–32 | Mild | Deckchairs, small plants, heavy litter becomes airborne; minor damage to sheds. More serious dislodging of tiles, slates, chimney pots. Wooden fences flattened. Slight damage to hedges and trees |
| T2 B12 | 73–92 | 116–147 | 33–41 | Moderate | Heavy mobile homes displaced, light caravans blown over, garden sheds destroyed, garage roofs torn away, much damage to tiled roofs and chimney stacks. General damage to trees, some big branches twisted or snapped off, small trees uprooted |
| T3 B14 | 93–114 | 148–184 | 42–51 | Strong | Mobile homes overturned/badly damaged; light caravans destroyed; garages and weak outbuildings destroyed; house roof timbers considerably exposed. Some of the bigger trees snapped or uprooted |
| T4 B16 | 115–136 | 185–220 | 52–61 | Severe | Motor cars levitated. Mobile homes airborne/destroyed; sheds airborne for considerable distances; entire roofs removed from some houses; roof timbers of stronger brick or stone houses completely exposed; gable ends torn away. Numerous trees uprooted or snapped |
| T5 B18 | 137–160 | 221–259 | 62–72 | Intense | Heavy motor vehicles levitated; more serious building damage than for T4, yet house walls usually remaining; the oldest, weakest buildings may collapse completely |
| T6 B20 | 161–186 | 260–299 | 73–83 | Moderately devastating | Strongly built houses lose entire roofs and perhaps also a wall; windows broken on skyscrapers, more of the less-strong buildings collapse |
| T7 B22 | 187–212 | 300–342 | 84–95 | Strongly devastating | Wooden-frame houses wholly demolished; some walls of stone or brick houses beaten down or collapse; skyscrapers twisted; steel-framed warehouse-type constructions may buckle slightly. Locomotives thrown over. Noticeable de-barking of trees by flying debris |
| T8 B24 | 213–240 | 343–385 | 96–107 | Severely devastating | Motor cars hurled great distances. Wooden-framed houses and their contents dispersed over long distances; stone or brick houses irreparably damaged; skyscrapers badly twisted and may show a visible lean to one side; shallowly anchored highrises may be toppled; other steel-framed buildings buckled |
| T9 B26 | 241–269 | 386–432 | 108–120 | Intensely devastating | Many steel-framed buildings badly damaged; skyscrapers toppled; locomotives or trains hurled some distances. Complete debarking of any standing tree-trunks |
| T10 B28 | 270–299 | 433–482 | 121–134 | Super | Entire frame houses and similar buildings lifted bodily or completely from foundations and carried a long or large distance to disintegrate. Steel-reinforced concrete buildings may be severely damaged or almost obliterated |
| T11 B30 | >300 | >483 | >135 | Colossal | |

potential for damage so, again the scale is neither linear nor logarithmic. It is used in the United States and parts of the Caribbean.

In 1974 Terence Meaden of Oxford University in England proposed extending the Beaufort scale to include tornados, using the 3/2 Power relationship to develop a scale T0 to T10, with T0 equal to B8. He organized the Tornado and Storm Research Organization (TORRO) which has been active in documenting and cataloging storms in Britain and on the European mainland.

The relationship between the two scales is

$$B = 2(T + 4) \quad \text{and conversely}: T = (B/2 - 4).$$

Table 2.3 shows the TORRO scale and the terminology associated with it. The descriptions of damage at increasing intensity levels are in delightfully colloquial English.

## 2.9  Calendars and Time

Next to counting and keeping track of objects and wealth, the most important use of numbers in the ancient world was probably to keep track of time. The ultimate basis for time is astronomical: such as the length of a day, the phases of the Moon (the length of a month), and the length of a year. The need to track longer periods of time led to the development of calendars. In early human history the most important times to keep track of were years, seasons, and days. Smaller units of time were not needed until societies become more complex.

The most important thing to keep track of was the seasons, to be able to anticipate the changes in climate though the course of the year. The seasons are caused by the changes in energy received by different parts of Earth during the year due to the 23.5° tilt of its axis of rotation relative to the plane of its orbit around the Sun (Fig. 2.6). The hemisphere tilted towards the Sun will receive more direct sunlight and heat (photons) and therefore be warmer, the hemisphere tilted away from the Sun will receive photons spread over a larger surface area and so be cooler. Seasons are not, as some people seem to believe, caused by the varying distance of the Earth's orbit from the Sun; if this were the case then both hemispheres would experience the same season at the same time of the year rather than at opposite times of the year as they do in reality. Figure 2.6 shows how this produces the seasons and explains the meaning of the terms 'solstice' and 'equinox.'

Here it worth mentioning that the longest-lived civilization, that of the Egyptians, which lasted for about 3,500 years, only recognized three seasons, not four. At the latitude of the Sahara and Arabian deserts, the temperatures do not vary greatly throughout the year, and rainfall is a rare occurrence and not seasonal. The Egyptian calendar had a year that was 365 days long, with 12 months, each having 30 days followed by five extra days at the end of the year. Each month was divided into 3 weeks of 10 days each. The big event was the annual flooding of the Nile. The three Egyptian seasons were *Akhet* (June–September), the season of inundation by flooding of the Nile; *Peret* (October–February), the season after the floodwaters receded, leaving behind a fresh layer of rich, black soil which was then plowed, planted and crops grown, and *Shemu* (March–May), when the crops were harvested.

The source of the Nile is far to the south, about 3,500 km (2,100 miles) away, in the region of Lake Victoria, and unknown to the ancient Egyptians. Lake Victoria wasn't known to the outside world (Europeans) until its discovery by Richard Burton's companion explorer John Speke in 1858 on their expedition to find the fabled lakes of eastern Africa. The rainfall there is highest in March through May, about 3 months before the flooding of the Nile in Egypt. To the ancient Egyptians, the flooding of the Nile was a miraculous annual event.

The construction of the Aswan Dam in the 1960s has been one of the 20th century's great experiments in geo-engineering. It would generate electricity and regulate the flow of the river, but it has had unintended regional consequences. The sediment carried by the river is now trapped behind the dam. The annual flooding of Egypt with its delivery of fresh fertile mud that could be plowed into the soil has ceased. So now the farmers use fertilizers, some of which runs off into the now nearly stagnant river causing blooms of algae that kill the fish. The supply of Nile sediment that built the highly populated delta has been entirely cut off, so that coastal erosion is now a major problem. Because Lake Nasser, created by the dam, is in one of the Earth's regions of highest evaporation, with a rate of more than 2 m (about 7 ft) per year, the flow of the Nile into the Mediterranean Sea has almost ceased. Without the Nile outflow, the salinity of the Eastern Mediterranean has been increasing, severely impacting the fisheries as far away as Turkey. Geoengineering can have unexpected repercussions.

But back to calendars. Some calendars, such as the Gregorian calendar used throughout the world today, are based on the solar cycle. Others, such as the traditional Islamic calendar, are based on lunar cycles. Still others, such as the Hebrew, Hindu, and Chinese calendars are based on both. The solar and lunar cycles closely coincide every 18.6, usually rounded to 19 years (see Fig. 2.7).

The Egyptians knew about the north–south motion of the Sun through the course of the year, probably using poles and then obelisks as sundials, as discussed below and shown in Fig. 2.8.

Now we need to define what we mean by "year." The one you are probably thinking of is the 'tropical year' (or 'equinoctial year' or 'solar year'). It is defined as the time from one March equinox, when the Sun is over the Equator moving northward, to the next. The mean value is 365.2422 days. Not to alarm you, but there are other kinds of 'year' that take the slow changes in Earth's orbit into account. We will get to those later in this book; they account for the changes in our planet's climate that resulted in the ice ages.

But there is also a 'lunar year' based on the cycles of the Moon. A synodic month is the time from one new Moon to the next. It roughly coincides with the menstrual cycle of women, which averages about 28 days. The synodic month ranges from 29.27 to 29.84 days with a mean of 29.53059 days. There are 12 synodic months of 29.53 days, giving a lunar year 354.36708 days, so a lunar year of 12 lunar months is about 10.88 days shorter than a solar year. Figure 2.7 shows the difference between solar and lunar years.

Starting dates of calendars differ: the Hebrew calendar is based on their traditional date for the creation of Adam. The

**Fig. 2.6** The seasons are due to the 23.5° tilt of Earth's axis of rotation relative to the plane of its orbit around the Sun. The solstices are the times when the Sun is at its maximal northern or southern position in the sky. The equinoxes are when the Sun is over the Equator and the lengths of day and night are equal. Nomenclature is opposite in the two hemispheres. The *white dashed lines* are, from north to south: Arctic Circle (66.5° N), Tropic of Cancer (23.5° N), Equator (0°), Tropic of Capricorn (23.5° S), and Antarctic Circle (66.5° S)

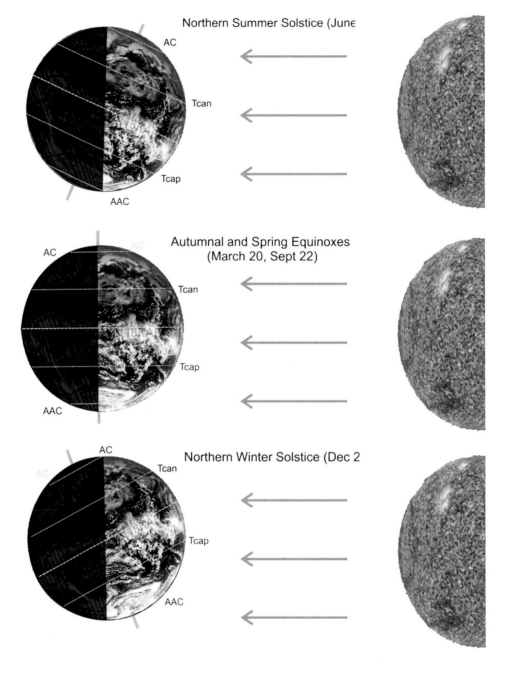

Northern Summer Solstice (June

AC

Tcan

Tcap

AAC

Autumnal and Spring Equinoxes
(March 20, Sept 22)

AC

Tcan

Tcap

AAC

Northern Winter Solstice (Dec 2

AC

Tcan

Tcap

AAC

Gregorian year 2015 is the year 5775 on the Hebrew calendar. The Hindu calendar begins with an alignment of the planets in 3102 BCE; Gregorian 2015 is Hindu 5117. The Islamic calendar starts with the Hijra (Hegira), the flight of Mohammed from Mecca to Yathrib, now known as Medina; Gregorian 2015 is Islamic 1436. But to really understand what has happened with our measurement of time we need to go back a couple of thousand years, to ancient Rome.

The Romans had a year of 355 days. The original calendar was attributed to Romulus, the (mythical) youth who founded Rome on April 21 of the year 1 AUC. AUC stands for *ab urbe condita,* meaning the time 'from the founding of the city' (Rome). 1 AUC corresponds to 753 BCE.

The original Roman calendar was divided into 10 named months: *Martius, Aprilis, Maius, Junius, Quintilis, Sextilis, September, October, November,* and *December* plus two unnamed months representing the winter (when not much happened back then). The year began on *Martius* (March) 1. Eventually, the Romans settled on names for the two winter months, inserting *Januarius* and *Februarius* before *Martius.* Now you know why December (literally the tenth month) is the twelfth month of the year. Good cocktail party stuff.

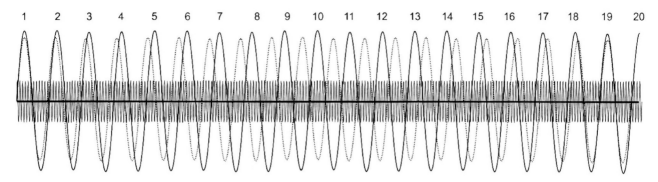

**Fig. 2.7** Comparison of tropical (solar, *solid line*), lunar (*dotted line*) years, and lunations (lunar months, the short cycles shown). The solar year is 365.242189 days; lunar year is 354.36708 days; average lunation (lunar month) is 29.53059 days. Topical and lunar years closely coincide every 18.6 years

The year of 355 days meant that the seasons gradually shifted relative to the months. So an extra month (*Intercalaris*) was inserted from time to time to keep the calendar in sync with the seasons. There was no specific rule for when to do it; just whenever it seemed like a good idea.

By the year 707 AUC (46 BCE), the seasons had drifted so far off from the months that the Roman dictator Julius Caesar took the bull by the horns and revised the calendar. By then it was known that the solar year was about 365.25 days. To correct the problem with the seasons, Caesar declared that 707 AUC would be 445 days long. The new Julian calendar went into effect the next year (708 AUC or 45 BCE). This calendar used 365 days for 3 years, followed by a fourth year with 366 days. The months alternated between 31 and 30 days in length, except for *Sextilis* which had 31 days, and *Febuarius* which had 28 days for 3 years and 29 days the fourth year. The year was reset to begin on *Januarius* 1. This system was to continue in perpetuity. The seventh month (*Quintilis*) was renamed *Julius* (July) to honor the creator of the new calendar. The middle of each month was called the *Ides*. It was on the Ides of March 709 AUC (44 BCE) that Julius Caesar was assassinated by a conspiracy of senators who opposed his dictatorial ways.

Julius Caesar had no legitimate children. In his will, he named his adopted son, Gaius Octavius, to be his successor. There ensued a series of disagreements and intrigues over the succession, involving, among others, Mark Anthony and Cleopatra, the inspiration of many novelists and playwrights. It all makes for great entertainment. Octavius became known as Octavian, and in 726 AUC (27 BCE) he became the first Emperor of Rome assuming the lofty name of Augustus. He thought the Julian calendar was such a good idea that he let the eighth month (*Sextilis*) be renamed after himself, *Augustus*, which became our August.

**Fig. 2.8** Using an obelisk as a sundial. By plotting points each day the solstices and equinoxes would become evident

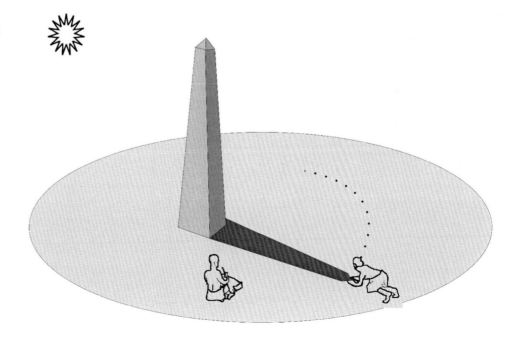

In Julius Caesar's time the winter solstice (*solstitium*, from *sol*, meaning Sun, and *sister*, meaning to come to a stop and stand still) was on December 25. The winter solstice is the time (in the northern hemisphere) when the Sun reaches its southernmost point in the sky and starts its return northward. Everyone in the northern hemisphere celebrated the solstice in ancient times. For the Romans it marked the time when the Sun god Sol was reborn. The week preceding the solstice, December 17 through 23 was a wild week of debauchery known as the *Saturnalia*. Over time it was moved to end on December 25, coinciding with the winter solstice. The week after the solstice continued the period of celebration and revelry so there was plenty to repent for at the beginning of the New Year, On January 1; everyone could make resolutions to not drink so much the next time, making them feel better about themselves. Julius Caesar had been a clever leader.

Eventually, came the Christians. They needed a date to celebrate the Resurrection of Christ. At the Council of Nicaea in 325 CE the Resurrection was set as the Sunday following the first full Moon after the northern hemisphere vernal (spring) equinox. The equinox is a time of equal lengths of day and night (*aequinoctium*, from *aequus* [equal] and *nox* [night]). The newly designated day, sometime after the middle of March, became independent of the Hebrew Passover. It was named Easter, after the pagan goddess of spring, Oestara.

To simplify determining the dates of the movable feasts the church adopted a new Ecclesiastical Calendar, a calendar based on the lunisolar cycle. In it the vernal equinox and the full Moon were not determined by astronomical observation but the vernal equinox was simply assumed to fall on the 21st day of March, and the full Moon (known as the Ecclesiastical Full Moon) was fixed to be 14 days after the beginning of the Ecclesiastical Lunar Month (known as the Ecclesiastical New Moon) The relevant Ecclesiastical New Moon (the Pascal New Moon) falls on or next after March 8. Depending on weather conditions and how good your eyes might be, the new Moon (total darkness) can vary by as much as 4 days. To make matters more confusing, the New Moon was determined not by when it reflects no sunlight, but upon the first appearance of a thin crescent. Easter falls on the Sunday after the Ecclesiastical Full Moon, not necessarily the real full Moon.

Then, in 350 CE Pope Julius I set December 25th, the traditional day of pagan festivities celebrating the winter solstice—as the day of celebration of the birth of Christ.

The Roman week had 8 days. The Babylonian week, adopted by the Hebrews, had seven. It was the Roman Emperor Constantine, a convert to Christianity in 330 CE, who introduced the 7 day week in the fourth century. The custom spread throughout much of Europe.

In about 525 CE Pope John I figured the birth of Christ to have been in 753 AUC, which he renamed 1 AD (Anno Domini). Earlier years have then become designated as BC (Before Christ) and there was no year 0, and unfortunately the Pope's calculations were wrong. According to the Bible, Christ was born during the reign of Herod the Great which ended in 749 AUC (now 4 BC). Most modern-day biblical scholars place the time of the nativity sometime between 747 and 749 AUC (6 and 4 BC/BCE).

Keeping time went more or less well for over a thousand years. However, the Julian calendar year of 365 days and 6 h was too short by 11 min. By the 16th Century this tiny error had made the calendar a mess. The dates of the solstices and equinoxes were off by 10 days.

In 1582 Pope Gregory cleaned up the mess by deleting 10 days from the calendar and proposing omitting a leap year every 400 years. This revision became known as the Gregorian calendar. In most Roman Catholic countries in 1582, Thursday, October 4th, was followed by Friday October the 15th. Other countries retained the Julian calendar.

Leaving out 10 days had unexpected consequences. It moved January 1, the beginning of the New Year, leading to confusion over the year in which something happened. The change to the Gregorian calendar did not occur until almost two centuries later in the United Kingdom. In London in 1757, Wednesday, September 2nd, was followed by Thursday, September 14th. Dates in Britain and its colonies before then, using the Julian calendar are referred to as "Old Style;" corrected to correspond to the Gregorian calendar as "New Style." Thus, the execution of Charles I was recorded by the Parliament as happening on 30 January 1648; the date is now referred to as Old Style. In modern accounts this date is usually shown as "30 January 1649 (New Style)." Similarly, Isaac Newton died on the 20th of March, 1726 (Old Style) which corresponds to the 31st of March, 1727 (New Style). Can you figure out how old he was when he died?

In Russia, the Great October Socialist Revolution of 1917, also known as the Bolshevik Revolution, was the seizure of state power by armed insurrection in Petrograd (which then became Leningrad, now known as St. Petersburg). In Russia this historic event happened on October 25, 1917, but in London, it was recorded as occurring on November 7. Britain was using the Gregorian calendar while Russia was still using the Julian calendar. Russia switched to the Gregorian calendar in 1918. Greece adopted the "new" calendar in 1923, Turkey in 1927. It was finally air travel that caused the Gregorian calendar to become accepted worldwide.

Needless to say, because calendar reform took place at different times in different countries local records for temperature, rainfall, and other weather/climate conditions need to be checked to determine the proper date.

It should be obvious that history based solely on dates without knowing what calendar was being used can be quite a mess. This applies to weather events as well as everything else.

At about the same time that Pope Gregory was convening his council of experts, there was another attempt at calendar reform underway. Joseph Justus Scaliger (1540–1609) was a Calvinist Protestant (Huguenot) cleric who had just missed being caught up in the St. Bartholomew's Day massacre in 1572 when as many as 30,000 Huguenots in France lost their lives. Sensing that trouble was afoot, he had fled to Geneva. When he returned he moved into the castle of his friend Louis de Chasteigner, Lord of La Roche-Posay, in western France staying there from 1574 to 1594. Scaliger was interested in ancient history and wanted to produce a calendar that would be consistent through time to allow him to understand the timing of historic events. He was aware of the problem that some years had different lengths, as well as the problem that there was no year 0 in the Christian version of the Julian calendar. He wanted his calendar to go back in time far enough to include all of the history of human civilization. He published his eight volume work proposing the term Julian Period, *Opus novum Emendatione Temporum*, in 1583.

Scaliger used three cycles to determine the length of one grand cycle of time which he called the Julian Period:

(1) The 19 year cycle for coincidence of solar years and lunations.
(2) The 28 year cycle for repetition of days of the week throughout the year.
(3) The 15 year Roman census cycle—The so-called 'Roman Indiction.'

Multiplying these together ($19 \times 28 \times 15$) Scaliger got 7980 years or 2,914,695 days. Scaliger rather arbitrarily set the starting time for the Julian Period at noon on January 1, 4713 BCE, which he thought was old enough to include all of human history, but he did not specify the geographic location for this noon. There were no time zones in those days. Until recently GMT (Greenwich (England) Mean Time), which was established in 1675, was used as the international standard for Scaliger's time system. GMT has now been replaced by UTC (Coordinated Universal Time) as the primary standard by which nations around the world consistently record and regulate time. The UTC system corrects for the tiny variations in GMT caused by changes in Earth's rotation rate, and other factors. UTC is sometimes abbreviated Z and called "Zulu time."

Because of the length of year problem, the time count in UTC is in days. This time is called the Julian date reflecting Scaliger's naming convention. Sciences requiring exact timing of events, such as astronomy, use this calendar. Thus the time when human beings (Neil Armstrong in particular) first set foot on the Moon is recorded as 2:56 UTC on July 21st, 1969. The Julian date for this event is 2440423.622222.

Yet another time convention is to use the UTC time and day number of the year. In this system (used by NASA among other institutions). Armstrong's Moon-touching date is 1969201 2:56. You can see these dates, sometimes abbreviated to just the day in the titles used on NASA's space photographs.

There have been additional, not-so-successful attempts at calendar innovation. In 1789 French revolutionaries tried to carry their enthusiasm for the number 10 to timekeeping. Their new calendar consisted of twelve 30 day months: *vendémiaire, brumaire, frimaire* (autumn); *nivôse, pluviôse, ventose* (winter); *germinal, floréal, prairial* (spring); *messidor, thermidor, fructidor* (summer). Each of the months consisted of three 10-day 'weeks' called *décades*. The names of the 10 days of the *décade* were simply numbered: *primidi, duodi, tridi, quartidi, quintidi, sextidi, septidi, octidi, nonidi, décadi*. The extra 4–6 days needed to fill out a year become holidays added at the end of the year. The day was divided into 10 h each with 100 min and the minutes divided into 100 s each. This innovative but deeply flawed calendar did not last long. The rest of the world continued to prefer the ancient Babylonian system of using 60s as the basis for timekeeping.

What about units of time smaller than the day? The day is naturally divided into times before the Sun is at its highest point in the sky, noon, and after: morning and afternoon. As societies developed in complexity it became useful to have smaller units of time. This can be done simply by following the shadow of a vertical pole on the ground. The vertical pole or other shadow-casting object is called a "gnomon." The Babylonians and Egyptians divided the day into 12 units, six for sunrise to noon, and six from noon to sunset. These units are now our hours.

We define a day as the length of time from one midnight to the next. This is a solar day, 86,400 s. But it is not a 360° rotation of the planet, but 360° 59′ 9.6″. This is because Earth's motion in its orbit around the Sun is the same as the direction of its rotation on its axis, counterclockwise as seen from above the North Pole. A full 360° rotation, a sidereal day, takes only 86,164 s.

If you follow the path of the tip of the shadow of the gnomon over the course of the year, you can determine the days of the solstices and equinoxes as the Sun moves north and south. This is how all these things were worked out in ancient times. An obelisk makes a very good gnomon, the taller the better, and Egypt, with its often cloudless skies, was able to develop a very fine system of time keeping. The gnomon can even be used for timekeeping at night. At night you can position yourself so that the North Star, Polaris, is directly over the gnomon (or in Egypt, right on the tip of the obelisk) and mark time through the night by the motion of the other stars around it.

I once asked George Swenson, an astronomer and close friend when I was at the University of Illinois, how the ancients were able to make good astronomical measurements without a telescope. The answer was simple, if you don't have a lens to magnify things, you can use a longer baseline to get the same result. Sighting on a star's position over the tip of an obelisk on the opposite side of the Nile Valley would allow the observer to make very fine measurements of its motion. You can imagine that with such a long baseline the observer would have to keep moving to keep up with the motion of the star.

The Sun shadow method of keeping time developed into the sundials, used by the Egyptians, Greeks, Romans, and throughout Europe until the invention of mechanical clocks in the 17th century.

To make measurements of units of time shorter than an hour, some sort of a clock was needed. A clock needs two things: (1) a regular, constant, repetitive process to mark off equal units of time, and (2) some means of keeping track of the accumulated number of units of time.

The earliest of these devices were "water clocks" using a steady flow of drops of water to fill or empty a basin. Keeping track of the time was done by means of marks such as scribed lines on the side of the basin. The trick was to ensure that the rate of the drops was constant, a problem never completely solved. The accuracy of water clocks was limited.

A pendulum is a much better time keeping device, because its period depends only on its length and not on the size of the arc though which it swings but because of friction, the pendulum needs to be resupplied with energy to keep it working. This is done by a mechanism called an 'escapement.' The escapement transfers energy, often from suspended weights or a coil spring, to the pendulum. It has a forked pallet that alternately checks and releases a toothed wheel imparting the energy in pulses.

Galilei Galileo conceived the idea of a pendulum clock with an escapement around 1637, but died before it could be built. A working pendulum clock was invented by Dutch mathematician and physicist Christiaan Huygens in 1656. Later the pendulum was replaced by a spring with an escapement, making pocket and wristwatches possible. Since the 1950s timekeeping has been done with 'atomic clocks,' using the electromagnetic spectrum of atoms as frequency standards. There will be more about clocks in the discussion of longitude in the next chapter.

## 2.10 Summary

Mathematics is a way of keeping track of things and organizing information. It took thousands of years for systems of easily manipulated numbers to be developed. The simple mathematical operations we know as arithmetic had their origins many thousands of years ago. We used to think that it was the Greeks in the first few hundred years BCE who developed mathematics beyond arithmetic, but recently we have learned that the more complex manipulations we call algebra were developed in Mesopotamia over 4000 years ago. The number zero, a symbol which stood for 'nothing,' was a special case. Although essential for solving many mathematical problems it was not introduced into Europe until about a thousand years ago.

Scientists deal with things that range in size from the universe to parts of atoms and use shorthand expressions, exponents and logarithms to keep written numbers manageable.

The need to keep track of time, especially the changing seasons through the course of the year, led to the development of calendars, but there were many obstacles and pitfalls along the way. Julius Caesar modified the older Roman calendar recognizing the year to be about 365¼ days long. The Julian calendar worked well for almost 1500 years, but its year was too short by 11 min. By the middle of the 16th century it was clearly out of sync with the seasons. Pope Gregory reformed the calendar in the 16th century, but many countries were slow in accepting the new calendar. It was air travel that forced its acceptance as a worldwide standard. And although used throughout the world today, the Gregorian calendar has a flaw, it has no year zero.

Accurate clocks were not developed until the middle of the 17th century. Precise timekeeping opened new vistas not only for scientific investigations but for determining the location of ships at sea as you will learn in the next chapter.

A Timeline for this chapter:

| Year | Event |
| --- | --- |
| Way back there | Humans devise a system of numbers with base 10 |
| ~4000 BCE | Sumerians devise a system of numbers with base 60 |
| ~1900 BCE | Babylonians using base 60 number system determine that $\sqrt{2}$ is about 1.41421296 |
| ~1800 BCE | Egyptian textbook shows how to do arithmetic |
| ~350 BCE | Aristotle defines mathematics as "the science of quantity" |
| 753 BCE | Rome is founded on April 21 of Year 1 of the AUC (*ab urbe condita*) calendar |
| 46 BCE = 707 AUC | Julius Caesar revises the Roman AUC calendar to have 365 days in a year and to add a day every fourth year. To get the seasons back in sync the year 707 AUC had 445 days. The revision is known as the Julian calendar. He renames the seventh month *Julius* (July) |

(continued)

(continued)

| Year | Event |
| --- | --- |
| 44 BCE | On the Ides of March, Julius Caesar is assassinated |
| 27 BCE | Caesar's successor takes the name Augustus and renames the eight month in his honor—*Augustus* |
| 325 | The Council of Nicea sets the day of the Resurrection of Christ to be the Sunday following the first full Moon after the northern hemisphere vernal equinox, becoming independent of the Hebrew Passover. It was named Easter, after the pagan goddess of spring, Oestara |
| 350 | CE Pope Julius I set December 25th, the traditional day of pagan festivities celebrating the winter solstice—as the day of the birth of Christ |
| 525 | Pope John I reckoned the birth of Christ to have been in 753 AUC, which he renamed 1 AD (Anno Domini). Earlier years then became designated as BC. There was no year 0 |
| 825 | Algebra is introduced by Baghdad mathematician Abu Ja'far Muhammad ibn Musa al-Khwarizmi in his book on equations *Al Kitab Fi Hisab Al Jabr Wa Al Muqabala* |
| 1202 | Leonardo Bonacci (Fibonacci) introduces zero to Europe |
| 1582 | Pope Gregory, in order to get the seasons back in sync, deletes 10 days from the Julian calendar. The new calendar is called the Gregorian calendar |
| 1583 | Joseph Scaliger publishes *Opus Novum Emendatione Temporum* introducing the Julian Period which began on January 1, 4713 BCE, and counts in days. Used by astronomers today |
| 1614 | John Napier publishes *Mirifici logarithmorum canonis description* with tables of logarithms |
| 1620 | Joost Bürgi publishes *Arithmetische und Geometrische Progress Tabulen* but forgets to include instructions on how to use the tables of logarithms |
| 1633 | Galileo Galilei's trial for the inconvenient truth that the Earth orbits the Sun |
| 1636 | Pierre de Fermat devises a three dimensional coordinate system but doesn't publish it |
| 1637 | René Descartes publishes his two-dimensional coordinate system |
| 1637 | Galileo Galilei conceives the idea of a pendulum clock but dies in 1642 before it could be built |

(continued)

(continued)

| Year | Event |
| --- | --- |
| 1656 | Christiaan Huygens builds a pendulum clock |
| 1665 | (The Plague Year) Isaac Newton dreams up calculus, not published until 1735 |
| 1675 | Leibniz develops his calculus and publishes the notation we use today |
| 1752 | Britain and its colonies adopt the Gregorian calendar by dropping 12 days from the Julian calendar |
| ∼1800 | Form of 'Arabic' (actually European) numerals finally becomes stabilized |
| 1805 | Francis Beaufort establishes a wind scale based on the winds effect on canvas sails |
| 1812 | Napoleon spreads the metric system through Europe |
| 1846 | Thomas Robinson invents the four-cup anemometer |
| 1884 | Giuseppe Mercalli formalizes an earthquake scale (MM) based on observations and damage |
| 1906 | A major earthquake devastates San Francisco—recently rated as 7.9 $M_W$ |
| 1906 | George Simpson publishes the 3/2 Power Law for the Beaufort wind scale |
| 1926 | Thomas Patterson invents the three-cup anemometer |
| 1935 | Charles Richter proposes an earthquake scale appropriate for California ($M_L$) |
| 1946 | The Beaufort wind scale is extended to B17 to include the winds in topical cyclones |
| 1954 | Paul Niggli publishes *Rocks and Mineral Deposits* |
| 1958 | My friends and I try our hand at making absinthe |
| 1971 | Tetsuya Fujita and Allen Pearson, introduce a scale for rating the intensity of tornados in the United States |
| 1972 | Herbert Saffir and Robert Simpson developed a scale for alerting the U.S. public to the danger posed by hurricanes according to the intensity of their winds |
| 1974 | Terence Meaden proposes a tropical storm wind scale, TORRO—T0—T10 based on the 3/2 Power Law, with T0 = Beaufort Scale B8 |
| 1979 | Seismologists Thomas C. Hanks and Hiroo Kanamori publish an earthquake scale based on the total amount of energy released: The Moment Magnitude Scale ($M_W$) |

(continued)

(continued)

| Year | Event |
|------|-------|
| 1988 | Cesare Emiliani publishes *The Scientific Companion* and predicts that doomsday will occur in 2023 |
| 1991 | Cesare Emiliani publishes *Avogadro's number: a royal confusion* in the Journal of Geological Education |
| 1992 | Emiliani publishes *Planet Earth Cosmology, Geology, and the Evolution of Life and Environment.* He concludes that Mathematics must be the work of the Devil |
| 1998 | James Rodger Fleming publishes his book *Historical Perspectives on Climate Change* |
| 2010 | A Mw 7 earthquake hits Haiti causing the deaths of over 200,000 people |
| 2010 | A Mw 8.8 earthquake near Maule, Chile, results in 577 deaths |
| 2011 | A Mw 9 earthquake and subsequent tsunami devastates areas of eastern Japan |

If you want to know more:

Adhikari, S. K., 1998. Babylonian Mathematics. Indian Journal of History of Science, v. 33/1, pp. 1–23.—A free copy of this fascinating article has been available on the Internet at http://www.new1.dli.ernet.in/data1/upload/insa/INSA_2/20005a5d_1.pdf

Fuller, R.B. & Applewhite, E.J., 1975. *Synergetics: Explorations in the Geometry of Thinking.* Macmillan, 876 pp.—A mind bender.

Fuller, R.B. & Applewhite, E.J., 1979. *Synergetics 2: Further Explorations in the Geometry of Thinking.* Macmillan, 592 pp.—For further bending of the mind.

Emiliani, C., 1992. *Planet Earth: Cosmology, Geology, and the Evolution of Life and Environment.* Cambridge University Press, 736 pp.—A great, stimulating introduction to our planet and its history.

Emiliani, C., 1995. *The Scientific Companion: Exploring the Physical World with Facts, Figures, and Formulas*, 2nd edition. Wiley Popular Science, 268 p.—An overall introduction to science and the scientific method.

Joseph, G.G., 2011. *The Crest of the Peacock—Non-European Roots of Mathematics*, 3rd edition. Princeton University Press, 561 pp.—A voyage into the unknown for those who enjoy mathematics.

Music for contemplation of what you have read: J.S. Bach: The Brandenburg Concertos, any or all.

Libation: Absinthe, of course. I recommend Absinthe Ordinaire. It is named not because it is in any way ordinary, but because it uses the formula created by Dr. Joseph Ordinaire in rural France in 1789, just in time for the Revolution. It is supposed to turn you into a wild-eyed liberal. These have included Ernest Hemingway, James Joyce, Charles Baudelaire, Paul Verlaine, Arthur Rimbaud, Henri

**Fig. 2.9** Edgar Degas (1834–1917). L'Ansinthe (1876)

de Toulouse-Lautrec, Amadeo Modigliani, Pablo Picasso, Vincent van Gogh, Oscar Wilde, and Erik Satie (Fig. 2.9).

It tastes of licorice with a hint of wormwood. Tradition requires that it be poured very slowly into a glass over a cube of sugar resting on a slotted spoon.

The ban on Absinthe started in Switzerland in 1908 after an investigation sponsored by the Swiss Winegrowers Association. They had become concerned that Absinthe was replacing wine as an Aperitif, a pre-dinner drink. In 1905 a Swiss farmer, Jean Lanfray, had murdered his family and attempted to take his own life after drinking absinthe. The fact that he was an alcoholic and had consumed a purported nine bottles of wine earlier in the day was overlooked and the Winegrowers Association succeeded to getting a ban on Absinthe written into the Swiss Constitution. Most of Europe and the United States, but not Australia, followed suit.

Investigations into the purported hallucinogenic properties of Absinthe were conducted in the 1980s and since; none were found. The ban was lifted in the U.S. in 2007. The French Senate voted to repeal the prohibition in 2011.

**Intermezzo II. Undergraduate Study—Three Universities and One Degree in 4 Years**

I graduated from a private high school (St. Mark's School of Texas) in 1951 when I was 16 years old. I was the youngest member of my graduating class. There was never any question where I was going to college, our family had close ties to Southern Methodist University, and my father was a Trustee. However, when I applied to SMU there was some concern that I was too young to enter, but that problem got solved. I might even have received an award for financial aid because I had been Valedictorian of my high school graduating class, but that was ruled out by my father. As far as he was concerned, financial aid was for those who could not afford to attend otherwise. Anyway, I would be living at home, and so all we had to worry about was tuition and 'incidental expenses.'

I started off taking a pre-med curriculum; that was regarded as the most rigorous set of courses for a biology major. During my freshman year I joined the photography club, and here is where serendipity enters the scene. One of those in the club was an exchange student from Germany, Heinz Just. He was considered a junior at SMU, and had come from the University of Munich. He wanted to learn how to develop film and make prints, and I already knew all about this because my brother and I had transformed the hall closet in our house into a darkroom so small that only one person could fit in at a time. SMU had a fine darkroom available for the photography club. I taught Heinz not only how to develop film, but the tricks of how to

make prints even when the negative wasn't all that good. We would talk, and I learned that his family had lived in the Sudetenland, which before the war was a largely German-speaking part of Czechoslovakia. During the war he had been sent to Bavaria, far from anyplace that might be bombed. At the end of the war the German population was expelled from the Sudetenland. In the chaos at the end of the war, Heinz and his parents had to search for each other. I introduced him to my family, who were interested to discover that the people we had been fighting were, after all, humans too. I tried to find out what 'exchange student' really meant and found out that Senator Fulbright had set up a program, but it was only open to Americans for graduate study. Again serendipity steps in. When my father was a boy, the Hay's used to vacation in Arkansas, and they stayed at the Fulbright's rooming house, so dad and the Senator were childhood friends. My father figured that if Senator Fulbright thought these exchanges were a good thing, they must be. Heinz went back to Germany at the end of my freshman year.

In those days, if you were interested in what was happening in Europe and particularly the inside history of World War II, SMU was a very interesting place to be. Robert Storey, who had been Executive Trial Counsel for the United States at the Nuremberg Trial of major Axis War Criminals, was the Dean of the Law School. Hans Bernd Gisevius was also on the campus. He had been a surreptitious member of the German Resistance, and as Vice Consul in the German Embassy in Zurich had acted as intermediary between Allen Dulles and the Resistance in Germany. The Resistance was responsible for the July 20th, 1944 attempt to assassinate Hitler. There were interesting and insightful lectures from time to time.

During my sophomore year, I met another student from Germany, Peter Trawnik, through the German Department's club. Peter had been invited to come for a year by his uncle who lived in Dallas. The Trawniks lived in Gräfelfing, just outside Munich, a village untouched by the war. But he described one night in 1944 when there was a devastating air attack on Munich. The Trawniks climbed up onto the roof of their house to watch the city burn.

Halfway through my sophomore year the German Department at SMU announced an interesting opportunity. Wayne State University in Detroit intended to reopen their "Junior Year in Munich" program (hereafter JYM) which had been shut down in the 1930s. The first group of students would go for the 1953–54 academic year. Serendipity strikes again. That was the year I would be a Junior, and I already had two friends in Munich. I asked my parents if I could apply. My mother was skeptical, but my father thought it was a great idea. We realized that living conditions would be very different from the peace and quiet of Dallas. And the cold war was heating up. There was always concern about whether it might turn into another war in Europe. Peter Trawnik convinced me that the most interesting way to travel to Europe was on a freighter. Before I left, I received some really good news. Heinz had managed to get me into the first student dormitory built for the University of Munich—a brand new building, with all the comforts of a dormitory in the US. Well, almost, anyway. It turned out we had warm water once a week.

I sailed from New York in June, 1953, on a freighter which took almost 2 weeks to reach Hamburg. There were five other passengers and no entertainment other than reading. We had meals with the Captain. Everything went well until we reached the North Sea and encountered huge waves. It was while I was at sea and 'out of touch' that the June 17th workers uprising in East Germany occurred. It was a huge event in the US, considered by many a prelude to war. Fifty years later I was to learn that our Junior Year in Munich was almost canceled because half of the students had withdrawn for fear they would get caught in a crisis. For me ignorance was bliss, and fortunately my family never expressed any concern.

Heinz had a friend in Hamburg, Horst, but I had been told that he worked all day. I was to go to his family's home and would stay there for a couple of days. When the freighter docked, I took a cab to the address. It looked rather bleak, and I can still remember hauling my suitcase up the decrepit stairway in a building that still showed many of the scars of war. I knocked on the door and was greeted by Horst's mother, who made me instantly feel at home. After unpacking I went out for a walk. Hamburg had been virtually destroyed in air raids, and many of the houses were simply shells. For someone who had never seen the destruction of war, it was a very sobering experience. That night we went to Planten un Blomen a huge park near the city center. There fountains and lights played to classical music provided by a symphony orchestra. I had never seen anything like it. Here in the midst of rebuilding from the destruction was a new unique art form—water, light and music. The next few days I wandered about the city fascinated by the modern buildings built (or rebuilt) of dark red bricks and at the ruins which still remained.

I took a train (changing successfully several times), to Neuburg an der Donau, where Heinz met me at the station and took me to his parent's apartment. After a couple of days I went alone to Bayreuth for the Wagner festival. I had managed to get tickets before I left the US. It was the year that there were radical new stage settings by Wieland Wagner. Again I had never seen anything like it. The Festival Opera house in Bayreuth was built especially for Richard Wagner's operas and had many special features. For one thing, you cannot see the orchestra pit; it is under the stage, and no light comes out from it. I had tickets for a complete performance of "The Ring"—four operas. The first "Das Rheingold" is performed without intermission and starts in late afternoon. You go in, take your seat, the lights go out, and it is pitch dark. The music begins in total darkness. The first part of *Rheingold* takes place in the Rhine River. In it. Rhine maidens are swimming about, singing. You only see these ladies appearing from time to time in the gloom toward the front of the auditorium. You can't see where the stage is, but gradually the faint rippling of light coming through the water covers the whole front of the space. After the performance, I had dinner in an outdoor restaurant on top of the nearby hill. The next day's performance was "Die Walkürie." It is in three acts, with an hour or more intermission between each, so you can go to one of the restaurants and have a meal spaced out over the evening. *Walküre* ends with Wotan putting Brünnhilde to sleep surrounded by a ring of magic fire. Those were the biggest, most impressive flames I have ever seen on a stage. I think it was done with a motion picture of flames projected onto a wall of steam. However it was done, it seemed unbelievably real. I was astonished at the originality of everything. The third opera, "Siegfried," was performed the next day, again with long intermissions. The fourth day was a pause, for 'light entertainment' in the Festival Hall—Beethoven's 9th Symphony conducted by Bruno Walter. And finally, on the fifth day the performance was, "Die Götterdämmerung," the 'Twilight of the Gods' complete with the rainbow bridge leading to Valhalla. At this point I was seriously considering scrapping the idea of becoming a paleontologist and thinking that stage set design would be a lot more fun.

The next day Heinz arrived about noon with the Volkswagen beetle we had rented for our grand trans-alpine tour. That evening we saw *Tristan und Isolde*. It was when going to dinner between the second and third acts, that I had an unexpected encounter with one of Hitler's more valuable deputies. Heinz was walking in front of me, and I was looking down at the path to the restaurant and thinking about what I had just seen, when I walked straight into an elderly gentleman, very nearly knocking him completely down. I caught him and helped him back up. Heinz had a shocked expression on his face. As the gentleman walked away with his friends, Heinz asked me "Do you know who that was?" I turns out I had bumped into Hjalmar Horace Greeley Schacht. Of course the name meant nothing to me. Heinz explained to me. Hjalmar Schacht had been Director of the Reichsbank under the Weimar Republic and continued in that position under Hitler. He was responsible for rebuilding the German economy and keeping it functioning during the war. He had been tried for war crimes at Nuremberg and found not guilty. It was concluded that he had just been doing what bankers do. If I had known who he was, I could have at least asked him how he got his name.

On Friday the 14th of August we drove to Nürnberg. Much was still in ruins, and there were large areas of the old town that were simply open spaces with rubble. But from what had been rebuilt it was already clear that the ancient center of this great medieval city was going to be reconstructed as close to its original form as possible. On Saturday we went swimming in the Main River and visited one of the Baroque palaces outside Bayreuth. In the evening we saw *Parzifal*. I've seen several performances since, but nothing quite as magical or mystical as that one in Bayreuth 1953. The Holy Grail was represented by a bowl, which began to glow and then gradually emitted intense brilliant light. I have no idea how that effect was achieved in those days.

The next day we were off on what was the greatest tour of Europe I have had. Following Tom Cranfill's advice I had obtained tickets to the Salzburg Festival. We went by way of Munich, my first chance to see the city I would be living in. At the end of the war the central part of Munich was more than 90 % destroyed and overall damage was over 60 %. That meant that 3 out of 5 buildings were either gone or in ruins. A lot of rebuilding had been done in the 9 years since the war, but there were still ruins everywhere. It seemed that the only buildings to survive the bombing were those Hitler had built: the two Nazi Party Headquarters buildings on Königsplatz (after the war one became the America House, the other became part of the Hochschule für Musik und Theater), the Gestapo Headquarters on Ludwigstrasse (then the US

Consulate), and the Haus der Kunst (originally the Haus der deutschen Kunst, for Nazi-approved art; after the war it was used to house the highlights of the fabulous Bavarian collections of art; the major galleries, the Pinakotheken, were all still in ruins). In contrast to some of the other cities which had been largely destroyed, the decision was made to rebuild Munich much as it had been before the war. I spent a sabbatical in Munich in 1988, and found that almost all physical traces of the war were gone. Interestingly, the students at the University then didn't seem to know that all of the streets lined with 18th and 19th century buildings and houses had been burned out and were ruins 40 years earlier.

We went for lunch at the Biergarten of the Löwenbräukeller on Stiglmaierplatz. It had been reconstructed just the way it had been before its destruction in a bombing raid in 1944. I had my first encounter with one of Munich's most famous specialties, Leberknödelsuppe (liver dumpling soup). A few hours later I was sick to my stomach and I blamed the soup. But we made it in good time to Salzburg and Heinz gave me the traditional remedy: a sugar cube soaked with several drops of Underberg bitters. It worked within a few hours, but I was still under the weather and decided to skip the performance of Handel's *Julius Cesar* that evening. It was many months before I tried Leberknödelsuppe again. Now I regard it as a delicacy, not to be missed on a visit to Munich.

The next day I had recovered and that evening saw the world premiere of a new opera, Gottfried von Einem's "*Der Prozess*" ("*The Trial*"). The opera is based on Franz Kafka's book. Josef K is arrested without being told why, tried for an unnamed crime, and convicted. It is a dark and surreal story, and very depressing. Von Einem's music is influenced by Stravinsky and Prokofiev, and was well suited to the theme. During the intermission I saw a small man standing alone looking at the audience. I walked over to him and asked how he liked the opera. He told me he would like to know what I thought of it. After some discussion we introduced ourselves. He was Gottfried von Einem. I liked *Der Prozess*, but it has not made its way into the repertoire of major opera houses. Little did I suspect that almost two decades later a close friend, Pavel Čepek, would be caught up in a real trial in Prague, very much like that in Kafka's play.

The next day we drove through the spectacular lake and mountain scenery of the Salzkammergut. And in the evening saw a memorable performance of "*Don Giovanni*" at the Festival Hall. The circum-alpine tour of the next month, through the Austrian Alps, northern Italy, southwestern France and Switzerland was literally the experience of a lifetime. In those days, you could stay in youth hostels and small hotels and eat in local restaurants for a pittance. On the advice of Tom Cranfill we discovered the 'other' Riviera, the one between Genoa and La Spezia which is still a gem. But perhaps the highlight of the trip for me was to Chamonix to see Mont Blanc and then on to Zermatt to see the Matterhorn. In 1953 there were none of the cableways that exist today; there were long and sometimes difficult hikes to see the sights.

Returning to Munich, I moved into the Red House, Studentenheim am Biederstein, where Heinz and I shared a tiny room. A few years ago Heinz, after having visited an exhibition on cubist art in Ferrara, reminded me that I covered the walls with prints of paintings by Braque. I think I was more fortunate that the other members of our American group in that I instantly had about 30 friends, all with interesting stories to tell. With one exception all had been too young to serve in the military. The exception was our wise Papa Wagner, in his twenties. He had been a fighter pilot in the Luftwaffe, and although he was less than a decade older than the rest of us, he was our local father figure.

Our Junior Year group started the semester early, in September, with special courses in German Language and European History. For the University of Munich (its real name is Ludwig-Maximilian Universität, better known simply as LMU) the semester began in October. We were able to enroll in regular courses in the University. The German students do not take exams until they are ready to finish their studies after 4 years at the University. If we wanted to attend University lectures and get JYM credit for a course, JYM provided tutors and we would get special exams arranged for us. A peculiarity of the University system in Germany (and much of Europe), is that the students keep their own records. The University issued each student a 'Studienbuch.' You would write in it what lectures you were attending. After one of the lectures, sometime during the semester, you would go up to the Professor and get his signature in the Studienbuch. Of course, that meant that in reality you only needed to attend one lecture (or give your Studienbuch to someone else to get the required signature). A complete honor system. Basically, you could learn the material any way you wanted. What counted was how well you did on the end of your studies when you took all the exams, usually over about a 2 week period. For the typical German student this meant that the first year was spent in confusion trying to figure out what you were doing. Year 2 was pure fun, having parties and general good times. Year 3 was a time for moderate panic, and year 4 was spent in total panic preparing for the exams. You also need to realize that German Universities (and most in Europe) do not have tuition. Higher education is free to those who are qualified. You pay a small fee and get the right to get a pass on public transportation, greatly reduced prices for tickets to concerts, opera, plays, museums, etc. The last time I was in Munich, the University was trying to figure out what to do about the perpetual 'students' who never seemed to finish but took advantage of these amenities, an estimated 20,000 out of 80,000 enrolled students.

My friends at the Studentenheim am Biederstein advised me that there were certain lectures you just had to have in your Studienbuch. I was able to sign up for them without asking for credit from the JYM. One absolute must was Advanced Sanskrit; no one ever seemed to take Introductory Sanskrit or to attend more than the one lecture required to get the Prof's signature. Another absolute must was to attend at least one of Professor Theodor Dombart's lectures on Egyptology. Dombart looked the part of a proper Egyptologist of the time. In his sixties, he had a triangular beard that extended down almost to his waist. German sentences usually have the verb at the end. In lecturing Dombart was able to devise very long and complex sentences that could go on for a minute or more, and then forget the verb at the end, leaving you up in the air as to what it was about. Or sometimes he would remember the verb and follow it with a 'nicht' which negated everything he had just said. And then there were the lectures on Art History by Hans Sedlmayer, open to all students in the University. The audiences always completely filled the AudiMax, the Auditorium Maximum, the largest lecture hall in the University, with students sitting on all the steps as well. The lectures were great, but you could tell when the end was coming because Professor Sedlmayer would start putting on his coat and shawl while still lecturing and give a final "Danke schön für Ihre Aufmerksamkeit" (Thank you for your attention) as he put on his hat and headed out the door.

The main building of the University on the Ludwigstrasse had opened, but some of the roof was missing, and parts of the building were still in ruins.

When winter came it was cold, and most of the classrooms were not heated. I wore a heavy winter coat and gloves—not the easiest gear for taking notes.

I called back to the US a few times to talk to my parents. It wasn't easy. Germany had been allowed to keep a single transatlantic cable at the end of the war. To place a call to America you went to the post-telephone office and made an appointment for the call, usually 10 days or more in advance. When the day came you would wait until you were called and assigned to a booth. Then you would wait in the booth until the connection had been made, usually a few minutes. Then you had maybe 5 min to exchange pleasantries before you were cut off so the next person could make his call.

Nevertheless, that year in Munich was the highlight of my life. I went to concerts, operas or plays an average of 4 days each week. Tickets for students always seemed to be available and were incredibly cheap. There was only one large concert hall, the Herkulessaal in the Residenz, but already two Opera houses. Munich already had four symphony orchestras (it now has five) and you could hear anything from music by Orlando di Lasso (16th century) to the concert series Musica Viva, which never performed anything older than 10 years. Serious opera was in the Prinzregententheater, a near duplicate of the Festival House in Bayreuth, and also built for Wagner's Operas. One of the highlights was Carl Orff's "*Carmina Burana*." I sat in the last row. Orff himself was in the box immediately behind me. Whether you know it or not, you have heard at least part of *Carmina Burana*; the themes have been used in so many films and TV shows. It is sometimes called the first piece of rock music. It consists of secular songs in vulgate Latin sung by monks. In that 1954 staging the 50 or so monks making the chorus were dressed in blue and red robes and were sitting around a huge round table that occupied most of the stage. Soloists would appear on the table. I've seen a number of performances since, none with such grand staging. Another memorable performance was a Stravinsky ballet with stage sets designed by Paul Klee.

The Theater am Gärtnerplatz, which burned out in one of the air-raids in July, 1944, had been rebuilt and was in operation every day. It did light opera, operettas and musicals. Again, I was fascinated by the staging which sometimes involved movies projected onto the background.

Our Junior Year program gave us the regular German University vacation periods off, three weeks over Christmas/New Year and two months between the winter and summer semesters. So I was able to spend several weeks in Vienna, Rome, and Paris. In Rome I attended a concert in the Palazzo Farnese. Much later I realized that the Palazzo Farnese was the setting for Act 2 of Tosca, where she murders Scarpia. A few years ago a friend going to Tosca as his first opera once asked me "what is it about?" I had to respond "Well, false imprisonment, torture, rape, murder, betrayal and suicide—the usual things."

Fasching is Munich's version of Mardi Gras, only instead of 1 day it lasts at least 2 months, from just after New Year until the beginning of Lent. It is the darkest coldest part of the winter, and to make up for it there are parties. Big Parties, costume parties called Balls. Almost all were open to the public for the modest price of a ticket. Attendance was usually between one and five thousand revelers. I believe their real purpose is to make sure you have something to repent during Lent. On weekends there might be as many as 30 costume balls each night. The festivities would start about 9 PM, but most students didn't arrive until much later, and the balls would end at dawn. The early morning streetcars would be full of people in every sort of outlandish costume. It was common to go with your girlfriend but to leave with someone else's. The art was removed from the Haus der Kunst, and the building was transformed into a painted cave with fantastic lighting. But for me the best parties were in the underground air-raid bunkers of the Regina Hotel underneath Lenbach Platz and the surrounding area. The bunkers, ordinarily sealed but opened for Fasching, were a labyrinth. There were many little rooms and three large ones where there would be orchestras and dance floors. Once into the maze you could have the devil of a time finding your way back out again. Most people wore masks or had their faces painted so well you couldn't even recognize friends. You would soon lose those you had come with and be having a good time with total strangers without ever finding out who they were. You might even wind up going together to a breakfast of Weisswurst and beer in one of the early-opening restaurants. In retrospect, it was remarkably innocent fun and good therapy for those who had experienced the war first hand. Now it is much more formal, much less exciting and fun: today many of the Munich students go to Venice for Carnival instead.

Although that year in Munich was the most exciting time of my life, I am amazed when I look back at the pictures I took that it was still a city largely in ruins.

And this was 8 years after the war had ended. Although I never participated directly in war, the experience of living in a bombed-out city left me with a real contempt for those who regard it as a reasonable way of solving problems.

My last year at SMU was spent partly planning on going to graduate school and more intensively planning on how to get back to Europe, maybe the next time to somewhere that had not been so affected by war.

I'M NOT SURE WE GOT THE NUMBER OF DIMENSIONS RIGHT

*All science requires mathematics. The knowledge of mathematical things is almost innate in us. This is the easiest of sciences, a fact which is obvious in that no one's brain rejects it; for laymen and people who are utterly illiterate know how to count and reckon.*

Roger Bacon (1214–1294)

To solve problems we first have to establish some standards to work from, give precise definitions to the words we will use, figure out how to locate things, learn a little about geometry and the way things grow, and explore how we can convey what we have learned to our friends.

## 3.1    Measures and Weights

Here I should note that the United States is the only country in the world which still adheres to an archaic system of weights and measures. It is a real impediment to public understanding of science. Even the last of the other holdouts, Myanmar and Liberia, gave up the ghost a few years ago and finally adopted the metric system. But there is a dark, well-kept secret: units used in the United States are now defined in metric terms. Take care, don't let the Congress know or they will make it a first order of business to try to repeal it.

But where did the units still used in the United States come from? We have to go back more than 2,000 years.

The Romans spread their system of weights and measures throughout the empire, replacing a myriad of older regional schemes. The basic Roman unit of length was the *pes* ('foot'), and it had a standard scribed on the base of a monument, now accepted to be equal to 296 mm (0.971 ft). The *uncia* (inch) was 1/12th *pes*; the *cubitus* (cubit) 1½ *pes*; the *passus* (pace) 5 *pes*; the stadium (furlong) 625 *pes*, the *mille passum* or *millarium* (mile) 1000 *pes*, and the *leuga gallica* (league, as used in Gaul) 1500 *pes*. Now you know where the words 'mile' and 'league' come from—perfect topics for cocktail party conversation. You can ask if someone knows how far 20,000 leagues under the sea (Jules Verne's story) might actually be—30,000 miles.

The Roman standard of weight was the *libra* (pound), equal to 328.9 g (11.60 oz or 5076 grains). It was divided into 12 units also called *uncia* (ounce). You will notice that the number 12 has crept in here, both in lengths and weights. We still use it. We buy eggs by the dozen; there are 12 in to the foot.

After the fall of the Roman Empire in the 5th century CE, almost everyone in Europe went their own way. Here we follow what happened in Britain, and how we acquired our current American system. As the Romans withdrew from England during the 5th century, new settlers arrived from the continent, the 'Anglo-Saxons.' They had their own system of measures. The new system of both lengths and weights was based on the size of grains of barley, the barleycorn. It was a standard that was easy to obtain and had the huge advantage that it was a major ingredient used in making ales and whiskeys. Here are some of the terms; 1 in (*ynch*) = 3 barleycorns laid end to end; 1 ft = 12 in But the Anglo-Saxons introduced the North-German foot of 13.2 in (335 mm) for routine measurements while the Roman foot continued to be used for construction of buildings and other structures. This confusing system seems to have been accepted by William the Conqueror in 1066, but in the late 13th century the modern English foot (=304.8 mm) was introduced to end the confusion. It is exactly $^{10}/_{11}$ of an Anglo-Saxon foot. Makes sense, doesn't it?

Continuing: 1 yd = 3 ft or 36 in; 1 fathom = 6 ft; 1 furlong ("One plough's furrow long" i.e. the distance a team of plough horses could be driven before they need to rest) = 40 *rods* or ten *chains* or 600 Anglo-Saxon feet = 660 'modern' (13th century) feet; 1 mile was originally the Roman mile of 5000 ft, but in 1592 it was extended to 5280 ft to make it a whole number (8) of furlongs.

The foot used in the North American colonies turned out to be equal to 0.3048006096 m whereas that in England was found to be 0.3047990 m. The difference was insignificant for most purposes, but leads to errors in measuring large distances. The whole matter was settled in July 1959 when

© Springer International Publishing Switzerland 2016
W.W. Hay, *Experimenting on a Small Planet*, DOI 10.1007/978-3-319-27404-1_3

the United States joined with the countries of the British Commonwealth of Nations to define the length of the international yard as exactly 0.9144 m. Well, almost; the United States still uses its older foot for certain purposes.

There two systems of weight in the English system: Avoirdupois (French for 'goods by weight'), is used for most materials, and Troy weights are used for precious metals (the origin of the name is unknown, but might refer to the town of Troyes in France). Weights in the English system were also based on the barleycorn. The standard unit for both systems is the 'grain,' the weight of a single barleycorn (0.06479891 metric grams). In Avoirdupois, 1 oz (oz) = 437.5 gains. 1 pound (lb—derived from the Roman *libra*) = 16 oz = 7000 grains; 1 stone = 14 lb; 1 hundredweight (cwt) = 112 lb; 1 ton = 2000 lbs. In Troy, 1 pennyweight (dwt) = 24 grains; 1 oz (oz t) = 20 dwt = 480 grains; 1 pound (lb t) = 12 oz t = 5670 grains. Not exactly the easiest system to remember, is it?

If you think all this is confusing, remember that in the English monetary system devised after the Norman Conquest (1066) there were 4 farthings to the penny, 12 pennies (pence) to the shilling, 20 shillings to the pound, and 21 shillings (1 pound and 1 shilling) to the guinea. A penny started out as one pennyweight of silver. A pound sterling was one pound of sterling silver, equal to 960 farthings. It was a wonderful system, great fun. To the dismay of accountants, the British changed to a dismally simple decimal monetary system in 1971.

## 3.2   The Nautical Mile

An oddity in all these measurement systems is the unit of length called the nautical mile (1852 m; 6067.115 ft). It is the oldest Earth-related measure, and still very useful today. The unit has had many names through the ages. The English term 'Nautical Mile' was introduced in 1730 by James Harris in his book *A Treatise on Navigation*.

It is generally assumed that this unit of measure goes back to the work of Eratosthenes of Cyrene (c. 276–c. 195 BCE). Cyrene was a Greek city in what is now Libya. He was a classical 'polymath' interested in many areas of science and mathematics. He became Chief Librarian at the fabled Library of Alexandria, a city founded by Alexander the Great on the western side of the Nile delta. The library is purported to have held as many as 400,000 manuscript scrolls. It was located on the edge of the harbor. Everything was lost when in, 46 BCE, Julius Caesar, besieged in Alexandria, set fire to his ships. One or more drifted into the Library and set it ablaze.

Eratosthenes is considered the founder of the science of geography, including the terminology used today of locating things in terms of longitude (360° of lines around the Earth passing through the Poles, similar to segments of an orange) and latitude (circles from north to south, parallel to Earth's Equator). There is some evidence that the idea of an Earth-size related system of measurement may go back to Babylonian times, but Eratosthenes is usually credited with its invention. But remember, he had access to that huge library and its store of knowledge.

He is credited with having calculated the circumference of the Earth. At local noon on the summer solstice in the city of Syene (now known as Aswan) on the Tropic of Cancer, the Sun would be at the zenith, directly overhead. The shadow of someone looking down a deep well in Syene would block the reflection of the Sun at noon off the water at its bottom. He measured the Sun's angle of elevation at noon on the solstice in Alexandria, and found it to be 1/50th of a circle (7° 12′) south of the zenith. He assumed that the Earth was spherical, and that Alexandria was due north of Syene, concluding that the distance from Alexandria to Syene must therefore be 1/50th of a circle's circumference. Surveyors had determined the distance between Swenet and Alexandria to be 5,000 *stadia*. He proposed a value of 700 stadia per degree, so that the circumference of the Earth would be 252,000 stadia. Of course, he did not know that the Earth is not a perfect sphere, but is flattened at the poles (oblate); his measurement was of a meridional (i.e. through the poles), not an equatorial circumference. If he used the Egyptian *stade* of 157.5 m in modern metric terms, his value for the circumference of the Earth translates to 39,690 km, an error of 1.0 % relative to the average meridional circumference of the Earth known today (40,008 km). Eratosthenes established the idea of using the size of the Earth for a unit of length of corresponding to 1 min of arc measured along a meridian—now known in English as the Nautical Mile.

One of the streets in Boulder, Colorado is 'Baseline Road' so called because it is laid out along the 40th parallel, i.e., the 40th degree of latitude. How far is Boulder from the Equator? There are 60 min in a degree and 40° to the Equator, so Boulder is 60 × 40 or 2,400 nautical miles from the Equator. Multiply by 1.852 and you have the distance in kilometers: 4444.8 km, or multiply by 1.1508 and you get 2761.9 US 'statute' miles. Mariners always use nautical miles; 'statute' miles are for landlubbers.

The nautical mile is the basis for measurement of a ships speed at sea, the knot; 1 knot = 1 nautical mile per hour. The term has its origin in the days of sailing ships. Speed at sea was measured using a chip log, a wooden panel attached by a line to a reel on the stern of the ship and weighted on one edge so that it floated perpendicular to the surface of the water. This provided substantial resistance to movement through the water so that it would stay in one place. The chip log was cast over the stern of the vessel and the line allowed to pay out as the ship moved away. One sailor allowed the line to pass between his fingers; another had a sand-glass timer. For the early US Navy the line had knots spaced 8 fathoms (48 ft) apart and the sand ran through timer in 28 s. The sailor with the timer determined the rate at which the knots were being passed through the other sailor's fingers. This method gave a value for the knot of 20.25 in/s, or 1.85166 km/h. Precision has been increased and the knot is now defined precisely as 0.514444 m/s = 1.852 km/h = 20.25372 in/s = 1.687810 ft/s = 1.150779 mph.

## 3.3   The Metric System

Scientists and, for that matter, everyone outside the United States, use the metric system. In the metric system, everything is ordered to be multiples of the number 10. No more 12 in to the foot, 5,280 ft to the mile, 16 oz to the pound, or 2,000 pounds to the ton.

The metric system originated in France. It had been suggested by Louis XVI but it remained for the Revolution of 1798 to begin the real organization of standards of weights and measures. The originators were ambitious; the meter was intended to be one ten-millionth of the distance along the surface of the Earth between the Equator and the North Pole on the meridian passing through Paris. However, they were wise enough to realize that maybe they did not know that distance exactly, so they made the length of a metal bar in Paris the standard. (They figured that the Earth's meridional circumference, the circle going through both poles and, of course, passing through Paris, was 40,000 km; that was back in 1790—they were off by 7.86 km).

All metric units are in multiples of 10; a 1000 m is a kilometer, a tenth of a meter is a decimeter, a hundredth is a centimeter, and a thousandth is a millimeter. The metric system makes everything very easy to calculate. The system was spread throughout Europe by Napoleon. Almost everyone seems to have thought it a good idea, except of course those who were not conquered by Napoleon, the British and their 'American cousins' (although the French helped us in our War of 1812 with the British).

After the Napoleonic Wars, The British and Americans retained their older systems of measure, or so they thought. In fact, things were not quite the same. Some of the terms still in common use in Britain are different and almost unknown in the United States. Britain's imperial fluid ounce

is about 5/6 of an American fluid ounce, so when an American buys a pint of ale in London, he gets a bit less than he would in Boulder, Colorado. If you diving through the English countryside ask how far to the next gas station and are told it is about four furlongs, you might be confused as to whether you can make it on an empty tank. Your weight in Britain is given in stones (14 pounds per stone), and remember that a hundredweight is not 100 but 112 pounds. The metric system is now in wide use in Britain, although highway signs still show miles as well as kilometers for speeds and distances. And there are still those who prefer to make their purchases in Guineas rather than Pounds.

There was one major change, however: in 1786 the American Continental Congress adopted a decimal monetary system, using the 'Dollar' (derived from the German 'Thaler') as its base.

Conversion to use of the metric system takes time. Young people adjust to it much more quickly than older people. "Metrication" began in the United Kingdom with use of certain terms and definitions for specific purposes in the late 19th century. In 1965, at the request of industry, the President of British Board of Trade announced that the metric system would be adopted with a target of completion within 10 years (i.e. 1975). The Commonwealth and other countries decided to follow Britain's example. Teaching of the metric system in schools was required after 1974. Most of the Commonwealth counties completed conversion by 1980, but the process is still not complete in Britain and the time-table originally agreed on has now been rejected.

Realistically, it requires at least one generation for the conversion from an older non-decimal system to the metric system to be completed and accepted by the public. The United States has not started to switch to the metric system, and has no plan to do so.

## 3.4   Temperature

The metric system applied not only to the weights and measures, but also to temperature. For the Celsius (sometimes called the centigrade) temperature scale fresh water freezes at 0 °C and boils at 100 °C when under the atmospheric pressure found at sea-level.

Numerical measurement of temperature began in 1654 in Italy. Ferdinando II de' Medici devised a sealed glass tube partially filled with alcohol which would expand or contract with temperature change independent of atmospheric pressure.

Daniel Gabriel Fahrenheit (1686–1736) substituted mercury for alcohol in 1714 making it possible to make more precise measurements. His original temperature scale (published in 1724, had three reference points: 0 being the coldest temperature, measured in a saline solution; 30 being the freezing point of fresh water, and 90 being the temperature of his wife's armpit. The degree sign came later. The problem of having three reference points was soon recognized, and several different calibration points were tried before accepting the freezing and boiling point of water as references. The degree sign was introduced for the units, and F added to indicate it is Fahrenheit's scale. On the modern Fahrenheit scale the freezing and boiling points are 32 °F and 212 °F, 180° apart.

In 1742 Anders Celsius (1701–1744) a Swedish astronomer, devised a decimal temperature scale, with 100° being the freezing point and 0° the boiling point of fresh water at sea-level atmospheric pressure (and no, I have not written those the wrong way around). It was not long before the calibration was reversed, so that 0° became the freezing point of water instead. By the time of the French Revolution it was referred to as the "centigrade scale" because it is defined by two points 100 units apart. The temperature is given as °C, which could be an abbreviation for either Celsius or centigrade. The scale was spread by Napoleon, but he lost the Battle of Waterloo, so Britain and the young United States retained the Fahrenheit scale. Britain began switching to the Celsius scale in the 1970s, although Fahrenheit temperatures are still published in newspapers. The United Sates is the only country that still uses the Fahrenheit scale.

In 1848 Lord Kelvin published his 'absolute thermometric scale' with temperatures now denoted simply by the letter K. At absolute 0 (–273.15 °C) all molecular motion ceases. This is the scale used in science.

## 3.5    Precisely Defining Some Words You Already Know

There are a number of words we use every day with vague meanings: density, acceleration, momentum, force, work, energy, pressure, and power. We use these general terms all the time in everyday life, but there are others like 'watt' which we also use every day, usually without knowing they have very specific definitions. Most of the time we can understand things using our intuitive feeling for what the word means, but sometimes we need to attach numerical values to them and then it is useful to know exactly what they mean. Figures 3.1, 3.2, and 3.3 illustrate a number of these in graphic form. Only metric units are used in these figures since the more complicated terms like 'watt' are defined in terms of metric units. The equivalent term to

'watt' in the English/American system would be 'horsepower.' If you really hate the metric system, try asking your local hardware store manager for a 0.13 hp (=100 W) lightbulb. Incidentally, there is no standard agreed-upon definition of horsepower. I think it depended very much on the horse.

First, something very basic: there are two kinds of values in science, scalars and vectors. Scalars are quantities that are described by a magnitude (or numerical value) alone. Vectors are quantities whose description requires both a magnitude and a direction (Fig. 3.1).

The term 'mass' is often defined as "the amount of material an object is made of," whatever that means. Better defined, it refers to the resistance of a body to forces acting on it. Weight is the description of the gravitational attraction a body has on the mass. Being Earthlings we have defined these terms, and we can use 'mass' and 'weight' almost interchangeably. An object with a mass of 100 kg (220 lbs), weighs 100 kg, but if you took an object with a mass of 100 kg to the Moon, it would weigh 16.6 kg (36.6 lbs) but, because it still has 100 kg of mass (resistance to forces acting upon it) it would still hurt just as much if you kicked it.

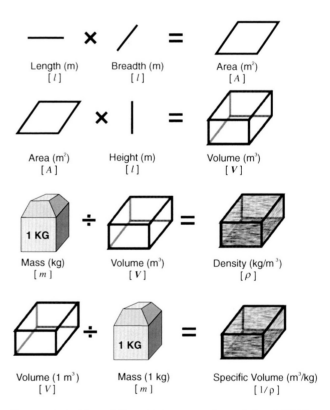

**Fig. 3.1** Definitions of Area, Volume, Density, and Speed (=Velocity). All of these are scalars. Symbols in square brackets are the abbreviations commonly used to represent these concepts in physics and mathematics. The symbol for density is the Greek letter rho

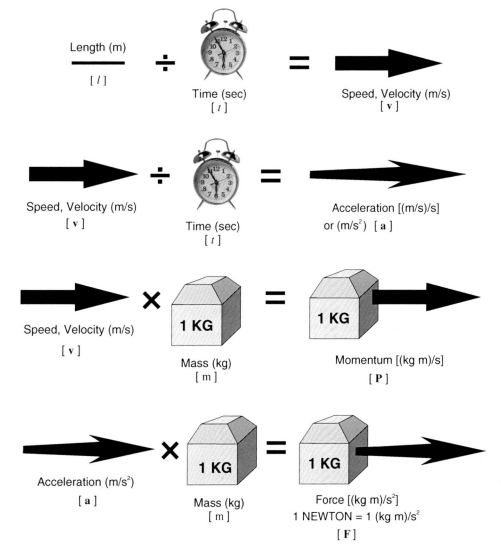

**Fig. 3.2** Definitions of Acceleration, Momentum, and Force. All of these are vectors. Symbols in square brackets are the abbreviations commonly used to represent these concepts in physics and mathematics

Actually, your weight will change if you move around the Earth. Because the planet is rotating and slightly flattened, you will weigh most at the pole, and least at the Equator. The difference for a 100 kg person is about 0.5 kg or about 1 lb for a 220 lb person.

In the next chapter we will discuss the historical development of our knowledge of chemistry, atoms, and related stuff, but here is a preview/refresher of things you probably learned in school. We use three terms for tiny particles of matter: protons are particles having a positive electrical charge; neutrons are particles with no electrical charge; and electrons have a negative electrical charge. Atomic nuclei are made of protons (particles having a positive electrical charge) and neutrons (particles having the same mass as a proton, but no electrical charge). The positively charged nuclei are surrounded by a negatively charged cloud of

electrons. The mass of the electron cloud is tiny compared with that of the nucleus. Hence, the number of protons plus neutrons determines the 'atomic mass.' The number of protons determines the element an atom belongs to. Carbon has six protons, oxygen has eight. The name of an element is abbreviated to a capital letter or a capital followed by a lower case letter: H stands for hydrogen; C stands for carbon; O for oxygen; He for helium, etc. Atoms that have the same number of protons but different numbers of neutrons are called 'isotopes' of the element (they remain the same element because they have the same number of positively charged protons; neutrons only change the mass). Hydrogen atoms consist of a single proton, no neutrons. There are three isotopes of hydrogen; they are so important that in studies of ancient climate they have different names. The atom with no neutron is 'hydrogen,' one with one neutron is 'deuterium,'

**Fig. 3.3** Definitions of work, pressure, and power. All of these are again scalars because they do not involve motion in a specific direction. Symbols in square brackets are the abbreviations commonly used to represent these concepts in physics and mathematics

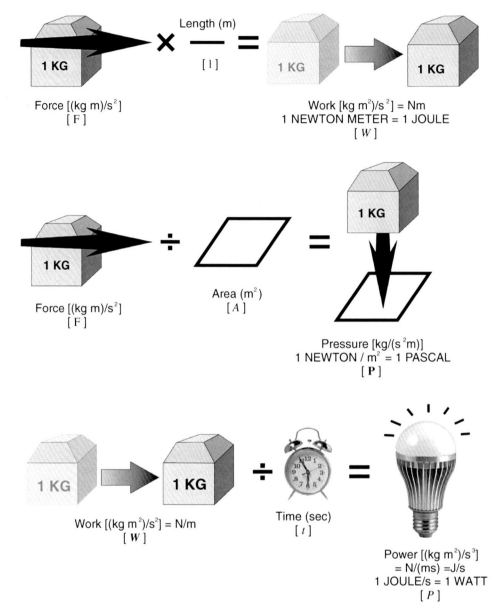

and the third, with two neutrons, is 'tritium.' Both carbon and oxygen also have three isotopes. The isotopes are indicated by a superscript number representing the total of protons and neutrons just before the letter abbreviation of the element: $^{16}O$, $^{17}O$, and $^{18}O$ are the three isotopes of oxygen. They differ by having 8, 9, and 10 neutrons in the atomic nucleus. The 'atomic weight' of an element is the average mass of all the isotopes of an element relative to $^{12}C$ (carbon 12) taking into account their relative abundance. In this system, $^{12}C$ is defined to have a mass of exactly 12 and all other elements and their isotopes are scaled to it. The atomic weight of hydrogen is 1.0078, carbon is 12.0107, etc.

Atoms are often bound together into molecules by sharing electrons. The number of atoms in a molecule is indicated by a subscript after the abbreviation for the element. For example, water is $H_2O$ because it has two atoms of hydrogen and one atom of oxygen. Common table salt is NaCl because it has a single atom of sodium (Na) and a single atom of chlorine (Cl), and methane ('swamp gas') is $CH_4$, with a single carbon atom, C, surrounded by four hydrogens.

How do we keep track of the ratios of one element to another? In contrast to other areas of science, geologists cite the proportions of elements to one another in percent by weight. The idea goes back a couple of centuries, to the beginnings of modern chemistry, when chemical analysis was new. John Dalton in England and Antoine Lavoisier in France were figuring out how to determine the proportions of different elements in a chemical compound. For example, it was found that common table salt, sodium chloride or NaCl, is 39.4 % sodium (Na) and 60.6 % chlorine (Cl) by weight. That may be nice to know but it is of very little use if you want to know how many atoms of Na there are for every atom of Cl. You have to know the atomic weights of Na and Cl in order convert weight percent to relative abundance of atoms in a compound. Sorting all this out took the better part of a century. We now know that the atoms are in the proportion 1 to 1, as the chemical formula NaCl indicates, but that the atomic weights of the two kinds of atom are different.

For reasons beyond my understanding, geologists have continued the tradition of reporting the composition of rocks and minerals in weight percent. When I was a student in Zurich in 1956 I missed hearing the famous lectures by Paul Niggli (1888–1953) on crystallography and mineralogy at the Federal Institute of Technology. He had died 3 years earlier, just 2 years after publication of his classic book *Rocks and Mineral Deposits*. But I did take courses from one of his successors, Conrad Burri. Burri was the expert on the use of polarized light to investigate minerals and wrote the classic book on that topic.

Niggli had realized that giving compositions of rocks and minerals in weight percent was useless in trying to understand the Earth. Instead, we needed to know how many atoms of this there were compared to atoms of that; that is, we needed to work with 'molar' proportions. It is molar proportions that are evident in a chemical formula, and if you want to understand reactions, that is all that counts. It took me about 20 years to understand this and realize its importance. Now, 50 years after I took the course in Zurich, this idea is gradually creeping into geology.

Back to the question: How do scientists keep track of how many atoms are in something? They use two scary terms, 'Avogadro's number' and 'mole.' They are scary because textbooks use incredibly complicated and sometimes downright incomprehensible ways of defining them. You may have run into them in a high school class and sworn off ever wanting to know anything more about science. Cesare Emiliani made a compilation of definitions of these terms in an amusing article entitled "Avogadro's number: a royal confusion" in the Journal of Geological Education in 1991. So, following Cesare, here is what they really mean. First, we have a standard number, the Avogadro; it's like 'dozen,'

only bigger. An Avogadro, named for Amedeo Avogadro, the 19th century Italian who figured all this out, is $6.02 \times 10^{23}$ things. It is just a number. It happens to be defined as the number of $^{12}C$ atoms in 12 g of carbon that only contains the isotope $^{12}C$, but it is the number itself that is important. Usually it refers to atoms or molecules, but you could conceivably refer to an Avogadro of avocados, which would make a very, very large guacamole, or as Cesare suggested, if you really want to terrify yourself, imagine an Avogadro of lawyers. Although they completely dominate the U.S. Congress, for some reason or other, lawyers are not very popular in America. In his Devil's Dictionary Ambrose Bierce defined 'lawyer' as One skilled in circumvention of the law.

Then there is the term 'mole,' sort of short for molecules, not the animals in the ground. A mole is the weight of an Avogadro of things. A mole of $^{12}C$ is 12 g (0.012 kg). A mole of overweight lawyers, at 100 kg each, is $6.02 \times 10^{25}$ kg. The mass of the Earth is only about $6 \times 10^{24}$ kg. So a mole of lawyers would be equal to the mass of ten Earths, a truly horrifying thought. The composition of many things is given as so many % of this and so many % of that, usually by weight or volume. That does not give you any idea of the relative proportions of different atoms. If something contains 1 mol of these atoms and 10 mol of those, you know that the proportion of atoms is 1 of these to 10 of those.

Think, if you listen to a politician's speech and someone tells you it is 10 % red herrings and 40 % hot air, how could you compare such things? One is a solid and one is a gas. On the other hand if your friend told you the speech contained 1 mol of red herrings and 4 mol of hot air you would know exactly what your friend meant.

For many more complicated physical concepts there is an internationally agreed-upon terminology: SI units. SI is the acronym for '*Système international d'unités*' or International System of Units. As usual, the French were ahead of the curve in these matters.

In the discussion below the symbols used for the various physical properties are shown in square brackets.

'Conservation' means that some measurable property of a closed system remains constant as the system changes. Such properties are said to be 'conserved.' Things that are conserved are momentum, angular momentum, and energy as a whole.

'Inertia' [*I*] was introduced into the English language by Isaac Newton in the context of physics. It is the property of matter by which it continues in its existing state, whether at rest or in uniform motion in a straight line; it is the resistance an object has to a change in its state of motion. The mass of an object is solely dependent upon the inertia of the object. The more inertia an object has the greater its mass and the greater its tendency to resist changes in its state of motion. In broader usage it simply refers to a tendency to do nothing or to remain unchanged.

'Moment of inertia' [also $I$] refers to objects going around in a circle. It is a quantity expressing the object's tendency to resist the angular acceleration that would increase or decrease its speed of rotation.

'Momentum' [P] was also introduced by Newton. It is the quantity that describes a moving object's resistance to stopping; it is shown in Fig. 3.2. It is the product of an object's mass times its velocity. It is the total of a force integrated over time.

'Angular momentum' [$L$] is the product of an objects moment of inertia and its angular velocity [T], the speed at which it goes around expressed as a angle covered per unit time.

'Energy' [$E$] is another common word we use, often without a very clear understanding of what it is. In science it is the ability of one physical system to do work [$W$] on one or more other physical systems. Energy as a whole is conserved, but it can be distributed among the different forms discussed below so they are not specifically conserved (Fig. 3.3).

Remember that work [$W$] is a force [$F$] acting over a distance, so energy represents the ability to push or pull something or alter its state. The SI unit of energy is the Joule (J), which is a newton meter. However, the very small amounts of energy associated with atoms and electrons are usually given in 'electron volts' [eV]. An electron volt is the energy acquired by a single electron accelerated through an electrical potential of one volt. 1 electron volt (eV) = $1.60217657 \times 10^{-19}$ joule (J); conversely, 1 J = $6.24150934 \times 10^{18}$ eV.

'Kinetic energy' [$K$] is the energy of an object in motion. It is equivalent to the total amount of work required to accelerate the object. It is a force integrated over distance. 'Potential energy' [$PE$] is energy that is sitting there, waiting to be used. A boulder sitting on a steep hillside, like one above my house, has potential energy. If it were to roll down the hillside, accelerated by the attraction of gravity, its potential energy would decrease as its kinetic energy increased. Some of its kinetic energy might be used to push over some trees, performing work, before it hits my home. 'Mechanical energy' [$E_{mechanical}$] is the sum of the potential and kinetic energies. 'Latent energy' or 'Latent heat' [$L$] usually refers to the energy that can be released by transformation of matter from one state to another, such as

freezing water to make ice. 'Chemical energy' is the energy stored in molecules that can be released by chemical reactions. The terms 'electric energy' and 'magnetic energy' refer to the energy in electrical and magnetic fields, which are actually inseparable. There are no specific symbols for these latter terms.

In any discussion of momentum or kinetic energy you will encounter the terms 'elastic' and 'inelastic' collisions. In physics these terms have meanings different from their everyday use. In physics, a perfectly elastic collision is one in which there is no loss of kinetic energy. Any collision between objects lager than the minute particles of matter called atoms and molecules will convert some of the kinetic energy into other forms of energy. No large scale impacts are perfectly elastic, but a good example of an (almost) elastic collision is when two billiard balls collide; neither is deformed and a tiny amount of the kinetic energy is converted into heat. An inelastic collision is one in which part of the kinetic energy is changed to some other form of energy during the collision. A good example is a collision between two automobiles. Much of the energy goes into deformation of the autos. Momentum is conserved in inelastic collisions, but one cannot track the kinetic energy through the collision since some of it is converted to other forms of energy.

One term we will use a lot in this book is 'radiation;' it is the 'radiant energy' transmitted as 'electromagnetic waves' which we will discuss in detail later. Other terms we use a lot are 'heat' and 'temperature.' Both relate to the motion of those minute particles of matter called atoms and molecules.

'Heat' is the random motion of atoms or molecules. The hotter something is the faster the particles are moving. The 'temperature' is a numerical measure of the momentum of the particles involved in this motion. If the motion of an atom or molecule is reduced to the point where it cannot be slowed any further, its temperature would be 'absolute zero,' which is 0 K, the base of the Kelvin temperature scale, or –273.15 °C, or –459.67 °F. This is a theoretical limit; nothing in the universe is this cold. Every object warmer than absolute zero emits and absorbs energy. The coldest temperature measure on Earth was in the Antarctic, about –90 °C (−130 °F); the hottest temperature record is a matter of dispute, but is around 58 °C (135 °F).

Two other terms involving 'heat' are especially important in climatology: 'latent heat of fusion' and 'latent heat of vaporization.' We know of water in three forms: a liquid, a solid (ice), and a gas (vapor). These different forms are called 'phases' and going from one to another consumes or releases lots of energy. We will discuss the properties of these phases in some detail in Chaps. 4 and 5.

When I was in school the energy involved in these phase transformations was all very easy to remember: a calorie was the amount of energy required to raise the temperature of 1 g of water 1 °C. You will notice that 'calorie' was spelled with a small 'c.' It is 1/1000th of the Calorie (capital 'C') used to describe the energy content of food. When the metric terminology was cleaned up, the calorie worked out to be 4.180 J. In the old days, 80 calories were required to transform 1 g of ice at 0 °C to 1 g of water at 0 °C, and 533 calories were required to transform 1 g of water at 100 °C to 1 g of vapor at 100 °C. Now we have to remember the numbers as $0.334 \times 10^6$ and $2.26 \times 10^6$ J/kg respectively; progress does not always make things easier to remember. These numbers are the 'latent heat of fusion' and 'latent heat of vaporization' respectively. What they mean is that it takes the same amount of energy to melt ice at 0 °C as it does to then raise the temperature of that water to 80 °C. If you freeze water to make ice, the same amount of energy is returned to the environment. It could be termed 'latent heat of freezing' but we use 'latent heat of fusion' for the amount of energy no matter which way the transformation goes.

It takes 5.3 times as much energy to transform water to vapor as it does to raise the temperature of the water from the freezing point to the boiling point. That is the 'latent heat of vaporization.' This is why the water being heated for the spaghetti just sits there on the stove and refuses to boil. The same amount of energy is returned to the environment when water precipitates as cloud droplets or rain. It could be called 'latent heat of condensation.' But again, we use 'latent heat of vaporization' for the energy involved in the transformation no matter which way it goes. The name of the transformation is always given from the perspective of starting with water in its liquid form; from there it either fuses into a solid or vaporizes into a gas. It is all part of the natural perversity of water to behave as it should, and can be blamed on the hydrogen atoms not being 180° apart.

In the Thesaurus I use, 'perverse' and 'evil' are close together in meaning, but I had not thought of water as being particularly evil until now. Anyway, good or evil, these phase transformations of water play a huge role in the way the atmosphere transports heat from one place to another and helps to even out Earth's climate.

Table 3.1 provides the relationships between some of the terms introduced above and other familiar things.

**Table 3.1** Relationships and values of some of the terms used in discussing climate

F stands for *Fahrenheit*, the oldest temperature scale still in use, but only in the United States and Belize

C stands for *Celsius*, the metric temperature scale where water freezes at 0° and boils at 100°. This is sometimes called the *Centigrade* scale

K stands for *Kelvin*, the absolute temperature scale; at 0 K there is no motion of molecules, at least in theory. There are 100 K between the freezing and boiling points of water. Degrees Kelvin are K, no degree symbol is used. 0 K is −273.15 °C

The '*calorie*' is now obsolete, replaced by the *Joule*, but some of us remember that 1 calorie is the amount of energy required to raise 1 g of $H_2O$ 1 °C. 1 calorie = 4.184 J

BUT you also need to remember that 1 Calorie (with a capital C, still in use to describe the energy in food) = 1000 calories

$4.157 \times 10^3$ J/kg/°C is the *specific heat* of water

0 °C = 273.15 K = 32 °F = the *freezing point* of water at sea level on the surface of the Earth
100 °C = 373.15 K = 212 °F = the *boiling point* of water at sea level on the surface of the Earth

Your body temperature is 37.1 °C or 310.25 K or 98.6 °F; each day you perform about $10 \times 10^6$ J of work

A *Watt* is 1 Joule per second (1 J/s). There are 86,400 s in a day, so if you perform $10 \times 10^6$ J of work in a day, your wattage is 115.74

Actually, your energy flow at rest is about 75 W, but can rise to 700 W during violent exercise

The average temperature of the Earth for the period 1940–1980 is usually taken to have been about 15 °C = 288 K = 59 °F. Now it is about 0.6 °C = 1.08 °F warmer

The average temperature of the surface of the Sun is 5780 K = 5510 °C = 9,900 °F

At 0 °C, the *latent heat of vaporization (or condensation)* of water is about $2.23\ 10^6$ J kg$^{-1}$; at 100 °C it is $2.26 \times 10^6$ J kg$^{-1}$

The general relation for latent heat of vaporization of water is $\approx (2.23 - 0.00235\ C) \times 10^6$ J kg$^{-1}$; (where C = temperature Celsius)

At 0 °C, the *latent heat of fusion (or freezing)* of water is about $0.334 \times 10^6$ J kg$^{-1}$

If ice sublimates to vapor at 0 °C, the latent heat of sublimation is about $2.83 \times 10^6$ J kg$^{-1}$

The solar constant, the amount of energy received by a panel in space facing the Sun at the average distance of the Earth to the Sun, averages 1353 watts per square meter (=W/m$^2$, =Wm$^{-2}$) as measured by satellites, but it varies slightly. This is usually rounded up to 1360 Wm$^{-2}$. To get the amount received in a day by 1 m$^2$ at the top of the atmosphere of our almost spherical Earth, you need to divide by 4 (the surface of a sphere is four times the area of a disc of the same diameter). Using 1360 Wm$^{-2}$ as the solar constant, that works out to be a convenient 340 Wm$^{-2}$, which everyone cites as though it were the real value everywhere, but a lot happens between the top of the atmosphere and Earth's surface

(continued)

**Table 3.1** (continued)

| |
|---|
| The average insolation (amount of energy received from the Sun) at the Earth's surface in the equatorial region is about 200 W/m$^2$ |
| The average insolation at the Earth's surface at Boulder, Colorado is about 200 Wm$^{-2}$ in summer, but drops to 100 Wm$^{-2}$ in winter |
| The geothermal heat flux from the interior of the Earth is about $60 \times 10^{-3}$ Wm$^{-2}$ (60 mWm$^{-2}$) mW = milliwatt. The geothermal heat flux is about 1/22,550th of the solar constant |
| For atmospheric pressure a variety of terms are still in use: In Thomas Jefferson's day atmospheric pressure was measured using the *barometer*, which had been invented in the 17th century. Early barometers expressed pressure in terms of the height of a column of mercury (Hg) it would support. In the US it was given in inches (in) of mercury. Sea level pressure was 29.92 in Hg For the metric system, the average sea-level pressure at the bottom of the atmosphere was supposed to be 1 bar. In the US today, atmospheric pressure is usually given in thousandths of a Bar, millibars. Unfortunately, when the Bar was established it was determined based on measurements of atmospheric pressure in Europe and North America, in a zone of low pressure. The global average is actually 1.01325 bar or 1013.25 mbar. The metric system now uses Pascals, rather than millibars, and if you are in another part of the world the weather report will give you the pressure in hectopascals (hundreds of Pascals) which, happily, happen to be equal to millibars |
| An Avogadro, named in honor of Lorenzo Romano Amedeo Carlo Bernadette Avogadro di Quaregna e Cerreto (1776–1856), is $6.02 \times 10^{23}$ things. You will hear a lot more about him later |
| A mole (mol) is the mass of an Avogadro of things. 1 mol $= 6.02 \times 10^{23}$ things expressed in kg |
| A hectare is 10,000 m$^2$ (a square 100 m on each side) and is equal to 2.47 acres. An acre is 43,560 ft$^2$ |

## 3.6 Locating Things

Fundamental to science, no matter what measurement systems are used, is a system for locating things in space and time. We'll start with space. You are going to meet a lot of interesting characters in this book, so our first one was sick, lying in bed and looking at a fly near the ceiling when he had an interesting idea.

Renatus Cartesius (Fig. 3.4), better known as René Descartes, was born in 1596 in La Haye en Touraine, south of Tours in the Loire Valley region of France. The town is now named La Haye-Descartes in his honor. He is now described as a philosopher and mathematician, and he wrote about many scientific subjects as you will read later. In 1618 he moved to the intellectually freer Dutch Republic. He died in 1650 in Stockholm. He is the person who in 1637 wrote "Je pense donc je suis" later Latinized as "Cogito ergo sum" and translated into English as "I think, therefore I am."

One of the most important things Descartes did was to watch the movements of a fly on a wall near the ceiling in a corner of his room as he lay in bed recovering from an illness. He figured out that he could describe the motions of the fly in space in terms of a two dimensional coordinate system. The system would have one axis, going left-right, and another axis going up-down. It worked very well as long as the fly did not leave the wall, and has become known as the system of Cartesian coordinates. Figure 3.5a shows his view of the fly and Fig. 3.5b the coordinate system he devised.

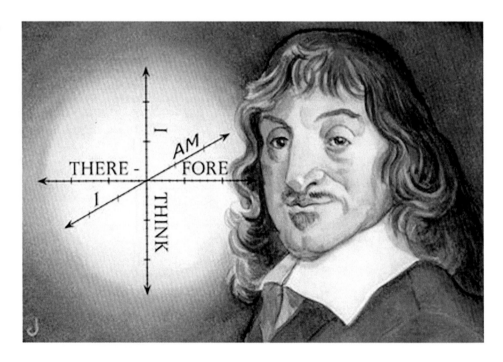

**Fig. 3.4** Renatus Cartesius a.k.a. René Descartes (1596–1650), in his Dutch finery

**Fig. 3.5 a** The fly in the corner near the ceiling in Descartes' room. **b** The location of the fly on the wall, in two-dimensional Cartesian coordinates. **c** The location of the fly in three dimensional coordinates (x, y, z = 2,0,4). **d** The Earth's coordinate system of latitude and longitude, which goes back to Eratosthenes (ca. 200 BCE)

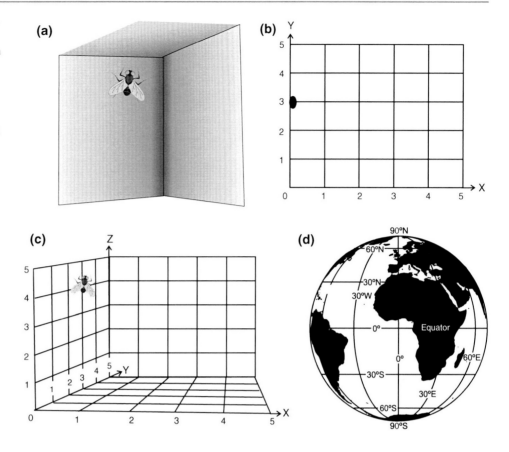

It became the basis for the area of mathematics we now call analytical geometry. He wrote about this in his 1637 work *Discourse on Method*. In the two dimensional system the x-axis goes left to right, and the y-axis goes upwards from the base. Adding another dimension, as the fly left the flat surface meant that the position of objects in space could be simply described in terms of three axes oriented at right angles to each other. Another Frenchman, Pierre de Fermat (1601–1665) of Toulouse, described today as an amateur mathematician, had actually described the three-dimensional coordinate system in a manuscript in 1636 but it was not published until after his death. Priority of publication of your ideas is important in science, so Descartes gets the credit. As you will find out later on, René Descartes was very cautious about publication of his ideas, having been intimidated by the ugliness of Galileo Galilei's recent trial for heresy in 1633.

The two dimensional coordinate system in use today has the two axes labeled as described above and shown in Fig. 3.5b. You will see diagrams using this coordinate system over and over again in this book. The values along the axes increase in these directions, as shown for the first version in Fig. 3.5c.

In reference to the Earth we fit the grid to a sphere and use the terms latitude for the lines parallel to the Equator (hence also called 'parallels') and longitude for the lines that go through the poles (also called 'meridians'). These coordinates are in three-dimensional space. The Earth is almost spherical, but its rotation causes it to be slightly flattened at the poles; it is an "oblate spheroid." The representation of the Earth as a globe that you may have on your desk is very nearly correct. However, the Earth is most often represented as a map on a flat piece of paper, and therein lies the problem. In most map projections the lines of latitude and/or longitude are not straight, and as you will see later, distortion is inevitable.

## 3.7  Latitude and Longitude

As discussed above, the idea of using circles through the poles and parallel to the Equator to provide a reference grid for locating things on the surface of the Earth is attributed to Eratosthenes. The convention of using 360° in a circle, with each degree further divided into 60 min (60′) and each minute into 60 s (60″) of arc goes back at least to the Babylonians. Circles parallel to the Equator are called lines of latitude or parallels. Circles through the poles are called lines of longitude or meridians.

One nautical mile along a meridian (1 min of longitude) is 1,852 m (6076.12 ft); 1 s of arc along the meridian is then 30.87 m (101.28 ft). Although conceived more than two millennia ago, this is a kind of accuracy that could not be routinely achieved until introduction of the satellite-dependent Global Positioning System (GPS) a few years ago.

Determining latitude and longitude is very important for mariners. There are no landmarks on the open ocean; no reference points. One of my colleagues at the Rosenstiel School in Miami, geochemist Oiva Joensuu would come up on deck after a night's sailing, look around, and say "We were here yesterday."

However, at night there can be plenty to see. Stars. In the Northern Hemisphere, there is one star, Polaris, the North Star, that is almost directly over the North Pole. At the Equator it is on the horizon all night long and at the North Pole it is at the zenith, but visible to anyone who might happen to be there only during the polar night. When you sight it, its elevation above the horizon measured in degrees is your latitude. So determining latitude in the Northern Hemisphere is very easy. All you need is a device that can measure the angle of the star above the horizon. At Boulder, Colorado, on the 40th parallel, Polaris is due north, in the same spot, 40° above the horizon, all night long. All the other stars appear to move in a circle around it through the night.

In the Southern Hemisphere determining latitude is not so easy. There is no star over the South Pole, only a rather empty space. So you need to know where to look.

Determining longitude is a wholly different matter. To determine how far you are to the east or west of some other place, you need to know the difference in time between the two places. Local time in each place can be determined by the position of the Sun, noon being when it is at its highest point above the horizon. The easiest way to determine the distance in longitude would be to set a clock in one place, and then carry it to the other place and compare it with a clock set to local time there. Both clocks would have been set at noon in each place. Since the Earth make one full rotation in 24 h, and longitude runs through 360°, each hour of time difference corresponds to 15° of longitude.

The problem is with the clock. In the early 17th century the water clock was still the standard. It could not be carried from one place to another. The problem of determining longitude was becoming critical. Mariners needed to know where they were and how far they had to go to reach their destinations. The distance between Europe and the Americas was just an educated guess.

Enter the telescope. When Galileo Galilei acquired one of the newly invented telescopes and observed Jupiter he saw the four large moons and noted that they were regularly eclipsed by the planet, each with a different rhythm. He realized that he had discovered a clock in the sky. Galileo had only a water clock to keep time and soon realized that he needed a clock that was more accurate. He conceived of using a pendulum for a clock in 1637, but died before it could be built. Many people had recognized the need for an accurate time keeping device, and Christiaan Huygens built the first one in 1656. It could be (very carefully) transported from place to place on land. But it could not tolerate the motion of a ship at sea. The problem of location at sea had become so serious that in 1714 the British Parliament established a Board of Longitude. It offered a prize to the first person to devise a practical method for determining the longitude of a ship at sea. The problem was finally solved by British clockmaker John Harrison (1693–1776). It is a great story wonderfully recounted in Dava Sobel's book *Longitude: The True Story of a Lone Genius Who Solved the Greatest Scientific Problem of His Time.*

## 3.8  Map Projections

Representing the surface of a sphere on a flat piece of paper is a challenging task. For the Earth the problem is compounded by the fact that the planet is not actually a sphere, but is flattened slightly at the poles because of its rotation. The difference between the equatorial and polar diameters is 43 km (26.7 statute miles; a statute mile is 5,280 ft). The average diameter of the earth is 12,742 km (7,918 statute miles; 6,880 nautical miles). Multiply those by $2\pi$ and you get a circumference of 40,000 km or 24,854 statute miles or 21,598 nautical miles. The even 40,000 km is because the meter was originally defined as 1/10,000,000 of the distance from the Equator to the North Pole on the meridian passing through Paris, France—one fourth of the way around the world. The modern measure of 21,598 nautical miles is remarkably close to the 21,600 nautical miles if there were 1 nautical mile per second of a degree as the ancients assumed.

Map projections used today go back at least 2000 years. The simplest, and the one still used in many publications of computer simulations of climate, is the Equirectangular projection, shown in Fig. 3.6a, sometimes called the Equidistant Cylindrical projection. It was invented by

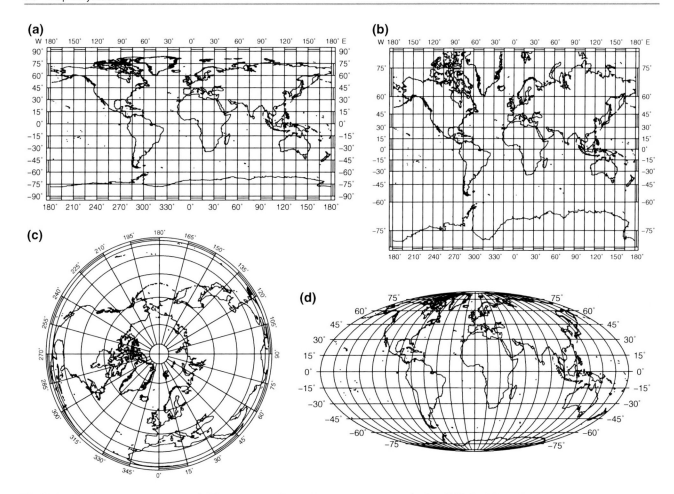

**Fig. 3.6** **a** Equirectangular projection. **b** Mercator projection. **c** Polar orthographic projection. **d** Mollweide projection

Marinus of Tyre in about 100 CE. The Earth is presented as a rectangle and the lengths of degrees of latitude and longitude are the same everywhere. Claudius Ptolemy (90–168 CE) who lived in Alexandria, Egypt and who produced an eight volume geography of the known world described this projection as the worst representation of Earth he knew of. When climate data are plotted on this projection it creates the false impression that the higher latitudes are much more important than they really are. As noted earlier, half of the area of the Earth lies between 30 °N and S, and the other half poleward of those latitudes. In this projection the area between 30 °N and S is 1/3 of the total on the map, and the area poleward of those latitudes is 2/3. Unless you are aware of the problem, the Equirectangular projection vastly overemphasizes the importance of the Polar Regions in global climate.

The map projection most of us are familiar with was designed by Belgian mapmaker Gerardus Mercator in 1569 (Fig. 3.6b). Some version of it hangs on the walls of many classrooms. It is often cut off at 60 °N and S to avoid the obvious distortions at higher latitudes. The version presented in Fig. 3.6b goes to 80 °N and S, and you will notice that on

it Greenland is larger than North America. Mercator's map was made for the specific purpose of navigation at sea. Lines of constant heading or course, also called rhumb lines, are straight lines on Mercator's projection.

Another way to look at the world is from above either of the poles. Figure 3.6c is a Polar Orthographic projection of the northern hemisphere. An orthographic projection is the view you would have if you were infinitely far away. Hipparchus (190–120 BCE), a Greek astronomer and mathematician, is credited both with the invention of this projection and of trigonometry. The polar projection shown here has the advantage that the shortest distance between two points is a straight line. This line is a segment of a 'great circle' passing through the two points and encircling the globe. Long distance air flights generally follow a great circle path, making adjustments to take advantage of the winds.

Finally, Fig. 3.6d shows my personal favorite, the Mollweide equal area projection. It preserves the correct areas of the lands and oceans at the expense of some distortion. It was invented by mathematician and astronomer Karl Brandan Mollweide (1774–1825) in Leipzig in 1805. Figure 3.7

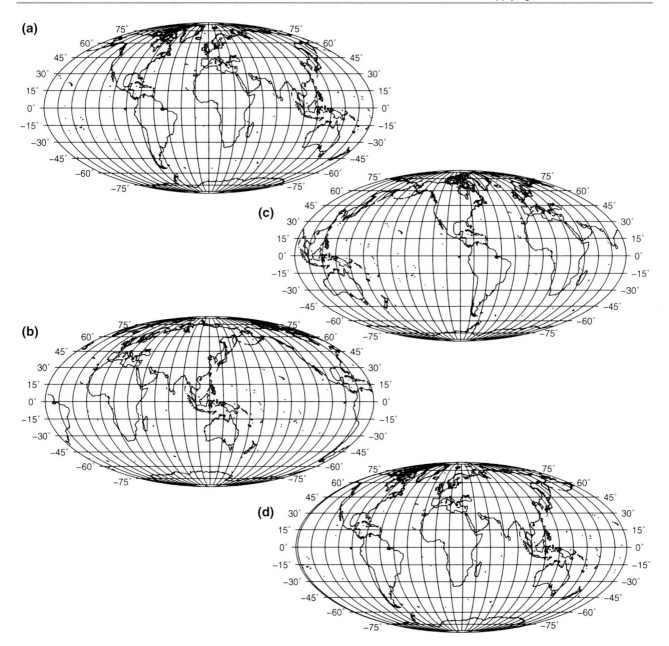

**Fig. 3.7** Four Mollweide Projections centered on the Equator but different lines of longitude. **a** Centered on 0° (Greenwich meridian). **b** Centered on 90 °W (Mississippi Delta)—the America-centric view. **c** Centered on 120 °E—near Beijing—the Chinese view **d** Centered on 38 °E (Moscow)—Many of the Russian and old Soviet maps use Moscow for the central meridian

shows several different versions of the Mollweide Projection centered at different longitudes, showing the distortions that depend on the meridian chosen for the center.

## 3.9  Trigonometry

Trigonometry originally started out as the study of geometrical objects with three sides—triangles. In school you certainly heard about Pythagoras' theorem: in a triangle, one of whose angles is a right (90°) angle, the area of a square whose side is the hypotenuse (the side opposite the right angle) is equal to the sum of the areas of the squares of the two sides that meet at the right angle (Fig. 3.8a). Pythagoras lived about 570–495 BCE and taught philosophy and mathematics in the Greek colonial town of Croton in Calabria, southern Italy. Although attributed to Pythagoras, this and many other mathematical ideas may be much, much older and were apparently known to the Babylonians who had figured it out by about 2,000 BCE.

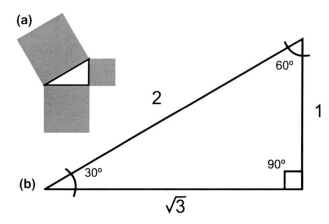

**Fig. 3.8 a** Pythagoras theorem: that the square on the hypotenuse is equal to the sum of the squares on the other two sides. **b** 30°–60°–90° triangle. For discussion, see text

Therein lies another important story. Why is a right angle 90°, and not, say, 100°? It goes back to the Babylonians and their method of counting. They figured the year was about 360 days long. With four seasons, marked by the time the Sun was in its lowest point in the southern sky (the winter solstice), its mid-point in returning north (vernal equinox), northernmost point in the sky (summer solstice), and mid-point in the return south (autumnal equinox) you divide 360 by 4 and get 90. Ninety degrees is one quarter of the way around a circle. The Babylonian mathematicians are also responsible for our timekeeping system. The 'normal' human pulse rate is 1/s. The Babylonians counted in 60s, so 60 s is a minute, and 60 min is an hour. They divided the day into 12 h of daylight and 12 h of night, or 24 h for the Sun to return to the same place in the sky each day. Amazing coincidence that it all works out.

Now remember that in Earth measure, a circle of latitude or longitude is 360°. Each degree is divided into 60 min, and each minute into 60 s. Latitude and longitude expressed to the nearest second will locate something on the surface of the Earth quite well; to within a square 30.92 m (101.45 ft) near the Equator, and with even greater precision toward the poles.

But back to trigonometry. There are a number of trigonometric functions that describe not only aspects of triangles, but many other things, including location and direction on the Earth. The simplest express the relations between the sides of a right triangle, shown in Fig. 3.8b. The sine (sin) of an angle is the ratio between the length of the side opposite the angle divided by the length of the hypotenuse. For example, the sine of 30° is ½ or 0.5; the sin of 60° is $\sqrt{3}/2 = 1.732/2 = 0.866$. The cosine (cos) is the side adjacent to the angle divided by the hypotenuse; cos 30° $= \sqrt{3}/2 = 1.732/2 = 0.866$; the cosine of 60° is ½ = 0.5. Sin 90° is the hypotenuse over the hypotenuse, or 1. There is another common trigonometric function, the tangent (tan).

The tangent is the ratio of the side opposite the angle to the side adjacent. It varies from 0 at 0° to become undefined as it approaches infinity (∞) at 90°. The values of sin, cos, and tan are shown in Cartesian co-ordinates in Fig. 3.9.

What does any of this have to do with the Earth? As discussed above, locations on the Earth are described in terms of coordinates: latitude and longitude. The latitude lines are parallel to the Equator and are numbered from 0° at the Equator to 90° at the poles, with +90° being the North Pole and −90° being the South Pole. The longitude lines meet at the poles. By convention, 0° longitude passes through the observatory in Greenwich, just east of London. The Earth has 360° of longitude, but it is usually divided into 180° east (or positive numbers) and 180° west (or negative numbers). The 180° line happens to conveniently run down the middle of the Pacific Ocean, so it (mostly) marks the International Date Line. Whatever the day is east of the line, say Sunday, it is the next day west of the line, Monday. This allows you to fly from Beijing back to the US and arrive before you left (but jet-lag will tell you otherwise). This problem of keeping track of what day it is was first noted when the one remaining ship of Ferdinand Magellan's fleet arrived back in Spain, on September 6, 1522, having completed the first circumnavigation of the globe in just over 3 years after the original flotilla of five ships had started out August 10, 1519. But there was a problem. The ship's log showed that it was September 5. They had lost a day by sailing westward around the world.

In thinking about the Earth, it is important to realize that the distance between the lines of longitude change with

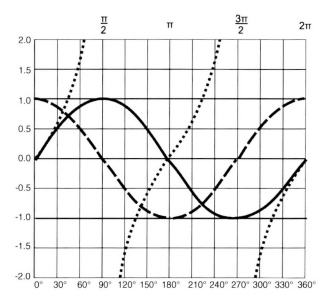

**Fig. 3.9** Values of the trigonometric functions with changing angle: sin = *solid line*; cos = *dashed line*; tan = *dotted line*. The top axis shows degrees expressed in terms of pi (π); the circumference of a circle (360°) is 2π

latitude. How? The distance is proportional to the cosine of the latitude. A degree of longitude at the Equator was originally intended to be 60 nautical miles, but it deceases towards the poles. The length of a degree of latitude remains constant. A minute of longitude at the Equator or a minute of latitude is 1 nautical mile. Like degrees, minutes and seconds, nautical miles are a form of measure that goes back to antiquity in one form or another. In other measuring systems, the nautical mile is 1,852 m or 6076 ft. The length of a degree of latitude is constant between parallels, but of course the length of a degree of longitude varies from 60 nautical miles at the Equator to 0 at the poles. That is where knowing the cosine of the latitude is important. The cosine is 1 at the Equator and 0 at the pole. Now can you guess the answer to the following question? Half of the area of the Earth lies between which latitudes? The cosine of 60° is 0.5, so half of the area of the Earth lies in the 60° of latitude that straddle the Earth between 30 °N and 30 °S. The southern margin of North America is about 30 °N. The part of the northern hemisphere poleward of 30 °N is 1/4 of the total area of the Earth. Here is the shocker. In climate studies it is often assumed that what happens in the 48 contiguous United States is typical of the Earth as a whole. Wrong: the 48 states occupy about 1.6 % of the area of the Earth and are not in any way a representative sample of the planet as a whole.

## 3.10 Circles and Angular Velocity

Circles are another fundamental geometrical form. Figure 3.10 shows the nomenclature associated with a circle. You learned most of these terms in grade school, except perhaps for four: the use of the Greek letter theta (θ) for angle, the letter 's' for the arc subtended by the angle θ, 'radian,' the length of a radius measured along the circumference, and 'chord,' a straight line connecting two points on a circle.

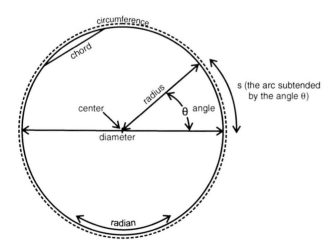

**Fig. 3.10** A circle and its nomenclature

There are two ways of describing the angle θ: either in terms of degrees (360° in a circle) or in terms of radians. You will recall that the ratio of the circumference to the diameter (d) of the circle is denoted by the Greek letter pi (π). The radius (r) is d/2. The ancients had a hard time trying to figure out the value of π. The Babylonians thought it was about 3.125. We now know that π is an irrational number, that is, it goes on forever starting with 3.14159265358979323846264 3383279502884197169399375105820974944459…, and the pattern of digits never repeats.

If you live in Indiana you may know that in 1897 the state legislature very nearly passed a law defining π simply as 3.2. The infamous Bill # 246 was based on the work of physician and amateur mathematician Edward J. Goodwin. He had suggested not one but three numbers for pi, among them 3.2. The Bill was introduced in February by Representative Taylor I. Record, and was initially sent to the Committee on Swamp Lands. The committee deliberated gravely on the question, decided it was not the appropriate body to consider such a measure and turned it over to the Committee on Education. The latter committee gave the bill a "pass" recommendation and sent it on to the full House, which approved it unanimously, 67 to 0. In the state Senate, the bill was referred to the Committee on Temperance. It passed first reading. However, a Professor of Mathematics from Purdue, Clarence Waldo, was passing through town the day the bill was introduced and shown a copy. He was asked if he would like to meet the author of the bill. He declined, remarking that he was "acquainted with as many crazy people as he cared to know." The bill died, saving Hoosiers the problems of roads that had offsets, collapsing buildings and bridges, and a set of measures unique to the state.

But back to the circle and "radians"—why does your computer ask for the angle in radians when you want to get the value of a sin or cos? The circumference of the circle is πd or 2πr. Consider finding the length of the arc subtended by the angle θ, s: If θ is in degrees, s = (2π/360) rθ, which could also be written (2πr)(θ/360); but if θ is in radians, s = rθ. Using radians you can forget the 2π/360 part of the calculation. It is just a lot simpler. If s = 1 rad, the angle θ = 57.2957795°.

Now consider motion around a circle. You know that velocity (v) is expressed in terms of meters per second (m/s or ms⁻¹). But if something is moving in a circle its velocity can be expressed in terms of the angle though which it moves. This is called the angular velocity, designated by the small Greek letter omega, ω. It is not expressed in m/s because it requires knowledge of the size of the circle, i.e. the length of the radius along the circumference, a radian. Angular velocity, ω, is radians per second, r/s. It is the angle (θ) swept out per second by a line from the center of the circle to its circumference. To convert this to m/s you need to know the size of the radius in meters. To keep the calculations simple, angular velocities are almost always expressed in radians.

If $\omega$ radians are swept out in 1 s, 1 rad will be swept out in $1/\omega$ seconds; $2\pi$ radians in $2\pi/\omega$ seconds.

The 'Period' (T) is one complete revolution. T = $2\pi r/v$ or T = $2\pi/\omega$.

Consider what all this means for the Earth. Earth's equatorial radius (r) is now known to be 6,378,137 m. Its circumference is $2\pi r$ which works out to be 40,075,017 m. The Earth rotates through 360° (one sidereal day) in 86,164 s. Earth's angular velocity, $\omega = 2\pi$ rad/86,164 s = 7.29212 × $10^{-5}$ rad s$^{-1}$. The tangential velocity at the Equator is $\omega r$ or 7.29212 × $10^{-5}$ rad s$^{-1}$ × 6,378,137 m rad$^{-1}$ = 465.1 m/s. At the Poles r = 0, so $\omega$ = 0. Because the radius of a circle of latitude varies as the cosine of the latitude, the velocity at 30° (cos = 0.866) is 402.8 m/s and at 60° (cos = 0.5) it is 232.5 m/s.

## 3.11 Centripetal and Centrifugal Forces

Go back and look at Fig. 3.2. Recall that for things going in a straight line 'acceleration' is a change in speed (=velocity), which is m/s. So acceleration is (m/s)/s or usually (and rather confusingly) written m/s$^2$. To change the speed of an object, to accelerate it, a force must be applied. A 'force' is acceleration times a mass: (kg m)/s$^2$. In everyday usage, we think of acceleration as being a change in speed of an object moving in a straight line. If you are in your car and speed up in the straightaway, your car is accelerating; its velocity is increasing. A force is being applied—more energy from the engine to make the car go faster, or fiction from the brake to make it go slower, to decelerate (which is simply acceleration with a minus sign in front of it). Acceleration requires that a force be applied.

However, in scientific parlance the word acceleration has a different meaning. Acceleration can involve an increase or decrease in speed, and/or a change in direction. If you go around a curve while maintaining constant velocity your car is also accelerating. Its direction is changing. A force is being applied—either as gravity from the slope of the roadway or from friction with the pavement or both. Application of a force to an object that has mass is critical to the definition of acceleration.

What are centripetal (from Latin *centrum* "center" and *petere* "to seek") and centrifugal (Latin *fugere* "to flee") forces? You have certainly heard of centrifugal force, but chances are you may not have heard the term centripetal force. These are pseudonyms describing real forces such as gravity, electromagnetic attraction, or the tension on a string with a weight on its end that you swing around your head.

The centripetal force is the force that keeps an object rotating about a center. The centrifugal force is an imaginary force exactly opposite to the centripetal force. It does not exist in the absence of the centripetal force. Physicists do not recognize the centrifugal force as an actual force of nature, but engineers treat it as one just for practical reasons, because for them it is useful in understanding the stresses in a rapidly rotating object, like a turbine.

The value of the centrifugal force depends on three factors: (1) the velocity of the object as it follows the circular path; (2) the object's distance from the center of the path; and (3) the mass of the object. Assume that the velocity of an object going around in a circle is constant. At any given moment the velocity of the object is constant and perpendicular to a line running from the object to the center of the circle; this is called its tangential velocity. If the centripetal force were to suddenly disappear, as it would if the string holding the ball flying around you head were to break, the ball would continue moving at the same speed in a straight line tangent to the circle at the point where the sting broke.

Centripetal force is what keeps things going in an orbit, like the Moon around the Earth or the Earth around the Sun. For these the real force is gravity ('gravitational attraction').

But what is this thing called the centripetal force?

Acceleration can be simply a change in direction without a change in velocity. For example, to change an object traveling 10 m/s forward so that it is instead traveling 10 m/s to the right, forward velocity needs an acceleration of −10 m/s and its velocity to the right accelerated by +10 m/s; if both of these happen instantly and at exactly the same time we have applied a total of 20 m/s acceleration to the object, yet at no point has its speed changed from 10 m/s.

Figure 3.12 shows a ball going in a circle around a center. It is going counter-clockwise, the same direction as the Earth rotating on its axis and going in its orbit around the Sun when viewed from the over the North Pole. This is also the direction of the winds in a cyclone or tornado in the Northern Hemisphere. Its velocity in its curved path is constant. In the following discussion we will use two symbols to express its velocity, $\omega$ if it is in radians, and $\Omega$ if it is in m/s. But the ball is constantly changing direction; it is constantly accelerating, and acceleration requires a force, and for things going around in a circle that force is called the centripetal force.

Now I will jump ahead and tell you that expressed in terms of the angular velocity $\omega$ of the ball about the center of the circle, the centripetal force ($F_{cp}$) is:

$$F_{cp} = mr\omega^2$$

where $m$ is the mass of the ball, $r$ is the radius of the circle, and $\omega$ is the angular velocity expressed in radians per unit time.

Expressed in terms of conventional velocity, $\Omega$, the centripetal force ($F_{cp}$) is:

$$F_{cp} = ma_{cp} = m\frac{v^2}{r}$$

where $m$ is the mass of the ball, $a_{cp}$ is the centripetal acceleration, $v$ is the velocity in m/s, and $r$ is the radius of the circle in m. Of course we could use velocity in mph and r in inches, but by now you realize that our archaic American system of measurement can make calculations quite difficult.

Note that in these formulas the speed is squared; this means that doubling the speed around the curved path requires that the centripetal force must be quadrupled to keep the ball moving along the same path. Note also that there is an inverse relationship with the radius of the circular path. Halving the distance of the ball from the center means that

**Fig. 3.11** Experiencing centripetal and centrifugal force

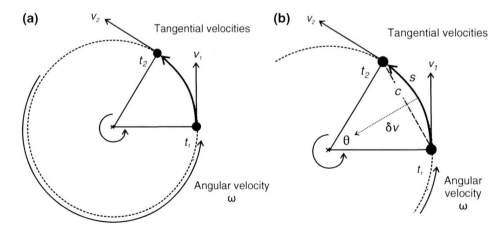

**Fig. 3.12  a** A ball going around in a circular path at constant velocity. At time $t_1$ it has tangential velocity $v_1$, at time $t_2$ it has tangential velocity $v_2$. **b** An enlargement of part of A showing more of terminology used in the discussion. $\Theta$ is the angle between the lines from the center to $t_1$ and $t_2$; $s$ is the arc between the position of the ball at times $t_1$ and $t_2$; $c$ is the chord connecting the locations of the ball at times $t_1$ and $t_2$; and $\delta v$ is the direction and magnitude of the centripetal force required to move the ball along the arc from its position at time $t_1$ to $t_2$

twice the centripetal force is required to keep it in the smaller circular path.

At this point there should be a little question in the back of your mind. Where did the squares of the velocities come from? This takes more than a bit of an explanation to understand. So here goes.

First we need to introduce some more terms with precise meanings in physics: scalars and vectors. Scalars are quantities that are simply described by a magnitude or numerical value alone. You have already run into scalars in the discussion of Earthquake scales, and the Beaufort and TORRO wind scales in the last chapter. Vectors are quantities that are described by both a magnitude and a direction.

Figure 3.12 shows our counter-clockwise rotating ball at two points in its trajectory, $t_1$ and $t_2$. The ball has the same instantaneous velocity (a scalar value, and here we are going to use the angular velocity $\Omega$, expressed in numbers like m/s) at each point, $v_1$ at time $t_1$ and $v_2$ at time $t_2$, but it is traveling in a different direction at each point (different vectors). The difference in time between the two points $(t_2 - t_1)$ is not shown, but it would be $\delta t$. The dotted line $\delta v$ is the force acting on the ball to keep it in its circular path. We use the expression $\delta v$ because it keeps changing its orientation as the ball moves. It is always at a right angle to the path, which means that it is pointing to the center of the circle. The angle $\theta$ is the angle between the two locations of the ball. Figure 3.13 shows you where $\delta v$ came from.

In Fig. 3.13a the two positions of the ball have been superimposed, showing the two vectors $v_1$ for time $t_1$ and $v_2$ for time $t_2$. Connecting the ends of the tangential velocity vectors is another vector, the dotted line; it is labeled $\delta v$ in Fig. 3.13b. The three vectors now form an isosceles triangle. Now remember that for an object going around in a circle, acceleration can be simply a change in direction while the ball moves at a constant speed. Acceleration requires a force. But recall that a force is a mass times an acceleration. In this case the mass is the ball itself; since it remains the same we will not show it in the equations below. The $\delta v$ is the acceleration required to change the direction of $v_1$ to that of

$v_2$. Keep the direction of this force in your mind as you look back at Fig. 3.12b. The $\delta v$ in Fig. 3.12 is the same $\delta v$ as shown in Fig. 3.11. It represents a force and it points directly toward the center of rotation of the ball.

Go back to Fig. 3.12b. The length of the arc between points $t_1$ and $t_2$, $s$ is the angle $\theta$ expressed in radians, $r$; we'll denote it as $\theta_r$. Now, in your mind move the ball at point $t_2$ back along the arc toward point $t_1$ until the distance between points $t_1$ and $t_2$ becomes very, very small. We will use the small Geek letter $\delta$ to denote a difference. The difference in time between time $t_1$ and time $t_2$, $\delta t$, becomes very short. The angle $\theta_r$ becomes very small, and the arc $s$ and chord $c$ become very short. As $s$ and $c$ become shorter, they ever more closely coincide, until when $s$ is infinitesimally short $c$ and $s$ have almost exactly the same length. In mathematical terms we say

$$\text{as } \delta t \to 0, \quad c \to s$$

Now the absolute values of $v_1$ and $v_2$ are the same, so we can call them simply $v$. Now, the difference in velocity $\delta v$ compared to the velocity $v$ is equal to a change in the length of the chord $c$ compared to the length of a radian $r$:

$$\frac{\delta v}{v} = \frac{c}{r}$$

But since we are talking about very small units of time, the length of the chord, $c$, is essentially equal to the length of the arc $s$, which is the velocity $v$ times the length of time $\delta t$: $v\delta t$. This can be substituted into the equation given above so that it becomes:

$$\frac{\delta v}{v} = \frac{v\delta t}{r}$$

And using simple algebra to rearrange the terms, this becomes

$$\frac{dv}{dt} = \frac{v^2}{r}$$

And remember that centripetal acceleration $a_c$ is a change in velocity with time ($\delta v/\delta t$)

$$a_{cp} = \frac{v^2}{r}$$

And remember that a force, $F$, is a mass times an acceleration; then the centripetal force $F_{cp}$ is

$$F_{cp} = ma_{cp} = m\frac{v^2}{r}$$

EUREKA! This is the second of the equations for centripetal force given at the beginning of this discussion. That is where the $v^2$ comes from!

The circuitous route to this conclusion led us though the exotic terrains of differential calculus and vector algebra and

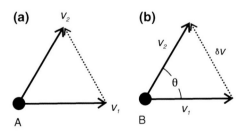

**Fig. 3.13 a** The two vectors superposed. The *dotted line* connects the ends of the two vectors and represents the force required to change $v_1$ into $v_2$. **b** Figure A relabeled to show terminology used in the discussion below

you are still alive. Congratulations, Leibniz and Newton would both be proud of you.

Now to convert from metric velocity $\Omega$ to angular velocity $\omega$, remember that distance $d$ = rate $r$ times time $t$ ($d = rt$). For our ball, going in a circular path at constant speed, the time $t$ for one revolution (or 'period,' $T$) means going around the circumference of the circle ($2\pi r$) in time $T$. Its angular velocity $\omega$ is then:

$$\omega = \frac{2\pi r}{T}$$

that is to say, the angular velocity is equal to one trip around the circle in a given unit of time.

Now, substituting $\omega$ for $v$ in the equation $a_{cp} = \frac{v^2}{r}$ gives us

$$a_{cp} = \frac{(4\pi^2 r^2)/r}{T^2} = \frac{4\pi^2 r}{T^2}$$

But $\frac{4\pi^2 r^2}{T^2}$ is $\omega^2$ so this expression for the centripetal force becomes simply

$$F_{cp} = mr\omega^2$$

EUREKA AGAIN! This is the first of the equations for centripetal force given at the beginning of this discussion. That is where the $\omega^2$ comes from!

Now for the really good news—the centrifugal force, $F_{cf}$, is simply the opposite of the centripetal force, so all you have to do is put a minus sign in front of the equations so they become

$$F_{cf} = -ma_{cf} = -m\frac{v^2}{r}$$

and

$$F_{cf} = -mr\omega^2$$

At the beginning of this section we mentioned the orbits of the Moon around the Earth and the Earth around the Sun. Unfortunately, as will be discussed in detail in later chapters, the orbits are not circular but elliptical, and their velocities in orbit are not constant but are faster when they are close to each other and slower when they are further apart. This means that the calculations for centripetal force are somewhat more complicated.

## 3.12  Exponential Growth and Decay

Exponential growth and decay are concepts very useful in understanding both why we are on the verge of a climate crisis and how we know about the ages of ancient strata.

First, let us consider what these terms mean. They can be expressed by simple algebraic equations, well worth knowing if you are trying to keep track of your savings account and investments, and are planning for the future. The expression is

$$y = a \times b^t$$

where a is the initial amount, b is the growth or decay constant, and t is the number of time intervals that have passed. If b is larger than 1, y will increase with time, or grow. If b is less than 1, y will get smaller with time, it is said to decay away. If the growth or decay is a constant percent, the equation can be written

$$y = a \times (1 + b)^t \text{ for growth, or}$$
$$y = a \times (1 - b)^t \text{ for decay.}$$

Growth can be steady through addition or subtraction of the same number; or it can be geometric, or exponential, as shown in Fig. 3.14.

What does all this mean in practical terms? If you put $100.00 into a savings account with 3 % interest per year for 10 years, the equation and its solution looks like this:

$$y = \$100 \times (1 + 0.03)^{10} = \$100 \times (1.03)^{10}$$
$$= \$100 \times (1.3439) = \$134.39$$

The '1.3439' is the result of multiplying 1.03 by itself ten times. However, you may also wish to know how long it will take to double your money at 3 % interest. This can be done by dividing the *natural logarithm* of 2 (ln2), which is 0.693147 by the interest rate (0.03), and the answer comes out to be 23.1 years.

Now here is a little trick you can use to be able to do these calculations in your head. A close approximation of the natural logarithm of 2 (ln2), is simply 0.70, so you can simply divide that number by 0.03. But most of us would have to write it down to make sure we get the decimal point in the right place, so here is the trick: multiply both sides by 100 so you don't have to convert 3 % to 0.03. Then it becomes a matter of dividing 70 by 3 = 23.33 years, a close approximation to the exact answer. You will immediately realize that if you were getting 7 % interest, the time required to double your money would be only (about) 70/7 or 10 years.

On the other hand, if you wanted to know what interest rate you would need to get to double your money in, say, 7 years, you would simply rearrange the equation to divide the natural logarithm of 2 (ln2 = $\sim$0.7) by the number of years, and the answer would be 10 %.

Where, you will ask, did the *natural logarithm* come from and why is it 'natural'? In school you learned about logarithms, but you probably only used logarithms to the base 10 (as we did in the section on orders of magnitude). The only

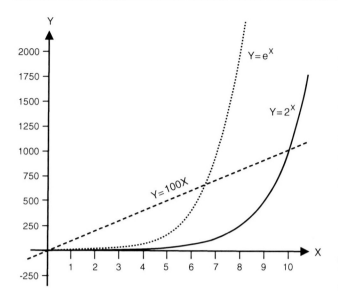

**Fig. 3.14** Growth scenarios. The x-axis is time; the y-axis the amount of growth. The three versions are linear (Y = 100X), geometric (Y = 2$^X$), and exponential (Y = e$^X$). If you invert these or put a negative sign on the exponent, they become the decay *curves* shown in Fig. 3.15

thing natural about 10 is that we have 10 fingers and 10 toes. The natural logarithms have as their base an 'irrational number,' specifically 2.71828182845904523536… (The dots mean the number is infinitely long, which is why it is 'irrational'). The number is represented by '*e*' in equations. It was 'discovered' by Swiss mathematician Jacob Bernoulli in the late 17th century, when he was exploring the effects of compound interest, but it actually seems to go back to money lenders in Babylonian or older times. It crops up as an important number all through science, mathematics, and nature. It is the basis of the spiral used by many animals and plants, such as snails and sunflowers.

Incidentally, when I was a postdoctoral fellow in Basel, Switzerland I worked in the Geological Institute, which is located in a building called the 'Bernoullianum,' named to honor Basel's mathematicians. One of my good student friends was Daniel Bernoulli. For the history-of-science buffs, I lived in an apartment in Eulerstrasse. Leonhard Euler was another Basel mathematician. Among other things, he was the one who named that supremely important irrational number '*e*' and calculated its value to 18 decimal places. Today *e* is called the Euler Number.

The root cause of most environmental problems can be traced to the exponential growth in the human population. 'Civilization' and particularly the 'industrial revolution' made it possible for more food to be grown to feed a larger population.

Surely you have heard of the Reverend Thomas Robert Malthus (1766–1834). He was an English scholar now remembered chiefly for his theory about population growth. His book *An Essay on the Principle of Population* went through six editions from 1798 to 1826. He argued that sooner or later population gets checked by famine and disease. He did not believe that humans were smart enough to limit population growth on their own.

Then, during the middle of the 20th century, the spread of antibiotics to prevent disease and extend life, and pesticides and fertilizer for protecting crops and growing more food made it possible for the growth rate of the human population to rise to 2.19 % per year. That is a doubling time of 31.65 years. Indeed, Earth's human population in 1961 was about 3 billion, and in 2000 it was about 6 billion. When Cesare Emiliani published his book *The Scientific Companion* in 1988 he calculated that at its rate of growth at the time, doomsday would occur in 2023. Doomsday was defined as the date when there would no longer be standing room for the human population on land. He had a second, really bad doomsday that would be reached only a few decades later, when the Earth is covered by human bodies expanding into space at nearly the speed of light. Fortunately, by 2001 the global human growth rate had slowed to 1.14 %, for a doubling time of 61 years. In 2011 the global growth rate has dropped to 1.1 %, a doubling time of 63 years.

Professor Al Bartlett at the University of Colorado has said that "The greatest shortcoming of the human race is our inability to understand the exponential function."

Today the global population is 7 billion. If nothing happens there will be 14 billion people on Earth in 2074. Most projections of the world's future population suggest an increase to 9 billion in 2050, with growth slowing or declining thereafter. Others project the population of 14 billion to be reached around the year 2200, but I believe that none of these will come to pass. Famine and disease will intervene. The current US population growth rate is 0.833 %, below the world average, but the highest of any industrialized country. European countries and Japan have negative population growth, while the Middle East and Sub-Saharan Africa have replaced Latin America in having the highest rates of population growth. In 1979 China became the first country to establish a policy of zero population growth. The goal is expected to be achieved by 2030. The consumption of natural resources both renewable, such as water, food, and wood, and non-renewable such as coal, natural gas, petroleum and metals increases both as the population grows and as living standards rise.

Similarly, decay can be steady through addition or subtraction of the same number; or geometric, or exponential, as shown in Fig. 3.15. Examples of decay include Carbon-14, used in carbon 14 dating, one of many forms of radiometric dating.

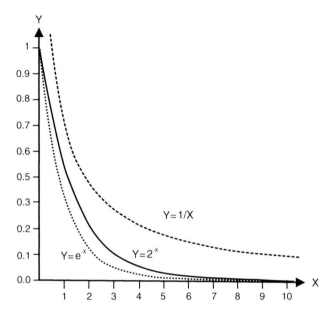

**Fig. 3.15** Decay scenarios. The x-axis is time; the y-axis the amount of decay. Again, the three versions are linear (Y = 1/X), geometric (Y = 2$^{-X}$), and exponential (Y = e$^{-X}$)

In exponential decay, the 'b' in the above equations is less than 1. It has a special Greek letter expression in mathematics, $\lambda$ = lambda. There are three terms commonly used to describe exponential decay: mean lifetime, half-life, and the 'decay constant' $\lambda$. In this book you will first encounter these in the discussion of radioactivity as a tool for determining the ages of things. So we will jump ahead a bit and use radioactive decay as our example. An atom of a radioactive element loses something from its nucleus with time, changing into either a different isotope of the same element (a different number of neutrons) or into a different element (a different number of protons). This transmutation is a random process; you can't tell which atom will decay next, but if you have a lot of them, say $6.022 \times 10^{24}$ or so, as you would have in a kilogram of uranium, the randomness goes away, and you can determine how fast they are decaying. The 'mean lifetime' (usually denoted by the Greek letter $\tau$ = tau), is intuitively obvious; it refers to the average lifetime of an atom. The average lifetime is when 1/e, or about 36.8 %, of the original number of atoms remain. The 'half-life,' $t_{1/2}$, is the length of time it takes for half of the original number of atoms to decay. The 'decay constant,' $\lambda$, is the inverse of the mean lifetime (1/$\tau$). Since this describes an exponential decay you can guess that the magic number '$e$' is going to be involved somehow, this time as the base of the natural logarithm ln.

The three terms are related as follows:

$$\lambda = 1/\tau = \ln2/t_{1/2} \quad (\ln2 = 0.69314)$$

Exponential decay crops up not just in discussing radioactivity, but all over the place: (1) the decrease of atmospheric pressure with height decreases approximately exponentially at a rate of about 12 % per 1000 m; (2) if an object at one temperature is placed in a medium of another temperature, the temperature difference between the object and the medium follows exponential decay; (3) the intensity of light or sound in an absorbent medium follows an exponential decrease with distance into the absorbing medium (important to know if your teenager has brought home a boom-box with strobe lights and you are trying to figure out how much insulation you will need to be able to sleep).

## 3.13   The Logistic Equation

Is there any way to predict the slowing of population growth? A method was found by Belgian mathematician Pierre François Verhulst (1804–1849). In 1838 he published his first version of what is now known as the 'logistic equation:'

$$P(t) = \frac{1}{1 + e^{-t}}$$

where $P(t)$ represents the population as a proportion of the maximum possible number of individuals at time $t$ and $e$ is the Euler Number, 2.71828183. If the population is 1, i.e. the maximum possible number of individuals, this is said to be the 'carrying capacity' of the environment (Fig. 3.16).

Exponential growth is characteristic of something that is 'out-of-control' such as the spread of plague or an infestation of undesirable organisms. If aliens were examining the Earth, they would probably conclude it is currently undergoing an infestation by humans. Infestations are usually restricted to particular regions, and are self-correcting when food supplies run low or disease runs rampant. The current explosion of human population appears to be a unique event in Earth's history. Never before has a single species spread so ubiquitously and so rapidly. How the correction will be made, and what it will look like are unknowns.

Another thing to note: There was a tiny flaw in Verhulst's idea about logistic growth. Unless you use just the right numbers that nice constant population on the right side of the equation won't last long. As you will see in Chap. 11, it can easily turn into what are called chaotic solutions. Each solution is different for even tiny changes in the initial numbers you use. Some solutions produce wild fluctuations (as is well known from rabbit populations); other solutions show total collapse of the population, which means extinction.

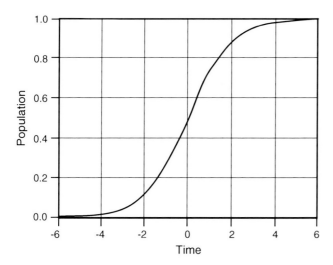

**Fig. 3.16** The 'logistic curve' devised by Verhulst to describe the growth of a population through time until it reaches stability. The population increase approximates exponential growth until it reaches $t = 0$, then the growth rate begins to decline. At population 1, the population is stable; the birth rate equals the death rate (or starvation rate as Malthus would put it)

## 3.14 Graphs

Graphs are a means of visualizing the relationship between different sets of data, often referred to as parameters. The most common format for a graph is with classic Cartesian coordinates. You may run into some jargon in the discussion of a graph. The horizontal axis is called the abscissa (from Latin *abscissa linea* "a line cut off; the vertical axis, called the ordinate (from Latin *ordinatus*, "arrange, set in order').

The growth of the human population over the past 12 millennia is an excellent example. Figure 3.17 shows the data plotted on an ordinary graph with both axes using simple numerical progressions. Other than showing that there has been an explosion of the human population since the Industrial Revolution, it is singularly uninformative.

Here I must introduce what may seem an odd convention for representing time that will be followed throughout this book. In most of the plots of a parameter against time in years that you will see elsewhere, the oldest will be on the left and the youngest on the right. The calendar numbers of years on the ordinate axis of the Cartesian coordinate system increase to the right. However, geologists, like me, we think of time as positive backwards; we think in terms of age. For us, the age is a positive number and so age increases to the right. Throughout this book youngest is on the left, oldest on the right. Don't worry, you will get used to it.

But a graph does not need to use the same numerical scale for both axes. Figure 3.18 shows the same data plotted on semi-logarithmic graph paper. The horizontal axis is the same simple numerical age progression, but the vertical axis is a logarithmic progression. It shows the changing major trends of population growth and decline over the millennia. The effect of wars cannot be seen. The population declines are related to a single cause—pandemics.

Here it is useful to note that the growth of the human population is the root cause of the climate and other changes Earth is currently undergoing.

**Fig. 3.17** Growth of the Earth's human population. Linear plot. *Solid line* is based on census data since ca. 1900 and estimates for the more distant past, *dotted line* is United Nations' current projection for the future

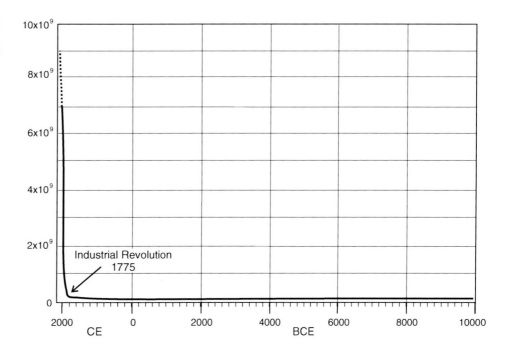

**Fig. 3.18** Growth of the human population based on the same data as given above using a semi-logarithmic plot showing the effect of agricultural innovations and pandemics. The large increase over the past few centuries is due to a variety of effects of the Industrial Revolution. The semi-log plot presents much more information in a smaller space

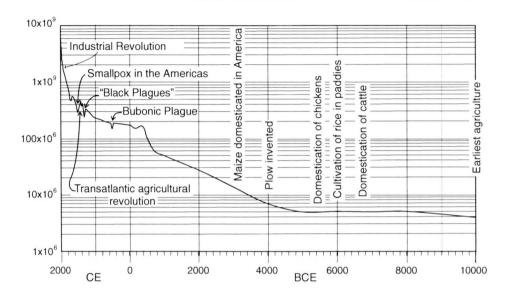

## 3.15   Statistics

Statistics is defined as the branch of science or mathematics concerned with the systematic collection and arrangement of numerical facts and their analysis and interpretation. There are a number of terms used to analyze and describe a set of data: (1) the average or mean is the sum of the data divided by the number of data entries; (2) the median is the datum in the middle of the numerically ordered list of data. (3) The standard deviation (SD) (represented by the Greek letter sigma, $\sigma$, describes the amount of variation in a set of data values. A standard deviation close to 0 indicates that the data points tend to be very close to the mean while a high points tend to be very close to the mean while a high standard deviation indicates that the data points are spread out over a wide range of values.

An example of a statistical analysis is shown in Fig. 3.19. It shows the NASA GISS (NASA = National Aeronautics and Space Administration; GISS = Goddard Institute for Space Studies) data compilation for the global average annual temperature of planet Earth from 1880 through 2014. The data are presented in terms of the anomaly (difference) from the average temperature from a base period: 1951–1980. The data have been handled very systematically but because of the uneven distribution and activity of weather stations the exact temperature of the planet for the 1951–1980 base period can only be determined to within a few

**Fig. 3.19** Scatter plot of the NASA GISS data set for Earth's global mean annual temperature from 1880 to 2014. Fitted through the data is a regression line. The slope on the line is about 0.7 °C (1.26 °F) per century

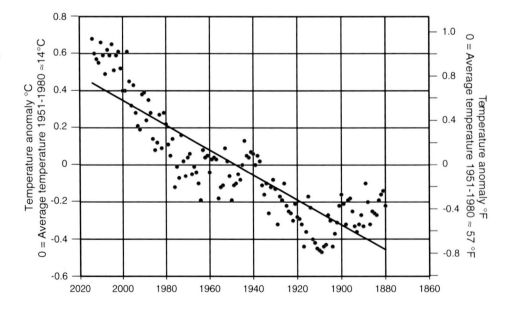

tenths of a degree, but is taken to be 14 °C, equal to 57.2 °F. The squiggly equal signs in the figure should be read as 'approximately.' Chapter 29 deals with the temperature record in detail.

Do you recall earlier I used the label "Global warming denier" rather than "Global warming skeptic"? A global warming denier would be likely to look at this graph and agree there is a trend in the data showing a change in the average temperature of the Earth, and that it clearly shows the Earth is getting cooler. That is, until you point out the fact that the time scale is going the opposite way to what they might be familiar with, having the most recent temperatures on the left and the oldest (greatest age) on the right. If at this point they claim there is no trend then they are denying evidence which clearly shows their position to be false, which is why they are called "deniers" rather than "skeptics". Try it!

## 3.16  Summary

If you managed to get through this and the last chapter you must be congratulated. It may seem like a lot to remember, but this background is essential to discussing and understanding what is going to happen in the future as we perform the grand experiment on planet Earth. If you get confused in later chapters, you can always return here to get the basics of scientific literacy.

Now we go on to the learn how it was discovered that our planet has a long history, and that the climate of the past 6,000 years, when humans developed 'civilization,' has been very unusual, having been both much warmer than most of the ice-age climate of the past few million years, and quite stable compared to the other brief warm intervals. All that is about to end.

A Timeline for this chapter:

| Year | Event |
| --- | --- |
| ~1800 BCE | Babylonians discover the relations among the sides of a triangle attributed to Pythagoras |
| ~500 BCE | Pythagoras develops the proof of his theorem of the triangle |
| ~200 BCE | Eratosthenes measures the circumference of the Earth and develops the latitude-longitude system for mapping the Earth |
| ~140 BCE | Hipparchus devises the Polar Orthographic map projection |
| 100 | Marinus of Tyre invents the Equirectangular map projection |

(continued)

(continued)

| Year | Event |
| --- | --- |
| ~150 | Claudius Ptolemy of Alexandria declares the Equirectangular map projection to be the worst representation of the Earth he knows of |
| 476 | Fall of the Roman Empire |
| 1066 | Norman Conquest of Britain |
| 1346 | The Black Plague reaches Europe; as much as ½ of the population dies |
| 1492 | Columbus ships carry smallpox to the Americas; having no immunity more than ½ of the native Americans will die |
| 1522 | Ferdinand Magellan's fleet completes the first circumnavigation of the globe |
| 1569 | Gerardus Mercator devises a new map projection to be used for navigation at sea |
| 1610 | Galileo Galilei observes the moons of Jupiter and realizes they might be a clock in the sky |
| 1633 | Galileo Galilei's trial for the inconvenient truth that the Earth orbits the Sun |
| 1636 | Pierre de Fermat devises a three dimensional coordinate system but doesn't publish it |
| 1637 | René Descartes publishes his two-dimensional coordinate system |
| 1654 | Ferdinando II de Medici devises a sealed glass tube thermometer |
| 1690 | Jacob Bernoulli 'discovers' the mathematical constant e |
| 1714 | Daniel Fahrenheit substitutes mercury for alcohol in a thermometer |
| 1714 | Britain establishes the Board of Longitude and offers a prize to the first person to devise a practical method of determining longitude at sea |
| 1724 | Fahrenheit develops his first temperature scale; it has three calibration points. |
| 1737 | John Harrison produces his first sea-going chronometer |
| 1742 | Andes Celsius publishes his temperature scale based on the freezing and boiling points of water |
| 1765 | John Harrison receives the final payment for his chronometer from the Board of Longitude |
| 1775 | James Watt invents a practical steam engine, starting the Industrial Revolution |
| 1786 | U.S. Continental Congress adopts a decimal monetary system |
| 1798 | Thomas Malthus publishes *An Essay on the Principle of Population* |
| 1805 | Karl Mollweide devises the equal area projection that bears his name |
| 1812 | Napoleon spreads the metric system through Europe |

(continued)

(continued)

| Year | Event |
|------|-------|
| 1815 | Napoleon defeated at the Battle of Waterloo |
| 1848 | Lord Kelvin (William Thomson), publishes *On an Absolute Thermometric Scale* |
| 1880 | Globally consistent temperature records begin |
| 1897 | Bill # 246 defining the value of pi as 3.2 is introduced into the Indiana Legislature |
| 1952 | Paul Niggli publishes *Rocks and Mineral Deposits* |
| 1958 | The National Aeronautics and Space Administration (NASA) is created |
| 1965 | British announce they will change to the metric system over 10 years |
| ~1970 | British start using Celsius temperature scale |
| 1974 | British schools start teaching the metric system |
| 1979 | British change to a decimal monetary system |
| 1991 | Cesare Emiliani publishes *Avogadro's number: a royal confusion* in the Journal of Geological Education |
| 1998 | James Rodger Fleming publishes his book *Historical Perspectives on Climate Change* |
| 2015 | Britain is still working on going metric |

If you want to know more:

Sobel, D., 1996. *Longitude: The True Story of a Lone Genius Who Solved the Greatest Scientific Problem of His Time.* Penguin, 192 pp.—An engrossing account of solving the longitude problem.

Music for contemplation: J.S. Bach: The Goldberg Variations (Glen Gould's interpretation is my favorite).

Libation: Give a toast to Napoleon and the spread of the Metric System with another glass of Absinthe. Or, if you consumed the rest of the bottle as you worked your way through the section on centripetal and centrifugal forces, try glass of the Pastis de Marseille; it is similar in taste to Absinthe but was never banned. The French National Anthem, the Marseillaise was composed in 1792 by Claude Joseph Rouget de Lisle in Strasbourg, a city in beautiful Alsace on the west bank of the Rhine River. It was a call-to-arms to protect the French Revolution of 1789 from Austrian and Prussian threats. Originally titled "Chant de guerre pour l'Armée du Rhin," (War Song of the Army of the Rhine). It acquired its nickname after being sung in Paris by volunteers from Marseille marching on the capital in support of the Revolution. The French National Convention adopted it as the Republic's anthem in 1795.

As you savor your Absinthe or Pastis you may wish to hear a rousing version of the French anthem, the Marseillaise at https://www.youtube.com/watch?v=CtaPWBzqEa4. If you have another glass of your apéritif you may wish to tour Versailles through the centuries at https://www.youtube.com/watch?v=tAIhuS7P24E.

## Intermezzo III. Graduate Study: Four Universities and Two Degrees in Four Years

My program of graduate study was certainly strange by today's standards. As I have said before, so much in life depends on serendipity, chance meetings that turn out to change the direction you were headed. As I described above, when I started as an undergraduate at Southern Methodist University, I told my advisor I wanted to be a paleontologist and to be a teacher in a college or University. It turned out he was a biologist, very much interested in evolution. He advised me that in order to understand fossils one first needs to be familiar with biology. I should be a biology major and then go into paleontology in graduate school. I followed his advice. My undergraduate degree was in Biology with a minor in Chemistry. Somehow long-term planning was in my blood.

My closest friend in high school had an uncle who taught at the University of Texas. Tom Cranfill was professor of English, specializing in Shakespeare. I had met Tom many times when he was visiting in Dallas. By 1952 he had already been to post-war Europe several times. I knew that Tom was on the Graduate Admissions Committee for the University of Texas, so during my senior year at SMU I asked him for advice on how to get into graduate school. What he told me then is still very good advice today.

He advised me to figure out what I wanted to do, find out who the most outstanding academics in the field were, contact them, and apply to their institutions. He also noted, "you can always change your mind after you're admitted and get started." It is surprising that even today few applicants for undergraduate or graduate study have a very clear idea of what they want to do, and even fewer ever bother to contact or meet the faculty they might work with.

Following Tom's advice, during my senior year at Southern Methodist University, I had discussions with Bob Perkins, the Geology Department's paleontologist. I told him I wanted to become a micropaleontologist and teach in a University. He wisely suggested I get as broad a background as possible, and study more than one group of microfossils. We went through lists of who was where, and decided on a plan which I then followed. The two major groups of microfossils being studied at that time were the foraminifera, single-celled amoeba-like animals that produce a huge variety of tiny shells, mostly made of calcium carbonate, and ostracods, tiny shrimp-like animals, each producing two shells, right and left, which are mirror images of each other. There are over 100,000 recognized species of foraminifera, but only a few thousand species of ostracods. Ostracod shells are sometimes quite elaborately ornamented, and you would never guess from the shell what the animal that lived inside looked like. The plan was that I would start by studying ostracods with an eminent US micropaleontologist, Harold Scott, at the University of Illinois in Urbana for my Master's degree. Then I would go to Stanford University and study the foraminifera, under Hans Thalmann, a professor from Switzerland. All of this was, of course, planned without my having applied or been accepted at either institution.

Step one was to apply to Illinois, and in my application to Illinois I spelled out who I wanted to work with and what I wanted to do. However, I also wanted to see if I could get in another year of study in Europe, so I went through the Institute of International Education in New York to apply for a year of study in Switzerland. Amazingly, although I was missing most undergraduate geology courses, I got a Fellowship offer from Illinois. From Switzerland I was offered a Fellowship to the University of Zürich sponsored by the 'Swiss Friends of the USA.' I called the Department Head at the University of Illinois, George White, and told him of my dilemma. George White didn't hesitate to tell me to accept the offer from Switzerland. I could fill in my undergraduate geology courses there and he suggested that if I reapplied I would then almost certainly be at the top of their list for Fellowship offers the next year. It was the long-distance beginning of an extraordinary friendship with another extraordinary person.

During the summer of 1955 I attended the University of Illinois' Geology Field Camp in Sheridan, Wyoming. It was my first real experience in seeing and studying geology in the field guided by expert faculty. It left me convinced I had made the right choice in choosing geology for a career.

In Switzerland I was officially enrolled at the University of Zürich, but the Geology Department was actually part of the Eidgenossische Technische Hochschule, the Swiss Federal Institute of Technology. My instructors included some of the best geologists in Europe. It was the first year of his teaching career for Rudolf Trümpy, who was to become one of the most distinguished geologists of the 20th century. We were a group of about 8 students taking the courses, and I was the only foreigner. There were field trips in the Alps, with lots of hiking and camaraderie with overnighting in alpine cabins made more comfortable by bottles of wine we carried along, but there was a problem—I couldn't understand what they were talking about.

One day in the fall of 1955 I remember well was Saturday, November 5. The peace treaty between the Allies and Austria had been signed on May 15, 1955, and had gone into effect on July 27. The withdrawal of Allied troops took place on October 25. But for many the great culminating event was the re-opening of the Vienna Opera. It took place on Saturday, November 5. I had taken that weekend for a trip to the Bernese Alps and was staying in the village of Mürren. It sits on a narrow ledge of pastureland on the vertical western wall of the Lauterbrunnen Valley, directly across from the Jungfrau. My little room had a magnificent view across the valley, and a radio. That evening there was a clear sky and beautiful Moon. I listened to the performance of Beethoven's Fidelio under the direction of Karl Böhm. One could hardly escape the significance for the Austrians; Fidelio is about prisoner's being released from confinement.

Somehow I had expected that since Zürich was in the middle of the 'German-speaking' part of Switzerland I would be able to converse with my fellow students. The Swiss speak a number of dialects of German that date back to the Middle Ages; they are literally unintelligible to outsiders. Switzerland is in many ways like a giant Club; if you speak one of the dialects you are a member of the Club, and if you don't you are a foreigner and treated politely as such. I found I couldn't understand anyone unless they made an effort to speak to me in 'high German,' a foreign language to them. From September, when I arrived, until December, except for the field trips, I walked around in a fog and very lonely—no friends. Then, one day it happened. I finally understood what all those funny sounds were. The great revelation came when I discovered that the Swiss word for 'was' (Past participle of to be) was 'gsi' (in German it is 'gewesen'). Within a couple of months I had mastered the Zürich dialect and suddenly had a lot of Swiss friends. I was accepted. The field trips in the spring and summer were even more exciting than those in the fall, and the evening discussions among faculty and students over interesting Swiss wines were an educational experience never to be forgotten.

It was also in Zürich that I was invited, for the first time, to a faculty member's home for dinner. Paleontology was taught in the University, and Emil Kuhn-Schnyder was our instructor. He was a vertebrate paleontologist, working on fossil reptiles 250 million years old. They occurred in rocks in a quarry at Monte San Giorgio in Ticino, the southern, 'Italian-speaking' Canton of Switzerland. These fossils occur in solid limestones, and Kuhn-Schnyder's team would quarry blocks and then X-ray them to see if there was a vertebrate fossil inside. If there was it would be carefully and painstakingly exposed using dental drills. As a result the Zürich institute had a fabulous collection of fossil Triassic marine reptiles. Kuhn-Schnyder invited his students to his home for a dinner of Ticinese cuisine prepared by his wife, and appropriate wines from Ticino. Again, the discussions went on till midnight. It was a custom I adopted during my teaching career. There is so much to be learned that cannot be passed on in the classroom.

Speaking of things that are difficult to be passed on from teacher to student: in a European university there are usually no prerequisites or requirements for courses. I happened to notice in the University of Zürich's catalog a course entitled 'Bewegungsgruppen der Kristallographie.' It was given by Professor Johann Jakob Burkhardt in the Department of Mathematics. To his great delight I was one of four students who showed up at the appointed time and place. I was a geology student; the others were majors in mathematics. He explained that it was all going to be very simple, the mathematical derivation of the 230 space groups in crystallography through motions in Euclidean space. All you needed to know was several things I had never heard of, including group theory, set theory, Bravais lattices, and wallpaper groups. The introductory lecture was totally beyond my comprehension. Nevertheless, I had a class in the same building the hour before, so I thought that maybe I might eventually catch on to what this was about. I even bought Burkhardt's book Bewegungsgruppen der Kristallographie. It was no help. After the fourth lecture or so, Professor Burkhart asked me to come to his office. I thought he was going to ask whether I was understanding anything at all. But no, he had apparently mistaken the baffled look on my face for one of total comprehension. He was so pleased that I was so interested in his lectures. Apparently no one from the Mineralogical Institute had ever come to his lectures before. And he had a favor to ask of me. His own derivations had been for three dimensional space. He had received a book from a Russian colleague who had derived the crystallographic space groups for four-dimensional space and he needed someone to read it so he could have a separate opinion about the validity of the mathematics. He pushed the book into my hands cheerfully telling me that in case I wasn't too fluent in Russian it was no problem; it was almost all mathematical expressions anyway. I politely accepted the book, quietly thinking 'how the hell am I going to get out of this.' Sure enough, that book could

have been written in Sanskrit as far as I was concerned. I skipped the next class and then decided I had better not be seen in the Department of Mathematics ever again. I slipped the Russian book under his door late one afternoon after he had left. Later I was to learn that all this wasn't really so esoteric after all. If you know the drawings of M. C. Escher, you have seen a variety of rotational motions in Euclidean space in practice (especially the series called Symmetry). I don't think that Escher tried to show them in four-dimensional space, but then maybe that's what some of the topological effects in the stranger Escher drawings (such as Ascending-Descending and The Waterfall) are about.

It was while I was in Zürich that my father wrote me that my mother had been diagnosed with cancer, and was to have a major operation. In those days transatlantic travel was a major undertaking; at least 5 days on a ship each way. Air travel was prohibitively expensive. A return home to visit her was not practical so I was to stay on until the year of study was finished in July. I planned to return via the shortest route, Zürich to Genoa, then on an Italian ship to New York, and by train to Colorado where my mother would be recuperating during the summer. The pride of the Italian fleet, the *Andrea Doria*, had a sailing on July 17th, scheduled to arrive in New York on the 26th; the timing was perfect for me. I went down to the travel agency, Reisebüro Kuoni, in the early spring, made a reservation for that crossing, and paid the deposit on my ticket. The remainder was to be paid in June. When I went to the Reisebüro in June to pay and collect my ticket, I happened to glance over the list of sailings to the US. The *Andrea Doria* was listed as fully booked, with a waiting list, but there was a special offer for another ship, the *Mauretania*, which was due to sail from Southampton a few days later but also arrive in New York on the 26th. Passage on the *Mauretania* was significantly cheaper. When I went to pay for my *Andrea Doria* passage, I was asked if I might like to change to the Southampton sailing, with the price to include rail and ferry transportation to England, and a one or two night stopover in London. Apparently someone on the wait-list for the *Andrea Doria* was willing to pay a premium to get on that sailing. It was a hard decision, the *Andrea Doria* was supposed to be a beautiful ship, just launched a few years earlier, while the *Mauretania* (not the original, which was sister ship of the *Lusitania*, but the second ship to bear this name) was much older, of pre-World War II vintage. I decided it was best to save the money, and a couple of days in London would be fun, so I took them up on the offer and switched to the *Mauretania*. That crossing of the *Andrea Doria* ended with its collision with the *Stockholm* and sinking before it reached New York. On the *Mauretania*, we steamed slowly through the floating wreckage a few hours after the *Andrea Doria* had gone under, got a good look at the Stockholm, which had lost its bow in the collision, and saw the *Ile de France* which had taken many of the survivors on board. Looking at the deck chairs and other debris in the water, it occurred to me that perhaps my letter telling my parents of my change of plans might not have arrived yet, and that morning of July 26th, 1956, I sent a telegram to my mother in Estes Park with the simple message: "Am fine, aboard the Mauretania, docking later today." Telegrams from ship to shore were expensive.

After visiting my mother in Colorado, I went back to our home in Dallas, then on to Illinois. The University of Illinois had awarded me a Fellowship, just as George White had suggested would happen. I finished my Master's degree at the University of Illinois in Urbana-Champaign at the end of the spring semester in 1957. My thesis was on Tertiary ostracods from southwestern France. You will remember that ostracods are tiny shrimp-like animals that distinguish themselves by building elaborate shells of calcium carbonate. I had picked out these fossil shells from samples available in the collections at the Swiss Federal Institute of Technology in Zürich the previous year. I could have gone on to a Ph.D. at Illinois, but my long-range plan called for me to go on to Stanford for that degree. George White was fascinated by my bizarre career plan and told me that when I got my Ph. D. and was looking for a position to be sure and apply to Illinois.

While at Illinois I applied to Stanford to work with Hans Thalmann. I was awarded an "Honorary Fellowship" to Stanford, which means 'we would be pleased to have you but you will not receive any money.' Fortunately, my grandfather had set up a fund for my education which we had hardly touched, so off I went to California via Route 66. And after I got to Stanford they discovered they needed a teaching assistant, and there I was. Hans Thalmann was one of the world's foremost experts on foraminifers. He was a Swiss who had spent much of his life before World War II working in Indonesia. He and his family had been in the US when the war broke out and Indonesia was invaded, and they stayed here. He worked part time as a consultant to oil companies, and held the interesting position of "Permanent Visiting Professor" at Stanford. He had only a part-time salary and that allowed him the freedom to carry on with his consulting practice from time to time although he was rarely away from campus. He had worked for Petroleos Mexicanos after the war, and still had good connections with them. Most of his students were from Latin America and he was delighted to have a Swiss-speaking American. Another of his students was Augustin Ayala, a faculty member of the University in Mexico City. In discussing possible topics for the Ph. D. Hans suggested a study of the foraminifera of the Velasco Formation, the geologic unit just above the Cretaceous-Tertiary boundary in the coastal plain of Mexico. Some of the foraminifera had been described 30 years earlier. Before going further I should explain to you what these tiny fossils were used for. In places like the coastal plain of Mexico there are thick sequences of shales and the calcareous shales called marls. When drilling a well for oil it is important to know where you are in the sequence of strata. You need to know if you approaching or have passed through the layer that might contain petroleum. The drill bit pulverizes the sedimentary rock down to millimeter-sized pieces. To keep track of where the well is in the stratigraphic sequence, a geologist 'well sitter' examines the drill cuttings. It is easy to tell where you are if the rock sequence consists of rocks of different types. But when it is all the same homogenous shale, you need something else, and the cuttings are just about the size of the fossil foraminifera. It was in the coastal plain of Mexico in the 1920s that it was discovered that these tiny fossils could be used to assist petroleum exploration by determining the age of the sediment. Since then they have become widely used all over the world to keep track of the age of the sediments being drilled in the exploration for petroleum. In Mexico it had been especially important to 'know where you were' because the impermeable shales overlie the reservoir rock that contains the petroleum, the El Abra limestone. The El Abra petroleum reservoirs were supercharged with natural gas. A well could easily get out of control if one suddenly, unexpectedly drilled into the limestone reservoir without being prepared. A well out of control is a 'gusher' which looks like a great way to discover an oil field in Hollywood movies, but is a disaster in practice because the oil is wasted and it may be very difficult to get the well back under control.

Hans wasn't the only micropaleontologist on the faculty at Stanford, Joseph J. Graham held the regular faculty position for that topic, But Graham was neither as knowledgeable or as well-known as Hans. He had only one graduate student working with him, and tried to talk me into switching to him as my advisor. I refused. I was happily taking courses during my first quarter at Stanford when Joe Graham called me into his office. He had been going over my transcripts and saw that I had never had a physics course in college. SMU had simply accepted my physics course under Doc Nelson at St. Mark's as fulfilling their requirement. Joe insisted I take freshman physics. So I signed up for a course taught in the spring quarter of 1958.

The introductory physics course was taught by Albert Baez. He was one of the best lecturers I ever encountered. He always started the class with some music and his lectures were exciting stories. The physics laboratory experiments were fun, and I was teamed up with a Chinese-American, Ming, who was really good at physics. One of the labs was especially memorable. It was in March, 1958, and we were going to conduct experiments on measuring radioactivity. It was a rainy day, but that was not a problem under the arcades of Stanford. We got everything set up properly and set the Geiger counter to filter out the background noise. However, something was wrong, we couldn't get a signal from our radioactive source. Ming was mystified. We called over that lab instructor and soon realized that everyone was having the same problem. The diagnosis was that the sources we had were old and had gotten very weak. A few days later we learned that something had gone wrong with one of the nuclear tests in Nevada and a plume of radioactive material had drifted over the San Francisco Bay area, where much of it had rained out. Our radioactive sources had been fine; they had just been masked by the radiation from the fallout. The fallout radiation in the Bay area was greater than considered permissible at the time, so the Atomic Energy Commission simply raised the permissible level so that no one would be alarmed.

A few weeks later Professor Baez invited some of us to his home for a barbeque. It was a great evening, and he asked me what I was doing in his course since I was obviously not a freshman. I explained my situation, and he told me to come see him later in the week. Then came the highlight of the evening. His teenage daughter Joan played the guitar and sang folksongs for us. Yes, that was the Joan Baez who would later become famous. A few days later Professor Baez gave me a quiz in his office, and wrote out a memorandum

to the Geology Department that I was proficient in Physics, that I should not be taking the introductory course, and I had earned an A in his course. The Geology Department accepted it and Joe Graham was very disappointed.

The thesis topic sounded great to me. Something strange seemed to have happened at the end of the Cretaceous. Hans knew that there was a marked change in the assemblages of foraminifera across the boundary, but it had not yet been well documented. I could get help in the collecting samples in the field from geologists at Petroleos Mexicanos. Augustin thought it was a good idea too, and offered to have his laboratory prepare the samples for me, if I would leave splits of my samples for him in Mexico City. Preparation of samples to extract the fossil foraminifera from the rock could be a messy business, involving soaking the samples in kerosene or carbon tetrachloride. In those days no one took precautions about inhaling the fumes, but it was a messy business. Augustin could also provide the paperwork necessary for me to be able to export the samples. So we made a deal, I would collect samples from the Mexican coastal plain and the edge of the Sierra Madre Oriental, during June and July, and then go to Mexico City, leave the samples at the Geology Department of the Universidad Autónoma de México where his people would split and prepare them, I could go out to the Ciudad Universitaria from time to time to check on things and perhaps start examination of the material.

I had not been to Mexico since I was in grade school, when my family visited Monterey. Then I found out that some friends would be in Mexico City that summer too. I knew that Tom Cranfill from the University of Texas was fascinated by Mexico and often spent his summers there. I called him in Austin for advice on how to find a place to live in Mexico City. He recommended a hotel for a short visit but suggested renting an apartment for staying a couple of months. I found out that he would be there during the summer of 1958 too.

Then, in the spring of 1958 my mother passed away. Efforts to control cancer in those days were primitive, and not often successful. I was able to fly back to Dallas for the funeral. Her death was not unexpected, but still a tragic loss to me.

JAMES HUTTON LOOKS ON AS JOHN PLAYFAIR STARES
INTO THE ABYSS OF TIME AT SICCAR POINT, SCOTLAND

*I was puzzled by the odd '4.' Why not a nice round 4000 years? And why such a very recent date? And how could anyone know?*
Arthur Holmes

One of the greatest areas of confusion about climate change has to do with the time over which changes occur. Our knowledge of past climates comes from geologic evidence. However, I find that unless you have spent most of your life thinking about the geologic past, you have no concept of the depths of time. As humans, we have a good idea of the length of a day, of a month, and of a year, because these are associated with light and darkness, the phases of the Moon, and the seasons. However, there are no natural cycles to mark decades or centuries, and these longer times become a blur. To a child waiting for a birthday party, time seems to run at a snail's pace; when you retire, you suddenly wonder whatever happened to all those years that flew by.

## 4.1    Age of the Earth—4004 BCE, or Older?

Many people have heard of James Ussher (1581–1656), Anglican Archbishop of Armagh in Northern Ireland, and Primate of All Ireland (Fig. 4.1). His precise dating of the moment of the Creation as occurring at nightfall proceeding Sunday, October 23, 4004 BCE, is included in the margins of fundamentalist Bibles. Published posthumously in 1658, *Annals of the World* is an extraordinary work attempting to give an account of major events in history from 4004 BCE to 1922 BCE, and from then an almost year by year account up until 73 CE. It was an attempt to trace back the genealogies in the Bible in literal terms and interweave this with Greek and Roman history. He was a great scholar, and well aware of the problems with the calendars. For the basis of his calculations he used Scaliger's Julian Period. It worked out that the Creation had taken place exactly 4000 years before the birth of Christ. Ussher also believed that the Earth would come to an end 6,000 years after it had been created.

Incidentally, the English translations of Ussher's book incorrectly translate the original Latin as Julian calendar rather than Julian Period. So when you discuss the age of the Earth with a creationist friend, you can ask whether the 4004

**Fig. 4.1** James Ussher (1581–1656), Anglican Archbishop of Armagh in Northern Ireland, and Primate of All Ireland

BC date is based on the Gregorian or Julian calendar. You might even mention the problem of the 455 day years the Romans used. You can leave them completely baffled when you tell them it is actually calculated from the Julian Period used by modern astronomers.

Ussher's was neither the first nor the last attempt to determine the age of the Earth from the Bible. Estimates had been made by the Venerable Bede in 723 (3952 BCE) and John Lightfoot in 1644 (3929 BCE), and later by Johannes Kepler (3992 BCE), and Isaac Newton (4000 BCE) among others.

One of the early challenges to the 'Biblical' date was made by Georges-Louis Leclerc, Comte de Buffon

**Fig. 4.2** Georges-Louis Leclerc, Comte de Buffon (1707–1788)

(1707–1788), a French nobleman and all-round naturalist (Fig. 4.2). In 1774 he concluded that the Earth was probably about 75,000 years old. His estimate was based on experiments on the cooling rate of iron. He also proposed that the planets had originated when a comet had struck the Sun. But Buffon's most important work was in biology, published as the *Histoire naturelle, générale et particulière* in 36 volumes between 1749 und 1788. His observations anticipated the idea of evolution. His age of the Earth was rejected by the authorities at the Sorbonne.

## 4.2    The Discovery of the Depths of Time— Eternity

The recognition of the immensity of geologic time seems to have occurred one clear day in the summer of 1788, when three people who considered themselves 'geologists' sailed along the Scottish coast east of Edinburgh and landed at a place called Siccar Point. James Hutton (1726–1797) is now known as the 'Father of Geology,' Sir James Hall of Dunglass (1761–1832), President of the Royal Society in Edinburgh, and mathematician John Playfair (1748–1819),

Hutton's close friend. There they saw steeply inclined beds of slate, covered by gently sloping red sandstone (Fig. 4.3).

Their thoughts that afternoon were recorded by Playfair in his eulogy to Hutton published in the *Transactions of the Royal Society of Edinburgh* in 1805:

> We felt ourselves necessarily carried back to the time when the schistus on which we stood was yet at the bottom of the sea, and when the sandstone before us was only beginning to be deposited, in the shape of sand or mud, from the waters of a superincumbent ocean. An epocha still more remote presented itself, when even the most ancient of these rocks, instead of standing upright in vertical beds, lay in horizontal planes at the bottom of the sea, and was not yet disturbed by that immeasurable force which has burst asunder the solid pavement of the globe. The mind seemed to grow giddy by looking so far into the abyss of time.

In a paper read before the Royal Society in Edinburgh later in 1788, *Theory of the Earth; or an investigation of the laws observable in the composition, dissolution, and restoration of land upon the Globe,* Hutton (Fig. 4.4) stated "we find no vestige of a beginning, no prospect of an end." The Earth might be eternal.

In 1795, Hutton published his magnum opus: *Theory of the Earth; with proofs and illustrations.* The book is a bit of a disaster. Hutton's writing is convoluted to say the least, and many of the drawings that were to be figures in the book were misplaced by the printer and did not appear in the published work. They were discovered in Edinburgh in the 1970s and finally published as *James Hutton's Theory of the earth: the lost drawings* in 1978. Better a couple of centuries late than never. John Playfair, realizing that Hutton's book was so difficult to read, essentially rewrote it as *Illustrations of the Huttonian Theory of the Earth,* published in 1802.

Hutton, Hall, and Playfair were important figures in what is termed the 'The Scottish Enlightenment,' which included such other well-known intellectuals as Frances Hutcheson, Adam Smith (*The Wealth of Nations*), David Hume, and Robert Burns. One of Hutton's close friends was Sir Joseph Black, a physicist and chemist famous for his work on latent and specific heat (concepts very important in climatology that we shall discuss in more detail later) and the discovery of carbon dioxide.

In 1997, to commemorate the bicentenaries of the death of James Hutton and birth of Charles Lyell, the Geological Society held a meeting that started in London and ended in Edinburgh. In London we gathered around Charles Lyell's tomb in Westminster Abbey for a special ceremony. I looked down and realized I was standing on the grave of one of his more famous adherents, Charles Darwin. On an excursion from Edinburgh we visited Siccar Point, where one can relive that day in 1788 and stare into the abyss of time.

The Scots had described one of the most spectacular features in geology, an 'angular unconformity,' with strata

**Fig. 4.3** Sir James Hall's sketch of the strata (layered rocks) at Siccar Point, on the coast of Berwickshire, in Scotland east of Edinburgh. The two sets of sloping strata represent original horizontal deposition on the sea floor, then tilting to a very steep angle through mountain building, erosion of the mountain range to allow the upper set of strata to be deposited horizontally on the sea floor, followed by uplift, tilting and erosion. This is what Hutton, Playfair, and Hall saw on that day in 1788, when they realized they were looking into the abyss of geologic time

tilted steeply by movements of Earth's crust, subaerial erosion planning off vast amounts of rock, and flooding with subaqueous deposition of sediments on the new sea floor. Moreover, Siccar Point shows these processes to have been repeated more than once.

## 4.3 Geologic Time Punctuated by Revolutions

A year after Hutton, Playfair, and Hall visited Siccar Point, on July 14, 1789, the great prison in Paris, the Bastille, was attacked and torn down. The French revolution was under way. One of the revolutionaries was Jean Léopold Nicolas Frédéric Cuvier (Fig. 4.5). He was 20 years old when the French revolution started. Since childhood he had been interested in animals and plants, and he became one of the great authorities on comparative anatomy of vertebrates. At the age of 30 he was appointed Professor at the Botanical Garden in Paris (formerly the Royal Gardens), which came to house the Natural History Museum. He was an avid collector of fossils, and an observant geologist.

The sedimentary deposits around Paris that Cuvier was familiar with were very different from those on the Scottish coast. Paris is situated in the middle of a geologic basin, and ringed by progressively older strata, many of which are rich in fossil mollusks and even vertebrates. The more distant strata were indeed tilted more than the deposits laid down in the center of the basin. What captured his attention was the effect of sequential flooding of the land as sea levels rose, followed by return to subaerial conditions as sea levels lowered. Geologists refer to these movements of the shorelines as transgressions and regressions. Based on what he saw, Cuvier noted that the fossil faunas, mostly mollusks and vertebrates in the marine strata, were the same from top to bottom. The retreat of the sea and subsequent flooding is often represented by thin strata or even bedding planes between the strata. Cuvier concluded that the transgressions and regressions occurred very rapidly in response to catastrophic geologic movements. The word 'catastrophe' is derived from the Greek and means turning things upside down. A catastrophe is defined as a sudden event with dire consequences. Cuvier held to the idea that the Earth was about 6000 years old, so catastrophes were required to fit in

**Fig. 4.4** James Hutton (1726–1797); the 'Father of Geology' in a formal pose. It would be nice to have a picture of him playing whist with his friends Joseph Black, Adam Smith, and David Hume at their club

all that had happened since the planet had formed. Certainly, for the French aristocracy, the 1789 Revolution was a catastrophe, as was the Reign of Terror of 1793–94. Cuvier survived all that as an observer, but the idea of catastrophic change was in the air. He became a close friend of Napoleon, and convinced him to send back to Paris the natural history collections from cities and towns in the areas he conquered. These greatly enhanced the collections already in Paris and

made the Musée d'Histoire Naturelle one of the world's greatest museums.

Cuvier assumed that the new replacement fauna that came in with each transgression had come from distant areas unaffected by the catastrophe. His ideas were modified by others to take on a religious aspect. Instead of the replacement faunas coming from distant regions, they were considered evidence of new Creations by God. There had been not one

**Fig. 4.5**  Jean Léopold Nicolas Frédéric Cuvier, (1769–1832) a.k.a. Georges Cuvier

'Creation,' but a number of successive 'Creations' of animal life, each with creatures more like those of today. Each of these 'Creations' had ended with a catastrophe like the supposed Biblical Flood. From what can be observed in the Paris Basin, it is a reasonable idea. The fossils are the same from the bottom to the top of each of the fossil-bearing strata; there is no evidence of gradual change. However, if you go further afield, you can find the intermediate forms that one would expect from evolution. We now know that the fossiliferous rocks in the Paris Basin are a very incomplete record; most of the geologic time is represented by 'hiatuses,' lengthy gaps in the record represented only by thin sediments or surfaces between the major rock units. Cuvier had thought these boundaries represented very short periods of time; today we know they represent much more time than do the deposits. The misinterpretation of Cuvier's ideas of two hundred years ago has been recently resurrected as 'Intelligent Design.'

Cuvier's idea of Catastrophism was first published in the preface to a treatise on Fossil Quadrupeds in 1812; it was presented more thoroughly in his *Discours sur les Révolutions de la Surface du Globe* published in 1825 and in its many subsequent editions. He had become such an integral part of the establishment that from 1819 on he was known as Baron Cuvier. His ideas were soon challenged by a young British geologist, Charles Lyell (1797–1875).

## 4.4  Catastrophism Replaced by Imperceptibly Slow Gradual Change

During the 1970s and 80s I made frequent trips to Europe on professional matters. Whenever possible, I would stop over in London and take a day to visit the library of the Geological Society in Burlington House on Piccadilly. The Geological Society (affectionately known as the Geol Soc) was founded in 1807. Its early members were 'gentlemen' interested in the Earth and its history. Many were educated as clergy, lawyers, or medical doctors, but few actually had to work for a living. In many ways it started out as a debating society. Until recently, the Society's lecture room was set up as a miniature version of the British parliament. Those who intended to speak in support of the presentation being made sat on one side, those who wished to raise objections on the other. It was a living example of the give and take that scientific research is all about.

In the Society's library I could often find obscure articles that were on my list of things to read but had been unable to locate elsewhere. The Geol Soc library has an extraordinary collection of older literature and is managed by a knowledgeable and efficient staff. By going up a narrow spiral staircase from the main reading room one reaches a narrow balcony where there are some remarkable treasures of the 18th and 19th centuries. There I stumbled on the first edition of Charles Lyell's *Principles of Geology*, in three volumes published from 1830 to 1832. Every geologist knows of it as the first "modern book" on geology, but few have ever read it. I took it down to the reading hall and spent the rest of the day in rapt amazement.

More than a third of the first volume of Lyell's work is devoted to the problem of climate change over the ages. He was convinced that the ferns and other trees that had formed the coal beds of England were tropical. Very methodically he tried to determine how the climate might have changed

**Fig. 4.6**  Charles Lyell (1797–1875)

from then to now. He knew the coal beds were very old (about 300 million years old, we know now) and that much had happened to the Earth since they were formed.

Lyell (Fig. 4.6) was, of course, quite familiar with the chalk of England. Chalk is that very special white limestone that makes up the white cliffs of Dover and the ridge called the South Downs, south of London (not far from Lyell's home). It is used to make the chalk for writing on blackboards. In 1837 Lyell visited Denmark and Sweden and saw chalk there as well. Because it was a form of limestone, and limestone is made of calcium carbonate—typically a warm-water deposit—he concluded that warm conditions must have extended much further poleward in the Cretaceous. It was not until 1853 as samples of the sea bed in the North Atlantic were being taken preparatory to laying the first transatlantic telegraph cable that the true nature of chalk became known. Today it is deep-sea sediment, ooze made up of the minute calcium carbonate fossil remains of coccoliths and planktonic foraminifera.

After considering a variety of possibilities for the cause of climate change, he concluded that it was probably a different distribution of land and sea that had been responsible for England and Europe's warmer climates in ancient times. At the time there was little known about the true nature of continents and ocean basins, and Lyell assumed that there was nothing permanent about either—with time, land could become sea or ocean and vice versa. He thought that the land areas were places where internal forces were pushing the surface up, and oceans areas where the crust was sinking; these forces would move from place to place with time. He reasoned that if land were concentrated in the Polar Regions, the Earth would be colder, and if the Polar Regions were

*Bless the baby what a wally he have a-made*

*Cause and Effect*

**Fig. 4.7** Henry de la Beche's cartoon (circa 1830) lampooning Charles Lyell's idea that processes that have shaped the Earth throughout its history are the same as those acting today. Colors and lettering enhanced. The (almost illegible) inscription from de la Beche to Buckland is in the *upper left*

water and most of the land was in the equatorial region, the Earth would be warmer. It was an extraordinary insight into the way the climate system works. Remember that when Lyell wrote the *Principles of Geology*, both the Arctic and Antarctic were largely unknown. The discovery that the Arctic was an ocean basin and not an ice-covered landmass did not come until the 20th century.

That Lyell's book could be considered the first modern introduction to geology is an understatement. Although some of the wording is archaic, it is so well written and the material so logically presented that if one didn't know it was written over 175 years ago, it could be read as a modern textbook. It might even get onto a bestseller list. Incidentally, it has recently been reprinted by the University of Chicago Press and is again available. Charles Lyell helped popularize the notion that the processes that acted to change the Earth in the past were the same as those we can observe today, a principle often stated as "the present is the key to the past." Lyell called this idea of slow change 'gradualism.' Unfortunately, this view of Earth History has come to be known by the more ponderous term 'Uniformitarianism' created by William Whewell in 1832. It was intended to contrast with Cuvier's 'Catastrophism.' Lyell argued that there was no need to invoke catastrophes, although some of his contemporaries made fun of him for what they believed to be a naive view. Lyell came to the conclusion that geologic processes were so slow that the Earth must be billions of years old.

A portrait of Lyell and a nice cartoon lampooning his idea is on the World Wide Web at www.strangescience.net/lyell.htm. Figure 4.7 shows an unflattering opinion of Lyell's idea of gradualism. It is a cartoon drawn by Henry de la Beche after Lyell's book had appeared and given to Oxford Professor William Buckland, Lyell's teacher. Titled 'Cause and Effect,' it shows a little boy urinating into a giant valley, and his nurse behind him exclaiming 'Bless the baby, what a walley (valley) he have a-made !!' The little boy is thought to be Buckland's son, Frank. Henry de la Beche was the founder of the British Geological Survey and loved to make cartoons lampooning his colleagues. This particular cartoon has an interesting history: long known only from descriptions, it was recently discovered among Buckland's papers in the archives of the University. It was much too racy to publish in the 19th century and finally appeared in print in the June 12, 1997 issue of *Nature*.

Lyell's 'uniformitarianism' gradually replaced catastrophism, and after publication of Darwin's 'Origin of Species' in 1859 it became geological dogma. When I defended my Ph.D. thesis, I presented the idea that something very drastic had happened to the plankton of the oceans about 65 million years ago at the end of the Cretaceous period. There had been a very rapid mass extinction, followed by a long slow gradual rebuilding of the assemblages of planktonic foraminifera to again form robust communities. I squeaked through on the understanding that I should downplay my idea of radical change when the work was published. However, the student defense following mine in the same room with mostly the same faculty members was by Roger Anderson, a paleobotanist working on the floras of the Cretaceous-Tertiary boundary in New Mexico. He presented evidence that there had been a very abrupt change in the nature of the floras, with forests replaced by ferns. At a reception later Professor Siemon Muller pulled me aside and said that maybe, just maybe, we were on to something.

## 4.5  The Development of the Geological Time Scale

During the 19th century, the abyss of time was organized into a sequence of major units. As can be seen in Fig. 4.8, it was very much a 'work in progress' throughout the century, and it continues to be modified even today as we learn more and more. However, most of the names still in use for the major intervals of geologic time were introduced over a hundred years ago. The geologic sequence is based on 'superposition,' the simple idea that the rock layers at the top of a sequence were deposited after, and are hence younger than, those at the bottom. The simplest system was devised during the middle of the 18th century by Giovanni Arduino of the University of Padua. It was intended to order sediments and rock from the Po River basin and southern Alps, calling the rocks, from oldest to youngest, 'Primary,' 'Secondary,' 'Tertiary' and 'Volcanics.' As luck would have it, the terms 'Primary' and 'Secondary' have dropped out (except in France), and 'Volcanics' has been replaced because most of the youngest sediments are not volcanic in origin. This leaves us with only 'Tertiary' still in use. The 'Quaternary' was added by French geologist Paul Desnoyers for the loose, relatively undisturbed sediments in the valley of the Seine River representing the youngest part of Earth's history. During the 19th century, geologists interested in paleontology substituted the terms 'Paleozoic' and 'Mesozoic' for 'Primary' and 'Secondary' and 'Cenozoic' for 'Tertiary' and 'Quaternary' as shown on Fig. 4.8.

Most of the advances in geology in the 19th century concentrated on establishing the sequence of layered rocks, particularly those containing fossils. The sister science of Paleontology worked out the general sequence of life forms that have inhabited our planet. Strangely, the person who realized the importance of establishing the sequence of rock layers and who discovered that fossils could be used to correlate the beds over long distances was not a geologist. It was William Smith, an engineer working on construction of canals in southern England, who made a map of the geology of southern England in 1815. He is now regarded as the 'Father of Stratigraphy'—the study of stratified rocks—but he didn't have the proper credentials to be able to join the Geological Society in London.

The legend for Fig. 4.8 gives the full names, birth and death dates, nationality, and occupations of those who helped to create the time scale we have today. Note that many of these were contemporaries and Fellows of the Geological Society (London) (FGS). Preston Cloud's 'Hadean' is unique in that, by definition, no rocks of that age are still in existence; it refers to the time when the planet was still molten. William Harland's 'Pleistogene,' as a parallel to Paleogene and Neogene, just hasn't caught on.

The four major units, which are now called 'Eras,' were divided into 'Periods.' Many of these have names that are now familiar because they pop up in discussions about past life on Earth and even in sci-fi films. The Paleozoic contains the Cambrian, Ordovician, Silurian, Devonian, Carboniferous, and Permian Periods. The origin of the Period names is diverse. In this case the first four refer to places (Cambria = Roman name for Wales; Devon = Devonshire) or ancient Celtic tribes in Wales (Ordivices, Silures). The 'Carboniferous' refers to the coal beds of Britain. Mapping out of the coal beds occurred early on because of the need for coal to fuel the industrial revolution after James Watt's invention of a practical steam engine in 1765. The Permian refers to a province of Russia, Perm, where British geologist and fox hunter Sir Roderick Impey Murchison found fossiliferous strata representing an age that had until then been poorly known. The Mesozoic contains three periods, the Triassic, Jurassic, and Cretaceous; again the names are of diverse origin. Triassic refers to the three-part subdivision of the strata which had been long recognized in central Europe; Jurassic to rocks in the Jura Mountains on the Swiss-French border; and Cretaceous to the special rock 'chalk' so characteristic of deposits of that age that ring Paris and form the White Cliffs of Dover. The Periods (usually called Epochs) of the Cenozoic were mostly established by Charles Lyell on the basis of the number of mollusk species still extant: Eocene, Miocene, and Pliocene. The Oligocene was inserted later between the Eocene and Miocene because there were unusual land plants in central Europe that didn't seem to fit Lyell's scheme. Although geologists and paleontologists at the time didn't know it, that strange Oligocene was the time when the Earth changed from a globally warm planet to one with ice in the Polar Regions. The Paleocene was added later to account for strata between the Cretaceous and Eocene.

By the beginning of the 20th century, it was becoming evident that the Earth had a long and complex climate history, having been at times much warmer than today and at other times much colder. More and more scientists were becoming interested in climate change, but the geological community was solid in its acceptance of Lyell's 'gradualism' and believed that climate change occurred only very slowly. Unfortunately nobody had any real idea about what 'slow' meant because no one knew how to get a handle on geologic time. The next chapter presents an overview of

| Volcanic Arduino 1759 | Quaternary Desnoyers 1829 | Pleistogene Harland 1989 | Recent Holocene | Lyell 1833 Gervais 1885 |
|---|---|---|---|---|
| | | | Pleistocene | Lyell 1839 |
| Tertiary Arduino 1759 | Cenozoic J. Phillips 1844 | Neogene Hörnes 1859 | Pliocene | Lyell 1833 |
| | | | Miocene | Lyell 1833 |
| | | Paleogene Neumann 1866 | Oligocene | von Beyrich 1854 |
| | | | Eocene | Lyell 1833 |
| | | | Paleocene | Schimper 1854 |
| Secondary Arduino 1759 | Mesozoic J. Phillips 1840 | | Cretaceous | d'Omalius d'Halloy 1822 |
| | | | Jurassic | von Humboldt 1795 |
| | | | Triassic | von Alberti 1815 |
| Primary Arduino 1759 | Paleozoic Sedgwick 1835 | | Permian | Murchison 1841 |
| | | | Carboniferous Coal Measures | Conybeare & W. Phillips 1822 Farey 1807 |
| | | | Devonian | Murchison & Sedgewick 1840 |
| | | | Silurian | Murchison 1835 |
| | | | Ordovician | Lapworth 1879 |
| | | | Cambrian | Sedgewick 1834 |
| | Precambrian Salter 1864 | | Proterozoic | Emmons 1887 |
| | | | Archaean | Dana 1872 |
| | | | Hadean | Cloud 1972 |

◄ **Fig. 4.8** The geologic time scale. Names of the units are given with the author and date of publication. Terms in *italics* have fallen out of favor and are no longer used. Arduino = Giovanni Arduino (1714–1795), 'Father of Italian Geology,' University of Padua (Veneto, Italy); Cloud = Preston Cloud (1912–1991), American geologist with U.S. Geological Survey; Coneybeare = William Daniel Coneybeare, English Curate, geologist, FGS, FRS; Dana = James Dwight Dana (1815–1895), Mineralogist/Geologist, Yale University; Desnoyers, = Paul A. Desnoyers, French geologist; Emmons = Samuel Franklin Emmons (1841–1911), American geologist with U.S. Geological Survey, FGS; Farey = John Farey, Sr. (1766–1826), English geologist, friend and supporter of William Smith; Gervais = Paul François Louis Gervais (1816–1879), French paleontologist, zoologist; Harland = Walter Brian Harland (1917–2003), Professor of Geology, Cambridge University, England, FGS; Hörnis = Moritz Hörnis (1815–1868), German geologist; Lapworth = Charles Lapworth (1842–1920), Professor of Geology, University of Birmingham FGS; Lyell = Sir Charles Lyell, Baronet, (1797–1875), English lawyer, geologist, FGS; Murchison = Sir Roderick Impey Murchison (1792–1871), English army officer, fox hunter, geologist, FGS; Naumann = Georg Amadeus Carl Friedrich Naumann (1797–1873), German crystallographer, mineralogist, geologist, Professor of Geognosy, University of Leipzig; d'Omalius d'Halloy, = Jean Baptiste Julien d'Omalius d'Halloy (1783–1875), Belgian geologist; J Phillips = John Phillips (1800–1874), English geologist associated with University College, London, FGS; W. Phillips = William Phillips (1775–1828), English mineralogist and geologist, FGS; Salter = John William Salter (1820–1869), English geologist and paleontologist, English Geological Survey, FGS; Sedgewick = Adam Sedgewick (1785–1873), English gentleman, Professor of Geology Cambridge University, FGS; Schimper = Wilhelm Philipp Schimper (1808–1880), Alsatian botanist/paleobotanist, Professor of Geology and Natural History at the University of Strasbourg; von Alberti = Friedrich August von Alberti (1795–1878), German geologist, salt technician; von Beyrich = Heinrich Ernst von Beyrich (1815–1896), Professor of Geology, University of Berlin; von Humboldt = Friedrich Wilhelm Heinrich Alexander Freiherr von Humboldt (1769–1859), German naturalist and explorer

climate change over geologic time, but first it is important to get a grasp on the immensity of geologic time.

Two terms were introduced in the middle of the 19th century which many of us wish could be changed: the Paleogene includes the Paleocene, Eocene and Oligocene, and the Neogene includes the Miocene and Pliocene. These terms were created before we knew very much about the geology of the Polar Regions. Now we know that a big change in Earth history took place at the end of the Eocene, when the Antarctic continent and Arctic Ocean became covered with ice.

There were older rocks that contained no fossils, simply called 'Pre-Cambrian,' later shortened to 'Precambrian.' The detailed study of them had to await new methods for determining their age; that took over a century.

The real basis for breaking up the geologic record into discrete units did not become clear until the middle of the 20th century. The rocks available for study in the 19th century were all on the continents, and those easiest to study are on the platform areas where the strata have not been greatly deformed. The sequences reflect long-term changes in sea level. The continents are alternately flooded and emergent. For this reason, a lot was known about the middle parts of the sequences and relatively little about what conditions were like during the time represented by the boundaries between the sequences.

Interestingly, geologists and paleontologists got along quite well without knowing anything other than 'relative ages,' that is, knowing that something is older or younger than something else. They knew that the Cambrian strata were older than those of the Ordovician, for example, and that both had been deposited a very long time ago. You could learn a lot by working out the sequence of events in Earth's history without knowing 'numerical ages' expressed in years. Even today most geologists think in terms of relative ages; to get the age in years they usually need to look it up.

Even before publication of Lyell's books, disturbing new evidence of climate change had come to light. Around the Alps, in northern Europe and in parts of the British Isles there were strange landforms—U-shaped valleys, long ridges of sand, gravel and boulders, and occasional huge blocks of rock resting on the soil. These features were attributed to the Biblical Flood; the deposits were called "drift" and their age given by Lyell as "Diluvian," the time of the deluge, that is, the Biblical Flood. Some of these things were hard to explain. The boulders, for example, were thought to have been transported by icebergs in the flood waters—but then came the question: how did icebergs get into the flood waters?

## 4.6    The Discovery of the Ice Age

In Switzerland, however, naturalists were discussing the similarity between the deposits of the Swiss midlands, between the Alps and the Jura Mountains, with the materials surrounding the glaciers in the Alps. The city folk were discovering what the farmers in the alpine valleys already knew. Glaciers were 'living things.' They advanced and retreated, and their deposits could be found far down the valleys they then occupied. In 1826 Swiss engineer and naturalist Ignace Vernetz (1788–1859) proposed to a meeting of the Swiss Natural History Society that the glaciers had been much larger and covered the whole region of the Swiss midlands between the Alps and the Jura Mountains. Word passed from one scholar to another until Louis Agassiz (1807–1873; Fig. 4.9) heard of it. At the time Agassiz was a Professor at the Lyceum of Neuchatel, and a paleontologist who specialized in studying fossil fish. He thought the idea was ludicrous, but after a field trip into the Alps with Jean de Charpentier, one of the hypothesis' proponents, he became convinced it might be right and spent several years gathering further evidence. Agassiz met Professor William Buckland of Oxford

**Fig. 4.9** Louis Agassiz, around 1840

University in 1838 at a meeting of naturalists in the German city of Freiburg-in-Breisgau on the Rhine not far from the Swiss border. Buckland, originally a catastrophist, was gradually changing his mind about how the Earth had evolved. He had been Lyell's teacher (and had been given Henry de la Beche's cartoon ridiculing Lyell's ideas a few years earlier). Agassiz invited Buckland to come with him into the Jura Mountains to look at the evidence that glaciers had been there. Buckland was not convinced, but displayed an open mind.

When Charles Lyell heard about the idea of widespread glaciation, he felt it was a challenge to his notion of gradual change, almost a return to Catastrophism. Other British geologists also rejected the idea. Then, in 1840, Agassiz attended a meeting of the British Association in Glasgow, Scotland. After the meeting he traveled through Scotland with Buckland, and Roderick Murchison, the Scottish gentleman geologist who had recently returned from his visit to Perm in Russia. With Agassiz's interpretation of the Scottish landscape as having a glacial origin, everything suddenly made sense. The U-shaped valleys had not been carved out by the tiny streams in them today but by moving ice; the strange longitudinal hills of rock detritus were glacial moraines; and those odd boulders scattered about the landscape had been transported by the glaciers. Buckland became an enthusiastic supporter of the glaciation hypothesis; Murchison remained unconvinced. Charles Darwin accepted Agassiz's ideas (1842) and wrote about glacial landscape features in Wales. Buckland eventually converted Lyell (1863).

In 1840 Agassiz published his first book on the glacial theory, *Étude sur les glaciers*, and in 1847 a more extensive work, *Système glaciare*, appeared. In the meantime Agassiz had moved to the United States in 1846 and found evidence for extensive glaciation in North America too. In 1848 he became Professor at Harvard University. Introduced to the US by Agassiz, the glacial hypothesis gradually gained adherents and by the end of the 19th century was generally accepted. Also, by the end of the 19th century evidence was beginning to be found that there had been more than one glaciation in the geologically recent past.

But there was more to this idea of large ice sheets on Earth. In 1856, evidence of a much older episode of glaciation of late Paleozoic age was found in India and in 1859 similar evidence was found in southern Australia. Glaciation had not been restricted to the relatively young geologic past in the northern hemisphere, but it had apparently occurred in the equatorial region and southern hemisphere in more ancient times.

At the beginning of the 20th century a burning question for geologists was to determine when the glaciation that had covered much of Europe and North America with ice had occurred? How long did it take for the ice to melt, and for how long had it been gone? Many of our cities were built on the deposits left by those ice sheets; when had the ice left the area? A few years ago I had the pleasure of translating Gerard de Geer's great 1910 paper on this topic into English. It had been presented at the 8th International Geological Congress in Stockholm in 1908. (Major discoveries don't always get into print right away). De Geer was a Swedish geologist who discovered that thin beds of sediments in the lake deposits of Sweden were in fact annual layers, and could be traced from one site to another to build up a history of the melting of the ice cap that had existed over Scandinavia. It was the first time anyone had ever been able to state 'this layer is 9,254 years old.' What was so wonderful about de Geer's paper was his sheer joy and enthusiasm in describing how, over the previous 25 years, he and his wife and students had been able to put together this first geologic record with an annual time scale extending back 12,000 years. Although he wasn't able to get back to the time of the height of the glaciation (after all, the ice had completely covered Sweden and the Baltic region then), He was able to document the changes that had occurred during the later part of the melt-off of the Scandinavian ice sheet.

Then, in 1992 the geological world was stunned as Joseph Kirschvink of Cal Tech presented evidence that in the Late Precambrian the entire planet had been covered with ice; the 'Snowball Earth' episodes. The idea had been suggested before, but always discounted for lack of convincing evidence. It is a current hot topic of interest to a wide audience. More about it later.

## 4.7   The Discovery of Past Warm Polar Regions

Just as the idea of large ice sheets during the Earth's recent and ancient past was becoming accepted, evidence was found that Earth had also been much warmer in the past. Adolf Erik Nordenskiöld (1832–1901; Fig. 4.10) was born in Finland, but later emigrated to Sweden. He was a geologist and Arctic explorer. In 1870, in a search for the Northwest Passage, he landed on western Greenland's Nûgssuaq (Noursoak) Peninsula (70.5° N), north of Disco Island. There in the Cretaceous strata he found a fossil flora rich in tropical plants, including the remains of breadfruit trees. It was evident that during the Cretaceous, the Period when chalk had been deposited in Europe and North America, tropical conditions had extended to what is now the northern polar region.

**Fig. 4.10**  Adolf Erik Nordenskiöld (1832–1901)

## 4.8 Throwing a Monkey Wrench into Explaining Climate Change

James Dwight Dana (1813–1895; Fig. 4.11) was an American mineralogist, a Professor at Yale, and, best known for his book *Dana's System of Mineralogy*. It has been through many editions, and anyone interested in minerals has run into it. He also published a *Manual of Geology* (1863) in which he introduced a concept that was to become both a tenet of American geology and a monkey wrench in development of any understanding of climate change—the permanence of continents and ocean basins. In contrast to Lyell's idea that land could become ocean and vice versa, Dana believed that the continents and ocean basins had formed when the Earth solidified. The more dense rock formed the continents and the less dense rock was to form the basement of the ocean basins. His idea was that the more dense rock did not contract as much as the less dense rock, hence as the Earth cooled, the less dense rock formed basins which collected water. These different types of basement rock had remained fixed in size and position ever since. His idea replaced that of Charles Lyell, that lands could sink beneath the sea and the sea floor could rise to make land. Curiously, Dana's whole idea was hypothetical. There was neither any geological nor experimental evidence to back it up.

There was, of course, that inconvenient observation of Late Paleozoic glaciation in India, almost on the Equator, but India was far away so that was a problem for the British to worry about. In fact there was a lot of information on Earth's past climates that made no sense, but not very many American geologists had traveled abroad or worked in other parts of the world. Dana stressed in his *Manual* that the ideas were based almost entirely on the geology of North America, particularly that of the United States, which he regarded as uniquely suited to understanding global geology. He introduced the idea that continents typically have mountain ranges on either side and a lowland interior. Unfortunately for making a general extrapolation, this situation is unique to North America. In other continents mountain ranges are mostly in the middle, not along the edges. Dana's permanence of continents and ocean basins quickly gained acceptance and became dogma in the United States for almost 100 years.

Incidentally, the America-centric idea that the 48 contiguous US states, occupying less than 1.6 % of Earth's surface, are representative of the planet as a whole is still around. Several recent discussions of climate trends assume that changes affecting the 48 states are typical of the entire planet. In particular, 1934 is often cited as the hottest year on record in the United States, hotter than 1998 by 0.02 °C. That has been trumpeted as proof there is no *global* warming. Actually, as you might guess, the *global* temperature in 1998 was warmer than the *global* temperature in 1934.

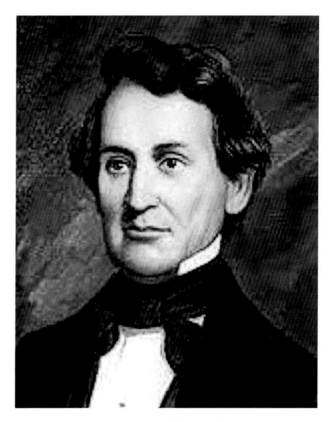

**Fig. 4.11** James D. Dana (1813–1895) American mineralogist whose idea of the permanence of continents and ocean basin dominated geology thought in the US for almost 100 years

I suppose that this could be considered a special case of P. T. Barnum's (or Abraham Lincoln's) observation 'you can fool some of the people all of the time, and all of the people some of the time' for the American public. But I guess George W. Bush did that with his famous variant 'you can fool some of the people all the time, and those are the ones you want to concentrate on.'

## 4.9 'Crustal Mobility' to the Rescue

A possible solution to the problem of the Late Paleozoic glaciation in India being near the Equator along with other climate anomalies was proposed by the German geologist Damien Kriechgauer in 1902. He suggested that possibly the Equator had a different position in the Late Paleozoic. Not that the Equator itself had moved, but that the continents had moved relative to it. It would be 70 years before the notion of 'true polar wander' became a respectable topic of discussion.

A more radical alternative view to that of Dana was proposed by Alfred Wegener (1880–1930), a German meteorologist/climatologist, in 1912. Wegener had earned his Doctorate in Berlin in 1906. While a student he had become enamored of the sport of hot-air ballooning, and in 1906 he set a record of 52 h for continuous time aloft, drifting from Bitterfeld (north of Leipzig) north to the Jutland peninsula, then southwest to land near the Dutch border. He also experimented with the use of tethered balloons to gather meteorological information. Because of his meteorological expertise, he was invited to join a 3-year Danish expedition starting in 1906 to explore northeastern Greenland. The frozen island had fascinated him since boyhood, and he instantly accepted. When he returned in 1909, he was hired as a lecturer at the University of Marburg, where he worked with Vladimir Köppen, who many considered the greatest climatologist of the time. Köppen had devised the system of climate classification based on vegetation as well as physical parameters that is still in wide use today. Wegener fell in love with Köppen's daughter, Else, and they were married in 1913. Wegener and Köppen plotted up maps showing the distribution of the then-known climate sensitive sediments and fossils. They came to a remarkable conclusion. The Late Paleozoic glacial deposits of India, Australia, South Africa and South America could better be explained if these continents were simply fragments of a much larger southern hemisphere continent, Gondwana. Furthermore, the similarity of the deposits in North America and Eurasia could be readily explained if there had been no Atlantic Ocean separating them, so that they formed a single northern continent, Laurasia. The idea was not new. It went back at least to the time of Sir Francis Drake (1620). Gondwana and Laurasia fit together nicely to make a Late Paleozoic supercontinent, Pangaea (meaning 'all earth'). There would have been a giant ocean basin 'Panthalassa,' the parent of the modern Pacific on the other side of the planet (Fig. 4.12).

I suspect that it was Wegener's experience with the pack ice around Greenland that gave him the idea that the continents had once been together and then broken into fragments and drifted apart. The climate sensitive sediments and fossils, at least from the Late Paleozoic on made latitude parallel bands around the globe, moving equatorward or poleward with time. Lyell's idea that land at the pole makes for a colder Earth was right! Suddenly, it all made sense. Except, there was something missing—a cause. Wegener guessed it was the different force of the tides on the continental blocks, due to Earth's rotation. Geophysicists were able to show that could not possibly be the case.

'Continental drift' seemed a simple enough idea, but had come from a meteorologist, not a geologist. American geologists remained not only skeptical, they became actively hostile. In 1922 Alfred Wegener was invited to attend a meeting of the American Association of Petroleum Geologists to discuss his idea, but he was not given an opportunity to speak. Instead he had to sit through a series of tirades ridiculing his idea. Apparently, he quietly sat there and smoked his pipe, said 'thank you' at the end, and returned to Germany.

Alfred Wegener was determined to understand more about the climate of Greenland. He died in the winter of 1930 trying to resupply the camp at Eismitte (mid-ice) on the Greenland ice-cap.

Across the Atlantic, in Britain, the reception was different. A young British geologist, Arthur Holmes (1890–1965), about whom you will hear much more later, took up the idea. He explained how it would work if there were convection in the Earth's mantle and proposed that the ridge running down the length of the Atlantic was the site where mantle rocks were rising, forcing the Americas apart from Eurasia and Africa.

The dogma of 'permanence of continents and ocean basins' finally died a painful death in the United States in about 1972, replaced by the theory of 'plate tectonics.' We took a poll of the audience of a session at the meeting of the American Association of Petroleum Geologists and found that the overwhelming majority favored the 'new' idea. In plate tectonics, the Earth's outer solid shell, the crust, is made of a number of large rigid plates. A few of the plates consist solely of oceanic crust but most consist of ocean and continental crust fused together. The crust making up the ocean floor grows by addition at the mid-ocean ridge. Where the edges of the plates opposite the mid-ocean ridges meet, one gets pushed over the other. The continents are passively attached to the ocean crust and move with it.

## 4.10 The Return of Catastrophism and the Idea of Rapid Change

The first reintroduction of Catastrophism into modern geologic thinking came in 1969. Digby McLaren was a prominent Canadian geologist and paleontologist. In his Presidential address to the Paleontological Society that year, he postulated that an asteroid impact was responsible for the great extinctions near the end of the Devonian Period, about 360 million years ago. I was in the audience and excited to hear about another time of sudden extinction and gradual recovery, although the mechanism of change seemed a bit fanciful. Needless to say, the idea received mixed reviews from the scientific community at large. However, in 1980 Walter Alvarez and his father, Luis, presented evidence that an asteroid had struck the Earth at the end of the Cretaceous period, 65 million years ago, causing the extinction of dinosaurs and many other forms of life. Skeptics that such a thing could happen had second thoughts when the fragments

**Fig. 4.12** Alfred Wegener's original figure showing continental drift (1912). Ideas of the general pattern have not changed greatly, but the ages have changed

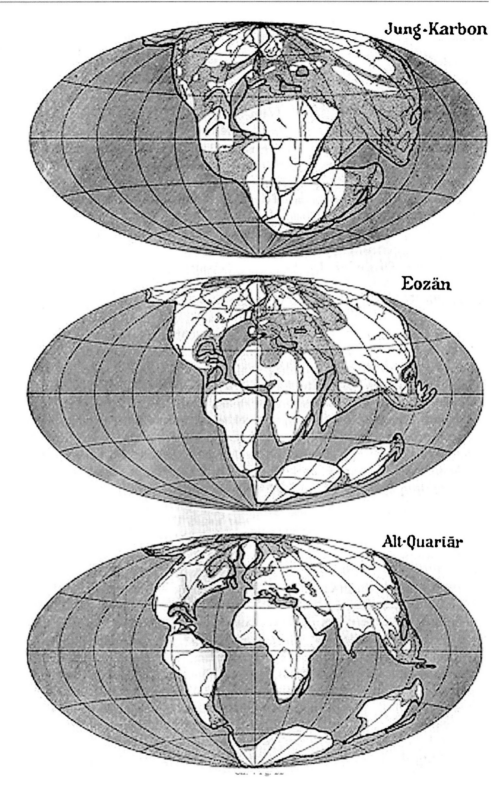

of the comet Shoemaker-Levy began crashing into Jupiter on July 16, 1994 (Fig. 4.13). The disturbances in Jupiter's atmosphere were larger in size than planet Earth itself. It was a sobering experience for everyone who thought that catastrophes were fantasy.

## 4.11   The Nature of the Geologic Record

As the sequence of layered rocks was being worked out in the 19th century, geologists became aware of the fact that most of the continents are covered by the younger

**Fig. 4.13** Impacts of fragments of the comet Shoemaker-Levy on Jupiter on July 25, 1994. Each of the impact marks is almost the size of planet Earth. Infrared image made by the Calar Alto Observatory, Spain

sedimentary rocks. Not just the deposits left behind by the glaciers in North America and Europe, but much larger areas are covered by Tertiary and Cretaceous rocks rather than older rocks. There were studies of the thickness of the rocks of different ages, and approximate age scales were devised following Lyell's principle of uniformitarianism and assuming that deposition rates had been constant through time, but in the absence of any method of actually assigning dates to the rocks, this was just guesswork. In the middle of the 20th century an alternative explanation was advanced by the German geologist Hans Stille. He proposed that the rate of mountain building, what geologists usually refer to as 'orogeny,' was increasing through time. More mountains mean more erosion, and that means more sedimentary rock being deposited. It made sense and the idea was widely accepted until 1969. That year Jim Gilluly, a geologist with the US Geological Survey, made the problem the topic of his Presidential Address to the Geological Society of America. He stated what should have been obvious all along: most sediments are produced by the erosion of older sediments. Older sedimentary rocks are continuously eaten away by erosion, moved to a new location by rivers and ocean currents, and redeposited as new sediment. Geologically speaking, the layered sedimentary rocks of the Earth's surface are a cannibalistic system. Figure 4.14 shows what exists today and what existed in the past but that has subsequently been eroded to make young sediment.

## 4.12  The Great Extinctions and Their Causes

We have learned a lot about the evolution of life on Earth from the fossil record. However, it is wise to remember that only a small fraction of the life forms become fossilized, and the record is strongly biased in favor of animals and plants that produce hard parts, such as shells. Also, the vast majority of fossils that have been collected are from marine deposits. The diversity of marine fossils appears to increase with time, but this is largely because young deposits are much more widespread than older deposits and have been more extensively collected. At the present time interest is focused on the record of land animals and plants, and our knowledge of those groups is growing rapidly.

From the fossil record paleontologists recognize five great extinction events during the Phanerozoic, shown in Fig. 4.15. The first was at the end of the Ordovician, about 445 million years ago and is thought to have been the result of climate change from a warm Earth into a glacial episode. The second was a multiple event near the end of the Devonian, 360 million years ago, and is now thought to have been the result of not one, but several asteroid impacts. The third was at the end of the Permian, ending the Paleozoic era about 250 million years ago. It is often referred to as 'the great dying.' Its cause is currently a topic of much discussion. Current ideas favor a massive release of greenhouse gases into the atmosphere and consequent global warming. The fourth was at the end of the Triassic, 200 million years ago, and seems to have involved changes in seawater chemistry associated with deposition of salt. The fifth was at the end of the Cretaceous, ending the Mesozoic Era 65 million years ago, and is the result of an asteroid impact.

The sixth great extinction is ongoing. It began about 40,000 years ago in Australia upon the arrival of humans. For hunting, the ancestors of the Australian aborigines set fires. In the dry climate they burned out of control and greatly affected both the animal and plant life resulting in many extinctions. Among the animals that became extinct in Eurasia were wooly mammoths, wooly rhinoceros, the huge Irish elk, saber-toothed tiger, scimitar cat, cave lion, giant ostrich, and the Neanderthal humans. The extinction of large mammals in both North and South America occurred shortly after the arrival of humans from Asia. It reached its maximum about 10,000 years ago. In the Americas, horses, camels, the stag-moose, the ancient- and long-horned bison, the giant beaver, the giant ground sloths, glyptodonts, saber-toothed tigers, the giant polar bear, the American lion, the dire wolf, mammoths, and a number of other large mammals vanished. We know very little about the extinctions of smaller animals and plants.

**Fig. 4.14** The amounts of Phanerozoic sediment originally deposited (*dashed line*) and remaining at the present time (*solid line*), expressed in terms of original global deposition rate. The difference between the *dashed* and *solid line* is the amount of sediment removed by subsequent erosion. The *dotted line* is a schematic fit of an exponential decay to the data (see Sect. 2.6 for a review of exponential decay)

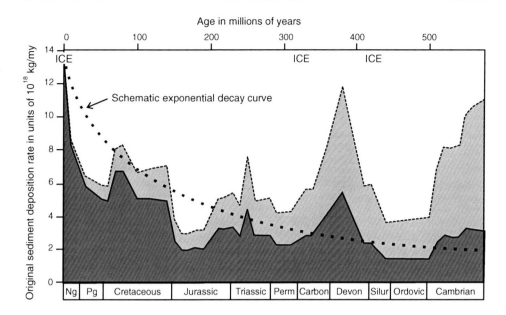

There are four major hypotheses to explain the 'terminal Pleistocene' extinctions: (1) Natural climate change; (2) Impact of an extraterrestrial object; (3) Virulent disease ('hyperdisease') introduced by humans; (4) Hunting by humans. Natural climate change refers to the deglaciation, the melt off of the Northern hemisphere ice sheets over the period from 25,000 to 11,000 years ago. The problem with this hypothesis is that there had been at least eight similar melt-offs during the past million years that had no analogous extinctions. The possibility that an extraterrestrial object impacted Earth on the ice sheet over Canada 12,900 years ago was proposed by nuclear centrist Richard Firestone and colleagues in 2007. It has attracted considerable attention, with both support and opposition from geologists and anthropologists. It doesn't seem to have affected the Paleo-Indians of North America and does not really coincide with the extinctions. It would not explain extinctions elsewhere. The hyperdisease hypothesis supposes that humans and perhaps their domesticated animals brought with them virulent disease organisms that spread through the other animal populations. The trouble is that this doesn't seem to be the way extinction of different species can be explained. Disease can jump from one species to another with evolution of the bacteria or virus, but they don't jump simultaneously to many different unrelated species at the same time. As we will discuss later, there is a very good possibility that disease (smallpox) introduced into the Native American population by the arrival of the southern Europeans may have caused a massive die-off of the American human population, which itself could have had global climatic implications. The final

possibility, hunting by humans, often called the 'overkill hypothesis,' involves not just hunting, but habitat modification. Once humans had invented the spear, they had a great advantage, even over animals much larger than themselves. But one of the widely used methods of hunting was to set fires to drive the wildlife into the open. I believe the great extinctions were caused by hunting, and much of today's ongoing extinction is caused by habitat destruction as we clear forests and use the land for agricultural production.

There is strong 'moral' objection to the overkill hypothesis because after all, humans are 'good' by definition, our own definition, that is. Population growth is 'good'—just ask your neighbor and they will proudly tell how many children and grandchildren they have. However, if you asked Mother Nature, I expect she would say that humans are a most unfortunate infestation, destroying the diversity of life on planet Earth.

## 4.13 Summary

So now we have a history of the Earth as a sequence of events. It is analogous to describing western history as Egyptian empire, Trojan War, Greece, Roman Empire, Charlemagne, Reformation, American Revolution, French revolution, World War I, and World War II, but not knowing the dates or number of years associated with these events. And if you try to fit in the Han, Tang, and Ming dynasties of China, you might get the order confused. Much of the 20th century was involved in putting the global geologic

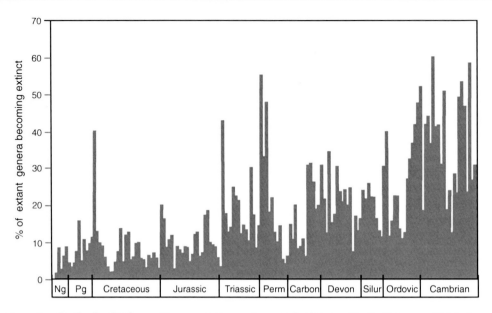

**Fig. 4.15** Extinctions of marine fossil animals over Phanerozoic time. The overall decline in extinction rates is probably an artifact of the greater knowledge of diversity of species in younger strata. The end Paleozoic (3), end Triassic (4), and end Cretaceous (5) extinction events stand out clearly from the background of 'normal' extinctions. The larger extinction event in the Paleogene (Pg) is the end of the Eocene, when polar ice appeared on Earth, and the last large one in the Neogene (Ng) coincides with the development of Northern Hemisphere ice sheets

sequence into a numerical framework. Lengths of periods of stability and rates of change matter.

But next, let's discuss Earth's climate, what it is and what controls it. You will discover that although we know a lot about climate, we don't know much about climate change.

A Timeline for this chapter:

| 723 | The Venerable Bede estimates the Creation to have taken place in 3952 BC |
| 1644 | John Lightfoot estimates the Creation to have taken place in 3929 BC |
| 1658 | James Ussher publishes *Annals of the World* with the famous date of the Creation as "at the start of evening preceding the 23rd day of October, 4004 BC" |
| 1759 | Giovanni Arduino (1714–1795) coins the terms *Primary*, *Secondary*, and *Tertiary* |
| 1774 | Georges-Louis Leclerc, Comte de Buffon estimates the Earth to be 75,000 years old |
| 1776 | Adam Smith's *The Wealth of Nations* is published |
| 1788 | James Hutton, Sir James Hall of Dunglass, and John Playfair visit Siccar Point |
| 1788 | Hutton reads his paper *Theory of the Earth* before the Royal Society in Edinburgh "we find no vestige of a beginning, no prospect of an end" |
| 1789 | July 14, Paris, the Bastille was attacked and torn down |
| 1793–94 | The reign of terror in Paris |
| 1795 | Hutton's *Theory of the Earth* is published |

(continued)

(continued)

| 1795 | Alexander von Humboldt gives the name *Jura* (*Jurassic*) to rocks exposed in the Jura mountains and southern Germany. |
| 1802 | John Playfair's *Illustrations of the Huttonian Theory of the Earth* is published |
| 1805 | In his eulogy to Hutton, Playfair describes 'staring into the abyss of time' at Siccar Point |
| 1807 | The Geological Society is founded in London |
| 1807 | John Farey, Sr., coins the term *Coal Measures* |
| 1812 | George Cuvier publishes his idea of multiple Creations separated by catastrophes |
| 1815 | Friedrich August von Alberti coins the term *Triassic* |
| 1822 | William Daniel Coneybeare and William Phillips coin the term *Carboniferous* |
| 1822 | Jean Baptiste Julien d'Omalius d'Halloy gives the name *Crétacé* (*Cretaceous*) to the chalk (French = *craie*) |
| 1826 | Ignace Vernetz proposes that Swiss Alpine glaciers had been much larger and had covered the Swiss midlands between the Alps and the Jura Mountains |
| 1829 | Paul Desnoyers coins the term *Quaternary* |
| 1830–32 | Charles Lyell publishes *Principles of Geology*; volume I is largely devoted to climate change |
| 1830 | Henry de la Beche draws a cartoon showing a glacial valley and making fun of Lyell's idea that processes that have shaped the Earth throughout its history are the same as those acting today. He gives it to William Buckland. |

(continued)

(continued)

| 1832 | William Whewell gives Lyell's idea that processes in the past were the same as those today the name Uniformitarianism (as opposed to Cuvier's Catastrophism) |
| 1833 | Charles Lyell coins the terms *Eocene*, *Miocene* and *Pliocene* for parts of the Tertiary |
| 1834 | Adam Sedgewick proposes the term *Cambrian* |
| 1835 | Sir Roderick Impey Murchison coins the term *Silurian* |
| 1835 | Adam Sedgwick coins the term *Paleozoic* |
| 1837 | Louis Agassiz proposes that the Earth had experienced an ice age |
| 1837 | Charles Lyell visits Denmark, sees the chalk there, and concludes that the Cretaceous climate was much warmer than that of today |
| 1839 | Charles Lyell coins the term *Pleistocene* |
| 1840 | Sir Roderick Impey Murchison and Adam Sedgewick coin the term *Devonian* |
| 1840 | Louis Agassiz, William Buckland, and Roderick Murchison tour Scotland; Buckland accepts the idea that the landscape reflects glaciation |
| 1840 | Agassiz publishes his first book on the glacial theory, *Étude sur les glaciers* |
| 1840 | John Phillips coins the term *Mesozoic* |
| 1841 | Sir Roderick Impey Murchison coins the term *Permian* |
| 1842 | Charles Darwin describes glacial features in Wales |
| 1844 | John Phillips coins the term *Cenozoic* |
| 1847 | Agassiz publishes a more extensive work on past glaciation: *Système glaciare* |
| 1854 | Wilhelm Philippe Schimper coins the term *Paleocene* |
| 1854 | Heinrich Ernst von Beyrich coins the term *Oligocene* |
| 1859 | Charles Darwin publishes *Origin of Species* |
| 1863 | Charles Lyell finally accepts the idea of an ice age |
| 1863 | James Dwight Dana introduces the idea of permanence of continents and ocean basins |
| 1864 | John William Salter coins the unimaginative term *Precambrian* |
| 1870 | Adolf Erik Nordenskiöld discovers Cretaceous tropical plants on the Nûgssuaq Peninsula of western Greenland (70.5° N) |
| 1872 | James Dana coins the term *Archaean* |
| 1879 | Charles Lapworth coins the term *Ordovician* |
| 1887 | Samuel Franklin Emmons coins the term *Proterozoic* |
| 1906 | Alfred Wegener makes his first trip to Greenland |
| 1908 | Gerard de Geer describes a numerical time scale for the deglaciation based on annual layers of lake sediments (varves) |
| 1912 | Alfred Wegener proposes the hypothesis of continental drift |

(continued)

(continued)

| 1922 | Alfred Wegener invited to attend a meeting of the American Association of Petroleum Geologists to hear his idea of continental drift ridiculed |
| 1930 | Alfred Wegener dies trying to resupply Camp Eismitte on the Greenland ice cap |
| 1934 | Hottest year on record for the US—until 2006 |
| 1959 | In my Ph.D. thesis defense I suggested that from my study of planktonic foraminifera in Mexico it appeared as though some catastrophe had occurred at the end of the Cretaceous; Roger Anderson's Ph.D. defense followed mine and he suggested that plant fossils in New Mexico indicated that the Cretaceous had ended with some sort of catastrophe. Our committees were skeptical in public, encouraging in private. |
| 1968 | An informal poll at the end of a session at the annual meeting of the American Association of Petroleum Geologists revealed that the overwhelming majority rejected the hypotheses of sea-floor spreading and plate tectonics |
| 1969 | Digby McLaren proposes that the extinction at the end of the Devonian was the result of an asteroid impact |
| 1969 | James Guilluly introduces the idea of cycling of Earth's sediments |
| 1972 | By an informal poll at the annual meeting of the American Association of Petroleum Geologists attendees finally accept the ideas of sea-floor spreading and plate tectonics, vindicating Wegener's hypothesis of continental drift |
| 1972 | Preston Cloud coins the term *Hadean* for the age between formation of the Earth and the oldest rock record |
| 1980 | Walter and Luis Alvarez present evidence that the extinction at the end of the Cretaceous was due to an asteroid impact |
| 1994 | Comet Shoemaker-Levy impacts Jupiter with spectacular results |
| 1997 | I stand on Charles Darwin's gravestone in Westminster Abbey to hear a eulogy celebrating the 200th anniversary of the birth of Charles Lyell |
| 1997 | I visit Siccar Point, east of Edinburgh, Scotland—the place where John Playfair had stared into the abyss of time |
| 1997 | Henry de la Beche's cartoon from 1830 ridiculing Charles Lyell's ideas is published in *Nature* |
| 2007 | Richard Firestone and colleagues propose that the late Pleistocene mammal extinctions of North America were caused by impact of an extraterrestrial object over the Canadian ice sheet. |

Suggested Reading:

Ogg, J. G., Ogg G., Gradstein, F., 2008. *The Concise Geologic Time Scale*. Cambridge University Press, Cambridge, UK, 184 pp.—An update of the latest state of the geologic time scale written for geologists.

Stanley, S. 2004. *Earth System History*. W. H. Freeman, New York, N. Y. 608 pp.—An excellent introduction to geology with half the book devoted to Earth's history.

Music: Since Scotland was the birthplace of geological science, then music of Alexander Mackenzie (1847–1935) would be appropriate. He is best known for the three *Scottish Rhapsodies* and I especially like the *Scottish Piano Concerto*.

Libation: A single malt Scotch Whiskey, and a toast to James Hutton.

A bit of history of Scotch whiskeys is in order here. Making and distilling whiskey in Scotland goes back to the end of the 15th century. The word 'whiskey' comes from Scottish-Gaelic word 'uishgy.' Whiskey is made from fermenting barley; this involves using the local water, and yeast for fermentation. When the alcohol content reaches a proper level, it is purified and concentrated by distilling. Unlike gin and most other stronger alcoholic concoctions, whiskey is then put in casks to age and mellow. Scotland has many different environments, and as a result each of the distilleries produces a spirit with unique flavors and aromas.

Single malts were largely unknown outside Scotland until the 19th century. The locals had their own preferences. However, the larger market in the cities wanted something that would always be more or less the same, so merchants began to blend whiskies from different sources to achieve stability and reliability. The first was a grocer named Johnnie Walker who had his shop in Kilmarnock in the Ayrshire district. As export to England developed more blends were made, and their whiskeys took on the names of the companies that blended them: Chivas, Ballantines, Dewar, etc.

It was not until after WWII that a few of the distilleries began to market their products outside the local area. The term 'single malt' means that the whiskey comes from a single distillery. Single malts soon captured the fancy of those who like Scotch whiskey. Because of the variations of the local waters, climate, fermentation and distillation processes, nature of the casks and the length of the aging process, single malts differ greatly from one another, providing a rich spectrum of tastes.

For now, the closest distillery to Siccar Point is Glenkinchie but its elixir may be hard to come by. It is a 'lowland' single malt, described as 'soft' and suitable for relaxing after a walk in the hills. (For more information check out Michael Jackson's 1994 Malt Whiskey Companion).

If you are a purist and don't want to dilute your precious single malt Scotch with water, you can have it 'on the rocks'—literally. Put some nice clean stone pebbles from a nearby stream into the deep freezer for a few hours and substitute them for ice cubes.

**Intermezzo IV. Graduate Study: Four Universities and Two Degrees in Four Years—Part II. The Summer of 1959: Field Work in Mexico; Completing My Ph.D. Degree**

After my first year of graduate classes at Stanford, I drove off to Mexico via Dallas to visit my father. My mother had passed away from cancer the year before. My first stop in Mexico was Monterey to talk with the geologists at Petroleos Mexicanos office there. They were very helpful in suggesting places where I might find outcrops (exposures) of Late Cretaceous and Early Tertiary rocks that would contain foraminifera. They explained that there was one important site on the banks of the Rio Panuco that was important, but not easy to reach, so they would meet me in Tampico on a certain date with a jeep and we would go there together. Most of the localities I needed to visit were cuts along the roadsides, so I could do those alone.

In the evening we went to an outdoor restaurant where entire goats and even cows were roasting, rotating on giant spits. I had never seen anything like it. I was introduced to a special drink, the "Petrolero" which consisted of roughly equal parts Tequila and Maggi. Maggi is a Swiss flavoring somewhat like Worcestershire Sauce. The drink does indeed look and taste like petroleum.

However, they gave me a tip about working alone in the field. In those days if you left your car alongside the road and were out of sight, it was considered abandoned, and windshield wipers, tires and anything else that could be easily removed from the car might be taken by passing motorists. The way to avoid this was to hire someone to watch the car while you were away. I asked how I could do this if I was out in the middle of nowhere. They told me to start with my sample collecting near the car, and soon someone, usually a little boy, would appear and ask if I needed him to watch the car. "You give him a peso or two, depending on how long you will be gone, and he will watch the car and everything will be perfectly safe." To my amazement, this worked, over and over again.

Another tip. Never ask if this is the road to so-and-so. In Mexico at that time it was considered very impolite to answer a question with 'non' (no). So the answer would always be 'si' (yes) even if so-and-so was off in a completely different direction. You always had to ask the question in such a way that the answer could not be simply yes or no. So you asked questions like "where does this road go" or "what is the name of the next town." I was already learning Mexican Spanish on the fly.

One obscure place I wanted to get to was a village named 'El Mulato.' There were supposed to be Cretaceous and Tertiary rocks exposed on a hillside just east of the village. It was shown on the maps, located about 40 km (24 miles) to the east of Ciudad Victoria. However, the map indicated that there was only a very rough road leading to it and I had no idea how I would get there. I was sure that my 1956 Plymouth would not be able to make it. I found a hotel on the Plaza in Ciudad Victoria, and in the evening, as I expected, it filled with people sitting on the benches and watching the boys and girls in their late teens/early twenties circling around the Plaza. The boys would go single file one way, the girls single file the other way. This was how they would look each other over. In Mexico at that time, a 'date' US-style was unheard of, at least in small towns. I casually struck up a conversation with some men on one of the benches, asking if they had ever heard of a place called El Mulato. It didn't take much time before I was directed to one of the more prosperous-looking gentlemen. He knew El Mulato well, he had ranch lands near there. I asked how I could get there. He said he needed to go to check on some things there; if I would share the gas costs we could go tomorrow. Early the next morning we started out. The 'road' was simply two ruts through the countryside. Before long we came to a river, which we crossed by simply driving through it. The water came up to the floorboards but that didn't seem to bother my friend at all. I guess the distance was about 35 or 40 miles, and we reached El Mulato about noon. The 'village' consisted of a few grass-walled huts with earthen floors. I saw the hillside just a short distance to the east. I explained to the people who had gathered around that I wanted to go there and collect some rocks. That was fine, but first they wanted to offer me lunch. We were invited into one of the huts. The 'stove' was a mound of clay with a depression in the middle, where some coals were burning. A flat piece of metal served as the only cooking utensil. A little goat meat, some frijoles (beans) and tortillas, and it was a great meal. After lunch my new friend/driver went off promising to be back in a couple of hours. I went to collect my rocks accompanied by all the children in the village. Everything worked out fine. The jeep appeared at the appointed time, and about 3 PM we started back. The sky had become cloudy and it started to rain. It soon changed into a downpour. Before long the road turned into slick mud. The ruts trapped the tires, and sometimes we went sliding

sideways forward. My friend/driver took it all in stride; he had apparently done this many times before. I have never seen such driving since. Sometimes going forward, sometimes sliding sideways, we made progress until we came to the river which was now a torrent. I thought we were there for the night. My friend got out, looked at the rushing water, and opened a bag on the back of the jeep. It contained a large sheet of canvas. We wrapped the engine in canvas, and drove through the river in water well over the hood and up to our chests. There was enough air trapped inside the canvas that the engine kept going until we were out of the water. We unwrapped the canvas, let the water drain from the jeep, and after a few minutes we were on our way back into Ciudad Victoria again. We arrived about 9 PM; the trip back had taken twice as long as the trip to El Mulato. He dropped me off at the hotel, and after he left I realized I didn't know his last name or have his address—he was certainly the best jeep driver in those parts.

Much of my collecting was done between the Sierra Madre Oriental and the Gulf of Mexico coast. I had originally planned to use Tampico as a base of operations, but I soon found out that in the summer it is unbearably hot and humid. Few of the possible accommodations had air conditioning and even fewer restaurants. I found a motel on the north side of town where there was a room air conditioner that worked part of the time but would go off unexpectedly. I found myself driving back to the edge of the Sierra Madre each day because that is where most of the exposures of the rocks I was seeking were to be found. Then I discovered, just where the mountains meet the plains, a local paradise, Hotel Taninul. Taninul was a huge hacienda with over 100 rooms, few of them occupied in the summer. Rates were less than in Tampico, the food was excellent, and it was cool at night, thanks to the air descending from the Sierra. There was a large natural spring which served as a swimming pool, with a bar in the middle where you could sit on submerged stones and sip tequila. This became my base of operations. There was another American guest staying there, Lon Tinkle, Professor of History at the University of Texas. He was a fascinating character, doing research in Mexico on the war with the Texans. His book *Thirteen Days to Glory* is a classic.

Finally the day approached when I was to meet the geologists from Petroleos Mexicanos in Tampico. I drove back and spent the night in the semi air-conditioned motel. We met the next morning and drove in a small convoy out the northern road from Tampico to Ciudad Mante. After a while we turned off onto a dirt track, which I was able to follow for several miles. When the going got rough, we decided to leave my car, found a boy to watch it, and went on by jeep. It was a trip of about 30 miles through jungle-like vegetation on small tracks. After a while we came to a large chain link fence and gate blocking the way. The gate was padlocked. No one around.

My friends from Petroleos Mexicanos called out "Juan, Juan" several times, as loud as they could. Most men in that area are named Juan, so this was to find out if anyone was nearby. No one appeared. Rather than smash the padlock, we dismantled the gate, taking it off the hinges. When we laid it down beside the road we looked back and there were at least 10 people watching us. My friends asked why they hadn't let us know they were there. They said no one was named Juan. We explained that we were from Petroleos Mexicanos, and PM personnel, as officials of the government, were allowed to go anywhere. They told us that the fence enclosed property of an expatriate American who was gone for the summer. We went on our way and eventually reached the Rio Panuco and the place called 'El Limón.' We found the exposures of rock we were looking for in a small clearing in the tropical vegetation. If you didn't know where it was, it would be difficult to spot from the air. We spent most of the afternoon collecting samples.

On the way back we put the gate back together again, good as new. Then, as dusk was falling and we were still on dirt tracks, we hit a rock, and the jeep soon came to a halt. One of the Petroleos Mexicanos geologists crawled underneath and found that the rock had torn a hole in the fuel line. I thought we were there for at least the night if not several days before someone happened along. But, this was Mexico, and after a few minutes, a boy appeared. Out of the jungle. We explained what we needed. He disappeared, and about 15 min later reappeared holding a replacement segment of fuel line. We thanked him and paid him for his trouble and drove on back to my car, where the boy was still dutifully watching it. And then drove back to the highway. There we parted company, the Petroleos Mexicanos geologists headed back to Tampico, and I, with the samples, back to Taninul and several Margaritas.

A few days later I left Taninul and drove to Mexico City in 1 day. It didn't look far on the map, but a short distance south of Ciudad Valles the highway starts up the Sierra Madre Oriental toward the Central Plateau. The road became a narrow two-lane highway on high steep mountainsides with spectacular drop-offs. Heavy truck traffic made it slow going. My first view of the

Valle de Mexico and its city was from the north in early evening. I should have known, but I really had no idea Mexico City was so large. And it is in a truly spectacular setting, in a valley at 7,400 ft with the two great volcanoes, Popocatépetl and Ixtlacihuatl, both about 17,000 ft high dominating the skyline to the southeast. I made my way down into the city, and eventually found the Paseo de la Reforma and the hotel just off it where I had reserved a room, the Maria Christina. The next day I called Tom Cranfill and found he was living in an apartment not far away, at Rio Tigris 24. We met, and Tom noted that there were apartments in the same building that could be rented by the week or month, saving me a lot of money. I moved into my own Mexico City apartment a couple of days later.

What amazed me about Mexico City was that it was a typical European city in almost every way. Streets that didn't run at right angles, and spectacular architecture spanning almost 500 years, and traffic like in Rome. After getting settled in I drove out to the University (UNAM = Universidad Nacional Autonoma de Mexico). It was really easy to get to; down Rio Tigris, across the Paseo de la Reforma, down Genova to the Glorieta Insurgentes, and then straight out Insurgentes Sur until you found the University complex. I knew the UNAM was going to be a spectacular exhibit of modern architecture, but it passed my wildest expectations. I easily found the Geology building, Augustin Ayala's labs and his assistants and unloaded my samples. They told me they would need a few weeks to process them. That afternoon, after looking around the campus and a couple of hours trying to decipher Juan O'Gorman's murals on the library, I drove back into the city center. This time I found myself going round and round in the roundabouts, first at the Glorieta Insurgentes, and then at the even more formidable Angel de la Independencia on the Reforma. I didn't know the primary driving rule in Mexico City—you are not allowed to hit another car. So if you are trapped on the inside of one of these roundabouts, you just turn and aim across the traffic for the street you want. Other cars will come screeching to a halt fractions of an inch from you. It is considered great sport to scare the wits out of the other driver, but you are not allowed to hit them. Incidentally the same rules seem to apply at the Arc de Triomphe in Paris. Not knowing how the system worked, it took me ten or fifteen minutes to get out of each of the circles and onto finally Rio Tiber and eventually back onto Rio Tigris. I put my car in a garage and swore never to drive in the city again. I had been working with hardly any time off for 2 years and I was ready for some vacation.

Tom was working on two projects, gathering a series of essays from Mexican authors for a special issue of the *Texas Quarterly* and purchasing art for the UT art museum. They had a grant for acquisition of art from Mexico, but it had two stipulations—no oil paintings, and nothing costing more than $100. He had a contact, Inés Amor, who ran the Galeria de Arte Mexicano. Inés had been responsible for promoting many well-known artists, and had received the highest award Mexico offers for service to the country—freedom from paying any taxes. A good friend of Tom's, Hans Beacham, was also living at Tigris 24. Hans was a photographer; he is especially well known for his portraits of Latin American writers and artists. During the summer he would photograph the essayists and artists to be featured in the *Texas Quarterly*. He was also working on a book on modern architecture in Mexico.

To make matters more interesting during the summer of 1958, the Mexican Presidential elections were coming up in October. In those days there was no real question who would win the election. The PRI (Partida Institutional Revolucionaria) always won, because the army counted the votes. The PRI candidate was Adolfo Lopez Mateos, and his name was everywhere. The protests against these rigged elections took the form of riots, and these would start in middle August. It was a traditional ritual.

I spent a few days learning my way around the city. There were three forms of public transportation, busses, peseros, and taxis. The busses were crowded but very cheap, the taxis were relatively expensive, and the peseros were cheap and convenient if you wanted to go where they went. Peseros were cars that would take as many passengers as they could hold. They had specific routes—I used those that went Chapultepec—Paseo de la Reforma-Alameda-Zocalo and back. Another of the major pesero routes was along the whole length of Insurgentes. Between those two routes you could visit almost all the major sights of the city. To get a ride in a pesero you stood on the side of the street and waved your hand and if they had room they would stop and you would squeeze in and hand the driver a peso. The driver would stop and let you out anywhere along the route. The ride was always cozy, usually at least four people on a seat that should accommodate three, but they were fast and efficient, and you could just close your eyes when going through the traffic circles. I've often thought we could take a giant leap toward solving public transportation problems in the US if we had peseros.

One of my first visits was to the Hilton Hotel on the Paseo de la Reforma. My father had given me a letter of introduction to the manager along with a note from the manager of the Hilton in Dallas. The purpose was to enable me to cash a check at the Hilton in case I needed to. I had heard that the hotel had been damaged in the earthquake that had hit Mexico City a year earlier. The Reforma Hilton was actually two buildings, an older one that had been a hotel with another name, and a second structure that enclosed it on three sides. The old part of the building was intact and being used. I had heard that the newer part had been damaged so badly it would need to be torn down. But when I arrived in Mexico City the Hilton was still seemingly intact. I went through the bustling lobby on the ground floor of the newer part of the building and figured that the reconstruction must have been completed. I found the manager's office and introduced myself. We had a chat, he set me up with the cashier so I could cash a check if I needed to (I didn't), and then he offered to show me something interesting. We went through some back doors and up a rough staircase. We were on top of the lobby. Above us was a totally empty space surrounded by walls about fifteen stories high. All of the structural steel had been removed with the intent that the new hotel would be built inside the old shell, and then the old walls demolished like an unveiling. To me those high, free-standing unsupported walls over the lobby full of people looked like a disaster waiting to happen. After that I never even walked on the sidewalk near the Hilton.

Earthquakes are a fact of life in Mexico City. They caused serious damage in 1957 and 1985. The central part of the city is built on a lake bed. The lake was still there when Hernando Cortes arrived in 1516. The sediments are very soft and filled with water. They have compacted because for many years wells drew out water for the local population. They also compact whenever a heavy weight is placed on then. The big opera house known as the Palacio de Bellas Artes was built of marble and because of its weight has been sinking since completion of its construction in 1934. By 1958 it had sunk far enough so that the steps, instead of leading up to it, led down to it. The steps had rotated 90 degrees, so that the treads (the flat part of the step) had become the risers (the vertical part of the step) and vice versa. Another exciting building is the Cathedral on the Zocalo. Built in the 17th century, the high columns along the nave are alarmingly tilted in a variety of directions, but the building has withstood many earthquakes. And then there was the Hotel Bamer, on the Alameda. It had been tilting over the street for years, but the 1957 earthquake nudged it a bit further. It was fun to go to brunch on the roof and see that the railings slope downward toward the Avenida Juarez, and to look over the front and see that it was above the center of the sidewalk below. It has survived two subsequent earthquakes.

The sediment underlying the city shakes like jelly when there is an earthquake. There are little earthquakes all the time, and each day I would check the shelves and cupboards to make sure the glassware and dishes were not slowly marching toward the edges. There are some classic sites in Mexico showing the subsidence, and as a geologist I visited as many as I could. One well casing, originally below the ground surface, stood like a monument over 20 ft high.

Tom Cranfill often invited me to go along with Hans and him to visit with artists. Inés Amor had set things up for us, so the 30 artists she had contacted for us knew the rules. Tom already knew from previous experience what was going to happen and suggested we make a pool of funds for purchasing the drawings and etchings we might be offered. Later we could divide them up. Tom said that the younger and even most of the older artists would be very anxious to have their works shown in the US and published, and the prices would be very favorable. He was absolutely correct. Our money pool was divided up as to how much we would invest in each artist's work. I've never seen anything like it before or since. Although I was almost totally ignorant on such matters he would ask my opinion and have me help in the selections. We would be looking at the works, and the more pieces we became interested in, the more the prices sank. Finally, after Tom finished his negotiations, he and I would leave with the purchases, and Hans would stay behind to take his portraits.

One evening, over some long gin and tonics, Tom explained his rules for building a personal collection of art. His own personal collection was built around 19th century Americans. First, select an age, area, style, and type of media. Unless you are wealthy you had better select contemporary art. The area must be where you live or someplace you regularly visit. The style is a matter of personal taste. The type of media, whether lithograph, drawing, water color, or oil depends to a large extend on the funds available. Next, set an annual budget, whatever you can spare for support of the arts, because you become, in effect, a supporter and driver in the art world. Buy only from reputable galleries or directly from the artists. It is amazing what lurks in the back rooms of galleries and closets of artists. Auctions are OK but only after you

know what you are doing (Tom had once picked up some signed Picasso lithographs at an estate sale for a song because no one could read the name). Then he gave me the most important piece of information: spend your entire annual budget on one or at most two items. After 40 years you will be amazed at what you were able to accomplish. Then and there I decided to work on building a collection based on contemporary Mexican art purchased mostly from Inés Amor at the Galeria de Arte Mexicana. I was able to do it for a decade, but lost my contacts after Inés passed away. Later, as other contacts developed, I switched to underground art from Communist Czechoslovakia.

I also had the opportunity to look at architecture with a specialist, Hans, who was working on his book *Architecture of Mexico: Yesterday and Today*, published a decade later. I will only cite one architect, who particularly fascinated me: Felix Candela. Candela invented a new way of building, thin curving sheets of concrete strengthened by internal reinforcing rods. No vertical walls, no flat roofs, just gravity-defying concrete. Some of the structures are breathtaking. To understand Candela's architecture you have to take a piece of paper, fold it, and then see which areas will bend and which won't. In the 1950s neither mathematical analysis nor computational capabilities were far enough advanced to really understand the forces and stresses involved in his structures. Candela's architecture challenges all your preconceived notions as to what a building should look like.

By the middle of August things were getting unruly in Mexico City. There had been a tiny increase in bus fares, and some of the students at UNAM decided that this was unfair to the poor. They started hijacking busses, taking them out to the campus and burning them. Soon riding a bus became an uncertain affair, not because you might get hurt, but because you might wind up somewhere other than your destination. It got difficult to catch a pesero or taxi as they took up the slack. Walking became the major mode of transportation. Then, toward the end of the month there was a real riot in the Alameda. Windows along the Avenida Juarez were broken. The police arrested a group of tourists from the US who were standing on the steps of one of the hotels. The reason was obvious; it was a lot easier to arrest the tourists than the rioters. The next day I had an appointment with officials of Petroleos Mexicanos at their building on the Alameda. The sidewalks were still covered with broken glass, and workmen were covering the broken windows with cardboard and plywood. I went to one of the upper floors of the building to meet with senior geologists and thank them for their hospitality. There was not the slightest indication that anything untoward had happened the day before.

Then came a break, an architect friend and his wife from Dallas arrived, and we drove off together to the south, via Puebla, Oaxaca and Taxco. That part of the country south and west of Mexico City is very different from the northeast. It reflects much older cultures, both indigenous and European. Ancient temples and pyramids vie with Baroque and Plateresque (the elaborate Mexican version of Rococo) architecture. We visited one of the churches in Puebla in which an Aztec altar hides beneath the altar cloth. Then we went on to Oaxaca, visiting the spectacular Mayan ruins at Mitla and Monte Alban. Seeing these through the eyes of a modern architect was a major educational experience. On the 31st we went to the Iglesia de la Merced to witness the "Blessing of the Animals." Everyone brings their animals to the square before and, amidst the ensuing chaos, they receive the blessing of the priest. It is pure impromptu folk theater. In Oaxaca I learned to eat a southern Mexican delicacy, grasshoppers. The market has an extensive section devoted to different kinds of grasshoppers served on a variety of tortillas. Rumor had it that grasshoppers are a significant source of antibiotics, if true eating them would be a good way to ward off disease.

We returned to Mexico City by way of Taxco. The Thomases had made arrangements for us to stay at the facilities of the Alexander von Humboldt Foundation, a delightful colonial era hacienda not far from the city center. Taxco is a beautiful town and could just as well be in the hills of Spain.

As we drove back into Mexico City it became obvious that the troubles had started. There were no busses on the streets, and we saw groups of people listening to someone giving a speech. After getting back and settled into Tigris 24, I said goodbye to the Thomases as they took off driving back to Texas. Tom and Hans filled me in as to what had been going on in the city. The newspapers gave no indication that anything was unusual was going on, but there were incidents of confrontation and violence going on each day. They suggested I collect my samples at UNAM as soon as possible and that we drive together as a convoy back to Texas. Hans, who spoke excellent Spanish, agreed to go with me to the UNAM campus.

We drove out Insurgentes Sur but as we approached the University we were stopped at a barricade manned by soldiers. They told us the campus had been taken over by rebellious students and did not want to let us pass. Hans explained that we were American scientists working with the University and needed to pick up some samples that had been left at the Geology Department. Eventually we were allowed to pass through, but told to come back by the same route we went in. On the edge of the campus there was a sandbagged machine gun emplacement and we were stopped again. Again Hans talked us through and again we were warned to come back by the same route. They could see the Geology building, and I expect they watched us through binoculars as we made our trip in. When we arrived at the Geology building we found it to be a mess. It had been all glass on the ground floor, but someone had driven a bus right through the building breaking most of the windows and display cases. There was no one around. We made our way upstairs and very fortunately found the laboratory unlocked. The samples had not been touched since I had brought them in. Gathering them up, we got them downstairs and loaded them into the back of the car. Everything was inspected at the two barricades on the way out. It had become obvious that the situation was far more serious than we had realized in the city center.

We took one more day to pack up our things. Later in the day Mexican friends called Tom to tell him that there were increasing numbers of incidents and advised that we should leave as soon as possible. The next morning we took off, planning to use the 'central road' through Queretaro, San Luis Potosi, and Saltillo to the Texas border. We had expected that there might be barricades controlling access to Mexico City. Once out of the city things seemed much more normal. Everything went well for a while; then I noticed a problem with my car. We pulled into Queretaro and I stopped at an auto repair shop. I had too much weight in the trunk, and the rear leaf springs were giving way; another spring needed to be added. They offered to fix the problem in a day. To our sheer amazement, the next afternoon, the car was ready. They had actually made the necessary additional springs themselves from scratch on the spot. In Mexico miracles just happen.

The summer of 1958 was an eye-opening experience for me. I had become familiar with Europe, and to discover that we had a very similar culture south of our border was a revelation. I really had not known that Mexico City and its Aztec antecedent Teotihuacan were always the largest or nearly the largest cities in the world. I was impressed that the racial discrimination that characterized the United States at the time did not exist in Mexico. At that time the Mexican population was estimated to be about 10 % European, 10 % Indian, and 80 % 'mestizo' or mixed. And until I spent time in Mexico I had not realized that it was a thriving 'European-style' country with major cities on the American continent two centuries before the British colonies achieved any significance.

That fall I worked up my samples, washing the shales to get out the specimens of foraminifera. After spending many weeks picking out specimens and identifying them, I began to put together a picture of what had happened at the end of the Cretaceous in this region. The planktonic foraminifera, those that had lived near the surface of the open ocean showed something quite remarkable. Large specimens (more than 1 mm across) with heavy ornamentation characterized the Late Cretaceous deposits. Then, at what I thought must be the Cretaceous/Tertiary boundary, they were replaced by minute unornamented forms. It was as though something very dramatic had happened to life in the ocean. Hans Thalmann then showed me Russian publications by Martin Glaessner on the same transition in the Caucasus. Everything there was exactly the same as what I had found in Mexico. Whatever had happened, it was a global phenomenon that had wiped out much of the ocean plankton.

The members of the Geology Faculty at Stanford were very conservative. In my thesis defense they smelled the odor of early 19th century Catastrophism —something they were sure had been thoroughly discredited by more than a century of research. They gave me a hard time for presenting such heretical views. But they couldn't explain why the same thing should be observed, documented by a highly experienced micropaleontologist, on the other side of the planet. So they let me pass. Immediately following my defense was that of another Ph.D. Candidate, Roger Anderson. He described a very sudden change in the land floras of the Raton Basin in New Mexico, also at the Cretaceous/Tertiary boundary. Many of the same faculty attending my defense were serving on his committee as well. At the party afterward a few faculty suggested that perhaps we were on to something.

Twenty years later Walter and Luis Alvarez presented geochemical evidence that an asteroid had collided with earth at that time, resulting in catastrophic extinction of many life forms.

Niels Bohr's first meeting with
J.J. Thompson - "This Is Wrong"

**5**

*As there is not in human observation proper means for measuring the waste of land upon the globe, it is hence inferred, that we cannot estimate the duration of what we see at present, nor calculate the period at which it had begun; so that, with respect to human observation, this world has neither a beginning nor an end.*
James Hutton—1785

In the last chapter you learned how the geologic timescale had developed. It consisted of a series of names, but no numerical ages. Without having numbers for the ages we had no idea how long something had lasted and no idea of rates of change. Baron Cuvier's idea of a history punctuated by catastrophes implied that conditions had remained the same for long periods of time, and were then changed by very rapid, catastrophic events. Charles Lyell's 'uniformitarianism' implied that change had occurred at rates that were imperceptibly slow relative to human lifetimes. Geology desperately needed a numerical foundation to understand Earth's history. The knowledge would come in stages, from wholly unexpected sources. First we would learn about the immensity of geologic time, and much later we would figure out how to determine the rates of geologic processes.

The rate at which something happens is critically important in determining *what* happens. Perhaps the simplest example is to consider the difference in what happens if you drive your car into a concrete barrier at 1 mile/h versus 10 miles/h or 100 miles/h. In the first instance the driver might experience a slight discomfort and no damage would occur to the car. In the second instance you might be a little shaken up, and the repair bill for the car would be a few hundred dollars. In the third instance, you would very likely be killed, so you would no longer need to worry about the damage to your demolished car.

What is happening to our planet right now is change that is going at a rate between 200 and 400 times faster than anything in the geologic past, except for that accident 65 million years ago when an asteroid struck on the northern coast of Yucatan.

Fifty years ago we knew virtually nothing about the rates of geologic processes. It is only during the past few decades that we have acquired knowledge of processes that operate on time scales of less than a million years.

Understanding how we know about the rates of change in the past and how they compare with what is happening today is so important it deserves some attention. Along the way you will acquire the basic knowledge you will need in order to understand how greenhouse gases work. Well worth knowing since there is a sea of misconceptions out there. Here is the story, told as briefly as I dare.

## 5.1 1788—An Abyss of Time of Unknown Dimensions

When, in 1788, John Playfair stared into 'the abyss of time' at Siccar Point, he had no real idea of the lengths of time involved in laying down the sediments, compressing them to form a mountain range, eroding that mountain range to lay down the sediments which would then be uplifted and compressed to form yet another mountain range, and eroding that to produce the rocks and landscape he observed. You will recall that Charles Lyell, in 1830, guessed that the age of the Earth might be measured in billions of years, but his idea was based solely on intuition. As a matter of fact, it was not until the 20th century that the length of geologic time could be expressed in actual numbers of years, as it turned out usually *millions* of years.

## 5.2 1863—Physics Comes to the Rescue—Earth Is not More Than 100 Million Years Old

Scottish physicist and engineer, William Thomson, thought he could come to the geologists' rescue in 1863. Thomson taught at the University of Glasgow, and when he was recognized by the British Crown for his achievements he was given the title "Lord Kelvin"—the Kelvin is the river that flows through Glasgow. Today he is usually referred to simply as Lord Kelvin (Fig. 5.1). One of Thomson's specialties was investigating the nature of heat. He adopted the then controversial idea that heat was a reflection of the motion of molecules and proposed an 'absolute' temperature scale.

W.W. Hay, *Experimenting on a Small Planet*, DOI 10.1007/978-3-319-27404-1_5

The zero on his absolute scale would be the temperature at which there is no molecular motion at all. The temperature scale was later named for him, and the units are termed Kelvin (not degrees Kelvin), abbreviated simply as K. The base of the scale, sometimes called 'absolute zero' is $-273.15$ °C. One degree on the Celsius temperature scale is equal to one K.

A related area he investigated was the conduction of heat. Having figured that out, he calculated the length of time required for the Earth to cool from a molten state to its present temperature and determined it to be between 24 and 400 million years old. The broad range reflected the uncertainties concerning the heat flow from the interior of the Earth and the thermal conductivity of rocks. From another calculation, based on the assumption that the Sun drew its energy from gravitational contraction and the argument that the Earth could not have existed when the Sun's surface extended out to the Earth's orbit, he arrived at an age limit of 100 million years. As an incidental conclusion, Thomson argued that the climatic zones of the Earth (tropical, temperate and polar) had not appeared until quite recently because as the planet cooled, the heat flow from the interior of the Earth was greater than the heat it received from the Sun. Paleontologists of the period argued that there was evidence for climatic zones far back in time, making this a serious debate between paleontologists and physicists in the latter part of the 19th century. Curiously,

**Fig. 5.1** William Thomson, Lord Kelvin (1824–1907) (Beards were in style.)

the argument that heat flow from the Earth's interior affects the climate comes up even today. So, just for the record, except for a few volcanic regions, the heat reaching the Earth's surface from the planet's interior is less than 1/3,000th the amount received from the Sun. As we will see, it may affect the behavior of ice sheets over timescales of many thousands of years, but it has nothing to do with large-scale climate change.

On its journey through the air and along rivers, rain water gathers various substances (including salt) which end up being transferred to the ocean. The matter of the age of the Earth seemed settled in 1899 when physicist John Joly of Dublin determined the rate of delivery of salt to the ocean by rivers and concluded from its present salt content that the ocean was about 90 million years old.

Many geologists and even Charles Darwin were highly skeptical of such a young age for the Earth, feeling intuitively it was not long enough to accommodate everything that had happened or the evolution of life. However, the general geological community came to accept the idea that Earth was at most about 100 million years old. The simple fact is that humans have no concept of time longer than a few generations, and imagining how long it might take to erode away a mountain range was purely a guess. The slowness of geologic processes is perhaps best expressed by the rate unit, the Bubnoff. The unit was proposed by Al Fischer of Princeton University to honor Sergei von Bubnoff, the first geologist who really worried about defining process rates and made it a major topic of his 1954 book *Grundprobleme der Geologie*. The Bubnoff is 1 micrometer or 'µm' ($1 \times 10^{-6}$ m) per year, which equals 1 mm per 1,000 years or 1 m per million years. Until recently no geologist ever expected to see any sort of global change in a human lifetime, but that is exactly what is happening now.

## 5.3   What We Now Know About Heat from Earth's Interior

The Earth is not simply cooling down as Lord Kelvin had assumed in his calculation of the age of our planet. It has its own internal heat sources in the slow growth of its core and the radioactive decay of uranium, thorium and potassium. They produce heat at a rate of about $3.8 \times 10^{13}$ W globally. Most of these are concentrated in the granitic rocks of the continents. Plate tectonics involves large scale convection of Earth's mantle, with warm material upwelling along the Mid-Ocean Ridge system. Directly on the ridge system heat flow may exceed 300 mW/m² and in regions of old ocean crust it may drop below 40 mW/m². The highest geothermal heat flux on Earth is at Yellowstone, where it may reach 5.5 W/m². Figure 5.2 shows an estimate of the geothermal heat flux for the continents, oceans, and Earth as whole. The values are in milliwatts; the global average is 0.26 %

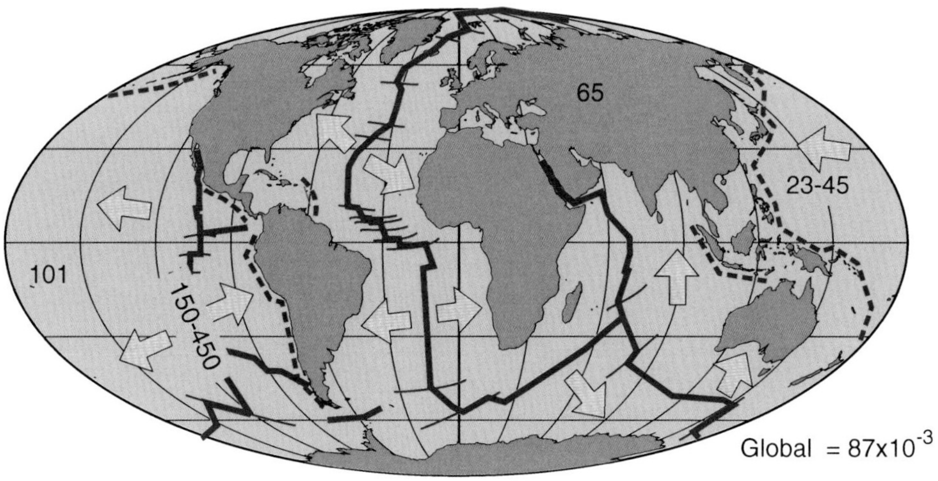

**Fig. 5.2** Average geothermal heat flows for the oceans, continents, and Earth as a whole. A new (2010) analysis of additional updated data suggests that these numbers may be about 5 % too low

(2.6 thousandths!) of the average energy received from the Sun. Even in Yellowstone heat from the Earth's interior has no impact on climate.

An interesting tidbit of information is that the variance in temperature due to seasonal changes only reaches about 1 m (3.3 ft) into soil or rock. Below that there is a steady rise in temperature with depth, typically between 15 and 35 °C/km on the continents, depending on the age and thickness of the sedimentary rocks. The influence of cold ice-age surface temperatures chilling the rock may extend down a kilometer or more and can be seen in temperature measurements from boreholes.

## 5.4 Some Helpful Background In Understanding 19th Century Chemistry

During the 19th century, the basic units of the structure of matter were 'atoms'—a Greek word meaning indivisible. But there were different kinds of atoms; these were the 'elements' of chemistry. However, before going further, it should be noted that the terms 'atom' and 'molecule' were essentially interchangeable during much of the 19th century, 'atom' by the English, 'molecule' by the French; in both languages for both single atoms and chemically bonded groups of atoms. The widespread strict use of the word 'molecule' as a group of atoms chemically bonded together and acting as a unit did not come until the 1880s.

An unlikely-seeming center for the advancement of science at the time was the soot-covered city of Manchester, England. Mark Twain commented: "I would like to live in Manchester, England. The transition between Manchester and death would be unnoticeable." The reason for the depressing conditions in Manchester was that it was at the forefront of the industrial revolution set in motion by James Watt's invention of the steam engine in 1775. A better understanding of the practical aspects of physics and chemistry was essential to the developing industries. The city fathers saw the need for local opportunities for higher education and the Manchester Academy was founded in 1786. It was one of several dissenting academies that provided religious nonconformists with higher education, including engineering and science. At that time the only universities in England were Oxford and Cambridge, and study there was restricted to Anglicans. Through the 19th century several other institutions were founded in Manchester to further the goal of technological development.

**Fig. 5.3** John Dalton (1766–1844) (Beards were not yet in style.)

In 1803, John Dalton (Fig. 5.3), then teacher of mathematics and natural philosophy at the Manchester Academy's 'New College,' made a presentation to the Manchester Literary and Philosophical Society (better known as the 'Lit and Phil') outlining a new way of looking at chemistry in terms of atoms. It was published in 1808 in his book 'A New System of Chemical Philosophy.' His 'atomic theory' dominated chemical thinking during the 19th century and sounds very familiar today. There were five main points in Dalton's theory: (1) elements are made of tiny particles called atoms; (2) all atoms of a given element are identical; (3) the atoms of a given element are different from those of any other element; the atoms of different elements can be distinguished from one another by their relative weights; (4) atoms of one element can combine with atoms of other elements to form chemical compounds—a given compound always has the same relative numbers of types of atoms; (5) atoms cannot be created, divided into smaller particles, nor destroyed in the chemical process—a chemical reaction simply changes the way atoms are grouped together. There were only about 37 elements known when Dalton wrote his book. He presented information on 20 of them. Dalton also thought that the simplest compounds, like water, were made of one atom of each of two different elements.

If the five rules sound familiar—you probably heard about them in school. Dalton's scheme was a synthesis of the ideas of the time concerning of chemical combination. There is a problem, however, in the rules as stated above. Developing methods originally devised by alchemists, elements and the compounds formed by combinations of different elements were distinguished by their relative weights for a specific volume. The weighing was done on a balance. Dalton believed that all atoms of an element were identical, and the weight of each was unique. He determined that the lightest element was hydrogen; it formed the base of the scale, with a relative weight of 1. Because all of the atoms of an element were thought to be identical, the relative weights of the elements became their 'atomic weight' and those of the molecules making up compounds became 'molecular weights.'

Unfortunately, figuring out an element's relative weight was a tricky business, especially for the lighter elements and Dalton and subsequent researchers made mistakes. What Dalton didn't know in 1808 was that the hydrogen gas he was weighing was actually $H_2$—two atoms of hydrogen combined. Furthermore, he thought that the simplest compound of two elements must be binary, formed from atoms of each element in a 1:1 ratio, so that his formula for water, for instance, was HO rather than $H_2O$. His weight for water

was 8, so by his reckoning the atomic weight of oxygen was 7. Despite the errors, Dalton's atomic theory provided a logical explanation of observations, and opened new fields of experimentation.

Dalton was unaware of research going being carried out on the other side of the English Channel. Joseph-Louis Gay-Lussac (1778–1850 Fig. 5.4 right) and Alexander von Humboldt (1769–1859 Fig. 5.4 left) had been working to determine the composition of air. They were convinced that for gases it was more accurate to compare their relative volumes rather than relative weights. They reported on their work at a meeting of the Académie des Sciences in Paris on the 1st of Pluviose in the year 13 (French Revolutionary Calendar). Their study was published that same year (1805) in the Journal de physique, de chimie et d'histoire naturelle but it was not widely circulated. They had found that air was 21.0 % oxygen, 78.6 % azote (nitrogen), and 0.4 % carbon dioxide. From a series of experiments they had also determined that water was composed of two parts of hydrogen to one part of oxygen.

In 1808 Gay-Lussac, just having been appointed Professor of Physics at the École Polytechnique in Paris, published his famous 'Law of Combining Volumes of Gases.' It received much wider distribution. He had discovered that when different gases reacted, they did so in small whole number ratios. He again noted that it took two volumes of hydrogen to react with one volume of oxygen to form $H_2O$. The same year Dalton's book was published it was shown that his rule that compounds were usually in a 1 to 1 ratio was found to be incorrect.

In 1811, Amedeo Avogadro, at the time a high school teacher in the town of Vercelli in the Piemonte west of Milan, formulated his famous hypothesis, that equal volumes of gases at the same temperature and pressure contain equal numbers of molecules. Avogadro's hypothesis stunned the scientific world. It was a highly controversial idea. In 1860 there was a Congress of Chemists held in Karlsruhe, Germany, to try to straighten out problems of nomenclature, methods of writing chemical formulae, etc. On the last day of the conference reprints of an 1858 set of notes from a course Professor Stanislao Cannizzaro had taught in Genoa were circulated. In it Cannizzaro showed how Avogadro's hypothesis, and the ideas that flowed from it, were correct. Members of the Congress were urged to adopt Avogadro's idea as a working hypothesis, and if serious problems were found, another Conference would be convened in 5 years to study the matter. The second conference was never called. Avogadro's revolutionary idea had been vindicated. Unfortunately he had died in 1856.

**Fig. 5.4** *Left* Alexander von Humboldt (1769–1859). *Right* Joseph-Louis Gay-Lussac (1778–1850)

## 5.5 Atomic Weight, Atomic Mass, Isotopes, Relative Atomic Mass, Standard Atomic Weight—A Confusing Plethora of Terms

In the 19th century scientists thought of atoms as being like marbles, all about the same size but having different weights. Of course, even in Dalton's day the expression should have been something like 'atomic mass' rather than 'atomic weight' because 'weight' is a measure of the attraction Earth's gravity has on a given 'mass.' Isaac Newton had made that distinction a century earlier. The 'mass' of something is the same on the Earth and on the Moon, but an object having a mass of 100 kg (220.5 lbs) weighs 100 kg on Earth but only 16 kg (35.3 lbs) on the Moon. However, the older term 'atomic weight' has stuck with us and is still commonly used by chemists, whereas scientists more concerned with physics use the more correct term 'relative atomic mass;' the two terms are interchangeable.

Here we need to introduce some special, often confusing, modern terminology used to describe different kinds of atoms. We now know that all atomic nuclei (except hydrogen) contain two kinds of particle, protons, which have a positive electrical charge opposite to the negative electrical charge of an electron, and neutrons, which are slightly more massive than protons and have no electrical charge. The number of protons is the 'atomic number;' it determines the element to which the atom belongs.

If the mass of a neutron is taken to be 1, the relative mass of a proton is 0.99862349 and that of an electron is 0.00054386734. Atoms with the same total number of protons plus neutrons are called 'nuclides;' such atoms can belong to different elements. Atoms with the same number of protons belong to the same element; atoms of an element having different numbers of neutrons are 'isotopes.' Thus, isotopes are a subclass of nuclides.

The 'atomic mass' ($m_a$) is the mass of an atomic particle, sub-atomic particle, or a molecule. The 'atomic mass unit' (amu) was established in the 19th century with oxygen as the reference standard. Confusion arose because the isotopic composition of samples of oxygen can vary significantly so different laboratories had different standards. The problem was resolved by international agreement in 1961, and the amu was replaced by the 'unified atomic mass unit' (u or dalton, Da) defined as 1/12 of the mass of a single atom of carbon having six protons and six neutrons at rest. That atom of carbon is designated $^{12}$C, where the superscript represents the total number of particles in the nucleus, six of which are the protons that define it as the element with the atomic number 6—carbon. About 98.9 % of all carbon atoms are $^{12}$C, but there are other isotopes of carbon $^{11}$C, $^{13}$C (1.1 %) and $^{14}$C. $^{11}$C and $^{14}$C are radioactive and present only in trace amounts. When the atomic mass of an atom is divided by unified atomic mass units (u) or daltons (Da), the ratio is a dimensionless number called the 'relative isotopic mass.' For example, the atomic mass of an atom of $^{12}$C is 12 u or 12 Da, but the relative isotopic mass of a $^{12}$C atom is simply 12.

'Atomic mass' refers to an individual species of atomic particle. Atoms of the same species are by definition identical, so atomic mass values are identical. They can include the masses of the electrons and are commonly reported to many more significant figures than relative atomic masses or atomic weights. The atomic mass or relative isotopic mass refers to the mass of a single particle. For example, the atomic mass or relative isotopic mass of $^{12}$C is (by definition) 12.000000 u; the atomic mass of the isotope $^{13}$C is 13.003355 u, and that of $^{14}$C is 14.003241 u. The relative atomic mass or standard atomic weight of carbon is usually cited as simply 12.11.

Note that the term atomic mass is fundamentally different from 'relative atomic mass' or 'elemental atomic weight' or 'standard atomic weight.' Those expressions are synonyms that refer to the mathematical averages of atomic mass values for samples of an element. Most elements have more than one stable isotope, so the relative atomic mass for any given sample depends on the mix of isotopes present. For many purposes a particular standard value for an element has been established; this may be referred to as a 'standard atomic weight.' As you will see later, it is the differences from these standards that allow us to determine the ages of materials, ancient temperatures, and other environmental parameters.

A 'standard atomic weight' is the mean value of atomic weights of a number of 'normal samples' of the element, where a normal sample is taken to be any reasonable source of the element or its compounds used by industry or science. Standard atomic weights are intended to be mean values compensating for small differences in the isotopic composition of elements in samples from different sources on Earth. They may differ from values from samples from unusual locations or from extraterrestrial objects. Standard atomic weights are cited in many textbooks and shown on charts of the Periodic Table as in Fig. 5.8 below.

Finally, a complication we won't worry about: the atomic mass of atoms, ions, or atomic nuclei turns out to be slightly less than the sum of the masses of their constituent protons, neutrons, and electrons, due to mass loss from the binding energy, related to Einstein's famous $E = mc^2$.

Because the 19th and 20th century literature on these topics uses these various terms in such a confusing way, I will use the expression 'relative atomic mass' most of the time, but its synonym 'atomic weight' when it seems appropriate in the following discussions.

Now back to our narrative: in 1815, William Prout, a physician living in London who was also an active chemical investigator, suggested that the relative atomic mass of every element is a multiple of that of hydrogen, and that the atoms of the other elements are essentially groupings of particles resembling hydrogen atoms. However, as an increasingly precise list of relative atomic masses was being developed it became apparent that it was not the simple set of whole numbers Prout predicted, but there were a lot of fractional values. His insight was discarded for a century. Nevertheless, the elements could be ranked by increasing relative atomic mass, and given a sequential 'atomic number' according to their position on the list.

The search for some sort of order among the chemical elements took a major advance in 1829 when a German chemist, Johann Döbereiner, discovered something he called 'triads.' He noticed that some elements formed groups of three which exhibited the same pattern. If you added the relative atomic masses of the first and third members of the triad you'd get a number very close to double the relative atomic mass of the middle member or, put another way, the average relative atomic mass of the first and third element approximately matched that of the middle element. The relative atomic masses he used for Cl and I were 35.47 and 126.47. The mean of these was 80.47, close to what the relative atomic mass of what Br was thought to be at that time, 78.383. The modern relative atomic masses for these elements are Cl = 35.5, Br = 79.9, and I = 126.9. Refining knowledge of the relative atomic masses has only improved the correlation. Other triads were calcium (Ca = 40.1), strontium (Sr = 87.6), and barium (Ba = 137.3); sulfur (S = 32.1), selenium (Se = 79), and tellurium (Te = 127.6); lithium (Li = 6.9), sodium (Na = 23), and potassium (K = 39.1); and several others. These triads of elements not only had similar chemical properties, but the strength of the bonds they formed with other atoms seemed to be proportional to their place in the triad. It would take another half century to really sort all this out.

Another critically important concept was introduced by another English chemist, Edward Frankland, in London. In 1852 he published the idea that the atoms of each element can only combine with a limited number of atoms of other elements. He termed this property 'valency' and in the US we call it 'valence.' Nobody had any idea what caused it; it was simply a property that atoms of each element possessed. However, by the middle of the 19th century the two principles of modern understanding of chemistry, the uniqueness

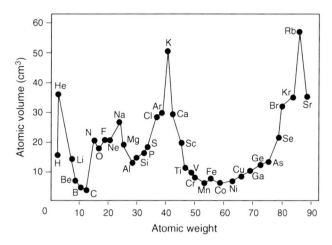

**Fig. 5.5** Lothar Meyer's diagram showing regular changes in the properties of elements. In this case 'atomic volume' is plotted against atomic weight (relative atomic mass). Atomic volume is the volume of a mole of the liquid or solid phase of the element. (After Meyer, 1870)

of atoms of each element and the limited number of atoms of other elements with which they could combine, were established and rapid progress was the result.

In 1863, John Newlands, an analytical chemist was just setting up his laboratory in London after having returned from fighting for Garibaldi in an effort to unify Italy (his mother was Italian). He gave a talk to the Chemical Society pointing out that there seemed to be a periodicity in the characteristics of chemical elements. In 1865, he named this the 'Law of Octaves.' Just as on a piano keyboard, every 8th element had properties that were much the same. The properties he was referring to were the ways the atoms of one element could combine with those of another, i.e. their valence. The idea was not very well received; Newlands was not part of the 'chemical establishment' of the time. Only his brief letters to the Society were published, because the editors of the Journal of the Chemical Society had a policy against publishing purely theoretical work as it would waste paper. England passed the honor of discovering the periodicity in the properties of elements to Germany and Russia.

Julius Lothar Meyer and Dimitri Mendeleyev (Fig. 5.6) had both studied with the famous chemist Robert Bunsen in Heidelberg, although they did not overlap. In the late 1860s Meyer was on the faculty of the University in Karlsruhe, and Mendeleyev, who had been born in a village in Siberia, had risen to become Professor of Chemistry at the State University in St. Petersburg. They both had the same idea about the periodic properties of the elements and both published diagrams showing this phenomenon. Meyer submitted his for publication in 1868, but it did not appear in print until 1870. Figure 5.5 shows his diagram of the elements hydrogen through to strontium plotted in terms of

atomic volume versus atomic weight (relative atomic mass), showing the periodic trends. Atomic volume is the volume of a solid or liquid phase of a mole of the element. A mole is the weight in grams corresponding to the relative atomic mass.

Dimitri Mendeleyev's publication appeared in the same year it was submitted, 1869, and he is credited as author of the first Periodic Table of the Elements. His table shows the elements arranged in octaves, with the rows representing the elements' relative atomic masses and the columns their valences (Fig. 5.7). The order of the elements was not strictly by relative atomic mass, but following his intuition he shifted the position of a few of them to fit with their valence properties. At the time there were 61 known elements, but places for 89 elements on the table. The empty places indicated that there were more elements to be discovered. At the time new elements were being discovered at a rate of almost one every year. As new elements were discovered, they fit neatly into Mendeleyev's table. The importance of the Periodic Table, which hangs on the wall in almost every chemistry laboratory and science lecture hall, is

**Fig. 5.6** Dimitri Mendeleyev (1834–1907). (Very long beards were in style)

| Reihen | Gruppo I. — $R^2O$ | Gruppo II. — $RO$ | Gruppo III. — $R^2O^3$ | Gruppo IV. $RH^4$ $RO^2$ | Gruppo V. $RH^3$ $R^2O^5$ | Gruppo VI. $RH^2$ $RO^3$ | Gruppo VII. $RH$ $R^2O^7$ | Gruppo VIII. — $RO^4$ |
|---|---|---|---|---|---|---|---|---|
| 1 | H=1 | | | | | | | |
| 2 | Li=7 | Bo=9,4 | B=11 | C=12 | N=14 | O=16 | F=19 | |
| 3 | Na=23 | Mg=24 | Al=27,3 | Si=28 | P=31 | S=32 | Cl=35,5 | |
| 4 | K=39 | Ca=40 | —=44 | Ti=48 | V=51 | Cr=52 | Mn=55 | Fe=56, Co=59, Ni=59, Cu=63. |
| 5 | (Cu=63) | Zn=65 | —=68 | —=72 | As=75 | Se=78 | Br=80 | |
| 6 | Rb=85 | Sr=87 | ?Yt=88 | Zr=90 | Nb=94 | Mo=96 | —=100 | Ru=104, Rh=104, Pd=106, Ag=108. |
| 7 | (Ag=108) | Cd=112 | In=113 | Sn=118 | Sb=122 | Te=125 | J=127 | |
| 8 | Cs=133 | Ba=187 | ?Di=138 | ?Ce=140 | — | — | — | — — — |
| 9 | (—) | — | — | — | — | — | — | |
| 10 | — | — | ?Er=178 | ?La=180 | Ta=182 | W=184 | — | Os=195, Ir=197, Pt=198, Au=199. |
| 11 | (Au=199) | Hg=200 | Tl=204 | Pb=207 | Bi=208 | — | — | |
| 12 | — | — | — | Th=231 | — | U=240 | — | — — — |

**Fig. 5.7** Mendeleyev's original Periodic Table (1869). Note the empty spaces that would be filled in as new elements were discovered

that it simplifies chemical properties and similarities so that one doesn't need to remember the details for each one separately. Indeed, I had a friend in the Chemistry Department of the University of Illinois who had such a bad memory for the names of the elements that he always referred to them by number. If you know the number you know where it is on the Periodic Table and from that you already know a lot about its properties.

Since that first version, the number of naturally occurring elements has risen to 92, and another 15 have been produced in atom-smashers. To accommodate this additional complexity, the Table has been modified so that row 1 has 2 elements (hydrogen and helium), rows 2 and 3 have 8 elements, rows 4 and 5 have 18 elements, and rows 6 and 7 have space for 32 elements. A modern version of the Periodic Table is presented in Fig. 5.8.

Five years before the end of the 19th century it was obvious to most scientists and the public that there wasn't much more to learn about chemistry or physics. All that remained to do was to refine the numbers for physical and chemical properties. Many 20th century scientists would be in their 20s when they made their most significant contributions, and many of them received Nobel prizes.

More importantly, with regard to understanding climate change, this first decade of the 21st century is very much like the first decade of the 20th was to understanding radioactivity and structure of the atom. Our knowledge is increasing at a staggering pace and new pieces to the puzzle are coming from unexpected sources.

## 5.6 1895–1913—The Worlds of Physics and Chemistry Turned Upside Down

Quite by accident, the world of physics and chemistry changed dramatically in the decade between 1895 and 1905. The events are worth recounting because this was a major revolution in the sciences of physics and chemistry. It profoundly affected geology, our knowledge of the history of the Earth and ultimately our understanding of how Earth's climate system works. The major players in this drama lived in France, Canada, England, Germany and the United States, communicating their results to one another through scientific meetings, rapid publication of their results, and postal correspondence.

A tip: if you can remember most or at least some of this account, you will be well equipped to hold your own at a cocktail party with scientists or engineers in attendance, and you will stun politicians. Nobody knows this stuff anymore. You can start by dropping the name 'Becquerel.'

## 5.7 Henri Becquerel and the Curies

In 1895 Antoine Henri Becquerel (Fig. 5.9) had been promoted to Professor of Physics at the Museum of Natural History in Paris. The Museum is not a single building as you might suppose, but an enormous complex of buildings on the left bank of the Seine, in the 12th Arrondissement in the

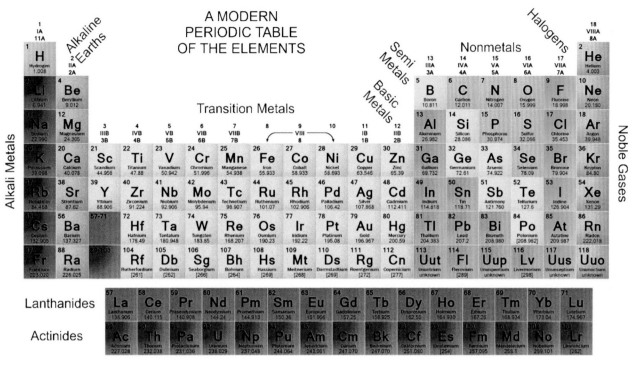

**Fig. 5.8** Periodic Table for 2015. Numbers *below* the element names are standard atomic weights. Elements *above* uranium (U) are not known to occur in nature but have been created in reactors. The logic behind the arrangement will become evident in Chap. 7

southwest part of the city. It became so large because French scientists, particularly the paleontologist George Cuvier, had encouraged Napoleon to bring back to Paris the collections of objects related to natural history from museums of all the lands he conquered. He did, and those specimens remained in Paris after the fall of Napoleon. The Muséum National d'Histoire Naturelle remains a great resource for scientific studies. Much of the space between the buildings is a botanical garden with a magnificent collection of plants from around the world. Today, just a few blocks down toward the river are the modern cubistic blocks of the Université Pierre et Marie Curie, devoted to the sciences. It is a part of Paris well known to many scientists but away from the usual tourist destinations.

As a physicist, Becquerel was especially interested in two phenomena, fluorescence, in which a material glows when excited by an energy source, and phosphorescence, in which the matter continues to glow for some time after the energy source is removed. You probably have some fluorescent lights in your house and some phosphorescent (glow-in-the-dark) materials as well. It was realized that both of these phenomena required that the material be 'excited' by some sort of external energy source, usually light. It was understood that no material could emit radiation on its own, without being excited by some external source.

In 1895 Becquerel heard about the discovery of Wilhelm Röntgen at the University of Würzburg in Germany of strange invisible rays that could penetrate aluminum, cardboard, and even human flesh and cause fluorescent paint on cardboard clear across the room to light up. Röntgen had recorded an image of his wife's hand, showing the bones and her wedding ring, on a photographic plate. Röntgen called them 'x-rays' because 'x' was the mathematical expression for an unknown. These strange rays emanated from a fluorescent spot on the end of a cathode ray tube. What is a cathode ray tube? Until recently, all our television sets and computer monitors were cathode ray tubes and took up a lot of space. Remember that even at the end of the 19th century there were no electric lights, and if you were a scientist experimenting with electricity you had to generate it yourself, usually by hand. Similarly, a cathode ray tube at that time was simply a glass tube, from which you could evacuate most of the air, with wires at either end connected to the two poles of your electrical generator or a battery. A visible ray passed down the length of the tube from the negatively-charged cathode to the positive anode. You could put different materials in the tube and see what happened as the electricity passed from one end to the other. Many minerals and shells would fluoresce brightly. Even the glass at the anode end of the tube would fluoresce (Fig. 5.10).

**Fig. 5.9** Antoine Henri Becquerel (1852–1908). A more modest beard

Physicians immediately jumped at the opportunity to see inside their patients using x-rays. Six months after Röntgen's discovery physicians were using x-rays to locate bullets in the bodies of soldiers injured on the battlefield and see the breaks in bones. However, in those days photographic plates were not very sensitive and the x-ray sources not very strong, so the patient had to remain motionless for many tens of minutes or even a few hours to get an image. In 1901,

Röntgen was awarded the first Nobel Prize in Physics for his discovery.

In 1896, Becquerel was trying to expand on Röntgen's experiments in order to understand the relationship between fluorescence and x-rays. He wanted to see if the rays produced by fluorescence included not only visible light, but x-rays. For these experiments he used a mineral known as Zippeite, a uranium salt (actually a hydrous sulfate of potassium and uranium if you are a chemist or mineralogist), from his father's collection. Zippeite is brilliantly luminescent, glowing bright yellow when excited and has a very short (1/100 s) persistence of luminescence once the source of excitation is removed. It comes from the mines at Joachimsthal (Jáchymov in Czech) in Bohemia. Those mines had operated for centuries, producing a lot of silver for coins that were called 'Jochimsthaler' or simply 'Thaler'—this is the origin of the word 'Dollar'. Incidentally, uranium was the heaviest known element at the time. It was extracted in larger quantities from the more common mineral 'pitchblende' found in the mines. Until World War I the mines at Joachimsthal were the only known sources of uranium, and they also supplied many rare elements used in making the colored glass for which Bohemia is famous. Glass with a trace of uranium is especially beautiful, with a delicate greenish hue.

Becquerel knew that the uranium mineral, when excited, was extremely fluorescent, and to excite it he planned to use the strongest possible source, sunlight. Of course, you would not be able to see the fluorescence in the bright sunlight, but if the fluorescence also included x-rays they would expose a photographic plate shielded from the light by being wrapped in black paper. The uranium salt was clamped onto the

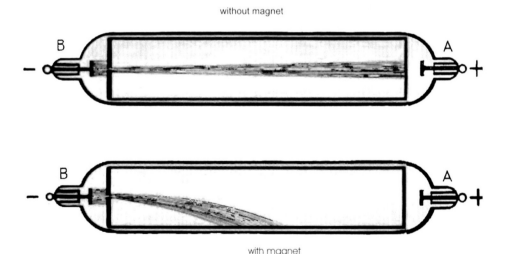

**Fig. 5.10** An early cathode ray tube showing the beam of negatively ionized particles moving from their source at the negative cathode toward the positive anode showing what happens when a magnetic field is introduced. The deviation of the beam when the magnetic field is applied means that the particles have an electric charge.

Your older television set is a more elaborate and complex version of this, but the idea that a shifting magnetic field could move the beam in such a way as to make a picture goes back to the 1870s, (Figure taken from Frederick Soddy's book '*Radio-Activity*' published in 1904)

wrapped photographic plate and set out to be exposed to the Sun. Unfortunately, it was late February and the weather was bad. The paper-wrapped plate with its uranium had been exposed only to diffuse daylight. Becquerel already knew that diffuse light would produce at best only a weak fluorescence. He put the photographic plate with its attached uranium salt away in a dark drawer. A few days later, on March 1, 1896, wanting to have some result to present at the weekly meeting of the Academie française, and probably thinking that perhaps he could show that even weak fluorescence might have produced x-rays, he developed the plate. To his great surprise there was a strong image of the uranium salt on the plate. It was evident that the image had been produced while the plate was in storage in the dark. He presented his result to the Academie the next day. It was considered an interesting result, but the question was raised whether the 'Becquerel rays' were the same as 'x-rays'? March 1, 1896 is often cited as the discovery of radioactivity, but at the time the real nature of the 'Becquerel rays' was not yet known. Henri Becquerel was 44 years old when he made his discovery.

Becquerel conducted a number of other experiments in the following weeks and found that the image on the photographic plate was just as intense after many days. It became evident that it was not the result of fluorescence or phosphorescence, which must decay away with time, but from some sort of radiation emanating from the uranium salts that did not require external stimulation. This was totally new to science.

Two co-workers in his laboratory, Marie (Fig. 5.11) and Pierre Curie took up the study of these strange rays. Among other things they found that the air around the minerals was more electrically conductive than normal. In 1898, they discovered that another element, Thorium, was also radioactive. Its radioactivity was independently discovered by an English chemist, Gerhard C. Schmidt the same year.

The Curies also found that the ore of uranium, pitchblende, produced even stronger radiation. They thought that this must be coming from an unknown chemical element that was a stronger emitter than uranium. Chemical analyses of the rocks had recorded every element present in proportions greater than 1 %, and they guessed that the new chemical must be present in very small amounts. By 1902, using sophisticated chemical techniques; they succeeded in isolating tiny amounts of the salts of not one, but two new elements from a couple of tons of uranium ore from Joachimsthal. The first they named 'Polonium,' after Poland where Marie had been born 35 years earlier, and the second 'Radium' which was so intensely radioactive as to produce an eerie green glow in a darkened room. Even more strange, they noticed that the cup containing it was warm. They coined the term 'radio-active' to describe the atoms which

produced the radiation. The amount of radium they had extracted was about 1/10 g (=0.0035 oz).

In 1903 Pierre Curie measured the heat emitted by the radium; it was not just a little, but enough to melt its own weight of ice in about an hour. Sir George Darwin (Charles' son, who had become an astronomer) and John Joly seized on this discovery and concluded that the heat generated by radioactivity was partly responsible for the heat flow from the interior of the Earth. The Earth had its own interior heat source from the decay of radioactive elements. One of Lord Kelvin's basic premises was knocked out from under the 100 million year age of the Earth.

The Curies could have patented their extraction procedure, but they did not. Instead they continued to isolate the highly radioactive polonium and radium and distributed samples of them to other scientists for their investigations. In 1903 Henri Becquerel and the Curies were awarded separate Nobel Prizes in Physics for their discoveries. Marie's was the first Nobel Prize awarded to a woman. In 1906, at the age of 47, Pierre was run over by a carriage in the streets of Paris and killed. In 1911 Marie Curie was awarded the Nobel Prize in Chemistry for isolation of these rare elements, becoming the first person to receive two Nobel Prizes.

To put things in context, it was in 1896, the year after the discovery of radioactivity, that Swedish scientist Svante Arrhenius published a paper entitled 'On the Influence of Carbonic Acid in the Air upon the Temperature of the Ground.' It was the first suggestion that the burning of fossil fuels might cause the planet's surface temperature to rise. It attracted some attention at the time, but the idea was almost forgotten for 60 years.

## 5.8 Nonconformists and the British Universities Open to All

I was an Honorary Research Fellow of University College in London from 1972 to 1990. In practical terms that meant that I spent a month in London each year, working with Professor Tom Barnard in the Micropaleontology Department. At University College, as in other British Universities, the title 'Professor' is restricted to the Head of the Department. University College, which opened its doors in 1826, was one of two Universities in England that were founded to be open to all students, regardless of their race, social position or religious beliefs. The other was a privately endowed institution in Manchester, Owens University (later to become Victoria University and later part of the University of Manchester). Cambridge and Oxford students had to be both wealthy and members of the Church of England. The main building of University College London is a neoclassical structure with a dome behind the portico entrance. It is

**Fig. 5.11** Marie Skłodowska
Curie (1867–1934)

in a courtyard opening off of Gower Street a few blocks
south of Euston Station (Fig. 5.12).

If you chance to visit the college, you will be greeted by
one of its benefactors, the jurist, philosopher and social
reformer Jeremy Bentham (Fig. 5.13). Jeremy died in 1832,
6 years after the College was founded. He believed in
individual and economic freedom, the separation of church
and state, freedom of expression, equal rights for women, the
right to divorce, decriminalization of homosexual acts, the
abolition of slavery, and animal rights. He opposed physical
punishment (including that of children) and the death penalty. He believed that mankind should strive for "the greatest
good for the greatest number of people." He had his body
mummified, and dressed in his finest clothes and great top
hat. He is seated on a chair in a glass box just to the right of
the main entrance. He had worked hard for the establishment

of the College and wanted to be able to greet the faculty and
students even after his death.

These nonconformist British Universities became centers
for laboratory experiments, while Cambridge and Oxford
specialized in mathematics and 'natural philosophy,' which
we usually refer to as 'physics.' In 1895, an English chemist,
Sir William Ramsay of University College in London, discovered helium on Earth—it was a gas trapped in a specimen
of uranium ore. Helium was about to become an important
piece of the radioactivity puzzle.

As mentioned previously, Manchester, considered the
world's first industrial city, was the site of the other institutions of higher education open to all. These were supported
by the 'nonconformist' business community, and included
the Mechanics Institute founded by James Dalton, also in
1826, and the non-sectarian Owens College, founded in

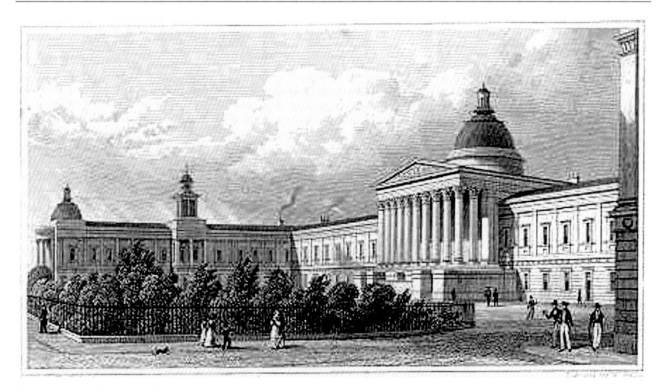

**Fig. 5.12** University College, London as it was in the mid-19th century. It looks the same today. It is not quite so grand on the inside

1851 as a result of an endowment by cotton merchant John Owens upon his death. Among Owens College's early undergraduate students was 14 year old John Joseph ('J.J.') Thomson (1856–1904) who went on to get his advanced degrees at Trinity College, Cambridge. The Mechanics Institute specialized in producing students in mechanics and chemistry, needed for the industrial revolution. During the latter part of the 19th and early part of the 20th century it became a factory of great scientists producing to date 18 Nobel Prize winners in Chemistry and Physics, and 5 in other fields. It also supplied Cambridge University's prestigious Cavendish Laboratory with much of its faculty. Manchester was clearly a great place to get an education in science and engineering, but not a very pleasant place to live.

## 5.9 The Discovery of Electrons, Alpha-Rays, and Beta-Rays

J.J. Thomson had become Cavendish Professor of Physics at Cambridge in 1884. In 1897 Thomson figured out that the mysterious 'cathode rays' were tiny particles with a negative electrical charge being emitted from supposedly solid, indivisible atoms. He called them 'corpuscles.' In 1891, George Johnston Stoney had suggested the name 'electron'

for the fundamental unit quantity of electricity, the name being derived from the Greek word for amber (ηλεκτρον). It had long been known that rubbing amber against fur is a very good way to generate 'static electricity' which is free electrons. Much to Thomson's distress, Stoney's earlier word won out over 'corpuscles' and to this day we call them 'electrons.'

One of Thomson's students at Cambridge was Ernest Rutherford (Fig. 5.14 *Left*), a 28 year old New Zealander. In 1898 Rutherford graduated from Cambridge and accepted a position at McGill University in Montreal, Canada. From his work in Thomson's laboratory at Cambridge, he concluded *"that the uranium radiation is complex and that there are present at least two distinct types of radiation—one that is very readily absorbed, which will be termed for convenience the alpha-radiation, and the other of more penetrative character which will be termed the beta-radiation."* (Philosophical Magazine, 1899).

In the meantime, a lot was happening in Paris. The search for new elements was continuing, and a young friend of the Curies, 26 year old André-Louis Debierne, discovered a new one in 1899. He named it 'Actinium.' The next year, Pierre Curie reported that radium radiation consists of two distinct types: rays that are bent in a magnetic field (β-rays), and rays that pass straight through (α-rays). Deflection meant that the β-rays had an electrical charge, and it was soon realized it

**Fig. 5.13** Jeremy Bentham
(1748–1832) seated just inside the
front door of University College,
London, greeting students faculty,
and guests

had the same mass as the electrons (corpuscles) discovered
by J.J. Thomson 3 years earlier. Another physicist in Paris,
Paul Villard, discovered in 1900 that there was a third type
of radiation coming from radium. It was similar to x-rays,
but able to penetrate much more material. We now know
these as gamma rays (γ-rays). It would be a long time before
the importance of this discovery would be recognized.

## 5.10   The Discovery of Radioactive Decay Series, Exponential Decay Rates, and Secular Equilibrium

In 1900 Rutherford's laboratory in Montreal had a strange
problem. His colleagues noticed that the count of alpha
particles from radioactive thorium varied according to

whether the window was open or closed. Rutherford guessed that one of the products might be a gas, 'Thorium emanation,' which we now know as radon. He was able to capture some of it, and found that it had a very short lifetime, after only a few minutes it was gone. But then the glass of the vessel that had held the gas had become radioactive. Some other 'decay product' had been formed and was adhering to the glass; it had a lifetime in the order of hours. It was the study of this short-lived gas and its decay product that gave rise to the idea of 'exponential decay' and the concept of the 'half-life'—the length of time required for half of a radioactive substance to decay.

It was also in 1900 that Rutherford acquired a colleague to help in his research. Frederick Soddy (Fig. 5.14 *Right*), a fresh 23 year old Oxford graduate who had been appointed as 'demonstrator' in the Chemistry Department at McGill. A demonstrator is one who sets up the equipment for the faculty lecturer. As Rutherford and Soddy worked together, they realized that something very odd was happening. If you had a pure sample of one of the uranium or radium salts it was soon contaminated by other heavy elements.

In 1902, using a strong magnetic field, Rutherford succeeded in deviating the path of the alpha rays. They were bent in the opposite direction as the beta rays, but to a much lesser degree. Pierre Curie's magnet had simply been too weak to show this effect. Rutherford realized that the 'α-radiation' must be particles with the same mass as the helium that had been found in uranium samples by Sir William Ramsay, but having a positive electrical charge. A giant puzzle was coming together. The uranium was undergoing 'spontaneous disintegration' with the loss of the alpha particles transforming uranium into other elements. They had discovered 'transmutation' of one element into another. Rutherford intensely disliked that term because it reeked of the alchemy of old. They soon figured out that uranium and thorium were the parents of 'decay series' resulting in new atoms with different relative atomic masses or valences. At the time chemists still followed Dalton's idea, that an atom of an element could have a single 'atomic weight' (relative atomic mass). Rutherford and Soddy were hesitant to assume that each of the radiation decay products was a new element, so they used terms like (U = uranium; Ra = radium) UX, UX2, UXII, etc., and RaD, RaE, RaF, etc. They were discovering what we now know as the uranium decay series. Their observations clearly didn't fit the accepted rules of the time. In this process they discovered something that was the key to dating ancient rocks as well as young sediments—secular equilibrium:

The normal or constant radioactivity possessed by thorium is an equilibrium value, where the rate of increase of radioactivity due

to the production of fresh active material is balanced by the rate of decay of radioactivity of that already formed. (Rutherford and Soddy, 1902).

Think about Rutherford and Soddy's statement for a moment. It means that, given enough time, matter will flow from the source through a series of reservoirs of different size, to an end reservoir where it accumulates. The source loses mass with time, while the end product gains it. With 'secular equilibrium,' given enough time, all the intermediate steps remain the same size. However, because their sizes vary greatly the rates of flow between them will vary greatly too. What this means is that if you can measure the masses of matter in the source and end reservoirs and the flow rate between any two of the smaller reservoirs, you can determine the flow rates between all of them! Now, for 'flow rate,' substitute 'half-life' or its close relative, 'decay constant.' This discovery allows you to calculate the half-life of uranium, which is measured in billions of years, from the half-life of radium, which is measured in thousands of years without waiting for millions of years!

## 5.11   The Mystery of the Decay Series Explained by Isotopes

In 1903 Soddy returned to Britain, first to University College in London where he worked as research assistant to Sir William Ramsay. Soddy verified that the decay of radium produced alpha particles, but after letting the experiment run for a long time helium was found in the container. The conclusion was that the alpha particles were the positively charged nuclei of helium atoms. In 1904, Soddy moved on to the University of Glasgow, as Lecturer in Chemistry. 'Lecturer' is roughly equivalent to Assistant Professor in a US University. Soddy was troubled by the results of his work with Rutherford and all those strange decay products they had found. Many of them could not be separated from one another by any known chemical means, but they seemed to have different relative atomic masses. He speculated in 1910 that there might be different kinds of atoms of the same element. He described this conundrum to a family friend, Margaret Todd, who was a physician in Glasgow and familiar with scientific questions. She suggested the term 'isotope' to denote these different forms of the same element. 'Isotope' means 'of the same place' and the word entered the scientific language in a paper by Soddy in 1913.

But in the first decade of the 20th century there was no good idea of what an atom might look like. In 1904 J. J. Thomson described the atom as being like a 'plum pudding' with the negatively charged 'corpuscles' (today's

**Fig. 5.14**  The new generation. *Left* Ernest Rutherford. *Right* Frederick Soddy. Beards had gone out of style

'electrons'—but J.J. hated the term and refused to use it) floating like plums within a positively charged cloud (the pudding). At least it was an improvement over the idea that atoms were relatively large solid objects.

## 5.12 The Discovery that Radioactive Decay Series Might Be Used to Determine the Age of Rocks

In 1905 Rutherford, in a talk at Harvard University, suggested that radioactive decay and the production of helium might be used to determine the age of the Earth. Soon after, he did just that and came up with an age of about 500 million years from a sample of pitchblende. Rutherford's helium ages were suspect because they assumed that the helium remained trapped in the rock, even though helium was known to slowly escape from almost any container.

That same year J.J. Thomson discovered that potassium was radioactive, but it would be many years before the nature of its radioactivity would be understood. It was not until late in the 20th century that the decay of potassium to argon became a major means of dating geological materials.

On his trip through New England, Rutherford also gave a talk at Yale. In the audience was a fresh young graduate in Chemistry, Bertram Boltwood. He looked into the matter and found that the ratios between uranium and lead in rocks from the same layer were the same, but that there was less uranium and more lead present in older samples. The conclusion was that lead was the end-member of the uranium decay series, and the age of the rock could be determined from the uranium/lead ratio. In the process he discovered what he thought was a new element, 'Ionium,' a new step in the decay chain which linked the chains of radioactive decay from uranium and radium. Boltwood figured that if one unit of radium decays to become lead for every 10 billion units of uranium in the parent rock, the age equals 10 billion times the ratio of uranium to lead. He dated 26 mineral samples, finding ages that ranged from 92 to 570 million years, but held back from publication. It was a good thing he did because these ages were based on the then current estimate of the half-life of radium of 2,600 years which was too large by a factor of about 150 %. The ages were soon corrected and Precambrian samples with ages over 1 billion years were included in his publication. The decimal point on the age of the Earth had moved over one place, from hundreds of millions to billions of years.

Lord Kelvin, still active in Glasgow, objected to all this radioactivity nonsense because it violated the principle of 'conservation of energy,' one of the basic tenets of physics. In his view, the energy released by radioactive decay simply appeared from nowhere—something absolutely forbidden by physics. Rutherford countered that the energy must have been put into the uranium by some event. There was also that nagging problem of what caused the Sun to shine and emit so much energy. If it wasn't gravitational contraction as Kelvin had proposed, what was it?

It was in 1905 that Albert Einstein formulated the relation between mass and energy, $E = mc^2$; his theory of special relativity, how mass and energy are interchangeable. These publications were in German, and that language had become very unpopular in the English-speaking world because of the First World War. In Britain, astrophysicist Sir Arthur Eddington translated and promoted Einstein's ideas. In 1919 he conducted an expedition to observe a solar eclipse and had found that during the eclipse the position of stars near the Sun appeared slightly out of place—the space the starlight had travelled through had been bent by the gravitational attraction of the Sun as had been predicted by Einstein's other breakthrough on General Relativity, which is a theory about gravity.

Back to the subject of the source of energy in radioactivity; Eddington knew that the major elements in the Sun were hydrogen and helium. The mass of hydrogen atom had been determined to be about 1.008 and that of helium to be 4.003. Eddington speculated that somehow four hydrogen nuclear particles might come together and fuse to become helium. If Einstein's equation $E = mc^2$ was correct, some mass would have been lost in this process and converted into energy. He speculated that this energy resulting from the fusion process could be what powered the Sun. However, he realized that the probability that the four particles could all come together at the same instant was very small and that the process of nuclear fusion must be more complex. A number of pieces to the puzzle were still missing.

It was several decades before a plausible 'event' could be identified. It came from a wholly different field, astronomy. In the first part of the 20th century the Milky Way, with its myriad of stars, was thought to be the 'Universe.' All of it. Then, in the early 1920s Edwin Hubble, using the telescope at Mount Wilson near Pasadena, discovered that many of the cloudy features of the night sky, called *nebulae*, were actually galaxies similar to but distant from our own Milky

Way. Like geologic time, the estimate of the size of the Universe had dramatically increased, by orders of magnitude. Hubble also observed that the light from these more distant galaxies was shifted towards red with increasing distance. This 'red-shift' indicated that they were all moving away from us and from each other. In 1931 Abbé George Lemaître of the Catholic University of Louvain in Belgium proposed the logical consequence of running Hubble's observations backwards—at some time in the distant past they were all together in one tiny place. He called it the 'hypothesis of the primeval atom.' The atom exploded in an event he called 'the big noise.' The term 'big bang' was substituted for 'big noise' by British astrophysicist Fred Hoyle in an attempt to ridicule the idea. Hoyle believed in an eternal unchanging universe. As a consequence, we now know Lemaître's theory today as the very descriptive 'Big Bang.'

Almost 20 years later, in 1950, Ralph Alpher and Robert Herman of the Johns Hopkins University proposed that this was the moment when the unstable, radioactive elements were created. If this were the case, they would all have the same 'age.' In 1957, Margaret Burbridge and her colleagues presented a very compelling argument that the synthesis of heavy elements, including those that are radioactive, was actually forged in the exploding stars known as supernovas. In this case, the heavy elements in different parts of the universe have different ages. Ordinary stars, like our Sun, fuse hydrogen atoms to make helium and release energy. As the hydrogen runs out, they fuse helium and hydrogen to make heavier elements, up to number 26, iron. Then they die. But if the mass of the star is large enough, at least 1.4 times as massive as our Sun, it can become a supernova. It is during these violent events that elements up to and even higher than uranium are forged. The supernova explosion also creates a shockwave that can induce further star formation from these new and heavier interstellar gases. Our own Sun has such heavier elements, which is how we know it is not a 1st generation star.

It has taken a while to get the dates all sorted out, but we now have some good milestones along the way. The latest estimate for the Big Bang is that it took place between 13.3 and 13.9 billion years ago. The supernova explosion that created the heavy elements in our Sun occurred about 4.57 billion years ago. It sent out a shockwave causing gases and other matter to begin to condense. The Sun formed shortly afterward, along with the solar system, about 4.56 billion years ago; the Moon formed about 4.527 billion years ago (Moon rocks have been dated) as a result of the collision of a large object with the Earth which would have been still molten. Finally, Earth's solid crust with continents and oceans was in existence by 4.4 billion years ago. The 90 million year cooling history Kelvin postulated did occur, but between 4.5 and 4.4 billion years ago, after this massive collision.

## 5.13   The Discovery of Stable Isotopes

In 1913, back at the Cavendish Laboratory in Cambridge, J. J. Thomson was investigating the effect of a magnetic field on a stream of positively ionized particles termed 'canal rays' or 'anode rays' because they are emitted from the positive anode rather than the negative cathode. These positive particles are boiled off the tip of the heated anode. This line of investigation was the beginning of what is now known as 'mass spectroscopy.' A mass spectrometer is analogous to the glass prism used to divide light into its different wavelengths, but it uses a magnetic field to spread out a beam of charged particles having different masses. Thomson concluded that those particles having a smaller mass were deviated more by the magnetic field than those having a greater mass. In an experiment with the then newly discovered gas neon, the photographic plate revealed not one, but two closely spaced sites where the ions had struck. Thomson realized that there were two kinds of neon, which differed in relative atomic mass. The stream of ions of the heavier form was bent less than that of the lighter form. He had discovered that a stable (non-radioactive) element had two isotopes, one with a relative atomic mass of 20, the other 22.

It would be another 40 years before the importance of this discovery was realized. Atoms and molecules are always in motion, but the heavier ones require more energy to alter their velocity than the lighter ones. It would later be discovered that there are stable isotopes of three very common elements, hydrogen, oxygen and carbon. The most common isotopes of these elements are $^1H$, $^2H$ (also known as 'Deuterium') (99.985 % and 0.015 % respectively), $^{16}O$, $^{17}O$, and $^{18}O$ (99.762 %, 0.038 %, and 0.200 % respectively) and $^{12}C$ and $^{13}C$ (98.9 % and 1.1 % respectively). Most water molecules ($H_2O$) have the $^{16}O$ isotope, but some have the $^{17}O$ and $^{18}O$ isotopes (known as 'Heavy Water'). However, when evaporation takes place the more rapidly moving water molecules with the lighter $^{16}O$ isotope escape into the atmosphere more frequently and the molecules with $^{18}O$ are left behind, resulting in isotopic fractionation. Fractionation also takes place in the formation of ice and even during the secretion of shells of marine organisms. As you will see later, the fractionation of oxygen isotopes gives us a powerful tool to make quantitative reconstructions of past climatic conditions. The carbon isotopes are, of course, involved with life processes, and they are fractionated as

different organic molecules are generated. The fractionation processes are so varied among different organisms that we can learn a lot about life from carbon isotope ratios.

In Thomson's day, the results of the mass spectroscopy were recorded on glass photographic plates. Later the records were on film. It would require major improvements in mass spectrometers before quantitative study of the distributions of stable isotopes could be carried out. A huge step forward occurred when digital detectors became available, and improvements are constantly being made even today.

## 5.14   Rethinking the Structure of the Atom

Now you may recall that a 100 years prior to these discoveries William Prout had proposed that the atoms of the different elements might be built up of units of hydrogen, but because the relative atomic masses of elements were almost never integers and always had decimal fractions, his idea was discarded. From Thomson's experiment it was evident that the relative atomic mass of neon, 20.18, could be the result of a combination of the two isotopes in a ratio of about 90 % $^{20}Ne$ and 10 % $^{22}Ne$ isotopes. (The '$^{20}Ne$' is the modern notation for the isotope of neon with an relative atomic mass of 20; the isotope is indicated as a superscript before the abbreviation for the element. You will run into a lot of this sort of notation in this book because a lot of our modern understanding of how the Earth works and process rates is based on isotopes, both stable and unstable.)

In 1907 Rutherford moved back to England to head up the Physics Laboratory at the University of Manchester (formerly Owens College). One of the post-doctoral students he inherited from his predecessor was Hans Geiger, who had received his degrees in Munich and Erlangen, Germany. Rutherford wanted to learn more about the alpha-ray particles, particularly how much matter they could penetrate, and how they were scattered as they passed through it. In those days the way this was done was to sit in front of a glass covered with fluorescent material and count and note the position of the scintillations as the alpha rays struck it. It was a task that drove Rutherford wild. He simply didn't have the patience. Geiger immediately devised a device that could do this automatically; this became the Geiger Counter we use to detect radiation today. With it he and Rutherford were able to demonstrate that the alpha particles had a double positive charge. Rutherford then co-opted a 20 year old undergraduate, Ernest Marsden, into helping out with what has become famously known as the 'gold foil experiment.'

Using their radium source, they used an electric field to direct a stream of the positively charged alpha particles toward the piece of very thin gold foil. The foil was surrounded by a circular sheet of zinc sulfide (ZnS) which would light up when hit with alpha particles. According to J. J. Thomson's plum pudding model, the alpha particles should have been deflected by no more than a few degrees as they passed through the gold atoms. However Geiger and Marsden found that a few of the alpha particles were deflected through angles larger than 90°. When they presented the results of the experiment to Rutherford he was astonished:

> It was quite the most incredible event that has ever happened to me in my life. It was almost as incredible as if you fired a 15-in. shell at a piece of tissue paper and it came back and hit you. On consideration, I realized that this scattering backward must be the result of a single collision, and when I made calculations I saw that it was impossible to get anything of that order of magnitude unless you took a system in which the greater part of the mass of the atom was concentrated in a minute nucleus. It was then that I had the idea of an atom with a minute massive center, carrying a charge. (Ernest Rutherford, Philosophical Magazine, 1911).

After graduating from Manchester, Marsden, on Rutherford's recommendation, became Professor of Physics at the Victoria University College in Wellington, New Zealand. Geiger went off to the Physikalisch Technische Reichsanstalt at Charlottenburg in Berlin and, after graduating, another of Rutherford's students, James Chadwick, followed him there. You will hear a lot more about this institute later on because this is where the data were located that were used by Max Planck at the turn of the century to develop the idea of 'quantum physics.'

Rutherford had started to develop our modern view of the atom, as having an incredibly tiny positively-charged nucleus surrounded by a negatively-charged cloud of electrons. The atom was mostly empty space. It could be thought of as a miniature solar system, but there was a problem with this simple concept. In Rutherford's atom the electrons should gradually lose energy and spiral into the nucleus. They didn't. The solution to this problem came from another of Rutherford's postdoctoral students, Niels Bohr.

Niels Bohr was a Dane and received his doctorate from the University of Copenhagen in 1911 when he was 26 years old. He was awarded a fellowship to spend a year abroad. He chose to go to Cambridge to work with J.J. Thomson. According to legend, their first meeting did not go well. Bohr walked into J.J.'s office with a copy of Thomson's book, opened it, pointed to an equation and said "This is wrong." Thomson was not amused. A few weeks later Bohr moved to Manchester to work with Rutherford. Bohr solved the 'electrons spiraling in' problem by introducing quantum physics into atomic theory. (More about the bizarre world of quantum physics later. It has a lot to do with how

greenhouse gases do what they do). The electrons were arranged in shells about the nucleus. In the current model, the innermost shell can have no more than two electrons, the next no more than 8, the next no more than 18, and subsequent shells no more than 32 electrons. Bohr's model of the hydrogen atom was published in 1913. The shells have names: K, L, M, N, O, P, and Q, or more simply 1, 2, 3, 4, 5, 6, and 7. The maximum number of electrons in each shell is K = 2, L = 8, M = 18, N = 32, O = 32 and P = 50; Q is not occupied unless the atom has been 'excited' (Fig. 5.15).

Meanwhile, from work carried out between 1911 and 1913, physicists in three different laboratories, Soddy in Glasgow, Kazimierz Fajans in Germany, and A. S. Russell in Oxford, all came to the same conclusion about the way in which the radioactive decay series took place. When an alpha particle was emitted, the atomic number decreased by two and the new atom belonged to the element two spaces to the left ('lower') on the periodic table. When beta decay took place the relative atomic mass stayed the same, but the element moved one place to the right ('upward') on the periodic table (Fig. 5.16). This has become known as the 'displacement rule' and its formulation was critical to the next step in understanding what is going on. Soddy, Fajans, and Russell all published their results within a few months of each other in 1913.

## 5.15   From Science to Science Fiction

H. G. Wells published *The Time Machine* in 1895, and *The War of the Worlds* in 1898; they were pure fiction. However, he followed the progress in physics closely, and in 1914 published The *World Set Free*, a novel that includes nuclear war with atomic bombs dropped from biplanes. The Prelude contains this remarkably prescient passage:

> And so,' said the professor, 'we see that this Radium … is an element that is breaking up and flying to pieces. But perhaps all elements are doing that at less perceptible rates. Uranium certainly is; thorium—the stuff of this incandescent gas mantle—certainly is; actinium. I feel that we are but beginning the list. And we know now that the atom, that once we thought hard and impenetrable, and indivisible and final and—lifeless—lifeless, is really a reservoir of immense energy. That is the most wonderful thing about all this work. A little while ago we thought of the atoms as we thought of bricks, as solid building material, as substantial matter, as unit masses of lifeless stuff, and behold! These bricks are boxes, treasure boxes, boxes full of the intensest force. This little bottle contains about a pint of uranium oxide; that is to say, about fourteen ounces of the element uranium. It is worth about a pound. And in this bottle, ladies and gentlemen, in the atoms in this bottle there slumbers at least as much energy as we could get by burning a hundred and sixty tons of coal. If at a word, in one instant I could suddenly release that energy here and now it would blow us and everything about

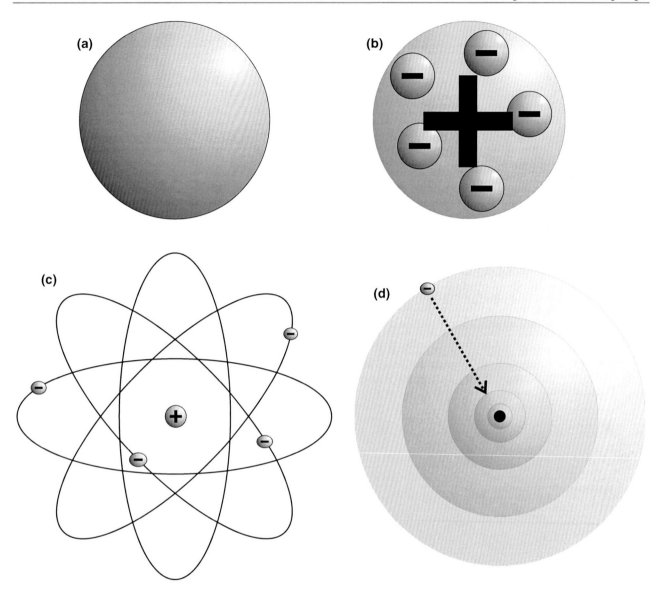

**Fig. 5.15** The development of our early understanding of the atom. **a** Dalton's (1808) 'billiard ball' model. **b** J.J. Thomson's (1904) 'plum pudding' model. **c** Rutherford's (1911) 'solar system' model. Although incorrect, this model has become incorporated into the logo of many organizations, such as the US Atomic Energy Agency. **d** Bohr's (1913) shell model of a hydrogen atom. In Bohr's model the electrons must be in one of the shells. External energy can move them immediately to a higher level; they can also lose energy in the form of a photon by dropping to a random lower energy level at some indeterminate point afterwards (as long as it that electron shell is not already full). A gradual buildup of small amounts of energy can, for example, eventually drop from the fifth level to the second level and emit a photon we perceive as blue light. This is why chemical substances glow different colors when you heat them in the chemistry lab

us to fragments; if I could turn it into the machinery that lights this city, it could keep Edinburgh brightly lit for a week. But at present no man knows, no man has an inkling of how this little lump of stuff can be made to hasten the release of its store. It does release it, as a burn trickles. Slowly the uranium changes into radium, the radium changes into a gas called the radium emanation, and that again to what we call radium A, and so the process goes on, giving out energy at every stage, until at last we reach the last stage of all, which is, so far as we can tell at present, lead. But we cannot hasten it.

Science fiction had become more science than fiction.

## 5.16   The Discovery of Protons and Neutrons

It was another 5 years before another major step forward came in our understanding of the structure of the atom. In 1919, Rutherford moved to Cambridge, succeeding J. J. Thomson as head of the Cavendish Laboratory. In June of that year he announced success in artificially disintegrating nitrogen into hydrogen and oxygen by alpha particle bombardment, proving that the hydrogen nucleus is present in

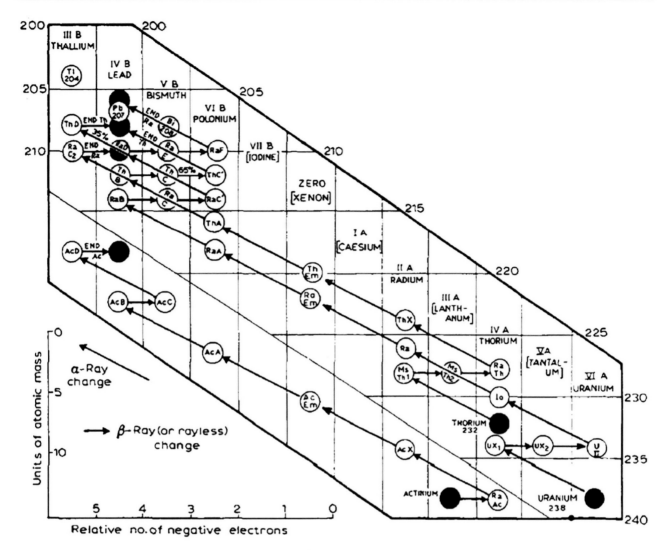

**Fig. 5.16** Frederick Soddy's 1913 diagram of the decay of uranium. The lighter elements are at the *top*, heavier at the *bottom*. The *sloping arrows* are α-decays, the *horizontal arrows* are β-decays. Don't worry if you are confused trying to read this diagram; it shows three different decay series the way they were represented at the time. It is upside down and backwards from the way this information is usually presented today

other nuclei. This result is considered to be the discovery of the proton. Rutherford named the 'hydrogen nucleus' the proton, either after the neuter singular of the Greek word for "first", or to honor William Prout, who had postulated its existence a century earlier, or both.

There were two nagging questions. If the particles in the nucleus were protons why didn't the atom have an electrical charge commensurate with its relative atomic mass? Worse, if protons carry a positive electrical charge, and positive charges repel each other, why doesn't the nucleus simply fly apart? It would be more than a decade before the first problem would be solved, and a half century before that problem about the protons repelling each other would be resolved.

James Chadwick had been interned in Germany during the war, and he returned to Britain in 1919 to work with his old mentor, Rutherford, in Cambridge. In 1923 he became Assistant Director of the Cavendish Laboratory. In 1932 he announced a major discovery, a final piece to the atom puzzle, one that could explain how the atoms of an element like neon might have two different masses. He reasoned that there was within the nucleus another kind of particle, one having essentially the same mass as a proton, but no electrical charge: the neutron. The number of neutrons is usually equal to or somewhat larger than the number of protons (with the exception of hydrogen, which is most commonly found in the form of 1 proton and no neutrons). Finally, everything made sense. Elements differ in the number of

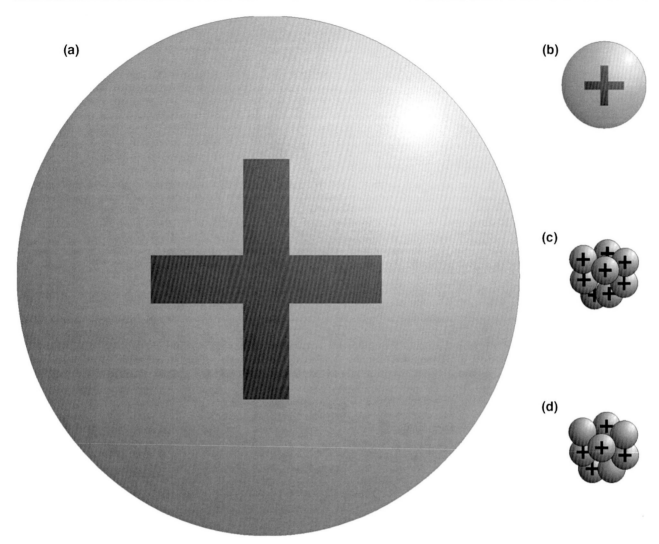

**Fig. 5.17** Evolution of ideas about the atomic nucleus. **a** Thomson (1904)—the 'plum pudding' nucleus is the size of the atom. The representation here is schematic compared with **b**, **c**, and **d**. We now know that the size of an atom as a whole is about 10,000 times the size of the nucleus. **b** Rutherford (1911)—the atom is mostly empty space; the nucleus is an extremely small object in its center. **c** Rutherford (1919)—the nucleus consists of positively charged particles called protons—and maybe something else. **d** Chadwick (1932)—the nucleus contains both positively charged protons and neutrons with no electrical charge. This finally provides an explanation for Soddy's, Fajan's, and Russell's (1913) 'displacement rule' and explains the nature of isotopes

protons in the nucleus, and the isotopes of an element differ from one another in having different numbers of neutrons. One can think of the role of neutrons as that to providing some separation between the protons and thus reducing the effect of the repellence of like electrical charges. Collectively, the protons and neutrons in an atomic nucleus are called nucleons (Fig. 5.17).

To summarize briefly what we have learned since then: An atom with at least one electron orbiting it is called a 'nuclide.' An atom with less than or more than a full complement of electrons is an 'ion,' positive if there are more protons than electrons, negative if there are more electrons

than protons (hence the seemingly back to front pH scale). Ions can be accelerated through an electrical field as they have a gross electrical charge that is not neutral. Alpha particles are helium nuclei, with two protons, two neutrons and no electrons. They have a 2+ electrical charge so they can be accelerated in an electrical field. They are released by radioactive decay and are the most used objects when investigating larger atomic nuclei.

Here are some statistics, as of 2010: There are 92 naturally occurring elements known on Earth. An additional 26 elements have been produced in particle accelerators (machines which use strong magnetic fields which make charged particles accelerate to high speeds and collide). The 92 natural elements have 339 nuclides or isotopes, 90 of the nuclides are completely stable, and 84 have been observed to undergo radioactive decay. The remainder appears to be stable but might decay under special circumstances. Thirty-three of the unstable nuclides have half-lives greater than 80 million years and have been present since the birth of the solar system. Fifty-four of the unstable nuclides have shorter half-lives and are intermediate products of a longer decay series or, like the $^{14}C$ used in dating objects less than 40,000 years old, are products of cosmic ray bombardment. Eleven of the 92 elements have no stable isotopes. Atoms up to element 20 (calcium) may have a stable nucleus with an equal number of protons and neutrons; above that number, the proportion of neutrons to protons gradually increases to a ratio of 1.5:1 in the heaviest nuclei. If the number of nucleons is 60 or less, fusion-releasing energy is possible; the fusion reaction is exothermic; it releases more energy than it takes to trigger the action. This can happen in ordinary stars. If the number of nucleons is greater than 60, the fusion process is endothermic (consumes energy). This means that the elements heavier than iron (Fe = 26), cobalt (Co = 27) and nickel (Ni = 28) can only be formed under very special energy conditions, such as during the explosion of a supernova.

**Fig. 5.18** The young Arthur Holmes

## 5.17   Arthur Holmes and the Age of the Earth

Back to geology. Arthur Holmes (1890–1965) (Fig. 5.18) was from Gateshead in the north of England. As a child he had heard about Archbishop James Ussher's determination that the Earth was created in 4004 BC. As I have mentioned before, the only thing from Ussher's book that ever gets cited these days is the date of the Creation, but the book is about trying to put numerical dates on historical events, such as when Julius Cesar was murdered, when the pyramids were built—stuff like that. It is a very long (almost 1,000 pages), very interesting scholarly work. Isaac Newton tried to figure out historical dates too, but didn't get back as far as the Creation with the details so he put it at a simple round number—4000 BC (Fig. 5.18).

As a youngster, it was that last 4 in the 4004 that troubled Arthur Holmes. It just seemed too precise, considering all the uncertainties. In his teens he was an undergraduate at the Royal College of Science (now Imperial College) in London, originally intending to study physics. He took a course in geology, and decided that was what he wanted to do. Maybe he could figure out the real age of the Earth. As an undergraduate he used the uranium to lead ratio in a mineral of Devonian age from Norway to date it at 370 Ma. In 1911, when he was 21 years old and had not yet finished his doctorate, he published a book, 'The Age of the Earth,' in which he discussed the various techniques that had been used before, including not only Kelvin's hypothesis, but estimates based on erosion rates as well.

At that time Holmes' preferred estimate of the age of the Earth was 1,600 million years. He argued that the future of the development of the geologic time scale hinged on the use of radioactive materials. It is difficult to imagine Holmes' audacity. He had not even been old enough nor had the credentials to join the Geological Society. However he gave a talk there and was astonished to discover he was the only one in the room who knew that there were different isotopes of lead. Most probably didn't even know what an isotope was. Holmes set about a lifelong career of refining the geologic timescale. In 1927 he published an updated version of his book, *The Age of the Earth, an Introduction to Geological Ideas*, which was read by a wider audience. In it he pushed back the age of the Earth to no younger than 3 billion years. In 1924 he joined the faculty of Geology at Durham, England, and in 1943 became a Professor in Edinburgh. In 1944 he wrote *Principles of Physical Geology* which became widely used as a textbook throughout Britain, and was updated with a new edition in 1965. In many ways, Arthur Holmes was the Charles Lyell of the 20th century.

## 5.18 The Development of a Numerical Geological Timescale

In 1917 the American geologist Joseph Barrell of Yale University used the framework of Boltwood's and Holmes' radiometric ages and his own estimates of the thicknesses of rock in each geologic system to make a guess about the ages of each of the geologic periods. Amazingly, as shown in Table 5.1, the framework he established is not greatly different from that used today but it was, of course, highly controversial at the time.

Already by the 1920s the uranium-lead isotope studies had made it evident to 'those in the know' that the Earth was billions of years old, although Lord Kelvin refused to believe that he could have been wrong. Interestingly, one of those individuals involved in developing the new radiometric techniques for measuring the age of the Earth was none other than John Joly, but since he knew from his salt calculation that the age of the ocean was about 90 million years he was mystified why the radiometric age determinations would give answers so different. In 1925 he wrote a book, *The Surface History of the Earth*, in which he concluded that something was wrong with the radiometric method of determining the age of the Earth, still insisting that the oceans were no more than 100 million years old. Old ideas sometimes die hard, and it was not until the late 1930s and 40s that the majority of geologists accepted the idea for an age of our planet measured in billions rather than millions of years.

The flaw in Joly's salt argument was that he thought that the salt content of the ocean had simply built up over time, when in fact there were many times when large amounts of salt were removed from the ocean water and deposited as rock. We now know that most of the salt in rivers comes from erosion of these ancient salt deposits; the rest is salt introduced into the atmosphere by the evaporation of spray from breaking waves. Knowing this, if we made a new calculation of the age of the Earth based on the salt content of the ocean, and its removal and redelivery, we would conclude that the oceans are infinitely old. In reality, the removal of salt into geologic deposits has caused the ocean to become fresher over at least the last half billion years. Amazingly, despite of the fact that Joly's method was shown to be wrong over a half century ago, it still shows up in textbooks and in television programs describing the history of the planet.

In 1935 it was discovered that there were actually two isotopes of Uranium that ultimately decay to two different stable isotopes of lead: $^{238}U$ to $^{206}Pb$ and $^{235}U$ to $^{207}Pb$ with half-lives of 4.47 billion years and 704 million years respectively. Today, about 99.28 % of the naturally occurring uranium is $^{238}U$, only 0.71 % is $^{235}U$, and 0.0054 % is $^{234}U$, an unstable decay product of $^{238}U$ with a half-life of only 245,200 years. Another radiogenic isotope of lead, $^{208}Pb$ is the final product of decay of $^{232}Th$ (Thorium), which has a half-life of 14 billion years. Unfortunately, the best uranium and thorium-bearing minerals do not occur in the sedimentary strata on which the relative age scale is based, so putting together a time scale was a tricky business.

**Table 5.1** Development of the numerical geological timescale

|  | Kelvin 1897 | Holmes 1913 | P and S 1915 | Barrell 1917 | S and D 1933 | Elias 1945 | Harland + 1989 | G and O 2004 |
|---|---|---|---|---|---|---|---|---|
| Quaternary |  | 1 |  | ? | ? | ? | 1.64 | 1.81 |
| Tertiary |  | 30.8 | 30 | 54 | 60 | 60 | 65 | 65.5 |
| Cretaceous |  |  |  | 119 | 140 | 135 | 145 | 145.5 |
| Jurassic |  |  |  | 154 | 175 | 175 | 208 | 199.6 |
| Triassic |  |  |  | 189 | 200 | 210 | 245 | 215.0 |
| Permian |  |  |  | 214 | 240 | 256 | 290 | 299.0 |
| Carboniferous |  | 340 | 140 | 289 | 310 | 319 | 362.5 | 359.2 |
| Devonian |  | 370 | 185 | 339 | 350 | 362 | 408.5 | 416.0 |
| Silurian |  | 430 |  | 379 | 380 | 389 | 439 | 443.7 |
| Ordovician |  |  |  | 469 | 450 | 460 | 510 | 488.3 |
| Cambrian | <98 |  |  | 539 | 540 | 538 | 570 | 542.0 |

Numbers are millions of years. Holmes 1913—*The Age of the Earth*; P & S 1915—Pirsson and Schuchert—*Textbook of Geology* (based on radiometric ages published before Holmes, 1913); Barrell 1917—*Rhythms and the measurements of geologic time*; S & D 1933—Schuchert and Dunbar *Historical Geology*; Elias 1945—*Geological Calendar*; Harland + 1989—Harland et al., *A Geologic Time Scale 1989*; G & O 2004—Gradstein and Ogg—*Geologic Time Scale 2004—why, how, and where next*

In addition to Holmes, Alfred Nier at Harvard and later at the University of Minnesota dated minerals on the basis of uranium, thorium, and lead isotope ratios. While I was a doctoral student at Stanford in the late 1950s I had a course taught by one of the grand old men of geology, Adolph Knopf, who had just retired from Yale. We spent a semester studying those heavy isotope dating methods in detail and examining the significance of each of the age dates then available. A framework for the geologic age scale had been established, but there simply were not enough reliable dates to allow one to calculate process rates. For me this was important because my thesis was on the nature of the Cretaceous-Tertiary boundary in Mexico, and I was finding out that something very dramatic had happened to the Earth in a very short time. We now know that the Cretaceous-Tertiary boundary event was an asteroid impact that changed the planet in a matter of hours and days. It is the only known instance of climate change being more rapid than what is happening today.

As wonderful as the development of the geological timescale was, something very important was missing. There were not enough dated points in the timescale to allow calculation of the rates of geologic processes occurring over less than a few million years. It now seems strange that when I was a student no one seemed to worry much about this gap in our knowledge, but Lyellian uniformitarianism ruled the day. We simply assumed that process rates in the past were the same as they are today. We were unaware that we didn't really know much about modern process rates. In this foggy landscape, Gerard De Geer's dating of the year-by-year history of the later part of the last deglaciation, discussed in Chap. 2, was the notable exception, but it was just the very youngest part of geologic history and no one cared that much about it. It would be another half century before more detailed time scales could be developed for the glacial epochs and younger parts of geologic time, and the extension of that more detailed time scale effort back to the more distant geologic past is just getting underway today.

To get geologic age dates close enough together to calculate rates when rapid change was occurring in the more distant geologic past would require new methods and about a half century more research.

## 5.19   Summary

The uranium, thorium and lead dating methods completely changed scientists' idea of the age of the Earth from about 90 million years to more than 2 billion years. The ages of the geological periods were established as values close to what we know today in only one decade, between 1907 and 1917. The values obtained in 1917 have been refined over the years, sometimes upward, sometimes downward, but never

to such a massive degree as the change from Kelvin's 90 million year estimate; the big advances in establishing a numerical framework for the geological time scale took place in a very short period of time. This advance was the first revolution in geologic knowledge of the 20th century. It was current when I entered college in 1951 intending to become a paleontologist. The second revolution would be the theory of plate tectonics which arrived in the 1960s.

As you will see, we are currently in something like one of these revolutionary times in science. In the field of cosmology we have discovered that there is more gravity in the Universe than we can account for with the atoms and other particles we know of, and there is more energy too. Scientists are now trying to understand what the 'dark matter' and 'dark energy' producing these effects might be.

Our understanding of the human impact on the planet and the resulting climate change is increasing day by day. A decade ago there could still have been some serious debate about whether humans could alter the planet, but since the beginning of the 21st century major new discoveries are made every few months revealing new ways in which human activities are unexpectedly and inadvertently changing planet Earth. The next chapters will outline what we know about how Earth's climate system works.

A Timeline for this chapter:

| Year | Event |
| --- | --- |
| 1775 | James Watt invents the practical steam engine; the Industrial Revolution begins |
| 1786 | Manchester Academy founded: a nonconformist academy providing religious nonconformists with higher education, including science and technology |
| 1788 | John Playfair—Stares into the abyss of time at Siccar Point |
| 1808 | John Dalton—'*A New System of Chemical Philosophy*' with the 'billiard ball' model of the atom and the idea that molecules are 1:1 combinations of atoms |
| 1808 | Joseph-Louis Gay-Lussac—publishes his "Law of Combining Volumes of Gases" which contradicts Dalton |
| 1811 | Amedeo Avogadro formulates his Law—equal volumes of gases at the same temperature and pressure contain equal numbers of molecules |
| 1815 | William Prout suggests that the atomic weight of every element is a multiple of that of hydrogen |
| 1826 | University College London founded—assisted by Jeremy Bentham |
| 1826 | The Mechanics Institute founded in Manchester by James Dalton |
| 1829 | Johann Döbereiner finds that elements come in groups of three ('triads') with similar properties |
| 1830 | Charles Lyell—Earth might be billions of years old |

(continued)

(continued)

| Year | Event |
| --- | --- |
| 1852 | Edward Frankland discovers that atoms of each element can only combine with a limited number of atoms of other elements (the concept of 'valence') |
| 1860 | Avogadro's 'Law' accepted by Karlsruhe Congress of Chemists as a working hypothesis. Avogadro had died 2 years earlier. |
| 1863 | William Thomson, Lord Kelvin, calculates on the basis of heat flow, and because the Sun draws its energy from gravitation, that the Earth is 100 million years old |
| 1863 | John Newlands—'Law of Octaves'—every 8th element has properties that are much the same |
| 1868 | Astronomers Pierre Jansen and Norman Lockyar discover a new element unknown on Earth, helium, in the spectrum of the Sun's atmosphere during a solar eclipse |
| 1868 | Julius Lothar Meyer—submits his paper on Periodic properties of the elements; publication delayed |
| 1869 | Dimitri Mendeleyev—submits his paper including a Periodic Table of the Elements; it is rapidly published |
| 1870 | Julius Lothar Meyer's paper on periodic properties of the elements is published |
| 1891 | George Stoney proposed the name 'electron' for the fundamental unit of electricity |
| 1895 | William Ramsay discovers helium on Earth |
| 1895 | Wilhelm Röntgen discovers x-rays, makes a photograph of the bones in his wife's hand |
| 1896 | Henri Becquerel, with his assistants Marie and Pierre Curie, discovers radioactivity of uranium |
| 1896 | Svante Arrhenius publishes 'On the Influence of Carbonic Acid in the Air upon the Temperature of the Ground' |
| 1897 | J.J. Thompson realizes that 'cathode rays' are tiny particles with a negative electrical charge—'corpuscles' (now called electrons) |
| 1898 | Marie and Pierre Curie discover that thorium is also radioactive |
| 1900 | Pierre Curie discovers β-rays (high speed electrons) |
| 1900 | Paul Villard discovers γ-rays (gamma rays) |
| 1902 | The Curies isolate two new radioactive elements, polonium and radium |
| 1902 | Ernest Rutherford discovers that 'α-radiation' must be positively charged particles with the same mass as the helium nucleus |
| 1902 | Ernest Rutherford and Frederick Soddy realize that radioactivity involves a decay series in secular equilibrium |
| 1903 | George Darwin and John Joly realize that radioactivity is an internal heat source for the Earth, knocking the base out from under Lord Kelvin's calculations |
| 1904 | J.J. Thompson proposes the 'plum pudding' model of the atom |
| 1905 | Rutherford suggests that radioactivity might be used to determine the age of the Earth |
| 1905 | J.J. Thompson discovers that potassium is radioactive |

(continued)

(continued)

| Year | Event |
| --- | --- |
| 1905 | Albert Einstein receives his Ph.D. from the University of Zürich with a thesis "A New Determination of Molecular Dimensions" and publishes articles on the photoelectric effect, special relativity, the $E = mc^2$ equation, and Brownian motion |
| 1910 | Frederick Soddy realizes that some elements have atoms with different atomic weights |
| 1910 | Gerard de Geer describes the history of the later part of the last deglaciation based on annual layers (varves) in lake deposits |
| 1911 | Rutherford proposes a 'solar system' model of the atom |
| 1913 | Soddy names the different forms of the same element 'isotopes' |
| 1913 | Niels Bohr proposes the shell model of the hydrogen atom |
| 1916 | Albert Einstein publishes his general theory of relativity |
| 1919 | Arthur Eddington promotes Einstein's Theory of Relativity and discovers that starlight is bent by the gravitational attraction of the Sun |
| 1921 | Astronomer Edwin Hubble discovers that many nebulae are actually galaxies |
| 1931 | George Lemaître proposes that the universe originated in an explosion he called 'the big noise.' He was ridiculed by Fred Hoyle who renamed the 'big noise' the 'Big Bang' |
| 1950 | Ralph Alpher and Robert Herman propose that the unstable, radioactive elements were created in the Big Bang |
| 1957 | Margaret Burbridge and colleagues propose that the heavy elements were not formed in the Big Bang but are synthesized in supernovas |

If you want to know more:

Gradstein, F., Ogg, J.G., 2004. Geologic Time Scale 2004—Why, How, and Where next! Lethaia, v. 37, p. 175–181.—A summary account. If you Google the title you will find a copy you can download for free.

Ogg, J. G., Ogg, G., Gradstein, F., 2008. *The Concise Geologic Time Scale*. Cambridge University Press, Cambridge, UK, 184 pp.—A detailed account of the present status to the geologic timescale, written for specialists.

If you Google 'Geologic Time Scale' you will find the latest updates.

Music: In honor of Marie Skłodowska Curie: something by her countryman Frédéric Chopin.

Libation: Why not settle in with a glass of a nice Californian Merlot.

## Intermezzo V. A Postdoctoral Year in Switzerland —and Europe—Part I

While working on my Ph.D. I gave serious thought about what I wanted to do next. As one of the measures taken to develop science in the United States after the launching of Sputnik in 1957, the National Science Foundation had announced a program of awards for Postdoctoral study. I was tempted to see if I could squeeze in one more year of learning before entering the teaching profession. I asked Hans Thalmann what he thought I should do. He suggested that if it were possible, it would be good for me to continue my studies in micropaleontology at an institute where many micropaleontologists had received their training—the tiny Geological-Paleontological Institute of the University of Basel in Switzerland. There were only two regular faculty members, Louis Vonderschmitt, geologist, and Manfred Reichel, micropaleontologist. Reichel's personal specialty was the study of 'larger foraminifera.' These fossils, abundant in Late Paleozoic, Cretaceous and Early Tertiary marine strata range up to several centimeters across.

In the meantime, I had been contacted by George White at the University of Illinois. He wanted me to apply for a position there. I wrote him that I might want to have a year of postdoctoral study if I could manage it. He thought that would be a good idea, but wanted me to apply anyway. I sent in applications to the University of Illinois and to the National Science Foundation. I received an offer from the University of Illinois and a postdoctoral award from the NSF at the same time. I called George White about my dilemma, and he said the matter was settled at Illinois. They wanted me, and could wait a year for me to come. I tell this because this is the way things were in those days. Lyndon Johnson, as leader of the US Senate saw to it that science was taking a massive upswing in the US. Academic positions became available as well as funding for research.

Two friends who attended our informal micropaleontology seminars at Stanford were Al Loeblich and his wife Helen Tappan. They had just published a major monograph on the planktonic foraminifera, which was immensely helpful to me in my research. Al worked for Chevron. When he found out I would be spending a year in Basel, he approached me with a proposition. He was very interested in obtaining sample material from the sites in Europe where planktonic foram species had first been described, known in the profession as 'type material.' Chevron would pay me $25 per sample; the samples had to be at least 1 kg. It turned out that there were more than a hundred localities involved, so it was a significant sum of money at the time. The possibility of having funds to travel around and see so much geology sounded great, but I needed approval from the National Science Foundation. To my great joy, they approved of the

arrangement, but with one proviso: I had to spend all of the money on travel expenses, every last cent had to be for expenses.

My father retired in 1959 as his company was sold to a larger insurance firm. He was in his early 60s and they had set him up with what at the time seemed a nice retirement plan. However, he remarked at the time that the executives at the new company had salaries that were almost double his. In his day, the executives worked to earn money for the stockholders; he feared that we were entering an era when executives would work for their own, not the stockholder's benefit, Only recently have I come to realize what great insight he had into economics.

Dad decided to make his first trip to Europe, with me as his guide, in the summer of 1959. We were going first class. We flew from Dallas to New York on one of the new American Airlines jets; they had been operating only 6 months. In New York we boarded the French Liner *Liberté* and sailed for France. My experiences in crossing the Atlantic had been on a freighter in 1953, and then in tourist class on the *Italia* in 1955 and *Mauretania* in 1956. I had bought a tuxedo, required for dinner, before leaving Palo Alto. Dad was immensely proud that I spoke some French, and I was in charge of negotiating the wine list.

Dad had an old friend, Larry Green, who had become very wealthy and now lived in tax-free Liechtenstein. Larry had made and lost several fortunes. His latest one had come about when he sold the French automobile company Renault on the idea of selling their cars in the US. Larry set up dealerships, and received a commission from every car sold. He had arranged accommodations for us in Paris, at a small hotel, the Raphäel, on the Avenue Klèbèr, just a block from L'Etoile with the Arc de Triomphe. We were greeted like old friends, M. Green was well known to the staff. We had a suite that Larry apparently kept reserved, in case he should want to visit Paris. Those were the most luxurious accommodations I have ever had in Europe (or the US for that matter). My father was impressed, and even a bit taken aback. A few days later we were to pick up a car at the Renault factory, but first we had a tour led by one of the executives. Even in 1959, the production line was robotized; the chief role of humans was to try to find bad welds and other imperfections, using hammers and chisels if necessary. It seemed way ahead of automobile production in the US.

After a week we left Paris and headed for Basel, where I dropped off a suitcase of winter clothing for storage at the Institute. All of the faculty and students were away on a field trip. We went on to Zürich were I showed dad my old haunts. Then to Liechtenstein and Larry's chalet. Liechtenstein is a small country, about 160 km$^2$ (62 miles$^2$) on the east side of the Upper Rhine valley between Switzerland and Austria. It is a constitutional monarchy governed by a Prince, who lives in a castle high on the mountainside. Larry's chalet was located about half way up to the castle and had a magnificent view over the town of Vaduz and the broad Rhine valley. In spite of its small size, Liechtenstein is an economic powerhouse. Many of the world's largest corporations have their home office there. The offices are actually mailboxes in a small office building in Vaduz.

From Vaduz we headed for the Alps, with our first overnight in Lucerne. There disaster struck. My father slipped in the shower and broke his leg. We got it set and put in a cast, and continued our trip, but not through the Alps, since he could hardly walk, but to Vienna, then Venice, then across northern Italy to Provence in the south of France. In Monaco we met Larry Green again. He had a yacht docked next to the harbor entrance. After that my father took the train back to Paris and flew back to the US. Later he said that his greatest mistake was to continue on with our trip rather immediately flying back to the US, and to come back to Europe later while I was still there.

From Provence I drove back to Basel, where my Postdoctoral year was to begin September 1. I went to the Institute to get my suitcase and learned that only a few days before, Professor Reichel had a heart attack and was in the hospital for an indefinite stay. Professor Vonderschmitt was very kind and introduced me to an Adjunct Professor, Hans Schaub, who worked on a special group of larger foraminifera, the Nummulites. The word 'Nummulite' stands for 'coin stone,' and these fossil forams are the size and shape of coins, but one species, found in the rock used to build the Great Pyramids in Egypt, reaches the size of a dinner plate.

Hans Schaub had worked on them for 20 years, carefully collecting them from layers of sediment in a deposit called the Schlierenflysch just inside the northern edge of the Alps a short distance south of the Lake of Lucerne. At that time, 'flysch' (pronounced fleesh) was a bit of a mystery. It was sediment formed as the Alps were being deformed in the early Tertiary and shoved onto what was then the southern margin of the European continent. We now know that the sands represent periodic submarine landslides down the slopes in front of the rising Alps while the shales are deposits formed by settling of particles through the water column during the normal process of sediment

accumulation. The sands represent deposits that formed in hours, the shales deposits that formed over thousands of years.

Everyone said that there were no more nummulite fossils to be found in the Schlierenflysch, Hans had taken them all. He had just recently put together a history of their evolution. He showed me his collection of samples and specimens. The Schlierenflysch is a sequence of alternating sands and shale. The nummulite fossils were in the sands, and the shales seemed to be devoid of fossils. Hans thought that maybe there could be planktonic foraminifera in the shales that had gone unnoticed.

Now a brief digression: while I was in my last year at Stanford, we had a brief visit from Milton Bramlette, an older geologist who had been with the US Geological Survey but was then at the Scripps Institution of Oceanography. He had gotten interested in extremely small fossils, 'nannofossils' that could only be seen with a high powered microscope. They were mostly a few microns (1 μm = 1000th mm) across. Some were elliptical 'coccoliths,' similar to the plates of *Emiliania huxleyi* you saw in Chap. 1 Others, 'discoasters,' looked like miniature stars. He had shown us how to extract them from shale by scraping off a bit with a knife and making a slurry in water.

The evening after Hans Schaub showed me his sample collection. I took a stereoscopic binocular microscope to look at pieces of the shale as he had suggested. If there were planktonic forams in the shale, some fragments should have been visible on the surface. But there seemed to be nothing there. I decided to try scraping off bits, making the slurry, and looking at it through a high powered microscope.

One of the meanings of the word 'epiphany' is 'a moment of great or sudden revelation.' I don't remember the exact date, but it was about 6 PM, and everyone had left the Institute. When I looked through the microscope, I saw myriads of nannofossils. I took another sample, from higher in the stratigraphic section. It too had lots of nannofossils, but they were different. I forgot all about getting dinner and worked through the night. By morning I had looked at many samples, each one different from the others. When Hans came in I, told him that his shale samples were full of nannofossils. Furthermore, I told him that if he gave me an unknown sample, I could probably tell him where it came from in the stratigraphic section. Hans was dumbfounded. He started giving me unknowns, and indeed I could tell him which layer it was. Louis Vonderschmitt came in and joined the party. This could be a major key to unraveling the history of deformation and uplift of the Alps. In one night I had made as much progress in deciphering the stratigraphy of the Schlierenflysch as Hans had made in 20 years.

Early in the fall I made a few collecting trips in Switzerland, now not just for planktonic foraminifera, but for nannofossils as well. Shipments were sent off to Al Loeblich at Chevron, and the fund for travel expenses began to grow.

Later, toward the end of October, Hans Schaub and I headed off to France and northern Spain to collect samples from a number of localities.

Our first stop was in Paris, to discover what was in the collections of the Musée d'Histoire Naturelle. Hans wanted to see the nummulites, and we were shown into a large room filled with storage cabinets. Each cabinet had many drawers with cardboard trays containing the fossils and a label telling what they were, and where they had come from. Hans was excitedly going through the drawers, making notes on what they contained. Then he opened one drawer, and it dropped out, spilling everything onto the floor. Fortunately he knew these fossils so well he was able, in a few hours, to get them all back into their correct trays. There was a note left over. It read something like 'On December 12, 1858, I opened this drawer, and it fell out of the case, spilling all the contents onto the floor. I have put everything back into the proper trays as best I can. I will ask the staff to have the drawer fixed. —D'Archiac'. 'D'Archiac' was Étienne Jules Adolphe Desmier de Saint-Simon, Vicomte d'Archiac (1802–1868). Hans roared with laughter on reading it. No one had looked in that drawer for a century, and the staff had not gotten around to repairing it. Hans left his own note in the drawer, presumably for someone in the 21st century'.

Paris lies in the middle of a geological basin, with Tertiary rocks underlying the city, and a ring of Cretaceous chalk surrounding it, and encircled by Early Cretaceous and Jurassic rocks further out. This was the region that had formed the basis for Baron Cuvier's idea of catastrophes and successive creations. Through his study of nummulites Hans Schaub had been able to show that the Tertiary deposits of the Paris Basin, that had been the crux of Cuvier's ideas, represented only moments of geologic time. Indeed, the time intervals represented by the Tertiary deposits were so short that no evolution was evident in them, just as Cuvier had described. Through his studies elsewhere Hans had been able to demonstrate that most of geologic time

was represented by the bedding plane surfaces separating the rock units.

Our first area of interest for sample collecting was in the Loire Valley and near the city of Le Mans. Then we headed south to the Bordeaux region. North of Bordeaux the rocks dip down to the south into what geologists call the Aquitaine Basin. After examining some sites along the edge of the Gironde Estuary, we headed to two famous places south of Bordeaux.

The first was the type section of the 'Burdigalian Stage.' The units of the geologic column presented in Chap. 3 are Systems; each is divided into smaller units, called Stages. Each stage has a reference section on which it was based, designated the 'type section.' The type section of the Burdigalian Stage (Lower Miocene) is in the forest behind an elegant home, known as Le Coquillat in the village of Léognan. We asked the owner of the home for permission to visit the exposure, and he was delighted to meet us and to show us the way. The locality is a pit in the forest, and the owner of the estate is very proud that he has such a famous treasure in his backyard. The name 'le coquillat' means 'the place with the shells' and it is indeed extraordinarily rich in fossil seashells, mostly mollusks and gastropods.

But we had an even more interesting experience visiting the 'type section' of the underlying Aquitanian Stage, not far away. It is the base of the Miocene Epoch. The type locality is an outcrop along the river in the 'vallon de Saint-Jean-d'Etampes,' (vallon = valley) at on old mill called the Moulin de Bernachon near the village of Saucats. We had already had a long day when we arrived at the Moulin. We could see the exposure of sediments outcropping across the river. A farmer lived in the old Moulin, and we explained that we were geologists and asked him if we could park our car here, on his land, next to the river while we visited the outcrop on the other side of the river. He too was delighted to have us visit; a number of geologists had been here over the years. The car was parked facing the outcrop. We started wading into the river; it wasn't flowing fast, but the water was well above my waist. Also we soon learned that there were deep holes, where the water was over our heads. But the farmer knew the river well, and from shore directed us where to go to avoid the holes. Evening was approaching, and we had our work cut out for us to measure the strata and collect samples before it got dark. We had about 100 pounds of samples each in our rucksacks when we were ready to cross back over. By now it was dark, and the farmer thought he could help by turning on the headlights of Hans's car. Now we were trying to cross back heavily weighted down and looking straight into the blinding headlights. The farmer again tried to help us avoid the holes, but now when he said 'to the left,' we didn't know whether he meant to his left or to our left. By the time we got back across, both of us had gone in over our heads, and we were thoroughly soaked. It was about 10 PM when we got everything loaded back into the car, and thanking the farmer for his help, started back to the main highway.

We had planned to drive on to Dax that evening, but that was several hours away, and we figured that all the hotels would be closed for the night. Bordeaux was closer so we decided to drive back into the city. We realized we were going to arrive about midnight. Now we had a dilemma. Our clothes were wet and partly covered with mud. Hans said 'We have two choices, we can go to a second class hotel, near the train station, and hope they will take us in, but then we will be awakened early by the trains—or we can go to a first class hotel and see how they react. We chose the latter approach; after all I had plenty of travel expense money to dispose of.

I got out my Guide Michelin and, using a flashlight, selected what was listed as the best hotel in Bordeaux, the Splendid. We arrived just at midnight. I walked in hoping my muddy boots didn't leave a trail across the impeccable red carpet. I introduced myself to the desk clerk as Dr. Hay and told him that Professor Schaub and I were geologists, had been studying the rocks and collecting fossils, and that we would like two rooms for the night if they were available. Academic titles are worth as much as nobility in Europe, and the clerk was pleased to have such distinguished visitors. He did a great job of not noticing any of my wet, muddy clothing. 'Of course,' and he handed me two keys. I asked "I don't suppose the restaurant is still open." "We usually close at midnight, but we will be happy to remain open if you can come down shortly." I assured him we would be down as soon as we changed our clothes. Then, the only indication he had noticed my muddy clothing. "Perhaps you would like to take the elevator from the garage directly to your room."

We showered and changed quickly and went down for dinner. We were the only ones in the dining room. I had seen that the specialty of the house was *Tournedos Bordelaise*, so that is what we ordered for a main course. The waiter went back to the kitchen, and the chef appeared. He was delighted with our order, and made suggestions for appetizers, salads, vegetables, and desert. Then the sommelier appeared with the wine list. It was very impressive. In those days Bordeaux wines were not nearly as expensive as they are today, and we ordered two bottles, a 1945 Chateau Lafite Rothschild, and a 1929 Chateau Margaux, both great and very rare vintages. As I recall, each bottle cost around $35.

By now the desk clerk, and waiter had joined us at the table, and Hans regaled them with stories of collecting fossils in the south of France and the glories of nummulites. The sommelier brought the bottles and decanted the wines using a candle. We insisted on sharing a glass of wine with each of our new-found friends, which they greatly appreciated. It was a meal never to be forgotten. After the tournedos were prepared, the Chef joined us at the table. But the highlight came when a fly that had been buzzing around made a straight shot down the neck of a decanter into the 1929 Margaux. The sommelier immediately disappeared and came back with a small sieve, and poured the wine into another decanter, removing the fly with the sieve. The meal and stories went on until about 3 AM.

There was another memorable culinary experience on that trip. We had ended another long day of sample collecting in the cemetery of the village of Minerve, east of Carcassonne. It was already dark, but Hans had some newspaper to make a torch, and we used that for illumination of the rock outcrop on the edge of the cemetery. I've often wondered what the people in the village must have thought seeing two strange figures working by torchlight, hammering away at the rocks in the cemetery. We finished about 10 PM and drove to Carcassonne.

Carcassonne is a modern city, but it has a magnificent, medieval, walled old-town on a hill, and we had decided to try our luck with a hotel there. Using the trusty *Guide Michelin*, we chose the Hotel de la Cité. Again we arrived about midnight. I tried my "I don't suppose the restaurant is still open?" "Non" the clerk replied, but he suggested that he could bring out a few cold things if we needed something to eat. We sat down in the darkened restaurant, and the clerk brought out huge trays of patés, cheeses, olives, tomatoes, and fruit. With good wine from the Corbières, it was a feast to rival the Splendid in Bordeaux.

A GEOLOGIST LOOKS BACK IN TIME

# Documenting Past Climate Change

*Geology is the science which investigates the successive changes that have taken place in the organic and inorganic kingdoms of nature; it enquires into the causes of these changes, and the influences which they have exerted in modifying the surface and external structure of our planet.*

Charles Lyell (1831)

You already learned that the first 'modern' geologist, Charles Lyell, was especially concerned about explaining past climates. He correctly believed that the coal beds of Europe and North America reflected tropical conditions. Without knowing that it was actually a deep sea deposit, he concluded that chalk, being limestone, had to have been deposited in warm waters. He suggested that in order for the Earth to have been warmer in the past, the distribution of land and sea must have been different.

Alexander Agassiz's 1840 demonstration that many of the unexplained features of the landscapes of Scotland were due to continental ice sheets in the not too distant geologic past made it clear that the Earth had been much colder than it is at present. The discoveries in 1856 and 1859 of evidence for late Paleozoic glaciation in India and Australia showed that the Earth had experienced cold episodes even further back in time. In 1870 the discovery of very ancient glacial deposits in Scotland showed that there had even been an ice age in the Precambrian.

Adolf Nordenskiöld's 1870 discovery of tropical plants, such as breadfruit trees, in the Cretaceous rocks of western Greenland at 70° N latitude removed any doubt among geologists that the Earth had been much warmer in the more distant past.

By the beginning of the 20th century it had become evident that the Earth's climate had undergone remarkable changes through time. To figure out what had happened, more information was needed from other parts of the world. Much of the puzzle has been filled in by geological exploration, and there have even been some major surprises. The first came in 1968 as the scientific Deep Sea Drilling Program proved both Alfred Wegener's concept of continental drift and Arthur Holmes' idea that the motion of continents might be driven by convection of the Earth's mantle. That led to plotting paleoclimate data onto plate tectonic reconstructions and everything made much more sense. At the same time it was found that there had been episodes when the deeper parts of the ocean had been completely deprived of oxygen; they had periodically become 'anoxic' during the Cretaceous. This simply cannot happen with today's ocean circulation, and it posed a major puzzle which still has not been solved to everyone's satisfaction.

## 6.1 What Is 'Climate'?

There is a famous adage: Climate is what you expect; weather is what you get.

Climate can be defined as the 'average weather.' In basic terms, it is how the Earth handles insolation—the radiation it receives from the Sun. In order to maintain constant conditions, the Earth must radiate the same amount of energy back into space that it receives. In order to do that it must redistribute the energy, and that involves moving it around by circulation of the atmosphere and the oceans. The total complex is the Earth's energy balance system.

'Climate' is the statistical description of all these factors. It is described in terms of the means and variability of the insolation, temperature, precipitation-evaporation balance, winds, and other relevant quantities that characterize the structure and behavior of the atmosphere, hydrosphere (oceans, seas, lakes, and rivers), and the cryosphere (ice and snow). It also includes information on the occurrence of extreme events. When the term 'climate' began to be used by scientists, nobody bothered defining the length of time over which the data were to be averaged, but in 1923 the International Meteorological Organization (IMO), parent of the World Meteorological Organization (WMO) defined the period for averaging to be 30 years.

The 30-year average used to define climate is intended to eliminate the effects of 11 year sunspot cycles and other shorter term multi-year variations such as El Niños. Unfortunately, if there were a rapid change of climate, we would need to be at least 15 years into it before it would be recognized. The current temperature rise began to emerge from the statistical average of the mid-20th century during the

1990s. We are 15 years into this temperature rise, and it is becoming obvious to (almost) everyone that something odd is going on.

Many of our American weather forecasters have a hard time accepting the idea of 'global warming.' As best I can tell, this is because when they were in college, they learned that the weather represents variations about a mean (the mean being the 'climate'). The idea that the 'mean' might change with time was not discussed. So perhaps it's difficult for someone with that background to deal with the current situation where not only the mean is changing, but the variability about the mean is also increasing. We see more and more new records for hottest day, coldest day, greatest precipitation amount, longest period without rain, etc. I'm waiting for the day when the forecaster will say: "Hey, wow, this is the hottest day on record for our town. And yesterday we had the heaviest rainfall ever. The climate is becoming unstable."

When geologists talk about climates of the past ('paleo-climates'), they have a different perspective. The problem comes in establishing the age equivalence of strata and their contained fossils from one place to the next. Geologists refer to this as 'correlation.' As you will see, it is possible to correlate annual layers in ice cores and some lake bed deposits that are up to a few tens of thousands of years old, but for all of the older deposits the age equivalence becomes less and less precise as we go back in time. This means that when we talk about the paleoclimate of the Cretaceous, for example, we are not talking about a 30-year average, but about conditions that existed over a time span of hundreds of thousands to even a few million years. There is another problem. The more precisely we can establish age equivalence of strata, the fewer places we can do it. The odd thing is that this problem of making precise age correlations doesn't seem to affect our understanding of past climate very much because for most of Earth's history the climate remained stable for very long periods of time. The oscillations between the extremes of glacial and interglacial clime of the last couple of million years are very unusual in the geologic past. Earth's climate seems to resist change.

What is the evidence for past climatic conditions? The thermometer wasn't invented until the 17th century, and the oldest record, that from central England, goes back to 1659. Few other areas have records older than 125 years. Precipitation records are even shorter. To learn about older climate conditions, geologists have used fossils, if possible with modern analog species. These are considered 'proxies' for climatological information. The best information comes from land plants. They are sensitive to the distributions of temperature, rainfall, and other factors during the different seasons of the year. Vladimir Köppen used vegetation to define modern 'regional climates' before detailed temperature and precipitation data were available for many areas. The distribution of fossil plants has been used to delimit climate zones on the Earth back into the Mesozoic. Very ancient plants are so unlike those of today that we can only make educated guesses about the conditions they lived under. Animal fossils are also useful but many of them, such as mammals and even dinosaurs, were warm-blooded, and independent of local climate conditions. Marine fossils are much more abundant, but the ocean is a great moderator of climate, so these fossils offer much less information about the seasonal variations. Often we can only say that they lived in warm, cool, or cold waters.

However, methods have been developed to use stable isotopes of oxygen and some other elements, as well as ratios of elements incorporated into fossil hard parts, to obtain numerical values for temperature and even ocean salinity. Much more about this later.

In addition to the information obtained from fossils, there are some sediments that form only under specific climatic conditions. Salt deposits ('evaporites') form only under warm to hot dry conditions. Coals are of two types: they represent either tropical swamp deposits formed under warm wet conditions or high-latitude peat bogs which form under cool wet conditions. Sand dunes are typical of deserts, and shallow water coral reefs grow in moderately warm, but not, as we have recently discovered, very warm waters. Glacial deposits, 'diamictites,' with boulders and cobbles in a matrix of finer-grained sediment, are very distinctive. 'Dropstones,' isolated pebbles in very fine grained sediment are often thought to have been dropped out of icebergs.

## 6.2  A Brief Overview of Earth's Climate History

When I began my professional career at the University of Illinois in 1960 it seemed as though geologists knew a lot about the world. We thought we knew about all that there was to know about the Earth's climate history. We assumed that the present day climate was Earth's normal state. The Quaternary ice ages had been deviations from that norm due to causes not fully understood. We really had no idea of the relative lengths of glacials ('ice ages') and interglacials when conditions were like those of today. We knew that the coal beds of the Late Paleozoic formed from plants and trees growing in tropical swamps. They had formed along the shores as a series of transgressions and regressions of the sea had over and over again flooded and laid dry vast areas of North America and Europe. There is a very specific sequence of sedimentary rocks associated with each of the coal beds, interpreted as sediments laid down in river beds on land, followed by development of a soil beneath the coal layer, followed by marine beds deposited in increasing water depths. One of my fellow faculty members at the University of Illinois, Harold Wanless, was the world's expert on these matters.

There were some troubling facts. The discovery of Late Paleozoic glacial deposits in India in 1856, and then in Australia in 1859, had suggested a possible cause for the repetitive sea level changes—buildup and melting of ice sheets in other parts of the world. But the Indian ice sheets would have been on the Equator, and that made no sense. Of course, 50 years earlier Alfred Wegener had offered an explanation with his theory of Continental Drift, but that idea was rejected out of hand by US geologists. As an alternative, one of Wanless' colleagues proposed that for some unknown reason North America and Europe had bobbed up and down like a cork, causing the repetitive sea level changes.

There was also the vexing problem of the fossils of tropical plants that had been found in Greenland in 1870. The plant fossils were in terrestrial deposits inter-bedded with marine rocks with typical Cretaceous fossils, so there was no question about their age. These fossils had convinced even American geologists that the planetary climate had been much warmer in the past. Two ideas for the cause of the warmer climate change had been proposed by Thomas Chrowder Chamberlin of the University of Chicago early in the 20th century: (1) changes in the atmospheric concentration of $CO_2$ (an idea he later abandoned), and (2) reversal of the deep sea circulation (nobody seemed to know what to make of this idea).

It had been discovered in the 18th century that the deep waters of the oceans were very cold, and at the beginning of the 19th century German naturalist Alexander von Humboldt had concluded that the cold deep waters off South America must have sunk into the ocean interior in the Antarctic. Today we know that cold saline waters sink into the ocean interior in both Polar Regions, but in Chamberlin's time that could only be surmised. Chamberlin reasoned that at times of global warmth the warm salty waters of the tropics could be more dense than the rest of the ocean. They could sink into the depths in the tropics and flow poleward, where they would return to the surface. It was a simple way to explain the warm poles.

In reality, during the first half of the last century geologists knew nothing about climatology, and 'atmospheric scientists' were almost all meteorologists concerned with predicting the weather. They knew nothing of geology and the history of the Earth. Only that outsider, Alfred Wegener, knew both fields.

There was also a problem in that we knew the geology of only a small part of the Earth as a whole. Until after World War II geologic research had been concentrated in Europe and North America. Knowledge of the rest of the world was sketchy. North American geologists concentrated on knowing about North America. I was very lucky to have joined the faculty of the University of Illinois where our Department Head, George White, thought that there was much to be learned from the rest of the world. He had hired a Swiss geologist, Albert Carozzi, onto the faculty, and arranged for a progression of visiting professors from abroad to come to spend a year with us in Illinois. It was a very stimulating, and for the US very unusual, atmosphere. Most importantly, nothing at all was known about two-thirds of the Earth—the ocean basins. It was simply assumed that they had always been deep, cold and dark—unchanging through geologic time. In retrospect, in the middle of the last century there just wasn't much to know about the Earth's history.

Then, in the mid-1960s everything changed. Over a period of only 5 years the idea of permanence of continents and ocean basins was discarded in favor of the concepts of sea-floor spreading and plate tectonics. The continents were mobile, as Alfred Wegener had proposed a half century earlier.

In 1956, Cesare Emiliani, a young Italian scientist who had come to the US to work with the isotope geochemistry group in Enrico Fermi's institute at the University of Chicago, reported his discovery that the oceans did have a history—the deep waters had once been warm; they must have had an interesting history. The next year Cesare reported his discovery that the deep-sea sediment record indicated that there had been many more than the four Quaternary ice ages accepted by land geologists at the time. He was instrumental in promoting the Deep Sea Drilling Project which recovered older sediments from the ocean basins. The DSDP operations began in 1967 and month by month its discoveries revolutionized our ideas about Earth's climate history. Cesare's work was based on using isotopes of oxygen; more about that in Chap. 12.

In Moscow, a small group of Russian geologists had been compiling all that was known about the geology of the USSR. Incidentally, that hammer in the hammer and sickle symbol on the Soviet flag is a geologist's hammer. The USSR had trained many geologists and set them about exploring their giant country searching for mineral deposits as well as general knowledge. With an area of 22,402,200 km$^2$ (8,649,500 miles$^2$), the USSR was almost as large as North America (24,709,000 km$^2$ = 9,540,000 miles$^2$). Detailed information on the geology of this huge area was published in a series of large atlases. By the late 1970s the geology of other new areas was becoming known. The Soviet scientists took on the monumental task of compiling the Geologic Atlas of the World for UNESCO. Starting in the 1970s Russian colleagues began to make truly global compilations of what was known of Earth's geology including information on ancient climates. These compilations were published as atlases in the late 1980s and 90s. They are so important that our group at GEOMAR in Kiel digitized them and have made them available on the Internet through the Global Earth Reference Model (GERM) database. In 2004, at the 32nd International Geological Congress in Florence, Italy, I asked Viktor Khain of the Lomonosov State University, one of the leaders in compiling global geology, "How did you come to expand your studies from the USSR to the whole world?" He replied, "Well, after you've put together the geologic information for an area as big as the USSR it didn't seem like that much additional effort to include the rest of the world." However, they were not alone: a group at the University of Chicago began to make similar compilations. Unfortunately, these were made under contract for the petroleum industry; so much of the information was not freely available.

So, during the latter half of the 20th century and up to the present, we have built up a rather detailed history of climate change on our planet over the Phanerozoic, the past 540 million years. In the early 1980s, Al Fischer at Princeton University realized that the Phanerozoic history of the planet could be divided into times with either of two climate states: one warm, without polar ice and relatively stable; the other cold, with polar ice displaying frequent expansions and contractions producing recurring glacial episodes interrupted by brief interglacials. He named these states 'greenhouse' and 'icehouse.' Both states showed internal variations which Fischer believed occurred on timescales of 100,000 years or less, but the climatic variations while the Earth was in its greenhouse state were much smaller than those of the icehouse state. His 1980s view of Earth history is shown in Fig. 6.1a. You will notice that he thought there were gradual transitions from one state to the other. His plots of two other important variables, intensity of volcanism and global sea level, are shown on the same diagram.

A decade later, in 1992, Larry Frakes of the University of Adelaide in Australia with his colleagues Jane Francis and Jozef Syktus made a more detailed analysis of the climate record and divided Earth history into warm and cool periods. These divisions are shown in Fig. 6.1 as a bar at the bottom (Fig. 6.1b). They do not correspond exactly to the greenhouse-icehouse episodes of Al Fischer, because during that decade much more had been learned about regional geology and more deposits had been accurately dated. They believed that the transitions from warm to cool states and back were abrupt.

Incidentally, some of the most spectacular evidence of the Late Paleozoic glaciation is found in city parks in the south of Larry Frakes' home town, Adelaide. I had opportunity to visit these a few years ago and was amazed at what I saw. There are grooves cut in the rock as the glacial ice carrying boulders moved across it. How could such features be preserved for more than 200 million years? The rock with grooves is quartzite, the hardest rock you can encounter in nature. But there is more: as you drive south from Adelaide (remembering to keep in the left lane), you see a glacial landscape, preserved by these very hard rocks. Looking at those glacial features in what is now a subtropical landscape that never sees snow makes an indelible impression of the magnitude of climate change.

In the meantime, while geologists and paleontologists have been sorting out the climate history of the Phanerozoic, geochemists have been hacking away at the problem of what conditions were like during the Precambrian. The amazing thing is that we now know that there have been oceans of liquid water and continents on planet Earth for 4.4 billion years. There is evidence for life on Earth back as far as 3.4 billion years ago. A best guess is that the climate history of the Precambrian probably looked rather like that of the Phanerozoic, with an alternation of warm and cold states. We know of two episodes of glaciation in the Proterozoic and one in the Archaean, so the icehouse state has been with us from time to time over most of Earth history.

In the 1980s a consensus was emerging that the forcing factor determining whether Earth was in a greenhouse or icehouse state was the atmospheric content of the greenhouse gas $CO_2$. How the $CO_2$ system works will be discussed in detail in Chap. 21; it is far more complex than originally thought. The idea was simple: during greenhouse climate times atmospheric $CO_2$ levels were several times higher than today; during icehouse times, they were similar to that of today, or somewhat lower during the ice ages. But the question of why there should be alternation of states with high and low concentrations separated by rapid transitions remained unclear. The last two decades have seen the simple alternation of greenhouse and icehouse worlds upset by the discovery of some extreme climatic events. The greatest of these came in 1992 with the discovery that during the late Proterozoic there were a series of episodes when the Earth seems to have been completely covered with ice, the 'Snowball Earth.' This extreme condition had been thought impossible. It opened a new perspective, the possibility of extreme climatic perturbations of the planet. This will also be discussed in Chap. 21.

At the same time, as new and better estimates of Cretaceous temperatures have been made, it has become evident that there were a number of major fluctuations during the greenhouse time. Some data indicate that tropical sea surface temperatures may have been as warm as 42 °C (=108 °F): they are rarely more than 28 °C (=82.4 °F) today. The flora represented by fossil leaves from the margin of the Arctic ocean indicate mean annual temperatures from 9 to 12 °C (=48–54 °F) with temperatures remaining above freezing even during the polar night. There is also evidence that geologically rapid falls of temperature may have occurred from time to time. These coincide with times when extensive areas of the ocean, particularly the young Atlantic, were devoid of oxygen, and large amounts of organic carbon were

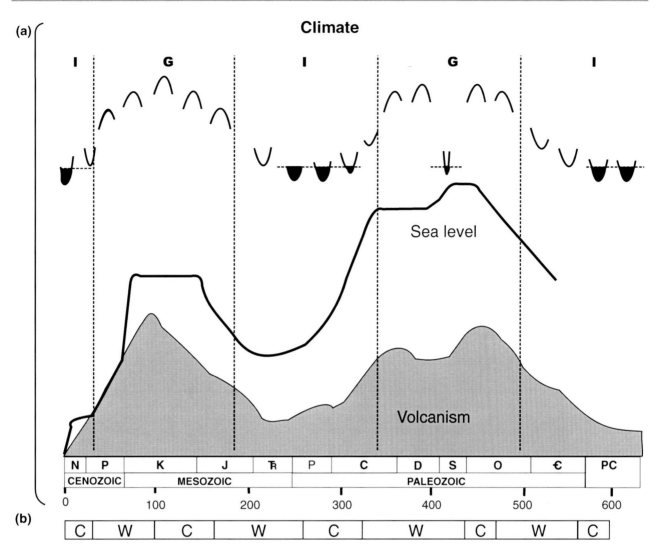

**Fig. 6.1** The pattern of global climate change as understood in the 1980s and early 1990s. **a** A replot of Al Fischer's 1982 diagram of an alternation of icehouse and greenhouse conditions together with his curves of changing rates of volcanism and variations of sea level. Both of the latter have been extensively revised since then. **b** The divisions of warm and cold Earth episodes proposed by Frakes, Francis, and Syktus (1992)

deposited on the sea floor. These changes of climate occurred over many tens to hundreds of thousands of years.

In 2010 David Kidder and Tom Worsley of Ohio University recognized that there were a number of times of extremely warm temperatures which represent a fourth major climate state for our planet: 'hothouse' climates. The hothouse intervals are geologically short, a few 10,000s–100,000s of years. Hothouse episodes are characterized by temperatures well above those of the Earths' normal greenhouse state. Kidder and Worsley believe that the hothouse intervals represent times when there was an active deep-ocean circulation opposite to that of today, with very warm salty waters sinking in the tropics and returning to the surface in the Polar Regions.

I have tried to make a more nearly complete account of the Earth's climate history using a diagram that includes changes by latitude with time, including the Snowball Earth and hothouse climate episodes (Fig. 6.2).

When examining Fig. 6.2, you will note that my view is that there have been apparently sudden transitions between the greenhouse and icehouse states, but that these states in themselves seem to be generally quite stable. A question is, assuming atmospheric greenhouse gas concentrations are responsible, why there should be just two dominant states, and why the extreme variants, Snowball Earth and hothouse states should be so rare and relatively short lived. It raises the question of how the atmospheric greenhouse operates; this is the subject of Chaps. 18–22. Table 6.1 shows typical temperature values.

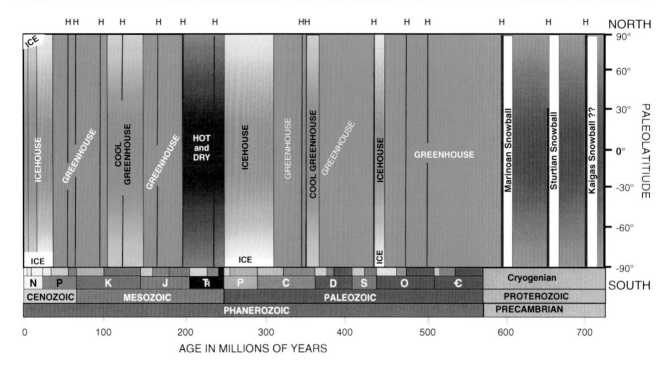

**Fig. 6.2** Major changes in Earth's climate over the past 750 million years. *Dark shades* warm, *white* cold (ice); *H* Hothouse episode. The pole to pole *white bars* at the *right* are the episodes of "Snowball Earth" conditions when the planet seems to have been completely covered by ice. There have been two major episodes of south-polar glaciation; one centered on 300 million years ago, the other covering the interval from 35 million years ago to present. The modern bipolar glaciation is started about 5 million years ago, although I believe that the Arctic Ocean has been ice-covered for the past 35 million years. We live in one of the brief interglacial episodes when the Northern Hemisphere ice is reduced to Greenland and the Arctic Ocean pack ice. The letters are the abbreviations for geologic intervals: *N* Neogene; *P* Paleogene; *K* Cretaceous; *J* Jurassic; **Ꞧ** Triassic; *P* Permian; *C* Carboniferous; *D* Devonian; *S* Silurian; *O* Ordovician; **Ꞓ** Cambrian. The Precambrian is the first 4 billion years of our planet's history; only its last 2 million years are shown here. We know very little about the climate during most of Precambrian time, but we do know that liquid water seems to have been present on the Earth's surface during all but the very earliest times

**Table 6.1** Guesstimates of mean annual temperatures at sea level for different latitudes for the four climate states

| Temperatures | Tropics | Mid-latitudes | Polar |
|---|---|---|---|
| Snowball | −30 °C (=−22 °F) | −40 °C (=−40 °F) | −50 °C (=−58 °F) |
| Icehouse | 22 °C (=72 °F) | 0 °C (=32 °F) | −30 °C (=−22 °F) |
| Greenhouse | 34 °C (=93 °F) | 20 °C (=68 °F) | 0 °C (=32 °F) |
| Hothouse | 42 °C (=108 °F) | 35 °C (=95 °F) | 16 °C (=61 °F) |

In the meantime, while geologists and paleontologists have been sorting out the climate history of the Phanerozoic, geochemists have been hacking away at the problem of what conditions were like during the Precambrian. The amazing thing is that we now know that there have been oceans of liquid water and continents on planet Earth for 4.4 billion years. There is evidence for life on Earth back as far as 3.4 billion years ago. A best guess is that the climate history of the Precambrian probably looked rather like that of the Phanerozoic, with an alternation of warm and cold states. We know of two episodes of glaciation in the Proterozoic and one in the Archaean, so the icehouse state has been with us from time to time over most of Earth history.

The lengths of time Earth has spent in its two major states, the 'icehouse' versus the 'greenhouse' are very different. Icehouse states account for only a very small fraction of Precambrian time. In the Phanerozoic, where we have good age control, Earth has been in the icehouse state for about 25 % of the time. However, the more intense icehouse state of the Quaternary, with both poles glaciated, represents only 0.3 % of geologic time, and the interglacials during that time, like the one we are currently in, represent about 10 % of the Quaternary. So the present climate is characteristic of only 0.03 % of Earth history. Further, the present interglacial, the Holocene, has been almost unique in having a remarkably stable climate for 8,000 years. It has been this climatic stability during an interglacial that has allowed humans to develop what we call 'civilization.' This condition is typical of only 1.7 % of the Quaternary, or 0.006 % of the Phanerozoic, or 0.00016 % of Earth's history. We are performing our grand uncontrolled experiment on planet Earth when it is in this exceedingly rare condition.

I find six of the Earth's climatic regimes particularly intriguing:

First, there is the present; the left edge of the diagram—an interval of time too short to be visible, thinner than the thinnest line you could see. The only climate state we know, and in which human civilization has developed, is characteristic of only the tiniest fraction of Earth history. If you think of Earth's history in terms of a 24-h day, all of human civilization has happened in the last 1/10th of a second.

Second, there were the ice ages that came and went over the past 3 million years; this is an interval too short to be shown clearly. Taken all together they would be a line at the left edge of the diagram. As recently as 19,000 years ago northern North America and northern Europe were covered by an ice sheet 3 km thick. The surface of these ice sheets was as high as the Tibetan Plateau today, but they were several times as large. The ice, which had required 80,000 years or more to accumulate, melted away in only 11,000 years. The dawn of civilization coincides with the disappearance of the ice sheets.

Third, there was the Cretaceous, the period indicated by the "K." It lasted from 140 to 65 million years ago. I have spent the last 30 years trying to understand it. There was no polar ice except possibly for brief episodes. Instead, polar temperatures were more like those of the tropics today. This was the time when the largest land animals, the dinosaurs, roamed the land. The modern flowering plants, the angiosperms, evolved during the Cretaceous. Deciduous trees, those that today drop their leaves and go into a winter sleep, evolved during the Cretaceous. However, freezing weather was not common in winter, and it is thought that the habit of dropping leaves and going into a resting phase was most likely a way of getting through the months-long night in the Polar Regions. During the older part of the Cretaceous the oxygen content of the oceans was very low and there were anoxic episodes when black sediment rich in organic matter accumulated. Throughout the Cretaceous sea level slowly rose to a peak thought to be 200–300 m higher than it is today. About a third of today's land area was flooded, and the world seemed more like a global ocean with large islands. Global temperatures rose to about 6–8 °C (10.8–14.4 °F) higher than they are today, and there may have been episodes when they were 14 °C (25 °F) higher than today.

Fourth, there was the Triassic, indicated by the special geological symbol ⚏. It lasted from 251 to 100 million years ago. It was a time when deserts were widespread throughout the vast interior of the remains of the supercontinent Pangaea. Many Triassic sediments are red, indicating that, although they were deposited by running water, they dried out completely after being laid down. If you had seen this planet from space, you might easily have mistaken it for Mars. The plants on land were mostly conifers, similar to modern pine and spruce; there were no 'flowering plants.' There were reptiles and dinosaurs, many bigger than modern crocodiles, but not nearly as large as those of the Cretaceous.

Fifth, there was the Carboniferous, indicated by "C." Lasting from 360 to 299 million years ago, this was the time of formation of the coal beds of the eastern United States, Western Europe, China, and many other areas. During the later Carboniferous there was ice on the giant continent of Pangaea, which included modern Antarctica, South America, Africa, India, and Australia that lay astride the South Pole. There were many glaciations and deglaciations, with rises and falls of sea level. At times of high sea level most of the present United States was under water (keep this in mind the next time someone tells you there is nothing to worry about, because the Earth has been warmer in the past). The coal swamps grew along the edge of the sea. The plants included ferns, but there were many other kinds, including large trees that became extinct and have no modern counterparts.

These first five examples represent the climatic extremes known through the 19th and most of the 20th century—oscillations between the greenhouse and icehouse states. As will be discussed later, there was good reason to believe that our planet had some sort of an internal thermostat that prevented more extreme variations in the climate. In the 1970s we figured out what it was—atmospheric $CO_2$. Then in the 1990s all that changed.

The 'Snowball Earth' condition, the sixth of my very interesting climate states, is a relatively new discovery, having been first proposed in 1992. It represents the most extreme climatic catastrophe the Earth has experienced since shortly after its formation. The youngest of these 'Snowball Earth' episodes occurred between 750 and 580 million years ago, before there were any higher forms of life. The Earth seems to have become entirely ice covered. The planetary thermostat had failed.

In just the past few years we have discovered several lines of evidence that there have been episodes of extreme warmth during the Cretaceous and Paleocene when temperatures rose to levels such that in the tropics photosynthesis by land plants may have shut down.

Each of these contrasting regimes represents something fundamentally different about the way the Earth's climate system operated. What is very peculiar is that the Earth seems to make transitions from one of these climatic states to another in very short periods of time, and then stays in one condition or another for long periods of time. The transitions from greenhouse to icehouse or vice versa take place very quickly, geologically speaking, and the transitions to the most extreme states seem to have occurred even more rapidly. Furthermore, the extreme states were relatively short-lived. This implies that 'positive feedbacks' were involved in climate change, as we shall discuss below. For relatively brief intervals these positive feedbacks overrode the stabilizing effect of the planetary thermostat. The last transition from greenhouse to icehouse climate, at the end of the Eocene, about 34 million years ago, is a good example.

## 6.3    The Cenozoic Climate 'Deterioration'

In the 1950s Cesare Emiliani had found that the few deep-sea benthic foraminifera available to him from sediments older than Quaternary indicated that oceanic deep water temperatures in the Miocene and Oligocene were warmer than today. One of the most intriguing results of the recovery of cores of deep sea sediments by the Deep Sea Drilling Project and Ocean Drilling Program was the production of a detailed account of the decline in deep water temperatures. Figure 6.3 summarizes what we have learned.

The evidence is that deep water temperatures were above 12 °C at the end of the Paleocene, and declined since the early Eocene. A sharp fall at the Eocene-Oligocene boundary about 34 million years ago is interpreted as the buildup of an ice sheet on Antarctica. The warmer deep water temperatures of the late Oligocene and early and middle Miocene are interpreted as reflecting shrinking of the Antarctic ice sheet, followed by a major expansion in the late Miocene. The decline in the Pliocene and Quaternary reflects the buildup of ice in the Northern Hemisphere.

Many of those working in this field thought that because today's deep water forms by sinking in the Polar Regions, the temperatures of the deep waters must reflect polar surface water temperatures as they do today. This is not necessarily true, because water dense enough to sink from the ocean surface can also be formed in regions with high evaporation, like the Mediterranean. However there is additional evidence, from fossil leaves of the plants that lived on the shores of the Arctic Ocean and fossil crocodiles found in the Canadian Arctic, that these estimates of warm polar temperatures are realistic. Equatorial ocean temperatures today are rarely higher than 28 °C, and polar waters in the Arctic are about −2 °C. The Equator to pole temperature gradient for surface ocean waters is about 30 °C. There is evidence that during the Eocene, tropical ocean temperatures were about 32 °C. The deep waters and fossils of coastal plants and animals indicate that the polar ocean temperatures were about 12 °C, so the meridional temperature gradient then was 20 °C. When you look at Fig. 6.3 what you are seeing is the increasing meridional gradient in ocean temperatures which is mostly a reflection of cooling of the Polar Regions.

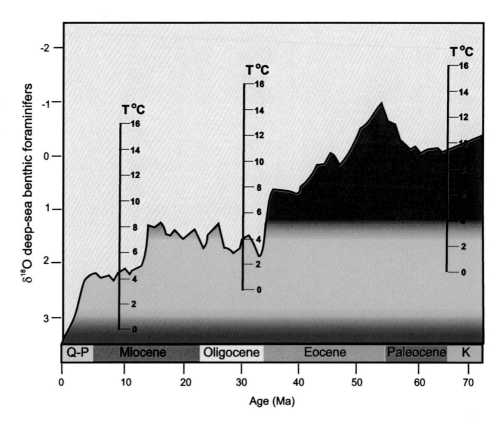

**Fig. 6.3**  The decline in ocean deep water temperatures during the Cenozoic, the last 65 million years of Earth history, based on the $^{18}O:^{16}O$ ratio in the shells of deep-sea benthic foraminifera. There are three temperature scales because the $^{18}O:^{16}O$ ratio in seawater changes as ice builds up on the continents. *Q-P* Quaternary-Pliocene, *K* Cretaceous

For the past 20 years I have participated in the discussion about the Cenozoic temperature decline. Something must have happened to cause the Earth's climate to cool. The 'usual suspects' are (1) a long term decline in the atmosphere's greenhouse gas content, particularly $CO_2$; (2) uplift of mountains, particularly the uplift of Tibet and formation of the Himalaya as India collided with Asia, and uplift of the North America's western Cordillera and Rocky Mountains, and South America's Andes; and (3) a reorganization of the circulation of the ocean.

Unfortunately, when the temperature graph was first published it was discussed as a 'deterioration' of Earth's climate. The warm period of the early Eocene has even been formally named the Early Eocene Climatic Optimum. Those who would like you to believe that global warming is nothing to worry about have seized on these terms. After all, what could be better than a return to the 'climate optimum'? I myself have given many talks to scientific groups about the Cenozoic climate 'deterioration' without it ever having occurred to me that the general public might think this means that the present climate could be improved on. I even have geologist friends in the petroleum industry who tell me that what was good for dinosaurs should be good for humans too. Climates are not good and evil as the terms 'optimum' and 'deterioration' would imply. The climate is simply what it is. If you happen to thrive in it you would consider it good. However, millions of years of human evolution have adapted us to live in ice age and interglacial climates, both much cooler than those of the Eocene. The Eocene Climatic Optimum would not have been optimal for modern humans —many areas simply would have been too hot for human habitation.

Remember that most of the animals and plants that thrived in those warm times have long since become extinct. Thirty five million years of gradual evolution have produced life forms adapted to the shifting glacial—interglacial conditions of the past few million years. For cold-blooded animals, like reptiles, the climate has definitely deteriorated. For mammals, which developed fur to keep them warm in winter, the colder climate has definitely been a plus. This has been the climate that resulted in the life forms we see all around us, including ourselves.

## 6.4    From Ages to Process Rates

One of the largest questions in geology is: how fast do things happen? The problem is that most geologic processes are so slow they cannot be observed by humans. We have a feeling for what things were like when our parents were young, and at least an idea of what things were like when our grandparents were young. But further back than that everything is in a fog. We can read accounts of life in the Middle Ages, but we really cannot grasp what it was like. The landscape you see when you are a child will look the same when you are old. De la Beche's cartoon, Fig. 4.7, is an excellent summary of the problem. Without numerical ages geologists could only guess at the rates of geologic processes.

One way of getting at rates of geologic processes was to find out how much sediment was being eroded and transported by rivers. Already in the 19th century geologists were collecting samples of river water, and measuring the amount of sediment in them. However, it was a bit of a hit or miss process, with lots of data from small rivers conveniently near universities and research institutions and only sporadic measurements of large rivers, especially those outside Europe and North America. Also, it seems that nobody liked trying to make measurements during a major flood, when the river is transporting abnormally large amounts of sediment and debris. In 1998 I wrote a review paper summarizing our state of knowledge on the erosion of detrital sediment (Detrital sediment fluxes from continents to oceans. Chemical Geology, 145, 287–323). One of the most interesting tables presented the information from a number of different sources. What was especially interesting was that it showed how often the data from older studies have been repeated over and over again. This is not any sort of plagiarism. It simply reflects the lack of any more recent data. In any case, already at the time the first measurements of sediment being transported by rivers were being made their drainage basins had been perturbed by human activity. By the time routine and regular measurements were made, farmers were plowing large areas of land, dams had been constructed, and the whole system had been completely altered. Hence, the data are of questionable value in trying to determine geologic process rates before human perturbation of the system. The question of how long it takes to erode a mountain range remained unsolved until quite recently.

Through the early decades of the 20th century many geologists were quite satisfied with the progress of the science. There was a solid framework for the geological timescale, albeit anchored by radiometric dates at only a few points. Geological exploration for mineral resources outside Europe and North America was giving the science a truly global aspect. The search for petroleum had created the new field of micropaleontology, the study of fossils too small to be ground up by the drill bit. These are the objects I studied at the beginning of my professional career. Far more abundant than larger 'megafossils,' the calcium carbonate shells of foraminifera, a kind of protozoan, were being found throughout sedimentary rocks of Late Paleozoic to Recent age. They offered the promise of much finer stratigraphic correlations than had been possible before. Many geologists were comfortable with the simple idea that it took a long, long time for a mountain range to grow and be eroded.

Very little thought was being given to the possibility of finding out about the actual rates of change in the past. The person who most eloquently argued for improving this situation became one of my heroes, but he remains almost unknown in the US: Sergei von Bubnoff (Fig. 6.4).

In 1931 von Bubnoff published '*Grundprobleme der Geologie—Eine Einführung in geologisches Denken*' = '*Basic Problems of Geology—An Introduction to Geological Thinking*.' Its third and last German edition appeared in 1954; it was finally translated into English and published in 1963 as '*Fundamentals of Geology*'. The English title sounds like it is just another textbook, so few people have read it. It is a treasure-trove of ideas. Von Bubnoff was born in St. Petersburg, Russia in 1888. His higher education was in Germany and Switzerland, and in 1929 he became Professor at the University of Greifswald, which you may never have heard of. Greifswald is a town in the northeast of the former German Democratic Republic (East Germany) just south of the shore of the Baltic, not far from the Polish border. Teaching began in 1436 and the University was officially recognized by the Pope in 1456. In 1950 von Bubnoff moved to the Humboldt University in (East) Berlin until his death in 1957. He emphasized that what was missing from our science, and what we really needed to know, were the rates of change.

As mentioned above, he has been honored by Al Fischer of Princeton University by having a geologic process rate, the Bubnoff, named after him. One Bubnoff is 1 m per million years. Physicists think in terms of the speed of light, 299,792,458 m/s; geologists think in terms of Bubnoffs, $3.1689^{-14}$ m/s.

Since the middle of the 20th century a number of techniques have been devised to refine the geologic time scale and get at the problem of geologic process rates.

## 6.5  Radiometric Age Dating in the Mid-20th Century

The problem with the uranium and lead dating that established the general framework of the numerical geologic timescale was that there were not enough dateable materials to determine rates of change of a million years or less. Worse, many of the dates were from mineral deposits that had no direct connection with the sedimentary rock sequence, so their placement in the geological timescale was uncertain. Scientific instruments of the day were crude by modern standards, and the experimental errors involved in making the measurements always introduced an uncertainty of a few million years. The youngest rocks that could be reliably dated using the uranium, thorium, and lead methods were a few million years old, older than the onset of the northern hemisphere glaciations. Before World War II there were only two geologists actively working on the geologic time scale, Arthur Holmes at the University of Durham, England, and Alfred Nier at the University of Minnesota. It just wasn't considered a very important topic.

To make progress in calibrating the geologic age scale other methods were needed. And to get at the problem of process rates, and rates of climate change older than those between 7,000 and 12,000 years that had been documented by de Geer in Sweden, very different approaches would be required.

Progress in this field was interrupted by the Second World War, but thanks to the research into atomic physics, the bathymetry of the oceans, and the Earth's magnetic field —all of which had military implications—there was a lucky series of accidental discoveries in the latter half of the 20th century.

**Fig. 6.4** Sergei von Bubnoff (1888–1957). The man obsessed with figuring out the rates of geologic processes

## 6.6 Potassium—Argon Dating

Potassium (K) is one of the most abundant elements; it is a major component of the feldspar minerals that make up most of the granitic rocks of the Earth's continental crust. It also occurs in lavas and volcanic ash.

You may recall that in 1905 J.J. Thomson had discovered that potassium was radioactive; not very, just a little bit. The original uranium salts used for studies of radioactivity often contained traces of potassium, so its radioactivity was masked by that of the uranium (and the radium it produced), but the nature of potassium's radioactivity remained a puzzle. The solution to the puzzle finally came in an unexpected way. In 1937 Luis Alvarez (1911–1988) (Fig. 6.5), a physicist at the Radiation laboratory of the University of California in Berkeley, discovered a wholly new process of radioactive decay—electron capture by the nucleus. His experiments had been on activated titanium and gallium. Alvarez received the Nobel Prize in Physics for this work in 1968. Incidentally, this is the same Luis Alvarez who, with his son Walter, showed in 1980 that the Cretaceous came to a cataclysmic end as an asteroid struck the Earth.

**Fig. 6.5** Luis Alvarez (1911–1988)

Electron capture had been considered theoretically possible by Italian physicist Gian-Carlo Wick in 1934. The German physicist/philosopher Carl Friedrich von Weizsäcker had proposed in 1937 that the strange radioactivity of $^{40}$K (Potassium 40) to produce $^{40}$Ar (Argon 40) could be an example of electron capture.

Now, let's back up for a moment. You remember that when Ernest Rutherford developed his 'solar system' model of the atom, one important question was: why didn't the electrons spinning about the nucleus eventually slow down and fall into the nucleus? This was explained by his Danish postdoctoral assistant Niels Bohr in 1913, who used the rules of quantum mechanics being developed by Max Planck to propose that the electrons orbited in shells at specific distances from the nucleus. We haven't discussed quantum mechanics yet, but basically it is a wholly foreign world outside our experience. Rather like what Alice found when she fell down the rabbit hole. So one way of thinking about this is that, although the electron is much smaller than the atomic nucleus, it occupies a space that is much larger than the nucleus, so it can't fall in. It is reminiscent of a lecture on a topic in physical oceanography I attended, which started out "Imagine a narrow, shallow channel that has no sides or bottom…" and then went into a series of equations.

If you are confused, don't worry, when you get down to the size of atoms, the world is very strange indeed. We will discuss this in greater detail later, in Chap. 12. It is important to have an understanding of how greenhouse gases work.

In electron capture by the $^{40}$K isotope, an electron from the innermost ('K') shell of the atom joins with a proton to make a neutron, but as it does so, a new particle is created and emitted, a neutrino. Neutrinos are electrically neutral elementary particles that usually travel at close to the speed of light. They have a very small, but non-zero mass. Furthermore they have a very weak interaction with atoms and other subatomic particles. They pass through ordinary matter almost undisturbed, which makes them extremely difficult to detect.

Potassium is element number 19 in the periodic table; it has three isotopes, two of which are stable and make up the overwhelming mass of the element: $^{39}$K (19 protons and 20 neutrons; 93.26 % of the K atoms), and $^{41}$K (19 protons and 22 neutrons; 6.73 % of the K atoms). The third isotope $^{40}$K (19 protons and 21 neutrons; only 0.012 % of the K atoms) is unstable. This rare isotope was discovered by Alfred Nier, while he was working on improving mass spectrometers at Harvard in 1935. It has a half-life of 1.25 billion years, but is very peculiar because about 88.8 % of the time it undergoes 'beta decay' ($\beta^-$) in which one of its neutrons emits an electron and an antineutrino (a subatomic particle with no charge) to become a proton. This changes the potassium atom's atomic number ($^{40}$K, with 19 protons and 21 neutrons) to that of a calcium atom ($^{40}$Ca, with 20 protons and

20 neutrons). The rest of the time one of the protons in the nucleus captures an electron, converting it into a neutron so that the atom becomes one of the noble gas Argon ($^{40}$Ar, with 18 protons and 22 neutrons). A 'noble gas' is an element with its outer electron shell filled so it will not combine with other elements (see Fig. 5.8).

When minerals crystallize from molten rock or seawater they do not incorporate argon into the mineral lattice. Once formed from radioactive decay of the potassium in the mineral, the radiogenic argon cannot easily escape from the crystal lattice. Many of the lavas and ash that volcanoes spew out contain these potassium-rich minerals. The volcanic materials become interlayered between the sedimentary rocks on which the classic geologic timescale is based so that their age dates can be directly integrated into the geology's relative age scale.

In 1948 Alfred Nier and his doctoral student L. Thomas Aldrich at the University of Minnesota reported that they had examined a variety of rocks for their radiogenic argon content, and found it in all of them. They suggested that this could be a new way of determining geologic age. Dating with potassium/argon began in 1950, and the technique improved over the years. It would appear to be easy: all you have to do is measure the ratio of the $^{40}$Ar to the $^{40}$K in the rock and determine its age. Unfortunately the measurements are not easy to make. The initial accuracy of the date was about $\pm 10$ %, and the margin of error has gradually been reduced over time. Since the middle of the last century $^{39}$K/$^{40}$Ar dating has become the workhorse of dating of sedimentary and volcanic rocks. A major advance was conceived in 1965—irradiation of the sample in a reactor to covert $^{39}$K into $^{39}$Ar, so that the $^{40}$Ar/$^{39}$Ar isotope ratio could be used to determine the age; the 'argon/argon' method. It was a great idea, but it proved to be far more difficult than originally anticipated. In the last decade it has become practical, although very expensive. The current accuracy (referred to in science as 'error') is thought to be as small as $\pm 1$ %, and it can be reduced to 1/10th of that. The most spectacular result has been the dating of ash from Pompeii, when it was analyzed in 1997 it was estimated to be 1925 years old ($\pm 94$ years for error). Using the Gregorian calendar, the age given by Pliny the Younger was 79 AD, which in 1997 would have made the ash 1918 years old (that's only 7 years difference). This means that the technique is now being used not only for ancient samples, but to date lavas and ashes in the younger geologic record to get a history of the eruptions of volcanoes. The latest version of the numerical timescale (Gradstein and Ogg 2004, see Table 5.1) avoided the older $^{39}$K/$^{40}$Ar dates and used over 200 of the newer $^{40}$Ar/$^{39}$Ar dates. These have provided fixed points between which other techniques could be used to refine the scale.

A major concern in $^{39}$K/$^{40}$Ar and $^{40}$Ar/$^{39}$Ar dating of lavas that flowed out on the deep sea floor is that under the huge pressure of the overlying water, Ar in the molten magma may have been forced into the mineral as it crystallized.

Other decay series used for dating include the less common unstable isotopes of rubidium and samarium. These have half-lives of 48.8 billion years and 106 billion years respectively, and are used mostly for dating very ancient Earth rocks and meteorites.

## 6.7 Reversals of Earth's Magnetic Field

Patrick Maynard Stuart Blackett (1897–1974) had begun his University studies after World War I under Ernest Rutherford at Cambridge. He had worked on observing the transmutation of nitrogen to oxygen under neutron bombardment, and became interested in cosmic rays. He joined the faculty of the University of Manchester and in 1953 he became Head of the Physics Department at Imperial College in London. After the Second World War he became interested in magnetic fields that might exist in space. This led him to the study of Earth's magnetic field and its regional variations.

He started collecting rock samples and examining the magnetism inherent in them, hoping to be able to decipher the history of Earth's magnetic field. He made a major discovery, that about half the rocks had magnetization corresponding to the Earth's present field, but half of them were magnetized in the opposite direction. Furthermore, his rocks from different continents showed that their latitudes had varied with time, in a way that supported Wegener's hypothesis of continental drift. In 1961, he published a paper entitled '*Comparison of Ancient Climates with the Ancient Latitudes deduced from Rock Magnetic Measurements.*'

Harry Hess (1906–1969) received his Ph.D. in geology from Princeton University in 1932. He joined the Princeton faculty in 1934 and pursued his special interest: study of arcuate island chains and their volcanoes. As World War II began he took leave of Princeton to become a naval commander operating in the Pacific, rising to the grade of Rear Admiral. His scientific curiosity unabated, he used the newly developed 'sonar' to make bathymetric profiling a routine operation on his vessel as it steamed from one site of action to another. He discovered the isolated seamounts we now know as 'guyots.' After the war, Hess returned to his faculty position in Princeton's Geology Department. Other research vessels started making bathymetric surveys of the ocean, and also began towing magnetometers on a routine basis. It was a natural progression of the technology, magnetometers having been used from airplanes to prospect for ore deposits over land, and from navigation problems it was known that there were variations in the magnetic field at sea.

Maurice Ewing, a pioneer geophysicist from Texas, had become professor at Columbia University in 1947. In 1949 he founded the Lamont Geological Observatory which emphasized marine geological studies. The Observatory had been the weekend residence of Thomas W. Lamont, a Wall Street banker. It was located in a romantic site on the Palisades of the Hudson, the bluff on the east side of the Hudson just north of New York City. Lamont had died in 1949, and his wife had given the estate to Columbia University. Ewing also acquired a ship, a former sailing yacht, which was converted into a research vessel, the *Vema*. He sent the *Vema* out to cruise the world's oceans, gathering geophysical information, including bathymetric, seismic, and magnetometer surveys, and taking cores of the ocean-floor sediments. He participated in many cruises himself. In those days, seismic surveying was done by lighting a stick of dynamite and throwing it overboard, something Ewing seemed to enjoy. He was also swept overboard in a rough sea, but succeeded in getting himself tossed back on board by another wave, albeit with a severe back injury. By the late 1950s Lamont Geological Observatory had become one of the world's most prominent oceanographic institutions. The data it had collected were instrumental in what happened next.

By the early 1960s it had become clear that much of the data supported Wegener's idea of continental drift. In 1928 Arthur Holmes had suggested that the mid-ocean ridge in the Atlantic—at that time only very poorly known—could represent the top of an upwelling convection cell forcing the continents apart. This idea appeared in his 1944 textbook 'Principles of Physical Geology' and its subsequent editions, but went largely unnoticed. In 1958, while still a graduate student at Stanford, I attended the annual meeting of the Geological Society of America in St. Louis, Missouri. The meeting was especially memorable because of the address given by the president of the Society, Raymond C. Moore, and entitled 'Stability of the Earth's Crust'. Ironically, a few hours after the talk St. Louis experienced one of its very rare earthquakes. At the meeting I met Bruce Heezen (1924–1977), a fresh young Ph.D. from the Lamont Geological Observatory in New York. He had given a talk on the similarity of the shape of the continental shelf edges in the east and West Atlantic. The fit of the edges of the continental blocks was much better than the fit of the shorelines that Wegener had used. His ideas had not been appreciated by his boss, Maurice Ewing, who firmly believed in permanence of continents and ocean basins, or by the audience who heard his talk. Bruce and I had dinner together and I told him about my student experience in Germany and Switzerland, where the idea of moving continents was far more acceptable. We remained lifelong friends.

A few years later, Harry Hess of Princeton and Robert Dietz of the Naval Electronics Laboratory in San Diego published papers suggesting that it looked very much like new ocean floor was being added along the mid-ocean ridge, pushing the continents apart. The process became known as sea-floor spreading. Incidentally, Bob Dietz had gotten his Ph.D. at the University of Illinois in Urbana in the late 1930s. He had proposed as his thesis topic 'The Geology of the Moon.' It was rejected because it was obvious to the faculty that there would never be any way to test his ideas. Bob and I got to know each other quite well over the years. In 1988, at the centennial meeting of the Geological Society of America, he was awarded the Penrose medal, the society's highest award. However, he was disappointed by the meeting because it turned out that the paper he had hoped to present in a technical session, on meteorite impacts, was one of the 10 % of papers that had been rejected. We had several drinks together to commemorate the event.

In the early 1960s Fred Vine was a graduate student under Professor of Geophysics Drummond Matthews at Cambridge. For his doctoral thesis he was examining the results of magnetometer surveys across the Reykjanes Ridge, the segment of the Mid-Atlantic Ridge just south of Iceland. Away from the ridge crest there were alternating strong and weak magnetic fields. It seemed as though some of the sea floor was magnetized in the same direction as the Earth's present field, reinforcing it, and adjacent strips of rock where the rocks were oppositely magnetized, counter to the Earth's present field and reducing its strength. As the story goes, he was leaving for the evening and folded the map down the length of the ridge. He realized that the pattern on the two sides of the ridge were mirror images of each other. The symmetrical magnetic lineations of the sea floor on opposite sides of the ridge supported the idea of sea floor spreading. Others tried this with data from other areas, and the result was always the same.

Then another young doctorate at Lamont, Jim Heirtzler, and his colleagues had an extraordinary idea. If one assumed a constant spreading rate for the sea floor and knew the age of one of the reversals, they could calibrate the timing of all of the other magnetic reversals and establish a paleomagnetic time scale. Suddenly, instead of having just a few numerical age reference points, there were many of them. Magnetic reversals occur apparently randomly with no set recurrence time. They are also recorded in many different parts of the globe, wherever the sediments or volcanic deposits carry a magnetic signature. The magnetized lavas can often be dated by $^{40}$K/$^{40}$Ar or more recently $^{40}$Ar/$^{39}$Ar, so that the ages ascribed to the individual reversals can be updated as new information becomes available.

As you can see in Fig. 6.6, some of the magnetic reversals occur close enough in time so that they might be used to determine rates of processes, but to be able to solve the sort

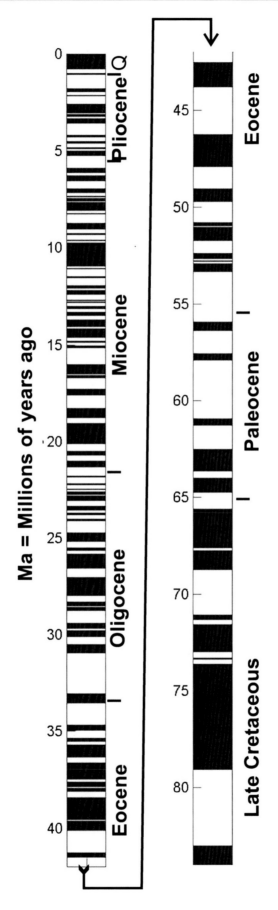

**Fig. 6.6**  A version of the paleomagnetic time scale from present back to 85 Ma. From Cande, S.C., and Kent, D.V., 1992. A new geomagnetic polarity timescale for the late Cretaceous and Cenozoic, Journal of Geophysical Research, v. 97, B10, pp. 13,917–13, 951. The longer time lengths of the reversals older than 45 Ma appear to be real, not just a sampling problem

of problems Serge von Bubnoff had posed, still finer resolution was required. It too, came from an unexpected direction.

## 6.8 Fission Track Dating

As will be discussed in greater detail in Chap. 12, one way in which radioactive elements can decay is through fission. In this process, discovered in Berlin in 1938, the radioactive atom, such as the 235 isotope of uranium, splits apart and the two pieces go flying off in opposite directions. If the atom is in a crystal, the pieces plow though the crystal lattice leaving a track that can be observed under a microscope. If you know the concentration of the radioactive element and know how often it undergoes fission you can count the number of 'fission tracks' and find out how old the mineral is. Well almost. One of the most useful minerals for this sort of study is apatite; chemically it is very similar to your teeth and bones. Apatite grains are usually tiny, but are widespread in both igneous rocks (those formed through the cooling and solidification of magma or lava) and sedimentary rocks. However, they have a peculiarity. If the apatite grain is warmer than about 100 °C the fission tracks become 'annealed.' They simply disappear. But eureka! The fission track age tells you the last time the rock was warmer than 100 °C, and if you measure the geothermal heat gradient (the rate of increase in temperature with respect to increasing depth into Earth's interior) you can determine when that mineral was at a given depth below the surface. Since the sample with the apatite grain was recovered at the surface, you can determine how long it has taken for the overlying material to be eroded and bring the grain to the surface. This rate at which material is removed is called the 'denudation rate.' It turns out, for example, that for the Alps it is about 1 m/1,000 years.

## 6.9 Astronomical Dating

A huge improvement in our ability to put precise dates on sediments came from the discovery 25 years ago that many sediments and sedimentary rocks record astronomical changes, particularly changes in the Earth's orbit around the Sun. We will explore the orbital cycles in detail in Chap. 14, but in short the ellipticity of our planet's orbit, the inclination of the Earth's axis of rotation, which causes the seasons, and the times of the year when the Earth is closest and farthest from the Sun change on time scales of tens to hundreds of thousands of years. These cycles cause changes in Earth's climate, and thereby leave a record in sedimentary rocks formed over time.

The story of astronomical cycles recorded in sedimentary rocks began almost a century and half ago, but it took a long, long time to mature. Today, orbital cycles can be calculated precisely and the cyclic record in the sediments can be dated to within a thousand years or less. The work on deciphering the geologic record in terms of astronomical cycles started in earnest only about 20 years ago, and was applied first to the youngest part of the geologic column. But the work has progressed so that we now have these records for much of the Cenozoic. As the research progresses we should be able to extend this very detailed record back into the Mesozoic.

## 6.10 Tritium, Carbon-14 and Beryllium-10

Three relatively short-lived radioisotopes are being formed continuously in the upper atmosphere by collisions between highly energetic cosmic rays and air molecules. Cosmic rays are not really 'rays,' like light, but protons and helium nuclei coming in toward Earth from outer space at very high speeds. The radioactive isotope of carbon, $^{14}C$, was discovered in 1940. It has a half-life of 5730 years. One way it is formed is when a speeding neutron collides with the nucleus of a nitrogen atom ($^{14}N$, 7 protons and 7 neutrons) displacing one of its protons so that the atom becomes one of carbon, with an atomic mass of 14 ($^{14}C$, with 6 protons and 8 neutrons). These heavier carbon atoms combine with oxygen to form carbon dioxide and are incorporated into plants during photosynthesis. Living plants continuously interact with the environment and so maintain a ratio of the different carbon isotopes that is the same as that of the atmosphere. Animals in the food chain eat the plants and acquire the same isotopic ratio. Once the organism dies it no longer incorporates new $^{14}C$ atoms. It becomes possible to determine the age of the plant or animal material by measuring the ratio between the remaining $^{14}C$ and the stable $^{12}C$ isotope. It has been used to date plant and animal remains, and even artifacts like the cloth of the Shroud of Turin which has turned out to have an age somewhere between 1260–1390 CE.

Although predicted in the 1920s, tritium ($^{3}H$) was not discovered until 1934. It is an unstable isotope of hydrogen with a half-life of 12.33 years. It is produced not only by cosmic rays, but also by hydrogen bombs. Bomb testing in the 1950s introduced a huge pulse of $^{3}H$ into the atmosphere. This 'artificial' pulse has been used by oceanographers to determine the rates of exchange of gases between the atmosphere and ocean, and to locate sites and rates of ocean deep water formation.

Beryllium-10 ($^{10}Be$) is actually formed both in the atmosphere and in the ground to a depth of about 1 m. It has a half-life of 1.5 million years. Within the last decade the techniques have been developed to use $^{10}$Be to determine long-term (thousands of years) erosion rates. Erosion is a discontinuous process, with most of the material being moved during extreme events like major floods or hurricanes. Beryllium-10 measurements integrate the erosion during these extreme events into the background of 'normal' erosion rates, and also let us evaluate the impact of human activities. This technique indicates that the most rapid erosion rates are 10 m/ky, in New Zealand and Taiwan. The Himalayas and Alps are eroding at a rate of 1 m/ky (confirming the fission track denudation rates). Denudation of the landscape in central Europe proceeds at a rate of 5 cm/ky, and the ancient shield areas, like eastern Canada, at a slow 1 mm/ky.

These isotopes only exist for relatively short periods of time, but they have been very important in establishing Sergei von Bubnoff's goal of determining geologic process rates. All will be discussed in some detail in Chap. 13.

## 6.11   The Human Acceleration of Natural Process Rates

Let's come back to the question: how long does it take for a mountain range to be eroded away? Much longer than you might expect, because it is not what you might expect from the denudation (erosion) rate. That is because the interior of the Earth behaves like a very thick fluid. The crust of the Earth 'floats' on this fluid like ice floats on water. Mountains are like a thicker block of ice; most of the mountain range is a mass of rock below the surface, displacing the more dense fluid material deeper within the Earth. As the surface of the mountains erode, the block becomes thinner and lighter, and the fluid material flows in beneath it and buoys it up. In fact, if a meter of rock is eroded off the surface of the mountain range, the remaining rock will rise about 80 cm, so the apparent change in elevation will only be 20 cm. Are any changes in the landscape perceptible in a human lifetime? Only where there are sudden mass movements, like landslides, or where humans have moved material on a massive scale.

The theory of 'plate tectonics,' developed during the 1960s and 70s, describes the surface of the Earth as a series of more or less rigid crustal plates, growing at one side and descending into the Earth to be destroyed at the other (Figs. 2.5, 6.7). The growth occurs along the mid-ocean ridge system that starts in the Arctic north of Europe, extends though the Atlantic and Indian Oceans and passes south of Australia into the eastern Pacific to end beneath western Canada. The ridge is broken into segments by huge cross-breaks called fracture zones or 'transform faults.' The sinking into the Earth's interior takes place along ocean trenches, such as those that surround the Pacific. The continents are carried along on these plates as they move. As a

consequence of the growth or spreading of the sea floor along the Mid-Atlantic Ridge, the Americas are moving away from Europe and Africa at a rate of 1 cm/year, or about half an inch a year, about the same as the rate of growth of your fingernails. This rate, originally deduced as a long-term average from the age of the ocean floor away from the Ridge, has actually been measured directly by precise measurements of the motion of islands of the Azores group on opposite sides of the Ridge.

But the plates do not necessarily move smoothly and continuously. Sometimes friction causes the motion to stop. Forces are built up until the friction is overcome and movement occurs in a few seconds or minutes, resulting in an earthquake. These are most spectacular when the relative motions of the plates are in the order of inches per year, as around much of the Pacific and beneath the northern Indian Ocean.

The accumulation of sediments in the deep sea is a relatively steady process, but occurs at rates of a centimeter or so per thousand years—a thousand times slower than the growth of your fingernails. A rate of one centimeter per thousand years is 10 Bubnoffs (Fig. 6.7).

The most rapid changes to our planet took place 65.5 million years ago, as an asteroid collided with it, striking the Yucatan peninsula. It resulted in catastrophic changes, which according to their nature, occurred in a matter of minutes, days or years. It is often said that this event caused the extinction of the dinosaurs, but most of them had died out in the last few million years of the Cretaceous when climates seem to have become unstable. It did result in a massive die-off of ocean plankton. Although there have been many other cases of asteroids impacting the Earth, that at the end of the Cretaceous, with its global effects, seems to have been a unique event during the past half billion years.

In 2005 University of Michigan geologist Bruce Wilkinson announced that over the past millennium humans have become the most important geologic agents on the planet. Human activity is now responsible for moving ten times more soil and rock than all rivers and glaciers combined. The role of humans as a geologic agent is directly related to population growth and the ability of our species to harness and use energy. But the role of humans goes back into the last glaciation, when Cro-Magnons made paintings in caves like Lascaux and Chauvet in France and Altamira in Spain. After seeing Altamira, Pablo Picasso is said to have remarked that everything since is decadent art.

We are in the midst of the greatest extinction since that asteroid collision; large animals seem to go extinct everywhere shortly after humans appear on the scene. About 9,000 years ago humans began to significantly modify the natural landscape by setting fires as a hunting tool and by breaking the land surface for agriculture. Both practices resulted in sudden increases in erosion rates. These changes

**Fig. 6.7** Plate tectonic sketch, showing the Mid-Ocean Ridge (spreading centers) as *broad black lines* and rates of sea-floor spreading from them, in cm/year. Multiply by 10,000 to get the rate in Bubnoffs. Some major Fracture Zones (transform faults) are shown as *thin black lines*, and major Ocean Trenches (subduction zones) as *dashed lines*. *Arrows* indicate general direction of plate motions

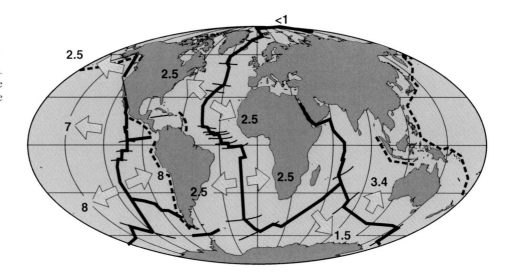

are recorded in the sediments supplied to the sea by rivers. The Yellow River of China is yellow because of the load of 1.6 billion tons of fine silt it carries from the Loess Plateau each year. The erosion is thought to have begun about 6,000 years ago as farming began on the Loess Plateau, the cradle of Chinese civilization.

Although humans have accelerated erosion by one order of magnitude, a factor of ten, our rate of consumption of fossil fuels is more than six orders of magnitude greater than their rate of formation. Over the last 500 million years, organic carbon has been buried at an average rate of $18 \times 10^6$ tons per year. Only a tiny fraction, about one hundredth of one percent, of this is available as coal, oil, or natural gas that humans can extract. Most is the finely disseminated carbon in sedimentary rocks that can never be recovered. However, we are currently burning this tiny fraction at a rate of 8 billion tons/year; that is over four million times faster than it formed!

The most rapid natural global changes occurred as Earth shifted from a glacial to an interglacial state, described in the next section. The modern average global temperature is about 15 °C; that of the last glacial about 9 °C. Our planet warmed by 6 °C over a period of about 10,000 years, or an average rate of 0.06 °C per hundred years. Global temperatures have risen about 0.6 °C in the last 50 years, about 20 times as fast as during the deglaciation. It may be hard to prevent an additional rise of 2 °C during the coming century. The average rate of rise of sea level during the deglaciation was 10 m/1000 years (1 mm/year). During the 7,000 years preceding the 20th century sea level was remarkably stable, and did not change globally. During the 20th century sea level rose about 0.15 m (1.5 mm/year). As the rate of melting of ice fields and glaciers accelerates, a sea level rise of 0.5–1 m (5–10 mm/year) is likely during the 21 century. But the greatest rate change is in the addition of carbon dioxide to the atmosphere. During part of the deglaciation its

concentration in the atmosphere increased at a rate approaching 1 part per million per century. It has increased by more than 100 parts per million since the beginning of the industrial revolution and is currently increasing at a rate of 195 parts per million per century. One might expect that the effect on the planet would be similar to driving your car into a concrete pillar at 1 mph versus 195 mph, assuming, of course, that there are no feedback mechanisms to exacerbate the effect.

The human effect on planet Earth is so extreme that ecologist Eugene Stoermer and atmospheric chemist Paul Crutzen have proposed a new Epoch for the geologic time scale, the Anthropocene. But now there is an argument as to whether it began in 1784 with the invention of the steam engine and start of the industrial revolution as Crutzen proposed in 2000 or, as my good friend Bill Ruddiman proposed in 2003, much earlier, when humans first began to affect the $CO_2$ content of Earth's atmosphere, about 8,000 years ago.

## 6.12   The Present Climate in Its Geologic Context

If you think that the Earth's climate has always been more or less like the one we enjoy today, you couldn't be more wrong.

From about 100 Ma (Ma = million years ago), in the Cretaceous (my special interest) until 50 Ma the Earth was generally very warm. There was an interruption 65 million years ago when an asteroid struck the Earth causing a brief, violent change of climatic conditions and the extinction of many forms of life. Although the event itself was short, it took many millions of years for animals and plants adapted to the new conditions to evolve. Except for the asteroid impact, and a few less catastrophic events, the Earth had experienced remarkably little variation in its climate. Fifty million years ago polar temperatures averaged about 15 °C, the average global temperature today. If there were glaciers anywhere, they were on high mountains. During the polar summer you could have gone for a comfortable swim in the Arctic Ocean, but you would need to keep an eye out for crocodile-like animals along the shore. Over the next 15 million years the temperatures gradually cooled. Then, 35 million years ago, something very dramatic happened. Ice sheets began to grow on Antarctica and the Arctic Ocean became covered with sea ice. The Earth had entered a different state.

It appears to have remained in this state, with some growing and shrinking of the Antarctic ice cover, until about 5 million years ago when Greenland became covered with ice. At 3.25 million years ago ice sheets began to appear on the Northern Hemisphere continents. These ice sheets would form over a period of about 30,000 years, and then rapidly melt off, leaving the continents ice free for about 10,000 years. Glaciation of Scandinavia and northeast Asia became pronounced at 2.75 Ma (Ma = million years ago), followed by glaciation of the mountains of Alaska and

western Canada at about 2.65 Ma and initiation of the huge Laurentian ice sheet over northeastern North America at 2.54 Ma. From then until 800,000 years ago, the cycles of glaciation-deglaciation had a 40,000 year periodicity.

Then something happened, because since 800,000 years ago, the period of the glaciation-deglaciation cycles changed to 100,000 years, with the ice gradually building up over about 80,000 years, and suddenly melting away in less than 10,000 years. The disappearance of most of the ice in the Northern Hemisphere was followed by an interglacial epoch in which the ice sheets were restricted to Greenland and perhaps Iceland and some of the islands of the Canadian Arctic, and there were glaciers in some mountainous areas (Fig. 6.8). During these interglacials the Arctic Ocean seems to have remained covered with ice. The interglacials typically lasted 10,000 years or less.

During the glacial episodes, the ice sheets gradually grew on land through the accumulation of snow that did not melt during the summer (Fig. 6.9). That snow came from water evaporated from the ocean. During the Last Glacial Maximum, the Northern Hemisphere ice sheets reached their maximum extent about 24,000 years ago. The global

**Fig. 6.8** The Northern Hemisphere in its *'abnormal'* Quaternary interglacial state. It has been in this state for only brief episodes totaling 10 % of the past million years

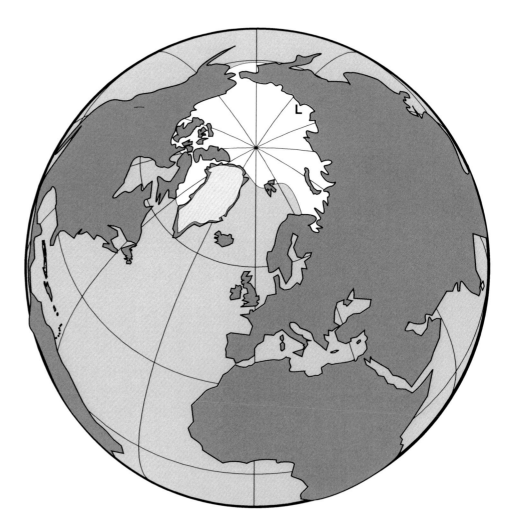

**Fig. 6.9** Earth in the *'normal'*
state it has had for 80 % of the
past million years, with ice sheets
growing over the northern
continents surrounding the North
Atlantic

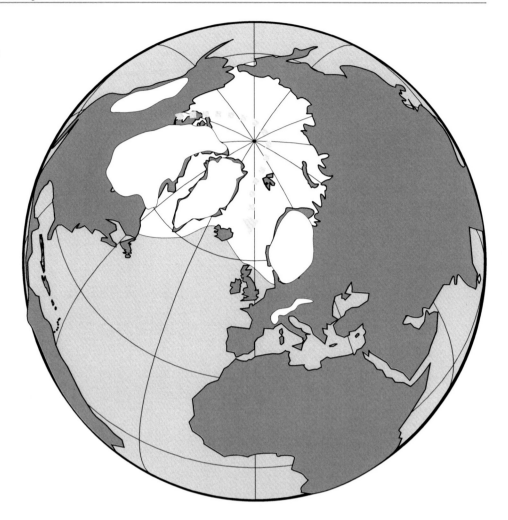

temperature then is estimated to have been about 5.6 °C colder than the mid-20th century average.

During the Last Glacial Maximum (Fig. 6.10) the ice over eastern Canada was about 3 km thick, but the top of the ice sheet was only about 2 km above sea level. This is because the 'solid' Earth is not really solid. At depths below 100 km or so beneath the surface, the rocks of Earth's mantle are hot and soft. They flow to accommodate the additional weight of the ice on the surface. The ratio of the density of the ice to the hot mantle rocks is about 1–3. The adjustment of the 'solid Earth' to the weight of the ice on the surface does not take place instantaneously and lags behind the growth and decay of the ice sheets. Today some areas, such as Scandinavia and the Hudson Bay region, are still rising in response to the unloading even though the ice was gone 6,000 years ago.

The transfer of water from the ocean to ice on land during the last glacial lowered sea level about 130 m. The coastlines were along the edges of the continental shelves. The shelves mark the edges of the continental blocks, where the continental slopes lead down into the deep sea. The shelves are largely submerged today but were exposed as land during glacial times. Starting about 20,000 years ago, the ice of the last glaciation began to melt. Sea level rose rapidly, sometimes by as much as 2 m or more per hundred years. The coastline along the eastern US was moving inland about 1.6 km (1 mile) per hundred years. About 12,700 years ago a sudden change in the climate occurred. There was a return to conditions like those of the Last Glacial Maximum. This drastic transition occurred in only one or two decades. Temperatures in Norway dropped by about 15 °C. This period of return to glacial conditions is known as the 'Younger Dryas' because of the widespread recurrence of *Dryas octopetalia*, a pretty little wildflower of the tundra. Its pollen and leaves occur in fossil floras of this age. It lasted about 1,200 years, ending 11,500 years ago as suddenly as it had begun.

Geologists refer to time since the Younger Dryas as the 'Recent' or 'Holocene.' Although there were still some remnants of the ice sheets on the northern continents, they were melting rapidly, and the early Holocene is taken as marking the end of the last ice age. The last remains of the great continental ice sheet of North America melted away about 6,000 years ago.

**Fig. 6.10** Earth in its *'extreme'*
Quaternary Glacial Maximum
(ice-age) state. It has been in this
condition for brief periods
totaling about 10 % of the past
half million years. This shows the
Northern Hemisphere,
24,000 years ago at the time of
the Last Glacial Maximum
(LGM). The names of the major
ice sheets are shown

Large areas of the northern continents are covered by fresh pulverized rock left behind by the ice. In much of eastern Canada this debris is just beginning to weather to form soils, a process that takes tens of thousands of years. Basins scooped out by the ice now hold lakes that trap the detritus from streams, preventing sediment from being transported onward to the sea. Some of the largest of these are the Great Lakes of North America. They mark the boundary between the hard crystalline rocks, granites and schists, of what geologists call the 'Canadian Shield' and the softer sedimentary rocks covering most of the eastern United States. As the ice flowed slowly south during the last glacial it was unable to pick up much of a load from the hard rocks of the Canadian Shield, but when it reached the softer sedimentary rocks, it scooped them up and left this load behind over the northern tier of the Midwestern states. The ice retreated into Canada about 12,000 years ago, and that 'ground moraine' spread over the northern Midwest has now weathered to form the rich farming soils that supply much of the wheat and corn in the US.

Forests and grasslands migrated rapidly northward during the deglaciation. Sometimes the climate changed so rapidly that some species of trees, which have lifetimes of many tens to hundreds of years, were not able to keep up. Some of the forests that grew during the deglaciation had mixtures of trees unknown today. Some remnant forests remain to this day; the 'Lost Maples' in west Texas are a good example.

The rapid sea-level rise shaped the present coastlines. The river mouths of glacial times were flooded and are now the estuaries of the eastern US and northern Gulf of Mexico. They trap sediment that would otherwise reach the sea. Coastal swamps and barrier islands characterize much of the rest of the low coastline.

Thus, instead of being a representative sample of the Earth's geologic history, the Holocene is one of the most peculiar of geologic epochs. The Holocene climate is characteristic of only brief intervals during the past 3 million years. We live in an interglacial episode when large ice sheets are restricted to the Antarctic and Greenland.

But that is not all that is peculiar about the Holocene. The Earth is about as warm as it typically gets during an interglacial. Since the ice sheets melted, climatic conditions have varied only slightly. The warmest time of the Holocene was between about 9,000 and 5,000 years ago, and is known as the 'Holocene climate optimum' (but remember that optimum means 'peak,' and not 'most ideal'). The actual warmer temperatures seem to have been restricted to Scandinavia and the northern part of European Russia. However, global temperatures may have been about 0.25 °C warmer than the mid-20th century average. The region of the modern Sahara was wet enough 9,000 years ago to support many kinds of vegetation and large animals. It gradually dried out to become desert by about 4,500 years ago. However, compared to older interglacials and glacials, the Holocene climate has been remarkably stable. Over the past few millennia, there was a warmer-than-average period called the 'Medieval Climate Optimum' that lasted from the 10th to 14th centuries. Temperatures were about 0.1 °C warmer than the mid-20th century average. It was followed by the 'Little Ice Age' that lasted until the late 19th century. Temperatures sank to 0.3 °C below the mid-20th century average. Both of these variations were largely confined to the Northern Hemisphere, perhaps only Europe. Even taking these perturbations into account the last 8,000 years of Earth history have had the most stable climate of the last million years. It is during this episode of climatic stability that societies were able to become organized and develop the partitioning of tasks leading to what we call civilization. We are now in the process of destabilizing the climate system and ending the conditions under which civilization developed.

Since the average length of an interglacial is 10,000 years, and we are already more than 11,000 years into this one, we might expect a return to glacial conditions soon. This is indeed what many scientists thought in the 1970s. But, as we shall soon see, there is good reason to believe that this is a completely unique interglacial, which, even without human intervention, would last for 70,000 years. With human intervention it may not end. There is also reason to believe that humans began to influence Earth's climate as early as 8,000 years ago. Humans had already steered our planet into uncharted waters long before the industrial revolution began. Thus, we cannot blame what is happening entirely on recent generations, but they are responsible for about 95 % of the effect.

If human interference with natural processes were tending to force the climate back toward a glaciation, we would have a much better idea of what might happen in the future. But greenhouse gas emissions are forcing Earth's climate toward warmer conditions—into a state the Earth has not seen for many millions of years. As we will see, human activity today is not giving climate a gentle nudge in that direction, but something rather more like a blow with a sledgehammer.

## 6.13   Steady State Versus Non-steady State

In chemistry, physics and climatology, a steady-state is a stable condition that does not change over time or in which change in any direction is soon balanced by change in the opposite direction. The variation around the average condition is called the variability of the system. In the 1970s we were making great advances in understanding both the chemistry of the ocean and the variability of the climate system, and how they maintained their stability. The overwhelming view of those studying the modern ocean and climate systems was that they were inherently stable. Ocean chemists, oceanographers, and atmospheric scientists were concerned with understanding the modern world. They knew very little about Earth's past history.

In a non-steady state the variability becomes uncontrolled, and there can be continuous change. Or there may be a 'change of state' in which the system moves to a new different steady state condition. Figure 6.11 shows these concepts graphically.

You can think of the Earth's different climatic states as depressions on a surface, and the Earth's climate as a ball lying in one of these depressions. Figure 6.11 shows the four common states. The two depressions on the left represent the icehouse states, with one or both poles covered by ice. As mentioned previously, over the past 600 million years the Earth has been in an icehouse state about 25 % of the time. In the icehouse state it alternates between glacial and interglacial states. During the glacial states, the ice extends well beyond the polar circles (about 66.5 °N and S). During the interglacials it is restricted to the higher polar latitudes. We live in an interglacial. In the 1970s as we were beginning to understand how the chemical systems of the ocean and atmosphere work, it was assumed that the interglacial we were living in was in a steady state. Few considered the possibility that any serious change might already be occurring now.

Over the past 3.5 million years, the Earth has spent about 90 % of the time in the glacial state, and 10 % in the interglacial state. The changes between the glacial and interglacial states correspond to changes in the Earth's orbit around the Sun, the 'Milankovitch parameters,' which we will discuss in detail later. The greenhouse states are shown on the right side of the diagram. There must be two of them, because we see regular alternations in the sedimentary rocks, but exactly what the difference is between the two states remains uncertain. Most likely it is a shift in the latitude of

**Fig. 6.11** Schematic diagram of the multiple states of Earth's climate system. The variations in Earth's orbit causing the alternation of glacial and interglacial states is discussed in Chap. 13. The effect of changing greenhouse gas concentrations is discussed in Chaps. 20–22. Over the past million years, atmospheric carbon dioxide concentrations have varied between 170 and 280 parts per million (ppm). In 2010, the atmospheric carbon dioxide concentration was 391 ppm and is increasing at a rate of about 2.2 ppm per year

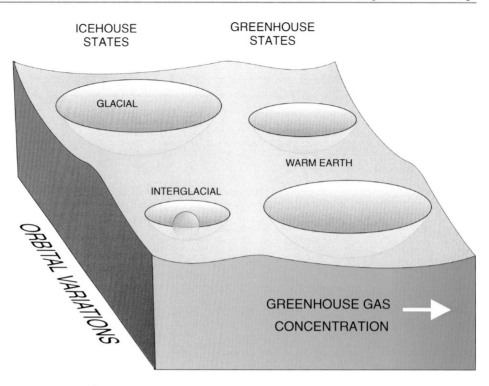

the climatic zones in response to those ongoing regular changes in Earth's orbital parameters. The difference between the icehouse and greenhouse states reflects the concentration of the major natural greenhouse gases, carbon dioxide and methane, in Earth's atmosphere. The greenhouse states are Earth's normal condition. It has been in the greenhouse state for 75 % of the last 600 million years.

The modern climate is represented by the ball in the interglacial depression. Variability of the climate can be thought of as the ball rolling around in the depression, going a little up the side of the depression, and the rolling back down. As we will see later on, the interglacial climate we are in has been extraordinarily stable; the last 7,000 years, during which all of modern civilization has evolved, have been the longest period of stable climate in at least the last half million years. Until very recently the ball has been very well behaved, and has stayed close to the bottom of the depression.

In the past 50 years the ball has been perceptively perturbed, largely by adding greenhouse gases to the atmosphere. It has started to roll up the side of the depression. If we could remove the perturbation, it would roll back down. On the other hand, if we continue to push it, it may roll out of the depression, and look for another one to fall into. As we will see, the changing orbital parameters are quite unfavorable for it to return to the glacial-state depression, and there is a big, deep greenhouse-state depression nearby. If it gets into that one, it may stay there for a long, long time.

## 6.14  Feedbacks

Feedback is defined as the process by which part of the output of a system is returned to its input in order to regulate its further output. Feedback can be either positive or negative. You know positive feedback as the screech from the loudspeakers when the microphone picks up the performing rock star's electric guitar, amplifies its sound, picks up that sound, amplifies it even more, and on and on until everyone is holding their ears. Unfortunately, the climate system is prone to positive feedbacks, so that a little perturbation by one factor leads to enhancement of that perturbation by another factor, and so on. Negative feedback is used to control systems. Negative feedback causes the effect of a perturbation to be reduced and may result in its eventual elimination.

As an example of the importance of feedback to climate, consider the following: On Earth, the solid form of water (snow and ice) is almost white. We perceive it as white because it reflects almost all of the Sun's light. From getting sunburn while skiing you probably know it also reflects the Sun's ultraviolet. It also reflects almost all of its infrared. In contrast, liquid water absorbs most of the energy from the Sun. The ultraviolet and infrared are absorbed at the surface. Red light is absorbed by about 10 m depth which is why red fish at that depth appear to be black to a scuba diver. Only blue light penetrates to a depth of 30 or so meters. The fact that cold snow reflects energy and warm water absorbs energy means that the snow stays cold while the water warms up. This results in a positive feedback. Early in the development of numerical models of climate it was realized that if the Earth has ice at the poles, it won't melt, and if the poles are ice-free, they cannot become ice-covered, unless the amount of energy received from the Sun changes.

Now consider conditions in another universe where snow and ice are black and water is white. What would be the temperature on the surface of a water-covered planet there? The cold snow would absorb heat and melt. The water would reflect incoming energy and cool. The surface temperature would be in balance at the freezing point of water. This would be an example of a negative feedback regulating the climate. If the amount of energy received from that planet's Sun changes, the negative feedback will prevent the temperature at the surface of the planet from changing. It will remain at the freezing point of water. One of my students named this hypothetical planet "Holstein" after the cows that have black and white markings.

In the section on Earth's climate history I noted that our planet seems to like to be in either a greenhouse or icehouse state. It doesn't seem to spend much time in between. This suggests that when conditions are right for change, positive feedback mechanisms kick in and ensure change with no possibility of return. As we will see later in this book, a surprising number of natural positive feedbacks are reinforcing the climatic changes induced by human activity.

## 6.15 Summary

Already in the 1830s Charles Lyell realized that there had been major changes in Earth's climate over time. Alexander Agassiz's championing of the idea of an ice age in 1840 introduced the idea that the climate had been radically different in the not too distant past. The discovery of evidence for 300 million year old Paleozoic glaciations in India and Australia, along with the discovery of 100 million year old Cretaceous tropical vegetation in Greenland made it clear that Earth has experienced major climate changes in the past.

'Climate' is defined as a 30 year average of temperature, precipitation, and related factors. Since data were sparse, Vladimir Köppen used vegetation to define modern 'regional climates,' and this practice has been extended to use fossils to determine climatic conditions in the ancient past.

The modern view is that Earth has two preferred states —'icehouse' and 'greenhouse.' In the icehouse state the Polar Regions and higher latitudes are covered by glacial ice. In the greenhouse state, ice is limited to high mountains. The planet spends about 75 % of its time in the greenhouse state, about 25 % in the icehouse state. We are currently in one of the recurring brief warm intervals, an 'interglacial' in the latest episode of Earth's icehouse state which began about

35 million years ago. Interglacials account for about 10 % of the icehouse time and are marked by retreat of the ice to the Polar Regions.

The last interglacial was about 110,000 years ago. It was followed by a gradual buildup of ice that lasted about 80,000 years. The Last Glacial Maximum, when the ice over Chicago was about a mile thick, was 24,000 years ago. The sudden melt-off started about 18,000 years ago and lasted about 10,000 years. We are now in an Interglacial, the 'Holocene.' Since 8,000 years ago the planet has had the longest period of climate stability in the last few million years. It is in this short interval of relatively stable climate that humans have developed what we call 'civilization.'

Geologic processes and past climate change are much too slow to be directly observed during a human lifetime. However, at present the human perturbation of the Earth system has accelerated these processes to two or more orders of magnitude, so that we can actually observe them. Our planet has entered a new phase of its history, the Anthropocene, when human activity has become Earth's major geological agent. We are currently destabilizing the climate system and pushing conditions on Earth in the direction opposite to what would happen in the natural course of events—toward a premature return to the greenhouse state.

We can now determine rates of climate change during the past few millions of years with great precision. The present rates of change are up to a couple of hundred times faster than anything that has happened to planet Earth before, with one exception—the impact of that asteroid 65 million years ago. Things that happen very fast, like driving your car into a concrete wall at a 100 miles an hour, tend to have very unpleasant outcomes.

A Timeline for this chapter:

| Year | Event |
| --- | --- |
| 800,000 BP (BP = Before Present) | Earth enters a period with approximately 100,000 year glacial cycles; glacial buildup lasting 80,000 years, deglaciation lasting 10,000 years, and interglacials lasting 10,000 years |
| 24,000 BP | Last Glacial Maximum (LGM) |
| 20,000 BP | Deglaciation begins |
| 12,700 BP | Deglaciation around the North Atlantic interrupted by a return to glacial conditions —'Younger Dryas' episode |
| 11,500 BP | End of the Younger Dryas—return to deglaciation |
| 11,000 | *Holocene* epoch (the current interglacial) begins |
| 9000 BP | Sahara is wet enough to support grasslands and large mammals |

(continued)

| Year | Event |
| --- | --- |
| 9,000–5,000 BP | Holocene 'climatic optimum' in parts of northern hemisphere |
| 8000 BP | Humans begin to be a major agent affecting Earth's climate |
| 6,000 BP | Major northern hemisphere continental sheets have melted away |
| 4,500 BP | The Sahara has dried out and become desert |
| 900–1300 CE | The 'Medieval Climatic Optimum' in Europe coincides with the development of Romanesque architecture |
| 1400–1800 CE | The Little Ice Age in Europe |
| 1784 | Steam engine invented; the Industrial Revolution begins |
| 1840 | Alexander Agassiz shows that the landscapes of Scotland were formed by glaciers |
| 1856 | Evidence of late Paleozoic glaciation is found in India |
| 1859 | Evidence of late Paleozoic glaciation is discovered in Australia |
| 1879 | Adolf Nordenskiöld discovers fossils of tropical plants in the Cretaceous of Greenland |
| 1923 | The International Meteorological Organization defines 'climate' as a 30 year average of the weather |
| 1931 | Sergei von Bubnoff publishes 'Grundprobleme der Geologie—Eine Einführung in geologisches Denken' |
| 1940 | The radioactive isotope of carbon, $^{14}$C, is discovered |
| 1948–50 | Alfred Nier and Thomas Aldrich devise the K-A dating method |
| 1950s | Tritium from hydrogen bombs is introduced into the atmosphere |
| 1956 | From study of deep-sea sediment Cesare Emiliani discovers that there had been many more than four Quaternary ice ages |
| 1961 | Patrick M. S. Blackett publishes 'Comparison of Ancient Climates with the Ancient Latitudes deduced from Rock Magnetic Measurements.' |
| 1963 | An English translation of Sergei von Bubnoff's book is published as 'Fundamentals of Geology.' Few people bother to read it. |
| 1968 | The Deep Sea Drilling Project proves both Alfred Wegener's concept of continental drift and Arthur Holmes' idea that it might be driven by convection in the Earth's mantle |
| 1969 | Al Fischer proposes the 'Bubnoff' as a rate of 1 m/10$^6$ my to describe geologic processes |

(continued)

| Year | Event |
| --- | --- |
| 1970s | Many geologists conclude that the next ice age is overdue |
| 1982 | Al Fischer divides earth history into 'greenhouse' and 'icehouse' states |
| 1992 | Larry Frakes and colleagues provide a detailed description of the succession of greenhouse and ice house states in Climate Modes of the Phanerozoic |
| 1992 | Joseph Kirschvink proposes the idea of a 'Snowball Earth' |
| 2000 | Paul Crutzen introduces the term Anthropocene for the time since the beginning of the industrial revolution |
| 2003 | Bill Ruddiman proposes that the Anthropocene actually began 8,000 years ago |
| 2005 | Bruce Wilkinson finds that humans are now the major geologic agent on planet Earth |
| 2010 | David Kidder and Thomas Worsley propose a new extreme climate state—the 'hothouse' |
| +70,000 | Present interglacial would end if it were not for human intervention |

If you want to know more:

Frakes, L.A., Francis, J.E, and Syktus, J.I. 1992. *Climate Modes of the Phanerozoic*. Cambridge University Press, Cambridge, UK, 286 pp.

Ruddiman, W.F., 2007. *Earth's Climate: Past and Future*, W.H. Freeman, New York, NY. 388 pp.

Music: If you have seen Walt Disney's 1940 film *Fantasia* you will undoubtedly remember the history of the Earth set to the music of Igor Stravinsky's ballet *The Rite of Spring* (*Le Sacre du Printemps*). The only thing that can compete with it is to see the ballet, which I did a few years ago in Berlin. If you have seen the film, you can just listen to the music and relive it; if you haven't seen it, you should seek it out.

Libation: So much of the work described in this chapter has been done by Americans, it's time for a sip of good Kentucky bourbon.

### Intermezzo VI. A Postdoctoral Year in Switzerland—and Europe—II

It took a few months for me to really appreciate Basel. The city lies on a broad bend in the Rhine River, where it changes from flowing to the west, out of the Lake of Constance, to the north, toward Frankfurt. Switzerland, France, and Germany come together at this point, and the metropolitan area is in all three countries. The south side of the bend is a high bluff, the north side much lower. The city on the south side is Basel proper; on the north side it is called Klein Basel (Little Basel).

The Romans founded a town, Augusta Raurica, a few miles upstream from Basel. The site there has beautifully preserved mosaic floors. Since I lived in Basel, excavations for building repair on both sides of the Rhine have exposed Roman mosaic floors beneath the medieval buildings, revealing that the Roman settlements were quite extensive. The city was destroyed in a major earthquake on October 18, 1356. Its magnitude is estimated to have been between IX and X on the Mercalli Scale, or between 6.7 and 7.1 on the $MM_W$ scale. One of the few buildings to survive, at least partially, was the Münster (cathedral), which had been started in 1019 and was finally finished in 1510. The earthquake holds the record as the largest to ever strike central Europe, and is commemorated in song and legend as the 'Tod zu Basel' (Death in Basel). The 'Totentanz' (Dance of Death) is a common topic in paintings of the time. There was a church with fine medieval paintings of the Totentanz near what is how Basel's Cantonal Hospital. For many years the tram stop next to the hospital was named Totentanz, until someone realized that was not very comforting for friends and relatives coming to visit a patient. Now the streetcar stop is an innocuous 'Kantonsspital.'

In the parts of the city on the bluff above the Rhine, on either side of the Münster, the construction dates on the houses are mostly from the late 14th and early 15th centuries. A notable exception is the Naturhistorisches Musem, the Natural History Museum, which dates from 1821, and which houses, among other things, very important collections of larger foraminifera and planktonic foraminifera.

From 1431–1449 Basel was the site of the Roman Catholic Council of Basel. The Council considered itself to have higher authority than the Pope. It concerned itself with the heresies of the followers of Johannes Hus, the reformer in Prague who preceded Martin Luther, and who had been burned at the stake in 1415. When, in 1437 Pope Eugenius IV, tried to move the Council to Ferrara, and change its membership to get it back under control, the Council in Basel simply elected an Antipope, Felix. After that church affairs got messy for a while.

Basel managed to come out of this mess with a University sanctioned and endowed by Pope Pius II in 1459. It opened its doors on April 4, 1460. I was able to see part of the elaborate ceremonies surrounding the 500th anniversary of this event. There was a memorable procession in the city with representative of all the major Universities of Europe, and a few of the older Universities of the Americas in their finest academic regalia.

Alas for the church, the University of Basel seemed especially attractive to progressives and liberals, like Erasmus of Rotterdam (1466–1536) the catholic priest who became better known as the Prince of Humanists, and Paracelsus (Philippus Aureolus Theophrastus Bombastus von Hohenheim, 1493–1541), a Swiss Renaissance physician, botanist, alchemist, astrologer, and general all-round occultist. Basel became, and remains, the center of Humanism in Europe. Humanism is variously defined as: (1) a system of thought that rejects religious beliefs and centers on humans and their values, capacities, and worth, or (2) an outlook or system of thought attaching prime importance to human rather than divine or supernatural matter. Humanism is the antithesis of religious dogmatism. This led the University of Basel to become a center for Mathematics (Leonard Euler, and the Bernoulli family), Science, Medicine (leading to Basel's present pharmaceutical industry) history (Jacob Burckhardt, famous for recognizing the importance of the Renaissance), and many others. My home in Basel was in the Eulerstrasse, and the Geological Institute was in a building known as the Bernoullianum.

Because of its, location, where Switzerland, France and Germany come together, Basel is inherently a bilingual city. Basler's don't actually speak High German, but have their own dialect, not unlike the Alemannic one hears in neighboring Swabia and Alsace. Fortunately, since I had become proficient in the Swiss dialect spoken in Zürich, it took me only a few months to learn 'Baseltüütsch.' So by December I was able to understand the discussions around the morning coffee table at the Institute. Fortunately, my French was good enough to handle that side of things as well. French is the language used in restaurants and clothing stores in Basel. German is used in hardware stores, the automobile repair shops, and most everywhere else. Dinner table conversation can switch back and forth depending on the topic. Of course, today most everyone also speaks English, but that marks you as an outsider. Remember that Switzerland is like a giant club. I discovered that if you spoke 'Baseltüütsch,' you didn't need your passport to go to dinner in neighboring Alsace (France) or the Black Forest (Germany).

Basel has a unique festival, 'Fasnacht,' which starts on the first Monday after Ash Wednesday—at 4:00 AM. All city lights go out at 3:59 AM, and then begins the most interesting part of the festival, 'Morgestre-ich,' which lasts until dawn. Groups of costumed, masked Basler's (called 'cliques') carrying fantasti-cally painted lanterns on poles and accompanied by fife and drum corps playing tunes from the time of the Napoleonic Wars march through the city center. There is no set route for the cliques, each goes its own way, and the scene is one of total chaos. The streets are filled with spectators who must get out of the way when one of the cliques comes marching along. I re-member getting crushed into doorways as they went by. The lanterns are new each year and make fun of political and current events. The only way to appre-ciate what it is like is to google Basler Fasnacht.

Professor Reichel had recovered enough so that he was able to come in for the morning coffee hour. He had some suggestions for collecting sites in Italy. A particularly interesting stratigraphic section had been studied in the 1930s by one of his doctoral stu-dents, H.H. Renz. However Renz had taken only hard limestone samples that could be studied in thin sec-tion. A thin section is made by grinding a flat surface on a fragment of the rock, gluing it to a glass slide, and then carefully grinding away the other side until the slice glued to the slide is so thin light can pass through it. It is a painstaking process and takes a lot of time and effort. Reichel said that the limestone layers, which were only a few centimeters thick, were interbedded with marl that might also contain micro-fossils. I asked why Renz had not also taken samples of the marl layers—it seemed so much easier to wash the microfossils out of it. But to do that, you need good brass sieves. Reichel pointed out that in the 1930s the Institute had not been able to afford sieves.

In the early spring, I started out on sample col-lecting again. This time the idea was to start in the south, Sicily, where it was warm, and work my way back up the Italian peninsula. One of the students in the Institute, Giovanni Lorenz, was a resident of Milan, and I invited him to go along with me and help out finding the localities, since I didn't know much Italian. Giovanni spoke not only Italian, but also German and the Basler dialect. If you have read Stendhal's *The Charterhouse of Parma* you will be familiar with the fact that Lombardy (where Milan is located) was part of the Austro-Hungarian Empire before the unification of Italy, and German was widely spoken there. Giovanni's family was bilingual.

In 1854 Christian Gottfried Ehrenberg (1795–1876) published a two volume work '*Mikrogeologie*' in which he presented beautiful drawings of the

microscopic fossils found in rock samples he had obtained from colleagues in Europe, Asia, Japan, Indonesia, and the Americas. Ehrenberg worked at the Museum of Natural History in Berlin, and became well known for his studies of microscopic organisms. However, the '*Mikrogeologie*' was unique. It opened a whole new field of science, which we now call Micropaleontology. Unfortu-nately, the localities from which his samples came were not very precise. For example, he described microfossils from North America as coming from samples taken 'west of the Mississippi.' One par-ticularly beautiful suite of microfossils had been described as coming from Cretaceous chalk near the village of Cattolica-Eraclea in Sicily. Another from marl thought to be of Cretaceous or Early Tertiary age from the Greek island of Aegina. The figures in his book are magnificent, and the specimens of planktonic foraminifera and nannofossils from these sites are among the most beautiful. There was just one small problem—they didn't look like any Cre-taceous microfossils I had seen. I wanted to recover these localities and started with the one in Sicily.

We drove south to Naples and took a ferry to Palermo. I had heard that it could be a dangerous city for foreigners (in Sicily at that time, 'foreigners' included Italians from the mainland). We stayed at one of the new 'Jolly Hotels' that were being built to open up southern Italy and Sicily to tourism. In the hotel we were warned it would not be a good idea to go out at night. The next day we did some sample collecting around Palermo and the following day headed off to the village of Cattolica-Eraclea across the island. We had no idea whether we could find the place which had provided the sample to Ehrenberg, but it was remarkably easy. There was only one fossiliferous layer near the town, marked as Pliocene 'Trubi' on the geologic map. All of the rest of the rocks were gyp-sum, now known to have been part of the salts deposited throughout the region when the Mediter-ranean Sea dried up at the end of the Miocene. But it was in the hunt for the fossiliferous layer that Gio-vanni discovered that Sicilian was an almost unintel-ligible version of Italian.

Our next place for collecting was near the Straits of Messina, and to get there, you have to go around Mount Etna. We started out spending a night in Agrigento, about 40 miles east of Cattolica-Eraclea. I knew that Sicily had once been a Greek colony, but I was unprepared for the magnificent Greek temple ruins just south of Agrigento. Even more interesting, the building stone is full of fossil shells, so the Greeks must have had some thoughts about geology.

On the east cost of the island we spent a night in the resort town of Taormina. Taormina has a Greco-Roman amphitheater, located so Mount Etna is in the background over the stage. It would have been an incredible site for the performance of a Greek tragedy with the volcano erupting in the background. I repeated much of this same itinerary in 2001, taking more time to enjoy the island. From the amphitheater in Taormina, Etna had recently erupted and there was a red glow from it at night. Sicily today is a very open and friendly place to visit.

We left Sicily taking a ferry across the Straits of Messina. There were a number of geological localities to collect in Calabria, the southern part of the peninsula. Finding them was often a problem, and Giovanni was getting frustrated that no one seemed to speak proper Italian. We worked our way gradually back north, recovering fossil localities that had been described in the 19th century. Finally we came to H.H. Renz's section, near Gubbio. It was a spectacular sequence of alternating thin beds of limestone and marl. We collected representative samples to take back to Professor Reichel. The Gubbio material was later worked up by students at the Institute and found full of well-preserved planktonic foraminifera. Gubbio has turned out to be a classic section across the Cretaceous-Tertiary boundary.

Now there remained that other Ehrenberg sample site, on the Greek island of Aegina, that had such beautiful microfossils. Enough travel money had accumulated in the fund for Giovanni and me to go to Greece and try to find it. According to the geologic map of Greece at the Institute, there were indeed Cretaceous rocks on Aegina, but there was no detailed map of the island. We flew to Athens and visited the geoscientists at the Museum there. They didn't know much about Aegina. So, without really knowing where we were going, we took a boat from the port of Athens, Piraeus, to Aegina. The island is one of those in the Saronic Gulf, about 17 miles south of Athens. We landed about noon, and decided to have lunch at an outdoor restaurant while we figured out how we were going to find the site which had provided Ehrenberg with his sample. We spoke no Greek and it seemed that no one on the island spoke any other language. During lunch I noticed a shop with pottery across the street. If it was made locally, the clay used for its manufacture might also have been what Ehrenberg had obtained. After much confused discussion, the shop owners thought we were looking

for a vineyard, we found that the pottery came from the village of Mesagros in the interior of the island. There was a bus that stopped at Mesagros every hour. We took the bus, and sure enough, on the hillside south of the town were clay pits in bluish marl. We selected one with the best-looking exposures and found its owner just ready to send a load to a local pottery. Looking at the clay with a magnifying glass, it was clear that it was rich in microfossils. They turned out to be of Pliocene age, like those at Cattolica-Eraclea.

Both of these rock samples represent deep-sea deposits. How they were uplifted to become part of these islands would become evident many years later when the history of the Mediterranean became known.

As spring arrived, the collecting trips turned to areas to the north. I invited Herbert Hagn, micropaleontologist in the Paleontological Institute at the University in Munich, to accompany me to hunt out classic localities around Paris, and in Belgium and the Netherlands. We planned to hunt down not only the sites where planktonic foraminifera had been described, but also to visit the type localities of Stages of the Tertiary and Cretaceous. He was delighted, and we started off by driving to Paris to consult with colleagues I had already met at the Musée d'Histoire Naturelle. To our distress we were told that the classic localities in the Paris Basin had all been lost. Deteriorated, built over, gone. Fortunately we also had scheduled a luncheon meeting with a priest, Abbe l'Avocat who worked on vertebrate fossils. He told us that the people in the Museum simply had never visited the classic localities. They were all still there. He put us onto an old guidebook that described them. We succeeded in locating every one of the old Stage type localities. Some were indeed off the beaten track, and one turned out to be in a World War I trench that was filled with leaves.

We had a lot of good luck and in Antwerp we stumbled into an area where new port facilities were being dug, exposing geologically young shell beds.

Later in the spring and early summer, I sought out localities in southern Germany and Austria. I made many geological friends that spring and summer; they went on to important academic and industrial positions, and we have remained in close contact ever since.

In his retirement my father was following one of his old interests. He had frequently been to Washington, doing what would now be called lobbying for the life insurance industry. He had been president of the American Life Convention. The 1960 election was coming up and Lyndon Johnson, whom he had known for many years, asked him to work for his nomination. Dad had been looking for a wife to fill the vacuum after my mother had passed away. Now he was dating Perle Mesta, the well-known Washington hostess, and enjoying the Washington social scene. They went to the Democratic National Convention in Los Angeles in July together.

Dad told me that as soon as he arrived in Los Angeles, he discovered that John Kennedy had already tied up the nomination. The Johnson supporters were amateurs compared with Kennedy's organization. Lyndon Johnson was offered the Vice-Presidency. Dad said that Lyndon called him that night to ask what he thought about it. My father's advice was for him to decline the offer. Lyndon Johnson was, after all, the most powerful person in the Congress. Johnson didn't take my father's advice, but they remained good friends.

The 21st International Geological Congress was organized by the five Nordic Countries, with the meetings held in Copenhagen during August, 1960. I attended, along with many of the faculty from the University of Illinois. Before the Congress I traveled to Sweden and Norway, and I participated in a Pre-Congress excursion to Spitsbergen, my first experience in the Arctic. George White had kept one faculty slot in the University of Illinois' Department of Geology open for a Visiting Professor each year. Two of the Congress organizers had been Illinois Visiting Professors. The Illinois Alumni party of the 21st IGC was the largest, and I was duly impressed with my soon-to-be home Institution. After the Congress I participated in a post-Congress Excursion to see the Cetaceous rocks of southern Sweden. It was on that excursion that I met Czech micropaleontologist Vladimir Pokorný, and his student Pavel Čepek, about whom you will hear more later. The ever patient George White had arranged for me to arrive a week late for the start of classes.

I sailed back to New York on the USS *United States*. The most memorable thing about the trip back was that the ship battered its way through every wave. A walk on deck of the *United States* was more of a series of lurches. It turns out that it was built to serve as a troop ship in an emergency, and it was the wrong length for operating in the North Atlantic. My father met me as the *United States* docked in New York the morning before Hurricane Donna struck. That evening we braved the storm, and I met the woman he was going to marry, Nadine Depuisieux. He had gone into Bergdorf-Goodman to buy a bottle of perfume for his former secretary. Nadine was the saleswoman behind the counter. She had just become a US citizen after emigrating from France. She had been born in Russia and their family had emigrated to France after the October Revolution. A day later I was at last on my way to the University of Illinois in Champaign-Urbana.

Over the years I have learned that for a geologist what you have read is not as important as what you have seen. That Postdoctoral year provided me not only with a broad knowledge of Europe and its geology, but with friendships in many countries that have lasted over the years. The mental flexibility to accept new ideas, and cast away old ones, comes from broad experience and encounters with things that are new, different, and unexpected.

THE MUSIC OF WAVES

# The Nature of Energy Received from the Sun—The Analogies with Water Waves and Sound

*The world is full of poetry. The air is living with its spirit; and the waves dance to the music of its melodies, and sparkle in its brightness.*

James Gates Percival

Earlier we said that Earth receives energy from the Sun—**insolation**. Our planet also receives energy from other sources in outer space: moonlight, starlight, and cosmic rays. However, the amounts of energy these bring to the planet are trivial compared with the insolation. The surface of the Earth also receives energy from the planet's interior, but it only averages about $1/10$ W/m$^2$; that is about 1/3,400 the energy received from the Sun, so we don't need to worry about it in our discussion of climate.

We referred to light and infrared 'radiation' as something everyone knows about. We have everyday experience with both; sunlight, mixing all the colors of the rainbow to appear to us as white light, and the heat we feel from a fire transferred through the air to us by invisible rays beyond the red end of the visible spectrum. We know that going to the beach on a summer day can result in 'sunburn.' You learned that it is the balance between incoming and outgoing radiation that determines the temperature of the planet and that greenhouse gases cause the surface to be abnormally warm. Now it is worth some time to understand what radiation is, how you see what you see and how rays of light are converted into heat, which you will remember is motion of atoms or molecules. How all this came to be understood is one of the great triumphs of human intellect and imagination.

Humans often reason by analogy. If we don't understand something, we think of some phenomenon that may be related and that we know something about. Then we explore the possibility that the two processes are related. The analogies with water waves and sound were a critical first step.

In the course of this discussion you will be prepared for an introduction to a world completely different from the one we are familiar with. You may recall that in Lewis Carroll's Alice in Wonderland, Alice's adventure began as she was lounging on a river bank, about to doze off, when she noticed a white rabbit with pink eyes. What caught her attention was that the rabbit withdrew his watch from his waistcoat pocket and said "Oh dear! Oh dear! I shall be late!" Having never seen a rabbit with a waist-coat before, much less one with a watch in his waist-coat pocket, her curiosity was aroused. She followed the rabbit to his hole, entered, and soon found herself falling down a long passage lined with cupboards and bookshelves.

'Lewis Carroll' was the pseudonym of Charles Lutwidge Dodson (1832–1898). Educated at Oxford, England, he became a tutor in mathematics and logic there. The Alice story was developed as an entertainment for the three children of Henry Liddle, the new Dean of Christ Church College, Oxford, during a boating trip on the river on July 4, 1862. Alice Liddle, age ten at the time, was so taken by the story that Dodson wrote it up as a manuscript which he gave to her in 1864. It was further developed as a 'children's book' and published in 1865. There are probably as many ideas about the meaning of the book as there are readers of it. My personal guess is that there was enormous ferment in mathematics and physics at the time (as you will soon see). Faculty of the colleges at Oxford usually dine together (they still do), and I expect Dodson was intimately involved with what was going on in many areas of science and philosophy through the dinner table conversation. It was a time when the scientific community was being flooded with information and definitive answers were missing.

Before arriving at the bottom of the rabbit hole, let's review what we know about waves and particles. We'll start with water waves as something we are all familiar with. They are, after all, the only waves we can observe in nature.

## 7.1 Water Waves

From common experiences we have some definite ideas about waves and particles. The most common waves we can see in everyday life are those on water. Humans have been familiar with waves throughout history. However, being familiar with something does not mean you understand it, and water waves are actually something quite special; they are on the interface between water and air. As you will see, the Pacific Islanders already knew a lot more about waves than their European counterparts during the time of the great

© Springer International Publishing Switzerland 2016
W.W. Hay, *Experimenting on a Small Planet*, DOI 10.1007/978-3-319-27404-1_7

voyages of discovery in the 14th, 15th, and 16th centuries. The first person to make a systematic investigation of them was Sir Isaac Newton in the 17th century. Although he is best known in this field for his work on figuring out the tides, he also discovered the relations between wavelength and wave speed among other things.

If you put some water in the bathtub and then perturb it using your hand or a board, you can generate a wave that will travel down the length of the tub, be reflected off the end and come back. After a few moments, the wave will take on the shape of a smooth curve. It looks very much like a plot of the trigonometric function known as a sine curve, shown in Fig. 7.1

There are three terms not introduced in Fig. 7.1: *Period, T*, is the length of time that elapses between the passage of one wave crest or trough and the next. *Frequency, f*, or Greek letter $v$ (nu), is the number of wave crests or troughs that pass a given point in a unit of time, usually 1 s. A frequency of one wave per second is called '1 Hz,' after Hermann Hertz—more about him later. The period ($T$) is the reciprocal of the frequency ($f$ or $v$); $T = 1/f$ or $1/v$, and vice versa. Another term that crops up is *Wave number, n* or $k$; it is the number of troughs (or crests) in a given distance. It is the reciprocal of the wavelength: $n$ (or $k$) $= 1/L$ or $n$ (or $k$) $= 1/\lambda$. This use of '$n$,' appropriate as it might seem, has been falling out of favor because there are already so many other $n$'s in mathematics. When we come back to wave number near the end of this chapter, I will use $k$ to avoid confusion.

Your bathtub experiment will also demonstrate another property of water waves: they can be reflected off a surface. You may even be able to observe another phenomenon, interference. You will need to use a board to set up the waves, and you can adjust their frequency by moving the board up and down more or less frequently. It won't take long to find the frequency where the wave reflected off the end of the bathtub meets one you just generated and they 'constructively interfere.' The wave doubles in height, and you get a face full of water plus lots of water on the bathroom floor.

If you put your rubber duck in the bathtub, you will discover that it just rides up and down, staying nearly in the same place as the waves move past. The water is not going anywhere. The wave is a disturbance in the water causing it

to move in a circle so that an individual speck in the water appears to move up and slightly forward, then down and slightly back. The motion is passed on from one parcel of water to the next. The wave is 'energy' moving through the water.

To make a wave, you require a 'disturbance' and a 'restoring force.' The disturbance perturbs the material and the restoring force brings it back to its original condition. Let's say you are on the edge of a pond on a calm day when the water surface is smooth, and you disturb the water by dropping a small pebble into it. The disturbance will produce small water waves. If the wavelength is less than 1.7 cm, these will be 'capillary waves;' the restoring force will be the surface tension of the water. Capillary waves also arise when a puff of wind disturbs the surface of smooth water. They roughen the surface and make it possible for the wind to 'get a hold' on the surface of the water and generate larger waves. You can also do this in your bathtub by letting the water become perfectly still and then letting a drop of water fall onto the surface. To see the wave group phenomenon, you need several feet of space, and you might want to damp out the waves reflected from the sides of the tub by using bath-towels draped over the edge of the tub and extending below the water surface.

Figure 7.2a shows a group of capillary waves. If you are very observant when you drop a pebble into the water, you will notice that about 8 waves will form and move away from the impact site in concentric circles. If you watch them you will see a remarkable phenomenon: the waves are a 'group' and if you keep your eye on it, you will notice that the wave at the back of the group moves though it to the front where it disappears. Shortly after that wave left the back of the group a new wave formed there. The number of waves stays the same, but individual waves form at the back and move through the group to die out at the front. The speed of the individual waves is about twice that of the group as a whole. Water waves are really special.

The waves you see on the ocean or a lake are generated by the wind. As the wind blows, disturbing the water and making larger waves, the restoring force becomes gravity, and water waves with a crest-to-crest distance of more than 1.7 cm are termed 'gravity waves.' As the wind blows harder, they lose the sinusoidal shape shown in Fig. 7.1, and take on a special 'trochoidal' shape shown in Fig. 7.2b, becoming pointed at the crests and more smoothly curved in the troughs. This is 'chop' and although it makes for a rough boat ride, it is very important to the way the ocean and atmosphere exchange gases. These pointy waves create spray and the droplets evaporate, leaving behind minute grains of salt in the air that can later serve as nuclei for raindrops. When the wind blows the top off these waves, they trap and carry bubbles of air down into the seawater, perhaps to a depth of 2 m (6 ft) or more. There the gases in

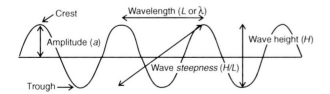

**Fig. 7.1** Terminology for describing water waves

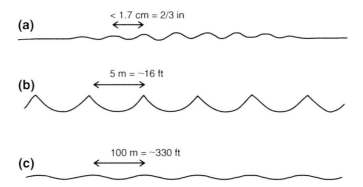

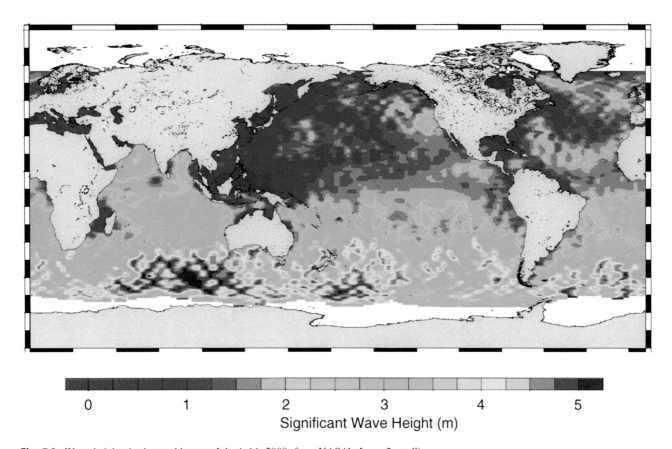

**Fig. 7.2** Water waves. **a** Capillary waves resulting from a disturbance of the water, such as a pebble dropped into it off the *left side* of the diagram. The restoring force is surface tension. **b** With strong disturbance by the wind, the waves become larger and may take on the trochoidal form shown here. The restoring force is gravity. **c** Ocean swell with long wavelengths and low amplitudes. Tsunami waves generated by earthquakes or submarine landslides have this same form but the wavelengths in the open are many kilometers. Again, the restoring force is gravity

the air are under pressure and dissolve more readily in the water. For this reason, oxygen is actually supersaturated in the surface layer of the ocean. These waves, like the capillary waves, travel in groups, but often there are multiple sets of waves intersecting each other to produce a confused 'sea.' Fig. 7.3 is a map of typical wave heights in the world ocean in July.

Major storms generate large waves. As these waves spread out over the ocean, their wavelength increases and heights decrease, as shown in Fig. 7.2c. In deep water, the speed of a wave increases with wavelength, and these waves, better known as 'swell,' can fan out across entire ocean basins. They outrace waves with shorter wavelengths. The Pacific is a special case. Storms around the Antarctic

**Fig. 7.3** Wave heights in the world ocean July 4–14, 2008, from NASA's Jason-2 satellite

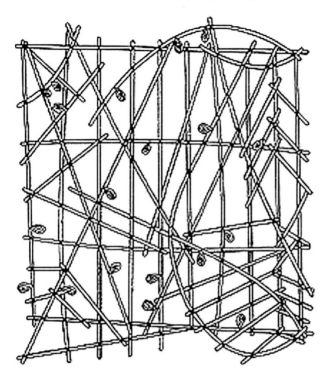

**Fig. 7.4** A Polynesian stick map. The *sticks* represent ocean swell; the shells islands (From National Geographic Society Map: Discoverers of the Pacific, 1974.)

generate swell that goes all the way to the northern coasts. In particular, the coastline of Kamchatka is linear enough that the Antarctic swell is reflected off it and moves out as swell with a new orientation across the entire basin again. If you know what you are doing, you can look at the different swells and orient yourself. Figure 7.4 shows a Polynesian 'stick map' of part of the Pacific; the sticks represent swell and the shells islands. Maps of this sort were used to navigate across vast distances of the Pacific. The bend in the Atlantic around Africa prevents Antarctic swell form getting into the North Atlantic, so until very recently Europeans and their North American cousins did not understand how to use this system of navigation.

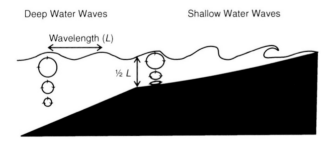

**Fig. 7.5** Motion of the water in deep and shallow water waves. In deep water waves, the motion of the water is circular with the diameter of the *circles* decreasing with depth. Shallow water waves are those where the water depth is less than ½ wavelength, and the water motion becomes elliptical or simply back and forth on the sea bed

Water waves can be classified into two kinds: deep-water waves and shallow-water waves. If the water is deeper than half a wavelength, it is a deep-water wave, and its speed increases proportionally to the wavelength. If the water depth is less, it is a shallow-water wave; its speed is controlled by the water depth and decreases as the water shallows. As shown in Fig. 7.5, deep-water and shallow-water waves behave quite differently. As an ocean wave approaches the shore, the transition from deep- to shallow-water wave affects the wave trough first. It is affected by the bottom before the crest is. The trough slows down, while the crest keeps its original speed. As the trough slows more, the water in a trough is pushed underneath the following crest, making it higher. This is a positive feedback effect, so that the wave becomes higher and steeper. Finally, the crest outruns the trough and the wave 'breaks.'

## 7.2  Special Water Waves—Tides and Tsunamis

There are other waves on the ocean you cannot see directly—the tide and tsunamis. Because of their great wavelengths both the tides and tsunamis are shallow-water waves. The tides are generated by the gravitational attraction of the Moon and Sun and have wavelengths of thousands of kilometers. The larger component of the tide is caused by the Moon and has a period of 12.4 h, half the time it takes for the Moon to return to the same place in the sky. The solar component has a period of 24 h. The tide waves interfere with one another, constructively and destructively, are retarded as the water shallows, and can be reflected if conditions are right.

Tsunami is a Japanese word. Tsunamis (often incorrectly called 'tidal waves' even though they have nothing to do with the tides) are generated by earthquakes or submarine landslides and have wavelengths of a few hundred kilometers and periods of five minutes to an hour or more. However, since the ocean is several kilometers (a couple of miles) deep, tsunami wave groups travel 500–700 km (300–400 miles) per hour. Again, their speed depends on water depth, and their progress across the Pacific becomes a complex interference pattern. Today, the arrival times and magnitudes of major Pacific tsunamis are shown on TV or the Internet as they occur, so that the public can take precautions. Their amplitude in the open ocean is usually only a meter or less, but when they enter shallower areas, the wave slows down and piles up on itself as the bottom of the wave moves slower than the top. Because of this effect, they can be 10s of meters high when they come on shore. The tsunami generated by the Great East Japan Earthquake (also known as the Tōhoku or Sendai earthquake) of March 11, 2011, had a maximum height of 38 m (125 ft). The wave trough arrives first, and the water seems to be sucked away from the shore, only to be followed minutes later by the crest which can run several kilometers (a mile or more) inland. Because of the long wavelength, the surge of water just keeps coming in for a long time; it can last 15 min or more. It used to be thought that the first tsunami wave was always the largest, but from the effects of the 2011 Japan tsunami in distant parts of the Pacific, we know that is not true. The second or even the third wave can be the largest. The 2011 Japan tsunami was not the highest ever recorded. An earthquake generated landslide into Lituya Bay, Alaska, during the night of July 9, 1958, generated a local tsunami 524 m (1,720 ft) high. Fortunately, the tsunami was largely trapped in the enclosed bay.

The tsunami warning system for the Pacific works well because the bathymetry of the ocean basin is well known, hence the speed of the wave from the earthquake source and its arrival times can be readily determined. Figure 7.6 shows the predictions for the 2011 Japan tsunami.

Remote sensing by satellites now detects the height of the wave in the open ocean, allowing refinement of the prediction of how large it will be when it comes onshore. None of this information was available for the December 26, 2004 tsunami in the Indian Ocean, which came as a surprise to many areas and resulted in up to 230,000 deaths. Tsunamis in the Indian Ocean are less frequent than those in the Pacific, so it was not until after the 2004 disaster that an effective prediction and warning network was set up.

Why are tsunamis often called 'tidal waves' in English? Simply because they have such long wavelengths, they are more like the tides than ordinary water waves.

Previously, we noted that what is being transported by waves is energy. Although the wind that produces waves may also induce some surface current flow in the same direction, that is a different matter. What is the energy in the waves? It is in two forms: kinetic energy, which is the circular or, in shallow water, elliptical movement of the water;

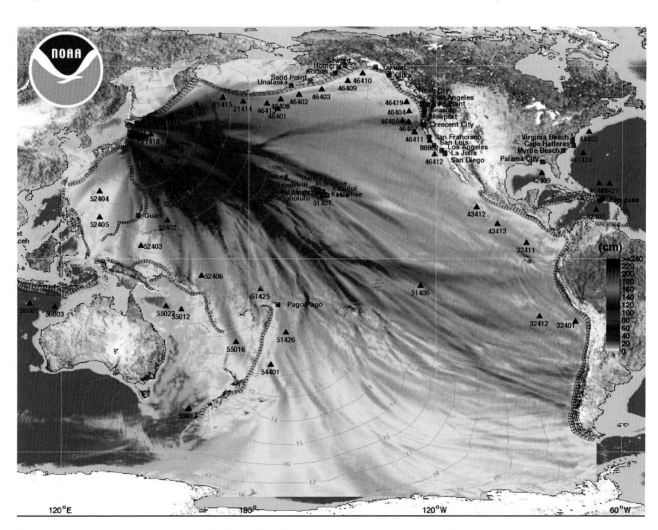

**Fig. 7.6** Wave height forecast for the Pacific Ocean from the Great East Japan Earthquake (also known as the Tōhoku or Sendai earthquake) of March 11, 2011. Because of its great wavelength the tsunami behaves as a shallow-water wave, affected and deflected by the sea-floor topography. *Black triangles* are sea-floor wave height monitoring stations. *Red squares* are onshore monitoring stations. These stations allow the predictions to be corrected as the tsunami proceeds. Ocean trenches are also shown

and potential energy, which is the displacement of the water from its mean position. The potential energy for a unit area of a wave is

$$E = \frac{1}{8}\rho g H^2$$

where $E$ is the Energy, $\rho$ (Greek letter *rho*) is the density of the water, $g$ is the acceleration due to Earth's gravity (9.8 m/s$^2$), and $H$ is the wave height. Obviously, in a given situation neither the density of the water or the acceleration due to gravity are going to change significantly, so the wave energy turns out to be proportional to the *square* of the wave height. That is, a wave with a height of 4 units (meters or feet or whatever) has 16 times as much energy as a wave with a height of 1 unit. This is why big waves coming on shore, such as during a storm, or a tsunami, are so destructive. While Hurricane Katrina was still in the Gulf of Mexico, it produced waves 19 m high in a region where waves 2 m high or less are normal. Those waves carried 90 times the energy of normal waves in the region, and individual waves were much higher. We know this because a few years earlier the Gulf of Mexico had been instrumented with pressure sensors on the sea floor.

Now think about this. If a tsunami wave in the open ocean is only a meter or so high, how can it carry so much energy? The equation above left out one important term. It described the energy in a wave of 'unit area.' The area of the wave is its length times its width. An ordinary wave coming on shore has a wavelength of 10 m ($\sim$30 ft) or so, but a tsunami has a wavelength of 100–500 km (60–300 miles) so the energy carried by it is 10,000–50,000 times as much as an ordinary ocean wave.

## 7.3  Wave Energy, Refraction and Reflection

How does the transport of energy work? You will recall that waves travel in groups, with new waves being formed at the rear of the group, moving through it, and dying out at the front. The energy travels with the group, at half the speed of an individual wave.

Water waves can be reflected off a smooth vertical surface, like a channel bulkhead, with almost no loss in energy, but more and more of the energy will be dissipated as the slope of the surface becomes flatter or the surface rougher. Ship channels are usually bordered by huge rocks to prevent reflection of the waves, but many waterfront communities have channels with vertical bulkheads and lots of DEAD SLOW signs.

Now suppose that a wave approaches the shore obliquely, as most ocean waves do. As the wave enters shallow water it slows, and bends, as shown in Fig. 7.7. The shallower the water, the slower the wave, so that at the shore itself, most waves arrive parallel to the shore. The slight oblique motion

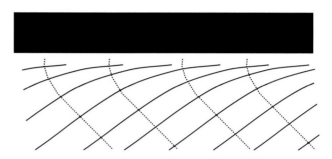

**Fig. 7.7** Waves approaching a shore obliquely, demonstrating gradual refraction

they still have means that they run up the beach at a slight angle, but recede straight down the sloping beach. As they do this, they gradually move the sand grains along the beach, causing longshore drift. The energy that was in the wave goes into moving the sand grains. This can be a phenomenon of great importance in shaping shorelines. The quartz sand beaches that extend along the east coast of Florida all the way south to Miami are made of quartz sand that was originally delivered to the Georgia coast from the Appalachians and carried south over 600 km (360 miles) by longshore transport. Today the longshore transport on the east coast of Florida is interrupted by a number of artificially cut inlets to harbors. The waves move the sand along shore to the inlet, and there they move it inward in the channel to fill the harbor. To those who live along such a coast, this is an example of the perverseness of nature.

As shown in Fig. 7.8, water waves demonstrate a feature that other kinds of waves also show. If you look at the support pilings of a pier or dock when there are only ripples on the water surface, you will see that the pilings block their progress, although they may curve in and come together again a short distance after passing them. On the other hand, if the waves are longer, they will seem to pass straight through as though the pilings were not there. If a piling is less than half a wavelength in diameter, the wave doesn't seem to know it is there.

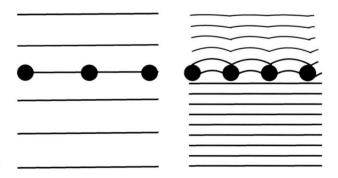

**Fig. 7.8** Interaction of water waves and pilings. If the piling is less than ½ wavelength across, the wave does not know it is there

In conclusion, water waves are instructive and wonderful to watch, but, as we shall see, they are very, very different from either sound or light.

## 7.4  Sound Waves

Now, let's consider how sound makes its way through the air. It had long been obvious that 'air' was something physical. You could feel it when the wind was blowing. Leonardo da Vinci is generally credited with the discovery that 'sound' is compression/rarefaction waves in air. In his *Notebooks*, written in the later part of the 15th century, the section on acoustics describes the effect of a cannon, with the explosion obviously compressing the air around it, and the effect of the vibrations of a bell after it has been struck being transferred to the air. Sound is waves of compression and rarefaction of the air spreading out along spherical fronts from the source. The waves were thus a longitudinal phenomenon. The pitch is related to the frequency of the waves. In the early days of making organs it became obvious that the pitch of an organ pipe was directly related to its length. Later it was discovered that the sound's wave- length was double the length of the pipe. By the 17th century it was known that the character of the sound of an organ pipe, the timbre, can be modified by different kinds of obstructions placed at the end of the pipe. Most of the great old European organs have an enclosure concealing the tops of the pipes so that these details could not be seen and copied.

Musical instruments do not generate pure sounds of a single wavelength. The note you hear is the fundamental frequency, the one with greatest amplitude; i.e. the loudest, but a musical instrument also produces waves of other frequencies. Most of them are integer multiples of the fundamental frequency, and are called harmonic frequencies. The resulting combination is called the timbre, or tone, or simply the sound quality of an instrument. Different instruments produce different levels of harmonic frequencies depending on what they are made of and their form, so they sound different. Brass instruments tend to produce odd-integer harmonics while woodwind instruments tend to produce even integer harmonics. String instruments are special because the sound depends on whether the string is plucked (guitar, harpsichord), hit with a hammer (piano), or continuously vibrated by another string on a bow (violins, cellos, etc.).

As you will see later, in the next chapter and Chap. 15, 'light' that interacts with molecules of air, water, or solids, can produce ancillary wavelengths different from the original.

## 7.5  Sound Waves and Music

The fact that vibrating strings would produce sound was known since ancient times. By the early 15th century the rules of sound related to vibrating strings and how the sound could be modified by a sound box were very well understood. The heyday of violin-making in Cremona, Italy—with Nicolò Amati (1596–1684), Antonio Stradivari (1644–1737), and Bartolomeo Giuseppe Guarneri (1698–1744)—was in the 17th and early 18th centuries. Furthermore, composers understood that sound took time to go from the source to the ear, and that it would be reflected off flat surfaces. Their music took reverberation times in large structures, such as cathedrals, into account. Mozart's Coronation Mass (Mass No. 15 in C major, KV 317) has wonderful deliberate pauses for the reverberation in the Salzburg Cathedral to become an integral part of the music. Composers also understood how sounds of different frequencies could combine to produce both pleasant and unpleasant effects (see Fig. 7.9).

Incidentally, in the middle of the 17th century Robert Boyle had built an air pump and discovered among other things that sound did not travel through a vacuum, although light did. Light passed unimpeded though the vacuum in the top of Torricelli's glass barometer tube. Sound and light were clearly different.

That sound traveled at a finite speed was evident from the experience of seeing something happen and hearing it a short time later. Even Aristotle (384–322 BCE) had concluded that sound involved some sort of motion of the air, but he also concluded that high notes traveled faster than low notes. I have a recording of a reconstruction of what Greek music of the time may have sounded like. It is rather strange to modern ears, to say the least, and may make it easier to understand why he made this mistake. Again, nobody's perfect.

If sound waves of different frequencies traveled at different speeds, symphonic music would be an incredible mess, with the high-pitched sounds from the violins arriving before the lower sounds from the cellos and basses. Imagine what one of Wagner's operas would sound like with the soprano bellowing out her high notes and the rumble intended to accompany it coming seconds later. As music developed, it was evident that all sound travels at the same speed, regardless of pitch.

The 'noise-canceling headphones' that many of us use in airplanes work by electronically sensing the frequencies of the sound waves reaching us and producing a set of sound waves offset exactly ½ wavelength, so that there is destructive interference and the external sound is canceled

**Fig. 7.9** Schematic representation of sound as compression/rarefaction waves. **a** A high frequency note (Middle C if the record is 1/10 s long). **b** A lower frequency note with a wavelength about twice as long (this would actually be C# in the next lower octave). **c** A combination of the two, showing the phenomenon of 'beats' as the two sets of sound waves alternately constructively and destructively interfere with one another

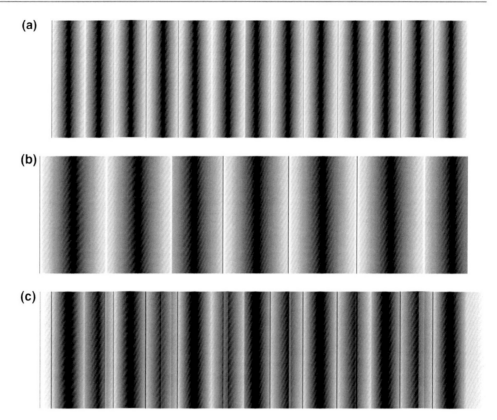

out. The same principle is used by baffles alongside high-speed highways in the United States and particularly in Europe. The sound bouncing off the baffles partly cancels out the noise of passing automobiles. In Europe some of these baffles are partly arched over the roadway, so the sound is reduced even on the highway itself.

There are many different musical scales. Doubling of the frequency of a sound produces a note an 'octave' higher. It is called an octave because in western music it is the 8th note that is double the frequency of the 1st note. Other musical scales have different numbers of notes between the doubling frequencies, but the doubling of the frequency is something the human brain seems to recognize as a reference.

The only musical note which has a more or less internationally recognized frequency is the 'A' above 'Middle C' on the piano. Its more or less accepted standard is 440 cycles per second (=440 Hz), but in fact different orchestras use different frequencies as their standard. In 1884 Giuseppe Verdi tried to get the Italian Music Commission to settle on 432 Hz for A. That makes Middle C an even 256 Hz. It is thought that this was the tuning used by Mozart, and certainly by Verdi. In 1939 Nazi Propaganda Minister Josef Goebbels tried to get international agreement on the more popular 440 Hz, but that failed. Swiss orchestras tune to A = 442, most German and Austrian Orchestras to A = 443 and Herbert von Karajan had the Berlin Philharmonic tune to A = 444. You may ask, what difference does it make?

Strange to say, it changes what one might call the mood of the music. I attended the Norbert Brainin/Günter Ludwig C = 256 concert in Munich on December 12, 1988, with the same pieces played first in A = 440, then the Verdi tuning of A = 432; C = 256. A recording of the concert can be found on the Internet. The difference is remarkable. The leader of the new movement to return to A = 432 is none other than perennial US Presidential candidate Lyndon LaRouche!

But wait. Organ pipes cannot be tuned; the wavelength they produce is half the length of the pipe. In organs Middle C is generated by a 2 ft pipe (measured in imperial feet from the mouth to the top of the pipe). Two feet is half the wavelength of the note that is generated. At 20 °C (= 68 °F) the frequency of the note is 256 Hz, so the A is 432, the Verdi tuning. The standard lengths of organ pipes were established before the metric system was introduced, and the length of the foot was slightly different in different countries; and of course, the frequency (pitch) will depend on the temperature and humidity. Table 7.1 shows the frequencies and wavelengths of sound waves for some notes over the ten octaves most humans can hear.

With a speed of sound at 340.29 m/s, the wavelength of waves with frequency of 440 is 0.773 m (2.536 ft). But the speed of sound changes a lot with temperature, slightly with humidity, and not at all with changes in air pressure. The often cited 'speed of sound at sea level' of 340.29 m/s assumes a temperature of 14 °C (57.2 °F), a relative

**Table 7.1** The common Western musical scale in terms of notes and their frequencies, with wavelengths based an A = 440 Hz, calculated for the speed of sound being 340.29 m/s

| Octave | Note | Frequency (Hz) | Wavelength (m) | Wavelength (ft) |
|---|---|---|---|---|
| 0 | C | 16.351 | 20.812 | 68.281 |
| 0 | A (Lowest note on piano) | 27.500 | 12.374 | 40.597 |
| 1 | C | 32.703 | 10.405 | 34.137 |
| 1 | E | 41.203 | 8.259 | 27.096 |
| 2 | C | 65.406 | 5.203 | 17.070 |
| 2 | E (Lowest note on guitar) | 82.407 | 4.129 | 13.547 |
| 3 | C | 130.813 | 2.601 | 8.533 |
| 3 | G (Lowest note on violin) | 195.998 | 1.736 | 5.696 |
| 4 | C ('Middle C') | 261.626 | 1.301 | 4.268 |
| 4 | C#/Db | 277.183 | 1.228 | 4.029 |
| 4 | D | 293.665 | 1.159 | 3.802 |
| 4 | D#/Eb | 311.127 | 1.094 | 3.589 |
| 4 | E | 329.628 | 1.032 | 3.386 |
| 4 | F | 349.228 | 0.974 | 3.196 |
| 4 | F#/Gb | 369.994 | 0.920 | 3.018 |
| 4 | G | 391.995 | 0.868 | 2.848 |
| 4 | G#/Ab | 415.305 | 0.819 | 2.687 |
| 4 | A (Tuning reference) | 440.000 | 0.733 | 2.536 |
| 4 | A#/Bb | 466.164 | 0.730 | 2.395 |
| 4 | B | 493.883 | 0.689 | 2.260 |
| 5 | C | 523.521 | 0.650 | 2.133 |
| 6 | C (high C for a soprano) | 1046.502 | 0.325 | 1.066 |
| 7 | C | 2093.005 | 0.163 | 0.535 |
| 8 | C (highest note on a piano) | 4186.001 | 0.081 | 0.266 |
| 9 | C | 8372.018 | 0.041 | 0.135 |

This spans roughly the full range of human hearing, although that varies from individual to individual

humidity of 53 %, and a pressure of 101.325 kPa (= 29.921 in. of Hg, = 1 standard atmosphere). The speed of sound is very sensitive to temperature, so at 35 °C (=95 °F) and the same relative humidity and pressure it is 353.65 m/s. The corresponding frequency produced by a foot long organ pipe is 455, about halfway up to A# or B$^{b}$. The temperature effect is why orchestras tune their instruments just before, and sometimes during, a concert.

If Chinese, Japanese, Indian, Indonesian, and Arab music sounds odd to you, it is because they use different scales and may have greater or lesser differences in frequencies between notes.

As noted previously, sound is compression and rarefaction of the air. As compression occurs, more air molecules strike one another, and the air warms; as the decompression occurs, the opposite happens. The process is largely 'adiabatic,' that is heat is neither gained from nor lost to the environment. But while this is true for the individual sound waves, the overall sound is gradually dissipated, and its energy is lost to the air as increased random motion of the molecules. This is always a problem with energy; some of it is always being lost as random motion imparted to the atoms and molecules of the environment. It can never be recovered as useful energy again. This loss of energy to random motion is what scientists call 'entropy,' a great term to throw into the middle of cocktail party conversation. Entropy is simply a measure of the amount of disorder in a system. On the grandest scale one can say that the amount of entropy in the Universe as a whole is always increasing.

Another interesting property of sound is the change in pitch due to the Doppler Effect. If a car or train is coming toward you, honking its horn or blowing its whistle, the pitch is higher than when it passes you, and as it goes away the pitch is lower. The effect was described in 1842 by Austrian physicist Christian Doppler (1803–1853) who had just been appointed to a position in the Prague Polytechnic Institute. He published a paper entitled *Über das farbige Licht der Doppelsterne und einiger anderer Gestirne des*

*Himmels = On the colored light of the binary stars and some other stars of the heavens.* He argued that the color of a star depended in part on whether the star was moving toward us or away from us. This would later be termed the 'red shift' and ultimately used to measure the distance to distant objects in the sky. The application to sound came in 1845, when Dutch geologist—later chemist, meteorologist and physicist—Christophorus Henricus Diedericus Buys Ballot (1817–1890), better known just as Buys Ballot, proved that the Doppler Effect also applied to sound waves. He had a group of musicians play a specific note on a train between Utrecht and Amsterdam and observed the sound by standing alongside the track. What happens is that as the sound source moves toward you the waves are compressed, that is the wavelength shortens and you hear a higher pitch. When the sound source moves away the waves are stretched out, the wavelength lengthened and the pitch lowered. The speed of the sound does not change.

## 7.6   Measuring the Speed of Sound in Air

You might think it would be easy to measure the speed of sound, but you need to be able to see the sound source, know how far away it is, and you need a device that can measure short intervals of time accurately. For reference in the following discussion, the modern value for the speed of sound in dry air at sea level with a temperature of 20 °C (68 °F) is 343.2 m/s (1,126 ft/s). This works out to be 1,236 km/h (768 mph), or approximately 1 km in 3 s or 1 mile in 5 s. This is useful information if you wish to know how far away a storm is, When you see a lightning flash count the seconds until you hear the thunderclap an multiply by the speed of sound (on a warm day: ~350 m/s or 1,150 ft/s). A few minutes later you can repeat this and determine whether the storm is moving toward or away from you.

After explosives were introduced to Europe from China, it was possible to see the flash of the powder and then hear the explosion. Although Galileo seems to have attempted to determine the speed of sound, the only readily available timing device for short intervals was the human pulse rate. For a normal, healthy fit person at rest, it averages about 60/min. This was probably the origin of the time unit 'second' in Babylonian times. Remember that the Babylonians used a sexagesimal number system (based on the number 60). This is why there are 60 s in a minute and 60 min in an hour.

Portable clocks using a spring regulated by an escapement mechanism were developed in the 16th century, but were not very reliable; they might gain or lose an hour or two or three per day, and they only had an hour hand. Gentlemen wore them around their necks as a sort of jewelry.

The first measurement of the speed of sound is attributed to Pierre Gassendi in Paris in 1635. His estimate was 1473 French pieds/second, (pied = French, not English feet) which works out to be 478.4 m/s (1569.6 ft/s). Another French scientist Marin Mersenne (1588–1648) (Fig. 7.10), often considered to be the father of 'acoustics,' revised this to about 450 m/s (1476.4 f/s) as published in his 1636 work on music: *Harmonie Universelle*. Finally, in 1656 Giovanni Alfonso Borelli (1608–1679) of Pisa and Vicenzo Viviani, who had been a student of Galileo and Torricelli, made a very precise measurement of the speed of sound based on the flash from a canon as equivalent to 350 m/s (1148 ft/s), quite close to the modern value given above.

Incidentally, Marin Mersenne condemned astrology and was a defender of Galileo and of science in general (rather dangerous positions for a priest to take at the time).

There could be a number of reasons why these early estimates of the speed of sound were so different: air temperature, humidity, the wind, and other physical factors. But the biggest problem was probably the assumption that the pulse rate is the same for different individuals and that it doesn't vary except with extreme exercise. Because portable spring-loaded timekeeping devices were so unreliable, the human pulse rate was used as a measure for short periods of time well into the 19th century. Many of the older

**Fig. 7.10** Père Marin Mersenne, who wrote the important treatise on music *Traité de l'harmonie universelle*, published in 1636, which includes a lot of information about the mathematical relationships between notes

**Fig. 7.11** Sir Isaac Newton (1643–1727). His own hair, no wig. He is honored with the modern unit of force, the 'newton.' One newton is the force required to accelerate a mass of 1 kg at a rate of 1 m/s (kg/ms$^2$)

realized that as the sound wave compressed the air, its temperature would rise; and as the rarefaction occurred, the temperature would go down. Taking this adiabatic process (adiabatic = without gaining heat from or losing heat to the surrounding environment) into account, he figured that Newton's value needed to be corrected to 337.7 m/s (1108 ft/s). Many English scientists of the time were furious about this simple solution to an old problem and even ridiculed Laplace—he was, after all, French, and the French had gone through that awful revolution, so how could they be trusted to get things right?

Table 7.2 summarizes data for the speed of sound in air at different temperatures, humidities, and pressures which, of course, vary with elevation. Note that the temperature effect is by far the largest, and the elevation (pressure) effect is negligible. However, temperature usually varies with elevation, and is called the 'lapse rate'—typically declining about 1 °C with every 100 m increase in elevation when the air is dry. The lapse rate will be discussed in detail later. The air pressure at my home, at an elevation of 8,200 ft (2,500 m) is typically 2/3 of that at sea level. The average annual temperature is 36.3 °F (2.4 °C). Washington D.C. lies at about the same latitude but is almost at sea level. There the average annual temperature is 57.5 °F (14.2 °C). If the air

observations of cannon fire were made during actual combat in warfare (gunpowder was expensive), so it is likely that the observer's pulse rates may have varied a lot depending on whether they were doing the shooting or were being shot at.

Isaac Newton (Fig. 7.11) in the second volume of his *Philosophiæ Naturalis Principia Mathematica*, published in 1687, produced an analytical derivation of the speed of sound taking into account the compression/rarefaction of the air. He also made observations of the time of return of an echo in one of the cloisters at Trinity College in Cambridge. Like his other work, it was a very sophisticated mathematical approach to the problem, but his answer, 968 ft/s (295 m/s) was wrong. In the second edition of the books, published in 1713, this was revised to 979 ft/s (298 m/s), still wrong. The problem of making a match between the analytical derivation of the speed of sound and the experimental data vexed the scientific community for an entire century.

In 1821, Pierre-Simon Laplace (Fig. 7.12), another French scientist, solved the problem. Newton had used Boyle's Law in making his calculations, but had overlooked that little phrase 'other things remaining constant.' Laplace

**Fig. 7.12** Pierre-Simon, Marquis de Laplace (1749–1827). His name is honored in mathematical terminology

**Table 7.2** The speed of sound in air at different temperatures, humidities, and elevations (pressures)

| Temp (°C) | Temp (°F) | Humidity (%) | Elevation (m) | Speed (m/s) | Speed (ft/s) | Where |
| --- | --- | --- | --- | --- | --- | --- |
| −20 | −4 | 60 | 5,000 | 319.21 | 1047.3 | Air–Mountain |
| −20 | −4 | 60 | 1 | 319.09 | 1046.9 | Air–Sea level |
| 0 | 32 | 60 | 1 | 331.64 | 1088.1 | Air–Sea level |
| 10 | 50 | 60 | 1 | 337.85 | 1108.4 | Air–Sea level |
| 20 | 68 | 60 | 1 | 344.12 | 1129.0 | Air–Sea level |
| 30 | 86 | 60 | 1 | 350.55 | 1150.1 | Air–Sea level |
| 10 | 50 | 0 | 1 | 337.46 | 1107.2 | Dry air–Sea level |
| 20 | 68 | 0 | 1 | 343.36 | 1126.5 | Dry air–Sea level |
| 30 | 86 | 0 | 1 | 349.15 | 1145.5 | Dry air–Sea level |

The air pressure at 5,000 m is half that at sea level. Note that the temperature effect is by far the largest, and the elevation (pressure) effect is negligible

were dry in both places the temperature difference would be greater.

## 7.7 Measuring the Speed of Sound in Water

The speed of sound in water was first measured in 1826 by two Swiss, Jean-Daniel Colladon (1802–1893), a physicist, and Charles-François Sturm (1803–1855), a mathematician. They were experimenting with sound in the waters of the Lake of Geneva. They contrived to ring a bell under water. They noted that they could hear the bell until they were about 200–300 yards away from it. Beyond that they could no longer hear it. It seemed that the sound was being reflected back down into the water from the water-air interface. Then they built a device that rang the bell under-water while simultaneously firing off some gunpowder above the water. This device was set on one boat, while one of them became an observer on another boat about 15 km (9 miles) away, listening through a tin cylinder pushed down into the water (Fig. 7.13). The cylinder was 3 m (yards) long, 20 cm (8 in.) in diameter, and closed off at the bottom, so that it remained filled with air. The observer had a stop-watch calibrated to 1/4 s. The water temperature was 8 °C (46.4 °F). They calculated the speed of sound to be 1435 m/s (4708 ft/s). They also discovered that the nature of the sound had changed, describing it as more like two knife blades hitting each other.

Table 7.3 shows the speed of sound in waters of different salinities and at different depths. However, the distances sound can travel in the sea depend on the wavelength and what is in the water. The shorter wavelengths can be scattered by sediment, plankton, and fish in the water and die out sooner than the longer wavelengths. The longer wavelengths can travel many thousands of kilometers, which is why low-pitched whale songs can travel so far.

## 7.8 The Practical Use of Sound in Water

During World War II much more was learned about sound in the ocean. A typical value for the speed of sound in seawater is 1500 m/s (∼4921 ft/s), but it increases with the temperature, the salinity (saltiness) of the water, and with the increasing pressure with depth. The structure of the ocean will be covered in much more detail in Chap. 25, but in the low and mid-latitudes there is a relatively thin layer of warm surface water representing about 10 % of the ocean volume, and a very large mass of cold deep water with a salinity of 34.5 ‰ (parts per thousand) making up about 80 % of the ocean volume. Between these two is the ocean thermocline, with the temperature transition between the warm surface waters and the cold deep water. Figure 7.14a shows a temperature-depth profile typical of the mid-latitude ocean. Figure 7.14b shows how the speed of sound varies in such a profile.

One of the major discoveries was that there is a region of minimal sound velocity (∼1480 m/s; ∼4856 ft/s) usually between 500 and 1000 m below the surface (1630–3280 ft; 280–560 fathoms). The speed of sound increases with depth in the warm water layer mixed by the winds above this depth. In this surface mixed layer, the pressure effect dominates. Then, in the thermocline, the temperature effect dominates, so that the speed of sound decreases to a minimum near its base. In the cold deep ocean, the speed of sound simply increases with depth due to the increasing pressure. Figure 7.14b shows these changes in sound velocity with depth.

Because the speed increases both above and below the minimum near the base of the thermocline, sounds generated at that level that go toward the surface or ocean depths are refracted back into it. It acts as a waveguide and sound trapped in it can travel enormous distances, even from the Atlantic into the Pacific Ocean. Known as the SOFAR

**Fig. 7.13** Jean-Daniel Colladon and Charles-Francois Sturm measuring the speed of sound in water in the Lake of Geneva, 1826

**Table 7.3** Comparison of speeds of sound in air, fresh water (lakes) and seawater under varying conditions of temperature, humidity, salinity and elevation/depth

| Temp (°C) | Temp (°F) | Salinity (‰) | Depth (m) | Speed (m/s) | Speed (ft/s) | Where |
|---|---|---|---|---|---|---|
| 30 | 86 | 0 | 1 | 1,507 | 4,944 | Fresh water |
| 20 | 68 | 0 | 1 | 1,481 | 4,859 | Fresh water |
| 10 | 50 | 0 | 1 | 1,447 | 4,747 | Fresh water |
| 0 | 32 | 0 | 1 | 1,403 | 4,603 | Fresh water |
| 4 | 39.2 | 0 | 200 | 1,425 | 4,675 | Fresh water |
| 20 | 68 | 5 | 1 | 1,488 | 4,882 | Coastal lagoon |
| 20 | 68 | 10 | 1 | 1,493 | 4,898 | Marginal sea |
| 20 | 68 | 20 | 1 | 1,505 | 4,938 | Marginal sea |
| 20 | 68 | 35 | 1 | 1,521 | 4,990 | Ocean |
| 5 | 41 | 35 | 1,000 | 1,487 | 4,879 | Ocean |
| 5 | 41 | 35 | 2,000 | 1,503 | 4,931 | Ocean |
| 5 | 41 | 35 | 3,000 | 1,521 | 4,990 | Ocean |
| 5 | 41 | 35 | 4,000 | 1,539 | 5,049 | Ocean |

(Sound Fixing and Ranging) Channel it has been used for underwater communications. Whales have known about this for millions of years, and often sing their songs in the SOFAR channel, shown in Fig. 7.15.

Late in the 20th century oceanographers realized that if the Earth is warming, this will eventually be reflected as a rise in the temperature of the deep ocean waters. Because of its size and huge thermal capacity it is the most stable climate monitor on the planet. Any change in the temperature of the deep ocean water would be extremely significant. It is impossible to deploy enough thermometers to detect a change in temperature, but a change in the velocity of sound could be determined. But the deep ocean waters are not still; there are deep ocean currents and eddies. To overcome this 'noise' the velocity of sound is being measured along a

number of different paths across the ocean basins. The experiment is called ATOC, for Acoustic Thermometry of Ocean Climate. The prototype used a low-frequency sound source off Kauai in the Hawaiian Islands and receivers near the continents in the North Pacific. It immediately ran into problems because humpback whales were disturbed by the signals. It operated from 1996–2006, and has been replaced by more refined systems, including one in the Arctic Ocean. Analysis of the decade of data showed that the noise problems were larger than anticipated but suggests warming in the order of a few hundredths of a degree. That may sound like a small number, but the heat capacity of the ocean, that is, its ability to absorb and retain heat, is more than 1,000 times that of the atmosphere. So a tiny change in the temperature of the deep ocean reflects a major change in the

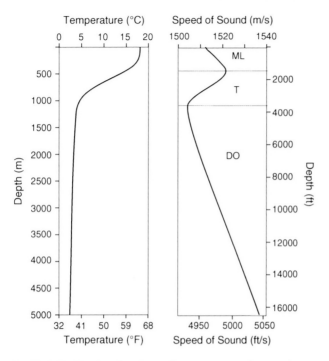

**Fig. 7.14** Profiles showing changes in temperature and speed of sound at an ocean site in the mid-latitudes. **a** Temperature versus depth. **b** Changes in the speed of sound with depth. *ML* surface ocean 'mixed layer' (water mixed by the winds). *T* Thermocline (the region of sharp decline of temperature with depth). *DO* Deep Ocean

Earth's climate state. We used to think that the temperature of the deep sea did not change appreciably from glacial to interglacial climates, but we now know that during the interglacials the deep sea warms about 4.6 °C (=8.3 °F).

Finally, if you are an older person, you may have put your ear to the rail of a railway line to listen if the train is coming. The speed of sound in an iron rail is about 5120 m/s (3.18 miles/s), 15 times as fast as in air, so you get plenty of advance notice.

In summary, sound is a compression-rarefaction wave. It requires an elastic medium to travel through. This includes solids, liquids, and gases if they are not highly rarefied. The speed of sound increases strongly with temperature. Although in air the speed of sound varies little with pressure, the opposite is true in the ocean. Sound can be reflected from surfaces, and refracted at interfaces or as the density of the medium changes. The different wavelengths of sound all travel at the same speed.

## 7.9 Summary

In this chapter we have covered two topics, which may have seemed unrelated to our goal of understanding climate change. We started by looking at water waves. They are the only waves that we can see and experience. We learned that the wave is something that moves through but does not actually transport the water. Even stranger, the thing that moves through the water is not an individual wave, but a wave train, with each individual wave having a finite and rather short lifetime. It turns out that the speed of a water wave depends on both the wavelength and water depth. Water waves are very special; nevertheless much of what we have learned from their behavior has been useful in understanding other waves.

Sound waves are compression waves in the air, water, or solids. In contrast to water waves, all wavelengths/ frequencies travel at the same speed, and their speed depends on the temperature and density of the medium though which they travel. The first accurate measurement of the speed of sound in air was not made until 1821. The speed of the compression-rarefaction waves in air is strongly dependent on temperature, somewhat on humidity, and very slightly on pressure. The velocity of sound in water was first measured in 1826. In water, sound velocity depends strongly on temperature, less strongly on pressure, which is a function of depth, and in seawater, on salinity. In solids, speed depends on density and temperature.

It turns out that both of these analogs are useful in understanding radiation.

**Fig. 7.15** Paths of sound from a source in the thermocline

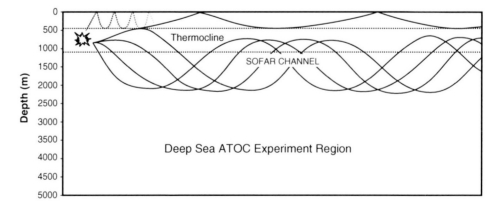

A Timeline for this chapter:

| ca. 340 BCE | Aristotle argues that sound is a disturbance of the air |
|---|---|
| 1620–1740 | Nicolò Amati (1596–1684), Antonio Stradivari (1644–1737), and Bartolomeo Giuseppe Guarneri (1698–1744) make great violins in Cremona, Italy |
| 1635 | Pierre Gassendi measures the speed of sound (478.4 m/s) |
| 1636 | Marin Mersenne publishes *Harmonie Universelle* and revises the speed of sound to 450 m/s |
| 1656 | Giovanni Borelli and Vicenzo Viviani make precise measurement of speed of sound using flash from a cannon—350 m/s (modern value at 20 °C and sea level is 343.2 m/s) |
| 1660 | Robert Boyle invents a vacuum pump and discovers that sound does not travel in a vacuum, although light does. |
| 1687 | Isaac Newton publishes *Philosophiæ Naturalis Principia Mathematica*; his Law of Gravitation explains the tides; he determines speed of sound in air to be 295 m/s |
| 1713 | Newton publishes second edition of the *Principia*; corrects speed of sound to 298 m/s |
| 1812 | Napoleon introduces the metric system, spreading its use through the lands he conquers |
| 1821 | Pierre-Simon Laplace makes an adiabatic correction to Newton's data, revising his estimate of the speed of sound to 337.7 m/s. |
| 1826 | Jean-Daniel Colladon and Charles-François Sturm measure the speed of sound in water (Lake of Geneva) to be 1435 m/s |
| 1865 | Lewis Carroll publishes *'Alice in Wonderland'* |
| 1884 | Giuseppe Verdi proposes that Italian Parliament should codify in law that 'A' on the musical scale is 432 cycles per second (Hz) |
| 1939 | Josef Goebbels, Nazi Minister of Propaganda, tries to get international agreement that 'A' on the musical scale is 400 Hz |
| 1996–2006 | ATOC experiment tries to use the speed of sound to measure the temperature change in ocean deep water |
| 2004 | Indonesian (Sumatra-Andaman) Earthquake generates Indian Ocean tsunami, killing 230,000 |
| 2011 | The Great East Japan (Tōhoku, Sendai) Earthquake generates a devastating tsunami |

If you want to learn more:

Open University, 2000. *Waves, Tides, and Shallow Water Processes*, Butterworth-Heinemann, Oxford, UK, 228 pp. There are many technical books on sound, but I like one that is fun to read:

Parker, B., 2009. *Good Vibrations: The Physics of Music*. Johns Hopkins University Press, Baltimore, MD, 288 pp.

Music: What else: Claude Debussy, La Mer.

Libation: 'Bordeaux' means 'at the edge of the water.' In that part of southern France, the soils are light and sandy and the winds blow in from the sea. The name usually conjures up visions of great red wines, but there is another treasure: white Bordeaux wines made from *sauvignon blanc* grapes. They are not unreasonably expensive, and a delight for an evening on the terrace.

## Intermezzo VII. New Horizons and the Berlin Wall

One of the prerogatives of an academic life is that you have the summer free of duties. Of course, you don't get paid either, but many academics arrange their salaries to be paid over 12 months rather than the 9 months they are at work. In reality, many academics, especially those in scientific or technical fields, have grants for research that help them out by providing summer salaries as well as support for undergraduate and graduate assistants to help in their research.

My first academic summer was that of 1961. I did not have funding, but I had saved up enough so that I was able to coax an undergraduate senior, Don Marszalek, to act as assistant to collect fossils with me in Europe in return for covering his expenses. We traveled via one of the especially cheap Icelandic Airlines flights for faculty and students headed to Europe. It was a long flight, with a propeller driven plane, a landing for refueling in Reykjavik, and finally arriving in Luxemburg. I had arranged to purchase a car, a little Renault, which I later had shipped back to the US and which served me well for a number of years. It got better gas mileage than almost anything on the road in the US today.

Early on, I said that my life has been strongly influenced by serendipity. Events that summer changed the direction of my research and allowed me to be a witness to history. Don and I started out collecting samples for microfossils in the Paris Basin, a classic area for geology. I had visited much of the area during a postdoctoral year in Basel, Switzerland in 1959–1960, but had not been able to see everything I wanted. From there we went south to Biarritz, on the Atlantic just north of the Spanish border.

We were walking along the promenade one evening when we passed a shop with snorkeling and SCUBA gear. Don asked if I had ever been snorkeling. I had not. The next day we came back and bought masks, fins, and snorkels. Don also bought a very nice Italian spear gun. Most spear guns at that time were complicated arbolette-type guns powered with one or more sets of rubber surgical tubing, with the external spear attached to the gun with a line. The Italian gun was metal and a very clean design. It was powered by a spring inside the barrel; it was basically a tube with a pistol grip and a nylon line attached to the protruding spear tip.

My first experience snorkeling was around the 'Rocher de la Vierge' in Biarritz. The Rocher is a big rocky promontory, jutting out into the sea from the center of Biarritz. Several people were walking among the rocks in shallow water—they were prying limpets and snails off the rocks and eating them. We went into the water, and in a few minutes Don had taught me what I needed to know and I discovered a new world. The rocky shoreline was covered by sea urchins, and a wealth of undersea life. Swarms of fish abounded. The undersea world was wholly different from what I expected.

Don, interested in trying out his new spear gun, swam out a considerable distance. In water depths of 6 or 8 m he came across several large boulders. Under an overhang he saw a guitar-shaped ray about 60–70 cm long. He knew they were edible and that someone on the beach would appreciate a fresh fish so he shot it—the gun worked perfectly. It was a long swim for him back to the beach, in part because he was towing the large flat fish several meters behind him. When he finally reached the beach, he stood up in the ankle-deep water—I couldn't see the fish but the spear was protruding in the air along with the wet nylon line attached to the spear gun he was holding. He suddenly jumped and the spear gun went flying in one direction and Don in the other. He yelled "What happened, I just got two strong electric shocks." One of the beach-combers identified the ray as a Torpille, also known as an Atlantic electric ray. They can produce electric shocks of 170 to 240 V. Using considerable care some of those on the beach gathered it up and took it off for dinner. Ray cooked in red wine is a local specialty.

During the course of the summer we went snorkeling in sites along the French and Italian Mediterranean coasts, and I realized that I wanted to learn a lot more about the marine world.

Don and I were in the south of France on August 1. Everyone in France has a month long vacation during the month of August; virtually everything shuts down and most of the country tries to visit resorts on the Atlantic and Mediterranean coasts. We had just come north from collecting samples in Spain and tried to find accommodations starting in the late afternoon, first in Perpignan, then in Montpelier, and finally around midnight in Nimes. Not a single room was available anywhere. Resigned to sleeping in the car, we drove into the hills, and came upon the place where the 2,000 year old Roman aqueduct built to bring spring water to Nimes crosses the river Gard, the Pont du Gard. We used coats and clothing to make beds and slept on the ground. I will never forget waking up in the middle of the night with the moonlight shining on the great structure. We spent part of the next day climbing around on it.

You can't sleep under the Pont du Gard anymore. The area is now a park, and there is an excellent museum about the aqueduct, its construction and even how it was financed. You can see the lead plumbing fixtures used to provide running water for homes in Roman Nimes. And the stone arches remain the largest ever constructed.

The other thing that happened that summer was an unplanned visit to Berlin. Don and I were in Munich, where I was visiting friends at the University. We also met some old friends of mine from Dallas, a German sculptor, Heri Bert Bartscht who I had met at an art exhibition at the little gallery in Dallas. Heri and his wife, Waltraud, or 'Wally' as we called her had just returned from Berlin. I had not been paying any attention to the news, and Heri told us that the East Germans were sealing off their part of the city from the west, and planned to build a wall. He said 'It's really interesting, you should go see it if you can.'

The next morning, August 15, Don and I drove the short distance to Munich's Rhiem airport, and simply went up to the counters asking if there was space on a plane to Berlin. There was space late that afternoon on an Air France flight, and we got on it. I jokingly told Don that the airline had earned the nickname Air Chance. The weather got bad as we approached Berlin, and the plane finally landed after three tries to get down on the runway. The nickname Air Chance no longer seemed so funny. We stayed at a Hilton hotel not far from Potsdamer Platz. The boundary ran through this famous old square, which once was the commercial center of Berlin but was a desolate wasteland of ruins in 1961. About 9 PM we left the hotel and started to walk the few blocks to Potsdamer Platz. We were stopped by an American soldier who told us we should not go any further; the border could be a dangerous place.

Frustrated, I remembered I had a micropaleontologist friend in Berlin, at the Technical University. Many of the buildings of the TU are near the famous 'Zoo' railway station, and the next morning we went there. Klaus Müller was in his office and delighted to see us. I asked if he could show us the border. "Absolutely, I'll not only show it to you, we'll cross it as many times as we can today. It's fascinating, the situation is chaotic, border crossing requirements are different everywhere."

Armed with our US passports and Klaus with his West Berlin identity card, we set off. As I recalled we crossed at least four different places that day. We said we had appointments at the Humboldt University and at the Academy of Sciences. The East German guards did not know what to do, so each time we were allowed to pass. At each border crossing station, construction on the wall had begun. It was a rather feeble thing at first, just concrete blocks cemented together. We did go to both the Humboldt University and the Academy of Sciences to visit paleontologists I had met earlier at conferences. We were having coffee in the Academy of Sciences when our host casually asked "By the way, how did you get in here? This building is supposed to be off limits to foreigners." Klaus said, "Well, I knew where your office was, so we just walked past the guard." Our host actually lived in West Berlin, but worked in the east. This was not uncommon before the wall was built. He said "When we finish our discussions I will give you an hour to get to the Friedrichstrasse crossing point (later to become 'Checkpoint Charlie'); then I am going to call security and denounce the guard. Then I will leave and go home to my place in West Berlin and never come back." We got to the Friedrichstrasse crossing point in half an hour, but leaving East Berlin had gotten more complicated. The guards wanted to know where we had been. Klaus talked us through.

This had been my first experience in a Communist country; I found it fascinating. When, a few months later I was asked to serve as a reviewer on a doctoral thesis at the Charles University in Prague, I happily accepted.

But now back to the story of how I got involved in marine research.

I resolved at the end of that summer that I wanted to study modern shallow water marine environments. To my amazement, the Department Head at the University of Illinois, George White, found this a great idea. Since Urbana, Illinois is not exactly on the seashore I needed to find a location where I could work. I decided to tour the marine laboratories of the Gulf Coast and Florida during the Christmas break after spending the holiday with my family in Dallas.

To make a long story short, the seawater around the laboratories in Texas and Alabama is muddy, and it is cold in winter, so I was much happier when I visited Miami. The University of Miami's Institute of Marine Sciences was filled with interesting characters, perhaps the most intriguing a geologist named Cesare Emiliani. Miami seemed like a good place to work, but I had one more place to visit on my list, a field station of the American Museum of Natural History, the Lerner Marine Laboratory. The Laboratory was located on the island of Bimini, Bahamas, just across the Florida Straits from Miami. After visiting the Institute in Miami, I flew with Pappy Chalk himself, owner of Chalk's airline, in a seaplane to Bimini. There I found a paradise for someone who wanted to study marine environments. Crystal clear waters, a huge variety of environments, including some small coral reefs, and an incredibly hospitable staff. I worked out of the Lerner Marine Laboratory for over 8 years.

AN INEFFECTIVE ATTEMPT TO MEASURE
THE SPEED OF LIGHT

# The Nature of Energy Received from the Sun— Figuring Out What Light Really Is

**8**

*For the rest of my life I will reflect on what light is.*
Albert Einstein

The analogs, water and sound waves, discussed in the previous chapter stimulated a lot of thought about the nature of light. However it was also realized that light had a number of peculiarities that set it apart. It was a mystery wrapped in an enigma.

## 8.1 Early Ideas About Light

The nature of light was long a mystery. In some ways, it still is. It was discovered that it has many strange properties: reflection, refraction, scattering, diffraction, interference, and polarization. It was obvious from daily experience that light came from the Sun, but the idea that rays from the eyes illuminated objects was also considered a possibility (the modern analogy would be Superman's laser emitting eyes). Euclid, the 3rd century BCE Greek famous for geometry, realized that light traveled in straight lines and described reflection: if reflection is from a flat surface such as a mirror, the angle of incidence of the light ray is equal to the angle by which the reflected ray leaves the surface. Claudius Ptolemaeus (we call him Ptolemy in the English speaking world), in the 2nd century CE, wrote a five-volume treatise on optics, but only one volume survives as a medieval Latin translation of a lost earlier, possibly inaccurate, Arabic translation of the missing original. In it he described the phenomenon of refraction, whereby the ray of light is bent as it passes from one material to another. He was unable to figure out the rule exactly. Even in those days there was dispute whether light consisted of particles or was some sort of wave. We really have no idea what the ancients knew about light—so many of the documents have been lost.

One of the most obvious properties of light is that, like water waves hitting a vertical bulkhead, it can be reflected. If the reflection is like that off a mirror, it is termed 'specular reflection' (Fig. 8.1a, b). This corresponds everywhere to the well-known rule: the angle of incidence of the light rays is equal to the angle of reflection. However, if the surface is uneven at the scale of the light waves or larger, the reflected rays still follow the rule, but go in different directions, producing diffuse reflection as shown in Fig. 8.1c. A good example is white paper.

A special case of reflection produces the colors we observe in objects. Color is the combined wavelengths of light that are not absorbed by the surface but diffusely reflected. For example, we associate land plants and photosynthesis with the color green. Green consists of the wavelengths of light that are *not* absorbed and *not* used for photosynthesis. When you buy a light especially to foster the growth of your house plants, it seems to have a strange reddish-bluish color, and if it is the only light, the plants do not look very green.

Clean, clear air transmits all of the wavelengths in the visible spectrum, but water does not. The red end of the spectrum is gradually absorbed by water, and blue penetrates deepest. Red is entirely gone at a depth of 10 m (30 ft) in seawater, so that a red fish appears black. Underwater photographs taken in the ambient light are a monotonous blue. That is why underwater photographers take lights with them, even when there is bright daylight at the surface.

Experiments show that not all animals see colors as do humans. Birds, for example, can see the ultraviolet that is invisible to us. Cats see only blue and green. Deer cannot see red. And not all humans can see all colors; some are 'color-blind.'

## 8.2 Refraction of Light

The bending of light rays as they passed from one medium to another was obvious to any observant person. You do not need a clear solid, like glass, to see the effect; water will do just fine. If you are out rowing, in a boat, you will notice that the oar seems bent where it enters the water. The toothpick

© Springer International Publishing Switzerland 2016
W.W. Hay, *Experimenting on a Small Planet*, DOI 10.1007/978-3-319-27404-1_8

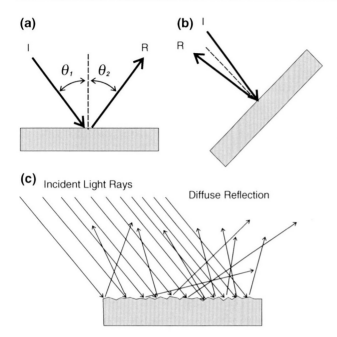

**Fig. 8.1  a** and **b** Light rays incident (I) on a smooth surface are reflected (R) at an angle the same as that of incidence ($\theta_1 = \theta_2$). **c** Light rays incident on a rough surface follow the same rule but are diffusely reflected in many different directions

in your olive in a Martini shows the same effect, perhaps less dramatically.

The problem of finding the law describing the bending of light as it passes from one medium to another ('refraction') that had frustrated Ptolemy was solved by several scientists in the early 1600s. They had no way to formally publish their results, so there is confusion over who was first. However, in 1621, Willebrord Snellius (Fig. 8.3) of the University of Leiden in Holland devised the mathematical form of the rule we still use. 'Snell's Law' states that the ratio of the sines of the angles of incidence and of refraction is a constant that depends on the media involved. An important descriptor, the refractive index of a medium, is simply the ratio between the velocity of light in a vacuum (or more practically, air) and its velocity in the medium, shown in Fig. 8.2.

Snellius' Law (English speakers have de-Latinized his name to simply Snell, so for most of us in the U.S. it's Snell's Law. Snell's Law states that the angles of incidence and refraction to the velocities of light and refractive indices are three simple ratios:

$$\frac{\sin \theta_1}{\sin \theta_2} = \frac{v_1}{v_2} = \frac{n_1}{n_2}$$

where $\theta_1$ is the angle between the incident ray and a vertical to the interface between the media; $\theta_2$ is the angle between the refracted ray and the vertical; $v_1$ and $v_2$ are the different velocities of light in the two media; and $n_1$ and $n_2$ are the refractive indices of the two media.

You might wonder how anyone in 1621 could know the speed of light. A few thought it was instantaneous; others thought it was just very, very fast. Of course, you do not need to know the absolute value of the speed of light to calculate the refractive index; it is just a ratio derived from the sines of the angle of incidence and refraction. The ratio has no units, like meters per second, it's just a ratio. Here are a few typical refractive indices, relative to the speed of light in a vacuum: Air at 0 °C at Earth's surface—1.0003 (so small a difference nobody bothers to make the correction between speed in air and speed in a vacuum); Water—about 1.335, which means that light travels 1/1.335 or 75 % as fast in water as in air; Window glass, about 1.457 or 69 % as fast as in air; Diamond—2.417, so the velocity of light is only about 41 % of what it is in air—that is why it bounces around between the facets and makes the jewel sparkle.

Until after the middle of the 17th century there were no scientific journals where one could publish the results of their work, so word was spread through letters to colleagues and pamphlets or books. Thanks to Google Books and the Gutenberg Project many of these rare items are now available for download on the Internet. The oldest science Journal, the *Philosophical Transactions of the Royal Society* was started in 1665, 20 years after the society was founded. Remember that in those days, what we call science was 'natural philosophy;' the word *science* did not appear in English until 1840.

Although it would be some time before it was formally published (and not by him), Snellius' idea got around. One of those who knew about Snellius' discovery was Renatus Cartesius, better known as the René Descartes, of 'Cartesian Coordinates' fame. In about 1615 Descartes wrote a treatise

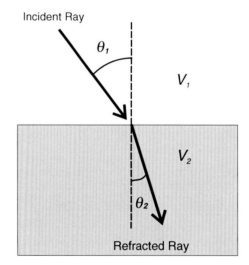

**Fig. 8.2** Refraction of light. $\theta_1$ is the angle of the incident ray; $\theta_2$ is the angle of the refracted ray, both measured with respect to a vertical to the interface between the media; $v_1$ and $v_2$ are the different velocities of light in the two media

**Fig. 8.3** Willebrord Snellius (born Willebrord Snel van Royen) (1580–1626)

covering optics, meteorology and a number of philosophical topics. However, the trial of Galileo Galilei took place in 1616. Although he was in France and a long way from Rome, Descartes was cautious about presenting his new ideas. He published a summary of his ideas in a rather informal manner in *Discours de la Méthode pour bien conduire sa Raison et chercher la Vérité dans les Sciences* (Discourse on the method of using Reason and searching for Truth in the Sciences), but he was so cautious that upon reading it, I found it almost unintelligible. It certainly would not have attracted an Inquisitor. His proper exposition of his ideas, *Le Monde ou Traité de la Lumiére* (*The World or Treatise on Light*), did not appear in print until 1664. In it he went off the deep end, proposing that light from the Sun reached Earth instantaneously. He explained refraction by mechanical analogies, concluding that light passes more easily, i.e. faster, through a denser medium than a less dense medium. This echoes what was becoming known about sound, which does travel faster in a denser medium. He regarded light as made of minute particles. Descartes did not believe that a true vacuum—space with nothing in it—was possible. Instead he proposed that there was a universal substance that filled all space, even the space between the particles of air and the atoms of apparently solid objects: 'æther.' He proposed that the æther formed swirling vortices, and it was these vortices that kept the planets in their orbits around the Sun. Descartes was a brilliant thinker, but many of his ideas were vestiges of thinking of the Middle Ages. Descartes' æther was with us for 250 years.

Refraction is responsible for twilight, where the Sun's rays are bent by the atmosphere: for the twinkling of stars, the

result of turbulence in the atmosphere; and for mirages, where hot air near the ground bends the light so strongly that objects below the line of sight or the horizon are brought into view.

## 8.3  Measuring the Speed of Light

When I started looking into the development of knowledge about light, I thought that no one knew much about the speed of light until the middle of the 19th century. Surprise, there were some very clever people around in the 17th century.

First, Johannes Kepler (1571–1630) (Fig. 8.4) had formulated his Laws of Planetary Motion

Kepler's Laws:

First Law (published in 1609), the orbit of every planet is an ellipse with the Sun at one of the two foci of the ellipse.

Second Law (also published in 1609), a line joining a planet and the Sun sweeps out equal areas during equal intervals of time. This has important implications for climate, because it means that if the orbit is highly elliptical, the planet will spend most of its time far from the Sun. If you remember the expressions 'caloric winter' and 'caloric summer' you will realize that with an increasingly elliptical

**Fig. 8.4** Johannes Kepler (1571–1630)

orbit 'caloric summer' becomes shorter while 'caloric winter' becomes longer.

Third Law (published in 1619), the square of the orbital period of a planet is directly proportional to the cube of the semimajor axis (i.e., half of the long axis) of its orbit. This is an amazing leap of insight—no wonder it took him 10 years to figure it out.

Although this third law sounds complicated, it is incredibly important because it is the key to the size of the solar system. What it means is that if you know how long it takes for a planet to orbit the Sun, you can figure out how far apart they are. If you do not know anything else, you can figure out how far it is relative to how far the Earth is from the Sun. This gave rise to the Earth–Sun distance being defined as 1 AU. Even if you do not know how far apart the two bodies are in terms of feet, miles, meters, or stadia, the distance is 1 AU.

Half a century later Giovanni Domenico Cassini (Fig. 8.5), an Italian astronomer working in Paris, figured out a way to determine how far away the Sun is. He used the phenomenon of parallax to solve the problem. Parallax is defined as the apparent displacement of an object as seen from two different points. Simple demonstration: close one eye, hold out your arm and put your thumb over some object in view. Close that eye and open the other. Your thumb no longer appears to be over the object. The angle you have to move your arm is a measure of the parallax.

In 1672 Cassini sent his colleague Jean Richer to Cayenne, French Guiana, in South America while he stayed at the observatory in Paris. They made simultaneous observations of Mars at its closest approach to Earth. From the measured parallax using Paris-Cayenne as a baseline and Kepler's laws, they determined the distance of Mars from Earth. It was not as simple as it sounds because at that time, determining the difference in longitude between Paris and French Guiana was a tricky business. You needed to know the difference in time between the two places. In other words you needed clocks in both places set to the same time.

In 1612 Galileo had proposed using the eclipses of the moons of Jupiter as a clock in the sky, but at that time the best timekeeping device was the water clock (based on dripping water), which was not very accurate. The far more accurate pendulum clock was invented in 1656 by Christiaan Huygens, solving the timekeeping problem—so long as you did not move the clock. Cassini is credited with being the first person to successfully use Galileo's 'moons of Jupiter' method so that synchronous measurements could be made in two different places. The difference in time between two places established the difference in longitude. The other key to the puzzle was to know the size of the Earth. Amazingly, Eratosthenes, in about 240 BCE, had calculated the circumference of the Earth to be (in modern measure) 40,000 km. Today we know that the equatorial diameter is actually 40,075.16 km; not much of an improvement.

**Fig. 8.5** Giovanni Domenico Cassini (1625–1712). Among other things, he discovered the large gap in the rings of Saturn, which is now named for him

However, Eratosthenes' estimate was not widely accepted. Other estimates had varied somewhat, and Christopher Columbus made a number of mistakes in using ancient data to estimate the circumference to be only about 30,000 km. It is possible that his errors were deliberate to make the proposed voyage seem more reasonable.

In 1670, another of Cassini's colleagues in Paris, Jean Picard, had determined the meridional (through the poles) circumference of the Earth to be the equivalent of 40,036 km. The modern value is 40,009.152. From the data they had gathered, Cassini arrived at an average Earth–Sun distance of 140,000,000 km (~87 million miles). This was a major correction over Kepler's guesstimate of 15,000,000 km (~9.32 million miles). But remember, using Kepler's third law, once you knew the Earth–Sun distance you knew the distance of all the other known planets from the Sun.

From about 1660, a Danish astronomer, Ole Rømer (Fig. 8.6), had observed the timing of eclipses of Jupiter's Moon Io at many different times as Jupiter's distance from the Earth changed. He made a prediction, which turned out to be correct, that when Jupiter was at its maximum distance from Earth the eclipse of Io would be 10 min late because the light had to travel a greater distance. He went to Paris and worked with Cassini to determine the Earth–Jupiter distance at the different times for which he had Io eclipse data. From

**Fig. 8.6** Ole Christensen Rømer (1644–1710). Those long wigs were back in style. He was a student of University of Copenhagen Professor Erasmus Bartholin. Bartholin published his discovery of the double refraction of a light ray by Iceland Spar in 1668, while Rømer was living in his home

this he worked out the speed of light to be 225,000 km/s (~ 140,000 miles/s). The modern value is 299,792 km/s (186,282.1 miles/s). Now we know that he underestimated the speed by 25 % but, after all, that was in 1676, a century before the American Revolution! They were incredibly clever people. Of course, there was no 'meter' or kilometer in those days; just about everything was measured in *Toise*, the French standard, which was exactly 6 *pieds* which turned out to be 1.949 m. Remember, the metric system was not introduced until 1812, by Napoleon.

The investigation of the properties of light was about to enter a new phase. The arguments over whether it was some sort of wave or particles would continue. It would be discovered that more than visible light was reaching Earth from the Sun.

## 8.4 The Discovery of Double Refraction or 'Birefringence'

In 1668 a remarkable deposit of a transparent mineral was discovered at Helgustadir in eastern Iceland. The mineral is calcite ($CaCO_3$), and at this locality it breaks into rhombs, as shown in Fig. 8.7. It was called 'Iceland Spar.' The rhombs

from Helgustadir are often nearly flawless, of a quality suited for optical instruments. The crystals have an unusual property—when you look through them you see a double image. As you rotate the crystal, one image stays in place while the other goes around it. How this happened would remain a mystery for a century.

A specimen of this remarkable mineral soon found its way to Erasmus Bartholin in Copenhagen. In 1669 he published a booklet describing the double image. Apparently the light was taking two separate paths through the crystal. Two different paths meant that it must be traveling at two different speeds. Crystals of Iceland Spar became popular objects of investigation.

In 1690 Christiaan Huygens (Fig. 8.8) published his *Traité de la Lumiére* (*Treatise on Light*), which includes a special section on '*l'étrange refraction du Cristal d'Islande*' (the strange refraction of Iceland Spar). Huygens argued that light is a form of wave, analogous to water waves, but wisely noting that the latter occur on an interface between air and water. It is important that he rejected the idea that light might be a compression-rarefaction wave-like sound, but must be some sort of a transverse wave. He also noted that whereas sound does not travel through a vacuum, as had been shown in Torricelli's barometer and Robert Boyle's vacuum chamber, light does.

Investigation of this strange property of Iceland Spar was made by making a dot on a piece of paper and placing the crystal on top of it. The double image appears in every orientation of the crystal. In discussing the double image, Huygens noted that one of the ray paths followed Snellius' Law whereas the other did not. Then he noted that as he rotated a crystal of Iceland Spar, one of the images stayed in place while the other went around it in a circle. Huygens named the image that stayed in place the 'ordinary ray' and the one that went around it the 'extraordinary ray.' He proposed that the 'ordinary ray' was propagating along a spherical front, whereas the 'extraordinary ray' must be

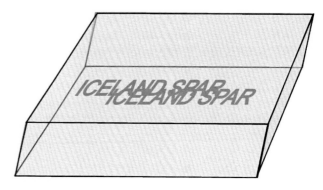

**Fig. 8.7** A rhomboidal crystal of clear calcite showing double refraction. Specimens from Iceland are rare today; most 'optical calcite' now comes from Mexico

**Fig. 8.8**  Christiaan Huygens (1629–1695)

propagating along an ellipsoidal front. He concluded that there must be something very special about the crystal that made the light take two paths through it and that they must be traveling at different speeds. Then he described the effect of looking at the image formed by one crystal through a second crystal of the same size set on top of it. When two crystals had the same orientation, the distance between the two images doubled. But when you rotated the upper crystal, things got really complicated. Sometimes he saw four images, as one might expect, but as the second crystal was rotated some of the images disappeared. Huygens found this quite strange and could offer no explanation. Subsequently examining more crystals of different minerals, Huygens and others found that double refraction occurred in many of them, but the separation of the two rays was much less than in Iceland Spar and just had not been noticed.

In those days, the words used to express the degree of refraction shown by a substance were 'refrangibility,' or 'refrinigbility,' terms hardly used today. But in modern scientific parlance the term 'double refraction' has been largely replaced with 'birefringence.' The phenomenon has no counterpart in water waves or sound and remained a curiosity without an explanation for over a hundred years Many scientists got themselves a couple of Iceland Spar crystals to play around with. But it would be more than a century before anyone figured out what was going on. There was something very strange about light.

Huygens also noted that if you took a knife to a piece of Iceland Spar, you could break it into smaller rhombs that had exactly the same angles between the faces as the original. He made a little sketch showing how the mineral could be thought

of as being made up of tiny ball-like pieces in a rhomboidal arrangement. This was rediscovered by accident a century later in 1784. René Just Haüy, an ordained catholic priest at Lemoine in France, was examining a beautiful piece of Iceland Spar belonging to a friend, when he accidentally dropped it on the floor. It broke into a number of pieces, each with a shape identical to the original. He had the brilliant thought that if you were to keep on breaking it down into smaller and smaller pieces, you would eventually come to one which was no longer divisible, and which represented the arrangement of the atoms of which the mineral was composed. This became the basis for the science of crystallography.

## 8.5  Investigating the Dispersion of Light

Dispersion involves separating the different wavelengths. As you will recall, dispersion happens with waves on the ocean. Because the speed of deep-water waves increases with their wavelength, long wavelength waves travel faster and outrun short wavelength waves. Fortunately, this does not happen with sound, which is what makes music possible to enjoy. But it does happen with light, which makes the colors of the spectrum.

During the Middle Ages, and perhaps earlier, it was noted that sunlight passing through a triangular prism of glass produced a spectrum with colors ranging from violet through blue, green, yellow and orange to red. Many thought that the glass of the prism colored the light but no one understood quite how that happened. The phenomenon of separation of the colors of light is called 'dispersion.'

In the later part of the 17th century, Isaac Newton tried out a new way of looking at the spectrum of colors. He made a tiny hole in his window shade and placed a prism in front of it. The spectrum of colors appeared on the wall opposite, much more clearly defined than just setting the prism in the sunlight. Newton then tried the experiment of using a card with a slit in it to select one of the colors and to allow this light to pass through a second prism. The color remained unchanged. Clearly, the light was not being colored by the prism. Then he tried inverting the second prism and allowing the entire spectrum produced by the first to pass through it. The colors were recombined to produce white light.

Newton concluded that white light was the combination of all the colors. In his book *Opticks* (not published until 1704 but reporting on experiments he had carried out many years earlier) he concluded that because of these properties, light must consist not of waves, as shown in Fig. 8.9a, but of particles ('corpuscles,' shown in Fig. 8.9b), and that color must be related to the velocity or mass of the particles. Further, he supposed that the phenomenon of double refraction or birefringence must be due to the corpuscles having different shapes, perhaps even facets, as suggested in Fig. 8.9c

**Fig. 8.9** 18th century ideas of the nature of light. **a** Huygens' idea of light as transverse waves. **b** Newton's idea of light as a stream of minute corpuscles. **c** Newton's idea of faceted corpuscles as a possibility to explain double refraction. **d** Young's idea of longitudinal waves

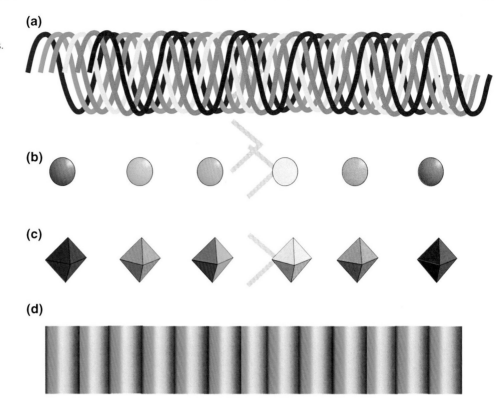

In 1800 William Herschel (Fig. 8.11), the English astronomer (and composer of music) reported to the Royal Society on the results of an experiment he had performed almost by accident. Herschel had wondered whether different parts of the Sun's visible spectrum were bringing different amounts of heat to the Earth. He contrived to set up a series of thermometers in different parts of the spectrum produced by a prism. The bulbs of the thermometers were painted black so

they could not reflect any of the incoming light. He found that the thermometers on the red end of the spectrum registered higher temperatures. But then, to check the results, he moved his thermometer array over, so some of it was beyond the red end of the spectrum. To his amazement, those beyond visible red light recorded even higher temperatures. He had discovered the 'invisible rays of the Sun' that existed beyond

**Fig. 8.10** William Herschel's diagram of the visible spectrum, *open curve* on the *right*, and the temperature spectrum, *shaded curve* on the *left*. From the *Philosophical Transactions of the Royal Society*, 1800. Note that in the diagram the *shorter wavelengths* are to the *right*, the *longer wavelengths* to the *left*, opposite to most modern diagrams

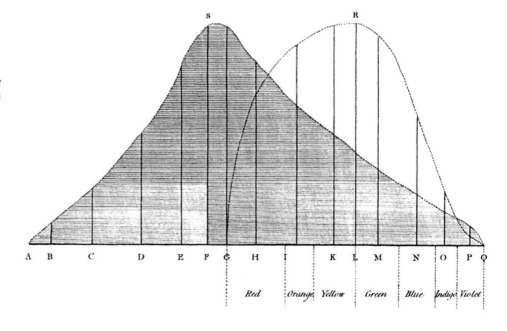

**Fig. 8.11**  Sir Frederick William Herschel (1738–1822)

red, the infrared. His original diagram showing the results of the experiment is reproduced in Fig. 8.10.

Another important discovery, again almost accidental, occurred in 1801. Johann Wilhelm Ritter (Fig. 8.12) was a

graduate student in pharmacy at the University of Jena in Germany. He was experimenting with silver chloride, a chemical which was known to turn black when exposed to sunlight (which later became the active ingredient in photographic film). He had heard that exposure to blue light caused a greater reaction than exposure to red light. He measured the rate of change when silver chloride was exposed to the different colors of light, using the spectrum of sunlight coming through a glass prism. The silver chloride in the red part of the spectrum showed almost no change, but became increasingly dark toward the violet end of the spectrum. Then placed silver chloride in the dark area beyond the violet end of the spectrum. He was surprised to find that the silver chloride showed an intense reaction beyond the violet end of the spectrum, where no visible light could be seen. There were invisible rays coming from the Sun beyond the violet end of the spectrum. Because of their capacity to induce chemical changes, he called them 'Chemical Rays'.

In only 2 years it had been discovered that visible light is just part of the overall spectrum of rays coming from the Sun. There were invisible rays beyond each end of the visible spectrum. The strange properties—that of inducing warming in the rays beyond red and that of causing chemical reactions in the part of the spectrum beyond violet—were a mystery. Herschel was 62 when he made his discovery; Ritter was 35.

We now know that the energy transmitted as visible light is just under half of that coming from the Sun. Something over a quarter of the energy is in the infrared part of the spectrum, and the remainder is in the ultraviolet.

**Fig. 8.12**  Johann Wilhelm Ritter (1776–1810). The relative wealth of Herschel and Ritter at the time of their discoveries is indicated by the quality of the portraits

## 8.6  Figuring Out the Wavelengths of Different Colors of Light

In 1801 Thomas Young (Fig. 8.13), an Englishman with very broad interests, gave a lecture to the Royal Society on how the eye works. He described the crystalline lens, and how it might change shape in order to change the eye's focus. He also came to the extraordinary conclusion that although the different colors of light appeared to be continuous, only three types of nerve receptors, one for red, another for green, and a third for violet, were required for color vision. The discovery of the cones and rods in the eye that are the receptors Young postulated came later, but in 1801 he had got it essentially right!

It is one thing to believe that light is a series of waves, but another to figure out the wavelengths of the different colors of light. It goes back to the 18th century, when Robert Hooke and Isaac Newton began to explore the colored rings that can be seen when you place a convex lens on a flat sheet of glass. These rings, now known as 'Newton's rings,' are an interference phenomenon brought about by the fact that different colors of light travel at different speeds through glass. Actually, you can often see distorted Newton's rings when you put two sheets of glass one on top of another. The surfaces are not usually completely flat and where there is space between them, the colored bands appear. The speed of all wavelengths of light is the same in a vacuum ($\sim 3 \times 10^8$ m/s). However, in anything other than a vacuum, the speed is proportional to the wavelength, so red light travels faster than blue light. In glass the difference in speeds is appreciable.

In 1802 Thomas Young used the Newton's rings interference phenomenon and a then state of the art estimate of the speed of light (500,000,000,000 ft in 8 1/8 min = 312,711.6 km/s; modern value is 299,792.458 km/s) to figure out the wavelengths of different colors of light, getting their values right to within about 1 % of the modern values. The modern values are given in Table 8.1.

It is often stated that the visible spectrum extends from 400 to 700 nm. We distinguish seven major colors. This is in slightly less than a doubling of the frequency, so we see less than an 'octave' of light. By comparison, remember that we hear nine octaves (doublings of frequency) of sound. The regions adjacent to the visible spectrum are the ultraviolet and infrared.

**Table 8.1** Frequencies and wavelengths of different colors of light

| Color | Frequency (THz = $10^{12}$ Hz) | Wavelength (nm = $10^{-9}$ m = $10^{-3} \mu$m) |
| --- | --- | --- |
| Violet | 668–789 | 380–450 |
| Blue | 631–668 | 450–475 |
| Cyan | 606–630 | 476–495 |
| Green | 526–606 | 495–570 |
| Yellow | 508–526 | 570–590 |
| Orange | 484–508 | 590–620 |
| Red | 400–484 | 620–750 |

## 8.7 Diffraction

The phenomenon of diffraction of light had been described in rather confusing terms in the 17th century. The first description is attributed to the Italian scientist Francesco Maria Grimaldi (1618–1663). He said that if a knife blade partially covered a beam of sunlight coming into a darkened room though a small slit, the shadow was not sharp, but appeared diffuse because some of the light had been bent in passing close to the blade. There seems to have been a lot of confusion about what was being described.

Others experimented with light passing through a narrow slit. It does not produce an image of the narrow slit but spreads out into a band, much like an ocean wave entering a narrow harbor entrance. But if you look at the band carefully you can see that it is brightest in the middle and becomes dimmer to each side. If you look really carefully you see that it does not just gradually become dimmer. As you go away from the center it dims, then becomes a little brighter, then dims again and again becomes brighter, and so on until it is too dim to see.

What happens if, instead of one, you make two slits? The pattern shows up again, only this time the contrasts are much more pronounced, and the image appears as a series of light and dark lines. This was such an important discovery it has become known in physics as the 'Two Slit Experiment.'

In 1803 Thomas Young described his experiment with light passing through two slits and forming an interference pattern. His original illustration of his idea of the diffraction of light through two slits is reproduced in Fig. 8.14 taken from the *Philosophical Transactions of the Royal Society* for 1800. It cast doubt on the particle theory, and he concluded that light must be some form of waves. He thought the waves were longitudinal, like the compression–expansion waves of sound, although later he suspected there might be a slight transverse component. His experiment was based on a comparison of observations of how water waves from two

**Fig. 8.13** Thomas Young (1773–1829)

**Fig. 8.14** Thomas Young's illustration of his idea how light passing through two narrow slits (*A* and *B*) would recombine to produce a diffraction pattern with alternating bright and dark areas (*C*, *D*, *E*, and *F*)

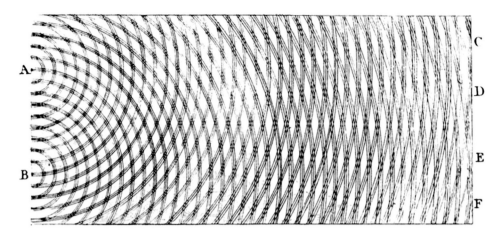

different sources could interfere with each other, producing reinforcement in some areas and disappearance in others.

As discussed above, the first person to investigate the spectrum produced by light passing through a glass prism was Isaac Newton, who performed his experiments in the late 17th century. But, as far as we know, Newton was concerned only with the spectrum of the Sun which he saw as a continuous spread of colors.

## 8.8  Polarization of Light

The discovery of polarization of light is attributed to Etienne-Louis Malus (1775–1812) (Fig. 8.15). He was an officer in Napoleon's army, engineer, physicist, and mathematician. He had participated in Napoleon's invasion and 'liberation' of Egypt. His diary of that expedition is one of the most horrifying accounts of man's inhumanity to man in wartime.

Like everyone else who had tried to figure out what light was, Malus had some specimens of Iceland Spar. Back in Paris after returning from the Middle East, he was examining one of the crystals in his apartment at 19 rue d'Enfer, just across the street from the Luxembourg Palace where Napoleon Bonaparte lived. The name 'rue d'Enfer' is curious; literally the Street of Hell. It drew its name from the city gate at the end of the street to the southwest, the Barriére d'Enfer, and that probably drew its name from the fact that it is located where the entrance to the catacombs of Paris is located. This particular 'rue d'Enfer' no longer appears on maps of Paris; that part of the street where Malus lived became the Boulevard Saint-Michel (the Boul Miche of the locals). To have a good view of the Luxembourg Palace, number 19 must have been near the intersection with the

modern rue Royer-Collard. The École des Mines (School of Mines, a haven for geologists) was (and is) across the street, a little farther on. Malus noted that the reflection of light from the setting Sun off one of the windows became brighter and darker, almost vanishing, as he looked at it while rotating his crystal. The light of the Sun itself showed no

**Fig. 8.15** Etienne-Louis Malus (1775–1812)

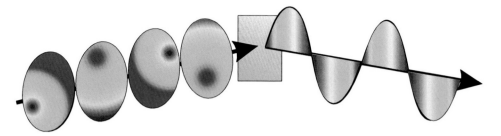

**Fig. 8.16** Normal light with wave motions in different directions on being reflected off a glass window becomes polarized, with the wave motions all having the same orientation. The intensity of each polarized ray varies according to the intensity and orientation of the incident ray, following Malus' Law

such effect, only the reflected light off the flat glass surface. He realized that instead of being a longitudinal wave with a small transverse component, as Thomas Young had proposed, the reflected light must be a transverse wave that was being blocked in some orientations of the crystal. Iceland Spar had revealed another of its secrets.

Borrowing the expression for division into sharply opposing factions, such as political groups, he described the reflected light as being polarized (Fig. 8.16).

He went on to formulate an expression for the changing intensity of the light passing through a polarizer:

$$I = I_o \cos^2 \theta_i$$

In words, this expression, now known as Malus' Law, says that the intensity of light $I$ observed through the polarizer is equal to the original intensity $I_o$ times the square of the cosine of the angle $\theta_i$ between the polarization of the light and the polarizer. If the polarizer is oriented at 90° to the direction of polarization of the light, the cosine of 90° is 0, and the light is completely blocked. If they are parallel to each other, all of the light passes through the polarizer. Because the intensity decreases as the square of the cosine, the intensity of the light drops off rapidly as the polarizer is rotated out of the parallel orientation.

On December 12, 1808, Malus reported his discovery at a meeting of the Institut de France. The very next week Simon Laplace announced that Malus had confirmed Huygens's explanation for double refraction of Iceland Spar. Again, as is so often the case in science, one thing leads to another.

In 1821 Augustin Fresnel (Fig. 8.17) produced a detailed theory of wave diffraction. He concluded from his investigation of polarization that light, unlike sound, must be exclusively a transverse wave. It was still generally assumed that these were waves in an unknown substance which filled all the spaces between atoms and molecules, and outer space (Rene Descartes' *æther*). It seemed inconceivable that waves of any sort could be transmitted through nothing at all (empty space!) The stage was set for the discovery of the remarkable nature of light.

## 8.9    Eureka!—Light Is Electromagnetic Waves

Electromagnetic waves are something very different. Back in the 18th century when electricity and magnetism were first being investigated scientifically, it became evident that they were somehow related. It was discovered that passing a metal wire though a magnetic field will induce a flow of

**Fig. 8.17** Augustin-Jean Fresnel (1788–1827). He is honored by a number of terms in optics and mathematics. You may have heard of 'Fresnel lenses' being used in lighthouses. Compared to a conventional spherical lens, it greatly reduces the amount of glass required by dividing the lens into a set of concentric annular lens segments known as 'Fresnel zones'

electricity through the wire. Similarly, passing an electrical current along the wire will cause a magnetic field to appear. It was found that it was possible to have a negative electrical charge in one place and a positive electrical charge in another. Further, the two charges would attract each other. If given the opportunity electricity, then thought to be some sort of mysterious fluid, would flow from the positive pole to the negative one to neutralize the difference. We now know that electricity is not a fluid, but a pool of discrete negatively charged particles, and the flow of electrons is from negative to positive. Magnetism was different. Again there were two poles, positive and negative, or north and south. One pole would attract its opposite or repel it if it were the same. But no matter how finely you ground up the magnetic material, the two poles could never be separated. Both were in every particle. To make a material magnetic these 'dipoles' needed to be lined up in the same direction. This happens readily in iron metal and some of the iron minerals such as magnetite. The dipoles line up in regions called 'domains.' Although small, magnetic domains in iron are visible with a microscope. To make a strong magnet, the metal is placed in a strong electrically induced magnetic field, and the domain size increases.

A lot of scientific effort was spent on trying to understand the relationships, but it remained for James Clerk Maxwell (Fig. 8.18) to make the great breakthrough in 1861. Incidentally, Maxwell's Great Uncle was John Clerk of Eldin, who had been a good friend of James Hutton. He was another major figure of the Scottish Enlightenment. It was John Clerk of Eldin who made many of the sketches of outcrops of rock included in Hutton's book "*Theory of the Earth.*" Interest in science ran in the family.

Maxwell worked on a mathematical formulation of the relations between electricity and magnetism, developing what have become known as the Maxwell Equations. These are developed in a paper "*On Physical Lines of Force*" published in three parts during 1861–1862 in the *Philosophical Magazine*. The paper starts out by describing the lines of magnetic force outlined by iron filings on a sheet of paper over a magnet. It then develops his ideas—too complicated to go into here—and finally concludes that there must be waves having both electrical and magnetic properties.

The paper discusses the term $V$ which appeared in his equations, finding it to be equal to the term $E$ of Wilhelm Weber and Rudolf Kohlrausch in a paper they had published in 1857. The $E$ in their paper, found through experiments with static and dynamic electricity that they had conducted at the University of Cologne in Germany, was the number of units of positive or negative electricity moving past a point

**Fig. 8.18** James Clerk Maxwell (1831–1879)

in a wire in one second. $E$ turned out to be 310,740,000 per second. Both Maxwell's $V$ and Weber and Kohlrausch's $E$ were speeds, and they were the same speed. Then Maxwell's paper refers to the unrelated 1849 measurement of the speed of light reported by Armand-Hippolyte-Louis Fizeau (1819–1896). Fizeau made his measurement by shining a light between the teeth of a rapidly rotating 720 tooth wheel. The measurement was made between a house on the Mont-Valérien in Suresnes, a suburb just west of Paris, and the hill of Montmartre, a distance of 8633 m. A mirror reflected the beam back through the same gap between the teeth of the wheel. The flash occurred with the wheel turning at 25.2 rotations per second. From this he determined that the speed of light was 313,000,000 m/s. This was essentially the same number as $V$ and $E$.

The startling conclusion is best described in Maxwell's own words, in part III of his famous 1864 paper "*On Physical Lines of Force*:"

The velocity of transverse undulations in our hypothetical medium calculated, from the electromagnetic experiments of MM. Kohlrausch and Weber, agrees so exactly with the velocity of light calculated from the optical experiments of M. Fizeau, that we can scarcely avoid the inference that **light consists in the transverse undulations of the same medium which is the cause of electric and magnetic phenomena**.

Maxwell's astonishment at the concordance of the results of these different investigations is beautifully expressed in his 1865 paper "*A Dynamical Theory of the Electromagnetic Field*" published in the *Philosophical Transactions of the Royal Society*:

> By the electromagnetic experiments of M. Weber and Kohlrausch v = 310,740,000 meters per second is the number of electrostatic units in one electromagnetic unit of electricity, and this, according to our result, should be equal to the velocity of light in air or vacuum.

> The velocity in air, by M. Fizeau's experiments, is V = 314,858,000; according to the more accurate experiments of M. Foucault, V = 298,000,000.

> The velocity of light in the space surrounding the earth, deduced from the coefficient of aberration and the received value of the radius of the earth's orbit is V = 308,000,000.

> Hence the velocity of light deduced from experiment agrees sufficiently well with the value of V deduced from the only set of experiments we as yet possess. The value of V was determined by measuring the electromotive force with which a condenser of known capacity was charged, and then discharging the condenser through a galvanometer, so as to measure the quantity of electricity in it in electromagnetic measure. The only use made of light in the experiment was to see the instruments. The value of V found by M. Foucault was obtained by determining the angle through which a revolving mirror turned, while the light reflected from it went and returned along a measured course. No use whatsoever was made of electricity or magnetism.

> The agreement of the results seems to show that light and magnetism are affections of the same substance, and that light is an electromagnetic disturbance propagated through the field according to electromagnetic laws.

The 'M. Foucault' Maxwell mentions was Jean Bernard Léon Foucault (1819–1868) who improved on Fizeau's device by using a spinning mirror rather than a toothed wheel to make a more precise measurement. This time the measurement experiment took place not across the city, but in a room, the Salle de la Méridienne in the Paris Observatory. The light path was only 20.2 m. Foucault's device shone a light onto a rotating mirror. The light bounced to a remote fixed mirror and then back to the first rotating mirror. But because the first mirror was rotating, the light from the rotating mirror finally bounced back at a slightly different angle. By measuring this angle, it was possible to measure the speed of the light. Foucault continually increased the accuracy of this method over the years. The speed cited by Maxwell was published in 1862. His final measurement, made in 1862, determined that light traveled at 299,796,000 m/s (only 458 m/s slower than the modern value).

Maxwell's conclusion was that light is a transverse wave with electrical and magnetic fields oriented at right angles to each other and to the direction of propagation, as shown in Fig. 8.19. The discovery that light is electromagnetic waves was revolutionary; but Maxwell also thought that the waves had to be moving through something, and that something, of course, must be René Descartes' *æther*, which now became the 'luminiferous æther.'

Maxwell's equations predicted that if light were electromagnetic waves, there must be other, perhaps unknown electromagnetic waves. The effort to produce artificial or man-made electromagnetic waves in the laboratory was taken up by Heinrich Rudolf Hertz (1857–1894).

Incidentally, it was while James Clerk Maxwell was working out the relation between electricity and magnetism and developing the idea of combined electric and magnetic fields, that another British scientist, John Tyndall, was measuring the effects of light and other radiation on different gases. He was generating the first numerical data on the greenhouse effect. He was providing the documentary evidence of the effect; an explanation had to come much later. It is one thing to know that something happens; it is something altogether different to understand why it happens.

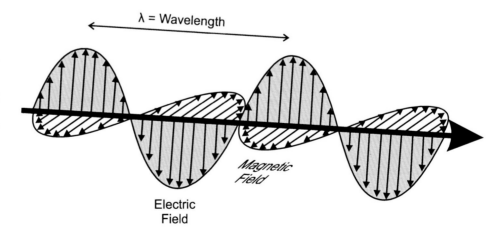

**Fig. 8.19** Electromagnetic waves. The electric and magnetic fields are oriented at right angles to each other and to the direction of propagation indicated by the *large arrow*. Now try to think of this not as a pencil-like ray, but as a band of light. If you can imagine that you are better than I am

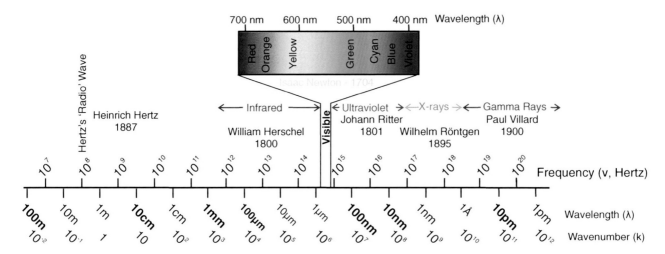

**Fig. 8.20** Discoveries of different parts of the electromagnetic spectrum. Isaac Newton (1704) showed that white sunlight was a combination of light of different colors. William Herschel (1800) discovered the invisible infrared rays of the Sun. Johann Ritter (1801) discovered the ultraviolet rays of the Sun, which he called 'Chemical Rays' because of their effect on silver chloride later used in making photographic plates. That was the state of things until three quarters of a century later, when Heinrich Hertz created artificial electromagnetic waves in the laboratory, the precursor of modern radio and TV. Experimentation with cathode rays led to the discovery of X-rays by Wilhelm Röntgen in 1895, and the examination of radioactive materials led to the discovery of gamma rays by Paul Villard in 1900, although it was Ernest Rutherford who gave them their name in 1903

## 8.10 A Review of the Discovery of the Invisible Parts of the Electromagnetic Spectrum

Figure 8.20 shows the discoveries of different parts of the 'invisible' rays of the Sun and other parts of the electromagnetic spectrum related to the development of cathode ray tubes, investigation of radiation from radioactive minerals, and Heinrich Hertz's experiments.

## 8.11 The Demise of the 'Luminiferous æther'

Nicolas Copernicus and Galileo Galilei had figured out that the Earth orbited the Sun, so, for a while, the Sun was the center of the Universe. But in the late 18th century our friend William Herschel, who was to discover infrared rays, was carefully measuring the motions of stars. From observation of seven bright stars he concluded that they appeared to be converging. He concluded that this meant that the Sun was moving away from them. His paper on this topic, published in 1783, was titled *"Motion of the Solar System in Space."* In the late 19th century it seemed obvious that if the Earth, in orbiting the Sun, was moving though the luminiferous æther, and if the Sun itself was also moving through the luminiferous æther, the velocity of light measured on Earth should change during the course of the year. Figure 8.21 illustrates the idea.

It was obvious that we could learn a lot more about the motion of our solar system through the Universe if we knew how fast we were moving through the luminiferous æther. Two American physicists set about to find out. The velocity of a point on the Equator, due to the rotation of the Earth, is 4,638 m/s (= 15,217 ft/s = 2.88 miles/s); this is much less than the instrumental errors involved in measuring the speed of light. However, its velocity in its orbit around the Sun is about 29,780 m/s (=18.5 miles/s), but this is still within the error of measurements using Foucault's measurement. What was needed to find the speed of the solar system through the luminiferous æther was the determination of the difference in the speed of light measured at different positions in the orbit, i.e. different times of the year.

Albert Abraham Michelson (1852–1931) had a bit of an obsession with measuring the speed of light. His parents had emigrated to the US from Prussia when he was a child. He had a career with the US Navy and taught at Annapolis, always bringing the speed of light into the discussion. He then spent some time back in Germany, and while there, in 1881, invented a new device intended to measure the differences in speed of light as the Earth moved through the luminiferous æther. The idea was to measure it in two directions at right angles to each other; as Earth moves through the æther, the velocity of light would necessarily be different in the two directions. The device used a special mirror to split a beam of light, send it in two different directions, and have the two beams bounce back from the mirror and be recombined. The device was called an interferometer, because since the two velocities would be different, constructive and destructive interference of the waves in the recombined beam should show a pattern of light and

**Fig. 8.21** The Earth and Sun moving through the streaming 'luminiferous æther'

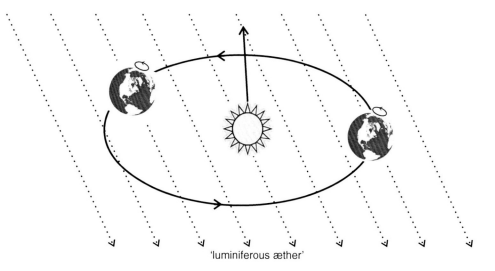

'luminiferous æther'

dark fringes. The effect would be similar to the fringes in Thomas Young's diffraction pattern. He calculated that if the æther were stationary relative to the Sun, then the Earth's motion in its orbit would produce a fringe shift 4 % the size of a single fringe.

The results of the experiments in Germany were inconclusive.

Michelson returned to the US, joining the faculty of the Case School of Applied Science in Cleveland, Ohio as Professor of Physics. The same campus also housed Western Reserve University, where one of the professors of chemistry was Edward Morley (1838–1923). In 1887 the two performed what was to become one of the most important experiments in physics.

They realized that they needed a very stable environment to carry out the measurements, so they set the equipment up in a closed room in the basement of Adelbert Dormitory, which was built of heavy stone. To reduce vibrations even further, the apparatus was set on top of a large block of sandstone, about a foot thick and five feet square. The sandstone block and the apparatus were then floated on a pool of mercury. (There was no OSHA—Occupational Safety and Health Administration—in those days).

The apparatus was similar to the one that Michelson had used in Germany, but had been improved by having the light reflected back and forth along the arms of the interferometer several times, so that the path length for each beam was increased to 11 m. They calculated that at this length the speed difference should be about 0.4 fringes on the diffraction pattern. With this redesigned instrument a difference of about 1/100th of a fringe would be detectable. So this time the experiment should work.

Since the sandstone block with the instruments was floating in the mercury pool, the device could be easily turned by hand, and it could rotate slowly in the 'æther wind'. Legend has it that Morley's main contribution to the

experiment was to provide the mercury. Observations and measurements could be made continuously. During each full rotation of the device, each arm would be parallel to the wind twice (facing into and away from the wind) and perpendicular to the wind twice. It was expected that a plot of the readings taken as the block rotated would show changes in velocity having the form of a sine wave with two peaks and two troughs each rotation. If the block was not rotated then the speed of Earth's rotation, which at the latitude of Cleveland is 3474 m/s (=11,398 ft/s, =2.16 miles/s), should produce the same pattern each day that would be clearly visible due to the increased precision of the instrument. Furthermore, the motion of the Earth around the Sun in the sea of æther should be readily detectable as a change in the speed of light over the course of a year.

Michelson and Morley made a number of measurements over the course of several months in 1887. To their utter distress, the speed of light turned out to be the same in all directions. The Michelson-Morley effort is sometimes referred to as the 'Great Failed Experiment,' even though accurate experiments that return unexpected results are not actually failures but successes. Despite the data, they continued to believe in the 'luminiferous æther,' others, among them Albert Einstein, realized that it no longer made sense.

## 8.12   Summary

The nature of light was long a mystery. The first measurement of the speed of light was made in 1676, only 25 % short of the modern value. In some of its properties, it seemed to be a wave phenomenon, but in others it behaved like a stream of particles. Light exhibits a number of properties analogous to water waves, such as refraction and diffraction. Although light of different wavelengths travels at the same speed in a vacuum, its speed differs in transparent solids, like glass, separating white light into a spectrum of colors. Light took two separate paths through the mineral Iceland Spar ('optical calcite'), mystifying scientists for over 200 years. Light can be polarized by reflection off a smooth surface. It still was not clear what light was but, finally, in the 1860s, James Clerk Maxwell discovered from his work on electromagnetic waves that they had the same velocity as that of light, leading him to conclude that light is an electromagnetic wave.

These clues were needed to make the huge leap from the classical physics of the 17th–19th centuries to the quantum physics of the 20th century. This would ultimately allow us to understand how greenhouse gases work to modify the climate system.

A Timeline for this chapter:

| ca. 300 BCE | In Euclid's *Optica* he proposes that light travels in straight lines and realizes that in reflection the angle of incidence equals the angle of reflection |
|---|---|
| ca. 240 BCE | Eratosthenes calculates circumference of Earth along a meridian to be about 40,000 km. Modern values are meridional (polar) circumference, 40,008 km (=24,860 miles); equatorial circumference 40,075 km (=24,901.5 miles) |
| ca. 150 CE | Claudius Ptolemaeus in *Optics* describes refraction of light |
| ca. 1490 | Christopher Columbus estimates equatorial circumference of Earth to be ~ 30,000 km—a serious mistake for planning his voyage to China and India |
| 1609 | Johannes Kepler publishes the first two Laws of Planetary Motion |
| 1612 | Galileo proposes using the moons of Jupiter as a 'clock in the sky' |
| 1619 | Kepler publishes the Third Law of Planetary Motion |
| 1621 | Willebrord Snellius develops 'Snell's Law' of refraction |
| 1656 | Christiaan Huygens invents the pendulum clock |
| 1660 | Robert Boyle invents a vacuum pump and discovers that sound does not travel in a vacuum, although light does |
| 1664 | René Descartes publishes *Le Monde ou Traité de la Lumiére* and proposes that light from the Sun reaches Earth instantaneously—through the 'æther' |
| 1668 | Iceland Spar discovered at Helgustadir in eastern Iceland |
| 1669 | Erasmus Bartholin describes the double image of Iceland Spar |
| 1670 | Jean Picard determines meridional circumference of Earth (through the poles) to be 40,036 km (modern value is 40,0009.152 km) |

(continued)

(continued)

| 1672 | Giovanni Cassini in Paris and Jean Richter in Cayenne, French Guiana make simultaneous observations of Mars' closest approach to Earth |
|---|---|
| 1676 | Ole Rømer and Cassini determine speed of light to be 225,000 km/s (modern value is 299,792 km/s) |
| 1690 | Christiaan Huygens publishes *Traité de la Lumiére* (*Treatise on Light*), with a special section on 'l'étrange refraction du Cristal d'Islande' (the strange refraction of Iceland Spar). He argues that light is a form of transverse wave |
| 1704 | Newton publishes *Opticks*, describing the spectrum of white light, and proving that the colors did not come from the prism; he proposes that light is particles ('corpuscles'), that color is related to velocity or mass of the particles, and explains double refraction as being related to the shapes of the corpuscles |
| 1800 | William Herschel discovers the 'invisible rays of the Sun' (infrared) |
| 1801 | Johann Ritter discovers the 'chemical rays' of the Sun (ultraviolet) |
| 1801 | Thomas Young concludes that the eye has receptors for only three colors: red, green, and blue |
| 1802 | Using the phenomenon of Newton's Rings, Thomas Young figures out the wavelengths of the different colors of light, getting them right to within 1 % of modern values |
| 1803 | Thomas Young describes his 'two-slit- experiment' and explains the diffraction patterns of light |
| 1808 | Etienne-Louis Malus discovers the polarization of light |
| 1812 | Napoleon introduces the metric system, spreading its use through the lands he conquers |
| 1821 | Augustin Fresnel publishes a detailed theory of wave diffraction, and concludes from investigation of polarization that light must be exclusively a transverse wave |
| 1849 | Armand-Hippolyte-Louis Fizeau measures the speed of light from Mont-Valérien to Montmartre as 313,000,000 m/s |
| 1857 | Wilhelm Weber and Rudolf Kohlrausch publish a paper showing that the number of units of electricity moving past a point in a wire in one second ($E$) is 310,740,000 |
| 1861–62 | James Clerk Maxwell publishes the first two parts of *On Physical Lines of Force* and concludes that there must be waves having both electrical and magnetic properties |
| 1862 | Léon Foucault measures the speed of light in a room in the Paris Observatory to be 299,796,000 m/s (modern value is 299,792,458 m/s) |
| 1864 | Maxwell, in part III of *On Physical Lines of Force*, cites Fizeau and Weber and Kohlrausch to conclude |

(continued)

(continued)

| | |
|---|---|
| | that 'light consists in the transverse undulations of the same medium which is the cause of electric and magnetic phenomena' |
| 1865 | Maxwell expands on his earlier work in *A Dynamical Theory of the Electromagnetic Field*, and cites Foucault |
| 1887 | Albert Michelson and Edward Morley conduct an experiment to determine the motion of Earth through the 'æther wind' and find that the speed of light is the same in all directions, independent of the motion of the observer |

If you want to learn more:
Zjonc, A., 1995. *Catching the Light. The Entwined History of Light and Mind*. Oxford University Press, 400 pp.

Music: Adolf von Henselt's Piano Concerto in E minor (Op. 16) is a delightful piece of music but rarely heard. Some of its themes have been copied for film music.

Libation: In honor of James Clerk Maxwell's Scottish heritage it would be appropriate to turn again to Scotch single malts. Try the Jura (from the Scottish Island of that name, not the mountains along the Swiss-French border).

### Intermezzo VIII. Back and Forth Through the Iron Curtain

At the International Geological Congress in Copenhagen in 1960, I gave a paper on the Cretaceous-Tertiary boundary in Mexico. The discussion then was not what had caused the extinction of so many animals and plants, but the rather esoteric question as to whether the 'Danian Stage' belonged to the Cretaceous or the Tertiary. Some members of the geological community were beginning to become aware that something dramatic had taken place at the end of the Cretaceous, and there were arguments about which rocks and fossils belonged to the Cretaceous and which to the Tertiary. The 'Danian Stage' had been described as the uppermost (youngest) strata of the Cretaceous. The concept of the Danian Stage of earth history is based on a very peculiar suite of rocks exposed in a quarry near Faxe and on the coast of the Danish island of Seeland at a place called Stevens Klint, just south of Copenhagen. If you are a paleontologist, the quarry is one of the strangest places you can imagine. It does not take long to figure out that it is a coral reef, but a very strange one, where fossils that are exceedingly rare elsewhere, such as crabs, are common. There are lots of corals, but they do not look like the ones you see on a coral reef today. These strange deposits sit on typical chalk that closely resembles that of the region around Paris. Separating the two deposits is an enigmatic, very thin layer of clay full of fish remains, the

'Fiskeler.' Even Charles Lyell had visited these localities and found them both fascinating and enigmatic.

A few years before the Congress a solution to part of the riddle had been found. Dredging off the coast of Norway had brought up corals like those found at Faxe, and it had been discovered that there is a wholly different set of coral reefs unlike those most of us know. These are what we now call 'deep-water coral reefs.' They grow in water depths of several hundred to a thousand meters. Unlike the corals of shallow water reefs, the coral animals of these deep-water reefs do not need sunlight.

The argument at the Congress was whether these deposits belonged to the Cretaceous period, named for the chalk deposits of France and Britain, or to the younger Tertiary Epoch. The Soviet view point, represented by Aleksandr Leonidovich Yanshin, was that the deposits were limestone; therefore they belonged to the Cretaceous. I was one of those arguing that the fossil foraminifera below the Fiskeler were like those of the Cretaceous, but those in the Danian deposits were like those of the Tertiary Velasco formation in Mexico. It sounds silly today, but it was the kind of argument geologists had in those days.

There was a young Czech micropaleontologist attending the Congress, Pavel Čepek, accompanying his Professor, Vladimir Pokornỳ. Pavel, Aleksandr Yanshin and I were together on a field trip after the Congress to look at Cretaceous deposits in southern Sweden. One evening Pavel was seated between Yanshin and me. I did not speak Russian, Yanshin did not speak English, but Pavel understood both, so he became the interpreter for a dinner-long argument. Yanshin and I had a great time, but Pavel was so busy translating he did not get much to eat.

On the field trip Pavel and I were looking at the chalk and I was telling him about how useful the nannofossils were in determining the age of rocks in the Alps. It turned out that Pavel had been looking at the nannofossils in the chalk deposits of Czechoslovakia. His thesis advisor, a very prominent micropaleontologist specializing in study of ostracods, Vladimir Pokorný, was on the field trip too. He inquired whether, if things could be worked out, I might be able to serve as an outside reviewer on Pavel's doctoral thesis. None of us knew what might be involved in having an American serve as reviewer on a Ph.D. thesis in Communist Czechoslovakia.

In the spring of 1962, I got an invitation to come to Prague during the summer to look at Pavel's work. The arrangement was the usual one for working with colleagues on the other side of the iron curtain; you paid for your trip to get there, and they paid your expenses while there. Since I was going to be in Europe anyway, it was an easy decision to make. I arrived in Prague in mid-August, coming on the train from Vienna. The border crossing into Czechoslovakia was interesting; there was a thorough examination of my passport and visa, and a look in the luggage. However, all this was not nearly as interesting as when leaving a week later. Pavel greeted me at the railway station in Prague.

Prague is one of the world's great cities. Essentially untouched by the war, everything in Prague was intact, but, like Vienna a decade earlier, it was a gray city. To get a feeling what these cities were like then, look at the film "The Third Man," filmed in Vienna in late 1940s. After the Peace Treaty with Austria was signed in 1956, Vienna blossomed back into color with the baroque houses mostly white or yellow. But the Communist government of Czechoslovakia did not believe in either repair or colorful building facades, and nothing had been done since before the war to bring them back to their original state. gray buildings and red banners with communist slogans printed in yellow, just like in the eastern sector of Vienna in the 1950s. I remember that no one, except Pavel, seemed to be smiling.

My hotel was on Václavské náměstí (Wenceslas Square) in the heart of the new city. Václavské náměstí is actually a rectangle a couple of hundred meters long, sloping up a hill and crowned by the Czech National Museum at the upper end. A dour clerk at the hotel took my passport; we put my luggage in the room, and went for a walk through the city. Prague lies astride a curve in the Vltava (Moldau) River, and that evening we walked down to the Charles Bridge. On the west and south sides of the river are high hills, that on the west, Hradčany, is capped by the magnificent Prague Castle with a curving unbroken baroque facade over half a kilometer long overlooking the city. The Vltava (Moldau) is a tributary of the Labe (Elbe). The rivers join about 40 km north of Prague and then flow north through sandstone canyons to Dresden and finally meet the sea north of Hamburg. Smetana's tone poem Má Vlast (My Country) contains the famous section known in English as "The Moldau" and is a beautiful musical description of the countryside and Prague.

Pavel was head of the Micropaleontology Section of the Czech Academy of Sciences. The offices and laboratory were on the upper floors of a building in the street called Spálená only a few blocks from Václavské náměsti. When we went there the next day,

I found it to be an island of warmth and cheerfulness in the sea of gray.

But before going any further, I have to explain that the Czech people are very special. In order to understand the Czech mentality, you have to read *The Good Soldier Švejk* (or *Schweik* in many translations) by Jaroslav Hašek, published in 1923. For me the best English translation is the Samizdat version available at www.zenny.com. Švejk is a Czech conscript serving in the Austrian Army during World War I. He carries out orders to the letter, with disastrous results. You are never quite sure whether Švejk is an imbecile, as he claims, or a genius at devising ways to frustrate authority.

One of the people I wanted to see on this first trip to Prague was another micropaleontologist studying coccoliths, Eva Hanzlíková. She worked in the Czech Geological Survey. Pavel accompanied me to the Survey at Klárov 3, a street near the Moldau. We went to the Survey building and told the Guard/receptionist at the front door that we wished to see Dr. Hanzlíková. He produced the 'Visitor's Book" and asked us to sign in. Pavel started to translate for me when the Guard interrupted to ask if I was a foreigner. Pavel said yes, and the Guard asked what country I was from. When Pavel said I was an American, the Guard picked up the telephone and started dialing. After a short conversation he handed the phone to Pavel. The person on the other end was the woman who was the Director of the Secret Political Section of the Survey. She immediately attacked Pavel: who was he to bring an American to this building without making prior arrangements. For such a visit an application must be made at least 3 months in advance. Pavel replied that I was a visitor to the Academy of Sciences and that he had only learned that I wanted to meet Dr. Hanzlíková a few days ago. She replied that he had broken the law and that there would be political consequences for this act. Further, I was not to be allowed into the Survey Building, and Pavel was forbidden to tell me anything about their conversation. We walked out the door.

Outside Pavel explained what had happened, and suggested we go to a Coffee House across the street. He called Hanzlíková from the coffee house and a couple of minutes later she joined us for a cup of coffee and a discussion of her studies and our work. There is a Czech expression for this kind of experience: 'The goat remains alive and the wolf is satiated.' As Pavel says, "Greetings from Schweik." Before that experience I really had no idea what Pavel thought about politics, but several comments he had made earlier indicated he did not like the regime. Now it was evident that he was a rebel.

I got a little introduction to daily life in Prague when we went out to get food for lunch. To make a purchase in a food store you needed first to stand in line to look and see what was available and how much it cost. Then you would get a slip of paper to take to the cashier, where you again stood in line. The cashier takes your money, and then gives you another slip of paper. After standing in line again you would give the slip of paper to the clerk who would then give you what you had purchased, if, of course, it had not sold out in the meantime. This was the routine for making purchases of any kind. There was an exception. If you had a foreign currency or a special voucher you could go to 'Tuzex' stores where you could buy western goods without standing in lines. But Tuzex stores did not accept Czech currency.

Two highlights of that first visit I remember very clearly were the garlic bread at a cellar restaurant, U Tomáše, and the Laterna Magica. The Czechs make garlic bread from day-old black bread which has dried out a bit. It is toasted by frying in a pan with a little vegetable oil and then spread with a layer of garlic about half a centimeter thick and sprinkled with salt. Guaranteed to keep witches, vampires, and most anyone who has not eaten it, at bay.

The Laterna Magica is one of the most remarkable theaters in the world. Performances are a combination of film and live acting or singing. We saw Jacques Offenbach's *Tales of Hoffmann* in a stunning performance. *Tales of Hoffmann* is about Hoffmann's three love affairs, all of which turned out badly. The first

woman he fell in love with was actually a mechanical doll who could dance ballet very well until she needed to have her spring wound up again. The second was an opera singer dying of consumption and being cared for by a certain Dr. Miracle. Needless to say, she dies. The third was a Venetian beauty, Giulietta, who had the unusual habit of singing duets with her reflection in the water. His brother, Schlemiel had no reflection in mirrors. Altogether a bad family to get mixed up with. The Venetian scenes at the Laterna Magica were amazing, with film of the city, floating gondolas on stage, some of the singing live and some on film.

On August 20th we made a trip into the countryside to see the chalk deposits he was studying. I had no idea that such beautiful chalks were found in Central Europe. The modern form of chalk is strictly a deep - sea deposit today, so seeing it in the middle of Europe set us both thinking about how the ocean plankton could have penetrated so far into the seas covering Europe. We made plans to see if we could keep on working together.

I left by train, and this time the border crossing into Austria was more complex. The train stopped in a village just before the border. Czech citizens traveling to this last stop had to have a special pass. I was alone in my compartment. Then the train pulled forward into an area surrounded by fences, and a thorough inspection took place. Border guards with mirrors on poles looked beneath the cars for stowaways, and dogs sniffed about. After about an hour, the train was underway again and soon stopped in an Austrian town, where lots of cheerful travelers climbed aboard.

ISAAC NEWTON DISCOVERS THE NATURE OF COLOR

*Every day sees humanity more victorious in the struggle with space and time.*
Guglielmo Marconi

How can you find out what something is made of if you cannot bring it into the laboratory and examine it? How do we know what the Sun is made of? How do we know what gases are in the atmospheres of other planets? How do we know what the glowing nebulae out there in outer space are made of? And why do they glow?

Nowadays the term 'remote sensing' is a household expression. There are two major kinds of remote sensing. Sound is a compression/rarefaction wave. Although high frequency sound waves are scattered, low frequency sound waves can travel thousands of kilometers through water and the solid Earth. Electromagnetic waves travel well through space and gases like the atmosphere, but are absorbed by liquids and solids. They penetrate only a few hundred meters through water and only a few millimeters at most through most solids.

In water, remote sensing uses sound to determine the depth to the sea floor without using a plumb line. We can even do swath mapping of the sea floor revealing great detail and even the nature of the seafloor deposits. Sound can be used to locate submarines (SONAR) and fish. Geologists use low frequency sound waves, seismic waves, to explore the interior of Earth's crust. They also make use of the even longer frequency sound waves (again called seismic waves) generated by earthquakes to explore the deep interior of our planet. Remote sensing in the sea does not involve electromagnetic radiation for one simple reason.

Remote sensing above the Earth's surface uses electromagnetic waves. They interact with the gases and airborne particles of the atmosphere. Just as important is the fact that the electromagnetic spectrum emitted by an object tells us about its temperature. That was the great contribution of Gustav Robert Kirchhoff's 'black body' investigations that started in 1875.

But electromagnetic radiation is also given off by chemical reactions, such as oxidation by burning something in a flame. It can also be given off by 'exciting' atoms or molecules with radiation. By passing electromagnetic radiation through liquids containing dissolved substances or even through solids, we can learn a lot about their chemistry. That is the science of investigation of the (electromagnetic) spectrum—spectroscopy.

Remote sensing did not begin with the satellite era; that is when it became a household expression. It actually began in the 17th century.

## 9.1 Spectra and Spectral Lines

As far as we know, the first person to investigate the spectrum produced by light passing through a triangular glass prism was Isaac Newton, who performed his experiments in the late 17th century. Newton's writings concern only the spectrum of the Sun which he saw as a continuous spread of colors, but it seems likely that he must have tried to observe the spectrum of the flames in the fireplace as well.

A half century later, in 1752, Thomas Melville, a physicist at the University of Glasgow in Scotland, noted that chemicals introduced into a flame often gave off colored light. You can check this out yourself, with common table salt (NaCl) which, in a gas flame, produces bright yellow light. Melville also observed that passing the light from the flame through a prism produced a spectrum, but unlike the apparently continuous colors—from indigo through violet to blue, green, yellow and finally red—seen in the spectrum of the Sun, it consisted of a few narrow bands of colored lines. It took another century before someone figured out what it meant. Progress in science is not always rapid.

In 1802 an English chemist, William Hyde Wollaston (1766–1828), described the results of a detailed examination of the spectrum of the Sun. He noticed something Isaac Newton had missed. It was not continuous as had been thought, but was interrupted by a number of dark lines. He had no idea what they meant.

In 1814, an optical instrument developer living in Bavaria, Joseph Fraunhofer (1787–1826), invented the spectroscope—a device for making precise measurements of spectra (Fig. 9.1). He independently rediscovered the lines

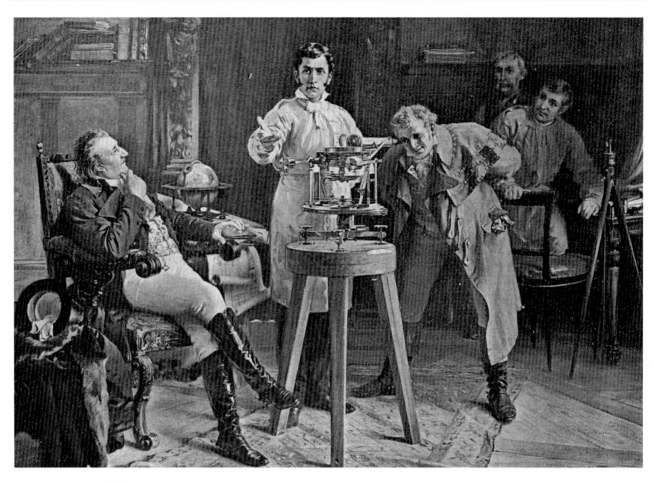

**Fig. 9.1**  Josef Fraunhofer demonstrating his new instrument, the spectroscope

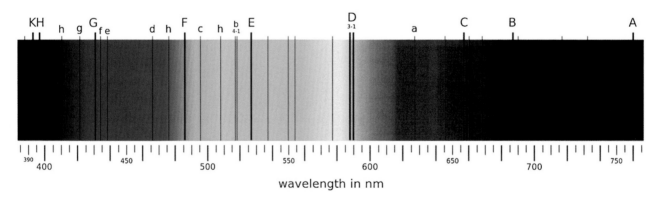

**Fig. 9.2**  The spectrum of the Sun, showing some of the *dark lines* Fraunhofer identified. The *lines* were later found to represent parts of the spectrum absorbed by gases in the Sun's atmosphere

Wollaston had seen. Fraunhofer began a systematic study of them and made careful measurements of the wavelength of more than 570 lines. He designated the major lines with the letters A through K, and weaker lines with other letters (Fig. 9.2). Still, no one knew what they meant. Although a number of other scientists observed and measured them, it would be almost another half century before their significance was recognized.

You will recall my mentioning Gustav Robert Kirchhoff (1824–1887), the father of the 'black body' concept. He is best remembered for his laws regarding electrical circuits in very practical use by electrical engineers and even

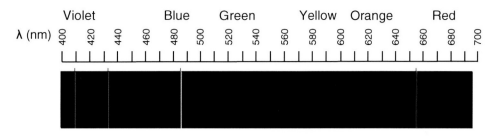

**Fig. 9.3** The emission spectrum of hydrogen. There are *four lines* in the visible spectrum, shown here. It was subsequently discovered that there are *four more lines* in the ultraviolet and infrared. The *blue line* at 486.1 nm dominates, so you would see light-emitting hydrogen as *blue*

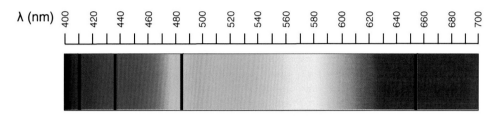

**Fig. 9.4** The absorption spectrum of hydrogen

electricians today. He formulated these in 1845 while still a graduate student at the University of Königsberg (in what was then East Prussia; now Kaliningrad in Russia). Later he was fascinated by the colors of light and the non-luminous rays that had been detected by William Herschel and Johann Ritter.

Serendipity, when events happening by chance lead to an unexpected, unusually good outcome, plays an important role in science. In 1852, Robert Bunsen, who had been Professor at Marburg, was hired as Professor of Chemistry by the University of Heidelberg. Part of the deal was that the University would build a new laboratory for him. The city was installing a gas lamp illumination system, and it was decided that gas lamps would provide lighting in the new building, and that gas would also be used for heating it. In 1854, while the building was still under construction, Bunsen realized that gas burners in the laboratory would be much more useful than the alcohol lamps previously used for chemical experiments. With the University mechanic, Peter Desaga, he designed a gas burner that used slits at the base to mix a small amount of air with the gas. At the top of the burner this produced an almost invisible flame—perfect for investigating the effects of chemicals in a flame.

Just at this time Kirchhoff came to work with Bunsen. Like Melville, they found that if chemicals are introduced into the almost invisible flame of the new 'Bunsen burner,' they produce brightly colored light. If the light is passed through a prism and spread out into a spectrum, these turn out to be very narrow bands of light representing very specific wavelengths. You can see the effect yourself if you put some

salt crystals into a gas flame. It emits a bright yellow light. This is called an emission spectrum. They also found that if white or colored light passes through a gas and is then split by a prism into a spectrum, black lines appear in some places. Specific wavelengths of the light are being absorbed by the elements in the gas, so this is called an absorption spectrum. More about what is actually happening to produce these phenomena later. Kirchhoff is regarded as the father of spectroscopy, a major technique in chemical analysis ever since. He formulated three laws of spectroscopy: (1) A hot solid object produces light with a continuous spectrum—a continuous emission spectrum; (2) A hot tenuous gas produces light with spectral lines at discrete wavelengths (i.e. specific colors) which depend on the energy levels of the atoms in the gas—a discontinuous emission spectrum (Fig. 9.3); and (3) A hot solid object surrounded by a cool tenuous gas produces light with an almost continuous spectrum which has gaps at discrete wavelengths depending on the energy levels of the atoms in the gas—an absorption spectrum (Fig. 9.4). Today, emission and absorption spectra are among the most fundamental tools of chemical analysis.

In the absorption spectrum you see the continuous colors of the spectrum interrupted by black bands corresponding to the emission lines shown in Fig. 9.4.

At the time, Kirchhoff could hardly have realized that investigation of these spectral lines would one day unlock the secrets to the structure of atoms and molecules. They would play a critical role in the revolution from classical Newtonian physics to Max Planck and Albert Einstein's quantum physics.

After laying the groundwork for spectroscopy during his work with Bunsen, Kirchhoff assumed the professorship that had been created for him in Berlin in 1875. There, he turned his attention to developing a theoretical explanation about how incoming and outgoing radiation operate, discussed in the next chapter.

## 9.2 The Discovery of Helium—First in the Sun, then on Earth

In the meantime a whole series of related discoveries were being made. Many years before, in 1868, two astronomers, Pierre Janssen in France, and Norman Lockyar in England had used a spectroscope while observing an eclipse of the Sun. The heart of a spectroscope is a triangular piece of glass, a prism, which spreads light into its component colors. Particular elements produce bright or dark lines in the spectrum, depending on whether they are emitting or absorbing energy. In the spectrum of the Sun's atmosphere, seen during totality, they found lines that indicated there was a chemical element present in the Sun but not known on Earth. Lockyar named the element 'Helium' from Greek *helios*, the Sun. In 1895, an English chemist, Sir William Ramsay of University College in London, discovered helium on Earth—it was a gas trapped in a specimen of uranium ore. Helium was about to become an important piece of the radioactivity puzzle.

## 9.3 The Discovery That Spectral Lines Are Mathematically Related

In 1885, two decades after Maxwell's discovery that light was an electromagnetic wave, another critical piece of the puzzle was fitted into place. Johann Josef Balmer was a mathematician who taught chemistry in a girl's school in Basel, Switzerland. When he was 60 years old, he discovered that there was a mathematical relation between the wavelengths of the spectral lines of hydrogen and the number 364.56.

To make hydrogen give off light, you need to enclose a small amount of the gas in a sealed tube with electrodes at both ends. At the time it was called a 'Geissler tube' or 'discharge tube;' when you apply a large voltage, such as 5,000–12,000 V, across the electrodes, the gas glows pink. If this sounds like some sort of exotic experimental device, it hasn't been, since in 1923 a Packard car dealership in Los Angeles installed two of the tubes filled with red-glowing neon gas. You see these many times a day. We now call them 'neon signs.' If the tube contains a tiny amount of mercury, it will be vaporized as the voltage is applied, and glow blue, but it also generates lots of ultraviolet radiation.

The ultraviolet radiation can be used to excite other compounds, 'phosphors,' and if the tube is coated with these you have a florescent light!

Balmer took advantage of the fact that the wavelengths of the hydrogen emission lines in the visible part of the spectrum (Hα, Hβ, Hγ, and Hδ) were well known at the time. The wavelengths had been measured quite precisely by Anders Jonas Ångström (1814–1874) in 1862 and 1871 as 6562.10, 4860.74, 4340.1, and 4101.2 Å. The Ångström unit, now simply the angstrom, is $10^{-15}$ m, so these are 656.210, 486.074, 434.01, and 410.12 nm, corresponding to red, blue, violet and indigo. These lines are now known as the Balmer Series. He found that all of the numbers reported by Ångström were related to a single value, 3645.6 Å. The expression he derived for the relation between these numbers wasn't simple, but it was exact:

$$\lambda = \frac{364.56 \times m^2}{m^2 - 4} \text{ or, without the numbers}$$
$$\lambda = \frac{bm^2}{m^2 - n^2}$$

where $\lambda$ is the wavelength in nanometers (nm). In the version without numbers, $b = 3.6456 \times 10^{-7}$ m (=364.56 nm), $m$ is an integer larger than 2 (such as 3, 4, 5, 6, etc.), and $n = 2$ (so $n^2$ is always 4). So if you make $m = 3$, the answer is 656.2; if $m = 4$, $\lambda = 486$; $m = 5$, $\lambda = 434$ nm; $m = 6$, $\lambda = 410$ nm; $m = 7$, $\lambda = 397$ nm (already in the invisible ultraviolet and not yet discovered when Balmer made his formulation). If you make m = ∞ (infinity), $\lambda = 364.56$ nm = b. One thing you will notice, the wavelengths keep getting closer and closer together as $m$ increases. Wonderful, but what could it possibly mean?

Meanwhile, in Sweden, physicist Johannes Robert Rydberg (1854–1919) was fascinated with periodicities among the chemical elements. Mendeleev's periodic table had been published in 1869, and new elements had been discovered which fitted into the blank spaces in the table. Rydberg also noted that there appeared to be periodicities in the spectral lines of the elements, and decided to pursue these. There was, after all, a wealth of data on spectral lines by the late 19th century. In 1888 he was working with the spectra of alkali metals. These metals, lithium, sodium, potassium, rubidium, cesium, and francium, are so reactive they do not occur in their elemental form in nature. They are always combined with something else. If you had a crazy chemistry teacher, like I did in high-school, he may have shown you what happens when you drop a little bit of metallic sodium into water. It explodes in a violent reaction that produces sodium hydroxide.

Rydberg was trying to figure out the relation between the spectral lines when he ran across Balmer's work. That progression of numbers being closer and closer together was something he had already noted in other elements. In the meantime, several more hydrogen lines had been discovered. Rydberg realized that the wavelengths Balmer had used belonged to some sort of special series (now known as the Balmer series). One additional emission line, in the infrared, fitted into Balmer's scheme, other newly discovered lines did not. In 1890, he revised the equation to cast it into a general form that would accommodate other series. Rydberg had been working with wave numbers rather than wavelengths, and this gave him a different perspective on the spectral lines. As you will remember from way back at the beginning of Chap. 7, the wavelength ($\lambda$) is the distance from crest to crest or trough to trough. The wavenumber ($n$ or $k$) is the reciprocal of wavelength ($n = 1/\lambda$); it is the number of waves in a given distance Rydberg first rewrote Balmer's equation in terms of wave number $k$ rather than wavelength, $\lambda$.

$$k = n_0 - \frac{4n_0}{m^2}$$

He realized that the '4,' which had been $2^2$ in Balmer's original formulation, was the fly in the ointment. Why couldn't it be any number? He rethought the whole problem and came up with a new equation for the spectral lines of hydrogen:

$$\frac{1}{\lambda} = k = R_\infty \left( \frac{1}{n_1^2} - \frac{1}{n_2^2} \right)$$

Here, $1/\lambda = k$ is the wavenumber, $R_\infty$ is the Rydberg constant, $1.097 \times 10^{-7}$ m$^{-1}$ (=109.7 nm) and $n_1$ and $n_2$ are integers (whole numbers), with the proviso that $n_2$ must be larger than $n_1$. Eureka! Now you can insert any integer for $n_1$.

Over the years, as spectroscopic techniques in the invisible parts of the spectrum have advanced, a number of other series in the hydrogen emission spectrum have been discovered. Table 9.1 shows six of the series discovered by the middle of the last century. There are even more now, but

they have not been given formal names. Why do additional series keep on being discovered? Because as $n_1$ increases, the series become fainter and fainter, and require ever more sensitive instrumentation to be detected. So, as the instrumentation continues to improve, new series are discovered. Another clue to the mystery.

Rydberg wasn't the only one investigating the meaning of spectra lines. A Swiss physicist, Walter Ritz (1878–1909), working at the University of Göttingen in Germany, also noticed something odd. He noted that if, for a given atom, there were spectral lines at two wave numbers, there was sometimes another spectral line at the precise sum of those two wave numbers. He published his idea in 1908, a year before his death from tuberculosis. This relationship became known as the combination principle. Whereas Rydberg's equation applied only to hydrogen and 'hydrogen-like' atoms, Ritz's combination principle had much broader application. The stage was set for someone to put all the clues together.

## 9.4 Heinrich Hertz's Confirmation of Maxwell's Ideas

The same year that one aspect of Maxwell's hypothesis—that the electromagnetic waves were transmitted through the luminiferous æther—was called into question by the Michelson-Morley experiment, Heinrich Hertz created electromagnetic waves in his laboratory, proving that Maxwell's hypothesis was essentially correct.

As a student Heinrich Hertz (1857–1894) (Fig. 9.5) studied with Kirchhoff and Hermann von Helmholtz in Berlin. Helmholtz (1821–1894) was an extraordinary individual with very broad interests. He was a physician as well as a physicist. Among other things, he was the person who took up Thomas Young's quest to understand how the human eye works and found the cones and rods that are sensitive to red, green, and blue—as Young had predicted in 1801. He inspired Hertz to explore the unexplored regions of physics. James Clerk Maxwell's ideas about electromagnetic waves were fascinating, but there was no experimental proof that they were correct. It was a major unexplored area in physics.

In 1885 Hertz became professor at the University of Karlsruhe. There he undertook experiments to confirm or refute Maxwell's ideas by generating electromagnetic waves in his laboratory, using an oscillating electric current. Maxwell had predicted that this could be done, and Hertz's goal was to prove or disprove Maxwell's ideas about electromagnetism. In effect, what Hertz built to send out an electromagnetic signal was what we would call an antenna. The detection device, the receiver, involved making a spark jump across the gap between two electrodes when the signal from the antenna was received. There was no physical connection

**Table 9.1** Series of emission spectral lines for hydrogen discovered through the middle of the last century

| $n_1$ | $n_2$ | Name | Discovered | Converges toward |
|---|---|---|---|---|
| 1 | $2 \rightarrow \infty$ | Lyman series | 1906–1914 | 91.13 nm (UV) |
| 2 | $3 \rightarrow \infty$ | Balmer series | 1884 | 364.51 nm (Visible + UV) |
| 3 | $4 \rightarrow \infty$ | Paschen series | 1908 | 820.14 nm (IR) |
| 4 | $5 \rightarrow \infty$ | Brackett series | 1922 | 1458.03 nm (IR) |
| 5 | $6 \rightarrow \infty$ | Pfund series | 1924 | 2278.17 nm (IR) |
| 6 | $7 \rightarrow \infty$ | Humphreys series | 1953 | 3280.56 nm (IR) |

*UV* ultraviolet; *IR* infrared. More have been discovered since

**Fig. 9.5** Heinrich Hertz (1857–1894). His name now represents the frequency of electromagnetic radiation. 1 Hz = 1 cycle (wavelength) per second

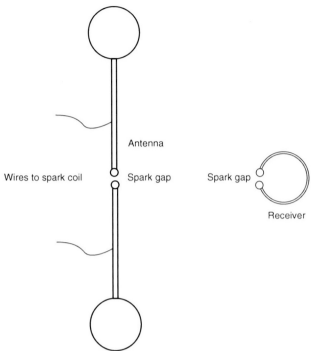

**Fig. 9.6** Schematic diagram of Heinrich Hertz's apparatus for generating and receiving an electromagnetic wave. The large balls at the ends of the 'antenna' store electricity until there is sufficient voltage to cause a spark to jump across the gap between the electrodes

between the generating and receiving devices. An annoyance in carrying out the experiment was that it was hard to see the spark. In 1887 Hertz concentrated his effort on observing the spark by enclosing the electrodes in a darkened box so that light from the spark was visible only through a glass window. He noticed something very odd. The spark was not as large as before. When he removed the glass window it got larger. He then experimented with a variety of different kinds of light sources. He discovered that invisible short wavelength ultraviolet light enhanced the spark effect. He reported the results of his experiment with ultraviolet light, but regarded it as a curiosity and did not follow up on it. However, others did, but more about that later.

In 1888 Hertz published his classic paper *Über die Einwirkung einer geradlinigen electrischen Schwingung auf eine benachbarte Strombahn* (=On the influence of uniform electrical oscillations on a neighboring electrical circuit). Hertz's apparatus, shown schematically in Fig. 9.6, generated a 3 m electromagnetic wave which was successfully transmitted across his laboratory. He had proven Maxwell's hypothesis to be correct. Oddly, it seems he did not realize the potential importance of his experiments to generate and receive electromagnetic signals. He never thought that it might be used as a means of communication. It was merely a curiosity of the physics laboratory.

Hertz died in 1894 at the age of 36. The cause of death was given as blood-poisoning—probably an infection. He had set

the stage for the development of a vast new technology, but as far as he was concerned, what he had been doing was just basic research. He had provided the experimental evidence that Maxwell's ideas concerning electromagnetic radiation were correct. That was his triumph. A collection of his papers was published posthumously in 1894.

## 9.5    Marconi Makes the Electromagnetic Spectrum a Tool for Civilization

One of those who got a copy of the 1894 book was a young Italian, Guglielmo Marconi (1874–1937) (Fig. 9.7). Marconi's father was Italian, his mother Irish. Marconi had been privately educated in Bologna, Florence, and Leghorn, Italy. He realized the practical implications of Hertz's studies. He began developing the 'wireless telegraph' in 1894, at his father's country estate at Pontecchio, in the hills just south of Bologna. There he succeeded in sending 'Hertzian waves' from one end of the house to the other, then from the house to the garden, and in 1895 over a distance of 2.4 km (1.5 miles). He approached the Italian government for financing to develop his 'wireless telegraph' but it showed no interest.

In 1896 with the encouragement of an Irish cousin, he took his apparatus to England and demonstrated it to the Chief Engineer of the British Post Office. He was soon granted a patent for his system of wireless telegraphy. He gave further demonstrations in London and on the Salisbury plain. He also made a transmission across the Bristol Channel. This new device might allow ships at sea to communicate with each other and with the shore; that would be a huge step forward in marine affairs.

Europe and America had been linked by a transatlantic undersea telegraph cable since 1858, but Marconi set out to determine whether it might be possible to send a wireless signal across the Atlantic. He had already discovered that the greatest distance the wireless signal could travel was proportional to the square of the height of the transmitting antenna. Common knowledge among scientists of the time was that electromagnetic waves travel in straight lines, but Marconi had the suspicion that for some unknown reason the transmission might follow the curvature of the Earth.

In 1901 Marconi set up a transmitting station at Poldhu, on the Lizard Peninsula, Cornwall, the southernmost point in England. The transmitter was 100 times more powerful than any previous wireless transmitter. The spark-gap transmitter fed a huge antenna array consisting of 400 wires suspended from 20 masts, each 200 ft tall, forming a circle. A similar station was set up on the American side of the Atlantic at South Wellfleet on Cape Cod. On September 17, 1901, a gale blew down the antenna array at Poldhu, and on the 26th of November the Cape Cod antenna was destroyed by a hurricane. A smaller 4 mast array was reconstructed in Poldhu, and the receiving station moved to what is now Signal Hill in St. John's, Labrador. The 500 ft (154.2 m) Labrador antenna was held aloft by a kite. The distance between the stations was about 3,500 km (2,100 miles). Around noon (local time) on December 12, 1901 Marconi reported receiving a signal of three Morse Code dots (the letter S) in Labrador. There is some second guessing as to whether this attempt at transatlantic transmission was really successful, because the wavelength was stated to be 366 m, corresponding to a frequency of 820 kHz. That is in the AM part of the modern radio dial, and we now know that AM transmissions are usually limited to a few hundred kilometers in daylight but go much further at night. Of course, it is possible that the frequency was reported incorrectly. Recent analysis suggests that the frequency might have been as high as 7 MHZ, which would have been much more likely to have been received. A much clearer, undisputed signal was detected in 1902, with the receiver located at Glace Bay, Nova Scotia.

Eleven years later it would be Marconi's wireless telegraphic transmission that was used to try to call ships to assist the sinking RMS *Titanic* on the night of April 15, 1912. The ship was about 350 nautical miles off St. Johns. Marconi's initiatives ultimately led to the transmission of audio radio (1919), FM radio and television (1937), and the many other wireless communication devices that are parts everyday life today.

**Fig. 9.7**  Guglielmo Marconi (1874–1937)

But how could the radio transmission follow the curvature of the Earth? It doesn't. It is reflected off a layer of ionized gas in the uppermost part of the atmosphere.

In 1902, American electrical engineer Arthur Edwin Kennelly (1861–1939) and British physicist Oliver Heaviside (1850–1925) independently predicted the existence of a layer of ionized air at the top of the atmosphere that could reflect radio waves. This became known as the Heaviside-Kennelly Layer or simply the Heaviside Layer. This layer of ionized gas extends from about 90–150 km (56–93 miles) above the Earth's surface. It reflects or refracts medium-frequency 300 kHz–3 MHZ radio waves, which includes what we know as the AM and shortwave broadcast bands. Because of this phenomenon radio waves can be propagated beyond the horizon, but for the AM radio frequency spectrum, the reflection occurs only at night. Already in 1902 it was evident that Earth was not simply surrounded by a gaseous atmosphere, but that free electrons were a part of its outermost layer. Radiation from the Sun was more than just light, infra-red heat, and some ultraviolet that might cause sunburn. Something else was producing very strange conditions far above the planet's surface.

If the name 'Heaviside Layer' seems familiar it is probably because you have seen or heard Andrew Lloyd Weber's musical 'Cats.' When Grizabella's time has come, Old Deuteronomy sends her

Up, up, up, past the Russell Hotel
Up, up, up, up, to the Heaviside Layer

In *Cats* the Heaviside Layer is a sort of heaven where cats go to be reborn. Grizabella sings the beautiful song 'Memories.'

Most of the lyrics of *Cats* are from T.S. Eliot's *Old Possum's Book of Practical Cats*, assembled from earlier poems in 1939, but the story of Grizabella is not among them. Eliot's wife, Valerie, thought it was too sad for children, and so it was not included. Mrs. Elliot attended the 1980 Sydmonton Festival concert where some of Andrew Lloyd Webber's musical settings of the Old Possum poems were performed. She gave him the paper with the fragment of Grizabella written on it. The rest is history. Andrew Lloyd Webber incorporated the Grizabella poem as a main theme in the plot of *Cats*.

## 9.6    Human Use of the Electromagnetic Spectrum for Communication, Locating Objects, and Cooking

Figure 9.8 shows the electromagnetic spectrum with detail for the part between 100 kHz and 10 GHz. It is in this part of the spectrum that humans now generate signals used for radio communications, television, radar, and microwave ovens. The original spark signals (the dots and dashes of Morse code) were replaced by more complex transmissions which allowed voice and music to be heard. The earliest of these involved modulation of the amplitude of the signal (AM radio, and shortwave radio), soon followed by using modulation of the frequency to carry the information (FM radio, television VHF and UHF bands). The first regular radio broadcasts started in London in 1922, and the first television broadcasts also started there in 1936. The first use of radar to detect ships was conducted on the Potomac by US Navy researchers in 1922. This idea was dropped until World War II.

During the Second World War it was found that signals in the 1–5 GHz range would bounce off large objects in the sky and on the ocean—radar became an important tool. In 1946, Percy Spencer, a self-taught engineer with the Raytheon Corporation, noticed that while he was testing a new vacuum tube called a magnetron used for radar, the candy bar in his pocket had melted. It was discovered that certain frequencies would not only soften chocolate candy bars but could even boil water, leading to the development of the microwave oven. Most commercially available microwave ovens use a frequency of 2.45 GHz. More about why they use that particular frequency and how a microwave oven works later. At higher frequencies (10–20 GHz) the signal bounces off raindrops, giving us weather radar.

It is interesting to note what a large part of the electromagnetic spectrum now has waves generated by human activity compared with the relatively narrow band representing visible light.

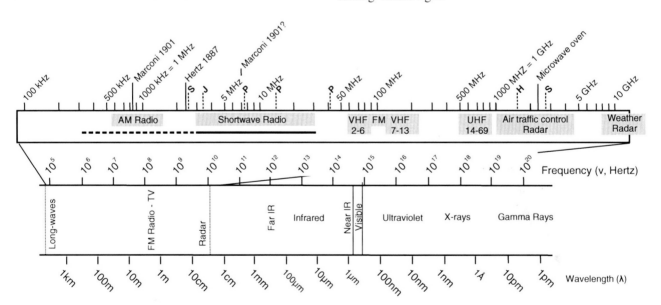

**Fig. 9.8** The electromagnetic spectrum, with detail showing the part now generated for human activities, shown in *gray*. The *dashed line* beneath 'AM Radio' is the part of the spectrum reflected by the ionosphere during night. The *solid line* beneath 'Shortwave radio' is the part of the spectrum reflected by the ionosphere during both day and night. VHF 2–6 are television channels 2 through 6; similarly VHF 7–13 and UHF 14–69 are television channels 7–69. In the frequency scale, *Hertz 1887* is the wavelength generated during his laboratory experiment. *Marconi 1901* and *1901?* are the wavelengths he may have detected in 1901. The *dashed lines* in the frequency scale are the frequencies of natural signals from outer space detected by radio astronomers: *S* from the Sun; the line at 2.8 GHz reflects sunspot activity; *J* Jupiter; the lines marked *P* are from Pulsars. No signals from other life forms have yet been detected, but the SETI Institute has been searching for such signals since 1985

## 9.7 Summary

In this and the preceding chapters we have covered many topics which may have seemed unrelated. We started by looking at water waves. They are the only waves that we can see and experience. We learned that the wave is something that moves through but does not actually transport matter. Even stranger, the thing that moves through the water is not an individual wave, but a wave train, with each individual wave having a finite and rather short lifetime. It turns out that the speed of a water wave depends on both the wavelength and water depth. Water waves are very special; nevertheless, much has been learned from their behavior.

Sound waves are compression waves in the air, water, or solids. In contrast to water waves, all wavelengths/frequencies travel at the same speed, and speed depends on the temperature and density of the medium though which they travel. The first accurate measurement of the speed of sound in air was not made until 1821. The speed of the compression-rarefaction waves in air is strongly dependent on temperature, somewhat on humidity, and very slightly on pressure. The velocity of sound in water was first measured in 1826. In water, sound velocity depends strongly on temperature, less strongly on pressure, which is a function of depth, and in seawater, on salinity. In solids, speed depends on density and temperature.

The nature of light was long a mystery. The first measurement of the speed of light was made in 1676, only 25 % short of the modern value. In some of its properties, it seemed to be a wave phenomenon, but in others it behaved like a stream of particles. Light exhibits a number of properties analogous to water waves, such as refraction and diffraction. Although light of different wavelengths travels at the same speed in a vacuum, their speeds differ in transparent solids, like glass, separating white light into a spectrum of colors. Light takes two separate paths through the mineral Iceland Spar ('optical calcite'), mystifying scientists for over 200 years. Light can be polarized by reflection off a smooth surface. But it still wasn't clear what light was. Finally, in the 1860s James Clerk Maxwell discovered from his work on electromagnetic waves that they had the same velocity as that of light, leading him to conclude that light is an electromagnetic wave.

It was discovered that black lines appeared on the spectrum of the Sun when it was examined in detail, producing what is now called an absorption spectrum. It was then discovered that substances can be caused to emit light under special circumstances, but the light is in very specific wavelengths. This is now known as an emission spectrum. It was found that spectra can be used to determine which elements are in a compound. Finally, it was discovered that the lines in the emission spectrum of the simplest element, hydrogen, bear an exact mathematical relation to each other.

These clues were needed to make the huge leap from the classical physics of the 17th–19th centuries to the quantum physics of the 20th century. This would ultimately allow us to understand how greenhouse gases work to modify the climate system.

A Timeline for this chapter:

| 1670–1672 | Isaac Newton water investigates the spectrum of colors created by passing white sunlight through a triangular glass prism |
|---|---|
| 1752 | Thomas Melville, in Glasgow, noted that chemicals introduced into a flame produced colors |
| 1802 | William Wollaston discovers black lines in the spectrum of the Sun |
| 1814 | Joseph Fraunhofer invents the spectroscope and discovers 570 lines in the spectrum of the Sun |
| 1845 | Gustav Kirchhoff, while still a student, formulates several laws concerning electrical circuits |
| 1853–1855 | Heidelberg installs gas-light illumination. A new Chemistry Laboratory Building being constructed for newly appointed Professor Robert Bunsen is outfitted with gas lines |
| 1854 | Technician Peter Desaga and Robert Bunsen invent the 'Bunsen Burner' |
| 1854 | Kirchhoff comes to Heidelberg to work with Bunsen, spends the next years developing the science of spectroscopy for chemical analysis |
| 1856 | James Clerk Maxwell discovers that light is an electromagnetic wave |
| 1858 | First transatlantic telegraph cable connects Britain and North America |
| 1862 | Anders Ångström determines the wavelengths of the hydrogen Hα, Hβ, and Hγ, spectral lines to be 6562.10, 4860.74, and 4340.1 Å |
| 1868 | Pierre Janssen (France) and Norman Lockyar (England) observe the Sun's corona during an eclipse and discover a new element—helium |
| 1869 | Dimitri Mendeleev publishes his Periodic Table of the Elements |
| 1871 | Anders Ångström determines the wavelengths of the hydrogen Hδ to be 4101.2 Å |
| 1875 | Gustav Kirchhoff becomes Professor at the University in Berlin, and proposes the 'black body' as a means of investigating absorption and emission of radiation |
| 1885 | Swiss high-school teacher Johann Balmer discovers the mathematical relation between the spectral lines of hydrogen |
| 1887 | Heinrich Hertz describes the effect of ultraviolet light on an electrically charged surface |
| 1888 | Hertz confirms Maxwell's prediction that there are unknown parts of the electromagnetic spectrum by creating and sending the first radio wave, with a wavelength of 3 m, across his laboratory |
| 1890 | Swedish chemist Robert Rydberg improves Blamer's equation to make it generally applicable |

(continued)

| 1894 | Hertz dies. A collection of his works is published as a book |
| 1894 | Guglielmo Marconi gets a copy of Hertz's book, devises a 'wireless telegraph' and sends 'Hertzian Waves' from one end of his father's house to the other end, then across the garden, then to a neighboring house |
| 1895 | Chemist Sir William Ramsay of University College, London, discovers helium on Earth—trapped in a specimen of uranium ore |
| 1895 | Marconi sends a 'wireless telegraph' message over a distance of 2.4 km |
| 1896 | The Italian Government not being interested in his device, Marconi takes his 'wireless telegraph' to England |
| 1901 | Marconi receives the first transatlantic wireless telegraph communication, three dashes for the letter "S". It is later questioned whether this was an actual message or merely noise |
| 1902 | Marconi receives an unequivocal transatlantic wireless telegraph communication |
| 1902 | American electrical engineer Arthur Edwin Kennelly and British physicist Oliver Heaviside independently predict the existence of a layer of ionized gas at the top of the atmosphere that can reflect radio waves, known as the Heaviside-Kennelly Layer or simply the Heaviside Layer |
| 1908 | Wilhelm Ritz discovers that if for a given element, there are spectral lines there is sometimes another spectral line at the precise sum of the wave number of the original two spectral lines |
| 1922 | First regular radio broadcasts, London |
| 1922 | First use of radar to detect ships on the Potomac |
| 1936 | First television broadcasts, London |
| 1939 | Research on radar revived |
| 1946 | Percy Spencer discovers that certain frequencies of microwave radiation can melt candy bars and even boil water |
| 1985 | SETI Institute starts listening for meaningful radio signals from outer space |

If you want to know more:
John D. Jenkyns 2009. *Where Discovery Sparks Imagination: A Pictorial History of Radio and Electricity*. The American Museum of Radio and Electricity, 224 pp.—A general overview with great pictures.

Music: John Williams wrote the music for the 'Star Wars' series and many other films. Great stuff. There are a number of recordings of various lengths.

Libation: A good Munich beer in honor of Joseph Fraunhofer! Franziskaner Weissbier is a favorite at the Gaststätte Fraunhofer in Munich, where you can hear brass music from 11 AM on Sunday mornings.

## Intermezzo IX. More Czech Adventures

After I returned to the University of Illinois, the fall was uneventful, but in the spring a friend in the Chemistry Department called and said that they had a postdoctoral fellow from Czechoslovakia. They knew I had been there, and thought that perhaps we might like to meet. So I met Jiři Jonas. Those first meetings with colleagues from the Soviet bloc were always a bit difficult because one never knew what the other person's political bent might be. However, after a few minutes together we found we had lots to talk about, and were soon laughing about all the rules and regulations one followed in Czechoslovakia. If you hadn't experienced life in the Communist bloc first hand, it was very difficult to imagine. We became very good friends.

I made another trip to Czechoslovakia in the summer of 1963. There was a lot of news. Pavel had divorced and remarried. His new wife, Jiřina, was a delight, always cheerful and bright. We continued to work on Cretaceous nannofossils at the office in Spálená, and this time all of Pavel's staff gave me warm welcome.

We made a longer field trip, partly for collecting rocks, but mostly for sightseeing. We visited Karlštejn, a magnificent medieval castle built by Charles IV. Then we took off on a trip through southern Bohemia and to Moravia where we finally found the little museum in the Augustinian Monastery in Brno devoted to Gregor Mendel. It was in the gardens of the Monastery that Mendel, working with different strains of sweet peas, discovered the rules by which traits are passed from one generation to another. His work laid the foundation for the modern science of genetics. In those days there were very few visitors to the little Museum, and I was able to see his notebooks and even tried on Mendel's eyeglasses.

Mendel was anathema to Soviet agricultural biologists, who were led by Trofim Denisovitch Lysenko, the biologist in charge of Soviet agriculture. Lysenko believed in the ideas of Ivan Vladimirovich Michurin, and thought that changes in plant species were brought about through hybridization and grafting, and that traits thus acquired could be passed on the offspring. This fitted in well with a Communist ideology fostered by Joseph Stalin and his circle: that all people were created equal and their differences as adults were purely a product of the environment in which they lived. This means that living under socialism would eventually result in a human population free of bourgeois tendencies. Mendelian ideas of heredity were heretical and anathema to the Communist Party. Lysenko was the symbol of ideologically correct 'science.' Visiting the Museum in the monastery in Brno was a sign of disrespect to Communist orthodoxy.

On the way back to Prague we found a field of corn. Pavel had never tasted roasted ears of corn before, so we stopped and decided to 'borrow' a few ears from the field. We got caught by the farmers, who asked what we were doing. Pavel explained that I was an American and wanted to show him how we eat corn in America. The farmer's couldn't seem to stop laughing. Then one of them opened an ear of corn for us. No kernels, just a cob. They were faithfully raising the corn from seed that state had provided following Lysenko's formulas. The farmers were doing exactly as they were told by their government, even though the result was no food. Good Soldier Švejks at work again.

The very next year Lysenko was out of a job due to massive agricultural failures in the Soviet Union, and Mendel was honored again as a Czechoslovak hero. Communist ideological pseudoscience was replaced by real science based on observations, research, and testing of hypotheses by experiment. It would be nice if that were the end of the story, but ideologically correct 'science' has reappeared in, of all places, the United States, often masquerading as religion or economic necessity. Surveys indicate that about 30 % of the US population believes that the Sun orbits the Earth, and only slightly over half believe in evolution, and even fewer believe in the 'Big Bang' origin of the Universe. The replacement of rational thinking by ideology is manifest in the current politicization of 'global warming.' And we even had our own Lysenko, Philip Cooney, originally a lobbyist for the American Petroleum Institute, was named Chief of Staff of the White House Council on Environmental Quality. He edited the reports on climate coming from NASA and the Environmental Protection Agency to remove references to global warming and climate change. It turned out that Cooney was a lawyer with a bachelor's degree in economics and no scientific training. He was hired by Exxon-Mobil immediately after he was exposed and resigned.

In the fall of 1963, Jiři Jonas was to return to Prague. He came to see me and told me he was very uncertain about his future life there. He had grown used to America and the freedom to pursue his research as he wanted, and didn't know what life would be like when he returned. We made a plan. If he felt he absolutely had to get out, he would send me a letter, and it would contain a reference to a fence around a piece of property I owned near Urbana. The mention of the fence meant he was desperate to escape.

A month or so after he returned I got a letter from Jiři telling me how well he had been received back into his institute, that everything was going very well and that he was very happy to be back in his homeland. He was even going to be attending a conference in Vienna to present the results of his research in a couple of months. I breathed a sigh of relief until I got to the last line of the letter: "Have you been able to finish the fence that we worked on around your property?"

I called Herb Gutowsky, the Chemistry professor who had been Jiři's advisor and host, and took the letter to his office and explained what it meant. He said "I have some contacts; I'll see what can be done." A couple of days later the local FBI agent came to my office and we discussed the matter. He complimented me on the subtle method of communication Jiři and I had used. He was passing on information to another agency.

However, I thought about what would happen if the US agencies didn't act. I got out my maps of Austria and Switzerland and made a plan. I knew the border area between the two countries rather well from visits to friends who lived very close to the border. I figured that if worse came to worse, I could go to Vienna, find Jiři at the meeting, and drive him to near the border, let him cross on his own, and pick him up on the Swiss side. I had friends in Switzerland who were well connected and I was sure we could arrange asylum there. The problem was I didn't know how the timing would work out. Then I did something that scared me out of my wits.

I picked up the telephone and called a friend in Vienna and said that I had a friend in Czechoslovakia who needed to leave the country. He would be coming to Vienna soon. Would it be possible for him to meet my friend and put my friend up at his house for a few days if necessary, until I could get to Vienna. The response was instantaneous. "Austria is a neutral country; we don't do things like that. How dare you call me and suggest such a thing." After a couple more nasty comments he hung up.

I sat there stunned. It wasn't exactly the response I expected from an old friend. It brought home to me the fact that aiding someone's escape from one of the Soviet bloc countries was an exercise not to be taken lightly. I also learned that in matters like this you should never trust anyone. Fortunately I had not mentioned Jiři's name or which Conference might be involved. My Viennese friend would almost certainly have assumed that I was talking about a Czech geologist or paleontologist. He might even have thought it was Pavel. Fortunately, Pavel wasn't headed to any conferences in Vienna that year.

A few days later the FBI agent walked into my office again to let me know that he had been told that things were moving along well.

In the spring of 1964, without warning, my office door opened, and there stood a smiling Jiři. "Hello Bill, I'm back."

After the shock of my telephone call to Austria, I was pleased that all this had happened without anyone, not even my Department Head, knowing about it. Then a couple of months later I was at a cocktail party where a faculty member from another Department, someone I hardly knew, introduced herself and said, "Oh, you're Bill Hay, the one who helped Jerry Jonas come back to the US." I realized that someone had 'let the cat out of the bag' and this was not good news; because I was to be on my way back to Prague a couple of months later.

ERNST PRINSHEIM AND OTTO LUMMER SHOW
WILHELM WIEN THEIR DEVICE FOR GENERATING
BLACK BODY RADIATION

*Progress is made by trial and failure; the failures are generally a hundred times more numerous than the successes; yet they are usually left unchronicled.*

William Ramsay

Through the 13th century no one in Europe seemed to have worried very much about why the Earth was such a comfortable place to live. After all, Augustine of Hippo (aka Saint Augustine, 354–430 CE) had proclaimed that 'we know all we need to know.' The centers of science and mathematics were in the Moslem world on the other side of the Mediterranean. During the Medieval Climate Optimum (ca. 950–1250 CE) temperatures in Europe were warmer than most of the 20th century. My own suspicion is that the heavy Romanesque architecture of the time with its thick walls and small windows provided cool retreats in the summer. They are pleasant places on a hot summer day today. Monasteries and castles did not have fireplaces for heating, only for cooking food.

These were the 'Dark Ages,' but they were not dark in terms of life being especially difficult. Food was plentiful, and the living was easy. They were 'dark' because the church believed that we knew all we needed to know. Intellectual curiosity was not valued. Dante Alighieri (1265–1321), in *The Divine Comedy*, written between 1308 and the time of his death, gives us an idea of values and politics at that time. Easier for us to understand is Umberto Eco's *The Name of the Rose*, a fascinating novel set at this time describing attitudes of the establishment and intelligentsia.

Then, in 1348 the Black Plague arrived in Europe, having come from Asia. Within a few years a third to half of the population had died. Suddenly the smaller population needed less food. Fields that had been used to grow food were abandoned and gradually replaced by forests. A century later the weather began to turn bad. Summers were not as warm and winters were sometimes very cold. The two historical events may not be unrelated. More about that later.

I have long had the suspicion that the Renaissance, which started in the middle of the 14th century, may have been in part a reaction to the changing weather in Italy. The idea that 'we know enough' was replaced by a new quest for knowledge. It was an intellectual revolution on a scale unknown since the Roman Empire. The Renaissance involved much more than architecture and painting—they were merely an outward expression of the new freedom of thought. People began to think about the world around them and to question old habits and ways of doing things. Leonardo da Vinci (1452–1519) showed us in his notebooks plans for all sorts of inventions, including many fascinating mechanical devices. Fireplaces began to appear outside the kitchen to heat the home during the colder winters.

## 10.1 What Is Heat?

Among the intellectuals there developed an interest in figuring out the nature of heat and cold.

You may recall that the ancient Greeks believed that there were four elements: *air, earth, fire,* and *water.* All four could change with time. Fire (=heat) was thought to be something physical like the other three. Aristotle had added a fifth element, *æther,* which he thought was incorruptible and unchanging. He thought stars were unchanging and therefore made of *æther.* In the 19th century æther will come back to haunt us in another form.

In 1667 German alchemist Johann Joachim Becher (1635–1682) proposed that fire was actually something which he called '*terra pinguis*' later renamed '*phlogiston.*' It was a substance without color, odor, taste, or mass contained within combustible bodies and released when they are burned. The idea was accepted for a century.

Antoine de Laurent Lavoisier (1743–1794) (Fig. 10.1) is regarded as the father of modern chemistry. Of noble birth, he worked successfully under both Louis XVI and the early Revolutionaries. Louis XVI was interested in producing a rational, uniform system of measurement, and Lavoisier took on the task. It would become the metric system, based on units that vary by powers of 10. After the Revolution it would be spread throughout continental Europe by Napoleon.

Lavoisier discovered that, although matter may change its form or shape, its mass remains the same. Oxygen had been

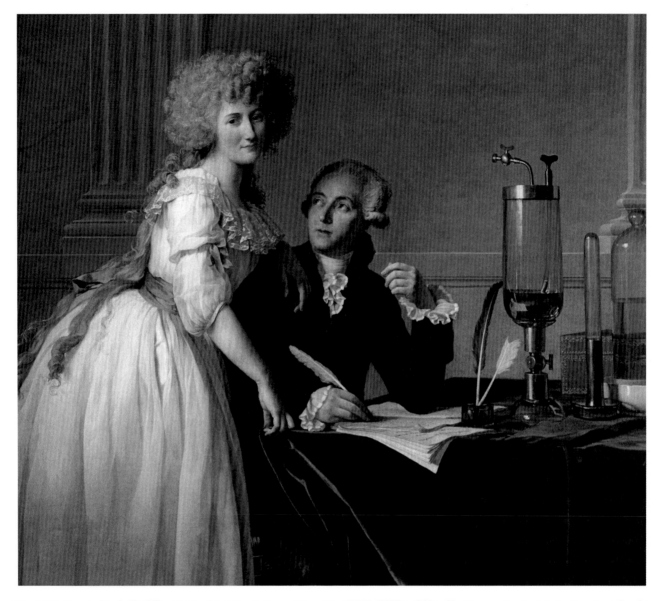

**Fig. 10.1**  Jacques-Louis David's portrait of Antoine de Laurent Lavoisier (1743–1794) and his wife. He was simply Antoine Lavoiser after the French revolution (1789). Guillotined during Robespierre's Reign of Terror, he is considered the father of modern chemistry

discovered independently by two chemists: Carl Wilhelm Scheele, in Uppsala, Sweden in 1773 or perhaps earlier, and Joseph Priestley in Wiltshire, in 1774. Priestley is given priority because his work was published first. In science, he who publishes first gets the credit. Although oxygen had already been discovered, it was Lavoisier who gave it its name in 1777. He figured out that burning was a chemical reaction combining oxygen with other elements. He determined that water was a molecule consisting of hydrogen and oxygen.

In 1783 he published a paper "*Réflexions sur le phlogistique*" in which he argued that the phlogiston theory was inconsistent with his experimental results. Instead, he proposed that there was a 'subtle fluid,' which he named '*caloric*' that was the substance of heat. He believed that the quantity of caloric is constant in the universe and that it was self-repellant, so that it flowed from warmer to colder bodies. Others, in the 1780s, believed that cold was also a fluid, '*frigoric*,' but Lavoisier's colleague Pierre Prévost argued that cold was simply a lack of caloric.

Lavoisier's experiments were cut short when he was branded a traitor by Maximilien de Robespierre during the Reign of Terror. He was executed by guillotine on May 8, 1779. He was exonerated posthumously a year and a half

**Fig. 10.2** Benjamin Thompson, a.k.a. Reichsgraf von Rumford (=Count Rumford of the Holy Roman Empire). Expatriate American

later and the French Government apologized to his wife for the mistake.

In 1804 the expatriate American colonist who had fought with the British during the Revolution, Benjamin Thompson (Fig. 10.2), announced the results of his experiments to the Royal Society of London. Thompson had acquired the title of Reichsgraf von Rumford (=Count Rumford of the Holy Roman Empire) by the then Prince-Elector of Bavaria, Karl Theodor. Rumford was the town in Massachusetts where Thompson had been born. He had been involved in manufacturing cannons at the arsenal in Munich. Curious about the frictional heat generated by boring cannon barrels, he designed an experiment: he immersed a cannon barrel in water and used a blunted boring tool to drill it out. He found that the water came to a boil in about two and a half hours. There was a seemingly inexhaustible supply of frictional heat. If the heat were a substance, like the caloric accepted at the time, it should not be inexhaustible. He also noted that the only thing that had happened to the barrel was that it had been subjected to friction by the motion of the boring tool. He compared the specific heats of the material machined away and the finished barrel and found that they were the same. There had been no physical change in the materials.

He concluded that heat was the motion of the atomic particles of which the matter composing the barrel and boring tool were made. The faster the motion, the hotter the object became. This discovery paved the way for our modern understanding of temperature. It was just at this time that the thermometer, which goes back at least to Galileo's time, was being developed and calibrated. Now there was an idea of what it might actually be measuring. Incidentally, Thompson's discovery was rapidly accepted in Britain, but 'caloric' lived on in other parts of Europe and in the United States. For the Americans, he was after all a traitor, and not to be trusted.

There was another seemingly unrelated discovery. In 1800 William Herschel found that there was more to the radiation coming from the Sun than just the visible spectrum. There were invisible rays beyond the red end of the spectrum that would heat thermometers. Somehow these non-luminous rays caused objects to warm. It was soon realized that similar non-luminous rays were being given off by warm objects on the Earth, like rocks heated by the Sun, or a sheet of iron placed in the back of a fireplace.

In 1824 French mathematician Jean-Baptiste Joseph Fourier (Fig. 10.3) made a remarkable suggestion:

> The earth receives the rays of the Sun, which penetrate its mass, and are converted into non-luminous heat; it likewise possesses an internal heat with which it was created, and which is continually dissipated at the surface; and lastly the earth receives rays of light and heat from innumerable stars, in the midst of which is placed the solar system. These are the three general causes which determine the temperature of the earth.

> The transparency of the waters appears to concur with that of the air in augmenting the degree of heat already acquired, because luminous heat flowing in, penetrates, with little difficulty, the interior of the mass, and non-luminous heat has more difficulty in finding its way out in a contrary direction.

From: *Remarques generales sur les températures du globe terrestre et des espaces planétaires* (Annales de chimie et de physique, 1824, Paris, 2me serie, 27, 136–167; English Translation by Ebeneser Burgess, 1837, American Journal of Science, 32, 1–20).

It was a succinct statement of what has come to be known as the energy balance model for determining the temperature of a planet, but it was just an idea. The data necessary to make a rigorous mathematical formation did not exist.

Based on the last sentence cited above, Fourier is now often cited as the father of the atmospheric 'greenhouse hypothesis,' but in reality he had no idea why the non-luminous heat had more difficulty in finding its way back into space. The greenhouse effect of specific gases is something that would be investigated and described by John Tyndall 25 years later.

**Fig. 10.3**  Jean-Baptiste Joseph Fourier (1768–1830)

The science of thermodynamics was developed to increase the efficiency of early steam engines. It can be traced back to French military engineer Nicolas Léonard Sadi Carnot (1796–1832) (Fig. 10.5) who believed that the efficiency of heat engines could be improved if the basics of heat transfer were understood. In 1824 he published a book, *Réflexions sur la puissance motrice du feu* ("Reflections on the Motive Power of Fire"). In it Carnot sought to answer two questions about the operation of heat engines: "Is the work available from a heat source potentially unbounded?" and "Can heat engines in principle be improved by replacing the steam with some other working fluid or gas?" He answered these in terms of 'caloric' which was still widely accepted in France as the basis of heat: "The production of motive power is therefore due in steam engines not to actual consumption of caloric but to its transportation from a warm body to a cold body." And: "In the fall of caloric, motive power evidently increases with the difference of temperature between the warm and cold bodies, but we do not know whether it is proportional to this difference." (From the Dover Press, 1960, translation of Carnot's book). In the book Carnot describes an engine (now known as the theoretical 'Carnot Engine') that makes use of both introduction

## 10.2  Thermodynamics

One of the pillars of understanding climate is the area of physics known as thermodynamics. The word is derived from the Greek words *therme*, meaning "heat" and dynamics, meaning "power." In effect it is the study of the conversion of energy between heat and mechanical work. It involves variables including temperature, volume and pressure.

Although its origins are ancient, the first practical steam engine was developed over the period 1763–1775 by James Watt (1736–1819) (Fig. 10.4) while working as an instrument maker at the University of Glasgow, Scotland. It was a great improvement over earlier versions because it used much less fuel (coal) to achieve the same result and was less likely to explode. His invention is credited with having initiated the Industrial Revolution. The steam engine is an external combustion engine, in that the fire is outside the engine itself. Most modern automobiles have an internal combustion engine, where the burning of gasoline takes place inside the cylinders of the engine. Steam power has been used for ships ('steamships'), railway locomotives, and even an automobile, the Stanley Steamer. It is still the major method used for generating electricity.

**Fig. 10.4**  James Watt (1736–1819). Technician at Glasgow University and later designer of the practical steam engine (1875). Contributor to the Scottish Enlightenment

and removal of heat, but notes that there is a problem because of the gradual loss of 'caloric.'

Sadly, Sadi Carnot died of cholera at the age of 36.

Carnot's ideas did not lead to an immediate improvement of steam engines, but were critical in the development of the diesel engine in 1893.

The next solid step in the formal development of the science of heat transfer was made by our old friend Lord Kelvin in 1851, early in his career. In an article "*On the dynamical theory of heat, with numerical results deduced from Mr. Joule's equivalent of a thermal unit, and Mr. Regnault's observations on steam*," published in the Transactions of the Royal Society of Edinburgh, he laid out much of what would become the mathematical basis of thermodynamics. In his paper he makes reference to the work of the young German physicist/mathematician Rudolf Clausius.

The father of formal thermodynamics is considered to have been German physicist/mathematician Rudolf Julius Emanuel Clausius (1822–1888) (Fig. 10.6). In 1850, just two years after completing his doctorate, he formulated what has become known as the Second Law of Thermodynamics. In 1865 he published an article with two famous statements: (1) The energy of the universe is constant; (2) The entropy of the universe tends to a maximum. Don't worry; I'll explain what entropy is below. By the way, that first statement

**Fig. 10.6** Rudolf Julius Emanuel Clausius (1822–1888). Looking quite stern

sounds a lot like Lavoisier's statement that the amount of caloric in the Universe is constant.

## 10.3 The Laws of Thermodynamics

Thermodynamics is the area of physical science that deals with the relations between heat and other forms of energy and the relationships between all forms of energy. You remember that heat is atomic or molecular motion. But let us go back and review what energy is: it is the ability of one physical system to do work on one or more other physical systems. Remember that work is a force acting over a distance. A force is an acceleration times a mass. An acceleration is a change in velocity, and velocity is expressed as distance covered in a length of time. Velocity is expressed in meters per second (m/s or ms$^{-1}$), and acceleration is meters per second per second: written as m/(s × s) or m/s$^2$ or ms$^{-2}$. Force is then kg ms$^{-2}$; if the units are 1 kg, 1 m, and 1 s, then the force, in SI units, is one newton (N). So energy represents the ability to push or pull something or alter its state. If the force acts over a distance (m) it becomes energy expressed as kgm$^2$/s$^2$ or kgm$^2$s$^{-2}$ or Nm. The SI unit of energy is the Joule, which is one newton meter. The unit is named after the English brewer of fine ale and beer and

**Fig. 10.5** Nicolas Léonard Sadi Carnot (1796–1832)

hobby physicist James Prescott Joule (1818–1889) (Fig. 10.7). The brewery was in Salford, a suburb of Manchester, so Joule, of course, knew the scientists at the institutes there. Since James' day, the brewery has moved to Market Drayton in Shropshire. Sadly it does not sell its ales and beer outside the local area, so to tip a pint of Joules in James' honor; you will just have to go there.

It was in 1843 that Joule published his very important paper on the relation between heat, energy, and work. The definition of energy includes an important word: 'ability,' and a mysterious clause 'alter its state.' This means that it includes potential energy, which is energy stored and waiting to be used, latent energy, which is inherent in a change of phase (such as solid to liquid to gas), chemical energy, which is inherent in the bonds between atoms, electronic energy, which involves changing the orbits of electrons in atoms and molecules, and nuclear energy, which involves fiddling with the atom's nucleus.

Although the SI unit of energy is the Joule, the very small amounts of energy associated with atoms and electrons are usually given in 'electron volts' (eV). An electron volt is the energy acquired by a single electron (its mass = 9.109383 $\times 10^{-31}$ kg or 1/1836 that of a proton) accelerated through an electrical potential of one volt. 1 electron volt (eV) = $1.60217646 \times 10^{-19}$ joule (J); conversely, 1 J = $6.24150974 \times 10^{18}$ eV. There are about 18 orders of magnitude difference when thinking about a baseball and an electron.

The "Laws of Thermodynamics" sound very impressive and you might think they are some sort of difficult abstract concepts. They are statements about things you already learned in high school. There are four laws:

**First Law**: Energy can be neither created nor destroyed. It can only change forms.

This is sometimes called the law of conservation of energy. It is another way of saying what Clausius said in 1865. To me it is something that is not intuitively obvious. One of the kinds of energy that changes form is chemical. When you burn wood in the fireplace, you are combining the oxygen in the air with the carbon in the wood in an 'exothermic' chemical reaction. Exothermic means the reaction releases heat. Most of the solid wood is converted into the gas carbon dioxide, and goes up the chimney, leaving the material that could not be oxidized behind as ashes for you to clean out.

When you rub two objects together the mechanical energy you put into rubbing them is converted into heat.

**Second Law**: If two isolated nearby systems, each in thermodynamic equilibrium in itself, but not in equilibrium with each other, are allowed to interact, they will exchange matter or energy and ultimately reach a mutual thermodynamic equilibrium.

**Fig. 10.7** James Joule pondering the nature of heat while patiently waiting for a pint of his fine ale to be drawn

That is, if you have an object that is cold and another that is hot and put them together, heat will be exchanged so that eventually both have the same temperature. The heat flows from the warmer object to the cooler object. Sounds like Lavoisier's idea of the flow of caloric from warm objects to cooler objects, doesn't it?

An important thing about the Second Law is that if you have something that is cold and something that is hot, you cannot take heat from the cold thing, making it colder, and give that heat to the hot thing, making it hotter without using energy from outside the two things.

Another way of stating the Second Law: Energy spontaneously tends to flow only from being concentrated in one place to becoming diffused or dispersed and spread out.

The Second Law introduces us to a critically important new concept: *entropy*. Entropy can be defined as energy that is no longer available for doing mechanical work. It is the diffused, dispersed and spread out energy of the preceding statement. Entropy is the random motion of atoms or molecules. That random motion is what we call 'heat.' It is waste heat; it cannot be used to do anything.

An example of the Second Law and entropy is your on-the-rocks Martini. To make the analogy simple we will omit the olive. The Martini has two systems which are not usually in thermal equilibrium (unless you keep the gin in the freezer compartment of the refrigerator), gin/vermouth and ice. The second law states that the ice will melt as the gin/vermouth cools. As the ice melts its highly ordered crystal structure is changed to the random motion of molecules in liquid water. The entropy of the Martini as a system has increased. Unfortunately the melting ice has also diluted what was a great drink.

By the way, the next time the waiter serves you a Martini that was sitting on the bar for a while before he brought it to you, you can complain: 'It's not the gin, it's the entropy.'

Another example. You have a child and have put all of his/her toys neatly away in the toy-box. You come back in an hour and find out that your darling got them all out and they are spread all over the floor. Order has given way to disorder. Your child has increased the entropy of the Universe. You can put them all back into the box, neatly arranged in order again, but that takes an external input of energy. Keeping order is a losing battle.

You can think of entropy as the tax Mother Nature collects on everything you do. The tax can be large or small depending on the efficiency of what you do in converting one form of energy to another. It is the loss due to friction, which is in the form of heat.

But there is a major implication in all this about the way things work. If entropy increases with time, it means that the Universe is gradually running down. The increase in entropy defines the arrow of time. This is a concept of great importance in physics where many of the equations work just as well backwards in time as they do forward in time. While it may seem obvious to you that time only goes in one direction, it is not so obvious to a theoretical physicist or, for that matter, to H.G. Wells when he wrote 'The Time Machine' in 1895. In a way, the Second Law of Thermodynamics is the key to unlock the way the Universe works.

**Third Law**: As temperature approaches absolute zero, the entropy of a system approaches a minimum.

Absolute zero is the temperature at which all atomic and molecular motion ceases. Entropy is random atomic and molecular motion, so if it ceases entropy goes to zero. Absolute zero does not exist in nature or the laboratory. It can only be approached, not reached.

In the 18th century there were lots of attempts to figure out what the coldest possible temperature might be ('absolute cold'). Other temperature scales were in use in those days, but in terms of the Centigrade scale in which the freezing point of water is 0 and the boiling point 100, estimates ranged from −240 °C (by the French physicist Guillaume Amontons in 1702), to −270° (by Swiss mathematician, physicist and astronomer Johann Heinrich Lambert, 1779), to between −1500° and −3000 °C by the much more famous French scientists Pierre-Simon Laplace and Antoine Lavoisier in 1780. John Dalton wrestled with the problem in his New System of Chemical Philosophy (1810) and concluded that it must be −3000 °C. The modern value for absolute zero, 0 K, is −273.15 °C. In most of the rest of this book we will simply cite absolute zero as −273 °C.

The first three laws of thermodynamics were developed in the 19th century, and have been stated in a variety of ways. As you might already have guessed, the person who really developed thermodynamics into a rigorous science was our old friend from the age of the Earth debate, William Thompson, Lord Kelvin. He is the one who figured out the temperature where molecular motion would stop, and the absolute temperature scale that today bears his name.

In 1848 Lord Kelvin proposed a scale of temperature based solely on thermodynamics. It used the intervals of the Celsius scale, with 100 steps between the freezing and boiling points of water. On the basis of the rate of contraction of air with declining temperature he figured that absolute zero must be about −273 °C. It has since been redefined and established as −273.15 °C. This absolute temperature scale is now called the Kelvin scale, and to distinguish it from other scales the temperature is cited simply as Kelvin, not °Kelvin.

In the 1930s another law was added. It is so fundamental to the science of thermodynamics it comes before the first law; it is the zeroth law:

**Zeroth Law**: If two thermodynamic systems are each in thermal equilibrium with a third system, then they are in thermal equilibrium with each other.

What this says, among other things, is that if you use a thermometer to measure the temperature of something, the same thermometer will register the same temperature for other things that are in thermal equilibrium.

Incidentally, some people believe that there is a Zeroth Commandment that comes before the other Ten that Moses brought down from the mountain. I learned it from Professor Hans Schaub's wife Eschte in Switzerland. It is quite simple, absolutely fundamental, and the Golden Rule of Politicians: "Lass Dich nicht erwischen" or "Don't get caught."

## 10.4   The Discovery of Greenhouse Gases

John Tyndall (1820–1893) (Fig. 10.8) was born in Leighlinbridge, County Carlow, Ireland. His early education emphasized technology and mathematics, and he became adept at land surveying. Between 1844 and 1847, he was lucratively employed in planning the construction of new railway lines in England. In furthering his education he moved to Germany, where laboratory experimental science was more advanced than in Britain. At the University of Marburg he studied under chemist Robert Bunsen and physicist Hermann Knoblauch whose specialty was

**Fig. 10.8** John Tyndall (1820–1893) reading a letter commenting on his experiments

magnetism. Tyndall's work on magnetism attracted the attention of Michael Faraday, who recommended him for a Professorship in Natural Philosophy (=Physics) at the Royal Institution in London in 1853. There Tyndall began experimenting with the interactions of radiation with gases.

He discovered that some gases were transparent to both 'luminous' and 'non-luminous' heat, while others were quite opaque to non-luminous heat. In a series of papers in the 1860s Tyndall reported on how he was able to measure the absorption of radiant heat by oxygen, nitrogen, hydrogen, water vapor, carbon dioxide, ozone, and even the fragrances of flowers. The first three had no capacity to absorb energy, but all of the latter did. He calculated that a single molecule of water vapor had a capacity to absorb 16,000 times as much radiant energy as nitrogen or oxygen molecules. He realized that water vapor was critical to regulation of the surface temperature of the Earth. He argued that more than 10 % of the radiation from the soil is stopped by vapor within ten feet of the surface. He speculated that if one could somehow remove the vapor from the air over a field at night, all of the plants would be frozen by morning. He found that perfumes, the fragrances of flowers, were particularly opaque to the radiant heat from the soil, and, in effect, suggested that they protected themselves with a blanket of warmth (Fig. 10.9).

It was almost another half century before the implications of Tyndall's experiments with gases were fully realized.

John Tyndall was also interested in the scattering of light, and used it to try to explain the blue sky and sunbeams. He was an avid mountaineer, and much interested in the motion of glaciers. At least three glaciers have been named in his honor. The smallest is in Rocky Mountain National Park, visible from my home.

## 10.5   Kirchhoff's 'Black Body'

The name Kirchhoff is not a household word, but it should be. Gustav Robert Kirchhoff (1824–1887) (Fig. 10.10) is best remembered for his laws regarding electrical circuits in very practical use by electrical engineers and even electricians today. He formulated these in 1845, as part of a report in a seminar he was taking as a graduate student at the University of Königsberg (in the then East Prussia; now Kaliningrad in Russia). In a seminar, the professor proposes the topic for discussion, and the students do the research. Often the students rediscover what the professor already knows, but if one is lucky new ideas will emerge. That brilliant work on electrical circuits became the basis for his doctoral dissertation which he completed in 1847. He then married the daughter of one of the professors who had conducted the seminar. They moved to Berlin, and a few years later Kirchhoff received an appointment at the University of Breslau in Silesia (now Wroclaw in Poland). In 1854 he was invited to work with Robert Bunsen at the University of Heidelberg. There he was fascinated by the colors of light and the 'non-luminous rays.'

In 1859, while at Heidelberg, he formulated what is now known as Kirchhoff's Law of Thermal Radiation: At thermal

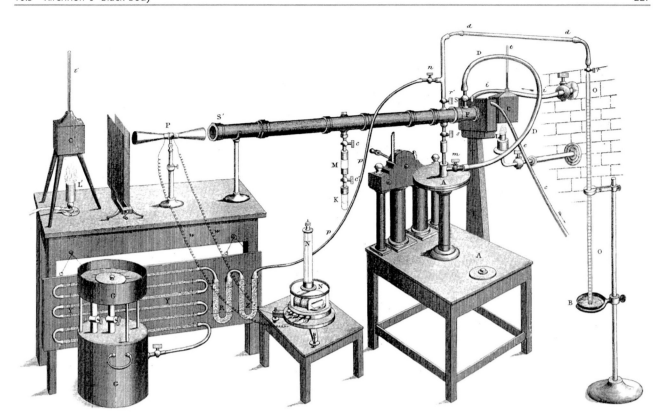

**Fig. 10.9** John Tyndall's apparatus for investigating the interactions between gases and radiation. (For the details you will need to look in Tyndall, 1861, *The Bakerian Lecture: On the Absorption and Radiation* *of Heat by Gases and Vapours, and on the Physical Connexion of Radiation, Absorption, and Conduction*. Philosophical Transactions of the Royal Society, v. 151, pp. 1–36.)

equilibrium, the emissivity of a body (or surface) equals its absorptivity. That is, for an object which neither warms nor cools, the radiation emitted by the object is equal to the radiation absorbed by the object.

In 1875 Kirchhoff was appointed to become the first Chair of Theoretical Physics at the University of Berlin (then called the Friedrich Wilhelm University; now the Humboldt University, named after its founder, the explorer/naturalist/scientist Alexander von Humboldt). There his research took a great leap. He attempted to develop a theoretical explanation about how incoming and outgoing radiation operate. It would ultimately lead to a method for determining the temperature of an illuminated object—like a planet—and for determining the role of greenhouse gases in climate.

He became the father of the concept of a 'black body.' A black body (or sometimes 'blackbody' in the US) is a hypothetical object that absorbs all the electromagnetic radiation incident upon it. Its absorptivity is 1. To maintain a specific constant temperature the black body re-radiates as much energy as it receives back into space. It is both a perfect absorber of energy and a perfect emitter of energy. Its emissivity is also 1. It absorbs the energy inherent in all wavelengths of visible and invisible light, and warms. To maintain a constant temperature it must radiate energy back out. This occurs at all wavelengths but most of the energy is

concentrated in a region of the spectrum that depends on its temperature. When it reaches about 798 K (=525 °C, 977 ° F) it is no longer black to the human eye, but begins to glow as a dull red. What is being seen is just the short wavelength tail of the mostly infrared radiation. As the temperature increases apparent color of the object gradually changes. It is an apparent color because it is a mixture of many wavelengths representing several specific colors.

Below 798 K the radiation is in the invisible infra-red; above 10,000 K it is mostly in the invisible ultraviolet, but a 'black body' would still be visible as having a pale blue hue. These colors and their associated temperatures hold true for virtually all materials, including metals and lava. The apparent color tells us the temperature. The temperature-color relation allows us to estimate the temperature of stars in the sky This is also how that temperature sensing device the doctor sticks in your ear works, only it is sensing the invisible infra-red radiation (Table 10.1).

Kirchhoff's black body is a hypothetical object. No such thing exists in nature. We now know that the energy distribution of the radiation emitted from a black body has the form of a skewed bell-shaped curve with a longer tail toward the longer wavelengths, as shown in Fig. 10.11. However in the 19th century the shape and magnitudes of the energy distribution was unknown. What was needed was a 'black

**Fig. 10.10** Gustav Robert Kirchhoff (1824–1887). Investigator of spectral lines and imaginer of the perfect absorber and emitter of energy, the Black Body

**Table 10.1** Temperatures of a glowing 'black body,' peak wavelength of the outgoing radiation, and apparent color of the 'black body'

| T in K | T in °C | T in °F | Peak $\lambda$ = nm = $10^{-9}$ m | Apparent color |
|---|---|---|---|---|
| 1200 | 927 | 1636 | 2415 | Red |
| 2000 | 1727 | 3076 | 1449 | Reddish-orange |
| 3000 | 2727 | 4876 | 966 | Orange |
| 4500 | 4227 | 7576 | 644 | Pale yellow |
| 6000 | 5727 | 10276 | 483 | White |
| 9000 | 8726 | 15676 | 322 | Pale indigo |

Note that the real color of the peak wavelength is not that of the apparent color perceived by the human eye, which is in fact a mixture of wavelengths

body' that could be used to make the necessary measurements. Although there are no true 'black bodies' in nature, some materials, such as soot and graphite, come close. The next step was to construct a device that emits radiation as though it were a black body so that detailed measurements could be made. This took almost a quarter of a century.

One of the problems that needed to be solved was the relation between the amount of energy a warm object radiated and its temperature. Attempts to figure this out went back to the early decades of the 19th century. At that time it

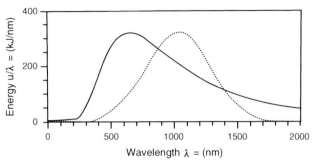

**Fig. 10.11** Comparison of a 'normal distribution' (*dotted curve*), with the skewed energy distribution of black body radiation (*solid curve*). A normal (or 'Gaussian') distribution is a bell-shaped curve symmetrical about the mean value. It is a first approximation to describe random variables that tend to cluster around a mean value. The distribution of wavelengths of thermal radiation from a black body is similar but is skewed, with a long tail toward the longer wavelengths (here to the right). The black body radiation curve for a given temperature has essentially the same shape as the distribution at any other temperature, except that the peak wavelength is displaced toward shorter values at warmer temperatures and longer values at cooler temperatures. At cooler values the tail to the right gets longer and longer

was recognized that there were three ways in which heat could be transferred: (1) conduction, as from the hot end of a metal rod to the cool end, (2) convection, as when warm water or air rises inducing circulation, and (3) through radiation. The first two were well understood, but how radiation worked was still a mystery.

In 1817 two French scientists, Alexis-Thérèse Petit (1791–1820) Professor of Physics at the École Polytechnique in Paris, and his student and successor Pierre Louis Dulong (1785–1838) had formulated an answer which was awarded a prize by the Académie Française. In modern terms, the answer was expressed in terms of watts per meter square radiated by an object at a given temperature. Unfortunately, their formulation didn't seem to always give the right answer as more and better measurements became available. Incidentally, there were a lot of problems in making precise measurements with thermometers in those days. Those based on the expansion of mercury from a bulb at the base up through a narrow tube varied according to whether the bulb was clear glass, silvered, or painted black. Other thermometers based on the expansion of air or another gas gave different readings. There was another major problem—a true absolute temperature scale had not yet been devised.

In the last quarter of the 19th century Berlin and Vienna became centers for documenting the relation between emitted radiation and temperature. Wilhelm Wien was a physicist who had studied under Kirchhoff. He worked at the Physikalisch-Technische Reichsanstalt in the Charlottenburg suburb of Berlin (analogous to our US Bureau of Standards). He used the data generated there and the hypothetical black body concept to formulate an empirical relationship between the two in 1893. It is now known as Wien's Displacement Law.

It states that the wavelength emitting maximum energy is inversely proportional to the absolute temperature.

$$\lambda_{max} = \frac{b}{T}$$

where $\lambda_{max}$ (lambda maximum) is the peak wavelength, T is the temperature of the black body in Kelvin, and b is a constant of proportionality which is now known to be =2,897,768 nm K. For example, the Earth's average temperature is about 15 °C (=59 °F = 288.15 K) so the peak wavelength emitted by our planet works out to be

$$b/T = 2,897,768 \, \text{nm K}/288.15 \, \text{K} = 10,054 \, \text{nm} = 10.054 \, \mu\text{m}.$$

The 10 μm wavelength is toward the middle of what is called the long wavelength infra-red (8–15 μm) and in the middle of the band of wavelengths absorbed by the greenhouse gases $CO_2$ and $H_2O$. At 0 °C (=32 °F) and 30 °C (86 °F) the peak wavelengths are 10.61 and 9.56 μm respectively. As you will see below, if Earth had no atmosphere with greenhouse gases, its average temperature would be about −18.6 °C (=1.5 °F) and the peak wavelength would be 11.40 μm, even more squarely placed for greenhouse gases to absorb energy. Earth's temperature just happens to be in the right range for greenhouse gases to play a big role in its climate.

## 10.6 Stefan's Fourth Power Law

In Austria, Josef Stefan (Fig. 10.12) was a Slovenian who studied at the University of Vienna, obtaining his doctorate in 1857. He joined the University faculty and later became Professor of Physics. He liked the formulation of Dulong and Petit but suspected that the data they had used might be inaccurate. He tried making corrections for the type of thermometer and most importantly for the conduction of heat through the vestiges of air remaining in the vacuum chamber Dulong and Petit had used. The best vacuum they had been able to achieve had been 2 Torr. Torr is an old expression for pressure meaning mm Hg (millimeters of mercury); 2 Torr = 0.00263 atmosphere = 2.67 mbar or hectopascals, or 2.7 thousandths of standard atmospheric pressure.

To start with, Stefan had only two measurements he trusted, but for them he realized that the radiance varied as the fourth power of the absolute temperature multiplied by a constant.

$$R = \sigma T^4$$

**Fig. 10.12** Josef Stefan (1835–1893)

where R is the radiance, σ (sigma) is a constant, and T is the absolute temperature (using the Kelvin scale). $T^4$ means that the temperature in Kelvin is multiplied by itself four times.

In science, basing a major hypothesis on only two measurements is generally considered to be highly risky, so Stefan searched the literature and found a few other measurements that were close enough to confirm his idea. This was an extremely simple formulation of the relationship. It was purely empirical, that is, it was simply a fit to the data. It had no theoretical basis. Finally published in 1879, it became known as Stefan's Law.

However, armed with the new law, Stefan tackled another problem where his intuition suggested that the Dulong and Petit Law had gotten the wrong answer—the temperature of the surface of the Sun. The solar constant had been measured by French physicist Claude Matthias Pouillet (1790–1868) 40 years earlier. In 1838 he published the result of his many measurements, estimating it to be 1228 $Wm^{-2}$, remarkably close to the 1366 $Wm^{-2}$ measured by satellites today. Using the Dulong-Petit Law, Pouillet had calculated the temperature of the Sun to be at least 1461 K and possibly as high as 1761 K. Other estimates were really just guesses and ranged all over the place, some as high as 20,000 K. Using an updated estimate of the solar constant and a still imperfect estimate of the constant σ, Stefan (1879) got a value of 5700 K. The modern estimate is 5778 K. Remember that all this was done when it was thought that the Sun drew its energy from gravitational contraction. It would not be until 1920 that Arthur Eddington came up with the idea that the Sun was powered by nuclear fusion of hydrogen to make helium.

American astronomer, physicist and aviation pioneer Samuel Pierpont Langley (1834–1906) made observations of solar radiation on Mount Whitney in California in 1884. He revised the solar constant upward to 2903 $Wm^{-2}$, and this value, over twice the modern value of 1,366 $Wm^{-2}$, was often used and cited for the next few decades. It made no sense in trying to figure out how the Earth's climate system worked, but nobody seemed to notice. As Cesare Emiliani would say, nobody's perfect.

Incidentally, Langley's name came to be used, particularly in the US, for a unit of energy per unit area. One Langley (Ly) is one calorie per square centimeter, so solar irradiation would be measured in Langleys per second. It is not an SI unit and hardly anyone other than TV meteorologists uses it anymore.

In 1863 Josef Stefan had an incredible stroke of luck. A bright young student, Ludwig Boltzmann (Fig. 10.13), decided to work with him. Boltzmann was a good theorist and mathematician, and only 4 years after Stefan had published his fourth power law, Boltzmann had worked it out from purely theoretical considerations about a black body and thermodynamics. How did thermodynamics get into this investigation of radiation? When James Clerk Maxwell had formulated his ideas about electromagnetic waves he believed that they moved through that mysterious undetectable substance, 'luminiferous æther,' that was thought to fill all of space. Maxwell thought that the waves moving through the 'luminiferous æther' must create some sort of pressure. This gave rise to the idea that the laws of thermodynamics, originally formulated for material things such as solid objects and air might also apply to the 'luminiferous æther.' Boltzmann used Maxwell's ideas and applied thermodynamics to the problem to arrive at the fourth power law without needing any data. The paper, published in 1884, is only four pages long, and even I can follow the math!

Ever more precise data have been used to get the modern value of σ, now known as the Stefan-Boltzmann constant: $5.670400 \times 10^{-8}$ $Wm^{-2}K^{-4}$.

What did all this have to do with the greenhouse effect? The simple $R = \sigma T^4$ equation can be used to evaluate a planet's energy balance. Take the Earth as an example. The Earth is a sphere, not a disc, so the solar constant, $\sim 1360$ $Wm^{-2}$ must be divided by 4 since the area of a sphere is 4 times that of a disc of the same diameter. This means that the energy received by each square meter at the

top of the atmosphere is 340 W. If you plug in 340 $Wm^{-2}$ for R on the left side of the equation and solve for T you get T = 278 K = 5 °C = 41 °F for the average temperature of the Earth. It sounds reasonable until you realize that the Earth is not a black body but reflects about 30 % of the incoming energy from the Sun back into space. That means that the energy from the Sun actually absorbed by the Earth is only 238 $Wm^{-2}$. If you substitute this for R, then T = 254.5 K = −18.6 °C = 1.5 °F; the average temperature on Earth would be well below freezing. In fact, that is the planetary temperature registered by a satellite far from Earth. But at the Earth's surface the average temperature is about 288 K = 15 °C = 59 °F. Obviously, something in the atmosphere acts to make the surface warmer: greenhouse gases; and their cumulative effect today is to make the planet's temperature about 33.6 °C = 60.5 °F warmer than it otherwise would be.

The importance of Stefan's Law in determining the relative importance of incoming solar radiation, planetary albedo, and the greenhouse effect on the climate and surface conditions on both the Earth and other planets cannot be overemphasized.

Table 10.2 shows our present state of knowledge: A. the relative distances of the planets and our Moon from the

**Fig. 10.13** Ludwig Boltzman (1844–1906). He appears to be scowling at someone who doesn't realize how simple and straightforward all this fourth-power stuff is

**Table 10.2**  Energy balance climatology of the planets of our solar system and Earth's Moon

| | Parameter | Mercury | Venus | Earth | Moon | Mars | Jupiter | Saturn | Uranus | Neptune |
|---|---|---|---|---|---|---|---|---|---|---|
| | Symbol | ☿ | ♀ | ⊕ | ☾ | ♂ | ♃ | ♄ | ⛢ | ♆ |
| A | Mean distance from Sun (AU) | 0.387 | 0.723 | 1.000 | 1.000 | 1.524 | 5.203 | 9.555 | 19.218 | 30.110 |
| B | Solar constant (W/m$^2$) | 9076 | 2599 | 1360 | 1360 | 586 | 50 | 15 | 4 | 2 |
| C | Average insolation (W/m$^2$) | 2269 | 650 | 340 | 340 | 146.4 | 12.56 | 3.72 | 0.921 | 0.375 |
| D | Temperature (K) if IR = OR | 447.3 | 327.2 | 278.3 | 278.3 | 225.4 | 122.0 | 90.0 | 63.5 | 50.7 |
| E | Temperature (°C) if IR = OR | 174.3 | 54.2 | 5.3 | 5.3 | −47.6 | −151 | −183 | −209.5 | −222.3 |
| F | Temperature (°F) if IR = OR | 345.7 | 129.6 | 41.5 | 41.5 | −53.7 | −239.8 | −297.4 | −345.1 | −368.1 |
| G | Albedo | 0.06 | 0.72 | 0.30 | 0.12 | 0.16 | 0.70 | 0.75 | 0.90 | 0.82 |
| H | Temperature (K) with albedo | 440.4 | 238.0 | 254.5 | 269.5 | 215.8 | 90.3 | 63.7 | 35.7 | 33.0 |
| I | Temperature (°C) with albedo | 167.4 | −35.0 | −18.5 | −3.5 | −57.2 | −182.7 | −209.3 | −237.3 | −240.0 |
| J | Temperature (°C) with albedo | 332.3 | −31.0 | −1.3 | 25.7 | −71.0 | −296.9 | −344.7 | −395.1 | −400.0 |
| K | Surface temperature (K) | 706–111 | 733 | 288 | 380–120 | 218 | 165 | 140 | 57 | 57 |
| L | Surface temperature (°C) | 427 to −168 | 460 | 15 | 107 to −153 | −55 | −108 | −133 | −216 | −216 |
| M | Surface temperature (°F) | 800 to −270 | 860 | 59 | 225 to −243 | −67 | −162.4 | −207.4 | −356.8 | −356.8 |
| N | Greenhouse effect (K, °C) | None | 495.0 | 33.5 | None | 2.2 | 74.7 | 76.3 | 21.3 | 24.0 |
| O | Greenhouse effect (°F) | None | 891 | 60.3 | None | 4.0 | 134.5 | 137.3 | 38.3 | 43.2 |

If the 'Greenhouse effect' is listed as none, the object has no atmosphere. Perusal of these data suggests that Earth is the only planet fit for human habitation

Sun in terms of Astronomical Units (1 AU = mean distance of Earth from the Sun (=149,597,870.7 km; or 92,955,807.3 miles). B. The solar constant, which is the mean amount of energy received from the Sun by a flat panel oriented perpendicularly to the rays from the Sun at the planet's average distance, in watts per square meter. C. The average insolation at the top of the planet's atmosphere (if it has one); it equals 1/4 of the solar constant since the planet is spherical and not a flat panel (remember the area of a circle or disc is $\pi r^2$ and the area of a sphere is $4\pi r^2$), and only half of it is illuminated at any given time. D, E, and F are the expected temperatures in K, °C, and °F, calculated using Stefan's fourth power law by assuming radiation balance; that is, the incoming radiation (IR) from the Sun is re-emitted by the planet as outgoing radiation (OR). If the two are not balanced, then the planet will either warm or cool. G is the albedo; it is the fraction of the incoming radiation reflected by the planet. H, I, and J, are the expected temperatures in K, °C, and °F calculated from Stefan's Law but reducing the average insolation received by the planet to account for the amount reflected, its albedo. K, L, and M are the surface temperatures of the planets in K, °C, and °F. For the inner planets (Mercury through Mars) these are well known, having been measured by probes or orbiters. For the outer planets they are still a bit of guesswork. The difference between the temperature calculation taking albedo into account and the actual surface temperature is the 'greenhouse effect' of the planetary atmosphere. For the two objects that lack an atmosphere, Mercury and the Moon, the surface temperature range is extreme, representing night and day temperatures. The table is worth perusing, especially if you are considering buying real estate on another planet. Figure 10.14 shows the planets distances and sizes.

By the way, if you use Samuel Pierpont Langley's value of 2903 Wm$^{-2}$ for the solar constant, you get 725.75 Wm$^{-2}$ for the irradiation at the top of the atmosphere, and with an albedo of 0.3 that reduces to an average of 508.03 Wm$^{-2}$ at the surface. Following Stefan's Law, the average temperature of the planet would be 307.7 K (=34.5 °C = 94.1 °F). That fourth power has a big effect. Since in the 19th century nobody knew what the average temperature of the planet might be, and knowledge of greenhouse gases was restricted to laboratory experiments, Langley's value for the solar constant seemed quite plausible. There was no need for any greenhouse effect to explain our planet's climate.

## 10.7   Black Body Radiation

The Stefan-Boltzmann Law allowed for the total emission from a black body at a given temperature to be determined. But a black body is a hypothetical concept that has no real counterpart in nature. Going further required measurements

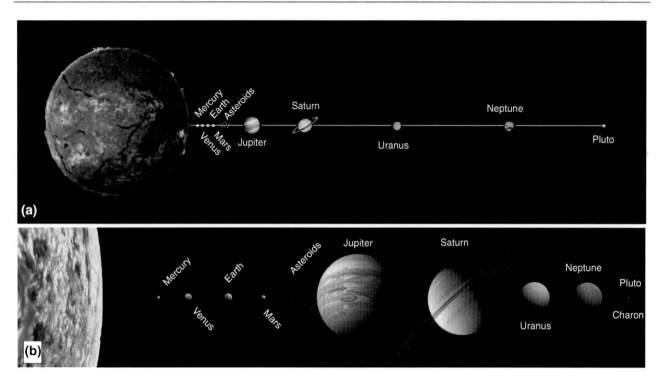

**Fig. 10.14**  **a** Planets distances from Sun to scale (sizes not to scale). **b** Planets sizes to scale (distances not to scale)

on something real that would be very close to acting as a black body. The answer is sure to surprise you; it is not an object at all, it is a cavity in an object (Fig. 10.15).

The idea, developed over a number of years, was that emission from a small hole in a copper globe or box blackened by a coating of platinum chloride and surrounded by a temperature-stabilizing liquid or gas very closely approached the emission expected from the hypothetical back body. The assumption is that the radiation comes from some sort of oscillating objects ('harmonic oscillators') inside the cavity without specifying whether they might be atoms, molecules or something else. A harmonic oscillator is something that when perturbed experiences a restoring force proportional to the displacement. Confused? Think of a pendulum. Or a weight bobbing up and down on a spring. Ah yes, but remember that both of these will gradually slow down and stop because of friction in the system. The friction produces waste heat = an increase in entropy. Over a decade or so enough measurements had been made to outline the shape of the radiation energy-wavelength distribution. A modern graph of the distribution for some high temperatures is shown in Fig. 10.16.

Berlin and Vienna became centers for documenting the relation between emitted radiation and temperature. In the Berlin suburb of Charlottenburg the Physikalisch-Technische

Reichsanstalt had been established in 1887 to make the measurements needed for the advancement of science. Wilhelm Wien had gotten his doctorate at the University of Berlin under Hermann von Helmholtz, Kirchhoff's successor. On graduation Wien was appointed as a physicist at the new Physikalisch-Technische Reichsanstalt. He used the data being generated and the concept of the hypothetical 'black body' to formulate an empirical relationship between the two now known as Wien's Displacement Law (1893).

This identified the wavelength of maximum energy, but did not describe the shape of the distribution away from its maximum. It states that the wavelength emitting maximum energy is inversely proportional to the absolute temperature

$$\lambda_{\max} = \frac{b}{T}$$

where $\lambda_{\max}$ is the peak wavelength, T is the temperature of the black body in Kelvin, and b is a constant of proportionality, now known as Wien's Displacement Constant, and is 2,897,786 nm K (=2,897.786 μm K).

Wien's displacement law identifies the wavelength of maximum energy, but does not describe the shape of the distribution away from its maximum.

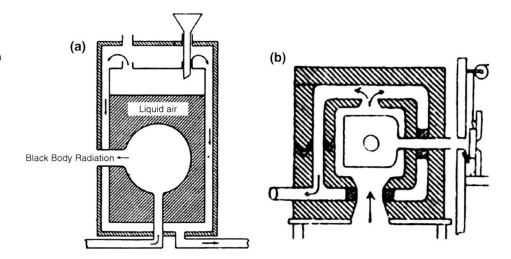

**Fig. 10.15** Black body-simulating devices consisting of boxes with cavities (a, b). The boxes are immersed in fluid baths, commonly liquid air, in order to be held at a constant temperature. These were devised by Otto Lummer and Ernst Pringsheim in the 1890s at the Physikalisch-Technische Reichsanstalt in Berlin/Charlottenburg

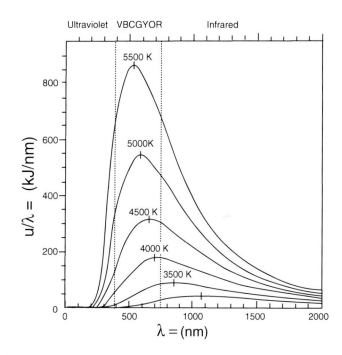

**Fig. 10.16** Plot of black body radiation energy (u) per wavelength ($\lambda$) plotted against wavelength ($\lambda$) for several absolute temperatures. 'VBCGYOR' are the wavelengths of the spectrum visible to humans (V = *violet*; R = *red*). The lowest curve is the emission curve for an incandescent lightbulb (T = ~2200 K); note that only a tiny fraction of the emission is in the visible light range, most is in the infrared

A few years later, in 1896, Wien derived an empirical approximation describing the shape of the spectrum of thermal radiation (frequently called the black body function). Sometimes called Wien's Distribution Law, it accurately describes the short wavelength spectrum of thermal emission from objects. However, it fails to fit the experimental data for long wavelengths emission.

In Britain others were at work on the problem. In 1900, Lord Rayleigh, using classical physics, derived an expression that fitted the experimental data for long wavelength emission. A few years later (1905) he and Sir James Jeans improved on the original equation, but noted that when this was applied to the shorter wavelengths of the ultraviolet, the answers diverged toward infinity. This problem soon became known as the 'ultraviolet catastrophe.' The failure of Wien's approximation in the longer wavelength part of the spectrum was labeled the 'infrared catastrophe.' Clearly, a better solution was needed; that would become the work of Max Planck.

Nevertheless, with the Wien and Rayleigh-Jeans approximations, the black body concept had enormous implications for understanding how the Sun-Earth system operated. It was an essential step in understanding Earth's energy balance. And the concept of harmonic oscillators would be crucial to understanding atmospheric greenhouse gases. However, none of this was obvious to anyone at the time.

Let's take a moment to review the state of knowledge about physics at the end of the 19th century. Until then everything had been going along hummingly. Sir Isaac Newton's view of the world completely dominated the field. Everything could be thought of in terms of mechanical processes. Arguments based on physics had shown that the Earth was about 100 million years old. Maxwell had explained the nature of electromagnetic waves, which included not only light but waves of much higher and lower frequency. They were perturbations of the 'luminiferous æther' that filled all of space, traveling at the speed of light. There was nothing mysterious, well almost nothing, left to investigate. The future would be devoted to refining the 'natural constants' that appeared in equations. Physics had become a boring field, nothing of any significance left to do. Basically, we were back to believing that we knew everything there was to know.

## 10.8   Summary

The nature of heat was a puzzle. Antoine Lavoisier, the father of modern chemistry, thought it was a fluid, which he named 'caloric.' Expatriate American, Benjamin Thompson, a. k. a. Count Rumford, figured out that it was actually the motion of the 'atoms' of which all matter is constructed. In spite of Thompson's argument, the idea of 'caloric' persisted for a few decades.

In 1824 Jean-Baptiste Fourier formulated the energy balance model for planet Earth hypothesizing that the incoming energy from the Sun must be balanced by outgoing radiation from the Earth, and postulating the existence of some sort of modulation by the atmosphere and ocean.

In 1775 a practical, efficient steam engine had been perfected by James Watt, initiating the Industrial Revolution, but the science behind it had not been worked out. In 1824 Sadi Carnot examined the physics behind the steam engine, and began a line or reasoning that would lead to the field of thermodynamics. In 1850 Rudolf Clausius began formulation of the Laws of Thermodynamics. In 1865 he published the famous statements: (1) The energy of the universe is constant; (2) The entropy of the universe tends to a maximum. These have become known as the first two Laws of Thermodynamics.

Also in the early 1860s John Tyndall conducted his experiments on the absorption of radiant energy by gases, discovering that, among others, water vapor is very important as a greenhouse gas.

Also about this time, Gustav Kirchhoff began to develop the concept of a 'black body'—a hypothetical object that absorbs all energy incident on it and emits an equal amount of energy; the balance determines the temperature of the 'black body.' This was a powerful new area of physics to explore. Over the next quarter century scientists at the new Physikalisch-Technische Reichsanstalt in Berlin worked on ways to simulate black body radiation.

In 1874 Josef Stephan announced his discovery that the radiance of an object varies as the fourth power of its temperature expressed in Kelvin. His hypothesis, based on limited empirical data, was confirmed in 1884 by Ludwig Boltzmann purely on the basis of theoretical considerations. This discovery was critical to understanding the energy balance between our Sun, Earth, and the planets. Before the end of the century we would have the first energy balance models for planet Earth.

At the Reichsanstalt in Berlin, Wilhelm Wien figured out a mathematical formulation to determine the wavelength of the maximum energy, and the spectral distribution of the radiation from the black body, but it only worked for the shorter wavelengths. In Britain, Lord Rayleigh and Sir James Jeans also devised a relationship for the spectral distribution, but it only worked for the longer wavelengths. Something was wrong somewhere.

However, the stage had been set for an understanding of how the Earth and Sun interact, and how the climate system works.

A Timeline for this chapter:

| ca 420 CE | Augustine of Hippo (Saint Augustine) declares " We know all we need to know" and ushers in the 'Dark Ages' |
|---|---|
| 950– 1250 CE | Medieval Climate Optimum |
| 1308– 1321 | Dante Alighieri writes the *Divine Comedy* |
| 1348 | The Black Plague arrives in Europe |
| 1348– 1351 | A third to half of the population of Europe dies of the plague |
| 1401 | The Renaissance begins with Lorenzo Ghiberti winning the competition for the bronze doors for the Baptistery of the Cathedral in Florence, Italy |
| ca. 1488– 1519 | Leonardo da Vinci records his ideas in his Notebooks |
| 1667 | German alchemist Johann Joachim Becher proposes that fire is something he called '*terra pinguis*' later renamed '*phlogiston*' |
| 1752 | Thomas Melville discovers that chemicals introduced into a flame produce specific colors |
| 1773 | Oxygen discovered by Carl Wilhelm Scheele, in Uppsala, Sweden |
| 1774 | Oxygen discovered independently by Joseph Priestley in Wiltshire, England |
| 1777 | Antoine Lavoisier gives oxygen its name |
| 1783 | Antoine Lavoisier proposes that there is a 'subtle fluid,' '*caloric*' that is the substance of heat |

(continued)

(continued)

| 1800 | William Herschel discovers the non-luminous rays beyond the red end of the spectrum |
|------|----------|
| 1804 | Benjamin Thompson, a.k.a. Count Rumford of the Holy Roman Empire, discovers that heat is the 'motion of atomic particles' |
| 1817 | Alexis-Thérèse Petit and Pierre Louis Dulong develop a formulation for the amount of energy given off as radiation by a heated object |
| 1824 | Jean-Baptiste Joseph Fourier publishes *Remarques generales sur les températures du globe terrestre et des espaces planétaires* outlining the basics of planetary energy balance |
| 1824 | Sadi Carnot publishes his book: *Réflexions sur la puissance motrice du feu* ("*Reflections on the Motive Power of Fire*") |
| 1838 | Claude Matthias Pouillet estimates the solar constant to be 1228 $W/m^2$ |
| 1843 | Beer and ale brewer and hobby physicist James Joule publishes *On the Calorific Effects of Magneto-Electricity and the Mechanical Value of Heat*—a contribution so important that the SI unit of energy is named for him |
| 1848 | Lord Kelvin writes about the need for an Absolute Thermometric Scale |
| 1854 | Robert Bunsen and Peter Desaga design a gas burner for the new chemistry laboratory at the University of Heidelberg; Gustav Robert Kirchhoff comes to work with Bunsen and, using the new burner, develops the science of spectroscopy for chemical analysis |
| 1859 | James Joule takes time off from brewing ale to publish *On some thermo-dynamic properties of solids* |
| 1861– 1864 | John Tyndall publishes several papers on radiation and gases, discovering the role of water vapor as the primary greenhouse gas |
| 1865 | Rudolf Julius Emanuel Clausius formalizes the first two laws of thermodynamics |
| 1875 | Kirchhoff is appointed to the Chair of Theoretical Physics at the University in Berlin and develops the concept of a 'black body' as an absorber and source of radiation |
| 1879 | Josef Stefan publishes his 'fourth power law'; he estimates the temperature of the surface of the Sun to be 5700 K |
| 1884 | Samuel Pierpont Langley (incorrectly) determines the solar constant to be 2,903 $W/m^2$ |
| 1884 | Ludwig Boltzmann derives Stefan's fourth power law purely from the laws of thermodynamics, not needing any data |
| 1890s | Ernst Pringsheim and Otto Lummer develop Black Body simulating devices at the Physikalisch-Technische Reichsanstalt in Berlin/Charlottenburg, producing lots of data for Wilhem Wien to study |

(continued)

| 1893 | Wilhelm Wien develops his Displacement Law describing the empirical relationship between temperature of a black body and the wavelength of maximum radiation |
|------|----------|
| 1893 | Based on Carnot's ideas, Rudolf Diesel perfects the Diesel Engine |
| 1896 | Wien devises his Distribution Law describing the shape of the spectrum of thermal radiation from a black body |
| 1900 | Lord Rayleigh uses classical physics to derive a mathematical expression for the long wavelength emissions from a black body |
| 1905 | Lord Rayleigh and Sir James Jeans improve on the earlier formula but discover that it doesn't work for short wavelengths (the 'ultraviolet catastrophe') |
| 1971 | The joule is formally adopted as the SI unit for energy |
| 1980 | Umberto Eco publishes *Il nome della rosa* (*The Name of the Rose*) |

If you want to know more:
Pais, A., 1986. *Inward Bound. Of Matter and Forces in the Physical World*. Oxford University Press, 666 pp.—A very readable account of discoveries in physics.

Music: Obviously Gustav Holst, Opus 32, *The Planets*.

Libation: So much of this work was done in Vienna, it seems appropriate to celebrate their work with an Austrian wine, perhaps the delicious white wine of the region, 'Grüner Veltliner.'

### Intermezzo X. Getting Used to the East

This time I arrived in Prague by air. Stairs were rolled out to our plane. We were told to remain in our seats for a few minutes; an official was coming on board to take someone off the plane. You can imagine my immediate concern and then utter surprise when Pavel came striding down the aisle with a huge grin on his face. Very officiously he said "Come with me please." We left the plane before any of the other passengers could disembark. We walked over to the immigration counter, and I was immediately ushered through. I had no idea what was going on, but kept my mouth shut. We picked up my luggage, and got in his car. He was laughing. "Pavel, what was that about? How did you do that?" He showed me his little red Academy of Sciences pass book. He told me it looked just like the ones used by the Secret Police, so he had simply flashed it at the guards and immigration people at the airport. He was just having fun to see what he could get away with.

During my visit to Czechoslovakia in the summer of 1964, the Micropaleontology Section of the Czech Academy seemed like a home away from home. I had become determined that I was going to learn Czech. Not as easy as it might sound. First there are 42 letters, some with very special sounds. indicated by the letters with little check-marks over them: Č is ch, Ď is dyu, Ě is ye, Ň is nye, Š is sh, and Ř is erzh (as in Dvořak) and involves tricks with the tongue that only Czechs can manage. It is often said that Czech is the only language where you can actually break your tongue. This is not true; I later learned that Slovak is just as bad. At the Micropaleontology office I would get help with learning such senseless expressions as: *Strč prst skrz krk* (which, translated means 'stick your finger through your throat'). Note, no vowels. Or *Vlk zmrzl, zhltl hrst zrn* (translation -the wolf froze, he swallowed a handful of grains) Again, no vowels. Or the classic *Třistatřiatřicettři stříbrných křepelek přeletělo přes třistatřiatřicettři stříbrných střech*, Three hundred and thirty three silver quails flew over three hundred and thirty three silver roofs. Pavel was especially concerned that I learn that last one. If you can say that properly, you can speak Czech. I was never successful, but I provided a lot of amusement for those in the office.

One of Pavel's assistants, Peter, had a problem that he hoped I could help solve. His uncle in Chicago had invited him to visit. Using his uncle's letter he had made application for a visa from the US consulate, but one necessary document was missing. He needed a statement from someone saying that they would cover his expenses in the US. His uncle would do this, but had not known he needed to include it in his letter of invitation. I typed out University of Illinois letterhead, and wrote out a letter to the US consulate stating that I would cover his expenses. Pavel told me, "You know he isn't coming back." I suspected as much, so now I was involved with assisting the departure of another Czech to the US.

One evening a few days later, Pavel came to my hotel to take me to dinner. He said quietly that we needed to talk. He picked up the telephone and pointed at the bottom, later explaining to me that telephones in the rooms for foreigners usually had microphones attached to listen to anything being said in the room. As we walked along the Vácslavské námĕsti, Pavel explained that although he was not a member of the Communist Party, friends kept him informed on what was going on. There was a movement toward reform being organized. Lots of young people were joining the party intent on changing it from within. There was a lot going on underneath the surface in Prague.

The summer of 1964 included more than just the stay in Prague. We made a trip to see the Cretaceous rocks in Slovakia, the eastern part of the country. Our first day was a drive to Bratislava, where we met Josef Salaj, a micropaleontologist with the Slovak Academy of Sciences. Josef was our guide as we explored the western end of the Carpathian Mountains. Slovakia was different from Bohemia and Moravia, with less of a western background. The farm houses were different, and in general it looked neglected. The towns still had loudspeakers throughout them, which at noon and a few other times during the day, would play martial music followed by slogans or even a speech from some communist official. You couldn't get rid of the propaganda simply by turning off your radio.

In the evenings we stayed in simple country inns. Westerners were unknown here, so I was quite a novelty. The food was surprisingly good, after all we were in a major farming area, and the farmers kept the best things for themselves and the locals. Then there was "The Book." Each inn (or hotel, restaurant or any other establishment for that matter) had a Book. You could ask for the Book and write comments in it. The Book was ordinarily used for making complaints. It would be reviewed periodically by party functionaries and the operators of the establishment would get lectured on their shortcomings or even reprimanded. We almost always asked for the Book, looking very serious. Then Pavel, Josef and I would write long complimentary notes, praising the staff for the excellent food and courteous service, ending with our signatures and titles. While we were doing this, the staff would be cowering on the other side of the room. Then, looking very serious, they would come and take the closed Book away. A few minutes later we would hear happy voices as they read what we had written. Glasses of Slivovitz and perhaps another bottle of wine would appear, and our stay would be a most pleasant one.

Each evening we would have 'map count.' We had detailed topographic maps to help us locate the rock exposures. And each evening Pavel and Josef would check off the maps against a list to be sure we had all of them. I finally asked what this little ceremony was about. They explained that the maps were secret, and no foreigner was ever supposed to see them. These were accurate maps, not like those you could buy in the stores. Maps available to foreigners had 'errors' on them, such as having the road or a town shown on the wrong side of a river. They were meant to confuse invading troops. I pointed out to Pavel and Josef that I was a foreigner and I was looking at the maps. The answer was purely from the Good Soldier Schweik: "When we checked out the maps they didn't ask us if there were any foreigners going with us on our field trip."

When we got back to Prague, I found we were not going to be able to finish off all the work before I was scheduled to leave. I needed to get a visa extension. This involved going to an office that handled visas for foreigners. Pavel accompanied me. We explained what we wanted. A clerk in a military-like uniform took my passport. About half an hour later she returned and handed me my passport back. There was no visa extension inside; instead "The Commissarin wishes to speak to you." I left Pavel behind as I was led down a long whitewashed corridor and shown a bench where I could sit. All I could think was "uh oh, this doesn't seem good." I wondered which of the Czech laws I had broken in the past few months they had found out about. I waited about a half an hour, and then the door opened and another woman in military dress appeared, "The Commissarin will see you now."

I was led into the Commissarin's room, and the door closed behind me. The Commissarin was a large, middle-aged very stern looking lady. She looked at me for a few minutes without saying anything. You can imagine how I felt; I was sure I was going to be led off to a prison cell. Then she said "You want a five day extension on your visa?" "Yes." "You are an American here on a visa; you are free to leave the country whenever you want. Why on earth would you want to stay in this damned country 1 day longer than necessary?" I was so surprised I didn't really know what to say, but I sputtered out that I was working with colleagues at the Czech Academy of Sciences and we needed a few more days to finish the work. She sort of shook her head as though I were nuts, put the visa extension stamp in my passport and handed it back to me. "There, enjoy your work, and have a good trip back home."

There definitely seemed to be some cracks appearing in the Communist monolith.

I had seen a lot of Czechoslovakia, and we discussed whether, if I were able to get it, Pavel might be able to accept a postdoctoral fellowship at the University of Illinois for a year. If he could come over, we could look at the Cretaceous of the US together in detail. But that would mean he would be away from his section of the Czech Academy for a whole year. He checked it out and found that it could be done. Shortly after I got back to Urbana, I started making inquiries about how to go about applying for the postdoctoral fellowship. It was not a straightforward procedure if the candidate was to be from a Communist country.

That fall I had a phone call from Peter who had just arrived in Chicago. That was the last I heard from him until a few years later when I met him at a Convention of the American Association of Petroleum Geologists and learned he was now employed in the US petroleum industry.

My trip to Prague in the summer of 1965 was relatively short. I had good news for Pavel: the University had approved a Postdoctoral Fellowship for him for the 1966–1967 Academic year. Pavel had bad news for me. He had joined the Communist Party to be one of those working from the inside for change. He thought this would probably prevent him from getting a visa to the US. I expected that this would certainly complicate matters, but all we could do was wait and see what happened.

In the meantime, Pavel had received an invitation to work at the West German Geological Survey (Bundesanstalt für Bodenforschung) in Hannover. He spent 8 months there in 1966, returning to Prague at Christmas time. His Postdoctoral appointment at the University of Illinois was to start on January 1, 1967. However, when he went to apply for his visa for the US, there was a problem. Pavel got caught up in the Kazan-Komarek affair.

Vladimir Komarek was born in Czechoslovakia, sent to Germany during the war as a slave laborer, escaped back to Prague in 1944, and served as a translator for the Allied forces in Czechoslovakia in 1945. He escaped to the West after the 1948 communist coup. He was then recruited by the Deuxième Bureau, the French intelligence agency, as an agent. He built up a network of spies in Czechoslovakia, crossing the Iron Curtain four times in the next 3 years. He left intelligence work in 1951 and took on a new identity, using the surname Kazan.

As Vladimir Kazan he worked as a travel agent in France, where he met an American woman and married her. In April, 1955, he moved to the United States, starting a travel agency in Cambridge, Massachusetts. He changed his name again, to Kazan-Komarek, and became a U.S. citizen in November 1960. Everything about his earlier activities seemed forgotten. He even got involved in arranging travel to the USSR using the Soviet Travel Agency, Intourist.

In the meantime, Czechoslovak and Soviet authorities had become aware of his activities as an agent for the French and decided to set a trap for him. Vladimir Kazan-Komarek, thinking that everything had been forgotten, innocently accepted an invitation from the Soviet Travel Agency Intourist to attend a travel conference in Moscow in the fall of 1966. On October 31, as he was flying out of the USSR on Aeroflot, the plane made an unscheduled stop in Prague, citing engine trouble. Vladimir Kazan-Komarek was arrested as a spy and charged with high treason and murder.

When Pavel applied for his visa, the US Consulate informed him that it was not issuing any visas, except for family visits and duties at the United Nations, until Kazan-Komarek was free. Kazan-Komarek was tried by the Czechoslovak authorities in January and sentenced to 8 years in prison. On February 3, 1967 he was granted amnesty and given 48 h to leave the country. Shortly thereafter Pavel got his visa and arrived to start his postdoctoral year in March, 1967. There was still one problem. The papers he was to sign to receive his stipend from the University had a loyalty statement that 'I am not now nor have I ever been a member of the Communist Party.' Pavel was not about to commit perjury. The University official responsible for these matters told me 'Send him over to my office; I'll give him the papers to sign. He won't sign that line; I won't be looking. I'll take the papers, put them in the file. Done.' Nothing was ever simple in those days, but somehow everything worked.

We had a year to work together on sorting out the distribution of Cretaceous coccoliths. We made several field trips to collect samples of rock. I would make out a detailed itinerary of the trip, including information on who we would meet with, and send it to the State Department for approval. In about a month I would get back a letter of official permission for Pavel's travel. The longest trip was to Miami, then through the Gulf Coast states, to Los Angeles, then to San Francisco, and back through Nevada, Utah, Colorado, Kansas, and Missouri to Illinois. We visited a number of micropaleontologists along the way, and they helped us with field collecting. I later found out that we were followed by a team of FBI agents who interrogated almost everyone we had met with. Micropaleontologists love to talk shop, and those agents must have learned a lot about the stratigraphic distribution of *Lithraphidites quadratus*, the paleobiogeography of *Watzaueria barnesae*, and other such fascinating topics.

Travel restrictions from the Czech side relaxed somewhat, and Pavel's wife Jiřina was able to come to Illinois for a few months.

Pavel returned to Prague in March, 1968, taking the Italian ocean liner *Leonardo da Vinci* back to Europe. Czechoslovakia was now in the midst of Alexander Dubček's internal communist revolution. When Pavel arrived back he was assigned a special duty in the Geological Institute of the Czech Academy of Sciences. The first free secret ballot of the members of the Institute selected five members to form a 'Rehabilitation Commission.' Pavel received the largest number of votes, and was thereby named its Chair. The purpose of the Commission was to examine the files of the Secret Police to determine whose careers had been damaged on a political basis. Officials thought that his experience outside the country made him especially well qualified for this task. It was a big job, and he was still working on it when I visited in the summer of 1968.

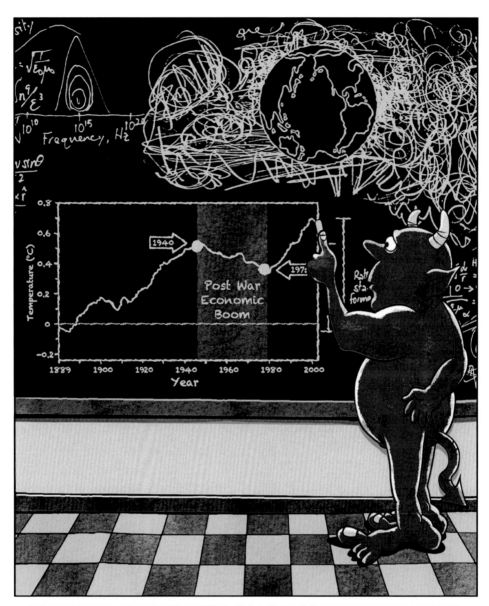

THE DEVIL INTRODUCES CHAOS INTO THE WEATHER

*The climate is what you expect; the weather is what you get.*
Mark Twain

## 11.1 An Introduction to the Climate System

It was just at the end of the 19th century that a true understanding of how the climate system operates began to develop. Stefan's Law was a critical component, but knowledge of Earth's reflectivity (albedo) and the role of greenhouse gases were also critical. As best I can tell, the first person who was able to take these major components into account to develop a comprehensive view was Swedish physicist/chemist Svante Arrhenius (1859–1927) (Fig. 11.1). He received his doctorate from Uppsala University in 1884. It contained 56 theses concerning chemistry and physics; most would still be accepted today. However, the most important idea in the dissertation was an explanation of the fact that pure salts and pure water are both poor conductors of electricity but solutions of salts in water are good conductors. His idea was that when introduced into water salt dissociates into electrically charged particles, the 'ions' that Michael Faraday had discovered during his experiments with passing electricity through water/salt solutions in 1834. Faraday's belief had been that ions were produced in the process of electrolysis; Arrhenius proposed that even in the absence of an electric current the salts in solution were in the form of charged ions. He concluded that chemical reactions in solution were reactions between ions. He believed that acids were substances which produce hydrogen ions ($H^+$) in solution and that bases were substances which produce hydroxide ions ($OH^-$) in solution.

After obtaining his doctorate, Arrhenius received a travel grant from the Swedish Academy of Sciences that allowed him to visit, among others, Ludwig Boltzmann in Graz, Austria.

In his famous 1896 paper *On the Influence of Carbonic Acid in the Air upon the Temperature of the Ground* he became the first scientist to speculate that changes in the levels of carbon dioxide in the atmosphere could substantially alter Earth's surface temperature. He speculated that these changes might have been the cause of the ice ages. The data available to him were different from what we have today. To estimate the effects of absorption of infrared radiation by 'carbonic acid' (his term for $CO_2$, carbon dioxide) he used the infrared observations of the Moon that had been made by Frank Washington Very and Samuel Pierpont Langley at the Allegheny Observatory in Pittsburgh. He had to conduct a thorough analysis of the data to discover clear trends. But from it he was able to formulate his 'Greenhouse Law.' In its original form, it reads as follows: "if the quantity of carbonic acid increases in geometric progression, the augmentation of the temperature will increase nearly in arithmetic progression." This simple expression is still widely used today as a first approximation of the relation expressed as follows: a doubling of atmospheric $CO_2$ results in 1 °C global warming.

Arrhenius' 1896 paper is often cited as the source of the idea that emission of $CO_2$ from burning of fossil fuels would lead to global warming, but that idea was actually introduced a decade later in a book he wrote for the general public: *Världarnas utveckling* (1906), German translation: *Das Werden der Welten* (1907), English translation: *Worlds in the Making* (1908). In it he suggested that the human emission of $CO_2$ would be strong enough to prevent the world from entering a new ice age, and he concluded that a warmer earth would provide conditions favorable to feeding a rapidly increasing population.

Arrhenius' 1896 paper includes discussion, in one form or another, of all of the elements of a modern global energy balance model; the shorter wavelength energy received from the Sun and the longer wavelength infra-red energy emitted by the Earth, modified by absorption by carbon dioxide and water vapor, the scattering of radiation, and albedo determined by the sea surface, land, snow, and clouds. How these ideas and the data required for their analysis changed over the 20th century is an epic story, but beyond the scope of this book. The advent of satellite observations in the 1960s and 70s initiated a revolution that continues today.

© Springer International Publishing Switzerland 2016
W.W. Hay, *Experimenting on a Small Planet*, DOI 10.1007/978-3-319-27404-1_11

**Fig. 11.1** Svante Arrhenius (1859–1927). Father of Physical Chemistry and investigator of the climatic effects of atmospheric $CO_2$

This chapter presents the classic mid-20th century view of the climate system. It uses pre-satellite estimates of the incoming and outgoing energy fluxes. Although it assumes that the climate is in a steady state, it recognizes that there may be some variability about the mean state, possibly due to changes in radiation from the Sun with the 11 year sunspot cycle. It was for that reason that the '30 year average' was adopted as a standard for defining the climate state. When this description of the system was developed, starting

in the middle of the 20th century, there was little thought given to the possibility of changes in greenhouse gas concentrations, land use, or other parameters inducing long term trends that would negate the steady-state assumption. This is what every older adult who studied climate science learned. It is in every textbook. It is a gross oversimplification and marginally relevant to the future.

## 11.2    Insolation—The Incoming Energy from the Sun

The Earth receives energy from the Sun. This is termed 'insolation.' Earth must return this energy back to space if it is not to heat up, as shown in Fig. 11.2. But the Earth is rotating; during the daytime it receives energy and during the night it can radiate this energy back into space. The Earth is also nearly spherical, so that most of the energy received from the Sun is around the Equator, where the Sun is overhead. In the Polar Regions the Sun is close to the horizon in summer and below the horizon in winter, so there the energy can be more effectively radiated back into space.

If the Earth were just a giant rock, the side facing the Sun would become very hot during the day and very cold at night. Earth's Moon is just such an object. Its average distance from the Sun is the same as the Earth's. However, the average temperature of the Moon is 29.5 Earth days (about 709 h). During the day the average temperature of the Moon's surface is 107 °C (=225 °F), and where the Sun is directly overhead the temperature reaches 123 °C (=253 °F), well above the boiling point of water on Earth. During the lunar night the average surface temperature drops to −153 °C (=−243 °F)

**Fig. 11.2** Earth receives a flux of short-wavelength radiation from the Sun ($Q_S$). Part is reflected, the rest absorbed. The Earth must reradiate a flux of energy ($Q_L$) equal to that received, but it is at long wavelengths. Energy can be transported on Earth by the atmosphere ($Q_A$) and ocean ($Q_O$), and the reradiation process is moderated by clouds and greenhouse gases

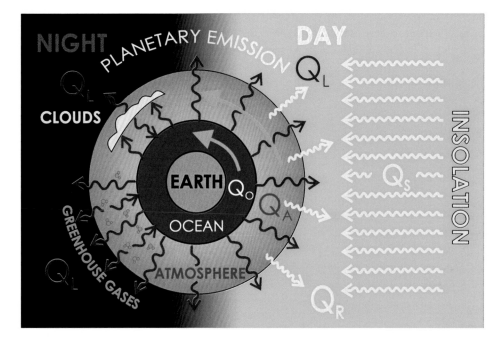

with some areas falling to −233 °C (= −387 °F). These troubling facts escaped the producers of the 1984 film "*2010 —The Year we Make Contact*" which shows the comfortable accommodations of our colony on the Moon.

If the Earth were just a rock it would also experience day-night temperature extremes, although they would be somewhat less because Earth's day length is 24 h, so that during the day there would be less time for the heat to build up and at night less time for the heat to dissipate. The greatest diurnal temperature ranges recorded at the Earth's surface are in the Sahara (50 °C = 122 °F) and Siberia (−67 °C = −88.6 °F). In the tropics, the night-day temperature range can be as small as 1 °C (= 1.8 °F).

Fortunately, the Earth is not just a rock. Two fluids cover it: liquid water (the ocean, seas and lakes), and a mixture of gases (the atmosphere). These fluids can circulate, and they redistribute the heat to reduce temperature contrasts on the Earth's surface. Some of the gases in the atmosphere absorb and reradiate energy. Transitions between the three phases of $H_2O$, ice, water, and vapor play a large role in redistribution of energy on the Earth.

The incoming energy from the Sun is in the form of "short wave" radiation. Earth dissipates this back into space as "long wave" radiation, shown in Fig. 11.2. To determine how much and what wavelengths of radiation will be emitted by an object, physics uses Stefan's Fourth Power Law and Kirchhoff's "black body" concept. Remember that a black body is a hypothetical construct; it is an object that absorbs all the electromagnetic energy that falls onto it and transmits none of that energy. It is an ideal source of thermal radiation. In the real world there are no perfect 'black body radiators,' but the Sun and the tungsten filament in an electric light bulb come close. If the temperature of an object is below about 700 K (427 °

C = 800 °F), the wavelengths emitted by it cannot be sensed by the human eye, and it appears black. However, if the object has a higher temperature we see it as colored. The Sun has a surface temperature of about 5778 K (= 5505 °C = 9941 °F). Most of the energy received from the Sun is in that part of the electromagnetic spectrum we call 'light,' which ranges from shorter wavelengths we call violet or blue, to the longer wavelengths we call red, as shown in Fig. 11.3. Some of the radiation is of even shorter wavelengths, the ultraviolet, and some is in longer wavelengths, the near infrared. Human eyes have evolved to be sensitive to the most intense radiation from the Sun. We cannot sense ultraviolet, probably because most of it does not reach the Earth's surface. We can sense the infrared, which we feel as warmth or heat.

If you read the labels on light bulbs or do much color photography, you are already familiar with this temperature-radiation effect. Particularly, when you replace your incandescent bulbs with the fluorescent or LED bulbs that use much less energy, you will find the color of the light emitted by the bulb described as some temperature Kelvin. For a warm candlelight feel you want about 4000 K, for a daylight effect you want 5700 K, of course, and for garish blue–white light you want 7000 K. There are some good sites on the web that will show you what the color of the light emitted from the bulb will be. In fluorescent bulbs, it is not a heated filament that emits the light; the fluorescent bulb is no 'black body' but uses a wholly different principle to emit light—that's why it uses so much less energy and why it doesn't get hot. You can even buy fluorescent bulbs that will work with a dimmer. Their color temperature stays almost the same as they are dimmed; they just emit less light. If you dim an incandescent light bulb it goes from white to orange to red.

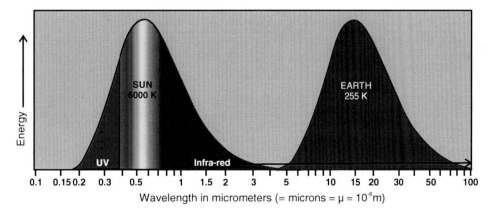

**Fig. 11.3** Radiation from the Sun and Earth by wavelength. The total amount of energy received from the Sun is equal to the total amount of energy reradiated by Earth. *UV* = ultraviolet

## 11.3   Albedo—The Reflection of Incoming Energy Back into Space

Figures 11.4 and 11.5 show what happens to the incoming energy. About 30 % of it is simply reflected back into space through three processes: backscattering (the blue sky), reflection from clouds, and reflection from the Earth's surface. The reflectivity of the planet is called its 'albedo'. Reflecting 30 % of the energy incident upon it, the Earth is said to have an albedo of 0.3.

Backscattering is produced by the interaction of the incoming light from the Sun with the gas molecules in the air.

This produces the blue of the daytime sky. It is what makes images of Earth from space look like a big blue marble. It can be affected by pollutants. Once I was in a barbershop in Santa Monica. The sky was orange-yellow. My eyes were watering. I asked the barber "Is it like this often?" He responded "No, we don't get such a clear day very often."

Clouds are the most important (and least understood) reflectors. Clouds are made of droplets of water. There are many different kinds of clouds, all having different reflectivities. Their formation depends on the amount of moisture in the atmosphere, how much the air is convecting (moving up and down) and whether there are tiny particles in the air

**Fig. 11.4** Generalized scheme the mounts of incoming solar energy reflected and by Earth. Numbers are approximations to give an idea of the relative importance of different processes

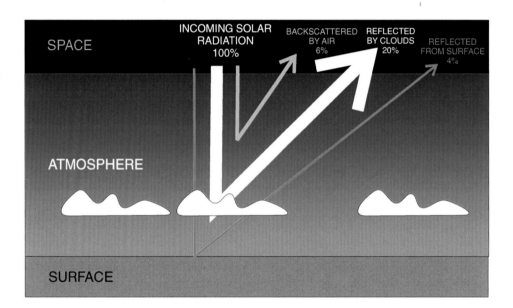

**Fig. 11.5** Generalized scheme of amounts of incoming solar energy absorbed by Earth. Numbers are approximations to give an idea of the relative importance of different processes

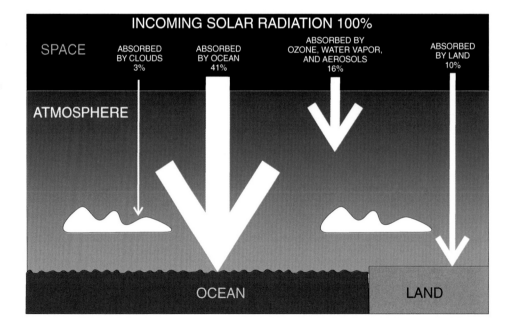

onto which the water can condense. Some clouds (contrails) are man-made.

Reflection from the surface varies greatly. Seen from directly overhead in space, the ocean appears black because water absorbs almost all the radiation it receives. Fresh snow reflects almost all the radiation it receives. Forests, farmland, and deserts all reflect different amounts of the incident radiation.

Some of the radiation coming to the Earth is absorbed by ozone in the stratosphere, by the greenhouse gases—those gas molecules with more than two atoms—in the air, and by dust. Some is absorbed by the water droplets in clouds. This absorbed energy heats the atmosphere, but because the greenhouse gases, dust, and clouds are not evenly distributed, the heating is uneven.

Finally, about half of the incoming energy is absorbed by the ocean and land. The ocean is a fluid. It can absorb the energy in one place and transport it to another. The energy absorbed on land is the only part of the energy absorption system that cannot be moved around.

## 11.4 Reradiation—How the Earth Radiates Energy Back into Space

How does the Earth get rid of this energy it receives in order to maintain a net zero energy balance? The Earth radiates the energy it receives from the Sun back into space, but its radiation is all in the far infrared, invisible to our eyes. Figure 11.3 shows the Earth's radiation spectrum assuming our planet's temperature to be 255 K (= −18 °C = −0.4 °F). This is our planet's average temperature as measured from far away in space. You notice that this is well below the freezing point of water (273 K = 0 °C = 32 °F). If the Earth were a perfect 'black body,' its temperature would be a more comfortable but still cold 278.3 K (= 5.3 °C, = 41.5 °F). However, the Earth is not close to being a 'black body' as we know from views from space. Like most objects in nature, it is a 'gray body' reflecting part of the radiation incident upon it. Our planet reflects about 30 % of the incoming energy; that is why we can see it from space. That is why its temperature as seen from space is 255 K. But in climatology its 'average temperature' (meaning surface temperature) is usually given as about 288 K (= 15 °C = 59 °F). That is its average *surface* temperature. Note that in climatology the term 'surface *temperature*' refers to the temperature of the air 2 m (= 6.6 ft) above the ground.

Why is this '*surface temperature*' higher than the average temperature for the planet measured from space? Because of the greenhouse effect. The greenhouse effect raises the surface temperature 33 K (= 33 °C = 59.4 °F). As far as we know, except for a few episodes 600–700 million years ago, the Earth has always had greenhouse gases in its atmosphere. If they weren't there Earth would be a frozen wasteland. Believe it or not, you can find websites that will tell you that the Earth as a whole is not warming: the overall planetary temperature is not changing. So don't worry, it is only the surface that is warming; the outermost atmosphere is cooling. I guess I do worry because along with about 6.8 billion other people, I happen to live on the surface of this planet.

Figure 11.6 shows the complicated way in which the Earth solves this reradiation problem. About half of the

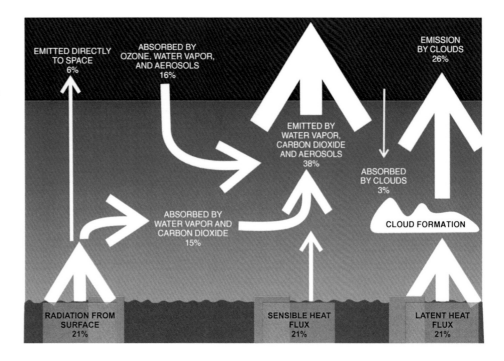

**Fig. 11.6** Reradiation of energy from Earth back into space. Absorption of incoming energy within the atmosphere is also shown. Note the important roles played by greenhouse gases, dust, and the latent heat flux. Again, the numbers are approximations to give an idea of the relative importance of different processes

EMITTED DIRECTLY TO SPACE 6%

ABSORBED BY OZONE, WATER VAPOR, AND AEROSOLS 16%

EMISSION BY CLOUDS 26%

EMITTED BY WATER VAPOR, CARBON DIOXIDE AND AEROSOLS 38%

ABSORBED BY CLOUDS 3%

ABSORBED BY WATER VAPOR AND CARBON DIOXIDE 15%

CLOUD FORMATION

RADIATION FROM SURFACE 21%

SENSIBLE HEAT FLUX 21%

LATENT HEAT FLUX 21%

energy that reaches the surface of the Earth is radiated directly back into space at infrared wavelengths. This is most effective at night when there is no incoming radiation (except from the Moon and stars). Of the other half, some of the energy that heated the land and surface ocean heats the lower atmosphere. This part of the system is called the sensible heat flux, sensible because we can feel it. But the largest share of the return of energy from the surface is as the latent heat in water vapor. This is water that evaporated from the surface of the oceans, seas, and lakes, and the water that is 'transpired' by plants. Transpiration is the process whereby plants take in water from the ground and it is evaporated though the leaves and stems. The phase transformation from liquid water to vapor involves a huge amount of energy, more than five times as much as raising the temperature of water from its freezing to its boiling point, so a little water goes a long way in Earth's energy balance.

Once in the air, some of the energy is absorbed by greenhouse gases, dust, and clouds. These reradiate the energy in all directions, down as well as up and to the sides. Eventually the energy goes back to space, but the process of reradiation is rather like going through a maze. Earth's surface and lower atmosphere are kept warm. As we will see later, if it were not for the greenhouse gases, the Earth would be an uninhabitable ball of rock covered by ice. There is just such an ice ball in our solar system, Jupiter's Moon Europa, but in its case, the ice may be resting on a very deep ocean.

By now you should be able to guess that the climate system is really very complicated. Any change in any part of the system will reverberate through the whole system and cause changes everywhere else. Because the system is so complex and interlocked it is said to be chaotic.

## 11.5    The Chaotic Nature of the Weather

This is a good place to recall why we have computers. The first computers were mechanical devices made with gears, wheels, bars, etc. I suppose the first computer was the clock. Much more complex mechanical computers were devised to predict the tides, knowledge useful for managing seaports. One of the most elegant of these is in the Deutsches Museum in Munich. But there was something else that would be much more important to be able to predict the weather.

If you could predict the weather far in advance, say a few years or more, you would be able to predict how much wheat would be produced, when to harvest strawberries, and determine how big the crop of oranges would be. You could predict the prices of commodities. More importantly, you could manage food production to meet needs. Since you know the components of the climate system, you should be able to express their relations in mathematical equations, program all this into a computer, and voila! You are king of the commodities market and can even manage the world food supply.

The first attempts to do this were made after World War II. Edward Lorenz (1917–2008), who had served as a weather forecaster for the Army Air Force during the war, went to the Massachusetts Institute of Technology, where he earned two advanced degrees and eventually became Professor of Meteorology. He was intimately involved in trying to build a computer to predict the weather. The transistor had not been invented, so these very early computers used electrical circuits and vacuum tubes. Can you believe it; most kids today have never even seen a vacuum tube? Lorenz made a hypothetical weather prediction run starting with a particular set of numbers for each of the elements of the climate system. These are called the 'initial conditions.' He got his result, and then, to make sure everything was working correctly, he decided to replicate it. He figured he didn't need to go through the whole process again, so he took a number generated half-way through the sequence, entered it, and left it to run to the end again. The results were very different from the first run. When he started the second run, he did not enter the whole number from the first run, which was 0.506127, but he had rounded it off to 0.506, thinking this wouldn't make any appreciable difference. But it made a huge difference in the results. With more experiments on his primitive computer, Ed Lorenz found that if you make a tiny change in one of the elements of the climate system, you get a completely different answer. At first this was attributed to instability of the vacuum tubes or possibly short circuits somewhere in the wiring. But then he came to an awful conclusion: the climate system is inherently chaotic. You cannot predict the weather far in advance. His 1963 paper *Deterministic Nonperiodic Flow* is a landmark in science.

Actually, the discovery of chaotic behavior in some physical systems goes back to the work of French mathematician Henri Poincaré in the early 20th century. In 1885 Poincaré had entered a contest to honor the 60th birthday of King Oscar II of Sweden, which was to be on January 21, 1889. One of the contest questions was to prove that our solar system is dynamically stable. That is, we don't need to worry about our planet simply wandering off into space by itself. Contestants had 4 years to work out the sort of 'birthday present' any scientist would like to have. Poincaré reduced the question to one of simply proving that you can predict the motion of one body moving in the gravity field of two others, say the Earth in the presence of the Sun and Jupiter. This has become known as the three-body problem. It sounds simple enough, but the solution involves solving nine simultaneous differential equations—a job unsuited for ordinary mortals. Poincaré thought he had solved the problem, showing that the orbits of the three bodies were stable. He submitted his solution for publication in Acta

Mathematica and was awarded the prize in 1899. However, one of the peer reviewers of the manuscript had a couple of little questions. In trying to answer the questions, Poincaré discovered to his great dismay that tiny differences in the initial conditions produced large differences in the outcomes. This phenomenon has become known as the "pathological sensitivity to initial conditions." He found that the orbits of the three bodies were not stable; he had discovered chaos in physical systems.

> A very small cause which escapes our notice determines a considerable effect that we cannot fail to see, and then we say that the effect is due to chance. If we knew exactly the laws of nature and the situation of the universe at the initial moment, we could predict exactly the situation of that same universe at a succeeding moment. But even if it were the case that the natural laws had no longer any secret for us, we could still only know the initial situation approximately. If that enabled us to predict the succeeding situation with the same approximation, that is all we require, and we should say that the phenomenon had been predicted, that it is governed by laws. But it is not always so; it may happen that **small differences in the initial conditions produce very great ones in the final phenomena**. A small error in the former will produce an enormous error in the latter. Prediction becomes impossible, and we have the fortuitous phenomenon. (H. Poincaré, 1903, Science et Méthode, Paris).

It is for this reason that the weather predictions are usually restricted to 5 day forecasts. In the description of 'climate' a 30 year average is taken as a benchmark, to eliminate the effect of the chaos inherent in the weather. You are sure to hear the argument "if we can't predict the weather next week, how can climate scientists possibly predict the climate a 100 years from now." The weather and climate are two different things. The weather operates on very short time scale; the climate is a description of the weather integrated over 30 years. Climate models differ in their predictions over long time spans both because of differences in the initial conditions and differences in computing methods. It is when the results of many different models generally agree, although they differ in the details, that cause and effect become plausible. And today, the results of climate models that differ significantly in their inputs and variables all have the same general conclusion—a warmer Earth is in our future.

As luck would have it, this problem of chaos in its astronomical context comes back to haunt us if we want to look at the long term history of insolation and how it may have affected Earth's climate in the distant past. How that problem has been solved is one of the great triumphs of collaboration between astronomers and geologists.

At this point it may be useful for you to see what mathematical chaos looks like. In Chap. 2 you encountered the 'logistic equation' for the growth of a population to the environment's 'carrying capacity.' This was Malthus' idea that the growth of a population has limits, and Verhulst's mathematical formulation of growth as a self-limiting

process. Good idea but rabbits don't know about it. Rabbit populations (and most everything else as well) experience large variations with time, not a stable level. In 1976, biologist Robert May published an equation that is more realistic:

$$X_{t+1} = a X_t (1 - X_t)$$

where $X_t$ is a number between 0 and 1 representing the ratio of the population existing at year t to the maximum possible population. Thus $X_0$ represents the initial ratio of the population at year 0 to the maximum possible population; $a$ is a positive number representing the combination of the rate of reproduction and death rate of a population. Sometimes called the biologic potential, it is an estimate of the maximum capacity of living things to survive and reproduce under optimal environmental conditions.

What you might expect of this is an initial rapid then slowing growth of the population as it reaches the environmental carrying capacity. What you get is something quite different. By the way, the title of May's paper was '*Simple mathematical models with very complicated dynamics.*'

Figure 11.7 shows six solutions to the equation, all starting with a population 30 % of its maximum possible size and carried out through 40 iterations. You can think of the iterations as generations. If the number for $a$ is 2, the birth and death rates are equal, and the population will remain the same size. If $a$ is less than 2, as shown in Fig. 11.7b, the death rate is greater than the birth rate, and the population will eventually die off, as shown in Fig. 11.7a. If a is between 2 and 3, the population will grow and oscillate about some value (0.6428) with the oscillations getting smaller and smaller. Above $a = 3$ the oscillations grow larger with time, but the population finally ends up oscillating back and forth between two values, as shown in Fig. 11.7c, or three or four or more values. Figure 11.7d shows the pattern repeating after every eighth oscillation. But when the value of a passes 3.6999 the solutions become chaotic and are unpredictable. Figure 11.7e–f shows this chaotic behavior. Figure 11.7e rather resembles the pattern of the ellipticity of Earth's orbit over time. Sometimes in these chaotic solutions there will be 'islands of stability' when the same oscillations keep repeating; but then suddenly and unexpectedly chaos will set in again. With $a$ greater than 4, the oscillations make huge fluctuations but crash to 0, extinction of the population, in short order.

Henri Poincaré noted that a small difference in the initial condition made a huge difference in the outcome. In Fig. 11.7 the initial condition was that the population was 30 % of the maximum possible size. In Fig. 11.8 the initial population is set at 50 % of the maximum possible size. With $a = 0.9$, the population dies off, a bit more slowly. With $a = 2.8$, the oscillations are a bit more pronounced at

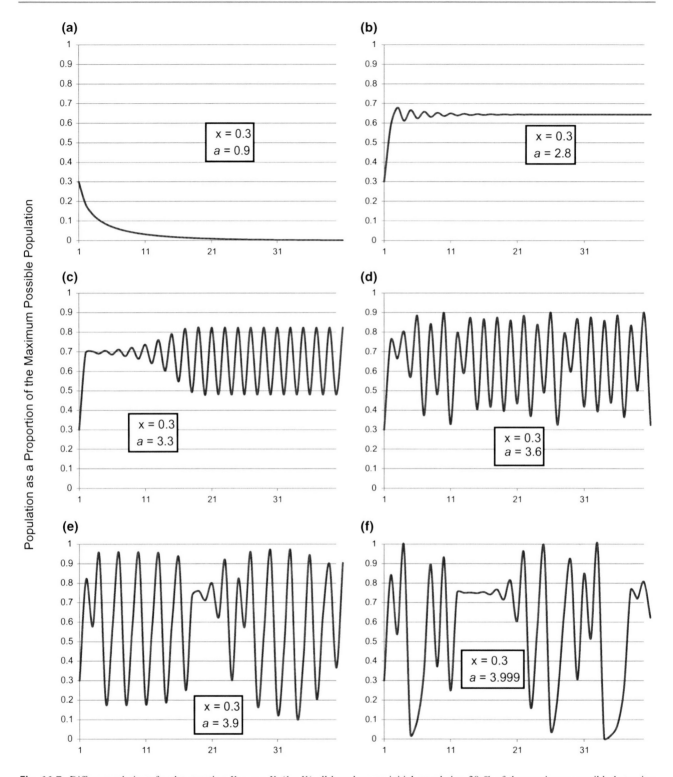

**Fig. 11.7** Different solutions for the equation $X_{t+1} = a\, X_t\, (1 - X_t)$ all based on an initial population 30 % of the maximum possible but using different values for the ratio of birth to death rates

first, but the end result is the same as with the 30 % initial population. But above $a = 3$ the solutions simply oscillate between two values. Again, chaos sets in above 3.6999.

This is the kind of behavior that Ed Lorenz found in his solutions for much more complicated equations in trying to predict the weather. The weather is inherently chaotic.

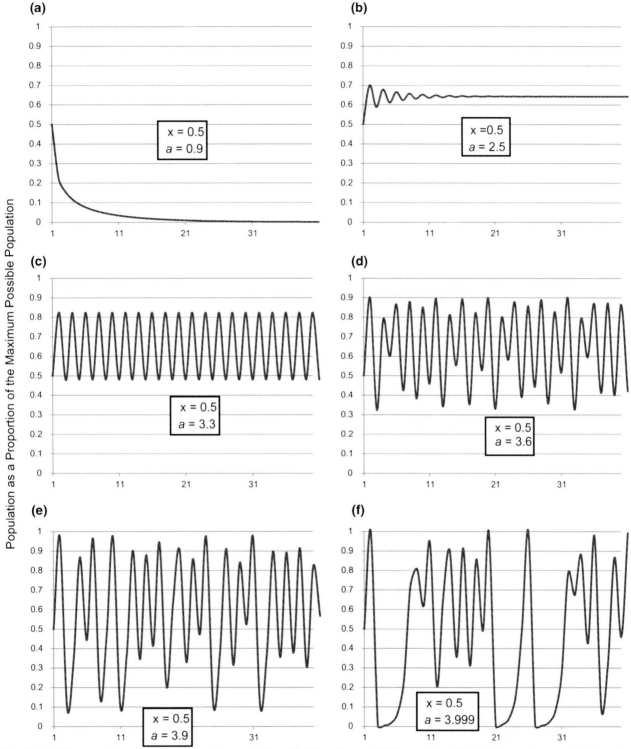

**Fig. 11.8** Different solutions for the equation $X_{t+1} = a X_t (1 - X_t)$ all based on an initial population 50 % of the maximum possible but using different values for the ratio of birth to death rates

You may hear some meteorologists say 'if we can't predict the weather a week in advance, how could anyone hope to predict the climate a hundred years in advance?' Remember that the climate is a long term average of the weather; 30 years is usually accepted as the averaging period. In running climate simulations with numerical models, there is a period from the beginning, the 'run up,' usually a few hundred years of calculations, that are discarded. Then, averages of 30 years are taken as typical of the 'climate.' In looking at Figs. 11.7 and 11.8 you will see that the first few years are special. They are the run up. But then if you average the next 30 years, and the 30 years after that, and so on, the averages will be very close. In our analogy, they are the 'climate.'

Different numerical climate models use different initial conditions, different equations, and different numbers of equations in attempting to take feedback processes into account. Their results will be different in detail. What is important is that in projecting a future with increasing atmospheric greenhouse gas levels, they all agree: Earth will be getting warmer and the weather will be getting more chaotic.

Always remember: the weather and the climate are two different things. The climate is what you expect; the weather is what you get.

Later chapters will examine each of the major components of the climate system to try to determine how they may have changed in the past and what effect they have had on the climate system as a whole. But first, here is an introduction.

## 11.6  The Earthly Components of the Climate System: Air, Earth, Ice, and Water

Empedocles (circa 450 BCE) got it almost right for climatologists. He thought there were four major elements: air, earth, fire, and water. Substitute ice for fire, and you have the four earthly components of the climate system. Add the fire back in as a fifth element and you have the geologist's world.

The energy that reaches Earth is first intercepted by the atmosphere and then a surface of land, water or ice. The different parts of the Earth's surface are termed the hydrosphere (ocean, seas, lakes, and rivers), the lithosphere (sediment and rock, mostly covered by soil on land), and the cryosphere (ice sheets, glaciers, and floating sea ice). Each of these has very different characteristics and behavior in response to the energy input from space. It is useful to know how big each of these is and how much energy they can store. Their ability to store incoming energy and then release it later moderates temperature changes from day to night, season to season, year to year and longer timescales. The mass of something per unit volume is called its 'specific gravity' or 'density.' The amount of energy required to heat up a unit mass of something is called its 'specific heat.' It varies slightly with the temperature. The heat capacity of something is the amount of heat required to change its temperature by 1 K (= 1 °C = 1.8 °F). It is a measure of how much energy can be absorbed, stored, and retrieved for a part of the climate system. Table 11.1 shows the relative importance of these parts of the climate system in terms of stored energy.

The atmosphere is the part of the system most directly involved with incoming and outgoing radiation. However, its heat capacity is less than 1/1000th that of the ocean. Because ocean water is a highly mobile fluid, it is in constant motion moving energy from one region to another. Most of the Earth's water is in the ocean. There are two other major reservoirs of $H_2O$, the cryosphere (ice and snow) and ground water in the pores of rock; both are about the same size. The amount of water in lakes and rivers is a fraction of 1 % of the total and the amount in the atmosphere as vapor is 1/1000th of 1 %. The cryosphere interacts directly with insolation, the atmosphere, and the oceans. In terms of area, the perennial sea-ice cover of the Arctic and ice on land are about equal, but the volume of floating sea ice is less than 1 % that of the ice on land. Under natural conditions, ground water does not respond rapidly to changes in climate, although it may contribute to sea-level changes on time scales of tens of thousands of years. In the past century, humans have been 'mining' groundwater at an unprecedented rate. This has had unexpected consequences for both regional climate and sea level. Pore water is included in the estimates of density and specific heat of soil and rock. The

**Table 11.1** Properties of the Earthly components of the climate system

| | Area ($10^6$ km$^2$) | Thickness (km) | Volume ($10^6$ km$^3$) | Density (kg/m$^3$) | Mass ($10^{18}$ kg) | Specific heat ($10^3$ J/kg) | Heat capacity ($10^{21}$ J/K) |
|---|---|---|---|---|---|---|---|
| Atmosphere | 510 | 8.3 | 4250 | 1.2 | 5.1 | 1.0 | 5.1 |
| Ocean | 361 | 3.8 | 1368 | 1039 | 1420 | 4.2 | 5964 |
| Cryosphere | 30 | 0.8 | 26 | 920 | 24 | 2.0 | 48 |
| Land-soil | 134 | 0.002 | 0.27 | 2000 | 0.536 | 1.5 | 0.8 |
| Land-rock | 134 | 0.5 | 67 | 2400 | 160.8 | 0.9 | 145 |

The heat capacity is the amount of energy required to raise the temperature of the component 1 K (= 1 °C = 1.8 °F). Air is compressible, so the true thickness of the atmosphere is measured in 100s of kilometers. The thickness given here is what it would be if air were incompressible and always had the same number of molecules per unit volume as it does at sea level. Ocean water is almost, but not quite, incompressible, so the average density of ocean water used here is that at mid-depths in the ocean. The thickness of rock involved with the climate system is estimated here to be 500 m. Below that level the flux of heat from the interior of the Earth dominates. That heat is still being generated by the ongoing radioactive decay of heavy elements such as uranium and thorium usual area given for land on Earth is $149 \times 10^6$ km$^2$. But of that $15 \times 10^6$ km$^2$ is covered by ice, so I use $134 \times 10^6$ km$^2$ as the area of land that is ice free. Seasonal temperature differences on land are smoothed out at 1 m depth, so in most places you can get the mean annual temperature by using a probe thermometer 1 m long. I've tried it. It works. Rock is an excellent thermal insulator. Heat propagation through rock or brick is very slow. If you have been around a firing kiln for pottery, you know that the inside can be glowing hot and the outside only warm.

From Table 11.1 it is obvious that the great thermal regulator of planet Earth is the ocean. Enormous amounts of energy are required to change the temperature of the ocean as a whole. It is the great stabilizer. If the temperature of the ocean starts to change it means that the planet is heating up or cooling down. It is the great climate decider.

## 11.7 The Atmosphere

The atmosphere is made up of three major gases: 78 % nitrogen ($N_2$), 21 % oxygen ($O_2$), and 1 % argon (Ar); and some other 'trace' gases: such as water vapor ($H_2O$), carbon dioxide ($CO_2$), methane ($CH_4$), ozone ($O_3$) and others. In case you were not aware of it, all of the $O_2$ has been produced by plants, and the amount in the atmosphere has varied over time, but there is a feedback mechanism to prevent it from getting too high. If the $O_2$ content were to increase to 35 %, trees and other living things would simply burst into flame. You may recall that on January 27, 1967, three astronauts were killed while training in the Apollo spacecraft filled with a pure $O_2$ atmosphere. An electrical spark was the cause, and items that were non-flammable in air burned readily.

But here is what is really peculiar about these gases: energy passing as radiation through the atmosphere doesn't interact with gas molecules that have only two atoms. But if the molecule is made of three or more atoms, it can intercept energy and vibrate faster, colliding with the surrounding molecules and causing them to move faster. Molecular motion is heat, so the air becomes warmer. But rapidly vibrating tri-atomic or larger gas molecules can also emit some of their energy as radiation. They are 'greenhouse gases.'

The atmosphere has a mass of $5.1 \times 10^{18}$ kg. (I hope you are comfortable with the metric system by now.) It contains $142 \times 10^{18}$ mol of $N_2$, $33 \times 10^{18}$ mol of $O_2$, and $1 \times 10^{18}$ mol of Ar. Each of these has a different density, but altogether air at sea level has a density of about 1.2 kg/m$^3$, as given in Table 11.1. That is, a cubic meter of air at the surface weighs about 1.2 kg, but you can't put it on the bathroom scale and measure it because it is buoyed up by the surrounding air, just like you are by the water in a swimming pool. On the other hand, if you would put the cubic meter of air in a box, put it on the scale, and use a vacuum pump to pump the air out of the bathroom, the scale would read 1.2 kg. But don't try this, because it is likely that your house will implode before you get a good vacuum in the bathroom.

Specific heat is the amount of energy required to raise the temperature of one unit of mass by 1 K = 1 °C. Each of the gases has a different specific heat, but $N_2$ and $O_2$ are both close to $1 \times 10^3$ J/kg K. The specific heat of Argon is only half that, but since it is less than 1 % of the atmosphere it makes almost no difference in determining the specific heat of air. Combining the numbers, the specific heat of air is about $1 \times 10^3$ J/kg K. Multiplying the specific heat of air by the total mass of the air gives us the heat capacity of the atmosphere, $5.1 \times 10^{21}$ J/K. Remember that number, you will want to compare it with other heat capacities later. And it's a good factoid to drop into the middle of a discussion of global warming at a cocktail party.

If the entire atmosphere had the same density as air at sea level, it would be about 8.5 km thick and there would be a sharp boundary between the atmosphere and space. But gases are compressible, so the atmosphere is denser at the Earth's surface and gets progressively thinner and more tenuous with altitude. There is no top to the atmosphere, no real boundary between the atmosphere and space.

Remember what Benjamin Thompson, later Reichsgraf von Rumford, discovered: heat is motion of molecules. The temperature of the air, what experts call sensible heat because you can sense it and measure it with a thermometer, depends on the motion of the molecules.

In the 18th century Jacques Charles demonstrated that as air is compressed, its temperature rises; as it expands, the temperature goes down. However, he did not publish his discovery. The publication of this important finding was by Joseph Louis Gay-Lussac in 1802, who credited the discovery to Charles. It is now known as Charles' Law.

The pressure-temperature relationship is actually a rather dramatic effect. In the old days, before computers, I could give my students in Miami the (for them) familiar case of filling a SCUBA tank. If you want to get a properly filled tank, it has to be immersed up to the valve in water to keep it cool. As compressed air goes into the tank, it warms up and if it is in the open air it can get quite hot. When it cools off, you find out that you have only about half as much air in it as you thought. Now that everyone has a computer, you are all familiar with the change in temperature with pressure because you have (or should have) a compressed gas cleaning duster to get the dust and cat hair off your computer. If you use it for more than a couple of seconds the can feels cool or even cold. You lowered the pressure inside the can, and the temperature went down. If you live in Florida, this is why the can of insect spray gets cold. As you go up in the atmosphere, there are fewer and fewer molecules in a given volume, and with fewer molecules moving around, their total kinetic energy is less and the temperature decreases.

So when air goes up, it cools, and when it comes down, it warms. Except, of course in the movie *The Day after Tomorrow*. In that film the air in the center of a gigantic hurricane is descending without warming, so that the −50 °C air from high in the atmosphere quick-freezes New York. There is another minor scientific glitch in that the air in the center of a hurricane moves up, not down. There is speculation that global warming might, paradoxically, lead to another glaciation (which we will discuss later), but it has nothing to do with defying the laws of physics as in *The Day after Tomorrow*.

Incidentally, the physics in most US science fiction films is so badly garbled and downright incorrect that our National Academy of Sciences has offered to provide advice to film directors on the subject.

Figure 11.9 shows the temperature changes in the atmosphere with altitude, starting with the global average surface temperature of 15 °C at sea level. On the right is the air pressure in hectopascals or millibars. Half of Earth's atmosphere is below 5 km, 75 % below 10 km, 90 % below 15 km, 95 % below 20 km, and 99 % below 30 km. At the top of Mount Everest, half of Earth's atmosphere is below you.

The lower part of the atmosphere, the troposphere, convects, with air rising and descending. Air warmed at the surface rises and cools, cool air descends and warms. At an elevation of 8–15 km the air temperature stops decreasing

with elevation and above that it starts to increase. The region where the temperature profile reverses is termed the tropopause. Above that is the stratosphere. Since the air gets increasingly warmer with elevation, it gets less dense than it would normally be, so the air is stratified. It cannot convect. Commercial jets usually fly at about 9–10 km elevation, in the tropopause or at the base of the stratosphere where the air is not convecting and is only 1/4 as dense as at the surface. The Kármán Line, named for Hungarian-American aeronautical engineer and physicist Theodore von Kármán (1881–1963), is considered the boundary to outer space. It is the altitude where the air is so thin that the velocity required to provide enough lift to keep an airplane aloft equals its orbital velocity.

Why does the temperature increase upward in the stratosphere? Because of a greenhouse gas, in this case ozone. Ozone specifically intercepts ultraviolet radiation from the Sun, and warms the surrounding air. The highest ozone concentration is in the lower part of the stratosphere.

## 11.8   The Hydrosphere

Most of the water on Earth is in the ocean. Like the atmosphere, the ocean is stratified into two layers. The upper layer is in contact with the atmosphere; winds blow across it and mix the water. This is called 'the mixed layer' or sometimes, the oceanic troposphere. The mixed layer is about 150 m thick. Below it is the 'deep water' or oceanic stratosphere. At its top there is a rapid temperature decline as

**Fig. 11.9** Vertical section through Earth's atmosphere, showing the nomenclature and temperature and pressure changes with altitude. The Kármán Line is commonly considered the boundary to outer space (After the US Standard Atmosphere, 1976)

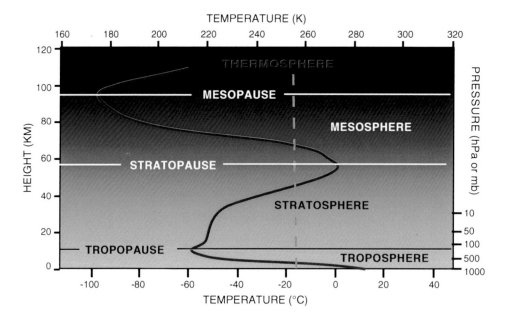

you go down, the 'thermocline' which is several hundred meters thick. Below this the water becomes homogeneous; it is cold and has essentially the same salinity (saltiness) everywhere. Unlike the atmosphere's stratosphere, which is always high above the surface, the ocean deep water comes to the surface in the Polar Regions. There it comes in contact with the air. If the air chills it, it may sink, but most of the time it simply convects, maintaining a constant cold temperature. The mixed layer of the equatorial region, tropics and subtropics is bounded by frontal systems at about 50° N and S that separate it from the polar waters. It is this low-latitude mixed layer that is what you know from bathing beaches, where the water is warm and wonderful. Only a few hundred meters down, the water is a frigid 2 °C. It may be hard to imagine, but only about 5 % of the ocean's water is in that warm mixed layer.

## 11.9 The Cryosphere

The cryosphere is ice and snow, but in reality it is all ice. Ice flows under its own weight. It can be thought of as a very viscous fluid. In terms of the comparative heat capacity shown in Table 11.1, it is about 1/100th that of the ocean. But that is not the whole story. Ice is frozen water; it represents a phase change. To convert the ice to water, a very large amount of energy is consumed, $335 \times 10^3$ J/kg, compared to $2 \times 10^3$ J/kg to change the temperature of ice below the freezing point by 1 °C. Table 11.1 gives the heat capacity of Earth's ice as $48 \times 10^{21}$ J, but that number assumes that only its temperature changes and none of it melts. Melting all the ice on Earth requires bringing it to the freezing point. The average temperature within the Antarctic ice sheet is below –20 °C, but we will use that as an average for all ice on Earth. Bringing it to the melting point requires $20 \times (48 \times 10^{21}$ J$) = 960 \times 10^{21}$ J. Melting it requires another $335 \times 10^3$ J/kg, multiplied by the mass of the ice, $24 \times 10^{18}$ kg, which equals $8040 \times 10^{21}$ J. Total energy required to get rid of all ice on Earth, $9000 \times 10^{21}$ J! That is almost enough to raise the temperature of the whole ocean 2 °C.

Melting ice doesn't change the temperature of anything. It just changes $H_2O$ from its solid to its liquid phase. It is a second great climate stabilizer, until you run out of ice. Think of that the next time you drink your gin and tonic—it was a great refresher until all the ice had melted.

## 11.10 The Land

The big differences between the ocean and land are that the land, soil and rock, have a much lower specific heat and does not move. It warms rapidly during the daytime and cools quickly at night. Diurnal and seasonal temperature contrasts are much greater than in the ocean.

Human modification of the landscape has been going on for at least 8,000 years. Humans have burned off forests to create better conditions for wild game as regrowth occurs. Humans have created vast shallow swampy areas for the cultivation of rice. Forests have been cleared for agricultural production of grains, fruits and vegetables and to provide grazing land for cattle, sheep, and other domesticated animals. Rivers have been dammed to produce large lakes. And vast areas have been covered by buildings and roadways. The area of paved roadways and buildings in the United States has been estimated to be equivalent to the area of the state of Ohio.

About 46 % of Earth's land area remains largely unaltered by humans. Most of that is in inhospitable regions: the tundra, the taiga, the remaining Amazonian rain forest, the Tibetan Plateau, the Australian outback, the Sahara and Gobi deserts and ice-covered Greenland and Antarctica.

After the atmosphere, land is the least stable part of the climate system.

## 11.11 Classifying Climatic Regions

The first global classification of climates was made by German–Russian scientist Wladimir Köppen (1846–1940) in the late 19th century. At the time there were no systematic measurements of temperature and precipitation, and for many parts of the world there were no data at all. By 1900 Köppen had recognized that plant communities provided a record of climatic variables on a time scale of many years. In 1918 he published a classification based on the multi-year temperature, precipitation and seasonality information provided by plants. His map of the 'Climates of the Earth' is reproduced here as Fig. 11.10.

Köppen's classification scheme is hierarchical, with categories represented by letters. The major categories were mostly defined by temperature: A = Tropical; B = Arid (defined by both temperature and minimal precipitation); C = Temperate; D = Continental. Each of these major

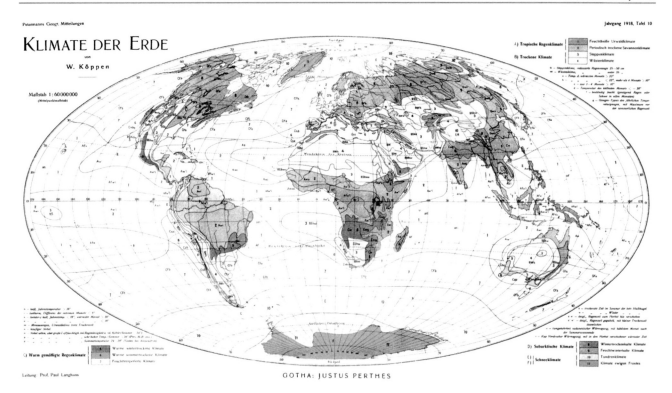

**Fig. 11.10** Wladimir Köppen's map of the climates of the Earth, published in *Petermann's Geographische Mitteilungen* in 1918

climate types is then subdivided by the nature of the precipitation, indicated by a second letter: W = Desert; S = Steppe; f = fully humid; s = summer dry; w = winter dry; m = monsoonal. These units can then be further subdivided using temperature, indicated by a third letter: h = hot arid; k = cold arid; F = polar frost; T = polar tundra; a = hot summer; b = warm summer; c = cool summer; d = extremely continental.

In 1923, the International Meteorological Organization defined 'climate' as the average of the weather over a 30 year period. Many trees and shrubs have a lifetime longer than 30 years, and Köppen's choice of plant communities as a base for climate classification was recognized to have been a brilliant idea. He continued to work on refining his climate classification scheme until his death in 1940. Other more complex schemes to classify climate based on more detailed analyses of temperature and the precipitation/evaporation balance have been proposed but most are not globally applicable.

Figure 11.11 shows a map of the climates of North America according to the Köppen classification with some modifications, from a paper by Edward Ackerman in 1941. All of Köppen's major classes and many of the original subclasses appear on the map, along with some additional

modifications of the subclasses. The purpose of this kind of climate classification was not only to be able to extrapolate from the areas where temperatures and precipitation are measured, but to make it much easier to visualize what the physical measurements mean. Everyone is familiar with terms like 'tropical rainforest,' 'desert,' and 'alpine,' and it is easy to visualize what they mean without knowing the numerical values of the temperature and precipitation ranges they represent.

After WWII, in 1954 Rudolf Geiger and Wolfgang Pohl published a wall map for classrooms that has become a classic with worldwide distribution: 'Wandkarte der Klimagebiete der Erde nach W. Köppen's Klassifikation (Wall Map of Climate regions following W. Köppen's classification). It brought things up to date before the modern episode of climate change began recently updates of the Köppen scheme have appeared, and it has become the most widely used climate classification in use today. A recent update of the world map is shown in Fig. 11.12.

By the way, don't forget that Wladimir Köppen was Alfred (Continental Drift) Wegener's father-in-law and for a while they all lived together. There must have been great conversations around the dinner table.

**Fig. 11.11** The climates of North America according to the Köppen classification (from Ackerman, 1941, Geographical Review, 35, 105–111)

## World map of Köppen-Geiger climate classification

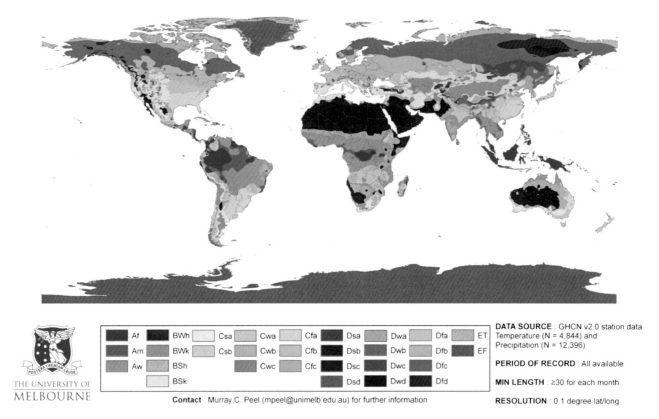

**Fig. 11.12**  An updated (2007) version of the Köppen-Geiger climate classification based on vegetation (Peel et al. 2007, Hydrology and Earth System Sciences, 11, 1633–1644.)

**Fig. 11.13**  Uncertainties in Earth's mean annual global energy balance; values in Wm$^{-2}$. The incoming solar radiation has been measured by satellites, correcting the older value of 340–342 Wm$^{-2}$. All other values are less certain. The ranges of other values are shown, with the most generally accepted values in **boldface**. A change in any one of the flux values means that compensating changes must be made in one or more of the other flux values. The Kármán Line, taken as the effective top of the atmosphere, also separates what we know relatively well from what we still know little about

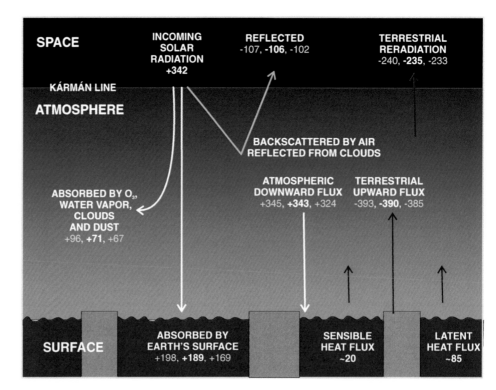

## 11.12    Uncertainties in the Climate Scheme

At the beginning of this chapter it was noted that the scheme presented above is that found in most textbooks. However, as more data from different sources are gathered, and different methodologies are used, there are uncertainties in the values of each of the processes. A recent evaluation of these is shown in Fig. 11.13.

## 11.13    Summary

What we have learned from this discussion is that the atmosphere is the part of the climate system most susceptible to outside influences. It is the most unstable part of the system, followed by land. In contrast, there are two major stabilizers fighting climate change, the ocean and the cryosphere. Unlike the other parts of the system, the cryosphere can absorb energy without changing temperature, so that things can be changing dramatically without your being aware of it.

This classic description of the climate system tells us *how* it works, but not *why*. Without knowing why something happens, predictions of what will happen if we change one of the parameters have no basis. Understanding the interactions between electromagnetic radiation and the molecules of air, land, and water requires a major leap in science: from classical physics to quantum physics. That is the topic of the next chapter.

A Timeline for this chapter:

| | |
|---|---|
| 450 BCE | Empedocles identifies the four major elements as 'air, earth, fire, and water' |
| 1879 | Josef Stefan publishes his Fourth Power Law based on empirical data |
| 1884 | Ludwig Boltzmann derives Stefan's Fourth Power Law from theoretical considerations alone |
| 1885 | Henri Poincaré enters contest to honor the 60th birthday of King Oscar of Sweden—the objective is to prove that our solar system is dynamically stable (the 'three body problem') |
| 1889 | Poincaré submits his solution to the problem and is awarded the prize, but a reviewer had a couple of small questions about the manuscript. In making the necessary revision, Poincaré discovers that the solution is chaotic |
| 1896 | Svante Arrhenius publishes *On the Influence of Carbonic Acid in the Air upon the Temperature of the Ground* |
| 1903 | Poincaré discovers chaos in natural systems and publishes the statement '*small differences in the initial conditions produce very great ones in the final phenomena*' |
| 1962 | The US Standard Atmosphere is established |

(continued)

(continued)

| | |
|---|---|
| 1963 | Ed Lorenz publishes *Deterministic Non periodic Flow* demonstrating that weather is chaotic |
| 1967 | Three US astronauts are killed while training in Apollo spacecraft filled with a pure $O_2$ atmosphere. |
| 1974 | Sherry Rowland and Mario Molina, chemists at the University of California at Irvine, warn about the dangers of chlorofluorocarbons on the ozone layer |
| 1976 | Robert May publishes '*Simple mathematical models with very complicated dynamics*' |
| 1976 | The US Standard Atmosphere, version 2, is published |
| 1987 | The Montreal Protocol on Substances that Deplete the Ozone Layer, an international agreement to phase out ozone-destroying chemicals, is signed |

If you want to learn more:

Barry, R.G. & Chorley, R.J., 2009. *Atmosphere, Weather and Climate*, 9th ed., Routledge, London, 536 pp.—A classic general text; lots of great pictures and figures; no complicated mathematics.

Peixoto, J.P. & Oort, A.H., 1992. *Physics of Climate*. American Institute of Physics, New York, 520 pp.—My personal standard reference for understanding modern climate. Lots of math but it can be skimmed over. Excellent figures.

Music: Good to get back to Johann Sebastian Bach. I recommend the Goldberg Variations played by Glenn Gould.

Libation: In honor of Wladimir Köppen, a nice Riesling from the Rheingau would be in order.

### Intermezzo XI. George W. White—A Most Unusual Department Head

I have already mentioned the then Head of the Department of Geology at the University of Illinois several times. I had managed to impress him as a graduate student simply because I had studied abroad in Germany and Switzerland. He was also impressed with my plan to study with Hans Thalmann at Stanford for my Ph.D., and already suggested before I left Illinois that he would like to have me come back as a faculty member. He was most patient when I let him know that I wanted to do a year of Postdoctoral work in Switzerland, funded by a National Science Foundation grant. He made sure I knew there was a faculty position waiting for me at the University in Champaign-Urbana.

It wasn't until years later that I realized what an extraordinary person he was, certainly the most innovative Department Head I have ever known He was born in 1903 in North Lawrence, Ohio, the son of a minister. He received his baccalaureate degree from Otterbein College in 1921 and the M.S. and Ph.D.

degrees in geology from Ohio State University in 1925 and 1933. He was associated with the University of New Hampshire from 1926–1941, where he began as an instructor and moved through the academic ranks to Professor and Head of the Department of Geology; he also served as Dean of the Graduate School. White returned home to Ohio as a professor of geology at Ohio State University from 1941–1947, and, for the last of those years, he served also as State Geologist of Ohio before moving to Illinois in 1947 as Head of the Department of Geology.

The University of Illinois was very well funded by the state, and George White was able to carry out some extraordinary schemes. He was able to hold open the faculty position in Structural Geology to be used for visiting faculty. At the time structural geology dealt with folding and faulting of rocks, largely an exercise in geometry. In the early and middle 1960s American geologists were firm believers in the permanence of continents and ocean basins fixed in their present positions. However, George White was clever and worldly enough to know that many Europeans believed in mobility of the continents. Widespread American acceptance of the ideas of plate tectonics and sea-floor spreading with all their complexities would not come later, in the early 1970s.

George once told me that the position in structural geology was an ideal one to hold open because any geologist could teach structural geology, so one could invite any geologist and not worry about their field of specialization; anyone could teach the undergraduate course. But they would be expected to also teach courses in their special field for advanced undergraduate and graduate students.

George looked abroad, to Europe, for geologists to invite for his open 'Visiting Professor' position. He sought out someone who was at a position in their career when they could benefit from an offer of a visiting professorship at a prestigious American University. He was spectacularly successful, Some of those were Frank Rhodes from Swansea in England, who later went on to become president of Cornell University, Albert Carozzi, who then became a permanent faculty member at the U of I, Theodore Sorgenfrei from Denmark, who later organized the 1960 International geological Congress in Copenhagen, Kingsley Dunham from Durham in England, who afterward became the Director of the British Geological Survey, and one of my recommendations: Hans

Laubscher (possibly the only real structural geologist to occupy the position) from the University in Basel, Switzerland. Hans went back to Basel to become Department Chair there.

George White instituted some very clever policies. He did not allow undergraduates to continue as graduate students. They could return for doctoral work if they had gotten a Master's degree at another University, but it was mandatory that they had to leave the U of I after getting their Baccalaureate. This was very important in preventing 'inbreeding' and assuring that the students were exposed to other faculty and their ideas. While George was Department Head, the University of Illinois at Champaign-Urbana was at or near the top in terms of Ph.D.s produced each year in U.S. Universities. About half went into industry, and the other half was split between employment in Universities and Research Organizations, such as Geological Surveys. Each year George would remind those in academia where they had gotten their advanced degree, and inquire whether they might have students they would like to send to the U of I for graduate study. This ensured a steady flow of excellent students into the graduate program.

George was passionate about building the collections of the University's library, especially the literature relevant to the history of geology. Wherever he went he sought out second hand bookstores and searched them for old geological literature. The University of Illinois Library, already one of the great libraries in the United States, had a special fund for the purchase of old and rare books. Over the years George built up a fabulous collection of the early geological literature, probably the best in the United States. It has become a major research tool for history of science, and the Geology Library is now named in his honor.

George's own field of expertise was Pleistocene Geology—the study of the glacial deposits so widespread in the central northeastern United States. Pleistocene geology tends to be something of an orphan. Many geologists consider the deposits to be simply stuff that covers up the real geology of the older rocks. But George maintained strong ties to the Illinois, Ohio, and Pennsylvania Geological Surveys, the Geological Survey of Canada, and several of the Canadian provincial surveys in Saskatchewan and Alberta—all areas with extensive Pleistocene deposits. He had an extraordinary group of graduate students; their names read like a Who's Who in the field. He authored 150 publications on glacial, economic, and engineering geology as well as his very special interest, the history of geology.

White stepped down as Head to become a Research Professor in 1965 and Research Professor Emeritus in 1971. His research continued through a very productive period of 1973–1981 as a consultant to the Ohio Geological Survey. He was awarded honorary degrees by Otterbein College, the University of New Hampshire, and Bowling Green State University.

It is only in retrospect that one realizes what an extraordinarily creative person he was. The open faculty position used for visiting professors was one of the best ideas in academia at the time. Sadly this possibility hardly exists in American universities any more.

FALLING DOWN THE RABBIT HOLE INTO A NEW WORLD

*Your theory is crazy, but it's not crazy enough to be true.*
Niels Bohr

The 19th century was in many ways the heyday of science, particularly physics. But in its last decade there were signs that something was amiss. It was as though Lewis Carroll's Alice had started to follow the rabbit and started the fall down its hole. Everything seemed almost normal as she fell down, but the world at the bottom of the hole was unlike anything she had known before.

In 1900 Lord Kelvin presented a lecture at the Royal Institution of Great Britain entitled 'Nineteenth-Century Clouds over the Dynamical Theory of Heat and Light.' He noted that the knowledge of physics was almost complete, but identified two small clouds that were troubling: (1) the problem in understanding the black body spectrum, which seemed to just keep getting more and more mysterious and (2) the 1887 Michelson-Morley experiment intended to measure the motion of the Earth relative to the luminiferous æther by showing that the speed of light varied in different directions. Instead it showed that the speed of light was the same no matter whether you were moving toward or away from its source. Although he didn't mention it, he was also troubled by the discovery of 'radioactivity,' which seemed to violate the conservation of energy prescribed by the First Law of Thermodynamics. Then there were those few pesky geologists and biologists who argued that 100 million years seemed too short a time to account for everything that had happened to planet Earth and life on it. We were about to fall down Alice's rabbit hole.

The two clouds Kelvin identified would undermine the classical physics of the 19th century—the first leading to quantum mechanics, the second to the theory of relativity. But a lot more was happening in the last 5 years of the 19th century: X-rays were discovered by Roentgen in 1895, radioactivity by Becquerel in 1896, the discovery that cathode 'rays' were actually 'corpuscles' (later named electrons) by J.J. Thompson in 1897, and the isolation of two new radioactive elements, radium and polonium, by the Curies in 1898. The nature of electromagnetic radiation was no longer the 'hot topic.'

As Kelvin noted, the failure to find a single mathematical expression to describe the skewed curve representing black body radiation was a troublesome factor that needed fixing. However, few scientists were aware of the problem. But Kelvin overlooked another phenomenon that needed explanation: the photoelectric effect. First, let's discuss the solution to the black body problem—from Max Planck—that was to start a revolution in physics.

## 12.1  Max Planck and the Solution to the Black Body Problem

Max Karl Ernst Ludwig Planck (Fig. 12.1) was born in Kiel, on Germany's Baltic coast. He obtained his doctorate at the University of Munich in 1879, working under Professor Philipp von Jolly. Jolly had actually advised him against pursuing a doctorate in physics because there was so little left to be done. But Planck loved the rigor of theoretical physics and was obsessed by the Second Law of Thermodynamics. His dissertation was on the Second Law and for the next two decades he continued to explore the implications of the Law and of the corollary of increasing entropy. The philosophical implications of the Universe just sort of winding down with time were enormous.

Max Planck liked perfection in developing an explanation of observations. So after Wien published his empirical Law of Distribution in 1896, Planck was concerned that it lacked a solid theoretical foundation. When it became evident that it matched the experimental data only in the shorter wavelength range, it was clear that it was only an approximation. Accounts of how Planck arrived at his solution to the problem vary greatly, and many are considered more mythology than history. The one I like the best is that Planck realized that Wien's Distribution Law was unsatisfactory at longer wavelengths and that Lord Rayleigh's alternative, based in classical physics, was also unsatisfactory because it led to solutions approaching infinity at shorter wavelengths.

Since both were good approximations for parts of the black body spectrum but failed in other parts, Planck set out to make a combination of the two that would be based solidly in physical theory and that would fit the entire spectrum. Unfortunately, beautiful as this story is, it is not true. Planck didn't seem to know much about Rayleigh's 1900 analysis, and nowhere did he discuss trying to make a combination of the two approximations. Max Planck was a purist. He sought a theoretically correct solution to Wien's distribution.

To understand what Max Planck did, we need to consider how people thought about the world at the end of the 19th century. Then as now everything in the human experience seems continuous. This is how our minds work. We see transitions in everything, not discontinuities. Nevertheless, the 19th century saw the general acceptance of the atomic theory, the idea that everything was made of tiny indivisible objects so small they could not be seen with any microscope. On the other hand there was overwhelming evidence that light has waves analogous to those on the surface of the sea, moving through the 'luminiferous æther.'

In 1900 he gave a series of presentations to the Academy of Sciences in Berlin, first explaining what he was trying to do—understand black body radiation—and then reporting on his experiments, the first four of them failures. He made the assumption that the electromagnetic radiation emanating from the hole in the box at the Reichsanstalt is produced by 'resonators' (the 'harmonic oscillators' mentioned above), presumably the atoms or molecules making up the walls of the box. Space does not permit me to go into the logical and mathematical problems he encountered, but on the 14th of December 1900 he presented his solution. He had come to the conclusion that under certain conditions, energy such as that carried by electromagnetic waves could not take on just any value, but must be some multiple of a very small quantity. Different levels of energy could not be envisioned as a continuous spectrum, but had to be 'discretized.' Energy comes in discrete pieces that differ by a minute amount. The pieces are like 'integers,' numbers like 1, 2, and 3. Fractions are not allowed. Planck designated the difference from one piece to the next as a constant, $h$. It came to be called a 'quantum'. It is exceedingly small, $6.626 \times 10^{-34}$ Js (Js = Joule seconds). This means that the lines showing the energy in electromagnetic radiation of different frequencies in Fig. 10.15 are actually a series of dots, but they are so close together you cannot see them, even with a microscope.

After the lecture, Planck's friend Heinrich Rubens went home and spent the night comparing the data he and his colleague Ferdinand Kurlbaum had produced at the Reichsanstalt with Planck's equation. The next morning he told Planck there was a perfect fit. But oddly enough, no one else seems to have paid much attention to Planck's idea that energy came in small packets. Even Planck himself wrote nothing further on the matter until 5 years later.

**Fig. 12.1** Max Planck (1858–1947). His name is immortalized in Planck's constant, $h$. $h = 6.62606896(33) \times 10^{-34}$ Js, or $4.13566733(10) \times 10^{-15}$ eVs

Now on to another troublesome problem, the photoelectric effect.

## 12.2  The Photoelectric Effect

You will recall that in 1885 Heinrich Hertz became professor at the University of Karlsruhe. He undertook to test James Clerk Maxwell's hypothesis that light and other forms of radiation were electromagnetic waves. He set up an experiment to generate electromagnetic waves using an oscillating electric current. Maxwell had predicted that this could be done. Hertz's experiment was designed to prove or disprove Maxwell's ideas about electromagnetism. Hertz built a device to send out an electromagnetic signal generated by a spark jumping across a small gap. Today we would call this an antenna. The receiver was a loop of wire, also with a gap for a spark, located several meters away in another part of the laboratory room—far enough so that there could not be an 'induction effect' due to the electric field of the antenna. The idea was very simple. If the spark generated by the antenna caused a spark to jump across the gap on the receiver, it meant that an electromagnetic wave had been sent from the antenna to the receiver.

It turned out that it was hard to see the spark on the receiver, so in 1887 Hertz concentrated his effort on better observing the spark by enclosing the electrodes in a darkened box. The light for observing the spark gap on the receiver was admitted only through a glass window. He noticed something very odd. The spark on the receiver was not as large as before. When he removed the glass window, it was larger. He then experimented with a variety of different kinds of light sources to illuminate the spark gap on the receiver. He discovered that invisible short wavelength ultraviolet light markedly enhanced the spark effect. He reported the results of his experiment with ultraviolet light, but regarded it as a curiosity and did not follow up on it. It was merely a troublesome curiosity on the way to a successful proof of Maxwell's ideas.

Spurred on by Hertz's discovery of the effect of ultraviolet light on electrical charges, a number of other investigators explored the effect of light on electrically charged metal surfaces. They got lots of different results—mostly attributed to the freshness and cleanliness of the metal surface. Nevertheless, the data began to show that the energy of the electrons released from the illuminated surface depended on the wavelength of the light, not on its intensity. This was the opposite of what James Clerk Maxwell's studies had predicted. He had thought that the energy would be directly related to the intensity of the light, not to its wavelength.

The solution to the problem of the photoelectric effect came from a young Albert Einstein (1879–1955) (Fig. 12.2). He was 12 years old when Heinrich Hertz discovered the photoelectric effect, and he would be the first person to figure out what caused it.

In 1905, Albert Einstein was 26 years old as he graduated with a doctorate from the Swiss Federal Institute of Technology in Zürich, perhaps better known as the ETH (Eidgenossische Technische Hochschule). When I was a student at the ETH in 1955–56 there was an exhibition for the fiftieth anniversary of that event. They included letters of recommendation for him as he sought a job. I remember one that had the phrase 'does not readily accept new ideas' in it. After a number of rejections, he was finally able to land a job in the Swiss Patent Office in Bern. But to continue, his dissertation was on the size of molecules, but he published several important papers on other topics that year. One concerned the photoelectric effect, which he explained by making the assumption that light was not waves but particles (quanta) which later (in 1926) came to be called photons.

Einstein built on Max Planck's theory of black-body radiation and proposed that each quantum of light was equal to the frequency multiplied by a constant, h (Planck's constant). He argued that if the light quantum had energy above some threshold it could eject an electron from an atom. The electron would be expelled with a given energy. Its energy depended on the energy inherent in the light striking the surface, which in turn depended on its wavelength. The intensity of the light determined how many electrons could be ejected from at metal surface, not their energy.

In 1921 Einstein received a Nobel Prize for this work. Most people are surprised to learn that Einstein's Nobel Prize was for the explanation of the photoelectric effect, not for his work on 'relativity' which had also been published in 1905. It

**Fig. 12.2**  The young Albert Einstein (1879–1955)

can be argued that Albert Einstein is the true father of quantum physics because Max Planck didn't seem to recognize the significance of his discovery. Incidentally, it was Max Planck who gave Einstein's 'Theory of Relativity' its name.

Max Planck's idea of energy coming in tiny pieces rather than as a continuum (quanta) and Albert Einstein's explanation of the photoelectric effect simply baffled most physicists. The ideas were so different from everything we know by human experience that they just didn't make any sense.

## 12.3   The Bohr Atom

You will recall from Chap. 5 that at the end of the 19th century there were two competing ideas about what an atom looked like. John Dalton's 'billiard ball' model of 1808 had satisfied most of the scientific community through much of the century. In 1904 J.J. Thomson had proposed that the atom was more like a 'plum pudding'—a sphere with electrons embedded in it. Ernest Rutherford, on the basis of the Geiger-Marsden Gold Foil experiment conducted in 1909, had concluded that the atom must have a very tiny central mass, the nucleus, surrounded by orbiting electrons. It was the 'solar system' model. However, there was an obvious problem with Rutherford's model. The nucleus has a positive electrical charge, and the electrons have a negative charge. Positive and negative charges attract each other. Why don't the electrons simply spiral in and join the nucleus?

Now it may be that although there was brief mention of Niels Bohr and his view of the atom in Chap. 5, you already knew of him though *The Simpsons.* In the eighteenth episode of the thirteenth season, *I am Furious (Yellow)*, Homer Simpson is about to watch one of his favorite TV shows only to find it preempted by the program *The Boring World of Niels Bohr.* Homer is so upset that he clutches an ice-cream sandwich, aims it at the screen like a remote control, squeezes out its contents, and it splatters all over Niels Bohr's image. Unfortunately, that is probably the only encounter most Americans have had with Niels Bohr. His world was anything but boring.

Ernest Rutherford had the good luck that Niels Henrik David Bohr (1885–1962) (Fig. 12.3), fresh from receiving his Ph.D. at the University of Copenhagen in 1911, had been granted a postdoctoral fellowship to work in Britain. Bohr had intended to work with Sir J.J. Thomson at Cambridge, but he was too brash for Thomson's liking, and so Bohr moved to Manchester to work with Rutherford.

Now here is one of the great lessons about academia. You probably think that a professor teaches the student. At the undergraduate level that is generally true, but there are exceptions. I had such a student in my Oceanography class at the University of Illinois, Marc Mangel. As a junior, Marc wrote a term paper that became a classic two-page paper in chemical oceanography: *A treatment of complex ions in seawater.* Marc went on to have an extraordinary career, founding the Center for Population Biology at the University of California, Davis in 1989 and then in 1996 he moved to the University of California, Santa Cruz where he is Distinguished Research Professor in the Department of Applied Mathematics and Statistics and directs a training program to teach young scientists the methods of quantitative population biology required for sustainable fisheries. I had the opportunity to visit him in Santa Cruz a few years ago. His recent book *The Theoretical Biologist's Toolbox: Quantitative methods for ecology and evolutionary biology* (Cambridge, 2007) is great fun reading for a scientist.

Although at the undergraduate level it is usually the professor who imparts knowledge to the students, at the graduate level the reverse can be and is often true—the student teaches the professor. This, of course, depends on the personalities involved. In many instances, the graduate student learns a technical trade from his professor, such as how to make measurements with a mass spectrometer and how to interpret the results. Many of them earn a Ph.D. and become what my old friend Bill Benson, a program officer with the National Science Foundation, called 'journeymen scientists.' There is nothing wrong in this; we need scientists who make careful measurements, generate data, and produce plausible interpretations. But there are some other academics, like Rutherford, who give their students or post-doctoral fellows problems to solve that they themselves can't figure out.

Older academics carry a lot of baggage in the form of old ideas that younger people are not burdened with. They

**Fig. 12.3** The young Niels Bohr

consider it especially gratifying when a young colleague shows them that their ideas were wrong. In my experience, academics of this latter sort are very valuable but not very common.

In Manchester, Niels Bohr thought about that problem of the orbiting electrons. He chose to make the problem as simple as possible by first considering only the hydrogen atom. At that time, if one believed Rutherford, the idea was that the hydrogen atom has a single tiny positively charged nuclear particle and a single negatively charged electron in orbit around it. The situation is rather like the Earth orbiting the Sun or the Moon orbiting the Earth. If the force of gravity were to suddenly disappear, the Moon would continue to move at the same speed, an average of 1.023 km/s (=288.8 mph), but in a straight path away from the Earth. Gravity makes the Moon keep falling towards the Earth, but its forward momentum is such that it exactly offsets the fall and the Moon simply keeps orbiting the Earth.

It is a question of angular momentum. Momentum, you will recall, is velocity times mass—the velocity being along a straight path. Angular momentum is velocity in an orbit times mass. Angular momentum, like energy, gets conserved. But it is not quite as simple as that. It is the angular momentum of the system that is conserved. Consider the Earth-Moon system. If the rotation speed of the Earth slows down, the planet loses angular momentum. But the angular momentum of the Earth-Moon system must remain constant, so the Moon increases its angular momentum by increasing the diameter of its orbit. In fact, the Earth rotates faster when there are ice caps on the poles, just as a skater spins faster by pulling in his/her arms. When the Earth rotates faster, the Moon comes closer, and that means that the tides get larger. There is evidence from fossils to suggest that in the

Devonian, some 400 million years ago, the Earth was rotating faster so that there were about 410 rather than 365 days in the year. This would mean that the Moon was closer and the tides higher. It is thought to be the friction of the tides that has slowed the Earth down.

Bohr was also concerned about the emission spectrum of hydrogen. Why were the lines so sharp? If electrons could spin in toward the nucleus the lines in the emission spectrum should show this by being broader and fuzzy. It is rare in science when one can identify the moment when a radical new idea is born. But we know from letters Niels Bohr wrote to Rutherford and George Hevesy, another postdoctoral fellow in Rutherford's institute, in late January and early February 1913, that he was not thinking about the meaning of the spectral lines. But on March 6, 1913, Bohr presented Rutherford with a draft of his paper on the structure of the atom. In recalling the events of those days later on, Bohr said "As soon as I saw Balmer's equation everything immediately became clear to me." It was a minor jump to incorporate Johannes Rydberg's generalization of Balmer's equation. Because energy came in quantum units, the electron could only orbit the nucleus at specific distances. The electron was behaving not like a particle or a planet. It was behaving like a standing wave of a specific wavelength.

What is a standing wave? Think of a guitar. If you pluck one of the strings in its middle, it vibrates back and forth between the two points where it is fixed; the bridge, and one of the frets on the neck. The sound has a specific wavelength, which you can change by adjusting the tension on the string. However, it is a single standing wave. But it can be split into two or more waves between the bridge and fret; these are the harmonics. They are exact even fractions of the original wavelength.

We usually think of waves as moving along straight lines, but in the case of the electron it is going around the atomic nucleus in some sort of orbit. The closest possible orbit was numbered n = 1. This came to be called the 'ground state.' But if exactly the right amount of energy were supplied by electromagnetic radiation, the electron could jump to a higher orbit. You might think of this as the first harmonic. In this state, the atom is said to be 'excited.' The dark bands on the emission spectra are where the energy of a photon is being absorbed by the atom and the electron is jumping to a higher orbit. But then the excited atom can radiate that energy back out as a photon of the same wavelength as the electron falls back to a lower orbit, producing the lines of the emission spectrum. These spectral lines are so well defined because the energies are quite specific. At this time Bohr was thinking that at any particular energy level the possible orbits of the electron would always be the same distance from the nucleus. The orbits did not need to be in the same plane, like the planets of the solar system, or the rings of Saturn as had been suggested by Japanese physicist Hantaro Nagaoka in

1904. Taking all the possibilities together, they would form a spherical 'shell.' The number of possible shells to which electrons can jump and fall back from corresponds to the number of emission lines in a spectral series.

A model of a hydrogen atom with the first four shells is shown in Fig. 12.4, taken from Bohr's 1923 paper *Über den Bau der Atome* (*On the structure of atoms*). The electron shells were initially given letter designations; K, L, M, N, O, and P, or numbered, from innermost to outermost, 1, 2, 3, 4, 5, 6, 7, 8. These shell numbers are now referred to as the 'principal quantum numbers.' We'll come back to that later.

Figure 12.4 shows the six possible transitions among the four inner shells of the hydrogen atom. Intermediate stages do not occur. Two of the jumps that correspond to visible spectral lines in the Balmer series are shown on the right, labeled Hα and Hβ. The Hα jump is the smallest and corresponds to a wavelength of 656.3 nm (pink); in terms of energy it is 1.89 eV. The Hβ jump is the next larger and corresponds to a wavelength of 486.1 nm (cyan) and an energy of 2.55 eV.

Many of the nebulae in the night sky are pink. That is the Hα emission line. Unfortunately, to see the color you need a good telescope, or even better, look up astrophotographs on the Internet by Googling 'pink nebula.' The images taken by the Hubble Space Telescope are especially impressive. If you want to see a beautiful example of both the pink and cyan emission lines together, Google 'Vela supernova remnant.' The picture taken by Marco Lorenzi is stunning. The hydrogen atoms that make up the nebulae are being excited by photons from nearby stars and glowing pink or cyan as the electrons fall back down to their ground state.

There is a similar phenomenon closer to home, the aurora borealis. If you live in the northern part of the US or in Canada or northern Europe, you have probably seen it.

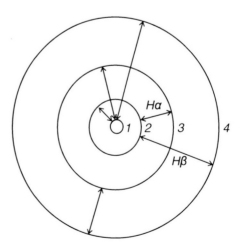

**Fig. 12.4** An illustration of Niels Bohr's model of the hydrogen atom, from his 1923 paper *Über den Bau der Atome* translated from a lecture in Danish by Wolfgang Pauli

It often looks like a green curtain hanging in the night sky, but it can also be blue or red. In this case it is not hydrogen that is being excited, but oxygen (green and reddish brown), and nitrogen (blue and red). More about this phenomenon in Chap. 14, where we discuss the atmosphere.

Only a few of the hydrogen emission lines are in the visible part of the electromagnetic spectrum. The transitions shown at the top of the diagram belong to the Lyman series and are jumps from the lowest shell. These are all in the ultraviolet. The transition in the lower part of the diagram is the α, shortest, jump of the Paschen series. It is in the infrared.

If the single electron orbiting the hydrogen nucleus is in the innermost orbit (1), it is said to be in the ground state. If incoming radiation has exactly the needed energy, that is, just the right wavelength, it is absorbed and the electron can jump to a higher orbit. The atom is said to be excited. Later, the electron can fall back down to a lower orbit, and as it does so, it emits a specific amount of energy in the form of electromagnetic radiation of a specific wavelength. When Niels Bohr first figured this out in 1913, only the spectral lines in the visible part of the spectrum, just a part of the Balmer series, were known. These are associated with jumps to and from the second shell. As the shorter and longer wavelengths of the spectrum beyond what we can see were investigated, other series were detected. The last one to be formally named is the Humphreys series discovered in 1953. Others have been discovered since and spectroscopic technology has improved. Each of the series is increasingly faint because the electron transitions are increasingly rare.

One means by which electromagnetic radiation and atoms, and ultimately molecules could interact had been discovered. The way lay open to figure out how greenhouse gases operate, but at the time that was a peripheral topic and would take many more years to sort out. First it would be necessary to figure out how things worked in atoms with more than one electron.

Figure 12.5 shows a more recent schematic model of the hydrogen atom with its shells. It shows the emission/absorption series that had been discovered by the middle of the last century, the wavelength (in nanometers, nm) of the electromagnetic radiation associated with each, and where these series of lines appear in the hydrogen emission spectrum. The emission spectral lines are the radiation emitted by the atom as the electron falls from a higher to a lower shell. In the absorption spectrum, these lines would appear to be black against the brighter background. In the diagram the arrows would be pointed in the opposite direction. Remember that the shorter the wavelength, the greater the energy involved.

The Lyman series involves jumps to and from shell 1, the ground state, with wavelengths entirely in the ultraviolet. The complete Balmer series is shown, including the line in the ultraviolet. The Paschen, Brackett and Pfund series for the hydrogen atom that were introduced in Chap. 9 are also shown on the atom model. Note that for each series, the longest wavelength (=lowest energy) is associated with the jump from one shell to the next lower shell. The wavelengths decrease (=energy increases) as the electron falls from successively higher shells. Also, note that the wavelengths become more closely spaced. When you look at the emission spectrum at the bottom of the diagram you will see that the Lyman and Balmer series do not overlap, but the spectral lines of the Paschen, Brackett, Pfund, and Humphreys series do. One can imagine that the possible number of shells is infinite, but in reality as the electron jumps sufficiently far away from the nucleus the attraction between the positive nuclear charge and the negative electron is so small that the electron can simply wander off. The resulting atom then has a positive charge; it becomes a positive ion. The electron may then attach itself to another atom which has space available in its incomplete outer shell, making that atom become a negative ion.

If you think of these jumps as being analogous to the harmonics of a vibrating guitar string, you will wonder why the wavelengths are not simple multiples or fractions of one another. That is because the vibrating string is linear, and the electron orbits are not really circles as shown in Fig. 12.5. They are more like the egg-shape of ovals. So that even the term 'shell' is misleading if you are trying to visualize what is going on.

These transitions of one or more electrons between different shells are called '*electronic transitions.*' If the electron leaves the atom, for whatever reason, the process is called '*ionization.*' The energies of photons required for ionization are in the ultraviolet range, and if these are the cause, then one speaks of '*photoionization.*' The electrons freed up by photoionization produce the photoelectric effect described by Heinrich Hertz in 1887 and explained by Albert Einstein in 1905.

One can imagine an infinite number of shells surrounding the nucleus, but only the first seven are utilized by the known elements for atoms in their ground state. Nevertheless you can calculate the wavelength or energy required to make the electron jump infinitely far away from the nucleus so that it leaves the atom altogether. For the series associated with the transitions between hydrogen shells shown in Fig. 12.5, the energy values for jumps to 'infinity' are 1.36, 0.340, 0.151, 0.085, and 0.038 eV. A complete list of the transition values between the inner shells of hydrogen in

**Fig. 12.5**  A 'modern version' of the Bohr model of the hydrogen atom showing the wavelengths associated with jumps of the orbiting electron from higher to lower levels. The ∞ hydrogen emission spectrum, with six of the series, is shown below. *H* Humphreys series

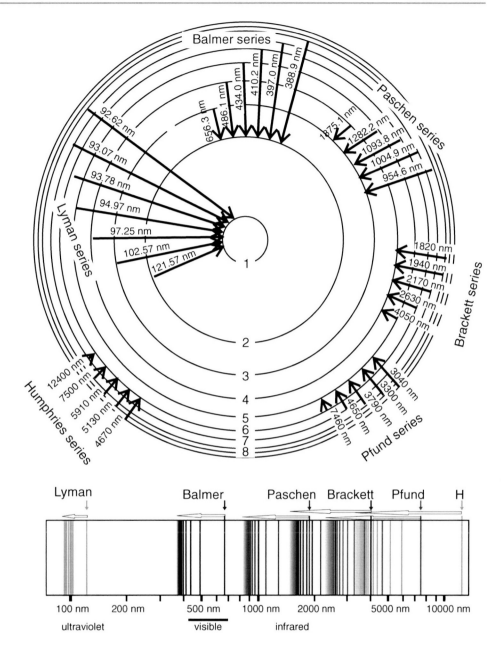

terms of wavelength and electron volts are given in Table 12.1

By inspection of the table you will see that none of the transitions have the same wavelengths or energies. Each transition is related to a different spectral line. How many are there? In the first 8 transitions there are 8! (8 factorial) possibilities. Numerically, 8! is $8 \times 7 \times 6 \times 5 \times 4 \times 3 \times 2 \times 1$, which, when you multiply it out is more than you might at first think, 40,320. If you consider the possibilities, which go out at least to an 11th shell, that number becomes 3,991,680. That is a lot of spectral lines. But remember that Planck's constant ($h$), which represents the energy difference between quanta, is, in terms of eV, $4.135667 \times 10^{-15}$. The quantum energy difference between wavelengths is so small that the

nearly 4 million possible wavelengths that might affect a hydrogen atom seem like a very small drop in the bucket of possible wavelengths.

## 12.4   Implications of the Bohr Model for the Periodic Table of the Elements

Once the Bohr model of the simple hydrogen atom was established, interest quickly turned to the more complex atoms with higher atomic weights. It was immediately discovered that applying Bohr's idea to more complex atoms was not easy. Remember that it was not until 1919 that

**Table 12.1** Wavelengths (λ, in nanometers) and associated energies in electron volts (eV) for the first eight series of transitions (Trans) of the single electron between the first eight of the many possible shells of the hydrogen atom

| Lyman | Trans | 1 ↔ 2 | 1 ↔ 3 | 1 ↔ 4 | 1 ↔ 5 | 1 ↔ 6 | 1 ↔ 7 | 1 ↔ 8 | 1 ↔ ∞ |
|---|---|---|---|---|---|---|---|---|---|
| | λ | 122 | 103 | 97.2 | 94.9 | 93.7 | | | 91.1 |
| | eV | 10.2 | 12.0 | 12.8 | 13.1 | 13.2 | | | 13.6 |
| Balmer | Trans | | 2 ↔3 | 2 ↔ 4 | 2 ↔ 5 | 2 ↔ 6 | 2 ↔ 7 | 2 ↔ 8 | 2 ↔ ∞ |
| | λ (nm) | | 656 | 486 | 434 | 410 | 397 | | 365 |
| | eV | | 1.89 | 2.55 | 2.86 | 3.02 | 3.12 | | 3.40 |
| Palmer | Trans | | | 3 ↔ 4 | 3 ↔ 5 | 3 ↔ 6 | 3 ↔ 7 | 3 ↔ 8 | 3 ↔ ∞ |
| | λ (nm) | | | 1870 | 1280 | 1090 | 1000 | 954 | 820 |
| | eV | | | 0.66 | 0.97 | 1.14 | 1.24 | 1.30 | 1.51 |
| Brackett | Trans | | | | 4 ↔ 5 | 4 ↔ 6 | 4 ↔ 7 | 4 ↔ 8 | 4 ↔ ∞ |
| | λ (nm) | | | | 4050 | 2630 | 2170 | 1940 | 1460 |
| | eV | | | | 0.31 | 0.47 | 0.57 | 0.64 | 0.85 |
| Pfund | Trans | | | | | 5 ↔ 6 | 5 ↔ 7 | 5 ↔ 8 | 5 ↔ ∞ |
| | λ (nm) | | | | | 7460 | 4650 | 3740 | 2280 |
| | eV | | | | | 0.17 | 0.27 | 0.33 | 0.54 |
| Humphries | Trans | | | | | | 6 ↔ 7 | 6 ↔ 8 | 6 ↔ ∞ |
| | λ (nm) | | | | | | 12400 | 7500 | 3280 |
| | eV | | | | | | 0.10 | 0.17 | 0.38 |

Note that the wavelengths of the Lyman series are all in the ultraviolet. The Balmer series is mostly in the visible spectrum. Wavelengths associated with the other series are all in the infrared

Ernest Rutherford succeeded in disintegrating nitrogen into hydrogen and oxygen by alpha particle bombardment, proving that the hydrogen nucleus is present in other nuclei. This was the discovery of the proton. That immediately raised a question.

There was clearly a problem in that the relative atomic weights of different elements was not a simple progression of integers, like 1, 2, 3, 4, 5, …, but rather they went 1, 4, 7, 9, 11, …. Something was missing somewhere. It would not be until 1932 that James Chadwick (Fig. 12.6), another of Ernest Rutherford's students, would prove there was another kind of particle in the nucleus, one without an electrical charge, the neutron. The number of neutrons is equal to or slightly greater than the number of protons. The differing numbers of neutrons made the isotopes Soddy had recognized. The atomic weights were not integers, but fractional numbers because of the mixture of different proportions of isotopes. Nevertheless, the elements could be numbered by integers following the order of their increasing atomic weight. This integer was simply the element's number and had allowed Mendeleyev's Periodic Table to be developed into series of rows of boxes.

In spite of the lack of understanding of the dual nature of particles in the atomic nucleus, progress was quickly made in using the periodic table to figure out how many electrons could fit in each shell of the heavier atoms. If there were only two atoms in the first row of the table, two electrons must be the maximum number that could fit into the first shell. The next row had eight elements, so it made sense to suppose that the next shell could contain up to eight electrons. The third row also had eight elements. The fourth and fifth rows each had 18 elements indicating that theses orbital shells

**Fig. 12.6** James Chadwick (1891–1974). He was awarded the Nobel Prize in Physics in 1935 for his discovery of the neutron in 1932

could have up to 18 electrons. The sixth row had 32 elements, so that the sixth shell might have up to 32 electrons. The seventh row was incomplete, but if the pattern continued, it should also accommodate 32 elements representing a shell with 32 electrons. The highest number of the naturally occurring elements is 92, uranium, but the seventh row of the periodic table could accommodate 26 more elements. None are known to occur in nature. However, as of the time of this writing, 19 of these non-naturally occurring elements have been produced in laboratories, albeit sometimes in quantities of only a few atoms.

As Niels Bohr's model of the atom was developed, it became evident the number of the element corresponded to the number of electrons orbiting the nucleus, when the atom had no net electrical charge. Figure 12.7 is his 'natural system of the elements' from his 1923 paper—in effect, the periodic table turned on its side with lines connecting the elements with similar properties. This was nine years before Chadwick's discovery of the neutron.

Figure 12.5 showed how a number of wavelengths of radiation could interact with the simplest atom, hydrogen, to produce electronic transitions. As you might guess, the radiation/atom interactions get much more complicated very rapidly as the total number of electrons increases. Furthermore, most elements occur in compounds, bound together with other elements by 'ionic,' 'covalent,' or 'metallic' bonds. An ionic bond is one in which one or more electrons from one atom are removed or one of these free electrons becomes attached to another atom. The resulting positive and negative ions attract each other. Some common materials, such as table salt ($Na^+Cl^-$) are held together by ionic bonding. In a covalent bond one or more pairs of electrons are shared by two atoms. Molecular oxygen, $O_2$, is an example. In its atomic form, each oxygen atom has eight electrons, two in the inner shell and six in the outer shell. It is as though atoms strive to have their outer shells completely filled with electrons. In the case of oxygen this is achieved by the two atoms sharing two of the electrons. Metallic bonds are very special. Metals have so many bonds sharing electrons that they can be thought of as positive ions in a sea of electrons. This gives them special properties, such as high heat and electrical conductivity.

**Fig. 12.7** Bohr's 1923 version of the 'natural system of the elements.' I have brought some of the symbols for elements up to date and indicated how the increasing numbers of electrons in the groups of elements correspond to the shells, using both the number and letter systems

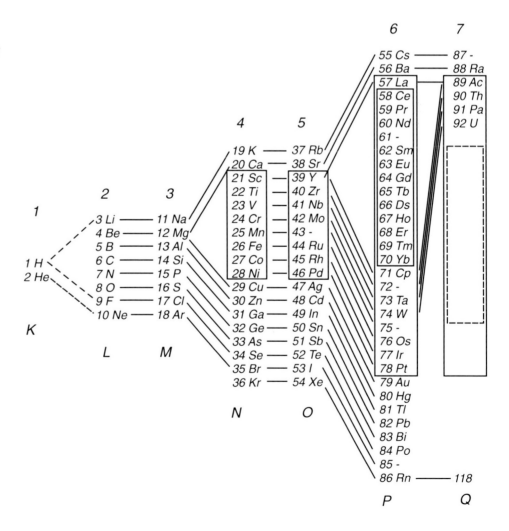

After Niels Bohr's model of the atom was introduced, a lot of things began to make sense. The first shell could accommodate only two electrons. And, strangely, the helium atom is, with a diameter of 0.98 Å, smaller than the hydrogen atom which has a diameter of 1.58 Å. Lothar Meyer had gotten this wrong on his diagram of atomic weights versus atomic volumes because neither had been liquefied at the time and he had been forced to guess their volumes. Hydrogen was first liquefied by James Dewar (1842–1923) in 1898, and helium, which had been discovered in the spectrum of the Sun but not discovered on Earth until 1898, was first liquefied by Heike Kamerlingh-Onnes (1853–1926) at the University of Leiden in The Netherlands in 1908. The helium atom is smaller because the increased attraction of the double positive charge of the nucleus pulls the electrons in closer. The electrons, of course, repel each other, so there is no more room for additional electrons in this innermost shell. The atom is happy; its outer shell is complete. It does not react with other elements to make compounds. It does not even react with itself to make a diatomic molecule like oxygen or nitrogen. Atoms with complete outer shells are the 'noble gases,' helium (He), neon (Ne), argon (Ar), krypton (Kr), xenon (Xe), and radon (Rn). The noble gases seem to be particularly happy and satisfied with themselves. They are aloof and do not combine with other atoms. Other atoms seem to have inferiority complexes and try to combine in such ways as to mimic the complete outer shells of the noble gases.

Adding another proton and electron to the helium atom (making it lithium) starts a new shell, which can potentially hold eight electrons and represents the beginning of a new series, $n = 2$. It is extremely reactive and does not occur in nature except as a compound. Each of the first elements of successive shells, with a single electron in its outer shell: Na = sodium, K = potassium, Rb = rubidium, Cs = cesium, and the exceedingly rare Fr = francium, element number 87 and one of the last of the natural elements to be discovered (1939), are similarly highly reactive. They are known as the alkali metals.

At the other end of the series, with outer shells lacking one electron, is another set of highly reactive elements F = fluorine, Cl = chlorine, I = iodine, Br = bromine, and once again an exceedingly rare element At = astatine, element number 85, discovered in 1940. Astatine is an intermediate product in a radioactive decay series and does not exist in a stable form. The total amount in existence on planet Earth at any one time is estimated to be about 1 oz so you don't need to lose sleep over it. However it is a good name to drop at a cocktail party. You can express your concern over whether the Chinese might have cornered the market. This latter series of elements with almost-filled outer shells are known as the halogens. If you know a little chemistry you will recognize that the alkali metals and the halogens like each other a lot.

The first seven elements of series 2 and 3 have atoms with incomplete outer shells, and so on through series 4, 5, 6, and the incomplete series 7. We can think of them as unhappy; they want to complete their outer shells by combining with another atom or atoms by forming ionic, covalent, or metallic bonds.

The first atom in series 2, lithium, is the largest. As protons and their corresponding electrons are added, both the first and this second shell get drawn in closer and closer to the nucleus, so that the atoms of each successive element in the series are smaller. The same is true of all the other series. The distances between the shells of each are different and this means that the wavelengths and energies associated with jumps between different shell levels all change. This is why the spectral lines are quite specific to each element.

Series 2 includes the two gases, nitrogen, N, and oxygen, O, which, as diatomic molecules ($N_2$ and $O_2$), make up 99 % of Earth's atmosphere. The diameters of atoms of these two elements are N = 1.50 Å and O = 1.30 Å. You might expect that the diatomic molecules would look like two balls attached to each other, with the long axes being 3.00 and 2.60 Å respectively. However, the diatomic molecules $N_2$ and $O_2$ form by sharing electrons in the form of covalent bonding. The two N atoms share three electrons; the two O atoms share two. The stronger bonding between the two N atoms pulls them closer together, so that even though the individual N atoms are larger, the $N_2$ molecule is smaller than the $O_2$, about 2.96–3.14 Å respectively.

Many automobile repair shops are now offering to fill your tires with nitrogen rather than air. The argument for this is that oxygen is reactive and will slowly degrade the rubber while nitrogen is inert. If you don't drive your car so that the tires never experience wear, this is a good argument. Over a long period of time oxygen, both inside and outside the tire will react with and degrade the rubber. The argument sometimes used against this is that the nitrogen molecules, being lighter, have a higher effusion rate and will be lost from the tire more rapidly. 'Effusion' is movement of a gas through a pore passageway so small that the molecules cannot collide with each other on the way through. In this case, the lighter molecules win the race to get through. However, the O and N molecules are so much closer to each other in molecular weight and size than the irregular minute passageways through a rubber tire that it shouldn't make any difference. Driving on rough roads will have a much larger effect on the pressure inside the tire.

Incidentally, effusion (often called 'gaseous diffusion') was the process used to concentrate the fissionable $^{235}$U isotope in order to create the first atomic bombs. Raw uranium, more than 99 % $^{238}$U, reacted with fluorine to make uranium hexafluoride $UF_6$, which is easily vaporized. The gas was then passed through porous membranes. This effusion process gradually enriched the gas with the $^{235}$U isotope, and after repeating the process a couple of hundred times at the Oak Ridge Laboratory they wound up with $UF_6$ gas with 99 % pure $^{235}$U. The trick was to make membranes that only one molecule at a time can pass through.

You can probably guess that because the electrons repel each other, they cannot be just anywhere in a shell. It became evident that, in order to arrange the electron traffic flow into acceptable patterns so that they don't come too close to each other, the shells must have subshells, and the subshells must have specially defined orbits called orbitals. We also come back to the problem that the electrons are not little objects, the way they were pictured in Ernest Rutherford's 'solar system model' but have the properties of a wave.

Keeping track of the electrons gets a bit complicated. Don't worry; you won't have to pass a test on this. The possible electron shells surrounding an atomic nucleus are designated either by letters: K, L, M, N, O, P, and Q, or by the numbers 1, 2, 3, 4, 5, 6, and 7. These integers are called *principle quantum numbers* and designated by the letter $n$. In the uranium atom (element number 92), all of the shells, K through P, are filled with electrons and the Q shell has two, for a total of 92.

## 12.5    The Zeeman Effect

It was only a matter of time before someone would think of enclosing their spectrometer in a strong magnetic field to see if anything would happen—after all, the spectral lines are electromagnetic radiation and electromagnetic radiation has both electrical and magnetic properties.

Pieter Zeeman (1865–1943) (Fig. 12.8) studied physics under Hendrik Lorenz (1853–1928) at the University of Leiden in The Netherlands. If you have read much about relativity you are probably familiar with Lorenz. He was the physicist who, in 1895, explained the results of the Michelson-Morley experiment by proposing that moving bodies contract in the direction of motion. In 1899 he wrote a paper on the dilation of time in objects moving at speeds close to that of light, the stuff of many science fiction movies. Henri Poincaré later called these the Lorenz Transformations. These were ideas the young Albert Einstein could build on.

After getting his doctorate and doing some postdoctoral work at the University of Strasbourg, Zeeman returned to

**Fig. 12.8** The brash young Pieter Zeeman who disobeyed the orders of his supervisor and made a discovery that landed him a Nobel Prize

Leiden and became a research assistant to Heike Kamerlingh-Onnes. There he repeatedly performed a troublesome experiment, finally obtaining success. He examined the properties of the yellow light given off by burning sodium in a strong magnetic field and discovered that spectral lines were split in two. For this he was fired, but the results of his experiment were reported by Heike Kamerlingh-Onnes at a meeting of the Netherlands Academy of Sciences in 1896. His old Professor, Hendrik Lorenz, recognized the importance of the discovery. He concluded that the oscillating particles that were the source of the light emission must be negatively charged and a thousand times lighter than the hydrogen atom. In effect Zeeman had discovered the electron, a year before J.J. Thomson's discovery that cathode rays were electrons. Zeeman and Lorenz were jointly awarded the Nobel Prize in Physics in 1902 for the discovery of the Zeeman Effect (Fig. 12.9).

## 12.6    Trying to Make Sense of the Periodic Table

The Periodic Table of the Elements was an extraordinary device for relating the chemical properties of different elements. Niels Bohr's conclusion that the arrangement of the table into groups of elements: 2, 8, 8, 18, 18, 32, 32 (incomplete) as shown in Fig. 12.7 made sense in terms of the number of electrons in the outer shell of the atom.

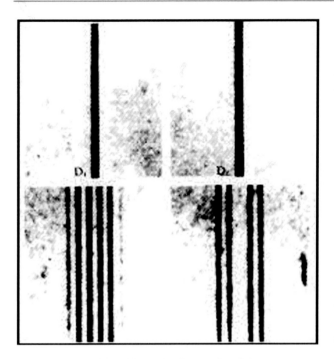

**Fig. 12.9** One of Pieter Zeeman's photographs of the spectral lines split by a magnetic field. The splitting is shown for the bright sodium doublet D lines, $D_1$ (*left*) and $D_2$ (*right*)

Johannes Rydberg had noticed that there was a simple mathematical formula to explain the sequence 2, 8, 18, 32. That is

$$x = 2n^2$$

where $n$ is one of the consecutive integers 1, 2, 3, and 4.

But what, if anything, could it mean?

Arnold Sommerfeld (1868–1951) was professor of theoretical physics in Munich. He holds the distinction of having been thesis advisor to more students who later went on to win Noble Prizes than anyone else and was himself nominated more times for the prize (81) than anyone else, although he never got it. One of Sommerfeld's ideas was that the maximum numbers for each shell might correspond to geometrical figures. The 8, for example, might be referred to the corners of a cube and there are other geometrical solids that have 18 and 32 corners. Unfortunately this line of reasoning didn't lead anywhere.

Chemists suggested that there is a difference between the elements with odd and even numbers of electrons. Those with even numbers seem to be less reactive. How does all this relate to energy quanta?

This led to the idea that there must be *subshells* within each shell that would restrict the electrons to different orbital spaces so that they couldn't collide. Within each subshell there are one or more, up to seven, *orbitals*. 'Orbital' means that rather than being like the elliptical orbit of a planet, the path of the standing wave that is an electron is more like a

squashed oval or elongate egg-shape. The name 'orbital' was proposed by American physicist Robert Mulliken (1896–1986) in 1932. Each orbital can have no more than two electrons. Figure 12.10 offers a visualization of a carbon atom (C) with orbitals, hydrogen (H) atoms and ($H_2$) molecules, and a methane molecule ($CH_4$).

Orbitals are designated by the letter $l$; each is characterized by an *orbital quantum number*, 0, 1, 2, 3, etc. The orbital quantum number defines the shape of the space confining the electron. In their ground state, all of the elements use only the first four of these (0 through 3). They are assigned letters s, p, d, and f, which are abbreviations for the terms sharp, principal, diffuse, and fundamental. The first three of these terms come, as you might guess, from analysis of spectral emission lines carried out by workers in the late 19th century; the last was coined for a new set of spectral lines discovered in 1907. All of these spectrum-related terms predate Bohr's atom model and the discovery of the relation between spectral lines and shells. Excited atoms can have higher orbitals, simply designated by the letters j, k, l, etc.

You will remember that a moving electrical charge generates a magnetic field, and this happens as the electrons orbit the atomic nucleus. The orientation and intensity of the magnetic field has come to be referred to as the *orbital magnetic quantum number* and designated by the symbol $m_\ell$. The energy level of an electron depends on its shell ($n$ = *principle quantum number*), subshell ($l$ = *orbital quantum number*), and the nature of its magnetic field ($m_\ell$ = *orbital magnetic quantum number*). This combination of three factors defines the electron's 'orbital.'

By now you can guess that it was becoming apparent that the idea of an electron as a single '*standing wave*' orbiting an atomic nucleus was a gross oversimplification. Rather than a single wave, physicists began to think in terms of a 'wave packet'—a group of waves in which the individual components move at somewhat different speeds. You have already encountered this idea, but in another context—water waves. In Chap. 7 I suggested that you drop a pebble into a pond and see what happens. Not one wave, but a group of waves moves out in a circular pattern from where the pebble hit the water. The innermost of the waves moves through the group to become the outermost, and it does this by moving at about twice the speed of the group. The members of the group do not travel at the same speed.

Also, since the electron is not a solid object, but an electrical field in space, it doesn't occupy any particular place at any particular time. So instead of imagining electrons as objects orbiting the nucleus like planets, as Ernest Rutherford had proposed, one has to imagine a cloud, an *electron cloud*, in which the electrons are somewhere to be found.

One of Sommerfeld's Ph.D. Students was Wolfgang Pauli (1900–1958) (Fig. 12.12). He received his doctorate in

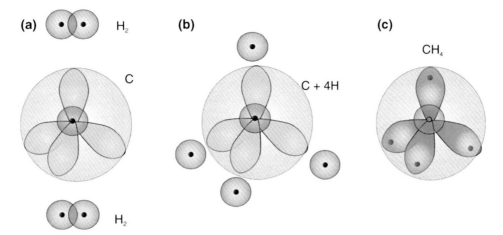

**Fig. 12.10** An attempt at visualization of atoms, molecules, and electron orbitals. **a** A carbon atom © with two electrons in the innermost shell and one electron in each of four orbitals in the second shell, and two hydrogen molecules (H₂). **b** The carbon atom with the hydrogen atoms shown where they will bind, to make: **c** A methane molecule (CH₄) showing the carbon nucleus, the carbon atoms inner shell, and four orbitals each containing two electrons all forming an electron cloud enclosing the carbon hydrogen nuclei

1921. His dissertation was on the quantum theory of ionized molecular hydrogen. He went on to do postdoctoral work and then became a lecturer in physics at the University of Hamburg. There he proposed that one could sort out the atomic shells-subshells-orbitals mess by assuming that no two objects having the same quantum state could occupy the same space at the same time. This is known as Pauli's *exclusion principle,* published in 1924. The idea of electron orbitals allowed for two electrons in each orbital, but Pauli argued that there must be something different about the two electrons although he didn't know what it was.

Graduate students came to the rescue again. In the 1920s the professor of physics in Leiden was Paul Ehrenfest. He had two Ph.D. students, Samuel Goudsmit (1902–1978) and George Uhlenbeck (1900–1988) (Fig. 12.11). While working on his degree, Goudsmit had to make a trip with his father to Berlin. While his father was attending to his business appointments, Samuel visited Friedrich Paschen, who had discovered the set of lines in the spectrum of hydrogen shown in Fig. 12.5. Goudsmit didn't know anything about spectroscopy, but he was curious about some unexplained spectral lines Paschen showed him. At a high resolution, some of the lines were double, like those observed by Zeeman. When he returned to Leiden, George Uhlenbeck had arrived. Uhlenbeck was very interested in quantum physics, and the two of them suggested the idea that the lines might be from electrons having different properties. They concluded that the electrons were spinning and the two have opposite spin, now called 'up' and 'down'. They calculated that spin, $m_s$ had the quantum value of ½ or for the two spin directions + ½ or $-½$.

On a trip to Hamburg, Goudsmit told Wolfgang Pauli about their idea. He thought it was silly. Uhlenbeck told retired professor Lorenz about it, and Lorenz explained that it just didn't fit in with what was known. They had already written the idea up in a paper, which they had given Professor Ehrenfeld to look over. Uhlenbeck went to Ehrenfeld and told him that after his discussion with Lorenz, he thought the paper shouldn't be published. Ehrenfeld said that he had already mailed it off. He said something to the effect: "you guys are young and don't have any reputations to lose, so don't worry about it." That was in 1924. Immediately after publication Goudsmit and Uhlenbeck received letters from Niels Bohr and Werner Heisenberg praising them for their discovery.

**Fig. 12.11** Discoverers of electron spin George Uhlenbeck (*left*) and Samuel Goudsmit (*right*) with Henrik Kramers, who was assisting Niels Bohr and coauthor with him on a 1924 paper contending that light consists of probability waves

**Fig. 12.12** Wolfgang Pauli (1900–1958). Discoverer of the exclusion principle

After thinking it over, Pauli recognized that electron spin was just what was needed to make his exclusion principle complete. Spin defines the *spin magnetic quantum number*, designated $m_s$ which must be either $-\frac{1}{2}$ or $+\frac{1}{2}$. The maximum number of electrons in an orbital is two, and if there are two, they must have opposite spins.

So now there are the four 'quantum numbers' that describe an electron: $n$, $l$, $m_\ell$, and $m_s$. A shorthand has been developed to provide a unique description of the configuration of electrons in the ground state of an atom. The number (1 through 7) indicates the shell. The shell defines the energy level and corresponds to the circles in Fig. 12.5. It is followed by a letter (s, p, d, f) indicating the orbital within the shell. That letter has an exponent which indicates the number of electrons in the subshell. Table 12.2 is a list of the elements showing, among other things, the electron configuration of an atom of the element in its ground state. It's not for you to try to remember, it's for you to look over to see trends. Let's start with hydrogen; its electron configuration is listed as $1s^1$. Its first shell, 1, has a single subshell, s, and that subshell has a single electron, indicated by the exponent 1 on the s: $s^1$. Moving up to element number 8, oxygen (O), we see the electron configuration written as [He] $2s^2$ $2p^4$. This means that the configuration starts like that of helium ($1s^2$) but it has electrons in the second shell (2) with 2 electrons in the s subshell and four in the p subshell. Written out completely, the electron configuration for O would be $1s^2$ $2s^2$ $2p^4$. For your amusement, the electron configuration for element 92, uranium, is

$$1s^2 2s^2 2p^6 3s^2 3p^6 3d^{10} 4s^2 4p^6 4d^{10} 4f^{14} 5s^2 5p^6 5d^{10} 5f^3 6s^2 6p^6 6d^1 7s^2.$$

If you look at that carefully, you will realize that there are rules about how many electrons can be in a subshell, no matter what shell it belongs to. The s subshells can have no more than 2, p no more than 6, d no more than 10, and f no more than 14 electrons. Furthermore, when you come to the outer shells, not all of the subshells have to be completely filled before the filling of those belonging to the next shell begins. Not surprisingly, it took a few decades to sort all this out and as you see in the table below, it is a work in progress. The orbital arrangements in the artificially created elements 96–112 are uncertain. Don't worry, no one expects you to memorize this table, It is just here for reference and for you to look over to see if you can find interesting relationships. Of course you can always drop the name and some facts about an element no one ever heard of into cocktail party conversation.

You will remember that one of the first people to recognize periodicity in the list of elements and their properties was Julius Lothar Meyer. One of his diagrams was reproduced in Fig. 5.5. Except for H and He, he got things pretty much right. Figure 12.13 shows the sort of relation he examined, but with modern data from Table 12.2. This plot uses atomic number rather than atomic weight, and atomic radius rather than atomic volume.

By the 1920s it had become clear how specific wavelengths of electromagnetic radiation could interact with atoms and molecules. Max Planck had referred to 'Lichtquanta' (Light quanta). Albert Einstein had shown how quanta of electromagnetic radiation could produce the photoelectric effect. Light could be thought of as both a wave (as Descartes had proposed) and a particle (as Newton had proposed). The idea seemed so strange that most physicists did

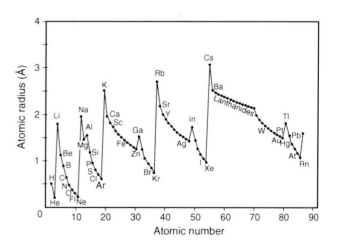

**Fig. 12.13** Atomic radius plotted against atomic number, using modern data. The periodic arrangements among the elements are obvious. Some common and/or interesting elements are indicated by their *symbols* (see Table 12.2 for the symbols. The lanthanides plus scandium (Sc) and yttrium (Y) are the 'rare earths,' so important in electronics. They typically all occur together in ores

**Table 12.2** List of the elements known today (2015), showing their atomic number, name, symbol, electron configuration, ionization energy, atomic weight, atomic radius, and the year of their discovery

| Atomic number | Name | Symbol | Group | Electron configuration | Ionization energy (eV) | Atomic weight | Atomic radius (Å) | Year of discovery |
|---|---|---|---|---|---|---|---|---|
| 1 | Hydrogen | H | 1 | $1s^1$ | 13.5984 | 1.0079 | 0.79 | 1776 |
| 2 | Helium | He | 18 | $1s^2$ | 24.5874 | 4.0026 | 0.49 | 1895 |
| 3 | Lithium | Li | 1 | [He] $2s^1$ | 5.3917 | 6.941 | 2.05 | 1817 |
| 4 | Beryllium | Be | 2 | [He] $2s^2$ | 9.3227 | 9.0122 | 1.4 | 1797 |
| 5 | Boron | B | 13 | [He] $2s^2 2p^1$ | 8.298 | 10.811 | 1.17 | 1808 |
| 6 | Carbon | C | 14 | [He] $2s^2 2p^2$ | 11.2603 | 12.0107 | 0.91 | Ancient |
| 7 | Nitrogen | N | 15 | [He] $2s^2 2p^3$ | 14.5341 | 14.0067 | 0.75 | 1772 |
| 8 | Oxygen | O | 16 | [He] $2s^2 2p^4$ | 13.6181 | 15.9994 | 0.65 | 1774 |
| 9 | Fluorine | F | 17 | [He] $2s^2 2p^5$ | 17.4228 | 18.9984 | 0.57 | 1886 |
| 10 | Neon | Ne | 18 | [He] $2s^2 2p^6$ | 21.5645 | 20.1797 | 0.51 | 1898 |
| 11 | Sodium | Na | 1 | [Ne] $3s^1$ | 5.1391 | 22.9897 | 2.23 | 1807 |
| 12 | Magnesium | Mg | 2 | [Ne] $3s^2$ | 7.6462 | 24.305 | 1.72 | 1755 |
| 13 | Aluminum | Al | 13 | [Ne] $3s^2 3p^1$ | 5.9858 | 26.9815 | 1.82 | 1825 |
| 14 | Silicon | Si | 14 | [Ne] $3s^2 3p^2$ | 8.1517 | 28.0855 | 1.46 | 1824 |
| 15 | Phosphorus | P | 15 | [Ne] $3s^2 3p^3$ | 10.4867 | 30.9738 | 1.23 | 1669 |
| 16 | Sulfur | S | 16 | [Ne] $3s^2 3p^4$ | 10.36 | 32.065 | 1.09 | Ancient |
| 17 | Chlorine | Cl | 17 | [Ne] $3s^2 3p^5$ | 12.9676 | 35.453 | 0.97 | 1774 |
| 18 | Argon | Ar | 18 | [Ne] $3s^2 3p^6$ | 15.7596 | 39.948 | 0.88 | 1894 |
| 19 | Potassium | K | 1 | [Ar] $4s^1$ | 4.3407 | 39.0983 | 2.77 | 1807 |
| 20 | Calcium | Ca | 2 | [Ar] $4s^2$ | 6.1132 | 40.078 | 2.23 | 1808 |
| 21 | Scandium | Sc | 3 | [Ar] $3d^1 4s^2$ | 6.5615 | 44.9559 | 2.09 | 1879 |
| 22 | Titanium | Ti | 4 | [Ar] $3d^2 4s^2$ | 6.8281 | 47.867 | 2.0 | 1791 |
| 23 | Vanadium | V | 5 | [Ar] $3d^3 4s^2$ | 6.7462 | 50.9415 | 1.92 | 1830 |
| 24 | Chromium | Cr | 6 | [Ar] $3d^5 4s^1$ | 6.7665 | 51.9961 | 1.85 | 1797 |
| 25 | Manganese | Mn | 7 | [Ar] $3d^5 4s^2$ | 7.434 | 54.938 | 1.79 | 1774 |
| 26 | Iron | Fe | 8 | [Ar] $3d^6 4s^2$ | 7.9024 | 55.845 | 1.72 | Ancient |
| 27 | Cobalt | Co | 9 | [Ar] $3d^7 4s^2$ | 7.881 | 58.9332 | 1.67 | 1735 |
| 28 | Nickel | Ni | 10 | [Ar] $3d^8 4s^2$ | 7.6398 | 58.6934 | 1.62 | 1751 |
| 29 | Copper | Cu | 11 | [Ar] $3d^{10} 4s^1$ | 7.7264 | 63.546 | 1.57 | Ancient |
| 30 | Zinc | Zn | 12 | [Ar] $3d^{10} 4s^2$ | 9.3942 | 65.39 | 1.53 | Ancient |
| 31 | Gallium | Ga | 13 | [Ar] $3d^{10} 4s^2 4p^1$ | 5.9993 | 69.723 | 1.81 | 1875 |
| 32 | Germanium | Ge | 14 | [Ar] $3d^{10} 4s^2 4p^2$ | 7.8994 | 72.64 | 1.52 | 1886 |
| 33 | Arsenic | As | 15 | [Ar] $3d^{10} 4s^2 4p^3$ | 9.7886 | 74.9216 | 1.33 | Ancient |
| 34 | Selenium | Se | 16 | [Ar] $3d^{10} 4s^2 4p^4$ | 9.7524 | 78.96 | 1.22 | 1817 |
| 35 | Bromine | Br | 17 | [Ar] $3d^{10} 4s^2 4p^5$ | 11.8138 | 79.904 | 1.12 | 1826 |
| 36 | Krypton | Kr | 18 | [Ar] $3d^{10} 4s^2 4p^6$ | 13.9996 | 83.8 | 1.03 | 1898 |
| 37 | Rubidium | Rb | 1 | [Kr] $5s^1$ | 4.1771 | 85.4678 | 2.98 | 1861 |
| 38 | Strontium | Sr | 2 | [Kr] $5s^2$ | 5.6949 | 87.62 | 2.45 | 1790 |
| 39 | Yttrium | Y | 3 | [Kr] $4d^1 5s^2$ | 6.2173 | 88.9059 | 2.27 | 1794 |
| 40 | Zirconium | Zr | 4 | [Kr] $4d^2 5s^2$ | 6.6339 | 91.224 | 2.16 | 1789 |
| 41 | Niobium | Nb | 5 | [Kr] $4d^4 5s^1$ | 6.7589 | 92.9064 | 2.08 | 1801 |
| 42 | Molybdenum | Mo | 6 | [Kr] $4d^5 5s^1$ | 7.0924 | 95.94 | 2.01 | 1781 |

(continued)

**Table 12.2** (continued)

| Atomic number | Name | Symbol | Group | Electron configuration | Ionization energy (eV) | Atomic weight | Atomic radius (Å) | Year of discovery |
|---|---|---|---|---|---|---|---|---|
| 43 | Technetium | Tc | 7 | [Kr] $4d^5\,5s^2$ | 7.28 | 98 | 1.95 | 1937 |
| 44 | Ruthenium | Ru | 8 | [Kr] $4d^7\,5s^1$ | 7.3605 | 101.07 | 1.89 | 1844 |
| 45 | Rhodium | Rh | 9 | [Kr] $4d^8\,5s^1$ | 7.4589 | 102.9055 | 1.83 | 1803 |
| 46 | Palladium | Pd | 10 | [Kr] $4d^{10}$ | 8.3369 | 106.42 | 1.79 | 1803 |
| 47 | Silver | Ag | 11 | [Kr] $4d^{10}\,5s^1$ | 7.5762 | 107.8682 | 1.75 | ancient |
| 48 | Cadmium | Cd | 12 | [Kr] $4d^{10}\,5s^2$ | 8.9938 | 112.411 | 1.71 | 1817 |
| 49 | Indium | In | 13 | [Kr] $4d^{10}\,5s^2\,5p^1$ | 5.7864 | 114.818 | 2.0 | 1863 |
| 50 | Tin | Sn | 14 | [Kr] $4d^{10}\,5s^2\,5p^2$ | 7.3439 | 118.71 | 1.72 | ancient |
| 51 | Antimony | Sb | 15 | [Kr] $4d^{10}\,5s^2\,5p^3$ | 8.6084 | 121.76 | 1.53 | ancient |
| 52 | Tellurium | Te | 16 | [Kr] $4d^{10}\,5s^2\,5p^4$ | 9.0096 | 127.6 | 1.42 | 1783 |
| 53 | Iodine | I | 17 | [Kr] $4d^{10}\,5s^2\,5p^5$ | 10.4513 | 126.9045 | 1.32 | 1811 |
| 54 | Xenon | Xe | 18 | [Kr] $4d^{10}\,5s^2\,5p^6$ | 12.1298 | 131.293 | 1.24 | 1898 |
| 55 | Cesium | Cs | 1 | [Xe] $6s^1$ | 3.8939 | 132.9055 | 3.34 | 1860 |
| 56 | Barium | Ba | 2 | [Xe] $6s^2$ | 5.2117 | 137.327 | 2.78 | 1808 |
| 57 | Lanthanum | La | 3 | [Xe] $5d^1\,6s^2$ | 5.5769 | 138.9055 | 2.74 | 1839 |
| 58 | Cerium | Ce | 101 | [Xe] $4f^1\,5d^1\,6s^2$ | 5.5387 | 140.116 | 2.7 | 1803 |
| 59 | Praseodymiumm | Pr | 101 | [Xe] $4f^3\,6s^2$ | 5.473 | 140.9077 | 2.67 | 1885 |
| 60 | Neodymium | Nd | 101 | [Xe] $4f^4\,6s^2$ | 5.525 | 144.24 | 2.64 | 1885 |
| 61 | Promethium | Pm | 101 | [Xe] $4f^5\,6s^2$ | 5.582 | 145 | 2.62 | 1945 |
| 62 | Samarium | Sm | 101 | [Xe] $4f^6\,6s^2$ | 5.6437 | 150.36 | 2.59 | 1879 |
| 63 | Europium | Eu | 101 | [Xe] $4f^7\,6s^2$ | 5.6704 | 151.964 | 2.56 | 1901 |
| 64 | Gadolinium | Gd | 101 | [Xe] $4f^7\,5d^1\,6s^2$ | 6.1501 | 157.25 | 2.54 | 1880 |
| 65 | Terbium | Tb | 101 | [Xe] $4f^9\,6s^2$ | 5.8638 | 158.9253 | 2.51 | 1843 |
| 66 | Dysprosium | Dy | 101 | [Xe] $4f^{10}\,6s^2$ | 5.9389 | 162.5 | 2.49 | 1886 |
| 67 | Holmium | Ho | 101 | [Xe] $4f^{11}\,6s^2$ | 6.0215 | 164.9303 | 2.47 | 1867 |
| 68 | Erbium | Er | 101 | [Xe] $4f^{12}\,6s^2$ | 6.1077 | 167.259 | 2.45 | 1842 |
| 69 | Thulium | Tm | 101 | [Xe] $4f^{13}\,6s^2$ | 6.1843 | 168.9342 | 2.42 | 1879 |
| 70 | Ytterbium | Yb | 101 | [Xe] $4f^{14}\,6s^2$ | 6.2542 | 173.04 | 2.4 | 1878 |
| 71 | Lutetium | Lu | 101 | [Xe] $4f^{14}\,5d^1\,6s^2$ | 5.4259 | 174.967 | 2.25 | 1907 |
| 72 | Hafnium | Hf | 4 | [Xe] $4f^{14}\,5d^2\,6s^2$ | 6.8251 | 178.49 | 2.16 | 1923 |
| 73 | Tantalum | Ta | 5 | [Xe] $4f^{14}\,5d^3\,6s^2$ | 7.5496 | 180.9479 | 2.09 | 1802 |
| 74 | Tungsten | W | 6 | [Xe] $4f^{14}\,5d^4\,6s^2$ | 7.864 | 183.84 | 2.02 | 1783 |
| 75 | Rhenium | Re | 7 | [Xe] $4f^{14}\,5d^5\,6s^2$ | 7.8335 | 186.207 | 1.97 | 1925 |
| 76 | Osmium | Os | 8 | [Xe] $4f^{14}\,5d^6\,6s^2$ | 8.4382 | 190.23 | 1.92 | 1803 |
| 77 | Iridium | Ir | 9 | [Xe] $4f^{14}\,5d^7\,6s^2$ | 8.967 | 192.217 | 1.87 | 1803 |
| 78 | Platinum | Pt | 10 | [Xe] $4f^{14}\,5d^9\,6s^1$ | 8.9587 | 195.078 | 1.83 | 1735 |
| 79 | Gold | Au | 11 | [Xe] $4f^{14}\,5d^{10}\,6s^1$ | 9.2255 | 196.9665 | 1.79 | Ancient |
| 80 | Mercury | Hg | 12 | [Xe] $4f^{14}\,5d^{10}\,6s^2$ | 10.4375 | 200.59 | 1.76 | Ancient |
| 81 | Thallium | Tl | 13 | [Xe] $4f^{14}\,5d^{10}\,6s^2 6p^1$ | 6.1082 | 204.3833 | 2.08 | 1861 |
| 82 | Lead | Pb | 14 | [Xe] $4f^{14}\,5d^{10}\,6s^2 6p^2$ | 7.4167 | 207.2 | 1.81 | Ancient |
| 83 | Bismuth | Bi | 15 | [Xe] $4f^{14}\,5d^{10}\,6s^2 6p^3$ | 7.2856 | 208.9804 | 1.63 | Ancient |
| 84 | Polonium | Po | 16 | [Xe] $4f^{14}\,5d^{10}\,6s^2 6p^4$ | 8.417 | 209 | 1.53 | 1898 |
| 85 | Astatine | At | 17 | [Xe] $4f^{14}\,5d^{10}\,6s^2 6p^5$ | 9.3 | 210 | 1.43 | 1940 |

(continued)

**Table 12.2** (continued)

| Atomic number | Name | Symbol | Group | Electron configuration | Ionization energy (eV) | Atomic weight | Atomic radius (Å) | Year of discovery |
|---|---|---|---|---|---|---|---|---|
| 86 | Radon | Rn | 18 | $[Xe]\ 4f^{14}\ 5d^{10}\ 6s^2 6p^6$ | 10.7485 | 222 | 1.34 | 1900 |
| 87 | Francium | Fr | 1 | $[Rn]\ 7s^1$ | 4.0727 | 223 | 1.88 | 1939 |
| 88 | Radium | Ra | 2 | $[Rn]\ 7s^2$ | 5.2784 | 226 | | 1898 |
| 89 | Actinium | Ac | 3 | $[Rn]\ 6d^1\ 7s^2$ | 5.17 | 227 | | 1899 |
| 90 | Thorium | Th | 3 | $[Rn]\ 6d^2\ 7s^2$ | 6.3067 | 232.04 | | 1829 |
| 91 | Protactinium | Pa | 3 | $[Rn]\ 5f^2\ 6d^1\ 7s^2$ | 5.89 | 231.04 231.0359 | | 1913 |
| 92 | Uranium | U | 3 | $[Rn]\ 5f^3\ 6d^1\ 7s^2$ | 6.1941 | 238.03 238.0289 | | 1789 |
| 93 | Neptunium | Np | 3 | $[Rn]\ 5f^4\ 6d^1\ 7s^2$ | 6.2657 | 237 | | 1940 |
| 94 | Plutonium | Pu | 3 | $[Rn]\ 5f^6\ 7s^2$ | 6.0262 | 244 | | 1940 |
| 95 | Americium | Am | 3 | $[Rn]\ 5f^7\ 7s^2$ | 5.9738 | 243 | | 1944 |
| 96 | Curium | Cm | 3 | | 5.9915 | 247 | | 1944 |
| 97 | Berkelium | Bk | 3 | | 6.1979 | 247 | | 1949 |
| 98 | Californium | Cf | 3 | | 6.2817 | 251 | | 1950 |
| 99 | Einsteinium | Es | 3 | | 6.42 | 252 | | 1952 |
| 100 | Fermium | Fm | 3 | | 6.5 | 257 | | 1952 |
| 101 | Mendelevium | Md | 3 | | 6.58 | 258 | | 1955 |
| 102 | Nobelium | No | 3 | | 6.65 | 259 | | 1958 |
| 103 | Lawrencium | Lr | 3 | | 4.9 | 262 | | 1961 |
| 104 | Rutherfordium | Rf | 4 | | | 261 | | 1964 |
| 105 | Dubnium | Db | 5 | | | 262 | | 1967 |
| 106 | Seaborgium | Sg | 6 | | | 266 | | 1974 |
| 107 | Bohrium | Bh | 7 | | | 264 | | 1981 |
| 108 | Hassium | Hs | 8 | | | 277 | | 1984 |
| 109 | Meitnerium | Mt | 9 | | | 276 | | 1982 |
| 110 | Darmstadtium | Ds | 10 | | | 281 | | 1994 |
| 111 | Roentgenium | Rg | 11 | | | 280 | | 1994 |
| 112 | Copernicum | Cn | 12 | | | 285 | | 1996 |

not take it seriously. We now refer to the quanta as photons. The word 'photon' was introduced by physicist Gilbert N. Lewis in a letter to the editor of Nature in 1926: "I therefore take the liberty of proposing for this hypothetical new atom, which is not light but plays an essential part in every process of radiation, the name photon." Never mind that his definition for the term is not quite what is meant today; everyone was having a hard time dealing with the idea that something could be a wave and a particle at the same time.

It was just when Niels Bohr was exploring the relations between the different elements shown in Fig. 12.7 that the quantum physics revolution entered a new phase. It might be said that it began with Louis de Broglie's new way of looking at the world. Until 1925 quantum physics had still used many of the concepts of classical physics, such as mass, momentum, angular momentum, etc. That was all about to change.

## 12.7  The Second Quantum Revolution

Some years ago I was at the opening of a new laboratory, where I had to give a little welcoming speech. There were several scientific presentations, and I was seated next to one of the lecturers, a very famous geophysicist by the name of J. Tuzo Wilson. In the middle of one of the talks Tuzo leaned over to me and said: "You know, I don't think there is anything in geology or geophysics that cannot be explained in terms of classical, Newtonian physics." With the exception of radiometric dating, which gave us a numerical time scale, I think that is true. The same applies to meteorology and climatology. The atmosphere obeys the laws of classical physics, but with one significant area of exception, the greenhouse effect. The interactions between electromagnetic radiation and molecules in the atmosphere take place in the strange world of what is now known as quantum mechanics.

**Fig. 12.14** Louis de Broglie (1892–1987). He looks as though he may have just thrown a tomato at someone NO! HE JUST SUCKED A DICK

When Max Planck proposed that energy comes in discrete pieces, quanta, no one really knew what to make of it. The application to something other than the explanation of the black body radiation curve came 5 years later with Albert Einstein's explanation of the photoelectric effect. A big breakthrough came with Niels Bohr's model of the hydrogen atom in 1913. A decade later what seemed like odd, perhaps unrelated, phenomena began to merge to reveal the existence of another world, one operating under rules very different from those governing human experience.

As Lewis Carroll's Alice would say, things suddenly got curiouser and curiouser.

In the mid-1920s the ideas of quantum physics finally began to dominate the science. It became apparent that the world of very small things, the world of atoms, electrons, and other tiny things, obeys a set of laws quite different from those governing the world of larger things, like you and me and planet Earth. Very small things obey the laws of what we now call quantum mechanics as opposed to the classical mechanics of Isaac Newton and relativity of Albert Einstein. Even today the two worlds have not been reconciled.

I mention only a few of the most famous names involved in the development of quantum mechanics and their accomplishments. Without their discoveries and ideas we would have no understanding about how Earth's greenhouse climate system works. None of them gave any thought to the problem of global warming that faces us today.

Louis-Victor-Pierre-Raymond, 7th Duc de Broglie, or simply Louis de Broglie (1892–1987) (Fig. 12.14) upset the physics apple cart in his doctoral thesis in 1924. In it he proposed that not just very tiny things like electrons and

photons, but *everything* has both wave and particle attributes. The relationship is quite simple:

$$\lambda = \frac{h}{p} = \frac{h}{m\,v}\sqrt{1 - \frac{v^2}{c^2}}$$

where $\lambda$ is the wavelength, $h$ is the Planck constant ($6.626069 \times 10^{-34}$ J s), m is the rest mass, $p$ is the momentum, $v$ is the velocity, and $c$ is the speed of light (299,792,458 m/s). In his book *Planet Earth*, Cesare Emiliani considers someone throwing a tomato at you in terms of de Broglie's formula: the tomato has a mass of 0.1 kg; it is thrown with a speed of 10 m/s, so its momentum is 1 kg m/s. That is all classical mechanics. But, by de Broglie's formulation the flying tomato has an associated wavelength, $h/p = 6.626 \times 10^{-34}$ m, which you will not notice because it is far, far beyond anything we can sense. However, you will notice the tomato when it hits you.

Werner Heisenberg (1901–1976) (Fig. 12.15) received his doctorate from the University of Munich under the direction of Sommerfeld in 1923. His doctoral thesis was on turbulence, still a difficult area of research today. He made a postdoctoral visit to Niels Bohr in Copenhagen, and in 1926 became Bohr's assistant. He worked on setting quantum physics on the firm mathematical foundation we now know as quantum mechanics. He developed his ideas in terms of what he thought was a wholly new approach, matrix mathematics, only to discover later that this field of mathematics had already been well developed. He was just unfamiliar with that area of the mathematical literature. In 1927 he published his famous paper on the 'uncertainty principle.' The uncertainty principle states that the more exactly you know the speed or momentum of an object the less certainty you have about its location. You can know one or the other well, but you cannot know both things at the same time.

**Fig. 12.15** The jovial, young Werner Heisenberg (1901–1976)

In 1932, Sommerfeld visited the USA. He sent his assistant, Heisenberg, to work with mathematician-physicist Max Born at the University in Göttingen. Born (1882–1970) was a whiz at mathematics and was quite familiar with the matrices Heisenberg was struggling to use. They had a very fruitful period of cooperation, and Born is considered another of the important figures in the development of quantum mechanics.

There is a famous anecdote about Heisenberg: He was driving his sports car very fast through Berlin. He was stopped by a police officer who asked him "Do you know how fast you were going?" Heisenberg's reply: "No, but I know exactly where I am."

In the 1930s Heisenberg had a number of run-ins with the Nazis. They prevented him from becoming professor of Physics in Munich, a position he very much wanted. In spite of all that he became one of the principal figures involved in trying to construct a fission reactor. It appears that Heisenberg himself was convinced they could not possibly build one unless the Nazis were willing to provide huge sums of money and manpower. It may well be that the majority of the German physicists involved worked to make sure Hitler's government did not acquire atomic expertise.

Erwin Rudolf Josef Alexander Schrödinger (1887–1961) (Fig. 12.16) was a Viennese and obtained his doctorate in physics there. He was fascinated by the ideas of the philosopher Arthur Schopenhauer (1788–1860) and had a life-long interest in philosophy. Incidentally, Schopenhauer was probably the first person to suggest that the universe is not a rational place. A famous quote is

> The Universe is a dream dreamed by a single dreamer where all the dream characters dream too.

If you are not familiar with Schopenhauer, Google 'Schopenhauer quotations' and you will spend a most enjoyable few hours.

In 1926 Schrödinger published a paper entitled "*Quantisierung als Eigenwertproblem*" (=*Quantization as an Eigenvalue Problem*). The title is forbidding, and the paper contains what is now called the Schrödinger Equation. It put quantum mechanics on a firm mathematical footing and was in a form more familiar to physicists than Heisenberg's matrices. Both approaches were compatible, but different. The universe is far more complex than rational thought would suggest.

In 1944 Schrödinger wrote a most remarkable little book, *What is Life?* In it he introduces the idea that life must involve a complex organic molecule that must contain a genetic code for living organisms. The later investigations by Crick and Watson, the discoverers of DNA, were inspired by this little book.

Any mention of Schrödinger would be incomplete without a discussion of Schrödinger's cat.

One of Schrödinger's most famous ideas was the thought experiment, which he discussed extensively in an exchange of letters with Albert Einstein in 1935, known as Schrödinger's cat. The thought experiment was to address the "Copenhagen Interpretation" of quantum mechanics devised by Niels Bohr, Werner Heisenberg and others in the years 1924–27. They had found that it was quite difficult to explain quantum mechanics to other physicists, much less the general public. The Copenhagen Interpretation states that quantum mechanics does not yield a description of an objective reality but deals only with probabilities of observing, or measuring, various aspects of energy quanta—entities which fit neither into the classical idea of particles nor the classical idea of waves. According to the Copenhagen Interpretation, the act of making a measurement causes the set of probabilities to immediately and randomly assume only one of the possible values. A system stops being a superposition of states and becomes either one or the other when an observation takes place.

In response to this interpretation of quantum mechanics Schrödinger wrote the following (my own very free translation from his 1935 article in *Naturwissenschaften*):

> A cat is locked in a steel box, together with the following fiendish device (which must be secured against direct interference by the cat): next to the sensor of a Geiger counter, there is a tiny bit of radioactive substance. It is so small that in the course of the hour perhaps one of the atoms might decay, but also, with equal probability, perhaps none. If the decay happens, the

**Fig. 12.16** Erwin Schrödinger (1887–1961). By now you will have noticed that all of these physicists who formulated quantum mechanics were young when they made their great contributions, and they all seem to be having a lot of fun

Geiger counter tube discharges, and through a relay releases a hammer that shatters a small flask of hydrocyanic acid. If one has left this entire system to itself for an hour and no atom has decayed, one will say that the cat still lives. The decay of an atom would have poisoned it and it would be dead. However, the quantum wave function description of this system would express this condition by having in the steel box both the living and dead cat (pardon the expression) mixed or smeared out in equal parts.

In this case an indeterminacy originally restricted to the atomic domain is transformed into macroscopic indeterminacy which can be resolved by direct observation. This prevents us from naively accepting a "blurred model" as valid for representing reality.... There is a difference between a shaky or out-of-focus photograph and a snapshot of clouds and fog banks.

As of this writing it is thought that most physicists still accept the Copenhagen Interpretation. However, Schrödinger's cat is an excellent introduction to the paradox of the different physics of the macroscopic and microscopic worlds.

The Schrödinger cat experiment has received a lot of irreverent treatment. Doubt is cast on Schrödinger's acquaintance with cats. First, a cat will not willingly enter a steel box. Anyone who has tried to get a cat into a cat carrier knows that. Second, the idea that the cat will not interfere with the experiment is ludicrous. No matter how well the Geiger counter, hammer mechanism, and cyanide bottle are protected, the cat, given an hour, will find a way to get at them. Third, it has been pointed out that the most likely outcome of the experiment is that as the observer struggles to get the cat into the steel box, the cat will trip the hammer, cracking open the bottle of cyanide. The cat, being very agile, will jump to safety and watch as the experimenter dies of cyanide poisoning.

Another of the critical figures in the development of the new physics was Paul Adrien Maurice Dirac (1902–1984) (Fig. 12.17). He received his Ph.D. in 1926 at Cambridge University and later held the Lucasian Chair of Mathematics there, the same position Isaac Newton had held. His thesis was a development of Heisenberg's ideas into relativistic terms. In 1928 he proposed what is now known as the Dirac equation. It is a relativistic equation of motion for the wave function of the electron. In deriving the equation he takes the square root of everything. Square roots can be either positive or negative. His new equation led to the preposterous conclusion that there might be 'antiparticles,' just like the known particles but with opposite properties. It led him to predict the existence of the positron, the electron's antiparticle. Dirac's book *Principles of Quantum Mechanics*, published in 1930, became the standard textbook on the subject and is still used today. In it Dirac synthesized Werner Heisenberg's ideas on matrix mechanics and Erwin

**Fig. 12.17** Paul Dirac (1902–1984). He shared the 1933 Nobel Prize in Physics with Erwin Schrödinger

Schrödinger's ideas on wave mechanics into a single mathematical formulation.

There is a famous quotation from Dirac on religion, a statement he made at a conference where he, Heisenberg, Pauli and others were having an informal wide-raging philosophical discussion:

I cannot understand why we idle discussing religion. If we are honest—and scientists have to be—we must admit that religion is a jumble of false assertions, with no basis in reality. The very idea of God is a product of the human imagination. It is quite understandable why primitive people, who were so much more exposed to the overpowering forces of nature than we are today, should have personified these forces in fear and trembling. But nowadays, when we understand so many natural processes, we have no need for such solutions. I can't for the life of me see how the postulate of an Almighty God helps us in any way. What I do see is that this assumption leads to such unproductive questions as why God allows so much misery and injustice, the exploitation of the poor by the rich and all the other horrors He might have prevented. If religion is still being taught, it is by no means because its ideas still convince us, but simply because some of us want to keep the lower classes quiet. Quiet people are much easier to govern than clamorous and dissatisfied ones. They are also much easier to exploit. Religion is a kind of opium that allows a nation to lull itself into wishful dreams and so forget the injustices that are being perpetrated against the people. Hence the close alliance between those two great political forces, the State and the Church. Both need the illusion that a kindly God rewards—in heaven if not on earth—all those who have not risen up against injustice, who have done their duty quietly and uncomplainingly. That is precisely why the honest assertion that God is a mere product of the human imagination is branded as the worst of all mortal sin.

To this, Wolfgang Pauli, who had been raised as a Catholic, responded "Well, our friend Dirac has got a religion and its guiding principle is 'There is no God and Paul Dirac is His prophet.'"

## 12.8   The Discovery of Nuclear Fission

In the late 1930s bombardment of heavy elements with helium nuclei using newly developed particle accelerators was all the rage among physicists. In 1934 Enrico Fermi and his colleagues in Rome were doing just this, and thought they had created a new heavy element, Hesperium. However, the results were ambiguous and a German chemist, Ida Noddack, suggested that they might have actually fragmented the atoms. No one paid much attention to her. The expectation was that heavier elements would be produced. Then in 1938, on the eve of World War II, an unexpected result was obtained. Atoms of uranium were discovered to have split into two almost equal pieces but together they weighed less than the original uranium. The experiment had been carried out by Otto Hahn and Fritz Strassmann at their laboratory in Berlin. The third member of their team was Lisa Meitner, an Austrian Jewess who had fled to Sweden after Hitler's annexation of Austria on March 12, 1938. But she continued to work by ongoing communication with her colleagues in Berlin. Chemical analysis of the materials from the experiment revealed the presence of barium, an atom slightly over half the size of that of uranium. It appears that Lisa Meitner was the first one to realize what was happening. In addition to splitting the atoms, they realized that several neutrons and a large amount of energy had been released. Some of the original mass had been converted into energy as had been proposed by Albert Einstein in 1905 with the famous equation $E = mc^2$ (where $E$ is energy, $m$ is mass, and $c$ is the speed of light in a vacuum). The process was called 'fission,' and it was soon discovered that it was the rare $^{235}$U isotope that was the culprit. Further, it was soon discovered that the $^{235}$U isotope underwent spontaneous fission in nature.

Niels Bohr carried the news of the Berlin experiment to the United States in 1939. In the meantime, Enrico Fermi had moved to New York. His wife was Jewish and Mussolini was adopting Hitler's anti-Semitism. The fission experiment was repeated in a laboratory at Columbia University. Meanwhile another refugee, Leo Szilard from Hungary, realized that if two or more neutrons were produced by the fission process, the result could be the chain reaction he had postulated some years earlier. President Roosevelt was warned of the possible consequences.

Fermi moved to Chicago, and an atomic pile named CP-1 was built in racquet-ball courts beneath the abandoned west stands of the University of Chicago's Stagg Field. It used unenriched uranium, 40 tons of it, mostly in the form of pressed uranium oxide spheres, but including 6 tons of uranium metal cylinders. To control the reaction very pure graphite rods, which absorb neutrons, were inserted into the pile. The first sustained chain reaction was achieved on December 2, 1942 and lasted for 28 min. The race to produce an atomic bomb was on, but what was needed was a way to enrich the $^{235}$U from its ordinary concentration of 0.71–3 % so that the chain reaction could be more effectively sustained. It is a fascinating story, but much too complicated to relate here. Soon almost all the atomic physicists in the US were preoccupied with the 'Manhattan Project,' and no progress was made in radiometric dating until after World War II.

In Germany the Nazi's distrusted many of their atomic scientists because of their Jewish connections, while at the same time many of the most competent German physicists had grave moral concerns about Hitler's policies. German progress toward atomic weapons was, fortunately, glacially slow.

After the war, Fermi's laboratory became the University of Chicago's 'Institute for Nuclear Studies' and included not just those interested in nuclear and quantum physics, but also individuals like Willard Libby, Harold Urey, and later Sam Epstein and Cesare Emiliani who were interested in Earth history. You will hear more about them later. The Institute became a center for, among other things, geochronology and the use of stable isotopes for paleoenvironmental studies.

But it was not just in Germany and the US that fission was being investigated. A subway station doesn't sound like a likely place for atomic research, but it was. In 1940 the spontaneous fission of $^{235}$U was discovered by Soviet physicists Georgy Flyorov and Konstantin Petrzhak. They were concerned that fission in surface laboratories might be caused by cosmic rays, which we will discuss in more detail in this chapter. To avoid this possibility, they made their observations of uranium in the Moscow Metro systems Dynamo station, 60 m (200 ft) underground.

Some 35 years later, in 1975, it was discovered that the fragments resulting from spontaneous fission leave tracks in the minerals that contain the $^{235}$U isotope. These tracks are visible under a microscope. They can be used to date the time when the rock cooled below a certain temperature. The annealing temperature that destroys the track depends on the mineral. Fission track dating is now an important technique in dating archeological materials and in determining the rates of erosion and denudation of landscapes.

Determining the age of the Earth has been a tricky business, since none of the original crust is preserved. A solution to the problem was found in 1956 when Claire Patterson dated a primeval piece of the solar system, the Canyon Diablo meteorite from Meteor Crater Arizona, using lead isotopes. The age, 4.55 billion years, fixes the age of the solar system. Subsequent studies have supported this date, with very minor modifications. However, recent dating of mineral grains in ancient sedimentary rocks gives an important date for the Earth itself. They show that the continents and oceans were already present 4.4 billion years ago.

## 12.9    Molecular Motions

In 1827 Scottish botanist Robert Brown (1773–1858) made the observation of apparently random motions of microscopic particles suspended in a liquid. He concluded that they were being jostled around by even smaller particles, the polyatomic objects we now know as molecules. It was a century before the implications of molecular motions became apparent and another quarter century before real understanding developed.

Molecules are 'polyatomic' objects, that is, they are composed of two or more atoms of the same or different elements. Like atoms, they can undergo electronic transitions, but these occur only with the energetic photons of the ultraviolet part of the electromagnetic spectrum. As you can guess, the number of possible electronic transitions possible for a molecule increases rapidly and soon becomes astronomical. But molecules have other more complex interactions with less energetic parts of the electromagnetic radiation.

As molecules gain energy by interacting with electromagnetic radiation, they are said to become 'excited.' The term 'excitation' includes the processes of (1) having electrons jump to a higher shell or energy level to fall back later as described above, or (2) changing the orientation and strength of the shared orbitals, thereby altering the motions between the atoms joined in the molecule causing it to vibrate or rotate, or (3) by imparting kinetic energy thereby increasing the translational motion of the molecule in space. The vibration frequencies are typically between $10^{12}$ and $10^{14}$ Hz, corresponding to the infrared range of the electromagnetic spectrum. The rotational frequencies are slower, typically of the order of $10^9$–$10^{10}$ Hz. Figure 12.18 puts all this in the context of the electromagnetic spectrum you already know.

The excited state does not last very long. A long-lived excitation lasts a few nanoseconds (1 ns = $10^{-9}$ s = 1 billionth of a second). This is the length of time during which the molecule can vibrate a million to a billion times.

When the molecule returns to its unexcited state, it emits energy. The emitted energy is always slightly lower than the excitation energy. The difference often goes into motion of the molecule in space, the motion we call heat. Mother Nature's tax. A good example is a florescent light. The electrical charge given the light tube excites atoms of mercury. The pause between the time you turn on the light and the light appears is the time it takes for the mercury to be vaporized from liquid droplets into gaseous form. The gradual increase in intensity of the illumination which usually lasts a few seconds reflects the increasing amount of mercury vapor being excited. The excited mercury vapor emits ultraviolet radiation, which you cannot see. But the ultraviolet radiation then excites 'phosphors,' chemicals coating the inside of the tube. The phosphors emit energy at a longer, less energetic wavelength, in the visible light spectrum. Depending on what phosphors are used, you can get any spectral color you want. Some of the energy involved in the transformation of ultraviolet light to visible light is lost as heat, but it is very small compared to an incandescent light bulb which loses 90 % of its energy as heat. The neon sign is even simpler. Electrodes at the ends of the tube excite the gas in the tube, and it emits colored light. The color depends on the gas.

A re-emitted photon can go in any direction. For a greenhouse gas molecule in the atmosphere, it is just as likely to be emitted downward toward the Earth's surface as upward toward outer space or sideways into the atmosphere where it could be absorbed by another greenhouse gas molecule. It is as though they are semi-transparent mirrors. Clearly, this process will slow down the loss of heat from Earth's surface and atmosphere through radiation.

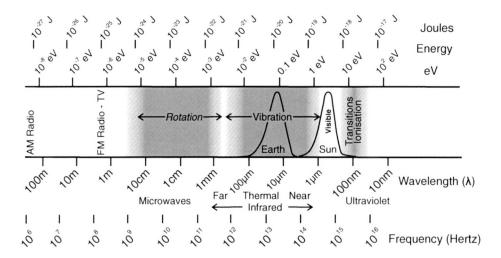

**Fig. 12.18** The parts of the electromagnetic spectrum involved in ionization, electronic transitions, and molecular vibration

Now here is where we, as inhabitants of Earth, are very lucky. The diatomic gases nitrogen and oxygen and the monatomic gas argon, which make up 99.9 % of 'dry' air, absorb hardly any of the wavelengths of either the incoming light from the Sun or the outgoing infrared radiation from Earth. Molecules that do not interact with photons are said to be transparent to the wavelength of the photons. Air is almost completely transparent to visible light. The common air molecules are non-participants in the greenhouse game. But every more complex gas molecule, those with three or more atoms, absorbs energy in either the ultraviolet ($O_3$ = ozone), or the infrared ($H_2O$ = water vapor; $CO_2$ = carbon dioxide, $CH_4$ = methane, $NO_2$ = nitric oxide, and a host of other less abundant gases).

Molecules become 'excited' by absorbing photons. Depending on the energy of the photon, the excitation can result in any of three modes of molecular motion: vibration for energy levels in the infrared, rotation for energy levels in the microwave range, and translation using whatever is left over. Vibration is a to and fro motion of parts of a molecule. Rotation involves rotation about the molecule's center. Both of these forms of molecular motion are quantized. There is not a continuum of vibrational or rotational motion; they occur in quantum steps. Translation involves movement of the position of the molecule in space. This is kinetic energy. It is not quantized. There is a continuum of translational motion. Translational motion is what we perceive as heat. The more rapidly the molecules are moving about, the 'hotter' we consider something to be.

The excited states of vibration and rotation last only for brief moments before the molecule emits a photon and returns to a lower energy state; but the kinetic motion, Mother Nature's tax, can last longer and be transferred to other molecules through collisions.

The specific ways in which electromagnetic radiation, from the Sun and other sources, interacts with the gases in our atmosphere are topics discussed in Chaps. 15 and 16.

## 12.10 Summary

In 1900 Max Planck found that the solution to the description of black body radiation required the assumption that energy is not a continuous spectrum but comes in small pieces, or quanta. In 1905 Albert Einstein realized that quanta of ultraviolet light impinging on a metal surface could free electrons from the metal, explaining the 'photoelectric effect.' In 1913 Niels Bohr explained that the spectral lines of the hydrogen atom were caused by the orbiting electron jumping from one quantum energy level to another.

Between 1923 and 1926 the world changed. It was realized that the world of the atom is one of wave-particle duality. Electrons do not occupy a specific point in space, but only have a probability of being in one place or another. The electron shells originally proposed by Niels Bohr were found to have subshells, and electrons were restricted to orbitals, but only two electrons can occupy the same orbital, and then only if they 'spin' in opposite directions.

During the middle of the last century the strange world of very small things was becoming better known, and a physical/mathematical basis for understanding it was being developed. Observations of spectral lines played a critical role in figuring all this out. But as spectroscopy developed, it became evident that not all of the spectral lines were associated with electronic transitions. There were a host of other lines in the lower wavelength parts of the spectrum waiting to be understood. These had to await the development of better technology that would allow them to be explored in detail.

Perhaps the best summary of the transition from classical to quantum physics is to quote from Wolfgang Pauli's Nobel Prize lecture:

> I shall only recall that the statements of quantum mechanics are dealing only with possibilities, not with actualities. They have the form 'This is not possible' or 'Either this or that is possible' but they can never say 'That will happen then and there.'

The world we experience on a human scale is a world of determinacy. The world of atoms is a world of probabilities.

A Timeline for this chapter:

| | |
|---|---|
| 1808 | John Dalton proposes the 'billiard ball' model of the atom |
| 1827 | Robert Brown observes the random motion of microscopic particles |
| 1865 | Charles Lutwidge Dodson, alias Lewis Carroll, publishes *Alice in Wonderland* |
| 1868 | Lothar Meyer submits his study of the periodic properties of elements for publication |
| 1869 | Dimitri Mendeleyev publishes his *Periodic Table of the Elements* |
| 1870 | Lothar Meyer's paper *Die Natur der chemischen Elemente als Funktion ihrer Atomgewichte* is published |
| 1879 | Max Planck obtains his doctorate from the University of Munich. His thesis is on the Second Law of Thermodynamics |
| 1885 | Johann Jakob Balmer publishes the mathematical relation between spectral lines of hydrogen in the visible light range |
| 1887 | Heinrich Hertz discovers that ultraviolet light enhances the intensity of a spark between two electrodes—the photoelectric effect |
| 1887 | The Michelson-Morley experiment shows that light travels at the same speed independent of whether you are moving toward or away from its source |

(continued)

(continued)

| 1895 | Hendrik Lorenz proposes that moving bodies contract in the direction of motion |
| 1895 | Wilhelm Röntgen discovers X-rays |
| 1896 | Pieter Zeeman discovers that a strong magnetic field can split spectral lines, the Zeeman effect; Hendrik Lorenz realizes the importance of this discovery |
| 1896 | Wilhelm Wien publishes his Law of Distribution of Black Body radiation—it does not work for longer wavelengths |
| 1896 | Svante Arrhenius publishes *On the Influence of Carbonic Acid in the Air upon the Temperature of the Ground* |
| 1896 | Henri Becquerel and the Curies discover radioactivity |
| 1897 | Sir Joseph John Thomson discovers that 'cathode rays' are 'corpuscles' (electrons) |
| 1898 | James Dewar liquefies hydrogen |
| 1898 | The Curies discover two new elements, radium and polonium |
| 1898 | Thomas Chrowder Chamberlin publishes 'The Influence of Great Epochs of Limestone Formation upon the Constitution of the Atmosphere' arguing that carbon dioxide steers climate |
| 1899 | Hendrik Lorenz proposes that time dilates in objects moving near the speed of light |
| 1900 | Lord Rayleigh publishes a different Law of Distribution of Black Body radiation—it does not work for short wavelengths (the 'ultraviolet catastrophe') |
| 1900 | Lord Kelvin presents his famous lecture *'Nineteenth-Century Clouds over the Dynamical Theory of Heat and Light'* |
| 1900 | Max Planck in analyzing Black Body radiation concludes that energy comes in small discrete packets—*Lichtquanta* = Light quanta |
| 1900 | Knut Ångström finds that $CO_2$ and water vapor absorb energy in the same spectral regions and concludes that $CO_2$ is not an important factor in climate |
| 1902 | Pieter Zeeman and Hendrik Lorenz receive the second Nobel Prize to be awarded in physics for the discovery of the Zeeman effect |
| 1905 | Albert Einstein receives his doctorate from the ETH in Zürich |
| 1905 | Einstein publishes his explanation of the photoelectric effect citing Planck's quanta |
| 1905 | Henri Poincaré groups Lorenz's equations as the Lorenz Transformations |
| 1905 | Einstein publishes his theory of relativity |
| 1905 | William Jackson Humphreys demonstrates that a column of $CO_2$ 50 cm long completely absorbs any infrared radiation entering it and concludes that changing amounts of atmospheric $CO_2$ have no effect on climate |
| 1906 | Theodore Lyman discovers a new line in the spectrum of hydrogen, then finds an entire new series in the ultraviolet |
| 1906 | Sir Joseph John Thomson is awarded the Nobel Prize in Physics for his investigations on the conduction of electricity by gases |

(continued)

(continued)

| 1908 | Heike Kamerlingh-Onnes liquefies helium |
| 1908 | Friedrich Paschen discovers a new set of lines in the spectrum of hydrogen |
| 1909 | The Geiger-Marsden gold foil experiment demonstrates that the nucleus of an atom is exceedingly small |
| 1911 | Wilhelm Wien is awarded the Nobel Prize in Physics for his discoveries regarding the laws governing the radiation of heat |
| 1912 | Niels Bohr wins a postdoctoral fellowship and goes to Cambridge to work with J.J. Thompson—they do not get along together |
| 1912 | Niels Bohr moves to Manchester to work with Ernest Rutherford |
| 1913 | Heike Kamerlingh-Onnes is awarded the Nobel Prize in Physics for his investigations on the properties of matter at low temperatures, including the liquefaction of helium |
| 1913 | Bohr presents Rutherford with his new model of the atom —it is published rapidly |
| 1918 | Max Planck is awarded the Nobel Prize in Physics for his discovery of energy quanta |
| 1919 | Ernest Rutherford disintegrates nitrogen into hydrogen and oxygen by alpha particle bombardment proving that the hydrogen nucleus is present in other atoms—this is considered the discovery of the proton |
| 1921 | Albert Einstein receives the Nobel Prize in Physics for his work on the photoelectric effect |
| 1922 | Frederick Sumner Brackett discovers another set of lines in the spectrum of hydrogen |
| 1922 | Niels Bohr receives the Nobel Prize in Physics for his investigation of the structure of atoms and of the radiation emanating from them |
| 1923 | Niels Bohr publishes *Über den Bau der Atome* (*On the structure of atoms*) presenting a new version of the periodic table |
| 1923 | T.C. Chamberlin concludes that water vapor, not carbon dioxide, is the gas controlling Earth's climate |
| 1924 | August Herman Pfund discovers another set of lines in the spectrum of hydrogen |
| 1924 | Wolfgang Pauli publishes his *exclusion principle* |
| 1924 | Samuel Goudsmit and George Uhlenbeck propose that electrons have spin |
| 1924 | Louis de Broglie proposes that everything has both wave and particle attributes |
| 1926 | Erwin Schrödinger publishes his paper *Quantisierung als Eigenwertproblem*, one of the basic foundations of Quantum Mechanics |
| 1927 | Werner Heisenberg publishes his *Uncertainty principle*, another of the basic foundations of Quantum Mechanics |
| 1929 | Louis de Broglie is awarded the Nobel Prize in Physics for his discovery of the wave nature of electrons |
| 1932 | James Chadwick discovers the neutron |
| 1932 | Robert Mulliken proposes the term 'atomic orbital' |
| 1932 | Werner Heisenberg is awarded the Nobel Prize in Physics |

(continued)

(continued)

| | |
|---|---|
| 1933 | The Nobel Prize in Physics is awarded jointly to Erwin Schrödinger and Paul Dirac for the discovery of 'new productive forms of atomic theory' |
| 1935 | James Chadwick is awarded the Nobel Prize in Physics for his discovery of the neutron |
| 1945 | The Nobel Prize in Physics is awarded to Wolfgang Pauli for his formulation of the Exclusion Principle |
| 1953 | Curtis J Humphreys discovers another set of lines in the spectrum of hydrogen |
| 2002 | In the TV show *The Simpsons*, Homer's favorite program is replaced by *The Boring World of Niels Bohr* |

If you want to know more:

Pais, A., 1986. *Inward Bound—Of Matter and Forces in the Physical World*. Oxford University Press, New York, 666 pp.—Again—A fascinating account of the development of modern physics.

Gamow, G., 1966 *Thirty Years that Shook Physics—The Story of Quantum Theory*. Doubleday, 224 pp.

Music: Something resembling quantum music might be György Ligeti's *Atmosphères* (1961). It was used in Stanley Kubrick's film *A Space Odyssey—2001*.

Libation: Perhaps it is time to discuss Gin. Originally developed by the Dutch as Genever, it is a grain alcohol flavored with juniper berries and other herbs. It was originally used as a pharmaceutical cure for almost all ailments. The British adopted it, modified it, and London became a source for the varieties now known as London Gin or Dry Gin (Fig. 12.19). One of my favorites is Beefeater's, named after the guards of the Tower of London. I often use it as the major component of a Martini. Other favorites are Tanqueray, Bombay, and Bombay Sapphire. For a Gin and Tonic it is best to use a less expensive gin, since the flavor of the tonic water largely masks that of the gin.

**Fig. 12.19** William Hogarth's 1751 sketch: 'Gin Lane' in London

**Intermezzo XII. Illinois and Miami**

Before I left Miami for my summer trip to Jerusalem and Prague we had a farewell lunch with a few friends in Miami. One was Ruth Trencher. Her position in the Institute never seemed to be really defined, but she worked on helping the scientists prepare proposals for funding. She also acted as a sympathetic ear, advisor, and offered encouragement when a proposal was turned down. As we walked outside Ruth turned to me: "Why don't you move down here?"

I had just received notice from the University of Illinois that I had been promoted to full Professor, and the thought of a move had never occurred to me. Ruth said, "I know the faculty would like to have you here; why don't we go speak to Fritz Koczy." Fritz was head of Physical Sciences in the Institute. Ruth led me straight to his office. She led the conversation, in her inimitable way: "Dr. Koczy, I've been talking to Professor Hay, and I think he would consider a move to our Institute." Fritz didn't pause. "Wonderful, we would be so pleased to have you here. I know you are headed off to Israel and Europe. I'll be in touch when you get back."

If you are a young faculty member reading this today, it will sound very, very strange, but that is the way things were done in the 1960s. No complicated rigmarole, no submitting a long complicated vita, no search committees. The United States was trying to build up its competence in the sciences, and there was a shortage of expertise in oceanography as well as many other fields.

It all happened so fast I didn't know what to think. However, an outside offer was always a good thing to have in your pocket. There was, of course, one small detail. At the Institute of Marine Sciences in Miami you had to find your own salary, generally from a federal agency. The University supplied only a minimal amount in return for teaching a course or two for graduate students.

When I got back to Urbana from Europe, a letter from Fritz Koczy was waiting for me. I was offered a Professorship in Marine Geology and Geophysics at a competitive (if somewhat imaginary) salary.

I took the letter to George White and explained that I didn't really want to move. I had my laboratory at Illinois very well set up and equipped with an electron microscope, and I had graduate students. I also explained that the salary was a bit of fiction. But I also suggested that the connection with Miami could be very beneficial to me and to my graduate students, offering new possibilities for research. To find out whether the Miami offer was realistic, I took a trip to Washington to talk with my program managers at the National Science Foundation. I learned that if I kept on writing good research proposals salary finding would not be a problem.

Within a few months we had an arrangement I would never have imagined. I found myself holding two full-time professorships, at the University of Illinois in Urbana, and at the University of Miami in Florida. To be at one institution I had to be on leave of absence from the other.

Now a few words about oceanography and the University of Illinois. Urbana Illinois sounds like the last place one might find scientists interested in the oceans. However, this is where the US's most famous marine geologist started his career. Working in the Geology Department since 1922, Francis Shepard served as Professor from 1939 to 1946, when he moved to the Scripps Institution of Oceanography in La Jolla, California. His family was from Massachusetts, and they were moderately wealthy. He used the family yacht to take samples off the New England coast. He also studied Lake Michigan, making one of the first bathymetric maps of the lower end of the lake. It was done with a lead line and a rowboat. One of my luncheon table partners at the Illini Union was Seward Stehli who had been Head of Physical Education before his retirement. Seward had rowed the boat while Francis took his measurements and a student, Jack Hough, retrieved samples of the bottom sediments. But earlier Francis had had an incredible stroke of luck. During the depression, I think it was in 1933, two youngsters rode the rails of a freight train into Champaign, Illinois. They wandered over to the University in Urbana to see if there was any work available. They ran into Fran Shepard, and he offered them jobs. Shepard had a little money of his own, and paid for his own research. That was the way it was done in those days. Those two boys were Robert S. Dietz and Kenneth O. Emery. Both got their degrees

under Shepard and both became eminent marine geologists.

Dietz's proposal for his Ph.D. dissertation was to work on the geology of the Moon. The idea was rejected by the geology faculty, because of course there would never be any way to check his results. Dietz went on to work with the Navy, Scripps, NOAA, and in 1977 became Professor of Planetary Sciences at Arizona State University. He was one of the first proponents of the idea of sea-floor spreading. When I knew him he had developed a particular aversion to Creationism shown in his 1987 book with artist John Holden: *Creation/Evolution Satiricon: Creationism Bashed—Did the devil Make Darwin Do It?* A great read if you can find a copy.

Ken Emery went on to become a much less flamboyant leader of the marine geology program at the Woods Hole Oceanographic Institution in Massachusetts.

When I arrived in Urbana, Jack Hough was the resident 'oceanographer,' and we became good friends. Jack had worked on developing sonar during World War II, using the facilities at Navy Pier in Chicago. He had become the expert on the history of the Great Lakes. When he moved to the University of Michigan's Great Lakes Research Institute in 1966, I inherited the role of resident oceanographer. My inexperience was embarrassing, but the connection with Miami during my sabbatical the next year gave me some credibility. All that was soon to change.

After the shock of the Russian invasion of Czechoslovakia I spent some time making inquiries about how I might be able to help my good friend Pavel Čepek. I knew he would need to get out of the country before the borders were closed down completely. I used my contacts to inquire whether the CIA might be interested in helping him leave the country. I got a quick rejection. But then I got word that he had been invited to accept a position at the German Geological Survey (Bundesanstalt für Geowissenschaften und Rohstoffe) in Hannover. Later, when the collapse of the Soviet Union came as a great surprise to the CIA, I wondered if they would not have been well served by having inside information on how the Dubček era had come to pass in Czechoslovakia, and how change in the 'communist empire' might occur. All the signs were there, just not recognized by our CIA.

In the fall of 1968, I suddenly became an important figure in oceanography. A couple of years earlier, a colleague at Princeton University, Al Fischer, had sent one of his graduate students, Dave Bukry, to work with me in Illinois. Dave came to learn about the calcareous nannoplankton, those tiny marine fossils that had become my specialty, and to use our electron microscope facilities. I should mention that the petroleum industry had already discovered their value in exploration, and several of my doctoral students had already gone into industry. I was almost alone in producing specialists in this new field.

The Deep Sea Drilling Project (DSDP) got underway in September 1968, and Dave was selected to be a micropaleontologist on the first Leg of the program. Much more about the DSDP in the next Intermezzo, but let me note here that Leg 1 was a learning experience for all concerned. The recovery of cores of deep-sea sediments by drilling proved more difficult than anticipated, and at first all that was recovered were some smears of the sediment on the core barrel. Not enough to do anything with, or so the Chief Scientists Maurice Ewing and Joe Worzel of Lamont Geological Observatory, thought. The smears were so small that there were none of the millimeter-sized fossils of planktonic foraminifera everyone expected to be the means of dating the sediments. Dave would take a bit of the smear, make a microscope slide, and after a few minutes announce the age of the sediment. It was regarded as some sort of a miracle, and since I had helped him learn the business, I was now regarded an expert on dating deep-sea sediments. As soon as the joint appointment agreement was signed in 1968, the Institute of Marine Sciences (IMS) appointed me as their representative on the Planning Committee for the DSDP. I started off my first semester as Professor in Miami by sailing on Leg 4 of the DSDP.

The next 5 years were a wonderful time. I spent the fall semester at the U of I in Urbana, teaching among other things, the oceanography course for undergraduates, which always had about 50 students. A few from each class went on to careers in ocean science. One of those students, John Southam, was just getting his Ph.D. in Physics. He became very interested in ocean problems, and after his degree was awarded he accepted an appointment in the Division of Geology and Geophysics in our Institute in Miami. We worked closely together for many years, with John solving many mathematical problems that were well beyond my meager expertise.

After a snowy Christmas with friends in Illinois, I would head south to spend the spring semester in Miami. Meetings of the whole Institute faculty were very different from those in Illinois. Walton Smith, the founder and Director of the IMS, would reserve a room at the *English Pub* on Key Biscayne, and our discussions were lubricated with beer and ale. It was marvelously informal and most of the discussions were about ideas for new projects.

I would make regular trips back to Urbana to work with my graduate students there. Even more exciting, I would lead a field trip to Bimini in the Bahamas for Illinois undergraduates over their spring vacation. That is, I would be in Miami, and the students would organize themselves into cars and make the trip south a nonstop overnight experience. Their second night out I would have eighteen students sleeping on the floor in my Miami apartment before leaving for Bimini the next day, using the Lerner Laboratory's vessel.

The Gulf Stream crossings were usually rough, and that would make for the first experience with sea-sickness for the prairie dwellers. At Bimini we filled the Lerner Marine Laboratory's dormitory facility, and made excursions to go exploring the marine life, wading on the shallow flats, snorkeling, and then going further afield using the Lab skiffs. We would find a few students who had experience with outboard motor boats on lakes, and turn them loose on the open sea.

A couple of memorable incidents from those field trips are worth recounting. Once on a trip around the island we were on the Gulf Stream side when a huge manta ray appeared. It was larger than our 15 foot skiffs, and probably weighed a ton. I stopped my boat and yelled to the others—"Stop and go into the water with the ray—it's a once in a lifetime experience." Mantas are plankton feeders, and the only danger is having them inadvertently bump into you. It was the highlight of the trip for the students that year.

Another interesting experience was when one of our skiffs learned about standing waves the hard way. The harbor entrance at Bimini is narrow, and the outgoing tide can reach a speed of several knots, the same speed as waves from off the Gulf Stream trying to enter the harbor. The result can be a standing wave up to 6 ft high, staying in one place like a wall. To get over it a skiff needs to be traveling at a fairly good speed. If too slow, the bow of the skiff simply pokes into the wave, and the boat fills up with water in a few seconds. One of our skiffs was going too slow, and it filled and capsized. The students all swam the short distance to shore, but their gear was on the bottom. We recovered the skiff, and during low tide the students were able to recover most of their things. It was a great lesson in wave dynamics.

In the late 1960s two events happened which eventually changed my career path. In 1967 the University of Illinois got a new Head for Geological Sciences, Fred Donath, and in 1969 the Institute of

Marine Sciences received a major endowment gift from Lewis Rosenstiel, and was renamed the Rosenstiel School of Marine and Atmospheric Sciences.

Then another odd thing happened. I had first met Tom Barnard, Professor of Micropaleontology at University College in London while a graduate student at Stanford. Tom wanted to add work on the calcareous nannofossils to the opportunities available for his students at University College. What was just a brief visit in 1969 turned into a long term relationship whereby I would spend a month each spring in London. We would work with Tony Rood, one of Tom's assistants and an expert electron microscopist, all day. Then I would take advantage of the city and attend plays and concerts in the evenings. The College and Geological Society libraries were a great resource for my research.

My joint arrangement with the two Universities continued until Walton Smith retired and the Rosenstiel School got a new Dean, Warren Wooster, a physical oceanographer from Scripps. Warren decided it was time to replace Cesare Emiliani as Chair of what was now the Division of Marine Geology and Geophysics, and wanted me to take over. Cesare had always been our leader by consensus, but he was now spending much of his time on the University's main campus working on introductory science courses for undergraduates. Cesare had become very concerned that the general public and especially politicians in the US had so little science background. They did not understand the implications of unlimited population growth and its impact on resources. Nor did they understand the threat posed by changing the composition of our planet's atmosphere by adding $CO_2$ from burning fossil fuels. He wrote several excellent books touching on these topics, one aimed at the general public, *The Scientific Companion: Exploring the Physical World with Facts, Figures, and Formulas*, and another aimed at college undergraduates, *Planet Earth: Cosmology, Geology, and the Evolution of Life and Environment*. Both are great reads.

Being Division Chair in Miami meant that I had to resign my appointment at Illinois. It was a sad time for me.

MILANKOVICH ORBITAL CHANGES
MAKE ALL THE DIFFERENCE

# Energy From the Sun—Long-term Variations

*In my scientific vocation I have found a pleasant shelter, for I was protected by it from many turbulences that shook the world.*
Milutin Milankovitch

The Earth's surface and atmosphere receive energy from four sources: the Sun, the Moon, Stars, and Earth's interior. Insolation is the energy received from the Sun, and currently averages about 340 W/m$^2$. Energy received from the Moon is about 0.002 W/m$^2$. Energy from the Stars is negligible ($1 \times 10^{-6}$ W/m$^2$), but there has been speculation that the explosion of a supernova in our near vicinity would make a very significant, possibly even catastrophic, contribution to the radiation balance. There is no geologic evidence that this has happened in the past half billion years. Energy from the Earth's interior averages less than 0.1 W/m$^2$.

The energy that Earth receives from the Sun is not constant over time. Our Sun is a normal everyday 'main-sequence' star. These stars obtain their energy by fusion of hydrogen nuclei (protons) and electrons to make helium nuclei. Their luminosity increases as they age until they run out of hydrogen fuel and start fusing helium into heavier nuclei until they run out of things to fuse, expand as a red giant, like Betelgeuse, and finally collapse into a dying dwarf star. In this process, the energy emitted by the star evolves over time, starting out relatively dim and then becoming brighter with time until it finally dies out. For main-sequence stars this process takes many billions of years. For a star of the mass of our Sun this whole life process takes about 10 billion years. Our Sun is about 4.6 billion years old, not quite half way through its lifetime.

## 13.1 The Faint Young Sun Paradox

The total energy received each year by Earth will change if the luminosity of the Sun changes. If the Sun emits more energy, the Earth must warm. From investigations of the evolution of stars we know that those like our Sun become brighter with time. Our Sun is now about 30 % more luminous than it was 4.6 billion years ago just after it had formed. Incidentally, that works out to a change of 0.000000007 % per year or an increase of the solar constant of 0.0000003 W/year.

When climatologists like Mikhail Budyko began to think about the long-term perspective of global climate change in the 1970s, they stumbled on the 'faint young Sun' paradox. If the Sun radiated only 70 % of the energy it does today, and Earth had the same albedo (reflectivity), the planetary temperature would have been about −40 °C (=−40 °F). If there were the same greenhouse gas content in the atmosphere then as today, it would have raised the average surface temperature to −7 °C (=19.4 °F), still below the freezing point of water, even salty ocean water. The Earth would have begun to freeze over, and the albedo would increase to perhaps 0.9, pushing the planetary temperature down to −130 °C (=−202 °F) and leaving it hopelessly frozen forever. Obviously, this catastrophe did not happen. Geologic evidence is that there was liquid water on the surface of the planet since shortly after its formation.

The Earth formed between 4.6 and 4.5 billion years ago. The oldest known 'rocks' are actually mineral grains that date to 4.4 billion years, and they indicate that liquid water, albeit very hot, was already present. The oldest sedimentary rocks yet discovered are at a place called Isua in western Greenland. They are more than 3.8 billion years old, and were laid down under water. They also contain evidence that life had originated by that time. The presence of liquid water in spite of the weak insolation from the faint early Sun means there must have been a greenhouse effect much greater than that of today. Until the 1990s we thought our planet had always maintained a moderate temperature throughout its history. This would require regulation of the temperature as the Sun became brighter over time by reducing the greenhouse effect by just the right amount. It was concluded that the Earth had a thermostat. More about that when we come to discuss greenhouse gases.

In the 1990s geologists discovered that our planet did not have such a stable climatic record. In fact, it appears that it did become ice covered from time to time, a condition that has come to be known as the 'snowball Earth.' There were times when the thermostat got out of whack. How this could happen and how the Earth could recover is a fascinating story, but I am going to postpone the story until we come to discuss greenhouse gases.

## 13.2    The Energy Flux from the Sun

Our Sun is a rather ordinary star in the 'main sequence.' These stars obtain their energy by fusing simple hydrogen nuclei (1 proton) to make helium nuclei (2 protons +2 neutrons) through a process called the proton–proton chain during which mass is converted to energy. If you are following everything in this book closely, you will remember that neutrons are slightly heavier than protons, so how can changing some protons to heavier neutrons release energy? That is because the process also uses electrons stripped from those hydrogen nuclei and also produces the antimatter positrons Paul Dirac had predicted. When positrons meet electrons both are annihilated. Both of their masses are converted to energy, so there is a mass loss in the process as a whole.

My father was on the Board of Directors of Texas Utilities, and I remember one evening in the early 1950s when he came home very excited. Over dinner he explained that he had heard a presentation about the energy source of the future—nuclear fusion. The same process that produces the Sun's energy would be available here, probably within a decade. I expressed my concern about having a miniature sun located in west Texas; wouldn't the fusion reactor generate a lot of waste heat? It is now a half century later, and the goal seems further away now than it did then. Part of the problem lies in creating on Earth the conditions that exist in the core of the Sun where the fusion reaction takes place: compressing the hydrogen to a density about 150 times that of water, denser than Earth's core, and heating it up to 15,000,000 K. We know how to make the fusion reaction operate for milliseconds in a hydrogen bomb, but getting it to go on longer than that is a lot more difficult. Don't hold your breath for this technology to come online anytime soon.

When you look toward the Sun, what you are seeing is the 'photosphere,' a glowing, apparently outer surface of the gas ball that is the Sun. That surface has a temperature of about 5778 °K (=5505 °C, =9877 °F). Following Kirchhoff's Law, that temperature is what determines the spectrum of radiation we see from Earth. But during a total eclipse you can see that the glowing part of the Sun is surrounded by a thin transparent 'atmosphere,' the corona. We now know that temperatures in the rarefied gas of the corona reach several million Kevin.

The flow or 'flux' of energy Earth receives from the Sun was designated 'Q$_S$' in Fig. 11.2 (the subscript S stands for 'short

wavelengths'). It is measured in watts (see Chap. 2, Sect. 2.8 if you have forgotten what a watt is). A flat solar panel 1 m$^2$ (roughly 3 1/3 ft on a side, =an area of 10.76 ft$^2$) directly facing the Sun at the average distance of the Earth from the Sun (149,597,870.691 km =92,955,807.23 statute miles, or if you are an astronomer, 1 AU or Astronomical Unit), receives about 1,360 W. This number is called the 'solar constant' and it was determined before satellite measurements were available. The energy flux actually measured by a panel on a satellite is 1,366 W/m$^2$. Most everyone still uses the good old 1360 W/m$^2$, and we will use it in our discussion below.

You can already guess that at the top of the atmosphere over the Equator on noon of the day of the equinox, when the Sun is directly overhead, Q$_s$ would be 1,360 (actually 1,366) watts. But by 6:00 P.M., as the Sun sets, the amount of energy received from the Sun goes to zero watts. And it stays at zero until Sunrise at 6 A.M. At the equinox the Sun's rays just graze the surface at the poles all day long. So how can we figure out how much energy is received by planet Earth?

Instead of our one m$^2$ solar panel, imagine a disc the size of the Earth. The area of the disc is $\pi r^2$, where r is the radius of the Earth (6.371 × 10$^6$ m = 3,959 miles). The area is 127.5 × 10$^{12}$ m$^2$ (=49,240,321 miles$^2$). The total energy received at the top of the atmosphere is then 173.4 × 10$^{15}$ W, or 173.4 PW. But the Earth is not a disc, it is a rotating sphere. The area of a sphere is 4 $\pi r^2$, so Earth's surface area works out to be about 510 × 10$^{12}$ m$^2$ (=510 × 10$^6$ km$^2$ =197 × 10$^6$ miles$^2$). Since the energy that would be received by a disc is spread over a surface four times as large, the average insolation is 340 W/m$^2$.

Armed with this piece of information we can use basic physics to calculate what the temperature of the Earth should be. In this instance, the 'energy balance' equation is

$$Q_S = Q_E$$

where Q$_S$ is the energy input from the Sun and Q$_E$ is the energy output from the Earth. The Earth must radiate the same amount of energy as it receives back into space, otherwise it would heat up or cool off. So we know that Q$_E$ is 340 W/m$^2$ (=340 Wm$^{-2}$ as a physicist would write it). You will recall that in 1879 Jožef Stefan found a simple equation that tells us how warm an object must be to radiate that amount of energy. The equation is

$$R = \sigma K^4$$

where R is the energy radiated, σ (the small Greek letter sigma) is the 'Stefan–Boltzmann constant' named for our old friends from Chap. 10, Jožef Stefan and Ludwig Boltzmann, who sorted all this out, and K is the temperature in Kelvin, the absolute temperature scale where 0 °C is 273.15 K (=32 °F). The Stefan-Boltzmann constant has the value 5.670400 × 10$^{-8}$ Wm$^{-2}$ K$^{-4}$. If you plug in 340 Wm$^{-2}$ for R, the answer comes out to be 278.3 K or 5.3 °C (=41.5 °F), which sounds sort of reasonable. But wait, that equation is for a 'black body.' A black body is a theoretical object that is a perfect absorber of incoming energy and a perfect radiator of outgoing energy. It would look black to us because it absorbs every wavelength of incoming energy. Our 'blue marble' Earth is not a black body. Its albedo (reflectivity) is 0.3, meaning that it reflects 30 % of the incoming energy; more about that in Chap. 15. So this means that the amount of energy from the Sun actually absorbed by Earth is only an average of 238 W/m$^2$. If we substitute this number for R, the answer comes out to be 254.5 K or −18.5 °C (=1.3 °F), which implies that the Earth should be frozen solid. In fact, 254.5 K is what you get if you take the Earth's temperature from deep space. However, most of us are concerned with the Earth's surface temperature, which happens to be near a more comfortable 288 K or 15 °C (=59 °F). The difference between the planetary temperature of 254.5 K and the planet's surface temperature, 288 K, is caused by some minor gases in the atmosphere, known as 'greenhouse gases.' They are the topic of Chaps. 20 through 22. If it were not for greenhouse gases, Earth would have been frozen over for the past 4 billion or so years.

These calculations make the assumption that the Earth–Sun distance does not change, when in fact it does change during the course of the year. This is because the Earth's orbit is elliptical, not circular. The changes in insolation received when the Earth is closest to and most distant from the Sun can be significant, as you can surmise from Fig. 13.1. The insolation varies as the square of our distance from the Sun. The energy emitted by the Sun changes too, during sunspot cycles. Now that we have spacecraft to measure such things we know it varies by about 1.3 W/m$^2$ between sunspot maxima and minima. More about that later.

It is the annual variations in insolation resulting from the tilt of the Earth's axis relative to the plane of the Earth's orbit around the Sun, not the changing Earth–Sun distance, which produces the seasons: summer, fall, winter, and spring. Figure 13.2 shows this schematically. It also shows two important features, the Tropics and the Polar Circles. The tilt of the Earth's axis varies with time, as we will discuss later, but today it is 23° 26′ 21.44″. This is usually rounded to be 23.5°. The Tropic of Cancer is 23.5° N of the Equator. It lies between Key West and Havana, Cuba. At noon on the day of the summer solstice, June 21, the Sun is directly overhead of the Tropic of Cancer. Its southern hemisphere counterpart is the Tropic of Capricorn, 23.5° S, which runs through the northern suburbs of São Paolo, Brazil. The Sun is directly overhead at noon on the winter solstice, December 21. These are facts important to know when planning your vacation.

More important for our discussion of the climate are the Polar Circles. The Arctic Circle is 23.5° from the pole, at latitude 66.5° N or 66.56° N to be more exact. People who

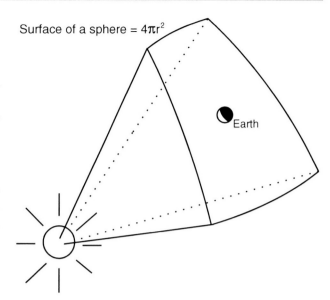

**Fig. 13.1** Energy emitted by the Sun spreads out as it extends further into space. The amount of energy received by planet Earth varies as the square of the Earth–Sun distance

live there tend to be quite exact about the position of the Circle because of the day-night cycle. The Sun does not rise on the Arctic Circle on the Winter Solstice, December 21, and it is above the horizon all day on the Summer Solstice, June 21. North of the Arctic Circle there are more days of 24 h sunshine in summer and a long period of darkness in winter. The Arctic Circle passes just north of the main island of Iceland, through southern Greenland, along the northern margin of the eastern Canadian mainland, through the Northwest Territories and through Alaska north of the Seward Peninsula. In Europe it is south of the town of Bodø, Norway, and well south of the university town of Tromsø, Norway. I had a postdoctoral student from Tromsø who informed me that the Sun sets there on November 21, and rises again on January 21. He told me that the 2 months of darkness become one long period of partying, reaching its climax when the Sun comes up again.

At a given distance the energy is everywhere the same, except during brief solar flares. The simplest calculation is to consider how the energy spreads out in space as a spherical surface. R is the distance from the Sun. However, because Earth's orbit is elliptical, the Earth–Sun distance varies during the course of the year, from 147.1 × 10$^6$ km (88.0 × 10$^6$ miles) to 152.1 × 10$^6$ km (94.5 × 10$^6$ miles). This causes the 'solar constant' to vary from 1382.7 to 1337.3 W/m$^2$ during the course of the year. In contrast, the average solar constant for Venus (108.2 × 10$^6$ km = 67.2 × 10$^6$ miles from the Sun) is 2599 W/m$^2$ and that for Mars (277.9 × 10$^6$ km = 172.7 × 10$^6$ miles from the Sun) is 586 Wm$^2$.

In terms of the energy received from the Sun, the Northern Hemisphere maximum is on the summer solstice,

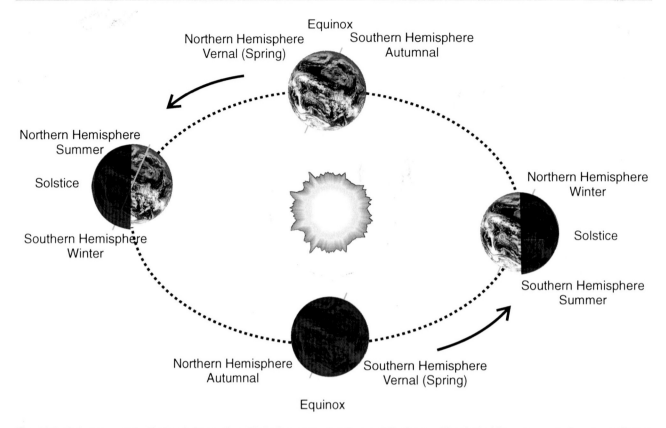

**Fig. 13.2** Orientation of the Earth relative to the orbital plane at the Solstices and Equinoxes. The *dashed lines* represent, from *top* to *bottom* Arctic Circle, Tropic of Cancer, Equator. Tropic of Capricorn, Antarctic Circle

June 21, and the minimum is on the winter solstice, December 21. However, these are not the warmest and coldest days of the year. It takes time for the Earth to heat up and cool down. The warmest and coldest days occur about 2 months later. When I was young, summer was when it was warm, and winter was when it was cold, and fall and spring were in between. However, we now have a rigid definition for the seasons. We now say that summer begins on June 21 and lasts until the Fall equinox when the Sun passes over the Equator headed south, September 23. Fall begins then and lasts until the winter solstice, winter is from then until the Spring equinox when the Sun crosses the Equator headed north, and then it is spring until the summer solstice.

Figure 13.3 shows the insolation at the top of the atmosphere received by different parts of the Earth during different times of the year. The diagram may look a bit odd at first. Latitudes are shown along the bottom axis: south is on the left, north on the right. You will note that the latitude marks are unevenly spaced. This is because the diagram is designed to show the relative areas of the Earth's surface in different latitudes. Half of the Earth's surface is between 30° N and 30° S. The other half is between 30° N and S and the poles. Along the top of the diagram is the sine of the latitude, a function you will remember if you took trigonometry or read Chap. 2 carefully. You will notice that these numbers are evenly spaced. The sine of the latitude is proportional to the area of the Earth's surface at that latitude.

The horizontal dashed line is the average insolation received by the Earth, the 340 W/m$^2$ we worked out above. Of course, in reality the Polar Regions do not receive as much energy as the equatorial regions, so there is a curved solid line, labeled 'mean annual insolation' that shows how the total energy received over the course of a year is distributed.

Now we come to the interesting part. There are two solid lines, one labeled NDJ (for November, December, January) and the other labeled MJJ (for May, June, July). These show the average energy received by Earth during those 3 month periods around the solstices. Look closely. You would think that these lines should be symmetrical, but you will notice that they do not cross at the Equator as you would expect. Further, you will also notice that the NDJ line is higher than the MJJ line. This is because the Earth's orbit around the Sun is not circular, but elliptical, and not centered on the Sun. At present the Earth is closer to the Sun in Northern Hemisphere winter than in Northern Hemisphere summer.

**Fig. 13.3** Insolation at the top of the atmosphere (W/m²) at different latitudes averaged over a year and averaged over the months NDJ (November, December, January) and MJJ (May, June, July). The global average insolation (340 W/m²) is shown as a *dashed line*. *AAC* Antarctic Circle, *TCAP* Tropic of Capricorn, *EQ* Equator, *TCAN* Tropic of Cancer, *AC* Arctic Circle

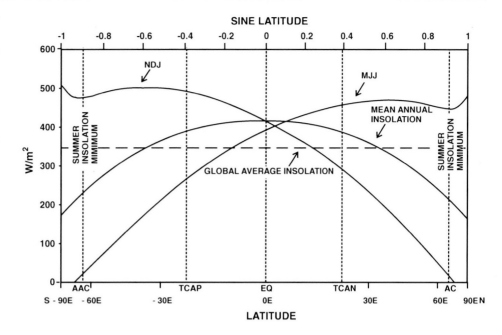

Our closest approach to the Sun occurs on January 3. Primarily because of the increase in day length during the summer, the Polar Regions receive more insolation over the 3-month season than other areas. During those months around the solstices, the Sun never sets on the regions poleward of the Polar Circles, which are presently at 66° 33′ 38″ (=66.56°) North and South latitude. The energy from the Sun just keeps on coming in 24 h a day beyond the polar circles, so the energy received by Earth at those high latitudes increases. If it were not for the 24-h sunshine in the Polar Regions, the curves would decline toward the poles. Another thing that I found very odd when I first plotted these diagrams is that the NDJ and MJJ curves do not rise smoothly toward the pole. Summer insolation has a secondary maximum at about 40° as a result of both the increase in day length and the elevation of the Sun above the horizon. Most importantly, there is a significant depression at the Polar Circles, about 66.56° N and S.

The insolation minimum at 66.56° N is closely associated with the growth and decay of the Northern Hemisphere ice sheets. Snow which accumulates during the fall, winter and spring is most likely to remain where the solar insolation is minimal during summer. We now know that the initial Ice Age glaciation of the Northern Hemisphere began on mountains in southern Greenland, at 66° N, about 5.5 million years ago. The largest Northern Hemisphere ice sheets, the Laurentian (eastern Canada and northeastern U.S.) and Scandinavian ice sheets, appeared later, but both initially nucleated on mountains at 66° N.

If you look at the older literature you will find that the latitudes of the Polar Circles are rounded to 65° N and S. This is because before the days of computers, when we used

pencil and paper for such things, it was easier to calculate isolation over the whole Earth in 5° latitude increments. The Earth's axis of rotation wobbles around like a child's gyroscope and, over the course of one of its 41,000 year cycles, the latitudes of the Polar Circles range between 67.9° and 65.5°.

The energy received by Earth also varies with our planet's distance from the Sun. Earth's orbit is not circular, but elliptical, so that the Earth–Sun distance varies during the course of the year. The ellipticity of the orbit changes with time as a result of the gravitational attraction of the other planets; sometimes it is elongate, at other times it briefly becomes circular. These cyclic orbital changes occur on timescales of tens to hundreds of thousands of years.

## 13.3 The Orbital Cycles

Today we know that the Earth's orbital parameters vary with time. The cause is the changing gravitational attraction of the Sun and other planets. There are four major orbital variables, shown in Fig. 13.4:

1) The ellipticity of the orbit changes on timescales of about 100,000 and 400,000 years (the variation is called the 'eccentricity');
2) the long axis of our planet's elliptical orbit precesses on a time scale averaging about 21,700 years;
3) the tilt of Earth's axis relative to the plane of its orbit around the Sun (the ecliptic), varying between 22° and 24.5° (currently 23.5°), changing on a 41,000-year timescale;

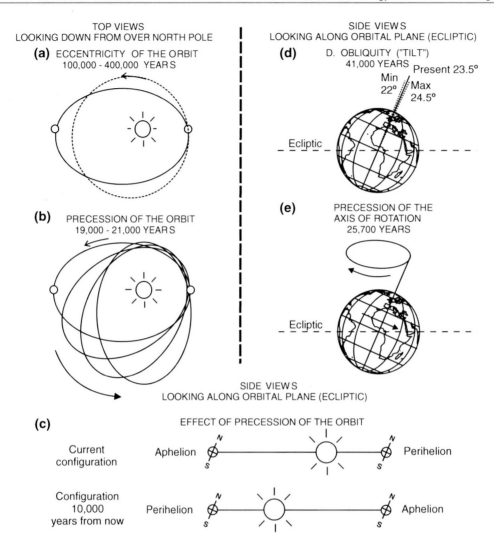

TOP VIEWS
LOOKING DOWN FROM OVER NORTH POLE

**(a)** ECCENTRICITY OF THE ORBIT
100,000 - 400,000 YEARS

**(b)** PRECESSION OF THE ORBIT
19,000 - 21,000 YEARS

SIDE VIEWS
LOOKING ALONG ORBITAL PLANE (ECLIPTIC)

**(d)** D. OBLIQUITY ("TILT")
41,000 YEARS  Present 23.5°
Min          Max
22°          24.5°

Ecliptic

**(e)** PRECESSION OF THE
AXIS OF ROTATION
25,700 YEARS

Ecliptic

SIDE VIEWS
LOOKING ALONG ORBITAL PLANE (ECLIPTIC)

**(c)** EFFECT OF PRECESSION OF THE ORBIT

Current configuration     Aphelion     Perihelion

Configuration 10,000 years from now     Perihelion     Aphelion

**Fig. 13.4** Milankovitch orbital parameters. **a** Eccentricity or Ellipticity of Earth's orbit which can vary from 0 (circular) to 0.07 on 100,000 and 400,000 year timescales (present eccentricity is 0.0167); **b** Precession of the elliptical orbit (19,000 to 21,000 years); **c** Effect of the elliptical orbit on seasonal insolation. At present Earth is closest to the Sun ('at perihelion,' Greek for 'near the Sun') on January 4th, just after the Northern Hemisphere winter solstice and farthest from the Sun (at 'aphelion,' Greek for 'away from the Sun') on July 4th, just after the summer solstice (see Current configuration); **d** Precession of the elliptical orbit. **e** Change in the tilt ('obliquity') of the Earth's axis of rotation relative to the plane of the Earth's orbit around the Sun (the 'ecliptic'), varying from 22° to 24.5°, currently 23.5°; E: Precession of Earth's axis of rotation

4)  the axis of rotation precesses on a 25,700-year timescale. The rotation of the planet, and its motion in its orbit are (like almost everything else in the solar system) counterclockwise when viewed from above the North Pole. The precession of the axis of rotation is clockwise when viewed from above the North Pole.

Early in the seventeenth century Johannes Kepler discovered that the motions of the planets could be understood if their orbits were elliptical, not circular. Kepler also realized that the Earth moved faster through its orbit when it was close to the Sun. He formalized this, stating that the Earth sweeps out equal areas of the ellipse in each equal unit of time. It was discovered much later that the ellipticity of the orbit changes with time, and that the axes of the elliptical orbit rotate ('precess') with time, as shown in Fig. 13.5. This has important climatological significance because it means that the Earth–Sun distance during the seasons changes over time. We are currently close to one of the extremes. The Northern Hemisphere winter solstice occurs on December 21; on that date the Northern Hemisphere reaches its maximum tilt away from the Sun. On January 3, the Earth, following its elliptical orbit, makes its closest approach to the Sun (it is at 'perihelion,' from the Greek peri = near, and helios = Sun). Because

**Fig. 13.5** Precession of the elliptical orbit of Earth around the Sun. 16,500 years ago, aphelion occurred on September 22; 11,000 years ago it was on December 21, coincident with the Northern Hemisphere winter solstice; 5,500 years ago aphelion was on March 20 and the vernal equinox was on December 21; at present aphelion is on July 4, just after the Northern Hemisphere summer solstice on June 21

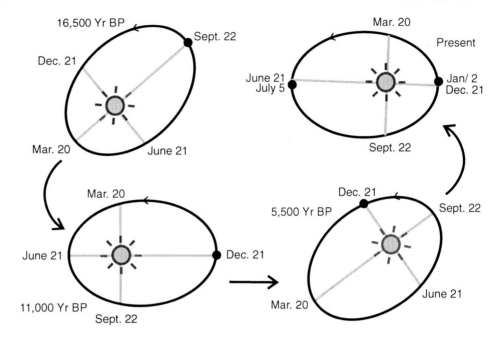

it is closest to the Sun, it receives more energy than at any other time during the year. Astronomers call this 'caloric summer.' So Northern Hemisphere winter and caloric summer coincide. The opposite is true near the time of the Northern Hemisphere summer solstice. What this means in practical terms is that at present the contrast between summer and winter temperatures in the Northern Hemisphere is at a minimum. The situation in the southern hemisphere is just the opposite; at present the contrast between summer and winter temperatures is at a maximum.

Knowledge that there is something odd about the way the stars appear to revolve around the Earth over the course of a year goes far back into history, probably to the time of the Babylonians and Egyptians. It is not just the annual parade of constellations, but the position of the constellations on a particular date changes slowly, about 1 day every 71 years. That would be over the course of a long human lifetime, so no individual could have noticed this, but astronomical observations made over long periods of time would have revealed it. Every 2,100 years a different one of the twelve constellations of the Zodiac is overhead at midnight on a particular date. Discovery of this phenomenon, called the 'precession of the equinoxes,' is attributed to Hipparchus in the second century BC, although his writings on the subject are not preserved. Presumably they were one of the losses when the great library in Alexandria, Egypt, burned. We know of Hipparchus through Claudius Ptolemaeus (Ptolemy) who presented the astronomical knowledge of his time (second century AD) in the *Almagest*.

At the time, it was thought that the Sun, Moon, planets and stars were attached to transparent spheres and rotated

around the Earth. The day-night lengths on the equinoxes are equal, but there were no accurate clocks to measure time. However, on the equinox the Sun rises exactly in the east and sets exactly in the west, and this could be determined. The long-term motions of the stars were as though the Earth's axis of rotation were slowly moving around in space.

After Nicolaus Copernicus discovered, in the middle of the sixteenth century, that the Earth orbits the Sun rather than being the center of the universe, the 'precession of the equinoxes' took on new meaning. Early in the seventeenth century Johannes Kepler discovered that the motions of the planets could be understood if their orbits were elliptical, not circular. The precession of the equinoxes can be represented differently, as shown in Fig. 13.5. Kepler also realized that the Earth moved faster through its orbit when it was close to the Sun. He formalized this, stating that the Earth sweeps out equal areas of the ellipse in each equal unit of time.

Another motion is a change in the tilt of Earth's axis of rotation relative to the 'ecliptic,' the plane in which it orbits the Sun. It is commonly referred to as the 'obliquity' of Earth's axis. At present the angle is about 23.5°, defining the latitudes of the Tropics and Polar circles. The obliquity of Earth's axis of rotation changes on a cycle of about 41,000 years from a maximum tilt of 24.5° to a minimum of 22.0°. These changes significantly redistribute insolation in the Polar Regions, but the effects can be detected down to the Equator.

It is important to realize that the precession of the equinoxes and the changes in obliquity do not change the total amount of insolation received by the Earth, but simply redistribute it to different latitudes and seasons. However, the changes in eccentricity actually do change the total insolation received during the year because the Earth moves faster in its orbit when it is close to the Sun and slower when it is further from the Sun. In reality this turns out to be a very small difference. However, the combination of an eccentric orbit and caloric summer or winter makes a big difference at the latitude of the Arctic Circle. At extremes of the ellipticity/precession cycle the difference in June insolation at 65° N is more than 20 %.

Also shown in Fig. 13.4 is the precession of Earth's axis or rotation, with a period of about 25,700 years. This is responsible for the phenomenon known as the 'precession of the equinoxes' by which the constellations of the Zodiac slowly cycle through the sky. It combines with the other motions to cause variations in their periods.

It has recently been suggested that another orbital variation may be responsible for the 100,000 cycle of glaciations and interglacials, the inclination of the Earth's orbit. Based on additional calculations physicists Richard Muller and Gordon MacDonald proposed in 1995 that the inclination of the Earth's orbital plane, the ecliptic, varied relative to the plane of symmetry of the solar system over a period of 100,000 years, approximately the same as the eccentricity cycle. They suggested that there might be a concentration of dust in the plane of symmetry of the solar system, and as the inclination of the Earth's orbit decreases relative to the plane of symmetry of the solar system and enters this dust band, the insolation received by the Earth would be reduced. They suggested that this might be the cause of the 100,000 year glacial cycles. Unfortunately, the inclination minima precede the glacial maxima by 33,000 years, implying that something else must be going on to cause such a long lag between cause and effect. Their idea attracted much attention at first but has lost favor because of lack of evidence for either the proposed concentration of dust, or a mechanism to cause the lag.

There is yet another complication in this whole business —the change in speed of the Earth through its orbit as its distance from the Sun changes. When in the sixteenth century Nicolas Copernicus proposed that the Earth and other planets orbited the Sun and not vice versa, he assumed that the orbits were circular. At the beginning of the seventeenth century, Johannes Kepler realized that to account for the observed motions of the planets, their orbits must be elliptical, not circular, and that the Sun must be at one of the foci of the ellipse. Further, he concluded that the planets must be moving faster when they are closer to the Sun and slower when they are further away. It was near the end of the seventeenth century when Sir Isaac Newton explained why all this was so, by introducing the idea of gravity. What all this means is that detailed calculations of the total amount of energy received by the Earth over different seasons must take into account not only the change in the solar constant related to the Earth–Sun distance, but the length of time Earth spends at that distance.

## 13.4   The Rise and Fall of the Orbital Theory of Climate Change

Joseph Adhemar was the first person to propose that there was a relation between astronomical considerations and climate, suggesting in 1842 that it was the precession of Earth's orbit that had caused the ice age. Adhemar's idea was promptly rejected by the eminent German scientist and explorer, Alexander von Humboldt. He noted that all that the precession did was to make for colder winters and warmer summers, and that the ice that formed in winter should have been melted in the warmer summer. The idea of precession affecting climate was dead for almost 50 years.

It seems as though no respectable scientist wanted take up the idea again, so it remained for someone well out of the mainstream to pursue the idea. James Croll (Fig. 13.6) was the son of poor Scottish farmers. He was self-taught, and had enormous curiosity about the Earth and astronomy. In 1859, when he was 38 years old, he was hired on as a janitor to the museum of Anderson's University in Glasgow so that he could gain access to the library. Five years later he was exchanging letters with Charles Lyell on variations in Earth's orbit and ice ages. To the idea of a precession effect, Croll added the long-term changes in the ellipticity of Earth's orbit, based on calculating the effect of the other planets (Fig. 13.7). He also realized that cool summers, when the snow would not melt completely, were an important factor. Charles Lyell recognized that he was not your everyday janitor, and helped arrange for him to have a position with the Scottish Geological Survey in Edinburgh. In 1875 he published 'Climate and Time, in Their Geological Relations.' The relation between changes in Earth's orbit and climate had become a serious topic of discussion.

Although Croll's book stimulated discussion, it did not convince many geologists. For one thing, the orbital changes had been repeated over and over, yet there was only one glaciation.

In the latter half of the nineteenth century the idea that the Earth had experienced a glaciation just before the modern era had been generally accepted. The problem then was what had caused this climate deterioration. For Lord Kelvin, the answer was obvious—cooling of the Earth. For others it was more likely to be changes in the amount of energy received from the Sun.

Then a new complication appeared. At the end of the nineteenth century geologists examining these young deposits became convinced that there had been more than one glaciation. One of these geologists was particularly important, Thomas Chrowder Chamberlin, the person who had suggested that scientists should have multiple working hypotheses. In 1882 he published a paper describing a second glaciation, and in subsequent papers expanded the number. In the United States the evidence was in the 'terminal moraines' and the soils that had developed on the glacial deposits. The youngest deposits had very fresh, rich soils. These extended from the Great Lakes halfway down to the Ohio River (Fig. 13.8). The youngest glaciation was termed the 'Wisconsinan.' A series of low hills across Ohio, Indiana and Illinois was identified as terminal moraines marking the southern limit of the Wisconsinan. South of these moraines were more weathered glacial deposits that extended down to the Ohio valley. The soil formation process had obviously acted on them for a much longer time, and they were attributed to an 'Illinoisan' glaciation. Skeptics were convinced when the remains of a forest were found sandwiched between the two glacial deposits. Soon it was realized that the Ohio River marked the edge of the older ice sheet. We now know that both the Ohio and the Missouri Rivers flow in valleys that once marked the edge of the continental ice sheets. Older more deeply weathered glacial deposits were identified to the west and were named the Kansan and Nebraskan. It was considered firmly established that there had been four glaciations in North America.

In Europe finding traces of the older glaciations was not so easy. They were recognized in the form of terraces along the river valleys north of the Alps. The deposits forming the terraces were laid down when the ice was eroding vast amounts of material from the Alps, and during the interglacials the rivers eroded these deposits leaving some of the material behind as the terraces. As in North America, four glaciations were recognized, although the penultimate one appeared to have two episodes. The problem of the glaciations had become twofold—why did they occur at all, and what caused them to come and go?

The next person to take up the investigation of the possible role of Earth's orbital variations in causing glaciation was a Serbian scientist, Milutin Milankovitch (Fig. 13.9). More precise astronomical data on them had been gained since Croll's time, and Milankovitch recalculated everything. Quite painstakingly, I might add, for there were no computing devices and all of the calculations had to be done by hand. It was a big job. Milankovitch realized that insolation at latitude 65° N was critical (see Fig. 13.3 to recall why it is so important). The initial publication of his result was in Serbian and, needless to say, did not attract many readers. However, a translation into French in 1920 was more widely read, and in 1924 his insolation curve was included in a book on paleoclimates by Vladimir Köppen and Alfred Wegener. They realized that it was the cool summers at 65° N, when the winter snow did not completely melt, that were critical to the growth of ice sheets.

Alfred Wegener deserves special attention. Born in Berlin, he received his school education there and after attending the Universities in Heidelberg and Innsbruck, obtained his doctorate in astronomy in Berlin in 1904. He immediately switched interest to the new field of meteorology. He

was already interested in polar exploration and particularly in Greenland. He was an avid balloonist, already having set records for time aloft as a graduate student. One of his goals was to reach the North Pole by balloon, and he made several flights over the ice. His first visit to Greenland was on a 2-year Danish expedition that started in 1906.

Back in Germany at the University of Marburg, he noticed the similarity of the coastlines on the two sides of the Atlantic. It seems likely that his idea that continents were drifting apart leaving behind the ocean basins must have originated with his views of the polar ice from above. In 1912, he published his thoughts on continental drift, organized another expedition to Greenland, and married the daughter of the most famous climatologist of the time, Wladimir Köppen.

I have mentioned the conservative nature of scientists and of geologists in particular, but the history of Wegener's idea of continental drift is an extraordinary example. His idea that the Americas had drifted away from Eurasia and Africa was taken seriously by a few geologists in Europe, although not accepted by most. Arthur Holmes, Professor at Durham and later at Edinburgh, included Wegener's idea in the many editions of his famous textbook *Principles of Geology*, albeit with the improvement that he thought the continents were riding on thermal convection cells in the Earth's mantle. In the United States, Wegener's ideas were roundly rejected. In 1926, the American Association of Petroleum Geologists even held a symposium on his idea, but almost all of the papers presented were arguments, often sarcastic, against it. Wegener was, after all, a meteorologist, not a geologist, so it was easy to reject his ideas as those of a dilettante scientist guilty of venturing outside his field. Wegener died in 1930, in an attempt to resupply a camp on the Greenland ice cap he had set up to investigate winter conditions.

But back to Milankovitch's insolation curves. The problem was clear. The match between the crude curves and the glaciations seemed contrived. Moreover, the orbital variations just seemed to go back in time indefinitely, but there had been only four glaciations. Something else had caused the glaciations. Virtually every other element of the climate system was examined as a possible cause, and each has had its proponents.

When I was an undergraduate in college in the early 1950s everyone knew there had been four glaciations. But when I started graduate study at the University of Zurich in Switzerland in 1955 I took a course in glacial geology in the Geography Department. There the story of the four glaciations was presented in detail, but there was some new information. Professor Büdel at the University of Wurzburg in Germany was arguing that there had been more than four glaciations, possibly as many as twenty. There was also a discussion of the work of Milutin Milankovitch, the Serbian academic who had promoted the idea that the glaciations were the result of changes in Earth's orbit around the Sun.

Upon my return to the United States the next year to work on my Master's degree at the University of Illinois in Urbana I found that the Illinois Geological Survey on the south edge of the campus housed many of the experts on the Quaternary. There were four glaciations. Period. No one had ever heard of Büdel. The idea that Earth's orbital variations had much to do with the glaciations was not popular. Not only was there no credible evidence for earlier glaciation, it had just been shown that the Milankovitch orbital cycles did not match the data. The search for a cause focused on very young mountain building, 'orogenic uplift' in 'geologists' parlance, the positive climatic feedback caused by the reflectivity of the ice sheets, and the effect of adjustments of Earth's surface in response to the loading of continental crust by glacial ice sheets. There were lots of convoluted arguments, none very satisfying to the casual observer.

A death knell for the orbital theory sounded when a new technique, carbon-14 dating, showed that the last Glacial Maximum occurred 18,000 years ago. According to Milankovitch's calculations, the minimum solar insolation in the Northern Hemisphere had occurred 24,000 years ago.

You will remember that Carbon-14 dating has become one of the most widely used techniques for determining the age of many materials. One of the early assumptions in $^{14}$C was that the rate of its production in the upper atmosphere had remained more or less constant over time. Using this assumption, by 1955 samples of wood representing forests overrun by the ice at its southernmost extent had been dated at 17,000–18,000 years old. This threw the Milankovitch hypothesis into serious doubt. Although not in my special field of interest, these exciting research results were coming in while I was working on my Master's degree at the University of Illinois.

Of course it turned out that the assumption of a constant rate of production of $^{14}$C in the upper atmosphere was wrong. But it has been possible to calibrate the $^{14}$C age scale to calendar years using other information. Knowledge of the changes in rates of production of $^{14}$C has been used to infer changes in solar activity, as we shall see later. We now know that the Last Glacial Maximum did in fact occur about 24,000 years ago.

I did not think much about the younger part of Earth history while working on my doctorate at Stanford. California had not been covered by ice, and the geology there is very different from that of the Midwest. However, a picture in the hallway of the Geology Department building on the northwest corner of the Stanford Quadrangle did remind me of Louis Agassiz's importance. It showed his statue toppled headfirst into the walkway adjacent to the Zoology building, a result of the 1906 earthquake (Fig. 13.10). Stanford President David Starr Jordan, who, like Agassiz was a paleontologist specializing in the study of fossil fish, later

recalled that someone, on seeing the fallen statue remarked 'Agassiz was great in the abstract but not in the concrete.'

My doctoral field work was concerned with the Cretaceous-Tertiary boundary in the Coastal plain of Mexico, far from the world of glaciers. The Cretaceous, which lasted from 145 to 65 million years ago, was a time of global warmth. Dinosaurs ruled the world. If you saw "Jurassic Park" you were seeing mostly Cretaceous, not Jurassic dinosaurs. I guess "Cretaceous Park" just doesn't sound as exciting. It had long been known that the end of the Cretaceous, 65 million years ago, was a time of major mass extinction but no one was sure what had caused it. I did not figure it out either, but I was able to show that the assemblages of foraminifera in the marine rocks of Mexico were more like those of the Caucasus, on the other side of the world than those of nearby Texas.

After finishing my doctorate I had a year of postdoctoral study at the University of Basel, in Switzerland, where my work continued on the Cretaceous and Early Tertiary. In the fall of 1960 I returned to the University of Illinois, this time as an Assistant Professor. Again I became involved in discussions of the ice age world, albeit peripherally.

One of the courses I taught each year as a young faculty member at the University of Illinois was Historical Geology, and I always spent an hour discussing Wegener and his idea of continental drift. The year before I came up for tenure review a friendly faculty member gently warned me that tenured faculty did not believe in Wegener, and advised me to avoid the topic in my classes.

However, only a few years later, in the mid-1960s, Wegener's ideas were subsumed into the theory of plate tectonics. I chaired a symposium on the results of the fledgling Deep Sea Drilling Project at a meeting of the American Association of Petroleum Geologists in 1969. Materials recovered by the DSDP in the South Atlantic in 1967 had provided what many considered obvious proof of the plate tectonic theory, and we asked how many in the audience, mostly petroleum geologists, accepted the idea. Eighty percent rejected it. A similar poll at an AAPG meeting 2 years later had 95 % in favor of it. It is typical for a revolutionary idea in science to be rejected for decades, then, with new evidence, to be accepted in only a few years. You will see this pattern over and over again in climate studies.

## 13.5    The Resurrection of the Orbital Theory

Over the Christmas vacation in 1961–62 I traveled to Florida and the Bahamas in search of a marine station I could use as a base for looking at shallow water sediments. My interest in this new field had been piqued by learning to snorkel in the Mediterranean while I was in Europe. I had found the undersea world to be wholly different from what I had imagined.

I visited the University of Miami's Institute of Marine Sciences and the Lerner Marine Laboratory on Bimini, operated by the American Museum of Natural History. I selected the Lerner Laboratories the site for future field work, but I enjoyed my visit with faculty at the Institute in Miami.

The most intriguing character in Miami was an Italian-born scientist by the name of Cesare Emiliani. Cesare had started out as a micropaleontologist at the University of Bologna. However, he immigrated to the United States in 1951, obtained a second doctorate from the University of Chicago, and in the early 1950s was a research assistant working for Harold Urey in Enrico Fermi's Institute for Nuclear Studies at the University of Chicago. You may recall that it was in a converted squash court beneath the University stadium that the first nuclear chain reaction took place. For students passing through, the area was known as the Metallurgical Project. Harold Urey was interested in the possibilities of using isotopes for scientific research in new areas.

One of Harold Urey's ideas was to use isotopes of oxygen to determine the temperature at which shells were formed. There are three common stable (that is, nonradioactive) isotopes of O. The most abundant (99.8 % of all O atoms) is $^{16}O$, the next most abundant (0.2 %) is $^{18}O$; the other stable isotope $^{17}O$ is half an order of magnitude less abundant than $^{18}O$. The relative abundances of these isotopes can be measured using an instrument called a mass spectrometer. Urey reasoned that the heavier isotope, $^{18}O$, might be preferentially incorporated into carbonate shells of clams grown at lower temperatures. Laboratory experiments, using clams grown in aquaria, showed that this was true. If this technique could be used on clams grown in the laboratory, it should also work on fossils. There was a way to determine the temperature of the water in which the fossil shell grew. The paleothermometer was born.

Strange to say, I had heard about this work from one of the participants in the research as I returned from my Junior Year in Munich aboard the Queen Mary in 1954. In the evening of the last day before landing in New York, I was outside looking at the water. I noticed that the water was no longer blue, and that the distance between the waves had gotten much shorter. I commented to a fellow passenger standing next to me—"looks like we are over the continental shelf now." We started to talk about the sea level rise that must have taken place at the end of the last glaciation, wondering whether this would have been land thousands of years ago. The passenger introduced himself as Ralph Buchsbaum. Startled, I immediately recognized him as the author of a book I knew well: *Animals Without Backbones*. He told me about work he had been doing with Harold Urey on determining temperatures of shell formation using oxygen isotopes, and later sent me a reprint of their paper. He also said that a young Italian scientist was working with Urey on determining the temperature of the ocean in the

**Fig. 13.6** James Croll (1821–90). Janitor in the museum of Anderson's University in Glasgow

past. It was an exciting prospect; little did I guess that I would meet that scientist, Cesare Emiliani, 7 years later.

Figure 13.11 shows a diagrammatic version of those early experiments with oxygen isotopes that Ralph Buchsbaum had been involved with. The isotopic analyses were performed by Sam Epstein on mollusks grown in aquaria held at different temperatures. The oxygen isotopic composition of the water and the shell were measured, and the differences induced by the different temperatures were expressed as the ratio between the abundance of atoms of $^{18}O$ and $^{16}O$. If there are 1,000 atoms of O, there will be, on average, 998 atoms of $^{16}O$ and 2 of $^{18}O$. This ratio will vary with several factors, as we will see later, but at the time these experiments were performed it was assumed that the temperature effect would dominate. The idea was this—otherwise identical molecules containing atoms of different weights will move at different speeds at the same temperature. You can think of it as a 'momentum' effect. Momentum is a function of speed times mass. At a given temperature, all of the molecules would have about the same momentum, but those with the heavier $^{18}O$ atoms would be moving more slowly. Since they are slower, they have less chance of becoming involved in chemical reactions. They tend to get 'left behind.' The shells

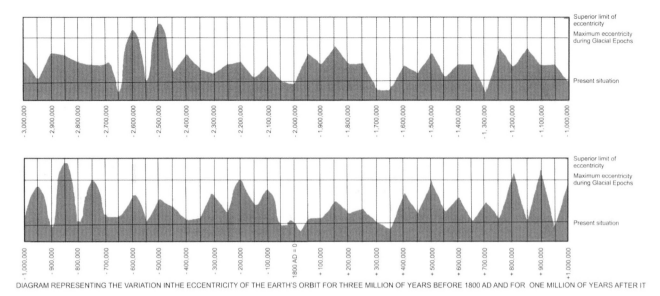

**Fig. 13.7**   James Croll's calculation of insolation changes due to eccentricity of Earth's orbit, from 3 million years ago to 1 million years in the future

of the mollusks are made of calcium carbonate minerals, calcite and aragonite, both of which have the chemical formula $CaCO_3$. The oxygen atoms in the $CaCO_3$ came from the dissolved bicarbonate ion ($HCO_3^-$) in the water. The ions that were moving more rapidly had a better chance of getting involved in the chemical reactions; those were the bicarbonate ions with $^{16}O$ atoms.

The accepted way of expressing the changes in relative abundance of $^{18}O$ atoms is in terms of numbers per thousand (or 'parts per thousand' = ppm) relative to a standard. There are two common standards in use, one is 'PDB,' a fossil 'belemnite' from the Cretaceous PeeDee Formation of South Carolina (which somebody happened to have handy when all this started); the other is Standard Mean Ocean Water (SMOW), which is in a jar in Copenhagen. The belemnite specimen is a piece of calcite that looks rather like the business-end of a very large caliber bullet. It was part of a squid-like animal, *Belemnitella americana*, that lived 80 million years ago. Sam Epstein and friends used the belemnite as their standard. It is now the universal reference for measurements on things made of calcium carbonate. The actual abundance of $^{18}O$ in the PeeDee belemnite is 0.0020672. This becomes the standard value, and analyses are reported in terms of the variation from this value. The difference from the standard is cited as $\delta^{18}O$ (delta $^{18}O$). 'Delta' is the scientist's shorthand for 'difference.' The abundance of $^{18}O$ in SMOW is 0.0020052; this becomes the standard for water, ice, and some other materials. Needless to say, the original standards would have been quickly used up as measurements were made, so a substitute source was set up by the International Atomic Energy Agency in Vienna, Austria to serve the scientific community.

The specimen shown in Fig. 13.11 is a mollusk known as the Queen Conch. It is a big snail which can be a foot long. Its scientific name is *Strombus gigas*, and if you have been to the Caribbean you may have brought back one of these beautiful shells. It is a food staple for many of the smaller islands. It can be used raw to make 'conch salad,' or cooked in oil to make 'conch fritters.' Both are very tasty. It is said that you can survive indefinitely on a desert island in the tropics because you can collect conchs by wading, and along with coconuts you have a completely nutritious diet.

Sam Epstein also used abalone, another large edible snail, for his experiments. Abalone shells are lined with 'mother-of-pearl' which is made of aragonite. He was interested in determining if there was a differential fractionation of the isotopes between the two species and between the two mineral forms of $CaCO_3$, calcite and aragonite. There are some differences, called the 'vital effect,' but once you know what they are, they can be taken into account and you can tell the water temperature from the $\delta^{18}O$. The ideal thing would be to make measurements in such a way that any 'vital effect' could be eliminated. That is exactly what another member of Harold Urey's group, Cesare Emiliani, did (Fig. 13.12). Cesare realized that if you made all the measurements on a single species the 'vital effect' would always be the same and could be neglected.

I spent a great afternoon with Cesare in January 1961. It was the beginning of a lifetime friendship. He had been studying the variations in oxygen isotopes in the carbonate shells of planktonic foraminifera. My doctoral thesis at Stanford had been done on these beautiful objects. The planktonic foraminifera are one-celled amoeba-like organisms that produce millimeter-sized carbonate shells that look

IDEAL MAP
of
NORTH AMERICA
during the
ICE AGE
—
T.C.Chamberlin
1894

**Fig. 13.8** T. C. Chamberlin's map of North America during the Ice Age, published in the 3rd edition (1894) of Archibald Geike's book *The Great Ice Age*

**Fig. 13.9** Milutin Milankovitch. Portrait from the NOAA (National Oceanographic and Atmospheric Agency) NGDC (National Geophysical Data Center) Paleoclimate Archive

like a spiral of hollow beads. Amoeba-like is perhaps a misnomer, because although they collect food by engulfing it like an amoeba, they look very different, more like a tiny living cobweb. A few years later, at the Lerner Marine Laboratory, I had opportunity to go hunting for these tiny organisms with some other researchers. We went out into the Gulf Stream. Using SCUBA gear we descended to about 10 m, and just waited for our eyes to adjust to the lower light level. It is a strange experience to dive in the open ocean. The bottom is thousands of meters below; when you look down you simply see blackness. After a few minutes you become aware of the many transparent organisms in the water, some of really indescribable beauty. And finally you see what you were looking for. Floating in the ocean, a planktonic foramifer looks like a gray sphere of tiny thread-like spines. The whole organism is about a centimeter in diameter with the tiny shell at the center. When the foraminifer dies (or reproduces) the shell drifts down to the ocean floor. Over much of the ocean floor almost half of the sediment is shells of planktonic foraminifera. The other half

is coccoliths. Much of our knowledge about the oceans in the past comes from study of these shells.

Cesare had gotten samples from cores of sediment taken in the Atlantic, the Caribbean, and the Pacific. Before World War II, cores of deep-sea sediments had been taken by simply dropping a length of pipe and letting it stick into the bottom. Needless to say, this technique did not recover more than a meter or so of sediment. After World War II, Börje Kullenberg (1906–1991) in Sweden improved the technique by including a piston inside the pipe. The pipe, or core barrel, is driven into the seafloor by having a great weight on its top. The piston stays at the level of the sea floor.

The Swedish Deep Sea Expedition of 1947–49 used this new device to recover ocean cores of unprecedented length, from a few to tens of meters. Incidentally, the person in charge of examining the deep-sea sediments collected by the Swedish expedition was Gustav Arrhenius, later Professor at Scripps Institution of Oceanography in La Jolla, California. Gustav was the grandson of Svante Arrhenius, the chemist who, at the end of the nineteenth century, had suggested that rising $CO_2$ concentrations might lead to global warming. The piston corer was used for other deep-sea expeditions. Under the direction of Maurice Ewing, the Lamont Geological Observatory of Columbia University had its research vessel *Vema* using similar piston corers to obtain samples of deep-sea sediments from all of the oceans, 320 days a year, from 1953 until its retirement in 1981.

The samples Cesare originally analyzed came from the Swedish Deep-Sea Expedition cores. The cores were sampled at 10 cm intervals, and the shells of foraminifera were picked out (Fig. 13.13). Making measurements on these shells, he found variations in the ratio of $^{18}O$ to $^{16}O$. He published his results in 1955 in a paper entitled simply *Pleistocene temperatures*. At the time it was known that the isotopic ratio could be due to three factors: (1) the organism itself might have some way of influencing the ratio through its metabolic processes; this was called the 'vital effect' and can be avoided by always using the same species for making measurements; (2) the temperature at which the shell was deposited—this is what Cesare was after; (3) the ratio of $^{18}O$ to $^{16}O$ in the ocean water. Cesare realized that this would change as ice sheets grew on land because the snow from which the ice sheets grew was water that had evaporated from the ocean. When water evaporates, the $^{16}O$ preferentially goes into the vapor leaving the $^{18}O$ behind in the ocean. That means that there is more $^{18}O$ to be incorporated into the shells. But if there were no change in the $^{18}O$ to $^{16}O$ ratio in the water, $^{18}O$ would be preferentially incorporated into the shells that formed at lower temperatures. In the 1950s and early 1960s no one knew how much of the isotopic signal was due to the temperature and how much reflected the amount of ice on land.

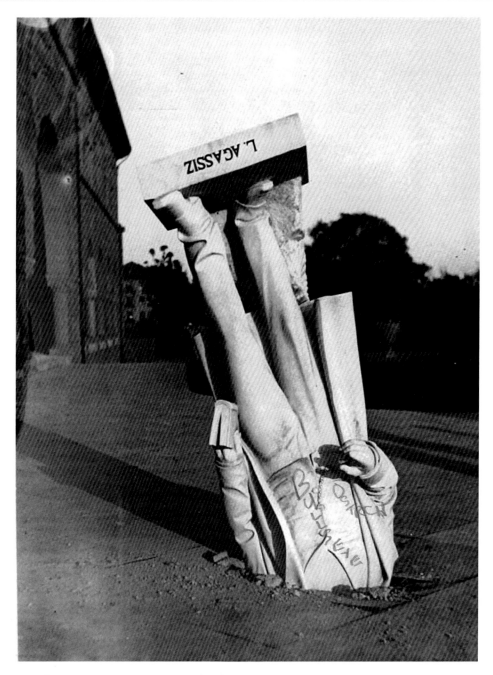

**Fig. 13.10** Statue of Louis Agassiz at Stanford university after the 1906 earthquake

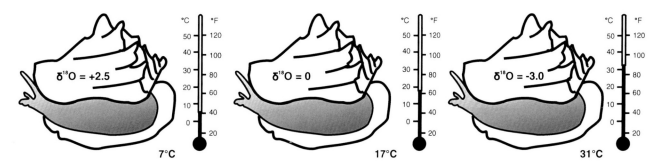

**Fig. 13.11** The difference in abundance of the $^{18}O$ isotope in the calcium carbonate shell of a mollusk grown at different temperatures

**Fig. 13.12** Cesare Emiliani. Collage with penguin prepared by Monica Abbott and Eloise Zakevitch. Cesare did not like to be photographed, and when I asked for one of him some years ago, I was given this picture which shows not only what he looked like but captures something of his remarkable sense of humor as well

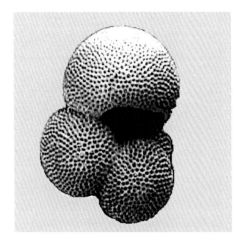

**Fig. 13.13** The shell of *Globigerinoides sacculifer*, one of the species of shallow-dwelling planktonic foraminifer used by Cesare Emiliani in his attempts to determine the temperature of the ocean in the past

Cesare guessed that 60 % of the signal was due to the temperature effect, 40 % to the ice effect. He concluded that equatorial and tropical ocean surface temperatures had been several degrees cooler during times of glaciation. At that time he believed that the cyclic glaciations were related to changing insolation (Milankovitch cycles), but thought that orogenic uplift, ice-albedo feedback, and the effect of isostatic adjustments to the loading of continental crust by glacial ice sheets—all topics still being actively discussed today—must also have played a role. Cesare concluded that the 41,000 obliquity cycle was the main driving force behind the waxing and waning of the ice ages.

Figure 13.14b shows the composite paleotemperature curve published by Emiliani and his colleague Johannes Geiss in 1958. The age scale was based on radiocarbon dates for the young part of the curve, back to 18,000 years. There was one older radiocarbon date that had been published. It had been made using an experimental, and as it later turned out, incorrect, technique, but was used to assign the second temperature minimum an age of 65,000 years. Emiliani and Geiss thought that this was the penultimate glaciation and concluded that the 41,000 year obliquity cycle was responsible for the timing of the glaciations. It seemed like a reasonable conclusion because the obliquity cycle has its greatest effect at high latitudes. The ages of the rest of the curve were extrapolated from these two dates. Figure 13.14a shows a version of Milankovitch's original solar insolation curve from the 1920s, and Fig. 13.14c is a newer solar insolation curve for 65° N, interpolated from the calculations of André Berger published in 1978. Berger's more detailed calculations were possible only after the development of the computer. It must be remembered that the age scale on Cesare's curve was largely guesswork, whereas that on the insolation curve is more rigidly fixed by astronomical motions. Also, Cesare's curve was based on a limited number of data points. Although the time scales are different, the similarity of the Emiliani and Berger curves is impressive, especially for the last glacial cycle. One wonders how science might have progressed if Berger's calculations had been available when the first paleotemperature measurements were being made. I have no doubt that the astronomical time scale would have been used to determine the ages of the variations in the $^{18}O:^{16}O$ ratios.

The Emiliani and Geiss paper ended with "The glacial-interglacial cycle was repeated several times during at least the last few hundred thousand years, and may continue for some million years in the future." In the 1960s the best evidence was that interglacials lasted about 10,000 years, and we were already about 11,000 years into the current interglacial. There was a lot of informal discussion about when the ice would start to accumulate over northeastern Canada again.

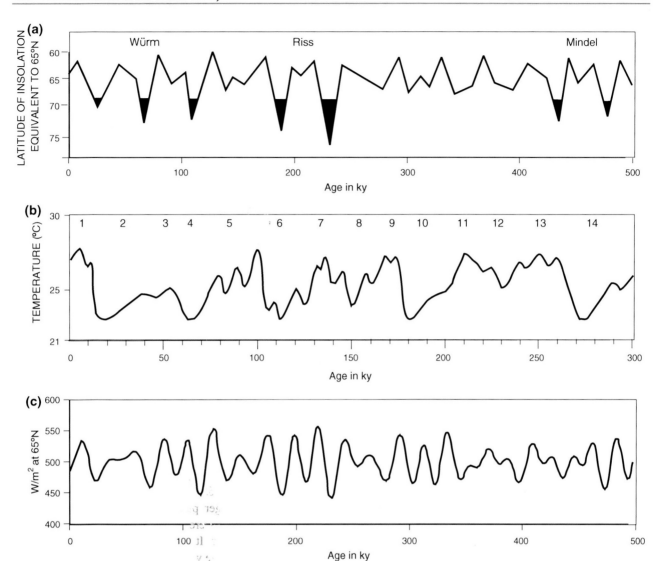

**Fig. 13.14  a** Milankovitch's insolation curve, from Köppen and Wegener (1924); note the crudeness of the calculations, made by hand. **b** The paleotemperature curve of Emiliani and Geiss (1955) for the tropical Atlantic based on oxygen isotopes. The age scale was a guess based on one calibration point at 65,000 years ago.

Kyr = kiloyear = 1,000 years. Numbers are their "Marine Isotope Stages" (MIS's) with odd numbers for warmer episodes and interglacials and even numbers for colder episodes and glacials. **c** Insolation curve for 65° N calculated from André Berger's (1978) orbital parameters

In 1972, Cesare sounded a warning: "Oxygen isotopic analysis and absolute dating of deep-sea cores show that temperatures as high as those of today occurred for only about 10 % of the time during the past half million years. The shortness of the high temperature intervals ("hypsithermals") suggests a precarious environmental balance, a condition which makes man's interference with the environment during the present hypsithermal extremely critical. This precarious balance must be stabilized if a new glaciation or total deglaciation is to be avoided." Similar warnings are found throughout the scientific literature of the 1970s, although most geologists thought a new glaciation was on its way.

## 13.6  Correcting the Age Scale; Filling in the Details to Prove the Theory

It was almost a decade before the Emiliani-Geiss age scale was corrected. It was done using ratios of the radioactive isotopes of Thorium ($^{230}$Th) and Uranium ($^{234}$U) in the skeletons of corals found a few meters above modern sea level in the Pacific, Florida Keys, and Bahamas. In about 1965 Wally Broecker and colleagues at Lamont Geological Observatory in New York and Washington University in St. Louis recognized that the coral terraces represented three interglacial sea-level highstands, and dated them at 85,000, 130,000, and 190,000 years. At about the same time,

Emiliani and Elizabeth Rona used a different pair of radioactive isotopes $^{230}$Th and Protactinium 231 ($^{231}$Pa) to date levels in two deep sea cores from the Caribbean which had excellent oxygen isotope records. They got ages centered on 91,000 years for the last interglacial (Isotope Stage 5 in Fig. 13.14b). As you might guess there developed a rivalry between the two groups over the correct timing of the glacial–interglacial events, but they were now more readily compared with the astronomical insolation record.

A few years later coral terraces on the Caribbean island of Barbados were studied, and it was concluded that they represented highstands of sea level during the interglacial between the last and the preceding glacial maximum. These were all included in Stage 5 of the Emiliani and Geiss paleotemperature curve. The coral terraces have been dated many times since then using this and other techniques, and three ages emerge: 80,000, 105,000, and 122,000 years. These are the three big humps on the Emiliani and Geiss curve shown in the middle box of Fig. 13.14. It became an open question whether or not the glaciations coincided with Milankovitch cycles, and if so, which ones? Evidently it wasn't the 41,000 year obliquity cycle Emiliani and Geiss had assumed.

Furthermore, studies of the assemblages of the planktonic foraminifera from different parts of the ocean resulted in different estimates of the amount of temperature change in the tropics. Louis Lidz, a graduate student working with Cesare studying a core from the Caribbean, concluded that the changes in species abundance indicated significant cooling, confirming Cesare's interpretation. Unfortunately Lou, one of the brightest students I have ever known, had contracted leukemia, and his life was cut short. His paper on the subject in *Science* was unnoticed. A large group project, CLIMAP, on the basis of samples from many parts of the world, concluded just the opposite, that the temperatures in the tropics had not changed appreciably during the glaciation. Into this controversy came a new figure, Nick Shackleton.

In the early 1960s, Nicholas Shackleton (1937–2006) (Fig. 13.16) was a young physics student in Cambridge, England. He became very interested in Emiliani's work in advancing knowledge of the glacial epoch. The Geology Department at Cambridge has always had an outstanding group of scientists, and the morning coffee hour is a place for stimulating discussion. It doesn't make much difference what field of Earth Science one is in, you will eventually wind up visiting a colleague in Cambridge. In the 1960s the leader of the Quaternary group was a botanist, Professor Harry Godwin, who had done some of the pioneering work on understanding Quaternary climate change from plant remains. Although not himself a physicist, Godwin had realized the importance of the developments in $^{14}$C dating and paleothermometry using oxygen isotopes.

As Nick began work toward a doctorate in the Institute of Quaternary Studies, he realized that the most important advance would be to improve the mass spectrometer used for $^{18}$O:$^{16}$O analysis so that smaller samples could be used. This would allow many more determinations to be made, speeding up the process of advancing knowledge. He also had an idea about how to settle the debate which had developed over whether the $^{18}$O:$^{16}$O variations in the deep-sea cores reflected water temperature or ice volume. Cesare had made his measurements on the shells of foraminifera that live passively floating near the surface of the ocean. Nick compared the $^{18}$O:$^{16}$O ratios of surface dwelling foraminifera with those that lived on the deep-sea floor. The deep water in the oceans sinks in the Polar Regions and is very cold. Its temperature can change only slightly from glacial to interglacial. Nick found that the glacial–interglacial range of isotopic values in both sets of shells were very similar, and concluded that most of the isotopic signal reflected changes in global ice volume rather than changes in water temperature. He received his Ph.D. in 1967.

Figure 13.15 summarizes his idea, brought up to date. He reasoned that the $\delta^{18}$O of ice sheets, which had by then been determined to be about −40, was very different from that of seawater ($\delta^{18}$O = 0). This is because as water evaporates, the molecules with $^{18}$O are left and those with $^{16}$O go preferentially into the vapor. This means that the ocean would have become enriched in $^{18}$O as the ice sheets grew. Now comes the confusion factor: a colder water temperature means that the shells of planktonic foraminifera become enriched in $^{18}$O, but the ocean water itself also became enriched in $^{18}$O as the ice sheets grew, so the shells would be enriched in $^{18}$O even if the water temperature stayed the same. Cesare Emiliani had guessed that the effect was 60 % temperature, 40 % ice volume. Nick Shackleton proposed it was largely ice volume.

Nick was one of the most memorable characters in this drama. Although he might easily have been mistaken for a long-haired hippie, usually wearing sandals and an old T-shirt, he became a distinguished faculty member at Cambridge, and was later Director of what became the Godwin Institute for Quaternary Research. In 1998 he was knighted for his contributions to earth science. All of his friends enjoyed addressing him as Sir Nicholas, to his occasional annoyance.

But Quaternary research was not Nick's only interest. He was an accomplished musician, playing the clarinet and other wind instruments. We had some wonderful discussions about early English music, and he introduced me to the work of the sixteenth century composer, Thomas Tallis. I once had the opportunity to hear Nick play in a concert in King's College Chapel, that masterpiece of flamboyant gothic architecture, in Cambridge. He was very much interested in the development of woodwinds and horns and over the years built up a valuable collection of historical instruments. Upon his death in 2006 he left the collection to the University of Edinburgh, where the instruments are used in concerts.

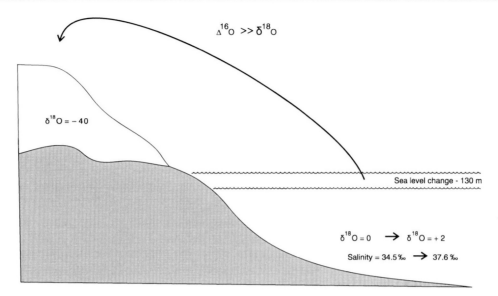

$$_\Delta{}^{16}O \gg \delta{}^{18}O$$

$\delta{}^{18}O = -40$

Sea level change - 130 m

$\delta{}^{18}O = 0 \quad \rightarrow \quad \delta{}^{18}O = +2$

Salinity = 34.5‰ $\rightarrow$ 37.6 ‰

**Fig. 13.15** How the isotopic composition of seawater changes as ice sheets build up on land. The ice is highly depleted in $^{18}O$, so that it becomes enriched in the ocean. Although the planktonic foraminifera living near the ocean surface would record this change in their calcium carbonate shells, they would also record changing temperature of the water. To avoid this problem, Nick Shackleton made measurements of the $\delta$ $^{18}O$ in benthic foraminifera living on the deep-sea floor. It was thought that the glacial–interglacial temperature changes there were negligible. As you might guess, that turned out to be not quite true, but a close approximation

**Fig. 13.16** Nick (later Sir Nicholas) Shackleton: Quaternary geologist and clarinetist

In 1976 Shackleton was co-author of a paper entitled '*Variations in the Earth's Orbit: Pacemaker of the Ice Ages*' published in *Science*. Jim Hayes, John Imbrie and Nick analyzed in detail the measurements of Nick's $^{18}O:^{16}O$ record in two cores from the southern Indian Ocean using 'spectral analysis.' Spectral analysis is intended to detect cycles in a series of measurements. One of the problems with spectral analysis is that you need a long record, one that should include at least seven of the cycles. The cores from the southern Indian Ocean were long, representing most of the last half million years of Earth history. They were also taken far from land so that the sediments should represent changes in the ocean alone. The spectral analysis of the $^{18}O:^{16}O$ record revealed periodicities at 106,000, 43,000, 24,000, and 19,000 years. These closely approximate the Milankovitch periods for eccentricity, obliquity, and the two major precession frequencies. It was evident that the Milankovitch cycles were controlling the frequency of ice advances and retreats even if the insolation changes alone did not seem large enough to produce the climatic changes Earth had undergone. It was also evident that the major glacial–interglacial cycles of the past half million years or so were occurring on a 100,000 year time scale, closely following the Earth's eccentricity cycle. But, they stated, the eccentricity cycle is the weakest in terms of changing the insolation, so there must be other factors working to amplify the signal. The question as to how the changes in eccentricity are related to glaciation was to be a source of argument for 20 years.

The Hayes, Imbrie, Shackleton paper ended with: "It is concluded that changes in the Earth's orbital geometry are the fundamental cause of the succession of Quaternary ice ages. A model of future climate based on the observed orbital-climate relationships, but ignoring anthropogenic effects, predicts that the long-term trend over the next several thousand years is toward extensive Northern Hemisphere glaciation."

Global temperatures had risen steadily from 1910 through the 1940s, and then declined during the 1950s through the 1970s. By the late 1970s many climatologists thought this might be a sign of the onset of a new ice age.

Strange to say, during those days no one seems to have bothered to project the Milankovitch cycles into the future. André Berger and Marie France Loutre of the Université Catholique de Louvain in Belgium not only did detailed calculations of the orbital cycles millions of years into the past, they also projected them into the future. They realized that eccentricity of the orbit is decreasing and will approach 0 (meaning that the Earth's orbit becomes almost circular) in about 25,000 years. After that, the eccentricity will increase again. This makes the insolation curve for the future look very different from that for the past. The present interglacial is unique among those of the past few million years and, without human interference, is likely to last 70,000 years. We will discuss this more later.

During the 1970s and 1980s many new analyses of the $^{18}O{:}^{16}O$ records were carried out on deep-sea samples from many different parts of the world. There were small differences from one place to another, and getting an age handle on the sediments was not very easy. For many of the cores studied, it was simply assumed that variations in the $^{18}O{:}^{16}O$ ratio were indeed a reflection of the Milankovitch orbital cycles, and they were used to establish the ages. This, of course, means that the data cannot be used to test the Milankovitch hypothesis, since it had already been assumed to be correct in establishing the age of the samples. Nevertheless, everything seemed to be going very well until the early 1990s. A composite 'average' record of the $^{18}O{:}^{16}O$ ratio in the shells of benthonic (bottom dwelling) foraminifera in marine cores was developed as a standard, and is shown in Fig. 13.17. It was thought that it was essentially

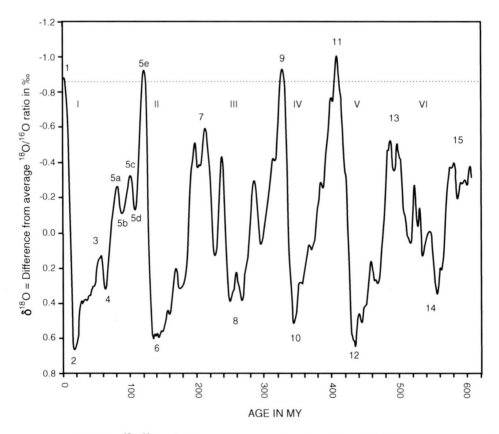

**Fig. 13.17** A composite record of the $^{18}O{:}^{16}O$ ratio in deep-sea benthic foraminifera (the 'Specmap Stack'), scaled to the average value. Note that the *left-hand* scale, the value of $\delta^{18}O$, is 'upside down,' with the values increasing downward, following the convention established by Emiliani. This allows us to visualize the data as a temperature curve, where up is warm and down is cold. When this composite was constructed it was estimated that about 75 % of the signal shown here was due to ice volume. Odd numbers are interglacials; even numbers are glacial episodes, following Emiliani's system. Roman numerals mark the 'Terminations' of the glacial episodes, the sudden changes from glacial maxima to interglacials. The *dotted horizontal line* is the present condition. It is evident that climatic states as warm as that of today are very rare

a reflection of the volume of ice on land, with some minor component of the signal coming from changes in the temperature of the ocean's deep water.

We also learned a lot more about how much ice was where during the last glaciation. The great ice sheets were all around the North Atlantic, covering northeastern North America and northern Europe, but not Siberia. There was a western North American ice sheet over the Cordillera, but not much in Alaska. It is thought that the great ice sheets were concentrated around the North Atlantic because the water there was relatively warm and could supply huge amounts of water vapor to the atmosphere to make snow.

In the early days of investigation of the Milankovitch hypothesis it was thought that glaciation of the Northern and Southern hemispheres must alternate. This is because of the precession cycle. When the Earth is closest to the Sun in the Northern Hemisphere winter, as at present, it provides extra heating for the southern hemisphere summer, and vice versa. It seemed obvious that if conditions were right for glaciation in the Northern Hemisphere they would be right for an interglacial in the southern hemisphere. However, in the 1970s and '80s investigations in the southern Andes and in Antarctica showed that this was not true. Glaciers advanced and retreated in both hemispheres almost simultaneously.

We now believe that this is because of the distribution of land and sea in the southern hemisphere. The continent of Antarctica is almost circular, centered on the South Pole. Almost all of it lies south of the Antarctic Polar Circle, and it is almost completely covered with ice up to 4.5 km thick. Only the northern tip of the Palmer Peninsula, that extends north toward South America, is crossed by the polar circle. Southern South America is narrow and mountainous. It does have glaciers, some very large, but the total ice area there is tiny compared with Antarctica. The problem for the Antarctic is that the ice is already so thick that there is no room for more unless the sea level goes down. The ice cap is so high, and the temperatures in the atmosphere over it so cold, that the air can hold almost no moisture. It is a desert. However, sea level falls when ice sheets build up in the Northern Hemisphere and then the ice-covered area on Antarctica can expand. As it does, it cools the neighboring tip of southern South America, and glaciers advance in the Andes too. How much extra ice builds up on Antarctica? About 10 % more than the 23.5 million cubic kilometers that is already there. So it works out that the glaciations in both hemispheres are synchronous even though the insolation is opposite.

The sawtooth nature of the 'paleotemperature' curve shown in the original Emiliani and Geiss figure and in the composite '75 % ice volume' curve shown in Fig. 13.17 has been a puzzle until recently. What it means is that ice builds up over the northern continents gradually, advancing and retreating, with the advances each time a bit greater than the last. It reaches a maximum size where the ice sheets extend southward almost to 40° N latitude. Then very suddenly the ice retreats and disappears, except in Greenland. After a relatively brief 'ice-free' interval the process starts over again. When you look closely at Fig. 13.17 you will see that each of the glacials has had a somewhat different history. The last interglacial was particularly interesting. It was warmer, and the ice seems to have had several false starts in ending it. However, you don't need to look at that diagram very long to realize what a strange, unique time we live in. It is usually stated that about 10 % of the time Earth is in an interglacial, but this diagram shows that it is even less. The Earth's 'normal state' for the past half million years or so is to have large ice sheets on the continents around the North Atlantic.

If you want to get a feeling for what conditions north of the Mason–Dixon line were like most of the time during the past half million years, I recommend a trip to Iceland. It is not one of the most favored tourist destinations, but if you really want to get away from it all, Iceland is the place. Some years ago I attended a geologists meeting in Reykjavik and took ten days off afterward, intending to drive the Ring Road around the island. The Ring Road is about 1350 km long. It doesn't sound like that far if you are used to driving the US Interstates or on the European Autobahns. As a tourist you can't easily drive across Iceland. There are no roads, but only tracks for 4-wheel vehicles, which are advised to travel in groups, preferably with someone who knows the way. I later learned that you can take a 'tourist bus' across the wild central area, which sounds like a great idea. One thing to remember is that when you rent a car, you will find that windshield replacement is not usually covered by the insurance. This is because paved roads exist only around the major towns (and now, I understand, for only part of the Ring Road). All of the 'highways' are gravel, with some cobbles inadvertently mixed in and deep gravel on the sides. Sometimes they are just wide enough for cars coming in opposite directions to pass, but if you pull to the side and get into the deep gravel on the side it's a bit like a slalom. It is recommended that you slow down when you pass cars coming in the other direction. Trucks seem to know how to grab those cobbles with their tires and throw them into the air just as you pass them. On the open road, the half-life of a windshield without cracks is measured in days. Mine went on the second day. Once out of the city the roads are fine. Fine means that the speed limit is 80 km/h, but you will be lucky to go 50 km/h before trouble sets in. There are occasional, rather unexpected, sharp turns where the roads go around boulders which are thought to be the homes of trolls. There are Icelanders who specialize in identifying which rocks trolls live in. It is very bad to disturb a troll's home, so roads go around them. As I understand it trolls only come out in the dark because they get turned to stone if hit by sunlight, so as long as you drive in the daylight you are OK. A lot of the rock formations are considered to be former trolls who made the mistake of staying out too late and getting caught in the sunlight.

The landscape is both beautiful and desolate. There are a few trees in Iceland but the landscape is mostly lava, tundra, ice-covered mountains in the distance, and sometimes the blue-white line of one of the ice caps on the horizon. The rocks are dark, the tundra greenish yellow, and the sky an incredible blue when there are no clouds or fog. There are broad sandy areas barren of plants. These are the outwash plains from the ice caps, appropriately called 'sanders.' It is about as different from the modern US Midwest as anything could be without actually having ice underfoot. After 5 days of driving and tourism I had reached Husavik, a delightful town on the north coast. Hearing that there might be delays due to flooding east of Reykjavik, I turned around and tried to retrace my route. I missed a turn and added another few hundred beautiful kilometers to my trip. You can often see a town a few kilometers ahead, only to discover that the road has to go around a fjord thirty kilometers long to get to it. So what looked like a short drive may turn out to take a few hours.

Imagine eastern North America as it was most of the time before the present interglacial. Today northern Canada is tundra, without trees. It is similar in many ways to Iceland, except that there are no roads at all. To the south of the tundra are the "North Woods" which starts with dwarf spruce trees around the latitude of Churchill on Hudson Bay (58.5° N). Larger evergreen trees continue south to near the Great Lakes. Then follows a forest of deciduous trees that extended down to the southern border of Tennessee until it was cut down in the 18th and 19th centuries to use the land for agriculture. Further to the south, nearly to the Gulf Coast, is a largely intact forest of pines. Along the Gulf Coast is another forest of warm-loving deciduous trees. Now imagine all those bands of vegetation compressed so that the edge of

the tundra is along the US-Canada border but the Gulf Coast forests are still intact. Now imagine those bands of vegetation expanding and contracting as the Laurentide ice sheet over eastern Canada and the northern US grows and shrinks. Most of the detailed evidence has long since been destroyed, but there is still a good record of what happened in the last deglaciation. We will discuss that later.

In addition to cores of deep-sea sediment, new records of the ancient climate were being uncovered by coring ice. Iceland has three small ice caps, but those on Greenland and Antarctica are much larger and thicker. The idea of drilling and coring those ice caps originated in the mid-1960s and by the 1980s a number of projects were underway. It was also discovered that caves, with their beautiful dripstone formations of stalactites, stalagmites and columns, could be another source of paleoclimate information. The $^{18}O{:}^{16}O$ in these carbonate deposits tell us about the temperature and rainfall of the past.

Investigation of these required a better understanding of how the $\delta^{18}O$ works with regard to evaporation, precipitation, and transpiration by plants. We now know a lot about this, largely thanks to Willi Dansgaard of Copenhagen. Figure 13.18 summarizes how the system works.

Figure 13.18 shows that average $\delta^{18}O$ of modern ocean water is 0. The average salinity of the ocean is 34.5 ‰. (Salinity is the amount of salt dissolved in the water, expressed in parts per thousand, shown by the symbol ‰). Evaporation selectively removes water molecules with the lighter isotope $^{16}O$ so that the $\delta^{18}O$ becomes positive; it also increases the salinity. Precipitation back onto the ocean surface enriches the surface ocean waters in $^{16}O$, so that the $\delta^{18}O$ becomes negative; at the same time the addition of fresh water decreases the salinity. However, the raindrops tend to collect the water vapor molecules containing $^{18}O$,

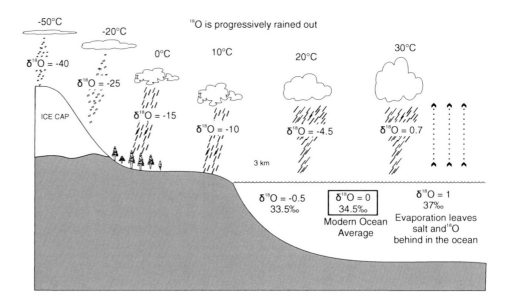

**Fig. 13.18** Schematic diagram of how evaporation and precipitation produce different ratios of $^{18}O{:}^{16}O$. See text for discussion

so that the remaining vapor becomes progressively 'lighter,' that is, the $\delta^{18}O$ becomes more negative. The depletion in $^{18}O$ becomes more extreme with colder temperatures. Over land the water is continuously being recycled as rain falls, plants take up the water and transpire vapor back into the air. At each step the molecules containing $^{18}O$ are retained, so that the water vapor in the air becomes progressively enriched in $^{16}O$ with distance from the coast. If all this sounds terribly esoteric, it is. However, knowing all this, you can tell from cave deposits whether the rainwater that fell in Tibet 15,000 years ago originally came from the Indian Ocean or the South China Sea! That tells you how strong the different monsoons were then. Cave deposits are powerful means of gaining insight into past climates.

However, as a monkey wrench to throw into the works, nothing could have been more appropriately named than 'Devil's Hole.' Devil's Hole is a hot spring in Death Valley. Carbonate rock has precipitated along the side of the spring, and since the 1980s cave deposits, particularly stalactites and stalagmites, have been investigated as possible records of the regional paleoclimate. They usually reflect the local rainfall and temperature patterns. When the record at Devil's Hole was investigated in the early 1990s, it showed an amazing $^{18}O{:}^{16}O$ record that looked almost exactly like that found in deep-sea cores, extending from 60,000 years ago back to about 550,000 years ago. Controversy arose because the investigators were able to use the Uranium/Thorium dating method to determine the end of the penultimate glaciation to be 145,000 years, rather than the 128,000 years previously accepted. As Cesare Emiliani was fond of saying, "nothing is ever perfect."

The Uranium/Thorium dating method takes advantage of the fact that one isotope of uranium, $^{234}U$, radioactively decays by releasing an alpha particle (the alpha particle consists of two neutrons and two protons; it is essentially the nucleus of a helium atom) to become a radioactive isotope of thorium, $^{230}Th$. $^{234}U$ has a half-life of 244,500 years, and $^{230}Th$ has a half-life of 75,400 years. In reality, both of these isotopes are part of a long decay series that starts with uranium 238 and leads through a series of 14 radioactive intermediate isotopes to end up as lead 206 ($^{238}U$), $^{206}Pb$, which is stable and doesn't decay. Many of the intermediate isotopes are used for age dating in one way or another. The half-life of the whole series is 4.5 billion years, roughly the same as the age of the Earth, so it was very important in finding out how old our planet is. $^{234}U$ and $^{230}Th$ are incorporated into carbonate when it is deposited, and measuring the ratio between them can date rocks formed over the past few hundred thousand years.

The problem was that the Devil's Hole record indicated that the last interglacial began 145,000 years ago, not 128,000 years ago as the marine records had indicated. The Devil's Hole record also indicated that the last interglacial was

20,000 years long rather than 10,000 years long. Just when most of the scientific community was coming to accept the role of orbital cycles in governing Earth's climate history over the past million years or so, these findings indicated that climate change was clearly out of phase with the Milankovitch cycles. It was soon argued that Milankovitch cycles had nothing to do with the Quaternary climate changes. To make a long story short, there has been much discussion about whether the Devil's Hole record upsets or verifies the Milankovitch theory. Something seems to be wrong somewhere, but my personal opinion is that if I were to look for a record of global climate over the past half million years, a hot spring in Death Valley would be about the last place I would look. Since the Devil's Hole controversy, hundreds of additional marine cores, and ice cores in the Antarctic and Greenland, confirm the same pattern of climate change as had been reported by Hays, Imbrie and Shackleton. Devil's Hole is an interesting anomaly. All it shows is that Milankovitch cycles do not rule the climate everywhere on Earth at all times.

The drilling on ice began to produce spectacular results in the late 1980s and '90s. A lot had been learned since the first ice cores were taken in the late 1960s. Recovering an ice core is a complicated business. The friction in drilling produces heat that would tend to melt the ice, and once recovered, the cores must be kept frozen. However, ice cores have some enormous advantages over deep-sea sediment cores. Many show annual layers that can be counted so that there is little question about the age of a particular level. Layers of volcanic ash can be analyzed to determine which eruption they represent, and thus also provide precise dates. They contain air bubbles that allow past concentrations of greenhouse gases to be measured. The trapped $CO_2$ can be dated back to 40,000 years or so. The oxygen isotopes of the ice yield a curve that looks very much like that obtained from the shells of foraminifera. But since the ice is $H_2O$, the ratios of the stable isotopes of hydrogen, $^1H$ (hydrogen) and $^2H$ (deuterium) can be used to determine the air temperature when the snow fell. As you can guess, the heavier water molecules, those with a deuterium atom or the $^{18}O$ isotope, are preferentially incorporated into snowflakes.

Paleoclimatologists were amazed and overjoyed when in 1994 a Russian-French-American group reported the results of a joint project, recovering continuous ice cores down to a depth of 3,623 m near the Russian Vostok station in central eastern Antarctica. Vostok station is located at 78.45° S, 106.87° E, at an elevation of 3420 m. The coring stopped 120 m above the surface of Lake Vostok. Lake Vostok is a huge subglacial lake about the size of the state of Connecticut. The coring stopped in lake ice that had frozen onto the ice sheet from below. The ice core contains a record of conditions near the South Pole extending back to 420,000 years ago. Lake Vostok has been isolated beneath the ice for at least that long. There is great curiosity about what it might contain. Are there living things down there? Organisms that have never known sunlight or temperatures above freezing? It has been agreed that penetration the last 120 m to Lake Vostok will not be done until someone understands how to do it with sterile tools so bacteria cannot accidentally be introduced into it.

Figure 13.19 shows the Vostok polar temperature record based on the proportions of two isotopes of the element hydrogen in the water molecules that make the ice. Unlike most other isotopes of elements, those of hydrogen have names. First is hydrogen itself, which has only a single proton in its nucleus, and deuterium, which has a neutron added to the proton in the nucleus, and tritium, which has two neutrons in the nucleus. Tritium is unstable and decays radioactively with a half-life of only 12.3 years. Tritium was very rare in nature until the hydrogen bomb tests of the mid-1950s.

In the same way as the oxygen isotopes vary with temperature, the deuterium/hydrogen ratio in water molecules varies with air temperature where snow formation takes place. This record has much more detail than is known from deep-sea cores, but the overall shape of the sawtooth curve is strikingly similar to that known from the deep sea. Because this record is based on evidence wholly different from that of the deep sea it is proof that similar climatic changes affect both Polar Regions simultaneously. Remember that the deep water in the ocean originates by sinking in the Polar Regions. Even more interestingly, close examination of this record reveals that it also reflects the pattern of variations in the Milankovitch orbital parameters at 65° N, but not at 65° S. This is a very important confirmation of the hypothesis that it is the changes in insolation at the Arctic Circle that have driven climate change over the past half million years

The temperature recorded by the deuterium/hydrogen ratio is warmer than the air temperature at the surface of the Antarctic ice cap. This is because it reflects the temperature

where water vapor condenses from the air to form precipitation, which is snow in the Antarctic. The air column over the Antarctic has a permanent and stable 'temperature inversion.' Ordinarily, the temperature of the air declines with altitude because of the decrease in pressure. But over the Antarctic it increases with altitude over the first few to 500 or so meters. This means that the air keeps getting less and less dense upward and the heavier air below is unable to rise and convect as it does most everywhere else. The temperature at the surface is very cold (typically −80 to −60 °C) and the air at the top of the inversion is typically 25–30° warmer (−40 to −35 °C).

In the meantime, another problem with the interpretation of the $^{18}O$:$^{16}O$ signal in deep-sea cores had raised its ugly head. The original assumption in using the $^{18}O$:$^{16}O$ ratio in deep-sea benthic foraminifera as a record of the ice ages had been that the temperature of deep-ocean waters had not changed very much. It was thought that 75 % of the signal was ice volume. It seemed like a reasonable assumption because the ocean's deep waters have an age of a hundred to a few thousand years; much shorter than the length of an ice age or interglacial. The time scales differ from place to place because the deep water sinks only at a few places and once inside the ocean interior it takes a while for it to penetrate to all the nooks and crannies.

But there would be plenty of opportunity for changes between the time of the last Glacial Maximum, 22,000 years ago, and today. Dan Schrag at Harvard University made a very clever attempt to solve this problem by looking at the $^{18}O$:$^{16}O$ ratio in pore waters in sediments of the age of the Last Glacial Maximum and comparing it with the $^{18}O$:$^{16}O$ ratio in the bottom water at a site in 3000 m water depth in the tropical Atlantic. The pore waters are trapped in the sediment as it is deposited, but as more and more material is deposited the sediments compact and the pore water is driven out, that is upward. This means that this technique will only work in relatively young sediment, before compaction

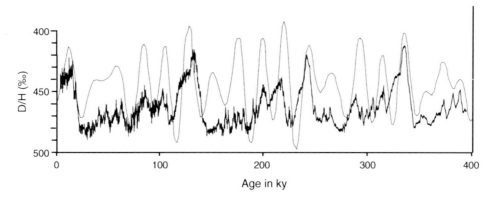

**Fig. 13.19** The Vostok ice-core record of the Deuterium/Hydrogen ratio interpreted as the paleotemperature of precipitation formation over Antarctica since 400,000 years ago. The Milankovitch insolation curve for 65° N is shown as a *dotted line*. Note that, like the plots of $\delta^{18}O$, the D/H scale is 'upside down,' with the numbers increasing downward. Again this allows us to visualize the curve as representing temperature with warm up and cold down

has taken place. Sure enough, the pore waters were different from the bottom waters, indicating that the temperature of the deep ocean water off northern South America was about 4 °C colder during the Last Glacial Maximum than it is today. He concluded that about 60 % of the $^{18}O{:}^{16}O$ ratio was due to change in ice volume. Nick Shackleton used a much more complicated but more precise method to arrive at essentially the same figure. Since we have two wholly different methods arriving at the same answer, it is fairly safe to conclude that this matter is largely settled.

In 2000 Nick Shackleton used this information to produce a diagram of the ice-volume variations in terms of both the $^{18}O{:}^{16}O$ ratio and sea-level change. Sea-level change is an easier way to visualize the ice volume changes because for most of us volumes of tens of millions of cubic kilometers of ice have no meaning. I present a modified version of his diagram in Fig. 13.20.

Comparing the dotted curve in Fig. 13.20, which represents the insolation at 65° N and the solid black ice-volume curve, it is evident that there are many similarities. If you look closely you will see that the changes in insolation occur a few thousand years before the ice volume responds. Buildup and decay of ice sheets takes time, so this lag is to be expected.

André Berger and Marie France Loutre of the Université Catholique de Louvain in Belgium devised a program to calculate hypothetical ice volume changes from the Milankovitch orbital variations. The results of their calculations, which are for both the past and the future, are shown in Fig. 13.21. Incidentally, the Université Catholique was also home to the Abbe George Lemaître, who in 1932 proposed the idea of the Big Bang for the origin of the universe. So much for the 'conflict between science and religion' in that part of the world.

Figure 13.21 shows the declining eccentricity of Earth's orbit. It will reach a minimum 26,000 years from now. This will be the most nearly circular orbit our planet has had in the last million years. When all the orbital motions are taken together, the insolation pattern at 65° N for the next 60,000 years looks different from that in the past. Until about 60,000 years ago, the insolation at 65° N showed wide swings. Then came an odd episode from 60,000 years ago until 30,000 years ago when the curve takes on a peculiar plateau shape. This is followed by a return to the broad swings, the downward trend marking the last glacial maximum and the upward trend forcing the ice to melt and leading into the present interglacial. The insolation at 65° N has been decreasing since 10,000 years ago, and this was what led to the belief of many in the 1970s that we were headed toward another ice age. But it turns out that we are only 2,000 years from the bottom of the declining insolation curve and then it will start to increase again. The next significant decline in insolation at 65° N will not occur until 60,000 years in the future. The lower diagram shows Berger and Loutre's hindcast (a hindcast is 'prediction' for the past) and forecast of Northern Hemisphere ice volumes based on the insolation and projections of future levels of carbon dioxide. The solid line represents the condition if atmospheric $CO_2$ concentrations had continued to be the same as they were during the last glacial–interglacial cycle. The dashed line indicates the condition if $CO_2$ levels rise to 750 ppm, double the present concentration and about three times the preindustrial concentration. We will discuss these and other aspects of this diagram in later sections of this book.

In 2004 Jacques Laskar and his colleagues at the venerable Institut de Mécanique Céleste et de Calcul des Éphémérides in Paris made even more detailed calculations of the orbital variations for the past and into the future. For the next 100,000 years the results are essentially identical to those of Berger and Loutre.

I should note that in reviewing the literature while writing this book, I came to the realization that almost everyone who worries about these things seems to be concerned with what

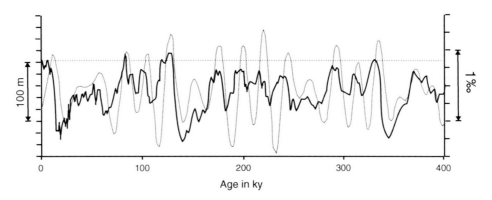

**Fig. 13.20** History of changes in ice volume on Earth over the past 400,000 years, shown as sea level change (*left scale*) and difference in the $^{18}O{:}^{16}O$ ratio (*right scale*). The insolation at 65° N is shown as a dotted squiggly curve in the background. The present condition is marked by the *dotted horizontal line*

**Fig. 13.21** Variations in eccentricity of Earth's orbit, June insolation at 65° N, and a hindcast/forecast of ice volume/sea level change from 200,000 years ago to 125,000 years in the future. The 0 sea level assumes no significant Northern Hemisphere ice (i.e. no ice on Greenland) and is 6 m higher than today. The *vertical dashed line* is 'today.' The development of civilization from the Roman Empire to the present fits within the width of that line. After Berger and Loutre (2002)

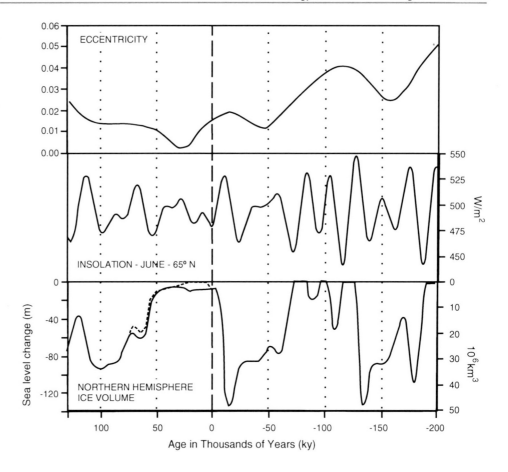

causes glaciation. In reality, glaciation, with larger or smaller ice sheets on northern Europe and North America, has been the normal condition for more than 90 % of the past million years. The real question is what causes interglacials, the condition we have now, when only Greenland has an ice sheet. Interglacials are the abnormality in an otherwise icy world. The ends of the glaciations are quite abrupt; Wally Broecker named them "Terminations." Each of the terminations seems to be associated with a combination of the Milankovitch parameters that has a maximum of insolation at 65° N coinciding with an obliquity maximum. Why the Earth stays in a glacial condition most of the time has to do with feedbacks in the climate system: albedo, changes in circulation of the atmosphere as the ice sheets thicken and block the winds, and the dynamics of the ice sheets themselves. All are topics we will discuss below.

## 13.7   The Discovery that Milankovitch Orbital Cycles Have Affected Much of Earth History

When I was a graduate student at the University of Illinois in the 1950s I took a course in 'Pennsylvanian' geology. 'Pennsylvanian' refers to the geological epoch between 318

and 299 million years ago when the coal beds of the eastern US, Britain, and Western Europe were formed. In most of the rest of the world this is the Late Carboniferous. These were the deposits that Charles Lyell had interpreted as evidence of a tropical climate. The Pennsylvanian rocks in the eastern US are remarkable in that they consist of a sequence of different rock types that are repeated over and over again. Each sequence starts with river deposits in channels cut into the underlying rocks. Above these is a fossil soil with tree roots. The fossil soil is very rich in the mineral 'kaolin,' which is used to make porcelain, so it is mined in several states. The soil is overlain by the coal bed, which is compacted fossil plant material. Above this a remarkable change occurs. The coal bed is overlain by black shale which is often rich in fossils of marine fish. Above this are shales and limestone that contain marine fossils. At the top is the erosion surface with river channels filled with sand and gravel. This sequence is repeated well over one hundred times.

In the 1950s the origin of these sequences was still controversial. It was agreed that they reflected repeated episodes of sea level change, but there were two ideas about what caused the sea level change. One group believed that the North American continent had bobbed up and down due to some unknown cause. The other group believed it had something to do with glaciation which had affected southern

South America, southern Africa, Australia, and India. The problem with the glacial hypothesis was no one understood how there could have been many repeated glaciations in such widely separated areas, some far from the pole. Also, it was common knowledge at the time that there had only been four Quaternary glaciation cycles, whereas in the Pennsylvanian there had apparently been many.

It was just a year after I had taken the courses in Quaternary geology and Alpine geology in Zurich, and recalled the discussions about Milankovitch cycles and the possibility that there had been many more than four glaciations in the Quaternary. I wrote a term paper suggesting that maybe the Pennsylvanian coal cycles might reflect a long repetition of southern hemisphere glacial/deglacial cycles on Milankovitch time scales. The site of the glaciation would have been the continent of Gondwana, an agglomeration of South America, Africa, Australia, Antarctica, and India hypothesized by Alfred Wegener as part of his theory of continental drift. Although my idea was considered quite radical (both Milankovitch cycles and continental drift were heretical ideas in the United States at the time), I got an "A" on my paper. In those days it was a rare instance of these heretical ideas being considered seriously by a member of the US geological establishment.

Now, to go on to the early 1960s, Cesare Emiliani had another idea that was to bear extraordinary fruit. He realized that the ocean had an undiscovered history and thought it might be very interesting to find out more about it. But to understand what was known in the 1950s we need to go back two centuries.

In the middle of the eighteenth century, British explorer Henry Ellis made the first deep soundings of temperature in the tropics, finding cold water below a warm surface layer, indicating the water came from the Polar Regions. The realization that the deep water in the oceans must come from the Polar Regions is usually attributed to 'Count Rumford' about 1800. Count Rumford was born Benjamin Thompson in 1753, in Woburn, Massachusetts. He worked as a teacher in Rumford, Massachusetts. He was fascinated with observing nature, particularly eclipses of the Moon, and tried to build a perpetual motion machine. At the time of the American Revolution he was suspected of being a royalist.

In fact, he was. He left America for England, to return as commander of the Queens Horse Dragoons fighting against the rebel colonists. During this time he became very interested in following the motions of the Moon, learning about gunpowder, and investigating the heat of the cannon barrel as it was fired. It was an interesting mix of curiosity about astronomy, chemistry and physics, being pursued in the midst of battle (Figs. 13.22 and 13.23).

Returning to Europe he went to Bavaria. That is where I encountered him, in spirit. In Munich there is a huge park, extending like a piece of pie, from the royal Residence near the city center out to the distant outskirts. It is known as the 'Englischer Garten,' so called because it is designed in the 'English style'—an informal park with curving pathways,

**Fig. 13.22** The *Glomar Challenger*, the scientific drillship that carried out operations for the Deep Sea Drilling Project from 1967 to 1983

**Fig. 13.23** Layered sedimentary rocks in Punta di Maiata, southern Sicily, with banding faithfully recording the orbital cycles. These sediments were laid down on the floor of the Mediterranean and have been uplifted since as part of the island. The distinctive patterns can be matched to different astronomical orbital calculations to determine which is correct. Using stratal patterns like these and the cyclic variations in oxygen isotope ratios, the geologic time scale has been calibrated back to about 15 million years ago and further studies are expected to extend the calibration back into the Cretaceous. (Photo courtesy of Lucas Lourens.)

huge trees, grassy areas, and lakes. It was the brainchild of Benjamin Thompson, who in 1792 had somehow gotten to be the Bavarian Minister of War. It was at the munitions works in Munich that Thompson observed the heating of cannon barrels as they were being bored out. It finally led to his conclusion that heat is molecular motion. At the time the prevailing idea was that heat was an invisible fluid, called caloric. Thompson's idea was rejected for many years as too radical. The Bavarian Prince, Carl Theodor, liked Thompson so much he gave him the title of Count Rumford of the Holy Roman Empire. I don't know whether the people of Rumford, New Hampshire (now Concord, NH) knew about his honor or what they would have thought of it. Later on he returned to England, installed a glass door in the front of his fireplace so he could observe the flames more carefully, and redesigned fireplaces to be tall and shallow so the bricks in the back would reflect and radiate about 10 % of the heat into the room. Unfortunately, this discovery has been largely forgotten. Most modern American fireplaces are designed in such a way as to suck heat out of the room and send it up the chimney. He is also credited with inventing central heating, the smokeless chimney, the kitchen oven, thermal underwear, and the pressure cooker. Because of his opposition to the American Revolution, he is cited in the American accounts as having been both a scientist and a scoundrel.

When I was a student in Munich in 1953–54 I lived in a dormitory on the edge of the English Gardens, and ever since whenever I visit Munich I take a long walk through Count Rumford's legacy. For me he will always be the person who thought a large part of any city should be parkland.

At the beginning of the nineteenth century, Admiral Ivan Fedorovich Kruzenshtern (1770–1846; the one whose statue is on the north embankment of the Neva in St. Petersburg) made the first scientific circumnavigation of the globe, measuring surface and deep water temperatures, and finding that the entire ocean was filled with cold deep water. Incidentally, although it took place after her death, Kruzenshtern's voyage had been ordered by Catherine the Great.

Curiously, except for Lord Kelvin, no one seems to have considered the possibility that there might have been different conditions in the ocean in the past. When I was in college the deep-sea environment was known to have been cold, dark, and unchanging throughout Earth history.

In 1954 Cesare Emiliani had published the results of $^{18}O{:}^{16}O$ paleotemperature measurements on three deep-sea samples known from their fossils to be older than Quaternary. One was from the Pliocene, one from the Miocene, and another which he thought was Oligocene, but has since been found to be Eocene. Each of the older samples showed a progressively higher temperature, and the 'Oligocene' one indicated that the deep waters were about 11 °C. That is the temperature of the waters off New York today. Needless to say, this caused quite a stir in the scientific community, but no one knew what to think of it. Everyone knew that the deep water of the oceans sinks in the Polar Regions, so it must mean that in the past the Polar Regions had been much warmer than today.

As recounted in more detail in Intermezzo XV, a major objective in American science in the early 1906s was 'Project Mohole,' the effort to drill a hole to the Mohorovičić Discontinuity, the surface separating the Earth's crust from mantle. Cesare Emiliani, however, was convinced that much more could be learned from recovering long cores which would record the history of the ocean. As the cost projections for "Project Mohole" escalated, it was finally canceled.

In 1963, Cesare submitted a proposal termed "LOCO" for Long Cores to the U.S. National Science Foundation (NSF). I heard about this on one of my frequent stopovers in Miami on the way to Bimini. A suitable ship, the *Submarex*, was chartered for test drilling of cores on the Nicaragua Rise in December of that year. Instead of dropping a core barrel to the bottom on a wire, cores were taken through the drill pipe that extended from the ship down to and into the sea floor. I was one of those who received samples. The results were exciting and ultimately led to the development of the Deep Sea Drilling Project (1967–1983) (Fig. 13.22) and the Ocean Drilling Program (1984–2003). As I describe in the Intermezzi, I became intimately involved in scientific ocean drilling and served in a series of advisory roles to these drilling projects from 1965 until 2003.

Now, to get back to the orbital theory. At first the cores taken by the Deep Sea Drilling Project (DSDP) were often distorted and incomplete. Nevertheless, the information from them was always new and exciting. Then, as the technology developed, the quality of the cores improved. In parallel, Nick Shackleton's mass spectrometer that could turn out lots of $^{18}O$:$^{16}O$ measurements was cloned, improved, and introduced into many laboratories. Thousands and thousands of measurements were made. Finally, a major breakthrough: in 1979 a new method of coring was introduced to the DSDP. The 'hydraulic piston corer' is the drill pipe version of the piston corer that had been used by oceanographic vessels. Since then hundreds of beautiful undisturbed cores of the soft upper sediments of the ocean floor have been recovered. Almost all of them showed the rhythmic changes of the sediments reflecting the changes in ocean circulation in response to the buildup and decay of ice sheets. In 1995 the effort to recover more sediment cores recording the younger part of Earth history was enhanced when the French launched the research vessel *Marion Dufresne*, named for the eighteenth-century French explorer Marc-Joseph Marion du Fresne. The *Marion Dufresne* has a giant piston corer that can take a weight of ten tons on the top of a large diameter core barrel 60 m long. With these resources we have accumulated information that gives us real insight into the history of the oceans. The $^{18}O$:$^{16}O$ measurements on the cores show that the Milankovitch cycles in deep-sea sediments go far back in time.

The idea that the orbital cycles might be reflected in the layering of sedimentary rocks goes back to a great American geologist of the nineteenth century, Grove Karl Gilbert. One of the characteristics of sedimentary rocks is that they are usually 'stratified,' that is, they are layered. It is the layers you see in the sketch of the rocks as Siccar Point in Fig. 4.3. The layering means that something about the environment in which the sediments were being deposited must have changed. Perhaps it was the supply of sediment, the effect of storms throwing the sediments back into suspension, mudslides down the slope of the basin, or possibly changes in the climatic regime affecting the region where the sediments originated. It took a long time before anyone came up with good ideas about what the layering represents, and probably there are many causes. In 1895, 20 years after Croll had suggested that orbital cycles might leave a geologic imprint, Gilbert suggested that the rhythmic bedding in Cretaceous rocks in Colorado might represent the 21,000 year precession cycle. It took almost a century before the idea was taken up again. In 1955, just as Emiliani was publishing the results of his oxygen isotope analyses, Gustav Arrhenius wrote that he suspected the variations in the carbonate content in the cores recovered by the Swedish expedition might reflect Milankovitch cycles. Thirty years later the idea that some strata appear to occur in cycles, probably Milankovitch cycles, blossomed everywhere. In 1982 Walter Schwarzacher in Belfast and Al Fischer at Princeton proposed that many sequences of alternating limestone and shale might reflect Milankovitch cycles. Almost immediately Milankovitch cycles were seen everywhere, mostly in Cretaceous rocks, but also in Carboniferous, Permian, Triassic, Eocene, Miocene and Pliocene strata. You did not need to make measurements to discover the cycles; you could see them in the exposures of rock. For a while everyone seemed to agree that they were definitely orbital cycles reflected in the sediments, but there was dispute over which cycles they were: eccentricity ($\sim$100,000 years), obliquity (41,000 years), or precession (21,000 years).

The University of Utrecht in the Netherlands had long had a reputation for excellence in micropaleontology and for improving the geologic time scale. The Paleontologic Institute, which I would visit from time to time in the 1960s and 70s, was located in a picturesque setting on the Oude Gracht (Old Canal) very close to the city center. In the 1970s and '80s a new campus for the University was built just outside the built-up area. The geological departments moved into beautiful new quarters. There are not many rocks to study in the Netherlands, so the favorite areas for research have come to be along the northern margin of the Mediterranean, from Spain to Turkey. In the 1980s Professor Johan Muelenkamp, the young head of the paleontologists, had the idea that the rhythmically-bedded early Quaternary and Pliocene rocks of

Calabria in southern Italy and on Sicily might reflect astronomical cycles (Fig. 13.23). The rocks contained abundant fossil planktonic foraminifers, and the species in the assemblages varied with the bedding. There had been a lot of study of the cycles in sediments less than a million years old, and Johan figured the time was ripe to see if the cycles could be traced further back. Nick Shackleton had used his isotope curves in an early effort to trace the record back. Utrecht graduate student Frits Hilgen took up the challenge, and completed his doctoral work on the Calabrian-Sicilian sequences in 1991. He was able to show a close correlation between the bedding and the astronomical cycles calculated by Andre Berger and Marie France Loutre at the Catholic University in Louvain, Belgium. It was truly amazing: one could take the astronomical curves and see them repeated in the rocks. More importantly, the details were there to see. Berger and Loutre had always made a series of different computations for the cycles going back in time, because, as you will remember from Chap. 11, small differences in the initial conditions have a big effect on the outcome. To project the astronomical calculations back in time they must be checked against some sort of evidence. Hilgen found out that it was possible to use the sedimentary exposures in Italy to determine which of the astronomical calculations was correct. Hilgen traced the correlation back to the early Pliocene, 5.5 million years ago.

However, there was a problem. The astronomical time scale did not agree exactly with the relatively few ages of the sediments based on the decay of radioactive elements. Frits was a graduate student. Those who doubted his work were experts representing the established view. There was one great comment in a public lecture to the effect that 'it would be tragic, considering all the work put into radiometric dating, if the astronomical calibration turned out to be correct.' After a few years an improvement in the technology of the radiometric dating solved the problem. The astronomical solution was correct, well almost.

The next step was taken by another graduate student, Lucas Lourens: to try to relate the changes in bedding to the changes in insolation. After all, this is the Mediterranean, not the Arctic. At first there still appeared to be a problem. None of the Berger and Loutre calculations fit the geologic data exactly. However, in the meantime a French astronomer, Jacques Laskar, had done a set of more detailed calculations of the astronomical variables taking into account the changes in the Earth's distribution of mass as the ice sheets built up and decayed, and the exact shape of the Sun (it is not a sphere, but is flattened at the poles because of its rotation, like the Earth). It turned out that these seemingly tiny corrections were critically important, and the entire astronomical time scale was revised. The corrections for the young part were very small, but then the older part fell into line,

and the correlation with geological data fit exactly, back to 5.5 million years.

The Mediterranean is located in a critical area for climate change. At present the Mediterranean lies in an arid region, and with its narrow, restricted connection to the Atlantic, it is saltier than the adjacent ocean. However, during glacial times it received more fresh water from rivers, and the evaporation rate was lower. Its surface waters periodically became fresher than those of the Atlantic. The alternations of the climate along the northern margin of the Mediterranean and changes in the salinity of its waters produced the rhythmic bedding and changes in the assemblages of planktonic foraminifera. It was an ideal area for recording climatic change for at least 15 million years.

With such a beautiful record, it was inevitable that the Utrecht group would take the next step and try to extend the record further back, before 5.5 million years. Now the record extends back into the Oligocene, about 30 million years ago, and fragmentary pieces are also in place for the Late Cretaceous-Paleocene, around 65 million years ago, and older.

The University of Utrecht invited me to be Guest Professor in the fall of 1993 to teach a course in paleoclimatology. However, I think I learned more from the Utrecht faculty and students than they learned from me. It was an incredibly stimulating experience as Frits and Lucas Laurens were setting one of the most fundamental areas of geology, the study of stratified rocks, on an astronomically calibrated base. This spectacular advance in our science has happened so fast that many geologists and almost no climatologists are aware of it.

In the meantime, one of Professor Michael Sarnthein's graduate students at the University of Kiel, Ralf Thiedemann, had written an extraordinary doctoral thesis. With thousands of analyses he had produced an amazing 8 million year record of the $^{18}O$:$^{16}O$ ratio in benthic foraminifera in a sequence of cores taken by the Ocean Drilling Program off northwest Africa. Alongside this was a record of variations in the dust coming from the North African deserts. The $^{18}O$:$^{16}O$ ratio largely reflects the volume of ice at high latitudes. The dust reflects climate change at low latitudes. The records indicate that climate changes over the past 8 million years have affected not only the Polar Regions, but the entire Earth. Figure 13.24 shows the last 5 ½ million years of the $^{18}O$:$^{16}O$ record.

The record shows some remarkable features. Before about 3.5 million years ago there were significant cycles, but these represent changes in the ice on the Antarctic. The glaciation of Greenland began about 5 million years ago, but the amount of ice on a fully ice-covered Greenland is only about 10 % of that on Antarctica. Until 3.5 million years ago, the present climate, indicated by the upper dotted line,

**Fig. 13.24** Cycles of $^{18}O:^{16}O$ ratios over the past 5 ½ million years. The *upper dotted line* is the present inter-glacial condition; the *lower dotted line* is the Last Glacial maximum

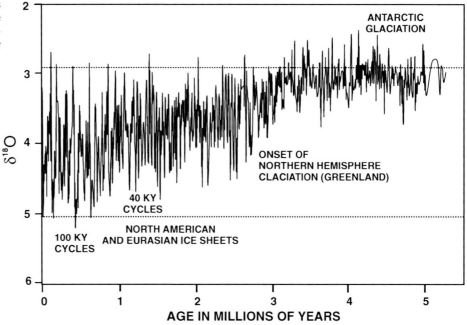

seems to have been about average for the Earth. Around 3.1 million years ago ice sheets began to grow and decay at 40,000 year time scales on the northern continents surrounding the North Atlantic. They reached a larger size around 2.74 million years ago, continuing to form and disappear on the 40,000 year time scale. Around 800,000 years ago the ice sheets became larger and the length of the cycles changed to about 100,000 years.

There has been a lot of discussion about what caused the change from 40,000 glacial–interglacial cycles prior to 800,000 years ago to 100,000 year cycles since. The arguments involve the eccentricity cycle, greenhouse gas concentrations, and the dynamics of ice sheets.

The conclusion of this section is quite simple. On time-scales of tens to hundreds of thousands of years, the Milankovitch orbital variations have had a great impact on our planet's climate. They have induced almost continuous climate change. It has been very unusual for the global climate to remain stable for as long as it has during the current interglacial.

## 13.8   Summary

Our Sun is a star about halfway through its lifetime of about 10 billion years. In its early stages it was less luminous than it is today. In the 1970s the problem of the 'faint young Sun' was considered a paradox. Why did the Earth not freeze over, become highly reflective and remain always in a frozen state? The answer was found in regulation of its temperature by greenhouse gases.

Under the influence of the other planets Earth's orbit is not circular but elliptical. The ellipticity of the orbit changes with time on a $\sim 100,000$ year time scale and the long axis of the ellipse precesses on time scales of the order of 21,000 years. The Earth's axis is tilted and also wobbles, by a few degrees on a time scale of about 41,000 years.

In 1875 James Croll published the first calculation of changes in insolation due to changes in Earth's orbit. It suggested that there might have been more than the one ice age known at the time. In 1882 Thomas Chrowder Chamberlin discovered that there had been an earlier ice age in North America. Shortly thereafter another, and then another even earlier ice age was discovered. Early in the twentieth century four ice ages were known.

In 1920 Milutin Milankovitch's early calculations of changes in insolation due to Earth's orbital changes, previously published in Serbian, were translated into French. A more complete set of his calculations was published in German in 1941, and finally translated into English in 1969. Vladimir Köppen and Alfred Wegener built on his ideas, realizing that it was the cool summers at 65° N, when the winter snow did not completely melt, that were critical to the growth of ice sheets. The problem with Milankovitch's hypothesis was that it suggested that there had been a large number of glaciations, but in the middle of the last century only four were known. Furthermore, Milankovitch's calculations indicated that the insolation minimum, and hence the maximum of the last ice age, should have been 24,000 years ago. The carbon-14 dates from the 1950s and 1960s indicated it was 18,000 years ago. The orbital cycle hypothesis was dead, but not for long. It was discovered that the rate of production of $^{14}C$ in the upper atmosphere varies with time. Over the past half century the carbon-14 scale has been calibrated using tree rings, and it now gives dates around 24,000 years for the Last Glacial Maximum.

In 1955 Cesare Emiliani showed that temperature record provided by the changing ratios of $^{18}O{:}^{16}O$ in fossil foraminifera in deep-sea cores indicated that there had been more than four glaciations. He had to guess at the length of the cycles, and thought, incorrectly, that they were the 41,000 year cycle of tilt of Earth's axis. A decade later dating of coral terraces representing the sea-level highstands of the interglacials suggested that the glacial–interglacial cycles were about 100,000 years long.

Emiliani realized that what was needed was a better record from the deep sea. His efforts led to the development of major programs to recover cores of deep-sea sediment: the Deep Sea Drilling Project (DSDP: 1967–1983), its successor, the Ocean Drilling Program (ODP: 1983–2003), and most recently the International Ocean Drilling Program (IODP).

In England Nick Shackleton improved the mass spectrometer used to measure oxygen isotopes so that many more measurements could be made. He argued that the oxygen isotope signal was more a reflection of the amount of glacial ice than changing ocean temperature. By the 1970s six glaciations had been well defined on the basis of deep-sea cores with no end in sight.

By 1976, it was established that the rhythm of glacial cycles followed the 100,000 year changes in orbital ellipticity, but it was recognized that other factors must amplify the signal. The signal had a peculiar saw-tooth shape. Apparently the northern hemisphere glacial ice built up gradually, over about 80,000 years and then the ice sheets melted off rapidly, in 10,000 years or so. These sudden transitions have been named 'terminations.' The intervening interglacials lasted about 10,000 years. We were already 11,000 years into the present interglacial. It was obvious to many geologists working in the field that a new ice age should be on its way.

New paleoclimate data were coming from ice cores in Greenland and Antarctica. The Vostok core from near the South Pole provided a 400,000 year record of Antarctic climate that was very similar to, but much colder than, the records from deep-sea cores.

More detailed projections of the orbital insolation patterns into the future show that the present interglacial is unusual. Orbital ellipticity is near a minimum, and without human intervention the present interglacial might last for another 60,000 years.

Examination of older geological deposits on land suggested that Milankovitch cycles pervade the record. Recovery of longer deep-sea records has shown that the cyclicity seems to go back without end. But there is a problem in fitting a timescale to the older orbital cycles. The changes in

the orbital parameters are inherently chaotic, and the calculations diverge as one goes further back in time. Frits Hilgen and Lucas Lourens at the University of Utrecht studied geologic sections on land to decide which of the possible mathematical solutions for insolation is the correct one. Step by step, the very detailed astronomical dating scale based on orbital parameters is being pushed into older and older ages. Deep-sea cores provide a detailed history of northern hemisphere glaciation to its onset almost 4 million years ago. We now know that the earlier glaciations followed at 40,000 year cycle, but 800,000 years ago the rhythm switched to a 100,000 year cycle. Its cause is still uncertain.

A Timeline for this chapter:

| ca. 200 BCE | Hipparchus describes the 'precession of the equinoxes' |
|---|---|
| ca 200 CE | Claudius Ptolemaeus cites Hipparchus 'precession of the equinoxes' in his *Almagest* |
| 1543 | Nicolas Copernicus' book *De revolutionibus orbium coelestium* arguing that the Earth orbits the Sun is published 1 year after his death |
| 1610 | Johannes Kepler publishes his first two laws of planetary motion |
| 1619 | Kepler publishes his third law of planetary motion |
| ca. 1800 | Count Rumford (Benjamin Thompson) postulates that the deep waters of the ocean come from the Polar Regions |
| 1803–1806 | Russian Admiral Kruzenshtern makes the first round the world voyage collecting scientific oceanographic information |
| 1842 | Joseph Adhemar proposes that precession of Earth's orbit caused the ice age |
| 1843 | Alexander von Humboldt rejects Adhemar's idea |
| 1864 | James Croll writes to Charles Lyell suggesting variations in Earth's orbit might be the cause of ice ages |
| 1875 | Croll publishes *Climate and Time, in Their Geological Relations* |
| 1882 | Thomas Crowder Chamberlin discovers that there had been more than one glaciation |
| 1895 | Grove Carl Gilbert suggests that Cretaceous rocks of North America may record orbital variations |
| ~ 1902 | Alfred Wegener makes balloon flights over sea ice |
| 1906 | After the earthquake, David Starr Jordan, President of the Leland Stanford Junior University remarks 'Agassiz was great in the abstract but not in the concrete' |
| 1906–07 | Wegener makes his first trip to Greenland as a member of a Danish expedition |
| 1912 | Alfred Wegener publishes the first of his paper on continental drift |
| 1920 | Milutin Milankovitch publishes *Théorie mathématique de phénomènes thermiques produit par la radiation solaire* |

(continued)

(continued)

| 1924 | Köppen and Wegener include Milankovitch's insolation curve in their book *Die Kimate der geologischen Vorzeit* |
|------|---|
| 1926 | The American Association of Petroleum Geologists invites Wegener to a special session; almost all the speakers sarcastically denounce his idea of continental drift |
| 1944 | Arthur Holmes devotes a chapter of *Principles of Physical Geology* to Wegener's hypothesis of continental drift |
| 1941 | Royal Serbian Academy of Sciences publishes Milutin Milankovitch's *Kanon der Erdebestrahlung und seine Anwendung auf das Eiszeitenproblem*; it will be translated into English as *Canon of insolation and the ice-age problem* in 1969 |
| 1947–49 | The Swedish Deep-Sea Expedition recovers the first long cores of deep-sea sediment using the Kullenberg piston corer |
| 1953–81 | The Lamont Geological Observatory's research vessel *Vema* recovers cores from throughout the world's oceans |
| 1954 | Cesare Emiliani publishes *Temperature of Pacific bottom waters and polar superficial waters during the Tertiary* showing the ocean bottom temperatures were warmer in the past |
| 1955 | Emiliani publishes *Pleistocene temperatures* showing variations in the ratio of $^{18}O$ to $^{16}O$ reflecting ice ages |
| 1958 | Emiliani and Johannes Geiss publish a composite paleotemperature curve indicating that there had been more than four glaciations each lasting about 41,000 years |
| 1958 | The idea to drill to the Mohorovičić Discontinuity in the Pacific Ocean basin is conceived at a backyard party in La Jolla California. The group names itself the 'American Miscellaneous Society" (AMSOC) |
| 1961 | Project Mohole drills five holes in 3,500 m (11,700 ft) of water off Guadalupe Island, Mexico, penetrating to 183 m (601 ft) below the sea floor. |
| 1962 | Mikhail Budyko publishes *Polar ice and climate* in the Proceedings of the Academy of Sciences of the USSR emphasizing the role of ice in planetary albedo |
| 1963 | Cesare Emiliani proposes a project to drill long cores in the sediments of the ocean basins (Project LOCO) |
| 1963 | The *Submarex* dills a test hole for LOCO on the Nicaragua Rise: I get samples to examine for nannofossils |
| 1965 | Wally Broecker and colleagues date coral terraces representing interglacial sea-level highstands to be 85,000, 130,000, and 190,000 years old |

(continued)

(continued)

| 1965 | A consortium, JOIDES = Joint Oceanographic Institutions for Deep Earth Sampling, is formed to carry out the goals of LOCO |
|------|---|
| 1966 | JOIDES' first drilling program takes place on the Blake Plateau off northeastern Florida |
| 1967 | Nicholas Shackleton receives his doctorate at Cambridge. He had improved the O isotope mass spectrometer and measured ratios in more cores, concluding that the $^{18}O/^{16}O$ isotope signal reflects primarily ice volume rather than temperature |
| 1967–83 | JOIDES' Deep Sea Drilling Project collects drilled cores in the world's ocean basins using the drill ship *Glomar Challenger* |
| 1969 | A poll taken at the meeting of the American Association of Petroleum Geologists found that 80 % rejected the hypothesis of seafloor spreading and plate tectonics |
| 1971 | A poll taken at the meeting of the American Association of Petroleum Geologists found that 95 % accepted the theory of seafloor spreading and plate tectonics |
| 1972 | Emiliani warns that a new ice age might be coming soon |
| 1975 | JOIDES becomes international |
| 1976 | Jim Hayes, John Imbrie, and Nick Shackleton publish '*Variations in the Earth's Orbit: Pacemaker of the Ice Ages.*' The paper ends with "ignoring anthropogenic effects … the long-term trend over the next several thousand years is toward extensive Northern Hemisphere glaciation" |
| 1978 | André Berger publishes detailed solar insolation curves |
| 1979 | The hydraulic piston corer is developed for the DSDP, greatly improving core recovery and quality |
| 1982 | Walter Schwarzacher in Belfast and Al Fischer at Princeton proposed that many older sequences of alternating limestone and shale may reflect Milankovitch cycles |
| 1983–2003 | JOIDES continues to collect cores from the ocean basins as the Ocean Drilling Program (ODP) |
| 1991 | Frits Hilgen in Utrecht traces orbital cycles in the rocks of southern Italy back to 5.5 million years ago |
| 1991 | Ralf Theidemann uncovers the glacial record of the Northern Hemisphere in an ODP core taken off northwestern Africa |
| 1992 | The isotope record at Devil's Hole Nevada throws the Milankovitch hypothesis into doubt |
| 1994 | The Vostok ice core project in Antarctica recovers 3,623 m of cores providing an climate record extending back 400,000 years |
| 1995 | France launches the research vessel *Marion Dufresne* with the capability of taking large, long piston cores |

(continued)

(continued)

| 1995 | Richard Muller and Gordon MacDonald propose that the Earth's orbit varies relative to the plane of symmetry of the solar system on a 100,000 year cycle |
| 2000 | Nick Shackleton revises estimates of sea-level change during the last 400,000 years |
| 2002 | Dan Schrag and colleagues devise a new technique for determining the temperature of ocean bottom water during the LGM |
| 2002 | André Berger and Marie France Loutre calculate hypothetical ice volume changes from the Milankovitch orbital variations for both the past and the future |
| 2003- | Recovery of long drill cores in the oceans continues as the International Ocean Drilling program (IODP) is headquartered in Japan |
| 2004 | Jacques Laskar and his colleagues make even more detailed calculations of the orbital variations into the future |

If you want to know more:
No real up-to-date summary I know of, but
Imbrie J. and Imbrie, K.P., 1979, *Ice Ages—Solving the Mystery*. Harvard University Press, 224 pp—is an entertaining read

Music: It seems obvious: Maurice Ravel's *Bolero*

Libation: To honor Milutin Milankovitch a Serbian wine: Prokupac is a heavy red, and dates back to the Middle Ages. A real touch of history. Not expensive, but hard to find. Slivovitz is a brandy made from plums to finish off the meal

## Intermezzo XIII. The Most Memorable Summer of My Life was 1968: Paris

Looking back I realize that sometimes events wholly unrelated to your career can have a huge influence on the path your life takes. For me it was the summer of 1968. During the spring of 1968 Vietnam War protests had grown to a fever pitch. Eugene McCarthy and Bobby Kennedy were challenging the leadership of Lyndon Johnson, running against him in the primaries. On March 31, Johnson announced that he would not run for a second full term as President. Then came a real shock: Martin Luther King was assassinated on April 4 in Memphis, Tennessee. In less than 5 years since the assassination of John F. Kennedy in Dallas, another American leader had fallen to an assassin's bullets. The country was in turmoil.

As a geologist or paleontologist and, I suspect, in many other academic fields in the middle of the twentieth century, the development of a career followed a simple plan. You were intended to become an expert in something. Preferably, the world's leading expert in something. In my field there were a lot of micropaleontologists, and each had to find some group of fossils they could specialize in. Planktonic foraminifera were

very important to industry, so there were lots of people working on them. Since industry people don't publish, they were safe. But those in the academic world were expected to publish papers with the results of new investigations. It was, and still is, important to produce and publish new data. In a crowded field these often turned into polemics against your colleagues, something still popular today.

But once you had selected your field of specialization you did not leave it. Paleontologists were expected to describe new species. Geologist colleagues often described this as stamp collecting. I was really lucky to have changed my research during my postdoctoral year to specialize in calcareous nannofossils, because the field was new. There were lots of new species to describe and most were useful in determining the age of rocks.

For me, that special summer started in Paris in May. I had spent the academic year 1967–68 at the Institute of Marine Sciences and had no teaching obligations. There was a meeting, the 'Colloque sur l'Éocène' on the geology of the Paris Basin. This is a very classic area, a nursery of geology. It had become a standard reference for geologists everywhere, but was in need of reexamination after a hundred years of neglect. We were meeting to report on new studies, trying to separate fact from fiction. France was in turmoil, and a general strike was threatened. The students of the Sorbonne had shut down the University. The Metro was not running. Getting a taxi was hopeless. You went everywhere by walking. Nevertheless, the geological field trips before the colloquium worked very well.

One of the sites we visited outside Paris had spectacular fossil oyster beds, looking good enough to eat. It was there that I learned about the fine points of consuming live oysters from a paleontologist who knew all about them - both fossil and recent. Henryk Stenzyl worked for a petroleum company, and oysters were his thing. As we looked at the beautiful fossil specimens being dug out by our colleagues he asked me, "Do you know why there are so many different types of oysters on the French menus?" I did not. If you have been to France and have gone to a seafood specialty restaurant, you have seen that there may be two or more pages of oysters listed, with greatly varying prices. Belons, Claires, Speciales, Fines de Claires, Portugaises and so on, with numbers after each ranging from 5, for the smallest size through 1 for a medium size to the large oysters in sizes 0, 00, 000, 000, 0000, and finally 'pied de cheval,' meaning as big as a horse's hoof. The prices range from reasonable to astronomical. There are also indications of where the oysters grew. There are "Crus" of oysters, just as there are for wine. And believe it or not, all of those different oysters have subtly different tastes. The position in an estuary or bay where the oysters were grown determines how salty the water was, and that affects the taste of the oyster. Some years later, on a geological excursion to the shores of Normandy to see the sediments there, I also saw some of the oyster farms in the Bay of Mont Saint Michel. The oyster spat, the young larvae, settle on solid objects, and the French make sure that at this stage of their life the only solid object they can find is a piece of rope. Once the spat have started to grow the ropes are wrapped around posts, and there they grow up. When it comes time to go to market, the ropes with their oysters are unwrapped and taken to be sold to the restaurants. They are opened just before you eat them, and a few drops of lemon juice on them will elicit a contraction so you know they are still alive. If you ask for the red hot sauce we use here as a dip for raw oysters, you will receive a sometimes courteous visit from the manager of the restaurant who will show you how to use the very delicate clear sauces made of vinegar and shallots to enhance the flavor.

In the US our oysters are dredged from muddy sea bottoms. Our *Crassostrea virginica* is what its name sounds like, a powerfully flavored cousin of its French counterpart.

After the field trips we were expecting to hold our meetings in one of the big, formal lecture halls in the Sorbonne. However, I got word to come instead to the side entrance of the Geological Institute at 9 AM on the first day of the conference. I showed up and was promptly ushered inside by a student waiting at the door, and shown to a classroom. Our French hosts explained that we could meet here as long as the other students did not find out about it. Also, the bridges over the Seine were going to be closed at 2 PM each day until further notice, so we had to be finished by 1 PM each day so those of us with hotels on the Right Bank could get across.

Almost all of us had hotels on the Right Bank, across the river. I was staying at the Regina, on the Place des Pyramides. I had discovered the Regina with help from some friends a few years earlier, and always stayed there. It was a truly classic hotel. At the end of your stay the desk clerk would ask if you had enjoyed your room. The first time I was there I had a room overlooking the Rue de Rivoli. The Rivoli is a racetrack for cars, trucks and every other form of motorized transportation. There is a stoplight where it crosses the Place des Pyramides. So on the Rivoli side you got treated to the noise of braking cars stopping for the traffic light and then the noise as they all started off again. You have to remember that in those days the main food markets for all of France were at Les Halles in the city center. There was an endless progression of trucks carrying everything edible to Les Halles starting about midnight and an endless progression carrying them away starting about 3 AM. I told the desk clerk the room was noisy. He said "Next time you might prefer the room opposite." I did not even know then that there would be a next time, but sure enough I was back at the Regina the following summer. When I walked in the same desk clerk greeted me by name, told me I did not need to register, and handed me the key to the room opposite. It had a pleasant view out over the rooftops of Paris, and it was absolutely quiet.

This was long before the days of computers. At the Regina they kept a record of every guest, what room they had occupied, what they had had to drink at the bar. When you made a reservation, they looked you up, and they were waiting for you when you arrived. When you went to the bar the waiter would ask if you wanted a Hennessy, like you had had last time. That is class. Today, all hotels keep track of their guests with computers, but over the years I know of only two that see to it you always have the same room, the Regina in Paris, and more recently my 'Stammhotel' in Munich, the Leopold.

After a couple of days of meeting it was becoming obvious that the half day sessions were not enough for us to get our work done. So we decided to break into small groups and find a nice restaurant where we could meet in the afternoon, work and continue our discussions through dinner. On May 24 the micropaleontologists were going to meet at the Restaurant Chartier, in the Rue de Faubourg Montmartre, just off the Boulevard Montmartre, an address easy to remember. One of our group had a room in a hotel in the Boulevard Poissonniére, not far from the restaurant. The Boulevard Poissonniére is the eastward extension of the Boulevard Montmartre, which is the eastward extension of the Boulevard Haussmann. Together they form a major east-west thoroughfare.

In case you did not know, the Grandes Boulevards of Paris were designed by Baron Georges-Eugène Haussmann to open up the city. Carried out in the 1850s and 1860s under the reign of Napoleon III, they had another purpose—to permit the authorities to control the sometimes unruly population. The Parisians had developed a nasty habit of coercing the government by means of riots and popular uprisings from time to time. The Boulevards were laid out in straight lines cutting right through the neighborhoods to connect strategic points. The demolition destroyed most of medieval Paris, and required new facades be built along the streets. The Mansard roofs, which had been designed by Francois Mansart in the 1600s, set the universal style. That is why Paris today has such a homogenous "Parisian" look. Facades the same, roofs the same. The effort at crowd control gave the city its special character. But Haussmann hadn't counted on the invention of the chain saw which, as you will see, has made a mess of the grand plan for crowd control.

Maria Cita had suggested we meet at her hotel at about 3 PM to get started. We were having vigorous discussions when a roaring noise came from the street. The room was on the fourth floor and had a balcony overlooking the Boulevard. A convoy of police vans was racing down the Boulevard headed east, off to control some crowd. The vans were heavily armored and had only tiny slits, wide enough for a gun barrel, for windows. People who had been on the sidewalks were running into the side streets. We did not know for sure what was going on, but moved our chairs onto the balcony to watch. After the police vans disappeared several workers in blue coveralls appeared with chain saws. They calmly proceeded to cut down the trees lining the boulevard, felling them into the roadway and blocking it. There would be no more police vans racing along this street. In half an hour it was a mess of fallen trees. We saw the police vans reappear some distance down the Poissonniére, and the police came running up the street firing tear gas canisters. We discovered that tear gas is heavier than air, and so we had ringside seats for the action below, with the police chasing anyone who appeared. By 6 PM the Boulevard was empty of people and police, and some of the trees had been pushed to the sides. We decided the time was ripe for a dash to the restaurant.

The restaurant Chartier was just around the corner, and we were shown to our large table as though nothing at all had happened. We were well into the dinner and had consumed liberal amounts of wine when rioters and police appeared in the street just outside the restaurant. Here we were, eating dinner and watching a riot just outside, where people were getting hit with billy clubs. One of the waiters had locked the door, but was standing by just in case more guests should arrive through the mayhem. The street filled with tear gas, and this time it worked its way into the restaurant just as our dessert was arriving. The management graciously suggested that we might be more comfortable in a private dining room in the basement.

We moved downstairs along with our desserts, and were served brandy. The tear gas, being heavier than air, followed us down and accumulated there. One of our party went up to check on how things were going, and came back to report that the street outside seemed to be clear. It was agreed that we should all try to make it back to our hotels 'par les petites rues' (by way of little streets). I crossed the street outside and turned left the few steps to the Boulevard Montmartre. I had gone about a half a block before I looked up. There seemed to be some smoke in the air, and through the haze down the street I saw a phalanx of police marching slowly toward me. I turned around and saw a phalanx of people in blue overalls carrying large red flags, marching slowly toward me. I decided that this was not a good place to be, crossed the Boulevard and slipped into a side street, the rue Vivienne, where several cars were ablaze.

Going past the cars I came to a large square with a magnificent building. Smoke was coming out of it. I soon realized it was the Bourse (the Paris Stock Exchange). It was very strange; no one was around, no spectators, no police, and no firemen. I just stood and watched the smoke coming out of the building. Then I noticed a group of police running into the square from one of the side streets. I reached into my jacket pocket and took out my passport for identification and started to walk to the policeman in the lead. About fifty yards away, he looked at me, raised his baton and started running toward me. He did not look at all friendly, and I ducked around a corner as fast as I could. There was a construction site there, and an arm waved me into a gap in the fencing. Someone had been watching me from inside the site. When the policeman rounded the corner he did not see where I had gone, and decided the fire in the Bourse was more important. The man who had waved me in never spoke a word; we were as quiet as we could be and communicated by hand gestures. By now lots of police had arrived, and cordoned off the square. With gesturing we made a deal that he would watch one street and I another for an opening where we might be able to get away. I often wondered since then if this was the person who set fire to the Bourse. It was at least an hour before an avenue of escape opened up. We were able to work our way through the construction and go out behind the police cordon. It was now about 11 PM, and my next problem was getting across the Rue de 4 Septembre, another of the broad Boulevards. The streets were now deserted, but the occasional police van would go racing by. It seemed as though a curfew had been imposed. Choosing the proper moment I dashed across the broad street and worked my way through the 'petites rues' to the Boulevard de l'Opera. Here it was not so quiet, there was a fracas going on in the direction of the Opera, I got across the Boulevard and ran the rest of the way down the Rue de Pyramides to the Hotel. The doorman was waiting for me, unlocked the door, handed me my key, and wished me a good night.

I went up to my room. Ordinarily the maid would have turned down the bed early in the evening when I would have been at dinner. Instead I found a note and a nice piece of chocolate. The note simply read "en gréve" (on strike). The Regina had more class than I had realized. I felt lucky to be alive and in one piece. It had taken me over three hours to get from the restaurant to the hotel. I got into bed and was finally dozing off, when there was a huge explosion outside. I did not know what it was, but I pulled on my pants, threw on a shirt and went downstairs. Others had gathered there. The concierge was explaining it was nothing to worry about, just the police firing off concussion grenades to clear the streets. Another went off nearby. Concussion grenades are designed to make a huge noise but do no damage. I went back to my room. The grenades kept going off until morning.

The next day I learned from the newspaper that the Sûreté believed that the fire at the Bourse was started by American agents, probably the CIA. Running from the police that night was one of the luckiest things I ever did. I went out to look around. The French are meticulous rioters. Government buildings had broken windows, but I did not see a single broken shop window. Gasoline had not been available for weeks and there were very few cars still running.

A day later we had a post-conference field trip to other geological localities outside Paris. I have no idea how it was arranged since all the gas stations in France had been closed for weeks. It was eerie to drive across railways and see the rusted rails. No trains had run for a couple of weeks. We visited the geological localities as promised, and then deviated from the schedule to go to a reception at the home of the conference organizer, Charles Pomerol, outside Paris. Charles apologized for the inconveniences, and we ended the conference appropriately, consuming many bottles of wine. He wished us a good voyage home. The bus took us back and dropped all of us off near the Arc de Triomphe. It was a long walk back to the Regina, and there was a musty odor in the streets. Since the metro was not running and the garbage collection had stopped, the citizens were using the underground stations as garbage dumps. They were getting full. Leaving France was not as easy as it might sound. There were no trains, no planes, no busses, no taxis, and no gasoline. The only possibility was busses that would drive in from Belgium and take a load of passengers out. They could refuel in Belgium. They would come to the Champs Elysees and, when they had a full load of passengers, leave for Brussels. It sounded like a dubious plan, but I trundled my suitcase the mile or so to the Champs Elysees, and lo and behold there were busses waiting. A few hours later I was in Belgium. From there I got across the Channel, and on May 30 flew from London back to Chicago.

ISIDOR ISAAC RABI ON THE DISCOVERY OF THE
'HEAVY ELECTRON' (LATER NAMED THE MUON):
"WHO ORDERED THAT?"

# Solar Variability and Cosmic Rays

**14**

*Before I came here I was confused about this subject. Having listened to you I am still confused. But on a higher level.*

Enrico Fermi

In the last chapter it was proposed that the gradual increase in solar luminosity since the Sun's formation has generally been compensated by declining greenhouse gas concentrations in the atmosphere. However, it was noted that between 800 and 600 million years ago there were several instances of extreme change that resulted in an episodically ice-covered Earth. It is thought that the ongoing gradual buildup of greenhouses gases in the atmosphere eventually brought each of these snowball Earth episodes to an end.

The redistribution of energy received by Earth at different seasons of the year caused by cyclic changes in its orbital parameters are reflected in the geologic record, and most spectacularly in the alternation of glacial and interglacial states over the past few million years.

The question whether shorter term solar variability might also affect Earth's climate has been a topic of discussion for over a century and a half and remains uncertain.

## 14.1 Solar Variability

The total energy received each year by Earth will change if the luminosity of the Sun changes. If the Sun emits more energy, the Earth must warm. Unfortunately we have only a very brief direct history of changes in solar radiance. The question sometimes arises—could our Sun be a variable star, with large changes in its energy output over time? The 'variable stars' astronomers speak of change in brightness over periods of hours to days to a few years. We do not have a long enough astronomical record to know how they might vary over centennial or longer timescales.

On the other hand, we do know about variations in our Sun's radiance on decadal and centennial time scales from direct evidence. There are also some indications that it may vary on thousand-year time scales. Finally, there are climatic variations on millennial time scales that are attributed to the Sun without any real evidence for a link. Let's start with what we actually know—the longest instrumental record of temperature.

Figure 14.1 shows the longest instrumental temperature record we have. The collection of the data started shortly after the invention of the first sealed-liquid thermometer. It allowed consistent measurements from place to place. The stations that contribute information to this record lie in the English countryside in a triangle bounded by Bristol, London, and Lancaster. The wide swings in the record reflect the phenomenon that a warm year is usually followed by a colder year, and vice versa. The lower temperatures from 1660 to 1710 reflect the later part of what is called the "Little Ice Age." In the 19th century it was thought that sunspots might account for climate trends.

Sunspots have been observed off and on for about 2800 years. They are dark spots which appear in pairs on the surface of the Sun. We now know that they are giant magnetic storms. Although you might think they would cause a decrease in the brightness of the Sun, its radiance actually increases because they are surrounded by brighter-than-normal areas called faculae (Latin for torches). The total change in solar radiance between times when there are no sunspots and when they reach their maximum abundance is about 0.1 %, less than 0.5 W/m$^2$ (you will recall that the average insolation at the top of the atmosphere is 340 W/m$^2$). This corresponds to a planetary temperature difference of 0.1 °C. However, in one part of the spectrum we cannot see, the ultraviolet, the change can be as much as 50 %.

The Chinese were the first to keep records of sunspots, presumably observing them when dust storms darkened the sky. Systematic Chinese records go back to the second century BCE. Sunspots were more frequently observed after the telescope was invented at the beginning of the 17th century, and after Galileo got hold of one and started to observe objects in the sky. His observation of sunspots was another of those things that got him in trouble with church authorities. Church dogma insisted that the Sun was perfect.

© Springer International Publishing Switzerland 2016
W.W. Hay, *Experimenting on a Small Planet*, DOI 10.1007/978-3-319-27404-1_14

**Fig. 14.1** The Central England
temperature record, measured
with thermometers. The *thin line*
traces mean annual temperatures
since 1659. The *thick line* is an
11 year moving average. The
sunspot record is also shown (for
greater detail see Fig. 14.2).
A close correlation between the
two records is not obvious

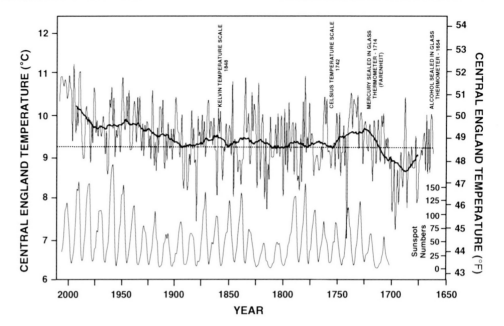

It couldn't have spots. But Galileo's trial for heresy in 1633 was not over sunspots, the existence of which the Vatican astronomers had confirmed, it was over whether the Sun orbited the Earth or vice versa. Perhaps fortunately for astronomers of the time there was a period during most of the rest of the 17th century when there were no spots on the face of the Sun. This period is now known as the Maunder Minimum. Nothing to report, no one burned at the stake.

It was some time before others took up the practice of making regular observations, but in 1826 Samuel Heinrich Schwabe, pharmacist and amateur astronomer in Dessau,

started to observe and count sunspots systematically. His 1843 paper asserting that they came and went in approximate 11 year cycles started a search for some effect on the Earth. Figure 14.2 is a recent compilation of sunspot numbers and cycles observed since 1690.

If you compare Figs. 14.1 and 14.2 you might think you see some similarities, but if you inspect them closely about the only clear correlation is with the rise in temperature at the end of the 17th century and the rise in sunspot numbers after the Maunder Minimum. Until quite recently it was thought that this especially cold episode in the Little Ice Age

**Fig. 14.2** Annual sunspot
numbers since 1690. Sunspots
were very rare between 1654 and
1715; the interval called the
'Maunder Minimum,' and a
second episode of low sunspot
numbers occurred in the early
1800s, the 'Dalton Minimum'

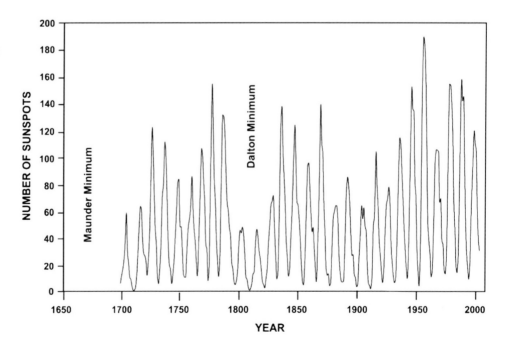

was caused by the lower solar radiance during the Maunder Minimum. As we will see later, there might be other possibilities as causes of the Little Ice Age.

Perhaps the most intriguing of these searches for a climatic effect related to the sunspot cycles was the apparent close correlation of the level of Lake Victoria, one of the sources of the Nile, with sunspot cycles during the period 1895–1922. Over this period of about 2½ cycles lake levels seemed to rise and fall with the number of sunspots. Unfortunately, after the paper describing this correlation was published, lake levels no longer followed the sunspot cycles. This story is often cited as an example of the danger of short-term apparent correlations. Then, amazingly, the correlation reappeared about 1970 and has continued to present. A more extensive investigation of the relation appeared a few years ago, when sediment cores from Lake Victoria were analyzed to obtain a record of lake levels extending back to about 1000 AD. There does often appear to be a correlation between sunspots and lake levels, but there is a problem. Since 1800 sunspot minima coincide with low lake levels, but before 1800 they coincide with high lake levels. The suspicion is that somehow the tiny variations in solar luminosity associated with sunspot cycles affect tropical rainfall. Why the effect reversed 200 years ago remains a mystery.

**Fig. 14.3** Franz Arthur Friedrich Schuster (1851–1934). Physicist, sunspot researcher, and connoisseur of the fine Rhine wines known in Britain as 'Hocks.'

Other efforts to find correlations seem more fanciful. Attempts to relate sunspot numbers to the financial markets go back to the 1700s, but without much success for investors. Then came a breakthrough. Franz Arthur Friedrich Schuster (1851–1934) (Fig. 14.3) was born in Frankfurt-am-Main, Germany and became a distinguished physicist at Owens College in Manchester, England. In the late 1870s he determined that good vintages of the German Rhine wines grown just downstream of Frankfurt occurred at approximately 11 year intervals, the average length of the sunspot cycle. This was practical, useful information for both the connoisseur and the bon vivant. Other attempts to relate sunspot cycles with other agricultural products and the commodities markets were less successful. The most extensive set of correlations were performed by Harlan True Stetson (Fig. 14.4) in the 1930s. He had been Assistant Professor of Astronomy at Harvard and research associate at MIT. He found correlations not only with wine vintages but noted that the stock market runup of 1928–29 lagged a sunspot maximum by 1 year, and the bottom of the great depression in 1932 matched a sunspot minimum. He also found a relationship between sunspot activity and automobile sales, rabbit populations, the growth of trees, and building contracts. Incidentally, in making his correlations he only had one sunspot cycle to work from, so practically any multi-year trend could be correlated with it one way or another. But he did make one real correlation—between sunspots and radio reception.

The cause and effect relationship between sunspot numbers, climate and the activities of animals and humans has remained elusive. In 2000, British meteorologist E.N. Lawrence noted that the Titanic sank (April 14, 1912) during a time of maximum numbers of icebergs in the North Atlantic that coincided with a sunspot minimum. I cannot help but note that the Monica Lewinsky scandal also coincided with a sunspot minimum!

When my dear friend, the late Stephen Schneider, (Fig. 14.5) author of many very interesting books including the fascinating *"Coevolution of Climate and Life,"* was at the National Center for Atmospheric Research in Boulder, he analyzed Stetson's work and tried to find correlations between sunspot cycles and other apparent cycles without appreciable success, although the correlation with rabbit populations intrigued him. He noted that the lack of any obvious causal relationship between two trends being correlated grates on the nerves of a scientist. In fact, Harland Stetson's work is often cited in classes in statistics as an excellent example of the misuse of apparent correlations.

Since the first clear association of sunspots with climate concerned the quality of wine, it would seem to warrant further investigation. I participated in one of those investigations with Stephen and friends in Boulder. Starting with a

**Fig. 14.4** Harland True Stetson (1885–1964). Correlator of sunspot activity with all sorts of things

selection of different vintages of a good California wine, we tasted and rated them according to our personal evaluations of quality, then tried to find a correlation with the sunspot cycle. The results were inconclusive. Of course, California wines are not supposed to vary much from year to year so the sunspot effect might be very difficult to detect. Nevertheless, this is the sort of investigation one can really get into. First, the quality of a given wine is a personal evaluation, there is no absolute scale. However, I must say that among his many talents Stephen Schneider had a vocabulary for describing nuances of taste that was second to none. Second, you need to remain unaffected by the wine as you proceed from tasting one to another. In the US we tend to swallow what we are tasting rather than following the European tradition, in serious wine tasting, of spitting it out onto the (hopefully straw-covered) floor. As a result of our inhibitions about arousing the ire of the hostess, the precision of judgment tends to decline as the tasting continues. The results become inconclusive requiring that the group

meet again and perform the tests all over again, perhaps starting off with a bottle from a different year as they seek to establish that elusive correlation between number of sunspots and wine quality. So far no one seems to have come up with a relation as clear as that described by Arthur Schuster, but maybe we just need to keep on trying. You don't need to be a climate scientist to join the quest to establish that subtle relationship.

Since 1978 orbiting satellites have monitored the luminosity of the Sun in detail on a regular basis. Figure 14.6 shows the record from 1978 through most of 2012, including much of three sunspot cycles. It shows the increase in variability from day to day as the sunspot maxima were reached in 1980, 1989, and 2000, and the much more constant output from the Sun during the sunspot minima of 1986 and 1996, and the unusually long minimum of 2006–2010. You may also detect a very slight overall decline, about 0.1 W/m$^2$ over the entire time these satellite observations were made.

We now know that sunspot cycles come in pairs; in the second cycle the magnetic polarity of the sunspots is reversed relative to the first. These 22 year cycles are termed Hale cycles. Longer cycles have been detected, one about 80–88 years (Gleissberg cycle) and another about 200 years, and another about 2,000–2,500 years (the Hallstadt cycle). I must confess that none of these are so obvious that they 'jump out at you' but the search for a direct relation between changes in solar luminosity and climate goes on. In spite of the fact that solar luminosity has been decreasing over the past decade, it has been blamed for global warming. Even more absurd, it has been claimed that this natural decrease as part of the sunspot cycle is a harbinger of global cooling. The cycle reached its minimum in July 2008, and solar luminosity increased over the next 5–6 years.

So in this sense, our Sun is a variable star although its variability would be hard to detect from another solar system. Its output changes by only about 0.1 % (1/1,000th) over the course of the 11 year sunspot cycle.

To go back before there were systematic sunspot records or to investigate the possibility of variability unrelated to sunspots, we must use something that would respond to changes in the Sun's activity. Two possibilities have been explored: the rate of production of Carbon-14 and the rate of production of Beryllium-10. Both are radioactive isotopes produced in the upper atmosphere by cosmic rays. To understand how these isotopes can tell us something about the variability of the Sun in the more distant past you need to know something about how these isotopes are formed, and that involves the nature of cosmic rays, the solar wind, and Earth's magnetic field. We will also explore whether cosmic rays affect Earth's climate.

**Fig. 14.5** Stephen Henry
Schneider (1945–2010).
Climatologist, visionary, and
connoisseur of fine wines. He
gave us early warning of global
warming, but many refused to
listen

## 14.2    The Solar Wind

The flux of these particles that reaches the top of Earth's atmosphere is strongly influenced by the solar wind. The idea that something other than light was leaving the Sun goes back to the solar flare of 1859. It caused deflection of compass needles and produced remarkable electrical effects on Earth's fledgling telegraph system. More about that below. The irregularities in the Sun's corona seen during a total eclipse, the relation between auroras and solar activity, and the observation that the tails of comets point away from the Sun all suggested that more than simple electromagnetic

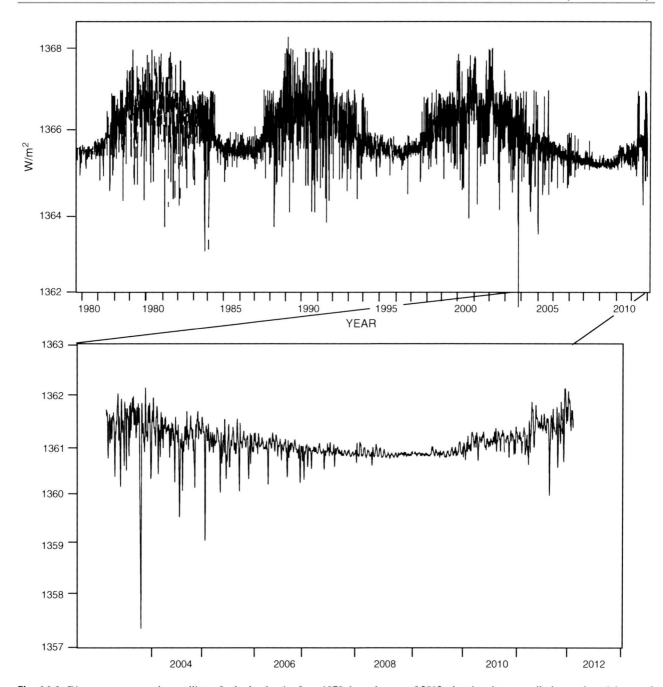

**Fig. 14.6** Direct measurement by satellites of solar luminosity from 1978 through most of 2012, showing the unusually long solar minimum of the mid-2000s

radiation was leaving the Sun. The term 'solar wind' was coined by Eugene Parker (Fig. 14.7), Professor of Astrophysics at the University of Chicago in 1958. Shortly thereafter, in 1959, the Soviet Luna I satellite made the first direct measurements of the solar wind.

The solar wind emanates from the Sun's corona. It is a plasma; that is, it is entirely composed of ionized matter, mostly hydrogen ions (protons) and electrons. At the very high temperatures that prevail in the rarefied plasma of the corona, the average speed of the protons is about 145 km/s

(=90.1 miles/s). The escape velocity for the corona, that is, the velocity an object must have to escape the Sun's gravity at the altitude of the corona, is about 618 km/s (=384 miles/s). Electrons in the corona, having a much smaller mass than the protons, have much higher velocities and create magnetic fields that can boost the speed of some of the ionized hydrogen nuclei so that they can leave the Sun and fly off into space as the solar wind. About 6.7 billion tons of this matter leaves the Sun per hour. It sounds like a lot of mass being lost from the Sun by this mechanism, but over

**Fig. 14.7** Eugene Parker (1927–). He named the 'solar wind' and realized it varied with the sunspot cycle

103 h (4.3 days). The fast particles travel almost twice as fast, about 750 km/s. They include the nuclei of helium atoms (α particles) that are present in the Sun's photosphere, and so seem to have a different source.

In 1958 Eugene Parker already noted that there was evidence of an 11 year cycle in the frequency of cosmic rays. It coincided with the 11 year sunspot cycle. If the solar wind varied with sunspot cycle, that would explain the change in the number of cosmic rays observed.

## 14.3   Solar Storms and Space Weather

Special events on the Sun, called 'solar storms,' can cause parts of the corona to be ejected into space. These coronal mass ejection events (CME's) send matter straight out in specific directions, almost like buckshot. Because they are so directional, the chance that some of the ejected matter will hit the Earth is small, but it does happen from time to time. The CMEs are often associated with spectacular sunspot eruptions known as solar flares. The most famous of these occurred on September 1, 1859. It has become known as the Carrington Flare for the amateur astronomer Richard Christopher Carrington (1826–1875) who observed and described it in detail. Two days after the flare appeared, things began to go seriously wrong with what few electrical devices there were in those days. Telegraph operators got an electrical shock when they tried to tap out messages. The telegraph wires were still active after they had been disconnected from their batteries. The magnetometer at Kew Gardens in London, set up to record the small changes in Earth's magnetic field, recorded a huge pulse. Auroras were seen down nearly to the Equator.

From what we know now about the speed of matter ejected from the Sun, these events should have happened about 4 days after the flare was observed. Later analysis showed that a few days prior to the giant flare of September 1 there had been another flare. It had caused a CME that cleared out a path for the CME associated with the Carrington Flare.

Records from ice cores now suggest that solar flares of the magnitude of the Carrington event occur perhaps once every 500 years. There is real concern, however, about the havoc such a stream of charged particles could wreak on the electronics of unprotected satellites, personnel in spacecraft, and even on the complex electrical grids through which we receive most of our power. What is now called 'space weather' is a matter of real concern for satellite operations and manned space stations. It is now carefully monitored. The SOHO satellite, launched in 1995, is devoted entirely to tracking the activity of the Sun and should give us at least a two day warning about trouble to come. Satellites and perhaps even the electrical grid could be shut down to avoid disastrous consequences to our technology-dependent society.

the course of its 4.7 billion year history it has lost only about 0.01 % (1/10,000) of its mass to make the solar wind.

Incidentally, the concept of an escape velocity goes back to Isaac Newton and his work on gravity and why Moons can orbit planets and why planets orbit the Sun. The escape velocity decreases with increasing altitude as the gravitational attraction weakens. Earth's escape velocity at the surface is 11.2 km/s (∼7 miles/s) but above the top of the atmosphere where air friction is not a problem it is about 10.9 km/s (6.8 miles/s). But you don't need to accelerate a rocket from 0 to that speed to put something in orbit. You can use Earth's rotation to give you a boost at the start. At the Cape Kenney Space Center it is 0.4 km/s, if you launch to the east. In Jules Verne's 1865 novel *From Earth to the Moon*, the launch site is Stone's Hill in "Tampa Town," Florida, very close to the modern Space Center launch site. Even though his works were science fiction at the time, Jules Verne knew a lot about hard science.

From satellites we now know that the solar wind can be divided into two components, slow and fast. The particles leaving the Sun are mostly ionized hydrogen (protons) and electrons. The slow particles have the same composition as the corona and travel through space at speeds of about 400 km/s (∼250 miles/s). They reach Earth's orbit in about

I always worry about once-in-500 or once-in-1000 year events. While I was working the problems associated with scientific drilling in the deep sea, we had an engineering firm trying to outguess what would happen to our drill rig in a once in a 100 year storm. At a meeting in late 1981, one of the fellows muttered, "I don't know about this, we've already had three once-in-a-hundred-year storms in Houston this year."

## 14.4   The Solar Neutrino Problem

During the 1980s there was a flurry of excitement, at least among the few who worry about such things, over what was called 'the solar neutrino problem'. Without going into any of the gory details, the Sun gets its energy from fusion of hydrogen nuclei to make helium. It is thought that this process is fairly well understood. The fusion takes place under tremendous pressures and temperatures at the center of the Sun. One of the by-products of this fusion process is neutrinos. Neutrinos are rather like electrons, but they have no electrical charge. They travel near the speed of light, and at that time were thought to have no mass. They pass right through 'solid objects' like the Earth. About 50 trillion of them go through your body every second. Since they go right though everything without interacting, they are very difficult to detect. The first attempt to detect those coming from the Sun was made by Raymond Davis and John Bahcall, who made an observatory consisting of detectors and a 100,000 gallon tank of cleaning fluid (carbon tetrachloride, $CCl_4$) at a depth of 4,800 ft in the Homestake Mine in Lead, South Dakota. If a neutrino collides with a chlorine atom in the cleaning fluid, it will produce a radioactive atom of argon, and that can be collected by bubbling helium through the cleaning fluid, and measured. The experiment was in operation from 1970 to 1994. They were able to detect neutrinos, but only 1/3 as many as should be coming from the Sun. The general conclusion of the scientific community at the time was that they had made a mistake in their calculations somewhere. Nobody could find the mistake. Bigger, better detectors were built in Japan, the Soviet Union, and Italy. Same result. After the results were confirmed by the Japanese, Davis and Masatoshi Koshiba received a Nobel Prize in 2002.

In the 1980s speculation rose that maybe the switch that turns the fusion reaction in the Sun on was being turned off. At that point I called my oracle for such esoteric matters, Cesare Emiliani. Cesare was, of course, intimately familiar with the experiments. The conversation went something like this: "Cesare I read that the neutrino flux from the Sun is only 1/3 what is should be. Are we likely to wake up one morning and find out it has gone black? And why isn't there already panic in the streets?" "First, there is no panic in the streets because nobody reads the scientific literature. Second,

I suppose it is possible the fusion furnace might oscillate, going on and off as they are suggesting, but we'll never know." "Why not?" "Because the photons which carry the energy from Sun to Earth are created in the interior of the Sun, but it takes a long time for them to work their way to the surface, several million years. They all go by different routes, so unless the oscillations of the solar fusion machine were on a timescale of millions of years we wouldn't know. Once the photons get to the surface, the flight time to Earth is only about 8 min." And then Cesare said something quite profound. "If there were significant variations in the output of the Sun we would know it from the geologic record."

Cosmologists have suggested many cosmic events that might destroy life on Earth, such as a supernova exploding in our vicinity. Yet the geologic record indicates that none of these has happened. There is good evidence for a few asteroids having collided with Earth causing major extinctions, but most changes in life on Earth seem to be 'home grown.' They have been caused by the evolution of new life forms that changed the way the biosphere operates or by changes in the distribution of land and sea in response to plate tectonics. Earth has not experienced any esoteric cosmic catastrophes for at least the last half billion years.

The solar neutrino problem was solved in 2002 in an unexpected way. In the 1970s and 80s it was thought neutrinos had no mass, but some physicists began to suspect they might have a very small mass. There are three kinds of neutrino (physicists call them 'flavors'): electron, tau and muon. As long as they were massless, they could not change from one to another. Then, when it was proven that they have mass, it was evident that they could oscillate between flavors. The early detectors were only sensitive to tau and muon neutrinos. Electron neutrinos weren't supposed to be there. When all three types could be detected, the flux from the Sun is what solar physicists had expected. We no longer need to worry; the Sun is stable but the neutrinos are not.

## 14.5   The Ultraviolet Radiation

We knew very little about the ultraviolet radiation from the Sun until recently. It is almost entirely absorbed in the upper part if the atmosphere. However, satellites have shown that it is more intense and variable than had been thought. Although the fluctuations of visible light intensity during the sunspot cycle are very small, that of the ultraviolet radiation may change by as much as 50 %. This should mostly be absorbed by ozone in the lower stratosphere. In principal that should enhance the stratification at the base of the stratosphere and make the differentiation between the stratosphere and troposphere more distinct. But the effect on the surface temperatures, weather or climate has not been detected with any certainty.

## 14.6    Cosmic Rays

You may recall from Chap. 9 that in 1902 Arthur Kennelly and Oliver Heaviside independently predicted that at the top of the atmosphere there was an electrically conducting layer of ionized air that could reflect radio waves allowing them to travel beyond the horizon. They thought the radio waves were being reflected from this ionized layer back down to the sea surface, which would then reflect them back up again —essentially forming a wave guide for electromagnetic waves in the short wave radio range ($\sim$3–30 MHz). The actual existence of this layer was finally demonstrated by English physicist Edward Appleton in 1924. It has been shown to be the result of ionization of oxygen molecules by ultraviolet light from the Sun, releasing free electrons. The air is so rarified that the electrons do not immediately find an ion to combine with. Because the phenomenon depends on ultraviolet rays of the Sun, the intensity of the ionospheric reflection varied between day and night, with the seasons, with the sunspot cycle, and with solar flares.

But during the first decade of the 20th century it was discovered that electrical phenomena were not restricted to the outer regions of the atmosphere. It was assumed that the electricity in the air at ground level was caused by the radioactivity (the 'β-rays') discovered by Henri Becquerel in 1896, and must come from minerals in the ground. In 1909, Theodor Wulf (1868–1946), a German physicist and Jesuit priest was investigating electricity in the air. He had built a sensitive electrometer, a device which could detect 'electricity'—ionized molecules or free electrons in air. He used a hermetically sealed bottle, so that detected electricity could come only from radiation from outside the bottle interacting with the air inside. To do this he made careful measurements at the base and top of the tallest structure in the world at the time, the Eiffel Tower in Paris built in 1889. He was startled to find that the intensity of the electrical signal in the air in the bottle at the top of the tower was greater than at the bottom. Initially his work was largely discounted by the scientific community, but others made the same sorts of measurements from balloons. Their results were the same. The electricity in the air was increasing upwards. But the density of the air in the lower part of the atmosphere is so great that ions and electrons can only exist for an instant. They could not be diffusing down from the top of the atmosphere. Something must be generating them in situ, and it wasn't radiation from the ground. The question was: could this effect be caused by radiation from the Sun?

One of the first to test Wulf's discovery by taking electrometers up in a balloon was Victor Hess (Fig. 14.8). In 1912 he took three of the devices on seven balloon flights to an altitude of 5300 m ($\sim$17,400 ft). The measurements indicated an increase of electrical activity by a factor of four. On April 17, 1912, he made use of the partial solar eclipse

**Fig. 14.8** Victor Hess preparing to go aloft in his instrumented balloon in 1912. He proved that ionizing radiation was coming from outer space, not from the ground

over Europe. The Moon had covered most of the Sun's disc, but the 'radiation' detected did not decrease. He reported "the results of my observation are best explained by the assumption that a radiation of very great penetrating power enters our atmosphere from above." The 'radiation' energy was not coming from the Sun but from outer space. Hess received the Nobel Prize in Physics in 1936.

The idea was that this electrical activity was being induced by some unknown form of electromagnetic radiation. One of those especially interested in this radiation was Robert Andrews Milliken (1868–1953). From its founding in 1921–1945 he was president of the California Institute of Technology (better known as Cal Tech). He became convinced that the radiation was an unknown part of the electromagnetic spectrum and in 1932 coined the term 'cosmic rays.' He was highly religious, and thought his cosmic ray photons were the "birth cries" of new atoms continually being created by God to counteract entropy and prevent the heat death of the universe.

On the other hand, Arthur Compton (1892–1962) at Washington University in St. Louis believed that the 'cosmic rays' were actually highly energetic particles. Compton also had a philosophical bent, and applied quantum theory to human activity.

A set of known physical conditions is not adequate to specify precisely what a forthcoming event will be. These conditions, insofar as they can be known, define instead a range of possible events from among which some particular event will occur. When one exercises freedom, by his act of choice he is himself adding a factor not supplied by the physical conditions and is thus himself determining what will occur. That he does so is known only to the person himself. From the outside one can see in his act only the working of physical law. It is the inner knowledge that he is in fact doing what he intends to do that tells the actor himself that he is free. (From "Science and Man's Freedom", in The Cosmos of Arthur Holly Compton, 1967, Alfred A. Knopf, New York, p.115)

In the first half of the 20th century, the investigation of cosmic rays depended largely on a device that revealed the presence of moving charged particles, the Wilson cloud chamber. Charles Thomson Rees Wilson (1869–1959) was a Scottish physicist. In 1894 he was investigating the formation of clouds and mist on Ben Nevis, the highest mountain in Scotland, when he experienced the phenomenon known as the Brocken specter. The Brocken itself is an easily accessible summit in the Harz Mountains of Germany where the 'specter' is often observed. The specter is a magnified image of the observer's shadow on the mist, surrounded by a halo of light. A black and white version of a recent photo of the Brocken specter on Ben Nevis is shown in Fig. 14.9.

Wilson began to experiment with condensation droplets and discovered that ions could serve as nuclei for their formation. By 1911 he had devised a 'cloud chamber' in which water saturated air could be made to condense into tiny droplets by suddenly cooling the air by expansion by means of a membrane. By the 1920s his device had been improved and was being used to trace the tracks of ionization left by α particles (helium nuclei) and β rays (electrons). The two could easily be distinguished because the tracks of the α particles are short and broad, and the β ray tracks are long and thin. Furthermore, in a magnetic field the paths curve in opposite directions. It was soon discovered that the cloud chamber could also detect cosmic rays.

Then in 1932, Carl David Anderson (1905–1991), working in Millikan's laboratory at Cal Tech photographed a long thin track that went the wrong way. He had discovered the positron that Paul Dirac, 4 years earlier, had postulated must exist. Dirac's hypothetical antimatter was real. Anderson received the Nobel Prize in Physics for 1936 for his discovery. The production and theft of antimatter for nefarious purposes is the topic of Dan Brown's book and the film *Angels and Demons*.

In 1934, things got more complicated. Bruno Rossi (1905–1993) at the University of Padua, Italy, was working on a study to document the 'east-west effect.' It had been discovered that more cosmic ray particles come from the west than the east. It was being recognized that this is due to deflection by the Earth's magnetic field, which rotates with

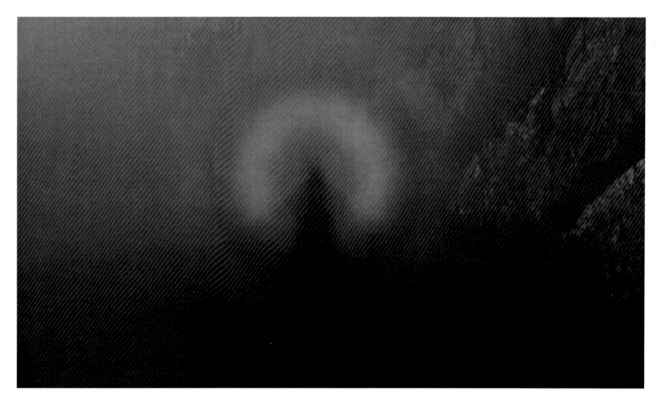

**Fig. 14.9** The 'Brocken Specter' seen on Ben Nevis in Scotland. From a photograph taken by the Knight Family of Ireland in 2009 and included in Scott Ramsey's TravelBlog 'Edge of Britain'

the planet. It also means that the incoming particles have a positive electrical charge. But Rossi noted something quite new: two widely separated Geiger counters recorded a simultaneous event.

Pierre Victor Auger (1899–1993) (Fig. 14.10) was a French physicist interested in cosmic rays. He had set up high-altitude observatories in the Alps, and in 1938 two detectors several meters apart recorded the arrival of high energy particles at exactly the same time. Along with Rossi's observations, this was the discovery of 'air showers' of elementary particles. Auger concluded that air showers were the result of a single event, collision of a high energy particle with atoms in the upper atmosphere resulting in a downward cascade of less energetic particles. He estimated that the energy of the incoming particle that could create large air showers must be at least $10^{15}$ electron volts (eV).

Gamma rays, the most energetic part of the electromagnetic spectrum, have energies of the order of $10^5$–$10^6$ eV. The cosmic ray particles reaching Earth's outermost atmosphere and causing air showers are a trillion times more energetic than gamma rays.

At the University in Kyoto, Japan, Hideki Yukawa (1907–1981) (Fig. 14.11) was worrying about why the protons in the atomic nucleus, all packed tightly together, but all having a positive electrical charge, didn't repel each other sufficiently to fly apart. In 1935 he proposed that there was a force, now known as the 'strong nuclear force' that

**Fig. 14.11** Hideki Yukawa (1907–1981). In 1935 he predicted the existence of mesons. They would not be found until 1947. He received the Nobel Prize in Physics in 1949 for his prediction

held them together. The strong nuclear force only operated over very short distances, the size of an atomic nucleus, so it obviously wasn't some form of electromagnetic radiation. He proposed that yet-to-be-discovered particles, which he called 'mesons,' were the glue that carried this force and held the protons together. He predicted that the particles would have a mass of about 100 MeV.

Elementary particle masses are often given in terms of energy, electron volts (eV). You can convert this to something more familiar by remembering Einstein's famous equation $e = mc^2$. That can be rearranged as $m = e/c^2$. In SI units that becomes $kg = J/(m/s)^2$. If you go back to Chap. 3 you will find that the J stands for joules, and a joule is a Newton meter or simply $J = kg\ m^2/s^2$. Now the equation becomes, in SI units

$$kg = \frac{kg(m^2/s^2)}{(m^2/s^2)}$$

or simply $kg = kg$, which is obviously true. Isn't math wonderful? The reason particle physicists use eV or MeV ($=10^6$ eV) or GeV ($=10^9$ eV) for mass is that it saves writing

**Fig. 14.10** Pierre Auger (1899–1993). The French physicist who began the systematic investigation of cosmic ray air showers

out a complicated number which turns out to be exceedingly small, having negative exponents. Remember that 1 electron V = 1.60217646 × 10⁻¹⁹ J (that is its energy), and the speed of light is 299,792,458 m/s. It was becoming clear that there was a lot more to learn about the world of very small particles. The use of eV for mass just makes it a lot easier to make comparisons.

When physicists were just beginning to feel comfortable with the idea of antimatter, a new elementary particle was discovered in the Cal Tech cloud chamber. In 1936 Carl Anderson and Seth Neddermayer (1907–1988) found the track of a particle that curved in the same direction as an electron but made a wider arc, indicating it was much heavier. It had a mass of 106 MeV, and they thought it was Yukawa's meson. They named it the 'mesotron.' It seemed so improbable that atomic physicist Isidor Isaac Rabi at Columbia University made the famous comment: "Who ordered that?"

When it was later discovered that it was not the particle Yukawa had predicted but a wholly new animal it was renamed 'muon'. A muon has a very short lifetime, about 2 μs ($2.197 \times 10^{-6}$ s), whereupon it decays into an electron and a couple of other things, that later turned out to be neutrinos. The short-lived muons were evidently being produced by energetic incoming particles still called cosmic rays.

As discussed in the last chapter, the period from the late 1930s until the end of World War II was a time when most physicists were preoccupied in the race to develop military weaponry. Basic research on subatomic particles and cosmic rays languished, but many perhaps unintentional discoveries were made anyway. Table 14.1 summarizes what was known about atoms and subatomic particles in 1939. This is what I learned about the world of particle physics when I was in high school at St. Mark's in Dallas in the late 1940s. I had an excellent teacher, "Doc Nelson" who kept up on progress in physics, so I even heard about the positron, although the idea of antimatter was really considered a fantasy world.

**Table 14.1**  The elementary particles known in 1939

| Particle | Electrical charge (e = 1.602 × 10¹⁹ C) | Mass (GeV = 1.8610 × 10⁻²⁷ kg) | Year discovered, *named* |
|---|---|---|---|
| e = electron | −1 | 0.0005 | 1897, *1891* |
| β⁺ or e⁺ = positron | +1 | 0.0005 | 1932, *1928* |
| μ = muon | −1 | 0.106 | 1936 |
| p = proton | +1 | 0.938 | 1919 |
| n = neutron | 0 | 0.940 | 1932 |
| γ = photon | 0 | 0 | 1901, 1905, 1926 |

Some had been postulated and acquired names before they were actually found and documented. The two discovery dates for the photon refer to Max Planck's 'energy quanta,' and Albert Einstein's more specific 'light quanta'

## 14.7  A Digression into the World of Particle Physics

After WWII better detectors and technology for determining the energy of cosmic rays were developed. Particle accelerators, called 'atom smashers' at the time, with ever increasing capabilities allowed discovery of a plethora of subatomic particles. Particle accelerators had been around since the latter part of the 19th century and by the 1950s there was one in almost every home in the US. The venerable cathode ray tube is a particle accelerator, accelerating electrons. Your old bulky TV set was a particle accelerator. But big accelerators did not exist before the mid-1950s. Then the technology developed for accelerating electrons and protons to near the speed of light. When they hit something going at that speed they can blow it apart into even tinier particles. Today there are more than 25 subatomic objects in what physicists refer to as the 'particle zoo.' Some were discovered during cosmic ray studies; some others were discovered in particle accelerator experiments and then noticed in cosmic ray detectors.

Before WWII the science of subatomic particles described them in terms of classic concepts, like mass and energy, albeit in the quantized form. Then 'spin' was added. Since the middle of the last century it has become evident that particles can also differ in other ways. Physicists have chosen a fanciful terminology to describe these differences. Particles may have 'strangeness,' 'charm,' 'flavors,' 'colors' (not the kind of colors we know), and so on. In this section I am only going to introduce those members of the particle zoo that are relevant to our discussion of climate and climate history, so you will miss the full flavor of modern particle physics. Pun intended.

I mentioned the use of cloud chambers in cosmic ray studies, but there was another, even simpler technology that was very useful. Gelatin-covered photographic plates were set out in light-tight holders on mountain peaks. The Pic du Midi de Bigorre in the Pyrenees and the Chacaltaya in the Andes were the major sites. After some days or weeks, the plates were recovered and developed. Cosmic rays had produced tracks on the plates. You can guess that by setting out the plate in a magnetic field, the particle tracks could be made to curve and indicate the particle's electrical charge.

In 1947 Cecil Powell (Fig. 14.12) and his colleagues at the University of Bristol, UK, reported on the tracks of new particles found in the photographic emulsion that had been placed at the Jungfraujoch in the Bernese Alps in Switzerland. The Jungfraujoch is reached rather conveniently by a cog railway inside the mountain. These new particles were the mesons Yakuda had predicted, and were given the name 'pion,' designated by the symbol π. It turns out that there are three kinds of pion, with positive, negative and zero electrical charges: $\pi^+$, $\pi^-$, and $\pi^0$. Actually there are five kinds of

**Fig. 14.12** Cecil Powell (1903–1969). Leader of the team that discovered pions

pions, because the $\pi^+$ and $\pi^-$ have antiparticles. Hideki Yukawa received the Nobel Prize in Physics in 1949 for his prediction, and Cecil Powell got the Prize in 1950 for his development of the photographic technique that revealed the existence of pions.

A couple of months after the discovery of pions, another particle was discovered in cloud chamber photographs taken at Manchester University. George Dixon Rochester and his postdoctoral assistant Clifford Charles Butler found a new particle. They called it a 'kaon' (K). Kaons have a new quantum property called 'strangeness.' Strangeness has to do with the way the particles decay. It turns out that there are six kinds of kaon, including the antiparticles. Rochester and Butler never received a Nobel Prize for what is considered one of the most important discoveries in particle physics.

One particularly important set of elementary particles is neutrinos. In 1930 Wolfgang Pauli had noted that the laws of conservation of energy, momentum, and angular momentum are violated in the $\beta$ decay of radioactive elements. He proposed that there was an unknown particle involved. He called it a neutron, but in 1932 James Chadwick used that name for the neutrally charged particle in atomic nuclei. In 1933

Enrico Fermi renamed Pauli's 'neutron.' It became the still hypothetical 'neutrino' emphasizing its small size. In 1956, at the Savannah River Research Center in South Carolina, Frederick Reines (1918–1998) and Clyde Cowan (1919–1974) finally found the neutrino, the particle that had the energy missing in $\beta$ decay. Small wonder it wasn't detected earlier. It has no electrical charge, and a mass so small it has not yet been determined, although estimates indicate that it is a very, very small fraction of an eV. It also has a 'weak interaction' with other particles; it is as though it does not know anything else is out there, and passes right through objects like planet Earth as though they weren't there.

Since WWII, cosmic ray observing devices have progressed from cloud chambers that fit on a table top, to photographic plates on mountaintops, to satellites, to the Auger Cosmic Ray Observatories—arrays of observation and data collection devices spread over a 35 × 35 km (21 × 21 mile) square area. The Southern Hemisphere Auger Observatory, located on the Pampa Amarilla of Argentina, has been operational since 2004, and the Northern Hemisphere Observatory is planned for construction soon, probably in Colorado. The Auger observatories are designed to investigate the less frequent high-energy cosmic ray showers.

We now know that beyond Earth's atmosphere the incoming particles are about 89 % protons (hydrogen nuclei), 10 % alpha particles (helium nuclei), and 1 % nuclei of heavier elements. That is a representative sampling of the abundance of elements in the Universe. But a lot happens as they approach our planet and encounter the atmosphere.

If you feel bewildered by all these strange particles with odd names you may be comforted to learn that you are not alone. In 1954, 10 days before his death, Enrico Fermi said "If I could remember the names of all these particles, I'd be a botanist."

## 14.8   How Cosmic Rays Interact with Earth's Atmosphere

Summarizing the above, 'cosmic rays' are not a form of electromagnetic radiation, but are highly energetic particles, most coming from outer space and others coming directly from the Sun. They are different from the photons of electromagnetic radiation because they have mass. Having mass, they cannot travel at the speed of light. The most common of these cosmic ray particles are single protons (hydrogen nuclei) and $\alpha$ particles (helium nuclei). The individual particles differ in their speed, and thus in their energy. Figure 14.13 is a compilation of data on the frequency of particles having different energies, based on data from observations at the ground.

You will note that the energies for cosmic rays shown in the lower part of Fig. 14.13 are all much higher than

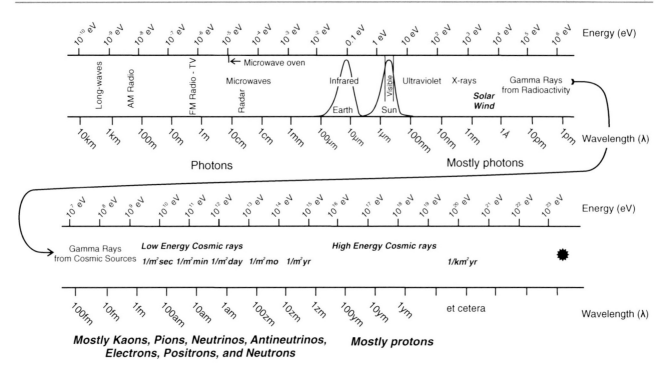

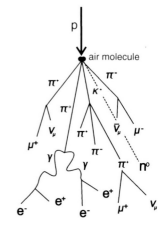

**Fig. 14.13** Comparison of the energies and wavelengths of electromagnetic radiation and cosmic-ray particles (*bold italics*). The fluxes for cosmic-rays particles are at the Earth's surface. The star is for the most energetic particle known, the Oh-My-God particle observed on the evening of 15 October 1991 over Dugway Proving Ground, Utah.

Its energy is estimated to have been to be approximately $3 \times 10^{20}$ eV. That is a kinetic energy equal to that of a baseball (142 g or 5 oz) traveling at 100 km/h (60 mph). You and your computer wouldn't want to get in its way

anything in the electromagnetic spectrum shown in the upper part of the diagram. What is important is the number of these highly energetic cosmic ray particles passing through a given area in a given length of time.

Because of their high energies, cosmic rays can cause 'soft errors' in computers. The flux of cosmic rays depends on altitude and location. In New York City, at sea level, the flux is approximately 140,000 n/m$^2$ ($= \sim 13,000$/ft$^2$) per hour. Near the Earth's surface the flux increases by a factor of about 2.2 for every 1000 m (1.3 for every 1000 ft) increase in altitude above sea level. Computers operated on the tops of mountains have an order of magnitude higher rate of soft errors compared to those used at sea level. The rate of upsets in aircraft may be more than 300 times the sea level upset rate.

Sunspot activity decreases the rate of cosmic-ray soft errors in computers. This is because the Sun produces few cosmic ray particles with the energies above the 1 GeV required to penetrate Earth's upper atmosphere and create particle showers. Remember that before they can reach Earth, cosmic rays must contend with the solar wind. The 'solar wind,' the general flux of particles from the Sun's corona, changes in strength with solar activity. It can deflect cosmic rays coming from sources outside the solar system. As they approach our planet cosmic ray particles encounter the Earth's magnetic field which can deflect those that carry an electrical charge. The increase in the solar wind during an

**Fig. 14.14** A cosmic ray air shower of particles generated by the collision of a fast-moving high energy proton (p) coming from outer space with a diatomic air molecule. The products include both particles and antimatter particles: kaons (K$^+$ is shown), pions ($\pi^+$, $\pi^-$, $\pi^0$), muons ($\mu^+$, $\mu^-$), neutrinos and antineutrinos ($\nu_\mu$, $\bar{\nu}_\mu$), electrons (e-), positrons (e +), gamma rays ($\gamma$) and neutrons (n)

active Sun period can reshape the Earth's magnetic field providing some additional shielding against higher energy cosmic rays and resulting in a decrease in the number of particles creating showers. However, the sunspot cycle effect is relatively small. It is thought to modulate the energetic neutron flux in New York City by about $\pm 7$ %.

Nevertheless, many of these cosmic ray particles reach the outermost levels of the atmosphere where they collide with the molecules of air and result in a cascade of secondary particles down to lower levels of the atmosphere, as shown in Fig. 14.14.

**Fig. 14.15** An artist's conception of a cosmic ay shower. Courtesy of ASPERA.com

Although only one is shown in Fig. 14.14, many of these particles produced through collisions are thermal neutrons. A thermal neutron is a free neutron with a kinetic energy of about $4.0 \times 10^{-21}$ J at room temperature. It has a characteristic speed of about 2.2 km/s. This has the same energy as the molecules of air near the Earth's surface. The name 'thermal' comes from the neutron's energy being the same as the room temperature gas or other material they are permeating. Neutrons generated from cosmic rays arrive at this energy level after ten to twenty collisions with atomic nuclei, provided that they are not absorbed.

Incidentally, your area, when standing, is less than a half square meter, but that still means that a particle with an energy of $10^{10}$ eV passes through you every second. But don't worry, an electron volt is a very small charge and anyway, most of these particles don't interact with anything we are made of. But a few do, and your brain is not totally unlike a computer. The next time you forget where you left your glasses or your wife's birthday you can blame it on a shower of cosmic rays.

The air shower produces billions of these exotic particles. Most of them have lifetimes of microseconds. The shower forms a cone a few degrees wide downstream from the original direction of the incoming particle. The incoming particle can come from anywhere in the sky; only a few come in from the zenith. The area covered by the shower as it reaches the ground can be quite variable depending on the angle of incidence of the original cosmic ray particle. It is typically about 16 km (10 miles) across. Whether such showers could have any significant effect on climate is uncertain, but it is thought that during turbulent storms when electricity is accumulating in the air, lightning may be triggered by the air showers from energetic cosmic rays (Fig. 14.15).

## 14.9   Carbon-14

The neutrons produced in the original and subsequent collisions lose kinetic energy (speed) through subsequent collisions as the air becomes denser. They become slower 'thermal neutrons,' traveling at about 2 km/s (=1.2 miles/s). When one of these thermal neutrons collides squarely with one of the atoms making up a nitrogen gas molecule, something very dramatic can happen. The common nitrogen

**Fig. 14.16** Left, Willard Frank Libby (1908–1980). He introduced carbon-14 dating. Right, Hans Suess. He found out what was wrong with carbon-14 dating and devised methods for making the necessary corrections. With Roger Revelle, Hans called attention to the great uncontrolled global experiment on which humanity had embarked

atom ($^{14}$N) is stable. It has 7 protons and 7 neutrons in its nucleus. A direct hit by a thermal neutron can knock a proton out of the nitrogen nucleus and replace it. The atom now has 6 protons and 8 neutrons in the nucleus and is $^{14}$C, Carbon-14. There are two stable isotopes of carbon, $^{12}$C (98.9 %) and $^{13}$C (1.1 %), and to these the collision has added a rare unstable isotope, $^{14}$C. The sole source of this unstable isotope $^{14}$C is its creation from nitrogen ($^{14}$N) by collisions with thermal neutrons. The $^{14}$C soon combines with oxygen to make a special radioactive form of $CO_2$, adding a minute amount to the concentration of $CO_2$ as a trace gas in the atmosphere. Carbon-14 has a half-life of about 5,730 years; through β-decay it becomes $^{14}$N again. But as carbon dioxide, it behaves like every other $CO_2$ molecule and is soon taken up by plants through photosynthesis and incorporated into plant tissue. This enables us to date wood and other plant products that are not too old.

Carbon-14 was discovered in 1940 by Martin Kamen and Sam Ruben at the University of California's Radiation Laboratory in Berkeley. However, development of the technology to use it for dating did not occur until after World War II. About 1949 Willard Libby (Fig. 14.16, Left), working in Fermi's Institute for Nuclear Studies in Chicago, figured out a simple way to do it. If you knew how much carbon was in the sample, and counted the number of beta particles emitted from the sample, and you also knew how much carbon was in the air today (as $CO_2$), and counted the number of beta particles emitted by an air sample, you could determine the age of the sample. Dating of samples of wood and other organic matter from Holocene and late glacial deposits began.

When $^{14}$C was introduced as a means of dating it was assumed that its rate of production in the upper atmosphere by cosmic rays was constant over time. However, even as the first $^{14}$C dates were published in the late 1940s it was recognized that something was wrong.

Hans Suess (Fig. 14.16, Right) was an Austrian physical chemist and nuclear physicist who was also working in Fermi's Institute. Incidentally, he was the grandson of Eduard Suess, the Viennese geologist who recognized that the Alps had been formed in a collision between Africa and Europe, anticipating Wegener's hypothesis of continental drift and our modern concept of plate tectonics. Hans Suess recognized that some of the $^{14}$C dates on Egyptian artifacts did not correspond to calendar dates. Either the Egyptologists' dates were wrong, or the rate of $^{14}$C production in the atmosphere was not constant.

In 1955, he saw another problem in $^{14}$C dating. The burning of fossil fuels since the industrial revolution had added $CO_2$ to the atmosphere, and the C in this $CO_2$ was 98.9 % $^{12}$C, 'dead carbon.' In other words, the atmosphere was already contaminated with fossil fuel carbon, and was different from the pre-industrial atmosphere. The original $^{14}$C had been diluted by fossil fuel $^{12}$C. To date samples more than a century old, a correction had to be made to remove the effect of the added 'dead carbon.' This correction became known as the 'Suess Effect.'

Remember that in 1958 Eugene Parker, also at the University of Chicago, had noted that the intensity of cosmic radiation varied with the sunspot cycle. The implication was that the rate of $^{14}$C production in the atmosphere might also have changed with the sunspot cycle and sunspot intensity.

To solve this problem the researchers in the Fermi Lab turned to tree rings. They had been extensively studied by Andrew Ellicott Douglass (1867–1962), better known as A. E. Douglass, an astronomer at the University of Arizona, where he had founded the famous Laboratory of Tree-Ring Research. Douglass's goal had been to investigate the cycles of sunspot activity which he thought would affect climate and be recorded by tree-ring growth patterns. Those results were ambiguous, but the collection of tree rings at the University of Arizona has become an international scientific resource.

By counting the rings, one could know the age of the wood precisely. Then, by measuring the ratio of $^{14}$C to the stable isotope $^{12}$C in the wood, one could determine whether there was too much or too little compared to the ratio of $^{14}$C in wood formed today or whether the wood formed before man's large-scale burning of fossil fuels introduced additional amounts of $^{12}$C into the system, the Suess Effect. By comparing how much $^{14}$C should be present with how much is present one can produce a 'corrected' $^{14}$C age scale. You might think that this would be limited to relatively young ages, but in fact there are trees in the Rocky Mountains that are thousands of years old. Until recently, the oldest living tree was thought to be a Bristlecone Pine in the White Mountains on the border between California and Nevada. Its age—4,767 years. Recently, an even older tree, supposedly 9,500 years old, has been found in Scandinavia. There are also logs that are even older. Recently, oak logs buried in bogs in Europe have been used to carry the record back much further, well into the time of the deglaciation. In the literature there are tables for converting the apparent age given by measuring the $^{14}$C to a calendar year age. This is an ongoing effort, refinements of the $^{14}$C scale are continuing, and the $^{14}$C-age/true-age tables are regularly updated.

We now know that at sunspot maxima, the rate of $^{14}$C production decreases and at minima it increases. This is due to the varying deflection of cosmic rays by the changing intensity of the solar wind. Unfortunately, there is a complication. The number of stable carbon atoms in the atmosphere varies with time as part of the global 'carbon cycle' which we will discuss later. The carbon cycle involves uptake of carbon by plants, the release of carbon by decay of organic matter, cycling of carbon through the ocean, etc. It is a complicated set of processes. So, unfortunately, the change in the rate of $^{14}$C is not solely due to solar activity deflecting cosmic rays. But the dating problem is alleviated by the direct comparison of $^{14}$C age dates with tree ring ages.

In the last two decades, a new technique, accelerator mass spectrometry, has made it possible to count the number of $^{14}$C atoms in the sample and compare it with the number of $^{12}$C atoms. The technique is costly in both time and expense, and there are relatively few laboratories with this capability. But it makes possible the measurement of much smaller quantities of radioisotopes and their decay products. Both precision and accuracy of the measurements has greatly improved. Now samples as old as 60,000 years can be dated.

## 14.10 Beryllium-10

In the ongoing search for radioactive isotopes for dating materials deposited during the last million years, other cosmic ray products have become useful. Galactic cosmic rays are really highly energetic particles. They can collide with atomic nuclei of $^{14}$N and $^{16}$O and simply break them apart. This process is called spallation.

For example, an oxygen atom, with 8 protons and 8 neutrons in the nucleus, can be split into a beryllium (Be) atom with four protons and six neutrons in the nucleus and some hydrogen, deuterium, or other small atoms. The $^{10}$Be atom is unstable, but its half-life is 1.51 million years, much longer than that of $^{14}$C. Beryllium is a light metal and has an affinity for slightly acidic water, so it is immediately absorbed into rainwater (or snow) and falls to the surface.

But beryllium-10 ($^{10}$Be) can be formed not only in the atmosphere but even in soils and rocks at the Earth's surface. Very high energy galactic cosmic rays can penetrate into these denser materials at ground level and cause spallation of atoms to a depth of about 1 m.

Beryllium-10 offers not only the possibility of dating ice cores and ocean sediments, both from areas where there can be no contamination from soil and rock ground cover, but also insight into the history of cosmic ray bombardment of the Earth. Beryllium-10 formed in soils and rock on ground can even tell us about long-term erosion rates.

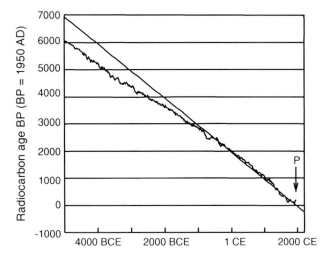

**Fig. 14.17** The Carbon-14 calibration. The straight solid line is the C-14 age determined assuming that the half-life is 5730 years and that production rate in the atmosphere has remained constant with time. The squiggly line is the age calibration obtained by using C-14 measurement of tree rings and other organic matter of known age. P is 'Present' = 1950 AD (=1950 CE)

Now, as with all good things, there is a problem. The stable isotope of beryllium, $^9$Be, is in most dust grains in the air, so the amount of dust that settles out could change the apparent Be concentration. To get around this problem, a group of French scientists, led by Edouard Bard, recovered an ice core from one of the least dusty spots on Earth, the South Pole. They used $^{10}$Be in the ice core to calculate what the $^{14}$C production rate over time should have been. This is possible because the half-lives of the two isotopes are so different. Their results are shown in Fig. 14.17. The arrows in Fig. 14.18 mark times of sunspot minima, when they are very rare for several decades. These minima seem to occur about every 200 years, so the next one is expected about 2100. The 'Medieval Warm Period' lasted from about 1000–1400; the 'Little Ice Age' from about 1500–1900. Of course,

one could say that the 200 year periods in the middle of each of these episodes were the real 'warm' and 'cool' times.

In conclusion, the decadal to centennial to millennial changes in solar radiation do not seem to have left an unambiguous global signal. The known variations in solar luminosity are by themselves capable of raising or lowering the planetary temperature by only about 0.1–0.2 °C. They might possibly have been responsible for the 'Medieval Warm Period' and the 'Little Ice Age' but they would need to be reinforced by some positive feedback mechanism.

## 14.11 Cosmic Rays and Climate

We have already discussed the role cosmic rays play in the production of $^{14}$C and $^{10}$Be, but there is also a possibility that they might have a direct influence on climate. In the 1990s a group in Denmark, led by Henrik Svensmark, a physicist at the Danish National Space Center in Copenhagen, found that there seemed to be a correlation in the intensity of cosmic rays bombarding the Earth and the formation of low-level clouds. I remember well the day in 1997 when one of my graduate students in Kiel brought me a copy of Svensmark and Eigil Friis Christensen's paper *Variation in cosmic ray flux and global cloud coverage—a missing link in solar-climate relationships,* which had just appeared. Among other things it was suggested that global warming was caused not by increasing $CO_2$ in the atmosphere, but by a change in the flux of cosmic rays.

The idea is that cosmic rays interacting with atoms in the upper atmosphere produce energetic particles such as muons, as described above. These then cascade down through the atmosphere and ionize (give an electrical charge to) molecules in the lower part of the atmosphere. These ionized molecules then serve as nuclei for the condensation of water droplets that ultimately form clouds. This was

**Fig. 14.18** Changing production rates of $^{14}$C in the upper atmosphere over the past millennium, derived from $^{10}$Be concentrations in an ice core from the South Pole. *Arrows* indicate sunspot minima indicated in the detailed younger record (since 1700) and the spotty older records. *Letters* indicate names of sunspot minima: *W* Wolf, *S* Spörer, *M* Maunder, *D* Dalton

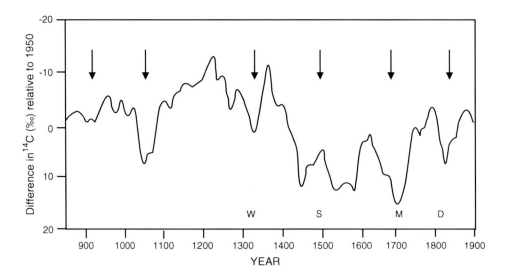

immediately accepted by the press in the 1990s as a certain correlation between cosmic ray flux and global warming.

To make a long story short, no one quite understands how the cascading particle flux would influence only low level clouds without affecting mid- and high-level clouds. A much more thorough comparison of the satellite-generated data sets on cloudiness, presented by the Norwegian Meteorological Office at the European Geosciences Meeting in Vienna in 2008, failed to find any consistent correlations of cloudiness with cosmic ray fluxes. Also, the correlation with low-level cloudiness disappeared after 1994 and doesn't seem to have come back since. Furthermore, the relation between cloudiness and Earth's surface temperature remains problematic. As we will discuss more extensively below, clouds reflect sunlight, and should thereby lower the surface temperature. But they absorb the long infrared radiation from Earth's surface, and should thereby cause warming. Everything depends on the kinds of clouds and where they are.

Nevertheless, Henrik Svensmark remains convinced that cosmic rays are one of the major driving forces behind climate change, and his 2003 book *The Chilling Stars: A New Theory of Climate Change* has attracted a lot of attention.

An old friend, Ján Veizer, who has also spent time oscillating between institutions in North America and Germany, and Nir Shaviv, of the Hebrew University in Jerusalem, thought that long-term changes in the cosmic ray flux might be responsible for the long-term episodes of glaciation that occur on our planet every 250 million years or so. You know that our Milky Way Galaxy looks like a pinwheel. What you might not realize is that the stars, like our Sun, are not stuck in one of the arms of the galaxy, but that the arms are waves of compression and rarefaction moving through the field of stars. Sometimes our Sun is in one of the arms; sometimes it is between arms. The idea is that the cosmic ray flux would increase as Earth passes through an arm, causing more cloudiness and global warming (or cooling). Other colleagues demonstrated that the evidence was too flimsy to raise this possibility out of the realm of speculation and that the hypothesis of greenhouse gas control rests on firmer ground.

This is perhaps the place to mention that some scientists in the 1950s and 60s thought that the arms of the galaxy were dustier than the space in between the arms. They suggested interstellar dust might be responsible for cooling the climate to allow the episodes of glaciation in the ancient past. This remains pure speculation.

## 14.12 Summary

The Earth receives energy from the Sun, insolation, at an average rate of 340 W/m$^2$ at the top of the atmosphere. The actual amount depends on latitude, season, and changing distance from the Sun. The Earth's orbit is elliptical, and at present our planet is closest to the Sun in early January and farthest in late June. The seasons are not caused by the changing distance from the Sun but by the inclination of the Earth's axis or rotation, currently about 23.5°. For the northern hemisphere the Sun reaches its greatest elevation above the horizon on June 21, and its lowest on December 21. These are the northern hemisphere summer and winter solstices. At the equinoxes, in September and March, the Sun is directly over the Equator illuminating both hemispheres equally.

Poleward of the polar circles, which lie 23.5° from the poles at latitudes 66.5° N and S, there is 24 h daylight during the hemisphere summer, with the Sun highest overhead during the summer solstice. The 24 h daylight means that the Polar Regions receive more insolation than the adjacent lower latitudes. There is a summer insolation minimum at the polar circle.

Because of the changing Earth-Sun distance, the northern hemisphere receives more insolation than the southern hemisphere during its winter and less insolation than the southern hemisphere during its summer.

Our Sun is a variable star, with sunspots appearing and disappearing on a roughly 11 year cycle. The Sun radiates more energy when sunspots are present than when they are not. This is because 'faculae,' bright spots among the dark sunspots, more than make up for the lowered radiation from the spots. The change in insolation between sunspot maxima and minima is about 0.5 W/m$^2$, a difference of less than 0.1 %.

There have been periods when there were no sunspots for several decades. The most recent of these was the Maunder Minimum which lasted from about 1645—1710. The Maunder Minimum coincided with an especially cold phase of the 'Little Ice Age' and it has been suggested that diminished insolation may have been the cause although other possibilities have been proposed recently. Other attempts to relate sunspot activity to global climate, wine vintages, and economics have proven inconclusive.

Satellite observations since 1978 have documented solar variability with a precision previously unknown and have shown that while the variations of the visible spectrum with sunspot cycle are very small, they may be as much as 50 % in the ultraviolet.

Solar storms are events on the Sun that produce flares and eject matter from its corona. The solar storm of 1859, known as the Carrington Flare, disrupted the then new telegraph lines and other electrical devices.

Cosmic rays are actually highly energetic particles, about 90 % protons, and the rest mostly helium nuclei, traveling at very high speeds. They come from outside our solar system. They can be deflected by particles steaming out from the Sun's outer atmosphere, the corona that forms the solar wind. The strength of the solar wind varies with the sunspot cycle, and induces an 11 year cycle in cosmic ray intensity. As they approach the Earth, the cosmic ray particles can also be deflected by the Earth's magnetic field.

In the 1930s it was discovered that cosmic rays produce showers of subatomic particles into the lower part of the atmosphere. These air showers are the result of the impact of highly energetic incoming particles with molecules in the upper atmosphere. Billions of exotic short-lived particles are produced in these collisions and rain down into the lower levels of the atmosphere. The investigation of cosmic rays and the advent of high energy particle accelerators after WWII led to the discovery of new kinds of subatomic particles: more kinds of muons, pions, kaons, and neutrinos. At present there are more than 25 kinds of subatomic objects in the 'particle zoo.'

Two of the byproducts of cosmic ray interactions are Carbon-14 and beryllium-10. Carbon-14 is formed from nitrogen-14 as one of the protons in its nucleus is replaced by a neutron. It has a half-life of 5730 years and has been widely used for dating materials back to 20,000 years; more recently, with the improved technology of accelerator mass spectrometry, back to 60,000 years. Carbon-14 dating is complicated by the fact that the rate of its production in the upper atmosphere varies with time and because of contamination of the atmospheric reservoir by 'dead' carbon introduced by burning fossil fuels. These problems have been overcome for the younger part of the record by comparing carbon-14 dates with tree ring ages.

Beryllium-10 is a spallation product produced by cosmic ray particles splitting atoms in the molecules of the air or materials in the ground. It has a half-life of about 1.5 million years, and has been used to determine the changes in the rate of carbon-14 production in the upper atmosphere.

There has been speculation that the ionization of air molecules by cosmic rays may serve to create nucleation sites for water droplets and influence cloud formation, and thus impact Earth's climate. This speculation was very popular for a brief period during the late 1990s and early 2000s. The evidence for correlations has not stood up to careful examination.

The energy from the interior of the Earth is only about one fourth of a percent of the insolation, much too small to have any effect on climate.

In conclusion, you have been very patient through this long discussion that so often seems to have led nowhere. We have determined that the effects of solar variability on climate are uncertain, but might be recognizable from time to time in the local climate. Perhaps is it time to get out the corkscrew and check out more wines. As far as I know there have not been systematic investigations of the influence of solar variability on southern hemisphere wines, so Australia,

New Zealand, Chile, Argentina, and South Africa are virgin territory for investigation. Keep a careful record of your findings.

A Timeline for this chapter:

| | |
|---|---|
| ca. 2750 BCE | A bristlecone pine starts growing in the White Mountains of Nevada |
| 364 BCE | Early Chinese records of sunspots |
| 28 BCE | Chinese record a very large sunspot on May 11 |
| 1612 | Galileo makes drawings of sunspots |
| 1633 | Galileo's trial begins, but it's not about sunspots |
| 1645–1715 | Maunder Minimum, no sunspots |
| 1826 | Heinrich Schwabe starts systematic observations of sunspots |
| 1843 | Schwabe announces discovery of the 11 year sunspot cycle |
| 1859 | Richard Carrington observes a huge solar flare on September 1; 2 days later electrical devices on Earth do strange things |
| 1865 | Jules Verne writes *From Earth to the Moon*. His spaceship launch is from a site near the present Kennedy Space Center in Florida. |
| 1875 | Arthur Schuster discovers the correlation between sunspot cycles and good vintages of Rhine wine |
| 1889 | The Eiffel Tower is built in Paris |
| 1894 | Charles Wilson sees the Brocken specter on Ben Nevis in Scotland |
| 1895–1922 | Level of Lake Victoria follows sunspot cycles |
| 1896 | Henri Becquerel discovers radioactivity |
| 1909 | Theodor Wulf finds that 'atmospheric electricity' increases with altitude, on the Eiffel Tower |
| 1911 | Charles Wilson builds his first cloud chamber |
| 1912 | Victor Hess proves that 'atmospheric electricity' increases with altitude, carrying his instruments aloft to over 5 km in a balloon |
| 1928 | Paul Dirac postulates the existence of antimatter |
| 1930 | Wolfgang Pauli notes that radioactive β decay violates the laws of conservation of energy, momentum, and angular momentum and proposes there is a missing particle, the *neutron* |
| 1932 | James Chadwick coins the name *neutron* for the neutrally charged particle in atomic nuclei |
| 1932 | Robert Milliken coins the term '*cosmic rays*' |
| 1934 | Carl Anderson discovers the first antimatter particle, the *positron* |
| 1934 | Bruno Rossi's separated Geiger counters record a simultaneous cosmic ray event |
| 1930s | Harland Stetson correlates sunspot activity with the Great Depression, automobile sales, rabbit populations, the growth of trees, and building contracts. |
| 1933 | Enrico Fermi renames Wolfgang Pauli's neutron to *neutrino* to avoid confusion |

(continued)

(continued)

| 1935 | Hideki Yukawa predicts the existence of *mesons* as the particles carrying the strong force holding atomic nuclei together |
| --- | --- |
| 1936 | Carl Anderson and Seth Neddermayer discover the '*mesotron*.' They think it is Yukawa's meson, but it was later determined to be a *muon* |
| 1938 | Pierre Auger's cosmic ray detectors record another simultaneous event—he postulates the existence of air showers of cosmic particles |
| 1940 | Carbon-14 is discovered by Martin Kamen and Sam Ruben at the University of California Radiation Laboratory in Berkeley |
| 1947 | Cecil Powell and colleagues at the University of Bristol discover *pions;* they are Yukawa's mesons. |
| 1947 | George Rochester and Clifford Butler at Manchester University discover another particle, which they name *kaon* |
| 1949 | Willard Libby figures out how to use Carbon-14 for dating organic matter |
| 1949 | Hideki Yukawa receives the Nobel Prize in Physics for his prediction of mesons |
| 1950 | Cecil Powell receives the Nobel Prize in Physics for figuring out how to find those mesons called pions |
| 1952 | It is apparent that some of the carbon-14 dates are wrong. The process of correcting the dates with tree rings begins |
| 1954 | My father tells me that fusion energy is coming soon |
| 1956 | Frederick Reines and Clyde Cowan find the neutrino everyone had been looking for |
| 1958 | Hans Suess discovers that Carbon-14 dates are contaminated by the 'dead' fossil carbon introduced since the industrial revolution—the Suess Effect |
| 1958 | Eugene Parker postulates the 'solar wind' and notes there is a correlation between the 11 year sunspot cycle and the varying intensity of cosmic rays |
| 1959 | Soviet satellite Luna 1 confirms the existence of the solar wind |
| 1978 | Satellites begin detailed monitoring of solar luminosity |
| 1981 | Houston, Texas experiences three once-in-a-hundred-year storms |
| 1982 | Stephen Schneider and I try to check out the correlation between sunspot cycles and wine vintages. Results were inconclusive. |
| 1993 | Henry Pollack and colleagues at the University of Michigan analyze the existing data on heat flow from Earth's interior and conclude that the global average is about 87 mW/m$^2$ |

(continued)

(continued)

| 1994 | Raymond Davis and John Bahall note that there are only about 1/3 as many neutrinos coming from the Sun as there should be, but there is no panic in the streets |
| --- | --- |
| 1994 | In response to my inquiry concerning the solar neutrino problem, Cesare Emiliani assures me: "If there were significant variations in the output of the Sun we would know it from the geologic record" |
| 1995 | NASA launches the European Space Agency's Solar and Heliospheric Observatory (SOHO) on December 2 |
| 1997 | Henrik Svensmark and Eigil Friis Christensen's paper *Variation in cosmic ray flux and global cloud coverage—a missing link in solar-climate relationships* is published |
| 2000 | E. N. Lawrence notes that the Titanic sank during a sunspot minimum |
| 2002 | Masatoshi Koshiba receives a Nobel Prize for confirming that there are missing solar neutrinos |
| 2002 | The solar neutrino problem is solved. Neutrinos can change 'flavor' after leaving the Sun; the 'missing' neutrinos simply weren't detected |
| 2003 | Henrik Svensmark's book *The Chilling Stars: A New Theory of Climate Change* appears |
| 2003 | Nir Shaviv and Ján Veizer's paper *Celestial Driver of Phanerozoic Climate* proposes that cosmic rays are the major forcing factor of climate |
| 2004 | Papers by Dana Royer et al. (*CO$_2$ as the primary driver of Phanerozoic climate*) and Stefan Rahmstorf et al. (*Cosmic rays, carbon dioxide and climate*) dispute the Shaviv-Veizer hypothesis and argue that atmospheric CO$_2$ is the main factor forcing climate during the Phanerozoic |
| 2004 | The Southern Hemisphere Auger Cosmic Ray Observatory is set up in Argentina |
| 2008 | Analysis of the satellite data by the Norwegian Meteorological Office fails to confirm Svensmark's cosmic ray—cloudiness hypothesis |
| 2010 | A new analysis of the expanded global heat flow data base by British scientists indicates that the global average is about 92 mW/m$^2$ |

If you want to know more:

Clark, S., 1999. *Towards the Edge of the Universe*. 2nd Ed., Springer, New York, 275 pp.—An outstanding review of modern cosmology.

Geisser, T., 1991. *Cosmic Rays and Particle Physics*. Cambridge University Press, Cambridge, UK, 296 pp.—A thorough, in-depth account of this fascinating topic.

Schneider, S. 1984. *Coevolution of Climate and Life*. Random House, Inc. 563 99.—A classic with lots of interesting ideas.

Music: Twelve tone music will remind you of the many kinds of strange particles generated by cosmic rays. Arnold Schoenberg's 1899 composition '*Verklärte Nacht*' (*Transfigured Ni*ght) is one of my favorites.

Libation: Here it is appropriate to honor my old friend Stephen Schneider with one of the red wines he liked so much: a Shiraz from Australia.

### Intermezzo XIV. The Most Memorable Summer of My Life Was 1968: Israel and Prague

I had returned to the US just in time to witness the assassination of Robert F. Kennedy on June 5. Although it was 2:15 AM in Urbana, I had awakened from my jet-lag nap and was watching the results of the California Primary. Bobby Kennedy has just given his brief victory speech and I was about to turn off the TV and go to sleep when the news came that he had been shot while leaving the ballroom through the kitchen. Another assassination barely 5 months after that of Martin Luther King. It seemed that assassination had become a routine part of the American Way of Life. The best hope for a successful successor to Lyndon Johnson and for an honorable end to the Vietnam War died a few hours later.

On the 9th of June I was on my way back across the Atlantic, changing planes in London before going on to Israel. The El Al flight landed at Ben Gurion Airport outside Tel Aviv in the early evening. I was met by my host Zeev Reiss. We drove up to Jerusalem, passing first through the towns of Lod and Ramla. Shortly after leaving Ramla it became very dark, no lights anywhere. Zeev explained that we were skirting a salient in the 1949 Armistice Line where there had been a Jordanian fortress known as Latrun. We came around a curve and there was the fort, darkened except for a very few lights. The road had been in no-man's-land, and for years sniper fire from the fortress had made it impossible to use. Capture of Latrun had been one of the major objectives in the 1967 war.

We arrived in Jerusalem late at night. I was welcomed into the Reiss's home where I would spend the next 6 weeks. My first impression of Jerusalem was that it had a wonderful climate. It was midsummer and it had been hot and humid at the airport. In Jerusalem located on the crest of the ridge that borders the Dead Sea Rift Valley on the west, it was cool and dry. The ridge is made of Cretaceous limestone, which is quarried as the major building stone for the city. The cool white of the limestone gives the city much of its uniquely homogenous character.

Zeev Reiss was one of the world's leading micropaleontologists and his wife Pnina was a wonderful cook. You can imagine my surprise when she served me my first breakfast in their home: ham and eggs. Of course they didn't eat pork, but she thought Americans had it every day for breakfast. I explained that I was grateful for her trouble, but I usually ate a very simple breakfast and after that we had more typical Israeli food. Zeev had immigrated to Israel from a village in Moldavia. Actually the area where he had grown up had originally been in Romania, but after the war it became part of the Soviet Republic of Moldavia. Zeev had played a very important role in the early development of Israel—he had developed the understanding of the groundwater geology of the country and his studies had been critical in ensuring a stable water supply for Israel's development.

Over the next six weeks I taught a course at the Hebrew University. It had been a year since the six-day 1967 war, and Jerusalem was calm except for the occasional explosion of land mines left over from the war. One day we drove down to Jericho and the Dead Sea, and stopped along the way to take pictures. I wandered off the roadway to get a better view. Zeev turned around and saw me. "Bill, don't move. Now walk back putting your feet exactly where they were when you walked over there." I did. "Land mines, they still haven't all been cleared, and you never know where they might be." I learned to stay on the pavement. I was able to visit many sites in the West Bank, the Golan Heights, and the Kibbutzim in northernmost Israel.

We also set up a laboratory in the Geological Institute for cultivation of live foraminifera. To get everything ready we needed to sterilize the aquaria. I had already set up a facility like this at the University of Illinois, and had learned that the best way to sterilize the aquaria was to fill them with a brine solution using kosher salt. I asked Zeev for Kosher salt. Zeev was not particularly religious, and he had never heard of it. He couldn't even imagine how salt could be kosher. After some inquiries of religious authorities who were equally mystified, we got in contact with a Professor Katz who had spent time in the United States. He explained that in the US most salt is actually a mixture of salt and clay. The clay coats the salt crystals, so that they are not affected by moisture. You have seen the motto 'When it rains it pours.' In fact, if you take ordinary US salt and try to dissolve it in water you will get a cloudy suspension of clay in water. Kosher salt is simply salt without the clay; only available the United States. In Israel ordinary salt would do just fine. The Katz's very kindly invited us to dine on the Sabbath (Friday evening). They were sufficiently religious that they did no work on the Sabbath, which included flipping on the light switches. I, as a gentile, was able to perform these duties.

To collect live foraminifera we drove from Jerusalem to Eilat on the Gulf of Aqaba (or Gulf of Eilat) on the Red Sea. The Marine Biology Laboratory in Eilat was close to the former border with Egypt (the Sinai was occupied territory), and was just getting refurbished. We were able to swim out far enough from shore to get some clean samples of the vegetation and sediment, and take them back to the lab to see if we had gotten any live foraminifera. After a few hours we found we had lots of them crawling up the wall to the aquaria. Zee was delighted. Imagine having spent much of your life using the shells of these little protozoans to determine the age of rocks and their environment of deposition, but never having seen the live animal. We celebrated at a sea-food restaurant near the beach in Eilat where both shellfish and fish were available, but of course those of our party who observed Jewish dietary laws could not eat anything that did not have fins.

In Israel in 1968 there was a lot of optimism about the future. There was a general feeling that now a lasting peace would come, and Zeev looked forward to a time when the Israelis and Arabs would work together to turn the Middle East into an industrial/agricultural society that would be the envy of postwar Europe.

From Jerusalem, I went to Prague. Pavel Čepek had returned from the US in March, and we were meeting in late July and August, 1968, to outline new projects. That was the year of the Prague Spring, with reformer Alexander Dubček in charge. It was fascinating. At least once a day, someone would show up at my desk in the Micropaleontological Institute of the Czech Academy of Sciences in Spálená and ask me (along with everyone else) to vote on questions being considered by the Parliament. We were always asked because we were very handy; the Parliament Building was only a few blocks away. The first time I was asked for my opinion I explained that I was an American and US citizens were not allowed to vote in foreign elections. The answer was quick: It's not an election; everyone in the country must express their opinion so the Parliament will know the will of the people. It was participatory democracy, Swiss style. A few days after I arrived we had a party at Pavel's tiny apartment. His wife Jiřina had prepared a buffet, and there was lots to drink, beer, wine, and Slivovitz. It was July 29th and

Leonid Brezhnev and Alexander Dubček were meeting at Čierná nad Tisou on the Czechoslovak/Soviet (Ukrainian) border. There was some concern that the Soviets might kidnap Dubček, so there was continuous news reporting. However, at the time their conference seemed to go reasonably well. Nevertheless, during the party someone went out on the balcony of the apartment and rushed back in shouting 'the Russians have invaded, tanks are in the streets.' We all went out to look. It was just a joke. Little did any of us realize that a few weeks later that was exactly what would happen.

The International Geological Congress was to start in Prague on August 21, but I couldn't stay for it. After my sabbatical year I had teaching duties at the University of Illinois and needed to get back.

I flew back to the US from Amsterdam on August 20, landing in Chicago. My father had flown up from Dallas to meet me, since I needed to go directly to the University for the start of the semester. We had a great dinner at Don the Beachcomber's and I filled him in on all that had happened. He was fascinated to hear about all the changes Dubček was bringing to Czechoslovakia. I told him that I expected Dubček's revolution to sweep through Eastern Europe and the Soviet Union, and would fit in well with the changes being demanded in France. I expected that in 6 months the world was going to look a lot different. We went back to our hotel room, turned on the TV and learned that the Russians were invading Czechoslovakia. The 4,000 + geologists who had arrived in Prague for the Congress were witnesses to the invasion. It was a summer to remember.

When I look back on it, the events of that summer made me realize that progress is made by exploring new ideas and breaking out of the old ways of thinking. Today we call it 'thinking outside the box.'

My friend Pavel was caught off guard by the invasion. On that day he was to give a talk in a session of the International Congress on our work using coccoliths to determine the age of Cretaceous rocks. He had set his clock radio for 6 AM to be sure he was up in plenty of time. He awakened to the news that the Soviets were invading Czechoslovakia and that Prague was being occupied. A few moments later he heard one of the huge Antonov transport planes from the USSR flying overhead, headed for the airport. There was another plane every four minutes.

One of the Swiss students I had mentored, Peter Roth, was staying with him. They walked toward the city center, to the office in Spálená. It was a complicated business, with Soviet troops blocking many streets. On the way Pavel tried to talk with a Russian Major, who simply took off the safety on his Kalishnikov and aimed it at him. Pavel retreated, and eventually succeeded in reaching the office. There he gathered together the Secret Police files and then took them to the apartment of Jiřjina's parents, where he spent three days burning them page by page. He was also able to meet with Hans Martini, the Director of the Bundesanstalt für Bodenforschung in Hannover, who was in Prague for the International Congress. Martini invited him to return to Hannover to continue his work at the Bundesanstalt.

Peter Roth had his own exciting adventures as he eventually found his way to a place where he could catch a ride on one of the transports taking foreigners to the border.

In the US, I made inquiry through the FBI whether we could accept Pavel and his family here. A few days later I got a negative reply. I was starting to think about other options when I got news that Pavel was already in Germany, back at the Bundesanstalt in Hannover, and would be joined by Jiřjina and Martin.

In 1974 Pavel was tried in absentia for unnamed crimes against the state and sentenced to 1 year in prison. In a real-life re-enactment of Kafka's *The Trial*, neither he nor his family were able to determine the nature of the charges against him. He returned, as a German citizen, to visit the other members of his family in 1982. He would not return again until after the 'velvet revolution' of 1989. His father had passed away, and it became necessary to settle the matter of the trial and sentence against him in order for him to be able to accept his inheritance. After the inquiry into the nature of the trial, the judgment against him was annulled.

Two decades later I came to wonder why the US had not been interested in someone who knew in detail what had been happening during the Dubček era. During the 1980s something very similar happened in the Soviet Union. Substitute the name Gorbachev for Dubček. The sequence of events was not exactly the same, but so close it should have been obvious to anyone who knew both countries what was happening. In the 1980s I became painfully aware that my contacts in the CIA knew almost nothing about that phase of Czechoslovak history, and were unable to understand the parallels to the developments in the Soviet Union.

REFLECTING SOME OF THE SUN'S RADIATION
BACK INTO SPACE

# Albedo

*We'd never have got a chance to go outside and look at the earth if it hadn't been for space exploration and NASA.*
James Lovelock

The discussion of solar insolation in Chap. 14 stated that the solar constant is about 1,360 W/m$^2$, but that is at the top of the atmosphere. As we discuss in this chapter, some of that is reflected back into space by clouds and dust before it reaches the surface, some is scattered to make the blue sky, and some is absorbed. When the Sun is directly overhead, the amount of electromagnetic energy reaching the Earth's surface on a clear cloudless day is a little more than 1,000 W/m$^2$. Most of this, about 527 W, is invisible infrared radiation. About 445 W is visible light. Most of the ultraviolet radiation is absorbed by ozone in the stratosphere and only about 32 W reaches the surface (still plenty to give you sunburn). We will explain scattering in Chap. 16 as part of the discussion of the interaction of air molecules with the Sun's electromagnetic radiation. The absorption by ozone will be treated in Chap. 16 in the introduction on greenhouse gases. In this chapter we discuss our planet's reflectivity to understand how much it differs from Kirchhoff's black body.

Albedo is a measure of the reflectivity of a planet. From considerations of black-body absorption and radiation of energy we know that albedo is a critical factor in determining the temperature of the planet. An albedo of 1 means total reflectivity and 0 means no reflectivity. With an albedo of 1, complete reflectivity, the temperature of the Earth's surface would be a function of the heat flow from the interior of the planet and would be close to absolute zero, 0 K (−273.15 °C, −459.67 °F). With a black surface absorbing all incoming radiation, the temperature of the surface would be 278.3 K (5.15 °C, 41.3 °F). However, the Earth has neither of these conditions. It is partially reflective, and its albedo can change with time.

As shown in Table 10.2, planets and other objects in our solar system have very different albedos. The Moon's albedo is 0.12, less than half that of the Earth. Yet the Moon does not look dark. Reflecting only 12 % of the sunlight incident upon it, the full Moon appears brilliant against the black night sky. Seen in the daytime, it is barely visible. The planet Mercury is even darker, with an albedo of 0.06. Mars' albedo is 0.16, similar to the Moon. But Venus shines brightly in the night sky because its albedo is 0.72.

## 15.1 Albedo of Planet Earth

Since early in the 1920s the Earth's albedo has been estimated to be 0.3, meaning that it reflects 30 % of the incoming energy. This estimate was made by measuring the 'Earthshine' on the Moon. You may have noticed that sometimes at the new Moon or in its early phases you can see the whole Moon as a faint disk. We see it because of 'Earthshine,' the light reflected from Earth. As more and more of the Moon becomes sunlit it becomes increasingly difficult to see the part illuminated only by Earthshine. As you might expect the reflectivity of the Earth changes with the seasons and distance of the Earth from the Moon and the Earth-Moon system from the Sun. For viewers in the northern hemisphere, Earthshine appears brightest in the spring. If you have never noticed it go outside and look for the dark "New Moon." If you can see the full disk of the Moon you are seeing it by Earthshine.

One current estimate of the Earth's average albedo is 0.297 ± 0.005, but of course it will vary slightly from season to season and year to year. The Earth's global temperature is strongly dependent on albedo. A change of only 1 % is a change in the average insolation received by the planet of 3.4 W/m$^2$. That is equivalent to a temperature change of almost 1 °C (Fig. 15.1).

What is it that reflects sunlight from Earth? Table 15.1 shows the albedos of different parts of the climate system. About 20 % of the reflection of sunlight is from the scattering of light that makes the sky blue. 60 % from high albedo clouds, 20 % from Earth's surface—most of that is from low albedo ocean water. Aerosols, tiny particles floating in the air, may be either reflective or absorbent of sunlight; they constitute at most a few % to Earth's albedo. It is a unique feature of planet Earth that its albedo is largely

**Fig. 15.1** NASA's *blue* marble. Earth seen from space

governed by the solid and liquid phases of water: ice and snow are white and highly reflective; the tiny droplets that form clouds are highly to moderately reflective; the larger volumes that form lakes, rivers, and oceans are highly absorbent of sunlight.

Water has special characteristics. The 'deep blue sea' is actually black when seen from space or when you look down while diving in the open ocean. It looks blue as you look along its surface because it is reflecting the blue sky. Ocean water absorbs all wavelengths of light, albeit in a specific

sequence as you go deeper in the water. Ultraviolet and red are absorbed first, then yellow, followed by green and finally blue. Plants on land are usually green because they are absorbing red and blue for photosynthesis and reflecting the unused green part of the spectrum. Only shallow water plants are green; those living at deeper levels are golden-brown. They use the green and blue light for photosynthesis. A fish that is bright red on the surface looks black at 10 m depth.

Water can have a mirror like surface. If there are no waves, all of the sunlight coming in at an angle of less than

**Table 15.1** Albedos of major features of the Earth's climate system. Albedos of aerosols is speculative, depending on their lifetimes, nature and extent

| Location | Surface | Minimum albedo | Maximum albedo | Present area |
|---|---|---|---|---|
| Earth's surface | Sand | 0.18 | 0.28 | 3 % |
| | Grassland | 0.16 | 0.20 | 3 % |
| | Green crops | 0.15 | 0.25 | 5 % |
| | Forests | 0.14 | 0.20 | 2 % |
| | Dense forests | 0.05 | 0.10 | 2 % |
| | Cities | 0.14 | 0.18 | 1 % |
| | Fresh snow | 0.75 | 0.95 | 7 % |
| | Old snow | 0.40 | 0.60 | 7 % |
| | Water | 0.05 | 0.80 | 70 % |
| Earth's atmosphere | Clouds | 0.30 | 0.80 | 60 % |
| | Scattering | 0.15 | 0.20 | 100 % |
| | Aerosols | 0.05? | 0.80? | 30 %? |

53° will be totally reflected. Fortunately, the ocean surface is roughened by waves so that the area of reflection of the Sun is much less than it would be if there were no winds. At the same time, calculating the albedo of part of the ocean surface becomes very difficult because it depends on the angle of the Sun, the size and steepness of the waves, breaking waves and whitecaps. In any case, when the Sun is only 20° above the horizon, almost all of the sunlight is reflected. When diving on reefs the light disappears so rapidly, usually within a few minutes; it is as though someone had turned down the light with a dimmer. There is still plenty of light to get back into the boat and go to shore before the Sun sets, but below the surface the night has already fallen.

With underwater lights, what happens next is amazing to behold. The number of fish on reefs is often double the number of "safe sleeping spaces." During this undersea twilight the day fish trade off sleeping spaces with the night fish. The night fish are completely different from the day fish. I am forever indebted to Roger Hanlon, now at the Marine Biological Laboratory in Woods Hole, Massachusetts, for having introduced me to this phenomenon when he was a graduate student in Miami. For anyone interested in marine life, Roger's film *'Aliens from Inner Space'* is a classic. The aliens are his favorite animals, squids.

Grasslands, croplands, and most forests have about the same albedo, about 0.15, but outside the tropics it changes seasonally as deciduous plants produce green leaves in the spring, grow during the summer, and lose their leaves in the fall.

Cities are similar, although internally they are warmer than the surrounding areas because they have many local heat sources. Urban sprawl is a special US problem related to our use and dependence on automobiles. The constructed area of the United States, buildings, roads, and other hard surfaces, is about 113,000 km$^2$, almost equal to the area of the state of Ohio. It has a regional effect but perhaps not as much of a global effect as one might imagine.

Figure 15.2 shows Earth's albedo as seen on March 5, 2005, near the time of the vernal equinox when the Sun is over the Equator so both poles are illuminated.

The brightest areas in Fig. 15.2 are the high ice sheets of Greenland and Antarctica. The next brightest are the areas of sea-ice in the Arctic and Antarctic, and the snow-covered areas of the northern continents coming out of winter. The tropical oceans are the darkest regions and bring the global average down to near the classic global albedo value of 0.3.

Recently, the CERES Science Team (CERES = NASA's Clouds and Earth's Radiant Energy System) reported Earth's

**Fig. 15.2** Earth's albedo as seen by the CERES satellite on March 5, 2005, near the time of the vernal equinox when the Sun is over the Equator and both hemispheres are equally illuminated. Image courtesy of NASA

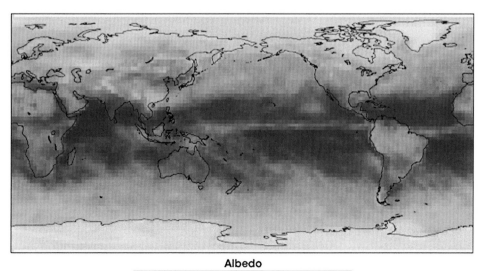

**Albedo**

0.0          0.5          1.0

shortwave albedo has been steadily declining since the instrument began making measurements in February 2000. From 2000 through 2004 the CERES instrument measured an albedo decrease of 0.0027. That translates to an increase of 0.9 W of energy retained in the Earth system. If everything else remained constant this would correspond to a temperature increase of 0.025 °C. It is uncertain what caused this decline in albedo. My guess as the most likely suspect would be declining snow cover. I base this guess on the work of our host in Leningrad, Mikhail Budyko, who in the 1960s realized how very sensitive our planet is to ice and snow cover.

It has long been known that Earth's albedo must have been higher when it was in its glaciated state, but it was in the late 1950s and 60s that Budyko published a series of papers attempting to quantify this effect and to determine whether this would be the feedback required to enhance the effect of the Milankovitch orbital variation to a level such that large-scale glaciation could occur. This was before the advent of computers, and he had to devise a clever mathematical approach. He assumed that polar ice has an albedo of 0.62, and that it is surrounded by barren outwash plains with some snow cover, having an albedo of 0.50. Using Milutin Milankovitch's calculations from 1941, he used a stepwise approach to explore the effect of starting with snow persisting through the summer at the critical 65° N latitude band. That changed the albedo both there and in the peripheral band further to the south. The increased albedo permitted the snow to expand to the south. He had to make guesses as to how long this would take, and then keep adjusting the insolation to the changing Milankovitch values. He concluded that the ice could advance to about 54° N in the Last Glacial Maximum and 2° further south in the preceding glacial maximum. These numbers correspond very well with what actually happened in Europe, but in North America the ice penetrated even further south. Budyko's computation methods were already too complex to allow him to take these geographical details into account. Also, the realization that the glaciation is largely a circum-North Atlantic phenomenon had not yet become evident.

Mikhail Budyko had not included any changes in cloud cover in his calculations. This was reasonable because the Northern Hemisphere ice sheets were thick and their surfaces were very high, between 2 and 3 km, like Antarctica today. White clouds over a white surface wouldn't make much difference anyway. However, in the 1970s we were trying to figure out why the Cretaceous was so warm, and one of the possibilities being considered was that the higher sea level at the time had expanded the area of the oceans so much that the Earth had become less reflective. In the middle Cretaceous, about 100 million years ago, sea level was about 300 m (= ~ 985 ft) higher than it is today. About 30 % of the present land area was under water in the Cretaceous, so a large area with an albedo about 0.15 would be replaced by water with an even lower albedo. At first it looked reasonable: this might be the answer to the problem of why the Cretaceous was so warm. However, the calculations did not include clouds. In effect, they were made for a hypothetical Earth-like planet that had no atmosphere. Was this right?

At the University of Miami's Rosenstiel School of Marine and Atmospheric Sciences we had a number of faculty who were working with the National Center for Atmospheric Research (NCAR) in Boulder, Colorado for their research. NCAR has been set up as a supercomputing center for academic research into weather and climate. At NCAR it was possible to run numerical simulations of climate. As a joint venture between our faculties in geology, atmospheric science, and physical oceanography, we decided to send a student to NCAR to work at on trying to solve the problem of the Cretaceous climate. Eric Barron agreed to be the guinea pig and went off to Boulder. We learned much more than we had hoped for. Eric became one of the nation's leaders in paleoclimate and climate change research, and is now President of Pennsylvania State University. In the 1970s the numerical climate models, although already extremely complex, were still relatively simple compared with those today. Eric and his NCAR advisor, Warren Washington, conducted a number of experiments, changing different 'boundary conditions' in the model one by one to see what their effect was on the climate. Boundary conditions are the assumptions that one makes about the distribution of land and sea, location and height of mountains, kinds of vegetation covering the land, greenhouse gas content of the atmosphere, insolation, etc. In a numerical model, you put in the boundary conditions, turn on the Sun, and, using all the appropriate basic laws of physics, the computer calculates the conditions at hundreds or thousands of points on the surface of the Earth. This is repeated over and over and over repeatedly recalculating the changing conditions that would have existed every few hours. Regional temperature differences set up winds, and the climate system develops. After 20 or 30 model years the calculations settle down, and the model starts 'simulating' the weather under the boundary conditions you had prescribed. To determine the 'climate,' many years of the model weather are averaged together. Eric and Warren conducted an experiment to see the effect of covering land with water, as it had been during the high sea level of the Cretaceous. In the simulation, clouds appeared over the seas and the overall albedo of the Earth changed very little. Apparently, flooding of the continents was not the reason the Cretaceous was so warm.

In the meantime, another problem was being explored at NCAR: if the northern hemisphere ice sheets start expanding, increasing the planetary albedo, why doesn't it just keep going and turn the planet into a snowball? It was found that as the ice proceeds further and further south it encounters ever increasing annual insolation, and with the present greenhouse gas concentrations, the global temperature at lower latitudes stays above freezing. Through the 1980s it was thought that, although there had been other episodes of glaciation far back in Earth's history, ice sheets have never advanced beyond 40° latitude. Even the faint young Sun discussed in Chap. 13 had not caused the Earth to turn into a snowball. It seemed as though we understood our planet's history rather well.

Then, in the 1990s, a startling geologic discovery was made. In the Flinders mountains of Southern Australia, finely laminated siltstones, rocks made of silty material, with 'dropstones,' pebbles that have been dropped from above, had been interpreted as having been deposited in water in front of glaciers. The dropstones would have fallen from icebergs. These rocks were deposited between 600 and 620 million years ago. It was already known that there had been glaciations in the Proterozoic, the Era of Earth history from 2,500 million years ago to 542 million years ago. But what was odd about these rocks was where they had formed. Using the magnetic properties of the rocks it was possible to establish that when they were deposited, they were near the Equator. Joe Kirschvink of the California Institute of Technology realized that ice sheets on Earth might once have been far more extensive than had been thought. Although most geologists were highly skeptical, others began to look into reports of glacial material in very ancient rocks. The glaciations occurred in two sets of episodes, the older ones in the early Paleoproterozoic, the older part of the Proterozoic, from 2,500 million to 1,600 million years ago, and the younger ones in the late Neoproterozoic (1,600–542 million years ago). The younger glaciations are concentrated in longer intervals, the Cryogenian (850–650 million years ago), and the Ediacaran (650–542 million years ago). There is good evidence that many of these glaciations resulted in ice cover right down to the seashore near or on the Equator. Even more interesting, the glacial deposits are intimately associated with carbonate rocks, limestones, that today form only in warm tropical waters. Apparently, the Earth went through a number of episodes when it became a snowball, and somehow managed to quickly recover.

During these episodes the Sun had not reached the intensity it has today. The solar constant 700 million years ago was about 324 W/m$^2$. The Snowball Earth albedo has been estimated to have been about 0.6, so the energy received by the Earth was about 130 W/m$^2$. This would correspond to a frigid planetary temperature of 241 K or −32 °C (−90 °F!). To find out how our planet got into this mess and how it recovered you have to wait until we discuss greenhouse gases.

## 15.2 Clouds

The greatest uncertainty in determining and predicting Earth's reflectivity is in evaluating the effect of clouds. At present they account for about 60 % of Earth's albedo. They can be very bright and reflective, or dark. They can appear, and grow or dissipate quickly. They can form a continuous layer or be very spottily distributed. They move, alternately masking ground surfaces of different albedos. They are the largest uncertainty in numerical climate modeling (Fig. 15.3).

**Fig. 15.3** Global albedo is dominated by clouds. Image by NASA's MODIS Atmosphere Science Team

Clouds form from the condensation of water vapor into droplets. This requires that the air be saturated or supersaturated with vapor. In the next two chapters we will go into controls on vapor saturation more extensively, but suffice it to say here that it is a function of both the amount of vapor in the air and temperature. Temperature decreases with altitude as the air pressure decreases. Remember that temperature is the kinetic energy of moving molecules. In air the molecules are flying around through space and colliding with each other and if there are fewer of them their overall kinetic energy, which is what we measure as temperature, is less.

There are two ways in which the water vapor content of the atmosphere is described. 'Specific humidity' is the amount of vapor in grams that is in 1 kg of air. 'Relative humidity' is the ratio of the amount of water vapor in the air and the amount that the air can hold when it is saturated. It is expressed in percent. It is a more useful concept for thinking about when and where clouds may form. The global average humidity at sea level is about 60 %. Of course, this varies greatly from place to place. The average air temperature at the top of the troposphere is about $-56$ °C ($=-69$ °F). At this temperature its ability to hold water is negligible; it is saturated with less than 0.1 g of vapor in 1 kg of air. The condensed droplet would be so small you wouldn't see it.

For droplets to form, the vapor needs a nucleation site, usually an aerosol, a minute particle suspended in the air. Aerosols range between 1 and 100 microns in size. Remember that a micron ($\mu$) is a common expression for the more formal term micrometer, one millionth of a meter ($1\ \mu = 0.0000394$ in). Water is a dipolar molecule: one side has a slight negative electrical charge, the other side a slightly positive charge. Don't worry, this will all be explained in detail in Chap. 17. Because of this property it will be attracted to any particle in the air that has an electrical charge. All of the major atmospheric gases, the molecules of $N_2$ and $O_2$ and atomic Ar are electrically neutral, but almost everything else in the atmosphere has some sort of a charge. There are three common natural aerosols that serve as nucleation sites: tiny salt grains, dust particles, and large organic molecules.

First let's consider the common sea salt, NaCl. It is the most common nucleation material over the ocean and for hundreds of miles inland. Much of the sodium chloride in river water is salt from the sea recycled through the atmosphere and rain. You may remember from Chap. 5 that the Irish physicist John Joly calculated the age of the Earth to be about 90 million years, based on the delivery of salt to the ocean by rivers. He didn't realize that most of the salt in river water is recycled from the ocean through the atmosphere.

How does salt get into the air? Wind ruffles the surface of the water and makes waves. As the wind increases and the waves get bigger, they 'break' and form whitecaps. Have you ever stopped to think about why 'whitecaps' are white? They are made of droplets of water.

Once a droplet is separated from its parent water, surface tension pulls it into a tiny round ball. The amount of light it will reflect depends on its size. The smaller the droplet the whiter it is. Something even stranger, as the droplet separates from its parent water, the surface tension shoots it forcefully away, like a tiny bullet. Once in the air, if the droplet is small enough, it is suspended in the air and does not fall back onto the ocean surface. You can experience the same thing in a hot shower or steam room. Small droplets evaporate readily if the air is not saturated. If they came from seawater, they contain some salt. As the water evaporates, it leaves the salt behind as a minute crystal in the atmosphere. Sometimes you can see the salt in the air as a whitish haze over the ocean. When the water vapor content of the atmosphere reaches saturation, that salt crystal will attract water molecules to form a droplet again. Incidentally, a typical cloud droplet is about 20 microns in diameter but a raindrop is typically about 2 mm in diameter, a factor of 100 different.

Dust is ubiquitous. In nature it is mostly small minute mineral grains or soil particles lofted by the wind into the air. In large quantities it can have an albedo effect of its own, but it is as important as nucleation sites for water droplets. Until humans began to farm the land using plows to expose the soil surface, dust sources were mostly in arid and semiarid regions where the water content of the atmosphere is low. Once in the atmosphere dust can travel great distances. From Joe Prospero, my colleague in Miami who studies such things, I learned that the dust that falls out in the rain over the Caribbean and southern Florida comes across the Atlantic from the Sahara. From another colleague, Gene Shinn of the US Geological Survey, I learned that the dust also includes microbes that can cause diseases affecting farm animals and crops.

Organic molecules are released by plants into the atmosphere as part of their exchange with the air. Light can cause these to combine and produce the large complex molecules that form a blue haze over forested areas and tropical jungles. They can serve as nucleation sites for droplets. Many years ago while I was at the University of Illinois I saw a demonstration of the development of the blue haze by Frits Went, who was then Director of the Missouri Botanical Garden in St. Louis. He had a large glass cylinder illuminated by an ultraviolet light. Over this he twisted a bit of lemon rind, like you might use for a dry vodka martini. For a moment nothing happened, and then as the invisible citrus oil vapors from the lemon rind settled into the cylinder it developed a beautiful blue hue. Frits had earlier been at Cal Tech and was well known for his studies of smog as well as being a great botanist.

As will be discussed in more detail below, 19th century impressionist painters were very much interested in the changing transparency of the air. Blue haziness and even brownish atmospheric tones play a big role in many of their paintings. Some of the accounts of those days refer to the air as being something visible, a property that had previously gone unnoticed. In fact, they were recording the pollution of the atmosphere by the expanding industrial revolution.

It is worth mentioning that human activities are creating some clouds where they would not occur in nature: water vapor injected into the upper troposphere and lower stratosphere by jet aircraft which we see as 'contrails.' Contrails (Fig. 15.4) are a recent contributor to Earth's albedo, becoming ever more significant since 1950. They may be one of the factors responsible for slowing global warming in the middle of the 20th century. The period from 1940 to 1980 is now referred to as the episode of 'global dimming' when a variety of anthropogenic effects reduced the transparency of the atmosphere and consequently the amount of insolation reaching Earth's surface. The global dimming was large enough to mask the greenhouse effect of increasing $CO_2$ levels.

Finally, you will recall from Sect. 14.11 he discussion of the possible climatic effects of cosmic rays that Henrik Svensmark thinks that ionization of the air by particle showers affects cloudiness. That is in fact the way Charles Wilson's cloud chambers work. The saturated air or in the cloud chamber condenses into droplets marking the path of a charged subatomic particle. If this really were an important process I would expect that we would from time to time see the cosmic showers producing beautiful downward displays of water droplets marking the paths of the particles. Perhaps it is in some science fiction film I missed.

The amount of clouds depends on the water vapor content of the atmosphere. It's an exponential function of temperature. The potential vapor content, 'saturation,' doubles with every 10 °C (18 °F) rise in temperature. Frigid air can hold almost no water vapor, so when it is very cold snowfall is minimal. At the freezing point (0 °C = 32 °F) a kilogram of air (about 1 m$^3$ = 10.8 ft$^3$ at sea level) can hold about 3 g of water as vapor, still not enough for a really good snowstorm. Heavy snow requires that warm air, with lots of vapor, be chilled. At 40 °C (= 104 °F), the temperature of good thunderstorm weather, it can hold 48 g of water vapor. There

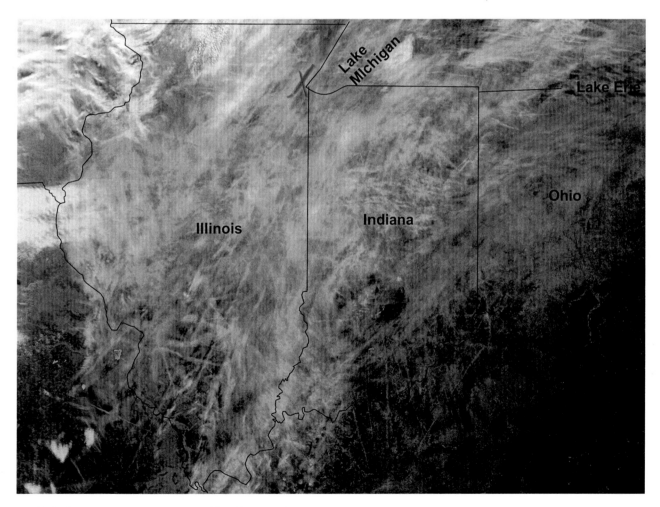

**Fig. 15.4** Contrails over the Midwestern United States on November 25, 2006

**Fig. 15.5** Some major cloud
types discussed in the text

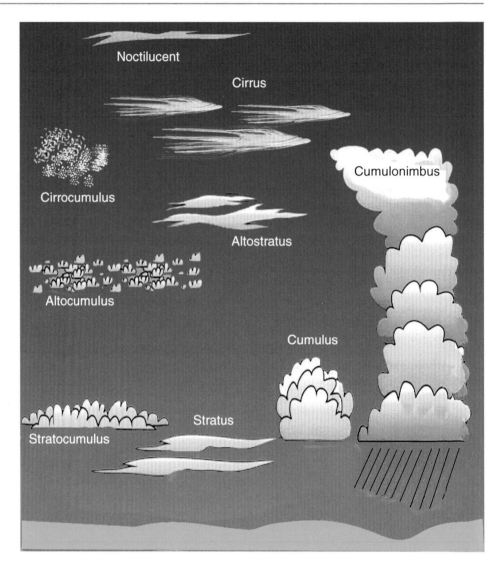

is little moisture in the air in the desert regions between 20°
and 30° latitude, and no clouds.

Figure 15.5 shows some of the major cloud types. They
can be grouped into two main categories, stratiform and
convective. Stratiform or layered clouds form blankets over
the landscape. They can even cover the ground surface; then
they are called fog. Fog banks can cover vast areas of the
ocean surface where the water is colder than the air. Convective clouds form when the air is vertically unstable, rising
in one place and descending in another. The rising air produces the puffy cumulus clouds seen on a warm day. The
clear areas in between the clouds are where the air is
descending. Each cloud marks an updraft and its usually flat
base is the elevation where the water vapor becomes saturated in the cooling air. The puffiness is the expression of the
turbulence in the convection. Different kinds of cumulus
clouds form at different elevations. The big fluffy ones are in
the lower troposphere. Smaller ones, because less moisture is
available, form in the mid-troposphere.

Clouds form where the air becomes saturated with water
vapor. The saturation occurs because as air rises it expands
and cools. The rate at which temperature declines with
altitude is called the lapse rate. More about it in the next
chapter, but here are some useful generalities: It is typically
about 6 °C/km ($\sim$3.3 °F/1,000 ft). This means that on a
pleasantly warm day (24 °C = 75.2 °F) the freezing point in
the atmosphere is about 4 km (2.4 miles = $\sim$12,700 ft)
above you. On a really hot day 40 °C (=104 °F) it would be
about 6.7 km (=4.16 miles = $\sim$22,000 ft) above you. At
higher altitudes the water droplets will be frozen into ice.

Both types of cloud, stratus and cumulus, occur at different levels in the troposphere, but their form varies with
altitude. Obviously, at the higher levels the clouds can
contain much less water. Stratus and cumulus are low level
clouds, usually found between a few hundred meters and
3 km (1.8 miles) above the ground. If they are supersaturated with water they produce rain and are called 'nimbus
clouds.' Clouds in the mid-level between 3 and 6 km

(1.9 and 3.7 miles) above the ground are termed altostratus and altocumulus. Those at altitudes between 6 and 12 km (3.7 and 7.5 miles) are called cirrus, a term or prefix meaning 'high.' Cirrus clouds often have the form of horse tails fanning out away from the direction of the upper level winds. Cirrocumulus clouds are small, dispersed high altitude versions of the lower level cumulus clouds. They often form what is called a 'mackerel sky,' often a forewarning of an approaching hurricane. These higher clouds are much less dense than the lower level clouds, and have lower albedos.

Cumulonimbus clouds form where there are very strong updrafts, and can extend through most of the troposphere. At the Intertropical Convergence, where winds of the northern and southern hemispheres come together near the Equator to form powerful updrafts, huge cumulonimbus clouds called 'towers' rise to altitudes of more than 18 km (=~ 60,000 ft = 11.2 miles).

Noctilucent clouds are a strange phenomenon, seen mostly at high latitudes. They are at altitudes in the range of 70–80 km. The name 'noctilucent' means 'night light.' They shine at night or for a while after twilight. They are so high that the Sun, below the horizon, shines off their undersurfaces long after it has set. How water vapor can be present at such great heights is still a mystery, but the most likely explanation is that it is produced in situ by the oxidation of methane to water and carbon dioxide under the influence of ultraviolet radiation. There is some anecdotal evidence that the occurrence of noctilucent clouds has increased since the beginning of the industrial revolution.

These different cloud types may have very different albedos, commonly ranging from 0.3 to 0.8 or even higher, as shown in Fig. 15.6. What controls the reflectivity of clouds when viewed from outer space? Droplet size and abundance. If the droplets are very small and abundant they are very reflective and make brilliant white clouds. As droplet size grows, they become less reflective and may appear gray. Ice crystals in the clouds tend to be larger than most droplets and are hence darker. The darker undersides of clouds are caused by the shadows of the droplets above. Remember that the albedo depends on the reflectivity of the upper surfaces of the clouds.

Another reason clouds are troublesome in understanding climate is that while they reflect light in the visible wavelengths, they are strong absorbers of the infrared from the Sun and the longer wavelengths radiated by Earth back into space. They are very important both in affecting Earth's albedo, and in acting as part of the overall atmospheric greenhouse system.

## 15.3 Could Cloudiness Be a Global Thermostat?

In the 1970s measurements of oxygen isotopes and study of fossil floras led to the conclusion that the Cretaceous tropical oceans had been 4–6 °C warmer and the polar oceans 15 °C warmer than today. As soon as these quantitative estimates of the warmer Cretaceous climate were published, they were questioned by some atmospheric scientists. It was argued that the Earth has a natural thermostat which would limit temperature increases. The amount of water vapor the air can hold doubles with every 10 °C increase in temperature. But the amount of energy required to convert water into vapor is very large, about $2.4 \times 10^6$ J/kg at these tropical temperatures. That is about 575 times the energy required to warm the water by 1 °C, $4.18 \times 10^3$ J/kg. As the energy of electromagnetic radiation is absorbed by water, a large part of it goes into turning some of the water into vapor. The kinetic energy of the water molecules is transformed into the latent heat of evaporation. This is effectively a heat storage mechanism. Atmospheric scientists proposed that there is a natural limit to tropical temperatures, imposed by these large amounts of latent heat involved in the evaporation process. They argued that a very large increase in insolation would be required to provide the energy necessary for the increased evaporation from a warmer sea surface.

In the early 1990s Veerabhadran Ramanathan and William Collins of the Scripps Institution of Oceanography in La Jolla, California proposed that the thermostat is more complex. Based on studies of cloudiness over the Pacific, they proposed that increased evaporation and atmospheric vapor content would induce formation of cirrus clouds high in the troposphere. These would reflect incoming solar radiation and limit surface temperature increases. However, this does not seem to be happening on a global basis.

The idea of an effective natural thermostat mechanism that would limit sea surface temperatures to approximately

**Fig. 15.6** Albedos of water and different cloud types

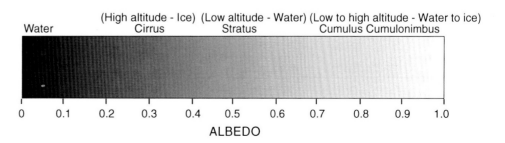

their present values is in conflict with our current understanding of the geologic evidence from the Cretaceous.

## 15.4    Volcanic Ash and Climate Change

In his work on changes in Earth's albedo, Mikhail Budyko concluded that there was another factor potentially as great as the ice-albedo feedback: volcanic aerosols in the atmosphere. We now know a lot more about volcanoes and their effect on climate than we did in the 1960s. Budyko thought the main culprit would probably be volcanic ash. However, volcanic emissions include not only ash, but gases, and particularly sulfur dioxide ($SO_2$) which attracts water vapor to eventually make droplets of sulfuric acid. Sulfuric acid has an extreme affinity for water, and forms condensation droplets. They are not as heavy as the ash and if injected into the stratosphere can remain aloft for long periods of time.

If these volcanic materials are lofted only into the troposphere, they will soon rain out. The effects will be strictly local. But if the volcanic explosion is violent enough, ash and other aerosols can be injected into the stratosphere and remain aloft for years or decades. The elevation of the troposphere-stratosphere boundary changes with latitude. Because cold air is denser than warm air, and because of the centrifugal force of the rotation of the Earth, the entire atmosphere is thicker at the Equator than it is at the poles. The troposphere at the Equator is about 20 km thick. At high latitudes it may be only half as thick. This means that the chance a volcano can loft material into the stratosphere depends on where it is. The volcanoes of Kamchatka, the Aleutians, and Iceland, all north of 50° N, are more likely to inject material into the stratosphere than those of Indonesia or the Philippines, between 20° N and S. Nevertheless, Mount Pinatubo on the Philippine Island of Luzon threw 20 million tons of sulfur dioxide into the stratosphere in 1991.

Undoubtedly the most spectacular volcanic event in historical times was the explosion of the Indonesian volcano Krakatau in 1883. It ejected 25 km$^3$ of pulverized rock into the atmosphere, with a column of ash rising 80 km into the sky, completely through the stratosphere and into the mesosphere. The ash spread around the entire planet, and lowered global temperatures in the next year by an estimated 1.2 °C. It is thought that the effects on the weather lasted for 5 years. Although these are spectacular events and may affect the global climate, they do not last long and do not occur frequently enough to cause permanent change in the global climate—or do they? Figure 15.7 shows the major eruptions over the past 1,000 years in terms of their effect on global insolation. The estimates of eruption intensity are based on sulfur dioxide ($SO_2$) preserved in ice cores in Greenland, so the suggested effects are probably biased

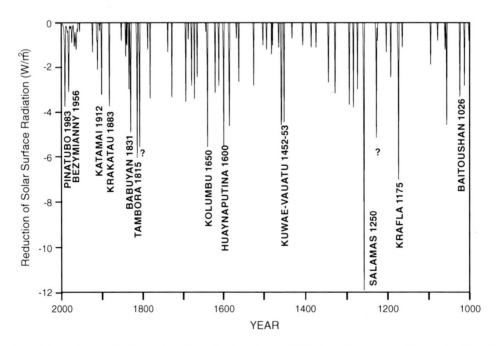

**Fig. 15.7** Reduction of insolation at Earth's surface by volcanic aerosols, mostly ash and sulfuric acid droplets (after Zielinski et al. 1994, *Record of Volcanism Since 7000 B.C. from the GISP2 Greenland Ice Core and Implications for the Volcano-Climate System*. Science, v. 264, pp. 948–952. Crowley 2000, *Causes of Climate Change Over the Past 1000 years*. Science, v. 289, pp. 270–277) Locations: Pinatubo–Luzon, Philippines; Bezymianny–Kamchatka; Katmai–Alaskan Peninsula; Krakatau–Indonesia; Babuyan–Philippines; Tambora–Indonesia; Kolumbu–Santorini, Greece; Huaynaputina–Peru; Kuwae-Vanuatu–South Pacific; Salamas–Indonesia; Krafla–Iceland; Baitoushan–China

toward eruptions in the Northern Hemisphere. In evaluating the climatological effects it is useful to know that a global decrease of insolation received by the Earth of 1 W/m$^2$ would lower the global temperature by about 0.27 °C.

Where are these explosive volcanoes located? If you recall the theory of plate tectonics, there are areas where one plate overrides another. We say that the plate being over-ridden is being subducted. On Earth's surface these zones of subduction are marked by the oceanic trenches, such as the Peru-Chile Trench bordering western South America; off western North America we don't see the trench because it has been overridden by the continent or is filled with sedi-ment. The oceanic crust being subducted is made of basalt, much of which has been altered by reaction with seawater percolating through it. The combination of altered basalt and water makes an explosive combination when the slab reaches a depth of about 100 km. Volcanoes form on the overriding plate above the slab; these are the volcanoes of Indonesia, the Philippines, and Japan. In South America these are the volcanoes of the Andes, in North America these are the Cascades. Almost all of the major eruptions shown in Fig. 15.7 are from subduction zone volcanoes. These explosive volcanoes are very different from the relatively quiet volcanoes like Mauna Loa in Hawaii, where the magma comes from sources in the mantle, and erupts though the plate with minimal infiltration of seawater releasing very fluid lava and very little ash.

The eruption that seems to have had the greatest impact on global climate in the past few centuries was that of Tambora in Indonesia in April of 1815. It immediately affected the weather in the tropics, but it was not until the next year that the ash and sulfur dioxide veil had spread throughout the stratosphere to higher latitudes. That year, 1816, is known as the 'year without a summer.' Snow fell and accumulated in Quebec City in early June. Frosts con-tinued throughout Canada and New England through most of the summer causing crops to fail. Cold and freezing temperatures alternated with extremely warm days. In Eur-ope the whole summer was cold and bleak.

In 1816 Lord Byron, John Polidori, Percy Shelly and his future wife Mary Godwin had rented a villa on the shores of the Lake of Geneva and expected a pleasant summer. Percy Shelly had separated from his first wife Harriet; she was to drown herself in the Serpentine Lake in Hyde Park, London, early in December 1816. The weather in Switzerland, as in all of Europe, was stormy, depressing and dreary. At the time the weather was blamed on sunspots. Unable to spend the time outdoors sailing on the lake as they had planned, they decided to try their hand at writing scary stories. Mary decided to write a book; in its preface she described the situation:

I passed the summer of 1816 in the environs of Geneva. The season was cold and rainy, and in the evenings we crowded around a blazing wood fire, and occasionally amused ourselves with some German stories of ghosts, which happened to fall into our hands. These tales excited in us a playful desire of imitation. Two other friends (a tale from the pen of one of whom would be far more acceptable to the public than anything I can ever hope to produce) and myself agreed to write each a story, founded on some supernatural occurrence.

But it was her story that has fascinated the public ever since. Her book, *Frankenstein*, was published in 1818. She married Percy Shelley on December 30, 1816. Percy drowned as his schooner was caught in a storm off La Spezia on the Italian Riviera in 1822.

The cold, nasty weather of 1816 caused more than depression of people's spirits. The crop losses caused famines and facilitated an epidemic of typhus.

One of the most significant eruptions in terms of history of civilizations was that of the volcanic island Thera in the Greek archipelago about 1626 BCE, also known as the Minoan eruption. It occurred at the time of the height of the Minoan civilization centered on the island Crete about 110 km (68 miles) to the south. The explosive eruption sent an estimated 100 km$^3$ (24 miles$^3$) of ash into the air. The site of the volcano is now a 12 by 7 km (7 by 4 mile) wide caldera partially surrounded by the semi-circular island of Santorini with volcanic ash deposits several hundred meters thick. It is thought that the explosion generated a gigan-tic tsunami that devastated the Minoan settlements on the north coast of Crete. The Minoans were seafarers, and must have had complete control over the eastern Mediterranean because their towns and cities, mostly along the coast, had no defensive walls. They appear to have been immune to attack by ships from the sea, but not to a tsunami 26 m (85.3 ft) high. Another increasingly popular idea is that pre-eruption Thera was Plato's legendary Atlantis.

Pyroclastic flows are suspensions of red hot ash in air. The deposit they form is called tephra. It was pyroclastic flows from Vesuvius in 79 CE that buried Pompeii and Herculaneum. The great Roman naturalist Pliny the Elder was fascinated by the eruption of Vesuvius, and was killed as he tried to observe a pyroclastic flow first hand. His son, who had stayed behind at their home on the other side of the Bay of Naples reported that the flows that had buried these towns continued out over the waters of the Bay and engulfed ships headed to rescue those on shore.

It was a pyroclastic flow from Mount Pelée, on the French Caribbean island of Martinique, that killed over 30,000 people in the town of St. Pierre on May 8, 1902. Melted glassware found in the houses was cited as an indication of the temperature of the flow, but it has recently been suggested that they might reflect the heat from burning wood set ablaze by the flow. Legend has it that the sole survivor in St. Pierre was a prisoner locked up in an interior cell in the town's jail.

The climatic effects of volcanic eruptions can be disastrous, but they generally do not last long. On the other hand, if you happen to live next to a volcano you might want to have a disaster plan. My favorite places for living on the edge would be: (1) in one of the suburbs of Naples on the slopes of Vesuvius where only narrow roads will serve as evacuation routes for your million or so neighbors; or (2) near Georgetown or one of the other picturesque towns on the northern end of St. Vincent in the Caribbean, under the shadow of the Soufrière Volcano (the name means 'The Sulfurer'); or (3) on the beautiful Shimabara Peninsula of Kyushu, Japan, where the Unzen volcano has a particular penchant for killing scientists studying it (43 to date), and where you must make sure there is a hill between your house and the volcano so you will not be in direct line of a pyroclastic flow.

At this point I need to tell you an interesting anecdote about the volcanism-climate story. At a meeting in the 1980s I attended a session on the relation between volcanism and climate. The speaker was trying to find correlations between eruptions and cooler temperatures in the same or the next year. Hugh Elsaesser, a climatologist with the US National Oceanic and Atmospheric Administration (NOAA), commented at the end of the talk that he thought you could make just as good an argument for cold weather being responsible for volcanic eruptions. He made it as a joke, but Hugh was thoroughly familiar with the data, and there are some good examples of the years before an eruption being colder than after. Examples: 1881 was cold, Krakatau erupted in 1883; 1901 was cold and Mount Pelée and the Soufrière in the West Indies and Santa Maria in Guatemala erupted in 1902. 1929–30 were cold; there was a major eruption of Azul Cerro in Chile in 1932. Hugh meant his comment as a joke, pointing out the fragility of the relationships found in the data. But a few years later I attended another symposium on this topic, and a meteorologist gave a talk relating snowfall to volcanic eruptions. Probably inspired by Hugh's comment he proposed that the weight of the snow caused the volcano to sink slightly, and opened up cracks for the lava to work its way to the surface. This was a case of not taking into account the orders of magnitude difference in the effects, and after discussions the idea did not make its way into print. However the idea wasn't totally forgotten.

It raised a very interesting point. Global surveys of the amount of volcanic ash in deep-sea sediments o indeed show that there have been increasing amounts of ash since the glaciation of the northern hemisphere began. Could this have something to do with the buildup and decay of ice sheets?

As the ice sheets grow and decay, their weight depresses the Earth's crust and it displaces the rock in the Earth's mantle. This rock is so hot and under such pressure that it can move as though it were a very viscous fluid. The density of ice is just slightly less than water and the density of the mantle rock is about three times as much as water. The ice sheets were typically 3 km thick. So in order to come into equilibrium, like the ice floating in your gin and tonic, an ice sheet will displace 1 km of mantle material. That has to go somewhere, and it flows toward the region not covered by the ice. You can imagine that as it flows away from the ice it might encounter one of those slabs where one plate is being forced beneath another. It might push against that slab as the ice builds up, and again from the other side to flow back as the ice melts away. This would cause the slab to wiggle a bit, and that might indeed open passages for magma to rise toward the surface. So glaciation might actually increase the rate of volcanic activity on Earth, while a few years of heavy snow would not have any effect.

Recently, concern has been expressed that the rapid melting of the ice sheets on Iceland and the accompanying acceleration of isostatic rebound might increase the frequency of volcanic eruptions there. The Icelandic volcanoes are unusual in that they are on a plate-tectonic spreading center, the Mid-Atlantic Ridge, where it is above water. The 2010 eruption through the Eyjafjallajökull ice sheet produced a plume of volcanic ash that reached the stratosphere and disrupted air travel in Europe for several weeks (Fig. 15.8). The media incorrectly reported that the eruption was introducing huge amounts of $CO_2$ into the atmosphere; it was sulfur dioxide ($SO_2$) that was released. Nevertheless, climate change deniers took advantage of the error, and claimed that the $CO_2$ from the eruption was much larger than any increase from burning fossil fuels. They even argued that almost all of the observed increase in atmospheric $CO_2$ must be due to volcanic activity. Afterward it was found the shutdown of air travel actually slowed the global rise of atmospheric $CO_2$. In May 2011 an eruption at Grímsvötn under the Vatnajökull glacier again spewed ash into the sky. In August 2014 eruptions at Bárðarbunga produced extensive lava flows but little ash. However, this eruption injected such large quantities of $SO_2$ into the atmosphere that, depending on winds, breathing conditions across Iceland were affected.

A detailed record of the history of Icelandic volcanic eruptions during the latter part of the deglaciation from the Last Glacial Maximum and through Holocene is potentially available in the ice cores from nearby Greenland. But identification of individual layers as originating from specific volcanoes requires detailed geochemical analysis. The work is ongoing and perhaps in a few years we will have a better understanding of the relation between ice sheet growth and decay and volcanic activity.

**Fig. 15.8** Ash plume from Eyjafjallajökull, April 15, 2010

## 15.5 Aerosols

Aerosols, tiny particles that are suspended in the air of the troposphere are an important transient component of the climate system. Aerosols include mineral grains ('dust'), sea salt, smoke, pollen, and sulfates from volcanoes and industrial operations. They are very unevenly distributed in the atmosphere, with different kinds dominating in different regions. Figure 15.9 shows NASA's 'Portrait of Global Aerosols.'

**Fig. 15.9** NASA's 'Portrait of Global Aerosols.' This image was produced by a GEOS-5 simulation at a 10 km resolution. Dust (*red*) is lifted from the surface, sea salt (*blue*) swirls inside polar cyclones, smoke (*green*) rises from fires, and sulfate particles (*white*) stream from volcanoes and fossil fuel emissions

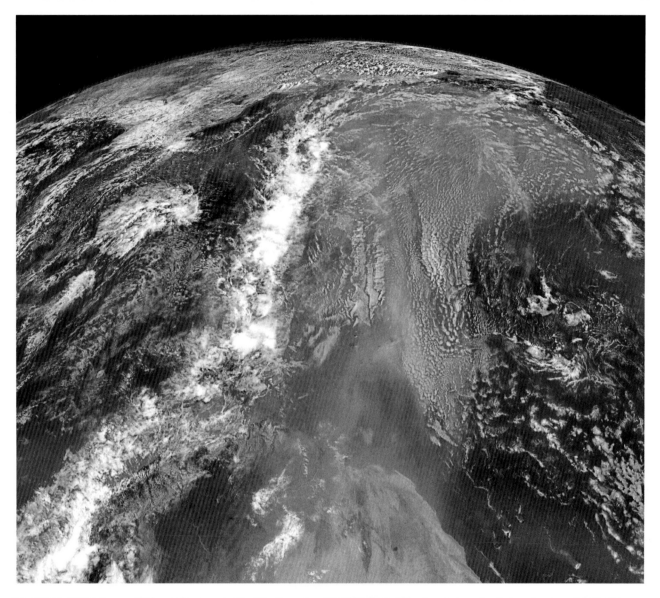

**Fig. 15.10** NASA image of Saharan dust crossing the Atlantic on June 24, 2014. West Africa is at *bottom*, South America at *top left*, Caribbean at *top center*. *White band* is the towering cumulus clouds of the Intertropical Convergence

Mineral dust, shown in red in Fig. 15.9 is generated in desert regions and area of drought. Carried by turbulent winds it can travel long distances. Few people realize that the dust that settles on the Caribbean Islands and Miami comes from Saharan Africa. The warm, dust-laden Saharan air rides over cooler air above the Atlantic and can cross the entire ocean (Fig. 15.10). In 1972 I led an expedition to recover sediment cores off West Africa. Near the Cape Verde Islands, we experienced a very peculiar phenomenon. There was so much dust in the air above us the Sun was just an orange disc in the sky. You could look at and see sunspots. Yet no dust settled on our ship. The dust was being carried westward in a layer of hot dry desert air overlying a thinner layer of more dense moist cooler air over the sea

surface. The radar showed a diffuse circle around the ship, the radar signal being reflected off the dust.

These dust storms are sporadic, but it is now thought that the unusually quiet Atlantic hurricane season of 2006 and some recent years have been due to dust reflecting and absorbing energy that would otherwise have warmed the ocean. Satellite imagery has shown that the dust content of the atmosphere decreased during the 1990s and has increased since then.

Dust was a much more significant part of the climate system during glacial epochs. There were much larger areas without continuous plant cover that could serve as sources. The ice sheets were surrounded by barren plains of fine rock ground up by the glaciers. Streams coming out of glaciers today are often milky white, loaded with 'glacial flour.' This is rock so finely ground up that the particles appear white. Winds blowing over the outwash plains loft the fine-grained sediment into the air, and can transport it for long distances. The dust can become a major sedimentary deposit of silt-sized particles in some areas. In the United States some of the cliffs along the east side of the Mississippi in Illinois are made of this wind-blown dust, which geologists call loess. But the most famous area is in China, where the 640,000 $km^2$ Loess Plateau extends from the Tibetan Plateau to the North China Plain. The loess is probably derived from the Tibetan Plateau, which was arid during glacials. Cultivation in the Loess Plateau began about 4000 years BP, and it has been very actively eroding ever since. The Yellow River flows through the Loess Plateau, and derives its name from the sediment load it picks up there; it is estimated to carry 1.6 billion tons of silt each year to the Yellow Sea. Examination of deep-sea cores taken in the Yellow Sea show that the river carried little sediment before the Loess Plateau was farmed.

Figure 15.11 shows dust concentrations in the Vostok ice core. Vostok is located in the sector of Antarctica south of Australia, about one third of the way between the Antarctic margin and the South Pole. The dust is clearly concentrated in the glacial intervals, which can have more than 100 times as much as the interglacials. The dust comes from Patagonia in southern South America. During the glacials, Patagonia had extensive barren outwash plains from the glaciers of the southern Andes.

## 15.6 Albedo During the Last Glacial Maximum

It was in the 1960s that Mikhail Budyko recognized the significance of changes of Earth's albedo as it alternated between glacial and interglacial state. His description of the ice-albedo feedback system was soon recognized as a major positive feedback mechanism enhancing the effect of changes in insolation resulting from alteration of the Earth's orbital parameters that had been described by Milutin Milankovitch. Budyko's hand calculations corresponded to the limits of glacial advances in Europe and North America known from geologic evidence.

The attempts by Budyko and others to reconstruct the planetary albedo during the Last Glacial Maximum (LGM) considered primarily the additional land areas covered by ice. However it soon became evident that there were

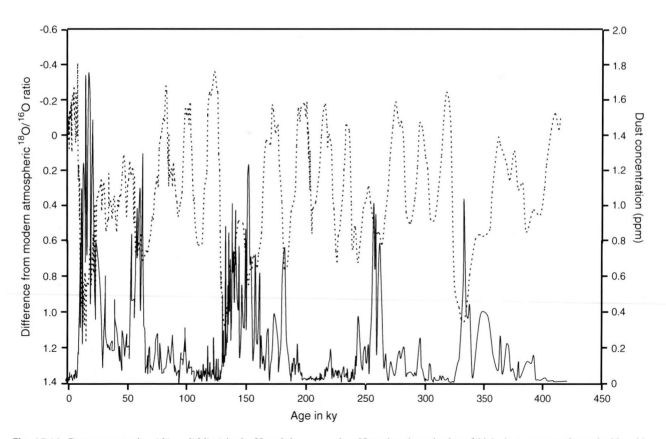

**Fig. 15.11** Dust concentration (*thin solid line*) in the Vostok ice core from Antarctica. The $^{18}O{:}^{16}O$ ratio curve, an index to the amount of ice on Earth, is shown by the *dotted line*; the larger numbers mean more ice. Note that the episodes of high dust concentration coincide with times of maximum glaciation. Data from NOAA's National Climate Data Center (NOAA NCDC)

other changes that would affect Earth's albedo: additional land areas were exposed as sea level fell exposing the continental shelves; there were changes in the vegetation on land; and there were increased levels of dust in the atmosphere. Then there is the problem of the nature of cloud cover on an Earth with expensive ice sheets. Different climate modeling facilities have used different patterns of these parameters. This led to the development of the Paleoclimate Modeling Intercomparison Project (PMIP), started in 1996 as an expansion of older numerical modeling comparisons. PMIP is an ongoing effort comparing the results of paleoclimate models from different institutions for specific ages and combining them as 'ensembles.' A recent comparison of the results of climate models at NCAR in Boulder, the Hadley Center in Exeter, England, and the University of Tokyo's Center for Climate System Research uses 0.304 for the albedo of the pre-industrial (PIND) Earth, and an average increase of 0.016 to 0.320 for the Earth's albedo during the LGM (Murakami et al. 2008, Journal of Climate, v. 21, pp. 5,008–5,033). Using the Kirchhoff/Stefan-Boltzmann scheme the planetary temperature would be 254.2 K (=−18.9 °C; =−2 °F) with the pre-industrial albedo, and 252.7 K (=−20.5 °C; −4.8 °F) with the LGM albedo,

The effect of the enhanced dustiness of the LGM was not included in the climate models cited above, for good reason. Dust in the atmosphere should have further increased the albedo especially at latitudes equatorward of the ice. However, dust over the ice will settle on to it and lower its albedo. We have already learned that as ice melts, the dust and soot particles in it accumulate on its surface. This results in absorption of more radiation, warming of the surface, and more rapid melting of the ice. To complicate matters further, as this layer of particles becomes thicker and thicker it begins to act as a shield slowing further melting of the ice. The darkening effect has become apparent on Greenland in recent years, and many of the Himalayan glaciers have such a thick cover of debris it is not certain where and how thick the ice is.

The difference due to changing albedo used in the models cited above is about 1/4 of the estimated global temperature difference of 6.5 °C (=11.7 °F) between pre-industrial times and the LGM. Because the insolation at the last glacial maximum was almost the same as today, the other 3/4 of the difference must come from changes in greenhouse gas concentrations and atmospheric aerosols. The increases in greenhouse gas levels recorded in the Antarctic ice cores from the Last Glacial Maximum (24,000 years BP) to pre-industrial time (1750 CE) were: $CO_2$: 182 to 285 ppm; $CH_4$: 340 to 667 ppb; $N_2O$: 200 to 275 ppb. In the Vostok core the mineral aerosol ('dust') levels changed from 1.8 ppm in the LGM to about 0.02 ppm in the pre-industrial Holocene, but this is a regional effect which cannot be extrapolated to Earth as a whole and was not included in the climate model experiments.

## 15.7  Changing the Planetary Albedo to Counteract Greenhouse Warming

There have been a number of suggestions about how one might counter the effect in warming due to increased greenhouse content of the atmosphere. Some are quite fanciful, others may be practical. None will solve the problem completely.

(1) The possibility of placing huge sunshades, like umbrellas in space has been explored. The costs would be enormous, and the lifetime limited by space debris.

(2) Injection of sulfur dioxide into the stratosphere, mimicking the effect of volcanoes, would be relatively cheap but would result in acid rain which would destroy most plant life on land, and increase acidification of the ocean.

(3) Using special ships to emit very tiny, highly reflective water droplets into the air over the oceans. The problem is that it changes the regional albedo with potentially unexpected consequences.

(4) Making the roofs of buildings white and using white aggregate in roadways to reflect more sunlight. This technique has already been used in some areas, such as Greece, to keep homes cool without air conditioning. This would have minimal cost impact, but again it would change the regional albedo with uncertain consequences.

## 15.8  Summary

Albedo is the reflectivity of the planet. The Earth's present albedo is about 0.30. During the Last Glacial Maximum it may have been closer to 0.32. The effect is dominated by clouds, one of the least well understood parts of the climate system.

Cloud albedos range from 0.3 to 0.8. The oceans have an even greater range of albedos, from 0.05 in the equatorial regions to 0.8 in the high latitudes where the incoming light is reflected off the surface. Albedo of the land areas average about 0.2 not varying greatly whether they are desert, grasslands, crops, forests, or covered by urban sprawl.

It has been suggested that cloudiness might be a global thermostat. Cloudiness would increase with the increased moisture available from a warmer sea surface. In the 1970s this appeared to be the case, at least for part of the eastern Pacific, but it has not been obvious as a more global phenomenon. The contrails produced by jet aircraft and the increased atmospheric pollution from cars and industry may have arrested the long-term rising temperature trend during the middle of the 20th century.

Volcanic eruptions can inject ash and sulfur dioxide into the stratosphere where it can remain for several years. This can affect the global albedo and result in lower surface temperatures. The most famous was the eruption of Tambora in Indonesia in 1815. It had global effects, making 1816 the 'year without a summer.'

Dust can also play an important role in raising the planetary albedo. There were episodes of greatly enhanced dustiness during glacial times, and these may have had a feedback effect reinforcing the colder climate. Enhanced dust transport from Africa over the Atlantic in recent years may have reduced the frequency and severity of hurricanes.

Changing the planetary albedo by geo-engineering has been suggested as a means of counteracting global warming due to greenhouse gas emissions, but the methods proposed are often prohibitively expensive and would produce uncertain results.

A Timeline for this chapter:

| 1940–1980 | Earth's temperature remains stable, later attributed to 'Global Dimming.' |
|---|---|
| 1941 | Milutin Milankovitch publishes his *Kanon der Erdbestrahlung und seine Anwendung auf das Eiszeitenproblem* tracing variations in insolation over time |
| 1952 | Toxic Fog engulfs London |
| 1969 | Mikhail Budyko publishes *The effect of solar radiation variation on the climate of the earth* |
| 1977 | We send Eric Barron from the Rosenstiel School of Marine and Atmospheric Sciences (RSMAS) (University of Miami) to the National Center for Atmospheric Research (NCAR) in Boulder, CO to work on modeling Cretaceous Climate for his Ph.D. dissertation |
| 1980 | Eric Barron's modeling confirms Mikhail Budyko's idea that $CO_2$ was the major factor controlling past global climate change |
| 1991 | Veerabhadran Ramanathan and William Collins publish *Thermodynamic regulation of ocean warming by cirrus clouds deduced from observations of the 1987 El Niño* proposing that clouds would act as a thermostat to control global warming |
| 1992 | Joe Kirschvink publishes *Late Proterozoic low-latitude global glaciation: The Snowball Earth* |
| 2000–2004 | CERES measures a decrease of Earth's albedo of 0.0027 |
| 2008 | Eric Barron becomes Director of NCAR |
| 2009 | Eric Barron becomes President of Florida State University |
| 2014 | Eric Barron becomes President of the Pennsylvania State University |

If you want to know more:

Again, Peixoto and Oort's book *Physics of Climate,* published by the American Institute of Physics has an excellent account.

Music: The classic 1968 film *2001—A Space Odyssey* uses Johann Strauss II's 'An der schönen blauen Donau' (*By the beautiful blue Danube*) as the graceful background music for space travel.

Libation: This would be a good time to breakout a bottle of champagne and toast NASA and the other space agencies that have contributed so much to our knowledge of Planet Earth.

**Intermezzo XV. The Deep Sea Drilling Project**
When I started my academic career, the tools used by a geologist in the field were a rock hammer, a 'Brunton Compass,' which had sights for determining directions, a tape measure, and a map. The motto of the International Geological Congress is "Mente et Malleo" (Mind and Hammer). The hammer in the Soviet Union's hammer and sickle flag was a geologist's hammer.

Occasionally geologists also obtained samples from wells being drilled. Oil companies used the cuttings from the drill bit that came up in the recirculating mud. I too used them as an academic researcher early in my career. But to obtain a larger sample of the rock from beneath the surface it was necessary to take a core. To do this while drilling a well, it was necessary to pull up the entire drill-string and substitute a core barrel for the drill bit. The drill-string consists of 30 foot long pipes screwed together. In 'pulling the string' they must be pulled up, unscrewed and stacked. This is the reason a drill rig has a tower. The whole string, now with a core barrel at its base, must then be reassembled and lowered back down the hole, and a core drilled out. Then it all has to come back up again. It is a very time consuming, and needless to say, expensive process, so cores were not taken often. The cores that were taken by industry were often considered proprietary because they were the source of information about the porosity and permeability of the rocks being drilled: information needed to determine whether a producing well might be profitable or not.

The oceanographic institutions took cores of deep sea sediments, but these were usually only a few meters or tens of feet long, and the age of the materials recovered was almost always Quaternary.

When the Soviet launch of Sputnik in 1957 awakened the US to the need to develop its scientific capabilities, most of the emphasis was initially directed at developing a space program. However, that fall a group of American Earth scientists led by Walter Munk of the Scripps Institution of Oceanography got together at a backyard barbeque in La Jolla and discussed whether this might not be a golden opportunity to make a major new advance in understanding our planet. Calling themselves the "American Miscellaneous Society" (AMSOC), they proposed to obtain a sample of the Earth's 'Mantle.'

The outermost layer of the solid Earth is called its 'crust.' It is about 30–50 km (18–30 miles) thick beneath the continents but only 15 km (9 miles) thick in the ocean basins. Its base is defined by a seismic discontinuity; a surface at which the velocity of seismic (sound) waves passing through the rock suddenly increases. The change in seismic velocity is known as the Mohorovičić Discontinuity (the Moho), named for Croatian meteorologist and seismologist Andrija Mohorovičić. The layer of rock beneath the discontinuity is called the 'mantle' and extends down to the Earth's core. In the 1950s we had a fairly good idea that the crust beneath the oceans was basalt, but no one knew for sure what kind of rock might make up the mantle.

AMSOC called its program 'Project Mohole,' identifying the Pacific Ocean basin north of Hawaii as the best site. There the MOHO is about 15 km (9.3 miles) down, and the first 4 km (2.5 miles) are water. They proposed it to the National Academy of Sciences, which was then able to obtain funding from the National Science Foundation to make a feasibility study. It is important to remember that at that time most American geologists considered the ocean basins to be permanent, ancient features. They also thought that the deep sea had always been a cold unchanging environment and it was assumed that the overlying sediments would contain only a monotonous, uninteresting four billion year history.

The technology required to drill a hole into the ocean floor in water 4–5 km (2.5–3 miles) deep did not exist. Drilling through water had been done in Lake Maracaibo in Venezuela, and in the Caspian Sea, but this was done from platforms. Some tests had been made of drilling from anchored barges, but that was it. To drill a hole to the Moho it would be necessary to hold the drilling platform nearly stationary for a long period of time. Anchors could not achieve this in several kilometers of water. The platform would need to be a vessel "dynamically positioned" by propellers. Further, it would be necessary to pull up the string of drill pipe, change the drill bit, and lower the bit and pipe back down to the sea floor and reenter the hole in the ocean floor. A lot of engineering needed to be done. Two feasibility tests were conducted 1961 using a drilling barge outfitted with outboard propellers and a system for positioning it relative to anchored buoys using radar. The deepest test was in water 3558 m deep (11,673 ft) off Baja California. It drilled through 170 m (558 ft) of sediment and 13 m (43 ft) of basalt. Only a fraction of the sediment and basalt were actually recovered, but it was evident that ocean drilling could be done. The engineering studies got into high gear. I heard a lecture about the planning for Project Mohole at the University of Illinois in 1963. I was impressed by the idea that the drilling vessel with its tower would be so tall that it could not pass beneath the Golden Gate Bridge. The cost estimates were becoming astronomical.

In the meantime, Cesare Emiliani's 1961 paper showing that the deep sea had been warmer in the past had appeared. That same year Bob Dietz's paper *Continent and Ocean Basin evolution by spreading of the sea floor* appeared. Harry Hess's ideas on ocean floor evolution were circulated in manuscript, and in 1962 his paper *History of Ocean Basins* was published. The basic ideas had already been presented in Arthur Holmes' 1944 textbook *Principles of Physical Geology* but had not been noticed in the United States. The theoretical underpinnings of project Mohole were becoming very wobbly.

In 1966 a young Congressman from Illinois, Donald Rumsfeld, conducted a review of Project Mohole. He questioned its budget and significance. He did not trust the contractor, Brown and Root, to be able to keep to the budget (Brown and Root is now a subsidiary of Halliburton). He concluded that the nation could not afford it, and the project was canceled.

But both Maurice Ewing, the Director of Lamont Geological Observatory, and Cesare believed that much more could be learned from the sedimentary record in the deep sea than by drilling to the Moho. In 1962 Ewing had approached several petroleum companies proposing a program of drilling and coring on a latitude-longitude grid. Emiliani argued that much more could be learned by drilling and recovering long sediment cores designed to investigate the history of the ocean and proposed a sediment drilling program, 'Project LOCO' to replace the single hole into the mantle. A LOCO advisory committee was established, with scientists from several oceanographic institutions. Then one of those things happened that could only have happened in those days. Bill Benson, program director handling the LOCO proposal in the US National Science Foundation, got word that a small shallow-water drilling vessel, the *D.V. Submarex*, was headed through the Panama Canal toward the Gulf of Mexico. He picked up the telephone and asked Cesare if he might be able to use this vessel, with only a few weeks' notice. Of course, Cesare jumped at the opportunity and in November and December of 1963 Project LOCO was initiated with the *Submarex* coring late Tertiary and Quaternary sediments on the Nicaragua Rise. The coring technique was improvised on the spot by Jim Jones of the Miami Institute, and consisted of dropping an oceanographic core barrel down inside the drill string. I was among those given the opportunity to work up the fossils from those cores for a publication.

The preliminary Mohole and *Submarex* cores demonstrated that the oceans had a history that was worth exploring, and the evidence for the theory of plate tectonics was rapidly emerging. The National Science Foundation urged four major oceanographic institutions to get together and submit a joint proposal for more sediment drilling and coring. In 1964 Lamont, Miami, Scripps and Woods Hole organized themselves as the Joint Oceanographic Institutions for Deep Earth Sampling—'JOIDES.' Their first drilling campaign was in 1965, with a transect of holes to depths of several hundred meters on the Blake Plateau east of Jacksonville, Florida using the drill ship, CALDRILL. Lamont was the operating institution and the cores were stored at Miami. A party of shipboard scientists from different institutions was aboard and carried out the initial analysis of the samples. This was to become the model for future operations. But the CALDRILL was too small for operations far from port.

JOIDES then established several science panels to plan a more extensive program. At first the Panels were regional. I was appointed to the Panel for the

Gulf of Mexico. In 1966 JOIDES formally proposed a program of ocean drilling as the "Deep Sea Drilling Project (DSDP)" to the National Science Foundation. It was for 18 months of drilling in the Gulf of Mexico and then along transects across the North Atlantic, South Atlantic, Caribbean Sea and Pacific Ocean.

More and more geophysical data were suggesting that the ocean basins were not permanent, but young dynamic features. Mapping of the Earth's magnetic field over the ocean south of Iceland and west of British Columbia and Washington showed 'magnetic stripes,' where magnetism of the sea floor alternately enhanced and suppressed Earth's magnetic field. It was already known that Earth's magnetic field had experienced a number of reversals in the past and the hypothesis was that ocean crust formed along the mid-ocean ridge system and then moved away as new crust was emplaced. The stripes reflected the alternations of earth's magnetic field. Ocean crust must be returning to the mantle along zones of convergence, such as oceanic trenches. The ocean floor must consist of a few large rigid plates, with new crust being added on one side, and older crust being lost on the other side. If these hypotheses of "sea floor spreading" and "plate tectonics" were true, the scientific assumptions that had gone into the design of Project Mohole had been quite wrong.

The engineering work that had been carried out for Project Mohole became important for the DSDP. The dynamic positioning system designed for the Mohole vessel became a reality, and the technology for re-entering the drill hole on the sea floor was further developed. In 1967 a ship already under construction for Global Marine Corporation intended for commercial drilling while anchored in shallow water, was modified to have thruster propellers installed in the bow and stern. These can move the ship sideways to help maintain position. The ship's location on the sea surface, relative to the position of the hole on the sea floor, would be determined not by anchors or by radar bouncing off floating buoys, but by using sonic beacons dropped onto the sea floor before starting the drilling procedure. The computer program to carry out the necessary calculations was the first of its kind. The ship was also equipped with the first satellite navigation instrument to assist in locating the site for drilling.

A rapid method of recovering cores was developed for use in the project, wholly different from the industry procedure but closer to that used on *Submarex*. "Wireline coring" involves dropping a core barrel down inside the drill pipe to latch into the drill bit. The core is taken through an opening in the center of the bit by drilling ahead. When the core barrel is thought to be full, a recovery tool is dropped down inside the drill-string on a wireline. On impact with the core barrel, this tool unlatches the core barrel from the drill bit so that it can be hauled back up to the ship. With the wireline system a core can be recovered in 1–2 h at most ocean depths.

The uniquely outfitted research drill ship was christened *GLOMAR Challenger* in honor of *HMS Challenger*, the British ship that collected samples from the ocean floor from 1872 to 1876. The first 18 month program of the Deep Sea Drilling Project revolutionized geology. In the North and South Atlantic the magnetic stripes were found to increase in age away from the crest of the Mid-Atlantic Ridge. Students I had trained were on two of the first three two-month legs. I was on the fourth.

The first Leg of the project upset everything we thought we knew. Sigsbee Knoll on the floor of the Gulf of Mexico turned out to be a salt dome, like those known on shore. Under pressure 'solid' salt flows, much like the ice of glaciers, but being lighter than the surrounding sediments, it tends to work its way toward the surface. But the core from Sigsbee Knoll also contained traces of petroleum. Petroleum in the deep sea? Unheard of. A 'Pollution Prevention and Safety' Panel was formed by JOIDES, with membership from industry to review the drilling plans. At the end of Leg 1 samples taken just east of the Bahamas showed that the Atlantic Ocean had been anoxic—devoid of oxygen—during the Cretaceous. In its first two months of operation the DSDP had turned geology upside down.

Leg 4 was Cesare Emiliani's dream and I was excited to be on it. It was to collect cores revealing the history of the area that served as the standard for studies of the Cenozoic in the oceans—the Caribbean. We expected to recover complete reference sections rich in the fossils, particularly planktonic foraminifera, used for global stratigraphic correlation. Most of the information on the stratigraphic distribution of these fossils had come from the margins of the Caribbean Basin, notably from Trinidad and Venezuela. The stratigraphic sections in the central Caribbean were sure to be even more complete. The shipboard party

consisted of a dozen geologists including Bill Benson of the NSF as a sedimentologist.

We were very surprised to discover that the stratigraphic sections in the Caribbean Basin were in fact much less complete than along the margins. Much of the sediment contained no calcareous fossils at all. This phenomenon had already been noted on Legs 2 and 3. Apparently, the depth in the Atlantic where calcium carbonate becomes undersaturated and below which carbonate is dissolved had moved with time. This had been known since the German South Atlantic Oceanographic Expedition of the 1930s. The dissolution horizon is called the 'calcium carbonate compensation depth ('CCCD'). At the time the CCCD was thought to be a pressure related phenomenon. On Leg 3 the suggestion had been made that the sea-floor itself had moved up and down as the CCCD remained at the same depth below the sea surface. But if that had been the case, there should have been simultaneous transgressions of the sea onto the continents as the water was displaced—and there weren't.

I spent many hours in the shipboard laboratory, mystified by the problem. There was not much else to do, my fossils were not there. I plotted up the ages and depths at different sites where the transition from carbonate rich to carbonate free sediment occurred. It was soon obvious that the planktonic foraminifera were more susceptible to dissolution than my beloved calcareous nannofossils. The dissolution of the nannofossils took place a few hundred meters below the level where the shells of planktonic foraminifera disappeared. It became evident that movement of the CCCD was not related to vertical motions of the sea floor but to changes in the chemistry of the water masses in the deep sea. I embarked on a decade of study of the relationships between carbon dioxide, carbonate and bicarbonate ion concentrations in the seawater, and how planktonic foraminifera and calcareous nannoplankton shells form and dissolve. A number of graduate students became involved.

Also, with time to spare, I gave thought to the problem of what happened when certain species of fossil, especially those required for determining the age of the sediment, might be selectively dissolved, thus confusing the record. This was my second excursion into probability theory. The first had been in working in the Bahamas, trying to figure out how large a sample of the biota must be in order to be sure that one knew which species were the most abundant organisms. For that effort in cooperation with my colleague John Dennison, then at the University of Illinois, I am frequently cursed as 'the guy who makes us count 300 specimens.' The two problems are not unrelated. But the idea of using probability to determine how well one knew the age of sediment from a sample of its fossils had major implications for the petroleum industry, and became a new field, known as 'probabilistic stratigraphy.' The age equivalence of strata is expressed as a probability, not as a certainty as had been the practice.

The problem is also related to evaluating how well you know anything at all. I decided that there were much larger questions to be resolved in Earth history than simply refining the methods for determining the age of sediments using fossils. What could have been 'wasted time' wasn't wasted.

I believe that it was on Leg 4 that we began to use the term 'paleo-oceanography' for the new field of science that was developing. We began to realize that Earth's past history was far more exciting than that revealed by the record on land alone.

DANIEL RUTHERFORD'S MOUSE AND CANDLE
EXPERIMENT (1772)

# Air

*One who neglects or disregards the existence of earth, air, fire, water and vegetation disregards his own existence which is entwined with them.*
Mahavira (599–527 BCE)

Our modern knowledge of air goes back to the middle of the 17th century. It had been recognized that there was a relationship between pressure and the volume of air, and Robert Boyle (1627–1691) quantified this through a series of experiments. He used a U tube 8 ft long on one side and 1 ft long on the other, with the shorter side sealed at its end. He poured mercury into the long side forcing the air in the short side to contract from the pressure of the mercury. Adding mercury a little bit at a time he determined that the pressure of a gas is inversely proportional to the volume it occupies. He published the results of his experiments in 1662. It is now known as Boyle's Law: at constant temperature for a fixed mass, the absolute pressure and the volume of a gas are inversely proportional. It can be expressed mathematically as

$$P \propto \frac{1}{V}$$

where $\propto$ means 'is proportional to.'

The second of the 'gas laws,' relating temperature and volume, was developed in France. French physicist and inventor Guillaume Amontons (1663–1705) realized there was a relationship in 1699, but an exact formulation was first given informally by Jacques Alexandre César Charles (1746–1823) in 1787. It was first published by French natural philosopher Joseph Louis Gay-Lussac in 1802 who credited the discovery to Charles. The law was independently discovered by British natural philosopher and chemist John Dalton about 1801, although his formulation was less thorough than Gay-Lussac's. Charles' Law can be expressed mathematically as

$$\frac{V_1}{T_1} = \frac{V_2}{T_2}$$

The investigation of the composition of air also began in the 17th century. Around 1639 Flemish scientist Jan Baptista van Helmont (ca. 1580–1644) identified a gas given off by burning wood and named it *gas sylvestre* ('wood gas');

today we know it as carbon dioxide. Van Helmont was the first person to understand that air is not a single gas, but a combination of gases.

However, this discovery was soon to receive a setback. In 1667, German alchemist Johann Joachim Becher (1635–1682) eliminated fire and air from the four classical elements: earth, air, fire and water. He substituted three forms of earth: *terra lapidea*, *terra fluida*, and *terra pinguis*. *Terra pinguis* was the key to combustion and was released when substances burned. In 1703 Georg Ernst Stahl (1659–1734), Professor of Medicine and Chemistry at the University in Halle, renamed Becher's *terra pinguis* to *phlogiston*. It was something inherent to combustible materials and was released when they burned. Its existence would be accepted as fact for three quarters of a century.

The major gases in the atmosphere were discovered about the time of the American Revolution. It all began in an odd way. Joseph Black (1728–1799) was born in Bordeaux, France. His father had originally lived in Belfast, but had moved to Bordeaux as a wine merchant arranging exports to Britain. What was a Scot doing in Northern Ireland? James VI of Scotland had become King of England in 1603, also owning the Kingdom of Ireland. In 1606 he began the Plantation of Ulster, confiscating land from the original Irish inhabitants, and moving in lowland Scots and English to help civilize the country. This was the start of the Northern Ireland problem, still with us today.

After his early home education in Bordeaux, Joseph Black was sent to live with relatives in Glasgow in 1740 to further his education. He entered the University of Glasgow in 1746, studying philosophy and languages. In 1748 he changed his major to medicine, studying under William Cullen. In addition to medicine, Cullen was investigating the evaporation of liquids at low pressure, making use of the air pump invented by Robert Hooke for Robert Boyle's experiments. In effect, Cullen's experiments related to refrigeration. In 1751 Black moved to Edinburgh to continue his academic work in pursuit of a doctoral degree. He became interested in finding alkalis milder than those in common use in that day for treatment of 'acid stomach' and urinary stones. His research was very methodical and involved weighting the components of an experiment as accurately and precisely as possible. His first academic position was in 1756 as Cullen's replacement in Glasgow. Cullen had accepted an appointment in Edinburgh.

In the 1750s Black continued his chemical experiments with alkalis and a gas he called 'fixed air' (carbon dioxide). In 1756, he proved that carbon dioxide occurred in the atmosphere and that it was in the breath exhaled by humans. He found that limestone, if heated or treated with acids, produced 'fixed air.' He determined that 'fixed air' was denser than normal air and a candle would not burn in it. When bubbled through a solution of lime it could precipitate out crystals that had the same composition as the limestone. Upon Cullen's retirement, Black moved from Glasgow to the University in Edinburgh in 1766. There he joined the Poker Club and played cards with James Hutton, Adam Smith, David Hume and other leaders of the Scottish Enlightenment.

In Edinburgh he asked one of his students, Daniel Rutherford (1749–1819), to investigate its properties further. In 1772 Rutherford performed a clever experiment. He kept a mouse in an air-tight container until it died. Then he burned a candle in the remaining air until it went out and then he burned phosphorus until it would not burn any further. The remaining air was then passed through a solution of lime to remove the 'fixed air' (carbon dioxide). The remaining gas did not support combustion, and a mouse could not live in it. Rutherford called it 'phlogisticated air;' we know it as nitrogen. At the time it was thought that there must be another element 'phlogiston' which was contained within combustible bodies and released during combustion. It turned out that one of the elements involved in burning was not in the combustible body itself, but already in the air.

The discovery of oxygen is variously credited to three people. Oxygen was first discovered by a Swedish chemist and pharmacist, Carl Wilhelm Scheele (1742–1786). He had produced oxygen gas by heating mercuric oxide and nitrates by 1772. He called the gas 'fire air.' It was the sole supporter of combustion. His account of this discovery was not published until 1777. In the meantime, in 1774, British clergyman Joseph Priestley (1733–1804) focused sunlight on mercuric oxide inside a glass tube, liberating a gas he called 'dephlogisticated air'. He found that candles burned brighter in this gas and that a mouse was more active and lived longer while breathing it. Priestley published his discovery in 1775 in a paper with the singularly uninformative title "*An Account of Further Discoveries in Air.*" Priestley published first, so he gets the credit for the discovery of oxygen. But the story is not over. Also in 1774 Antoine Lavoisier (1743–

1794) in Paris had started a series of chemical experiments. One of these involved heating tin and air in a closed container. He noted that there was no overall increase in weight of the closed container but that air rushed in when it was opened. He concluded that part of the trapped air had been consumed. The tin had increased in weight and that increase was the same as the weight of the air that rushed back in. His experiments on combustion were documented in his book *Sur la combustion en général*, published in 1777. He had discovered that air is a mixture of two gases; 'air vital,' essential to combustion and respiration, and 'azote,' derived from the Greek for 'without animal life,' which did not support either. Most importantly, what Lavoisier did was to demolish the phlogiston theory.

By 'air' we mean all gases that make up the atmosphere. We now know a lot more about its composition. For dry air, 99.964 % of these gases are $N_2$, $O_2$, and Ar. The remainder is the trace gases, mostly $CO_2$, $CH_4$, and $N_2O$, all greenhouse gases. In water-vapor saturated wet air, $H_2O$ can displace some of the dry air molecules and make up to 5 % of the air by weight.

The atmosphere is the thin layer of gas covering the surface of the Earth. The radius of the solid Earth is about 6370 km (= ∼3960 miles). As you may remember from Chap. 11, the thickness of the atmosphere is commonly taken to be 100 km (= ∼60 miles), defined by the Kármán Line. But the gases extend further out, becoming increasingly tenuous. The 100 km is about 1.5 % of the radius of the solid Earth. Think of a large peach. The thickness of the atmosphere would correspond to the skin of the peach. Unlike the atmospheres of other planets, ours is layered. Its mass is less than one thousandth that of the other fluid body on the planet, the ocean. It has a 'memory' of only a few weeks. To paraphrase Raymond Pierrehumbert of the University of Chicago, Earth's atmosphere is a sea of atoms, molecules and photons. To these we need to add the occasional subatomic particle, particularly thermal neutrons and muons. Let's start by considering the atoms and molecules.

## 16.1 The Nature of Air

The major gases of the atmosphere are nitrogen (78.084 %), oxygen (20.946 %) and argon (0.934 %). The three are shown in their relative sizes in Fig. 16.1.

The nitrogen and oxygen are both in the form of diatomic (=two atom) molecules, $N_2$ and $O_2$. Argon is present as a monatomic (=one atom) component of the atmosphere. It is a noble gas, which means that it has a full outer electron shell and does not combine chemically with any other atom. The molecules of $N_2$ and $O_2$ have the general form of two partially merged spheres or an elongate capsule with a medial constriction. Although the atoms of nitrogen are larger than

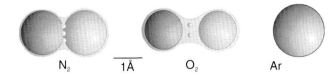

**Fig. 16.1** The three particles that make up 99.964 % of dry air. One Ångström is 0.1 nm or $1 \times 10^{-10}$ m

those of oxygen, the molecule is smaller in its longest dimension because the two atoms share three electrons, as indicated in the figure, whereas the atoms of the oxygen molecule share only two. To shorten the following discussion I will include the argon atoms in the term 'molecule.'

At room temperature at sea level, these three molecules fill about 0.1 % of the space they occupy; that is, 99.9 % of the space is empty. Once thought to be filled by the hypothetical all-pervasive æther, we now know it is vacuum—empty space. Well actually, space doesn't seem to be empty. Einstein's Theory of Relativity tells us that it can be distorted, bent, and stretched or compressed. It is hard to imagine all that happening to nothingness. Something seems to be there, but we don't know what it is. But as far as you and I are concerned, it's empty.

However, there are a lot of molecules in the almost empty space that is air. How many are there in one cubic meter of air at sea level? The mass of that cubic meter of air (35.3 ft$^3$) is about 1.295 kg (=2.855 lbs); the average molecular weight of dry air is 28.97, so there are 44.70 mol of air in that cubic meter. Multiply that by Avogadro's number ($6.022 \times 10^{23}$) and you get $269.2 \times 10^{23}$ which is about 27 septillion. A handy number to throw out at a cocktail party.

The atmosphere also contains some other 'harmless' gases in trace amounts: Neon (0.0018 %), Helium (0.0005 %), Krypton (0.0001 %), Hydrogen (0.00005 %), and Xenon (0.000009 %). All except hydrogen are non-reactive noble gases. The few molecules of diatomic hydrogen form by breakdown of water molecules by energetic ultraviolet and cosmic rays at the top of the atmosphere. The hydrogen molecules are very light, so that in order to have the same kinetic energy as the N$_2$ and O$_2$ molecules their velocities are very high. So high that they can escape Earth's gravity and be lost into space. Fortunately, as we will see, Earth's atmosphere is stratified in such a way that water molecules have a very hard time working their way to the top where they can be broken down. This means that to our great good fortune, our planet loses water very, very slowly. The stratification of the atmosphere has enabled our planet to retain its water as oceans, whereas Venus and Mars have lost the water they may have originally had.

Although very rare on Earth H$_2$ is the most abundant gas in the universe. There is about one molecule per cubic centimeter in 'empty' space. But H$_2$ also forms clouds, like the Orion nebula which contains up to a million molecules per cubic centimeter. That may sound like a lot, but it is a concentration about one quintillionth that of our atmosphere at sea level. The hydrogen in the Orion and other nebulae is excited by the radiation from nearby stars and glows with that beautiful pink 656.3 nm spectral emission line, making them some of the most glorious objects in the night sky—through a telescope.

## 16.2 The Velocity of Air Molecules

At sea level the molecules in air travel at an average speed of about 500 m/s. But this is an average; some go much faster, some much slower. In spite of their small size, these speeds (or, in technical talk, translational velocities), result in frequent collisions. The collisions are elastic; the molecules bounce off each other, like a tennis ball or ping-pong ball hitting a hard surface. The collisions between billiard balls hitting each other and bouncing off the cushions surrounding the playing area of the table are another good analogy. The only difference is that for the tennis, ping-pong or billiard balls, some of the energy of the collision is absorbed and becomes heat, so they don't bounce as high or go as far each time. That is Mother Nature collecting her tax of entropy. Molecules act as though they are highly elastic, so that kinetic energy can be transferred from one to the other. The mean time between collisions among N$_2$ and O$_2$ molecules and Ar atoms at sea level pressure is about $0.2 \times 10^{-9}$ s, which translates to 5 collisions per nanosecond or 5 billion collisions per second for each molecule. The mean free path, that is, the average distance traveled by a molecule before it collides with another molecule is about $10^{-7}$ m or 1000 Å. Remember that the size of an atom is about 1 Å and most air molecules are between 2 and 4 Å, so the spacing is like that of a couple of soccer balls on a soccer field. Or, you could think of it as being like a three dimensional billiard game. The molecules possess kinetic energy, and each can gain or lose kinetic energy and change velocity by the nature of the collision. It is their average kinetic energy and the number of these moving objects that we call heat. Measuring that average energy gives us the number we call 'temperature.'

## 16.3 Other Molecular Motions

Translational movement in space is only one of the motions of the molecules. It is the only motion possible for the monatomic (=single atom) gas argon. Most forms of energy are quantized, that is they come in the discrete tiny pieces first postulated by Max Planck. Translation is not quantized.

However, molecules also vibrate and rotate. Both are forms of kinetic energy and both are quantized. The possible ways in which molecules can vibrate is a function of the number of atoms in them. The diatomic molecules $N_2$ and $O_2$ have only one mode of vibration, symmetrical stretching and compression, shown in Fig. 16.2.

The vibration occurs as though the two molecules were connected by a spring—that is, as though it were a 'harmonic oscillator' (remember that term from the discussion of black body radiation in Chap. 10). A harmonic oscillator obeys Hooke's Law of Elasticity: that the extension of a spring is in direct proportion to the load applied to it. The energy involved in the vibration oscillates between kinetic (the actual motion of the molecules), and potential (the stored energy that reaches a maximum when the molecule is at either of the extremes of the vibration cycle). You will note that I said 'almost as though' because in reality the oscillation is not perfectly harmonic. If it were, it would completely cease at a temperature of absolute zero; but even at that temperature, at which all motion is supposed to cease, there is still some 'zero-point' vibration.

For this reason the vibration of molecules is said to be 'anharmonic.' The magnitude of the oscillation can be increased by adding energy to the molecule. This can happen when photons collide and are absorbed as described below. The transitions between energy levels are not continuous, but quantized. The rates of vibration are in the range of $10^{11}$–$10^{14}$ Hz, corresponding to the far infrared.

Incidentally, the Robert Hooke (1635–1703) (Fig. 16.3) of Hooke's Law was an English 'natural philosopher,' architect and 'polymath.' 'Natural philosophy' is what we today call science. A 'polymath' is a person whose expertise spans many different subject areas. He is perhaps best known for his book *Micrographia*, a best seller in 1665, with drawings of the minute things he observed through his microscope, like fleas, louses, gnats and other things too small to be seen clearly by the human eye. Hooke must have had a good sense of humor, because he first described his discovery of the nature of a coil spring in 1660 with the anagram "ceiiinosssttuv." It was 18 years before he published the solution to the anagram: "*Ut tensio, sic vis*" which translates as "As the extension, so the force." He did not get along with Isaac Newton, and even claimed that Newton had plagiarized his work in the *Principia Mathematica*. His polemics were one of the reasons that Newton became something of a recluse and did not attend the meetings of the Royal Society.

If you look up molecular motions in a book on physics or mechanics, you will immediately be confronted with two special terms: 'degrees of freedom' and 'degenerate.' In physics, a degree of freedom of a system is a parameter that contributes to the state of a physical system. That can cover a multitude of sins, so to speak. In the case of the molecules

**Fig. 16.3** Robert Hooke (1635–1703). Microscopist, physicist, and curator of experiments of the Royal Society. As Robert Boyle's technical assistant he had constructed the air pump required to create a vacuum. Later he was obsessed with trying to make light of the work of Isaac Newton. His original portrait in the Royal Society was lost; it is thought to have been destroyed by Newton. This is a modern portrait by Rita Geer, 2004, based on contemporary descriptions of his appearance

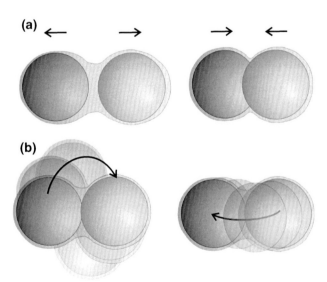

**Fig. 16.2  a** The vibrational mode of $N_2$, stretching and compressing. **b** The two rotational modes of $N_2$: they are 'degenerate' in that although they are different, they seem to be the same, depending on the viewpoint of the observer

listed above, the body can move independently in the three directions of space (remember the Cartesian coordinate system?). The momentum of a body consists of three components; each called a degree of freedom. A molecule with x atoms has $3x - 6$ normal modes of vibration, but a linear molecule, with the atoms in a straight line, has only $3x - 5$ because rotation about the linear axis doesn't count as vibration. For diatomic molecules like $N_2$ and $O_2$: $(3 \times 2) - 5 = 1$. 'Degenerate' means that although there might be two modes, as shown for rotation of $N_2$ in Fig. 16.2, you cannot distinguish between them because they depend on how you view the object. The two modes of rotation shown above are degenerate because if you observed D from the plane of this page, it would look like C. Your perception of things depends on your point of view.

There is another important thing about these particles that make up most of the air. They have no electrical polarity. Their electrical charge is neutral; they have the same number of protons and electrons, and the electrons are evenly distributed.

In addition to these gases, the atmosphere contains variable amounts of trace gases that readily interact with photons in the thermal infrared range. To do this the molecule must be 'dipolar,' that is it must have some degree of electrical charge, positive on one side, negative on the other. The concentrations of the gases are expressed in terms of parts per million by volume. Avogadro's Law states that, at a given temperature and pressure, equal volumes of gases contain equal numbers of molecules. The most variable and potentially most abundant trace gas is water vapor ($\sim 0$–40000 ppmv = 0–4 %, depending on temperature and availability). The other major trace gases are carbon dioxide ($CO_2$, 278 ppmv in 1750, 394 ppmv in 2011 = 0.0278 and 0.0394 % respectively), methane ($CH_4$, 0.700 ppmv in 1750, 1.808 ppmv in 2011), and nitrous oxide ($N_2O$, 0.270 ppmv in 1750, 0.323 ppmv in 2007). These are the major greenhouse gases affecting the temperature of Earth's surface. Schematic models of them are shown in Fig. 16.4.

Another 'greenhouse gas molecule,' ozone, interacts with photons in the ultraviolet; it is responsible for stratification of Earth's atmosphere, partitioning the lower atmosphere into troposphere and stratosphere. It is the gas responsible for keeping water from reaching the top of Earth's atmosphere where it could be dissociated into $H_2$ and O. It has a special story of its own to tell, so we will get to it later, in Chap. 19.

Three of these gases, water ($H_2O$), carbon dioxide ($CO_2$), and nitrous oxide ($N_2O$), have three atoms in the molecule; one, methane ($CH_4$), has five. Each is comprised of two different elements. One, $H_2O$ has an asymmetric bent form. Two have a linear arrangement of the atoms, but one of them, $CO_2$, is symmetrical, and the other ($N_2O$) is

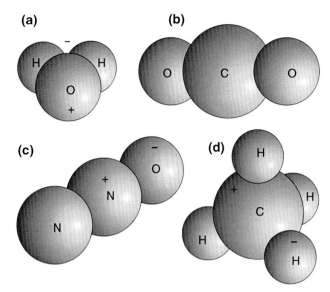

**Fig. 16.4** The four major greenhouse gases that interact with infrared radiation. **a** Water. **b** Carbon dioxide. **c** Nitrous oxide. **d** Methane

asymmetrical. The fourth, the methane molecule, resembles a tetrahedron with the C atom in the center and four H atoms at the corners.

Unlike the diatomic and monatomic gases that make up most of the atmosphere, two of these greenhouse gases are always electrically dipolar. That is, they have one side that is relatively negative while the other is positive. Water has the strongest dipolarity, a feature which gives it many unusual properties. It plays such an important role in the climate of the Earth that it is discussed in detail in the next chapter. Although $CO_2$ and $CH_4$ in their symmetrical ground states are not dipolar, they become momentarily dipolar when they vibrate asymmetrically. The asymmetric vibration of $CO_2$ is shown in Fig. 16.5. I leave it to your imagination (best after a Martini or snifter of brandy) to picture the nine asymmetrical vibration modes of methane; remember $(3 \times 5) - 6 = 9$.

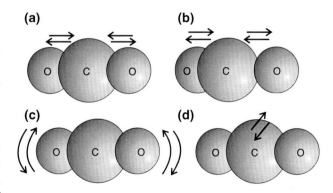

**Fig. 16.5** Modes of vibration of the carbon dioxide molecule. Each of these vibrations creates a momentary dipolar state

## 16.4  The Other Major Component of Air—Photons

The idea that electromagnetic radiation comes in tiny packets, 'quanta,' was the basis for Einstein's explanation of the photoelectric effect in 1905. The term 'photon' for these quantum objects as 'the carriers of radiant energy' was proposed by American physical chemist Gilbert Lewis (1875–1946) in 1926. Lewis is best remembered as the discoverer of the covalent bond in chemistry.

Photons fill the atmosphere day and night; photons in the visible light range coming from the Sun make 'daylight.' Before the late 19th century visible light at night was only from the Moon and stars, and from natural and human made fires, lightning, volcanic eruptions, the aurora, and meteor trails—Earth was dark at night.

Since the invention of the electric light, now in so many different forms, that has changed dramatically, as shown in Fig. 16.6. In the United States we seem to be obsessed with trying to illuminate the night sky as much as the ground, so that in many areas it is no longer possible to see the stars. In Europe and much of the rest of the world lights have reflectors so that all of their illumination is directed toward the ground, and you can see the stars. London and Paris are notable exceptions because many buildings are illuminated by lights shining upward on the facades from below. You will also notice that South Korea seems detached from Asia because of the lack of lights in North Korea. Putting reflectors on lights to direct the illumination toward the ground not only makes for a beautiful night sky, illuminating only the ground costs half as much. Our night lights confuse many insects, birds and other animals—the full impact on ecosystems remains to be assessed.

At least as important to our planet as the visible light are the invisible rays of the Sun, the ultraviolet and near infrared, and the infrared radiation coming from the Earth itself. The significance of Kirchhoff's 'black body' concept was that it allows us to understand the energy balance of a planet. The amounts of energy in the incoming radiation from the Sun, Moon, and stars must be equal to the energy radiated back into space by planet Earth. Otherwise the planet will warm or cool. You will remember from the introduction to the climate system in Chap. 11 that the Earth is not a perfect black body. As you learned in the last chapter, Earth does not accept all of the incoming radiation, but at present reflects about 30 % of it back into space.

It is easy to make a mental image of atoms and molecules as being like balls or collections of balls that can partially fuse into one another. We can think of atomic nuclei as being a collection of tiny balls, some with a positive electrical charge (protons) and others with no electrical charge (neutrons). We can think of electrons as sort of fuzzy clouds, each having a negative electrical charge, orbiting the nucleus. Even if these images are inaccurate, they are useful and easy to imagine. All submicroscopic particles have both wave and particle properties. A photon is both a wave and a particle at the same time. For example, a single photon can pass through two separated slits at the same time, and then

**Fig. 16.6**  The Earth at night; from NASA imagery

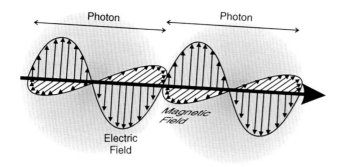

**Fig. 16.7** One way of imagining the unimaginable photon

strike an atom knocking an electron out of orbit. In passing through the two slits it was a wave, but when it hit the atom it was a particle.

I found the illustration in Fig. 16.7 useful in trying to imagine a photon in terms of both an electromagnetic wave and quantum object.

Remember that there are two major sources of photons that encounter Earth's atmosphere—those from the Sun and those from the Earth itself. The photons from the Sun are about 51.7 % in the near infrared, 38.3 % in the visible range, and 8.7 % in the ultraviolet. Absorption in the atmosphere is such that at ground level the distribution is 52.7 % near infrared, 44.5 % visible, and only about 3.2 % ultraviolet.

## 16.5  Ionization

The energy of ultraviolet radiation is sufficient to strip electrons from air molecules, break the molecules apart and ionize the atomic fragments; these processes produce free electrons. Most of the ultraviolet light reaching the Earth is removed by being absorbed, causing electronic transitions and ionization of air molecules in the outer atmosphere, the special region termed the 'Ionosphere,' which extends from 50 to 1,000 km ($\sim$30–600 miles) above Earth's surface. This is where the O ions required to form ozone are produced. This is also the region that interacts with long wavelength electromagnetic radiation to reflect radio waves allowing over-the-horizon transmission. It will be discussed in greater detail later in this chapter.

## 16.6  The Scattering of Light

Air is almost completely transparent to photons in the visible light range. These photons are not absorbed, but they can be scattered by air molecules. Scattering means that the electromagnetic wave interacts with the tiny electric field surrounding the molecule in such a way as to cause the wave to

change its direction. The molecule's electric field is generated by the cloud of electrons orbiting the atoms that make up the molecule. This is known as Rayleigh scattering, described in 1897 by the same the Lord Rayleigh who, you will remember, had fitted an equation to the long-wavelength side of the black body radiation curve.

Rayleigh scattering occurs when the electromagnetic wave encounters a molecule smaller than 1/10th wavelength. The light most readily scattered by $N_2$ and $O_2$ are at the blue end of the spectrum, with wavelengths less than 480 nm = 4,800  Å  (Å = Ångström = 0.1 nm = 0.0001  μm). The molecules of air are much smaller, between 2 and 4 Å. The scattering process is something like water waves encountering a series of pilings, except in this case the pilings are moving. Only a small fraction of the incoming blue light is scattered. Some photons are sent back into space, contributing to Earth's albedo. Others spread out sideways before being scattered downward. The result is the blue sky. The longer wavelengths of light are scattered as well, but to a lesser degree. The effects are only seen when the Sun is low over or on the horizon. At sunset rays of the Sun have a much longer passage through the atmosphere, about ten times as far as at noon, and the blue light is completely scattered out. The scattering of the longer wavelengths then becomes apparent and the sky appears red. Clouds take on the reddish hue because they are illuminated by the red rays. Figure 16.8 shows the percentages of the Sun's light of different wavelengths scattered when the Sun is overhead and on the horizon.

The wavelengths scattered depend on the size of molecules in the air. When the Earth was young, it is thought to have had a much higher concentration of $CO_2$ in the atmosphere. This not only compensated for the fainter Sun by increasing the greenhouse effect, but it means that the sky would have had a different color—yellow. Cesare Emiliani

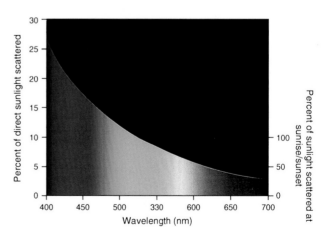

**Fig. 16.8** Rayleigh scattering of light of different wavelengths by air molecules when the Sun is directly overhead and when it is on the horizon

speculated that the reason land plants are green is because photosynthesis developed to take advantage of the most abundant wavelengths of light reaching Earth's surface. If yellow light is absorbed for photosynthesis, that leaves green to be reflected. Green light is not used for photosynthesis.

An especially beautiful example of the scattering effect occurred for a few years after the eruption of the volcano El Chichon in Mexico in April, 1982. It injected seven million tons of $SO_2$ into the stratosphere along with about three times as much ash. The ash gradually settled out, but the $SO_2$ remained for several years. The result was a special scattering of the Sun's rays to produce a beautiful indigo hue to the sky seen from high mountains and after sunset.

Aerosols are discrete particles in the air. They too scatter light, but by a different mechanism called 'Mie scattering.' Aerosols remain suspended in the air because they are constantly bombarded by air molecules. They are commonly the same size as or larger than the wavelengths of light. They include water droplets, dust, smoke and pollen. The Mie scattering process is different from Rayleigh scattering. All of the wavelengths of light are scattered together. The result is that the scattered light appears more or less white. Cloud droplets are Mie scattered, so that clouds appear white or gray.

## 16.7  Absorption of the Infrared Wavelengths

Investigation of the phenomenon of scattering took on new dimensions in 1928. When light is scattered most of the scattered photons are of the same wavelength as the incident light; this is the Rayleigh scattering described above. However, in 1928, Indian physicist Chandrasekhara Venkata Raman (1888–1970) at the University of Calcutta discovered that a small fraction of the scattered light had a different wavelength and that this difference depended on the molecules present in the sample. This is now known as the Raman Scattering Effect. He was awarded the 1930 Nobel Prize in physics. In the years following its discovery, Raman spectroscopy was used to provide the first catalog of molecular vibrational frequencies. However, the measurements were so difficult to make that Raman spectroscopy fell out of favor when commercial IR spectrophotometers became available in the 1940s. However, it was the introduction of the laser in the 1960s that changed everything. New, simpler Raman spectroscopy instruments with increased sensitivity along with improved infra-red spectroscopes have made it possible to investigate molecular motions in detail.

Single atoms interact with electromagnetic radiation by having orbiting electrons move to higher levels—these are quantized electronic transitions. However, atoms bound together as molecules have much more complex clouds of electrons orbiting them and offer many more possibilities for interaction. The number of possible electronic transitions, vibrations and rotations for diatomic molecules becomes enormous. Instead of a few spectral lines there are complex series of absorption/emission lines, as shown for a small portion of the $CO_2$ spectrum in Fig. 16.9.

In reality, the fine structure shown in Fig. 16.9 disappears. It is smeared out into continuous absorption bands both because of the changing population densities of the air molecules with the declining pressures and changing temperatures with altitude, and because of the Doppler Effect due to the different speeds of the individual molecules. An additional complication working to smear the lines together into bands is the differing isotopic compositions of the air molecules. There are three stable isotopes of O: $^{16}O$ (99.76 %), $^{17}O$ (0.004 %), and $^{18}O$ (0.20 %), and there are two stable isotopes of carbon: $^{12}C$ (98.93 %), and $^{13}C$ (1.07 %) as well as a trace of $^{14}C$. These combine together in 18 possible ways to produce molecules of $CO_2$ with 18 different molecular weights. Each of these has slightly different absorption wavelengths. To make matters more complex, the photon being absorbed does not have to have exactly the right amount of energy for vibrational or rotation excitation of the molecule. Remember that vibration and rotation are quantized. While the photon cannot have less energy than is required to change either of these quantum states, it can have slightly more energy. This excess will be converted into the kinetic energy of translation of the molecule in space. The speed of the molecule is increased. Remember that 'translation' = speed is not quantized. In short, it's a complicated mess, but Mother Nature only has to contend with the overall patterns of absorption.

Of course, the energy absorbed by the molecule is quickly re-emitted. But whereas the thermal radiation from the Earth intercepted by the gas molecule was directed upward toward space, the radiation re-emitted by the molecule goes in all directions, about half of it back downward toward Earth's surface. The radiation re-emitted by the molecules is just as complex as that absorbed. The decay of a vibrational or rotational state need not be the same as that produced by the original excitation but could occur in intermediate quantum steps.

**Fig. 16.9  a** The total absorption spectrum of $CO_2$. **b** The fine structure in the part of the $CO_2$ absorption spectrum between 13 and 17 microns (μm) through 1 cm (0.4 in.) of $CO_2$ at sea level pressure (1013.25 hPa = mb) and room temperature (296 K = ~23 °C = 73 °F). **c** The $CO_2$ absorption spectrum as it would appear if all the $CO_2$ in the atmosphere formed a layer at its base, at sea level pressure and room temperature. The layer would be about 334 cm (131.5 in. = 10 ft 11 ½ in.) thick. Data plots were made using Spectralcalc.com

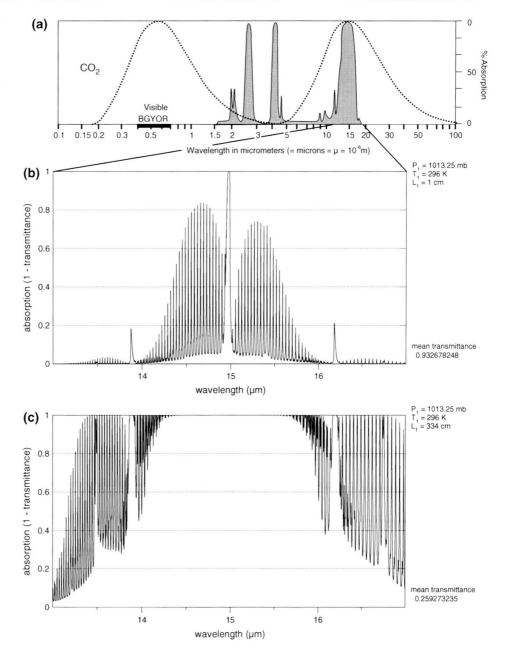

## 16.8   Other Components of Air: Subatomic Particles

Continuously passing through the air is a stream of subatomic particles generated by the cosmic rays, mostly high energy protons, colliding with air molecules in the upper atmosphere. Most interactions in the lower atmosphere are with 'thermal neutrons' having an energy of about 0.025 eV and traveling at an average velocity of 2.2 km/s. That is about four times the average speed of an air molecule. Most of these come from cosmic ray showers, but some also come from the ground as byproducts of radioactive decay. On average there is about 1 thermal neutron passing by for every 100 septillion air molecules. That is a ratio of $1:1 \times 10^{26}$ or written out: 1:100,000,000,000,000,000,000,000,000. With so many molecules out there it sounds like our thermal neutron might have a good chance of hitting one. But remember, those molecules themselves fill only 1/1000th of the space, and the nuclei of $O_2$ and $N_2$ molecules, which would be the thermal neutron's target, occupy only about 2.5 millionths of the space within the molecule. That means you can add another nine zeros to that number written out above. Further, the thermal neutron has no electric charge so it is not attracted to anything. Taking all this into account, the chance of a thermal neutron passing through the air actually hitting an atomic nucleus and combining with it to make an isotope would seem to be rather slight; and it is. But some of them collide with nitrogen molecules in the air in the region between 9 and 15 km above Earth's surface. With some luck, the collision will happen in such a way as to split the two nitrogen atoms apart and allow the thermal neutron to replace a nuclear proton in one of the nitrogen atoms. That transforms it into an atom of carbon-14. The original nitrogen atoms had seven protons and seven neutrons; the new carbon atom has six protons and eight neutrons and is unstable. It will combine with oxygen to make $CO_2$ and soon be taken up by a plant in photosynthesis. It will undergo radioactive decay, having a half-life of 5,730 years which will make it possible for us to date organic matter back several tens of thousands of years.

Also passing through the atmosphere are a lot of muons. The flux at sea level is about 1 muon per square centimeter per minute. They have the same electric charge as an electron, but have about 200 times its mass. Their average lifetime is about 2 μs. This is actually very long for one of these exotic particles created by cosmic rays. This would allow them to go about 600 m before decaying into an electron and two neutrinos. But because they travel at near the speed of light, they live in Einstein's relativistic world where time and space are stretched out. This means that they can penetrate the whole atmosphere and even travel far underground before finally decaying or interacting with something else.

As discussed in Chap. 14, there are a number of other subatomic particles produced in the upper atmosphere by collisions of the high-energy protons we call cosmic rays, but most of them decay before reaching the lower atmosphere. There are also some radioactive isotopes of elements produced by spallation of air molecules. One of these atomic fragments, Beryllium-10, with its long half-life of 1.36 million years, is soon adsorbed onto an aerosol and is then rained out of the atmosphere. As discussed in Chap. 14, its accumulation in glacial ice can be used to investigate past solar activity.

## 16.9 Summary

'Air' is the gases that make up the atmosphere. In dry air, $N_2$ (78.084 %), $O_2$ (20.946 %), and Ar (0.934 %) are 99.964 % of the total. The remaining 0.036 % are the trace gases, mostly $CO_2$, $CH_4$, and $N_2O$, all of which are greenhouse gases. In water-vapor saturated wet air, $H_2O$ can displace some of the dry air molecules and make up to 5 % of the air by weight. At room temperature sea level air molecules fill about 0.01 % of the space they occupy. They are moving at a speed of about 500 m/s and each molecule collides with another about five billion times a second. It is the total

kinetic energy of these molecules that we measure as 'temperature.'

In addition to flying around, technically called 'translation,' the molecules vibrate with the individual atoms moving back and forth and sideways to each other, and rotate. Rates of vibration are in the range of $10^{11}$–$10^{14}$ Hz, and rotation is in the range of $10^9$–$10^{11}$ Hz, corresponding to infrared wavelengths of the electromagnetic spectrum. Interaction of molecules with infrared radiation requires that they have, at least momentarily, a differentiation of its electrical charges so that one area is more positive, another more negative. This does not happen for gases with only two atoms, but does for all others. It is the interaction with infrared radiation that makes the tri-atomic and more complex gases 'greenhouse gases.' They can capture the energy of photons and become 'excited,' but they almost immediately reradiate it in any direction. Energy capture for vibration and rotation is quantized; translation is not quantized. Because the orbital motions of electrons in a molecule are so complex many different wavelengths can excite a greenhouse gas molecule. To excite the molecule, the photon must have at least the amount energy required for a vibrational or rotational transition. It can also have slightly more, and the extra energy then goes into non-quantized translation. Because there are so many molecules and because they are all moving at different speeds, there is a Doppler effect and the specific energies required for excitation become smeared together into spectral absorption bands rather than single lines.

A Timeline for this chapter:

| 1639 | Jan Baptista van Helmont identifies a gas given off by burning wood and names it *gas sylvestre* ('wood gas' = carbon dioxide); he realizes that air is more than one gas |
|------|------|
| 1660 | Robert Hooke publishes his Law of Elasticity with the anagram "ceiiinosssttuv" |
| 1662 | Robert Boyle publishes the results of his experiments showing that the volume of a gas is directly related to pressure |
| 1665 | Robert Hooke publishes his *Micrographia* with pictures of microscopic creatures |
| 1667 | Johann Joachim Becher eliminates fire and air from the four classical elements: earth air, fire and water, and substitutes three forms of earth: *terra lapidea*, *terra fluida*, and *terra pinguis*; the latter is released when objects burn |
| 1678 | Robert Hooke publishes the solution to his anagram: "Ut tensio, sic vis" = "As the extension, so the force" |
| 1699 | Guillaume Amontons discovers a relation between the temperature of a gas and its volume |
| 1703 | Georg Ernst Stahl renames *terra pinguis* to *phlogiston* |

(continued)

(continued)

| 1740–1790 | Heyday of the Scottish Enlightenment led by David Hume, Francis Hutcheson, Alexander Campbell, Adam Smith, Thomas Reid, Robert Burns, Adam Ferguson, John Playfair, Joseph Black and James Hutton |
| 1752–1754 | As a doctoral student in Edinburgh Joseph Black performs chemical experiments with alkalis |
| 1756 | Black becomes Professor of Medicine at the University of Glasgow, continuing his experiments with alkalis and 'fixed air' (carbon dioxide) |
| 1766 | Joseph Black moves to Edinburgh as Professor of Chemistry |
| 1772 | Black's student Daniel Rutherford removes oxygen and carbon dioxide from air and calls the remaining gas 'phlogisticated air' (nitrogen) |
| 1772 | Carl Wilhelm Scheele produces oxygen gas by heating mercuric oxide and nitrates; he calls it 'fire air' (oxygen) |
| 1774 | Joseph Priestley (1733–1804) focuses sunlight on mercuric oxide inside a glass tube, liberating a gas he calls 'dephlogisticated air' (oxygen) |
| 1774 | Antoine Lavoisier heats tin and air in a closed container and notes that although there is increase in weight of the container after air rushes in when it is opened; he concluded that part of the trapped air had been consumed and discovers that the tin had increased in weight by the same amount as the weight of the missing air; he calls the missing gas 'air vital' and the remaining air 'azote' |
| 1776 | The American Revolution begins |
| 1777 | Lavoisier publishes his book Sur la combustion en général; he demolishes the phlogiston theory |
| 1787 | Jacques Charles describes the relation between temperature and volume of a gas in detail but does not publish his discovery |
| 1802 | Joseph Gay-Lussac publishes details of the relation between the volume of a gas and temperature, crediting the discovery to Charles |
| 1871 | Lord Rayleigh explains the scattering of light by small particles |
| 1905 | Albert Einstein explains the photoelectric effect in terms of Lichtquanta |
| 1926 | Gilbert Lewis names the 'carriers of radiant energy' photons |
| 1928 | Chandrasekhara Venkata Raman discovers that a small fraction of scattered light has wavelengths different from the incident light |
| 1982 | El Chichon erupts in Mexico injecting $SO_2$ into the stratosphere and creating indigo skies after sunset for several years |
| 1960s–present | Space age investigations provide lots of new information about the atmosphere and the upper atmosphere in particular |
| 2011 | Raymond Pierrehumbert of the University of Chicago describes Earth's atmosphere as a sea of atoms, molecules and photons |

If you want to know more:

Gribbin, J. 2010. *Science: A History*. Penguin Books, 672 pp.—An excellent coverage of many topics, this among them.

Again, I recommend Peixoto and Oort's book, *Physics of Climate* for a more technical account.

Music: Beethoven's 6th (Pastoral) Symphony is very different from the others, and includes a great storm.

Libation: A light white wine, such as a Pinot Grigio from the Veneto in Italy goes well with the Beethoven.

## Intermezzo XVI. The History of JOIDES

In the last Intermezzo I gave you a brief account of how the Deep Sea Drilling Project (DSDP) got started—now I'll recount how I got involved in it.

When JOIDES was formed in 1965, I was asked to be on one of its geographic panels, that for the Gulf of Mexico. I was still full time at the University of Illinois, but had students studying the calcareous nannofossils and planktonic foraminifera of the Gulf Coast sediments in Alabama, Mississippi, and Texas. The Panel met during the Annual Convention of the American Association of Petroleum Geologists in New Orleans. Among the things we wanted to know: what was a strange hill, Sigsbee Knoll, doing on the very flat floor of the Gulf of Mexico?

In 1966, when the JOIDES proposal for an 18 month drilling campaign was sent out by the National Science Foundation for review, I was one of the recipients. I called Bill Benson, who was in charge of the review at NSF, and told him I had a conflict of interest. My conflict was that if the program was funded and cores were recovered, I would want samples to examine for calcareous nannofossils. I remember that conversation very well. Bill was delighted. He told me to write a review expressing my enthusiasm, and to note my conflict of interest. It was just the sort of strongly supportive review from outside the JOIDES institutions that was needed to be able to fund such a large project in geology. The program was approved in 1967, with the management office located at Scripps in La Jolla.

JOIDES had an Executive Committee, which consisted of the directors of the Oceanographic Institutions. But, of the four original institutional members, only Maurice Ewing at Lamont had a geological background. Because of the Executive Committee's lack of expertise in the field they relied on a Planning Committee for the actual design of the program. Each of the four institutions had one member on the Planning Committee. As soon as my joint appointment with the Institute of Marine Sciences in Miami was settled in 1968, I was appointed to be its representative on the JOIDES Planning Committee. That same year the University of Washington's Department of Oceanography joined JOIDES. The other Planning Committee members were Joe Worzel (geophysicist), representing Lamont, Art Maxwell (geophysicist) for Woods Hole, Bill Riedel (paleontologist) for Scripps, and Joe Creager (sedimentologist) for Washington.

My first meeting was very exciting. The DSDP had just started operations, and had drilled several holes but recovered very little sediment. Joe Worzel had been on Leg 1 and gave us a firsthand account. The Project Manager at Scripps, Ken Brunot, was an engineer from industry, where the objective was to 'make hole,' not recover cores. We were meeting at Scripps, and decided we needed to urge the Institution's Director, Bill Nierenberg, to replace Ken with a scientist to oversee the program. Since it had been Bill who had decided that an engineer should be in charge and had chosen Brunot, our meeting could turn into a difficult confrontation. Bless their hearts, my three senior colleagues elected me to the honor of telling the Director of the US's largest oceanographic institution that he had made an unwise decision.

We invited Bill to have lunch with us at a little restaurant in La Jolla. Pleasantries were exchanged, and then I had to jump in with all four feet to make our case. Bill was obviously startled; I was a youngster, and telling him rather forcefully that he had to do something. Perhaps it was because I was someone who actually used the materials recovered on a routine basis, he listened and understood. At the end of the meeting he told us he would pass the management of the program to a scientist, but it might take some time.

A few weeks later Melvin Peterson, a Scripps geochemist was appointed Project Manager, and Terry Edgar Chief Scientist.

After a few months had passed, when the Planning Committee met again, the other members dropped another bombshell on me. The project was so successful that a proposal for an extension, this time for 30 months, would probably be needed. With a straight face, Art Maxwell explained that these jobs always fell to the youngest member of the committee. I ended up writing the proposals not only for this first extension of the program, but also for next two as well.

To us it was the Deep Sea Drilling Project. To the National Science Foundation, thanks to Bill Benson, it was the Deep Sea Drilling Program, both abbreviated 'DSDP.' The Program was not funded by a grant, but by a contract to the Scripps Institution of Oceanography in La Jolla, California. There is an important distinction. For NSF, 'projects' funded by grants end when the grant period expires; 'programs' can be extended through amendments to the original contract. Bill had already foreseen that the DSDP was going to be a great success, and had worked to steer it down the right path from the very beginning. It could be continued through amendments to the contract.

For planning for the future I had to make a number of trips to the office at Scripps. Mel, Terry and I became close friends. The extension was made from taking proposals for drilling submitted by individual scientists or groups of scientists, having them evaluated and ranked by the Planning Committee, and then trying to design a ship track that achieved the largest number of high priority objectives with the least unproductive steaming time from one area to the next.

During this time, I was spending half of the year at the University of Illinois, and I made use of the best students in my oceanography class to help design the ship tracks. I've never told this story before, but here is how it was done: I would invite five students to dinner. They would come about 5 PM, but before we ate, which usually turned out to be closer to 9 PM, they had to solve a problem: design alternative ship tracks taking into account the proposals for drilling. The first extension was to a global circumnavigation. I had world maps for each student, and a little Texas Instruments calculator that could determine distances between latitude/longitude points. Some were told that they could go through the Panama Canal, others that they could not use the Canal but had to use the Drake Passage to get from the Atlantic to the Pacific. One could use both. They already knew the general pattern of ocean currents, and 'to go with the flow.' They had to start from a Gulf Coast port. I had even checked out books on global port facilities from the library to make sure they could handle the Glomar Challenger. Bottles of wine were opened, and by dinner time I had five different solutions to the problem.

I then took three of the solutions to the next Planning Committee meeting. Two efficient ones, but going to different sites, and the least efficient one. I figured that the five member Planning Committee would have a difficult time with three competitive routes, but they could handle two. Sure enough the Committee soon figured out which was the least efficient, and threw it out so they could concentrate discussion on the other two solutions. The global circumnavigation ship-track chosen by the Committee was the one the Illinois students had agreed was best after they had finished dinner. Terry Edgar, who attended the Planning

Committee meetings as a guest, was delighted at having this hurdle overcome, and when I met with Mel Peterson later, he was equally impressed. I used the Illinois students to help design the next extension as well, but till now, no one ever knew.

The Chairmanship of the Planning Committee rotated every 2 years, and I became Chair in 1972, just as the program was about to enter a new phase, internationalization. In reality, the program had been international from its inception, including shipboard scientists from many countries to take advantage of their expertise, and others onboard as observers from countries in whose waters we would be drilling.

Bill Nierenberg was deeply involved in Republican politics, and working with the National Science Foundation was able to persuade the Nixon administration that it might be possible to expand the funding for the project to include contributions from the Soviet Union, Japan, Germany, France, and Great Britain. By 1971 serious approaches were made to the possible foreign partners.

In the spring of 1972 I was walking down a hallway in the National Science Foundation, which in those days was located on 16th street in Washington, only a block away from the Executive Office Building and the White House Complex. A beautiful, tall woman came out of a door and looked at me. "You're Bill Hay, aren't you?" "Yes," I replied. "Come into my office, I have something to show you." She was Mary Ann Lloyd, Chief Counsel for the Foundation. She handed me the JOIDES agreement. "Have you ever read this?" I confessed I had not. She told me that it made every institution, including their parent Universities, and every individual involved in planning the program personally and collectively liable for damages should anything go wrong. "What if you make a miscalculation and have an oil spill off some country that depends on its offshore fisheries for its food. You've wiped out the entire nation's livelihood. Who pays? You folks have to incorporate before we can go any further with international funding possibilities. We need to be sure that the liability ends up with the Foundation and United States Government." I had a good idea of what she was talking about; as Chair of the Planning Committee I now served on the Pollution Prevention and Safety Panel and considering disaster scenarios was its business.

The next week was the Annual Meeting of the American Association of Petroleum Geologists in my home town, Dallas. I flew to Dallas and asked my father if I could use our family lawyer, Ed Smith, to check on a little matter. Monday morning I handed him the JOIDES agreement to look over. I told him that the National Science Foundation's Counsel thought it might have liability implications. I left it with him and went off to the Convention. A few hours later I was paged. Ed Smith needed me to call him right away. I returned his call. "Bill, you don't have anything to do with this organization, do you?" "Yes, I'm the Chairman of the Planning Committee that is responsible for the program." Silence. I'll never forget what he said next. "Bill, how long would it take for you to get out of that position?" His interpretation of the original JOIDES agreement was the same as Mary Ann Lloyd's. Everybody was liable for everything.

A few minutes later I saw Manik Talwani, who had replaced Maurice Ewing as Director of Lamont, and just happened to be the Chair of the JOIDES Executive Committee. I told him about the situation. He laughed, "Lawyers are always looking for something to do. I'm sure it's not a problem, but I'll have our lawyers take a look when I get back."

A week or so later Manik called a special meeting of the JOIDES Executive Committee, and explained the situation. Walton Smith asked me to represent Miami at the meeting. Manik explained that we had to have a corporation to relieve our institutions and ourselves of liability. Within a matter of weeks the corporation was formed: Joint Oceanographic Institutions (JOI, Inc.) to be located in Washington, D.C. JOI immediately contracted with JOIDES to provide the scientific advice for the drilling program. In case of a disaster, the buck stopped with JOI and NSF. The road was clear for making overtures to other countries.

WATER MOLECULES ARE GREGARIOUS

*Water is the driving force of all nature.*
Leonardo da Vinci

Compared to the other planets in our solar system, Earth is unique in its abundance of the chemical HOH (Hydrogen–Oxygen–Hydrogen), better known as $H_2O$, or 'water.' In this chapter I refer to it as HOH because that approximates its structure, and will keep reminding you why it is so odd. You may recall from Chap. 5 that in 1805 Joseph-Louis Gay-Lussac and Alexander von Humboldt found that it took two volumes of hydrogen to react with one volume of oxygen to form water, so its formula must be $H_2O$.

This apparently simple compound has some of the most complex properties of any molecule. On Earth, HOH occurs in three different states or phases: gaseous vapor, liquid water, and solid ice. Incidentally, it is quite unusual for any substance to have its gaseous, liquid and solid phases exist side by side, but that it just what we have in the case of HOH on the surface of our planet. We often use the term 'water' in two ways: as a synonym for HOH in all its forms, and just for its liquid phase. But that can lead to confusion, so, although it may seem awkward, I'll use 'HOH' for the chemical compound in all its forms, and 'vapor' for it as a gas, 'water' for the liquid, and 'ice' for the solid. Each of these phases has very unusual properties.

## 17.1 Some History

There were undoubtedly many investigations of the properties of water over the centuries, but it was Scottish medical doctor and chemist, Joseph Black who laid the groundwork for our modern ideas. You were introduced to Black as a chemist in Chap. 16. It was while he was Professor in Glasgow (1756–1766) that Black carried out experiments with phase transformations of gas to liquid to solid.

The thermometer had been invented in the 17th century, but the first standardized scales were those of Fahrenheit (1724) and Celsius (1742); much more about all this in Chap. 30. The thermometer had made it possible to determine that ice and the water surrounding it remained at the same temperature as the ice melted, and this took a long time. Black was able to determine that the amount of heat required to melt ice would raise the temperature of an equal mass of water from 32° to 140° F. He coined the term 'latent heat' for the heat involved in the phase transformation. He also coined the term 'specific heat' for the amount of heat required to change the temperature of a phase of a material by a given amount, determining that the specific heats of water and metals like iron were very different. Other experiments concerned the phase transformation of water to vapor. He presented these ideas at a meeting of the Glasgow University Philosophical Club in 1762. Although they were widely adopted Black never formally published them.

Black was fortunate to have a bright young technician to help him, James Watt (1736–1819)—the James Watt who went on to improve the steam engine and thereby start the industrial revolution.

By now you have probably begun to wonder why there was so much interest in heat and phase transformations in Glasgow, Scotland. This was after all, where William Thompson (1824–1907) would later become famous as the first Baron Lord Kelvin. He earned the title for his studies of heat. Well, it seems to have something to do with a major business in Scotland, distilling whisky. Distilleries abound through the Glasgow area and capturing the superb essences of malt whiskies is deeply involved with evaporation and condensation.

## 17.2 Why Is HOH so Strange?

The HOH molecule consists of two hydrogen atoms connected to a single oxygen atom. But because the hydrogen atoms are much smaller than the oxygen atom, the molecule is asymmetrical, as shown in Fig. 17.1; it is a classic dipolar molecule.

Each of the hydrogen atoms (H) shares an electron (−) with the oxygen atom (O). This linkage is known as a

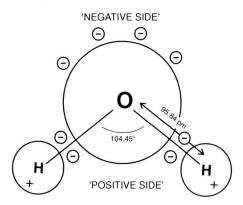

**Fig. 17.1** Schematic view of a water molecule. O is the oxygen atom. H's are positively charges hydrogen atoms, and the small circles with minus signs are electrons. In reality each of these parts of the molecule is in motion, the electrons forming a cloud around the nuclei of the oxygen and Hydrogen atoms. pm = picometer = $1 \times 10^{-12}$ m = 1 trillionth meter

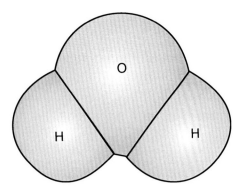

**Fig. 17.2** A Three-dimensional representation of a molecule of HOH

'covalent bond.' Because they have the same charge, the two hydrogens repel each other, but their charges are not strong enough for them to be on opposite sides of the oxygen. Instead they are 104.45° apart. The oxygen atom has four unshared electrons which are repelled by the positive electrical charges of the hydrogens and swarm in a cloud on the other side of the molecule. Figure 17.2 gives a better impression of what a molecule of HOH might actually look like with its electrons swarming around to make a cloud enclosing the atomic nuclei.

The asymmetry of the HOH molecule results in one side having a slight positive charge, the other side a slight negative charge. These forces are so small that in its gaseous phase as vapor they have almost no effect. However in the liquid phase, water, they are extremely important and make it behave in a thoroughly unexpected way. In water the molecules attract each other; they are 'sticky' and want to hang together. It is thought that they form clusters which last for a tiny fraction of a second, then disperse only to form clusters with other molecules an instant later. The tendency to form clusters decreases as the water becomes warmer.

This changes the viscosity of the water, as you may have noticed. Cold water seems to be slightly 'thicker' than hot water. In the solid phase, ice, these charges are integrated into the crystalline structure.

The atomic mass of a hydrogen atom is 1; that of an oxygen atom is 16, so the molecular mass of water is 18. Other molecules near this size and mass behave very differently with regard to the phase changes between vapor, liquid and solid. Consider a compound of similar molecular mass, methane. Methane has four hydrogens symmetrically surrounding a carbon atom. There are no 'positive' and 'negative' sides to the molecule. The atomic mass of carbon is 12, so the molecular mass of methane is 17. Under the atmospheric pressure at Earth's surface the freezing point of methane is −183 °C and its boiling point is −169 °C. Water is only slightly heavier, and we would expect its freezing and boiling points to be close to those of methane. Instead, at Earth's surface, water freezes at 0 °C and boils at 100 °C. Not only are the temperatures of the phase changes much higher, the temperature difference between the freezing and boiling points is much larger. Water is completely unique in this respect. Because of its stickiness in the liquid phase, it behaves as though it has a much higher molecular mass. Furthermore, the heat of fusion and heat of vaporization, that is, the amounts of energy required to melt ice to form water and to boil water to form vapor are much higher than for most other substances. Figure 17.3 is a graphical representation of the amount of energy consumed in heating ice from −100 °C to its melting point, transforming it to water, heating the water to its boiling point, transforming it to vapor, and heating the vapor to +200 °C.

Although the amounts of energy involved in heating ice and water are usually considered constant, as shown in Table 17.1, they are in reality variables, dependent on temperature. However, these variations are very small. Perhaps you can barely make out the gentle curve in the ice line in Fig. 17.3; the differences for water are too subtle to be visible in the diagram. If you study the numbers you can figure out that it takes the same amount of energy to melt a kilogram of ice as it does to raise the temperature of water from 0 to 80 °C! And look at how much energy it takes to boil water, converting it to vapor. No wonder "a watched pot never boils."

Table 17.1 compares some of the properties of water with other common Earth materials.

Comparing the different phases of $H_2O$ we note that the density of the gaseous phase, vapor, is about 1/1000th that of the liquid. This is a common relationship for these two phases in all sorts of substances. However, we also note that the solid phase, ice, is less dense than the liquid phase, water. This is very strange. This means that when water freezes, it expands. The only other common liquid that shares this property is acetic acid (the acid in vinegar). In all

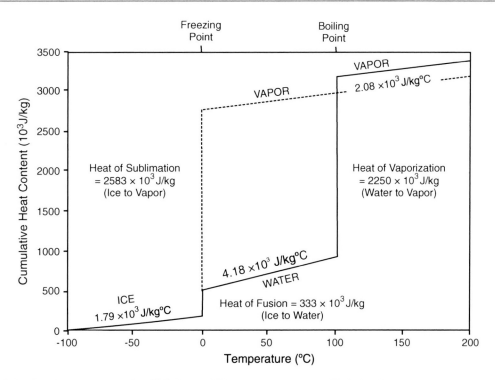

**Fig. 17.3** Cumulative heat involved in heating $H_2O$ from $-100$ to $+200\,°C$ at sea level, showing the varying amounts of energy (expressed as Joules per kilogram per degree C: J/kg °C) required to change the temperature of the three phases, ice, water, and vapor, and the very large amounts of energy involved in the phase transformations

from ice to water (fusion) and, especially, water to vapor (vaporization). Ice can also transform directly to vapor (sublimation); the energy involved in sublimation is the sum of that involved in fusion and vaporization. The energy consumed in the phase transformations can be returned by condensing vapor to water and freezing water to ice

**Table 17.1** Comparison of the physical properties of some materials common on Earth's surface

| Material | Vapor | Water | Ice | Air | Methane | Ethanol | Limestone | Salt |
|---|---|---|---|---|---|---|---|---|
| Composition | $H_2O$ | $H_2O$ | $H_2O$ | $N_2$ (78 %) $O_2$ (21 %) | $CH_4$ | $C_2H_5O$ | $CaCO_3$ | NaCl |
| Molecular mass | 18 | 18 | 18 | 30 | 16 | 46 | 100 | 58 |
| Phase | Gas | Liquid | Solid | Gas | Gas | Liquid | Solid | Solid |
| Density (kg/m$^3$) | 0.8 | 1000 | 920 | 1.3 | 0.68 | 789 | 2700 | 2300 |
| Boiling point (°C) | 100 | 100 | | $-194.3$ | $-161.6$ | 78.4 | 898 | 1465 |
| Freezing point (°C) | 0 | 0 | 0 | $-216.7$ | $-182.5$ | $-114.3$ | | 800 |
| Specific heat (10$^3$ J/kg °C) | 2.08 | 4.181 | 2.00 | 1.0035 | 2.226 | 2.44 | 0.84 | 0.92 |
| Thermal conductivity (W/m °C) | 0.016 | 0.6 | 2.3 | 0.025 | 0.032 | 0.14 | 3.5 | 5.9 |
| Latent heat of vaporization (10$^6$ J/kg) | | 2.25 | 2.58 | 0.20 | 0.51 | 0.88 | | |
| Latent heat of fusion (10$^6$ J/kg) | | | 0.33 | 0.02 | 0.06 | 0.11 | | 0.49 |
| Surface tension (10$^9$ N/m) | | 7.20 | | | | 2.23 | | |

Data are taken from many sources

other naturally occurring materials, the solid phase is more dense than the liquid phase. The fact that water expands when it freezes is very important in breaking down rock to make soils. Water can seep into cracks and pry the rock apart when it freezes.

Specific heat is the amount of energy required to change the temperature of a substance, and here too water is almost

unique. Liquid $H_2O$, water, has the highest specific heat of any common substance except ammonia. The specific heat of water is $4.18 \times 10^3$ J/kg °C. That is, it takes 4,180 J of energy to raise the temperature of a kilogram of water (which is 1 liter by volume) 1 °C. In contrast, the specific heat of an ordinary building brick weighing 1 kg is 1,000 J/kg °C, less than 1/4th that of water. The difference in

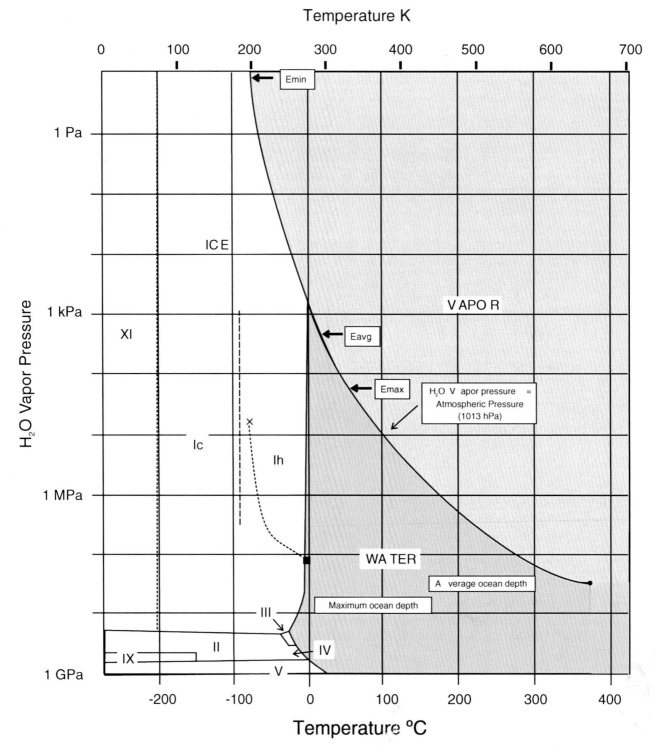

**Fig. 17.4** Phase diagram for HOH. Roman numerals indicate different crystal forms of ice. Eavg marks the average temperature at the surface of the Earth—note how close it is to the 'triple point' where all three phases can co-exist. Emin and Emax mark the minimum the maximum temperatures at the surface of the Earth. The dot where the line separating 'water' and 'vapor' ends is the 'critical point' where the two phases cannot be distinguished It is at 647 K = 374 °C = 705 °F and 22.064 MPa = 3200 lb/in.² or 218 atmospheres; roughly equivalent to a depth of about 2,180 m or 7,150 ft in the ocean. The point marked 'x' is the Antarctic surface at Vostok, and the dashed line below it is the temperature profile in the Vostok core which ends just above subglacial Lake Vostok shown as a black square

specific heat is due to that 'stickiness' of the water. It changes its structure as it is heated, and that change in structure absorbs energy.

You might be surprised to learn that all of the phases of $H_2O$ are good insulators. The ability to transmit heat is called 'thermal conductivity' and is expressed as watts per meter per degree. All gases are good thermal insulators because the individual molecules are far apart compared to a metal or other solid. But the thermal conductivity of water vapor is even less than that of dry air and half that of methane. Liquid water is a better thermal conductor than ethanol, but a much better insulator than liquid mercury (thermal conductivity = 8.3 W/m °C) and almost all solids. Ice has an unusually low thermal conductivity for a solid.

Stop! You will say. When I put a pot of water on the stove it all heats up evenly, more or less. But you are heating the pot from the bottom, making the water at the bottom expand and rise; you are making the water convect. If you heat the water from the top (scientists sometimes do strange things like this just for the hell of it) you will find that the water becomes stratified, hot on top and cold on the bottom. Now think about insolation—the Sun heats the water from above. Why doesn't it become stratified? Actually, it does— if there is no wind to make waves and mix it. I remember some very calm days in the Bahamas when there would be a thin surface layer of water that could be unpleasantly hot.

Electrical conductivity is a different matter; it is measured in terms of either 'conductivity' or 'resistivity'—the ease of or resistance to flow of an electric current. Conductivity is measured in 'Siemens' (S), a term less familiar than that used for resistivity, the Ohm ($\Omega$), but is simply the reciprocal, $1/\Omega$. An Ohm is defined as a resistance between two points of a conductor when a constant potential of one volt applied to these points produces a current of one ampere in the conductor. The electrical conductivity of water varies through 8 orders of magnitude depending on how much salt it contains. It is expressed in S per meter, = S/m or $Sm^{-1} = \sigma$. For pure water, $\sigma = 5.5 \times 10^{-6}$, meaning it is very resistant to an electrical current. However, the minor impurities in drinking water alter its conductivity by two to four orders of magnitude: $\sigma = 5 \times 10^{-4}$ to $5 \times 10^{-2}$. Salt water is two orders of magnitude more conductive, $\sigma = 4.8$ for sea water with 35 g of salt per kilogram of water (=35 ‰) at 20 °C. But the change in electrical conductivity with salt content is so pronounced that it has become the standard method of determining the salinity (saltiness) of seawater, replacing the old chemistry lab techniques. This property allows a probe to measure the salinity of sea water at any depth very precisely without the need to take a sample.

Figure 17.4 is a phase diagram for HOH. The phase diagram plots the boundaries between the different phases against temperature and pressure. Note that here pressure on the left axis is purely the pressure of the HOH.

Two of the phases, liquid water and ice, occur in relatively pure forms on the surface of the solid Earth. However, the salt in seawater modifies the phase diagram slightly; at sea-level seawater boils at about 103 °C rather than 100 °C and freezes at about −2 °C rather than 0 °C. Over 95 % of all the water on or near the surface of the Earth is saltwater.

The gaseous phase, vapor, is mixed with air. When discussing air, one can speak of the partial pressure of the different gases. You will recall that dry air is made of 78 % Nitrogen ($N_2$), 21 % Oxygen ($O_2$), and 1 % Argon (Ar). The other components, such as $CO_2$, and other trace gases make up much less than 1 %. The atmospheric pressure at sea-level is about 1013 hPa, so the partial pressure for $N_2$ is 790 hPa, for $O_2$ 213 hPa, and for Ar, 10 hPa. If the air were 'wet,' and saturated with water vapor at the Earth's average temperature of 15 °C, its partial pressure of the HOH would be 17 hPa. It would be displacing some of the regular air molecules. If you have HOH at sea level, and heat it to 100 ° C, its partial pressure is equal to that of the regular air molecules, it can displace all of them. Water at this temperature boils as the vapor pressure exceeds that of the atmosphere. This point is indicated on Fig. 17.4.

All of this has very practical implications. If you live near sea level you can believe what you read in a cookbook. All is well until you go up into the mountains and start cooking there. My home is at an elevation of about 8,000 ft (= ~ 2,440 m) and water boils at about 92 °C (197.6 °F). To achieve the equivalent of a 'three minute egg' cooked at sea level, you need to boil it for 6 min or so; there is some uncertainty because it seems to also depend on the weather. For a proper discussion of how to boil an egg, I refer you to Charles D.H. Williams (Professor of Physics, Exeter University, UK) 'The Science of Boiling an Egg' (http://newton.ex.ac.uk/teaching/resources/CDHW/egg). Making bread or baking cakes becomes an art, because not only the boiling point of water is lower, the gas produced by the yeast expands more rapidly.

But back to the phase diagram. The boundary separating the water and vapor fields marks the changing boiling point (if you go from water to vapor) or condensation point (if you go from vapor to water). It continues upward in the diagram as the boundary between the ice and vapor fields. There it marks the sublimation point (if you go from ice to vapor) or deposition point (if you go from vapor to ice). Along the line bordering the vapor field you will see Emax, the maximum temperature ever recorded on Earth's surface (57.8 ° C = 136 °F at Al Aziziyah, Libya in September 1922); Eavg, Earth's average surface temperature (15 °C = 59 °F); and Emin (−89.4 °C = −129 °F). The slope of the curve bounding the vapor field means that the temperature of condensation or deposition decreases with pressure. As we have already seen, the practical lesson of this is that the boiling point of water deceases with elevation. Continuing into the area where water no longer exists, we see that the temperature of deposition of ice from vapor also decreases with elevation.

You will notice that there is a point where the lines separating water from vapor, ice from vapor, and water from ice come together. This is called the triple point. It is as 0.01 °C and a pressure of 6.12 hPa. Under these conditions, the three phases, solid, liquid, and gas, or for HOH ice, water, and vapor can happily coexist. It just so happens that many places on Earth have these conditions as the seasons change. There are two very strange things about planet Earth: first, it happens to have an abundance of HOH, and second, it has surface conditions that hover around the triple point for this remarkable substance. As far as I know there is no comparable substance on any of the other planets of the Solar System.

Within the water field you will find the average depth of the ocean and the deepest point in the ocean (in the Marianas Trench) marked. A circle close to the ice boundary marks the temperature of the deep ocean, several degrees above the freezing point. The salt in the ocean depresses the freezing point of seawater, so actual freezing line for seawater is a few degrees to the left of the line shown here.

If you follow the boundary between vapor and water down to the right, you will see that it ends at a point before the right edge of the diagram. This is called the 'critical point.' It is at temperature of 374 °C (705 °F) and a pressure of 22.06 MPa, corresponding to a depth of about 2,150 m depth. At the critical point something very, very strange happens. Beyond the critical point it is no longer possible to make a distinction between liquid and gas. As the temperature rises, the liquid, in this case water, expands and becomes less and less dense. As the pressure rises the gas, in this case vapor, becomes more and more dense. At the critical point the densities of both are the same. They are indistinguishable. Both gas and liquid are fluids, but when they become indistinguishable they become something new, a supercritical fluid. For seawater, the critical point is at 407 °C and 29.8 MPa, corresponding to a depth of about 2850 m in the ocean.

At first this may sound very esoteric, but then you will recall that you have seen pictures of the black and white 'smokers,' vents for hot hydrothermal fluids on the mid-ocean ridge. The average depth of the mid-ocean ridge is about 2500 m, and the temperatures of the escaping fluids have been measured to be in the range between 350 and 450 °C, so one could suspect that at a hydrothermal field along the ridge some of the fluids might actually be supercritical. Of course, the surrounding ocean water is cold, so this bizarre condition cannot exist for any distance outside the vent. Just recently, in 2008, proof of supercritical conditions in the deep sea was reported by Andrea Koschinsky and a group of marine geologists from several German institutions. Using a deep-sea submersible they explored and sampled a hydrothermal field located at 5° S on the Mid-Atlantic Ridge at a depth of 3000 m. The venting fluids had temperatures as high as 464 °C. So not only does this planet have conditions over much of its surface near the triple point for HOH, the ocean contains sites where the critical point is surpassed.

Now examine the ice field on the phase diagram. You will immediately notice a number of subfields labeled with roman numerals. The common ice most of us know is Ih, ice with a hexagonal crystal form. It is the only kind of ice found in abundance on planet Earth. However, if there is a bare surface and it is very cold (below −80 °C), vapor may deposit directly on the surface as ice with a cubic structure (shown as Ic on the diagram) or as amorphous non-crystalline ice. This is about all we have under natural conditions on Earth. However, there are a number of more exotic forms of ice, each having its own crystallography. A few of these are shown on the diagram but there are still more that exist only at higher pressures. While these exotic forms of ice do not exist on Earth, they may exist on other bodies in the solar system, such as Jupiter's Moon Europa or Saturn's Titan. You can visit Steven Dutch's Ice Structure website at (http://www.uwgb.edu/dutchs/petrolgy/Ice%20Structure.HTM) to learn more about them.

The coldest temperature ever recorded, −89.4 °C, was measured at Vostok, the Russian Station near the center of East Antarctica. Vostok has an elevation of 3,488 m. it is shown by an X in the ice field. Beneath it there are about 3,750 m of ice, and below that a lake. Lake Vostok is about the size of Lake Ontario, and has some areas deeper than 1,000 m. It is indicated by a small black square on the phase diagram. The dotted line between these points is a guess at the temperature/ pressure profile through the Antarctic ice sheet.

## 17.3    The Hydrologic Cycle

Now, let's think about how much $H_2O$ there is on Earth and where it is. Table 17.2 summarizes the information. Be careful to note the exponents on those powers of ten.

We call the transfer of water among these different reservoirs the 'hydrologic cycle.' It has been obvious to observers since ancient times that water evaporates into the air, falls out again as precipitation, and flows down rivers back to the ocean. The hydrologic cycle was described in some detail by the Roman author Lucretius in his work 'On the Nature of Things' written about 50 BCE. He noted that although rivers kept flowing into the ocean, sea-level didn't change. He also observed that if you look at the ocean on a cold day you can sometimes see 'steam' rising from its surface and disappearing into the air. He concluded that this

**Table 17.2** Major reservoirs of water on or near the Earth's surface

| Reservoir | Volume (m$^3$) | Mass (kg) | Moles |
|---|---|---|---|
| Ocean | $1,371.3 \times 10^{15}$ | $1,371.3 \times 10^{18}$ | $76.12 \times 10^{21}$ |
| Ice | $26.3 \times 10^{15}$ | $24.5 \times 10^{18}$ | $1.36 \times 10^{21}$ |
| Active ground water | $8.4 \times 10^{15}$ | $8.4 \times 10^{18}$ | $0.47 \times 10^{21}$ |
| Inactive ground water | $50.2 \times 10^{15}$ | $50.2 \times 10^{18}$ | $2.79 \times 10^{21}$ |
| Lakes and Rivers | $0.2 \times 10^{15}$ | $0.2 \times 10^{18}$ | $11.1 \times 10^{18}$ |
| Atmosphere | $0.013 \times 10^{15}$ | $13 \times 10^{15}$ | $0.72 \times 10^{18}$ |
| Biosphere | $0.0006 \times 10^{15}$ | $0.6 \times 10^{15}$ | $0.03 \times 10^{18}$ |
| Bound in sedimentary minerals | $8.1 \times 10^{15}$ | $8.1 \times 10^{18}$ | $0.44 \times 10^{21}$ |

Active ground water is near the surface and flows slowly through the rocks and sediments. Inactive ground water is more deeply buried and flows little if at all. The volume of water in the atmosphere is given in terms of liquid water. Some water is bound up in sedimentary minerals as part of the crystal structure. There is thought to be more water in the Earth's mantle, but no one knows how much. Estimates usually range from one to several ocean equivalents

was what reappeared as clouds and rain. We now know that the rainfall on land that runs back as rivers to the sea is equivalent to a layer of water 10 cm thick (0.01 m) over the entire surface of the ocean each year.

Geologists like to use the term 'residence time' as an index to process rates. Residence time for water is the average length of time it would take for all the water in the ocean to cycle through the atmosphere, precipitate as rain, and flow back in. The average depth of the ocean is about 3800 m. So if 10 cm flows in each year, the residence time is 380,000 years. Of course, this calculation makes the assumption that each molecule of seawater goes through the cycle only once. Nevertheless it is a useful way of emphasizing that the water spends most of its life in the ocean; once every 380,000 years the molecule gets to spend a year cycling through the atmosphere, falling as rain or snow on land, and becoming part of a river flowing back to the sea. This is equivalent to your employer giving you 1 day off for every 1,041 years you work. The residence time for a water molecule in the air is only a few days.

By now I hope you are convinced that H$_2$O, one of the most common substances on the surface of our planet, is a very unusual substance. It is the main reason our planet is very different from the others of our solar system. Let's take a closer look at the properties of each of the phases. The behavior of the vapor phase is similar to that of many other gases, and will be discussed in detail in Chap. 20. However, water and ice are very special substances, in many ways unlike most other liquids and solids. They deserve special treatment.

sea level, a kilogram of air (a kilogram of air occupies about 1 m$^3$ at sea level) with a temperature of 21 °C (70 °F) can hold about 12 g of HOH vapor. The amount doubles with every 10 °C, so that at freezing, a kilogram of air is saturated with only 3 g of vapor.

If the air becomes supersaturated, the excess vapor will be precipitated out as water or snow. The latent heat involved in the phase change warms the atmosphere. In fact, the major means of energy transport by Earth's atmosphere is not by moving warm or cold air from one place to another, or by the force of the winds. It is through the 'latent heat' involved in the evaporation and condensation of HOH. As Fig. 17.3 shows, this conversion process involves more than five times as much energy as is involved in heating water from its freezing to its boiling point.

For the most part, HOH vapor behaves like any other gas molecule. But, as we will discuss further in this chapter, any atmospheric gas molecule that has more than two atoms intercepts and retransmits radiation. Water vapor is especially effective in trapping the outgoing long-wave radiation from the Earth. In fact, it is Earth's major greenhouse gas, but unlike the others its distribution is strongly dependant on temperature and elevation. It is so important that Chap. 20 is devoted exclusively to it.

The other oddity about HOH vapor is that its molecular mass is only about half that of the other gases in the atmosphere. This means that as its concentration increases, the air becomes less dense. If the air is warm, the density difference between dry and saturated air can be significant enough to produce atmospheric pressure gradients.

## 17.4 Vapor

Vapor is the gaseous phase of water, and it enters the air wherever water or ice is available. However, the amount of vapor that can be held by the air depends on temperature. At

## 17.5 Pure Water

Water is a truly strange substance. The amounts of energy it releases on freezing or absorbs in order to boil are prodigious, unlike any other natural substance found in any

**Fig. 17.5** Density of fresh water at sea level, plotted against temperature. Inset shows the density change between 0 and 10 °C. The maximum density, usually said to be at 4 °C is actually between 4 and 5 °C. Specific volume is the volume occupied by a given mass; it is the reciprocal of the density. Note that the units for density are in kg/m³, while those for specific volume are in L/kg (L = liter). 1 L = 1/1000 m³

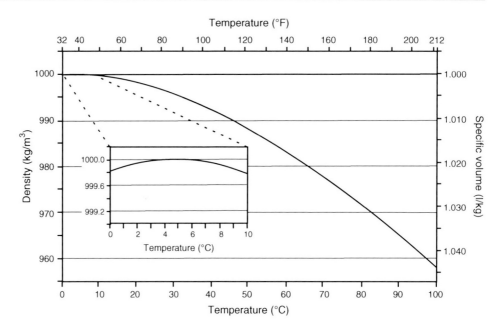

abundance on the surface of our planet. And water is truly abundant, covering a bit more than 70 % of Earth's surface.

Just as the transport of latent heat by water vapor is the dominant means of redistribution of energy by Earth's atmosphere on time scales of days to months, the thermal capacity of the surficial layer of the oceans is the dominant means of stabilization of climate on time scales of years to centuries. To appreciate how much energy the ocean can take up or release, consider its heat capacity. The heat capacity of something is the specific heat times the mass of the something. You can already guess that the 'heat capacity' of Earth's oceans is very large. If fact, the surficial layer of ocean water that changes temperature with the seasons is about 100 m thick, and has an area of $335 \times 10^6$ km². Taking the density of the water to be about 1000 kg/m³, that works out to be $33.5 \times 10^{18}$ kg. The area of land surface on our planet is about $149 \times 10^6$ km², but did you know that the seasonal temperature change only extends down about one m? That is because the thermal conductivity of rock, the ability for it to transmit heat, is quite low. Taking the density of the rock to be about 2500 kg/m³, the mass of rock that changes temperature with the seasons is about $0.37 \times 10^{18}$ kg. That is a difference in mass of about a factor of 100. Multiply that by the difference in specific heat, a factor of about four, and we come up with a heat capacity of the surficial layer of the oceans about 400 times that of the land. No wonder the temperature changes at sea and along the coast are so much less than in the interior of the continents! Earth's oceans and lakes are huge stabilizers moderating the effect of the seasonal changes in insolation.

The upper part of the ocean that changes with the seasons is called the 'mixed layer.' It is mixed by the winds and waves on time scales of days and weeks. It is less than 5 %

of the ocean by volume. The deeper layers of the ocean are in communication with the surface on much longer time scales of hundreds to perhaps a thousand or so years. They respond to longer term climate change. Most of the ocean's deep waters sank in the Polar Regions and are cold. Changes in the rate of overturning in the ocean can bring up greater or lesser amounts of this cold water in lower latitudes, and affect global climate. Convection, or overturning, depends on the upper waters becoming more dense than those below (Fig. 17.5).

Density is the mass of a given volume of something. Density is affected not only by what the substance is, but by the temperature and pressure at which we make the measurement. In the old days it was commonly the weight in grams of a cubic centimeter. Fresh water at sea level and 4 ° C has a density of 1 g/cc. Nowadays it is more common to express density in terms of kilograms per cubic meter, so that same fresh water has a density of 1000 kg/m³.

But water has a special peculiarity. In every other liquid we know, density increases as it becomes colder. For fresh water, this is true from its boiling point (100 °C) down to 4 ° C. But as it cools further toward its freezing point at 0 °C it expands and becomes less dense.

Why does water behave this way? Detailed investigations of the structure of this apparently simple substance have been going on for over a half century, using ever more sophisticated techniques. And the answer is: we are still not quite sure. Obviously, because it behaves as though it has a much higher molecular mass, the individual molecules must be clumping together into clusters, with more clusters when it is colder, fewer when it is warmer. Figure 17.6 is a schematic representation of warm water, showing a few loose clusters. We now know that the clusters have very

**Fig. 17.6** Schematic diagram of the molecular structure of warm water

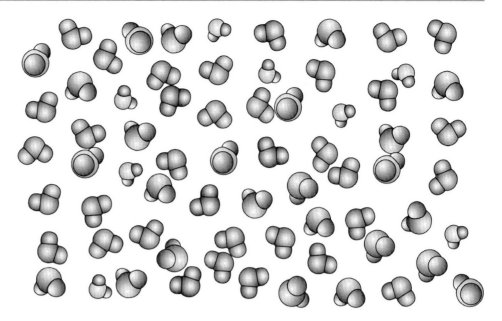

short lifetimes, measured in 'picoseconds' ($10^{-12}$ s = one trillionth of a second). The increasing number of clusters as the temperature declines makes it easy to understand the increase in density from 100 °C down to 4 °C. The more the molecules group together in clusters, the smaller the volume of the liquid, and therefore the higher its density.

Increasing the temperature of water reduces the stickiness, which we perceive as viscosity. So if you have ever had the suspicion that water coming to a boil seems to be 'thinner' or less viscous than it was when you put it in the pan, you're right.

But what happens between 4 and 0 °C? It is thought that this expansion is due to the formation of hexagonal clusters, as shown in Fig. 17.7. Note that the cluster has a hole in the middle, and occupies more space than would six non-clustered molecules. These hexagonal clusters are called 'ice rings' because they resemble the structural elements of crystalline ice.

Incidentally, I keep my Vodka in the freezer (in Russia they just put it on the window ledge in winter). The water becomes cooled below its freezing point but does not turn into ice because of the alcohol. But when you pour it out, it flows like syrup; the difference in viscosity is obvious.

Apropos Vodka, Russians like to flavor it. Just a bit of lemon peel, or a few peppercorns, or anything else you like will add flavor and make it very special. It takes about a day

**Fig. 17.7** Schematic diagram of the molecular structure of water about 4 °C, when 'ice rings' begin to form

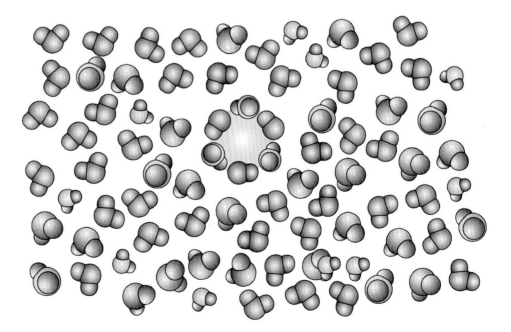

**Fig. 17.8** Schematic diagram of
the molecular structure of water
as it approaches the freezing
point, with multiple 'ice rings'

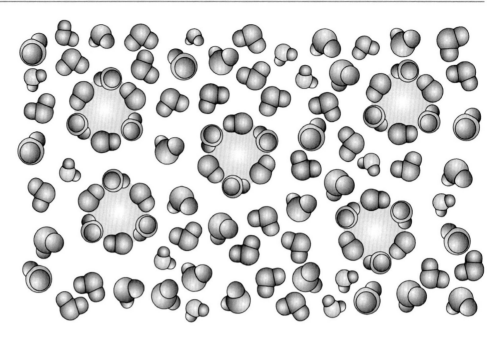

for the alcohol to extract the flavor. Beware of trying this with more than one dried chili pepper.

As the water temperature decreases, approaching the freezing point, more and more of these 'ice rings' form, and the water expands, as shown in Fig. 17.8. You can see from Fig. 17.8 that the amount of expansion is tiny, but it is enough to allow the colder water to float on the 4 °C water. This thin layer of frigid water on the surface of a lake or pond can then chill to the freezing point, and undergo the phase change to a solid, forming ice. Ice itself has a much lower density, only 920 kg/m$^3$, well below the range shown in Fig. 17.5. As we will see below, this is because all of the molecules have the hexagonal arrangement and occupy more space.

This peculiar property of having a maximum density above the freezing point has significant consequences for temperate and polar fresh water lakes. For a fresh water lake in the latitudes where winter temperatures are below 0 °C, this means that the surface waters become increasingly dense during the fall, and sink into the depths. Then in late fall, all of the water in the lake cools to 4 °C. All of the waters in the lake have the same temperature, and without any density contrast to stratify the waters, the entire lake can convect from top to bottom. This phenomenon, known as the fall turnover, can bring odoriferous deep waters, where organic matter has been decomposing, to the surface. As winter comes on, the surface waters become lighter, and float on the dense 4 °C waters in the bottom of the lake. Finally the surface waters reach the freezing point, and ice forms. As the winter ends, the ice melts, the surface layers warm to 4 °C again, and the spring turnover occurs. These sometimes unpleasant turnovers are familiar to anyone who lives along

the shores of the Great Lakes. Lakes that experience seasonal turnover are termed 'holomictic.'

The situation can be very different and much more dangerous when large bodies of water become stratified and do not turnover regularly; they are called 'meromictic' lakes. Lakes in the tropics never become so cold that their average temperature reaches 4 °C. Their deep waters form during cool spells, but are often stagnant. Sinking organic matter from the surface layers decompose and consume the oxygen on the deep waters. Under the pressures deep in the lake, $CO_2$, $CH_4$, $H_2S$ and other gases dissolve in large quantities. When the lake turns over it is like opening a bottle of soda pop. The gases come out. The methane ($CH_4$) could form an explosive mixture with the air, but over marshes it usually burns as flickering flames called will o' the wisp. The $CO_2$ is much more dangerous. It is colorless, odorless, and heavier than air. It suffocates people and animals. There is a famous fossil locality south of Frankfurt/Main in Germany, Messel. During the Miocene, about 20 million years ago Messel was a stagnant lake, producing large amounts of $CO_2$. Animals would come to the edge of the lake for water and suffocate. Their remains could not decompose because there was no oxygen, and even their soft parts became fossilized. But can you guess what the most common fossils are? Bats. Bats have a high metabolic rate, and when they unsuspectingly flew into the invisible layer of $CO_2$ they suffocated and fell into the lake. The most dangerous gas of all is hydrogen sulfide ($H_2S$). It is colorless, but in very small concentrations has the odor of rotten eggs. At higher concentrations it is odorless and highly poisonous. It has the nasty ability to go through rubber, so that it can kill you even if you are wearing a gas mask. Some of the sinkholes in Florida frequented by SCUBA divers have $H_2S$ in their bottom waters, making them dangerous sites for diving.

Many tropical lakes are in volcanic regions. They can be heated from below causing their deep waters to rise to the surface bringing with them noxious and poisonous gases. Lakes Monoun and Nyos, in Cameroon, Africa, are meromictic lakes in volcanic terrain. On August 15th, 1984, a 'limnic explosion' occurred at Lake Monoun. Vast amounts of $CO_2$ bubbled to the surface. 37 people died but their cause of death was not immediately understood. Then, on the 21st of August, 1986, Lake Nyos overturned, releasing a massive amount of $CO_2$. The heavy gas spread into the surrounding villages and suffocated over 1,700 people and 3,500 livestock. Lake Kivu, in the East African Rift Valley is much larger, with an area of 2,700 $km^2$. Its deep waters have very high concentrations of $CO_2$ and methane. It has a geologic record of overturn causing massive extinction of animal life in the surrounding area about every 1,000 years. The cause of the overturning is not known with certainty, but is thought to be most likely the result of periodic volcanic activity. Today more than 2 million people live in the Lake Kivu basin and a nearby volcano is showing signs of activity.

## 17.6   Natural Water

There is no 'pure water' in nature, only in the laboratory. Many compounds dissolve in water. It is sometimes called the universal solvent. The reason is, again, the asymmetric nature of the water molecule. The positive and negative sides of the molecule attract the positively and negatively charged ions in minerals, and cause them to go into solution. The most common dissolved mineral in river water is calcite, $CaCO_3$, from limestone. It is carried in solution as calcium ions, $Ca^{++}$, and bicarbonate ions, $HCO_3^-$. In rivers in volcanic regions the common dissolved salt is natrona, $NaCO_3$, carried in solution as the sodium ion Na + and the bicarbonate ion $HCO_3^-$. It often accumulates in playa lakes, and in some areas its presence is evident because the water 'feels slippery.'

Seawater contains ions of every known element, but the most abundant are sodium $Na^+$ and chlorine, as the chloride ion, $Cl^-$. These two ions make up about 86 % of the salt in seawater. It is interesting that almost all of the chlorine on Earth is either dissolved in the ocean or incorporated into salt deposits that formed from evaporation of seawater. It seems to have been in the ocean since it formed. Much of the chlorine in river water is recycled directly from the ocean. The water in sea spray from breaking waves evaporates and leaves salt NaCl behind in the air. These minute salt crystals can then serve as nuclei for raindrops falling on land. If the river water contains unusually high amounts of the chloride

ion, it means that there are evaporite deposits somewhere in the drainage basin. We will discuss the composition of seawater in more detail in Chap. 25, but one term that needs to be introduced here is 'salinity.' Salinity is the amount of solid mineral that precipitated out as seawater is evaporated to dryness. At least that was the idea. It turns out that some of the inorganic matter in seawater leaves as a gas before it is completely dried out and some of the water gets bound in the minerals, so it all turns out to be a horribly complicated mess. But in principle, salinity is the amount of 'salt' left behind, expressed as grams per liter. This is indicated by the symbol ‰. Today the average salinity of ocean water is 34.5 ‰. By the way, nobody bothers trying to measure the salinity by evaporating water anymore; today we measure the electrical conductivity of the water and convert that to salinity.

Seawater has another property that is very unusual. In contrast to fresh water, its maximum density lies below its freezing point. I learned this odd fact during my first year as an Assistant Professor at the University of Illinois. One of my colleagues, Professor Jack Hough taught a course in Limnology and Oceanography (Limnology is the study of lakes). Jack had been involved in developing sonar techniques for locating submarines in World War II. However, his base of operations was Navy Pier in Chicago, and most of his experience was in the Great Lakes. It was a good place for American submariners to learn their trade safely, but he recognized that lakes were different from the ocean. When I joined the faculty of the Rosenstiel School of Marine and Atmospheric Sciences in Miami in 1968, I found that one of my colleagues there, Frank Millero, had made the most detailed determination of the relationship between salt content, temperature and density of seawater. The relationship cannot be described exactly, but it can be approximated with a long mathematical expression, called the 'Equation of State.' Millero's version comes in two parts, one describing the temperature-salinity-density relation at the ocean surface. It has 15 terms. The second part describes relationships at different depths.

You might think that if you took 1 kg of water and added 35 g of NaCl salt to it, its weight would be 1.035 kg, and the volume the solution would occupy would be equal to the volume of the water plus the volume of the salt. You would be correct about the weight, but wrong about the volume. Let's assume that the fresh water is at 4 °C so that its density is 1000 $kg/m^3$. This means that 1 kg will have a volume of exactly 1 L or 1000 cc. Table 17.1 shows that NaCl is 2.3 times as dense as water, and 1 g of water at 4 °C occupies exactly 1 cubic centimeter (cc). So 35 g of NaCl will have a volume of 15.2 cc. The total volume should be 1015.2 cc, right? But it's not; instead it turns out to be 987.6 cc—less than the original volume of the water alone. What happened? The $Na^+$ and $Cl^-$ ions seemed to have slipped into the spaces

**Fig. 17.9** The relation between temperature, salinity, and density for fresh, brackish, and sea water. *Solid curved lines* are density at the sea surface; *dashed lines* are density at 40000 Pa (approximately 4 km depth). Lines of equal density are called 'isopycnals.' The *straight line* labeled MD is the maximum density; that labeled FP is the freezing point. More than 80 % of the water on Earth has the temperature and salinity indicated by the *gray circle*. Technically speaking, this diagram is "The Equation of State"—great for cocktail party conversation

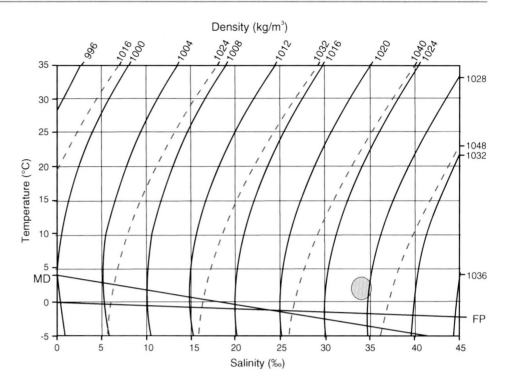

between the water molecules, and made the whole liquid contract. In many ways the Equation of State, which is the numerical description of this effect, is what we call 'counterintuitive.' Water is really strange.

You have probably heard that water is incompressible. Well almost, but not quite. Under pressure it occupies a slightly smaller volume, another way of saying that its density increases slightly. But again this is not straightforward, the degree to which the density changes is again dependent on temperature and salinity. The second part of the Equation of State, describing the compressibility of sea water with depth has 25 terms. It has military implications because it determines the speed of sound in the ocean, and sound is used to locate submarines. The complexity of the temperature-salinity-density-pressure relationship has a variety of unexpected consequences. I have spent many hours over the past forty-odd years thinking about the implications of these strange phenomena. It always helps to have a Martini first.

Figure 17.9 is a graphical representation of the relation of salinity, temperature, density and pressure. In this diagram I have plotted temperature on the vertical scale, rather than on the horizontal scale as in Fig. 17.5. That is because the shapes of the density lines look similar in both figures, they represent very different things. You won't need to look at Fig. 17.9 very long before you will notice some very odd things.

Figure 17.9 covers most of the major kinds of water you hear about, 'fresh' with salinities less than 0.5 ‰, 'brackish' (0.5–18 ‰), 'seawater' (18–50 ‰, but mostly in the range 30–38 ‰). More concentrated solutions are called 'brine.'

Sodium chloride begins to precipitate out when the salinity reaches 350 ‰. It should be noted that there are no universally accepted definitions for these terms. One very important point on the diagram is where the lines of maximum density and freezing point cross. That is at a salinity of 24.7 ‰. Waters with lower salinities behave like fresh water; water at the freezing point will float, and ice will form on the surface. Waters with higher salinities are different; the colder the water, the denser it is. Think about this for a moment: as seawater with a salinity greater than 24.7 ‰ cools toward the freezing point, it becomes denser and denser. At the freezing point it is at its maximum density and if it is being cooled by the air, it will sink from the surface. It would seem that there is no way ice could form on the surface of the ocean unless the whole ocean were chilled to the freezing point. But sea-ice does form. We'll discuss that later.

Now consider something else. Let's say you have two containers each containing the same amount of sea-water and both waters have the same density. But they have different salinities and temperatures; they lie along one of the density lines shown in Fig. 17.9. You mix them together. For this mixed water, both the temperature and salinity will be the average of the original waters. What about the density? If you are like 99.999 % of the population you will think the density must remain the same. But look carefully at Fig. 17.9. Connect two points on any density line with a straight line. We can call it a 'mixing line.' If there were equal proportions of both waters, the mixture will be at the midpoint of the line. What is the density at this point? Amazingly, it is always greater than the original density. You may think that makes

**Table 17.3** Numerical values for density of surface waters at different temperatures and salinities

| Temp (°C) | Salinity (‰) | | | | | | | | |
|---|---|---|---|---|---|---|---|---|---|
| | 0 | 5 | 10 | 15 | 20 | 25 | 30 | 35 | 40 |
| 0 | 999.84 | 1003.91 | 1007.95 | 1011.99 | 1016.01 | 1020.04 | 1024.07 | 1028.11 | 1032.15 |
| 5 | 999.97 | 1003.95 | 1007.91 | 1011.86 | 1015.81 | 1019.76 | 1023.71 | 1027.68 | 1031.64 |
| 10 | 999.70 | 1003.61 | 1007.50 | 1011.39 | 1015.27 | 1019.16 | 1023.05 | 1026.95 | 1030.86 |
| 15 | 999.10 | 1002.95 | 1006.78 | 1010.61 | 1014.44 | 1018.28 | 1022.12 | 1025.97 | 1029.83 |
| 20 | 998.21 | 1002.01 | 1005.79 | 1009.58 | 1013.36 | 1017.15 | 1020.95 | 1024.76 | 1028.58 |
| 25 | 997.05 | 1000.81 | 1004.56 | 1008.30 | 1012.05 | 1015.81 | 1019.57 | 1023.34 | 1027.13 |
| 30 | 995.65 | 999.38 | 1003.10 | 1006.81 | 1010.53 | 1014.25 | 1017.99 | 1021.73 | 1025.48 |
| 35 | 994.04 | 997.74 | 1001.43 | 1005.12 | 1008.81 | 1012.51 | 1016.22 | 1019.93 | 1023.66 |

These values represent the mass, in kilograms, of one cubic meter of fresh, brackish, or seawater. They correspond to the solid curved lines in the graphical representation of the Equation of State shown in Fig. 17.9

no sense, but you are not seawater. This is not a silly mind experiment; it is exactly what happens in the deep North Atlantic, where six different water masses, all with very similar densities meet and mix to produce a denser water mass known as North Atlantic Deep Water.

Here's another one. You take those same two water samples and let them sink toward the bottom of the ocean. They start off sinking at the same rate. Will they reach the bottom at the same time? Look again at Fig. 17.9. See the different slopes of the density lines at the surface and at 4,000 m depth? Whichever of the two waters is colder will become compressed faster. That means it becomes denser faster, and sinks more rapidly. It will reach the bottom first. Again, this is not a trivial result. It happens to be how ocean 'bottom waters' form.

You can ascribe all this to the natural perversity of nature.

## 17.7   Water—Density and Specific Volume

Another expression for density is 'specific gravity.' Both refer to the mass of a given volume, but specific gravity is simply the ratio of the density of something relative to that

of water at sea level and 4 °C. The reciprocal of the density (1/density) is called the 'specific volume.' It is the volume a particular mass of something will occupy. Table 17.3 shows specific gravities and Table 17.4 the specific volumes for waters at different temperatures and salinities.

An easy way to think about changes in specific volume is to consider what happens to 1000 kg of water in a container with a 1 m² base as the temperature changes. The numbers in the columns of Table 17.4 then represent the height of the water column in our container, expressed in meters for waters of different salinities as the temperature changes. So heating fresh water from 0 to 30 °C will cause the surface to rise from 1.00016 to 1.00437 m or 4.2 mm. If we heat typical ocean water, with a salinity of 35 ‰ from 0 to 30 °C, the height of the column goes from 0.97649 to 0.98233 m or 5.8 mm. That may not sound like much until you realize that the average depth of the ocean is 3790 m. As we will see in Chap. 25, almost 90 % of the world's ocean water is what is called 'deep water;' it has a temperature around 2 °C and a salinity of 34.7 ‰, and we now know that it is gradually warming. To make a crude calculation, let's assume that the average temperature of the whole ocean is 5 °C, its salinity is 35 ‰ and that the entire ocean warmed from 5 to 20 °C,

**Table 17.4** Numerical values for specific volume of surface waters at different temperatures and salinities

| Temp (°C) | Salinity (‰) | | | | | | | | |
|---|---|---|---|---|---|---|---|---|---|
| | 0 | 5 | 10 | 15 | 20 | 25 | 30 | 35 | 40 |
| 0 | 1.00016 | 0.99610 | 0.99211 | 0.98816 | 0.98424 | 0.98035 | 0.97649 | 0.97266 | 0.96885 |
| 5 | 1.00003 | 0.99607 | 0.99215 | 0.98828 | 0.98444 | 0.98062 | 0.97684 | 0.97307 | 0.96933 |
| 10 | 1.00030 | 0.99640 | 0.99255 | 0.98874 | 0.98496 | 0.98120 | 0.97747 | 0.97376 | 0.97006 |
| 15 | 1.00090 | 0.99706 | 0.99326 | 0.98950 | 0.98576 | 0.98205 | 0.97836 | 0.97468 | 0.97103 |
| 20 | 1.00180 | 0.99800 | 0.99424 | 0.99052 | 0.98681 | 0.98314 | 0.97948 | 0.97584 | 0.97221 |
| 25 | 1.00296 | 0.99919 | 0.99546 | 0.99177 | 0.98809 | 0.98444 | 0.98081 | 0.97719 | 0.97359 |
| 30 | 1.00437 | 1.00062 | 0.99691 | 0.99324 | 0.98958 | 0.98595 | 0.98233 | 0.97873 | 0.97515 |
| 35 | 1.00600 | 1.00227 | 0.99857 | 0.99491 | 0.99127 | 0.98765 | 0.98404 | 0.98046 | 0.97689 |

These values represent the volume, in cubic meters, of 1,000 kg of fresh, brackish, or seawater the expansion of the water would raise the surface by 0.0028 × 3,790 = 10.48 m (about 34 ft and 4 in.)! That is enough to drown all of the cities of the Gulf Coast, Florida, and east coast of the US. And that is without adding any water to the ocean, just warming it.

In fact, as we will see in Chap. 27, a significant part of the current sea-level rise is due to thermal expansion of the ocean. The warming of the deep waters is happening because the waters sinking into the interior of the ocean are now slightly warmer than they were in the past. Even if the major cause of global warming, addition of $CO_2$ to the atmosphere, could be stabilized immediately, the continued sinking of slightly warmer water would continue to warm the ocean interior and cause sea level to continue to rise for at least another century.

Just how compressible is seawater? First, a little note about how pressure in the sea is described. You may recall that 1 atmosphere of pressure was originally designated to be 1 Bar. It turns out that the pressure under 10 m of seawater is about equal to 1 atmosphere. This means that 1 m of depth in seawater equals 1/10 Bar or 1 dbar. When oceanographers describe pressures in the sea, they talk about decibars because the number of decibars is a close approximation to the depth in meters. Of course, it's not exactly equal because of the differences in density of seawater from place to place, but it is a very handy, easy-to-remember conversion.

This easy conversion means that a pressure of 2000 dbar is found at about 2000 m (2 km) depth. If oceanographers were really up to date, they would use kiloPascals (kPa). 10 kPa is one decibar, so the pressure at 2000 m would be about 20,000 kPa—not so handy a conversion. So oceanographers stick with decibars. Next time you are in a cocktail bar, instead of asking for three fingers of Scotch, you could ask the bartender for 1/20th decibar. If you are in Woods Hole, Massachusetts, home of Woods Hole Oceanographic, he bartender would probably be a graduate student, earning extra income on the side and he would know what you meant.

Questions that might come to mind: does the temperature of the water increase under pressure? Yes, but the temperature increase due to pressure is small. It does show up clearly in water temperature measurements made with a thermometer in deep-sea trenches where the water can appear to be 1 or 2° warmer than the water above it. To correct for this 'adiabatic' (=occurring without gain or loss of heat) temperature increase due to pressure, oceanographers use the term 'potential temperature.' The 'potential temperature' is the temperature the water would have if it were brought to the surface without gain or loss of heat. Second question: above it was stated that if two waters of equal density sink, the colder water will be more compressible and increase in density faster, but here we see that the warmer water becomes compressed more than the colder

water. What is going on? The waters we were discussing had the same *density* at the surface, not the same *salinity*. But if you thought of the questions on your own and they troubled you, you are well on your way to understanding just how strange water is.

## 17.8    Water—Surface Tension

Another of the special properties of water is its high surface tension. It is the property that causes the water surface to behave as though it were an elastic sheet. Water has the highest surface tension of any common liquid except mercury. Because the water molecules attract each other, they do not want to leave the group; so they form a distinct boundary surface. As a kid you probably made a compass by floating a magnetized needle on the surface of water. You may have also seen insects walking on the water. These things are possible only because of the special nature of the surface of the water. It is also surface tension that makes it possible for you to dry yourself with a Bath Towel after a shower. The fuzziness of the towel comes from many tiny loops of fabric. Because of its surface tension, water prefers the myriad of tiny surfaces in the towel to the smoothness of your body, and you can 'dry yourself off.' If you don't believe this try drying yourself with a smooth cloth, like a shirt or blouse. It turns out that this peculiar property of water has very important climatic effects.

First, it allows the wind to ruffle the surface of the water and to make waves. Waves become very important in mixing the surface waters of the ocean downward to assist in developing that 'surface mixed-layer,' that has such an enormous heat capacity. As the wind increases and the waves get bigger, they 'break' and form whitecaps. Have you ever stopped to think about why 'whitecaps' are white? They are made of droplets of water.

Once a droplet is separated from its parent water surface tension pulls it into a tiny round ball. The amount of light it will reflect depends on its size. The smaller the droplet the whiter it is. Something even stranger, as the droplet separates from its parent water, the surface tension shoots it forcefully away, like a tiny bullet. Once in the air, if the droplet is small enough, it is suspended and does not fall back onto the ocean surface. You can experience the same thing is a hot shower or steam room. Small droplets evaporate readily if the air is not saturated. If they came from seawater, they contain some salt. As the water evaporates, it leaves the salt behind as a minute crystal in the atmosphere. When the water vapor content of the atmosphere reaches saturation, that salt crystal will attract water molecules to form a droplet again.

Water droplets can condense from saturated air to form clouds. Now, this is where we get involved with one of the most important variables in Earth's albedo. As you learned in Chap. 15, clouds as a whole are responsible for about 60 % our planet's albedo. Yet, among themselves, clouds may have very different albedos, commonly ranging from 0.3 to 0.8 or even higher, as shown in Fig. 15.6. What controls the reflectivity of clouds? Droplet size and abundance. If the droplets are very small and abundant they are very reflective and make brilliant white clouds. As droplet size grows, they become less reflective and may appear gray. Ice crystals in the clouds tend to be larger than most droplets and are hence darker. The darker undersides of clouds are simply a reflection of the shadows of the droplets above. The albedo depends on the reflectivity of the upper sides of the clouds.

While clouds are responsible for 60 % of Earths albedo, the surface of the ocean accounts for about 70 % of the remaining 40 %. But the albedo of water is even more variable, ranging from 0.05 to 0.8. How is it possible for a single substance to have such a wide range? The albedo of water depends on the reflectivity of the water surface, shown in Fig. 17.10. The figure assumes a perfectly plane unperturbed water surface, which of course is extremely rare. The water surface is ruffled by the wind, and that is where surface tension enters into the picture. The water surface is also bent into waves, as discussed in Chap. 7. Figure 17.10 is simply a guide to give you a feeling of how this complicated system works.

When the Sun is directly overhead and out to an angle of 45° from the vertical (=45° above the horizon) the amount of the incoming solar radiation reflected off the water surface is very small. Almost all of the energy is absorbed by the water. The albedo is very low. The reflectivity increases as the Sun goes lower toward the horizon, and when it is greater than 75° from the vertical (=15° above the horizon) the amount of light reflected increases dramatically. If you have been SCUBA diving in the late afternoon, this is when 'the lights go out.' The amount of sunlight decreases very dramatically as the Sun approaches the horizon.

But there is a further complication. The Sun is directly overhead any given place only very briefly. Because of the t.. n the Earth's axis of rotation, the Sun makes an apparent .ion from south to north and back as our planet orbits the .. There are three vertical lines on the diagram. One ws the mid-day Sun angle at the polar circle at the sol- ice for that hemisphere. The next shows the mid-day angle of the Sun for the polar circle at the equinox; this is the same angle for the Sun at the pole at the solstice. The last line, at the right edge of the figure is the angle of the Sun seen from the ole at the equinox, when the Sun is just along the horizon.

What is happening now is that more and more of the Arctic Ocean is becoming ice free each year, and the reflectivity of the water surface during the summer is less than when it was ice-covered. Solar energy is being inserted

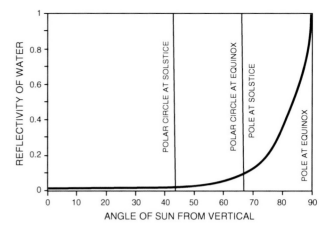

**Fig. 17.10** Reflectivity of a horizontal water surface plotted against the angle of the Sun from the vertical

into the northern polar region in a way that has not happened in the past. The amount, of course, will depend on the roughness of the water surface and whether the polar ocean is covered by clouds. We have no past experience with the situation to guide us. We will learn a little more each year as this drama unfolds.

## 17.9 Ice

The solid phase of HOH is what we call ice. As can be seen in the phase diagram, Fig. 17.4, it can form directly from vapor or water. Almost all of the ice occurring naturally on Earth is the crystalline form known as Ih, with the molecules ordered into hexagons adjoining each other to form sheets, as shown in Fig. 17.11.

The atoms forming the sheet do not all lie in a single plane, every other H sticks up or down out of the plane of the sheet and these connect with the overlying or underlying sheet to form a three dimensional structure shown in Fig. 17.12.

There are a number of other forms of ice having different arrangements of the HOH molecules but with one exception they do not exist under the conditions we have on Earth. They may exist on other objects in the Solar System. The other form of ice that may exist on Earth is called Ic, and it has a cubic arrangement of the molecules like those in a grain of salt. It can form when temperatures are below −80 °C. These can occur in the higher parts of the East Antarctic plateau and high in the atmosphere. However, it is 'meta-stable' and readily converts to the hexagonal ice Ih. It is also possible for HOH to freeze to a solid without any regular structure, called 'amorphous ice.' This can happen when vapor comes in contact with a very cold surface. But for all practical purposes ice Ih is the only form existing on our planet. Sadly, the chance of your seeing a single crystal of ice

**Fig. 17.11** *Top view* of a single layer of an ice Ih crystal. The HOH molecules alternate orientations to form the three-dimensional crystal lattice

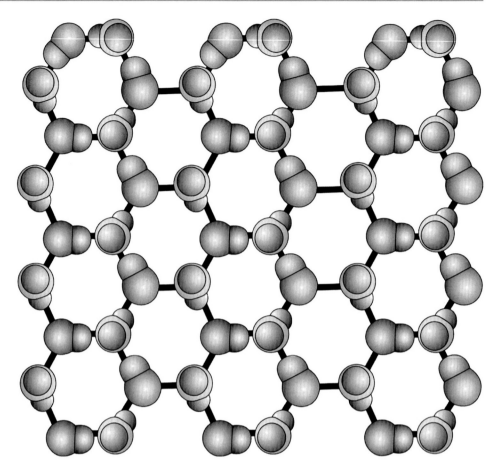

is exceedingly small. The basic form of a single ice crystal is a thin hexagon. I saw one in 1960, in a mine in Spitsbergen. It was a single, beautiful transparent crystal of ice about 30 cm across and a few millimeters thick. I have never seen anything like it since.

Much more common are the complex multiple crystals of ice that we know as snowflakes. They often show the hexagonal symmetry, and can be exceedingly beautiful. If you have never seen one in a magnifier or small microscope, it is well worth the effort. A good place to start is Kenneth G. Libbrecht's website SnowCrystals.com. Snowflakes form within the atmosphere, but where the vapor is in contact with a cold surface, it may deposit as frost, or if the crystals grow and become very large, hoarfrost. The climatological implications of making snow will be discussed in detail in the next chapter.

Early in the development of numerical climate models in the early 1970s it was discovered that if the Earth is free of polar ice, it is very difficult for permanent ice to accumulate there, and if the poles are ice covered it is very difficult to make them ice free. It was concluded that there are two stable states for our planet, one with polar ice, and one with ice-free poles. It was found that a very large perturbation of the climate system would be required to change from one state to the other. In 1977, a close friend at Princeton

University, Al Fischer, and a young Ph.D., Mike Arthur, published a classic paper entitled "Secular variations in the pelagic realm" documenting these two states in the geologic record and coining the terms 'greenhouse' and 'icehouse' for them (Fig. 6.1). Earth has been in the 'icehouse' state for 35 million years. It takes a very dramatic change in some aspect of the climate system to shift the Earth from one of these states to the other. The ongoing addition of $CO_2$ to Earth's atmosphere is forcing our planet toward its 'greenhouse' state and if allowed to proceed unchecked could result in a world very different from the one we know.

## 17.10 Earth's Ice

Although there are high-mountain glaciers at all latitudes, the Polar Regions are our planet's harbors of permanent ice. However, the two Polar Regions are very different, as shown in Fig. 17.13. For orientation, the Greenwich meridian, 0°, which passes through the Royal Observatory at Greenwich just east of London, is shown as a heavier line. It is the boundary between east and west longitudes, and extends through much of the Atlantic. The north-polar region has an ocean basin surrounded by land. The south-polar region has a continent surrounded by ocean.

**Fig. 17.12** *Top view* of the upper two layers of an ice Ih crystal showing the offset of one to the other

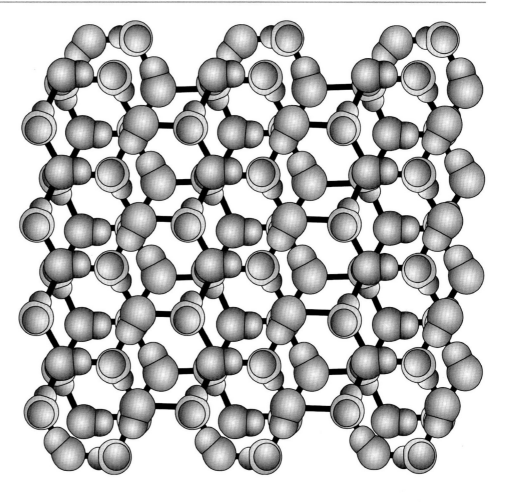

The greatest difference between the two Polar Regions is simply their geography. Beneath the sea-ice of the northern polar region is the Arctic Ocean Basin, averaging about 1 km deep. Beneath the ice of Antarctica is a continental block with an elevation only slightly above present day sea-level. Each polar region has an Ice Sheet, the Antarctic in the south, and Greenland in the north. The Greenland Ice Sheet is only about 1/10th the size of that of Antarctica. There are some smaller 'Ice Caps,' almost all in the Arctic region, covering islands or archipelagos, such as Baffin Island, Iceland, Svalbard, Franz Josef Land and Novaya Zemlya. Most of the world's larger glaciers are in the Alps, Himalaya, Andes, and northern Rockies.

As shown in Fig. 17.13, much of Earth's ice cover is floating sea-ice; its area varies greatly with the seasons. By the end of southern hemisphere winter in September, a vast area of sea-ice surrounds Antarctica with its margin extending well beyond the south polar circle. Its limit is fixed by the west-to-east flowing Antarctic Circumpolar Current which encircles the globe at 45°–55° S. In the northern hemisphere, late winter (March) ice covers Baffin Bay between Greenland and Canada, Hudson Bay, part of the Bering Sea between Alaska and Siberia, and much of the

Barents and Kara shelves north of Scandinavia and eastern Russia. Most of the sea-ice around the Antarctic disappears by late southern summer in March. Until recently, sea-ice covered most of the Arctic Ocean even into the late northern summer.

Table 17.5 shows that the total area covered by ice in the Antarctic is more than twice that in the Arctic. But the nature of the ice in the two Polar Regions is very different. In terms of mass, 90 % of Earth's ice is associated with Antarctica, 2 % of that amount is in the floating ice shelves of the Ross and Weddell Seas. Another 9 % is on Greenland, and the remaining 1 % is about evenly divided between other ice caps and glaciers, and sea-ice. The areas covered by winter sea ice are very similar in both Polar Regions, but that in the Arctic is thicker than that in the Antarctic. Note that the estimates of sea-ice area and thicknesses given in Table 17.5 are for conditions prior to 1980. Since then major changes have begun to affect both the Arctic and Antarctic regions, as will be discussed below and in more detail in Chap. 32.

Finally there is winter snow cover. It is mostly restricted to the northern hemisphere because the appropriate latitudes in the southern hemisphere are ocean. In the last century, until 1980, the northern hemisphere area of snow cover

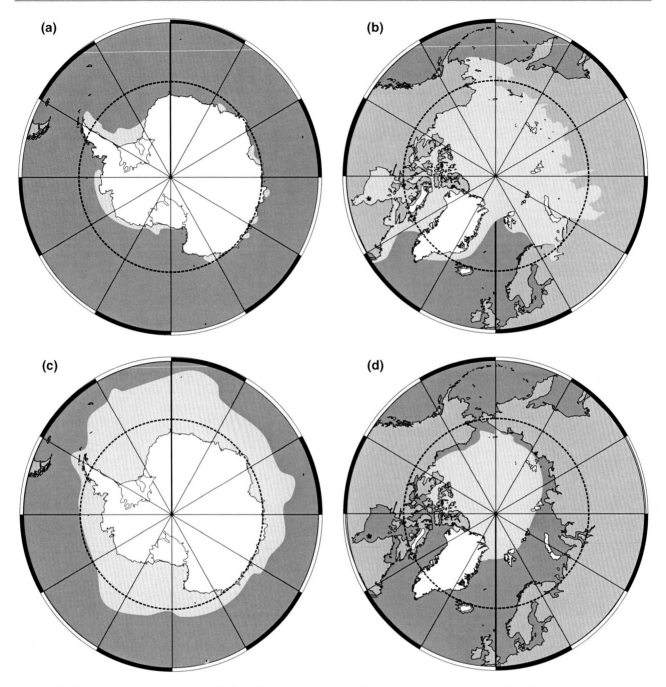

**Fig. 17.13** The southern and northern Polar Regions of Earth showing ice-covered areas varying with the seasons. Ice sheets and caps are in *white*. The two largest Antarctic ice shelves are *outlined*. Sea ice is shown in *light gray*. Continents and islands are in *darker gray*, and ocean basins in *dark gray*. A and B show typical late 20th century conditions for the month of March; C and D are for September. The Greenwich meridian, longitude 0°, is shown as a heavier line. The polar circles at 63.5° N and S are shown by dashes. All of the area of the northern hemisphere continents shown in figure B may experience snowfall during the winter

averaged about $37.5 \times 10^6$ km$^2$. Since 1980 it has averaged about $35.5 \times 10^6$ km$^2$. It is relatively thin, and because most of the snow is air, its water content is small. However, it is important to realize that almost all of Earth's major rivers have their origins in snow-covered mountains and through spring, summer, and fall are fed by meltwater.

Table 17.5 also shows the amount of energy required to melt the major ice masses. This includes both the energy required to warm the ice to the melting point, and the latent heat of fusion (melting) to transform the ice to water. A glance at the table shows that of the land-based ice, glaciers and small ice sheets are by far the most vulnerable, and

**Table 17.5** Major bodies of ice on Earth

| Region | Area ($10^6$ km$^2$) | Ice thickness (km) | Volume (ice) ($10^6$ km$^3$) | Mass (ice) ($10^{18}$ kg) | Energy to melt ($10^{21}$ J) |
|---|---|---|---|---|---|
| Ice grounded on land | | | | | |
| Antarctica | 13.58 | 2.45 | 29.32 | 27.18 | 8969.4 |
| Greenland | 1.73 | 1.52 | 2.90 | 2.69 | 887.0 |
| Glaciers and ice caps | 0.76 | 0.236 | 0.18 | 0.17 | 56.1 |
| Total grounded ice on land | 16.07 | 1.402 | 32.40 | 30.04 | 9912.5 |
| Floating ice (sea-ice and ice shelves) | | | | | |
| Arctic ocean winter | 15.0 | 0.003 | 0.045 | 0.041 | 13.53 |
| Arctic ocean summer | 7.0 | 0.003 | 0.021 | 0.019 | 6.37 |
| Arctic ocean seasonal difference | 8.0 | 0.000 | 0.024 | 0.022 | 7.15 |
| Southern ocean winter | 18.0 | 0.001 | 0.018 | 0.017 | 5.61 |
| Southern ocean summer | 3.0 | 0.001 | 0.003 | 0.003 | 0.99 |
| Southern ocean seasonal difference | 15.0 | 0.001 | 0.015 | 0.014 | 4.62 |
| Antarctic ice shelves | 1.6 | 0.475 | 0.732 | 0.671 | 221.43 |
| Total floating ice (winter) | 34.6 | | 0.795 | 0.729 | 240.57 |
| Total polar areas covered by ice in winter | | | | | |
| Arctic | 16.73 | | 2.95 | 2.73 | 900.57 |
| Antarctic | 33.18 | | 30.07 | 27.87 | 9196.44 |

Data are for conditions prior to 1980. Note that because of the lag time involved in melting or freezing ice, in the Arctic, the 'winter' maximum for sea-ice actually occurs in April and the 'summer' minimum in October. In the Antarctic the 'winter' maximum for sea-ice occurs in October and the 'summer' minimum in April

this is borne out by their recent history. The past few decades have seen retreat of mountain glaciers in all parts of the planet. The second most sensitive component is the Arctic sea-ice, and it too has shown dramatic and accelerating changes in recent decades. The amount of energy required to melt the Greenland ice sheet is two orders of magnitude larger. It may have largely disappeared during some earlier interglacials. The amount of energy required to melt the Antarctic ice sheets is yet another order of magnitude larger, and although it has certainly varied in size, it has probably not melted completely in the last 35 million years.

## 17.11    How Ice Forms from Fresh Water and from Sea water

Ice can directly form from freezing the cold surface of a fresh-water lake. The ice forms thin layers of superposed sheets of hexagonal HOH groupings. The successive sheets are offset from one another and held together by the hydrogen bonding forces. A view showing two superposed sheets is shown in Fig. 17.12. These layers can slip past one another, so that the ice is flexible, as all skaters know. If the freezing occurs on a very cold windless night, very large transparent sheets of ice can form. With wind and waves, air

is incorporated in the ice and it becomes gray or white. This is what we usually see. Ice serves as an insulator between the cold air above and the cold water below. This insulating effect usually prevents the layer of ice from getting very thick. As the ice thickens, water can freeze onto the underside of the ice layer, but remember that when freezing takes place, latent heat is released, warming the water, so that it becomes increasingly difficult for the ice layer to thicken. Lake ice doesn't need to be very thick, only 15 cm (6 in.), to support the weight of a human, and a slightly lesser thickness can support a moving skater. The maximum thickness that lake ice can achieve is probably about 5 m, observed on some of the Dry Valley lakes of Antarctica near McMurdo Sound where the mean annual temperature is about −20 °C. In the Great Lakes of North America winter ice in sheltered bays ranges from 45 to 75 cm, but on the open lake surfaces ice floes can pile up to a thickness of 1 m or more.

Making ice on the surface of the ocean, sea-ice, is a very different matter. Above, we noted that if seawater has a salinity above 24.7 ‰, it density increases right down to the freezing point. In principle it should sink from the surface before it can freeze. But sea-ice does cover the surface of the ocean in the Polar Regions. How is this possible? The usual explanation is that snow falling on the surface of the sea forms tiny areas where the salinity is below 24.7 ‰. There these tiny puddles of fresher water of water can freeze, and once they do, the ice grows further by attracting water molecules. However, ice cannot incorporate more than about 7 ‰ salt in its crystal lattice. So as the ice crystals grow, the salt is concentrated in brine droplets. The brine droplets are heavier than the ice, and gradually work their way down through the ice and are expelled out the bottom.

This is a good description of what happens around the Antarctic, where ocean salinities are generally above 34 ‰. However, the situation in the Arctic is very different. A number of major rivers, Canada's Mackenzie, and Russia's Kolyma, Yenisei, Lena, Ob, Pechora, and Severnaya Dvina empty into the Arctic basin. Although they flow only during the summer, they carry more than 10 % of the Earth's total runoff from land to the ocean. The Arctic Ocean accounts for only about 2 % of the total area of the global ocean. This seasonally concentrated freshwater inflow into a restricted ocean basin greatly reduces the salinities in coastal areas, particularly the Siberian shelf, so that as summer ends it is much easier for sea-ice to form. The broad shelf off the Lena River has been called the 'ice-factory.' As the ice begins to form in the fall, the continuing river flow loads it with sediment. The pattern of currents in the Arctic carries the ice out to sea, and eventually across the Arctic ocean toward northern Greenland. Russian colleagues suspect the process of sea-ice formation may be even more complex. They suggest that some of the sea-ice may actually form within the upper layers of the Arctic Ocean rather than at the surface. The sea-ice coming from the Lena River ice factory contains a great deal of clay, and as some of it melts, these clay particles sink. If the sea water below the surface is near the freezing point, it may use these clay particles as nuclei to grow as ice. The ice particles then grow by attracting more water molecules, becoming larger and lighter, and drifting up to the surface.

In the middle of the last century the average thickness of the Arctic sea-ice was about 3 m, and this served to insulate surface sea-water with a temperature of about $-1.8$ °C from overlying winter air temperatures of $-40$ °C. Over the last decades the situation has changed drastically, average sea-ice thicknesses are less than 2 m, and toward the end of the Arctic summer in September increasingly large areas are ice -free. The perennial sea-ice cover of the Arctic Ocean has begun to disappear. In 2007, at the end of the Arctic summer in September, the Arctic sea-ice cover had declined to just over 4 million km$^2$, only 1/3 of the ocean area. At the same time in 2008 the area of ice cover was slightly larger, but both of the 'Northwest Passages' from the Atlantic to the Pacific, that north of Siberia and that through the Canadian Archipelago, were open for the first time in recorded history. To me, these unexpected changes in Arctic sea-ice cover are the most serious effects of global warming. The future would

be more predictable if ice-formation in the Arctic followed simple rules. The details and implications of these drastic changes will be discussed in detail in Chap. 32.

Incidentally, you will sometimes hear that the area of sea-ice around the Antarctic is getting larger (that's true), and therefore the climate must be getting colder. Think about that for a moment. Ocean salinities around the Antarctic are about 34 ‰ or higher, so if it gets colder that water should just sink below the surface as it approaches the freezing point (Fig. 17.9). To make sea-ice, the surficial water must become less dense, and the only way that can happen is for it to become less salty. That means it must be freshened, and the only source of fresh water for the Southern Ocean in winter is snow. But the amount of snow that can fall depends on the temperature of the atmosphere and the waters further north. Even near freezing the amount of water vapor the air can hold doubles with every 10 °C increase in temperature. If both the atmosphere and ocean in the subtropics warm, more snow can fall further south, and more sea-ice will form. The growth of the area of sea-ice that forms during the Antarctic winter means that the thin surface layer of the Southern Ocean is getting fresher from more snowfall, and this is happening because the southern hemisphere oceans and atmosphere are warming.

In 2008, the rate of sea-ice formation on the Arctic Ocean during the fall increased dramatically over earlier years. Does this mean that the Arctic was getting colder than usual? No, it means that the surface waters of the Arctic Ocean are getting fresher. In fact, the summer inflow from rivers into the Arctic has been increasing in recent years, diluting the coastal waters and increasing the size of the 'ice-factory.' And the decline in summer sea-ice area means that as winter comes on, there are still large areas of open water that can provide moisture to the atmosphere which can fall as snow on the more saline offshore waters.

## 17.12  Snow and Ice on Land

The amount of energy released in converting vapor to snow is $2583 \times 10^3$ J/kg. That is an enormous amount of heat being returned to the atmosphere. A major snowstorm requires both large amounts of moisture in the air and cold temperatures. But, as we will discuss in greater length in the next chapter, cold air does not contain much vapor. A major snowfall occurs when warm moisture-laden air meets cold air. The cold air is denser and flows underneath the warm air lifting it up and cooling it. And that is why, when there is a major snowfall, there is so little change in the temperature of the air perceived by someone on the ground.

Although snow is one of the most important aspects of the climate system, we know very little about it. We know the areas of snow cover from satellite observations, but there is little information on snow depths, the equivalent water content, and what happens to the water. Measuring snow depth is a notoriously complex exercise. All you have to do is look outside after a snowstorm. Unless it was very calm, the wind has blown the snow into drifts and any average the thickness is very difficult to determine. Snowfall in the mountains is highly variable. I may have boot-deep snow at my house, and there will be none in the town a few miles away.

Also, snow doesn't necessarily melt to make water. It can simply evaporate directly back into the air. This phenomenon, known as sublimation, occurs in very dramatic fashion north of the Alps. In the winter, when the lowlands of Austria, Germany, Switzerland and France north and west of the Alps are covered with snow, a warm dry wind may blow from warm Mediterranean Coast over the Alps. The wind is called the 'Föhn' and the snow simply evaporates. It also dries out everything else, including humans. People become irritable and headaches are common. In Zürich, hospitals carry out only the most necessary operations. Incidentally, in German an electric hair dryer is called a "Föhn." In Colorado the eastern side of the Rockies experiences the same phenomenon, when the strong westerly winds, the 'Chinooks,' come over the Front Range. Since so many of our rivers depend on snow in mountain source regions for their summer flow, more information about this critical aspect of the hydrologic system is sorely needed.

Fresh snow is mostly air. The actual snowflakes make up only about 30 to 40 % of the volume of the snow. The density of the snow is only about 300 kg/m$^3$, a third that of water. Colorado is famous for its 'powder snow' which has one of the highest proportions of air. As the snow builds up, the weight of the upper layers begins to compress the deeper layers. However, up to a depth of 10 m, the air is able to circulate within the snow, and to equilibrate with changes in the atmosphere. The circulation is forced by winds over the snow.

By the time the snow has built up to a depth of 10–30 m, it makes up 60–65 % of the volume; the amount of air has dropped from 60–70 to 35–40 %. The density has increased to about 500–600 kg/m$^3$. The excess air has escaped by migrating up and out. The compacting 'snow' takes on a granular texture, like sugar, and is called 'firn.' The original snowflakes have fused together as larger ice crystals. At depths of 75–85 m the porosity, the space occupied by air, has been reduced to 10–15 % and the air is concentrated in 'bubbles.' The density has increased to about 800 kg/m$^3$. At these depths the bubbles of air become closed off, and air can no longer escape toward the surface. As more and more snow falls and the ice thickens, the air bubbles become compressed under the pressure of the overlying load. These bubbles are a sample of the atmosphere when the snow accumulated, but because of the mixing of the air that can take place in upper layers of snow the sample represents an average over several years. In central Greenland, the air samples contained in the bubbles typically represent a 6-year average; in the Vostok core near the South Pole the bubbles represent an average closer to 40 years. As more and more snow is loaded on and the pressure on the underlying ice increases, the ice becomes increasingly transparent.

The air bubbles come under greater and greater pressure as they are buried. Finally, near 1 km burial depth, the air molecules begin to escape into the ice. However, of the air molecules, only the rare gas helium is able to fit into the crystals of ice Ih. For the larger air molecules, the ice must recrystallize to form a cage around them. These cages of HOH molecules surrounding a gas molecule are called 'clathrates.' The HOH molecules rearrange themselves to form cubic, octahedral more complicated cage structures. Some are so complex they are called 'Buckyballs' after Buckminster Fuller, because they resemble the structure of his geodesic domes.

The air trapped in the ice has been used to reconstruct the past history of the atmosphere, with the records from Greenland extending back 250,000 years. In Antarctica the EPICA dome core extends back 800,000 years and there is a possibility that records back to 1.5 million years can be found.

Once the ice thickness is over about 10 m it begins to deform and flow. In a mountain glacier, the flow is downhill. However, in the case of an ice sheet, where the topographic relief is small compared to the area covered by the ice, the flow is out from the center toward the periphery. These differences are shown schematically in Fig. 17.14. The rate of flow depends on the thickness of the ice and its temperature. It flows by deformation of the bonds within and between the molecules of HOH. Slipping occurs along the plane of the hexagonal sheets and though rearrangement of the crystals. If the temperature is not very cold and pressure conditions are right, ice can melt and refreeze. This often happens at the base of mountain glaciers outside the polar regions. There, when the ice encounters obstacles it cannot move, it may melt on the uphill side and refreeze on the downhill side. Even as it flows, however, ice remains brittle, and where the gradient steepens it may break and pull apart forming crevasses. If the base of the ice sheet is warm, approaching the melting point, the ice can move over the substrate melting and freezing, and plucking pieces of rock from it. If the base is cold, it can become solidly frozen to the base and immobile.

What determines whether an ice sheet has a warm or cold base? The original temperature of the snow and ice formed from it can be very cold; in East Antarctica the mean annual

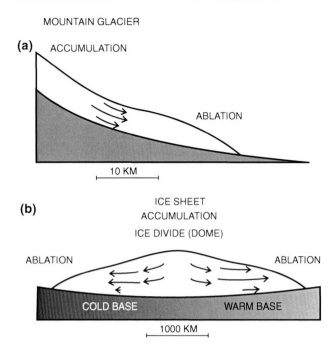

MOUNTAIN GLACIER

**(a)**  ACCUMULATION

ABLATION

10 KM

**(b)**  ICE SHEET
ACCUMULATION
ICE DIVIDE (DOME)

ABLATION                                                ABLATION

COLD BASE          WARM BASE

1000 KM

**Fig. 17.14** Comparison of a mountain glacier (**a**) and an ice sheet (**b**). Note the difference in scale between the two diagrams. *Arrows* indicate relative speed of motion at different levels within the ice. In the accumulation regions, the snowfall and buildup of ice exceeds melting; in the ablation region melting exceeds the supply of ice

temperatures are about −60 °C. There is no possibility that ice at these temperatures can melt. However, the solid Earth beneath the ice sheet is heated from below, by the 'geothermal' flux of heat from deeper within the Earth. Although this geothermal heat flux only averages about 50 mW/m², it gradually warms the base of the ice, and heat is conducted very slowly upward. If the ice at the base of the sheet is stagnant, it will eventually be heated to the melting point. If it is on a slope, it will begin to move downhill. However, in Antarctica there are many basins beneath the ice, and the meltwater stays in place as a subglacial lake. Where the warmed ice moves downhill, it will be replaced by colder ice from above. This may result in alternating pulses of motion followed by refreezing and stagnation. The time scale for these pulses and freeze-ups depends on the geothermal heat flux, which varies from place to place. It is generally on the order of thousands of years.

Recently a new phenomenon has been observed. As temperatures over the ice warm, the area of melting increases, and the water works its way down into the interior and finally to the base of the ice. There it acts as a lubricant, and allows the glacier or ice stream to flow much more rapidly. This phenomenon, first observed in some 'tidewater' glaciers that end in the sea, has now become more widespread. It is now known to be a very active process in Greenland and in West Antarctica, speeding up the flow of ice to the ocean.

The only other common solid on Earth that undergoes plastic deformation is rock salt, sodium chloride (NaCl). There are extensive deposits of rock salt in the Mediterranean Basin, Red Sea Basin, and beneath the US Gulf Coast and Gulf of Mexico. Along the American continental margin and deep Gulf of Mexico the salt is of the order of a kilometer or more thick; it was extracted from the ocean about 160 million years ago. The salt is less dense than the overlying sediment. As it has been loaded by younger sediment deposited on top of it, and heated by the geothermal flux from below, it has begun to flow. Under the Gulf Coastal Plain, the salt wells up through the overlying strata as 'salt domes.' Although dissolution by groundwater prevents the salt from reaching the surface, the salt domes are often represented by a low flat-topped hill. Perhaps the most famous of these is Avery Island, a hill surrounded by swampland in southern Louisiana. It is on this hill that a very special hot sauce, Tabasco, is manufactured. As far as I know, Tabasco is the most ubiquitous American export. Tabasco sauce can be found in grocery stores in almost every country on the planet. *THANKS FOR THE FLASH — GORDON !!!*

Along the continental margin the loading of the salt by sediment is forcing it to flow forward into the deep Gulf of Mexico, much like toothpaste being squeezed out of a tube. The effect of this flow can be recognized in a deep, steep cliff, the Sigsbee Scarp along the base of the continental margin off Texas and Louisiana.

But back to ice—in terms of area, most of the northern hemisphere ice is sea-ice, as shown in Table 17.5. The largest northern land-based ice sheet is on Greenland, but it is only about 1/10th the size of that covering the Antarctic. If the ice on Greenland were to melt, we would see a large island with a central lowland surrounded on its margins by hills and, especially in the south, mountains.

In the south, the Antarctic continent is covered by what appears to be a continuous ice sheet (Fig. 17.15). However, that view is deceptive. West Antarctica, the part of the continent that lies south of the Pacific Ocean, is mountainous and has a number of ice centers and ice-streams flowing slowly to the ocean. Its ice sheet contains about 10 % of the total ice on Antarctica. East Antarctica, which lies south of the Atlantic and Indian Oceans and Australia, is a continental plateau with some large upland regions. Although it has about 90 % of Antarctica's ice, it has only a few large ice centers, called 'domes' and a few very large ice streams leading to the ocean.

If we were to remove the ice, and restore the continent to its original configuration before it was covered with ice, we might not recognize it (Fig. 17.16). We would see a low continental area, East Antarctica, bordered on the west by a mountain range, the Transantarctic Mountains. They are bordered on the west by a shallow sea, the Ross Embayment,

**Fig. 17.15**  A contour map of the ice-covered Antarctic continent. Contours are in meters. The major ice divides are shown as *dashed lines*. The seaward edges of the ice shelves are indicated by *dotted lines*. The South Pole is marked by a *circle enclosing the letter S*

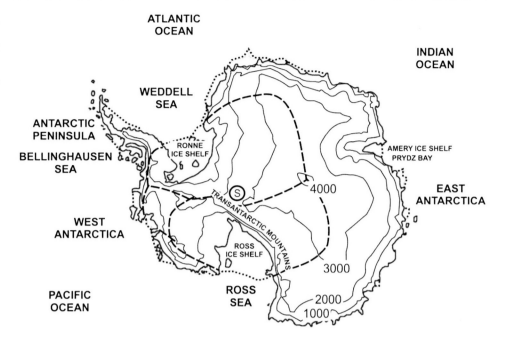

**Fig. 17.16**  Map of the surface of Antarctica beneath the ice. Areas below sea level are shown in shades of *blue*; areas above are in *yellow* and *orange*. Image produced by BEDMAP, British Antarctic Survey

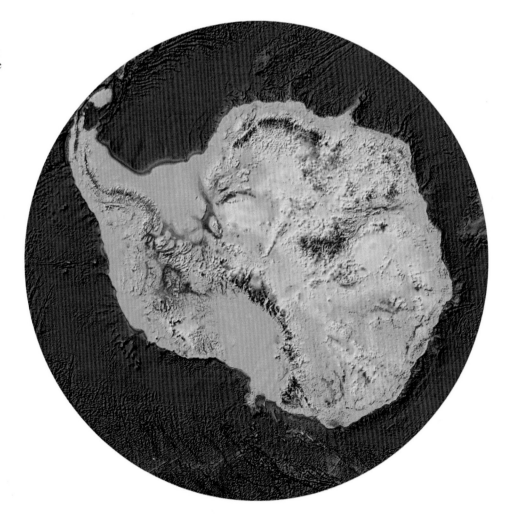

and lowlands. West Antarctica is a mountainous archipelago, with many islands. The most peculiar thing about Antarctica is its overall shape. Continents are made of granitic rock that is less dense than the basalt of the ocean floor. They resemble corks floating on water. Most continents have the general shape of a low dome rising sharply out of the ocean. The steep falloffs into ocean depths are the continental slopes. Today water laps onto the edges of the dome, forming the continental shelves. The peripheral areas not covered by water are the coastal plains The boundary between the gentle seaward slope of the continental shelves and the steeper slope into the ocean depths is the shelf-break. Today the shelf-breaks, the outer edges of the continental shelves, commonly have a depth of 200 m. During the Last Glacial maximum, when sea level was about 130 m lower than today, the outer edges of the Antarctic continent were just awash. How can a 130 m sea-level fall result in a 200 m deep edge of the continent being awash? There is a phenomenon called 'isostatic adjustment' that takes place. Beneath the solid continental blocks and ocean crust the rock becomes less solid and behaves like thick molasses, flowing sideways as the block is loaded or unloaded. So as sea level goes down reducing the load on the edge of the continental block, the edge rises. More about this in Chap. 27.

Antarctica is unique because unlike other continents its upper surface is saucer-shaped. The ice sheet has the shape of a giant lens, thicker in the middle (about 3.5 km) and thinner at the edges (1–0.5 km). The weight of the thicker ice in the center of the continent has depressed the rock surface, creating the bowl shape. Today the edges of the ice sheet do not extend to the shelf break, leaving the fringe of the continent ice-free. Beyond the edge of the ice sheet, the continental shelves slope upward from the ice-covered area to their outer edges. The shelf breaks around Antarctica are often about 200 m deep, but water depths at the edge of the ice sheet can be 600 m or more. During the Last Glacial Maximum the ice covered the continental block completely, extending out to the edge of the continental shelf.

The Antarctic and Greenland ice sheets differ not only in size, but in their nature. These are large areas of the Antarctic ice sheet where the ice rests on rock that is well below sea-level. Remember that the density of ice is about 920 kg/m$^3$ while that of the seawater is about 1027 kg/m$^3$. You will recall that you have often heard that 9/10ths of an iceberg is below water. That is because of this density difference. This also means that as long as the ice sheet is at least 10 % thicker than the water depth would be, it will rest on solid rock. However, if the ice sheet is thinner, it will float above the sea floor. This happens where the glacial ice moves seaward and becomes detached from the deepening rock floor. Such thick floating ice is well developed in several places around the Antarctic continent as 'ice shelves.' The largest of these are the Ronne-Filchner Ice Shelf in the Weddell Sea and the Ross Ice Shelf in the Ross Sea.

The ice shelves are glacial ice streams that have advanced into the deeper parts of the Antarctic continental shelves. The water is deep enough so that they float above the bottom, but are held in place by their sheer mass, and by islands and submarine rises. They are eroded from beneath by the relatively warmer ocean waters. Their outer edges are usually 200–300 m thick and the inner margins can be more than a kilometer thick. The moving ice may be forced over shallow areas of the sea floor forming humps on the surface of the ice shelves known as 'ice rises.' Being more or less locked in place, the ice-shelves block the flow of ice from the interior of the continent. However, in recent years, large pieces have broken off and drifted into the Southern Ocean. These losses are permitting grounded ice from the continent to flow more rapidly toward the sea.

## 17.13    Ice Cores

Geologists had been using the sequences of fossils in stratified sediments to interpret Earth history since beginning of the 19th century. Originally, the stratigraphic sections that provided this history were all on land. After WWII, cores taken in the deep sea added new vistas to this history.

The idea that the ice caps of Greenland and Antarctica might have similar stratigraphic records began with Ernst Sorge, a member of one of Alfred Wegener's expeditions to central Greenland. During the winter of 1930–31 Sorge dug a pit 15 m deep near his snow cave, at Station Eismitte (Eismitte = middle of the ice) and realized there was a stratigraphy and a history to be found in the ice.

After WWII, Henri Bader, who was then Research Professor of Mineralogy at Rutgers University, New Jersey, wrote a report on the state of polar latitude research in Europe. Bader had been born in Zürich, Switzerland, and was a student of Paul Niggli, the famous mineralogist at the Swiss Federal Institute of Technology. Bader's favorite mineral was ice. He had emigrated to the US before WWII, and became involved in the study of permafrost and ice in Alaska. After the war, the US Government realized that knowledge of the Polar Regions might be important in the future and created the Snow, Ice, and Permafrost Research Establishment (SIPRE) as part of the U.S. Army Corps of Engineers. Bader had some experience in coring ice in Alaska, and he convinced the Army and the US National Academy of Sciences that a lot could be learned if only we could drill cores of ice. A Third International Polar Year had been planned for 1957–58, and it was decided that this would become a larger program of global exploration, the International Geophysical Year. The US component was to include ice core drilling in both Greenland and Antarctica.

Coring is not as simple as it might sound. The ice must never be allowed to melt; conditions in the hole must always be kept cold. Worse, ice under pressure flows, as we have discussed above. This means that when you pause in the drilling procedure, the ice will try to close the hole. It has taken decades to for us to become proficient at coring ice. It was necessary to develop special drill bits and coring apparatus, and to find fluids that could keep the hole open and yet not freeze. Great precautions must be taken against contamination. It was discovered that ice recovered from great depths, where it had been under high pressure, tended to expand when brought to the surface; if it was disturbed it might suddenly crack into many fragments and become useless for study. Now the cores from great depths are allowed to 'rest' for up to a year after they are recovered so that they can slowly, gently expand. Like making a fine wine!

Selecting the sites for coring takes care. First, sites must be located where the temperature has always been well below freezing. If the ice were to melt, the record would by distorted or destroyed. The air in the bubbles in the ice would escape, some of it even becoming dissolved in the meltwater. Secondly, because at depth the ice deforms and flows, it is necessary to select sites where this effect can be minimized. If you look at Fig. 17.14b you will realize that the best place to recover a core with minimal disturbance is on top of an ice divide or dome. This is going to be in the center of the ice sheet, and, of course, the logistically most difficult place to reach and bring in the drilling equipment. However, there the snow compacts to ice in a straightforward manner and shows little evidence of deformation with depth until near the base. Many of the ice cores have the word 'dome' attached to their names. Third, the details of climate change recorded in the ice will depend on the rate at which it accumulated. The Greenland ice cores have provided a remarkably detailed record of the last 120,000+ years, back into the last interglacial. Some of the Antarctic cores were taken at near-coastal sites with high accumulation rates, and have provided a detailed record of the last deglaciation and Holocene. Central Antarctic sites have much slower accumulation rates, but a record that extends back almost 1 million years. Finally, there is the problem of hauling in the drilling equipment, constructing an on-site core storage facility (usually excavated as a cave to ensure cold temperatures), and arranging for transport of the cores back to the scientific laboratories where they will be studied. This explains why these projects are carried out by governments, often assisted by the military which has the ability to carry out operations in such hostile environments.

It has taken much time and effort to understand what the ice cores can tell us. Unlike the rock record, the things in which we are interested can migrate in the ice cores. We already discussed the circulation of the air within the firn layer, before it becomes a solid ice mass. Direct, rapid circulation with the overlying atmosphere is generally limited to the first few meters, but below that the air is pressed out as the firn compacts. The lighter molecules are able to migrate upward more rapidly than the heavier ones so there is some differentiation of the gases that must be taken into account. By the time the bubbles are closed off the air in them is decades or hundreds of years older than the ice surrounding them.

Determining the age of the ice is not always easy. Sometimes there are distinct annual layers and these can be counted. Figure 17.17 shows a meter-long segment from a core from the GRIP2 site in central Greenland (GRIP = Greenland Ice core Project) (http://www.globalwarmingart.com/wiki/Image:GISP2_Ice_Core_jpg). The core is from a depth of 1837 m. The layering represents the seasonal changes in the snow, larger snowflakes in summer, smaller in winter. The larger flakes trap more air, and the tiny air bubbles give the different transparency to the ice. In this core the age has been determined by counting the annual layers, and this segment is about 16,250 years old.

The age of the air in the bubbles can be determined using the $^{14}C$ in the carbon dioxide or methane molecules. This works back to 50,000 years. Because it has migrated upward, the air in the bubbles is usually a few hundred to a few thousand years older than the surrounding ice. By knowing the relation between the abundance of bubbles and the age difference to the ice, the age of the bubbles in the older parts of the cores can be estimated.

While counting the annual layers works well in many of the Greenland ice cores because they accumulated at a relatively rapid rate in the order of 10 cm/yr, it is difficult in Antarctica where the accumulation rates have been much slower. Fortunately, there are layers of volcanic ash in the cores. Volcanic eruptions produce ash that has distinctive characteristics, like a fingerprint. If the eruptions can be identified, and their date is known, these layers can be used

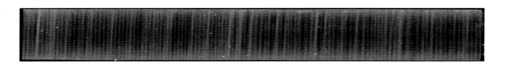

**Fig. 17.17** Segment from the GISP2 ice core, 1837 m depth, showing annual layering. GISP2 cores were taken at the highest point on the Greenland ice sheet (38° 28′ W, 72° 35′ N, 3208 m above sea level) on the "ice divide" of central Greenland

as age markers in the core. Between such markers the age of the ice can be estimated from knowledge of the way it compacts and deforms. Recently an elegant method for determining the ages in the long Antarctic cores and for correlation of the Antarctic and Greenland cores has been developed using the concentration of methane in the cores. It turns out that global methane production takes place in low latitudes, and reflects the intensity of the monsoons. The strength of the monsoons depends on the changes in insolation at low latitudes. The variations in methane content correlate very well with the changes in insolation at 30° N. These changes are recorded synchronously in the ice of both Polar Regions and have been very important in making global correlation of climatic events over the past 100,100 years.

The ice cores have produced an extraordinarily detailed history of much of the last million years of Earth history. The information comes from a wide variety of material incorporated into the ice, including aerosol sulfate and nitrate, dust, radioactive isotopes produced by interactions with energetic cosmic rays at the top of the atmosphere, and trace metals.

The foundation of much of the ice core research is the $^{18}O{:}^{16}O$ ratio found in the ice itself. The ratio varies with the isotopic composition and temperature of the ocean water where evaporation took place, and with the distance the water had to travel before it fell as snow. The pattern of the $^{18}O{:}^{16}O$ ratio in the ice cores closely resembles that recovered from benthic foraminifera in deep-sea cores, but in many cases offers much more detail than that in the deep-sea record. The ratio $^{1}H{:}^{2}H$ (Hydrogen/Deuterium) can be used to determine the temperature at which the snow condensed from the air. The varying concentration of dust in the cores is

an index to the changing areas of land where dust can be picked up by the wind as well as the strength of the winds. However, it is the record of $CO_2$ and other greenhouse gases trapped in the air bubbles that has attracted the most attention. The 450,000 year ice core record from Vostok, Antarctica shows a near-perfect correlation between $CO_2$ and temperature. The same close correlation is found in all of the other ice cores, and the consistency of the results from so many different places is quite remarkable.

In recent years ice cores have been taken on mountains, such as Kilimanjaro, in the Andes, Alps, and in Tibet. These records are generally much shorter than those from the polar regions, but offer new insights into the regional climate history.

## 17.14   Ice as Earth's Climate Stabilizer

A comparison of the solar energy received by the polar regions versus the Equator over the course of a year is shown in Fig. 17.18. During the polar summers the poles receive significantly more energy than the equatorial region. Regions poleward of the polar circles experience 6 months of darkness and 6 months of light. Although at the pole the Sun is at most 63° above the horizon, compared with almost directly overhead at the Equator, it does not set during the polar summer. The incessant sunshine results in a continuous energy input. If it were not for the ice, these would be the hottest places on the planet during their respective summers. Note that at present, the south-polar region receives more energy than the north. You will recall that this is because the Earth is closer to the Sun in January, the southern hemisphere summer, than in June.

**Fig. 17.18** Seasonal changes in insolation at the top of the atmosphere at the poles and equator, and annual averages for the Polar Regions

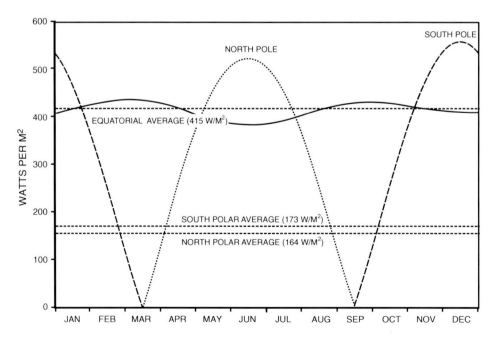

**Fig. 17.19** Detail of the boats in the painting by Giovanni Antonio Canal (Canaletto): The River Thames with S. Paul's Cathedral on Lord Mayor's Day (ca. 1747–1748). Courtesy of www.canalettogalley.org

The seasonal changes in sea ice cover are very important in stabilizing polar climates. In the north, when the Sun comes above the horizon at the time of the spring equinox, the entire polar region is covered by white snow and ice. These reflect the sunlight back into space, so that there is a long delay time, and the height of polar 'summer' is in September.

The ice shelves are glacial ice streams that have advanced into the deeper parts of the Antarctic continental shelves. The water is deep enough so that they float above the bottom, but are held in place by their sheer mass, and by islands and submarine rises. They are eroded from beneath by the relatively warmer ocean waters. Their outer edges are usually 200–300 m thick and the inner margins can be more than a kilometer thick. The moving ice may be forced over shallow areas of the sea floor forming humps on the surface of the ice shelves known as 'ice rises.' Being more or less locked in place, the ice-shelves block the flow of ice from the interior of the continent. However, in recent years, large pieces have broken off and drifted into the Southern Ocean. These losses are permitting grounded ice from the continent to flow more rapidly toward the sea.

## 17.15 Summary

Our planet differs from all other objects in our solar system in having both abundant HOH and a temperature very close to its phase triple point, so that it can exist abundantly in all three states, as a gas, vapor, as a liquid, water, and as a solid, ice. In its gaseous state it is the Earth's most important greenhouse gas; in its liquid state, water, the enormous heat capacity of the oceans serves to stabilize the planet against climate change; in its white solid state, it the most important control on Earth's albedo. The phase changes between these states involve huge amounts of energy. The conversion of water to vapor (latent heat of vaporization) and vapor to water (latent heat of condensation) is the major means of heat transport for the atmosphere. The strange condition that water is dark and absorbs incoming solar radiation while snow is white and reflects it is the most powerful positive feedback in the Earth climate system. The abundance of HOH makes our planet uniquely suited for the development of diverse life, but one more thing is needed—protection from the Sun's life-destroying ultraviolet radiation—the topic of Chap. 19.

A Timeline for this chapter:

| 1724 | Daniel Gabriel Fahrenheit proposes a standard temperature scale |
| 1742 | Anders Celsius proposes another, simpler, temperature scale |
| 1762 | Joseph Black coins the terms 'latent heat' and 'specific heat' in describing the phase transformations of water in a talk to the Glasgow University Philosophical Club |
| 1775 | Black's former instrument maker, James Watt, makes a huge success with his business making a steam engine that is efficient and safe—and launches the Industrial Revolution |
| 1805 | Joseph-Louis Gay-Lussac and Alexander von Humboldt repot that water is made of two volumes of hydrogen to one volume of oxygen |
| 1808 | In his *New System of Chemical Philosophy* John Dalton states that water is made of hydrogen and oxygen, and gives its formula as HO. |
| 1809 | Joseph-Louis Gay-Lussac publishes his classic paper *Sur la combinaison des substances gazeuse, les unes avec les autres* showing that it takes two volumes of hydrogen to react with one volume of oxygen to form water, so its formula must be $H_2O$. |
| 1930–31 | Ernst Sorge digs a pit 15 m deep near his snow cave, at Station Eismitte and realizes there is a history to be found in the ice. |
| 1957–58 | US scientists begin ice core drilling in both Greenland and Antarctica as part of the International Polar Year |
| 1960 | I first learn about the strange properties of seawater from Jack Hough at the University of Illinois |
| 1980 | Frank Millero and colleagues publish the classic *International one-atmosphere equation of state for sea water* |
| 1988–93 | The GISP2 ice core is drilled in Greenland |
| ca 2002 | Charles D.H. Williams (Professor of Physics, Exeter University, UK) publishes 'The Science of Boiling an Egg' on the Internet |

If you want to know more:

Discussions of the properties of water are scattered throughout the scientific literature, but I know of no comprehensive source on the whole topic. Somebody needs to write a good book about the properties of water. Perhaps, if you have a year or so to spare you might wish to do this.

Music: George Frideric Handel's Water Music. Composed in response to King George I's request for a concert to be played for his amusement on the Royal Barge on the River Thames (Fig. 17.19). It premiered of on 17 July 1717 and became a favorite to be performed on Lord Mayor's Day (now the Lord Mayor's Show, usually in mid-November of each year).

Libation: A chilled London Gin, such as Beefeaters or Tanqueray, clear as the purest water; taken straight, or, as I prefer, in the American style with an olive or two and a dash of white vermouth, as a 'Martini.'

## Intermezzo XVII. Internationalization of the DSDP

The Deep Sea Drilling Project had always had an international character. The shipboard science party consisted of ten to twelve specialists in different fields, and we always wanted to have the greatest expertise possible aboard. The members of the party were responsible for describing and making preliminary examinations of the cores. This involved studies of the sediments, their contained fossils, and even the chemistry of the pore waters.

Jerry Winterer of Scripps perhaps best described the experience of participation as 'attending the great traveling seminar.' In the laboratories and at mealtimes we discussed all sorts of topics. It was an incredible learning experience for all those who sailed on one or more of those 2 month Legs aboard the *GLOMAR Challenger*. The participants developed a camaraderie that extended across the old institutional barriers.

I was Chair of the Planning Committee from 1972 to 1974 when the plans for including international partners were being developed. It was also decided that we would expand the US membership, including Universities that had major oceanographic programs. Art Maxwell had become Head of the Institute of Geophysics at the University of Texas, and they had at-sea seismic operations underway, so they were to be included too.

Until now the job of the Chair of the Planning Committee occupied about a quarter to half of a scientist's time. It was not funded from the DSDP, but contributed by the institution that had the honor of having its member serve in that position. During the planning for the expansion and internationalization of JOIDES the job became full time plus. Instead of preparing five or ten copies of reports it was now necessary to make more like fifty copies so they could be sent to the prospective participants to keep them informed. Previously we had reported to only one funding agency, now we were going to have six. Little mistakes in the paperwork could cause confusion and a lot of trouble. About half way through my term I explained to my colleagues that things were getting out of hand, and at the inception of the new program we would need to have some support for the Planning Committee Chair. We also needed a more efficient way of communicating, not just among all the new members of the program but with all of our Panel members, past participants in the shipboard science parties and to aid in recruiting prospective future participants. I think I put a stack of reports and correspondence on the table to show what we were up against. We now had an observer from the National Science Foundation attending our meetings, and he too was impressed.

The result was the establishment of a JOIDES Office with a science coordinator and a secretary. The Office would move with the Planning Committee Chairmanship, and there would be a 'newsletter.' It immediately morphed into a little magazine, the *JOIDES Journal*. These ideas were presented to the Executive Committee, which immediately approved them. Our glorious days of cozy informality in planning and running what had become the most respected program of geological exploration of planet Earth were over.

The financial plan for internationalization was for the United States' National Science Foundation to provide the largest share, about 2/3 of the funding, thereby retaining ultimate control. The five other participating countries, the USSR, Germany, France, Britain, and Japan each provided equal sums, initially $ 1 million each. One institution in each country became a member of JOIDES. At the same time, the appropriate departments or institutions at five more US Universities, Hawaii, Oregon, Texas, Texas A & M and Rhode Island became members of JOIDES. All at once, the simple informal discussions with six old friends sitting around the table became a very formal operation with sixteen participants. Bruce Malfait, program officer at NSF once described it as like herding cats.

I was still Chair for the first meeting of the expanded Planning Committee. It was held at Scripps in La Jolla. This gave the new participants the opportunity of seeing the DSDP onshore facilities and to meet with the staff. It was largely an organizational meeting, and one of our first tasks was to find a name for the new program.

A few weeks earlier I had broken my ankle on a visit to London, and had a cast on my leg. A few days before the meeting Terry Edgar, the resident DSDP chief scientist had an accident while surfing, and one of his legs was wrapped in bandages. Both of us were on crutches. The two of us sitting at the head of the table with our legs propped up and our crutches made quite an impression on our colleagues. It was suggested that somehow the moment should be commemorated in the new name for the project. 'UniPOD' was suggested, for 'United Program of Ocean Drilling.' POD, of course, is Greek for foot, and the projects one-legged leaders would thus be immortalized. But we settled on the simpler 'IPOD' which stood for 'International Program of Ocean Drilling' but would

also recall our condition at this meeting. This is how the program was known until its end in 1983. To smooth the funding procedure inside the National Science Foundation, the name officially became the 'International Phase of Ocean Drilling of the Deep Sea Drilling Program.' But to everyone for the next decade, it was simply IPOD. We should have copyrighted the name.

Another of our tasks was to set up topical advisory panels. We identified the scientific topics of interest in a new way and established twelve panels to cover them, with a total membership of about 125. We also established a panel to provide liaison with industry.

All of those first meetings with expanded membership were truly exciting. I was still Chair of the Planning Committee at the first meeting of the new JOIDES Executive Committee. When it had been four, then five people the Executive Committee, the meetings were like a continuing soap opera. The Planning Committee Chairman was there to present the Committee's reports and requests for approval. Since only one member of the original Executive Committee, Maurice Ewing, was a geologist, much of the technical discussion was a dialog. But then the Planning Committee business would be finished, and the Executive Committee could get on to arguing about other things, such as who had the best plans for expeditions, the best faculty, who would carry out special studies, etc. These often became acrimonious, and rarely was anything in those regards decided. But the arguments were always carried over to the next meeting. Remember that two of the original four members of the Executive Committee, Walton Smith of Miami, and Maurice Ewing of Lamont were Founders of their institutions. Bill Nierenberg of Scripps and Paul Fye of Woods Hole and later Maurice Rattray of Washington were simply Directors. As far as the Founders were concerned, here was a big distinction between being a Founder and a Director. And Bill Nierenberg would always remind us that he had been involved in the Manhattan Project. Five cats in one room.

At the first meeting of the new expanded Executive Committee I made the presentations for the Planning Committee. But as had been the case in previous Executive Committee meetings, the soap opera began where it had left off last time: in the middle of some abstruse discussion of who should get the credit for what. Speaking German and French was a valuable asset to help in discreetly bringing those representatives up to date on what was going on. But I was worried about Soviet representative Andrei Monin, Director of the Shirshov Institute of Oceanology in

Moscow. He was accompanied by an assistant, Igor Mikhaltsev, who was more proficient in English. But proficiency in English was not much help in understanding the ongoing arguments among the original American members of the Executive Committee. Their disputes had more to do with inter-institutional politics than with ocean drilling. I sat next to Andrei and Igor, and quietly explained to them what was going on, and what the discussions/arguments were really about. It was something they greatly appreciated, and never forgot.

Incidentally, for history buffs, all issues of the JOIDES Journal are available at http://www.odplegacy.org/program_admin/joides_journal.html.

EARTH'S ATMOSPHERE IS STRATIFIED

*Science never solves a problem without creating ten more.*
George Bernard Shaw

The atmosphere is the gas surrounding our planet and held in place by gravitational attraction. The ancients were quite aware that the atmosphere was something tangible even if invisible. There were, after all, winds, sometimes warm, sometimes cold. Galileo Galilei is commonly credited with the invention of the thermometer, then known as a thermoscope, in 1607. Although temperature measurements were made, it would be over a century before standard temperature scales were developed to allow the measurements to be compared. The development of thermometers is discussed in detail in Chap. 30; it is an important topic in the documentation of global warming.

Systematic investigation of the atmosphere began with Evangilista Torricelli's invention of the mercury barometer in 1643. This was soon followed up, in 1648, by Blaise Pascal's discovery that atmospheric pressure decreases with altitude. He concluded that there must be a vacuum above the atmosphere.

Air is a compressible fluid, and its density at any given altitude depends on the weight of the total amount of air above it, i.e., the pressure on it, and its temperature. The density/pressure relation follows the relation between volume and pressure formulated by Robert Boyle in 1662 and known as 'Boyle's Law.' The change of temperature with decreasing pressure follows 'Charles' Law.' Jacques Charles discovered the relationship in the 1780s, but the first publication of it was by Joseph Louis Gay-Lussac in 1802, who graciously noted that the discovery should be credited to Charles. The law was independently discovered by British natural philosopher John Dalton by 1801. But in fact the basic principles had already been described a century earlier by Guillaume Amontons (1663–1705). Priority of discoveries in science can be a messy business.

For a century there were no major new developments of our understanding of the atmosphere. Then, in 1900 a program of unmanned balloon soundings of the upper atmosphere by the French physicist Léon Philippe Teisserenc de Bort (1855–1913) revealed a large region with constant temperature. It was wholly unexpected, and he named it the 'stratosphere.' Significant further investigation of the atmosphere had to await the space age.

## 18.1 Atmospheric Pressure

The Earth's atmosphere is held in place by our planet's gravitational attraction. However, if an object is moving fast enough it can escape Earth's gravity and sail off into outer space. Earth's escape velocity is 11.2 km/s (about 7 miles/s) at the surface. It decreases gradually with altitude. Fortunately, the $N_2$ and $O_2$ molecules and Ar atoms of air do not travel that fast. But some very light gases that are in the atmosphere in trace amounts, $H_2$ and He, do achieve kinetic velocities of that magnitude. As they work their way to the top of the atmosphere, they can escape Earth's gravitational attraction and fly away to be lost to space.

In reality, there is no 'top' to the atmosphere. It just becomes thinner and thinner with altitude and some trace of it extends out to 10,000 km. However, for practical purposes, the Kármán Line, at an altitude of 100 km above sea level is considered the 'top' of the atmosphere. You may remember from Chap. 11 that the Kármán Line is the altitude at which an airplane must travel as fast as a satellite in orbit in order to stay aloft. The thinner the air, the faster the plane has to go to generate enough lift to stay up. At 100 km above sea level that speed is about 7.8 km/s (1,746 miles/h).

## 18.2 The Structure of the Atmosphere

Air changes its temperature as it is compressed or allowed to expand. Rising air cools with expansion because of the reduced pressure from the thinning overlying atmosphere. The air is not losing any energy by becoming cooler, but it is converting the energy of heat into another form, 'potential energy.' Work was involved in raising the gas from a lower to higher altitude against the force of gravity. The energy for that work came from the sensible heat of the air; that is the heat we can measure with a thermometer. The sensible heat was converted to potential (=gravitational) energy as the air rises. This can be converted back to sensible heat when the air descends.

Unlike the atmospheres of other planets, ours is distinctly layered. It is layered because the temperature of the air does not steadily decrease with altitude, as one might expect, but at two different levels it increases with altitude. This creates a situation where warm air, with a lower density, overlies colder air with a greater density. Under these circumstances vertical convection cannot occur, and the air is stratified. This phenomenon is called an inversion.

Temporary inversions can occur at the base of the atmosphere. In Colorado the cold dense air from the high mountains can flow down into the local depression where Denver is located, sliding underneath warmer air. The local inversion prevents convection, and the pollution from automobiles is trapped below the temperature inversion where it can accumulate to form the city's infamous 'brown cloud.' The same thing happens from time to time in Los Angeles, where the cooler air from the ocean slides in underneath the hot air coming from the deserts to the east, causing a trap for pollution (Figs. 18.1 and 18.2).

In Chap. 11 you encountered the 'US Standard Atmosphere'—a schematic representation of the change in temperature with altitude and the atmospheric layers it defines. Figure 18.3 shows temperature changes with altitude for three areas of our planet in more detail. The different areas are the tropics, the mid-latitudes, and the Polar Regions.

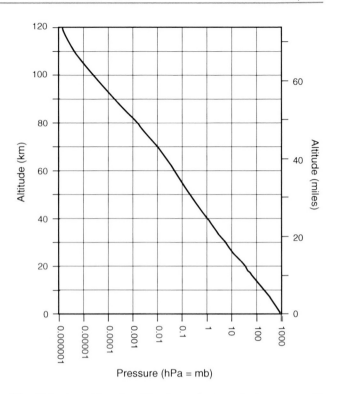

**Fig. 18.2**  A semi-log plot of the change in atmospheric pressure with altitude. The *line* is almost but not quite *perfectly straight* because of the effects of changing temperature with altitude. With sharp eyes you can see a little blip at the tropopause, about 18 km

There are two temperature profiles for the mid-latitude and Polar Regions, representing summer and winter. This diagram contains a wealth of information and you may want to spend some extra time exploring it.

**Fig. 18.1**  A NASA orbiter view of Earth's stratified atmosphere

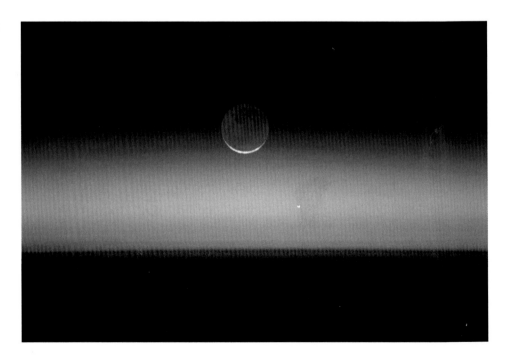

**Fig. 18.3** Changes in temperature with altitude for the topics (*solid line*), the mid-latitudes with summer represented by the *long dashed line* and winter by the *short dashed line*, and the Polar Regions with the summer represented by the *medium dashed line* and the winter by the *dotted line*. The names of the atmosphere's layers are shown opposite the appropriate parts of the temperature curves. Also shown are Mount Everest, and an airplane at the altitude flown by most long-distance jets, the heights reached by some instrumented helium-filled balloons, and the altitude where meteors burn up. A pressure of 1 Atmosphere = 1013 hPa (hPa = hectopascals = millibars)

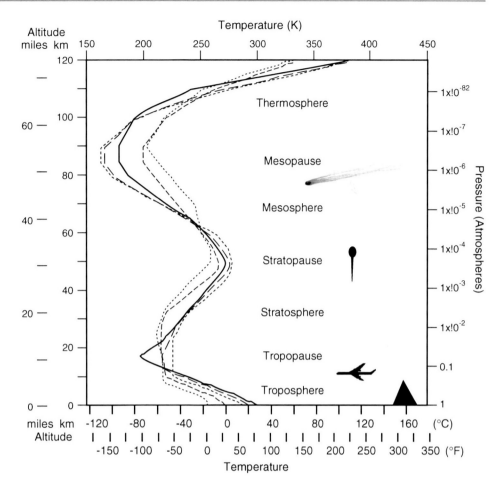

## 18.3   The Troposphere

Starting at the bottom of Fig. 18.3 you will see the troposphere, defined by the decline in temperature with altitude. If you have sharp eyes you may be able to see tiny changes in curvature of the lines at the very bottom of the diagram. These define the basal part of the troposphere which responds to the diurnal temperature changes from day to night. Over the oceans and lowlands the top of this layer is where the pressure equals 850 hPa, an elevation of about 1.5 km.

You can also see a change of the cooling upward trend at the base of the profile for the Polar Regions in winter. That is the polar inversion, which develops over Antarctica. It is a persistent feature, but rarely more than 100 m above the surface of the ice. Now follow the tropical temperature line. The temperature decreases steadily upward to about 18 km, and then sharply reverses into an increasing trend. That sharp reversal is the tropical tropopause. A similar sharp reversal of the temperature trend is not evident in the mid-latitude and Polar Regions. You will notice that the temperature decline ends at lower altitudes, about 9 km in

the Polar Regions. Away from the tropics the tropopause becomes progressively thicker, as a layer of isothermal temperatures almost 20 km thick in the Polar Regions.

## 18.4   The Stratosphere

The increase in temperature above the tropopause is caused by the capture of energy from the ultraviolet rays from the Sun. Thus is done by a very special greenhouse gas, ozone, the topic of the next chapter. The upward temperature increase makes it impossible for this layer of the atmosphere to convect. It is stratified. Hence its name, the stratosphere. It extends upward to about 50 km where the temperature trend gradually reverses forming the stratopause.

## 18.5   The Mesosphere

For the next 40 km the temperatures again decrease with altitude, forming the layer known as the mesosphere. The term mesosphere was proposed by Sydney Chapman in 1950 "for the layer between the top of the stratosphere and the major minimum of temperature existing somewhere below 100 km." This is where meteors and space junk falling out of orbit encounter enough resistance from the thin atmosphere to be heated to the point that they burn up.

## 18.6    The Thermosphere

The temperature trend gradually reverses again around 90 km, the mesopause. Above this the temperatures again increase with altitude. This is the thermosphere, which extends out to 640 km (400 miles). It too was proposed by Sydney Chapman in 1950 for "the layer of upward increasing temperature above… the mesosphere." The name 'thermosphere' derives from the Greek word for 'hot.' But for this region the word 'temperature' takes on another meaning from that at the surface of the Earth. A thermometer won't work at these heights because the energy imparted to it by the few colliding air molecules is less than the energy it radiates into space. A conventional thermometer would read a very low temperature. But for the outer parts of the atmosphere, including the thermosphere, 'temperature' refers to the total kinetic energy—translation, vibration and rotation—of the molecules. The temperature of the thermosphere varies greatly between day and night and with varying solar activity. It can reach up to 1,500 °C (=2700 °F).

Beyond the thermosphere is the 'exosphere' where air molecules become progressively more rare. Its top is usually considered to be about 10,000 km (6,200 miles) above Earth's surface, but in reality it varies greatly under the influence of the solar wind.

## 18.7    The Exosphere

The upper part of the atmosphere is very different from that below 100 km. It is held by the Earth's gravity but is very much under the influence of the stream of particles from the Sun known as the solar wind. Remember that the solar wind is a stream of charged particles, mostly electrons and protons, coming from the Sun's corona.

Its lower boundary, the exobase, is at the top of the thermopause, ranging through altitudes of about 500–1,000 km (310–620 miles) depending on solar activity. The base of the thermopause is about 250–500 km (160–310 miles). By analogy with the other layers, the thermopause is the region where the increase in temperature of the thermosphere ceases. It is effectively the region where the influence of solar radiation on the temperature of the

atmosphere begins. The exobase is typically defined as either the height above which there are negligible collisions between the particles or the height above which atoms or molecules are on purely ballistic trajectories—two ways of saying the same thing.

The major gases in Earth's exosphere are different from those of the lower atmosphere. They are the lightest gases, mainly hydrogen, some helium, and, near the exobase, atomic oxygen.

The upper boundary of the exosphere has been defined as the altitude at which the influence of solar radiation pressure on the velocities of hydrogen atoms exceeds that of the pull of Earth's gravity. In theory that would be 190,000 km (120,000 miles) above Earth's surface. That is half the way to the Moon. Satellites can see the exosphere from space as a geocorona extending out to at least 100,000 km (62,000 miles) from the surface of the Earth. The exosphere is commonly considered the transitional zone between Earth's atmosphere and interplanetary space. Table 18.1 summarizes the physical properties of the atmosphere.

## 18.8    The Magnetosphere

The term 'magnetosphere' was coined by British physicist Thomas Gold in 1959 as: The region above the ionosphere in which the magnetic field of the earth has a dominant control over the motions of gas and fast charged particles and is known to extend out to a distance of the order of 10 Earth radii; it may appropriately be called the 'magnetosphere.'

Incidentally, this is the same Tommy Gold who, to the consternation of geologists and geochemists, in the 1980s revived old ideas that natural gas and petroleum are abiogenic (not formed from biological materials) and came from the interior of the Earth, not from the decomposition of buried organic matter. As Cesare Emiliani would say, "Nobody's perfect."

We now know that Earth's magnetic field extends into space an average distance of about 20 times the planet's radius, about 128,000 km or 80,000 miles. However, because the solar wind acts like a flow streaming from the Sun, the magnetosphere is not spherical but bullet-shaped, extending out about 96,000 km (60,000 miles) on the side facing the Sun, and 160,000 km (100,000 miles) on the downstream side away from the Sun. It intercepts the charged particles of the solar wind and guides them in toward the planet. Nothing much happens until the charged particles encounter the molecules of the outermost atmosphere, the 'exosphere.'

**Table 18.1** Physical properties of the atmosphere

| Layer | $h$ | $T$ | | | $P$ | $\rho$ | $n$ | $v$ | $\mu$ |
|---|---|---|---|---|---|---|---|---|---|
| | km | K | °C | °F | Pa | kg m$^{-3}$ | m$^{-3}$ | s$^{-1}$ | |
| Tropo | 0 | 288.1 | 15.0 | 59.0 | $1.01 \times 10^{+05}$ | 1.23 | $2.55 \times 10^{+25}$ | $6.92 \times 10^{+09}$ | 28.8 |
| | 2 | 275.1 | 2.0 | 35.6 | $7.95 \times 10^{+04}$ | 1.01 | $2.09 \times 10^{+25}$ | $5.55 \times 10^{+09}$ | 28.8 |
| | 4 | 262.2 | −11.0 | 12.2 | $6.17 \times 10^{+04}$ | 0.819 | $1.70 \times 10^{+25}$ | $4.41 \times 10^{+09}$ | 28.7 |
| | 6 | 249.2 | −24.0 | −11.2 | $4.72 \times 10^{+04}$ | 0.660 | $1.37 \times 10^{+25}$ | $3.47 \times 10^{+09}$ | 28.8 |
| | 8 | 236.2 | −36.9 | −34.4 | $3.56 \times 10^{+04}$ | 0.526 | $1.09 \times 10^{+25}$ | $2.69 \times 10^{+09}$ | 28.8 |
| | 10 | 223.2 | −49.9 | −57.8 | $2.65 \times 10^{+04}$ | 0.414 | $8.60 \times 10^{+24}$ | $2.06 \times 10^{+09}$ | 28.8 |
| | 20 | 216.6 | −56.5 | −69.7 | $5.53 \times 10^{+03}$ | 0.0889 | $1.85 \times 10^{+24}$ | $4.35 \times 10^{+08}$ | 28.8 |
| Strato | 30 | 226.5 | −46.6 | −51.9 | $1.20 \times 10^{+03}$ | 0.0184 | $3.83 \times 10^{+23}$ | $9.22 \times 10^{+07}$ | 28.8 |
| | 40 | 250.4 | −22.8 | −9.0 | 287.0 | 0.00400 | $8.31 \times 10^{+22}$ | $2.10 \times 10^{+07}$ | 28.8 |
| | 50 | 270.6 | −2.5 | 27.5 | 79.8 | 0.00103 | $2.14 \times 10^{+22}$ | $5.62 \times 10^{+06}$ | 28.8 |
| Meso | 60 | 247.0 | −26.1 | −15.0 | 22.0 | 0.000310 | $6.44 \times 10^{+21}$ | $1.62 \times 10^{+06}$ | 28.8 |
| | 70 | 219.6 | −53.6 | −64.5 | 5.22 | $8.28 \times 10^{-05}$ | $1.72 \times 10^{+21}$ | $4.08 \times 10^{+05}$ | 28.8 |
| | 80 | 198.6 | −74.5 | −102.1 | 1.05 | $1.85 \times 10^{-05}$ | $3.84 \times 10^{+20}$ | $8.66 \times 10^{+04}$ | 28.8 |
| | 90 | 186.9 | −86.3 | −123.3 | 0.184 | $3.42 \times 10^{-06}$ | $7.12 \times 10^{+19}$ | $1.56 \times 10^{+04}$ | 28.7 |
| Thermo | 100 | 195.1 | −78.1 | −108.6 | 0.0320 | $5.60 \times 10^{-07}$ | $1.19 \times 10^{+19}$ | $2.68 \times 10^{+03}$ | 28.2 |
| | 120 | 360.0 | 86.9 | 188.4 | 0.00254 | $2.22 \times 10^{-08}$ | $5.11 \times 10^{+17}$ | 163.0 | 26.0 |
| | 140 | 559.6 | 286.5 | 547.7 | 0.000720 | $3.83 \times 10^{-09}$ | $9.32 \times 10^{+16}$ | 38.0 | 24.6 |
| | 160 | 696.3 | 423.1 | 793.6 | 0.000304 | $1.23 \times 10^{-09}$ | $3.16 \times 10^{+16}$ | 15.0 | 23.3 |
| | 180 | 790.1 | 516.9 | 962.4 | 0.000153 | $5.19 \times 10^{-10}$ | $1.40 \times 10^{+16}$ | 7.20 | 22.2 |
| | 200 | 854.6 | 581.4 | 1078.5 | $8.47 \times 10^{-05}$ | $2.54 \times 10^{-10}$ | $7.18 \times 10^{+15}$ | 3.90 | 21.2 |
| | 300 | 976.0 | 702.9 | 1297.2 | $8.77 \times 10^{-06}$ | $1.92 \times 10^{-11}$ | $6.51 \times 10^{+14}$ | 0.420 | 17.6 |
| | 500 | 999.2 | 726.1 | 1339.0 | $3.02 \times 10^{-07}$ | $5.21 \times 10^{-13}$ | $2.19 \times 10^{+13}$ | 0.0160 | 14.2 |
| Exo | 1000 | 1000.0 | 726.8 | 1340.2 | $7.51 \times 10^{-09}$ | $3.56 \times 10^{-15}$ | $5.44 \times 10^{+11}$ | 0.000750 | 3.9 |

$h$ height, $T$ temperature, $P$ pressure, $\rho$ mass density, $n$ particle density = number of particles per cubic meter, $v$ number of particle collisions per second, $\mu$ average particle mass relative to that of a proton = average molecular weight. Data from the University of Utrecht's Department of Astronomy

## 18.9 The Ionosphere

The term 'ionosphere' is an alternate term used for that part of the upper atmosphere where air is ionized by solar radiation. It was named by Sir Robert Alexander Watson-Watt (1892–1973) in 1926 "for the region in which the main characteristic is large scale ionization with considerable mean free paths." Sir Robert is considered by many to be the inventor of radar. The ionosphere corresponds to the uppermost mesosphere, the thermosphere and part of the exosphere, extending from about 60–1000 km (∼40–60 miles). In this outer region of the atmosphere the Sun's ultraviolet radiation interacts with air to cause ionization resulting in charged molecules, ions, and free electrons. Several different levels are recognized, as shown in Fig. 18.4. Each layer has different properties, absorbing or reflecting radio waves.

Solar X-rays and ultraviolet radiation become more intense as solar activity increases. The higher energy solar radiation is progressively consumed in interactions with the air so that only the lower energy non-ionizing wavelengths penetrate the mesosphere into the stratosphere and troposphere. The interactions of the high energy x-ray and ultraviolet rays with air molecules cause them to be ionized or broken down into individual atoms, freeing large numbers of electrons. It is the electrons that refract electromagnetic waves with frequencies between 3 kHz and 30 MHz—radio waves. The process is very much like the bending of sound waves in water, discussed in Chap. 7.

Because the ionosphere is reacting to radiation from the Sun there are large differences between day and night, with the seasons, with latitude, and over solar activity cycles. Some of the day-night changes are shown in Fig. 18.4.

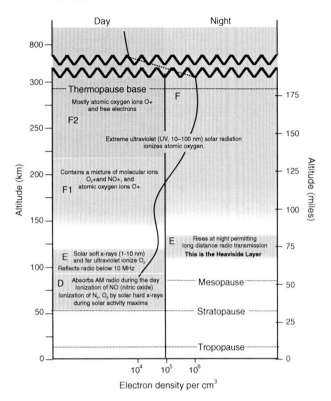

**Fig. 18.4** The ionosphere during day and night. The major layers and their characteristics are indicated in the diagram. Starting at the bottom, the ionosphere's D layer develops only during the daytime. It absorbs AM radio signals during the day; the signals travel further after sunset as the D layer decays. For very low frequency signals, the D layer and Earth can serve as a wave guide. Refraction begins with the E layer, which acts to bend radio wavelengths below 10 MHz back toward Earth. The E layer partly decays during the night and migrates upward. This allows the radio waves to travel further before they reach the refracting layer and are turned downward to again be reflected off the ground. The result is a major improvement in radio transmissions at night. The F layer is the main refractor of radio signals and becomes differentiated into two parts during the day, but becomes more homogeneous at night. These effects are summarized in Fig. 18.5

## 18.10    The Atmospheric Greenhouse Effect

The planetary greenhouse analogy came from the idea that the atmosphere acts like the glass in a greenhouse. But before starting this section I have to confess my sheer amazement at what I have found both on the internet and in

print, even from atmospheric scientists, about what a greenhouse is and how it works.

In the 17th through the late 20th century people knew what a greenhouse was: a structure made mostly of glass that would let in sunlight and trap the long-wave radiation from the ground, keeping it warm. Now I find that many people, who probably have never spent more than 5 min in a greenhouse, think it works by preventing the breeze from entering and removing the warmth that today is generated by electric heaters. There are many people who are unaware that glass is transparent to short-wave radiation and opaque to long-wave radiation. In case you are one of those, go to your heated glass-front oven or fireplace with a fire going, and hold your hand near the glass. If you turned on the light in the oven you know the short-wave visible radiation is coming through the glass. Open the glass door. Put your hand back in the same place. Get the message. Glass is transparent to short wave radiation and opaque to long-wave (infrared) radiation. If you are building a greenhouse, the Royal Horticultural Society website offers advice on what glass to use for maximum effect.

Any discussion of the greenhouse effect is supposed to start with an obligatory nod to the early 19th century French mathematician, Jean Batiste Joseph Fourier, as the original source of the idea. James Roger Fleming, in his *Historical Perspectives on Climate Change*, gives a delightful account of the profusion of confusion over what Fourier actually intended. His ideas are summarized in the following statement from his 1824 paper *Remarques generales sur les températures du globe terrestre et des espaces planétaires* (Annales de chimie et de physique, Paris, 2me serie, 27, 136–167; English translation by Ebeneser Burgess, 1837, American Journal of Science, 32, 1–20):

> The earth receives the rays of the Sun, which penetrate its mass, and are converted into non-luminous heat; it likewise possesses an internal heat with which it was created, and which is continually dissipated at the surface; and lastly the earth receives rays of light and heat from innumerable stars, in the midst of which is placed the solar system. These are the three general causes which determine the temperature of the earth.

> The transparency of the waters appears to concur with that of the air in augmenting the degree of heat already acquired, because luminous heat flowing in, penetrates, with little difficulty, the interior of the mass, and non-luminous heat has more difficulty in finding its way out in a contrary direction.

**Fig. 18.5** The effects of ionospheric layers on radio transmission

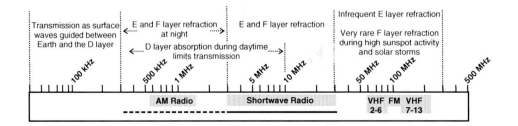

In order understand how ideas about the greenhouse effect developed, we need to know more about the atmosphere, its composition, and the abundance and importance of different greenhouse gases.

Let's briefly review what we know about the atmosphere. The Earth's atmosphere has a mass of $5.136 \times 10^{18}$ kg, which is equivalent to $178.7 \times 10^{18}$ mol of air. Remember that moles are avogadros of molecules; an avogadro is $6.02 \times 10^{23}$ things, so the number of molecules of mostly nitrogen, oxygen, and argon is $107.6 \times 10^{42}$, give or take a few. That may sound like a large number, but remember this: if the human population on planet Earth were to continue to grow at its late 20th century rate of 1.3 % per year, the entire land area will be covered by $153 \times 10^{12}$ people standing on it by about January 1, 2780. There will be one square meter of land for every person on our planet. Not even any room to lie down. And that will happen in the same length of time that has elapsed since the cathedral of Notre Dame de Paris was built. At the same rate of growth, the human population will be equal to the number of molecules of air in about the year 8140. That is a little more time than has elapsed since the Trojan War, and a little less than has elapsed since the founding of the city of Jericho. So when someone tells you that humans have no effect on the Earth, feel free to tell them they might enjoy jumping off the nearest cliff to see if the Law of Gravity is still in force.

To come back to the greenhouse gases, here is a handy thing to know: *equal volumes of gases at the same temperature and pressure contain the same number of molecules.* That is Avogadro's Law, proposed as a hypothesis by Amedeo Avogadro (Fig. 18.6) in 1811, and demonstrated in many ways to be true. By being tested and found to be correct it has become one of the most important laws of physics. It was step one in figuring out how to compare the number of molecules of one thing with the number of molecules of another.

As discussed in Chap. 16, the atmosphere is made of gases and the concentrations of the different gases are usually given in terms of percent by volume, or parts per thousand (ppthv or pptv), parts per million by volume (ppmv), parts per billion by volume (ppbv), or parts per trillion by volume (pptv). Note that pptv is used in two ways, so sometimes it can lead to confusion, but in atmospheric matters pptv almost always means parts per trillion. Almost always; nobody's perfect. Because the numbers are by volume, not by weight, they refer to the relative numbers of molecules (Avogadro's Law!). You don't need to go to a Periodic Table to figure out what is going on.

The atmosphere contains many gases that absorb radiation and pass on energy to adjacent molecules to raise the temperature, but all of them are present in only small quantities. As you will remember, the most abundant are water vapor ($H_2O$), methane ($CH_4$), carbon dioxide ($CO_2$), nitrous oxide ($N_2O$), and ozone ($O_3$). In all of these greenhouse gases the molecules are made of three or more atoms and have at least a transient dipolar electrical character so that they can absorb and reradiate electromagnetic radiation.

Table 18.2 shows both the absolute and relative amounts of these gases and their 'greenhouse potential.' The greenhouse potential (GP) is a measure of the ability of each molecule of the gas to intercept the long-wave radiation (the non-luminous heat of the 19th century scientists) from Earth's surface; it is shown relative to the GP of $CO_2$. To get a feeling for the contribution of each gas to the overall greenhouse effect you multiply the GP by either the 'Relative molar abundance vis-a-vis $CO_2$' or the total number of moles in the atmosphere.

**Fig. 18.6** Amedeo Avogadro and Avogadro's Law. Suddenly you realize you can read Italian

**Table 18.2** The atmosphere and major greenhouse gases: masses, molecular weights (MW), total moles, concentrations (where appropriate) in 1750 (before the industrial revolution) and in 2007, relative molar abundances and greenhouse potential (GP)

| Gas | Mass (kg) | MW | Moles | Conc (ppmv) | Relative molar abundance vis-a-vis $CO_2$ (1750) | GP | GP × RMA |
|---|---|---|---|---|---|---|---|
| Atmosphere (total) | $5.14 \times 10^{18}$ | 29 | $177 \times 10^{18}$ | | 261,000 | | |
| Stratosphere | $0.90 \times 10^{18}$ | 29 | $32 \times 10^{18}$ | | 45,760 | | |
| Troposphere | $4.24 \times 10^{18}$ | 29 | $147 \times 10^{18}$ | | 215,700 | | |
| $O_3$ stratosphere | $2.7 \times 10^9$ | 48 | $56.3 \times 10^9$ | 0.00181 | 0.0000008 | | |
| $O_3$ troposphere | $0.3 \times 10^9$ | 48 | $6.2 \times 10^9$ | 0.00043 | 0.00000009 | | |
| $CO_2$—1750 | $2.17 \times 10^{15}$ | 44 | $49.2 \times 10^{15}$ | 278 | 1 | 1 | 1 |
| $CO_2$—2011 | $3.08 \times 10^{15}$ | 44 | $70.1 \times 10^{15}$ | 394 | 1.417 | 1 | 1.417 |
| $CH_4$—1750 | $1.98 \times 10^{12}$ | 16 | $124 \times 10^{12}$ | 0.700 | 0.0025 | 23 | 0.0575 |
| $CH_4$—2011 | $5.12 \times 10^{12}$ | 16 | $320 \times 10^{12}$ | 1.808 | 0.0065 | 23 | 0.1495 |
| $N_2O$—1750 | $2.10 \times 10^{12}$ | 44 | $478 \times 10^{12}$ | 0.00027 | 0.00000097 | 296 | 0.0003 |
| $N_2O$—2011 | $2.52 \times 10^{12}$ | 44 | $572 \times 10^{12}$ | 0.00032 | 0.0000012 | 296 | 0.0004 |
| $H_2O$ | $12.9 \times 10^{15}$ | 18 | $716 \times 10^{15}$ | 4900 | 14.5 | 0.28 | 4.06 |

The GP is a measure of the relative ability of a pure sample of each gas alone to trap and reradiate long wavelength energy. GP × RMA = greenhouse potential times the relative molar abundance vis-a-vis $CO_2$ in 1750 (Compiled from various sources)

You don't need to actually do this to realize that the major greenhouse gas is water vapor, $H_2O$, followed by carbon dioxide, $CO_2$, then methane, $CH_4$, and finally nitrous oxide, $N_2O$. Greenhouse potential for stratospheric and tropospheric ozones, $O_3$, are not shown because the former is intercepting short wave radiation from the Sun, and the estimates for the role of ozone in the troposphere vary from one investigator to the next, largely because of its uneven distribution.

Each of these gases interacts with specific wavelengths of radiation, as shown in Fig. 18.7. As described in Chap. 16, the nitrogen and oxygen molecules that make up most of the air do not absorb energy, but they do interact with incoming solar radiation by preferentially scattering the blue part of the spectrum resulting in our planet's blue sky.

The greenhouse gases in Fig. 18.7 are arranged by the way they are distributed. Ozone is at the top since it is largely restricted to the stratosphere. Water vapor is at the bottom because its largest concentrations are in the lower troposphere. The other three gases are more evenly distributed but largely in the troposphere. A glance at the figure immediately shows that water vapor is the most effective greenhouse gas, intercepting both incoming and outgoing radiation. Carbon dioxide absorbs a broad group of wavelengths in the middle of the spectrum of outgoing radiation. The details of their distributions are our next topic.

## 18.11  The Distribution of Gases in the Atmosphere

Gases are not evenly distributed through the atmosphere. The heavier molecules are most abundant at the base of the atmosphere, the lighter molecules at the top. Figure 18.8 shows the relative concentrations of the most abundant gases.

At the right side of the diagram are the two gases that make up 99.9 % of the atmosphere, $N_2$ and $O_2$. The most abundant atmospheric gas, nitrogen, shows no change with altitude. However, the next most abundant gas, oxygen, shows a decrease in its relative concentration above 100 km. This is because at that altitude the air molecules are so far apart that gravitational settling begins to be a significant process. The average oxygen molecule has a molecular weight of 32, while that of nitrogen molecules is 28. Oxygen, being heavier, is settling out of the exceedingly thin air.

At the left on the bottom of the diagram is NO, nitric oxide. Like $N_2$ and $O_2$ it is a diatomic gas and has no dipole moment, so it is not a greenhouse gas. It forms from the decomposition of $N_2O$, and its abundance increases upward as the latter declines.

All of the other gases are greenhouse gases. Water vapor, $H_2O$, is the third most abundant gas in the lower part of the atmosphere. It is also Earth's major greenhouse gas. With a

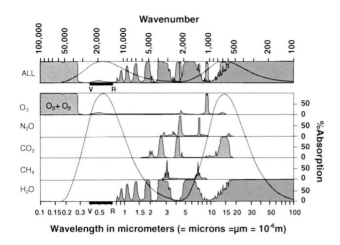

**Fig. 18.7** Absorption of different wavelengths of solar and Earth radiation by major greenhouse gases: ozone ($O_3$), nitrous oxide ($N_2O$), carbon dioxide ($CO_2$), methane ($CH_4$), and water vapor ($H_2O$). The % absorption amounts take into account both the greenhouse potential (GP) and the relative abundances of the greenhouse gases. Dotted curves show re-radiation from the Sun (*left peak*) and Earth (*right peak*). *Black* band along the bottom axis is the visible spectrum; *V* violet, *R* Red

molecular weight of only 18, it is much lighter than $N_2$ and $O_2$ and could be expected to diffuse upward toward the top of the atmosphere. There it would be broken down into its components by ultraviolet radiation, and the hydrogen would escape into space. This seems to be what happened to the water originally on Mars, but we are lucky that water is retained on Earth. This is because of two peculiarities, one of $H_2O$ itself; the other the layered structure of the atmosphere.

$H_2O$, or HOH, as I prefer to think of it, has the peculiarity that it can exist as a gas, vapor; or a liquid, water; or a solid, ice, at temperatures encountered from the bottom to the middle of the troposphere. However, the transition from vapor to water or ice requires some sort of non-gaseous surface, in the form of small aerosol particles on which the transition from vapor to water can occur. These are typically about 0.2 µm across, or 1/100th the size of a typical cloud droplet. If no cloud condensation nuclei are present, water vapor can be supercooled below 0° C (32° F) before droplets form spontaneously. One of the hypotheses for the cause of the crash of Air France Flight 447 over the central equatorial Atlantic on June 1, 2009, is that it entered a body of cold air supersaturated with vapor and the vapor immediately condensed on the solid surfaces, blocking the airplane's pressure-sensing devices and causing it to go out of control.

Cloud droplets range from 1 to 100 µm in size. Although they are much denser than air, the collisions with air molecules keep them aloft. As they grow they become rain drops

ranging from 0.5 mm to as much as 10 mm in diameter, but mostly in the 1–2 mm size range. At these sizes, gravity takes over, and the drops fall out as rain, carrying the cloud condensation nuclei with them. This process continuously purges the atmosphere of aerosols and pollutants.

Finally, the temperature structure of the atmosphere acts as an efficient cold trap for water vapor. Its concentration above the troposphere is only 1/1,000th the average at the surface and much of the vapor at high altitudes is probably formed in place by the oxidation of methane.

The fourth most abundant gas is $CO_2$. It is a greenhouse gas, and a critical part of the cycle of carbon, being consumed by plants for photosynthesis and reintroduced to the atmosphere through respiration of both plants and animals. It is the topic of Chap. 21. However, it is important to note here that it is well mixed upward in the atmosphere to about 100 km, where it begins to show the same gravitational settling effect as $O_2$.

The next gas is ozone, $O_3$. It is unstable and decays back to $O_2$ over time, so it must be continuously replenished. This occurs mostly through interaction with ultraviolet radiation in the upper atmosphere, but some ozone is generated by lightning in the troposphere and by human activities. Ozone will be discussed in detail in Chap. 19.

Methane, $CH_4$, and nitrous oxide, $N_2O$ are all greenhouse gases produced by sources on Earth's surface. Their concentrations decrease upward as they gradually decompose. The will be discussed in Chap. 22.

## 18.12   The Overall Effect of the Atmosphere on Solar Irradiance

Solar irradiance is the electromagnetic energy emitted by the Sun. Figure 18.9 summarizes the ways in which the radiation is absorbed and reflected by Earth's atmosphere, based on data from NASA and NOAA.

Note that the extreme ultraviolet is completely absorbed. Scattering, particularly of the blue end of the visible spectrum, and reflection of part of those wavelengths back into space accounts for much of the loss. Remember that scattering accounts for about 20 % of Earth's albedo and makes it the blue planet. The selective absorption of particular wavelengths in the infrared part of the solar spectrum is largely due to water vapor. Water vapor used to be essentially restricted to the lower troposphere, but since the 1950s jet aircraft have been introducing more and more of it into the stratosphere. Often the water exhaust shows up as contrails which then evaporate leaving the water in the lower stratosphere.

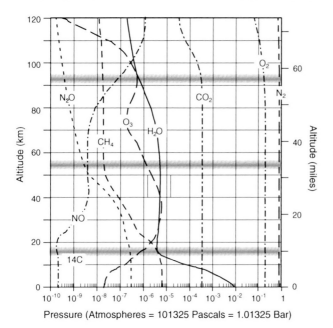

**Fig. 18.8** Relative amounts of the eight most abundant gases in Earth's atmosphere. Data plot made using http://spectralcalc.com. The tropopause, stratosphere and mesopause are shown as *horizontal bars*

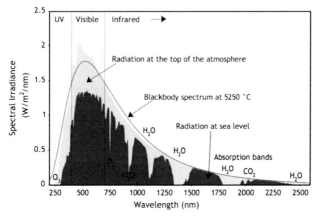

**Fig. 18.9** The loss of solar radiation through the atmosphere. *Yellow* area represents the extraterrestrial solar radiation at 1 AU (Astronomical Unit—the average Earth-Sun distance). The *right part* of the curve is smooth; that is the 'back-body' radiation from the hot surface of the Sun's photosphere. The *left part* is ragged because the emission is from excitation of the atoms of the Sun's plasma. The *red* area represents the radiation reaching the Earth's surface through an air thickness of 1.5 atmospheres. This corresponds to the Sun being 37° above the horizon. All of the *yellow* area is absorbed in the atmosphere. Based on the 2000 ASTM Standard Extraterrestrial Spectrum Reference E-490-00 and the ASTM G173-03 Surface Irradiance spectrum (*ASTM* American Society for Testing and Materials)

## 18.13   The Effects of Anthropogenic Atmospheric Pollution

Two kinds of pollutants have been added to the atmosphere by human activities, the invisible greenhouse gases discussed above, and visible and invisible particulate materials, anthropogenic atmospheric aerosols. Both can affect the passage of solar radiation through the atmosphere. The incoming solar radiation has wavelengths between about 200 and 3000 nm, most of it concentrated in the visible light region of 400 (violet) to 700 nm (red). The major gaseous absorbers of incoming radiation have their effect mostly outside the visible light region. Ozone ($O_3$) absorbs wavelengths in the ultraviolet and water vapor ($H_2O$) in the far red and infrared. The greenhouse gases being introduced by human activities, carbon dioxide ($CO_2$), methane ($CH_4$), and nitrous oxide ($N_2O$) are active in absorbing and emitting the longwave radiation from Earth back into space and do not affect insolation. Their cycles and effects are discussed in detail in Chaps. 21 and 22.

Atmospheric aerosols can have two effects: (1) they can reflect incoming solar radiation and increase Earth's albedo, or (2) they can absorb radiation and heat the atmosphere. To understand the effects of atmospheric aerosols we need measurement of the amount of solar radiation reaching the Earth's surface, but these have been developed only in the past few decades so there is no direct long-term record. Because this field of research is so new the terminology for describing insolation has not completely stabilized and reading the scientific literature can be a challenge. The direct and diffuse solar insolation reaching Earth's surface is variously called 'global radiation,' 'surface insolation,' 'solar/shortwave irradiance,' or 'surface solar radiation.' The use of the term 'global' in this context is also confusing. Often it means simply incoming solar radiation reaching some part planet Earth, or it may refer to the entire planet.

If I may digress for a moment: In a restaurant in France, when the waiter hands you the bill for the meal and you discover that the addition is wrong (almost always favoring the waiter), he will take it back with the explanation "Pardon monsieur, c'est une somme globale" and then make the correction. Again, the French are ahead of us in ambiguous use of the term 'global.'

Natural particulates in the atmosphere include the dust, sea salt, and volcanic ash discussed in Chap. 15, along with smoke from natural fires, pollen, spores, and bacteria. Currently these natural materials make up about 90 % of atmospheric aerosols. The particulates added by human activities include ash, soot, and smoke from industries, man-made fires ('biomass burning'), sulfur dioxide ($SO_2$)

from industrial activities such as coal-burning power plants. The SO$_2$ reacts with water vapor and other gases in the atmosphere to create a variety of sulfate aerosols. Sunlight causes photochemical reactions with automobile exhaust fumes producing complex particles we see as 'smog.'

Anthropogenic aerosols can affect transmission of energy from the Sun in very different ways. Those with light color, such as fine smoke particles can reflect light and increase the regional albedo. Larger smoke particles, such as 'black carbon' or soot, absorb energy and warm the atmosphere. These effects are called 'Direct Radiative Forcing' (DRF). In small concentrations anthropogenic aerosols can serve as nuclei for water droplets, forming white clouds, and increasing albedo. Such particles are called 'cloud condensation nuclei,' CCNs. This reduction of irradiance by reflection resulting from cloud formation by CCNs is called Indirect Radiative Forcing (IRF).

The differences between reflective clouds and clear air can promote convection. In large concentrations aerosols can form a dark layer, which then absorbs energy and forming a warm layer within the troposphere, an 'inversion,' which inhibits convection. Large concentrations of aerosols also inhibit cloud formation. These processes are so complex they are still not well understood and usually omitted from numerical climate models.

Scientists studying the atmosphere speak of "Aerosol Optical Depth." It is a measure of the extinction of the incoming solar radiation by dust and haze. It is a dimensionless number that tells us how much direct sunlight is prevented from reaching the ground by aerosol particles in the vertical column of atmosphere over the observation location. An Aerosol Optical Depth value of 0.01 corresponds to an extremely clean atmosphere, and a value of 0.4

corresponds to a very hazy condition with 40 % of the incoming sunlight blocked by particles in the air. Prior to the Industrial Revolution aerosol optical depths were close to 0.01 over much of the Earth Today average aerosol optical depths for the U.S. vary from 0.1 to 0.15.

Anthropogenic aerosol production goes back to primitive man's discovery of the use of fire, but it increased by orders of magnitude with the Industrial Revolution and subsequent explosion of the human population. Figure 18.10 shows the growth of human population and a guesstimate of how atmospheric pollution may have decreased the amount of energy reaching Earth's surface. The values for tropospheric aerosols are questionable because they do not include the effects of large scale burning of grasslands and forests as part of the hunting activities of pre-industrial populations. Today smoke from biomass burning to clear forests is a regionally significant, but not well quantified, component of the climate system.

The human population had grown only slowly over thousands of years, with significant reductions from time to time as a result of pandemics and possibly famines. A more detailed history was shown in Figs. 3.17 and 3.18. Exponential growth of the human population began with the Industrial Revolution which made possible division of labor and improved farming techniques. The result was more plentiful food and better nutrition. Population growth was further enhanced by the introduction of disease-reducing antibiotics and widespread use of industrially-produced nitrate fertilizers which increased crop yields; both advances occurred shortly after World War II. All of these activities added anthropogenic pollutants to the atmosphere.

It has been apparent since the late 19th century that something has been happening to the air. If you look at

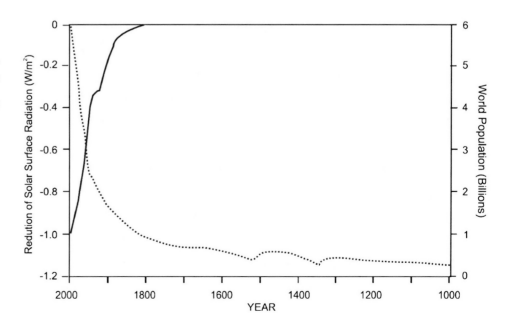

**Fig. 18.10** An estimate of the effect of anthropogenic atmospheric pollution with aerosols in the troposphere (*solid line*) and world population growth (*dotted line*). Tropospheric aerosol data after Crowley (2000) *Causes of Climate Change over the Past 1000 Years* Science, v. 289, pp. 270–277

**Fig. 18.11** John Constable
(1776–1837) Wivenhoe Park,
Essex, England, 1816. Now part
of the campus of the University of
Essex at Colchester, about
66 miles (110 km) northeast of
London. Note the crisp view and
clean, clear air

**Fig. 18.12** Albert Bierstadt
(1830–1902) Mount Rainier, ca.
1880. At the base of the mountain
there is a natural haze from the
forest

pe-19th century paintings with landscapes such as those by
Peter Paul Rubens (1577–1640; look past the naked ladies),
Claude Lorrain (1600–1682; who introduced landscape
painting as a genre), Thomas Gainsborough (1727–1788),
and John Constable (1776–1837), both famous English
landscape painters. Notice how sharp the hills and how blue
the skies are. The atmosphere was very clear. The same is
true of 19th century paintings of the wild American west,
like those by Albert Bierstadt (1830–1902) that hang in the
U.S. Capitol. In France and England things changed during
the 19th century. Painters like Joseph Mallord William ('J.
M.W.') Turner (1751–1851), Camille Pissarro (1830–1902),
Alfred Sisley (1839–1899), Claude Monet (1840–1926), and
Pierre-Auguste Renoir (1841–1919) documented the effects

of what we would call atmospheric pollution. To them the
atmosphere had become 'palpable.' It had 'substance' that
had not been noticed before; it was a fascinating, dramatic
phenomenon. The paintings are the best documentation we
have of this early rise of anthropogenic atmospheric
aerosols.

The Industrial Revolution is commonly thought of as
having begun about 1775, with James Watt's invention of a
practical steam engine. This made mechanization of many
manufacturing processes possible, speeding up production
many times faster than had been possible when everything
had to be done by hand. In its early phases the Industrial
Revolution was fueled almost exclusively by wood. Many of
the forested areas of Britain, Europe and the eastern United

**Fig. 18.13** Claude Monet painted a series of paintings of the Palace of Westminster, home of the British Parliament in London between the years 1900–1905. All the same size and viewpoint, a window at St. Thomas' Hospital overlooking the Thames, and all show the London fog

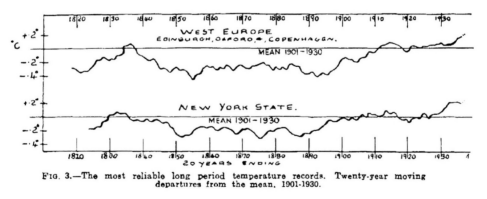

Fig. 3.—The most reliable long period temperature records. Twenty-year moving departures from the mean. 1901-1930.

**Fig. 18.14** Figure 3 from George Callendar's 1938 paper showing the changes in temperature in Western Europe and northeastern North America since the early 19th century. Note that oldest dates are on the *left*, youngest on the *right*

States were sacrificed to supply the demand. Then in the early to mid-1800s the new industries turned to coal as an energy source. Coal introduced not only carbon dioxide into the atmosphere, but the much more obvious pollutant, soot. The soot particles acted as nuclei for water, forming tiny droplets, resulting in increases in fog in industrialized areas. Claude Monet pained beautiful pictures of London fogs along the Thames.

Later some of these English 'fogs,' like that in London in the mid-20th century, became deadly. Over the years, soot accumulated on the surfaces of stone buildings tuning white surfaces black. When I first visited London and Paris, in 1954, all of the monumental buildings were blackened with soot. Lights shone on the facades at night, making them dark gray with shadows. Along with the soot came sulfuric acid, which etched the building stones and statues. In Europe, the acid in the air and rain of the previous 100 years had caused more 'weathering' of architectural works in one century that in the previous millennium.

The soot-laden atmosphere was obvious; it was considered a sign of progress. The introduction of greenhouse gases went almost unnoticed. However, you will remember that in 1894 Svante Arrhenius had suggested that increasing levels of $CO_2$ from burning coal might eventually cause global warming. His idea was soon being questioned. In 1900 another Swedish scientist, Knut Ångström (1857–1910), son of Anders Ångström for whom the unit of distance is named, showed that $CO_2$ and water vapor absorb energy in the same spectral regions. Over much of the Earth, water vapor is much more abundant in the atmosphere than $CO_2$, so he concluded that changing atmospheric $CO_2$ concentrations had no real effect on climate.

Then came the great blow to Arrhenius' idea. In 1905 German physicist Clemens Schaefer (1878–1968) reported in the *Annalen der Physik* (the same journal in which Albert Einstein published his three revolutionary papers that same year) that a column of $CO_2$ 40 cm long would almost completely absorb any infrared radiation entering it. American physicist and atmospheric scientist William Jackson Humphreys (1862–1949) discussed this in his textbook *Physics of the Air* published in 1920. Humphrys reckoned that any heating of Earth's surface by $CO_2$ would come from that in the stratosphere, where it would act like a semi-transparent mirror reflecting the heat back down. He calculated that the $CO_2$ content of the stratosphere would be roughly equivalent to the 40 cm of Schaefer's experiment. He concluded: "Now, according to the experiments of Schaefer, a layer of carbon dioxide 40 centimeters m thick is sufficient to produce very nearly full absorption, and therefore, no increase in the amount of carbon dioxide in the atmosphere could very much increase its temperature" and hence could have no effect on climate. Humphreys' textbook was used in most courses in Meteorology. It went through two more editions, the last in 1940, but it has been reprinted several times since, most recently in 2012. That statement from 1920 is still cited by some climate change deniers as the latest word on the temperature effect of atmospheric $CO_2$; an increase has no effect on the Earth's surface temperature.

T. C. Chamberlin of the University of Chicago, who had originally championed Arrhenius' idea by proposing that $CO_2$ had been a major control on climate throughout Earth history, became increasingly impressed by the argument that water vapor was the real greenhouse gas. In 1923 he rejected his earlier ideas, concluding that he had been fooled and that $CO_2$ was unimportant.

By the end of the first half of the 20th century, it seems that the only person who still thought that $CO_2$ was important in climate was a British steam engineer, George Callendar. In 1938 he published the graphs shown below.

Callendar's Figure 3 shows a general episode of cooler temperatures in from about 1835 to 1900. This is a pattern similar to that seen in 20th century records from 1950 to 1980 as a plateau, now referred to as an episode of 'global dimming.' But the compilations shown by Callendar in Fig. 18.3 are not global but specific small areas with intense human activity. It was during this time that Edinburgh acquired its nickname 'Auld Reekie' (Old Smokey) and soot turned its buildings black. In Oxford, coal heated the great University dormitories and it is close enough to London to be affected by industrial activity there. Copenhagen grew rapidly during this time. In America, New York State was a major center of industrial development. My guess is that the mid-19th century plateau in temperature for these areas reflects atmospheric aerosol pollution.

Figure 18.15 shows Callendar's attempt to expand his temperature graphs to the rest of the world. His Figure 4 shows slight late 18th century variations followed by a steady upward trend since about 1910. He attributed this to the effect of $CO_2$ as a greenhouse gas. No one paid much attention to Callendar; he was, after all, just an engineer working in industry. His official title was 'Steam Technologist to the British Electrical and Allied Industries Research Association.' He was not a meteorologist or physicist at a major institution. He was considered a dilettante.

In case you are an older person, like me, who thinks the sky was a more intense blue when you were young, you're right. For much of the world, the term 'sky blue' has changed from an intense blue caused by the scattering of sunlight by air molecules to a pale blue representing the effect of a variety of aerosols introduced by human activities. Figure 18.10 showed the increase in this effect over the last thousand years. It is a direct effect of the Industrial Revolution, population growth, the use of coal, and automobiles. The local effects were sometimes spectacular. You will recall the famous London pea-soup fogs that made detective stories of the 19th and early 20th century so exciting. Happily I missed the Great London Fog of December 5–10, 1952 that killed about 4,000 people, but I experienced one of the last major pea-soupers in the late 1950s as the pollution was being reduced. The earlier pea-soup fogs were yellowish but mine had already lost some of that color and was just a dense gray with the odor of burning coal. Many people still had coal stoves in their apartments. I was staying at a hotel in Mayfair, and had been at a theater on Shaftesbury Avenue. I walked back to my hotel with some difficulty. The fog was so thick you couldn't see across Piccadilly Circus, and crossing Regent Street was a real problem. Regent Street is a grand thoroughfare about 20 m (66 ft) wide and you couldn't see from one side to the other. Furthermore, although the cabs were going at a snail's pace, they couldn't

**Fig. 18.15** Figure 4 from George Callendar's 1938 paper showing a global rise in temperature since the beginning of the 20th century. Note that oldest dates are on the *left*, youngest on the *right*

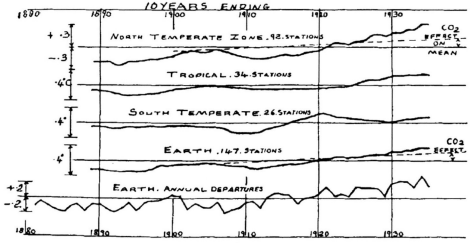

Fig. 4.—Temperature variations of the zones and of the earth. Ten-year moving departures from the mean, 1901-1930, °C.

see much of anything in the street except the red lights on the back of the cab in front. They would just suddenly appear as out of nowhere. In those days gentlemen wore dark clothing so you were invisible to the cabbies. We waited in groups until someone with a light or brightly colored coat came along and then we would cross together. A year or so later the anti-pollution measures were all in place, and although I've visited London many times since I never experienced another real fog. I noticed that the number of sunny days in London has markedly increased over the years. London has demonstrated that the problem of air pollution can be solved (Figs. 18.11, 18.12, 18.13, 18.14 and 18.15).

In the early 20th century Los Angeles had been famous for its clear skies and dry weather. These made its suburb, Hollywood, an ideal site for making motion pictures. But after WWII things began to change. The number of automobiles increased dramatically. The air had become visible; it was called 'smog.' By 1947 Los Angeles' smog problem had become quite apparent. Its source was a mystery, and no one blamed automobiles. Auto exhaust is clear and the Los Angeles smog was brown. It was a biochemist who had been studying the flavor of pineapples at the California Institute of Technology, Arie Haagen-Smit, who discovered what was happening. The smog was affecting the growth of his plants and he was curious about what it was. It took time, but he finally discovered that the clear hydrocarbon emissions from tailpipes, when exposed to sunlight and nitrogen oxides, turned into brown 'photochemical smog.' Photochemical smog caused eye irritation and a host of respiratory problems. It took more than a decade to convince politicians, regulators and industry that cars were the biggest smog culprit in Los Angeles. In the late 1950s and 1960s other cities began to experience the same problem. Californians began to experiment with anti-smog devices. In 1963, the U.S. Congress

enacted the first Clean Air Act, acknowledging that smog had become a national problem. Two years later, it called for the first national emissions standards for cars. In 1967 the Congress gave California permission to set even stricter emission standards than those imposed by the federal government. In 1969 the Justice Department sued automakers for conspiring to delay anti-smog devices, a lawsuit ultimately settled out of court. Then, Congress enacted the law that has set the framework for U.S. air pollution regulation, the Clean Air Act of 1970. The modern solution to the smog problem, the catalytic converter was introduced in 1974.

The catalytic converter had a problem that delayed its introduction—it was incompatible with the additive to make 'anti-knock' gasoline, tetraethyl lead. This additive was very popular in the United States. The lead is released into the atmosphere as through the automobile exhaust. Lead is a poison affecting the nervous system and the brain, especially in children, and for several decades Americans had been getting a high dose of it from the gasoline we used. Many members of the U.S. Congress grew up in the era of atmospheric lead pollution.

Ice cores from Greenland show that there was an earlier episode of lead pollution of the atmosphere. There are high concentrations of lead in ice dated between 200 BCE and 400 CE, the time of the Roman Empire. The Romans were especially fond of lead as a useful metal. It was easy to extract in smelters, and because it is soft, it is easy to work. The Romans used it for plumbing; their cities had water supply systems similar to our own, with valves in the homes for tuning the water on and off. It was all made of lead, and it is suspected that the decline of the Roman Empire may have been due in part to lead poisoning. The smelters must have produced a lot of lead-laden smoke to leave such an impressive record in Greenland.

Sulfur dioxide ($SO_2$) emissions are an important contributor to the anthropogenic atmospheric aerosol inventory. Figure 18.16 shows the increase and decline of $SO_2$ emissions over the past 150 years. The main source of $SO_2$ emissions is the coal used in power plants. In the atmosphere, $SO_2$ reacts with other chemicals to form a variety of sulfate compounds and with water to be a major cause of acid rain. The seriousness of the acid rain problem became apparent in the 1980s when lakes in the northeastern U.S. and Canada were found to be rapidly deteriorating and had become so acidic that they would no longer support fish such as Brook Trout and minnows. Serious efforts to reduce $SO_2$ emissions in the US began in 1989. The problem continues to plague many parts of the world, particularly China and Russia.

Figure 18.17 shows a modern compilation of temperature for the Northern Hemisphere. A good way to understand the development of ideas regarding the effect of $CO_2$ and global warming is to cover the left side of the diagram with a piece of paper so that at first you see only the record to 1940, just after Callendar had published his paper. What this segment shows is a series of large annual temperature fluctuations about the slow general increase Callendar had documented.

Now move your sheet of paper over to show the pattern up to 1980. You will see that from about 1940 to 1980 there is a plateau. The rise expected from the continued $CO_2$ increase had stopped. This was not apparent until the 1970s. Speculation about why this was happening included the idea that we were approaching an orbital solar minimum, and that there might be another ice age on the way. In the 1980s and 90s it was recognized there was no significant reduction in the energy output from the Sun. It was the amount of the Sun's energy reaching the Earth's surface that was declining. The effect was termed "Global Dimming." It is now thought that this episode of global dimming was caused when atmospheric pollution reached a level such that it masked the gradual temperature rise caused by increasing levels of greenhouse gases.

The suspicion was that the change in solar energy reaching the Earth's surface was more likely to be anthropogenic, due to human activity, rather than natural. Then came an unexpected experiment on the atmosphere. After the terrorist attack of September 11, 2001, all commercial planes over the United States were grounded for 3 days. There were no jet aircraft in the stratosphere over the US and the contrails they would usually leave behind were eliminated. The result was an increase in the daily temperature range over the US, with an increase in the maximum and decrease in the minimum temperatures. The atmosphere had become more transparent to both incoming and outgoing

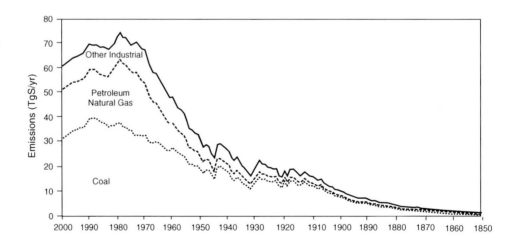

**Fig. 18.16** History of sulfur dioxide emissions in terms of teragrams ($10^{12}$) grams of sulfur

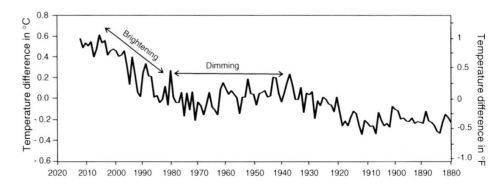

**Fig. 18.17** Changes in Northern Hemisphere annual average surface temperatures between 24° and 90° S from 1880 to 2014, shown as differences from the 1951 to 1980 average

radiation. To many this was a strong indication that the dimming effect had indeed been anthropogenic and largely related to contrails.

Now remove the paper and see what has happened since 1980. The temperature increase is proceeding, with the usual ups and downs but overall at a more rapid rate that in the early part of the 20th century. We now refer to these changes as episodes of 'brightening' and 'dimming,' referring them to Earth's surface temperature response to the amount of solar insolation reaching the planet's surface In other words, the temperatures observed at Earth's surface are a reflection of both the effect of increasing greenhouse gas concentrations and of the transparency of the atmosphere. Conclusion: the albedo of Earth's atmosphere responds to changes in the scattering of light, cloudiness, and to the amounts of suspended natural and anthropogenic aerosols. Keep in mind that Fig. 18.17 is only for the Northern Hemisphere between the Tropic of Cancer and the North Pole. This includes most of the industrialized nations, and is the region most likely to be affected by human activities.

Figure 18.18 shows the temperature record for the Southern hemisphere from the Tropic of Capricorn to the South Pole. There appears to be only a slow, rather even rise in temperature starting early in the 20th century. The episode of global dimming is barely detectable. The overall rise is less than in the Northern Hemisphere because the area of high heat-capacity ocean water in the Southern Hemisphere is much larger than in the Northern Hemisphere.

A more complete account of local, regional, and global temperature records, along with the details of how they are compiled is in Chap. 30.

Figure 18.19 compares emissions of $CO_2$ and S over the past 150 years. Neither of these is directly proportional to their competing greenhouse or albedo effect, and other aerosols contribute to enhancement of the albedo, but the departure of the trends of the two curves about 1980 offers strong support to the idea that the end of the episode of global dimming is a response to increasing atmospheric transparency.

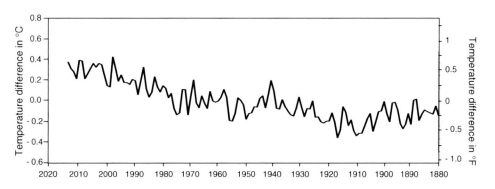

**Fig. 18.18** Changes in Southern Hemisphere annual average surface temperatures between 24° and 90° S from 1880 to 2014, shown as differences from the 1951–1980 average. The mid-20th century plateau is less obvious

**Fig. 18.19** Comparison of increasing emissions of $CO_2$ and $SO_2$ (as S) since 1850

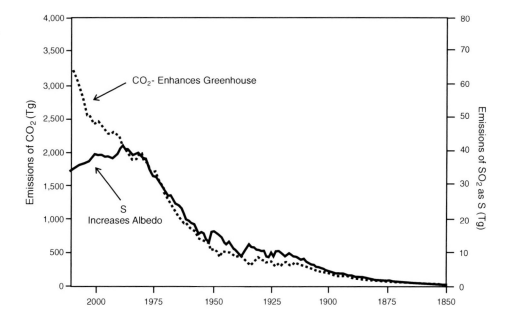

## 18.14   Summary

The atmosphere is divided into layers by temperature inversions. An 'inversion' is where light, warm air overlies denser cold air, preventing convection. Normally, the temperature of a gas declines as it expands, and this happens in the lower part of the atmosphere, the troposphere. The air in the troposphere convects and is homogenized by continuous mixing. But at an altitude ranging from about 20 km in the tropics to 10 km at the poles the trend reverses and the temperature increases with height. This is due to the absorption of ultraviolet radiation and conversion of the excitation energy into heat by the special greenhouse gas ozone. The site of the temperature reversal is called the tropopause, and the layer in which the temperature increases with altitude is called the stratosphere. In the stratosphere warmer air overlies colder air and convection cannot occur. The air is stratified. At a higher altitude, called the stratopause (about 50 km), the temperature trend again reverses, declining with altitude. This is the mesosphere, where atmospheric pressure is about 1/100,000th that at sea level. At about 90 km, the mesopause, the temperature trend again reverses. The kinetic energy of the molecules increases as they are hit by energetic ultraviolet photons. The molecules are traveling very fast, but there are so few of them that we cannot measure 'temperature' with a thermometer. The top of the thermosphere, the thermopause, varies in height from 500 to 1,000 km because of pressure from the solar wind, particles streaming out from the Sun. Above the thermopause is the outermost layer of the atmosphere, the exosphere. Only the lightest gases, hydrogen and helium make it this far out. It is the transition to interplanetary space and may extend half-way to the Moon.

The ionosphere is the region in the upper atmosphere where energetic ultraviolet radiation ionizes air molecules and frees electrons. These electrically charged particles interact with electromagnetic radiation at frequencies between 200 kHz and 30 MHz—radio waves—and refract them, permitting long distance over-the-horizon radio transmission.

Introduction of anthropogenic aerosols can both reduce and increase Earth's albedo, changing the amount on insolation reaching our planet's surface. It is likely that anthropogenic aerosols are responsible for the episode of 'global dimming' noted in temperature records between 1950 and 1980.

A Timeline for this chapter:

| 1607 | Galileo Galilei invents the 'thermoscope' |
| --- | --- |
| 1643 | Evangilista Torricelli invents the mercury barometer |
| 1648 | Blaise Pascal discovers that atmospheric pressure decreases with altitude and concludes that there must be a vacuum above the atmosphere |

(continued)

(continued)

| 1662 | Robert Boyle finds that the volume of a gas is in proportion to the pressure applied to it |
| --- | --- |
| 1699 | Guillaume Amontons investigates the relationship between pressure and temperature in gases but lacks a good thermometer |
| 1780s | Jacques Charles discovers the relationship between the temperature and pressure of a gas |
| 1801 | John Dalton also notes the relationship between temperature and pressure of a gas |
| 1802 | Joseph Louis Gay-Lussac formalizes and publishes Charles' Law |
| 1811 | Amedeo Avogadro publishes his idea that equal volumes of different gases at the same temperature and pressure contain the same number of molecules |
| 1824 | Jean Batiste Joseph Fourier publishes *Remarques generales sur les températures du globe terrestre et des espaces planétaires* |
| 1900 | Based on unmanned balloon soundings Léon Philippe Teisserenc de Bort (1855–1913) discovers a high altitude region with constant temperature and names it the 'stratosphere' and the layer beneath it the 'troposphere' |
| 1926 | Sir Robert Alexander Watson-Watt coins the term 'ionosphere' |
| 1950 | Sydney Chapman proposes the terms mesosphere and thermosphere for upper layers of the atmosphere |
| 1959 | Sir Thomas Gold proposes the term 'magnetosphere' for the region above the ionosphere in which the magnetic field of the earth has a dominant control over the motions of gas and fast charged particles |
| 1980s | Sir Thomas Gold argues for an abiogenic origin of petroleum |
| 1960s–present | Space age investigations provide lots of new information about the atmosphere and the upper atmosphere in particular |
| 2001 | After the 9/11 attack U.S. commercial aircraft were grounded The elimination of contrails resulted in an increase in diurnal temperature changes over the U.S. demonstrating that human activities were indeed affecting climate |
| 2011 | Raymond Pierrehumbert of the University of Chicago describes Earth's atmosphere as a sea of atoms, molecules and photons atoms, molecules and photons |

If you want to know more:

José Peixoto and Abraham Oort's *Physics of Climate*, mentioned before, is excellent on these topics.

Adam Voiland's excellent 2010 article 'Aerosols: Tiny Particles, Big Impact' is at http://earthobservatory.nasa.gov/Features/Aerosols/

Music: Sibelius Symphony #5—A song for swans. Listen for the swans flying overhead.

Libation: Finland makes great vodka too. Sip one flavored with lemon peel as you listen for the swans.

**Intermezzo XVIII. Becoming a Dean** I had moved full time to Miami in the fall of 1973, and became Chair of the Rosenstiel School's Division of Marine Geology and Geophysics in 1974. In the fall of 1976, I was attending a geological meeting in Honolulu, Hawaii when I got a telephone call from Dick Robins, Chair of the School's Division of Biology and Living Resources. Quite unexpectedly, our Dean since 1973, Warren Wooster, had resigned a few days earlier, effective immediately, and was leaving Miami. The School's faculty had met, and elected me to be Interim Dean. I think I said "Wait a minute how can ...?" and Dick responded, "It's done, and you're it. Don't worry, we'll all help you." Needless to say, it was quite a shock. Overseeing a multimillion dollar operation with several hundred employees is far different from helping to manage the affairs of a faculty of 12.

My father had retired some years earlier, and I called him for advice, since he had founded and been the President of a Life Insurance Company. He had some very simple advice: find out who your immediate subordinates are, what they are supposed to do, and whether they have been doing their jobs effectively. If need be, replace them, but be sure their duties are well defined and give them a free hand to do their jobs. Don't expect them to solve a problem the way you would and don't interfere with them unless there is a real problem. In other words, don't try to micromanage the operation.

When I got back to Miami, I called the 'managers' together. I had an Associate Dean, Warren Wisby, who apparently now had no duties but was there because he had served in that position under Walton Smith, the School's Founder. Jim Gibbons was in charge of operations, managing the land facilities and our vessels. Bill Muff was in charge of bookkeeping and accounting. I asked how the School was doing overall. Bill said we had one small problem: we were going to be bankrupt in a few months. "That's the bad news. The good news is that we have plenty of money. The problem is that we can't spend it."

What had happened was that Dean Wooster had tried to make the Rosenstiel School more like Scripps on the way it handled faculty salaries. Under Walton Smith salaries were closely related to productivity, where productivity is a reflection of the amount of grant funding and publications. Warren had capped the higher salaries and was bringing up the lower salaries. That might go well in an institution funded by the state, but the Rosenstiel School's funding was almost entirely from government agencies and donations.

Let me explain how the funding of oceanography and some other areas of science works in the US. Although originally founded with endowments from private sources, our major oceanographic institutions became wards of the government during World War II, and then in the late 1950s as our country realized the need to improve its capabilities in science. Funding of these institutions comes mostly from the National Science Foundation, the Office of Naval Research, NOAA, and NASA. Lesser amounts come from private foundations and donors. The grants or contracts from the government agencies include direct operating expenses for carrying out the proposed research and for faculty, technicians, and graduate student salaries along with their tuition. In addition to these direct expenses, there is a category called 'indirect costs.' These cover the operation and maintenance of the buildings and all the other expenses involved with the research enterprise, including the administration. Indirect costs generally range from 40 to 60 % of the direct costs, but depending on how the accounting is done may be less or even more. To recover the indirect costs the institution has to spend the money in the grant or contract; the funds dribble in as the work proceeds.

Potential faculty salary increases are included in the grant proposals. Whether these salary increases will actually be made depends on policies at the University. What had happened at the Rosenstiel School was that by holding down the salaries of researchers with large grants, the recovery of indirect costs used to finance the routine operations of the School had been reduced. What Bill Muff meant when he said we had lots of money was that the money was there in the grant accounts, but since salaries had been held back our indirect cost recovery had become less than the actual cost of operations. There was a way to solve our problem: to have significant salary increases for the highly productive faculty, bringing them up to the levels the government agencies had already approved. I had to do some serious negotiating with the university administration to get a major exception to its salary raise policy.

All of this may sound quite esoteric, but I wish the general public had a better understanding of how the funding of scientific research works in the United States. As I write this in 2015 there is a very strong anti-science movement being advanced by a number of politicians. In the 114th Congress (2015–2017) the Chairs of the committees overseeing science funding are religious fundamentalists who believe the Earth is less than 7,000 years old and that 'climate change' is

God's will. A majority of the Congress purport to believe that although the climate might be changing, it is not due to any human activities. Cutting science finding not only slows basic research, but it has a multiplier effect throughout the higher educational system. Some suggest we should concentrate only on applied research that will have a rapid economic return. That can work for a few years, but it is like cutting the top off a candle and expecting just the top to keep on burning for as long as the whole candle would. Fortunately for civilization, this is an American malady not shared by the other developed countries, and especially not by China. But back to my first days as a Dean.

I gave my Associate Dean, Warren Wisby, a job. Take care of handling faculty matters. Because of the salary policy over that past few years many of our best faculty were actively pursuing possibilities elsewhere. He was to lend a sympathetic ear to complaints and keep our best researchers from jumping ship. I told Jim Gibbons to keep the place running and not bother me unless he had a problem he could not handle. His response was a huge smile and "Aye, aye sir." Bill Muff was to work out a detailed plan to keep us solvent.

Walton Smith had retired from the University, but was just across the street (the 'street' being Rickenbacker Causeway) where he was President of the International Oceanographic Foundation, a charitable organization he had founded in 1953 to support research at what was then the Institute of Marine Sciences, but that now was operating a museum, Planet Ocean. Walton was delighted with my appointment as Interim Dean. He had actually made me a trustee of the IOF in 1974, the only member of the School to be granted such an honor. In my new role, he always addressed me as 'dear boy.'

We reinstituted one of Walton's old traditions. When the School's Doherty Marine Science Center was constructed in the 1960s, it included a dining facility and an auditorium. Walton managed to get a liquor license for the dining facility to fulfill one of his dreams—having an 'on-campus' pub where staff and students could gather after the work day. My predecessor had regarded it as an unnecessary expense and had closed it. I reopened it and resumed Walton's habit of going there after work. The kitchen was closed and there was no dinner service, but we had sandwiches and snacks for those who needed them. That way I was available for informal talks with anyone: faculty, staff, students, and all the rest who worked at the School. Walton, who had not often visited the School since his retirement, and his wife Mae, were regulars. Our pub soon became more widely known, and we had visitors from Key Biscayne who enjoyed mixing with the scientists and learning about our work.

I still had to make my presentation to the University officials to get permission for special salary raises that would both retain our best people and provide the overhead income that would allow us to continue operations. Fortunately, the university had a new Provost, Clyde Wingfield, who was just learning about how things worked in a private University and was receptive to innovative ideas. We made a presentation to him and his new staff, and, to make a long story short, after analyzing our situation they approved our plan.

We had a faculty meeting where everything was explained, and faculty were notified of their new salaries, which often made up for several years. Those who had become depressed were excited and began to think of new ways to seek outside finds for their research. The result was amazing. When I became Interim Dean, our grant funding was about $6 million per year. Two years later it had doubled. The gamble that Wingfield had taken paid off.

My term as Interim Dean was over at the end of the academic year in June, but I decided to be considered for a permanent appointment. I wanted to see our changes carried through. After interviewing some others, the faculty again selected me for the job.

SIGNING THE MONTREAL PROTOCOL (1987-89)

# Oxygen and Ozone—Products and Protectors of Life

*The evolution of oxygen-producing cyanobacteria was arguably the most significant event in the history of life after the evolution of life itself. Oxygen is a potent oxidant whose accumulation into the atmosphere forever changed the surface chemistry of Earth.*

David Canfield

Our planet is unique among those in our solar system in that its atmosphere contains two unusual gases, diatomic oxygen ($O_2$) which we call simply 'oxygen,' and triatomic oxygen $O_3$, known as 'ozone.' These gases are two of several allotropes of the element oxygen. Both intercept incoming ultraviolet radiation, and in addition, ozone acts as a greenhouse gas. Both react to combine with other elements or compounds; this process is called 'oxidation.'

## 19.1 Diatomic Oxygen—$O_2$—'Oxygen'

It may surprise you to learn that oxygen, so vitally important to life, is the most abundant element in planet Earth. It completely dominates other elements in Earth's crust and mantle of the Earth. In terms of mass, it is about 44 % of these parts of the planet. But almost all of it is firmly locked up, bound with silicon (Si) in silicate minerals. As a matter of fact, you can think of the Earth as made of an enormous number of silica tetrahedra, $SiO_4$ (shown in Fig. 19.1) with other elements in the intervening spaces. This silicate oxygen is unavailable for reacting with anything.

The second largest reservoir of oxygen is more accessible, it is bound with hydrogen as water. With enough energy, water molecules can be split apart into their component parts, as shown in Fig. 19.2. The hydrogen gas molecule, having a low molecular weight (MW) of only 2 will work its way to the top of the atmosphere where it is likely to have a translational velocity high enough to escape into interplanetary space. With a molecular weight of 18, $H_2O$ is lighter than the nitrogen gas in our atmosphere, $N_2$ (MW = 28), and should work its way to the top of the atmosphere where it would be destroyed. Indeed, if Earth's atmosphere were not layered in such a way as to act as a cold trap for $H_2O$ as described in Chap. 18, the planet would steadily lose its water. This seems to be what happened to Mars, which once had water but is now a desert.

The other way in which $O_2$ is produced is through photosynthesis by plants. Photosynthesis uses the energy of photons to build chemical compounds that store it as chemical energy. The plants' energy collection devices are large molecules called chlorophylls (green) or carotenoids (yellow). Their colors are the wavelengths of light they do not use and reflect. The wavelengths they do use are shown in Fig. 19.3, plotted against the energy in the spectrum of the Sun's radiation at Earth's surface. Most of the absorption of radiation is in the range from near ultraviolet though cyan (350–550 nm). There are two major chlorophylls, a and b; chlorophyll a is in almost all plants. It takes advantage of the spectrum coming through the atmosphere. Water absorbs part of the spectrum, so aquatic plants must adjust. Chlorophyll b is a supplement in some plants such as green algae. Carotenoids cannot directly carry out photosynthesis, but collect energy and pass it on to chlorophyll a. The brown algae known as kelp and the microscopic diatoms have their distinctive colors because of carotenoids. There are a number of other pigments in plants that can serve to gather energy.

Photosynthesis is a complicated multi-step process, but can be expressed simply by the chemical equation

$$CO_2 + H_2O \rightarrow CH_2O + O_2 \uparrow$$

or in words: carbon dioxide and water are converted to sugar and free oxygen. The energy required for the conversion comes from sunlight. The actual sugars produced have the formula $C_6H_{12}O_6$, but in many discussions '$CH_2O$' is used as a generic term for any carbohydrate produced by a plant. There are two sugars that have the formula $C_6H_{12}O_6$, glucose and fructose. They differ in the way the atoms are arranged. Glucose is a sugar found in much of our food. Sucrose has a similar formula ($C_{12}H_{22}O_{11}$); it is our 'table sugar' and comes from sugar cane or sugar beets. Fructose comes from corn, and is in corn syrup but you can also get it in granular form. Fructose is almost twice as sweet as sucrose to our taste

© Springer International Publishing Switzerland 2016
W.W. Hay, *Experimenting on a Small Planet*, DOI 10.1007/978-3-319-27404-1_19

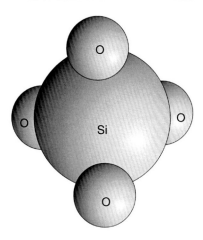

**Fig. 19.1** The silica tetrahedron. The structure common to most silicate minerals, and basic building block of planet Earth. The oxygen atoms are firmly locked in place

buds, and is used in commercial soft drinks. I have a Viennese cookbook that recommends using fructose rather than sucrose in recipes for pastries and candies because you get the same sweetness for half the calories.

Photosynthesis is not a very efficient process; only about 3–6 % of the light energy actually gets converted to chemical energy. Furthermore, it has an Achilles heel. A critical first step in the process involves an enzyme, 'Ribulose-1,5-bisphosphate carboxylase oxygenase,' commonly called 'RuBisCO' or simply RUBISCO. An enzyme is a substance produced by a living organism that facilitates a chemical process without being in the process. In the non-living world such a substance is called a catalyst. RUBISCO is also involved in the uptake of $CO_2$. Unfortunately, RUBISCO is

sensitive to heat stress. At elevated temperatures it begins to confuse $O_2$ with $CO_2$ and takes up the wrong gas. It starts a process that is the opposite of photosynthesis, 'photorespiration,' which adds to the burden of normal respiration carried out by the plant. The efficiency of RUBISCO declines at temperatures above 35 °C (=95 °F) and essentially ceases at 45 °C (=113 °F). This loss in efficiency can be counteracted by another enzyme, 'RUBISCO activase' produced by the plant in response to the stress, but only to a limited degree.

Further, more than one photosynthetic pathway has evolved. First of all, it is important to know that the plants take up $CO_2$ through special openings in the leaves, called stomata that can open and close. But when the stomata are open, the plant loses water through 'transpiration.' The classic sequence of events in photosynthesis is called the Calvin-Benson cycle. Plants that use it are called $C_3$ plants because one of the critical enzymes involved in $CO_2$ uptake has three carbon atoms. The stomata are open all day and photosynthesis takes place throughout the leaf. Other plants have developed a special method of handling the relatively low levels of atmospheric $CO_2$ that have persisted for the past 35 million years. They use a different photosynthetic pathway called $C_4$. These plants use another enzyme, called 'PEP Carboxylase,' for the initial uptake of $CO_2$. It works very quickly and delivers the $CO_2$ to RUBISCO for photosynthesis. This then takes place in specialized inner cells of the leaf; this special arrangement is called 'Kranz Anatomy.' Because of their ability to take up $CO_2$ rapidly, $C_4$ plants do not leave the stomata open as long, and thereby conserve water. In essence, $C_4$ plants can concentrate and store $CO_2$ for later use. There is a third pathway, called CAM

**Fig. 19.2** Dissociation of $H_2O$ by solar ultraviolet radiation

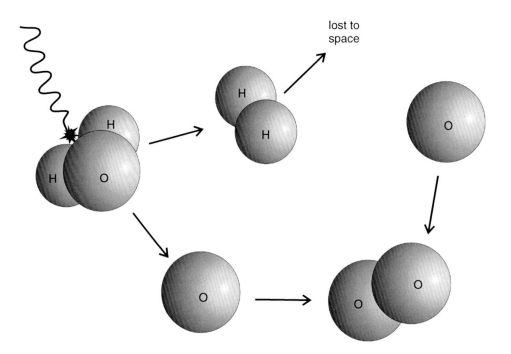

**Fig. 19.3** Absorption of different wavelengths of light by plant pigments associated with photosynthesis. The *squiggly purple curve* at the top is the energy of the Sun's spectrum at Earth's surface, calibrated on the right axis in terms of watts per square meter of Earth's surface for each nanometer (nm) of wavelength. The *green curve* is absorption by chlorophyll A; the *yellow curve* is absorption by chlorophyll B; the *orange curve* is absorption by carotenoids

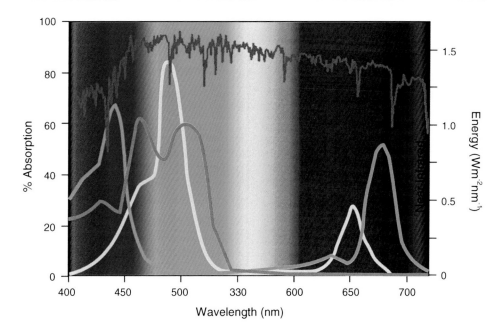

photosynthesis. CAM stands for 'Crassulacean Acid Metabolism,' named for a plant family using this route. In CAM plants, the stomata open at night when it is cooler and evaporation rates are lower and close during the day. $CO_2$ is taken up during the night, converted to an acid, and stored. During the day, the acid is broken down and the $CO_2$ is released to RUBISCO for photosynthesis.

Now, what is what? Most plants are $C_3$. They are more efficient than $C_4$ and CAM plants under cool and moist conditions and under normal light. Almost all plants that normally grow in the mid and high latitudes are $C_3$. $C_4$ plants photosynthesize more rapidly and grow faster than $C_3$ plants under high light intensities and high temperatures, but they prefer lower atmospheric $CO_2$ concentrations.

$C_4$ plants have a more efficient use of the available water, and have shallower roots than $C_3$ plants. $C_4$ plants include several thousand species in at least 19 plant families. Tropical grasses, which we have imported for our lawns, are $C_4$. Almost all fast-growing plants, like corn (maize), sorghum, sugar cane, and bamboo, all important to humans, are $C_4$. In fact, humans are the greatest friends $C_4$ plants could possibly have. We have spread them everywhere.

The CAM plants are more efficient in water use than either $C_3$ or $C_4$ plants. They live under arid conditions, and most desert plants such as cactus and succulents, and some orchids and bromeliads are CAM. Under unfavorable conditions they can even go into a holding pattern called CAM-idle. They leave their stomata closed day and night. The $CO_2$ given off in respiration is used for photosynthesis and the oxygen produced by the photosynthesis is used for

respiration. It sounds like this might go on forever, but Mother Nature takes her tax on energy, so it has its limits.

Why all this discussion about different plants and photosynthetic pathways? One of the most important aspects of our great uncontrolled experiment on planet Earth is almost never mentioned. How will plants react to the changing environment? $C_4$ and CAM plants appeared and expanded their rages between 10 and 5 million years ago. This was about 20 million years after $CO_2$ levels had dropped from high levels of 1,000 ppmv to the oscillations between 180 and 300 ppmv that have persisted since. The reason for the lag is unknown. But we are already approaching 400 ppmv; we already have higher atmospheric concentrations that Earth has seen for 35 million years. All life has evolved to adapt to the lower $CO_2$ concentrations. For at least the last 8 million years, $C_4$ plants have had an advantage of $C_3$ plants in lower latitudes because they are better adapted to the lower levels of $CO_2$. The advantage crossover is thought to be about 500 ppm $CO_2$. Above that level, $C_3$ plants gain the advantage again. That will happen in a few decades. We have no idea what will happen. It is a complicated world, too bad we do not yet fully understand it.

When a plant dies, the organic matter it produced through photosynthesis decays:

$$CH_2O + O_2 \rightarrow CO_2 + H_2O$$

The oxygen that was produced by photosynthesis is consumed in 'burning' the sugar to make carbon dioxide and water. The only way to accumulate $O_2$ in the atmosphere is to bury the organic matter so that it cannot decay. This is exactly what has happened over Earth's history. For every molecule of $O_2$ in the atmosphere there is a carbon atom in an organic molecule buried somewhere. Although the process is inefficient, production of atmospheric $O_2$ through photosynthesis completely dominates over any $O_2$ produced by dissociation of water vapor.

Here is another interesting thing. Oxygen is a reactive gas; its ability to react and 'oxidize' other material is dependent on its concentration and the temperature. With the present atmospheric oxygen concentration of 21 % the temperature required to start combustion of organic matter, such as dry grass, wood, or paper, is the 451 °F (=234 °C) made famous by Ray Bradbury's 1953 novel of that name. He imagines a world where books are outlawed and must be burned. As oxygen levels increase the combustion temperature decreases, so there is a natural limit on how high atmospheric $O_2$ levels can be. The 1967 fire in an Apollo space capsule that killed three astronauts occurred in a pure oxygen atmosphere at room temperature.

How much $O_2$ is where? Table 19.1 shows the major reservoirs, fluxes into and out of them and the residence time of a molecule in them. Residence time is simply the reservoir size divided by the flux rate. It is simply a measure of how sensitive the system might be to change.

Now an important question: can human activities change the concentration of $O_2$ in the atmosphere? Certainly not by burning fossil fuels. As noted above there must be an atom of carbon buried somewhere for every molecule of $O_2$ in the atmosphere. But the extractable, burnable fossil fuels (coal, petroleum, natural gas) amount to only about 0.015 % of the total amount of buried carbon. Most of it is finely disseminated in sedimentary rock. Perhaps it is worth mentioning here why the energy output of the different fossil fuels is different. Coal has a formula rather like $CH_2O$, natural gas (methane) has the formula $CH_4$, and petroleum is somewhere in between. You simply can't combine as much oxygen with coal as you can with methane, so for a given amount, methane yields more energy by burning. This is a good place to recall a comment by my fellow graduate student at Stanford, Keith Kvenvolden. Keith went on to work for NASA, looking for organic molecules in samples from the Moon and elsewhere, and to studying the distribution of organic carbon on Earth. Keith once said, "Petroleum is one of the marvels of nature; it contains so many

exotic organic compounds with unusual characteristics. Compounds that required millions of years to make, and that we cannot duplicate in the laboratory. And what do we do with this fabulous material? We burn it!"

There is however, another ominous possibility. We now know that at least half of the oxygen supplied to the atmosphere each year comes from phytoplankton in the ocean. These are minute single-celled plants. As will be discussed at greater length in subsequent chapters, part of the $CO_2$ from burning fossil fuels is absorbed by the ocean, gradually making it more acidic. At the same time the man-made fertilizers being used to increase food production on land are finding their way to the sea, further upsetting the natural balance. There is a clear trend toward depletion of the oxygen in the ocean. The phytoplankton are adapted to live in the ocean as it was for many millions of years. They are already responding to the changing ocean chemistry by changing the distributions of species and by the development of 'dead zones' where life has become impossible. The consequences of the anthropogenic perturbation toward a more acidic, less well oxygenated ocean are not well known. It is quite conceivable that they may result in the death of some or all of the ocean's phytoplankton. This would severely impact global oxygen production its fluxes into and out of the biosphere. The decay of organic matter resulting from a phytoplankton die-off would consume oxygen and produce carbon dioxide. That would be a strong positive feedback, accelerating acidification and oxygen depletion; a global catastrophe. It is rarely mentioned, but very real.

## 19.2   Triatomic Oxygen—$O_3$—Ozone

Ozone ($O_3$) is formed from diatomic oxygen ($O_2$) through interaction with energetic ultraviolet radiation in the upper atmosphere. The process is shown schematically in Fig. 19.4. An ultraviolet photon splits an $O_2$ molecule into its component atoms and these can then collide and combine with other $O_2$ molecules to make ozone, $O_3$.

This new molecule, with a molecular weight of 48, is heavier than $O_2$ and $N_2$, and works its way down through the mesosphere to settle into the stratosphere. At the base of the stratosphere it accumulates as the 'ozone layer.' The process of ozone formation consumes the energy of incoming short-wavelength solar ultraviolet radiation and protects life

**Table 19.1** Major reservoirs of oxygen

| Reservoir | Size (Moles of $O_2$) | Fluxes in/out (Moles of $O_2$ per year) | Residence time years |
|---|---|---|---|
| Atmosphere (as $O_2$) | $43.75 \times 10^{18}$ | $9.37 \times 10^{15}$ | ca. 5,000 |
| Ocean (as $O_2$) | $3.06 \times 10^{18}$ | – | – |
| Biosphere (as $CH_2O$) | $0.5 \times 10^{18}$ | $9.37 \times 10^{15}$ | ca. 50 |
| Lithosphere (as XO) | $9062.5 \times 10^{18}$ | $0.018 \times 10^{15}$ | ca. 500,000,000 |

Recall that a mole is one Avogadro ($6.022 \times 10^{23}$) of molecules

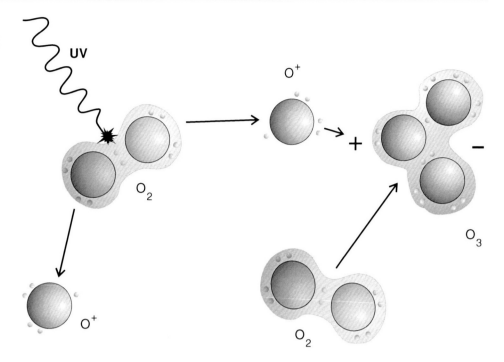

**Fig. 19.4** A solar ultraviolet photon splits an $O_2$ molecule; the fragments then collide with other $O_2$ molecules combining with them to make $O_3$

on Earth from its most harmful effects. In the stratosphere ozone absorbs longer wavelengths of ultraviolet radiation further reducing the intensity of that reaching Earth's surface.

The highest concentrations (about one or two molecules of ozone to one million molecules of air) are in the lower stratosphere, between 15 and 30 km above Earth's surface. The ozone there is responsible for the temperature inversion that allows the stratosphere to exist. At the inversion, the temperature of the air starts increasing upward, rather than decreasing upward as it does in the troposphere. Unlike the convecting lower part of the atmosphere, the troposphere, the stratosphere is stratified and stable. It is ozone that allows Earth to have this unique structure to its atmosphere. Remember that the stratosphere acts as cold trap for water vapor, preventing it from reaching the upper atmosphere where it would be broken down and the hydrogen lost to space. Ozone is responsible for ensuring that water remains on Earth.

Ozone is inherently unstable. It is a strong oxidizing agent and will break down and do its job if it encounters anything that can be oxidized. It is especially sensitive to chlorine (emitted by volcanoes), and bromine. Its oxidizing ability makes it a useful sterilizing agent. In Europe it is used instead of chlorine to ensure a clean water supply. Like chlorine, it kills bacteria and oxidizes other impurities, but it also breaks down naturally and leaves no taste in the water. It concentration in the stratosphere's ozone layer depends on its rates of production and destruction.

Ozone from the stratosphere works its way down into the troposphere. There it encounters many more pollutants in the atmosphere that are likely to destroy it. It is also produced

naturally by lightning. A number of human activities produce ozone in one way or another; in particular, combustion of fossil fuels and electrical sparks. It is also produced by energetic devices such as laser printers (you may have noticed the smell). Its concentrations are higher in populated areas where it is a component of smog. Amounts of tropospheric ozone are highly variable both in time and from place to place. Concentrations above 100 ppbv (parts per billion by volume) are considered unhealthy. It does have some absorption of wavelengths in the Earth's outgoing radiation and is a greenhouse gas. Its contribution to global warming is uncertain because of its uneven distribution and short lifetime, but it has been estimated to be responsible for as much as 1/3 of the warming observed through the 1990s. It contributes to the effect of cities being heat islands.

## 19.3 The Oxygen-Ozone-Ultraviolet Connection

The Sun radiates ultraviolet starting from end of the visible spectrum at 400 nm (nm) down 100 nm. This radiation is commonly classified as long, UV A (400–320 nm); medium, UV B (320–280 nm); short, UV C (280–100 nm); or very short (100–10 nm) with many other terms in this category. Figure 19.5 shows these categories and their absorption by $O_2$ and $O_3$.

Both $O_2$ and $O_3$ interact with the Sun's ultraviolet radiation, but in very different ways. As described above, diatomic oxygen molecules ($O_2$) are split by the incoming UV A radiation into single atoms or ions with an electrical charge. Before they can recombine, some these collide with

**Fig. 19.5** Major subdivisions of ultraviolet radiation in the rage 100–400 nm (=0.1–0.4 µm). *Gray area* is the extraterrestrial radiation at a distance of 1 AU (=Astronomical Unit = average Earth-Sun distance) from the Sun. The *black area* is the radiation received at Earth's surface at a Sun angle of 37° above the horizon (mid-morning or mid-afternoon). Data from Spectralcalc.com

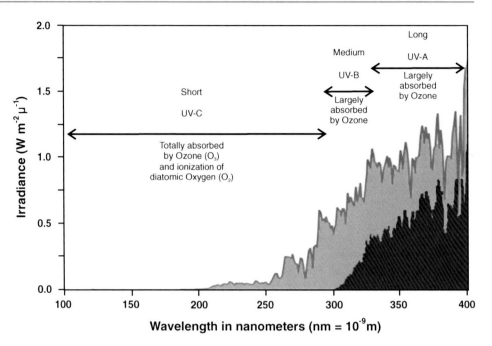

other oxygen molecules to produce ozone ($O_3$). Because ultraviolet, especially UV A and UV B, is effective in breaking down organic molecules and corrupting DNA, the presence of a protective ozone layer was essential to the development of life on land. UV C is inimical to life; in fact, it is used for sterilization. UV B, in small amounts enhances production of vitamin D in humans. It also enhances the production of melanin, the pigment which gives 'white' Caucasians a suntan. In larger amounts it causes sunburn, and can result in skin cancer. UV A is less dangerous. It is

visible to birds and many insects. Bug-zappers use UV A to attract insects. Figure 19.6 shows how the ultraviolet radiation is absorbed through the atmosphere.

You will notice that $O_2$ is four orders of magnitude less effective in absorbing ultraviolet radiation in this range than $O_3$, so you might expect it to be relatively unimportant. But the molecules of $O_2$ are five orders of magnitude more abundant than $O_3$ in the upper atmosphere (see Fig. 19.7), so in fact $O_2$ is the most important interceptor of the dangerous UV C radiation.

**Fig. 19.6** Portions of the spectrum from 200 to 325 nm absorbed by $O_2$ and $O_3$. The term 'Absorption Cross Section' on the *left* axis shows the relative intensity of absorption by the two gases. The absorption is expressed in terms of area, although in reality it is not an area. It is a quantum world thing, don't worry about it

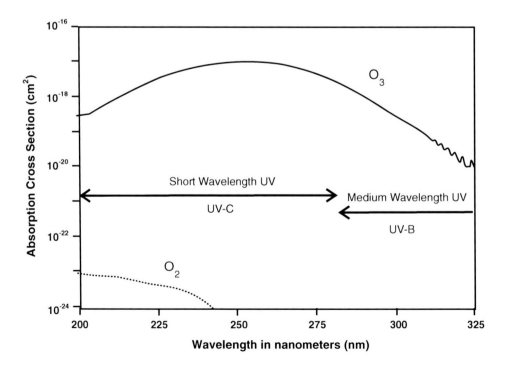

**Fig. 19.7** Ultraviolet radiation at the top of the atmosphere, at 30 km altitude, and at Earth's surface. Note that virtually all of the dangerous radiation in the 225–280 nm range is gone at 30 km

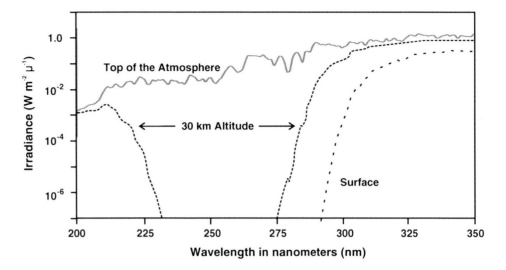

It is obvious that life on land would not be possible without $O_2$ and $O_3$ in the atmosphere, but $O_2$ and ultimately $O_3$ are products of life. When I was a student at SMU in the 1950s it was thought that almost all of the production of $O_2$ was by land plants. Those in the ocean were thought of as producing about the same amount as they consumed for respiration. Now we think both sources of $O_2$ are about equal.

## 19.4    The Oxygen-Ozone-Ultraviolet Conundrum

If $O_2$ has been introduced into our atmosphere largely through photosynthesis, how and when did it appear? Further, ozone could not exist if oxygen were not present in the atmosphere. Only very small amounts of oxygen could be produced in the uppermost atmosphere as radiation from the Sun's ultraviolet breaks down water into hydrogen and oxygen molecules. We know that today almost all of the oxygen in our atmosphere is produced by plants as a byproduct of photosynthesis. When the first space probes were sent to Mars to look for life, James Lovelock, author of the Gaia hypothesis, stated confidently that none would be found. The basis for his statement was that we already knew there is no oxygen in the Martian atmosphere.

The history of atmospheric oxygen has fascinated a number of us for years. We now know that the Earth formed as a solid planet about 4.5 billion years ago. There were already oceans of water on the planet by 4.4 billion years ago. There is also evidence from carbon isotopes that life forms were already busy replicating molecules 3.8 billion years ago. These early life forms did not use the energy from light to make their food supply, but they took advantage of chemical reactions. Some terminology: Organisms that produce their own food are 'autotrophs.' There are two kinds of autotrophs, those using chemistry, 'chemoautotrophs' and

those using light 'photoautotrophs.' Organisms that cannot make their own food and must consume other organisms or materials produced by them are 'heterotrophs.'

Much of our knowledge of chemoautotrophs has come in just the past decade from exploration of the life in deep sea sediments and in hot water vent systems ('hydrothermal vents') on the mid-ocean ridge. We have discovered forms of life we never imagined could exist. Most are previously unknown kinds of bacteria, but there are many higher organisms as well It has been estimated that as much as half of the planet's total mass of living organisms lives as bacteria and other primitive organisms in the pore spaces of ocean sediments and seafloor basalts. All of them are chemoautotrophs.

Recent geologic exploration has pushed back evidence for photosynthesis having evolved by about 3.5 billion years ago. But the photosynthesizing organisms had to live several meters down in the water to avoid being killed by the Sun's ultraviolet radiation. They were very gradually contributing $O_2$ to Earth's atmosphere. Then, about 2.4 billion years ago, something very important happened, the 'Great Oxygenation Event.' We are still discussing exactly what it was that happened, but somehow atmospheric oxygen levels rose to a few percent of what they are today. That is enough to start making ozone a significant gas and set in motion a positive feedback protecting photosynthetic life. This history is summarized in Fig. 19.8.

After the Early Proterozoic oxygenation event, oxygen levels dropped back, but then slowly rose so that by the beginning of the Phanerozoic, the time when fossils of higher forms of life appear, atmospheric oxygen was about 30 % of its modern value. Plants first appeared on land in the Ordovician, perhaps 450 million years ago. At first small plants limited to the edges of streams, they gradually spread onto the land surface. By the Devonian 380 million years ago they covered the entire land surface where water was

**Fig. 19.8** History of oxygen in Earth's atmosphere in terms of percent of the present level. After studies by James Walker and Robert Berner

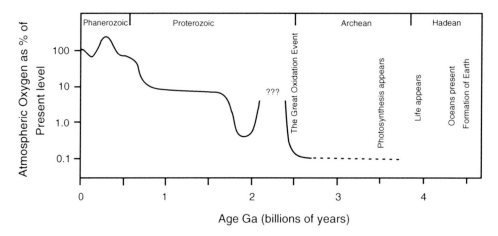

available, and were actively making soils. A few million years later they had developed the technology to elevate fluids high above the ground. Trees evolved and grew in swamps where their remains formed the coal deposits of the appropriately named Carboniferous. The enormous burial of organic carbon caused atmospheric oxygen levels to rise to as much as 30 % higher than today. Insects reached enormous size; there were dragonflies with a wingspan of 1 m (3 ft). By the end of the Paleozoic, fungi had learned how to decompose lignin, the decay-resistant component of wood, and this innovation meant the end of massive carbon burial and a lowering of atmospheric oxygen levels to near what they are today.

How all this worked out is a topic of ongoing research, but the early history of life on earth was probably facilitated by the faint young Sun. The temperature of the Sun's photosphere, its apparent surface, has been measure quite precisely to be 5778 K. From our knowledge of black body

radiation we can calculate that if, when it was young, it emitted only about 70 % of the energy it does today, its surface temperature would have been 5285 K. Figure 19.9 is my best guess as to what the faint Sun's ultraviolet spectrum may have looked like. Clearly there would have been less of the harmful short wavelength ultraviolet.

The early development of photosynthesis, oxygenation, and the interactions of life with the Sun's ultraviolet radiation are topics of great current interest and every year brings new ideas and understanding.

## 19.5  The Human Interference with Ozone

The idea that human activities might affect ozone goes back to 1970 when Dutch chemist-climatologist Paul Crutzen (1933) argued that nitrous oxide ($N_2O$) introduced into the

**Fig. 19.9** Comparison of black body radiation at 5778 K (*black line*) and 5285 K (*dashed line*). The *light gray area* is the present extraterrestrial radiation at 1 AU. The *back area* is a guess as to what the extraterrestrial radiation at 1 AU may have been 4.4 billion years ago. In both cases the actual radiation differs from the black body curve because of the complex interactions of energetic processes taking place in the plasma that is the Sun

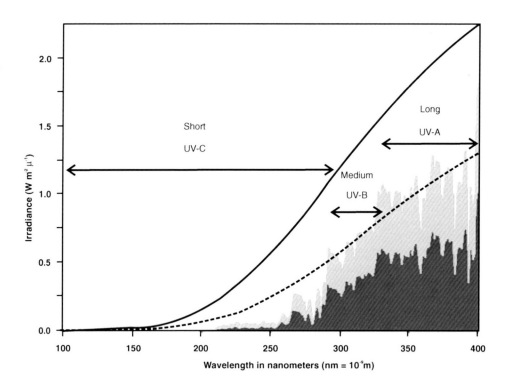

air from fertilizers would decompose ozone back to diatomic oxygen. Remember that man-made nitrogen fertilizers did not come on the market until after World War I, and exponential growth of the fertilizer industry did not begin until after World War II. It is another of mankind's great uncontrolled experiments. Crutzen's 1973 doctoral dissertation at the University of Stockholm was entitled "*On the photochemistry of ozone in the stratosphere and troposphere and pollution of the stratosphere by high-flying aircraft*". Measurements of stratospheric ozone showed that its concentration was declining.

On a cruise in the South Atlantic in 1971, James Lovelock (later of 'Gaia hypothesis' fame) found that almost all of the chlorofluorocarbons and bromofluorocarbons ('CFCs' commonly called 'Freon' and used in refrigerators and aerosol spray cans) that had been manufactured since their invention in 1930 were still present in the atmosphere. They were simply accumulating in the air. Then, in 1974 Frank Sherwood Rowland, Chemistry Professor at the University of California at Irvine, and his postdoctoral associate Mario J. Molina suggested that long-lived CFCs might also cause breakdown of ozone. The idea was that when the CFCs got into the stratosphere they would be dissociated by UV radiation releasing the Cl atoms which had already been shown to be even more effective than $N_2O$ at catalyzing the destruction of ozone.

The discovery of a hole in the stratospheric ozone layer over the Antarctic was made in the 1980s by British Antarctic Survey scientists Joseph Farman, Brian Gardiner and Jonathan Shanklin. In May 1985, they published a paper in *Nature* showing that ozone levels had dropped to 10 % below previous normals for Antarctica in its January summer.

It soon became evident that this feature was expanding rapidly. Holes in the ozone layer developed in both Polar Regions. Ultraviolet radiation was becoming more intense at high latitudes in the summer. Although the contiguous United States was not affected, northern Europe, Canada and Alaska were. In the early 1990s I was living in Kiel in northern Germany. Kiel is at 54° N, the same latitude as Churchill on Hudson Bay. Denmark, Sweden and Norway are further north. Kiel is on the Baltic coast. To the east is what we called the Baltic Riviera, where summer vacation beach towns have exotic names like California and Brazil. There would often be warnings of dangerously high UV levels, asking people not to stay out in the Sun long. It was more than annoyance. At those latitudes you spend long, dark cold winters indoors. When summer comes you want to enjoy the long daylight and warm days. When you are told that you can't stay outside long because the intense UV might cause cancer, it is quite depressing.

One of my colleagues in the University of Colorado and NOAA's 'Cooperative Institute for Research in Environmental Science' (CIRES), Susan Solomon, was instrumental in discovering what was depleting the ozone layer. In 1986 and 1987 she led expeditions to Antarctica. She was able to show that the abundance of chlorine dioxide, a product of the reaction of chlorine with ozone, was about 100 times greater than it should be under natural conditions. It was the reaction of ozone with chlorine and bromine from the decomposing CFCs that was the cause of the Antarctic ozone hole. Her work was critical in leading to the Montreal Protocol and the ban on compounds causing ozone depletion (Fig. 19.10).

The amazing thing is that most of the CFC chlorine that destroys ozone is not used up in the process. Much of it remain insidiously intact, so that after it destroy an ozone molecule it can move on to destroy another, and then another, and so on.

Needless to say, the industry making CFCs was outraged. The same folks, who are now arguing that there is no global warming or if there is warming it has nothing to do with burning fossil fuels, were then arguing that civilization would collapse without CFCs and that increased UV was a good thing.

However, in this case science won out over the need for quick profits. In 1987, representatives from 43 nations signed the Montreal *Protocol on Substances That Deplete the Ozone Layer*. The protocol is an international treaty designed to protect the ozone layer by phasing out the production of numerous substances believed to be responsible for ozone depletion. Meanwhile, the halocarbon industry (including DuPont in the US) had shifted its position and started supporting a protocol to limit CFC production (Fig. 19.11).

If nothing had been done, we would now be well on our way to sterilizing the land surface and upper waters of the Earth. Today CFCs have been almost completely phased out and the depletion of the ozone layer is being reversed. However, a strange phenomenon has developed in the Polar Regions. As ozone replenishment has begun, the global structure of the ozone layer has changed. Persistent holes on the ozone layer have formed so that the tropopause, the boundary between the troposphere and stratosphere, has become diffuse. This is affecting the high-latitude tropospheric circulation in both hemispheres, but is most strongly noticed around Antarctica. Figure 19.12 shows the Antarctic ozone hole at its maximum recoded extent, in September, 2006.

The economic disaster predicted by the halogen industry was replaced by a boom in innovation. Paul Crutzen, Sherry Rowland, and Mario Molina were awarded the 1995 Nobel Prize in Chemistry for their work on stratospheric ozone.

But the story is not over yet. The Montreal Protocol did not include the $N_2O$ from fertilizers. It has become the primary ozone destroyer of the 21st century.

**Fig. 19.10**  Susan Solomon and friends in the Antarctic

**Fig. 19.11**  Chlorofluorocarbons
and related compounds in the
atmosphere. The units are an
estimate of the total effective
amount of halogens (both
chlorine and bromine) that would
affect ozone in the stratosphere.
These compounds are also
greenhouse gases but highly
variable in their effect

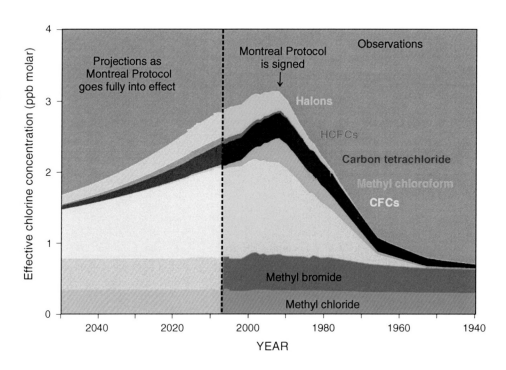

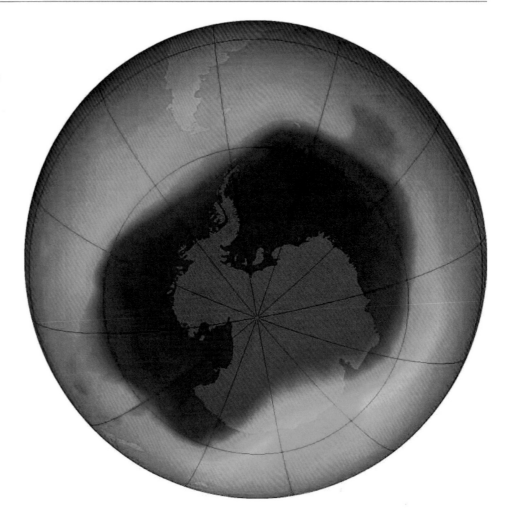

**Fig. 19.12** The Antarctic ozone hole in September, 2006; the largest extent it has had to date. *Shades* of *green* indicate normal levels of stratospheric ozone, *blue* indicates areas of depletion, *purple* indicates absence. Image from NASA

## 19.6  Ozone—The Greenhouse Gas

In the infrared spectrum, ozone is a classic greenhouse gas. It fills a gap in the effects of other greenhouse gases. Figure 19.13 shows its role in the overall scheme of things.

The absorption of the infrared in the stratospheric ozone layer causes it to warm and produces the upward increase in temperature that prevents convection. It is responsible for the structure of Earth's atmosphere, and the reason why our layered atmosphere differs from those of the other planets. The absorption region is rich in detail, as shown in Fig. 19.14. To me it resembles a spectrum of the sound of a crescendo in one of Anton Bruckner's symphonies.

The depletion of stratospheric ozone has caused the lower stratosphere to cool as the troposphere warms and the two effects have had some unexpected consequences. The ozone hole over the Antarctic has caused a reorganization of the winds that have protected the Antarctic from warming while at the same time causing greater than expected warming over the ocean surrounding the continent. The result has been more evaporation from the warmer water resulting in more snowfall on the ocean and an expansion of the circum-Antarctic winter sea ice. Conducting more than one uncontrolled experiment at a time is a messy business.

## 19.7  Summary

Diatonic oxygen ($O_2$) and triatomic oxygen ($O_3$, ozone) are very unusual gases to find in a planetary atmosphere. Although a very small amount of $O_2$ could be produced by ultraviolet radiation dissociating water molecules, it would be readily consumed in the weathering of minerals and oxidation of iron. The oxygen entering Earth's atmosphere is the result of photosynthetic organisms (plants) combining carbon dioxide and water to make sugars and free oxygen. The energy for photosynthesis comes largely from the blue part of the electromagnetic spectrum.

Decay of organic matter is the reverse of the photosynthesis reaction. Organic matter ('sugar') and oxygen from the atmosphere combine to produce carbon dioxide and water. The only way to leave oxygen in the atmosphere is to

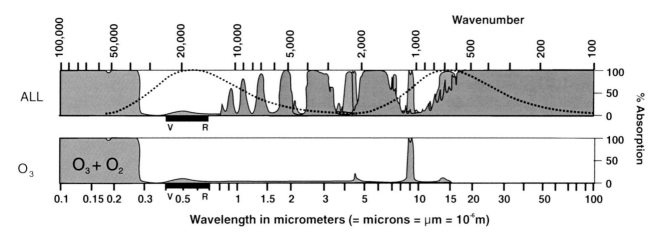

**Fig. 19.13** The role of oxygen and ozone in the overall pattern of greenhouse gas activities. Ozone fills in a narrow band between 8 and 10 μm (=800 and 1,000 nm)

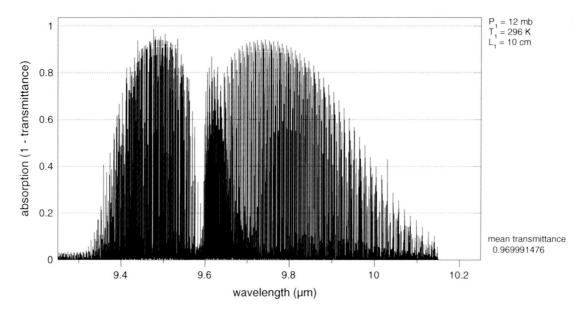

**Fig. 19.14** Part of the IR absorption spectrum of $O_3$ in the lower stratosphere. Calculated using Spectralcalc.com

bury the organic carbon (the 'sugar') so that it cannot decay. For every molecule of $O_2$ in the atmosphere there is an atom of carbon buried as organic matter. Once in the atmosphere, oxygen absorbs short wavelength ultraviolet radiation. This is critically important because short wavelength ultraviolet is inimical to life. It destroys DNA and other molecules essential for life. The UV radiation splits $O_2$ molecules into the individual atoms. Some of these then combine with $O_2$ molecules to make ozone, $O_3$. Ozone absorbs longer wavelengths of the ultraviolet. Together the two UV absorbers make the surface of the planet habitable.

Although photosynthesis evolved about 3.5 billion years ago, it took about 3 billion years for it to become effective enough to produce the levels of atmospheric oxygen we have today.

Ozone plays a double role. It is both an absorber of harmful ultraviolet radiation and an absorber of a part of the infrared spectrum not affected by other greenhouse gases. Ozone is formed near the top of the atmosphere and because of its greater molecular weight settles downward. It accumulates in the stratosphere which it heats by absorbing Earth's infrared radiation. It is responsible for the temperature inversion that allows the stratosphere to exist.

Human activities have interfered with the natural workings of this delicately balanced system. The man-made chloro- and bromofluorocarbons ('CFCs, Freons') widely used in the last century for refrigeration and aerosol spray cans break down in the stratosphere releasing chlorine and bromine. These act as catalysts enhancing the decay of ozone back to diatomic oxygen. If the ozone layer were to be lost, ultraviolet radiation reaching the surface of the Earth would be very destructive to life. The discovery of the ozone depletion in the 1970s and 80s led to an international agreement, the Montreal *Protocol on Substances That Deplete the Ozone Layer* in 1987. Production of CFCs has largely stopped and the stratospheric ozone layer is recovering. However, another ozone destroying chemical, nitrous oxide ($N_2O$) is released by man-made fertilizers and has become the major ozone destroyer of the 21st century.

A Timeline for this chapter (Ga = Billions of years ago; Ma = Millions of years ago):

| | |
|---|---|
| 4.54 Ga | Earth consolidates matter to become a solid planet |
| 4.4 Ga | Oceans exist of Earth |
| 3.8 Ga | Life exists on Earth |
| 3.5 Ga | Photosynthesis appears |
| 2.4 Ga | The Great Oxygenation Event |
| 500 Ma | Atmospheric oxygen reaches 10 % of its modern value |
| 450 Ma | Plants begin to invade the land, limited to streams at first |
| 380 Ma | Plants now cover the land surface |
| 350 Ma | Coal swamps develop; atmospheric oxygen reaches 130 % of its modern value |
| 250 Ma | Fungi learn to decompose lignin, atmospheric $O_2$ concentration declines toward the modern level |
| 1970 | Paul Crutzen argues that nitrous oxide lignin ($N_2O$) introduced into the air from fertilizers can cause ozone to decompose to diatomic oxygen |
| 1971 | James Lovelock discovers that almost all of the chlorofluorocarbons and bromofluorocarbons manufactured since their invention in 1930 are still present in the atmosphere |
| 1974 | Depletion of stratospheric ozone is becoming apparent |
| 1974 | Sherry Rowland and Mario J. Molina suggested that long-lived CFCs might also cause breakdown of ozone |
| 1985 | British Antarctic Survey scientists Joseph Farman, Brian Gardiner and Jonathan Shanklin show that ozone levels had dropped to 10 % below normal for the Antarctic summer |
| 1986–87 | Susan Solomon leads expeditions to Antarctica, discovering that the abundance of chlorine dioxide, a product of the reaction of chlorine with ozone, is about 100 times greater than it should be under natural conditions |
| 1987 | Representatives from 43 nations sign the Montreal *Protocol on Substances That Deplete the Ozone Layer* |
| 1995 | Paul Crutzen, Sherry Rowland, and Mario Molina are awarded the Nobel Prize in Chemistry for their work on stratospheric ozone |
| 2006 | Antarctic ozone hole reaches its maximum extent |

If you want to know more:

Stephen O. Andersen and K. Madhava Sarma, 2002. *Protecting the Ozone Layer: The United Nations History*, Routledge, 513 pp.

Music: Somehow Richard Wagner's Siegfried Idyll seems appropriate to celebrate the Montreal Protocol.

Libation: A Highland single malt Scotch with a rich peaty flavor. Consult Michael Jackson's book.

## Intermezzo XIX. The Cretaceous Problem

Before I start, let me explain something about the state of geology as a science in the 1960s. It was a descriptive science as opposed to an experimental science. Physics, chemistry, and biology are experimental sciences; you can perform experiments and observe their outcomes. Astronomy and geology are descriptive sciences; we can make observations, but we can't make experiments on the object we are observing. Of course, there are some small areas where once can perform experiments—in the study of igneous rocks one can make melts of different materials and let them crystallize to see what minerals form and in what order. And with sediments, one can experiment with wave tanks and flumes to see bow bedding features are produced.

When I got my Ph.D. paleontologists were called 'stamp collectors' by their colleagues. We collected fossils, identified them, named them if they were new species, and we used them to determine the age of the rocks that contained them. As best I can remember, only paleobotanists, the guys dealing with fossil plants, attempted to interpret the environments in which they had lived.

Our tools had always been simple, dental tools and scrapers to prepare the larger fossils, a magnifying glass or hand lens, and the trusty binocular microscope for microfossils. For the nannofossils we needed as high a magnification as one could get with an optical microscope. I was very lucky when I joined the faculty at the University of Illinois; Ken Towe, who later moved on to the U.S. National Museum in Washington, was an expert electron microscopist and helped me get started with that new technique. Then in the later 1960s another major improvement in technique became available, the Scanning Electron Microscope. My fellow paleontologist in the Illinois Geology Department, Phil Sandberg, and I helped the University acquire one of the first of these new instruments in the United Sates. It revolutionized the way we could look at our microfossils. We realized that there were a number of new things we could do with the fossils, but the main interest remained their use in determining age.

All of that changed in the mid-1960s but the revolution in using fossils to infer environmental conditions went almost unnoticed in the much larger revolution of plate tectonics and sea-floor spreading.

In the early part of the 20th century Thomas Chrowder Chamberlin at the University of Chicago had proposed two major ways by which the Earth's climate would have been different in the past. The first was, following Svante Arrhenius' idea that the concentration of $CO_2$ in the atmosphere might have varied through time. He thought that might explain the alternation between glacials and interglacials. The second was that there might have been a reversal of the deep sea circulation. At present, the deep waters of the ocean are cold, reflecting the fact that they sink in the Polar Regions. They return to the surface in the low latitudes. That the cold deep waters must originate in the Polar Regions had been known for over a hundred years. Chamberlin suggested that when the Earth was warmer, as during the Cretaceous, the saltier waters of the tropics might have sunk into the ocean interior and returned to the surface in the Polar Regions. They were interesting ideas, but few geologists had given them any credence.

It was on Leg 4 of the Deep Sea Drilling Project in 1969 that I began to seriously worry about the climate of the Cretaceous Period (145.5–65.5 million years ago). It had been known for almost a century that polar temperatures were warm enough to support tropical vegetation. Cretaceous rocks in North America were famous for their dinosaur fossils. You have seen marvelous animations of them in the film 'Jurassic Park.' The dinosaurs in the film are almost all of Cretaceous, not Jurassic age.

On Leg 4 I realized that the fluctuations of the calcium carbonate compensation depth might reflect changes in ocean chemistry. Although I didn't realize it, it was a revolutionary idea. It was published in the *Summary and Conclusions* section of the DSDP Initial Report for Leg 4, in 1970. Nobody read it. The consensus at the time was that the chemistry and salinity of the oceans had remained constant for billions of years. The chemistry of the oceans was thought to be purely the result of inorganic reactions between water and minerals. The classic paper on this topic, by Lars Gunnar Sillén had been published in 1967.

But then, in 1972, Wally Broecker published his great paper *A kinetic model for the composition of seawater*. Although you might not guess it from the title, the paper shows how marine organisms control the concentrations of many of the ions in seawater. It removed the straightjacket that had constrained our thinking.

What did we actually know about the Cretaceous? In Miami we worked on reconstructing the paleogeography, at least the positions of the continents. We

didn't know much about the distribution of mountains, but made some guesses. It was obvious that sea-level had been higher in the Cretaceous. A large area of the continents had been flooded, but we didn't know much about the general elevations of the continental blocks at that time. We made a lot of guesses. A Visiting Professor from Oregon, Bill Holser, introduced us to the idea that the salinity of the ocean had not been constant in the past, but must have changed. Over the Phanerozoic lot of salt had been removed from the oceans and stored on land as evaporite deposits—salt, gypsum and anhydrite. And we were just discovering through the DSDP that there was a lot more salt stored beneath the floors of marginal seas, like the Gulf of Mexico and Mediterranean. We were discovering that large areas of the oceans had become anoxic, devoid of oxygen. This simply cannot happen today because even if somehow every living thing in the ocean were suddenly to die, the decomposition of the dead organic matter would consume only half the oxygen in the ocean. In 1960 geologists had known so much about the Earth and its history that our science was concerned with making a few minor revisions to the story, sort of mopping up. By 1970 we were becoming aware of just how much we did not know.

The puzzle of a warm Earth with tropical temperatures in the Polar Regions gave rise to several questions. Overall, the temperature contrasts on the Earth were much smaller than today. The climate was said to be more 'equable.' First, how could the Polar Regions become so warm? The implication was that the heat that accumulated in the equatorial region was being transported poleward more efficiently. But the driving mechanism for the winds is the temperature difference from one place to another. Temperature controls the density of the air, and it is the density differences between different air masses that produce the pressure gradients that power the winds. With warmed polar regions, the winds should have been slower, and the transport of heat less efficient. This became known as the heat transport conundrum. But could it be that the ocean circulation was more vigorous, and it was the oceans that had carried the heat poleward? Unfortunately, the surface currents of the ocean are driven by the wind. They should have slowed too.

Could the Earth's albedo have been different? In Leningrad Mikhail Budyko had begun to explore the climatic effects of changing Earth's albedo and published several important papers on the topic in 1968 and 1969. The original publications were in Russian, but the ideas began to percolate through the iron curtain. One of my colleagues at Scripps did some calculations on the effect of removing the present reflective polar ice, and taking into account the larger area of dark ocean. It worked; simple calculations yielded something like a Cretaceous climate. Only one small problem, his Earth had no atmosphere, no clouds, and they are 60 % of today's albedo. Back to the drawing board.

When I started my joint arrangement between Illinois and Miami, I was able to take advantage of the expertise in physical oceanography and atmospheric science in Miami. Claes Rooth was a physical oceanographer who loved to speculate about how the ocean circulation might respond to different forcing factors, such as changing the speed of the winds and the evaporation-precipitation balance. Eric Kraus was an equally adventuresome atmospheric scientist who liked to think about how different atmospheric gas compositions might affect its structure and circulation. We were also able to organize short courses with well-known geochemists such as Dick Holland of Harvard and Bob Garrels of Northwestern as instructors. They told us about the chemical reactions involved in weathering rocks, and how life interacted with the inorganic world. These short-courses were largely funded by tuition charged the industry participants, with academics attending for free.

Our frustration in trying to understand the enigma of the 'warm Earth' grew until finally, as recounted in Chap. 15, we sent one of our best students, Eric Barron, off to the National Center for Atmospheric Research in Boulder to do some numerical climate modeling. It was in the late 1970s that a consensus began to form that the culprit in the 'warm Earth' drama was atmospheric $CO_2$. It was the only thing that answered all the questions. I remember well that I still didn't understand why just changing the content of $CO_2$ in the atmosphere would preferentially warm the Polar Regions and result in an 'equable' global climate. Warren Washington at NCAR explained to me that the Earth already has an even more important greenhouse gas, water vapor, but its concentration depends on temperature. So the Earth always has a greenhouse effect in the tropics. Figure 20.4 shows this effect. But if the poles are cold, there is no water vapor in the air. There it is the concentration of $CO_2$, the second most important greenhouse gas, that is important. If its concentration is low, the poles are cold; if its concentration is high, the poles are warm. Simple as that.

But then, once you begin to think about how the atmosphere and ocean might circulate on a warm Earth you realize that many of the things we take for granted are no longer true. My ideas on this are given in more detail in Sect. 6 of Chap. 25, but here is a preview:

Without polar ice Earth becomes a very different planet. Today both Polar Regions are cold, in winter and summer, because of the high albedo of the ice. But if you replace the ice with water or land, the Polar Regions will experience seasonal reversals of temperature. In the Cretaceous this is complicated by the fact that the Arctic is water, the Antarctic land. This means that the reversals are in the opposite sense in each hemisphere. The Arctic had a high pressure in summer, low in winter, while the Antarctic had a low in summer and high in winter. The same geographic situation exists today.

Not only today's atmospheric circulation, but that of the ocean as well, depend on the stability forced by polar ice. Without the ice as a stabilizing factor, the winds would shift latitude and become stronger or weaker with the seasons. If the winds shift, the ocean currents become unstable. The polar sites of deep water formation can shut down and be replaced by tropical sites. The entire vertical structure of the ocean changes. Today the ocean is highly structured, with a number of distinct layers. My best guess is that the Cretaceous ocean was completely different—very unstructured. Unfortunately, after 40 years of work we are still in the process of sorting things out.

ROMAN PHILOSOPHER TITUS LUCRETIUS CARUS
FIGURES OUT THE HYDROLOGIC CYCLE (CA. 50 BCE)

*Some scientists believe climate change is the cause of unprecedented melting of the North Pole, and that effects these very uncertain weather patterns. I think we should listen to those scientists and experts.*

Dalai Lama

It has been obvious to observers since ancient times that water evaporates into the air, falls out again as precipitation, and flows down rivers back to the ocean. The hydrologic cycle was described in some detail by the Roman author Lucretius in his work "On the Nature of Things" written about 50 BCE. He noted that although rivers kept flowing into the ocean, sea-level didn't change. He also observed that if you look at the ocean on a cold day you can sometimes see 'steam' rising from its surface and disappearing into the air. He concluded that this was what reappeared as clouds and rain. We now know that the rainfall on land that runs back as rivers to the sea is equivalent to a layer of water 10 cm thick (0.01 m) over the entire surface of the ocean each year.

Geologists like to use the term 'residence time' as an index to process rates. Residence time for water is the average length of time it would take for all the water in the ocean to cycle through the atmosphere, precipitate as rain, and flow back in. The average depth of the ocean is about 3800 m. So if 10 cm flows in each year, the residence time is 380,000 years. Of course, this calculation makes the assumption that each molecule of seawater goes through the cycle only once. Nevertheless it is a useful way of emphasizing that the water spends most of its life in the ocean; once every 380,000 years the molecule gets to spend a year cycling through the atmosphere, falling as rain or snow on land, and becoming part of a river flowing back to the sea. The residence time for a water molecule in the air is only a few days.

Water vapor, or simply 'vapor,' is the gaseous phase of water. It evaporates directly into the air wherever water is available, or enters through the transpiration of plants. It is the most important greenhouse gas, as shown in Fig. 20.1. It has absorption bands in the infrared part of the solar spectrum as well as throughout most of the longer wavelength part of Earth's thermal infrared spectrum. But there is a vapor 'window' in the Earth's thermal infrared from about 8–18 μ. This is very important to the way the Earth works as a planet because the only greenhouse gas that captures energy that passes through this gap is stratospheric ozone.

You can easily recognize the importance of water vapor as a greenhouse gas by comparing the day-night temperature difference in a dry area with that in a humid region. In desert regions the heat during the daytime may be almost unbearable, but at night it is cold. The Texas panhandle is not a desert, but it often has cloudless clear skies and low humidity. A day-night temperature difference of 55 °F (=28 °C) is likely to occur several times each winter. In the equatorial region and tropics day-night temperature differences are minimal, often only a degree or two, because of the high vapor content of the atmosphere.

The very uneven and variable distribution of vapor in the atmosphere affects the way Earth radiates energy back into space. We usually think of the Polar Regions as the most effective sites of re-radiation, but deserts are very important too. All of this unevenness in the radiation balance has a major impact on the weather.

## 20.1 The Behavior of Dry Air

Air, being a compressible fluid, changes its temperature with pressure. Rising dry air cools because it expands in response to the declining pressure of the thinner overlying atmosphere. Comparison of the right- and left-hand scales on Fig. 20.2 shows how the air pressure changes with elevation in the lower troposphere. The temperature of the air reflects its

**Fig. 20.1** The electromagnetic absorption/emission bands of water vapor compared with the overall greenhouse gas absorption and emission, and the details of vapor absorption/emission in parts of the Earth's thermal infrared spectrum. Note that it is a very effective energy absorber/emitter of many of the wavelengths longer than 20 μ

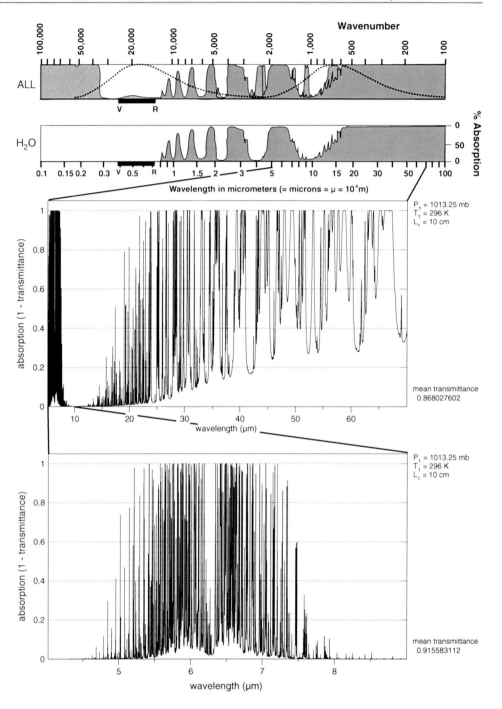

energy content. As the air rises its sensible heat, the temperature you can measure with a thermometer, is converted into what is called potential or gravitational energy. This is converted back into sensible heat when the air descends.

The rate at which the temperature declines with elevation is termed the "lapse rate." For completely dry air, the lapse rate in the troposphere is about $10.8\ °C/km^{-1}$, as shown in Fig. 20.2. If the dry air rises or descends without exchanging heat with its surroundings it is said to change adiabatically.

Its temperature change follows the almost straight diagonal lines in Fig. 20.2.

I live in the Rocky Mountains, where the air is usually very dry. A third of the atmosphere lies below the crest of Colorado's Front Range. My house is at an elevation of about 2,400 m ($\sim$8,000 ft). Boulder, Colorado, where the University is located, is about 1,650 m ($\sim$5,400 ft), a difference of 750 m. The expected temperature difference is about 13.5 °C (10.8 °F) and this is usually the case. But if I

**Fig. 20.2** Change in temperature as dry air rises or descends. Starting with its temperature at a particular elevation the rising or descending air changes temperature along the diagonal lines. Atmospheric pressure is shown on the left. Americans still often use 'millibars' (1000 mbar = 1 Bar = ~1 Atmosphere); the rest of the world uses 'hectopascals' which are numerically the same. Note that the spacing between units indicated on the left decreases toward sea level, reflecting increased compression of the air. Note that half of the air on Earth is below the Tibetan Plateau

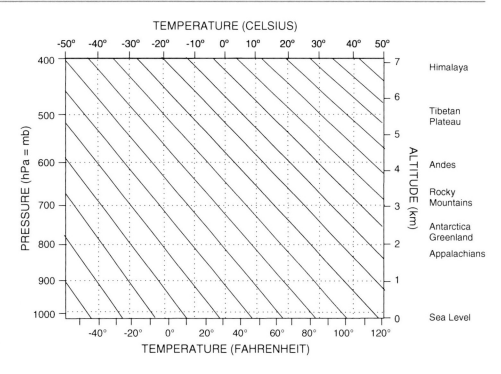

## 20.2 The Behavior of Wet Air

know a bit more I can tell whether I will run into snow as I drive up the canyon.

Air is never completely dry. Its ability to hold moisture approximately doubles with every 10 °C increase in temperature. Figure 20.3 shows the mass of the water vapor that can be accommodated in saturated air at different temperatures and elevations or pressures. At the surface of the earth, the moisture content of the air may vary from <0.05 g $H_2O$/kg at −50 °C (as in Antarctica) to >60 g $H_2O$/kg at a scorching 45 °C. Water vapor will condense as raindrops or snow when it becomes supersaturated. Because the temperature of the air goes down with elevation, most the vapor is in the lower half of the troposphere, and because the surface temperature decreases from the Equator to the poles, it is concentrated in the air at low latitudes. It is useful to know that wherever there are large expanses of water, the water controls the air temperature. The air takes on the temperature of the water.

There is a powerful positive-feedback greenhouse effect wherever water is available to evaporate. In the warm tropics the water evaporates putting large quantities of vapor in the air; this intercepts the outgoing radiation, further warming the air and water. In the Polar Regions, temperatures are so cold that almost no vapor is in the air, so its greenhouse effect is effectively nil. This is a very important point in understanding how Earth's greenhouse system works. Warm water produces a greenhouse, cold water does not. Like the

ice-albedo feedback, this is another classic example of a positive feedback in the climate system. It acts to keep the tropics warm and the poles cold.

This peculiar greenhouse gas distribution is shown schematically in Fig. 20.4. The greenhouse effect of vapor is essentially restricted to warmer regions and the lower part of the atmosphere. We can't refer to it as a 'vaporsphere' because its distribution is not spherical. I suppose we could refer to the 'vaporspheroid' or maybe even more precisely as a 'vaportoroid.' You can try dropping those terms at a cocktail party.

You will remember that an enormous amount of energy is involved in the transformation of water to vapor. This energy is called the latent heat of vaporization. 'Latent' means present or potential but not evident or active. The atmosphere transports energy from one place to another as sensible heat, as the kinetic energy in the winds, and as latent heat in the water vapor. The kinetic energy of the wind turns out to be only about 1 % of the total energy transport. But the transformation of water to vapor through evaporation and then back to water through precipitation is about 60 % of the atmospheric energy transport system. If you stop and think about it for a moment—most of the energy transport in our atmosphere is accomplished by a substance that doesn't exist in any abundance on any other planet in our solar system.

Because of the rapid increase in the capacity of warm air to hold water vapor, very large amounts of energy are required to saturate it. You will recall that in the 1970s it was argued that the large amounts of energy required for evaporation from the sea surface to saturate the overlying air with

**Fig. 20.3** Amount of water in saturated air. The diagonal lines are grams of vapor per kilogram of air. Note that the vapor content increases very rapidly with temperature

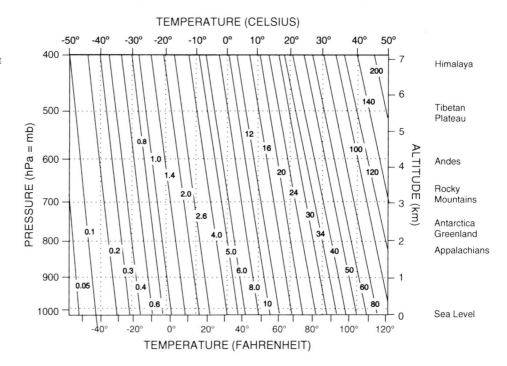

vapor would limit the temperature of tropical seas to less than 30 °C. It was proposed that this was a thermostat acting to keep the Earth from getting too warm. You will also recall that the geologic evidence is that tropical ocean temperatures in the Cretaceous were about 34 °C, suggesting that under the right circumstances the thermostat can be turned up. If you go on the internet at look at a map of sea surface temperatures today you will see large areas that are above 30 °C. This thermostat doesn't seem to be working very well.

As saturated air rises, the temperature decrease resulting from expansion reduces the amount of water that can be

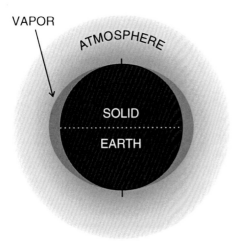

**Fig. 20.4** Schematic representation of Earth, showing its atmosphere and the distribution of the greenhouse gas vapor. Vapor forms a toroid around the equatorial region and tropics, but moves northward and southward with the seasons

carried as vapor. The excess water is transformed back into the liquid phase as droplets which appear as clouds. If the excess is large enough, the droplets increase in size and may fall out as precipitation. In the transformation from vapor to liquid, the latent heat of vaporization is released as sensible heat, warming the air. As a result, the lapse rate for vapor-saturated air is less than that of dry air. Rising saturated air that does not exchange heat with its surroundings follows the path indicated by the curved sloping lines ("wet" or "saturated adiabats") in Fig. 20.5. At very cold temperatures the amount of moisture in the air is negligible, and the lapse rate does not differ appreciably from that for dry air. However, at warm temperatures the lapse rate is very much less than that for dry air. For saturated air at 30 °C near sea level the lapse rate is less than 5 °C/km. At lower elevations and warmer temperatures the wet adiabats for vapor-saturated air are steep. At higher elevations, as the air cools and the saturation capacity lessens, the wet adiabats begin to slope from upper left of the figure to the lower right. Above the middle of the troposphere the moisture content of the atmosphere is so small that the wet adiabats approximate dry adiabats. The global lapse rate for planet Earth is usually given as 6.5 °C/km. This is an average of dry and saturated lapse rates in the mid-latitudes. However, this is the number I use to figure out whether I will run into snow as I drive up the canyon from Boulder to my home in the winter. If snow is about to fall, the temperature at my home, 750 m above Boulder, can be expected to be about 4.8 °C cooler than Boulder. If the temperature in Boulder is 2 °C I can expect to see the road covered with snow about half way up the canyon.

**Fig. 20.5** Change in temperature as air saturated with water vapor rises or descends

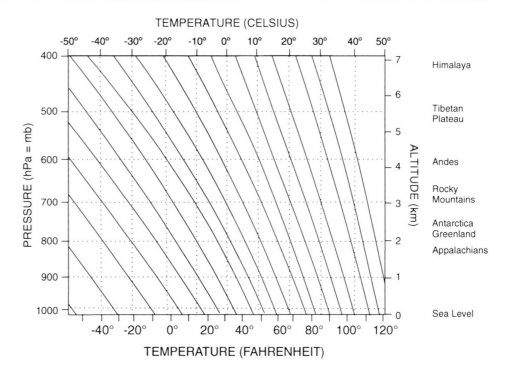

One aspect of the transport of latent heat by water vapor is to moderate temperature changes in the interiors of the continents. The moisture picked up over the oceans is precipitated as rain over the continents moderating the effects of seasonal changes. In summer, the effect is to heat the upper levels of the troposphere, while the rainfall into the warmer lower levels and onto the surface evaporates cooling them. In winter the primary effect is to raise temperatures in the lower troposphere and at ground level.

## 20.3   What Controls Atmospheric Water Vapor?

One of the larger red herrings in the greenhouse gas controversy is that water vapor is the major greenhouse gas, much more abundant than any of the others, and so it is the real problem. The others are not worth worrying about. This idea goes back to our old friends John Tyndall and Thomas Chrowder Chamberlin. Chamberlin had at first been enamored of the idea that changes in the concentration of carbon dioxide could be an explanation of the ice ages as had been suggested by Svante Arrhenius in 1896. However, after the turn of the century the importance of water vapor as a greenhouse gas became better known, and Chamberlin became one of its strongest proponents. Of course, he was thinking about natural, not man-made changes.

The problem with water vapor as a greenhouse gas is that its abundance depends on temperature, doubling with every 10 °C rise. This is a very strong positive feedback. If the amount of another greenhouse gas increases, it will receive massive reinforcement from water vapor. Is there any way human activity can affect the amount of vapor in the atmosphere?

Although the usual answer to this question is a resounding no, that is not strictly true. Humans have already inadvertently altered part of the hydrologic cycle. You can immediately imagine that the construction of artificial lakes might have an effect. Man-made lakes and ponds are estimated to have an area of about 259,000 $km^2$. That is about 1/1000th the area off natural lakes, rivers, and swamps, so if there is any effect it is likely to be very local. But there is an anthropogenic effect; one that not many people have considered. It is the effect of human alteration of plant communities.

Most of the land surface is covered by plants. Plants make their own 'food' through combining carbon dioxide and water to make sugar and free oxygen, which escapes into the atmosphere. You recall the general photosynthesis reaction:

$$CO_2 + H_2O \xrightarrow[Light]{} CH_2O + O_2 \uparrow$$

The carbon dioxide comes from the atmosphere, but the water comes from the soil. To obtain carbon dioxide, the leaves of plants have openings called stomata, which can be opened or closed as necessary for the photosynthetic activity of the plant. The oxygen produced by photosynthesis leaves the plant through the stomata, but at the same time, the plant loses some of its water as well. This loss of water as the plant breathes is called 'transpiration.' In the dark, when plants are not photosynthesizing, plants take in oxygen and lose water to the atmosphere by breathing. Just like animals, they must breathe to carry on life functions. Animals also lose water when they breathe; you can see the vapor condense when you breathe out in cold weather. So, the water that falls as rainfall soaks into the soil. From there it is taken up by the roots of the plants, and eventually finds its way into the leaves, where the photosynthesis factories are located. There it evaporates back into the atmosphere though the stomata.

The evolution and spread of plants onto land occurred in the Silurian, about 440 million years ago. For a brief period they were confined to the shores of streams and lakes, but after a few million years they had conquered hillsides. As they covered the land they spread rainfall into the continental interiors making conditions there more suitable for their colonization. They have had a great impact on weathering, erosion, and climate ever since. They radically altered the Earth's hydrologic cycle and by increasing the water vapor over the land, enhancing the greenhouse effect over large land areas. Plants have modified the climate in such a way as to promote their own existence.

Where I live in Colorado it is about 1,500 km to the nearest ocean. Some of the moisture in rainfall gets here directly from the ocean, brought by major storms. However, most of the rain that falls here is recycled though plants. It has fallen one or more times as rain, gone into the soil, up through the plants, and back into the air. Often the summer days are punctuated by the 3 PM rain, which lasts for an hour or so. It is water recycled by the plants and falling back to earth.

In western Amazon Basin, the rainfall is water that has been recycled six or more times. Clearly, the plants on land run the show as far as the long-term vapor content of the atmosphere over land is concerned. There are short-term increases in vapor content as major storms pass through, but the background level of vapor content is determined by plants.

The problem is, not all plants are equal when it comes to transpiration. This topic has already been discussed in the preceding Chapter, but it is so important in introducing vapor into the atmosphere that I'll take the liberty of going over it again.

The photosynthesis equation presented above is, as you might suspect, an oversimplification. Photosynthesis involves a number of steps, using complex molecules to facilitate the process. Most common plants go through a sequence of steps known as the 'Calvin cycle.' One of the molecules involved at a critical step in this cycle contains 3 carbon atoms. All plants using the Calvin cycle are called $C_3$ plants. Most trees and shrubs and herbaceous plants other than tropical grasses are $C_3$ plants. The leaves of the plants have openings, called stomata, which lead to internal chambers where atmospheric gases can be exchanged with the plant tissues. Although these plants take in $CO_2$ and through photosynthesis release $O_2$, they also respire, consuming $O_2$ and giving off $CO_2$. During the day photosynthesis dominates, and plants are able to grow, storing food. The situation reverses at night, only respiration goes on, not photosynthesis. During the day, the balance between photosynthesis and respiration is controlled by the atmospheric ratio of $CO_2$ to $O_2$, the temperature, and the brightness of the light. $C_3$ plants have their maximum efficiency at high levels of $CO_2$, low levels of $O_2$, temperatures between 15 and 25 °C, and medium illumination. Their efficiency decreases with decreasing levels of atmospheric $CO_2$, temperatures above 25 °C, and in bright light. $C_3$ plants typically have deep roots, and can tap the moisture in the deeper layers of soil. Because their stomata stay open during the day to take in $CO_2$, they transpire readily and return much of the moisture from rainfall to the atmosphere. They play an important role in recycling water over land areas. You may have noticed that when you walk into a forest, the air feels cooler. It is. Some of the sensible heat in the air went into evaporating water from the leaves of the trees, cooling the air. A curious feature of this transpiration effect is that if you measure the lapse rate (the decline in temperature with elevation) as you climb a mountain, you will find that it is different from the lapse rate measured by a weather balloon ascending directly from the valley below. This is because the plants are releasing vapor into the air, and it is more concentrated near the land surface creating a special near-surface 'microenvironment.'

Many grasses and sedges, and some other plants typical of grasslands, savannas, and semi-arid regions utilize a photosynthetic pathway which is markedly different from that of the $C_3$ plants. In these plants the photosynthesis process is more complex and the steps take place in different parts of the leaf. At a critical step the process involves a molecule with four carbon atoms, so that these plants are known as $C_4$ plants. The $CO_2$ taken in from the atmosphere is first concentrated by about a factor of ten, and then this air with concentrated $CO_2$ is passed along to another part of the leaf for the photosynthesis proper. These plants use the stored concentrated $CO_2$ to carry out photosynthesis during the daylight hours. Although energy is expended in concentrating the $CO_2$, the overall efficiency of photosynthesis is greatly enhanced and water loss is reduced.

It is not known when this alternative photosynthesis pathway evolved, but the spread of $C_4$ plants is thought to be an adaptation to the lowering of atmospheric $CO_2$ levels during the Cenozoic. They became widespread about 8 million years ago. The fastest growing plants, maize (corn), sorghum, sugar cane, switchgrass, amaranth, finger millet, and bamboo are all $C_4$ plants. Their maximum efficiency occurs at temperatures between 30 and 40 °C and under bright light. They are adapted to warmer, drier, brighter conditions than $C_3$ plants. They typically have shallow roots and remove moisture only from the upper layers of the soil. Because of their efficient use of water and inability to use water in the deeper soil layers, they have an important effect on the hydrologic cycle. They reduce the return of water to the atmosphere by as much as 25 % compared to $C_3$ plants. The water they do not use percolates downward into the groundwater system. Since they return so much less water to the atmosphere, they promote drier conditions downwind. They create conditions in which only they can survive.

One group which uses $C_4$ metabolism, the tropical grasses, is probably the most abundant land plants on earth. Grasses have several special adaptations—their leaves are produced from the central stem at ground level rather than from the tips of branches. Hence they can quickly recover from grazing. They are also capable of becoming brown and dormant during dry periods; again because of the way in which leaves are produced they can produce green leaves in a matter of days after a rainfall. They are able to choke out other herbaceous plants and survive drought. Their ability to become dormant and recover rapidly enhances unevenness in the rainfall. The development of $C_4$ grasses has played an important role in spreading aridity and desertification of the mid-latitudes over the past 8 million years. In the United States we have imported these grasses and use them for our lawns, especially in the South: Bermuda grass from the African savannah (not Bermuda), St. Augustine grass, native to southern North America, the Caribbean and West Africa, and, you guessed it, Crabgrass.

I first became aware of the role that the evolution of $C_4$ plants has played in Earth's history through some numerical climate experiments performed by Robert DeConto at the National Center for Atmospheric Research (NCAR) as he was working on his Ph.D. at the University of Colorado. I was his advisor. Rob had run a series of models of the Cretaceous testing the effects of different levels of carbon dioxide. It differentially warmed the Polar Regions as expected, but a very serious problem remained. The continental interiors remained very cold in winter. There is a famous (to those interested in the Cretaceous anyway) site in eastern Siberia along the Grebenka River, a tributary of the Anadyr, which has an extraordinary assemblage of fossil leaves from the Cretaceous. These leaves are from plants which do not tolerate freezing temperatures. It was at nearly the same latitude then as it is now. Then as now, it was about as far from any ocean as it is possible to be. Today the site has extremely cold winter temperatures. The problem was to find out how the interiors of large land areas could remain warm in the winter.

Rob's climate simulations used a very elaborate scheme for taking the effects of plants on climate into account: changes in albedo, transpiration, soil development, and other factors for 110 plant types. Jon Bergengren at NCAR had devised this scheme as an ancillary part of the climate computation. It is known as EVE, the Equilibrium Vegetation Ecology model.

EVE works like this: A uniform plant cover over all the continents is used as an initial condition. The climate model is run, generating monthly temperature and rainfall patters. Using this information, the plant cover is modified replacing the original plants with those appropriate to the new climatic conditions. The model is run again and new climate data are generated. Again the plant cover is modified to be compatible with the climate. After thirty or so iterations, the plant cover and climate come into equilibrium. The obvious intricacy of this part of the climate model gives some insight into how complex the models are, and why supercomputers are required to do the calculations.

Rob and Jon decided to go through the list of plant types and remove those which had not yet evolved in the Cretaceous. These included all of the $C_4$ plants. Running the climate model without the $C_4$ plants, the continental interior warmed enough to have its coldest winter temperatures near freezing. This was caused by the greater moisture content of the air, which had the effect of moderating both winter and summer temperatures.

Today, humans are $C_4$ plants' best friends. In warmer regions, such as the US Midwest, we replace $C_3$ plants with fast growing $C_4$ crops like corn. In recent decades large areas of tropical rainforest have been cut down to provide pastureland. These massive replacements of $C_3$ plants with $C_4$ plants have as a result the general drying out of the land downwind. You will recall that around the time of the American Revolution, it was thought that the climate was being improved by cutting trees and making the country more like Britain. In fact, without understanding the processes involved, Thomas Jefferson got it right with his intuitive feeling that the climate would be less humid and more tolerable if the forests were cut down. What no one suspected then and few do today, was that these agricultural changes set the stage for drought.

At the higher latitudes of northern Europe, the major grains are wheat, rye, and barley; all of these and other grains that grow well under cooler conditions are $C_3$ plants.

Several years ago I gave a lecture on this topic at a university in Germany. Afterward I was approached by some agricultural scientists in the audience who had not been aware of the water retaining properties of $C_4$ plants. They were exploring the possibility of genetically 'improving' these plants so that they could use $C_4$ metabolism and grow faster. They realized that there needed to be some serious consideration to the possible climatic consequences of this kind of genetic engineering.

The extreme in water conservation in the plant world uses what is called CAM metabolism. As you will recall, CAM stands for Crassulacean Acid Metabolism. These plants close the stomata during the daytime to prevent water loss and open then at night to take in $CO_2$ when it is cool and water loss can be kept to a minimum. Exotic as it sounds you may well have some of these plants in your home. Jade plants belong the Family Crassulaceae. Other plants using this special water conserving metabolism are the cacti, and the pineapple. CAM plants are beautifully adapted to dry conditions, but have little further effect on the regional climate.

## 20.4  Anthropogenic Effects

Another of man's activities is also having an effect on climate, but we know little about it. Jet aircraft often fly in the stratosphere. When jet fuel burns two of the products are $CO_2$ and $H_2O$. We see the water initially as man-made clouds—contrails. They soon evaporate introducing vapor into the stratosphere. Because the temperature decline with elevation in the troposphere causes water to precipitate out as rain or snow, vapor is caught by this 'cold trap' and does not ordinarily get into the stratosphere. Our planet protects its water with something no other planet has, the tropospheric-stratospheric 'cold trap.' If the cold trap were not there we would already have lost most of our water to space. Cosmic rays and energetic radiation from the Sun dissociate water molecules at the top of the atmosphere into hydrogen and oxygen. The hydrogen can easily escape into space. This dissociation phenomenon may explain why Earth seems to have been developing a stratosphere even before $O_2$ producing plants became common, but this is just speculation today, as intense research on this subject goes on. Today the vapor introduced into the stratosphere intercepts some of the infra-red radiation from the Sun as well as longer infra-red coming from below. It should act to reinforce the warming effect of ozone.

The use of groundwater for irrigation is another important anthropogenic effect. Not so much where the irrigation is applied directly to the plants, but where the water is shot into the dry atmosphere to fall back as artificial rain. In California and the US Midwest there are large rotary irrigation devices on wheels that spew water from wells into the air. When you fly over the area between the Mississippi and the Colorado Rockies you will see circular areas on the ground where these inefficient machines are at work. Much of the water they are using entered the groundwater system thousands of years ago.

## 20.5  The Changing Area of Exposed Water Surface

You might not think of it, but the area in which water is in contact with the atmosphere can change, introducing more vapor into the atmosphere. The place where the biggest change is possible is the Arctic. The Arctic has been covered by sea-ice for a long time. I wrote a paper a few years ago arguing that it has been continuously ice-covered for the past 35 million years. As in most of science I was looking for something completely different when I realized this. I won't go into the details here, but there is good evidence that the Arctic has remained mostly ice-covered even during the interglacials. As I noted above, if there is a large expanse of water, the air immediately above it takes on the temperature of the water. Ice, however, is a good insulator. The Arctic Sea Ice used to have an average thickness of about 3 m. The seawater below the ice has a temperature of $-1.5$ °C; the air above has a winter temperature of $-30$ °C and a brief summer temperature just above 0 °C. That is a winter temperature difference of over 30° in just three months. In the past 30 years about 1 m of the ice has melted, and the melting rate is accelerating (Fig. 20.6). When the ice is gone, the temperature of the air will take on the temperature of the water, jumping up 30 °C. The water vapor content will increase eightfold just because the ice is gone. It will continue to increase rapidly as the air warms, and voila!—we have a powerful positive greenhouse feedback in the Arctic caused by the water vapor. At the same time the Earth's northern polar region substitutes a dark ocean for its reflective ice surface, another positive feedback even taking into account the steep angle of incident sunlight. These were changes Mikhail Budyko had already anticipated and discussed with us in our meeting in Leningrad in 1982.

Let's think for a moment about how much energy is pouring into the Arctic to melt the ice. To melt 1 m in 30 years is equivalent to melting 0.03 m in 1 year. Ice has a density of about 931 kg/m$^3$, somewhat less than seawater at its freezing point, so it floats. The energy required to melt a layer of sea ice 0.03 m thick with an area of 1 m$^2$ is

$$\underset{\text{Volume}}{\left(0.03\,\text{m} \times 1\,\text{m}^2\right)} = \underset{\text{Mass}}{\left(0.03\,\text{m}^3\right) \times 931\,\text{kg/m}^3 = 27.7\,\text{kg}} \times \underset{\text{Heat of fusion}}{\left(0.335 \times 10^6\,\text{J/kg}\right)} = 9.36 \times 10^6\,\text{J}$$

Spread over a year, how many watts is this? One watt is a Joule per second. There are 31,556,943 ($31.6 \times 10^6$) seconds in a year, so the energy to melt the ice at this rate is

$$9.36 \times 10^6\,\text{J}/31.6 \times 10^6\,\text{s} = 0.296\,\text{J/s} = 0.296\,\text{W}$$

**Fig. 20.6** Melting Arctic sea ice. Image courtesy Michael D. Lemonick

The difference between having a stable sea-ice cover and melting sea-ice seems to be very small. However, over 30 years the energy required to melt a cubic meter of sea ice (913 kg) is $306 \times 10^9$ J. If that had been a cubic meter of water 1 at 0 °C, and you had put the same amount of energy into it without letting any of it evaporate, its temperature would be

$$\underset{\text{Latent heat of fusion of water}}{0.335 \times 10^6 \text{ J}} \quad / \quad \underset{\text{Specific heat of water}}{(4.18 \times 10^6 \text{J}/^\circ\text{C})} = 80\,^\circ\text{C} = 176\,^\circ\text{F!}$$

Of course the Arctic surface waters would not actually reach this temperature because they would mix with the deeper and surrounding waters. But some of the energy would go into evaporation, raising the $H_2O$ content of the Arctic air and further enhancing the greenhouse effect.

The good news is that melting sea ice does not cause sea level to rise. The ice is floating in the water, and it does not change the water level as it melts. It floats because ice is less dense than water; a given mass of ice occupies more space that the equivalent mass of water so it floats. Also, all that energy going into melting sea ice is preventing the water from getting any warmer. The summer air temperature over the ice is close to 0 °C because that is the melting point.

The bad news is that when the Arctic ice is gone, in a few decades, the Arctic Ocean will begin to warm. In the Arctic there is a layer of relatively low density water on top of much saltier denser water. The layer is about 30 m thick. If, over 30 years, the same amount of energy continued to be introduced to the Arctic, that layer would warm by about 8 °C, from −1.5 to 6.5 °C, still not warm enough for comfortable swimming, but on its way to becoming the Arctic Riviera. To understand why all this energy is concentrated into preferentially warming the polar region, you need to understand how $CO_2$ works. That's in the next chapter.

## 20.6   Summary

Water vapor is a powerful greenhouse gas. Its content in the atmosphere depends on water availability and temperature. Over the oceans and coastal regions it has a positive feedback, increasing temperatures where they are warm and having no greenhouse effect in regions that are cold. It is an amplifier of the effects of other greenhouse gases. Over land, the main source of atmospheric water is through transpiration by plants. Humans have changed the conditions that originally existed by promoting the spread of $C_4$ plants that return less vapor to the atmosphere than $C_3$ plants. In case you have just finished that second Martini, I will take the liberty of reminding you that $C_3$ plants are trees shrubs, and herbaceous plants other than tropical grasses—essentially all of the plants native to North America before the Europeans came. The $C_4$ plants are typically fast-growing species, and include maize (corn), sorghum, sugar cane, switchgrass, amaranth, finger millet, bamboo, and the tropical grasses such as Bermuda grass, St. Augustine grass, and the highly successful but unwanted crabgrass. The effect is to reduce rainfall over land while reducing the greenhouse effect. The most dramatic change that is occurring on Earth today is the melting of the sea-ice cover on the Arctic Ocean. It is too late to stop it, and its consequences are sure to be dramatic climate change. Unfortunately we do not know exactly what will happen

A Timeline for this chapter:

| 1896 | Svante Arrhenius suggest that $CO_2$ might be the a cause of the ice ages |
|------|---------------------------------------------------------------------------|
| 1899 | T.C. Chamberlin proposes that reductions of atmospheric $CO_2$ caused the ice ages |
| 1903 | Chamberlin begins to recognize the importance of water vapor as a greenhouse gas |
| 1923 | Chamberlin recognizes water vapor as the dominant greenhouse gas |
| 1950s | Jet aircraft begin to inject water vapor into the stratosphere |
| 1982 | Arctic sea-ice meltback begins, exposing polar water |
| 2007 | Major meltback of Arctic sea-ice meltback occurs |
| 2012 | A larger meltback of Arctic sea ice occurs |

If you want to know more:

I've already recommended my favorite books on atmospheric science but perhaps a change of pace is in order: Dale DeGroff, 2002. The Art of the Cocktail. Clarkson Potter Publishers, 230 pp.

Music: Benjamin Britten: The Four Sea Interludes from the opera 'Peter Grimes.'

Libation: At this point you may wish to make yourself a gin Martini and think about all this. Of course, you may wonder about how much vermouth to add to the gin. Noel Coward suggested that the best Martini is made by "filling a glass with gin then waving it in the general direction of Italy." If you can, use one of those conical glasses that fits into a bowl full of ice (Fig. 20.7). If you keep putting your glass back into the bowl, the Martini will stay at about the freezing point and not become diluted. Taste it from time to time. Unless you foolishly shook the Martini with ice rather than stirring it, thereby bruising the gin, it will continue to have the same taste. James Bond liked to have his Martini shaken because he preferred it made with vodka made from potatoes. Strictly lower class. Upper class vodka is made from grain—there is a world of difference. Anyway, once the ice in the bowl has melted, your Martini will warm and soon taste insipid. That is

the sea-ice melting effect on a scale a human can understand. Incidentally, you can't do the math properly here because the heat capacity of olives varies with the stuffing.

When you realize what all this Arctic sea-ice melting implies you may need a second Martini.

### Intermezzo XX. Fund-Raising

There was another aspect to being Head of an institution like the Rosenstiel School that many people do not think of: fund raising. It is a very important part of the duty of the administration of any private non-profit organization, like the University of Miami. The founder of Miami's Marine Institute, F.G. Walton Smith, had become a master at the art, but he passed the money through his International Oceanographic Foundation (IOF). He always had a fear that if he operated directly through the University, the 'main campus' in Coral Gables would claim a part of what he brought in. Further, if the main campus got directly involved with the donor it might completely divert the donation. Many non-profit institutions have separate Foundations to receive funds, so Walton was simply following a well-established path.

I never knew quite why, but in 1974, just after I had moved permanently to Miami, Walton put me on the Board of Trustees of the IOF. I was the only faculty member of the School to receive that honor. The other Board members were a very distinguished group of wealthy individuals. My role was simply to add credibility to the operation. I learned some of the fundamentals:

The first requirement in having a successful organization to support science, medical research, or environmental studies is to have a large membership of small donors. In the 1970s the annual dues to the IOF were $25.00. In return the members received a quarterly publication, *Sea Frontiers*, which had articles written for the general public on various aspects

**Fig. 20.7** Apparatus for pondering the climatic effects of melting Arctic sea ice. **a** Initial condition. **b** Final condition after two experiments—as perceived by the investigator

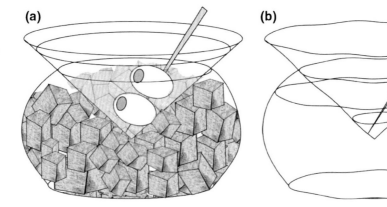

of marine science. The authors were often from the School. Walton's wife, Mae, was the editor. This was a break-even proposition. It costs as much to send out dues notices and the magazine as the membership dues bring in.

However, this opens the door to another level of donors. Anyone, or any charitable foundation, interested in making a large donation to such a non-profit will want to know how many small donors there are. It is the small donor base that determines how much a major donor will give. It is said that for many of these organizations about 20 % of the income is from thousands of small donors, and 80 % from a few large donors.

Walton had cultivated the right people and had built a formidable Board of Trustees, so that when he retired from the University he was able to call on them to fulfill one of his dreams—to build a museum to educate the public about marine science, *Planet Ocean.*

Interestingly, the 1968 gift of $12 million to the University of Miami by Lewis Rosenstiel had come out of the blue. About $8.8 million went to the Institute of Marine Sciences. In 1969 the Institute was renamed the Dorothy H and Lewis S Rosenstiel School of Marine and Atmospheric Sciences. The remainder of the gift was used to match federal funds to complete a building under construction at the University's Medical School which also bears their names. It was the largest single gift the University had ever received. Lewis Rosenstiel had founded Schenley, the whiskey company, and was reported to have had a lifelong interest in marine science.

For fund-raising I had a special problem because I did not want to compete with Walton's IOF operation across the street at Planet Ocean. But I too needed independence from the main campus fund-raising operation. Provost Clyde Wingfield and his staff saw to it that I got my own fund-raising officer, Al Veliky. Al began to actively pursue contacts, and made sure I got onto the Miami social circuit. Soon I was in a tuxedo three nights each week, attending balls and fund-raisers for other charities and institutions. I always paid my own way to those gatherings.

Before I knew it I was dating women who knew their way around through Miami's social circuit. My favorite companion was Evelyn Spitalny, who only the older readers of this book may remember. In the 50s and 60s 'Phil Spitalny and His All Girl Orchestra' were regulars on radio and TV. Their star was 'Evelyn and Her Magic Violin.' Evelyn was Phil Spitalny's

wife, but he had passed away some years earlier. She was not only an accomplished violinist, but had a vast knowledge of music and we had a great time together.

After his death in 1975, Lewis Rosenstiel's third wife, Blanka, spent part of each year in Miami, and I got to know her. She had been born in Warsaw, and emigrated to the United Sates in 1956. When she married Lewis in 1963 she was 32 and he was 72. Blanka founded the Chopin Foundation of the United States in 1977. I had always been a fan of Chopin, and she knew it, so I became a Trustee of the Foundation. The Foundation held a competition each year to help select young American pianists to go to Warsaw for the International Chopin Competition there.

Somehow the editor of the social page of the Miami Herald, Grace Wing Bonet, decided I was someone she wanted to promote. I became the 'dimpled Dean' and 'most eligible bachelor in Miami.' Many of the duties of the President of the University of Miami, Henry King Stanford, had been taken over by Provost Wingfield, so Henry's main function now was looking for donors. We were often at the same affair, and although in competition with each other, we got along wonderfully well.

Donations to oceanographic institutions do not always come in the form of money. Yachts are often offered as gifts. The donor will take a tax deduction for the estimated value of the yacht. As you might guess, the donor's estimate and the actual resale value of the vessel did not always match, and sometimes a boat needs work before it can be resold. It is customary for the institution to hold the vessel for a year or two, to show that it has had a 'use related to the mission of the institution.' I was in friendly competition with Walton for yacht donations. Both of us had 'related uses.' For a marine institution, one of the 'related uses' of a yacht is entertaining prospective donors.

Some of the yacht donations were quite valuable, in the $1 million range. But we would also get offered racing boats ('Cigarettes') that had probably been used for drug smuggling. We quietly did a lot of checking out of potential donors' credentials.

I had a very astute evaluator of the condition of potential donations in Jim Gibbons, our operations manager. I also found a close friend in the yacht brokerage business, Alex Balfe, who could give us estimates of the market value of what we were being offered. We never informed our potential donors of the results of our investigations, just an acceptance or rejection of the offered gift. When I later became

Director of the University of Colorado Museum I discovered that there is a parallel activity in potential gifts of paintings, artifacts and other objects to museums.

A number of the faculty at the School were from Europe, and we decided we would institute the tradition of Fasching/Carnevale/Mardi Gras with an annual all-night costume party at the School on the Saturday before Mardi Gras ending with breakfast at dawn. Our original intent was that it was just for faculty and students, but Al Veliky realized its potential and invited some prospective donors. We had three ticket prices, one for students, another for faculty and staff, and another for the general public—all priced out to cover costs, and for the outsiders, to provide us a nest egg for the next year's event. After our first Marine Mardi Gras, Grace Wing Bonet decided it should become a major event on the Miami social calendar,

but implied that one needed to make a significant contribution to the school to get an invitation, something we had not thought of. Outside participants in costume became as abundant as our students.

Working 8 AM to midnight most days was really more than I had anticipated when I first accepted the job. After all, I had started out as an academic scientist. My studies in micropaleontology had led to an interest in ocean history and understanding how the climate system works. I had never intended to become an administrator. Walton Smith explained to me that as Dean I would have very little time for my own scientific interests. He told me I would have to satisfy my scientific curiosity by following the research achievement of the School's faculty. I missed my laboratory.

But there was a major goal to be achieved—construction of the building the Rosenstiel gift had envisioned.

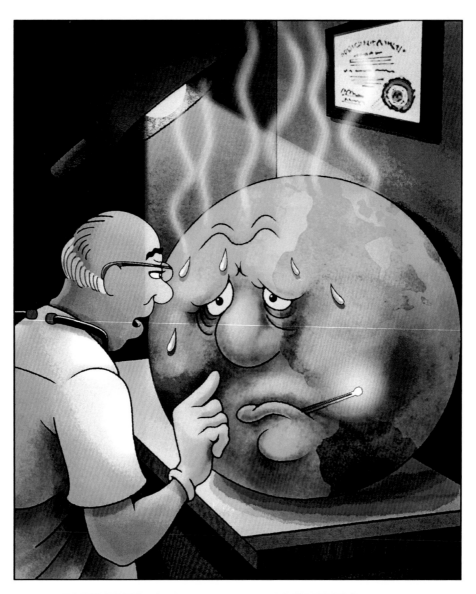

"TAKE TWO ASPIRIN, GET BACK INTO ORBIT,
AND LAY OFF THE CARBON DIOXIDE"

*We're running the most dangerous experiment in history right now, which is to see how much carbon dioxide the atmosphere can handle before there is an environmental catastrophe.*

Elon Musk

Now we finally come to something I know about from personal experience. Investigating the role of carbon dioxide in climate change is the line of research that got me invited to participate in the conference in Leningrad described in the first chapter. So let me tell you the convoluted story of how I got involved in this field, and then I will give you an over view of how the $CO_2$ system works. I hope to give you a feeling for the complexity of the system and how tightly it is interwoven with life, both in the ocean and on land. For reference, Earth's major reservoirs of carbon are shown in Table 21.1.

As you may recall, I started out my professional career as a micropaleontologist. It was in that role that I was a member of the Shipboard Scientific Party on Leg 4 of the Deep Sea Drilling Project in 1969. The project was organized into 2-month cruise legs using the ocean-going drill ship *Glomar Challenger*, and had begun in August 1968. Each leg drilled as many as 20 holes in the ocean floor, and recovered cores of the sediments. The first DSDP leg had made two incredible discoveries, oil-soaked sediment in the bottom of the Gulf of Mexico and black shales rich in organic carbon in the deep Atlantic just east of the Bahamas.

The second leg crossed the North Atlantic and discovered something very odd. In the middle of the Atlantic is a Ridge extending from Iceland to south of South America and Africa. The Mid-Atlantic Ridge had been discovered during the 1840s when telegraph cables were being laid between America and Europe, but it had not been mapped until after World War II. It consists of subsea lava flows, commonly called 'pillow basalts.' Basalts are made of feldspars in a matrix of volcanic glass formed when the lava contacted the ocean water. Submarine lavas cannot flow like rivers as they do on the surface, but typically advance as a series of balloon-shaped blobs, appropriately called pillows. The common deep-sea sediments on the flanks of the Mid-Atlantic Ridge are calcium carbonate 'oozes' and 'red clay.' The ocean margins are bordered by sands, silts and clays eroded off the continents. The carbonate oozes, which look (but do not taste) like coffee ice cream, are made of the shells of planktonic foraminifera and coccoliths produced by calcareous nannoplankton that lived near the surface. They accumulate at a rate of 1–2 cm per thousand years. The carbonate ooze blankets the slopes of the Mid-Atlantic Ridge down to a depth of about 4,000 m. Below that depth the sea floor is covered by red clay, which is mostly dust that has settled from the atmosphere into the ocean. It accumulates at a rate of a millimeter or so per thousand years. Of course, the dust settles everywhere, but when it is mixed in with 10–20 times as much calcareous material you don't notice it. The boundary between calcareous ooze and red clay marks a depth boundary in the ocean. It marks the boundary between waters sufficiently saturated with $CaCO_3$, so it doesn't dissolve, and deeper waters that are corrosive to $CaCO_3$ and dissolve it. It is not apparent in the sediment until the carbonate was dissolved away. In the Atlantic this boundary, called the Calcium Carbonate Compensation Depth (CCCD) is very sharp, only 100 m or so thick. In 1969 the conventional wisdom was that the boundary between carbonate sediments and red clay, the CCCD, in the Atlantic was caused by the increasing pressure of the water.

The Leg 2 Shipboard Scientists thought that when the ship drilled into red clay, it would extend right down to the basaltic submarine lava flows that floor the ocean basins. They still had difficulty recovering cores, but even with incomplete records they were startled to find that the red clays were underlain by carbonate ooze. The sequence should have been the other way round; carbonate oozes should be deposited in shallower water and should always be on top of red clays. It was evident that there was a lot we didn't yet know about the oceans.

Leg 3 crossed the South Atlantic, and by this time the drilling crew had figured out how to recover intact sediment cores on a routine basis. This was 1968–1969, and the ideas of sea floor spreading and plate tectonics were new and many geologists considered them untested, unproven hypotheses. Leg 3 drilled a transect of holes through the sediment to the basalt basement across the South Atlantic.

Using the fossil planktonic foraminifera and calcareous nannofossils to date the sediment just above the basalts, it was found that the age of the sea floor increased systematically away from the crest of the Mid-Atlantic Ridge—just as predicted by the hypotheses of sea-floor spreading and plate tectonics. In just 2 months these speculative notions had been supported by solid facts.

Leg 3 also found the same strange sediment sequences as in the North Atlantic. But here the sediments sometimes went from calcareous ooze to red clay and back again to calcareous ooze before reaching the basalt. Since the CCCD was supposed to be a pressure level in the ocean, it was proposed that the whole Atlantic Ocean floor must have moved up and down by as much as 1 km to account for the changes in the boundary between red clay and carbonate ooze.

I went aboard the Glomar Challenger in Rio de Janeiro for Leg 4. We drilled a few holes off Brazil and then went on to our major objective, the Caribbean. At that time most of what we knew about planktonic foraminifera came from studies in Trinidad, where the geology is extremely complicated. We expected the geologically simple Caribbean to have the most beautifully preserved fossils and most complete geologic record for the last 65 million years. My job was to identify the calcareous nannofossils in the cores and determine the age of the sediments. It really didn't take much time to do this, so I had time to do other things as well. Instead of the pure calcareous oozes full of pristine fossils that we had expected, our Caribbean cores were interlayered red clay and carbonate ooze. It was a great disappointment. I started plotting all the information that had been gathered so far. It was soon obvious that the ocean floor could not have moved up and down; if it had there would be records of the sea level changes by marine sediments covering large areas of the continents at those times. Before long it became evident that the boundary between carbonate ooze and red clay could not be due to pressure but corresponded to a balance between the rain of carbonate shells and coccoliths from the sea surface and the corrosive effect of carbon dioxide-rich deep waters. I found that worrying about all this was a lot more fun than identifying calcareous nannofossils and determining the age of the sediments. I was hooked on trying to understand carbonate deposition and $CO_2$ in the ocean. The problem turned out to be far more complex than I had anticipated. Over 40 years later I am still working to understand it.

## 21.1  Carbon Dioxide as a Greenhouse Gas

The absorption/emission bands for $CO_2$ are shown in Fig. 21.1. It has several absorption bands in the solar infrared, but these are unimportant for the greenhouse effect, In the Earth's thermal infrared spectrum $CO_2$ has very strong absorption/emission bands centered almost exactly on the

Earth's peak emission wavelengths, from about 8 to 18 μm. This is part of the spectrum where the effect of $H_2O$ vapor is relatively weak.

Unlike the spotty temperature and water-availability determined concentrations of water vapor, $CO_2$ is ubiquitous. Its distribution is latitude independent. It's concentration does vary with the seasons, by about 6 ppmv, or about 1.5 % of its concentration, as deciduous plants consume it for photosynthesis during the northern hemisphere summer and become dormant in the winter. The fact that global seasonal variations in atmospheric $CO_2$ correspond to the northern hemisphere seasons simply reflects the much greater area of land in the mi-latitudes in the northern hemisphere than in the southern hemisphere. Furthermore, $CO_2$ is well mixed through the troposphere, stratosphere, and mesosphere decreasing only in the thermosphere where its relatively greater molecular weight tends to prevent it from going higher (see Fig. 18.8).

## 21.2  The Carbon Cycle

Figure 21.2 shows diagrammatically how the carbon cycle worked before humans appeared on the scene. It also provides a brief review of plate tectonics. The tectonic plates that cover the surface of the Earth are formed by the addition of lava along the mid-ocean ridges. The volcanoes along these ridges are mostly underwater, and unseen except by deep-sea expeditions. A notable exception is Iceland, where the Mid-Atlantic Ridge rises above sea level. These volcanoes probably bring $CO_2$ from the mantle to the surface, but most is recycled as seawater percolates through the volcanic systems. The plates move away from the mid-ocean ridges as new material is added, a process called 'sea-floor spreading.' Some other volcanoes may appear within the plates. These 'intraplate volcanoes,' like Hawaii, have sources deep within the mantle and bring new, primordial $CO_2$ to the surface. As in the case of Hawaii and the Hawaiian-Emperor seamount train of older volcanoes trailing off to the west and north, these mantle 'hot spots' seem to be relatively stable, fixed in position as the plate moves over them. As it ages, the plate loses heat and becomes denser, until finally it sinks beneath an adjacent plate along a 'subduction zone,' usually marked by a deep-sea trench. Volcanoes on the overriding plate spew forth a combination of the melting rock from the descending slab and the overlying oceanic sediment. Some of the oceanic sediment being subducted is carbonate ($CaCO_3$) which breaks down to release $CO_2$ to rise through the volcanoes and be expelled into the atmosphere. The sediments also contain seawater in their pore spaces. This also provides chlorine and $CO_2$ to the overlying volcanoes.

The other long-term source for $CO_2$ is from the weathering of coal and sediments containing previously buried organic carbon, often referred to as 'kerogen'.

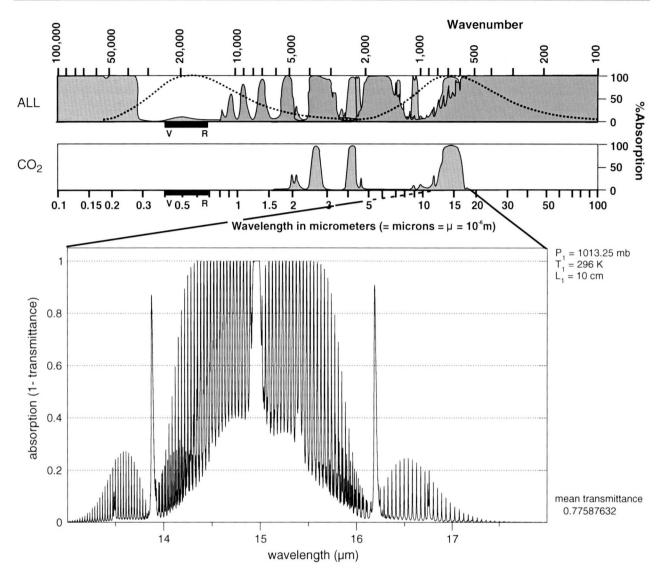

**Fig. 21.1** The absorption/emission bands of $CO_2$ at sea level and room temperature. Note that these bands are centered on the maximum of Earth's thermal infrared spectrum

## 21.3 The Very Long-Term Carbon Cycle

Where does $CO_2$ come from? When the Earth accreted 4.6 billion years ago, some of the pieces that fell into the young planet must have been similar to modern 'carbonaceous chondrite meteorites.' About 5 % of the known meteorites fall into this class. They are made of silicate minerals, which we will discuss below, carbonaceous material, including amino acids—one of the building blocks of life—and water. They are probably much more abundant than the 5 % would suggest because they do not look very distinctive and weather rapidly, so the chance of finding one is very small. They were part of the original material from which the solar system condensed. They must have brought both water and carbon to the Earth. Some of that water is in the oceans, but

more is probably still in Earth's mantle. Carbon from the mantle occasionally reaches the surface as diamonds, the high-pressure crystalline form of pure carbon. But most of the carbon remaining in the mantle is still slowly emerging through 'outgassing,' reaching the surface through volcanoes. The Earth's sedimentary rocks contain about $68 \times 10^{18}$ kg, about 86 % as carbonate rock and 14 % as organic carbon in sedimentary rocks. We do not know how much carbon remains in the Earth's mantle, but that in the Earth's sedimentary cover implies an average rate of outgassing of $15 \times 10^9$ kg C/year.

Carbon dioxide has long been suspected to be the greenhouse gas that controls Earth's climate on the longest time scales. Unlike $H_2O$, which is in high concentrations over warm water and very low concentrations over ice, $CO_2$

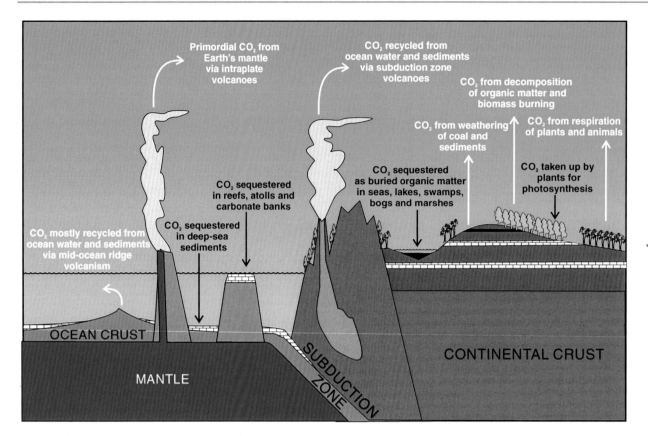

**Fig. 21.2** Major sources and sinks of carbon and carbon dioxide throughout most of the younger Phanerozoic time

is more or less evenly distributed throughout the atmosphere. It is an effective greenhouse gas for the entire planet.

In 1981 John Walker and P. B. Hays at the University of Michigan and Jim Kasting at the National Center for Atmospheric Research realized that $CO_2$ might be the Earth's thermostat. To understand how it works, you need to know about where the $CO_2$ outgassed from Earth's mantle goes. Forget about burning fossil fuels for the moment as we explore the natural system.

In case you hadn't noticed it, our planet is made up mostly of oxygen. It is locked up tightly with silicon in what are called silicate minerals, so it can't be used for anything. In most of these minerals each silicon atom is surrounded by four oxygen atoms, with the silicon at the center and the oxygen atoms at the corners of a tetrahedron enclosing it. If you were very tiny and could see all this it would be incredibly boring; a world made of silica tetrahedra repeated over and over and over again. On the other hand, if you were Buckminster Fuller you would realize that you are looking at the basic architecture of our planet.

The most abundant elements in the outer part of the Earth (the core maybe different) are O, Si, Al, Fe, and Ca. Fe is iron. This sequence of abundance of the elements in our planet can be easily remembered if you say "Oh see all the

fearsome cats." The Earth's mantle and crust are made of silicate minerals, perhaps the most common of which are feldspars. Feldspars are mostly silicon and oxygen with a few atoms of other common elements. They have the formula X (Ca, Na, or K) $Al_{1\ or\ 2}\ Si_{3\ or\ 2}\ O_8$. That is, they have one atom of a metal, either calcium (Ca), sodium (Na), or potassium (K), two or three atoms of aluminum (Al), two or three atoms of silicon (Si) and eight atoms of oxygen (O).

OK, back to the early Earth, as the first rains fell they encountered rocks that contained carbon or carbon dioxide. Aided by water these minerals broke down, releasing $CO_2$. Later volcanoes have continued to supply carbon dioxide to the atmosphere. Once free $CO_2$ was available, it began to combine with water to make carbonic acid ($H_2O + CO_2 \rightarrow H_2CO_3$). Carbonic acid is the fizzy stuff in soda pop. It is a weak acid, you can drink it. It slowly reacts with silicate minerals like feldspar. One of the common feldspars, anorthite, is $CaAl_2Si_2O_8$. It 'weathers' slowly, reacting with carbonic acid in what is appropriately termed the 'carbonation reaction:'

weathering reaction          residuum          in solution
$$CaAl_2Si_2O_8 + H_2O + 2H_2CO_3 \rightarrow Al_2Si_2O_5(OH)_4 + Ca^{++} + 2HCO_3^-$$
Anorthite   Water Carbonic acid   Kaolinite  Calcium Bicarbonate

It may look complicated, but isn't. It is a chemical reaction equation. The arrow means "goes to" and both sides of the equation must have the same number of each kind of atom. In the above equation both sides have one calcium atom, two aluminums, two silicons, six hydrogens, twelve oxygens, and two carbons. On the right-hand side, the pluses and minuses indicate that the atoms carry electrical charges; these are ions. They are dissolved in lots of water not shown in the equation.

The feldspar reacts with carbonic acid and water to make a clay ($Al_2Si_2O_5(OH)_4$) plus calcium ions and bicarbonate ions. The positive and negative ions are dissolved in river water and headed for the sea. The clay is something you have in your house. It is kaolin, the clay used to make fine porcelain. If 'kaolin' seems to have a Chinese ring to it, that's because it is from the Chinese—the inventors of porcelain. It is also what is on the grains of salt so that 'when it rains it pours.' If you dissolve a bit of commercial salt you will find that it makes cloudy water. The grains are coated with clay to prevent them from absorbing water from the air. I never knew this until I spent the summer of 1968 in Israel.

I was helping a paleontologist friend, Zeev Reiss, whom I had met in Basel, Switzerland, to set up aquaria for growing foraminifera in the Geological Institute of the Hebrew University. We were going to collect foraminifera in the Gulf of Aqaba off Eilat and bring them back to Jerusalem so students could study them. But first we needed to get the aquaria set up in Jerusalem. Everybody in the US knows that the best way to sterilize an aquarium is by washing it out with a strong solution of salt water made from kosher salt. I asked for kosher salt in Jerusalem. Blank stares from everyone. No one ever heard of kosher salt. We solved this mystery by going to a religious authority in the Hebrew University who fortunately had happened to have lived in the US. He explained that most US salt includes clay. Kosher salt is salt without the clay. Mystery solved. We went to Eilat; got the forams, brought them back and they thrived for many years in the laboratory in Jerusalem.

Back to the ions in solution. Once in the ocean the following reaction takes place:

$$Ca^{++} + 2HCO_3^- \quad \rightarrow \quad CaCO_3 + H_2O + CO_2 \uparrow$$
Calcium Ion Bicarbonate        Calcium Carbonate Water   Carbon Dioxide

The calcium and one of the bicarbonate ions combine to form one of two minerals, calcite or aragonite, both with the composition $CaCO_3$. The two minerals have the same composition but different structures or arrangements of the atoms. These are the minerals that make up limestone. You will notice that we started the weathering process with two molecules of $CO_2$, and ended up with one being fixed in a mineral, and one being returned to the atmosphere. The arrow going up means escape to the atmosphere. Over the

long run this is the way Earth disposes of $CO_2$ in the atmosphere, using the first equation to get the $CO_2$ into solution as bicarbonate, and the second to precipitate the bicarbonate as one of the components of limestone. It takes a long, long time.

How could the outgassing of $CO_2$ provide a thermostat for the Earth? It turns out that the rate of carbonation reaction, as with most other chemical reactions, depends on temperature. Like the solubility of water vapor in air, the rate of this reaction doubles with every 10 °C rise in temperature. That doesn't sound like much of a thermostat, but think about it. If the reaction runs at a rate of one unit of $CO_2$ consumed per unit time at 0 °C, then at 10 °C 2 units are consumed in the same length of time. At 20 °C 4 units are consumed, at 30 °C 8 units, at 40 °C 16 units, at 50 °C 32 units, at 60 °C 64 units, at 70 °C 128 units, at 80 °C 256 units, at 90 °C 512 units, and at 100 °C 1,024 units. Between the freezing and boiling points of water this reaction rate increases by a factor of over a thousand! So the idea is this—as $CO_2$ is released into the atmosphere by volcanoes its concentration in the atmosphere goes up. But then the temperature also goes up and the reaction rate increases, lowering the $CO_2$ concentration. This is a wonderful negative feedback mechanism regulating the temperature of our planet as long as the changes in $CO_2$ supply are not too fast. This is what will eventually happen to the present spike in atmospheric $CO_2$ resulting from burning of fossil fuels. Unfortunately the time constant associated with this process is in millions of years. Nevertheless, it has worked in the past to keep the Earth from freezing over or going above the boiling point of water. Or so we thought until the 1990s, when we learned that the Earth had turned into a snowball several times.

In order to turn into a snowball, the concentration of greenhouse gases in the atmosphere would have to become very low. Water vapor is just an amplifier of the greenhouse effect; it does what the other gases tell it to. If the major greenhouse gas was always $CO_2$ we would need to have some way of getting it out of the atmosphere in a hurry.

The second way $CO_2$ can be removed from the atmosphere is through photosynthesis. $CO_2$ is used by plants in order to produce organic matter. You remember the chemical equation

$$CO_2 \quad + H_2O \quad \rightarrow \quad CH_2O \quad + \quad O_2 \uparrow$$
Carbon Dioxide Water        Organic Matter   Oxygen

$CH_2O$ is our general expression for organic matter. To fuel the activity of the plant, or the animal that eats it, or upon death of the organism the organic matter is oxidized back to $CO_2$ and water.

$$CH_2O + O_2 \rightarrow CO_2 + H_2O$$

Only if the organic matter is buried, so it cannot decay, does the $O_2$ stay in the atmosphere. The $CO_2$ that went into the organic matter is buried with it. This process removes $CO_2$ from the atmosphere until the buried sediments containing organic matter are returned to the Earth's surface and are weathered. On the other hand, if we dig up the organic matter and burn it, $CO_2$ is immediately returned to the atmosphere. This is what humans do when we extract natural gas and petroleum and dig up coal and burn these fossil fuels.

Removal of $CO_2$ from the atmosphere by this process could only occur after organisms developed photosynthesis, and that was long, long after the origin of life. We think the earlier life forms, primitive bacteria, used a variety of 'chemosynthetic' pathways to obtain energy. Life existed on Earth for more than 2 billion years before photosynthesis was discovered as a way of obtaining energy. You have probably heard that life may have originated in the 'hydrothermal vents,' hot springs on the sea floor where light never penetrates. Many of these chemosynthetic bacteria are still around today, living inside the Earth's sediments and crustal rocks where there is neither light nor free oxygen. To most of them, oxygen is a deadly poison. The total amount of carbon in living things on the surface of the Earth and in the ocean is estimated to be about $600 \times 10^{12}$ kg. More than 99 % of it is in plants on land. Recently it has been estimated that there may be just as much carbon in these 'exotic' buried bacteria.

The Snowball Earth episodes occurred shortly after a supercontinent, containing almost all the modern continents broke up. Plate tectonics broke this Late Precambrian Supercontinent, Rodinia, into fragments which moved about during the Paleozoic to come back together as a single supercontinent again at the end of the Paleozoic, Pangaea. It was suggested that maybe there had been an enormous flux of nutrients to the ocean, massive blooms of phytoplankton had taken in $CO_2$ and converted it to organic matter, died and dropped it to the bottom of the ocean where it was buried before it could be decomposed. If this sounds far-fetched to you, it did to me too. You can imagine my joy one evening in Felix's Oyster Bar in New Orleans' French Quarter, between servings of platters of raw oysters, when Jim Kasting told me about another, more plausible, explanation. But for that you have to wait.

Using $CO_2$ to get out of the snowball state makes more sense. On a frozen Earth the volcanoes keep up their work, adding $CO_2$ to the atmosphere. As you will have discovered on opening a warm bottle of soda pop, cold water can hold more $CO_2$ than warm water. So where the ocean was not covered by ice it absorbed $CO_2$ from the atmosphere. The temperatures were so cold that the carbonation reaction had ceased. Without silicate minerals to weather, the $CO_2$ built

up and built up until the greenhouse was effective enough to melt the ice. Once the ice melted the oceans started to warm releasing more $CO_2$ into the atmosphere. Suddenly there was an episode of global warming raising ocean temperatures to the highest levels ever. The Earth went from one extreme to the other. There is good evidence that this is exactly what happened. The glacial deposits of the Snowball Earth episodes are overlain by what are called 'cap carbonates.' These are massive limestones. Paul Hoffmann of Harvard has described his amazement when he realized what these 'cap carbonates' really were—huge crystals of calcium carbonate that grew on the floor of the shallow seas. As the oceans warmed, the dissolved calcium carbonate was precipitated so rapidly it formed giant crystals.

There is another reaction involving $CO_2$ you may hear about—its reaction with limestone. Once the carbonate has been deposited in the sea as limestone, it can be uplifted above sea-level and be exposed to weathering on land. There are many exposures of limestone: the chalk that makes the white cliffs of Dover, the Austin Chalk of Texas, the Selma Chalk of Alabama, the Paleozoic limestones of Kentucky and Tennessee, the fantastic landscapes of China around Guilin, and so on. Most caves are formed by solution of limestone. Limestones are much more easily weathered and eroded by carbonic acid than are silicate minerals. Here is what happens. First carbon dioxide is absorbed by raindrops to make carbonic acid:

$$CO_2 + H_2O \rightarrow H_2CO_3$$

Then, the carbonic acid reacts with limestone to produce ions of calcium and bicarbonate in solution

$$H_2CO_3 + CaCO_3 \rightarrow Ca^{++} + 2HCO_3^-$$

The ions are transported to the sea, where the following reaction takes place:

$$Ca^{++} + 2HCO_3^- \rightarrow CaCO_3 + H_2O + CO_2 \uparrow$$

You will notice that one molecule of carbon dioxide was used in the first equation to make carbonic acid. The carbonic acid dissolved the limestone ($CaCO_3$), turning it into ions that can be transported by rivers to the sea. There the calcium and two molecules of the bicarbonate recombine as limestone, releasing one molecule of carbon dioxide back into the atmosphere. The carbon dioxide molecule that went in to weather the limestone came back out as limestone was precipitated in the sea again. There was no net change in atmospheric carbon dioxide concentration. Just one molecule was used temporarily to move the limestone from one place to another by transport in solution. All rivers on Earth are calcium bicarbonate solutions. If there is just a little, the water tastes especially good to most people. If there is a lot we call the water 'hard' and it interferes with using soap and detergents.

Carbonic acid on limestone is a very active weathering process, but it does nothing to change the concentration of carbon dioxide in the atmosphere. Some people, like me, used to think that the erosion rate of limestone would go up if the atmospheric concentration of carbon dioxide was higher. But that is not true for one simple reason. The weathering of limestone does not take place very rapidly where it is directly exposed to the air. There it is usually dry, so nothing happens. Most of the weathering takes place where damp soil covers the limestone. The $CO_2$ content of the air in soil is an order of magnitude or more higher than the concentration in the atmosphere. This is because the organic matter in soil is constantly decaying and releasing carbon dioxide.

The idea that the $CO_2$ in the atmosphere will be consumed by weathering limestone on land is a red herring. However, if the $CO_2$ can get into the ocean and work its way down into the deep sea, something wonderful happens. It can react to dissolve the carbonate deposits on the sea floor. These carbonate deposits consist of coccoliths and the shells of planktonic Foraminifera, and are called 'oozes' because of their texture. They look much like vanilla ice cream (but don't taste like it!). Depending on whether the dominant material is coccoliths or planktonic foraminifera, they are called 'coccolith ooze' or '*Globigerina* ooze' (after the most abundant genus of planktonic foraminifer. Both of these materials are made of calcite and do not readily dissolve above water depths of several thousand meters. In a few small shallower areas the carbonate is dominated by the tiny shells of pelagic snails, called pteropods, and are known as 'Pteropod ooze.' The shells of pteropods are made of aragonite, and are much more soluble, so that they are very rarely found below depths of 1000 m. The area of the ocean floored by carbonate oozes is about $128 \times 10^6$ km$^2$, and the average thickness is several hundred meters. That is larger than the area of land today not covered with ice ($121 \times 10^6$ km$^2$). Compared to this expanse of carbonate flooring the deep sea, the area of exposures of limestone on land is trivial. You can think of this deep-sea carbonate as a huge chemically reactive surface waiting to respond to changes in the amount of carbon dioxide in the atmosphere and dissolved in the ocean. The problem is that there is a long time lag between the injection of $CO_2$ into the atmosphere and its arrival in the deep sea where this reactive surface is waiting; and the dissolution process itself is not instantaneous. The time lags involved are of the order of hundreds to thousands of years.

How much carbon has been outgassed over the 4.5 billion years since our planet formed? How fast is new $CO_2$ from the mantle being added today? Unfortunately, the subsea volcanoes are influenced by seawater seeping into the rocks, and it is difficult to get an accurate estimate of the modern rate of addition, but one would assume that most of it came out early in the planet's history, so the modern rate should be much less than the long term average. The subduction zone volcanoes are recycling $CO_2$ from deep-sea sediments. In fact the average rate of carbon emissions (as $CO_2$) associated with all volcanic eruptions between 1800 and 1969 is thought to be about $1.8 \times 10^9$ kg C/year, but these are almost exclusively subduction zone volcanoes. In mountain ranges the sedimentary rocks can be subjected to enormous pressures and temperatures which causes carbonate rocks to undergo alteration, termed 'metamorphism' which releases $CO_2$. The amount and rate of these emissions is not well known. These are the bubbles in the fine mineral water from the Southern Alps, San Pellegrino. The natural decomposition of organic matter in sediment as it is eroded is another natural source of $CO_2$.

To the west of Cologne and Bonn, Germany is a volcanic field known as the Eifel. One of the most beautiful Romanesque abbey churches is located on the shore of a small crater lake, or 'Maar,' Maria Laach. It is the youngest of the Eifel volcanoes; its eruption 11,000 years ago was about 20 times the size of Mount Saint Helens. This crater is but one of many in the region and there are many springs with carbon dioxide-rich mineral water. These waters are bottled and sold world-wide. The water from the Appolinaris and Gerolstein springs are available even in our village in Colorado. A few years ago we had a small geological excursion to visit measuring facilities at the springs where geologists are attempting to determine how much $CO_2$ is bubbling out in the Eifel region. It is evident that much of the $CO_2$ emitted from volcanic regions comes out long after the eruptions. However, it is useful to note that the estimates of volcanic $CO_2$ entering the atmosphere are about three orders of magnitude (1,000 times) less that the $CO_2$ currently being introduced by the burning of fossil fuels.

## 21.4   Carbon Dioxide During the Phanerozoic

The idea that $CO_2$ is the greenhouse gas that has controlled Earth's climate over the eons can be traced back to Mikhail Ivanovitch Budyko's work in the early 1970s. He had been working with my argumentative friend from the 1960 excursion in Sweden, Alexander Leonidovich Yanshin, and Alexander Borisovitch Ronov of the Vernadsky Institute of the USSR Academy of sciences in Moscow. They thought they had found a key to the problem of the composition of Earth's atmosphere in the past. Alexander Borisovitch Ronov had started gathering data on the abundance of different rock types on the Earth in the 1950s. It had been a daunting task. In the 1970s his first global compilations were appearing. Of particular interest was the distribution of carbonate rocks, which had been very abundant until the

Tertiary, and in younger geologic times had become quite rare as deposits on the continents. They interpreted this as a long-term decline in atmospheric $CO_2$. They also noted that in case there might be something wrong with the argument that atmospheric $CO_2$ concentrations could be determined from the relative amounts of limestone on the continents, the abundance of volcanic rocks followed the same pattern. This work culminated in a classic book, *History of Earth's Atmosphere*, published in 1985, with an English edition in 1987.

In the 1970s several of us in the US were busy at work compiling data on the distribution of deep-sea sediments gathered by the Deep Sea Drilling Project. I realized that Ronov's compilations did not include the deep sea. It was also becoming apparent that the decline in limestone deposited on land was paralleled by an increase in carbonate oozes deposited in the deep sea. The abundance of limestone on the continents was not an index to the amount of $CO_2$ in the atmosphere, but the abundance of volcanic rocks might be. That was what got me invited to the Leningrad conference.

When we sent Eric Barron from Miami to work on modeling the climate of the Cretaceous at the National Center for Atmospheric Research in Boulder in 1976, we really had no very good idea why the Cretaceous had been warmer. Oxygen isotope measurements on fossils provided quantitative information it had been several degrees warmer, and everyone agreed that there was no ice on the poles. Study of fossil plants had made it clear that freezing temperatures were rare, even deep in the interior of the continents during the Cretaceous. The first problem to solve was why the Polar Regions should be warm. The usual suspects for climate change were the lack of high mountains, the high sea level and flooding of the continents, and with the advent of plate tectonics, the different configuration of the continents. A different level of $CO_2$ was added as a new possibility. But plate tectonics was 'en vogue,' and we thought it would provide the answer.

Figure 21.3 compares the geography of the present day with that of the middle Cretaceous, 90 million years ago. Today the land masses are mostly in the Northern Hemisphere. In the Cretaceous they were more evenly divided

**Fig. 21.3** The Earth today (0 Ma) and in the middle of the Cretaceous Period (90 Ma = 90 million years ago). In both cases there are circumglobal seaways. That today is around the Antarctic continent. In the Cretaceous the circumglobal seaway, the Tethys, was at a low northern latitude, and separated the northern and southern continents. The entire Earth is shown on these Mollweide map projections. Mollweide maps have the advantage that they are equal area projections. If you overlay a square on one part of the map it will have the same area anywhere else on the map, useful in recalling how small the Polar Regions are compared to the tropics

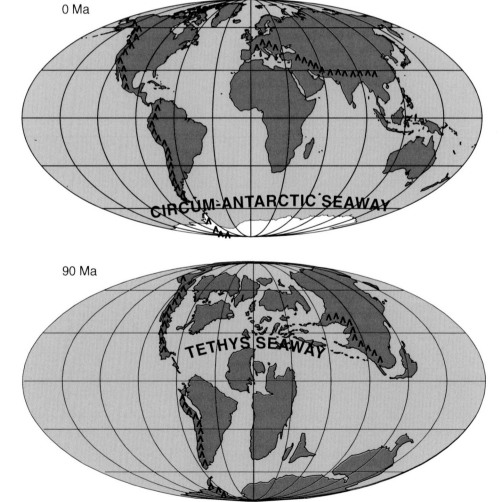

between the two hemispheres. The much greater extent of the oceans and seas in the Cretaceous is obvious. South America and Africa are readily recognizable on the Cretaceous map, but some of the other continents are not. Australia is separating from Antarctica. The triangular island east of South Africa is the Indian subcontinent, headed north to collide with Asia and produce the Himalaya about 50 million years ago. In the north, Western and Eastern North America, Greenland, Europe and Asia appear as islands separated by seaways. The east-west seaway between the northern and southern continents is known as the Tethys.

Eric performed a series of experiments with the climate model at NCAR, comparing the effects of each change in boundary conditions with the present. Removing mountains had little overall effect except for allowing the winds to blow more efficiently around the Earth. The higher sea level had almost no effect; instead of providing much more absorption of the incoming radiation by the water as expected, the model showed the seas to be covered with clouds reflecting the radiation back into space. The different positions of the continents also had only a small effect. The only thing that worked well to simulate the Cretaceous climate was to increase the concentration of atmospheric $CO_2$, and it specifically warmed the poles.

In 1978 I asked Eric Barron's NCAR advisor, Warren Washington, why $CO_2$ would be so effective at selectively warming the Polar Regions. The answer was simple. The Earth has a temperature-related greenhouse gas, water vapor. But it is missing where the surface temperatures are cold. In the tropics, the effect of $CO_2$ is small because the water vapor dominates the greenhouse effect. But when the Polar Regions are cold, $CO_2$ can be the sole greenhouse gas. It brings up the temperature and, as the region warms, the water vapor content of the atmosphere increases as a positive feedback. The effect of increased $CO_2$ is concentrated there where the Earth would otherwise be coolest, at the poles. I asked my question in 1978. Thirty years later we are seeing exactly this happening.

After Eric's work there was a specific problem left over: the warm interior of Siberia in wintertime. It would be solved a decade later by Rob DeConto and Jon Bergengren. As described above, they found that plants have evolved to exercise some control over the water-vapor greenhouse effect over the continents. $C_4$ plants now restrict moisture transport into the interiors of the continents, making for colder winters.

The early papers by Budyko and Ronov had started many geologists thinking again about the history of the atmosphere as a possible cause of climate change. The topic had almost dropped out of sight for 50 years. Ronov's global compilations of the amounts of different sedimentary rocks of each age were translated into English and appeared in the late 70s and early 80s. They made it possible, for the first time, to think of planet Earth as a geochemical system evolving over time.

The quest for a method of determining atmospheric concentrations of $CO_2$ in the past had begun. Bob Berner at Yale, Tony Lasaga at Princeton, and Bob Garrels at Northwestern University cooperated on an effort to develop a quantitative model of the carbon-$CO_2$ cycle. It includes introduction of $CO_2$ from volcanoes and from the weathering and erosion of sediments containing organic carbon. They calculate the removal of $CO_2$ through weathering of silicate rocks and deposition of carbonates, and burial of organic carbon. Volcanism plays a large role, and they assume that the rate of volcanism on Earth is closely related to the rate of sea-floor spreading and plate tectonic activity. We only have a record of sea-floor spreading rates for rocks younger than Jurassic (the last 145 million years). Because of this their first model, published in 1983, covered only the Cretaceous and Tertiary. Then it was realized that if sea-floor spreading rates are high, the mid-ocean ridge displaces more water, and sea level rises on the continents. The highest sea-level of the Phanerozoic, the last 542 million years of Earth history, occurred during the middle of the Paleozoic, about 400 million years ago. Using sea level as an index to sea-floor spreading rates, and by analogy to the rate of volcanism, they extended their model back to the beginning of the Phanerozoic. The model, originally called BLAG, after the names of the authors, is now known as GEOCARB. Over the past two decades it has been developed and refined. One of my colleagues at GEOMAR in Kiel, Klaus Wallmann, has further refined the model for the Cretaceous and Tertiary. The latest results for the Phanerozoic history of atmospheric $CO_2$ are shown in Fig. 21.4.

The model indicates that in the Early Paleozoic, before about 350 million years ago the atmosphere held about 15 times as much $CO_2$ as it does today. During this time the chief way of getting $CO_2$ out of the system was through the deposition of carbonate sediments in shallow seas. Large scale deposition of carbonate was accomplished by marine plants, corals, and shellfish found most abundantly in waters shallow enough for light to penetrate to the bottom. Organic matter could also be buried in these sediments. But during this time the amount of carbon incorporated into carbonate rocks was as much as 20 times that buried as organic matter.

The world changed as the first land plants appeared alongside streams in the Silurian (S) and spread over the land during the Devonian (D). They caused a massive drawdown of atmospheric $CO_2$, to levels similar to what we have today. Only six times as much carbon was incorporated in carbonate rocks as in buried organic matter. During the Carboniferous (C) vast amounts of carbon were buried as coal deposits. At that time there were no organisms capable of breaking down lignin, the material that wood is made of. Accompanying the $CO_2$ drawdown was a glacial episode on the southern continent of Gondwana, comprised of today's South America, Africa, India, Australia and Antarctica. At that time Gondwana was located over the South Pole. The glaciation lasted from about 345–270 million years ago and consisted of glacial and interglacial episodes very much like what Earth has experienced in the last 3.5 million years. In the Permian (P), fungi learned how to digest wood, and it became much harder to bury organic carbon. These fungi are still around today, causing fallen trees to rot away. Since wood became liable to decay and decomposition releasing $CO_2$, the levels of atmospheric $CO_2$ rose again.

A great mass extinction at the end of the Permian, and formation of the supercontinent of Pangaea, which included all the modern continents in one giant landmass, combined to make life on land very stressful. During the Triassic (Ṝ) deserts were widespread on land; the bright red sediments of these desert areas are widespread in North America and Europe. But offshore great reefs grew. Some of the most spectacular can be seen today as the mountains of the Southern Alps in Italy, the Dolomites. The carbonate rock, of which these mountains are made, gave its name to a carbonate mineral, 'dolomite,' $CaMg(CO_3)_2$, in which magnesium atoms replace half the calcium atoms. It is less easily eroded than limestone, and the reefs of the Dolomites stand like great monuments. In the Triassic the balance shifted again toward deposition of carbon in carbonate rocks, the ratio rising once more to 20.

Levels of atmospheric $CO_2$ hovered between 5 and 10 times that of the present until the Cretaceous (K), when another biological innovation changed the world. My beloved

coccolithophores and the planktonic foraminifera had evolved in shallow seas during the Jurassic (J). In the Cretaceous they spread throughout the oceans. Their coccoliths and shells settled to the ocean floor, forming a new kind of deep sea sediment, the calcareous oozes. Before the Cretaceous, carbonate sediment had been deposited only in shallow seas; now it was also being deposited in the deep ocean on a massive scale. The deep sea oozes are on ocean floor that is slowly moving toward the subduction zones to descend into the mantle. Carbon had become involved in the global plate tectonic cycle.

In looking at Fig. 21.4, the question must arise: with such high levels of atmospheric $CO_2$ in the Paleozoic, Earth's surface temperatures must have been very high. Here it is important to remember that the luminosity of the Sun increases with time, so that the solar constant, the amount of energy received by the Earth from the Sun, was less in the Paleozoic. Today it is about 1340 $W/m^2$; in the mid-Paleozoic the solar constant would have been about 1300 $W/m^2$. It is thought that the modern threshold level for atmospheric $CO_2$ required to initiate glaciation is about 500 ppmv, but that in the Paleozoic this threshold was about 3000 ppmv.

In 1975 I gave a talk on this drastic change in the carbon cycle at the annual meeting of the Geological Society of America. At the end I jokingly pointed out that these tiny organisms had set in motion a Doomsday Machine. If they continued to take carbon from the Earth's surface, and drop

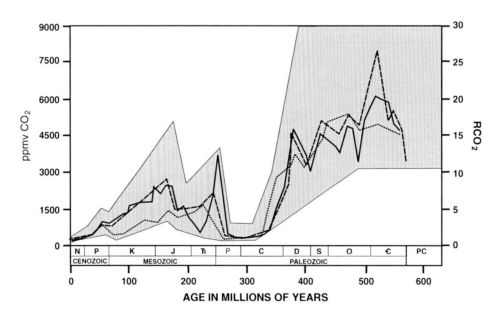

**Fig. 21.4** Levels of atmospheric $CO_2$ during the Phanerozoic, based on model calculations by Bob Berner and his co-workers. Two recent versions are shown, the latest as a *solid line*. $RCO_2$ means 'times present $CO_2$' whereby the 'present' value is arbitrarily taken to be 300 ppmv, the level of atmospheric $CO_2$ concentration in 1915. The letters are the abbreviations for geologic intervals: *N* Neogene; *P* Paleogene; *K* Cretaceous; *J* Jurassic; Ṝ Triassic; *P* Permian; *C* Carboniferous; *D* Devonian; *S* Silurian; *O* Ordovician; Ꞓ Cambrian; *PC* Precambrian

it to the sea floor where it could be returned to the Earth's mantle, and did it as at the same rate as today for another half billion years, there would be no more carbon on the surface of the Earth. It would all be in the form of diamonds in the mantle. And diamonds are forever. At the end of my talk a member of the audience came up to me and told me that the President of the Geochemistry Society had just given the inverse of my talk. He had discussed how, over the vastness of geologic time, plate tectonics had brought carbon from the Earth's mantle to the surface. It is as though the world has come full circle.

A more detailed view of the changes over the past 140 million years is shown in Fig. 21.5 based on both Berner's and Wallmann's model calculations, geochemical measurements, and study of fossil leaves.

The measurements shown in the diagram are derived from two unrelated techniques. The black squares represent measurement of a particular set of compounds found in marine phytoplankton (algae that live in the near-surface waters of the ocean). These compounds, alkenones, are sensitive to the amount of $CO_2$ in the water; their structure changes with the $CO_2$ concentration. Since the surface ocean

and atmosphere exchange gases and are in equilibrium, it should be possible to determine the atmospheric $CO_2$ concentration from the alkenones. When this technique was initially proposed by Mark Pagani in 1998, the idea was met with considerable skepticism for two reasons. First, it is a complicated line of reasoning, and perhaps some of the assumptions might be wrong. Secondly, the results were completely out of line with what everyone thought at the time. Mark's studies indicated that atmospheric $CO_2$ levels had been about the same as modern pre-industrial levels (around 270 ppmv) for the past 24 million years. When he published his work most geologists believed that the glaciation of the Northern Hemisphere had been brought on by a decline in atmospheric $CO_2$.

Then came evidence from a wholly different source, the density of stomata on leaves. Stomata are the pore-like structures on the undersides of leaves that open and close to admit $CO_2$ into the leaf for photosynthesis and regulate transpiration. The number of stomata per unit area varies with the concentration of $CO_2$ in the atmosphere. Wolfram Kürschner at the University of Utrecht in the Netherlands and Dana Royer, at Wesleyan University, have both made a number of

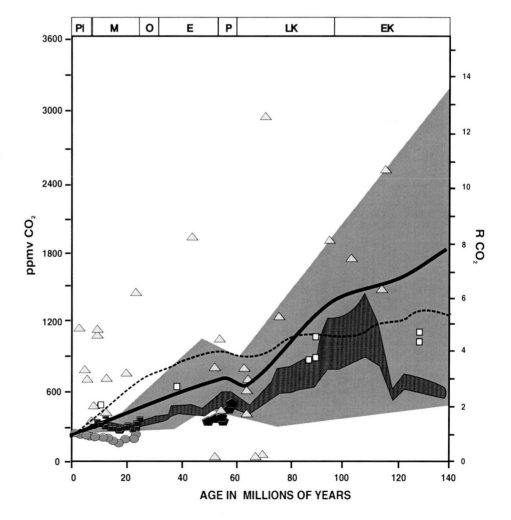

**Fig. 21.5** Atmospheric carbon dioxide concentrations calculated by Bob Berner and Zavareth Kothavala in 2001 (GEOCARB III) shown as a *black line* (the uncertainty of this model is shown in Fig. 21.4) and the field of results obtained by Klaus Wallmann (*grey field*). Also shown are estimates of $CO_2$ levels based on measurements of marine planktonic algae (*red squares*), the stomata of land plants (*purple pentagons*), boron isotopes in shells of planktonic foraminifera (*green ovals*), sedimentary porphyrins (*white squares*), and calcretes (carbonate concretions in ancient soils) (*yellow triangles*) which might represent short-term fluctuations

measurements of stomatal density on fossil leaves to learn about ancient $CO_2$ levels. Their results are close to those of Pagani, and the geological community has had to make a major adjustment in its thinking. Then, in 1999 British researchers Paul Pearson and Martin Palmer, using another, wholly different geochemical technique involving boron isotopes came up with similar results. Calcretes (concretions in paleosols, ancient soils) show a much greater variability for reasons as yet unknown. The consensus is that for the last 25 million years or so atmospheric $CO_2$ levels have remained close to those of the pre-industrial era. The current levels of atmospheric $CO_2$ are higher than any the Earth has seen for at least 25 million years.

The University of Utrecht has a large collection of extant plant material collected over the past centuries. Examining the stomata on these leaves and others from plants grown under controlled conditions Kürschner and his colleagues made the unsettling discovery that European birches and oaks stop adjusting to rising levels of $CO_2$ when the atmospheric concentrations reach 300 and 450 ppmv. Many of these plants have already reached their limit. I find this to be staggering news when you realize that none of the plants or animals around today has evolved under the levels of $CO_2$ we have today. The entire evolutionary process of the past 25+ million years has taken place under atmospheric $CO_2$ levels lower than today.

Do today's rising levels of $CO_2$ stress plants? One argument is that plants need $CO_2$, so the more the better. This is similar to the argument that humans need fresh water to drink, so if you are dropped off a boat in the middle of Lake Michigan you will be just fine. Realistically you need to remember that the $C_4$ metabolism of many of our fast-growing crops, like corn, evolved as a means of coping with low levels of $CO_2$. The $C_4$ plants do not grow as well under higher levels of $CO_2$. Some plants do thrive with higher levels of $CO_2$. In the Netherlands, greenhouses add $CO_2$ so that the atmosphere inside has double concentration of that outside. They are reported to produce about 25 % more tomatoes. Tomatoes are $C_3$ plants. Other $C_3$ plants do not do as well, and some studies suggest that the plants that thrive in higher $CO_2$ concentrations are what we call 'weeds.' Of course a weed is simply a plant that is out of place, but if you are raising crops they are a nuisance. The important thing to realize is that some plants may do well with higher levels of atmospheric $CO_2$, while others will have a harder time of it. It is likely that the structure of plant communities will change. When you remember that some plants are annuals and others, like many trees, require 50 years or more to reach full maturity, you can imagine that the response to increasing $CO_2$ is likely to very complex. A young tree starts its life under one set of conditions but reaches maturity under different conditions. This is what happened during the last deglaciation, resulting in plant communities 'with no modern analogs.' The problem we face is not just climate change, it is environmental change. I would expect that genetic engineering will cope with many of the problems for crops, but the vast number of plants that are of no special interest to mankind is sure to be impacted. The outcome is unpredictable.

Ideas concerning the onset of the Northern Hemisphere glaciation have had to be revised. Unfortunately, we do not yet have independent data, such as stomatal density indices, that would allow us to evaluate levels of atmospheric $CO_2$ during the interval from the Middle Miocene to the Eocene, but I will describe what I think is a reasonable scenario.

## 21.5 Rethinking the Role of $CO_2$ in Climate Change

To understand how our knowledge of climate change over the past 65 million years has progressed, we need to go back to the beginnings of modern geology, back to Charles Lyell. He recognized that a major change had taken place between the time of chalk deposition, the Cretaceous, and the younger rocks, known as the Tertiary Epoch of Earth history. The change from the white chalks of Dover to the brown sands and shales on which London is built is spectacular. We now know that this boundary was caused by the impact of a small asteroid in the northern Yucatan. It was one of the major mass extinctions for our planet.

But Charles Lyell's main interest was in the Tertiary sediments and the fossils of mollusks they contained. In his works in the 1830s he recognized four successive groups of strata, each with assemblages of the fossil shellfish that ever more closely resembled modern shallow water faunas. He named these, from oldest to youngest, Paleocene, Eocene, Miocene, and Pliocene. These names were derived from Greek and mean Ancient age, Dawn age, Few age, and More age, referring to the number of species of modern mollusk to be found in the deposits.

The names of most of the geologic Periods that make up the Phanerozoic are based on marine rocks and fossils. There are two exceptions, the Carboniferous, the time of deposition of the coal beds, and the Oligocene which was inserted into the middle of Lyell's classification by the German paleobotanist Heinrich von Beyrich in 1854. He was referring to some peculiar non-marine strata in Europe that contained fossil leaves very different from those found in non-marine Eocene beds. One of my friends from the 1960s in Europe, Claude Cavelier, is a paleobotanist among other talents. He wrote a monograph on the fossil plants of the Paris Basin pointing out that the Eocene-Oligocene boundary marks a shift from a tropical to a more Arctic flora.

In 1853 the Tertiary had been divided into two superepochs, the Paleogene and the Neogene. The Paleogene came to include not only the Paleocene and Eocene but also the

Oligocene; the Neogene includes the Miocene and Pliocene. These subdivisions are still around, but if it were to be done today, the Oligocene would certainly be part of the Neogene. Something very important happened at the Eocene-Oligocene boundary.

The first clue came from cores recovered by the Deep Sea Drilling Project in 1973. One of the Co-Chief Scientists, Jim Kennett, asked Nick Shackleton to work up the O$^{16}$:O$1^{8}$ record for cores recovered at Site 277, located halfway between New Zealand and Antarctica. Up to that time it was known that something odd was going on about then, but what Nick found was truly startling. The measurements indicated that the temperature of the deep sea had dropped from 10 °C to less than 6 °C in a very short period of time. This, tied in with other evidence gathered by the DSDP on that Leg, led to the conclusion that the initiation of glaciation of Antarctica occurred at that time. Once the ice had covered the continent the ice-albedo feedback prevented it's melting. It was a major step in the evolution of Earth's climate, and it had happened in an interval of time so brief that it could not be measured. Unfortunately this conclusion was accompanied by the idea that there was really much more to learn about the history of the Antarctic.

Here I should mention that I have never directly been involved in Antarctic research, but I have acted as the facilitator of such research on several occasions. When it was first proposed that we send the *Glomar Challenger* to the Antarctic, it was our little Planning Committee that had to make the decision. I was the only member of the Committee solidly in favor of the idea. It would be a risky business; the ship did not have the hull strength to be able to operate in ice. We were meeting at Lamont Geological Observatory in Palisades New York, in 1970, and one of the proponents was a Lamont scientist. I suggested that before we scrap the idea we have the proponent come give us a presentation. After some arguments, because we rarely invited proponents to give us presentations, it was agreed. Our committee listened intently. From the presentation it sounded like the project would just learn a little more about the local geology of the Ross Sea. Local geological problems were something we always avoided. As I looked around the room it was obvious no one was changing their mind. At the end of the presentation I started a vigorous questioning, asking about the implications of what we might learn in terms of Earth's history. I kept on hammering about whether this had real global significance. The proponent was getting very angry with me for raising all these objections, but answered them. He finally left the room in a huff thinking I had deliberately torpedoed his project. After he left the question was put to us. "Well?" Unanimously the members replied "Let's do it."

In the mid 1990s drilling in the Antarctic was again a major topic for discussion. The Ocean Drilling Program had become much more formal than the old DSDP. The proposals for drilling were now quite formal documents with elaborate justifications and detailed plans for the drilling program, and were sent out for independent reviews before the Planning Committee got a chance to discuss them. In the 1990s the JOIDES Planning Committee had 17 members and at least 5 observers, so you can imagine how difficult it was to achieve consensus on anything. Antarctic scientists were complaining that their proposals never seemed to work their way to the top. One of the problems was that they had submitted five competing proposals. Another objection was that there was nothing more to be learned in the Antarctic; it got covered with ice at the end of the Eocene, and that was that. Then there was the refrain "It's so far away." In 1996 I was contacted and asked to Chair a special Planning Group on Antarctic Drilling. I asked "Why me, I don't know anything about the Antarctic?" That was the reason I was chosen, I didn't know much about either the science or the personalities involved. But as soon as I started looking over the proposal I realized I wasn't the only one in the dark. Most of my acquaintances still thought not much had happened in the Antarctic since the Eocene.

The Antarctic community was a diverse group, consisting mostly of scientists from Australia, New Zealand, Britain, Germany, Italy, Russia and the US. Their previous work had documented episodes of ice buildup and decline, and had shown that there was a major mismatch in the geologic data. If the oxygen isotope data were correct, and ice had covered Antarctica at the end of the Eocene, there should have been an accompanying sharp fall of sea-level. Yet other geologic evidence showed that the sea-level fall didn't occur until 10 million years later. There was another example in the Miocene where there was an 8 million year discrepancy between the change in oxygen isotopes and a sea-level fall. The Antarctic had a much more interesting history than I or most of my 'unAntarctic' colleagues suspected. Our planning group was able to prioritize the proposals and suggest a way of accomplishing the work. Some of it was completed before the Ocean Drilling Program ended in 2003. More remains to be done in the future.

In the meantime, one of my former students, Rob DeConto at the University of Massachusetts was trying to figure out what had happened at the end of the Eocene. He and Dave Pollard at the Pennsylvania State University used coupled numerical models to simulate the development of the climate, oceans, vegetation, and ice cover in the Antarctic region over multi-million year time scales. They used Milankovitch orbital variations calculated for that time to determine the effect of varying solar insolation with different atmospheric CO$_2$ concentrations. What they found was that as CO$_2$ drops below three times its pre-industrial level (which they take to be 280 ppmv), certain orbital configurations will cause glaciers to form on the Antarctic highlands. These are ephemeral unless the CO$_2$ level drops to twice the pre-industrial level. Then the ice grows in stages to eventually cover the whole continent.

In the meantime, the oxygen isotope record across the Eocene-Oligocene boundary has been refined, using ocean drilling cores with a more detailed record. It is not a single sharp drop as had originally been thought, but occurs in a couple of stages. Amazingly, to me at least, this is what the DeConto-Pollard model predicts.

At GEOMAR I was interested in trying to learn more about the development of Earth's climate from the global distribution of calcareous nannofossils and planktonic foraminifers. We put together a plate tectonic program that can also plot the sites where particular species of fossils have been found. It is on the web at www.odsn.de. This whole business was worked out by one of my graduate students at GEOMAR, Emanuel Söding. As with all my graduate students I asked him to do something I couldn't. I asked him to set up a program to plot occurrences of fossils on our plate tectonic reconstructions. I told him where he could find the data files on the web. I expected to hear back in a couple of months. However, the next morning he showed up and suggested I try it out. I did. It worked. I drank several beers at lunch that day. Just another example of the adage that professors have been around for a while, they know what the problems are. Graduate students are smart, they solve the problems. It's a symbiotic relationship.

My goal in setting up this program was to determine how far the frontal systems in the ocean move as the climate changes. Frontal systems are places where currents bring water masses together and the water sinks. Sometimes they can be so sharp they look like a line on the sea surface. The warm tropical and cold polar oceans are separated by what are called the Subtropical and Polar fronts. Although the names might suggest they are very far apart, they are in fact quite close together, usually only about 5°–10° of latitude apart. That translates to 300–600 nautical miles. My guess was that they must have moved equatorward as ice built up on Antarctica, and I thought that plotting the migrations of ocean fronts would reveal other, more subtle changes in Earth's past climate, especially back in the Cretaceous. You can easily locate them today using only information on the distribution of ocean plankton. The geographical ranges of calcareous nannoplankton and planktonic foraminifera are controlled by these fronts. One big clue is that the calcareous nannoplankton do not live in the cold waters poleward of the Polar Front, and they do not live beneath sea ice.

Imagine my surprise to find that when I got back to the Eocene and the Cretaceous, there seemed to be no evidence that such frontal systems had existed in the ocean at all. It raised the possibility that maybe ocean circulation then had been very different from that of today. Our program only included data from the ocean drilling programs, but there was a lot more information available in the scientific literature. I started plotting on the occurrences of calcareous nannofossils from outcrops of rock exposed on land as well. I discovered that there was even a collection of data on their occurrence in the Arctic Ocean Basin. Here the calcareous nannofossils were not in the rocks in which they had originally been deposited, but they had survived erosion and were now in young sediments on the Arctic Ocean floor. As I started adding them to what we already knew I realized that they had lived in the Arctic Ocean throughout the Cretaceous, Paleocene and Eocene. But then they had disappeared. And no younger fossils were to be found. Apparently the polar front had developed and sealed off the Arctic at the end of the Eocene. If the polar front had been there it was very likely the Arctic had been covered by sea ice for the last 35 million years. Ice had covered both poles at the same time.

Here is where serendipity strikes again. As the Ocean Drilling Program approached its end in 2003, the marine geological community was thinking about how the work might go on. The Japanese had proposed a joint new initiative with the US. They intended to construct a giant drilling vessel that would be able to drill deeper and better than any scientific vessel before. In the meantime, that vessel, named the Chikyu, has been built and is at sea.

I was again Chairman of the JOIDES Planning Committee in 2000–2002 when the planning for future programs was going on. Only now I represented Germany at the table. The Europeans were more than a little upset at the possibility of being excluded from the new project since they had supported international ocean drilling for over 25 years. Helmut Beiersdorf of the German Geological Survey was the German member of the Executive Committee. On the train to a meeting in Strasbourg one day, he began to talk about the possibility of a European group continuing with its own scientific ocean drilling program, linked to the Japanese-American venture. Others in Europe had the same idea, and soon a new group, ECOD (European Consortium for Ocean Drilling) was born. Scientists in the US urged that so much had been learned from ocean drilling that the *JOIDES Resolution* should be refurbished and continue operations. In effect we now have three ocean drilling operations, each with very different capabilities and great prospects for learning more about the ocean.

A proposal to drill and core the Lomonosov Ridge, a long narrow rise that crosses the Arctic Ocean from Russia to Greenland, and passes close to the North Pole, had been floating around for a while. To go there the Ocean Drilling Program would have to hire an icebreaker drill ship and other icebreakers to protect it while on site. It would be very expensive and change obligations for the participating countries. But it was an ideal venture for the new European drilling consortium. I desperately wanted to see this project carried out in order to learn about the history of the Arctic.

My biggest concern was that we might send an icebreaker drill ship to the North Pole without proper support, and the venture might fail to achieve its goals. I got quite a reputation for being an obstructionist. However, my concerns were not unfounded. Our group at GEOMAR had had almost a decade of experience in Arctic operations. GEOMAR's projects were being conducted in cooperation with Russian scientists. The goals had been to understand processes on the Siberian shelf, in particular the Laptev Sea off the mouth of the Lena River. The scientific operations required help from several icebreakers each season. Fortunately, Russia has a fleet of large nuclear icebreakers, and we were able to rent a few each summer.

The expedition to the Lomonosov Ridge went off with fair if not total success. There were still problems with ice even though the ice-strengthened drill ship had the support of two icebreakers. The cores they recovered confirmed that the Arctic Ocean had been ice-covered back to the Eocene. Most startlingly, it discovered that, at times during the Eocene, the Arctic 'Ocean' had been covered by floating mats of a fresh-water fern, *Azolla*. The Arctic 'Ocean' has at times been a fresh-water lake, by far the largest that has ever existed on Earth.

Now we are beginning to zero in on what happened at the end of the Eocene. Atmospheric carbon dioxide levels had been declining since the middle of the Cretaceous, about 100 million years ago. I suspect that this was because the calcareous plankton were dropping their carbonate coccoliths and shells onto the ocean floor as deep-sea carbonate ooze. The process was interrupted by the end-Cretaceous asteroid impact, which essentially wiped out all the ocean plankton as well as the dinosaurs. But by the beginning of the Eocene, 10 million years later, evolution had done its work and there were new species of calcareous plankton back at work, making deep-sea carbonate ooze and lowering atmospheric CO$_2$ again. At the end of the Eocene, CO$_2$ levels had dropped to about 500 ppmv. When the Milankovitch orbital changes lowered insolation in the Polar Regions, ice covered both Antarctica and the Arctic Ocean. Next something really drastic took place.

Cold salty waters in the Polar Regions began to fill the ocean interior producing the condition we have today, where 90 % of the ocean is cold deep water that sank at high latitudes. CO$_2$ is much more soluble than any other gas in seawater because it is not just a little more soluble; it is a lot more soluble. Oxygen makes up about 21 % of the atmosphere, CO$_2$ about 0.03 % of the atmosphere. These gases are in equilibrium across the atmosphere/ocean interface. But the concentration of dissolved O$_2$ in the surface ocean is about 7 mg/kg. One mg is 1/1,000th of a gram, and to get the numbers into terms one can really compare, 7 mg of O$_2$ per kilogram of water is equal to 0.000219 mol of O$_2$ per kilogram of water. The concentration of CO$_2$ in seawater is

about 90 mg/kg. That translates to 0.002142 mol of CO$_2$ per kilogram of seawater. Think about that for a moment. For every molecule of CO$_2$ in the atmosphere there are 919 molecules of O$_2$. But there are ten times as many molecules of CO$_2$ in the ocean as there are molecules of O$_2$. Carbon dioxide is almost 10,000 times more soluble in seawater than oxygen. How can this possibly happen? Unlike oxygen, nitrogen and argon, the principal atmospheric gases, carbon dioxide reacts chemically with seawater to form carbonic acid (H$_2$CO$_3$) and the bicarbonate (HCO$_3^-$) and carbonate (CO$_3^=$) ions. This has major implications for life in the ocean, but more about that later.

As the polar oceans cooled at the end of the Eocene, the water became dense enough to sink into the depths. The deep ocean changed from containing relatively warm water to being filled with cold water. Cold water can absorb much more CO$_2$ than warm water so, as the Polar Regions cooled, most of the atmosphere's remaining CO$_2$ was sucked into the ocean.

Figure 21.6 shows the effect of temperature and salinity on the solubility of CO$_2$ in fresh and sea-water. Note that the units on the vertical axis are moles per kilogram (of water) per atmosphere of pressure. One atmosphere of pressure ($\sim$1000 hPa or 1 Bar) is equal to 10 m of seawater. CO$_2$ becomes much more soluble under the increasing pressure with depth in the ocean. Seawater saturated with CO$_2$ at the surface (under 1 atmosphere of pressure) would only be half saturated at a depth of 10 m. Each 10 m of water is equivalent to another atmosphere of pressure. You can conduct interesting experiments on your own, taking a bottle of carbonated mineral water into a swimming pool and opening it near the bottom to see whether any bubbles come out. In principle, you can do the same thing with the bottle of champagne you deserve for having read this far.

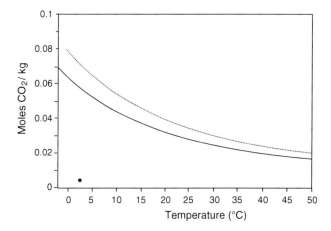

**Fig. 21.6** Solubility of carbon dioxide in fresh water (*dotted line*) and seawater with the average salinity of about 35 ‰ (35 grams of salt per kilogram of water). The dot is the average concentration of CO$_2$ in seawater at the present time; it is far below saturation

What happened at the end of the Eocene was a massive change in the state of Earth's climate system. It changed from what it had been to a climate much more like that of our present interglacial. I will describe what I think atmospheric and oceanic circulations were like prior to the Oligocene in the later chapters of this book devoted to atmospheric and oceanic circulation.

The deep sea is well undersaturated with $CO_2$, but some shallow waters can become supersaturated on warm days. For the entire ocean to become saturated with $CO_2$, the atmosphere would have to have about 20 times as much $CO_2$ as it does now (about 6000 ppmv). If you look back at Fig. 21.4 you will see that levels that high may have occurred during the Paleozoic, but probably not since.

## 21.6    The Ice Core Record of the Last Half Million Years

All of the discussion above is based on interpretation of the levels of atmospheric $CO_2$ from 'proxy data.' Proxy data are measurements of one thing, such as $^{18}O:^{16}O$, that tell us about something else, such as ice volume and temperature. But ice cores recovered from Greenland and Antarctica

actually contain samples of the atmosphere when the snow fell. Figure 21.7 shows the locations of ice cores recovered in Greenland and Antarctica that are mentioned in this book.

As the snow compacts to form ice, bubbles of air are trapped. The air bubbles can be analyzed, and giving us direct measurements of the amount of $CO_2$ in the atmosphere when the snow fell. Being lighter than the ice, the bubbles of air migrate upwards, so the ages of the air samples are slightly older than the ice that contains them. However, this was recognized early on, and it was discovered how to make the necessary corrections. There was also a question as to whether the $CO_2$ in the bubbles was really a sample of the atmosphere, or whether it might have been generated by decomposition of organic material in the ice. After all, there are algae and bacteria that live in snow banks and thrive when the temperature nears freezing. In the US west, you may have noticed the rose red color of some snow banks; the color is from single-celled algae. Fortunately, these organisms leave a distinctive signature in the isotopes of carbon as they produce $CO_2$, very different from the usual atmospheric $CO_2$. The isotopic composition of the $CO_2$ is routinely checked to make sure it is uncontaminated.

Organic matter can also be carried in with dust, but again, it has a distinctive isotopic signature that can be recognized.

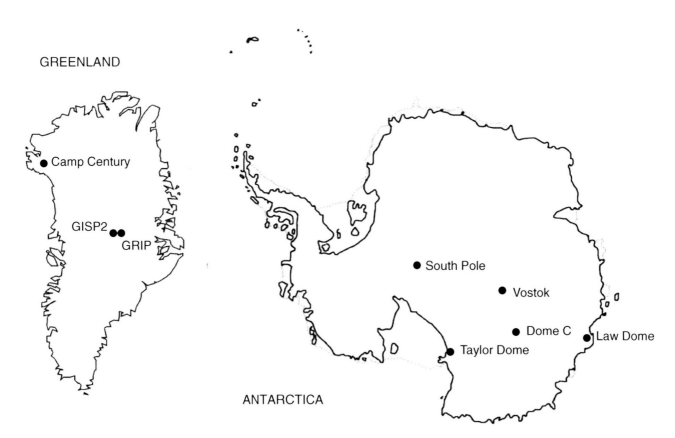

**Fig. 21.7**  Greenland (*left*) and Antarctic (*right*) ice coring sites discussed in this book

Unfortunately, dust contamination is a real problem in the ice cores from Greenland. The North American and Scandinavian ice sheets were surrounded by sandy outwash plains that served as a nearby source for dust. Contamination is much less of a problem in the cores taken in Antarctica. Those cores also contain some dust, most of which comes from Patagonia, but it is much less of a contaminant.

Figure 21.8 shows the $CO_2$ record for the Vostok core in Antarctica, plotted against the Milankovitch insolation curve for 65° N. Over the past 450,000 years levels of atmospheric $CO_2$ have varied between 185 and 300 ppmv, with the

average being 234 ppmv. The $CO_2$ variations follow the insolation closely, usually with a slight lag of about 1,000 years. The primary forcing of glacials and interglacials is from the changing Milankovitch orbital parameters. Once that starts the cycle, the $CO_2$ follows and soon becomes a significant positive feedback.

The relation of atmospheric $CO_2$ concentration to temperature in the Vostok record is shown in Fig. 21.9. Looking carefully you will see that the $CO_2$ levels rise after the temperature rises. The lag is of the order of a hundred to a thousand years. The lag during a temperature decline is even

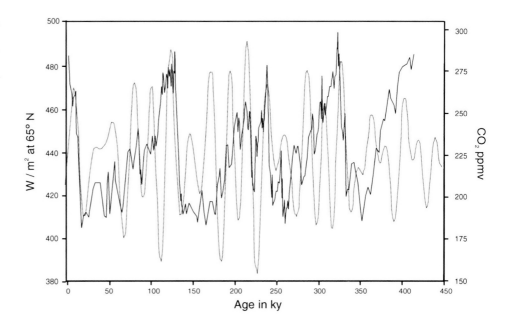

**Fig. 21.8** Past atmospheric $CO_2$ concentrations recorded by air trapped in the Vostok ice core in East Antarctica (*solid line*). The insolation at 65° N is shown as a *dotted line*

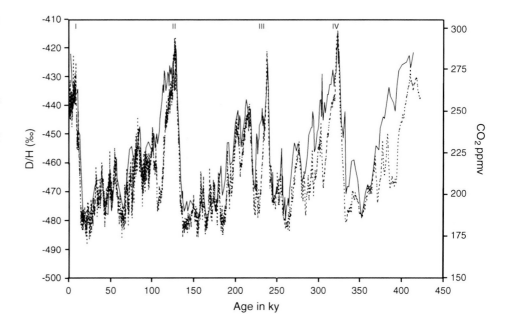

**Fig. 21.9** Past atmospheric $CO_2$ concentrations recorded by air trapped in the Vostok ice core in East Antarctica (*solid line*). The temperature variations, interpreted from the ratio of deuterium to hydrogen, are shown as a *dotted line*. The glacial-interglacial temperature difference is of the order of 10 °C

greater, up to almost 10,000 years. It is clear that it is not $CO_2$ changes that initiate the temperature change, but that as the temperature changes, something happens to cause $CO_2$ to change. Of course, as the $CO_2$ changes it reinforces the temperature trend, a typical positive feedback.

Before this information was available many of us speculated about how $CO_2$ might have a positive feedback to reinforce the effects of Milankovitch cycles. As Bob Dietz, one of the original proponents of sea-floor spreading used to say, "It's a good idea to develop your hypotheses before there are too many facts. Facts tend to confuse the issue." One of the ideas was that as sea level declined during the buildup of ice on land; the sediments on the continental shelves would be eroded, releasing nutrients into the ocean. This would cause ocean plankton to bloom and they would consume $CO_2$, drawing down the atmospheric levels and reinforcing the cooling trend. But you can just as well present an argument that the shelf seas are the most productive. That is, they have the most living plants and animals, and if sea level eliminates the space available for these organisms, they cannot remove $CO_2$ from the atmosphere, so atmospheric $CO_2$ would either stay the same or even increase. Here are several other variants to these arguments, all of them leading nowhere in particular because there simply was no way to know what had happened.

It was this relation between $CO_2$ and temperature on the glacial-interglacial scale that prompted Congressman Joe Barton's comment after Al Gore's testimony on global warming:

> In your movie, you display a time line of temperature and compared to $CO_2$ levels over a 600,000-year period as reconstructed from ice core samples. You indicate that this is conclusive proof of the link of increased $CO_2$ emissions and global warming. A closer examination of these facts reveals something entirely different. I have an article from Science magazine which I will put into the record at the appropriate time that explains that historically, a rise in $CO_2$ concentrations did not precede a rise in temperatures, but actually lagged temperature by 200–1,000 years. $CO_2$ levels went up after the temperature rose. The temperature appears to drive $CO_2$, not vice versa. On this point, Mr. Vice President, you're not just off a little. You're totally wrong.

Congressman Barton (Republican) represents the 6th Congressional District in Texas, which includes Corsicana. Probably unknown to its residents, the District contains some of the most classic sites of upper Cretaceous rocks rich in microfossils. These localities, which I was delighted to visit and sample many years ago, are now covered by urban sprawl.

The article Congressman Barton refers to is probably that by Nicolas Caillon et al. (2003) Timing of Atmospheric $CO_2$ and Antarctic Temperature Changes across Termination III: *Science* 299, 1728–1731. (See also E. Monnin et al. 2001, Atmospheric $CO_2$ Concentrations over the Last Glacial Termination: *Science* 291, 112–114). Remember that a

'Termination' is the episode of rapid deglaciation that marks the end of each of the glaciations. The ice builds up to a maximum over a period of about 80,000 years, and during the deglaciation it melts away over an interval of only about 10,000 years, called the 'Termination.' Terminations are designated by Roman Numerals, starting from the last one (Termination I), which lasted from about 18,000–6,000 years ago. Termination III was the antepenultimate deglaciation, lasting about 10,000 years centered about 240,000 years ago. It is true that in the ice cores the changes in $CO_2$ lag behind the changes in temperature, as one would expect in a natural system where the ultimate forcing is from changes in Earth's orbit. This was evident in the original description of the Vostok core (Petit et al. 1999, Climate and atmospheric history of the past 420,000 years from the Vostok ice core, Antarctica, *Nature* 399, 429–238). It is what one would expect if the ultimate cause of the termination is the change in insolation due to Milankovitch orbital variations. Most importantly, Congressman Baron confuses the effect on two very different timescales, that associated with geologic history, when these changes occur over thousands to tens of thousands of years, and that for recent human history, in which changes in atmospheric history are occurring on timescales of decades. In the first chapter of this book I mentioned the importance of timescales for processes. The Congressman's comment illustrates an inability to grasp the difference between processes acting over decadal and millennial timescales. Or perhaps it's just a desire to confuse the public. Incidentally, Congressman Barton is a religious fundamentalist who believes the Earth was created in seven days six thousand years ago, so all of this discussion of the past is pure fantasy as far as he is concerned.

Figure 21.10 shows the atmospheric $CO_2$ variations plotted with the oxygen isotope curve reflecting the global buildup and decay of ice in the Vostok core. The variations of $CO_2$ occur a few thousand years before the variations in ice volume, as one would expect. The massive ice sheets are slow to respond to climate change. The reasons for the $CO_2$ variations are due to a combination of factors involving changes in ocean temperature and circulation, changes in the areas of shelf seas, with their abundant life, changes in open ocean plankton populations, and changes in vegetation on land. This is a work in progress, and will be discussed more extensively later.

More recently, the European Consortium for Ice Coring in Antarctica (EPICA) has completed drilling at a site designated Dome C. This site is located south of Australia, about one third of the way from the Antarctic coast to the South Pole. Drilling there has recovered a record even longer than at Vostok, going back to about 750,000 years. It is still being analyzed. Figure 21.11 shows the $CO_2$ record from 400,000 to 650,000 years. Atmospheric carbon dioxide levels remained below 275 ppmv over this entire interval.

**Fig. 21.10** Past atmospheric $CO_2$ concentrations recorded by air trapped in the Vostok ice core in East Antarctica (*solid line*). Also shown is the oxygen isotope ratio curve for the air, representing the variations in the amount of ice on Earth (*dotted line*)

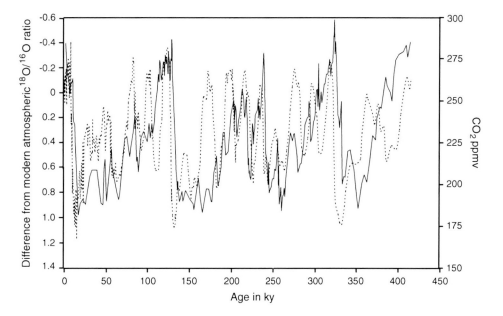

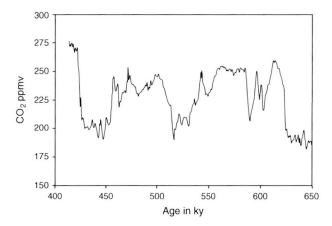

**Fig. 21.11** Atmospheric $CO_2$ levels for the interval from 400,000 to 650,000 years ago, from the DOME C ice core (East Antarctica)

The Taylor Dome ice cores have a detailed record of the last 70,000 years. The complete record is shown in Fig. 21.12. As the glaciation progressed toward the Last Glacial Maximum, atmospheric $CO_2$ levels gradually decreased from about 215–185 ppmv. The variations that can be seen correspond closely to the changing insolation at 65° N with only brief lags. However, you will notice that the two lines depart from one another at about 8,000 years ago. Insolation is declining but $CO_2$ is increasing. Bill Ruddiman of the University of Virginia has devoted most of his life to investigating the transition from the Last Glacial Maximum to our present condition. He noticed that this pattern of a continued rise of atmospheric $CO_2$ after the end of the deglaciation is unique to the latest Termination, as shown in Fig. 21.13. The penultimate deglaciation, Termination II, is complex with a series of rises and falls of $CO_2$ reflecting

frequent climatic shifts during the last interglacial. In spite of the swings in atmospheric $CO_2$ concentration averaging about 15 ppm during the Eemian interglacial which followed Termination II, the oscillations remain centered on 275 ppm. The earlier Terminations III and IV show distinct peaks in atmospheric $CO_2$ concentration at the end of the deglaciation followed by gradual declines through the later part of the interglacial until the next ice age begins.

Ruddiman attributes the post-8,000 year BP (BP = before present, where 'present' is usually taken to be 1950) rise in $CO_2$ to human activity. Early human activities involved burning of large areas as a hunting technique and clearing of land for agriculture. He pointed out that although it might seem as though these activities were on a very small scale, they acted over a long time. He noted that if the clearing of land started 8,000 years ago and proceeded at a rate of 0.04 Gt C/year ($0.04 \times 10^{12}$ kg of carbon per year) for 7,800 years, the cumulative total is 320 Gt ($320 \times 10^{12}$ kg) of carbon. Assuming that over the past 200 years the rate increased by a factor of 20 to an average of 0.8 Gt C/year, the cumulative total would be 160 Gt of carbon, half that of the preceding 7,800 years. Human activity may have been slowly introducing $CO_2$ into the atmosphere since the beginning of civilization. Until the industrial revolution, this gradual rise may have offset the decline in insolation at 65° N since its maximum 11,000 years ago. This gradual increase may have inadvertently helped to stabilize the climate avoiding the large variations that occurred during the earlier interglacials.

The Termination I through modern interglacial $CO_2$ record is shown in Fig. 21.13. It is based on the younger part of the Taylor Dome ice core data spliced together with $CO_2$ measurements from Law Dome ice core. The latter has a

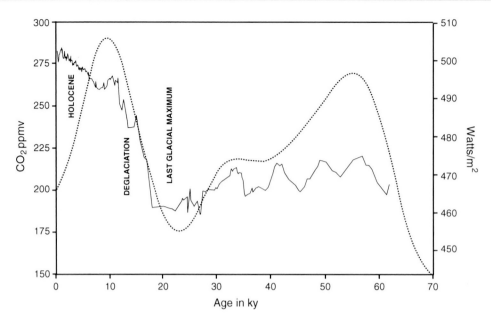

**Fig. 21.12** Atmospheric $CO_2$ concentrations over the past 70,000 years, recorded in the ice cores from Taylor Dome (East Antarctica). *Dotted line* is the insolation at 65° N

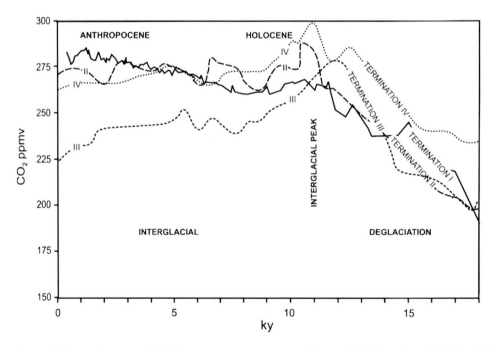

**Fig. 21.13** Comparison of changes in atmospheric $CO_2$ concentrations during and after four Terminations: I (the last deglaciation; from about 18,000 to 11,000 years ago), II (the penultimate deglaciation; from 116,000 to 136,000 years ago), III (the antepenultimate deglaciation; from 245,000 to 238,000 years ago), and IV (the preantepenultimate deglaciation; from 336,000 to 323,000 years ago). The unusual upward trend since 8,000 years ago is apparent. The rise over the past 200 years is not shown. The February 2016 $CO_2$ level of 404 ppmv would plot above the top of this figure

shorter but more expanded section because of much higher snow accumulation rates. The rise of $CO_2$ that started with the industrial revolution is evident. In the background of Fig. 21.14 is the $CO_2$ record for the last 650,000 years. Note

that the highest levels recorded in the ice cores for the past are never above 300 ppmv. As far as we know today, the February 2016 level of 404 ppmv, has not been seen for the past 35 million years.

**Fig. 21.14** Atmospheric CO$_2$ concentrations based on the younger part of the Taylor Dome ice cores (*thin solid line*) and the last 1000 years based on the Law Dome ice cores (*thick solid line*) connected to the November 2011 level by a *dotted line*. *Dotted lines* in the lower part of the diagram are the 650,000 year record of the Vostok and Dome C ice cores. Late Holocene CO$_2$ levels have been higher than 99 % of the earlier record. The present level is as much higher than Holocene levels as glacial CO$_2$ concentrations were below Holocene levels. All of these ice cores are from the Antarctic

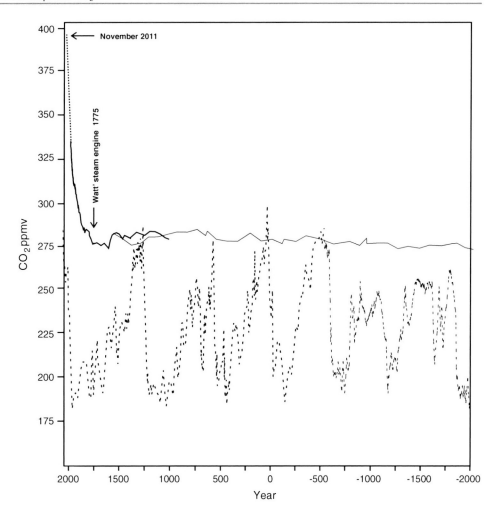

## 21.7   The Industrial Era Increase in Atmospheric CO$_2$

At the end of the 19th century, Svante Arrhenius had recognized than human activity was changing the levels of atmospheric CO$_2$ and that this was likely to have an impact on climate. He believed that the climate would slowly warm, and that for countries at high latitudes, like Sweden, that might be a good thing. It would make possible a longer growing season. He did not think about the other implications, nor did he expect that the amounts of CO$_2$ being introduced into the atmosphere would increase as rapidly as they did. As discussed above, water was recognized as a more important greenhouse gas, and during the first half of the 20th century, CO$_2$ received little attention.

A British steam engineer, George Callendar, was also an amateur meteorologist. He became interested in the increase of CO$_2$ and its radiative properties. He published one of the first diagrams showing the way in which the absorption of energy by CO$_2$ coincides with part of the long-wave infrared radiation from the Earth's surface. He collected data on the

rise of CO$_2$ concentrations in the atmosphere. He established that in the first 35 years of the century, CO$_2$ levels had increased about 10 % and that the rate of increase was itself increasing. He published several papers between the late 1930s and 1950s, including an overview in 1958 suggesting that this increase was a result of burning fossil fuels, and that it might result in a warmer Earth.

Hans Suess, who you will recall was one of the pioneers in $^{14}$C dating, discovered that the burning of fossil fuels was introducing a systematic error into the method. Fossil fuels contain no $^{14}$C, only $^{12}$C, so-called 'dead carbon'. The $^{14}$C dating method depended on determining the ratio between the two isotopes, and the addition of $^{12}$C since the beginning of the industrial revolution had to be taken into account. This error correction became known as the 'Suess Effect.'

Gilbert Plass was born in Toronto. During World War II he worked with Fermi's group at the University of Chicago as a physicist. He received his Ph.D. in physics from Princeton in 1947, and then held a number of positions, both industrial and academic, ending up at Texas A & M. His work was in the field of radiative energy transfer, and he is credited with developing a detailed account of absorption and emission of infrared energy by greenhouse gases, in particular $CO_2$ and water. In 1956 he published two important papers on carbon dioxide and climate. He recognized that humans were performing an uncontrolled experiment on the climate of planet Earth. He wrote:

> If at the end of this century measurements show that the carbon dioxide content of the atmosphere has risen appreciably and at the same time the temperature has continued to rise throughout the world, it will be firmly established that carbon dioxide is an important factor in causing climate change.

The pieces required for real understanding of the role of carbon dioxide and Earth's climate were falling into place. What was needed was someone to initiate a long-term study of cause and effect.

Roger Revelle was born in Seattle in 1909. He received his doctorate from the University of California at Berkeley in 1936, working with the Scripps Institution of Oceanography in La Jolla. He became a Professor at Scripps in 1948. He was concerned about how the rising levels of atmospheric $CO_2$ would affect the oceans, and wrote a paper with Hans Suess (also at Scripps then) terming the rise of $CO_2$ the 'Callendar Effect.' That article included a prophetic statement:

> Human beings are now carrying out a large-scale geophysical experiment of a kind that could not have happened in the past nor be reproduced in the future. Within a few centuries we are returning to the atmosphere and ocean the concentrated organic carbon stored in sedimentary rocks over hundreds of millions of years. This experiment, if adequately documented, may yield a far-reaching insight into the processes determining weather and climate.

From 1963 to 1976 Roger was at Harvard University, where he founded the Center for Population Studies. One of his students was Al Gore. Gore was inspired by him, and became one of the very, very few U.S. political leaders who had some background in science. In 1976 Roger returned to the University of California at San Diego as Professor of Science Technology and Public Affairs. I got to know him through studies being conducted by the National Research Council. As a 'peer reviewer,' I suggested editing out some of his more prophetic statements in the introduction to the 1990 National Academy Report *Sea level Change*. Unfortunately, he took my advice seriously and cut out some predictions which have since proven to be correct.

But to get back to the $CO_2$ story, before leaving Scripps Roger realized that what was needed was a series of regular measurements of atmospheric $CO_2$ at a site far from industrial sources of pollution. He found a willing graduate assistant, Charles David Keeling, and an appropriate site, the top of Mauna Loa volcano on the Big Island of Hawaii, and the measurements began in 1958. The air is sampled daily,

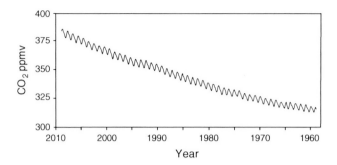

**Fig. 21.15** Atmospheric concentration of $CO_2$, measured at Mauna Loa observatory through 2007. Since then concentrations have passed 400 ppmv

and monthly averages computed. The record through July 2007 is shown in Fig. 21.15.

I think this figure is probably familiar to every reader, but it is usually shown the other way around, with the first year of measurement, 1958, on the left and the latest value on the right. Geologists usually make diagrams with the past on the right and the future on the left, so I have shown this diagram as a geologist would view it. The wiggles are the Earth's Northern Hemisphere 'breathing' with the seasons. When trees and other plants leaf out in the spring, they take in $CO_2$, causing the curve to go down. When they go into their winter sleep they are no longer net consumers of CO2 but respiration gives off $CO_2$ causing the curve to go up. We can say that this is the Northern Hemisphere breathing because that is where there are broad landmasses in the zone where large seasonal changes occur.

I was amazed to find on the Internet reports stating "atmospheric $CO_2$ levels have started to decline." What is being reported is the seasonal decline, but this is not made clear, and the impression is given that the overall increase has been reversed.

If you lay a ruler along the trend in the diagram you will see that it is not straight, but that the annual rate of $CO_2$ addition to the atmosphere is increasing. The overall rate for 1958 through 2007 is about 11.8 %. That means a doubling time of about 6 years. That is 6 years from now we will be putting twice as much $CO_2$ into the atmosphere as today.

In discussing $CO_2$ in the atmosphere it is useful to distinguish between modern and fossil carbon. Modern carbon is what is in living plants and animals, and the organic matter in soils. It contains the isotope $^{14}C$ generated by cosmic radiation at the top of the atmosphere. It is the carbon that forms the 'natural' part of the carbon cycle. Fossil carbon is carbon that was buried in sediments and is millions of years old. It is the carbon now being extracted from the rocks by human activity. It contains no $^{14}C$, sometimes it is referred to as 'dead' carbon. It is the un-natural part of the carbon cycle that humans have created.

Figure 21.16 shows the amounts of carbon dioxide from fossil carbon introduced into the atmosphere each year since 1750. Overall the curve has the shape of an exponential function; that is the rate of increase increases with time. However, since 1950, the rate of increase has been relatively stable, at 8.9 % per year, with a doubling time of a little over eight years.

Where does this fossil carbon come from? As shown in Fig. 21.17, the source of most of the carbon being added to the atmosphere is from the burning of fossil fuels. However,

about 3 % comes from are two other major sources and several minor ones. The other major sources are manufacture of concrete. The active agent in concrete is calcium oxide, made from roasting limestone:

$$CaCO_3 \rightarrow CaO + CO_2 \uparrow$$
$$\text{Heat}$$

To make concrete, the calcium oxide is mixed with silicate minerals or quartz. These react to form a solid material.

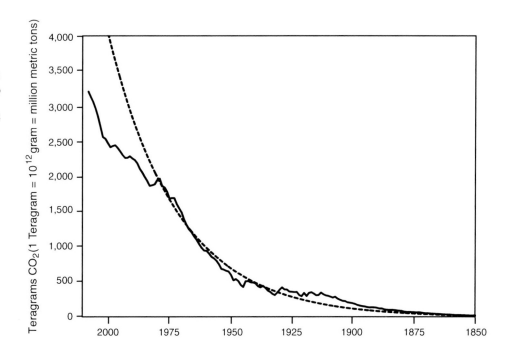

**Fig. 21.16** Cumulative amount of carbon dioxide from fossil carbon added to the atmosphere since the industrial revolution began (1775). The amount added from 1775 to 1850 is too small to show and is not included. An exponential curve fit to the data is shown as a *dashed line*. Data compiled by Tom Boden, Gregg Marland, and Bob Andres, Carbon Dioxide Information Analysis Center (CDIAC), Oak Ridge National Laboratory, Oak Ridge, Tennessee 37831–6290 You can test your knowledge of history by figuring out what the dips in the curve represent

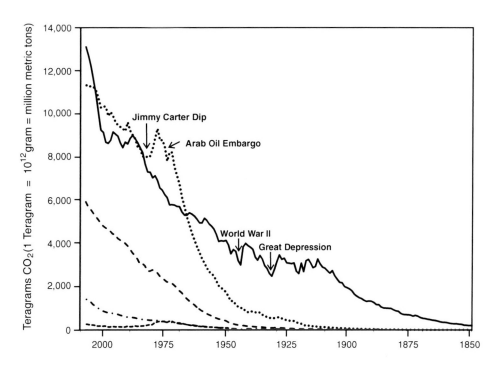

**Fig. 21.17** Cumulative amounts of carbon dioxide from fossil carbon added to the atmosphere since the industrial revolution began (1775) by type: *black line solid* (coal); *dotted line liquid* (petroleum); *long dashed line* gas (natural gas); *dash-dot line* cement manufacture; *short dashed line* flared gas. The amount added from 1775 to 1850 was from coal but is too small to show and is not included. Data compiled by CDIAC

Concrete has been made since ancient times. The Roman's perfected it to a degree not seen again until recently. Roman concrete has survived in excellent condition for over 2,000 years, and it has recently been discovered that the 'secret ingredient' used to make this very durable concrete was volcanic ash.

The other major anthropogenic source of carbon dioxide involves modern carbon. It comes from burning of trees and shrubs, to clear forested areas so they can be used for agriculture or as pastureland for cattle. This is frequently referred to simply as 'biomass burning.' Remember that Thomas Jefferson thought it was the clearing of forests that would improve the climate of the United States and make it more habitable. Bill Ruddiman estimated to long-term deforestation rate from 8,000 years ago until the industrial revolution to be $0.04 \times 10^{12}$ kg C/year, with a cumulative total of $320 \times 10^{12}$ kg of carbon having been added to the atmosphere over 7,800 years. However, the rate of deforestation has increased dramatically as world population has exploded, and it is estimated that over the past 200 years about $200 \times 10^{12}$ kg of carbon has been added through changes in land use. The total modern carbon added from human activity is then about $520 \times 10^{12}$ kg of carbon, $200 \times 10^{12}$ kg of it in the last 200 years. The effect of the addition in changes in land use prior to the industrial revolution was to prevent the level of atmospheric $CO_2$ from declining as it would have under normal conditions after a glacial termination. The level stopped declining at about 260

ppmv 8,000 years ago, and slowly climbed to the pre-industrial level of about 275 ppmv in 1750 (Table 21.1).

One of the other possibly important sources of modern carbon to the atmosphere comes from the use of the plow in farming. Where soil rich in organic matter is plowed, the $CO_2$ in the soil is freed and the organic matter exposed to oxidation. A possible countermeasure comes in the more recent practice plowing under the litter after the crop is harvested.

One possible danger of global warming is that $250 \times 10^{12}$ kg C, about 1/4 of the planets soil carbon, is stored in the peat bogs and frozen soils of the high Arctic. As the temperature warms this massive reservoir becomes susceptible to decomposition and conversion to $CO_2$, and could be the source of a massive release of both $CO_2$ and methane to the atmosphere.

To understand the relation between these masses of carbon and the amount of $CO_2$ in the atmosphere, you need to know how to convert a mass of carbon into ppmv $CO_2$ in the atmosphere. One ppmv $CO_2$ is approximately equal to $2 \times 10^{12}$ kg of carbon ($2.111 \times 10^{12}$ kg C to be exact).

If all of the anthropogenic $CO_2$ generated over the past 250 years were still in the atmosphere, the $CO_2$ concentration level would be about 520 ppmv, but it is only about 400 ppmv. The increment remaining in the atmosphere as a result of the industrial revolution and modern era amounts to about $200 \times 10^{12}$ kg C. This is the amount that has brought the concentration of $CO_2$ from 275 to 400 ppmv. Only about 40 %

**Table 21.1** Major reservoirs of carbon

| Reservoir | Mass of C (kg) | Moles of carbon | Residence time (year) | Form | % C |
|---|---|---|---|---|---|
| Carbonate rocks | $60 \times 10^{18}$ | $5 \times 10^{21}$ | $1 \times 10^{9}$ | $CaCO_3$ | 12 |
| Buried organic carbon | $12 \times 10^{18}$ | $1 \times 10^{21}$ | $260 \times 10^{6}$ | $CH_2O$ | 40 |
| Land plants | $0.60 \times 10^{15}$ | $50 \times 10^{15}$ | 50 | $CH_2O$ | 40 |
| Organic matter in soil | $1.22 \times 10^{15}$ | $101.6 \times 10^{15}$ | 500 | $CH_2O$ | 40 |
| Peat | $0.36 \times 10^{15}$ | $30 \times 10^{15}$ | 1000 | $CH_2O$ | 40 |
| Microbes | $0.025 \times 10^{15}$ | $2 \times 10^{15}$ | ? | $CH_2O$ | 40 |
| Organic matter in surface sediment | $0.15 \times 10^{15}$ | $12.5 \times 10^{15}$ | 100 | $CH_2O$ | 40 |
| Dissolved organic matter in ocean | $0.74 \times 10^{15}$ | $58.3 \times 10^{15}$ | 5000 | $CH_2O$ | 40 |
| Particulate organic matter in ocean | $0.005 \times 10^{15}$ | $0.4 \times 10^{15}$ | 50 | $CH_2O$ | 40 |
| Bicarbonate ion in ocean | $36 \times 10^{15}$ | $3000 \times 10^{15}$ | 2000 | $HCO_3^{-}$ | 20 |
| Carbonate ion in ocean | $2.4 \times 10^{15}$ | $200 \times 10^{15}$ | 2000 | $CO_3^{=}$ | 20 |
| $CO_2$ in atmosphere (1750) | $0.42 \times 10^{15}$ | $9.9 \times 10^{15}$ | 4 | $CO_2$ | 27 |
| ? Deep biosphere | $0.60 \times 10^{15}$ | $13.6 \times 10^{15}$ | $260 \times 10^{6}$ | $CH_2O$ | 40 |
| Methane in atmosphere (1750) | $2.12 \times 10^{12}$ | $0.13 \times 10^{15}$ | 10 | $CH_4$ | 75 |
| Methane clathrates | $11 \times 10^{15}$ | $916 \times 10^{15}$ | $1 \times 10^{6}$ | $CH_4$ | 75 |

The numbers are estimates and may well be off by 25 %. Buried organic carbon includes that concentrated enough to be considered fossil fuels (petroleum, natural gas, coal). They are less than 1 % of this total. The residence time for $CO_2$ in the atmosphere is only a few years because it continually exchanges with living organisms and the ocean. However, its lifetime in the system is of the order of hundreds of thousands to millions of years. Methane and methane clathrates will be discussed in the next chapter. "Form" is the compound containing the carbon. $CH_2O$ is used as a generic formula for organic matter, living and dead. % C is the percent of the molecular weight of the compound that is carbon of the carbon added to the atmosphere over the past few centuries is still there. What happened to the rest of the 700 CO$_2$?

## 21.8 The Short Term Sinks for Atmospheric CO$_2$

The remainder of $320 \times 10^{12}$ kg C added by humans to the atmosphere has been absorbed by the ocean and land in roughly equal proportions. Eventually it will be stored in sedimentary rocks once again, but in the meantime it is in temporary storage in seawater and in new plant growth, wooden structures, filing cabinets, and garbage dumps. First, let's consider the ocean.

The ocean and atmosphere maintain a balance of gases, interchanging them across the air-water interface. The changes can be monitored and measured, so this part of the cycle is relatively well known. However, what happens is fairly complicated and has enormous implications for life in the sea. It will be discussed more fully in the Chap. 26 on the biosphere, but can be summarized as follows: CO$_2$ from the atmosphere can be absorbed directly into the seawater, but only about 1 % of the absorbed CO$_2$ remains in the form of dissolved gas. The rest reacts with the water and the carbonate ion, CO$_3^=$ to be converted into the bicarbonate ion, HCO$_3^-$:

$$CO_2 + H_2O + CO_3^= \rightarrow 2HCO_3^-$$

About 91 % of total CO$_2$ in the ocean is already in the form of the bicarbonate ion, HCO$_3^-$, and 8 % is present as the carbonate ion, CO$_3^=$. This reaction causes the carbonate ion to be replaced by the bicarbonate ion, and makes it easier for seawater to dissolve the calcium carbonate shells of animals and plants or conversely, more difficult for organisms such as corals to secrete the carbonate framework on which they grow. Unfortunately, the ability of the ocean to absorb additional CO$_2$ from the atmosphere decreases as the atmospheric concentration rises. This is because as the availability of the carbonate ion decreases, more of the CO$_2$ remains in its dissolved gas form, and it becomes more and more difficult for the ocean to absorb more gaseous CO$_2$. At today's atmospheric concentrations, 400 ppmv, the ocean can only absorb about 60 % of the increase in CO$_2$ that it could when the atmospheric concentrations was 275 ppmv. The ocean will be able to absorb a decreasing proportion of the atmospheric CO$_2$ as atmospheric concentration levels rise. This means that as more and more CO$_2$ is added to the atmosphere, and the ocean able to absorb less and less of the increase, more will remain in the atmosphere.

The CO$_2$ that has been taken up by the ocean has already started to change its chemistry. It is thereby having an effect on marine life not directly related to global warming. This will be discussed below in Chap. 26.

The uptake of carbon on land is thought to be mostly related to the reforestation of some areas previously dedicated to agriculture. It has been estimated that if all of the land that has been deforested were to be turned back to forest again, it would consume about $90 \times 10^{12}$ kg C from the atmosphere, reducing the atmospheric concentration by about 45 ppm. On the other hand, if deforestation were to continue until all of the planet's forests had been converted into grassland, another $400 \times 10^{12}$ kg C would be added to the atmosphere, raising the level of the concentration by 200–400 ppmv. The uncertainty in these numbers results from estimates of the amount of wood in forests, their areas, and plant growth rates.

It has recently been suggested that significant reforestation has occurred, possibly several times during human history. The largest episode of reforestation occurred after the bubonic plague of the Middle Ages killed about a third of the human population of Europe and probably many other areas. This was followed by the inadvertent introduction of smallpox to the Americas. The disease came along with the European explorers in the late 15th and early 16th centuries. It wiped out all of the Caribs who inhabited the Caribbean islands, and much of the indigenous population of the continents even before direct contact with the explorers venturing into North and South America. The result of these massive reductions of population was a return of much agricultural land to forest, and a global drawdown of CO$_2$ of about 10 ppmv during the 17th, 18th, and 19th centuries as trees grew and forests spread. This can be seen as a small dip on the Law Dome CO$_2$ concentration curve shown in Fig. 21.14. That dip coincided with the 'Little Ice Age.' The suspicion is that the Little Ice Age had its origin in plagues that significantly reduced human populations. Similarly, the regrowth of forests on formerly cultivated land in the eastern U.S., and the largely successful efforts to prevent forest fires in the western U.S. during the 20th century may have absorbed significant amounts of CO$_2$ from the atmosphere, slowing the overall increase in concentration and reducing the rate of global warming so that it was not so obvious.

However, when all of the possible 'sinks' for carbon other than the ocean are added up there remains a significant unknown of the order of $2 \times 10^{12}$ kg C. It is going somewhere, but just where is not known. It may be as simple as being the wood used to construct houses in the United States. Our method of constructing wood-frame homes with wood or plywood siding sometimes covered with a decorative veneer of brick or stone is unique, and was developed to accommodate a rapidly growing population. It was originally called 'Chicago construction,' developed about 1832. It got a bad name after Chicago's Great Fire of 1871, so afterward it was called 'balloon framing' or 'balloon construction.' It has been largely replaced by a similar but less labor intensive method of building a house called 'platform framing' or 'stick construction.' In any case an American 'brick house' is really mostly wood. Another possible sink is the paper in company and bureaucrats' files. I made this suggestion once in a meeting, thinking it was a joke, but some in the audience took me seriously, and calculations suggest there may be a lot of carbon in all that paper. Imagine the effect the Xerox copier had on the amounts of paper going into files. I suspect another part of the missing carbon is the paper (the catalogs you get in the mail) and cardboard in landfills. It has been found that these do not decay as rapidly as was expected.

## 21.9 Carbon Dioxide Catastrophes

Carbon dioxide is significantly heavier than ordinary air, having a molecular weight of about 44 compared with those of the $N_2$ ($\sim$28) and $O_2$ ($\sim$32) that make up more than 99 % of air. The average molecular weight of air is about 29.

If vented into a low place, the colorless, odorless $CO_2$ can collect and suffocate animals. One famous site for this is in the Campi Flegrei, a volcanic caldera on the western side of Naples, Italy. There the Solfatara crater emits pungent sulfurous gases which are easy to detect, and since Roman times have been considered to have curative powers. However, the Solfatara is also a prodigious source of natural $CO_2$ with a mean emission rate of 1067 tons per day. That is seven times higher than today's rate for Vesuvius (151 ton/d). During a calm night with no winds, the concentration of $CO_2$ in the Solfatara can rise to 2,400 ppmv or more. There are anecdotal stories that dogs wandering into the crater have suffocated by the gas.

More serious is the problem with three volcanic crater lakes in Africa: Lakes Nyos and Monoun in Cameroon, and Lake Kiva in the Congo. The dormant volcanoes housing these lakes emit $CO_2$, and it accumulates dissolved in the bottom waters of the lakes. It is estimated that 90 million tons of $CO_2$ are trapped in the deep waters of Lake Nyos (Fig. 21.18). On August 21, 1986, a 'limnic eruption' (lake water eruption), possibly caused by a landslide within the lake, occurred at Lake Nyos. It triggered the sudden release

**Fig. 21.18** Lake Nyos, Cameroon; site of the August 21, 1986, limnic eruption of $CO_2$. The gas cloud poured over the lake outlet, seen on the far side in this picture (Courtesy Wikipedia)

of about 100,000–300,000 tons of $CO_2$. A 100 m (330 ft) fountain of water and foam is thought to have formed on the surface of the lake, producing a tsunami at least 25 m (82 ft) high that scoured the shore along one side of the lake. The invisible $CO_2$ gas cloud spilled over the northern lip of the lake into a valley and then spread into two other valleys branching off it. The heavier $CO_2$ displaced the air in these valleys for as far as 25 km (16 miles) from the lake, suffocating some 1,700 people, mostly rural villagers, and about 3,500 livestock. Older occurrences of such events are the topic of legends, but are not documented. Engineering technology to slowly vent the deep waters of the lakes is being developed.

## 21.10  Projections of Future Atmospheric $CO_2$ Levels

Projections of future levels of $CO_2$ depend on when and what measures are taken to reduce emissions. For the year 2100, the Official estimates range from 600 to 900 ppm. After stabilization of emissions, it is expected that there will be a period of stable atmospheric concentrations for one or a few thousand years, followed by a decline over the next several thousand years. In reality, the introduction of fossil fuel $CO_2$ into the atmosphere regularly exceeds the 'worst case' official scenarios.

Computer models of the initiation of glaciation on Antarctica suggest that there is an important climatic threshold between 700 and 800 ppmv $CO_2$. Above that threshold no stable Antarctic ice sheet can build up. The question arises whether the Antarctic ice sheets will remain intact if atmospheric $CO_2$ levels approach that threshold as southern hemisphere insolation continues at its current maximum. Complete melting of the Antarctic ice sheet at the current rate would require over 250,000 years, but the melting rate is likely to increase significantly during the 21st century. My best guess is that the future history of the Antarctic is too close to call.

## 21.11  Summary

The role of $CO_2$ as the climate-regulating greenhouse gas for planet Earth has become evident in recent years. Over the history of the planet it has operated to counter the effect of the changing luminosity of the Sun to regulate the temperature of Earth's surface. It may have failed in this role several times during the Precambrian, allowing the Earth to become completely ice-covered, but it was able to recover and bring the Earth back to the condition where liquid water covered most of the surface of the planet. Geochemical modeling indicates that the concentration of $CO_2$ in Earth's

atmosphere declined from about 6,000 ppmv 500 million years ago to 500 ppmv 35 million years ago. This gradual decline was interrupted by a major drawdown in the Late Paleozoic, from 325 to 270 million years ago, as major coal deposits formed. The lower $CO_2$ concentrations permitted ice sheets to develop on the southern continent of Gondwana. At the end of the Eocene, 35 million years ago, a threshold of about 500 ppmv was reached and polar ice appeared in both the northern and southern hemispheres. Filling of the deep ocean with cold water drew the atmospheric $CO_2$ levels down further to about 270 ppmv. They remained near this level, with variations of about 20 ppmv as the ice sheets grew and decayed until very recently. There is evidence that human changes in land use started a slow rise in atmospheric $CO_2$ levels 8,000 years ago. A major increase started with the industrial revolution in 1750, and became an exponential increase starting in 1850. The present level of 390 ppmv is the highest since the decline 35 million years ago. It can be attributed to two causes, the burning of fossil fuels and the large-scale clearing of forests for agriculture.

There is no timeline for this chapter.

If you want to know more:

David Archer, 2010. *The Global Carbon Cycle*. Princeton University Press, 216 pp.

Music: To celebrate the bubbles of $CO_2$: Johann Strauss II's Champagner Polka (Opus 211).

Libation: Any wine with lots of $CO_2$ bubbles: Champagne of course, but the term is an 'Apellation Controlé' restricted to those elixirs produced in the Champagne region, around Reims, France. You should also try Crémant, a bubbly wine made by the same method in Alsace, Burgundy, and other parts of France.

---

### Intermezzo XXI. Building a Building

When Lewis Rosenstiel made his gift to the University of Miami for the Marine Science institute, a significant portion was designated for the construction of a new building. The Trustees of the University were hesitant to approve any such project because it did not include funds for operation or maintenance. They were afraid it might become an unfunded liability. Also, new buildings always cost more than they were supposed to. Neither Walton Smith nor Warren Wooster had been able to convince them otherwise.

I had no idea how to accomplish this goal either, but I knew that the endowment was not growing as fast as construction costs were rising. A few weeks after I became Interim Dean, I attended a meeting of the JOIDES Executive Committee in my new role. The meeting was at Lamont-Doherty Geological

Observatory on the Palisades of the Hudson, just north of New York City. It occurred to me that perhaps someone I had met years before, an architect who seemed to know how to get things done, might be able to give me some advice. I called Max Abramovitz at his Manhattan office and asked if he might have an hour or so to talk with me after our meeting at Lamont was over. I briefly explained my problem. "Of course, come on into the city." I changed my flight back to Miami to an evening flight.

Now I need to digress and tell you how I had come to know Max. In the spring of 1962, when I was an Assistant Professor at the University of Illinois in Urbana, I was given an unusual duty. Each year the University hosted a state-wide high-school science fair. I was asked to be the University's organizer for the fair to be held in the spring of 1963. It was going to be held in a new building, the Assembly Hall which was still under construction. The fair would be held just as the construction was completed.

The Assembly Hall resembles a flying saucer that has just landed. Max Abramovitz was its architect, and Bill Poston was his resident architect on site overseeing construction. I had already met Bill because I had gone out to see the building when construction was just getting underway. We had run into each other at the site, and he explained the unique way in which this building, with the largest open interior space at the time, was being constructed. I had always had a great interest in architecture and was fascinated by the sheer audacity of this project.

The building is circular, 400 ft in diameter with no internal supports for a dome 128 ft above the center floor. A special horizontal-wheeled tractor wound 614 miles of 1/5" steel wire around the edge of the dome, circling it 2,467 times gradually squeezing it inward and upwards, so that the 800,000 ft² of wooden scaffolding that had supported the concrete when it was poured could be removed. The interior space has 16,000 permanent seats. It is surrounded by a quarter mile circular concourse with 24 bridges lit by 24 skylights. The concourse was to be the site of the science fair.

Bill and I had common interests in music, opera, and good food, and made trips to Chicago together to attend the Lyric Opera and the Chicago Symphony. Now Bill and I worked out plans to ensure that there would be enough electrical outlets, water, and all the other special things required by the science fair. Things were still being installed the night before the students arrived. The fair, in its new venue, was a great success. Max had come out to see how things went.

After the fair closed we celebrated at my tiny apartment with champagne and good wine.

Then a few years later, Max built the Krannert Center for the Performing Arts at the University of Illinois. It is a wonderful complex with a concert hall with excellent acoustics. It was based on studies his firm had made of the great concert halls in Europe as they were designing Philharmonic Hall for Lincoln Center in New York. The New York hall had to be much larger than their optimum designs, and could not incorporate wood, and its acoustics were initially not as good as had been hoped. The Krannert was the real thing. I would see Max from time to time as the construction proceeded and we enjoyed getting to know each other. The Krannert Center opened in the spring of 1969, when I was officially in Miami, but I returned for the occasion.

When I called him in 1976 I didn't know whether he would even remember me. When we met he listened carefully to my problem, told me it was not unique. He offered to come down to Miami to have a look.

A couple of weeks later I picked him up at the Miami airport and took him out to the School. He walked around the 'campus' and through all the buildings. As we walked through Marine Geology and Geophysics he smiled: "I hope the University didn't pay the architect the full fee for this one. I designed this for the Canal Zone Authority in Panama and the plans are available from the US Government Printing Office for $ 5.00." His reaction overall: "It looks like these buildings were dropped out of a passing airplane." The next day we got serious. Max told me, "This is hopeless, you need more space." We visited the Seaqarium next door, just to see if they had any unused space. None. Then Max asked about the very large public parking lot in front of our School. It was for fishermen using the catwalks along the bridge to Key Biscayne. It had several hundred parking spaces and there were about four cars parked there. "You need this piece of land." We went back to my office, and met with my management team. "First you need that get that parking lot out front from the county. You need it before you can do any new building project. Have someone start taking pictures of it each day to document how many cars use it. In some of your buildings, corridors take up a third of the space. But you don't have outsiders using the corridors; by removing the walls the hallway space could be incorporated into the laboratories and get you some needed space right away. That building with the running seawater system and seawater tank on top is going to become unsafe as the corrosion of the reinforcing rods in the concrete proceeds. To free up the money from the trustees you will need to show how you will pay for operations and maintenance not only of a new building, but of the ones you already have. You need to know how your overhead income will be generated. You need to have at least a 5 year plan for faculty replacements and hires. You need to know whether you are overtaxing some of your funding sources and ignoring other possibilities."

Max made it clear we were not ready to proceed with a new building. First we needed a detailed analysis of what space was needed—how much, and what kind. We needed a Planning Program that could formalize our needs. "This is a great opportunity; I hope I can be involved."

We certainly had our work cut out for us. We had just experienced what Max's understudy, Gerry Schiff, called the "Max Factor."

Before he left Max took me aside. "When you took on the job as Dean, you were given a list of your duties, a job description. If you do what is in your job description this new building will never be built. You need to delegate everything in your job description and make this your real job." And then he went on to tell me the same things about delegation of authority that my father had told me: "And another thing. Do you know the story of Fillipo Brunelleschi and the dome on the Cathedral in Florence?" "No." "Read Vasari's account. Everything that happened to Brunelleschi is going to happen to you."

The book is Giorgio Vasari's Lives of the Most Excellent Painters, Sculptors, and Architects, published in 1550. The entry is for Filippo di ser Brunelesco. At the beginning of the 15th century the Cathedral of Florence had been largely finished, but the crossing, the intersection of the nave and transepts, was still open to the sky. It was to be roofed over by a dome. There was a problem. The dome was to be octagonal, starting 52 m (171 ft) above the floor and spanning 44 m (144 ft) without the use of flying buttresses seen in gothic cathedrals. There was another problem. A dome of that size had not been built since 126 AD, in Rome: the Pantheon. Slightly smaller, and much closer to the floor, its dome is Roman concrete. It was poured on a wooden scaffolding that was later removed. There were not enough large trees in Tuscany to build the sturdy scaffolding required to use the same technique in Florence. And anyway, no one was quite sure just how the Romans had done it. The Florence story has all the elements of innovative modern construction: insider trading; the low bid coming from a contractor who didn't know how to do it but got paid anyway; the endless questioning of Brunelleschi's innovative solution; and endless problems of payment. Anyone who builds a building with public funds or with government overseers should be familiar with the story.

My entire management group thought Max was great. But the idea that we might be able to engage a well-known New York architect seemed out of the question. Everything the University of Miami had built had been done by local architects, most with ties to the Board of Trustees.

I went over to the Coral Gables campus to see the person in charge of such matters, Oliver Bonnert. I told him about Max's visit, and the need to acquire more space and to develop a detailed Planning Program. He was startled. I had certainly violated University protocol in such matters, but he was certainly intrigued that an architect would identify the need for additional space as the initial goal. Miami architects built buildings; they didn't tell you why you couldn't build a building.

I had already broken ground with the University trustees on the salary matter, and in December 1976 I made my request to hire Max to develop a Planning Program that would justify a request to the county for land. I found a friend among the trustees, a lawyer who was impressed with both the salary adjustment and planning arguments. He asked the hardest questions, and became my strongest supporter.

The firm Harrison, Abramovitz and Kingsland was hired to do the Planning Program. Gerry Schiff was their man on the spot. He interviewed literally everyone in the School to determine their needs and gather ideas. While this was proceeding, something new happened that changed everything.

Eric Kraus, our Chair of Atmospheric Sciences, had been exploring the possibility of forming a joint Institute with NOAA. It was to be on the lines of the University of Colorado's Cooperative Institute for Research in Environmental Sciences (CIRES). Eric's proposed new organization would be the Cooperative Institute for Marine and Atmospheric Science (CIMAS). The School's major contribution would be space to house the new Institute along with its administrative support. It wasn't that much space, just offices and meeting rooms for 7 faculty, 6 post-doctorates/graduate students, and their administrative support; a little over 2,000 ft$^2$. It was a small job and it had to fit into a small space and it also needed to be adjacent to the Marine Science Center where the physical oceanographers were located. Gerry identified a small parking lot at the end of the Marine Science Center as a possible site. I called Max to see if he might be interested in such a small project. He was.

Ollie Bonnert agreed that Max's firm could be among those competing for the CIMAS building design. When the presentations were made Harrison, Abramovitz and Kingsland hit a home run, coming in with ideas much more innovative than the others. It was a small project after all, and the Trustees approved. When the design was complete, the documents were sent out to possible contractors for bids. They came in under the estimates. The Trustees didn't know what to think; bids were always over and had to be negotiated down. Max later explained to Ollie Bonnert how this had been done. His firm had hired one of the best contractors in Miami to cost out the project; but that carried the stipulation that they could not bid on the contract. Everything, down to the kinds of screws to be used, had been in the bid document. The CIMAS project was built under budget and completed ahead of time.

In the meantime we had gotten permission to have Max's firm give us a general design for a new building to be built mostly on the county's parking lot in front of the School. Max sent Gerry Schiff around to each of the major oceanographic institutions in the US to look at buildings they had recently built and to interview faculty to learn what was good and what was bad about them. Working from the Planning Program a preliminary design was drawn up, in a large format brochure that could be distributed to all the interested parties, including the Dade County Commissioners.

I think we paid about $25,000 for a beautiful large model to be made. We also received a grant of planning funds from another foundation, but with the stipulation that they would expire after a year if the project were not moving forward. Then we had the County Commissioners over to the School for a visit and a serious discussion about that parking lot. Of course they were non-committal. Waterfront property in Miami is very valuable.

I waited to hear that our proposal for acquisition of the land was on the agenda of the County Commission. A few months went by. I called Max and told him nothing was happening. He told me "You need to find out who makes up the agenda for the County Commission. Then you need to contact him and ask for an appointment, explaining you have something to show him. And then give him or her a presentation and ask them to put you on the agenda."

I did just that, I gave the official our big brochure and showed him our plans. Before I was through with my presentation he said "This should be on the Commissions' agenda." I told him about the planning funds that would expire. "It's got to be on the agenda right away."

A couple of weeks later Warren Wisby and I attended the Commission meeting where our proposal would come up. The Commissioners spent most of the morning arguing about a new lock on a door that would cost $250. And a traffic light. They broke for lunch.

After lunch the first item: Transfer of 1.6 acres of County Land on Virginia Key to the University of Miami's Rosenstiel School; estimated value $2.5 million. Any objections? Hearing none, Approved.

METHANE - AN EVEN MORE POWERFUL GREENHOUSE GAS

*Every once in a while I feel despair over the fate of the planet. If you've been following climate science, you know what I mean: the sense that we're hurtling toward catastrophe but nobody wants to hear about it or do anything to avert it.*

Paul Krugman

To review what we have discussed so far. Any gas with more than two atoms in the molecule will absorb and re-emit radiation in the thermal infrared part of the electromagnetic spectrum. Water vapor is the most important greenhouse gas. Its concentration in the atmosphere depends on both the availability of moisture and temperature. It is very unevenly distributed, with high concentrations over the tropical oceans and very low concentrations over deserts and the Polar Regions. Carbon dioxide is the second most important greenhouse gas. It is intimately involved in the carbon cycle, and is a product of respiration. Its concentration varies with the seasons, declining with the growth of mid-latitude northern hemisphere plants in the spring and summer, and increasing as they become dormant in the fall and winter. Except for these seasonal variations, it is well mixed in the troposphere. It's concentration in the atmosphere has increased markedly since the beginning of the industrial revolution because of the burning of fossil fuels and cement manufacture.

There are two other major greenhouse gases, methane and nitrous oxide. Both are produced naturally, but have large contributions from human activities. There are also a large number of minor greenhouse gases, almost all man-made.

## 22.1 Methane—CH$_4$

Methane ($CH_4$) is a more powerful greenhouse gas than $CO_2$. As shown in the comparisons in Table 18.2, its greenhouse potential is 23 times that of $CO_2$. Although geologists regard $CO_2$ as the most important gas in regulating Earth's temperature over the younger part of the planets history, methane may have played a major role in its earlier history. This is related to a change in its lifetime in the atmosphere.

Methane has two absorption/emission bands, one at the overlap of solar and terrestrial infrared spectra, the other in Earth's shorter wavelength thermal infrared. The absorption/emission bands are broad and highly structured, as shown in Fig. 22.1.

The natural and anthropogenic sources of methane today are quite diverse, as shown in Fig. 22.2. The methane is produced mostly by anaerobic bacteria. Anaerobic means that they can only live in an oxygen-free environment. This environment occurs in wetlands, bogs, and in sediments rich in organic matter. From the sediments, they can escape into the water of oceans and lakes, and thence into the atmosphere. They also live in the digestive tracts of ruminant animals such as cattle, and in the past, dinosaurs. They also occur in the digestive tracts of termites, which use them to digest the wood they eat.

Today, methane introduced into the atmosphere reacts with the hydroxyl radical, $OH^-$ and atomic oxygen in the upper atmosphere to ultimately produce water and carbon dioxide: Note that the term ion is used for a charged atom or molecule in water; it is called a radical if it is the air. The high energy radiation coming from the Sun dissociates water and oxygen molecules high in the atmosphere, producing $H^+$, $OH^-$ and $O^=$.

$$CH_4 + 4OH^- + 2O^= \rightarrow 4H_2O + CO_2$$

This process is a major source for water vapor in the stratosphere and upper atmosphere. As you might guess, the concentration of methane in the atmosphere decreases upward. Although it is well mixed in the troposphere, its concentration declines with elevation above the stratopause.

Before the industrial/agricultural revolution, the concentration of $CH_4$ in the atmosphere was about 0.700 ppmv. Today it is about 1.750 ppmv. Because it is readily susceptible to decomposition in an oxygen-rich atmosphere, methane molecules in the atmosphere today have an average lifetime of only about 10 years. This contrasts sharply with

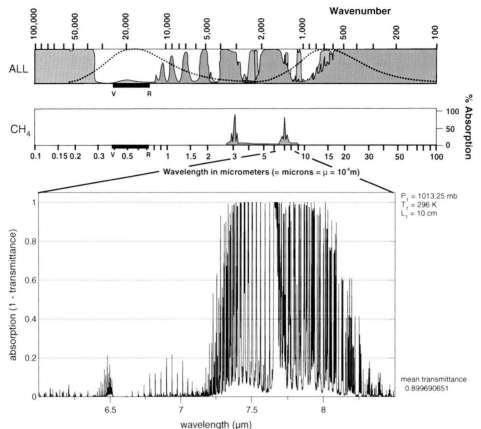

**Fig. 22.1** The absorption/emission bands of CH₄ at sea level and room temperature. Note that these bands are on the short wavelength side of Earth's thermal infrared spectrum

CO₂ which requires hundreds of thousands to millions of years to be neutralized by silicate weathering and finally removed from the carbon system.

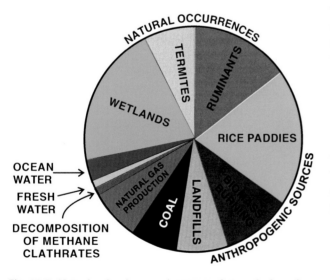

**Fig. 22.2** Natural and anthropogenic sources of atmospheric methane (*Source* Gavin Schmidt, NASA-GISS—Goddard Institute for Space Sciences)

## 22.2  The History of Methane in Earth's Atmosphere

During the first 2 billion years of Earth history conditions were very different from those of today. There was no oxygen in the atmosphere except for extremely small amounts produced by the dissociation of water molecules at the top of the atmosphere by incoming radiation. The abundance of atmospheric oxygen was about 1/100th of its present value of 21 %. And with lots of other molecules ready to be oxidized, the lifetime of one of those rare oxygen molecules was very short.

During the first 2.3 billion years of our planet's history, the atmosphere was anoxic, that is, without free oxygen. 2.3 billion years is half of Earth's lifetime since it accreted from debris orbiting the young Sun about 4.6 billion years ago. For the first 700 million years, Earth was bombarded with asteroids, one large enough to form the Moon, and others large enough to vaporize the young oceans. But life appeared early in Earth's history. There is isotopic evidence for the existence of life more than 3.85 billion years ago. Although this was not accepted by all geologists when first published, additional analyses seem to support this ancient origin.

Most larger modern life forms obtain energy by 'burning' organic matter using oxygen.

$$CH_2O + O_2 \rightarrow H_2O + CO_2$$

This was impossible for the first half of the planet's history, because there was no oxygen in the atmosphere. Even more importantly, it would have been impossible to build the complex organic molecules required for life in the presence of oxygen. The primitive organisms that inhabited the early Earth were a number of different kinds of bacteria. They used a variety of chemical compounds as sources of energy. For Earth's early life forms 'chemosynthesis' rather than photosynthesis was the major way of life. Chemosynthesis forms the base of the food chain today in deep sea communities associated with hydrothermal vents on the mid-ocean ridge. It is also the means of life for bacteria that live in sediments and rocks where there is no oxygen.

Methane was one of the materials that could be used as food. However, some bacteria used other compounds as food, and produced methane. These 'methanogens' provided a steady supply of methane to the atmosphere resulting in concentration levels 600 times that of today. In the absence of oxygen, the lifetime of methane molecules in the early atmosphere was in the order of 1,000 or more years, 2 orders of magnitude longer than today.

In the preceding Chapter we discussed the idea that $CO_2$ has been responsible for regulating Earth's temperature over its entire history. That idea arose in the 1970s. Thirty years later it was recognized that methane was also an important greenhouse gas in Earth's early atmosphere. The combination of $CO_2$ and $CH_4$ provided a more efficient greenhouse. Higher concentrations of $CH_4$ had the interesting property that when its concentrations became too high its interaction with incoming solar radiation resulted in formation of an organic haze which changed the Earth's albedo producing an anti-greenhouse effect and preventing the planet from becoming too warm. This was another temperature regulation mechanism, complementary to that of $CO_2$ but operating on a shorter time scale. To me, the combination $CH_4$–$CO_2$ greenhouse provides a more realistic and understandable idea of what the early atmosphere was like, and how Earth's temperature was regulated.

Ideas about the composition of Earth's early atmosphere have changed. In the middle of the last century it was thought that it would have been impossible for free hydrogen ($H_2$) to be present in any significant quantities in the early atmosphere. This is because today Earth's gravity is insufficient to keep these molecules from being swept away by the solar wind. Then as the implications of the faint young Sun were investigated, it became apparent that the top of the atmosphere would have been much colder than previously thought, and that the escape of hydrogen was unlikely. It became evident that Earth's early atmosphere could have contained large amounts of both $CO_2$ and free $H_2$. These gases would also have been dissolved in the water on the surface of the planet. There are ancient bacteria, methanogens, that can obtain their energy for life by combining those two gases to produce methane:

$$CO_2 + 4H_2 \rightarrow CH_4 + H_2O$$

These were probably some of the earliest life forms.

Recently Jim Kasting has estimated that the early atmosphere may have had $CH_4$ concentrations of 500–1000 ppmv, three orders of magnitude greater than its abundance in the atmosphere today.

As I have mentioned previously, it was one evening in 1995 in New Orleans, that Jim Kasting told me there was a better explanation of the cause of the Snowball Earth episodes than removal of $CO_2$ by a pulse of fixation of organic carbon by phytoplankton in the ocean. It involved an early atmosphere with methane as a major greenhouse gas. At that time I had never thought about the possibility that methane might also be involved in regulating Earth's temperature. Jim pointed out that photosynthesis did not become significant as a life process until about 2.3 billion years ago. As oxygen was introduced into the atmosphere, the lifetime of methane would have decreased dramatically, with methane being converted to $CO_2$. There is evidence that the level of $O_2$ in the atmosphere rose to at least 21 ppmv (compare that to the modern level of 210,000 ppmv). Since the greenhouse potential of $CO_2$ is much less than that of $CH_4$, there would have been a sudden drop in the magnitude of the greenhouse effect. This decrease in the greenhouse effect would have initiated the earliest of the major series of glaciations affecting the Earth about 2.3 billion years ago. These may have even been global Snowball Earth episodes.

More recently, Alexander Pavlov of the Center for Astrobiology at the University of Colorado, along with Jim Kasting, and others at The Pennsylvania State University have suggested that after the 2.3 billion year oxygenation event, methane levels returned to 100–300 ppmv. Unfortunately there is no sure way to determine $O_2$ levels during the period from 2.3 to 1.05 billion years ago but there is some evidence that it was between 10,500 and 37,800 ppmv (5–18 % of the present level). It is thought that these low oxygen levels promoted generation of methane by bacteria. Another oxygenation event occurred about 750 million years ago, and as oceanic and atmospheric $O_2$ concentrations rose, atmospheric $CH_4$ again declined sharply. This led to the several Snowball Earth episodes of the Late Precambrian. The violent climate swings associated with the Snowball Earth's were finally countered by another evolutionary advance. Worm-like creatures capable of ingesting the organic matter in sediment took away the 'food' supply required by methanogenic bacteria, and ended the role of methane as a major greenhouse gas. These are fascinating topics under active discussion today, it's a bit like an ongoing crime investigation—trying to figure out the cause of the very dramatic climate events that occurred 600–700 million years ago, and why they have never occurred since.

Unfortunately we do not yet have clear evidence of the abundance and role of atmospheric methane throughout the Phanerozoic. But we do have a record for the past half million years from the ice cores in Antarctica and Greenland.

Today the behavior of methane as a greenhouse gas is very different from what it was in the Precambrian. Because of the high levels of oxygen in the atmosphere today, the life of a methane molecule in the atmosphere is only about 10 years. Today the major sources of methane are primitive anaerobic bacteria that can only live in an oxygen-free environment. They live in bogs and in sediments rich in organic matter. They also live in the digestive tracts of ruminant animals such as cattle, and in the past, dinosaurs. They also live in the digestive tracts of termites, which use them to digest the wood they eat. In the cases of bogs and sediments the oxygen available in interstitial waters is used up in oxidizing some of the organic matter, but there is more organic matter left over that can be used by chemosynthetic organisms. The organic matter is broken down into methane and carbon dioxide.

$$2CH_2O \rightarrow CH_4 + CO_2$$

From the Antarctic and Greenland ice cores we know that atmospheric methane levels have gone up and down with the glacial-interglacial cycles. The atmospheric $CH_4$ concentrations are slightly different between the two hemispheres, the Northern Hemisphere being higher.

The cause of the glacial-interglacial changes in methane concentration was originally thought to be changes in the area of bogs along the Arctic Ocean, which would expand in the warmer interglacial climate. More recently it has become evident that the change is in response to the Asian monsoon. Recently, Bill Ruddiman and Maureen Raymo discovered that the variations in $CH_4$ correspond more closely to variations of insolation at 30° N, than to that at the Arctic Circle, as shown in Fig. 22.3. They suspect that the variations in methane may be related to the southwest Asian monsoon.

What is the southwest Asian monsoon, and why should it respond to solar insolation, and what does it have to do with methane? The term 'monsoon' derives from Arabic and refers to winds that reverse with the seasons. The very high Tibetan Plateau responds sharply to changes in insolation. It lies mostly around 5,000 m, and most of the atmosphere is below it. It becomes warmer in summer and colder in winter than the surrounding areas. As a result it develops a low atmospheric pressure system in summer and a high in winter. At the surface air flows toward low pressure and away from high pressure.

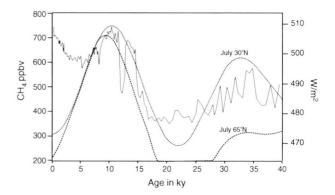

**Fig. 22.3** Fluctuations of atmospheric methane concentrations in the GRIP core from Greenland. *Dotted lines* are the solar insolation *curves* for 30° N and 65° N. During the last glacial termination there are two drops in methane concentration, one about 12,000 years ago, the other about 8,000 years ago. The 12,000 year drop is known as the Younger Dryas event, and marks a 1,000 year long return to glacial conditions in the North Atlantic. Its cause will be discussed in the Chapter on ocean circulation. Note that the methane concentrations start upward about 5,000 years ago, reflecting human agricultural practices, such as raising cattle and growing rice

Figure 22.4 shows the situation over the Indian Ocean, southern Asia and Tibet during the Northern Hemisphere summer. A high pressure system exists over the southern Indian Ocean about 30° S, and a low over Tibet, about 30° N. The moisture-laden air from the Indian Ocean tries to flow from south to north, but the west to east rotation of the Earth causes its path to deviate from a straight line. As it moves northward in the southern hemisphere, the west to east speed of the surface beneath it increases as it approaches the

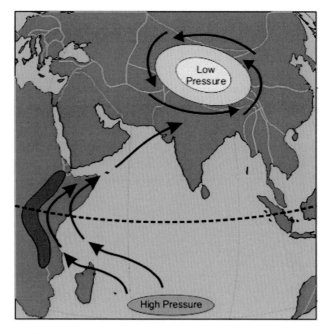

**Fig. 22.4** The Southwest Asian Monsoon

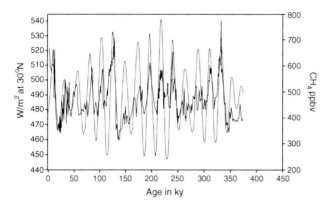

**Fig. 22.5** Atmospheric methane concentrations in the Vostok core (*solid line*) and insolation at 30° N (*dotted line*)

Equator. The result is that, seen from the ground, the air flows from the southeast to the northwest. It flows straight across the Equator, but as it moves further northward the speed of the ground beneath it decreases, and it flows from southwest to northeast. The deflection in direction is aided by the East African Rift, an uplift of eastern Africa about 2 km high. It drops some of its moisture over eastern Africa. By the time the air reaches Asia, it has picked up more moisture from the Arabian Sea and then it hits the wall of the Himalaya, and is forced to lose most of it moisture as the monsoon rains. This seasonal rainfall produces wet areas where plant material can decay under anaerobic conditions, and methane is generated.

In the winter the conditions are reverse, and the winds blow from the Tibetan region to the southeast, are again deflected by the East African Rift, and end up at the low pressure system that develops over the Equator. These winds, are, of course, dry. The effect of the monsoon circulation is a very strong seasonality of rainfall—a wet season and a dry season rather than a summer and a winter. In the days of sailing ships, the reversal of winds was used by sailors to cross the Arabian Sea from west to east in the summer and back from east to west in the winter.

It is evident that the intensity of the monsoon depends on the difference of the insolation over the Tibetan Plateau with time. That is why the pattern of atmospheric methane concentration coincides so closely with the insolation at 30° N.

Figure 22.5 shows the relation between atmospheric methane concentrations in the Vostok core and insolation at 30° N according to Ruddiman and Raymo's 2003 paper. They used the methane/insolation relation to adjust the time-scale for the Vostok core, arriving at conclusions very similar to those that had been made by Nick Shackleton. In the last few years methane variations have become an important tool for precise age correlation between the Greenland and Antarctic ice cores.

Note that the natural variations over the past 350,000 years have ranged between 350 and 750 ppbv. The trend over the

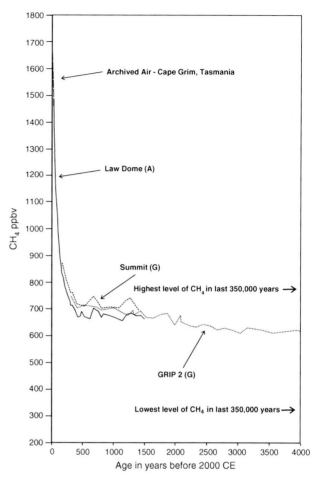

**Fig. 22.6** Changes in the concentration of methane in the atmosphere over the past 40,000 years. Data from the GRIP 2 and Summit ice cores from Greenland (G), the Law Dome ice core from Antarctica (A), and archived air from Cape Grim, Tasmania

past 4000 years is shown in Fig. 22.6. Note that it includes part of the gradual upward trend observed since 5000 years ago. Like the rise in $CO_2$, the sharp upward trend in $CH_4$ concentration began in the latter half of the 18th century.

The gradual increase from 5000 years ago to the mid-18th century is usually attributed to cultivation of rice, which takes place in wetland environments. There was a sharp rise in atmospheric $CH_4$ until about 1987. It is thought to have been mostly due to increased cattle ranching and farming. Some countries famous for their dairy products, such as the Netherlands and Denmark, have locally elevated $CH_4$ levels. The $CH_4$ increase slowed, with some small increases during volcanic eruptions such as that of Mount Pinatubo in 1991 and wildfires in Indonesia and the Arctic in 1997 and 1998. From then until now there has been almost no growth in the atmospheric concentration of methane. It is thought that this may be due more to reducing losses from pipelines and oil-field operations than to stabilization of the cattle industry.

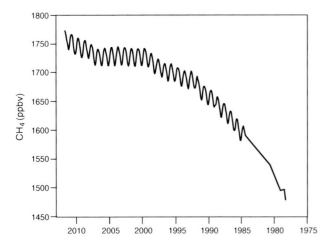

**Fig. 22.7** Changes in concentration of atmospheric methane since 1975 based on the archived air taken at Cape Grim, on the northwest tip of Tasmania. The seasonal variations have an amplitude of about 25 ppbv and are thought to be related to the Asian monsoons. The maxima occur in August–October. The plateau from 1999–2006 is thought to be due to reduction in leakage from gas transmission pipelines and oil-field operations. From 1978 to 1985 air samples were taken infrequently. The program of routine sampling several times a month began in 1985. Data from CSIRO Marine and Atmospheric Research and the Australian Bureau of Meteorology (Cape Grim Baseline Air Pollution Station). *CSIRO and the Australian Bureau of Meteorology give no warranty regarding the accuracy, completeness, currency or suitability of the data for any purpose"*

With a greenhouse potential about 23 times that of $CO_2$, the current levels of $CH_4$ are thought to contribute about 1/3 of the greenhouse effect. The biggest unknown in the methane equation is the future of the enormous amounts of methane stored as clathrates and vulnerable to release by global warming.

One of the most important air sampling and archiving programs to document the changes in chemistry of the atmosphere is being carried out by the government of Australia. The sampling site is at Cape Grim on the northwest corner of the island of Tasmania. This site is well away from any local sources of pollution. The program was started in 1970s but sampling was sporadic until 1991. Since then the samples of the air have been taken several times a month. The collection provides a representative sampling of the southern hemisphere atmosphere at 40° S. Figure 22.7 shows the results of the sampling program for the changes in concentration of atmospheric methane since 1975.

## 22.3   Methane Clathrates

What are clathrates? This is another of those topics where, by serendipity, I had the pleasure of being in on the ground floor. I first heard the term at one of those exciting meetings of the five members of the JOIDES Planning Committee in the early days of the Deep Sea Drilling Project. This one was in the fall

of 1969. Joe Worzel, the representative from Lamont-Doherty Geological Observatory came in very excited with a big roll of paper in his hands. "Look at this" he said as he unrolled a 'seismic profile' of across the outer ridge of the Blake Plateau, a deep sea topographic feature offshore northern Florida and Georgia. Marine geologists and geophysicists use the term 'seismic profile' for an acoustical image made my bouncing sound waves off the sea bottom and the underlying sediments, and recording the echoes on a roll of paper as the ship steamed forward. This as a low-frequency version of the 'ping' used to locate submarines or measure the distance to the sea floor. The sound waves used in making a marine seismic profile are audible to the human ear as a 'boom' and an 'air gun' is one of the devices commonly used to produce them. The air gun periodically releases a large bubble of air into the water, producing the booming sound. Incidentally, the term 'seismic profile' has nothing to do with 'seismic activity' or earthquakes except that both involve sound waves. Those generated by earthquakes have a much lower frequency and cannot be heard, although they can be felt. We had a minor earthquake at Stanford in 1958, when I was a graduate student. It felt like elephants were running through the building.

In those early days of Deep Sea Drilling the seismic profiles we had to work with were relatively crude. If the density of the sediments kept increasing downward and if there was layering, it would show on the seismic profiles. You could literally see the layering of the sediments beneath the sea floor. But the profile Joe Worzel showed us was different from any I had ever seen before. Several hundred meters below the sea floor was another reflection that was parallel to the sea floor and cut across the layers of sediment. Actually something similar is seen on many seismic profiles. The sound wave goes to the sea floor, bounces back, reflects off the sea surface, goes back down again, and bounces off the sea floor again. This is called a 'multiple' and can obscure some of the detail in the diagram. It is readily recognizable, however, because it looks like a vertically exaggerated version of the sea-floor topography. Joe's profile was different. The reflector inside the sediment was an exact duplicate of the sediment surface. There was no exaggeration of the relief. All of us looked at it in puzzlement.

"Anybody ever heard of clathrates?" Joe asked. None of us had. "Me either until a few days ago" Joe confessed. "They are ices made of gas and water. In Siberia they form a trap for the natural gas in the Arctic. No one knew until recently that they existed in nature. I think this is what we are looking at." Joe looked at me. "These things have been known since the 1930s in natural gas transmission pipelines. Under pressure the gas combines with water vapor to form an ice that clogs valves in the pipelines. Bill, you are in your 'Illinois semester' phase and there is a big engineering school there. Maybe somebody knows something about

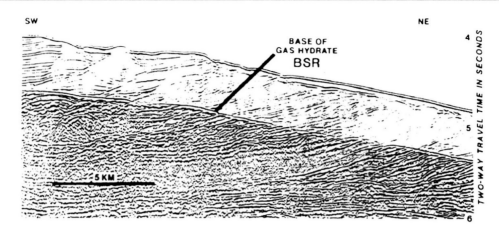

**Fig. 22.8** Seismic profile across the Blake Outer Ridge, showing the Bottom Simulating Reflector that is so prominent there. The vertical scale is 'two way travel time' (twtt). It is the length of time it takes for the acoustic pulse to travel from the ship on the sea surface down to the sediment surface or to layers within the sediment and then back to the listening devices on the ship. The speed of sound in seawater is about 1500 m/s. The top of the sediment on the left is about 4 s twtt, or 2 s deep; or 3000 m (3 km). The speed of sound in the sediments is greater, depending on their density. Near the sediment surface it is usually about 1750 m/s increasing downward. So a geophysicist has to make an educated guess how deep the BSR is, but we now know from drilling that this BSR is about 440 m below the sea floor

these things." We went on to do our regular work for the rest of the meeting, but always returning to the 'clathrate' discovery over coffee breaks. At the end of the meeting we agreed we would try to educate ourselves about clathrates to have an informed discussion next time we met.

Figure 22.8 shows a seismic profile across the Blake Outer Ridge, similar to the one I saw in 1969. You can see what is now known as a 'Bottom Simulating Reflector' or BSR for short.

It was the winter semester, and I went back to Urbana and tried to find someone who knew something about these things. Some people had heard about clathrates but nobody knew very much about them. However, the University of Illinois has a great library, and with a little searching I found some articles about gas clathrates and their role in plugging gas transmission lines. I copied the articles I could find and sent them off to Joe.

In one of those odd coincidences, my apartment-mate at Stanford, Kieth Kvenvolden, became the world's expert on clathrates. Kieth started out as a geochemist interested in the traces of organic matter in meteorites and on the Moon and Mars. He worked for NASA at the Ames Research Center near Palo Alto, California. When the natural occurrences of clathrates became known, Kieth jumped in to collect as much information as possible about them. We have had many discussions about them over the years.

Clathrates are a very special form of ice. The word clathrate comes from the Greek for cage. Water molecules form a three dimensional structure around a gas molecule. Most often that molecule is methane, but it might also be hydrogen sulfide (the gas that smells like rotten eggs and is extremely poisonous) or another gas, such as $CO_2$, found in

sediments. Figure 22.9 shows the simplest form of a methane clathrate molecule.

So technically, methane clathrate is an 'ice.' But what is very strange is that these clathrates form well above the freezing point of water. What is required for their formation is pressure, and not very much when you think about the pressures that exist at depth in the ocean. Remember that 10 m of water is equal to one atmosphere of pressure.

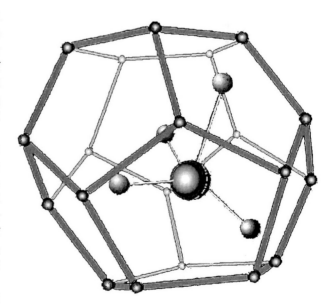

**Fig. 22.9** The simplest form of methane clathrate. The methane molecule ($CH_4$) is in the center, surrounded by a cage of water molecules locked together. The geometric form created by the water molecules is a pentagonal dodecahedron, that is, a twelve-sided enclosure having pentagonal faces. It is one of those geometric structures that Buckminster Fuller found so interesting

Figure 22.10 is a clathrate stability diagram. It may look a bit complicated at first, so let's go through it step by step.

The vertical axis on the diagram is the depth below the sea surface. The horizontal axis, marked at both the top and bottom of the diagram is the water temperature in °F and °C. In the grey area of the diagram dissolved methane gas will combine with water to make methane clathrate, a solid. The boundary between the grey and white backgrounds is called a 'phase boundary'. In the white area the methane and water are fluids (gas and liquid). Remember that seawater contains salt, so its freezing point is below that of fresh water. The coldest seawater can get before it freezes is about $-3$ °C. At that temperature, the phase change occurs at water depths slightly over 200 m. This is the condition in the Arctic, but it is not specifically shown on the diagram.

One question you might think of is whether methane clathrates can exist floating around in the seawater. The answer is probably yes, but of course if it was just one gas molecule with its cage of water molecules it would be far too small to see. However, the clathrate cages might clump together and form groups large enough to be seen. But then something else would happen. Methane clathrate is lighter than seawater, so the clumps would start to drift upward and as they got nearer the surface they would cross the phase boundary and turn back into water and methane. You can also bet that if there are clumps bacteria will find them and use that methane as food.

There are some places, such as on 'Clathrate Ridge' off Oregon, where methane clathrate lies on top of the sediment. It looks like opaque white ice, perhaps more like old snow. Erwin Suess, one of my colleagues at GEOMAR in Kiel, was investigating it in a submersible. He tried to pick some of it up with the mechanical arm of the submersible and put it in a basket. There is a beautiful film you may have seen on one of the Science Channels, showing this operation. Somehow everyone thought that if the clathrate was resting in the sea floor it must be heavier than water. But when he tried to put it in the collecting basket attached to the submersible it floated upward to the surface. Even at great depths, it is lighter than water. The clathrates on Clathrate Ridge are frozen onto the sediment surface.

Clathrates are usually found within the sediment, from the sediment surface down to a depth where the temperature is warm enough to prevent their formation. The diagram shows how this works. Figure 22.10 shows two profiles of the change in temperature of seawater with depth, one for the mid-latitudes (40°–45° latitude) and one for the tropics (10°–20° latitude). You will notice that both curves converge at about 800 m depth.

Let's assume that the sediment surface is at 800 m depth, as it is along the continental margins, for example. If methanogenic bacteria are producing methane from organic matter within the sediment it will combine with water to form methane clathrate, which is, remember, a solid. This solid will act as a glue to hold the sediment particles together. It the clathrates are abundant, they act like cement, holding loose clay, silt, and sand-sized grains together.

But as you go deeper into the sediment, it gets warmer. It is warmed from below by the heat coming from the interior of the Earth. You may recall that this is not very much energy compared with that coming from the Sun, but it is enough to produce a 'geothermal gradient' within the sediment. At some depth the geothermal heat will have raised the temperature of the sediment enough so the clathrates will reach the phase boundary again, and below that depth inside the sediment methane will exist as a gas and water will be in its liquid form.

The picture you get is that of a sort or armor plating on top of the sediment, with solid clathrate in the spaces between the grains cementing everything together. Now go back and look at Fig. 22.8. The BSR is the base of the clathrate zone. Between the BSR and the sediment surface clathrate is holding the sediment together, making it more solid. Below the BSR there is no clathrate. The contrast in solidity of the sediment above and below this boundary

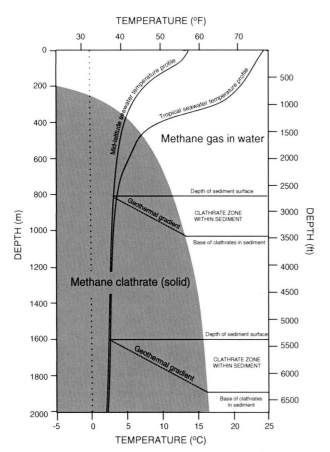

**Fig. 22.10** Methane clathrate stability diagram

makes the sound waves bounce off it, so it shows up clearly on the seismic profile.

How much methane clathrate is there and where is it? Hold on to your hat. When methane gets incorporated into the clathrate the volume it occupies diminishes greatly. When one cubic meter of methane hydrate is decomposed, it produces 164 $m^3$ of methane and only 0.8 $m^3$ of water! Estimates of the total amount of methane hydrate on Earth vary widely, but Kieth Kvenvolden believes that it is likely that there are more than $10^{15}$ kg of carbon stored in this form. To put it in perspective that is twice the total amount of carbon in recoverable and non-recoverable fossil fuels: coal, petroleum, and natural gas. It is 5,000 times the amount of methane currently in Earth's atmosphere. To many of us, it is like a ticking time bomb waiting to be set off by global warming.

If you look at Fig. 22.10, and imagine those ocean temperature profiles moving to the right, you will realize that as the ocean warms, methane clathrates will decompose and be released. This can have two important effects: (1) destabilization of the continental margins so that there may be large landslips into the deep sea, and (2) release of large amounts of methane into the ocean and eventually into the atmosphere as a positive greenhouse gas feedback.

How serious is this problem? It is clearly serious; it is already going on in the Arctic, where methane bubbling up through bogs as the permafrost melts has been observed in many areas. There are small craters on the floor of the Beaufort Sea off Alaska, where methane is emerging. Furthermore, in Siberia, where much of the methane release is occurring, there are those very large gas field areas, mostly methane, trapped beneath the permafrost.

What about the rest of the world? The problem is that we do not know that much about the distribution of methane clathrates. In order to make a BSR visible on seismic profiles it must be present in large quantities. But there are many where we know from ocean drilling that it is present but does not show on the seismic profiles. It is even sometimes tricky to tell whether it is present in the recovered cores because they warm as they are brought up from the deep sea, and they warm further as the 30 ft-long cores are removed from the core barrel, placed on deck, and then cut into 6 ft sections of manageable length. During the time between the cutting of the core and its examination in the onboard laboratories the clathrates may decompose. Imagine the surprise when you open a core and find white snow-like ice interbedded with the sediment. Everyone gathers around as a piece of the ice is taken from the core, and set aflame with a match. Burning ice, good stuff for pictures to remember the event.

I should mention that the JOIDES Resolution has managed to recover not only methane clathrates, but also the dangerous hydrogen sulfide clathrates, which required that those who might be exposed to the gases from the cores wear hazard suits with gas masks. Although $H_2S$ in very low concentrations has the odor of rotten eggs, it is odorless in the higher concentrations which are extremely poisonous. When gas clathrates are indicated on the seismic profiles there is no way of telling what the caged gas might be.

As you have already guessed, this is a large potential energy resource, just sitting there unexploited. Several efforts have been made to find out how these huge reservoirs of methane clathrate might be converted into natural gas and used for fuel. There are three problems. First, it takes a lot of energy to 'melt' the clathrate and release the methane. It is that same old nasty problem of the large amounts of energy required to make a phase change—such as, you remember, going from ice to water or water to vapor. Secondly, the clathrates are relatively thin and spread out over huge areas. To physically harvest methane clathrates you would have to dig up huge areas of the sea floor, particularly on continental margins. The third problem hardly needs to be mentioned, this would be another source of fossil fuel being converted to $CO_2$, and if added to the atmosphere it will promote more global warming.

Methane clathrates are thought to be the culprits in two major catastrophes: an episode of sudden global warming that occurred 55 million years ago, and the cause of a huge submarine landslide off Norway 8,100 years ago. Known as the Storegga Slide it caused a huge tsunami in the North Atlantic.

## 22.4 Methane Trapped in Permafrost

The Arctic is a special place because the ground us underlain by 'permafrost'; it is 'permanently frozen.' Permafrost means that the temperature does not rise above 0 °C for two or more consecutive years. About 1/4 of the Northern Hemisphere land, $25 \times 10^6$ $km^2$, is underlain by permafrost. The permafrost layer can range from a few meters to over 1400 m in thickness. The thickest permafrost is in Siberia. The ground became deeply frozen during the ice ages, especially where there was no glacial ice to protect the land from the deep winter cold. The permafrost contains methane clathrates, but we do not know for sure how much. It also trapped natural gas, mostly methane, being generated by anaerobic bacteria in the sediment, preventing it from escaping to the surface. Some of the major gas fields of Siberia are methane trapped beneath the permafrost. "Permanently frozen" has turned out to be a misnomer, because the permafrost is melting all around the Arctic as warming occurs. Construction in the Arctic, whether it be buildings or pipelines or whatever, is done on stilts to hold the structures above the land surface and prevent the warmth emanating from the structures from melting the permafrost. As the permafrost melts today these

**Fig. 22.11** Distribution of
atmospheric methane in the upper
troposphere in 2011, reflecting its
release from permafrost and
clathrates as the Arctic warms.
The volume mixing ratio is the
ratio of the number density of the
gas to the total number density of
the atmosphere (density is the
number of molecules per unit
volume) *Image* NASA GSFC
AIRS satellite program

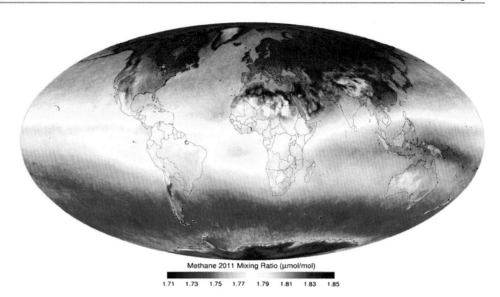

stilts find themselves resting on soft slush rather than solid
ice. It is like removing the foundation from underneath a
structure—an interesting sidelight to global warming, and a
costly one if you happen to live there.

Monitoring of methane emissions from the permafrost
has been going on since the 1980s but the monitoring sites
are irregularly distributed. Furthermore, they only measure
local concentrations, which can be highly variable. A better
estimate of methane emissions comes from satellite observations of methane concentrations in the upper troposphere,
where the gas has been mixed more evenly into the air.

In 2011 the levels of methane in the upper troposphere
increased dramatically, so much so that it was initially
questioned whether the satellite instruments had been damaged and were giving false readings. Then it was discovered
that there were indeed many new methane vents, some
producing craters in the Siberian tundra. Methane gas was
found as bubbles in Arctic sea ice, indicating that some of
the venting was occurring from the sea floor. A major new
source of greenhouse gas emissions has appeared, in what
Mikhail Budyko called 'the Achilles Heel of the climate
system'.

Figure 22.11 shows another very interesting result of our
grand uncontrolled experiment on planet Earth: the separation of the northern and southern hemisphere air masses. The
Intertropical Convergence Zone near the Equator is acting as
a barrier, keeping the Arctic-sourced methane trapped in the
northern hemisphere.

## 22.5  Nitrous Oxide—N₂O

Nitrous oxide ($N_2O$) is a colorless gas with a slightly sweet
odor. It has two major absorption/emission spectral bands,
one in the overlap of solar and terrestrial infrared spectra,

where energies are very small, and the other in the short
wavelength part of the Earth's thermal infrared, between 7.5
and 9 $\mu$. Figure 22.12 shows the distinctive pattern of these
bands.

Nitrous oxide is best known to the public as 'laughing
gas' and has been used as an anesthetic/analgesic in dentistry
and surgery. The most memorable use in this form was by
Inspector Jacques Clouseau (Peter Sellers) of the Sûreté as
he extracted the wrong tooth from his former mentor Chief
Inspector Dreyfus (Herbert Lom) in one of the funniest films
ever made, *The Pink Panther Strikes Again.*

In the atmosphere $N_2O$ is the most powerful natural
greenhouse gas, with a warming potential estimated at 296
times that of an equivalent amount of $CO_2$. Fortunately its
atmospheric concentration is much less, about 7/10,000th
that of $CO_2$. However, like $CO_2$ and methane, it's abundance
has increased markedly since the beginning of the industrial
revolution.

Its natural sources are primarily from bacteria in soils and
the ocean. The bacteria progressively remove the oxygen
from nitrogen-containing organic compounds, such as the
animal wastes. If you keep aquarium fish you are already
familiar with the process of developing an active denitrification system to allow the fish to lead a happy life. I kept fish
as a high-school student, and we didn't know much about
how to do this efficiently in the 1940s. It took months of
careful tending to get a 'balanced aquarium' that could go
for a week without a water change Now we have subsurface
filters, or more elaborate filtration systems where the denitrifying bacteria grow and process the waste from fish into
harmless compounds that can fertilize the plants. Now you
can purchase denitrifying bacteria in bulk, add them to a new
aquarium, and you are instantly off and running. Now I keep
African cichlids, and give partial water changes every 3
months or so. Denitrification involves stepwise removal of

**Fig. 22.12** The absorption/emission bands of N$_2$O at sea level and room temperature. Note that these bands are centered on the shorter wavelengths of Earth's thermal infrared spectrum

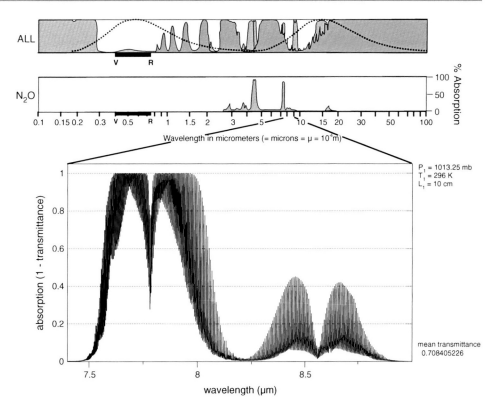

oxygen; it is used for metabolism of the bacteria. One of the waste products excreted by fish or decomposition of organic matter is usually ammonia, NH$_3$. Urea also decomposes into ammonia. In acidic water, with a pH below 7, ammonia dissolves to form the ammonium ion:

$$NH_3 + H_2O \rightarrow NH_4^+ + OH^-$$

The ammonium ion is harmless to living organisms. However, if the water is alkaline, with a pH above 7, free ammonia is produced. It is poisonous to aerobic life forms. You know it from liquids used for disinfecting dirty areas, and in the old days you might even have experienced it as 'smelling salts' used to revive women from a swoon. Judging from novels and plays of the time, women seemed to swoon a lot in the 19th century. They don't seem to swoon much anymore.

In the balanced aquarium, ammonium is oxidized to nitrite (NO$_2^-$) then further oxidized to nitrate (NO$_3^-$), a fertilizer that is taken up by plants in the aquarium. In your garden things work a bit differently.

Plants require three major fertilizer elements, nitrogen, phosphorous, and potassium. When you buy a bag of fertilizer for the plants in your garden you will see numbers on it that tell you the percentages of available nitrogen, phosphorus and potassium. A 10-10-5 fertilizer has 10 % nitrogen, 10 % phosphorous and 5 % potassium. In a 100-pound

bag, therefore, 10 pounds is available nitrogen, 10 pounds is phosphorous and 5 pounds is potassium. The other 75 pounds is filler that no direct value to the plants but prevents the shock of being exposed to large concentrations of the fertilizing elements. The nitrogen in a bag of fertilizer is often in the form of ammonium nitrate (NH$_4$NO$_3$), a white powder. Only part of the fertilizer you spread on your garden plants or lawn will be used by the plants. Some of it will be used by bacteria in the soil. This is harmless unless the soil is wet and becomes anaerobic, or if the fertilizer runs off into a wetland area.

In the absence of free oxygen in the environment, the oxygen in the oxides of nitrogen can be removed by bacteria. The last oxide to be produced, before pure nitrogen gas (N$_2$) is nitrous oxide, also a gas. If it escapes into the atmosphere it acts as a potent greenhouse gas.

| Nitrate to nitrite | $NO_3 + 2H^+ \rightarrow NO_2 + H_2O$ |
| Nitrite to Nitric oxide | $NO_2 + 2H^+ \rightarrow NO + H_2O$ |
| Nitric oxide to Nitrous oxide | $2NO + 2H^+ \rightarrow N_2O + H_2O$ |
| Nitrous oxide to Nitrogen | $N_2O + 2H^+ \rightarrow N_2 + H_2O$ |

The present natural and anthropogenic sources of N$_2$O are shown in Figs. 22.13 and 22.14. Some of this comes indirectly from the oxidation of NH$_3$ released into the atmosphere from industry as suggested by Fig. 22.14. It is thought that much of the release in agricultural soils comes

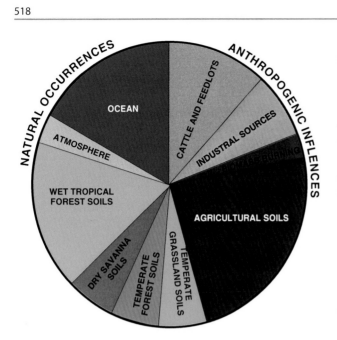

**Fig. 22.13** Natural and anthropogenic sources of N$_2$O today. Note that in the natural system, wet forest soils in the tropic account for about 1/3 of the source. Glacial-interglacial changes in atmospheric N$_2$O concentrations are related to the reduction of area of these soil types and to the cooler climate which slowed chemical reaction rates (*Source* Gavin Schmidt, NASA GISS)

from tropical and temperate regions where the available phosphorous has been depleted so that there is a large surplus of ammonium nitrate in the fertilizer that cannot be used by the plants.

We have no information about the abundance of N$_2$O in the atmosphere older than that contained in the ice cores, and even in those the record is sometimes incomplete. Figure 22.15 shows the record from Dome C in the Antarctic. Although the record is incomplete, it is evident that N$_2$O concentrations are about 40 % higher during the interglacials than during the glacials.

The younger part of the record, the past 1000 years is shown in Fig. 22.16. The increase is more gradual than that of CO$_2$ and methane. This is because the systematic manufacture of ammonia-based fertilizers did not start until after the discovery of the Haler-Bosch process in 1910, with full scale production being used during World War I to produce munitions. After the war the process was used to produce commercial nitrogen fertilizer. The use of manufactured fertilizers increased dramatically after World War II. Nitrogen fertilizers are relatively inexpensive to make, since the nitrogen requires in the manufacturing process comes from the air, and is free. Phosphate and potassium fertilizers require mining of deposits containing these elements, and such deposits occur only in limited areas.

If the nitrogen fertilizers are used where other nutrients, such as phosphate or potassium, are absent, most of it is decomposed and much eventually enters the atmosphere as N$_2$O.

This is often the case in tropical regions where the soils have been leached of nutrients.

The results of the Cape Grim, Tasmania, sampling program for N$_2$O are shown in Fig. 22.17. Note that there are

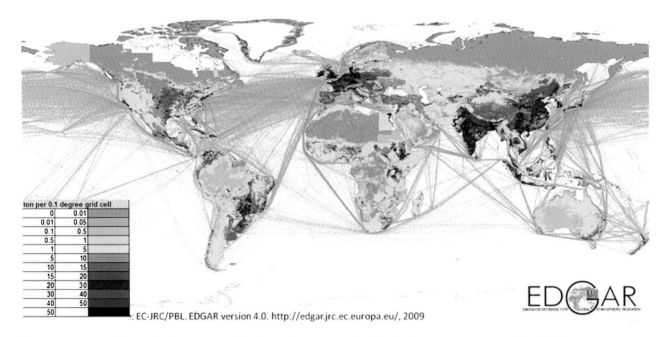

| ton per 0.1 degree grid cell | |
|---|---|
| 0 | 0.01 |
| 0.01 | 0.05 |
| 0.1 | 0.5 |
| 0.5 | 1 |
| 1 | 5 |
| 5 | 10 |
| 10 | 15 |
| 15 | 20 |
| 20 | 30 |
| 30 | 40 |
| 40 | 50 |
| 50 | |

: EC-JRC/PBL. EDGAR version 4.0. http://edgar.jrc.ec.europa.eu/, 2009

**Fig. 22.14** Sources of atmospheric N$_2$O, compiled from satellite data collected by the European Space Agency EDGAR Program

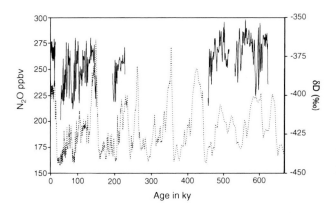

**Fig. 22.15** Concentration of N₂O in the Dome C ice core from the Antarctic (*solid line*). The *dotted line* shows the change in isotopic abundance of deuterium in the ice, an index to the temperature of the air where snow formation takes place. There is no temperature scale for the deuterium isotope ratio, but the familiar sawtooth curve of glacials and interglacials is evident

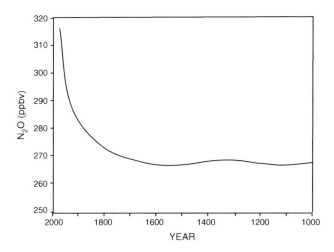

**Fig. 22.16** Atmospheric concentration of N₂O since 1000 CE. Sketched though data gathered from firn near the South Pole, ice cores, and the Cape Grim (Tasmania) air archive. Like the increases in CO₂ and methane, the increase in N₂O began as the industrial revolution proceeded and with global population growth

seasonal fluctuations, similar to those seen for $CO_2$. You can also see, if you put a ruler along the trend in the diagram, that the amount of $N_2O$ introduced into the atmosphere each year is increasing, although not as rapidly as $CO_2$.

## 22.6 Other Greenhouse gases

The three tropospheric greenhouse gases discussed so far, $CO_2$, $CH_4$, and $N_2O$ account for most of the increase in the greenhouse effect since the industrial revolution began 250 years ago. They currently represent about 80, 23, and 7 % of the current greenhouse forcing. However, there are a

number of other greenhouse gases, most of them anthropogenic.

Particularly important are the chlorofluorocarbons (CFCs) and related gases developed in the late 1920s as a replacement for the ammonia then in use in refrigerators, and the related Halons used in fire extinguishers. Although present in only minute amounts, measured in parts per trillion (ppt), they are extremely potent, with greenhouse effect potentials as much as 14,000 times that of $CO_2$ taken over a 100 year period and, indeed, they have atmospheric lifetimes of 100 years or more. Their concentrations in the atmosphere have been declining since the Montreal Protocol was ratified in 1987, as shown in Fig. 22.18.

There are some other gases which, although not greenhouse gases in themselves, affect the rates of formation or destruction of other greenhouse gases. Carbon monoxide (CO) reduces the abundance of the $OH^-$ radical in the stratosphere, slowing the decomposition of methane. It also assists in the production of ozone. CO is introduced both from natural and anthropogenic sources, probably in roughly equal amounts. The anthropogenic sources are deforestation, biomass burning, and fuels. It has been suggested that the effect of CO in slowing the destruction of methane may be as powerful a radiative forcing factor as the emissions of $N_2O$. The hydrogen ion ($H^+$) also reduces the abundance of $OH^-$, increasing the lifetime of $CH_4$ and some of the anthropogenic gases. It too has both natural and anthropogenic sources, again in roughly equal amounts. The anthropogenic source is mostly from burning of fuels.

There are a number of other exotic gasses used in manufacturing processes which have very high greenhouse potentials. Some could be catastrophic if even in trace amounts were released into the atmosphere. A discussion of these is beyond the scope of this book. It is important that

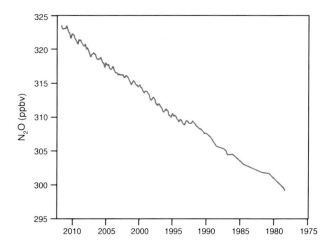

**Fig. 22.17** Atmospheric concentrations of N₂O at Cape Grim, Tasmania from 1977 to 2005

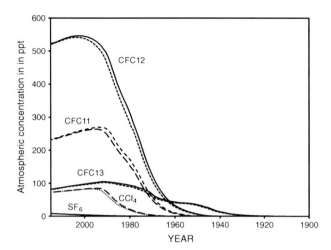

**Fig. 22.18** Annual mean tropospheric CFC-11, CFC-12, CFC-113, carbon tetrachloride ($CCl_4$) and sulfur hexafluoride ($SF_6$) concentrations in the atmosphere. In each case the *upper line* is for the northern hemisphere, the *lower line* for the southern hemisphere

they be identified and properly controlled before they enter the environment.

## 22.7　Summary

One greenhouse gas, ozone, is important and beneficial in both protecting the surface of the planet from ultraviolet radiation and in giving the earth's atmosphere a special layer, the stratosphere. The stratosphere restricts convection in the atmosphere to the layer between it and the planet's surface, the troposphere. It is a feature unique to Earth. Recognition that man-made chlorofluorocarbons were having a dramatic effect in enhancing the decomposition of ozone led to the Montreal Protocol of 1989, which mandated their elimination as soon as possible. The reversal of the trend toward destruction of the ozone layer is occurring at the present time.

The three tropospheric greenhouse gases having a major effect are $CO_2$, $CH_4$, and $N_2O$. Emissions of all of these have increased dramatically since the beginning of the industrial revolution and the explosion in population that it made possible.

In comparing the increases in atmospheric concentration of the three gases, it can be seen that the start of the rise is different for each. The first to increase is $N_2O$, undoubtedly related to the increase in land under agriculture. Natural fertilizers, such as manure, were used until early in the 20th century when industrial production of ammonia began. The rise in methane is related to the increase in cattle production, expansion of rice farming, biomass burning involved in clearing land for farming and pastures, coal mining, leaks associated with natural gas production and use, and from

decomposition of organic matter in landfills. The rise in $CO_2$ is largely associated with the use of fossil fuels, biomass burning, and manufacture of cement. As noted above, these three gases account for 7 %, 23 %, and 80 %, respectively, of the increase in greenhouse forcing since the beginning of the industrial revolution.

The seeds of climate change have already been planted. If we can summon the will to do it, we can take measures to slow the rates of greenhouse gas emissions and eventually reverse the trends. If all we had to worry about were gradual even warming of the planet, we could make careful plans for the future. We could make accommodations for gradual changes in the local climate, water supply, agricultural production, and sea level change. Unfortunately that is not what is going to happen. Instead of a predictable, slowly changing future climate we should expect increasingly chaotic climate change. As Hermann Flohn suggested in 1982 we can expect to see many records broken: the hottest day, the coldest day, the greatest rainfall, the longest drought, the biggest snowfall, the least snowfall, the biggest hurricane, etc. And of course, with the continuing growth of cities and further urbanization, we can expect larger floods, more destruction from tornadoes and winds, and other climate-related problems. Some of the possible changes in climate are even counter-intuitive, such as the possibility than some areas might see a return to glacial conditions. The problem is that, with our nudging, the Earth is making a transition from one climate state to another, and the next few hundreds to thousands of years may be a rough ride.

A Timeline for this chapter:

| | |
|---|---|
| 1910 | The Haber-Bosch process for fixing atmospheric nitrogen as ammonia is discovered—it is used intensively to make munitions in WWI and WWII |
| 1920s | Chlorofluorocarbons ('Freon') are introduced for use in refrigerators |
| 1946 | The Haber-Bosch process is used to begin to make fertilizer on a global scale |
| 1969 | Joe Worzel proposes that 'bottom simulating reflectors' seen on seismic profiles of the ocean floor might be clathrates |
| 1978 | Monitoring of greenhouse gases begins at Cape Grim, Tasmania, Australia |
| 1983 | Keith Kvenvolden and Leo Barnard describe gas hydrates drilled by the DSDP on the Blake Plateau's Outer Ridge, in the northwest Atlantic Ocean |
| 1987 | The Montreal Protocol begins the phase-out of chlorofluorocarbons |
| 1995 | Jim Kasting tells me that methane was probably the early Earth's greenhouse gas |
| 1995 | Timothy McVeigh uses ammonia-based fertilizer to make the bomb that destroys the Alfred P. Murrah Federal Building in Oklahoma City |

(continued)

| 1998 | Keith Kvenvolden publishes his paper *'A primer on the geological occurrence of gas hydrate'* in Geological Society, London, Special Publications, v. 137; p. 9–30 |
| 2003 | Bill Ruddiman and Maureen Raymo show that glacial-interglacial changes in atmospheric methane correspond to changes in insolation at 30° N and changes in the southwest Asian monsoon |
| 2011 | Methane venting from permafrost suddenly increases in the Arctic |

If you want to know more:
Intergovernmental Panel on Climate Change (IPCC), 2013. *Climate Change 2013: The Physical Science Basis.*

Available at https://www.ipcc.ch/report/ar5/wg1/ to understand this comprehensive survey of the state of our knowledge.

Music: Paul Dukas: L'apprenti sorcier (The Sorcerer's Apprentice).

Libation: Perhaps a cognac is in order. They are produced from grapes that grow on chalk around the town of Cognac in the northern part of the Aquitaine region of southwestern France. The grapes do not make a very good wine, so it is distilled into a fine 'eaux-vives'—a water of life. There are a number of distilleries and most offer cognacs of different ages. They get better with time, Never mix cognac with ice or water, and it should be presented in a 'snifter' a bulbous glass that has a somewhat narrower opening at the top. The aroma is to be savored first, then the taste. Consume very slowly, and warm it with your hands.

### Intermezzo XXII. An Unexpected Detour

'Jimmy' Carter won the Presidential election in 1976, just after I had become Dean. At that time the world was entering a period of 'stagflation,' with stagnant economies, inflation, and rising interest rates. He is perhaps our only President to have ever had a serious background in science. He had been commander of a nuclear submarine and knew a lot of the physics you have read about here. He was also a good systems analyst, and very early on he figured out that our country's dependence on oil and coal as its primary energy sources was leading us toward a crisis.

On April 17, 1977 he gave a televised "fireside chat" declaring that the U.S. energy situation during the 1970s was the moral equivalent of war. He recalled the effect of the Arab Oil Embargo of 1973–74, when gasoline was in short supply, there were lines at the gas pumps, and the price of fuel skyrocketed. He urged the citizenry to conserve energy, installed solar panels on the White House, and wore sweaters while turning down the heat. This earned him the intense dislike of the petroleum industry. He was a prophet before his time.

Carter's Head of the Office of Science and Technology Policy (OSTP) was Frank Press. Frank was an eminent Geologist/Geophysicist from MIT. He had gotten his doctorate under Maurice Ewing. The OSTP developed two initiatives: 1) explore the US continental margins as a new source of petroleum; 2) require the US automobile industry to develop 48 mpg cars by 1995.

On July 15, 1979 Carter gave his "Crisis of Confidence" speech, outlining a program for achieving energy independence: "On the battlefield of energy we can win for our nation a new confidence, and we can seize control again of our common destiny." As nearly as I can determine, that speech was the only event that actually resulted in a reduction of petroleum use, and that occurred several years later, after many Americans had bought smaller cars.

At the same time the idea of exploring our continental margins was being developed, the end of the Deep Sea Drilling Program was in sight. It had already gone on for a decade, and was scheduled to end at the latest in 1983. Frank Press thought that we might be able to have a successor program to conduct exploration of our continental margins. He envisioned a joint US Academic—Industry program of more sophisticated drilling, with funding to come half from the US Government and half from industry.

To understand what 'more sophisticated drilling' means, you need to recall how the scientific drilling from GLOMAR Challenger was carried out. The ship lowered a drill string with a bit on the end down to the sea floor, and drilled a hole. The bit had a hole in the center. To flush the drill cuttings out of the hole, seawater was continuously pumped down the drill pipe, through the hole and carried the cuttings up through the space that formed between the rotating drill pipe and the surrounding sediment or rock. Cores were recovered by dropping a core barrel down inside the pipe. On impact the core barrel attached itself to the bit. The bit then drilled ahead and sediment passed upward through the hole into the core barrel. After the core barrel was thought to be full, a retrieval device was dropped down on a wireline. On impact it latched onto the top of the core barrel and at the same time caused the core barrel to be released from the bit. That is, if all went well; and almost always it did. It was a very elegant technique, but not copied by industry because a bit with a large hole in the center does not drill as fast as one with many small holes.

Our DSDP drilling technique had one significant flaw. There would be no way to control a blow-out of gas or oil if we should encounter it. After the discovery of petroleum at Sigsbee Knoll on Leg 1, JOIDES established a Pollution Prevention and Safety Panel (PPSP) to insure that we did not drill anywhere that there might be a chance of encountering petroleum.

While I was Chairman of the JOIDES Planning Committee I was also on the PPSP. Hollis Hedberg, the Panel's Chair and a very highly respected petroleum geologist, would always start the meeting by passing around a photograph. It was an aerial shot of a shallow water drilling vessel entirely enclosed in a ball of flame. "Gentlemen, this is what we are here to prevent."

To drill where gas or petroleum is expected the operation must have well control. On land this control is achieved by pumping 'drilling mud' down the pipe. The drilling mud is heavier than the rock through which the hole is being drilled. The mud is heavier because it contains minerals like barite which are more dense than those making up most rock. It passes through the bit and returns up the annulus between the drill pipe and the rock. The pressure on the pore fluids in the rock, the 'lithostatic pressure,' is proportional to the weight of the overlying rock. The pore fluids, whether salt water, petroleum or gas cannot escape into the drill hole because the 'hydrostatic' pressure of the drilling fluid is greater. The upper part of the hole, where the sediments or rock are not as solid and might fail because of the pressure of the drilling mud, is usually cased with concrete pipe so the drilling mud cannot penetrate into it. If an 'overpressured zone' is encountered, where the pore fluids in the rock are higher than expected from the weight of the rock itself, the well can be controlled by increasing the weight of the drilling mud.

Drilling such a hole from a ship is a much more complicated process. Drilling mud must be used both to control conditions in the hole and to prevent a possible blowout, and to bring the drill cuttings out of the hole. In the DSDP operation, the drill cuttings simply accumulated on the sea floor around the hole. Drilling mud is very expensive because of the special minerals that give it its weight. It must be recycled, and this means that there must be return circulation, so that the mud pumped down the pipe returns to the ship where the cuttings can be filtered out and the mud reused.

Return circulation requires that there be a second pipe, actually a tube, from the sea floor back up to the ship. The drill pipe is inside this larger tube, which is known as a 'riser.' The drilling mud is pumped down the drill pipe and the mud and its contained cuttings return through the annulus, the space between the drill pipe and the riser. You might begin to guess that this is all going to get very complicated (and expensive) in a hurry. In 1979 almost none of the necessary engineering had been carried out. But this is exactly what would be needed to make an exploration for petroleum on the continental margins.

I was now serving on the Executive Committee of JOIDES, overseeing the very successful International Phase of the Deep Sea Drilling Program from a distance. Frank Press wanted to introduce us to the responsible executives of the US Petroleum Industry. I was invited to Chair this adventure. I have no idea why I was chosen, but it was probably because I was thought to have close ties to the industry. I had taught regularly in the American Association of Petroleum Geologists' "Petroleum Exploration Schools." At that time there were about 35 large petroleum companies in the US, and many of my students were working in the industry.

I explained that I would rather see a continuation of the DSDP, with international partners. We had made a lot of progress in understanding the history of the Earth, and were recognized as the most successful international science program in history. The argument was that for the future we needed to do something really new. A simple continuation of DSDP-type drilling would not be sent forward for approval by the congress. It was essentially 'take it or leave it.'

After a Welcome by Frank Press and Introduction by Phil Smith of the OSTP, I chaired the Scientific Presentations at the first meeting of potential academic participants and industry representatives at Rice University in Houston in early 1979. The review of what had been learned from the DSDP was an eye opener for many of the industry officials. Although their research staff were aware of the program, and some of them served on JOIDES Panels, the executives were just becoming aware of the impact the theory of Plate Tectonics was having on exploration. Of all the industry contacts made at that meeting, George Pichel of Union Oil was especially helpful. He told me that our presentations were good, but needed to be improved. I would need to go around to each company and make a presentation —a really professional presentation—'at a level oil company executives could understand.' Most of them were not geologists. He put me onto a company, Green Mountain Photo, on the west side of Denver that made slide presentations for petroleum industry personnel giving talks at national meetings. Remember, there were no personal computers, much less 'PowerPoint' at the time.

Accordingly I visited Green Mountain outside Denver. Its head, Lev Ropes, not only designed beautiful graphics, but he explained to me how people learn. It was a simple rule. First, tell the audience what you are going to tell them. Second, tell them. Third, tell them what you told them. An interesting human limitation is that we are very poor at remembering more than seven things at a time. So a presentation should never try to make more than seven points. All of this was new to me. The presentation slides were beautiful and got widely distributed.

Over the next months I visited over 20 companies giving my pitch. In the middle of this exercise, we had an additional request from Frank Press. He told us that the academic community needed a high-level person to be available in Washington not only to coordinate activities on selling this program to industry but to make sure the National Science Foundation was fully on board, and that the Congress was being brought along.

JOI, Inc.'s first president was Robert White. He had been Administrator of NOAA and had taken on the job part time when he left NOAA. He was an atmospheric scientist and was resigning to pursue other interests.

Since I was the one making overtures to industry, I was on the spot. At first I volunteered to become another part time President of JOI while continuing to be Dean in Miami. I didn't want to leave Miami, there was so much still to be done. So, for the better part of a year I commuted two or three days a week from Miami to Washington. I would catch a very early morning flight from Miami to National Airport, and be in the JOI Office in the Watergate Complex by 9AM, and take the 7PM flight back to Miami; it was a routine not to be envied.

In 1980, with many election year activities under way, we were informed by the OSTP that the President of JOI needed to be full time, reside in Washington, and be willing to travel a lot. JOI's Board of Governors offered me the job. There really was no alternative; so much of my life had been devoted to promoting scientific ocean drilling I felt I had to continue even if the proposed new program was not entirely to my liking. Somehow we had to devise a way to keep our international partnerships going.

On the late afternoon of May 16, 1980, I called JOI's Board Chairman, John Knauss, Dean of the University of Rhode Island's School of Oceanography, and told him I would accept the job and move to Washington. The next day a jury in Miami acquitted white Miami police officers who had been indicted for manslaughter in the death of a black man, Arthur McDuffie. The Miami Riot of 1980 began a few hours later. A pall of smoke hung over the city.

THE CORIOLIS EFFECT IN EVERYDAY LIFE

*The deflective force due to the Earth's rotation, which is the key to the explanation of many phenomena in connection with the winds and the currents of the oceans, does not seem to be understood by meteorologists and writers on physical geography.*
William Ferrel (1871)

The circulation of Earth's atmosphere and oceans would be very simple and straightforward if it were not for the planet's rotation. Aristotle, based on common sense observations of the movements of the Sun and stars across the sky, believed that Earth was the center of the universe and everything revolved around it. Nicolas Copernicus and Galileo Galilei argued that these seemingly sensible observations were an illusion, and they proposed that Earth both rotates on its axis and orbits the Sun. The search for effects of the planet's rotation began in the 18th century. It was easy to understand that a decline in rotational speed of Earth's surface from the Equator to the poles, due to the planet's decreased circumference, should affect the circulation of the atmosphere. However, scientists realized that there were also some unanticipated and mysterious effects at work. The discovery of these factors is a fascinating story.

This unanticipated effect of rotation was discovered by mathematicians and engineers in the early 19th century, but it was generally unnoticed or misunderstood by meteorologists and oceanographers until the mid-20th century. The Coriolis Effect (also known as the Coriolis Force although it is not a force of nature recognized by physicists), is in many ways the most mysterious thing about living on a rotating globe. It, like the centrifugal force has even been termed a "fictitious force," because it is seen by an observer on a rotating surface but not seen by a stationary observer. Yet it affects everything that moves on the planet—including the atmosphere, the oceans, rivers, trains, cars, and airplanes. Meteorologist Anders Persson, who has written a number of amusing essays on the Coriolis Force that are readily available on the Internet, calculated that a golf ball that is putted at a speed of 2 m/s on a frictionless green in central Europe will, after 15 m, have deviated 5 mm to the right—enough to risk missing the hole. By contrast, the deviation would be 4 mm in southern Europe and 7 mm in Scandinavia. These differences are due to the changing magnitude of the Coriolis Effect with changing latitude.

Most people are wholly unaware of the Coriolis Effect, and when they try to understand it, it seems to make no sense. The following discussion attempts to bring some clarity to this subject and to put it in historical and scientific context.

## 23.1 The Sun Does not Orbit the Earth, but Earth Rotates 1543–1651

In 1543, Polish monk Nicolaus Copernicus published his book *De revolutionibus orbium coelestium*, proposing that the Sun is at the center of the universe and that the Earth orbits the Sun—a concept known as "heliocentrism." This was a revolutionary idea intended to replace Aristotle's concept, from 350 BCE, that Earth is at the center of the universe, and the Sun and planets revolve around it. Aristotle's idea, known as "geocentrism," was based on simple observations of the sky. The Sun and stars move across the sky from one horizon to the other, so it makes sense that they must be circling the planet. But there were some problems with this idea—particularly the fact that the planets moved at different rates, and sometimes they even seemed to move backwards. To explain that, a series of nested transparent celestial spheres was proposed.

Copernicus discarded the ancient idea of moving celestial spheres to propose the concept of a rotating Earth. In 1588, Danish astronomer Tycho Brahe developed an alternative explanation for the movements observed in the sky—the Sun orbits the Earth, and the planets orbit the Sun. In Italy, Dominican friar Giordano Bruno went even further, proposing that the stars are actually suns like our own, and that they probably had planets orbiting them. Bruno was 400 years ahead of his time. In 1600, he was burned at the stake in Rome's Campo dei Fiori for this and other heresies.

No one seems to have paid much attention to the geocentrism/heliocentrism controversy until 1609, when

© Springer International Publishing Switzerland 2016
W.W. Hay, *Experimenting on a Small Planet*, DOI 10.1007/978-3-319-27404-1_23

Galileo Galilei, using the newly invented telescope, found support for Copernicus's idea of heliocentrism. Galileo observed the moons of Jupiter moving around that planet, and he watched the changing phases of Venus, which were much like those of Earth's Moon. The Roman Catholic Church was married to Aristotle's ideas, and this new cosmology was a highly inconvenient truth coming right in the middle the Protestant Reformation. For the Catholic Church to admit that it had erred on such a fundamental matter would have been catastrophic. Galileo was charged with heresy in 1614, but it was not until 1633 that he was formally put on trial before the Roman Inquisition. Galileo was convicted and sentenced to house arrest for the rest of his life. He died in 1642. 350 years later, in 1992, the Vatican formally and publicly cleared Galileo of any wrongdoing.

Interestingly, thinking people seemed to pay little attention to the Church's edict against Galileo. Before his trial the Vatican's own astronomers had actually shown that Galileo was correct. Soon the intellectuals of the time were using the heliocentric model to make all sorts of progress in science, including Ole Rømer's measurement of the speed of light in 1676. Nevertheless, some old dogmatic ideas are hard to shake. In 2014, a poll of 2,000 Americans chosen at random revealed that 29 % of the people in the United States still believe that the Sun goes around the Earth.

One of the obvious questions that arose to scientists contemplating living on a rotating sphere was: why don't we notice the motion? At latitude 40° N, we are moving eastward at a rate of 364 m/s (1,195 ft/s), yet it seems like we're going nowhere at all. In the 17th century, the search for a method of proving that Earth actually rotates was a hot topic.

The first experiment in this matter was performed in 1643 in the south east of France, in the region then known as the Dauphiné. Calignon de Perrins set up a large plumb-line pendulum with a heavy weight. He observed that the pendulum moved on its own, in synch with the rhythm of the nearby tides. No one knew what to make of this observation.

That same year, Pére Marin Mersenne attempted to demonstrate the rotation of Earth by firing artillery projectiles vertically upward from a site near Paris. The experiment was unsuccessful; the balls landed in all directions around the cannons. Some couldn't even be found and were assumed to have flown off into space. You might remember Mersenne from his 1636 treatise on sound and music, *Harmonie Univeselle*, discussed in Chap. 7. He was a "polymath," interested in all aspects of what we now call physics.

In 1651, Giovanni Battista Riccioli and Francesco Maria Grimaldi presented arguments for and against Earth's rotation in their encyclopedic astronomical treatise, *Almagestum Novum*. Book 9 of that massive collection is devoted to the question of whether the universe is geocentric or heliocentric. The authors presented 126 arguments concerning Earth's motion—49 for a moving planet and 77 against.

They preferred a modified version of Brahe's model of the universe. For Riccioli and Grimaldi, the Sun went around the Earth, as did Jupiter and Saturn, but Mercury, Venus, and Mars orbited the Sun. They argued that if Earth were rotating, a cannon ball fired due north should land east of its target. But cannons being what they were in those days, it was impossible to experimentally accept or reject that idea.

## 23.2   Robert Hooke and Isaac Newton 1674–1686

In 1674, Robert Hooke published *An Attempt to Prove the Motion of the Earth*, in which he describes the attraction that celestial bodies have for one another. At the end of the book, he makes statements that presaged Newton's Laws of Motion:

Rules of Mechanical Motions: …

First, That all Cœlestial Bodies whatsoever, have an attraction or gravitating power towards their own Centers, whereby they attract not only their own parts, and keep them from flying from them, as we may observe the Earth to do, but that they do also attract all the other Cœlestial Bodies that are within the sphere of their activity; and consequently that not only the Sun and Moon have an influence upon the body and motion of the Earth, and the Earth upon them, but that ♀, also ♀, ♂, ♄, and ♃ [Note—these are symbols for the planets Mercury, Venus, Mars, Saturn and Jupiter] by their attractive powers, have a considerable influence upon its motion as in the same manner the corresponding attractive power of the Earth hath a considerable influence upon every one of their motions also. The second supposition is this, That all bodies whatsoever that are put into a direct and simple motion, will so continue to move forward in a straight line, till they are by some other effectual powers deflected and bent into a Motion, describing a Circle, Ellipsis, or some other more compounded Curve Line. The third supposition is, That these attractive powers are so much more powerful in operating by how much the nearer body, wrought upon is to their own Centers.

In 1679, Hooke tried to engage Isaac Newton in a discussion of how the attraction between celestial bodies worked. Newton declined the invitation with the comment, "I have a fancy of my own." Newton was at his family's country home in Lincolnshire at the time, cleaning up estate matters after the death of his mother. There were apple trees in the garden. Newton probably saw apples falling from the trees and began to wonder about the paths they took to the ground. The prevailing idea was that they should fall straight toward the center of the Earth, but that idea did not take the rotation of the planet into account.

Newton speculated about what would happen if a falling apple did not encounter the ground and could keep falling into the Earth. See Fig. 23.1. He knew that because the tree was in motion as the Earth rotated, the apple would not fall in a straight line to the planet's center. His first speculation was that the apple would spiral in toward the center. But realizing that the spiral path required friction to slow the apple, Newton

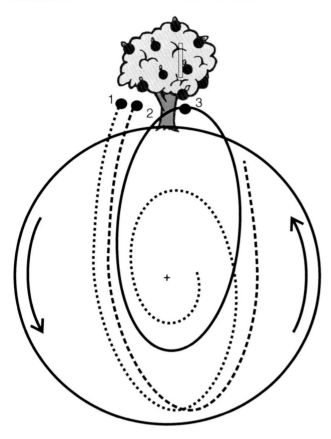

**Fig. 23.1** Isaac Newton's developing ideas about gravitational attraction. *1* The apple spirals in toward the center of the earth. *2* The apple follows a parabolic path. *3* The apple follows an elliptical orbit, like a planet according to Kepler's first law

soon replaced that idea with the notion that the apple's path of fall would be a parabolic curve. But that idea did not explain what happens when the apple returns to the height from which it started. It soon became obvious to Newton that the apple's path would have to be elliptical. Although Newton did not realize it until later, this elliptical path was the same as the orbit of a planet around the Sun, as described by Kepler's first law, formulated in 1609. Kepler's law had been based solely on analysis of astronomical observations; the idea of gravitational attraction had played no role in it. Kepler had assumed that the Sun was located at one of the foci of a planet's elliptical orbit. Newton eventually realized that it was not the Sun, but the common center of gravity of the Sun and the planet, that was at one of the orbit's foci.

In 1686, Isaac Newton presented a copy of the first volume of his *Philosophæ Naturalis Principia Mathematica* to the Royal Society at a meeting in London. It was formally published the next year. At the meeting, Hooke accused Newton of plagiarizing his ideas, and a bitter feud ensued between the two scientists that continued throughout the rest of their lives.

An apple may not have hit Newton in the head, but thoughts about falling apples helped lead him to develop his equation for gravitational attraction:

$$F = G \frac{m_1 m_2}{r^2}$$

In this equation, $F$ is the attractional force, $G$ is the gravitational constant (now known to be $6.67384 \times 10^{-11}$ m$^3$ kg$^{-1}$s$^{-2}$ = $6.6734 \times 10^{-11}$ Nm$^2$ kg$^{-2}$), $m_1$ is the mass of one body, $m_2$ is the mass of the other body, and $r$ is the distance between the two bodies.

Newton's ideas on gravity were major steps forward, but there was still the vexing problem of definitively proving that Earth rotates. It is not inherently obvious.

The controversy over how to prove that Earth is indeed rotating continued for over a century. Then a brilliant old idea for a test was revived and gained widespread acceptance among the intelligentsia. Because the top of a building is further from the center of Earth than the base, the top must be moving eastward faster than the base. If you were to drop a weight from the building, Earth's surface should turn under it as it falls. The original idea, proposed in the early 17th century, was that the weight would fall straight down and, since the planet's surface is moving eastward, the weight should land to the west of its starting position. It was Galileo who pointed out that this idea was wrong. He noted that the weight's drop site, being farther from the center of the Earth, already had a greater eastward velocity than the ground below it, so the weight should land to the east.

In Galileo's day, physics had not yet advanced to the point that a quantitative test could be carried out on this matter. In addition, greater heights and dropping distances were needed to carry out experiments. Robert Hooke had tried a dropping experiment in 1669, but his fall distance was only about 27 ft (8 m), making the results inconclusive.

## 23.3  Explaining the Trade Winds–Halley and Hadley's Ideas 1686–1735

In 1686, Edmund Halley, a polymath best known for his work in astronomy, produced a map of the trade winds, the winds that blow from east to west, both north and south of the Equator. They are called trade winds because sailing ships, including trading vessels, used them to cross the ocean from east to west. Sailors found these winds quite reliable. To get to the Americas from Europe, the instructions were simply to sail south until butter melts, then turn west, and the winds will take you across the Atlantic. Halley proposed that the winds resulted from the air constantly flowing toward a westward-moving barometric low-pressure area, which was created by the heating of the Sun as it moved to the west though the course of a day. This seemed like a good idea at the time—essentially saying that the air chases the low pressure created by the warmth of the apparently moving Sun. Halley was, of course, a heliocentrist, but his explanation would be plausible whether the Earth rotated or the Sun orbited the Earth.

In 1736, however, George Hadley, a Londoner who was a lawyer by profession but a meteorologist by avocation, attempted to explain why the trade winds should be so reliable and constant—even at night—and why they would blow from east to west. He proposed that the trade winds were caused by air descending in the latitudes of the tropics and flowing equatorward. He used Newtonian physics to argue that a moving object keeps moving in a straight line unless a force is applied to make it turn. The air was his moving object. The air isn't attached to the surface of the Earth, so as one approaches the Equator, the surface is moving from west to east beneath the air moving from north to south. To an observer on the surface, the winds would blow from east to west, becoming strongest just before the air rose to great heights again. His presentation to the Royal Society reads:

> For this reason it seems necessary to shew how these Phænomena of the Trade-Winds may be caused, without the Production of any real general Motion of the Air westwards. This will readily be done by taking in the Consideration of the diurnal Motion of the Earth: For, let us suppose the Air in every Part to keep an equal Pace with the Earth in its diurnal motion; in which Case there will be no relative Motion of the Surface of the Earth and Air, and consequently no wind; then by the action of the Sun on the Parts about the Equator, and the rarefaction of the Air proceeding therefrom, let the Air be drawn down thither from the N and S Parts. The Parallels are each of them bigger than the other, as they approach the Equator, and the Equator is bigger than the Tropicks, nearly in the proportion of 1000–917, and consequently their Difference in Circuit about 2083 Miles, and the Surface of the Earth at the Equator moves so much faster than the Surface of the Earth with its Air at the Tropicks. From which it follows, that the Air, as it moves from the Tropicks towards the Equator, having a less Velocity than the Parts of the Earth it arrives at, will have a relative Motion contrary to that of the diurnal Motion of the Earth in those Parts, which being combined with the Motion towards the Equator, N.E. Wind will be produced on this side of the Equator and a S.E. on the other. These, as the Air comes nearer to the Equator, will become stronger, and more and more Easterly, and be due East at the Equator itself, according to Experience, by reason of the Concourse of both Currents from the N and S, where its Velocity will be at the rate of 2083 Miles in the Space of one Revolution of the Earth or Natural Day, and above 1 Mile and ½ in a Minute of Time; which is greater than the Velocity of the Wind is supposed to be in the greatest Storm, which according to Dr. Denham's Observations is not above 1 Mile in a Minute. But it is to be considered, that before the Air from the Tropicks can arrive at the Equator, it must have gained some Motion Eastward from the surface of the Earth or Sea, whereby its relative Motion will be diminished, and in several successive Circulations may be supposed to be reduced to the Strength it is found to be of.

This elegant explanation of the trade winds was widely accepted and still appears in some textbooks and encyclopedias today. The speeds of rotation of Earth's surface at the equator and tropics are, respectively, 475.6 and 436.2 m/s. Assuming Hadley's idea to be correct, and neglecting friction, the velocity of the winds at the surface would be 39.4 m/s, which would be hurricane/typhoon force. The trade winds' actual velocities are typically 5–6 m/s. Hadley was shrewd in realizing that something was making them slower than the difference in surface speed between the topics and the equator—perhaps the friction of the air with mountains. However, friction cannot supply the entire answer. A more accurate explanation of the trade winds would not come for another century, and full acceptance of the explanation by the meteorological community would require another century after that.

Figure 23.2 shows Hadley's proposal and some of the things it didn't explain. In particular, Hadley did not discuss the prevailing westerly winds of the mid-latitudes, which were well known at the time and used to sail eastward across the oceans. He could have made the same argument to explain those winds, but he didn't—perhaps because his ideas would incorrectly require the westerlies to be much stronger than the trade winds. Earth's eastward velocity at 45° of latitude is 328.4–107.8 m/s slower than at the tropics. That is a latitudinal difference of more than 241 miles per hour, similar to the highest wind speeds ever recorded (103 m/s (231 mph) on April 12, 1934, on Mount Washington, New Hampshire; 113 m/s (253 mph) on April 10, 1996, on Australia's Barrow Island during tropical cyclone Olivia).

In fact, the westerlies are neither as strong nor as steady as the trade winds. The fastest consistent westerly wind speeds are in the "Roaring Forties" of the Southern Hemisphere, where they average 8–18 m/s (18–40 mph).

Voyages from Spain to Mexico used the trade winds for the westward crossing and the westerlies for the return crossing. Vessels sailing from England to North America took advantage of weaker high-latitude westerly winds for the westward crossing, and they used the stronger mid-latitude westerlies for their return.

Hadley's idea was so simple and easy to understand that it dominated meteorological investigations until late into the 19th century. Unfortunately it is wrong.

To understand the error in Hadley's reasoning, it is necessary to reconsider two familiar terms: momentum and inertia. You may remember from Chap. 3 that for objects traveling in a straight line, linear momentum (P) is mass × speed (kg m/s) (Fig. 3.2). Momentum is conserved. For example, when a collision occurs between two objects in an isolated system, the total momentum of the two objects before the collision is equal to the total momentum of the two objects after the collision. In other words, the momentum lost

**Fig. 23.2** A schematic view of Hadley's 1736 idea of winds on Earth, redrafted from Plate 1 in Matthew Fontaine Maury's 1855 *Physical Geography of the Sea.* This idea persisted until corrected by William Ferrel in 1856

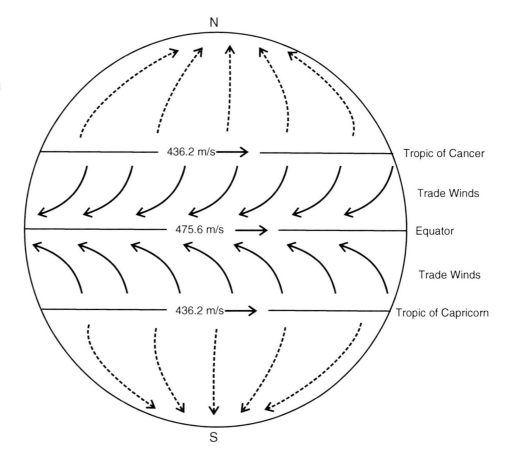

by object 1 is equal to the momentum gained by object 2. In air, there are many molecules of different weights traveling at different speeds and often colliding with one another. Although the momentum of a single molecule may change, the total momentum of the entire system is conserved and constant. Inertia ($I$) is the property of matter by which it continues in its existing state, whether at rest or in uniform motion in a straight line. Inertia is the resistance an object has to a change in its state of motion. The mass of an object is defined by the inertia of the object.

There are analogous terms for objects moving around in a circle. Angular momentum (also termed moment of momentum or rotational momentum) is the rotational analog of linear momentum. It is also a conserved quantity—the angular momentum of a system remains constant unless acted on by an external torque. A torque is simply a force that changes the speed of rotation. Moment of inertia ($I$) is a term referring to objects going around in a circle. It is a quantity expressing the object's tendency to resist the angular acceleration that would increase or decrease its speed of rotation.

Hadley's explanation works only for north-south motion, leaving other motions unexplained. Most importantly, the conservation principle regarding angular momentum on the rotating Earth is not the same as if it were linear momentum.

At any moment the linear momentum of an object going around in a circle is $\Omega v$, in which $\Omega$ is the velocity around a circle expressed in meters per second. You may remember from Chap. 3 that we use two symbols to express velocity around a circular path: $\Omega$ if it is expressed in radians (angular velocity) and $\Omega$ if it is expressed in m/s (conventional velocity). The object would have this linear momentum ($\Omega v$) if the centripetal force causing it to move in a circle were suddenly to vanish; it would fly off in a straight line tangent to the circle at the point of release. In contrast, because its direction is constantly changing, the angular momentum of an object going around a circle is $2\Omega v$, twice as large.

Hadley had incorrectly assumed the conservation of linear momentum. During an object's circular motion around Earth's axis, its angular momentum is conserved, but this is not the case with what its linear momentum would be. To understand what 'linear momentum' means in this context, think of an object being released from gravity and flying off into space. As shown in Chap. 3, its path would along the tangent to the point of release. If it broke away on one side of the Earth the object might have an eastward direction (the same as Earth's rotation); if the break occurred on the other side it would moving westward (opposite to Earth's rotation). As another example, consider an object moving in a circle on the surface of the planet, like air going counterclockwise around a stable low pressure system a few 100 km across in the northern hemisphere. Assuming that its velocity around the low remains constant, its angular momentum remains constant; i.e. it is conserved. However, its linear momentum at any point is its mass times its angular velocity around the low plus or minus the angular velocity of the planet, which will be higher at a low latitude and lower at a high latitude. Its linear momentum is constantly changing as hence is not conserved. This is what is happening to the air taking the curved path between the Tropics and the Equator, shown in Fig. 23.2 Why the air circulates around a region of low pressure rather than simply flowing in to even out the pressure gradient involves the mysterious Coriolis Force and will be discussed later in this Chapter.

Another mistake made by Hadley was his unrealistic assumption that there was some sort of an impulsive force pushing the air. Today, we think of the air as being accelerated down a pressure gradient from high to low.

Despite its inaccuracies, Hadley's idea was so simple that it seemed a perfect explanation for the trade winds. In the 19th century, it was championed by the famous German meteorologist Heinrich W. Dove (1803–79) so persuasively that it became known as the Dove-Hadley theory. It is still alive and well in some textbooks and encyclopedias.

## 23.4 Back to Proving that Earth Rotates 1791–1831

Toward the end of the 18th century and the beginning of the 19th century, several serious attempts were made to use falling objects to prove the rotation of Earth. But where could people of that time find suitably tall towers? There was the famous leaning bell tower, the Campanile, in Pisa, Italy, that Galileo is supposed to have used in an experiment; it is about 56 m (183 ft) high on the low side. But in northern Italy, there were many taller towers, and most had overhanging parapets at the top. These tall towers, most of which were actually the homes of wealthy citizens, were typically eight to ten stories (70–100 m = 230–330 ft) high. The unique design of these buildings—with one room on top of another and with stairs and doors inside—offered protection against intruders. The overhanging parapet at the top was a great place to drop stones on potential intruders; keep in mind that there were no police in those days, so you were pretty much on your own. In the town of San Gimignano, near Florence, a number of tower homes have survived, making for a wonderful place to see what a medieval Italian town looked like.

Bologna is thought to have had as many as 180 towers in the past, but only two survive (Fig. 23.3). Incidentally, Bologna's two towers were the architectural inspiration for the twin towers of the World Trade Center in New York

**Fig. 23.3** The Torre de Garisenda (*left*) and Torre de Asinelli (*right*) in Bologna, Italy

City, destroyed in the September 11, 2001, terrorist attacks. One of the Bologna towers, the Torre Asinelli, became famous as the site of falling weight experiments. The Torre Asinelli, finished in 1119, is 97.2 m (320 ft) high. Its neighbor, the Torre Garisenda, was built at about the same time, but its foundation had structural problems, and it soon began to tilt. The upper part of the Torre Garisenda was removed in the 14th century to prevent its collapse.

Over a period of several months in 1791 and 1792, Giovanni Battista Guglielmini (1763–1817) carried out a series of experiments at the Torre de Asinelli. The first task was to calculate the exact deflection that should happen during the fall of the ball used in the experiments. This was done in two steps: (1) determining the ball's fall time, and (2) determining the eastward distance that the ball would travel as it fell from the top of the Torre to the base.

The fall-time, disregarding the resistance of the air, was calculated following Newton's equation:

$$t = \sqrt{2h/g}$$

where $t$ is the time, $h$ is the fall distance, and $g$ is the acceleration due to gravity. If $h$ is 97.2 m and g is 9.80492 m/s$^2$ (as measured at Bologna in 2011), then $t$ would be 4.45 s.

Calculating the ball's expected displacement during its fall involved determining the difference in eastward velocities between the base and the top of the Torre. Earth rotates completely around (360°) in 86,164 s (the sidereal day). The angular velocity of any point on Earth's surface, ω, is 2π/86,164 s = 7.292 215 × 10$^{-5}$ rad s$^{-1}$. In discussing the conventional velocity (Ω) of a point on Earth's surface, we need to know the length of the radian at that point. A radian's length is the distance from the surface point to the planet's axis of rotation, as shown in Fig. 23.4. Thus, the length of such a radian varies with the cosine of the latitude (1 at 0°, 0 at 90°). Earth's rotation causes it to be shaped like an oblate spheroid, flattened at the poles, but this oblateness was not discovered until the late 18th century, and its role in Guglielmini's experiments were insignificant.

Today, Earth's dimensions are known with great precision. The equatorial radius is 6,378.1370 km (3,963.1906 miles), and its polar radius is 6,356.7523 km (3,949.9028 miles). Remember that a radian is the length of a radius along the circumference of a circle. The length of a radian at 40° is 4,885,936 km. The numbers used in 1791 were less precise, but more detailed than was actually needed for the experiments.

The traveled between an object 100 m above the surface of the Earth and the ground surface, which we will call Δv, is then

$$\Delta v = \frac{h}{r} v \cos(\varphi)$$

where $h$ is the fall distance, $\varphi$ is the latitude, $v$ is the angular velocity of Earth's rotation at the equator (464.6 m/s), and $r$ is the radius of the Earth (6,371 km). For the 97.2-m-high Torre

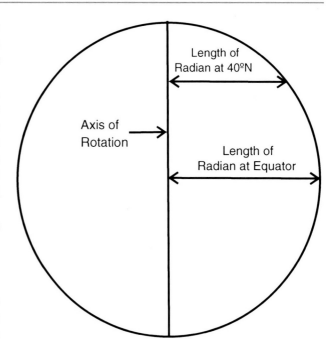

**Fig. 23.4** The length of a radian of circumference depends on its latitude on Earth

de Asinelli in Bologna, $\varphi$ is 44.5° (already determined in the 18th century by astronomical observations). At latitude 44.5° a radian is about 4549209.085 meters at sea level (modern measurement). We now know that the base of the Torre is 75 m above sea level. The eastward velocity of the base is 331.7338 m/s. The velocity of the top is 331.7409 m/s. Thus, the top of the tower is moving eastward at a velocity of 0.0071 m/s (7.1 mm/s) faster than the base. When a lead ball is dropped, if it retains the eastward velocity it had at the top of the drop throughout the 4.45 s of its fall, it should land 0.03156 meters (31.6 mm) east of the drop site (Figs. 23.5).

Guglielmini dropped 16 metal balls, each one inch in diameter, from the Torre. They were released from a cleverly constructed device built to try to ensure that the ball had no motion of its own before the drop (Fig. 23.6 left). A large wax tablet was placed at the base of the Torre to record their

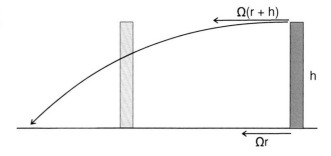

**Fig. 23.5** The expected path of a ball dropped from the Torre de Asinelli. The ball outpaces the eastward motion of the Torre by 31.6 mm during its 4.45 s of fall

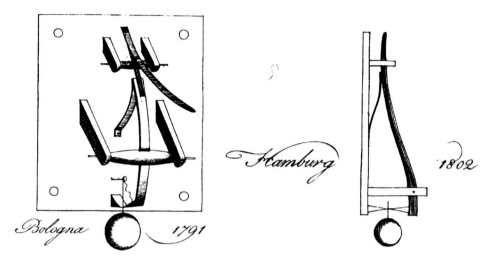

**Fig. 23.6** The devices designed to drop the balls in the fall experiments at Bologna and Hamburg, as shown in Benzenberg's 1804 report, *Vesuche über des Gesetz des Falls, über den Widerstand der Luft, und über die Umdrehung der Erde* (Experiments on the law of fall, on the resistance of the air, and on the rotation of the Earth). The Bologna device is shown in front view, the Hamburg device is shown in side view, making the minor differences difficult to see. Even with these complex mechanisms, it was difficult to prevent the devices from imparting some unknown lateral motion during release

exact impact points. In Guglielmini's later experiments, some of the balls were covered with a thin layer of mercury to help mark the exact point of impact, which was then determined with the aid of a microscope. The impact point was recorded relative to a reference point on the ground. The fall time was about 4.5 s. The main problem with these experiments was ensuring that the dropping mechanism itself did not impart some lateral motion at the moment of release. It took 50 years for another experimenter to find a solution to this problem.

Another problem was determining exactly where to mark the reference point for measurements on the ground beneath the release point of the balls. This turned out to be a far more complicated procedure than that for dropping the balls. It was accomplished in 1792 using a plumb line hung from the release point. A plumb line, sometimes called a lead line, usually has a lead weight attached to it (the Latin word for lead is *plumbium*). But experience had shown that if a heavy lead weight were used on such a long wire, it would behave like a pendulum. It would take days to slow its swing without ever coming to an absolute rest. Accordingly, Guglielmini used a wire line with a marble ball 6 cm in diameter as the weight. Even with the marble ball, the work required many hours on windless nights, because even the motion of people or carts in the streets nearby could perturb it.

The results showed that instead of the calculated landing location 31.6 mm east of the drop site, the average actual location of the impact points was 15.9 mm east and 10.6 mm south of the point determined by the plumb line to be directly beneath the release point. Something was wrong somewhere.

Why did the balls fall short by half the calculated distance? And why did they land south of the expected point? These

discrepancies inspired another attempt at the experiment in 1801 and 1802, this time by a German astronomer, geologist, and physicist named Johann Friedrich Benzenberg (Fig. 23.7).

**Fig. 23.7** Johann Friedrich Benzenberg (1777–1846). His stern expression reflects his determination to get the right answer

He used the tower of the church of St. Michael in Hamburg (Fig. 23.8). The original church having been destroyed by lightning in 1750, the new church was completed in 1786. It is located at latitude 53.5° N, as determined by astronomical observations. St. Michael's was one of the highest towers in Europe, and it offered several different sites for the drop. There was an open space in the center of the staircase spiraling up the tower, allowing the ball drops to be performed inside the structure. The fall heights for Benzenberg's experiments varied from 71 to 103 m (235–340 ft).

There were a few problems with using this site for the experiment. Two busy streets ran nearby the tower. Also, northern Germany is famous for its strong westerly winds, which caused drafts inside the tower. There was a possibility that the Sun might heat one of the walls more than the others, inducing a slight tilt of the tower during the daylight hours. And the church officials did not allow Benzenberg to work at night, as he would have preferred.

In preparation for these experiments, Benzenberg asked one of his friends, Heinrich Wilhelm Matthias Olbers (1758–1840), who was a physician and astronomer in Bremen, to work on the problem of determining the expected deviation. He also asked Olbers to explain the problem to two of the best mathematicians of the time, Carl Friedrich Gauss, in Brunswick, Germany, and Pierre-Simon de Laplace, in Paris. Gauss was a 24-years-old prodigy whose groundbreaking book on mathematics, *Disquisitiones Arithmeticae*, had just been published. By the middle of 1801, Gauss had already figured out the orbit of the asteroid Ceres, which had been discovered on January 1 of that year, a problem Laplace considered insoluble. Laplace was 53 years old and had made major contributions to the fields of mathematics, statistics, physics, and astronomy. The first volume of his *Mécanique Céleste* had appeared in 1799. It included his 1775 work on tides which involved a rigorous mathematical discussion of the motions of objects on the surface of the Earth.

Benzenberg's preliminary experiments began in October 1801. He started out using balls made of flint and a container with clay to mark the impact points. He found it was very difficult to get his plumb line to come to rest, so he had to guess where the midpoint of its motions might be. It seemed as though everything that could go wrong did go wrong. He performed 16 experiments over the next few weeks. Some of the balls fell to the west of the supposed vertical, some to the east, some to the north, and some to the south. He attributed the lack of success to the fact that he had performed his experiments during the daytime, not at night as Guglielmini had.

Benzenberg spent about 6 months making improvements on his experimental design. He modified the drop device (Fig. 23.6 right) and devised a new way of recording the impact sites. He also switched to using balls made of lead and zinc. From July through October of 1802, he performed 31 experiments. Most balls fell to the east and south of the drop site, but a few went to the west and north. The average eastward deflection was two-thirds of that calculated by Olbers, Gauss, and Laplace. The southward deflection was a mystery.

**Fig. 23.8** The tower of the church of St. Michael in Hamburg, Germany. Frontispiece in Benzenberg's 1804 report, *Vesuche über des Gesetz des Falls, über den Widerstand der Luft, und über die Umdrehung der Erde* (Experiments on the law of fall, on the resistance of the air, and on the rotation of the Earth)

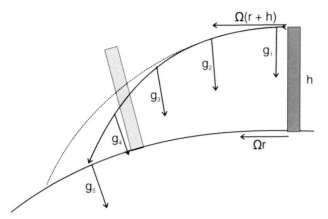

**Fig. 23.9** A revision of the path of a ball dropped from the Torre Asinelli, taking into account the curvature (here highly exaggerated) of Earth's surface. As the ball falls, the direction of gravity (g) is changing, indicated by the arrows labeled $g_1$, $g_2$, $g_3$, $g_4$, and $g_5$

It seems to have been Olbers who figured out why the balls did not land as far to the east as expected. The calculations all assumed the pull of gravity to be vertical during the course of the fall. Olbers finally realized that the direction of gravity changed during the fall, as shown in Fig. 23.9 (using the Torre Asinelli as an example). Gauss agreed, but they had no explanation for the deflection to the south.

In Fig. 23.5, the surface of Earth is shown as flat. In reality, however, it is curved. In the 4.45 s it took for the ball to reach the ground, the Torre Asinelli moved 1.4762 km eastward in an arc. Figure 23.9 shows the same scene with a highly exaggerated curvature of the Earth.

Although Fig. 23.9 offers an explanation for why the ball would fall short of the distance it would travel if gravity always pulled in the same direction, it does not explain why the fall deviated to the south. Something was still wrong.

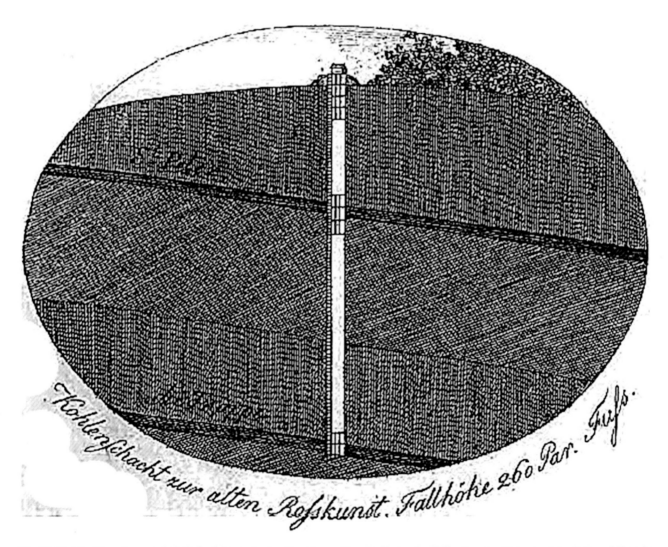

**Fig. 23.10** The 90-m deep vertical shaft of the coal mine Zur alten Rosskunst, near Schlebusch in Westphalia, Germany. The fall height is given as 260 Parisian feet. From Benzenberg's 1804 report, *Vesuche* *über des Gesetz des Falls, über den Widerstand der Luft, und über die Umdrehung der Erde* (Experiments on the law of fall, on the resistance of the air, and on the rotation of the Earth)

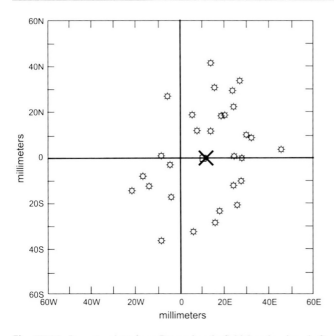

**Fig. 23.11** Impact points from Benzenberg's Schlebusch mine shaft experiments. The middle of the diagram is the site on the ground determined to be directly beneath the drop site. The large X to the right of this location is the calculated eastward deflection of the drop. The small square next to the X is the average of all impact points. This is a stunning example of the statistical principle that the average of a large number of measurements is likely to be closer to the true value than any of the individual measurements

While writing up his results from the tower of St. Michael, Benzenberg chanced to visit a coal mine (named "Zur alten Rosskunst") near Schlebusch in Westphalia, Germany (Fig. 23.10). There, he was shown a vertical mine shaft 90 m deep. He soon decided that this would be much better place to conduct his experiments. There was no wind problem, the plumb line could come to rest, and there was no possible tilting during the day. The only problem was the physical climb down the shaft and back up—and the dust, which got on everything.

In the meantime, both Gauss and Laplace worked out more detailed calculations of the deviations expected in three-dimensional space. They used different methods to deal with the fact that observations are being made on a rotating globe, not from a stable point in space. In their many pages of equations, both mathematicians incorporated two key terms—one for centripetal force and the other for the observer's position on a rotating frame.

The term for centripetal force was $\Omega \times (\Omega \times r)$, which could also be written as $r\omega^2$ if expressing angular velocity in radians ($\omega$) instead of conventional velocity in meters per second ($\Omega$). You may remember from Chap. 3 that centripetal force is determined by:

$$F_{cp} = mr\omega^2$$

The term that took into account the observer's position on a rotating frame was $2\Omega \times r$. It had already appeared in Laplace's original work on tides in 1775, and it is better known today as the "Coriolis term." This key factor would explain the southward deflection noted in the Hamburg experiments.

During October 1804, Benzenberg conducted 29 fall experiments in the vertical mine shaft using lead-zinc balls. The results are shown in Fig. 23.11. The scatter of the impact points is impressive, and it seems almost a matter of luck that the average point turned out to be so near the calculated deviation. The lack of any southward deviation, which was predicted, probably made Laplace and Gauss suspicious that the results were not as good as they seemed, but Benzenberg was pleased at the confirmation of the eastward deflection.

Yet another set of drop experiments was carried out in 1831. Ferdinand Reich (1799–1882) (Fig. 23.12), was a professor of physics at the venerable Bergbauakademie (Mining Academy) in Freiberg, Saxony. He used a 158.5-m vertical shaft in the Drei-Brüder-Schacht for his experiments. He reckoned that the major problem with both the

**Fig. 23.12** Ferdinand Reich (1700–1882), who proved that Earth rotates and who discovered iridium

**Fig. 23.13** Gaspard-Gustav de
Coriolis (1792–1843)

Guglielmini and Benzenberg experiments lay in their release
mechanisms for dropping the ball. He tried a number of
ideas using different kinds of balls and devices before
devising a very clever solution to the problem.

Reich designed an apparatus consisting of a short tube
with a diameter just large enough for the ball to pass through
when cold. First, he placed the ball in boiling water, so that it
expanded slightly, enough to prevent its passing though the
tube. Positioned on top of the tube, it would slowly cool and
then finally fall though. This was not an easy task to
accomplish, because the balls all needed to be perfectly
round and of exactly the same size. But the result was worth
it—a much smaller cluster of impact points.

Reich performed 106 drop experiments. The calculated
eastward deviation was 28.396 mm; the average of Reich's
experiments was 28.282 mm. The calculated southward
deviation was 4.374 mm; the average of the experiments

5.061 mm. The balls fell to the south as well as to the east,
just as Laplace and Gauss had predicted. Reich's experiments were soon accepted as proof of Earth's rotation.

## 23.5    The Ideas of Gaspard-Gustave de Coriolis 1829–1835

You may be surprised to learn that Gaspard-Gustave de
Coriolis (Fig. 23.13) never wrote a word about atmospheric
or oceanic circulation. He was an engineer interested in
mechanical devices.

His father, Jean-Baptiste-Elzéar Coriolis, had been in the
military service of King Louis XVI and was a member of the
Rochambeau Corps sent to America to fight with General
Washington against the British. After he returned to France
in 1784, he became part of the guard assigned to the king.

**Fig. 23.14** Émilie de Breteuil Marquise de Châtelet (1706–1749). Portrait by Maurice Quentin de La Tour

When the Revolution broke out in 1789, he suddenly found himself in deep trouble. In 1792, he escaped to Nancy, in northeastern France, where he established a business. Louis XVI was guillotined in 1793.

Gaspard-Gustave was born in Paris in 1792, but he grew up in Nancy. He was very bright, finding particular enjoyment in mathematics. Fortunately, new educational opportunities for young people interested in mathematics had recently opened up in France.

You may recall that Manchester, England, was the center of Europe's Industrial Revolution, and that Manchester Academy had been founded in 1786 to promote education in science and engineering and provide talent for the growing industries. A similar institution, the École Polytechnique, was founded in Paris in 1794. Napoléon Bonaparte was a great promoter of science, and after he proclaimed himself

emperor of France in 1804, he moved the École Polytechnique from a western suburb into central Paris' Latin Quarter near the recently completed Pantheon. Under Napoléon, the École Polytechnique was primarily a military academy devoted to the development of engineering skills. The emperor gave it the motto, "*Pour la Patrie, les Sciences et la Gloire*" (For the Nation, Science, and Glory). The faculty of those days included the most prominent mathematicians of the time. (Incidentally, the oldest institution of this type in the United States is Rensselaer Polytechnic Institute in Troy, New York, founded in 1824.)

Coriolis enrolled in the École Polytechnique in 1808. Two years later he graduated and moved on to study at the École des Ponts et Chaussees (School of Bridges and Highways), where he joined the Corps de l'École, which allowed him to work as an engineer on projects in eastern

France. In 1816, he returned to the École Polytechnique as a teaching assistant to Augustin-Louis Cauchy (1789–1857) and André-Marie Ampère (1775–1836). After realizing that students were having problems following the lectures, he wrote what we would call a textbook: *Du calcul de l'effet des machines, ou Considérations sur l'emploi des moteurs et sur leur évaluation: pour servir d'introduction à l'étude spéciale des machines* (On calculating the effects of machines, or Considerations on the use of motors and their evaluation to serve as an introduction to the special study of machines). The book, published in 1829, became a classic. In it, Coriolis defined two special terms in physics: *travail* (work; see Fig. 3.3) and *forces vives* (kinetic energy).

Here, it would be helpful to review several scientific and philosophical concepts as they were understood at the beginning on the 19th century. The Law of the Quantity of Motion was proposed by René Descartes in his *Pricipia philosophiae*, published in 1644. Descartes believed that when God created the universe, he included a certain amount of motion, and as part of God's perfection, the total quantity of motion remains constant forever. In other words, for the universe as a whole, the product of mass times velocity ($mv$) is conserved. According to Descartes' consideration of the collision of bodies, it is only the magnitude of the quantity of motion that is conserved, not the direction of motion. The magnitude is a scalar, a number that is always positive, and never a vector, a number which requires direction and can be negative. But remember that Descartes believed that all of space was permeated by the invisible æther, which must somehow impede the motion of visible objects. This meant that absolute equations and knowledge of what was happening was impossible.

In 1687, Isaac Newton had proposed three laws of motion: I. An object either remains at rest or continues to move at a constant velocity, unless acted upon by an external force; II. The sum of the external forces on an object is equal to the mass of the object multiplied by the acceleration of the object ($F = ma$); III. When one body exerts a force on a second body, the second body simultaneously exerts a force equal in magnitude and opposite in direction on the first body. He called the product of mass times velocity ($mv$) momentum (Fig. 3.2). Since the velocity ($v$) is the product of an acceleration ($a$) over time, the expression could be written as *mat* (mass × acceleration × time), or as simply as *Ft* (force × time, meaning force acting over time).

Taken together, Newton's three laws describe another law: the Law of Conservation of Momentum. This was discussed previously in relation to Hadley's ideas about the trade winds. The momentum in Newton's law is linear momentum; the idea of another kind of momentum, angular momentum, would come later. Newton's ideas were similar to what Descartes had proposed, but they can apply to things much smaller than the whole universe.

While thinking about colliding objects, Gottfried Wilhelm von Leibnitz suspected that the situation might be more complex than either Descartes or Newton had imagined. In his *Discours de metaphysique* (Discourse on metaphysics), published in 1686, Leibnitz distinguished between Descartes' quantity of motion without taking direction into account (here designated as $m|v|$), and Newton's quantity called momentum (designated by $mv$) which involves direction. He coined the terms *vis mortua*, for resting bodies about to begin motion, such a stretched spring at the moment it has stopped and just before it starts to recoil, and *vis viva*, associated with actual motion. *Vis viva* was soon better known by its French name, *forces vives* (living force) for the force one body might exert on a second body. We now know the *forces vives* as kinetic energy, but that term did not enter the English language until 1849, when it was coined by Lord Kelvin. Leibnitz proposed the expression $F \propto mv^2$, meaning that the *vis viva* ($F$) is proportional to $mv^2$ in some undefined way. There was no experimental proof of the relationship.

Now $v^2 = 2as$, where s is the displacement (velocity$^2$ = 2 × acceleration × displacement), so that $mv^2 = 2mas = Fs$ (mass × velocity$^2$ = 2 × acceleration × displacement = Force × displacement). This means that the *forces vives* ($Fs$) of Leibnitz is a force acting over space whereas Newton's momentum $Ft$ is a force acting over time. That is the difference between kinetic energy and momentum.

The next step came from Dutch lawyer and natural philosopher Willem's Gravesande, who enjoyed performing experimental demonstrations of classical mechanics. One of his experiments, conducted in about 1722, involved dropping metal balls into soft clay in an effort to determine the proportionality factor in Leibnitz' equation. It looked to him like it might simply be $F = mv^2$. He later wrote to an acquaintance in France, Émilie de Châtelet, telling her about his thoughts on what the relationships between force, mass, and velocity.

Gabrielle Émilie Le Tonnelier de Breteuil Marquise de Châtelet (1706–1749) (Fig. 23.14) was an extraordinary woman. Born into a noble family, she was 19 years old when she married the Marquis Florent-Claude Chastellet (later written as Châtelet), military governor of Semur-en-Auxois in Burgundy. He was often away on military business. In the 18th century, women of the nobility were expected to be largely decorative, attending to their husband's needs. They were also expected to have a few lovers on the side. Émilie loved classical literature and mathematics, and her lovers tended to be mathematicians and philosophers. Her most constant companion in the later part of her life was François-Marie Arouet, the writer, historian, and philosopher of the French Enlightenment better known by his *nom de plume* "Voltaire."

Émilie received the message from Willem's Gravesande while she was working on translating Newton's *Principia* into French. The letter intrigued her, prompting her to conduct a series of experiments dropping metal balls into soft clay from different heights. Her results supported the simple $F = mv^2$ relationship. That was in the mid-1740 s. Her translation of Newton's *Principia* was published in 1756, 7 years after her death. That book made Newton's new ideas about forces, masses and motions widely available and accessible in Europe.

It was almost a century later when Coriolis added his refinements to these topics.

What is 'work' (*travail*) in physics? Let's review a few things from Chap. 3. (It might be useful to refer to Figs. 3.1, 3.2, and 3.3 to refresh your memory.) Velocity ($v$) is distance ($d$) traveled per unit time ($t$), or $v = d/t$ expressed in metric terms, such as meters per second. Acceleration ($a$) is an increase in velocity with time, or $a = v/t$. This might be more readily understood if written as $v = (d/t)/t$ or, in metric terms, as (m/s)/s or m/s$^2$. Force ($F$) is acceleration ($a$) of a mass ($m$): $F = am$, or in terms of metric units, [(m/s)/s]kg or (kgm)/s$^2$ or newtons. Work ($W$) is a Force ($F$) operating over a distance ($d$): $W = Fd = amd$; in terms of metric units this is m[(m/s)/s]kg, which could be simplified to m$^2$kg/s$^2$ or Joules.

Now we will explore some new territory, as Coriolis did. With Galileo's inclined plane experiments, he had found that the velocity of an object undergoing constant acceleration (like that due to gravity) doubles with every unit of time.

We will designate the initial velocity $v_i$, the final velocity $v_f$, the average velocity $\bar{v}$ (pronounced v bar), the distance $d$, the time $t$, the acceleration $a$, the average acceleration $\bar{a}$, and the change in each quantity by $\Delta$ (Greek capital letter delta). Then the average velocity is

$$\bar{v} = \frac{\Delta d}{\Delta t} \text{ and the average acceleration is } \bar{a} = \frac{\Delta v}{\Delta t}$$

The symbol $\bar{v}$ can also be written as $\bar{v} = \frac{v_i + v_f}{2}$, where $v_i$ is the initial velocity, and $v_f = v_i + at$, the velocity acquired though acceleration over time $t$.

Then $d = v_i + \frac{1}{2}at^2 = v_f^2 = v_i^2 + 2ad$

Now, if the object starts from rest, $v_i = 0$ so $v_f^2 = 2ad$ or rewritten $a = \frac{v_f^2}{2d}$

Since Work $W = amd = mad$, it could be written $W = m\frac{v_f^2}{2d}d = \frac{1}{2}mv_f^2$.

Coriolis made a major contribution by adding in the ½, the proportionality factor that Leibnitz had suspected—though it had escaped his experiments, as well as those of Willem's Gravesande and Émilie de Châtelet. There had been problems with the clay that those researchers had used. Once that issue was resolved, and once work was precisely

defined, it was possible to figure out the energy requirements for starting a machine and then keep it going. Coriolis wrote of "travail moteur" (motive work) and "travail resistant" (resistance to work), which might initially be the inertia (resistance to change) of a waterwheel at rest, and later the friction involved in its operation.

He related work to energy. Remember from Chap. 3 that energy is the ability of one physical system to do work on one or more other physical systems. Potential energy is energy that is available, waiting to be used. Kinetic energy is the energy of something in motion. Coriolis realized that in any mechanical situation, the sum of potential and kinetic energy is constant—that is, it is conserved. As work is performed using kinetic energy, the reservoir of potential energy decreases.

Coriolis next described the forces involved in rotating parts of machines, such as those driven by waterwheels, which were very common in the early years of the Industrial Revolution. In 1832, he published a major paper on the topic: *Memoire: Sur le principe des forces vives dans le movement relatifs des machines* (On the principle of kinetic energy in the relative movements of machines) in the *Journal de l'École Polytechnique*. This paper addresses the centrifugal forces affecting the rotating parts of machines (Fig. 23.15).

Coriolis explored what happens when an outside force changes the path of an object moving on a two-dimensional disc. Figure 23.16 shows an object of mass $m$ moving at a velocity $v_0$ on a circular path around a center of rotation different from that of the turntable. It is acted upon by a force $F$ over a distance $ds$; $\theta$ is the angle between $F$ and $ds$. The object is set onto a new trajectory and accelerated to a new velocity $v_1$. Remember that 'acceleration' can mean either a change of speed or a change of direction, or both. Coriolis built on his earlier definitions of work and kinetic energy (*travails* and *force vives*) to show that, in this case, work is done by the force $F$ as some potential energy is transformed into kinetic energy. In equations, Coriolis showed that the relation between potential and kinetic energy for the body is the same in rotating and nonrotating systems. Figure 23.16 also shows the centripetal and centrifugal forces acting on the object, $F_{cp}$ and $F_{cf}$. Although noting them in his discussion, Coriolis observed that they are equal in magnitude and cancel each other out.

In his 1998 paper cited in Fig. 23.16, Anders Persson noted that although Coriolis was concerned with early 19th-century technology, his analysis could be related to the increase in speed and kinetic energy of a satellite falling toward Earth, where work is done by the gravitational force along the trajectory of the satellite.

Coriolis' next major undertaking was a work inspired by the commandant of the École Polytechnique, Henri-Alexis de Tholosé (1781–1853): *Théorie mathématique des Effets du Jeu de Billiard* (Mathematical Theory of the Effects of the

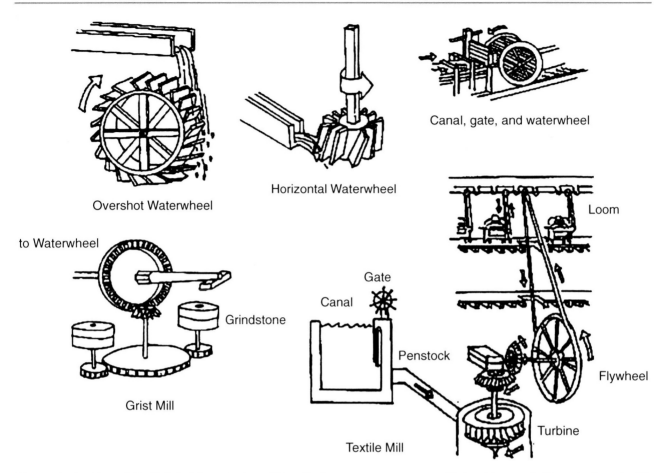

**Fig. 23.15** Machines of the kind Coriolis wrote about. Waterwheels were a common source of industrial power in France in the early 19th century. They were used to drive a variety of increasingly complex machines, transferring energy though rotating parts we would call gears

Game of Billiards), published in 1835. This may sound like an odd topic for investigation by an engineer. However, billiards—which had evolved in the 16th century from an outdoor version to an indoor game with balls on a table top being struck and set in motion by a cue stick—had a number of industrial analogies.

At that time, there was a great interest in determining how energy is passed from one object to another, or from one part of a machine to another. The discussion originated in the 17th century with disputes between Rene Descartes and Christiaan Huygens. Descartes believed that the universe was permeated by 'æther,' implying that even if air were removed, a moving object would still have to push æther out of the way. Since æther was a hypothetical substance with unknown properties, exact calculations regarding motion would be impossible

Christian Huygens put the matter on a more solid footing, and published his ideas in 1666. Several important ideas from that essay are listed here, in his own words:

- The quantity of motion that two hard bodies have may be increased or diminished by their collision, but when the

quantity of motion in the opposite direction has been subtracted there remains always the same quantity of motion in the same direction.

- The sum of the products obtained by multiplying the magnitude of each hard body by the square of its velocity is always the same before and after collision.
- A hard body at rest will receive more motion from another, larger or smaller body if a third intermediately sized body is interposed than it would if struck directly, and most of all if this [third] is their geometric mean.
- A wonderful law of nature (which 1 can verify for spherical bodies, and which seems to be general for all, whether the collision be direct or oblique and whether the bodies be hard or soft) is that the common center of gravity of two, three, or more bodies always moves uniformly in the same direction in the same straight line, before and after their collision.

It sounds as though Huygens was an avid fan of billiards and learned much about basic physics from the game. In retrospect, it seems that billiards may have been crucial to the development of ideas concerning kinetics in physics.

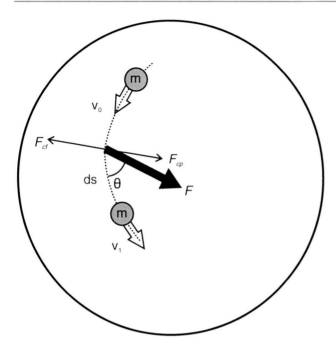

**Fig. 23.16** Schematic diagram illustrating Coriolis' 1832 discussion of a force acting on a moving body. After A. Persson, 1998, How do we understand the Coriolis Force. *Bulletin of the American Meteorological Society*, 79/7, 1373–1385

Billiards developed from the games played with a stick or mallet and balls on lawns—games such as croquet, golf, and trucco. Lawn games go back to the middle of the 14th century. Today, most American homes are surrounded by a lawn that requires frequent maintenance. The concept of a lawn originated as a place to play games, and when I was a child, that's what we did. We had a croquet court in the back yard and played almost every day if it wasn't raining. Tennis could also be played on a lawn. I haven't seen a croquet court or lawn tennis net for many years. Many Americans know that you are supposed to have a lawn and regularly mow the grass, but they have no idea why.

The word "billiard" probably comes from the French *billart* or *billette*, meaning "stick," or *bille*, meaning "ball." The ball is struck with a cue, a word derived from *queue*, the French word for a tail. It the good old days, players used the tail of a mace to strike the ball. It was a new and innovative use for an instrument originally designed to hit someone over the head. By the 18th century, the billiards had become so popular that almost every café in Paris had a table.

Key to the game was the ball. The earliest billiard balls were made of wood; then later from clay baked into a ceramic. However, balls made of ivory were favored from at least 1627 until the early 20th century. By the middle of the 19th century, elephants were being slaughtered on a massive scale for their ivory tusks, needed to keep up with the demand for high-end billiard balls. Only about eight balls

could be made from a single elephant's tusks. In the later 19th century public alarm over the slaughter of elephants led to a search for substitutes. The first material tried was a plastic, nitrocellulose, marketed as 'celluloid' in 1870. Other materials followed, leading to the more refined composition balls we have today. The last ivory billiard balls were made early in the 20th century.

Every billiards table is in effect a physics demonstration laboratory for the public. Many concepts that might otherwise seem quite abstract become common experience with billiards. Coriolis' book on the subject remains a classic.

In 1835, Coriolis published his work on the forces at play in machines with rotating parts: *Mémoire: Sur les équations du movement relative des systémes de corps* (On the equations of relative movement of systems of bodies) in the *Journal de l'École Polytechnique*. It is a fine example of technical writing at its most obscure. From the title you can hardly tell what it is about. The text is a further description of the forces involved in the rotating parts of machines—this time taking into account parts that are coupled in such a way that one rotation is related to, but different, from another rotation. One disc on a machine may be rotating or moving at one speed or in one direction, while another, attached disc may be rotating or moving in a different speed or direction.

Coriolis referred back to his 1832 paper and the engineering concepts of *force vives* and centrifugal force (to refresh your memory, review Chap. 3, Sect. 3.11). But in the 1835 paper, he explored the forces acting on an object moving independently on a rotating surface, describing a new concept that he called the *forces centrifuges composes*. This "composite centrifugal force" is what we now call the Coriolis Effect or Coriolis Force. Although the text makes many references to rotating platforms and the forces acting on them in terms of simple geometric figures, such as parallelograms, there are no illustrations. Coriolis uses mathematical equations to show that the *force centrifuges* acting on an object fixed to the rotating platform is:

$$F_{cf} = -mr\omega^2$$

And the *forces centrifuges composes* acting on an object on the disc but moving relative to it is:

$$F_{cfc} = 2m\omega$$

In effect, Coriolis introduced his ideas through what we would today call vector algebra, a topic of mathematics that did not exist at the time. Consequently, almost no one seems to have read the 1835 paper. Nevertheless, his work eventually led to his name becoming associated with the phenomenon so crucial to understanding atmospheric and oceanic circulation.

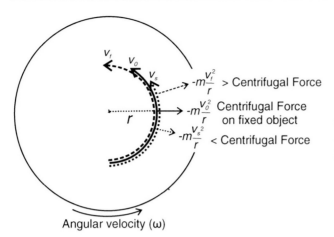

**Fig. 23.17** The Coriolis effect of objects on a turntable rotating at speeds different from that of the turntable

The following discussion, much of which is adapted from the items on the Coriolis Force published by Anders Persson, is presented in terms of motions associated with objects on a turntable. It is a loose account, in words and equations, of what Coriolis described in his work.

Figure 23.17 shows a turntable rotating counterclockwise with an object fixed to it at a distance $r$ from the center and having a velocity $v_0$. Two other objects move around at the same distance from the center, but one is faster ($v_f$) and the other is slower ($v_s$) than the fixed object. The object fixed to the turntable experiences a centrifugal force equal to its mass times the square of its velocity divided by its distance from the center. As indicated in Sect. 3.11, the object is held in place by a centripetal force designated with a positive sign, and it experiences an opposite centrifugal force of the same magnitude designated with negative sign. The object that is not attached to the turntable and is moving faster along the same path experiences a greater centrifugal force, while the

unattached object that is moving slower experiences a lesser centrifugal force. The path of each unattached object will be forced outward to the right. Both objects will eventually fall off the turntable, with the faster-moving object falling off sooner than the slower-moving object.

The centrifugal forces acting on these objects have the same radial orientation. directly away from the center of the turntable, but different magnitudes. The differences between the centrifugal forces acting on the fixed object and the moving objects produce the Coriolis Effect.

Now consider the situation with an object moving in a spiral inward on the turntable—compared to an object fixed to the turntable (Fig. 23.18a) and an object spiraling outward (Fig. 23.18b). The objects are spiraling around centers of rotation different from that of the turntable, and the centrifugal forces associated with them have orientations different from that of an object fixed to the turntable. Again, the difference between the centrifugal forces experienced by the fixed and moving objects is the Coriolis Effect.

Figure 23.19 shows the nature of the different forces, using the example of an outward-spiraling object with velocity $V$ relative to a turntable rotating with an angular velocity $T$, so that $V = Tr + V_r$ ($T$ is the angular velocity of the turntable, $r$ is the distance from the center of the turntable, and $V_r$ is the additional component of the velocity of the object that makes its path differ from a simple circle about the center of the turntable. The path of the object is shown by the solid curved arrow. The total force on the object is shown by the solid straight vector line drawn from its center of rotation. The total force can be decomposed into the centrifugal force related to the motion of the turntable, shown as a dotted straight vector from the center of the turntable. To make the discussion simpler, we will now (and throughout the rest of this chapter) consider the centrifugal force as having a positive value, rather than the negative value previously given to

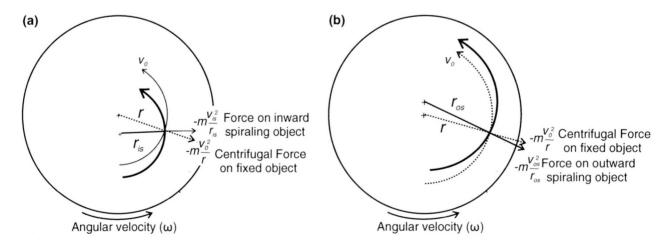

**Fig. 23.18** Forces acting on an object fixed to the turntable and one spiraling inward on the turntable (**a**) and one spiraling outward (**b**)

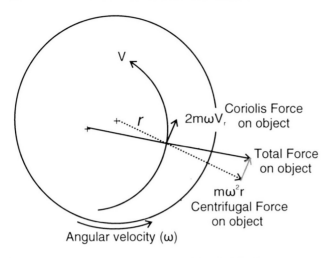

**Fig. 23.19** Evaluating the magnitude of the Coriolis force

it, so that it is simply $mT^{-2}r$. The other force is indicated by the short gray arrow connecting this centrifugal force vector with that representing the total force. It is also shown as a short black arrow at the intersection of the total force vector and the object's curved path. The value of this force was determined by Coriolis to be $2\,mTV_r$. You will notice that in this diagram, the Coriolis Force has a magnitude of about one-third that of the centrifugal force.

Figure 23.20 helps us understand where the 2 in the $2\,\omega V$ comes from. The figure shows a turntable on which an object moves from the center to the periphery with a velocity v. Seen from outside, the object moves in a straight line while the turntable rotates beneath it. Seen by an observer on the turntable, however, the object appears to take a curved path. While the turntable rotates though an angle $\alpha$, the object's path as it reaches the turntable's edge is at an angle $2\alpha$.

Coriolis was concerned with two-dimensional rotating discs and their closely associated objects, not with rotating three-dimensional objects like the Earth. It would be

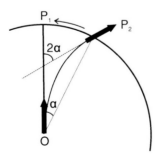

**Fig. 23.20** Path of an object moving from the center of a turntable (O) toward a point on the edge (P₁). As the turntable rotates though an angle α, the object appears to take a curved path and arrives at the periphery at point P₂. The direction of is motion at the edge is 2α. (After Fig. 3 in Persson 1998, How do we understand the Coriolis Force. Bulletin of the American meteorological society, v. 79, 1373–1385)

24 years before the possible significance of his work to processes on the Earth was suggested—as an explanation for the meandering rivers of Siberia.

German naturalist Karl Ernst von Baer (1792–1876), best known as the father of the science of embryology, proposed that the meandering of the south- to north-flowing Siberian rivers was due to the rotation of Earth. He argued that the rotation of the planet had no effect on rivers that flow from east to west. In considering his ideas at a meeting in 1859, the French Academy of Sciences had problems accepting them because of the obvious meandering of such east-to-west flowing rivers as the Seine and Loire. This may be why the French began to question whether Hadley's model was correct.

## 23.6 The Foucault Pendulum 1851

In 1841, French astronomer Jean Bernard Léon Foucault (1819–1868) reported to the Académie des Sciences on his development of a new kind of clock. The pendulum clocks developed by Christiaan Huygens split time into minutes with jumps in the minute hand. The pendulum kept track of time as it made an arc back and forth, and energy lost to fiction was replaced by a weight and an escapement mechanism, as discussed in Chap. 7. For refining astronomical observations, Foucault wanted a clock that could show smaller intervals of time. His solution was a "conical pendulum" in which the bob at the end of a wire goes around in a circle rather than back and forth. Huygens had tried to make a similar "pirouette" pendulum, but it did not work properly.

While experimenting with his invention, Foucault noticed that if the bob was swinging back and forth, it maintained its orientation even as the base of the mechanism was rotated. Then he noticed that the swinging bob gradually changed its orientation when the base was not rotated. A decade after constructing his conical pendulum clock, he realized that he had found a new way to prove that the Earth rotates.

Foucault reasoned that if you had a pendulum swinging back and forth at the North Pole, it would maintain its orientation as the Earth rotated beneath it, retuning to its initial position in one sidereal day (360° of rotation, or 23.9345 h). The pendulum's arc would remain stable with respect to the stars but appear to an observer to rotate clockwise, because in reality the Earth rotated counterclockwise beneath it. At the South Pole, the effect would be opposite, with the pendulum rotating counterclockwise. This means that at the equator, the path of the pendulum does not rotate at all. The sine of 90° is 1, and the sine of −90° is −1, so the period of the pendulum's rotation is a function of the sine of the latitude.

Foucault made an initial experiment at the Paris Observatory in February 1851, and then set up his famous "grand experiment" in the Pantheon on the hill above Paris' Latin Quarter. He used a pendulum with a wire 1.4 mm (0.055 in) in diameter and 67 m (220 ft) long supporting a 28-km (62.7-lb) brass-coated lead bob hung from the dome. The latitude of the Pantheon was determined to be 48° 50′ 49″ The sine of its latitude is 0.7529543. The plane of the swing rotated about 11° per h, making the complete circuit in 34 h, 47 min, 14.6 s. This was the first unequivocal proof that the Earth rotates.

Foucault pendulums were soon set up around the world in public buildings and science museums to prove to the public that our planet rotates. The "force" that causes the swing plane of the pendulum to rotate is what we now call the Coriolis Force. Foucault had never read—and was totally unaware of—Coriolis' 1835 paper.

## 23.7    The Earth is Neither a Disc nor a Sphere

Coriolis had been concerned with rotating discs. However, the Earth is not a disc, nor is it a sphere. Its rotation has slightly flattened it at the poles into an oblate spheroid. Earth's angular velocity, $\omega$, is 7.292 rad/s. Its equatorial diameter is 12,756 km (7,926 miles), and its polar diameter is 12,720 km (7,900 miles)—a difference of only 64 km (40 miles). One could consider each of the circles of latitude on the Earth to outline the edge of a disc and evaluate the forces of an object moving along it. Figure 23.21 shows the

eastward velocities of Earth's surface at different latitudes and the changes in relative strengths of the centrifugal and Coriolis forces. The centrifugal force varies with the velocity of Earth's surface and is a function of the cosine of the latitude. It is perpendicular to Earth's axis of rotation, not Earth's surface (except at the Equator). The Coriolis Force varies with the sine of the latitude.

On our rotating spheroidal Earth, gravity is a function of both the gravitational attraction of the planet's mass and the centrifugal force. Figure 23.22 shows the outline of a sphere as a dotted line and the surface of a spheroid as a solid line. $R_{eq}$ is the radius at the equator, $R_{lat}$ is the radius at some other latitude, and $F_{cf}$ is the centrifugal force at that latitude. The gravitational attraction of the planet is g*. The force of gravity noted by an observer is g; it is the resultant of the two forces g* and $F_{cf}$. A stationary observer finds the force of gravity g to be perpendicular to Earth's surface.

An object moving across the surface of Earth will experience different centrifugal forces and different directions and magnitudes of gravity as it moves from one latitude to another.

Figure 23.23 shows a decomposition into vertical and horizontal components of the forces depicted in Fig. 23.22. Here, gravity g is the difference between the vertical component of gravitation g*$_V$ and the vertical component of the centrifugal force $F_{cfV}$. The object is stationary, because the horizontal components of both gravitation g*$_H$ and the centrifugal force $F_{cfH}$ are equal, thereby balancing out. If the object moves eastward, the centrifugal force will increase,

**Fig. 23.21** Speed of the Earth's surface at different latitudes. Relative strengths of the centrifugal and Coriolis forces vary as the cosine and sine, respectively, of the latitude

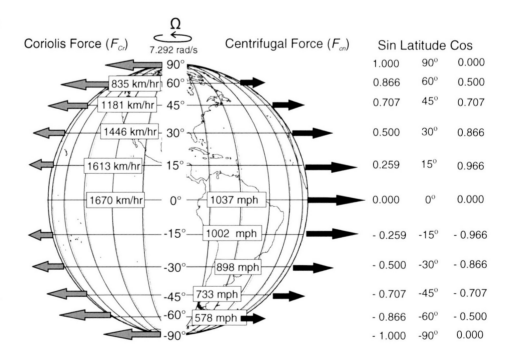

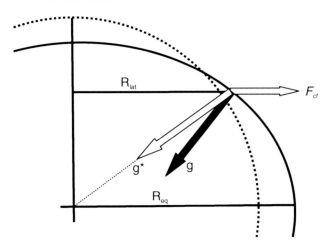

**Fig. 23.22** The gravity g observed at any place other than the poles is the resultant of Earth's gravitational attraction g* and the centrifugal force of the rotating Earth. $R_{eq}$ = equator, $R_{lat}$ = some other latitude

with its horizontal component becoming larger than that of the gravitational force, and the object will move equatorward (to the right). If the object moves westward, the centrifugal force will become smaller than the gravitational force, and the object will move poleward.

This raises an interesting question. What would happen if suddenly the gravitational attraction that keeps you on the surface of the Earth were to disappear? Most people might think that you would fly straight up into the air. But the centripetal force acting on you on the surface is not directed toward Earth's center, but toward the axis. Thus, if the gravitational force were to vanish, you would fly off the surface at the angle shown by $F_{cf}$ in Fig. 23.23—an angle that is 90° minus the latitude above the equatorward horizon.

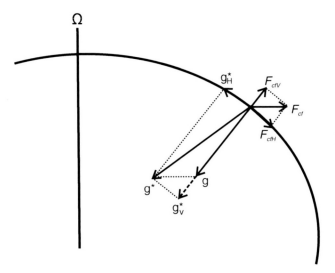

**Fig. 23.23** Decomposition of the forces shown in Fig. 23.22 into vertical and horizontal components. For a discussion, see the text

At the equator (0°), you would fly straight up, at 90° to the horizon. But at 30° N (along the U.S. Gulf Coast), you would fly off at an angle 60° above the southern horizon, and in Boulder, Colorado (40° N), you would go off at an angle of 50° above the southern horizon.

After becoming airborne, you would not circle the Earth. Rather, you would fly away into space, traveling in a straight line at the tangential velocity of your original latitude.

## 23.8   Inertial Circles

On a rotating sphere, the Coriolis Force acts to keep turning a moving object to the right in the Northern Hemisphere and to the left in the Southern Hemisphere. It acts to bring a moving object around in a circle and return it to the place where it started. We call this path an inertial circle. Its size is:

$$R = v_r / 2\Omega \sin \varphi$$

where $R$ is the radius of the inertial circle, $v_r$ is the velocity of the object going around the circle, and $\varphi$ is the latitude on Earth's surface.

Remember that during an object's circular motion, its angular momentum around Earth's axis is conserved, but not its linear momentum and that Hadley's error was in assuming conservation of linear momentum in a rotating system. As the object moves through a circle on the surface of an eastward-rotating spheroid like Earth, the object sometimes has an eastward direction (the same as the spheroid's rotation). In such circumstances, linear momentum will be greater than when the object is moving westward (opposite to the spheroid's rotation).

Figure 23.24 shows a global pattern of inertial circles for air moving at velocities between 50 and 70 m/s (110–160 mph). These velocities are unrealistically large but allow the general pattern to be clearly seen.

As you know from experience winds are not constant, but we can see their common cycloidal motions only when the pick up dirt or sand forming 'dust devils'. The variations in air speed over water are reflected by the production of rotating cycloids a few kilometers in diameter. Ocean water does not move in straight lines. Plots of the tracks of drifting buoys or floats suspended at particular depths in the ocean are often called 'spaghetti diagrams' because of the complex patterns. The inertial mini-eddies are grouped together into larger 'mesoscale eddies' 100 km or so across which are then in turn incorporated into the larger gyral patterns of ocean circulation shown on maps and globes.

For those of you who really want know how this works, consider an "object" to be an air or water mass, moving with speed $v$ and subject only to the Coriolis Force ($F_{cor}$). It will travel in a trajectory called an inertial circle. Because the

**Fig. 23.24** The Coriolis force produces inertial circles that tend to restore a body to its initial position. This hinders the geographical displacement of air masses

Coriolis Force is always directed at right angles to the motion of an object, the object will move with a constant speed around a circle whose radius $R$ is given by:

$$R = \frac{v}{F_{cor}}$$

where $F_{cor}$ is $2\Omega\sin\varphi$ ($\varphi$ is the latitude). The time $T$ required for the mass to complete a full circle is:

$$T = \frac{2\pi}{F_{cor}}$$

In the mid-latitudes, $F_{cor}$ typically has a value of about $10^{-4}$ s$^{-1}$. Hence, for a typical atmospheric speed of 10 m/s (22 mph), the radius of an inertial circle is 100 km (62 miles), with a period of about 17 h. For an ocean current with a typical speed of 10 cm/s (0.22 mph), the radius of an inertial circle is 1 km (0.6 mile). These inertial circles are

**Fig. 23.25** Inertial circles. **a** On a rotating turntable, where $F_{cor}$ is constant. **b** On a rotating sphere, where $F_{cor}$ increases with latitude ($\varphi$), showing the westward drift known as the $\beta$ effect

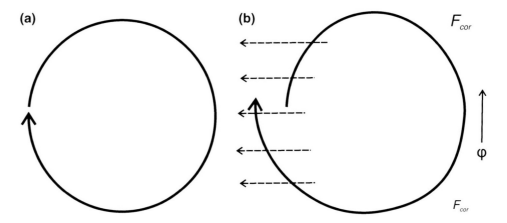

clockwise in the Northern Hemisphere (where trajectories are turned to the right) and counterclockwise in the Southern Hemisphere.

However, on our rotating planet, $F_{cor}$ varies with latitude, being larger at higher latitudes and smaller at lower latitudes. This means that the paths of objects will not form exact circles. On our eastward-rotating planet, this means that instead of returning to the original site of the start of the motion, the object will end up displaced to the west, as shown in Fig. 23.25. This result is known in meteorology and oceanography as the $\beta$ (beta) effect. Because the Coriolis parameter varies as the sine of the latitude, the radius of the oscillations associated with a given speed are smallest at the poles and largest as they approach the equator. This relationship ensures that eddy motions can never become stable unless some other factor, such as topography or bathymetry, comes into play.

### 23.9   William Ferrel Figures It Out 1856–1858

William Ferrel (1817–91) (Fig. 23.26) was born in southern Pennsylvania. He had little formal schooling while a child, but he educated himself by reading books about science, and he became a school teacher. He graduated from Bethany College in its first graduating class in 1844. It was while he was teaching in Missouri and Tennessee that he made a major contribution to the developing science of meteorology. During this time, he was studying three books—Pierre-Simon Laplace's *Mechanique Celeste* (five volumes originally published from 1799–1825, with English translations (Celestial Mechanics) published from 1829–1839); Matthew F. Maury's *Physical Geography of the Sea* (1855); and James Espey's *Third Report on Meteorology* (1851).

In 1856, Ferrel published *An Essay on the Winds and Currents of the Ocean* in volume 6 of the *Nashville Journal of Medicine and Surgery*. It was revolutionary. He proposed

that the circulation of the atmosphere is driven by four forces related to differential heating and vapor content, resulting in air masses of different densities and flows from regions of high to low pressure under the influence of Earth's rotation. It may well be the most concise account of atmospheric circulation ever written:

**The forces concerned in producing the motions of the atmosphere.**—There are four principle forces which must be taken into account in a correct theory of the winds. The first arises from a greater specific gravity of the atmosphere in some places than others, on account of a difference of temperature and

**Fig. 23.26**  William Ferrel (1817–1891), the man who finally got it right

of the dew point; for when it becomes heated or charged with vapor in any place to a greater degree than at others, it becomes specifically lighter, and hence the equilibrium is destroyed. There is a flowing together, then, of the heavier air on all sides, which displaces the lighter air, and causes it to rise up and flow out in a contrary direction. This is the primum mobile of the winds, and all the other forces concerned are dependent upon it for their efficiency. A second force arises from the tendency the atmosphere has, when, from any cause, it has risen above the general level, to flow to places of a lower level. These two preceding forces generally produce counter-currents. Again, when from any cause, a particle of air has been put into motion toward the north or south, the combination of this motion with the rotary motion of the earth produces a third force, which causes a deflection of the motion to the east when the motion is toward the north, and a deflection toward the west when it is toward the south. This is the same as one of the forces contained in La Place's general equations of the tides, the analytical expression of which is 2 n u r sin l cos l; l being the latitude, n the motion of the earth at the equator, u, the velocity of the particle north or south; and r, the radius of the earth. The fourth and last force arises from the combination of a relative east or west motion of the atmosphere with the rotary motion of the earth. In consequence of the atmosphere's revolving on a common axis with that of the earth, each particle is impressed with a centrifugal force, which, being resolved into a vertical and horizontal force, the latter causes it to assume a spheroidal form conforming to the figure of the earth. But if the rotary motion of any part of the atmosphere is greater than that of the surface of the earth, or, in other words, if any part of the atmosphere has a relative eastern motion with regard to the earth's surface, this force is increased, and if it has a relative western motion, it is diminished, and this difference gives rise to a disturbing force which prevents the atmosphere being in a state of equilibrium, with a figure conforming to that of the earth's surface, but causes an accumulation of the atmosphere at certain latitudes and a depression at others, and the consequent difference in the pressure of the atmosphere at these latitudes very materially influence its motions. This force is also expressed by one of the terms of La Place's equations, the analytical expression of which is 2 n u r sin l; u being the relative eastern or western velocity of the atmosphere (Fig. 23.26).

The last two forces are the east-west and north-south components of what we now call the Coriolis Force. However, in this first paper, Ferrel made an error that was probably due to a misunderstanding of Laplace's equations. Ferrel thought that the east-west force was smaller than the north-south force by a factor of the cosine of the latitude. In 1858. he published *The Influence of the Earth's Rotation Upon the Motion of Bodies Near Its Surface* in the widely read *Astronomical Journal*, offering a complete and correct mathematical derivation of the forces involved. It is in this paper that he stated what became known as Ferrel's Law: "If a body is moving in any direction, there is a force arising from the earth's rotation, which always deflects it to the right in the northern hemisphere, and to the left in the southern."

In the mid-19th century, it was thought that there were two atmospheric cells in each hemisphere, with air descending along the band of subtropical highs at about 28° N and S. It was believed that most of this air then flowed equatorward as

the trade winds, while the remainder flowed to Earth's poles, as shown by the diagram in Maury's book (see Fig. 23.2). In his 1856 paper, Ferrel noted that this belief did not make sense, observing that Arctic expeditions had discovered another set of high-latitude winds that flow outward from the pole all year long. Ferrel made the first sketch showing a three-cell-per-hemisphere atmospheric circulation, reproduced here as Fig. 23.27. Everything now made sense.

In 1858, Ferrel took up a full-time position as a meteorologist on the staff of the *American Ephemeris and Nautical Almanac* in Cambridge, Massachusetts. In 1882, he joined the U.S. Army Signal Service (which would become the U.S. Weather Bureau in 1891). He retired in 1887 and died in West Virginia in 1891.

## 23.10  Buys-Ballot's Discovery 1857

Christophorus Henricus Diedericus Buys-Ballot (1817–1890) was a Dutch meteorologist who was very much concerned with organizing meteorological observations so that larger patterns of air masses could be recognized and their movements traced (Fig. 23.28).

In 1857, Buys-Ballot published his famous *Note sur le rapport de l'intensité et de la direction du vent avec les écarts simultanés du barometer* (Note on the relation between the intensity and direction of the wind and concurrent spread of barometry) in the *Comptes Rendus* (Reports) of the French Academy of Sciences. In that highly influential work, he noted that if you stand with your back to the wind, there is a low-pressure system to your left and a high-pressure system to your right, as shown in Fig. 23.29. This observation has become known as Buys-Ballot's Law.

## 23.11  Flow Down a Pressure Gradient

To understand how Earth's rotation affects the circulation of the atmosphere and ocean, consider the flow of air from a region of high pressure to one of low pressure, as diagrammed in Fig. 23.30. The pressure gradient can be caused either by differential heating of the atmosphere over the planet's surface or by differing water vapor content of the air or both. On a non-rotating planet the parcel of air would simply accelerate down the pressure gradient. However, on a rotating planet, the accelerating parcel will be turned to the right by the Coriolis Force (in the Northern Hemisphere) until, as the two forces, the pressure gradient force and the Coriolis Force balance. Then the air simply flows along one of the pressure isobars. This is termed 'geostrophic balance' or 'geostrophic flow.'

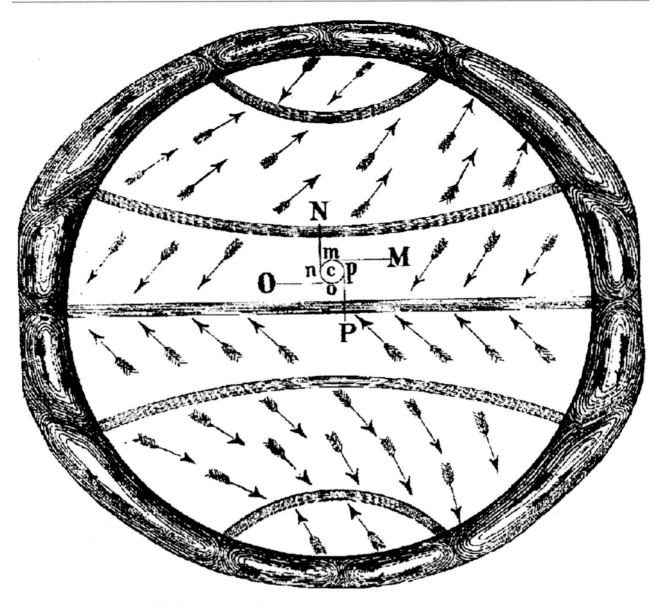

**Fig. 23.27** William Ferrel's 1856 diagram of the winds

The size and speed of rotation of Earth are the reasons that our planet has three atmospheric circulation cells in each hemisphere, as discussed in the next chapter.

An extreme case regarding planetary rotation in our solar system is Venus, with a period of retrograde (clockwise when viewed from over its north pole, opposite to that of Earth and all of the other planets) 360° rotation of 243 Earth days. That rotational period is close to the period of its orbit around the Sun, 224.7 Earth days—meaning that, as a practical matter, Venus does not rotate. Another extreme case is Jupiter, a gaseous giant planet that rotates completely around in only 9.84 Earth hours. Figure 23.31 shows the very different appearances of these two cloud-covered planets.

## 23.12 Introduction to the Use of "Coriolis" in Meteorology

Gaspard Gustave de Coriolis never wrote a word about the effects of Earth's rotation on the atmosphere or ocean. The first person to do that was Pierre-Simon Laplace, in 1775, as part of his theory of the tides published in the Mémoires de l'Academie royale des Sciences, and included in the more widely available *Traité de mécanique céleste* (1798, Tome I-II, Livre 4, pp. 171 ff.). The first use of Coriolis' name in the context of meteorology appears to have been in 1900, in Marcel Brillouin's book *Mémoires originaux sur la circulation générale de l'atmosphère* (Original memoirs on the general circulation of the atmosphere).

**Fig. 23.28** C.H.D. Buys-Ballot (1817–1890)

## 23.13 A Mathematical Derivation of the Coriolis Force

One of my favorite memories of London concerns an evening "London Walk" several years ago through the East End, focusing on the sites of the Jack the Ripper murders. At the site of the first murder, our guide, a delightful older lady, told us, "Now if you are squeamish, go over to the churchyard and wait for us there, but for those of you who want to hear all about the bloody gore, come closer."

For those who want to wait by the churchyard, here is the Coriolis Force in relatively simple English: For a given rate

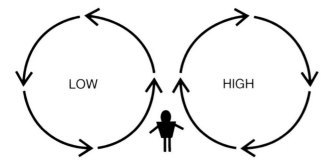

**Fig. 23.29** Buys-Ballot's law. If you stand with your back to the wind there is a low-pressure system to your *left* and a high-pressure system to your *right*

of rotation of the observer, the magnitude of the Coriolis acceleration of an object is proportional to the velocity of the object and the sine of the angle between the direction of movement of the object and the axis of rotation.

If you are a normal human, you should skip the next section and follow the advice of Henry Stommel and Dennis Moore in their 1989 book, *An Introduction to the Coriolis Force*:

> All professional meteorologists and oceanographers encounter this mathematical demonstration at an early stage in their education. Clutching the teacher's hand, they are carefully guided across a narrow gangplank over the yawning gap between the resting frame and the uniformly rotating frame. Fearful of looking down into the cold black water between the dock and the ship, many are glad, once safely aboard, to accept the idea of a Coriolis force, more or less with a blind faith, confident that it has been derived rigorously. And some prefer never to look over the side again.
>
> So, when they are asked for an explanation by a landlubber standing on the dock, who has never been carefully guided across the perilous plank, they find themselves singularly unable to explain the curious force. Incomplete explanations abound in popular books and magazines.

But for those of you who want to see an example of the whole gory business, the following derivation of the Coriolis Force is adapted from that available from NASA.

In this discussion, $\vec{R}$ is the vector distance from Earth's center to the wind, and omega (ω) is the rotation vector of the Earth; it lies on the axis and points to the North Pole.

Over a certain time period ($dt$), the distance ($d\vec{R}$) is the velocity of the Earth plus the velocity of the wind relative to

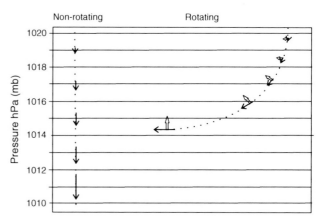

**Fig. 23.30** Acceleration of a parcel of air down a pressure gradient from a high- to low-pressure area on a non-rotating and rotating planet (*solid black arrows*). The *horizontal lines* are isobars, lines of equal pressure. On a non-rotating planet the air will simply flow from high to low. In the Northern Hemisphere of a rotating planet, such as Earth, the parcel of air will be turned to the right by the increasingly strong Coriolis Force (*hollow arrows*) as it accelerates, and it will eventually flow along one of the pressure isobars, never reaching the site of low pressure. It is then in geostrophic balance. A pressure difference of 10 hectopascals (hPa = millibars mb) is typical between Earth's subtropical high and equatorial low systems

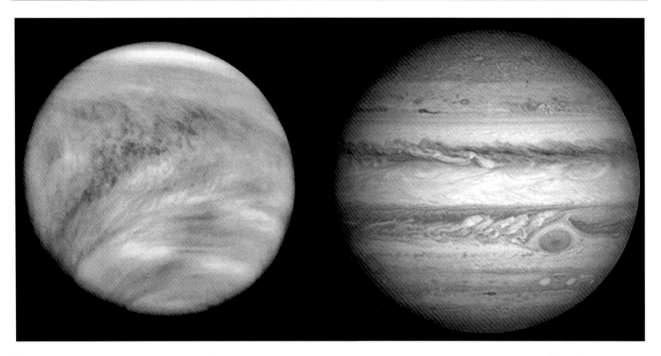

**Fig. 23.31** *Left* Venus has a retrograde (*clockwise*) rotation rate close to its orbital period and no obvious pattern of atmospheric circulation. *Right* Jupiter has a rotation rate almost three times faster than Earth and nine bands of directionally alternating winds in each hemisphere

the Earth. The velocity of anything is defined as the first derivative of its position:

$$\vec{v} = \frac{d\vec{R}}{dt}$$

so the wind velocity as viewed from space is simply the wind velocity as viewed from Earth plus the velocity of the Earth's rotation as viewed from space.

$$\frac{d\vec{R}}{dt_{Space}} = \frac{d\vec{R}}{dt_{Earth}} + \vec{\omega} \times \vec{R}$$

To find the acceleration acting on the parcel of air, we differentiate according to the normal rules of calculus:

$$\frac{d^2\vec{R}}{dt^2_{Space}} = \frac{d}{dt}\left(\frac{d\vec{R}}{dt_{Earth}}\right)_{Space} + \frac{d(\vec{\omega} \times \vec{R})}{dt}$$

After performing the differentiation and collecting terms, we arrive at:

$$\frac{d^2\vec{R}}{dt^2_{Space}} = \frac{d^2\vec{R}}{dt^2_{Earth}} + 2(\vec{\omega} \times \vec{v}) + \vec{\omega} \times (\vec{\omega} \times \vec{R}) + \frac{d\vec{\omega}}{dt} \times \vec{R}$$

where $2(\vec{\omega} \times \vec{v})$ is the Coriolis term and $\vec{\omega} \times (\vec{\omega} \times \vec{R})$ is the centripetal term.

There are three terms on the right, which we must add when we are looking at the winds from the ground. The first is the Coriolis term, the second is the centripetal term, and the third is the effect of a changing rotation rate on the winds if Earth's rotation rate changed. Fortunately, for meteorological purposes, only the Coriolis term is important, and the other two can be neglected. So the problem reduces to:

$$\vec{a}_{Space} = \vec{a}_{Earth} + 2(\vec{\omega} \times \vec{v}_{Earth})$$

where $2(\vec{\omega} \times \vec{v}_{Earth})$ is the Coriolis acceleration.

Because force is mass times acceleration, we can calculate the force by substituting $\frac{\vec{F}}{m_{Space}}$ and $\frac{\vec{F}}{m_{Earth}}$ for the acceleration terms $\vec{a}_{Space}$ and $\vec{a}_{Earth}$ in the last equation, resulting in:

$$\frac{\vec{F}}{m_{Space}} = \frac{\vec{F}}{m_{Earth}} + 2(\vec{\omega} \times \vec{v}_{Earth})$$

where the right-hand side is now what we call the Coriolis Force.

To get out of the vector notation, assume you are looking at only the horizontal component of the wind with respect to Earth's turning surface. Then, for any latitude lambda $\varphi$:

$$F_{Space} = F_{Earth} + 2|\omega|\sin(\varphi)$$

That's all there is to it.

## 23.14    Summary

The Coriolis Force, or Coriolis Effect, is in many ways the most mysterious thing about living on a rotating globe. It has even been termed a "fictional force." Yet it affects everything that moves over the surface of the Earth—including the atmosphere, oceans, rivers, trains, cars, airplanes, golf balls, and you yourself.

After Nicolas Copernicus introduced the idea in 1543 that the Earth rotates and orbits the Sun, an intense controversy regarding this matter raged for over 200 years. Efforts to prove the rotation of Earth led to the exploration of falling bodies and to Isaac Newton's theory of gravitation.

In 1735, George Hadley proposed a simple explanation for the trade winds: they result from the more slowly moving air of the subtopics moving toward the more rapidly moving equatorial region and, hence, appearing to blow from the east. This satisfied almost everyone for the next couple centuries.

At the end of the 18th century and the beginning of the 19th century, a number of sophisticated attempts were made to prove that Earth rotates by dropping objects from towers and down mineshafts. It took about 50 years of trials to get a satisfactory and convincing result. Pierre-Simon Laplace and Carl Friedrich Gauss made correct calculations for the expected deflection of dropped objects, though their calculations were still not proof of rotation.

During the 1830s, Gaspard Gustave de Coriolis was a mechanical engineer at the *École Polytechnique* in Paris concerned with the behavior of billiard balls, waterwheels, and other rotating machinery. He developed formal, scientific definitions for work, kinetic energy, and the deflective forces active on rotating objects. But he never wrote a word about the atmosphere or ocean.

It was Léon Foucault's pendulum experiment in 1851 that finally proved that the Earth indeed rotates.

Also in the middle of the 19th century, modern meteorology was born. William Ferrel was an American who finally sorted things out regarding atmospheric circulation. In his classic paper from 1856, Ferrel cited Laplace's tidal equations as well as his new observations of high-latitude atmospheric conditions to propose a three-cell-per-hemisphere atmospheric circulation. He realized that the main forces acting on the air are heat and humidity, resulting in higher or lower pressure; movement from regions of high pressure to low pressure; and forces causing air masses to turn to the right in the Northern Hemisphere and to the left in the Southern Hemisphere.

Dutch meteorologist C.H.D. Buys-Ballot observed that if you stand with your back to the wind, there is a low-pressure system to your left and a high-pressure system to your right.

Although Coriolis was not the first researcher to work out the forces acting on objects moving on a rotating globe, his name was introduced into the meteorological literature in 1900, and it is now used to describe the deflective force experienced by the winds and moving water.

A Timeline for this chapter:

| | |
|---|---|
| 1534 | Nicolaus Copernicus, in *De Revolutionibus Orbium Cœlestum*, proposes that the Sun is the center of the universe and that Earth orbits around it |
| 1589 | Galileo Galilei drops balls from the Leaning Tower of Pisa, showing that the rate of fall is independent of mass. He discovers the rule for acceleration due to gravity using a ball rolling down an inclined plane |
| 1674 | Robert Hooke publishes *An Attempt to Prove the Motion of the Earth* |
| 1679 | Robert Hooke and Isaac Newton exchange letters regarding falling objects |
| 1686 | Edmund Halley proposes his theory that the trade winds follow the Sun |
| 1686 | Newton presents his inverse law of gravitation to the Royal Society |
| 1686–89 | Hooke and Newton argue over who first conceived the inverse square law for gravitational attraction |
| 1686 | Leibniz coins the terms *vis mortua* and *vis viva*, which in French becomes *forces vives* |
| 1687 | Newton's *Philosophiæ Naturalis Principia Mathematica*, with the law of universal gravitational attraction, is published |
| 1722 | Willem 's Gravesande suggests that the proportionality factor in Leibnitz' *vis viva* equation might be $F = mv^2$. He writes to Émilie de Châtelet |
| 1735 | George Hadley proposes his theory of the trade winds |
| Mid-1740s | Émilie de Châtelet performs experiments proving the $F = mv^2$ relationship |
| 1756 | Émilie de Châtelet's translation of Newton's *Principia* is published |
| 1775 | Pierre-Simon Laplace proposes his theory of the tides |
| 1791 | Giovanni Battista Guglielmini performs his experiments dropping balls from the Torre de Asinelli in Bologna |
| 1801 | Johann Friedrich Benzenberg starts his experiments dropping balls inside the tower of St. Michael's church in Hamburg |
| 1804 | Laplace and Gauss calculate the deviations expected from Benzenberg's experiments |
| 1804 | Benzenberg uses the shaft of the coal mine Zur alten Rosskunst for his experiments |
| 1829 | Coriolis publishes his textbook, *Du calcul de l'effet des machines, ou Considérations sur l'emploi des moteurs et sur leur évaluation: pour servir d'introduction à l'étude spéciale des machines* |
| 1831 | Ferdinand Reich uses a new type of drop release for falling-ball experiments, leading to better results |
| 1832 | Coriolis publishes his *Memoire: Sur le principe des forces vives dans le movement relatifs des machines* |
| 1835 | Coriolis publishes *Théorie mathématique des Effets du Jeu de Billiard* |
| 1835 | Coriolis publishes his *Mémoire: Sur les équations du movement relative des systémes de corps*, describing mathematically what we now call the Coriolis Force |

(continued)

(continued)

| 1851 | Léon Foucault displays his pendulum experiment in the Pantheon in Paris |
| 1856 | William Ferrel publishes *An Essay on the Winds and Currents of the Ocean* in the *Nashville Journal of Medicine and Surgery* |
| 1857 | C.H.D. Buys-Ballot publishes his *Note sur le rapport de l'intensité et de la direction du vent avec lesécarts simultanés du barometer* |
| 1900 | The French expression for Coriolis Force is introduced into the meteorological literature in a review by Marcel Brillouin |

If you want to know more:

Persson, A., 1998. *How do we understand the Coriolis Force?* Bulletin of the American Meteorological Society, v. 70, pp. 1373–1385.

Persson, A.O., 2005. *The Coriolis Effect: Four centuries of conflict between common sense and mathematics, Part I: A history to 1885.* History of Meteorology, v. 2, pp. 1–24.

Music: Something especially appropriate for this complicated story is Ferruccio Busoni's Piano Concerto in C major, Op. 39 (BV 247). Most piano concertos have three movements. Busoni's has ten, tending to wander off in all directions.

Libation: At least a double of whatever you like most.

**Intermezzo XXIII. Course Correction**

Unfortunately the major events in one's life do not happen in an orderly sequence, so now I have to back up to 1978 and tell you about another series of events.

The Rosenstiel School belonged to a number of other organizations. We were a member of the Gulf and Caribbean Fisheries Institute, so as Dean I was put on their Board. I learned a lot about Fisheries, and met the man who had figured out how to peel shrimp using a device with mechanical rollers. It peeled the shrimp the same way he had done it with his toes when he was peeling them for himself on a Shrimper Boat.

We also belonged to the Gulf Universities Research Consortium which had its headquarters in Houston. I didn't know of anything we actually did with GURC, but one day in 1978 Jim Sharp, its President, showed up at my office in Miami. He knew that the DSDP would end in a few years and guessed that it might be looking for a new vessel to replace the GLOMAR Challenger. He wanted to come to Dallas to talk with an official of another company that might like to bid on the operations contract. He assured me that it would be a very interesting meeting.

A couple of weeks later I was in Dallas at the offices of South East Drilling Company (SEDCO) to meet with Dillard Hammett. The SEDCO offices were in an old renovated School Building. Dillard was a Dallas history buff, and when he found out it was my home town he asked if I was any relation to the Stephen J Hay who had been mayor. I explained that I was his grandson. The furnishings of much of SEDCO's offices were antiques from the Stephen J. Hay School that had recently been closed.

It was one of the strangest meetings I have ever had in my life. For the first part of the meeting he had a telephone to his ear. He was also negotiating with a shipyard in Singapore over the construction of a new drilling vessel. They would come up with a number, and he would have a question that sent them back to the drawing board to make new calculations. Rather than to try to place the call again, he just hung on the phone waiting for their answer.

In the meantime he explained that SEDCO was a leader in the offshore drilling business and would like to bid on any new contract for scientific drilling. He showed me pictures of their vessels and discussed the technology they had developed, including risers. I explained that a new program was several years off but that if I had anything to do with it I would certainly see to it that SEDCO could bid on the contract—along with any other companies that wanted to, of course.

Now back to the riots in Miami. That weekend of May 17 and for the next few days whole blocks of the city were burned, and it wasn't just near Liberty City and the northern area. In the evenings I used my view over Key Biscayne from my 26th floor condominium to spot fires being set on our island. I would call the fire department to report them. By late May real estate prices had dropped precipitously in Miami, and I realized I would be unable to sell my condo and buy a place in Washington.

So I decided to live at the Cosmos Club until things in Miami returned to normal. I had become a member of the Cosmos Club while I was at the University of Illinois, and had stayed there on visits to the city. The Cosmos Club is a private social club. It was founded by John Wesley Powell in 1878. Powell was the geologist/explorer who led the first party through the Grand Canyon in 1869. Its members have interests in science, literature and art. For a long time the Club was located on Lafayette Square across from the White House, but in 1952 moved to a mansion at 2121 Massachusetts Avenue near Dupont Circle. That part of Massachusetts Avenue is sometimes called 'Embassy Row.' The public rooms are quite grand. The accommodations for members are in contrast quite Spartan. My travel schedule was such that I was rarely in Washington for more than a few days at a time, so I would check out, leave my things in the storage room, and check in again a few days later. While there I would go out for dinner at one of the many restaurants nearby, but first have a cocktail in the lounge. Many distinguished academics and scientists stayed at the Cosmos Club on visits to Washington, and it was a great place to meet people. One would see famous newsmen informally interviewing politicians at lunch.

The Ocean Margin Drilling Program with industry was always touch and go. The NSF and industry set up a pool of about $16 million to prepare a series of Atlases of the geology of the US continental margins and to carry out engineering studies. The atlases were produced through contracts to the oceanographic institutions and were compilations of all the existing knowledge in the public domain, and sometimes what had been proprietary information from industry. The engineering studies were contracted out to Santa Fe International's offices in Houston.

Although I was nominally President, JOI's day--to-day operations were carried out by John 'Jack' Clotworthy. Jack had been Bob White's Director of Congressional Affairs at NOAA but had moved over to JOI with him. Jack stayed on with JOI after Bob left. Jack had plenty of experience in dealing with the Congress and science contracting, areas in which I had none. We made a good team, and I was able to be on the road most of the time cultivating our industry participants and performing the important function of liaison with our JOIDES partners. I would attend both the Planning and Executive Committees to keep them informed. It was clear to me that the overwhelming majority of the academic science community had little interest in exploring the petroleum potential of our continental margins. And some of them already knew something industry didn't seem to know: the Atlantic margin did not have the source rocks that would generate petroleum. The organic carbon buried in sediments of the Atlantic margin would generate gas but not oil. And the California margin was already producing. The Oregon and Washington margins did not have enough sediment to harbor significant resources. The international partners were hanging on in the hope that the industrial initiative would collapse and we could continue to learn about our planet's history. Many of my friends thought I was a traitor, but they were unaware of the threat that there would be no future program at all if we did not explore the OSTP's initiative.

Then came another unexpected surprise. We were offered a ship that, if refitted, could carry out riser drilling, the *GLOMAR Explorer*. The story of the *GLOMAR Explorer* is the stuff fantasy films are made of. To make a long, complex story short, here is what happened. In April 1968, a Soviet submarine, the K-129, was lost in 4.8 km (3 miles) of water north of the Hawaiian Islands. Aboard it was the code book for encoding messages between the submarine and the Soviet Admiralty and several nuclear missiles. Our Central Intelligence Agency wanted that device, and devised a subterfuge to get it. The eccentric millionaire Howard Hughes owned Global Marine Development, Inc., which built offshore drill rigs and drill ships. The CIA secretly paid Hughes about $350 million for the construction of the gigantic *GLOMAR Explorer*. It was supposedly going to mine manganese nodules from the sea floor. Manganese nodules grow very slowly on the sea floor where the sediment accumulation rates are low. The real goal of the *GLOMAR Explorer* was to recover the submarine and its contents. A salvage operation of this magnitude had never been carried out before. Now the story gets more interesting. No one in the metal business could understand why Hughes would go to such an expense to mine manganese nodules from the sea floor. There is plenty of manganese available cheaply from sources on land. The rumor was that manganese nodules contained other metals that were far more valuable but had been overlooked in chemical analyses. Soviet ocean scientists figured that if Howard Hughes was building such a huge ship to harvest them, there must be something about them they didn't know. They began a program of systematically photographing the floor of the Pacific Ocean to determine where the highest concentrations of the nodules were. The data archive they produced is a modern treasure. American scientists realized that if both Howard Hughes and the Soviets knew something about manganese nodules we didn't, we had better find out what it was. Through the NSF the US launched a major multi-institutional program of investigating manganese nodules. Nothing about the *GLOMAR Explorer* really made sense. But the ship was a marvel of engineering. Like other drill ships it had a derrick, but one much sturdier and taller than any other ship. Furthermore, the derrick rested on four huge ball bearings so it would stay vertical as the ship rolled and pitched in the sea. Instead of a giant vacuum cleaner-like tube to the sea floor, the *Explorer* had a drill string made of what were essentially cannon barrels screwed together. The heart of the operation was a giant grapple the size of the submarine, designed to enclose it and allow it to be brought to the surface. From the *Explorer*'s location and the nature of the operations, the Soviets soon figured out what it was after and changed their encoding system. It is known that the grapple failed at a critical time, and most of the submarine was lost, but we do not know what was actually recovered (although, presumably, the CIA does). We visited the Glomar Explorer at its anchorage in Suisun Bay east of San Francisco. It is very impressive, even though its derrick has been removed. SEDCO prepared estimates on the cost of conversion for continental margin drilling. They were about the same as constructing a new ship, but its operating costs would have been much greater. We had looked the gift horse in the mouth and didn't like what we saw.

In November 1980 Ronald Reagan was elected President. One of his first actions was to cancel the energy conservation initiatives Jimmy Carter had started. The solar panels on the roof of the White House survived until 1986, but the tax breaks for wind turbines and other renewable energy resources were rescinded. He encouraged Americans to indulge themselves by buying big low-mileage cars that would support our petroleum industry, although the oil they were producing and selling was mostly from foreign sources.

Reagan appointed Michel T. Halbouty to be interim Secretary of Energy. Mike Halbouty was a legend in the petroleum industry. He was a not-so-small independent and a strong believer that the competition of the free-enterprise system was what had made the US petroleum industry great. The son of Lebanese immigrants had received his advanced education at Texas A & M, and Mike was Texan through and through. I already knew him casually through the American Association of Petroleum Geologists. I went the see him in his new office in the Energy Department and made my presentation about the proposed academic-industry Ocean Margin Drilling Program. He listened intently and it seemed he might wish to recommend it to President Reagan. He would later change his mind.

A 15TH CENTURY VIEW OF THE NATURE OF THE WINDS

> *A bird maintains itself in the air by imperceptible balancing, when near to the mountains or lofty ocean crags; it does this by means of the curves of the winds which as they strike against these projections, being forced to preserve their first impetus bend their straight course towards the sky with divers revolutions, at the beginning of which the birds come to a stop with their wings open, receiving underneath themselves the continual buffetings of the reflex courses of the winds.*
>
> Leonardo da Vinci

## 24.1 Why the Earth's Atmosphere and Oceans Circulate

Whether you realized it or not most of the discussion up to now has treated the Earth as though it were a uniform body. Only in the discussions of insolation and albedo did we recognize that different areas receive and emit different amounts of energy. Earth is unlike Mercury, Venus, the Moon and Mars in having both an atmosphere and an ocean. Earth's overall energy fluxes are shown schematically in Fig. 24.1. Earth's atmosphere and ocean are fluids that redistribute the energy introduced by the Sun's insolation so that it can be more effectively reradiated into space.

These two fluid systems have very different characteristics: the density of air at sea level is about 1/1000 that of water. The specific heat of air is $1.0 \times 10^3$ J/kg/°C, less than 1/4 that of water ($4.2 \times 10^3$ J/kg/°C). The mass of the atmosphere ($5.1 \times 10^{18}$ kg), is 1/275th that of the ocean ($1.4 \times 10^{21}$ kg). The total heat capacity of the atmosphere is equivalent to a layer of ocean water only 3.2 m in thickness. Because of these differences, the atmosphere responds to external forcing (e.g., seasonal insolation) in a matter of weeks while the ocean responds only on a much longer time scale of years to decades. Most of the work in redistributing energy over the short-term (days to weeks) is done by the atmosphere. As Leonardo da Vinci noted, birds live in and contend with this very mobile environment.

As we found out in the chapter on insolation, the average amount of energy received by the Earth at the top of the atmosphere is 340 W/m². With an albedo of 0.3, the energy penetrating into the atmosphere and to the surface of the planet is 238 W/m². With a total surface area of $510 \times 10^6$ km² ($=510 \times 10^{12}$ m²), the total energy received by Earth is $121 \times 10^{15}$ W. In much of the literature, $10^{15}$ W is termed a

Petawatt (PW), so this is 121 PW. Most of this energy is reradiated from near the site of input, but the atmosphere and ocean carry some of it poleward to the higher latitudes where the input is less. The total poleward energy transport reaches about 5.5 PW at about 35° N and S. Because of the present uneven distribution of land and sea in the northern and southern hemispheres, the partition between atmospheric and oceanic transport is asymmetric and actually rather complicated, as shown in Fig. 24.2. The ocean component reaches a maximum of about 2 PW in the low latitudes of the Northern Hemisphere, and 1 PW in the low latitudes of the Southern Hemisphere. The total atmospheric transport reaches almost 4 PW in the mid-latitudes.

If you examine Fig. 24.2 carefully you will notice that the total poleward energy transport goes to 0, reversing direction not at the Equator, as you might expect, but a few degrees north of the Equator. This is because Earth's heat equator does not coincide with the geographical equator but is displaced into the northern hemisphere. The reason for this is that the South Polar Region is colder than the North Polar Region. To make the hemispheric temperature gradients more equal, the heat equator gets pushed north of the Equator.

To explain the confusion you see in Fig. 24.2 I need to jump ahead and give you a preview of how the oceans and atmosphere do their jobs. The oceans transport energy in three ways: as sensible heat (heat you can measure with a thermometer), kinetic energy (the energy of the moving water), and, in the Polar Regions, the latent heat involved in freezing and thawing of sea ice. The sensible heat carried by ocean currents completely dominates the other two which are so small they cannot be shown on the figure.

The atmosphere has four means of transporting energy. Two of them, sensible heat and potential energy (the energy gained by working against gravity to raise a parcel of air to a higher altitude) are intimately related, as was discussed in

© Springer International Publishing Switzerland 2016
W.W. Hay, *Experimenting on a Small Planet*, DOI 10.1007/978-3-319-27404-1_24

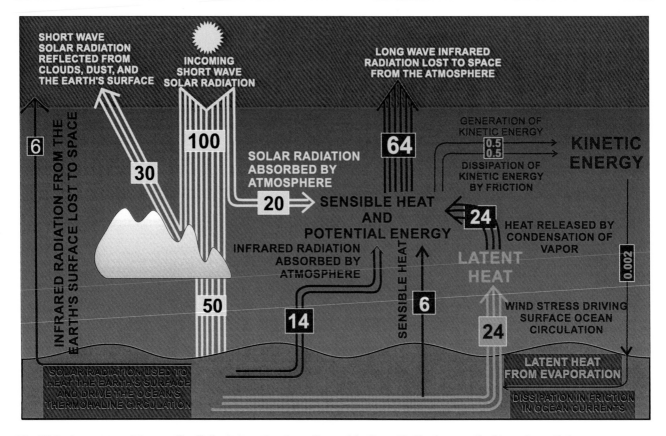

**Fig. 24.1** A summary of the overall radiation balance for planet discussed in Chap. 11. *Numbers* are % of incoming radiation

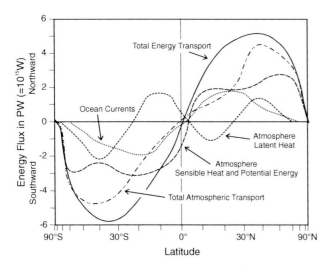

**Fig. 24.2** Earth's energy transport by the atmospheric mechanisms and the ocean's surface currents. The scale for latitude is a cosine function, making it proportional to area

Chap. 18. You will notice that the reversal of direction of the transport of sensible/potential energy is about 5° N of the Equator. This is the mean position of the Intertropical Convergence Zone (ITCZ) where the Trade Winds of the northern and southern hemispheres meet, and the warm air rises. A very large part of the overall atmospheric energy

transport is as the latent heat inherent in the water vapor-water phase transformation (gas to liquid cloud droplets and rain). But you will notice that at the lower latitudes the latent heat goes in the wrong direction; its energy is being transported toward the heat equator rather than poleward. This is because, as you will remember, the tropical trade winds blow toward the ITCZ from the southeast in the south and from the northeast in the north. At the ITCZ the vapor that had been picked up by these warm winds as they blew over the ocean is released as heat as the air rises. This is an important positive feedback stabilizing the ITCZ.

Before going any further, there are two things you need to know: (1) Winds are named by the direction from which they come; ocean currents are named by the direction in which they go. Thus a wind from west to east is a 'westerly' while an ocean current moving in the same direction is called 'easterly.' (2) The rotation of air or water, whether it be in a small eddy or a giant gyre is said to be either cyclonic or anticyclonic. Cyclonic rotation, as you might guess, implies that the rotation is around a low pressure area. Anticyclonic rotation is around a high pressure area. Because of the rotation of the Earth, cyclonic rotation is counterclockwise in the Northern Hemisphere, and clockwise in the southern hemisphere. Anticyclonic rotation is just the opposite, clockwise in the Northern Hemisphere and counterclockwise in the southern hemisphere. But there is more to it than just rotation. In the center of a cyclonic eddy or gyre the air is moving upward. A tornado is a very small, tight cyclonic eddy, a tiny low pressure system. The upward motion of the air in the middle can pick up objects and carry them aloft. On the other hand, air in the center of an anti-cyclonic eddy or high pressure system is descending, and hence will be dry. There will be no clouds to be seen.

Now let's see how the global atmospheric circulation system works.

## 24.2   The Geographically Uneven Radiation Balance

The incoming solar radiation is constantly changing with the seasons, as shown in Fig. 24.3.

Adding to the complexity, Earth's albedo is unevenly distributed, as shown in Fig. 24.4.

The outgoing longwave radiation is also very unevenly distributed, concentrated over the desert regions, as shown in Fig. 24.5.

This uneven distribution of the energy balance is constantly changing, and the atmosphere is constantly moving

heat in both its sensible and potential forms to try to maintain a constant overall radiation balance.

## 24.3   The 'Modern' Atmospheric Circulation

I use the term 'modern' here to distinguish the overall pattern of circulation during the present Interglacial, the Holocene, from the very different circulation that characterized the Last Glacial Maximum and Termination I, the period of deglaciation. As will be discussed later there is good reason to believe that the Holocene, the last 11,000 years, has been the longest period of climatic stability in at least the last half million years. Since the end of the 20th century that stability has begun to fade, so that today's 21st century circulation might be called 'post-modern.' The general pattern is shown in Fig. 24.6.

Until very recently, both of Earth's poles were covered by ice. There is still a large amount of ice cover on the Arctic Ocean, although it is melting more and more each year. Ice-covered poles are illuminated only half the year and have a high albedo so that they reflect much of the sunlight during their summer seasons. As a result they remain cold throughout the year. Cold air is denser than warm air, so this

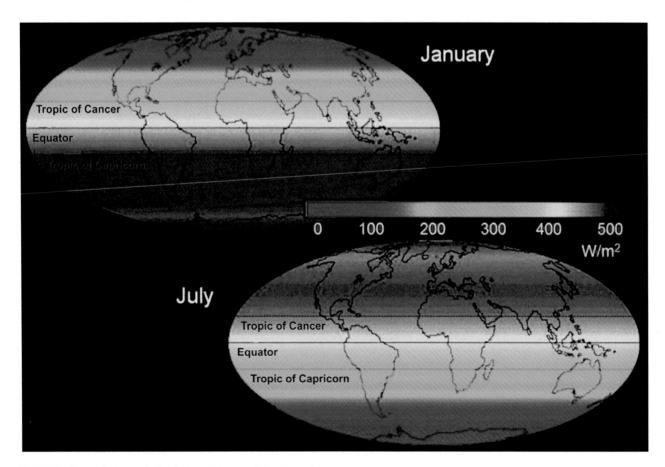

**Fig. 24.3**  Seasonal change in insolation at the top of the atmosphere

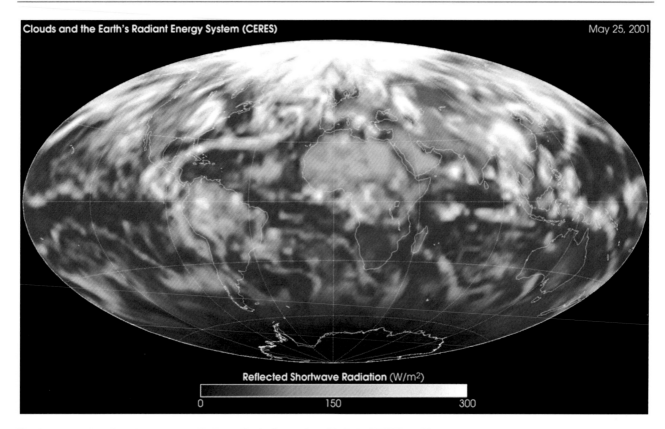

**Fig. 24.4**  Earth's reflected shortwave radiation—albedo. Image from NASA's CERES satellite program

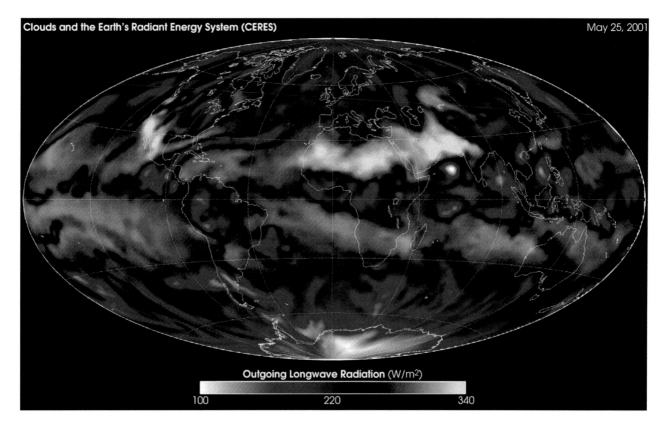

**Fig. 24.5**  Earth's outgoing longwave radiation. Image from NASA's CERES satellite program

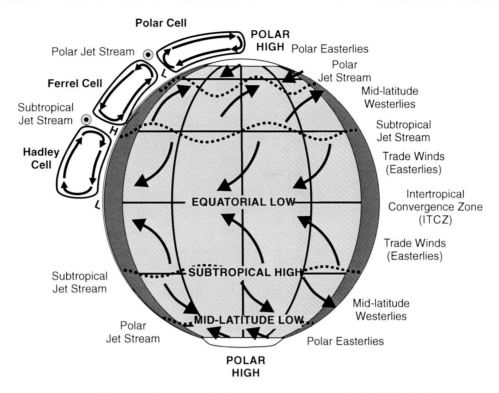

**Fig. 24.6** The general circulation of the atmosphere on an idealized Earth with polar ice caps, a pole-to pole ocean (*light gray*) between two pole-to-pole continents (*dark gray*). Regions of atmospheric low and high pressure, and the high-altitude jet streams along the cell boundaries, discussed below, are also shown

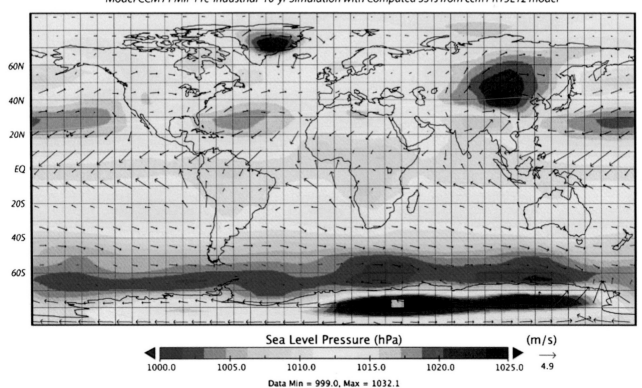

**Fig. 24.7** Mean atmospheric pressure and winds from the NCAR's CCM1 Global Circulation Model with computed sea-surface temperatures. The air pressure at the surface is shown in the *colors—blue* represents low pressure; *red* is high pressure. Also shown are the approximate average wind directions and strengths that would result from this map of pressure difference. The winds move from high pressure areas to low pressure areas, but they are bent by the Coriolis Effect. *Credit* David Bice

**Fig. 24.8** Variations in the
northern hemisphere jet stream in
June (**a**) and July (**b**) 1988,
Images from NASA's Scientific
Visualization Studio (Trent
Schindler)

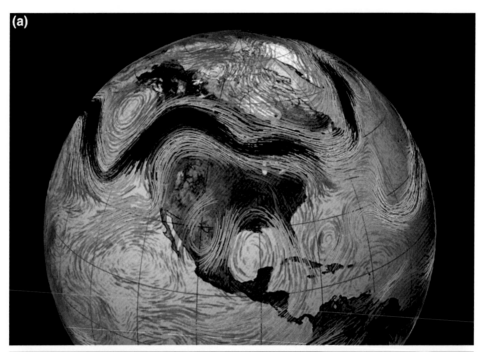

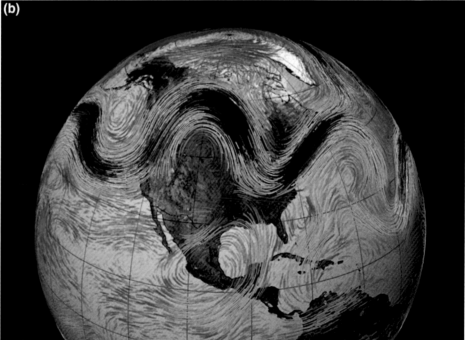

dense air has, until very recently, made permanent atmospheric high pressure systems at the poles throughout the year. This acted to stabilize the global pattern of atmospheric circulation. In contrast, the equatorial region experiences high year-round solar insolation resulting in permanent atmospheric low pressure there.

From the last chapter you will remember George Hadley, the London lawyer and hobby meteorologist who, in the 18th century, explained the Trade Winds as resulting from air descending in the tropics and then moving Equatorward over the increasingly rapidly eastward moving surface of the Earth. This tropical-to-equatorial circulation system has become known as the atmosphere's Hadley Cells. A century later American meteorologist William Ferrell realized that the movements of the air would produce spinning eddies in the mid-latitudes. This intermediate circulation, with westerly winds is now called the Ferrell Cells. These spinning eddies are the 'highs' and 'lows' that March from west to east across North America and produce our weather. They are transient features, but they obscure the overall pattern of

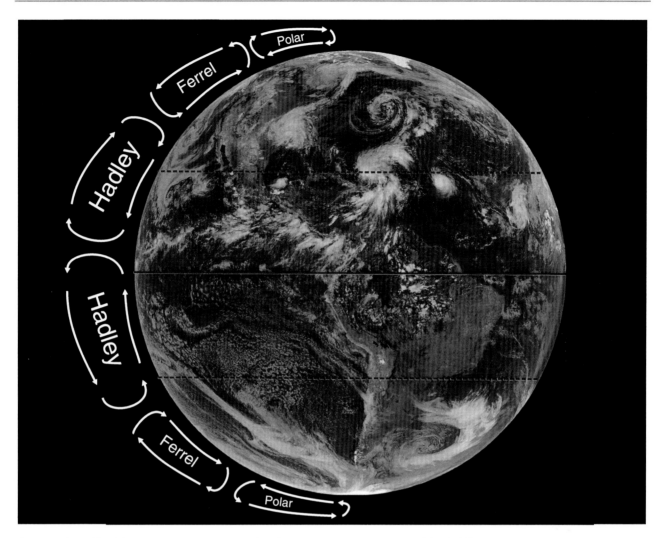

**Fig. 24.9** One of NASA's Blue Mable images showing the western hemisphere. The Equator and Tropics of Cancer and Capricorn are marked in *red*. The major atmospheric cells noted in Fig. 24.6 are indicated on the *left*, but their locations vary with longitude and, except for the arid zones marked by deserts, it is difficult to trace them across the globe

circulation. However, the ocean, with its slow response time, sees the general pattern of atmospheric circulation and the ocean waters move in accordance with it. The high latitude circulation is referred to simply as the Polar Cells. Figure 24.6 is a schematic representation of the Earth's long term atmospheric circulation.

Going from an idealized Earth-like planet to our planet as it exists today makes things more complicated, but the general pattern remains. A synthesis of the 'average modern' distribution of atmospheric pressure and winds on the Earth is shown in Fig. 24.7, compiled by researchers at The Pennsylvania State University using data from NCAR and NOAA.

The boundaries between the Polar and Ferrel cells are marked by narrow bands of rapidly flowing air in the upper troposphere, the Polar and Subtropical 'jet streams.' These exist because the adjacent air masses have very different temperatures and pressures, creating a sharp pressure gradient. Air flowing down the gradient from the higher to the lower pressure accelerates and is turned to the right in the northern hemisphere, to the left in the southern hemisphere by the Coriolis Effect. Wind speeds in the polar jet streams are typically 100 km (60 miles) per hour but can be double that. The northern polar jet stream typically has four to six north-south undulations, called Rossby waves. These waves move from west to east carrying with them the smaller high and low pressure systems that make the local weather. The north-south extent of the Rossby waves is small in summer and large in winter. The subtropical jet streams are much less intense, and may even disappear from time to time (Fig. 24.8).

In the southern hemisphere most of the area beneath the jet streams is water, so that there is less longitudinal temperature contrast; consequently they are not as strong and their north-south undulations are much less.

**Fig. 24.10**  Another NASA view of the Blue Mable showing the ITCZ as a band of towering cumulus clouds a few degrees north of the Equator

The thickness of the troposphere is very strongly affected by Earth's rotation. It can be up to 18 km (60,000 ft) thick at the Intertropical Convergence Zone, and as little as 7 km (23,000 ft) thick at the poles.

Figure 24.9 is one of NASA Blue Marble images showing Earth's western hemisphere from above the Equator with the major atmospheric cells indicated. Note that the Intertropical Convergence Zone making the boundary between the Hadley cells is north of the geographical Equator. Its moving location follows the seasonal march of Earth's thermal equator.

The Hadley cells contain about half of the mass of the atmosphere. The air moves toward the low pressure caused by warming at the Equator where it rises to the upper troposphere. From there it moves poleward and descends around 30° N and S latitude. The descending air is dry and heated by the compression it experiences during its descent. It defines the location of the Earth's major arid zones, from about 25°–35° N and S, including the Arabian, Saharan and Sonoran deserts. At the surface the air moves Equatorward as the Easterly Trade Winds. As it moves across the ocean

**Fig. 24.11** The northern hemisphere with a cloudless frozen Arctic. Image from NASA/Goddard Space Flight Center's Scientific Visualization Studio

surface, it absorbs large quantities of water vapor and its inherent latent heat. The surface circulation of the Hadley cells carries potential energy (as latent heat) toward, rather than away, from the Equator, concentrating the energy in the equatorial region. The Trade Winds north and south of the Equator meet along the Intertropical Convergence Zone (ITCZ). There the air rises as the ascending limbs of the Hadley cells. The air is cooled by expansion as it rises, and the water vapor condenses forming a band of towering

cumulus clouds (Fig. 24.10). This releases huge amounts of latent heat that warm the air and reinforce its tendency to rise. It is as though there is an internal furnace driving the air upward. This internal source of energy stabilizes the Hadley cells. The energy in the air is transported poleward in the form of potential energy. This potential energy is the energy used to counter the force of gravity. If you pick up a book off the table and hold it above the floor, you used energy to lift the book. As long as you hold onto it, it has potential energy.

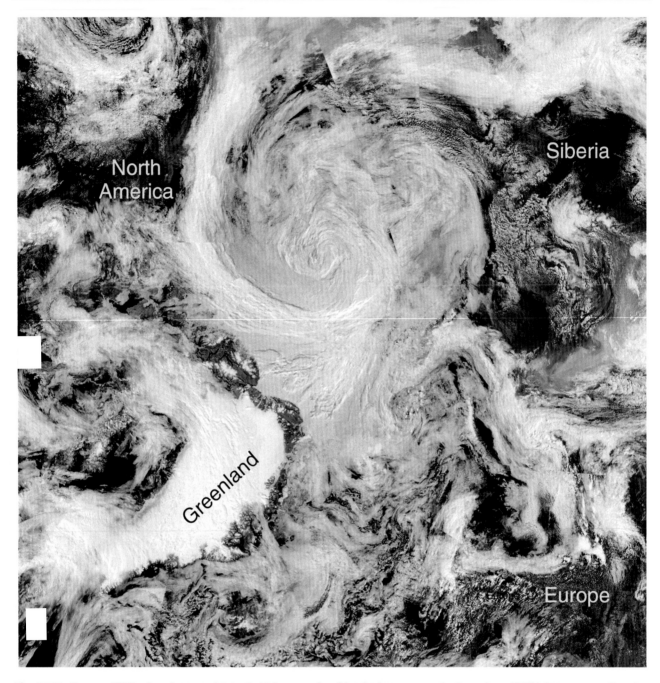

**Fig. 24.12** Summer 2012, when the perennial Arctic high was replaced by the low pressure Arctic cyclone. NASA image composites circa August 5, 2012

That potential energy can be converted to kinetic energy by letting go of the book so it falls to the floor. The air aloft has potential energy. At the descending limbs of the Hadley cells that potential energy is released in the form of the heat that warms the air.

Because the Hadley cells are inefficient at transporting energy poleward most of the low- latitude poleward energy transport is carried by the ocean. Models of the atmospheric circulation show that the Hadley cells are a persistent feature of

the atmospheric circulation whether the Earth has ice-covered poles or is ice free. However they do tend to expand poleward as the climate warms. This is happening today and the northern hemisphere is most strongly affected because of the instability induced by the loss of Arctic sea ice. We are seeing a northward expansion of the arid zones of northern Mexico and the southwestern U.S. and much of the Middle East.

The Polar cells contain less than 14 % of the atmosphere. The classical Holocene pattern has been that air sinks at the pole and flows equatorward, rising again at about 60° N and S latitude. At the surface this air flow is known as the Polar Easterly Winds. The uneven distribution of land and sea in the Northern Hemisphere makes these winds variable, but in the southern hemisphere they are more regular. Because the air of the Polar cells is cold, it cannot take up significant amounts of water vapor and hence latent heat transport within the Polar cells has been insignificant (Fig. 24.11).

However, what has been a stable pattern for most of the past hundreds of thousands of years is becoming unglued. In August 2012 the perennial high pressure system over the Arctic reversed to become a low, seen as the huge cyclonic storm shown in Fig. 24.12.

Between the Hadley and Polar cells are the third units, the Ferrel cells. These contain about 36 % of the mass of the atmosphere. The Ferrell cells are often said to be a passive response to the circulation of the Hadley and Polar cells. They can be thought of as driven by the two adjacent cells, but their northern and southern boundaries are best characterized as waves that move from west to east, circling the planet and carrying eddies of high and low pressure from west to east with them, as shown schematically in Fig. 24.13. The surface air flow is poleward, generating the mid-latitude Westerly Winds which are most stable at 45° N and S. The air is warm enough so that latent heat transport by water vapor is important. Both the sensible heat of the air and the latent heat of the water vapor are carried poleward by these cells. Hence it is in the latitudes between 30° and 60° N and S that the atmosphere is most important in poleward energy transport.

In the Northern Hemisphere the mid-latitude circulation is perturbed by the high elevations of the Tibetan Plateau and the Cordillera of western North America. The Tibetan plateau at an elevation of 5 km extends half way through the atmosphere; it is like a gigantic island around which the air must flow. Similarly, the Cordillera and Andes, with many elevations between 3 and 4 km also act as blocks; they force much of the air to move around them. In the Northern Hemisphere the spacing of the uplifts perturbs the air flow in such a way as to set up four or five planetary waves which move around the Earth. A 'planetary wave' is simply a wave so large as to be on the general scale of the Earth itself. These Northern Hemisphere planetary waves carry eddies, the high and low pressure systems of the television meteorologist, with them.

In contrast, the spacing of the continental and oceanic areas in these latitudes in the Southern Hemisphere results in a system of quasi-stable lows over the continents and highs over the oceans producing a very different sort of circulation. It was first described by Sir Gilbert Walker (1868–1958), a director of the British Meteorological Service in India in the early 20th century. What has become known as the Walker

**Fig. 24.13** Another representation of the atmospheric circulation. The *broad wavy line* in the Northern Hemisphere represents the boundary between polar and temperate air masses. The waves move from *left* to *right*, circling the globe in a week or two. The eddies that are high and low pressure systems are at the boundary between the polar and temperate air, and move along with them, as the waves circle the globe. In the Southern hemisphere the highs and lows are stationary, with highs over the cooler water and lows over the warmer land. This is the Southern hemisphere's "Walker Circulation"

Circulation' describes the air flow between stable high pressure systems located over the Indian, South Pacific, and South Atlantic Oceans, and low pressure systems located over southern Africa, Australia, and South America. Walker was trying to understand why the Indian monsoon (the 'Southwest Monsoon' discussed above) should fail from time to time, resulting in failing crops and sometimes famine. He didn't have enough information to solve the problem at the time, but his work provided a basis for understanding the 'El Niño' phenomenon.

If you refer back to Fig. 24.7, you will see that the greatest atmospheric pressure contrasts on Earth occur around the Antarctic, where the extremely cold air above the continent meets the warmer air over the Southern Ocean. The highest wind speeds recorded in Antarctica were at Dumont d'Urville station in July 1972: 91 m/s (327 km/h, 199 mph). A circumglobal wind system, the "Roaring Forties" develops over the Southern Ocean; average wind speeds are 8–12 m/s (28.8–43.2 km/h, 18–17 mph) (Fig. 24.14).

**Fig. 24.14**  A view of the southern Polar Region, showing Antarctica surrounded by sea-ice and the waters of the circumglobal Southern Ocean

## 24.4 The Indian and Southeast Asian Monsoons

The word 'monsoon' derives from the Portuguese *monção* which is in turn derived from terms used in the lands surrounding the Arabian Sea to describe the seasonally reversing winds of that region. These winds were critically important for trade using sailing vessels. During the summer the winds blow from west to east, so sailing vessels could easily travel from Arabia and western Africa to India. In the winter they reverse to east to west making for an easy return voyage.

The Indian Summer Monsoon, also known as the Southwest Monsoon is caused by air flowing from the atmospheric high pressure systems south of the Equator toward the atmospheric low that develops over the Tibetan Plateau in the summer, as shown in Fig. 24.15. The Southeast Asian Monsoon is (today) less strong but draws air from the seas off southeast Asia over the continent. There is a similar effect in North America, when summer lows over the

**Fig. 24.15** The Indian and Southeast Asian Summer Monsoons

High Plains and southern Rockies draw in air from the Pacific and Gulf of Mexico.

In the Indian Summer Monsoon the air from the high pressure system south of the Equator is first directed to the left by the Coriolis Effect, but as it approaches the Equator the Coriolis Effect vanishes and on emerging in the northern hemisphere the Coriolis Effect reappears and turns it to the right. The curve of the path is enhanced by the highlands of the East African Rift system. The winds pick up moisture from the ocean and then, upon arriving over India and meeting the Himalaya, the moisture falls out as torrential rains. In the winter the wind system reverses, and rainfall over India is rare.

We know from the deep sea record in the Arabian Sea that there have been large changes in the intensity of the Indian Monsoon during the last glacial and the deglaciation. However, their causes remain elusive. Because greenhouse warming is expected to be stronger over land this would increase the pressure differences shown in Fig. 24.15 and strengthen the monsoon. However, pollution of the atmosphere over India as industrialization proceeds is dramatically increasing and changing the regional albedo. This will tend to weaken the albedo, but some Indian climatologists predict that there will be no change at all. Other climatologists suspect that the Indian Monsoon has a chaotic behavior and may be completely unpredictable.

## 24.5   Cyclonic Storms

Cyclones are storms that develop from atmospheric low pressure systems. Responding to Earth's rotation, air at in the lower part of the atmosphere spirals in toward the center of low pressure but because of the Coriolis Effect it reaches a point where it simply flows in a circle along an isobar. It forms a gyre rotating counterclockwise in the Northern Hemisphere, and clockwise in the Southern Hemisphere. Remaining intact, it can pick up heat from the ocean and intensify. Intense cyclonic storms form both in the tropics and at high latitudes.

In major storms the winds flowing inward along the pressure gradient reach equilibrium with the Coriolis Force circling on the isobar at the edge of the cyclone's 'eye.' The eye has almost no wind and an open sky. It gives brief relief from the storm as it passes. But the strongest winds are at the inner edge of the 'eyewall' and return suddenly and with great force as the eye moves on. Figure 24.16 shows a beautiful example of these features in Typhoon Dolphin in May 2015.

Cyclonic storm systems are best known from the tropics where, in the Atlantic, they are called hurricanes, in the Pacific, typhoons, and, in the Indian Ocean, simply cyclones. Until recently their intensity was described solely by the

maximum sustained wind speeds. As discussed in Chap. 2, there are two major scales in use. One, developed in Britain as an extension of the Beaufort Wind Scale, is based on a regular increase in wind velocities. This TORRO scale uses the 3/2 power relationship of the Beaufort Scale to develop an extension with categories T0–T10. Table 2.3 shows TORRO scale.

The other was developed in the early 1970s by wind engineer Herbert Saffir and meteorologist Robert Simpson, Director of the U.S. National Hurricane Center. The SSHS = Saffir-Simpson Hurricane Scale, or SSHWS Saffir-Simpson Hurricane Wind Scale was designed to alert the public to the danger posed by a hurricane according to its potential for damage. The SSHWS uses the maximum sustained surface wind speed (defined as the peak 1-min wind at the unobstructed standard meteorological observation height of 10 m [33 ft]) associated with the storm as the basis of the scale. Because its categories are based on the perceived potential for damage the SSHWS is neither linear nor logarithmic. It is used in the United States and parts of the Caribbean (Table 24.1).

Figure 24.17 shows that there are far more tropical storms in the northern than in the southern hemisphere. This is because the Earth's thermal (heat) equator is about 10° north of the geographic Equator. Tropical storms originate in the eastern part of an ocean. In the North Atlantic, most start out off of West Africa on the eastern margin of the ocean basin. In the Pacific they tend to start in the middle of the ocean, but some originate on the eastern margin. In the Indian Ocean there are two groups, one in the Bay of Bengal starting off southeastern Asia, and another set in the Arabian Sea, starting off western India.

Tropical cyclones usually follow a more or less predictable path. In the northern hemisphere they move first to the west. They soon acquire a gradually increasing northward component, then move northward, and finally move more or less sharply eastward steered by the prevailing westerly winds. One notable exception was Hurricane Sandy, in 2012. Just as it was dropping down from being a hurricane to simply a tropical storm, Sandy suddenly turned westward to come onshore in New Jersey. That change of course happened just after the dramatic change in northern polar circulation shown in Fig. 24.12. The cyclonic system in the Arctic had forced the perennial high pressure system over cold ice-covered Greenland to the south. Its anticyclonic circulation (clockwise in the northern hemisphere) guided Sandy onto a westward course. This new change of course may happen more frequently as the summer Arctic sea-ice meltback becomes ever more intense.

It has become evident that maximum sustained wind speed is not a good representative of the real strength of a hurricane. NASA imagery has suggested that the overall size of hurricanes may be increasing. Figure 24.18 shows a typical 20th

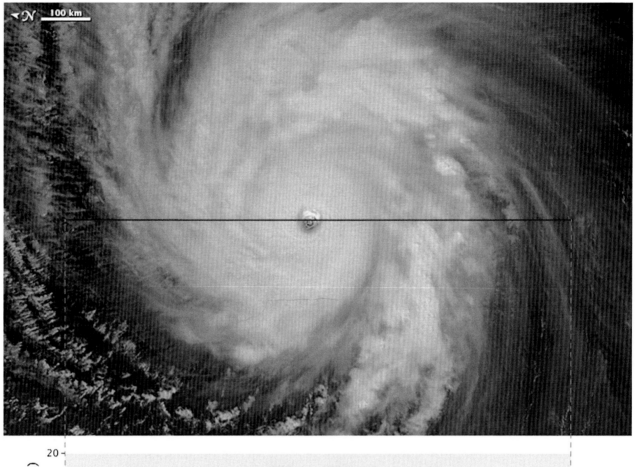

**Fig. 24.16**  NASA's portrait of Typhoon Dolphin in the western Pacific, May 16, 2015 with cross-section using reflective radar

century hurricane, Floyd as it approached Florida in 1999. I noticed while looking at satellite images of the 2005 hurricane Katrina that it seemed to occupy almost the whole of the Gulf of Mexico before coming ashore and devastating the coast from New Orleans to Alabama. I was really startled when I saw NASA's image of Hurricane Irene on August 27, 2011 (Fig. 24.19). Irene covered the eastern seaboard of the U.S., from southern Georgia to northern Maine.

Kerry Emanuel of MIT has pioneered use of a new index that would express the total energy evolved though the lifetime of the storm. His Power Dissipation Index (PDI) is defined as the sum of the maximum 1-min sustained wind speed cubed, at six-hourly intervals, for all periods when the cyclone is at least tropical storm strength. Comparison of

calculated PDI's for past hurricanes and sea surface temperatures, shown in Fig. 24.20 shows a good correlation and a sharp rise since 1990. Unfortunately, the complete PDI cannot be compiled until the hurricane has dissipated back down to tropical storm level. However, partial compilations while the hurricane is underway can give a better estimate of its potential for damage.

Polar cyclones are much less well-known, probably because so few people live along the Arctic margin. However, they are becoming more intense. Unfortunately we have no experience with an increasingly ice-free Arctic so, as Roger Revelle would say—if we keep good records we will learn a lot about what Mikhail Budyko called the 'Achilles Heel' of the climate system.

**Table 24.1**   The Saffir-Simpson hurricane wind scale (from NOAA's National Hurricane Center website)

| Category | Sustained winds | Types of damage due to hurricane winds |
|---|---|---|
| 1 | 74–95 mph<br>64–82 kt<br>119–153 km/h | **Very dangerous winds will produce some damage**: Well-constructed frame homes could have damage to roof, shingles, vinyl siding and gutters. Large branches of trees will snap and shallowly rooted trees may be toppled. Extensive damage to power lines and poles likely will result in power outages that could last a few to several days |
| 2 | 96–110 mph<br>83–95 kt<br>154–177 km/h | **Extremely dangerous winds will cause extensive damage**: Well-constructed frame homes could sustain major roof and siding damage. Many shallowly rooted trees will be snapped or uprooted and block numerous roads. Near-total power loss is expected with outages that could last from several days to weeks |
| 3 (major) | 111–129 mph<br>96–112 kt<br>178-208 km/h | **Devastating damage will occur:** Well-built framed homes may incur major damage or removal of roof decking and gable ends. Many trees will be snapped or uprooted, blocking numerous roads. Electricity and water will be unavailable for several days to weeks after the storm passes |
| 4 (major) | 130–156 mph<br>113–136 kt<br>209–251 km/h | **Catastrophic damage will occur**: Well-built framed homes can sustain severe damage with loss of most of the roof structure and/or some exterior walls. Most trees will be snapped or uprooted and power poles downed. Fallen trees and power poles will isolate residential areas. Power outages will last weeks to possibly months. Most of the area will be uninhabitable for weeks or months |
| 5 (major) | 157 mph or higher<br>137 kt or higher<br>252 km/h or higher | **Catastrophic damage will occur**: A high percentage of framed homes will be destroyed, with total roof failure and wall collapse. Fallen trees and power poles will isolate residential areas. Power outages will last for weeks to possibly months. Most of the area will be uninhabitable for weeks or months |

**Fig. 24.17** Historic tropical storm paths and intensity on the Saffir-Simpson scale

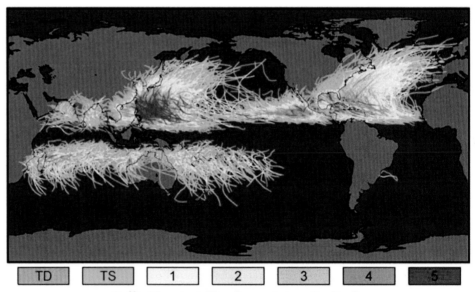

Saffir-Simpson Hurricane Intensity Scale

## 24.6   Tornados

Tornados are very intense tiny cyclones initiated and driven by the very large pressure differences that can build up under the highly turbulent conditions in major thunderstorms Wind speeds must be estimated after the fact by inspection of the damage.

In 1971 Theodore Fujita of the University of Chicago developed the Tornado Intensity Scale in use today. It is based on after-the-fact analysis of damage. Wind speeds estimates are actually guesses and have never been scientifically verified. Different wind speeds may cause similar-looking damage from place to place—even from building to building. Table 24.2 is an updated version of the Fujita Scale.

Information about tornados often depends on luck. There must be an observer who happens to be in the right place at the right time, and viewing conditions must be favorable. They

**Fig. 24.18** Hurricane Floyd approaching Florida, September 12, 1999

**Fig. 24.19** Hurricane Irene off
the eastern U.S. on August 27,
2011

can be detected by the reflections from Doppler radar of debris aloft. Tornados are often shrouded in rain and invisible. Although they are most frequent in the afternoon, many occur at night. Much can be reconstructed after the event, especially in built-up areas. Recently it has become possible, through satellite imagery to determine the path and size of the base of a tornado. However, the only good statistics concern the number reported each year, shown in Fig. 24.21.

**Fig. 24.20** Compilation of power dissipation indices for Atlantic hurricane since 1860 (*solid line*) compared with average sea-surface temperatures off the eastern U.S. (*dashed line*)

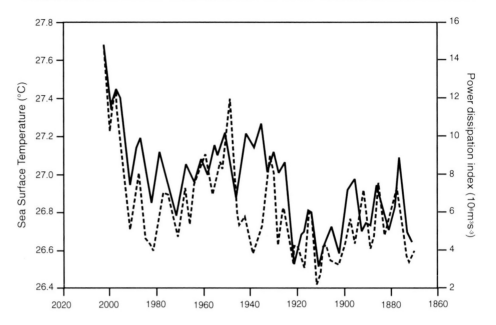

**Table 24.2** The Fujita Tornado intensity scale (from the NOAA website)

| Scale | Wind estimate (MPH) | Typical damage |
|---|---|---|
| F0 | <73 | **Light damage**. Some damage to chimneys; branches broken off trees; shallow-rooted trees pushed over; sign boards damaged |
| F1 | 73–112 | **Moderate damage**. Peels surface off roofs; mobile homes pushed off foundations or overturned; moving autos blown off roads |
| F2 | 113–157 | **Considerable damage**. Roofs torn off frame houses; mobile homes demolished; boxcars overturned; large trees snapped or uprooted; light-object missiles generated; cars lifted off ground |
| F3 | 158–206 | **Severe damage**. Roofs and some walls torn off well-constructed houses; trains overturned; most trees in forest uprooted; heavy cars lifted off the ground and thrown |
| F4 | 207–260 | **Devastating damage**. Well-constructed houses leveled; structures with weak foundations blown away some distance; cars thrown and large missiles generated |
| F5 | 261–318 | **Incredible damage**. Strong frame houses leveled off foundations and swept away; automobile-sized missiles fly through the air in excess of 100 m (109 yds); trees debarked; incredible phenomena will occur |

**Fig. 24.21** Number of tornados reported in the U.S. per year since 1970. The *black line* is a 5-year running average. Adapted from McCarthy, D., and Schaefer, J., 2004. Tornado trends over the past 30 years. Preprints, 14th Conf. Applied Climatology, Seattle, WA, Amer. Meteor. Soc., CD-ROM

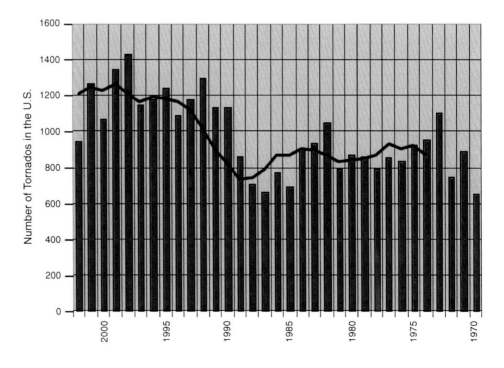

The United States is home to most of the world's tornadoes. Although tornadoes occur in other parts of the world, they are a relatively are weather phenomenon. Europe currently averages about 300 per year, but they are generally weak with short paths.

## 24.7 Summary

The circulation of the atmosphere and ocean try to adjust the regional and local inequalities of Earth's energy balance. The atmosphere responds much more rapidly than the ocean and is responsible for the day-to-day weather. It transports energy in the forms of latent heat of water vapor, kinetic energy of the winds, and potential energy. There are three broad bands of circulation in each hemisphere. The tropical Hadley Cells extending from the subtropical highs at about 30° N and S to Earth's thermal equator located about 10° north of the geographical Equator. The Ferrel Cells cover the mid-latitudes from the subtropical highs at about 30° N and S to the subpolar lows at about 60° N and S, They are the belt of westerly winds with imbedded high and low pressure systems. The northern hemisphere Ferrel cell circulation is perturbed by the Tibetan Plateau and the mountains of western North America, so that the circulation is in the form of a series of four to six N-S waves (Rossby Waves) that move eastward around the planet. The high-latitude Polar Cells, from about 60° N and S to the poles have long been stabilized by high-albedo polar ice which has forced perennial high pressure and anticyclonic circulation. The meltback of Arctic sea ice is causing instability of the northern polar high, and the Arctic became a site of cyclonic circulation in 2012.

Cyclones are intense low pressure systems that develop over the oceans. They can develop at both low and high latitudes. Tropical cyclones are known as hurricanes in the Atlantic, typhoons in the Pacific and simply as cyclones in the Indian Ocean region. Most occur in the northern hemisphere, usually forming just north of the heat equator, Their path is initially westward, gradually turning northward, and finally back eastward. Hurricane Sandy, in 2012, was an exception turning westward from its northerly course in response to the southward displacement of the Greenland high by the late summer reversal of the Arctic circulation that year.

Tornadoes are very intense, ephemeral mini-cyclones initiated and driven by the very large pressure differences that can build up under the highly turbulent conditions in major thunderstorms.

A Timeline for this chapter:

| 1735 | George Hadley publishes his theory of the cause of the Trade Winds |
| 1835 | Gaspard Gustave de Coriolis publishes *Sur les équations du mouvement relatif des systèmes de corps*, showing that Newton's laws of motion can be used in a rotating frame of reference if an extra force (now called the 'Coriolis acceleration' or 'Coriolis Force') is added to them |
| 1856 | William Ferrel publishes his first paper on the circulation of the atmosphere, outlining the three cell per hemisphere pattern a polar cell, an equatorial cell, and an intermediate cell |
| 1924 | Sir Gilbert Walker describes the peculiar atmospheric circulation of the southern hemisphere |
| 1971 | Theodore Fujita proposes an intensity scale for tornados |
| 1972 | NASA's Apollo 12 Mission makes beautiful photographs of Earth from space |
| 1973 | The Saffir-Simpson hurricane intensity scale is introduced to the public |
| 1975 | The TORRO cyclone scale is introduced in Britain |
| 1997 | NASA initiates CERES: Clouds in Earth's Radiant Energy System |
| 2002 | NASA initiates its 'Blue Marble' Earth imagery program |

If you want to know more, again I recommend:
Barry, R.G. & Chorley, R.J., 2009. *Atmosphere, Weather and Climate*, 9th ed., Routledge, London, 536 pp.—A classic general text; lots of great pictures and figures; no complicated mathematics.

Peixoto, J.P. & Oort, A.H., 1992. *Physics of Climate*. American Institute of Physics, New York, 520 pp.—My personal standard reference for understanding modern climate. Lots of math but it can be skimmed over. Excellent figures.

Music: Edward MacDowell Piano Concertos: No. 1 in A minor and No. 2 in D minor. MacDowell was an American composer of the late 19th century. The second concerto is the most well known, but the first is my favorite.

Libation: To go with MacDowell, I suggest a fine Kentucky Bourbon.

### Intermezzo XXIV. Exciting Times
In 1981 we were getting the results of the engineering studies we had commissioned from Santa Fe International. I recount some of the conclusions here because they are relevant even today in the light of the recent BP *Deepwater Horizon* disaster in the Gulf of Mexico.

The first major disaster in offshore drilling occurred on June 3, 1979, in the Gulf of Mexico's Campeche

Bay, about 100 km (62 miles) northwest of Ciudad del Carmen, Mexico. Mixtec 1 was an exploratory oil well being drilled by the SEDCO's semi-submersible drilling rig 135-F in waters 50 m (160 ft) deep. The well was equipped with blow-out prevention devices, but they were not activated until the well was no longer controllable.

Drilling for petroleum at sea is very different from drilling on land. On land a drilling fluid, heavy mud and water, is pumped down the drill pipe; it exits through the drill bit which is cutting up the rock into small pieces. The drilling mud is denser than the rock fragments so they float and are carried upward by the return flow of drilling mud in the drill hole outside the pipe. At sea the circulation from the top of the hole at the sea floor to the ship is through a 'riser,' a large pipe enclosing the string of drill pipe. This allows return circulation of the fluids in the well so that the expensive drilling mud is recycled and not lost. The connection at the sea floor is a complicated affair. There must be a solid base for the riser, and this is where the well must be shut off should anything go wrong. The shut off is accomplished by squeezing the riser and drill pipe closed. That requires a lot of force and is done by hydraulic rams, usually powered by explosive charges.

What can go wrong? Gas is generated by the decomposition of organic matter in the sediment. If it is produced in large amounts its pressure can exceed the lithostatic pressure, the weight of the rock above it. It can't escape because the rock is glued together by mineral cements. If there is 'overpressured gas' in the pore space of the sediment it can escape into the drilling mud. This makes the mud lighter, and the weight of the mud column decreases so that more gas flows in, and before you know it gassy drilling mud is coming up the annulus between the drill pipe and hole wall. The more gas gets into the well the faster the process goes. This is what caused 'gushers' back in the early days of petroleum exploration.

Downhole sensors can detect the changes in weight of the drilling mud column. The sensors are monitored on board the drillship, and if they show that something is happening the blowout preventers must be fired off. That means a loss of many millions of dollars; the well is lost. The problem is that the deeper the hole beneath the vessel, the smaller the danger signal from the sensors. It is my understanding that at Ixtoc, the crew on the ship needed permission from Petroleos Mexicanos Headquarters in Mexico City to approve firing the blowout preventers. It came much too late.

Here is the problem. The person monitoring the sensors gauges on the drillship has a very boring job. But it is a huge responsibility. That person will probably want to see a very clear signal before pressing the button. After all, by pressing it they are costing the company many millions of dollars. And, if the blowout preventer works as it should, one will never really know whether a blowout was about to happen. That employee will surely be blamed for over-reacting. We had long discussions at Santa Fe in Houston about this problem. It was thought that the monitors would need to attend a special school for training. I do not know whether such a school was ever established.

Exactly this sequence of events, with the delays in decisions, appears to have happened to the *Deepwater Horizon* on April 20, 2010. Only this time there was a further complication. Already, days earlier, bits of the rubber sealing rings of the blowout preventers had been seen in the returning drilling mud. Even if the blowout preventers had been fired off in time, the sealing of the hole would not have occurred. Incidentally, the *Deepwater Horizon*, being operated for British Petroleum, belonged to Transocean, which is the amalgam of Global Marine, SEDCO, and Santa Fe International.

Another aspect of the engineering was to ensure that the operations could be carried out even if there were to be a hurricane. This would mean that the drill ship should be able to disengage from the drill string and riser, but leave them held up by floats at a safe depth below the surface. The ship would be moved to calmer waters and return later when conditions permitted. We had to plan for a 100-year storm. Our meeting to discuss this was held in late September, 1979, after a series of very heavy rainstorms in Houston. Someone remarked, "We're supposed to plan for a once in a hundred years storm. I think we've already had three of those this year in Houston."

The engineering studies indicated that the proposed Ocean Margin Drilling Program might be ahead of its time.

The final meeting of industry, NSF, and JOI to decide on the future of the OMD took place in early October, 1981 on Catalina Island, California. Transport to island was by yachts or other vessels belonging to several of the companies. On the way out to the island, my confidantes in industry told me what was going to happen. They would no longer be able to support the program. There were two reasons: (1) they felt it would infringe on their competitive interests in economic exploration of the continental margin; (2) they had discovered that the composition of the US

Congress changes every 2 years and they believed it was risky to suppose that the US Congress could guarantee it would honor a 10 year commitment of support as they were being asked to do.

Then, as we approached the island, the Director of NSF, John Slaughter, took me aside. The Reagan administration had decided that the OMD program was really a means of subsidizing the US petroleum industry and could no longer support the program.

It was a wonderfully harmonious meeting. After flowery introductions of all involved, the head of Exxon Production Research Corporation, speaking for the industry participants, said "We can't support it" and John Slaughter said "Neither can we." We had a lot to drink and a great time celebrating the demise of this ill-formulated venture. Although I could not show it, I was delighted. It meant that our international program might have a future.

The Catalina conference lasted for a couple more days, where we simply enjoyed each other's company as we listened to technical reports and discussed matters over excellent food and wines. But on the side I learned a lot, in strict confidence. In fact, industrial wells had already been drilled at most of the prospective sites off the US East Coast. They had not found any petroleum but had encountered very highly overpressured gas, levels so extreme that drilling operations had been forced to cease. In fact, the proposed OMD drilling would not have been possible. Offshore gas production was not economical because of the cost of pipelines and other infrastructure required to bring it onshore.

The OMD was finished. Now it was essential to go back to our international partners to explore a future program of ocean drilling to carry on the DSDP tradition. A revolution had been brewing in the academic science community for months. With the proponents of OMD in the OSTP gone as the new administration took over, the time for action had arrived. Since the beginning of the year JOIDES had begun to make preparations for a challenge to the proposed OMD. The Catalina meeting had avoided a confrontation.

For the US community I was the villain in the OMD affair so I was careful to be a very inconspicuous observer at the conferences to organize the new program. A new program for Scientific Ocean Drilling beyond 1983 was prepared at a conference held in Austin, Texas in mid-November, 1981.

Some members of Congress had been following the whole affair, and NSF selected me to testify before the House's Committee on Science and Technology. It was like walking a tightrope to explain how all this had worked out without offending anyone. But the science community and NSF had a very supportive Chairman, Don Fuqua of Florida. If I recall correctly my exchanges were primarily with member George Brown of California who made sure that the right ammunition for a special appropriation for NSF was in the record. A young Al Gore was a member of the Committee.

My father had testified before Congress several times, and he called to congratulate me on having survived the experience.

1769: BEN FRANKLIN PRESENTS HIS MAP OF THE GULF STREAM

# The Circulation of Earth's Oceans

*There is a river in the ocean. In the severest droughts it never fails, and in the mightiest of floods it never overflows. Its banks and its bottom are of cold water, while its current is warm. The Gulf of Mexico is its fountain, and its mouth is in the Arctic Sea. It is the Gulf Steam.*

Matthew Fontaine Maury

Except in certain passages, like the Bosporus, the Straits of Messina, with its legendary whirlpools and Strait of Bab-el Mendab connecting the Red Sea with the Indian Ocean, it is unlikely that sailors were aware of the overall motions of the ocean's waters. Sailing ships were moved by the winds; and assistance from the moving water was generally unnoticed.

In 1513 Ponce de Leon had noted the northward-flowing current between Florida and the Bahamas but it was not until two centuries later that its northern extension was documented. In the middle of the 18th century the Board of Customs in Boston, Massachusetts, noticed that packets traveling between Falmouth, Massachusetts, and New York, New York, by sea took 2 weeks longer to arrive than merchants traveling from London to Rhode Island. This was very strange because on land Falmouth and New York were less than a day apart by road.

Benjamin Franklin decided to investigate this mystery. He consulted with a Nantucket sea captain, Timothy Folger, who had been collecting data relevant to the problem. They are credited with the discovery and documentation of the Gulf Stream. Ben Franklin's famous map of 1769 is shown in Fig. 25.1.

## 25.1 The Modern Ocean's General Circulation

In the open ocean, the winds drive the water. However, instead of going directly downwind, the surface waters move 45° to the right of the wind in the Northern Hemisphere, and 45° to the left of the wind in the southern hemisphere. This was worked out by Swedish oceanographer Vagn Walfrid Ekman in the first half of the last century. It means that on the surface, the major east to west or west to east currents flow parallel to the lines of latitude. Ekman also found that the motion of each layer of water forces the next lower layer to move slightly more to the right or left, depending on which hemisphere it is in. He found that the net motion of

the water in the layer affected by the wind is 90° to the right of the wind. This general motion is known as the 'Ekman Transport' in his honor. All of these surprising motions are the effect of Earth's rotation.

The circulation of the ocean in the lower and middle latitudes is dominated by the torque between the Easterly and Westerly Winds. This produces huge clockwise gyres in the Northern Hemisphere and counterclockwise gyres in the Southern Hemisphere. An idealized version of the ocean circulation driven by the winds is shown in Fig. 25.2.

You will remember from the last chapter that as the air moves upward in cyclonic and downward in the anticyclonic atmospheric eddies we call highs and lows, the huge anticyclonic gyres in the tropical oceans not only rotate, they pump water downwards. The cyclonic circulation in the Polar Regions pumps water upwards.

The ideal gyral circulation of the surface waters is most closely approached in the Pacific. However, there are barriers on both sides of the Pacific. The Pacific's North and South Equatorial Currents encounter the continental margin of Asia in the Southeast Asian Seas. Part of the flow is diverted to the north and part to the south as Western Boundary Currents. The remainder returns along the Equator just beneath the surface as the east-to-west flowing Cromwell Countercurrent. The Countercurrent is locked onto the Equator by the Coriolis Force. If the water starts to drift off to the south it is turned to the left, back onto the Equator; if it starts to go northward, it is turned to the right, back onto the Equator. The other part of the gyral system that is nearly ideal is the west to east flowing Antarctic Circumpolar Current, sometimes called the West Wind Drift, which circles the globe.

Another effect of the rotation of the Earth is that the gyral circulation is not symmetrical. The poleward-flowing currents on the western sides of the ocean basins, such as the Gulf Stream, are narrow and deep while the return currents on the eastern side of the ocean basins, such as the Canary Current, are broad and shallow. The amount of water carried

© Springer International Publishing Switzerland 2016
W.W. Hay, *Experimenting on a Small Planet*, DOI 10.1007/978-3-319-27404-1_25

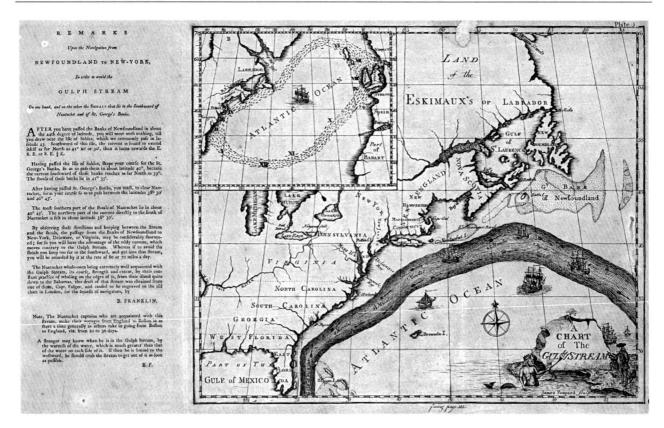

**Fig. 25.1** Benjamin Franklin's 1769 map of the Gulf Stream

**Fig. 25.2** The general
circulation of the ocean on an
idealized Earth with a pole-to
pole ocean (*light gray*) between
two pole-to-pole continents (*dark
gray*). The winds are shown in
*light gray*. The ocean currents
they drive are shown in *black*.
The *dashed lines* are fronts in the
ocean, where water masses come
together and descend

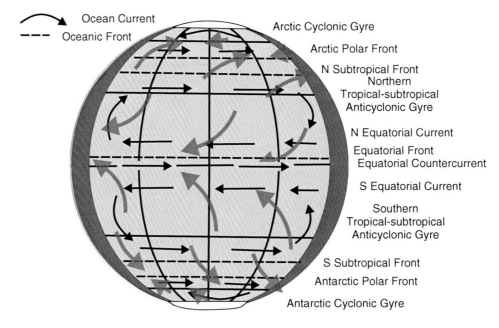

by the currents in measured in Sverdrups. One Sverdrup is
one million cubic meters of water per second. That is
approximately the flow of all rivers on Earth combined. The
Gulf Stream transport is about 30–40 Sverdrups.

The southern tropical anticyclonic gyre in the South
Atlantic is perturbed by a geographic peculiarity. The eastern

tip of South America splits the Southern Equatorial Current
into northern and southern flows introducing asymmetry into
the system. The North Atlantic has an anticyclonic gyre, but
it is surrounded by marginal seas that introduce saltier or
fresher waters modifying the flow. The Indian Ocean waters
are so perturbed by the climatic peculiarity of the seasonally

reversing winds of the Asian monsoon that the ocean currents reverse with the seasons.

The waters at higher latitudes are dominated by the cyclonic circulation induced by the Polar Easterly winds and the Mid-latitude Westerly winds. This happens poleward of 45° N in Pacific Ocean, where a cyclonic gyre forms south of the Alaskan Peninsula. Poleward of 45° S the waters of the circumglobal Southern Ocean are driven by the Westerlies as the Antarctic Circumpolar Current. The polar Easterlies coming off Antarctica force the east to west circulation of the Antarctic Coastal Current. These two ocean currents flowing in opposite directions create a band of cyclonic circulation around the Antarctic continent. These surface currents are shown in Fig. 25.3.

This is the pattern of ocean currents you will find on any map of the oceans, from National Geographic or in a World Atlas. However, this is an average of the way the ocean circulates taken over a period of months. In reality the winds are always somewhat unsteady and variable, and they generate 'mesoscale eddies' a 100 km or so across. It is a field of eddies that move as a current, as shown in Fig. 25.4. Only in narrow passageways, like the Florida Strait between Florida and the Bahamas, does the water flow as a "river in the ocean."

You may be surprised to learn that only about 10 % of the water in the oceans is involved in these surface currents. The other 90 % does something quite different.

The anticyclonic circulation of the tropical-subtropical gyres forces warm, less dense surface water downward onto the cold, more dense water underlying it, creating a sharp density gradient that forms the floor of the gyres. This boundary layer in the ocean is called the thermocline, because temperature of the water decreases sharply with depth. The thermocline forms the floor of the anticyclonic gyres. It generally lies about 200 m below the ocean surface. So although the huge tropical-subtropical gyres occupy almost 75 % of the ocean surface, they contain only 8 % of the ocean's water. The ocean most of us know from our vacations is the warm water of tropical subtropical anticyclonic gyres. Even Norway is bathed by a relatively warm water current headed for the Arctic. The other 92 % of the ocean volume is occupied by the cooler and cold waters which fill the ocean interior. They are exposed on the surface only at high latitudes, and only the residents of Greenland, northern Canada, and Arctic Siberia know them well.

The cyclonic circulation at high latitudes forces deep waters upward to the surface. This region occupies about 21 % of the ocean surface. With no wind-driven downwelling water to contend with, the rising deep waters are cooled by the atmosphere and sink again. As a result, the ocean convects from the surface to the ocean floor. The surface water temperatures are everywhere about the same, very close to the freezing point of seawater.

The oceans' low and high-latitude gyres are separated by frontal systems along which waters converge and sink to intermediate depths and flow equatorward through the ocean interior. This sinking region represents the outcrop of the ocean thermocline and intermediate waters underlying the tropical-subtropical anticyclonic gyres where they come to

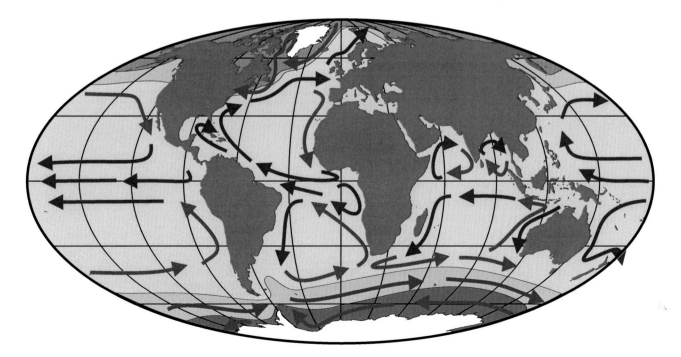

**Fig. 25.3** Idealized schema of the oceans' major gyre and frontal systems forced by the winds

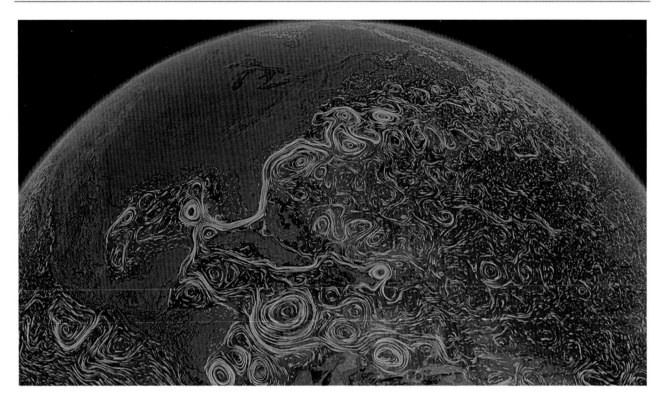

**Fig. 25.4** A snapshot of circulation in the northwestern Atlantic. This is a single frame from NASA's Perpetual Ocean animation (http://www.nasa.gov/topics/earth/features/perpetual-ocean.html#.VYbjN_m6fys)

the surface. The sinking region is bounded equatorward by the Subtropical Front at about 45° latitude and poleward by the Polar Front at about 50° latitude. Over the short distance of 500 nautical miles (926 km) the ocean temperature changes from warm to frigid. Figure 25.5 is a schematic section through the ocean from pole to pole, showing the major water masses and the deep water circulation.

These ocean frontal systems are critical to the structure of the modern ocean. They form beneath the strongest Westerly Winds. The winds drive the water equatorward with a force proportional to their velocity. The cooler polar waters sink rather than piling up over the more slowly moving waters driven by the lesser winds over the subtropical ocean. Figure 25.6 is a sketch showing how this works.

The formation of the deep interior waters of the oceans is a special process. Seawater is different from fresh water in that it reaches its maximum density below the freezing point. Fresh water has its maximum density at 4 °C. Below that temperature it becomes less dense. Temperate lakes, which experience cold temperatures in the winter, cool to 4 °C. The water in the lake convects until the entire lake has this temperature, top to bottom. Then as winter proceeds, the surface layer becomes colder. As it gets colder, it becomes lighter (less dense), making a surface layer that floats on the deeper 4 °C waters. The surface layer finally freezes when its temperature reaches 0 °C.

But as salt is added to the water, the temperature of maximum density decreases, and when the salinity is 27.4 ‰ it is at the freezing point of the water, about −1.5 °C. The ‰ symbol means parts per thousand, and salinity is the amount of salt, by weight, in a given volume of seawater. The density of the seawater depends on its temperature and salt content. It also depends on pressure, although this is a smaller effect.

At the average salinity of seawater, 34.7 ‰, the water just gets denser and denser until it reaches the freezing point, −1.8 °C for that salinity. This being true, one would expect that in order to make sea ice on the surface, the whole ocean would have to be chilled to the freezing point. Obviously this doesn't happen because both polar oceans have a sea ice cover that expands greatly in winter. Sea ice forms when snow falls on the surface and dilutes the surface water slightly. It is less dense, and can freeze. But sea-ice does not contain as much salt as the seawater from which it forms. The excess salt is expelled into the surrounding waters, making them saltier and denser. They become sufficiently dense that they overcome the tendency to rise, induced by the cyclonic circulation, and they sink to the bottom of the ocean. There, they spread out, flowing underneath the tropical-subtropical gyres, and filling the ocean interior. The deep waters of the ocean are remarkably homogenous, with a temperature of about 2 °C and a salinity of about 34.7 ‰.

**Fig. 25.5** An idealized meridional section through the ocean. The *solid lines* represent boundaries between major water masses. They also coincide with density surfaces (isopycnals) within the ocean. Densest water is in the deep ocean, but outcrops in the Polar Regions. The lightest water is in the subtropical-tropical gyres. If nothing were moving, the lines of equal density, isopycnals, separating them would be horizontal. It is the rotation of the Earth that causes them to deviate from horizontal

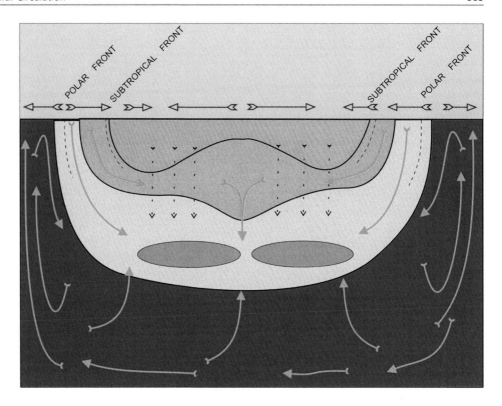

The sinking of water into the deep interior does not take place over broad areas; it happens in restricted regions where special conditions can exist. One of Henry Stommel's most interesting papers, published in 1962, is entitled *On the smallness of sinking regions in the ocean.*

Sea ice formation in the Arctic is different from that of the Antarctic because there is a special region that acts as an ice factory. The flow of the Eurasian and North American Rivers into the Arctic is almost 10 % of the world's entire river discharge into the sea. Most of it comes from the

**Fig. 25.6** The role of the westerly winds in generating and stabilizing the subtropical and polar frontal systems. This sketch shows the situation in the Northern Hemisphere, as it exists over the North Pacific Ocean. However, these frontal systems are best developed in the southern hemisphere where they encircle the Antarctic continent and are only 5° of latitude apart

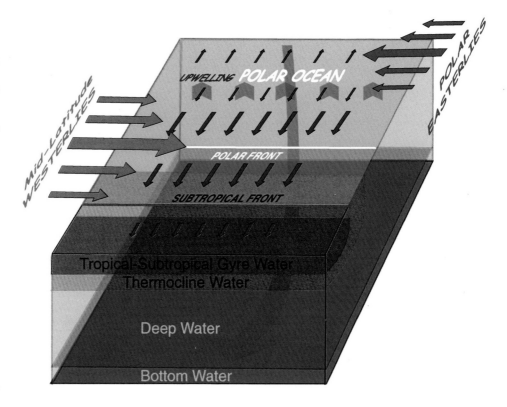

Russian, particularly the Siberian, rivers. This large inflow occurs in only a few months of the year. One of these, the Lena, empties into the broad, shallow Laptev Sea. Its waters spread over the broad shelf off northeaster Siberia creating a low-salinity surface layer. As winter comes on these low salinity waters freeze and the ice drifts out toward the center of the Arctic Ocean. Arctic sea-ice is a special product of this unique geographical situation. Over the past 30 years, the Arctic sea ice has been melting faster than it forms. The fabled 'Northwest Passage,' an ice-free route that would allow ships to go from the Atlantic to the Pacific though the Arctic first opened in September 2007. Estimates of the year when the Arctic will be ice-free in summer were 2100 a decade ago, 2080 five years ago, and 2050 three years ago. Now it is recognized that the melting is accelerating, going faster than anyone had anticipated. The Arctic could be ice-free in September 2030.

Another factor affecting the ocean is the difference between precipitation, river runoff, and evaporation over the ocean and marginal seas. This aspect of the hydrologic cycle is referred to as the freshwater balance. The freshwater balance for a sea is positive if precipitation plus runoff from land is greater than evaporation. If there is a positive freshwater balance, the outflow from the marginal sea into to ocean is at the surface and inflow occurs at depth. This condition is termed "estuarine." With a negative freshwater balance, evaporation exceeds the river runoff and precipitation. The sea receives surface inflow from the ocean and outflow occurs at depth; this condition is termed 'lagoonal.'

One of the modern Earth's geographic peculiarities is that 55 % of the water that evaporates from the world ocean and falls as rain on land flows back through rivers into the Atlantic and its tributary Arctic Ocean. The Atlantic Ocean complex with its marginal seas represents only 30 % of the total ocean surface area. You might expect that with such a disproportionate share of the world's river discharge the Atlantic would be fresher than the Pacific. But it is not. The average salinity of the Atlantic is 34.92 ‰ whereas that of the Pacific is 34.60 ‰. This is because of the odd shape of both the North and South Atlantic. Both are widest beneath the descending limb of the Hadley cells. The water vapor carried equatorward by the surface winds of the Hadley cells rises in the Intertropical Convergence Zone, but the ITCZ is not where you would expect it to be, over the Equator; it is usually between 8 and 10° N. This geographic peculiarity makes it possible for the Atlantic to be saltier than the Pacific. How does this work?

A special peculiarity of the Earth today, mentioned previously, is the hemispheric thermal contrast induced by the ice-covered continent at the South Pole, with an average elevation of 2300 m, and the ice-covered Arctic Ocean, with an average elevation of 1 m above sea level. The result is that in the middle of the troposphere, the pole to Equator

temperature gradient is greater in the Southern Hemisphere than in the Northern Hemisphere. It is this difference that displaces the Intertropical Convergence Zone from the Equator, where it should be, into the Northern Hemisphere. Its average location at 8–10° N crosses Central America in Costa Rica where the mountains are relatively low. As a result, large amounts of water are transported from the warm tropical Atlantic to the Pacific as vapor. The vapor precipitates as rain onto the Northern Equatorial Current and becomes incorporated into the North Pacific gyre, which has the lowest salinity of any open ocean area, 34.57 ‰. The Atlantic exhibits lagoonal circulation, and the Pacific is estuarine.

Figure 25.7 shows the temperature of the world ocean after the spring solstice. Note the very warm temperatures in the western Pacific and Indian Oceans. In the late 1970s it was argued that if global temperatures were to rise evaporative cooling would prevent sea surface temperatures from rising above 28 °C. Now there are large areas that have temperatures of 28–30 °C.

The weather over the ocean results in heavy precipitation along the ITCZ and the atmospheric lows at higher latitudes, while the regions under the subtropical highs experience strong evaporation. The result is regional differences in salinity of the surface ocean varying from 30 to 40 ‰ (parts per thousand), as shown in Fig. 25.8. Note the increasing progression of salinity from the North Pacific to the South Pacific to the South Atlantic to the North Atlantic and into the Mediterranean.

As discussed in Chap. 17, waters of different temperatures and salinities have different densities. Although the winds are the major drivers of ocean currents, density differences can also play an important role, especially when two bodies of water are separated by a narrow passageway, Fig. 25.9 is a temperature-salinity-density diagram showing the major water masses of the oceans.

This ocean structure and circulation system appears to have been stable during at least the past few million years, but they depend on the constancy of the wind systems, which in turn depend on the presence of ice at the poles. These features are shown schematically in Fig. 25.10. During the Quaternary alternations between glacials and interglacials the ocean structure remained the same although the areas covered by sea-ice expanded and contracted.

## 25.2    The Global Ocean's Great Conveyor

The interior waters of the ocean have a global circulation pattern very different from that of the surface waters. Beneath the surface waters, the ocean is stratified according to density, but the isopycnals, the surfaces of equal density, are not horizontal (See Fig. 25.5). They slope upward

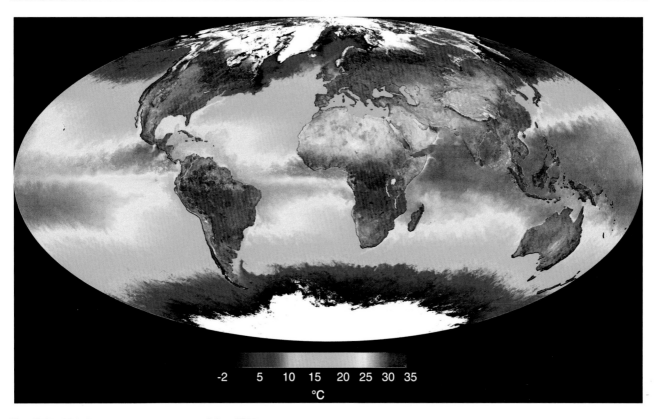

**Fig. 25.7** Global sea surface temperatures. May, 2001

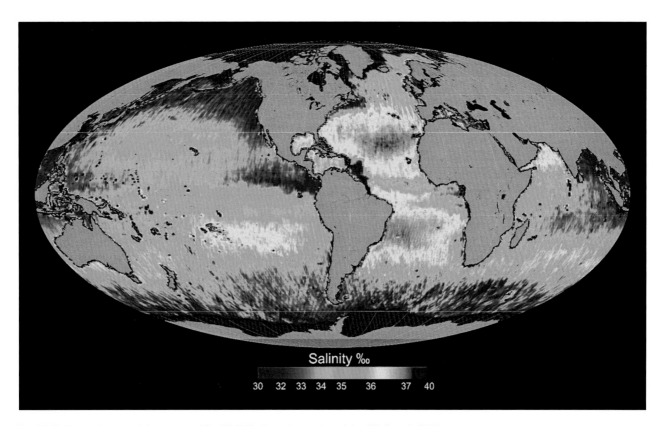

**Fig. 25.8** Sea surface salinities measured by NASA's Aquarius satellite, May 27–June 7, 2012

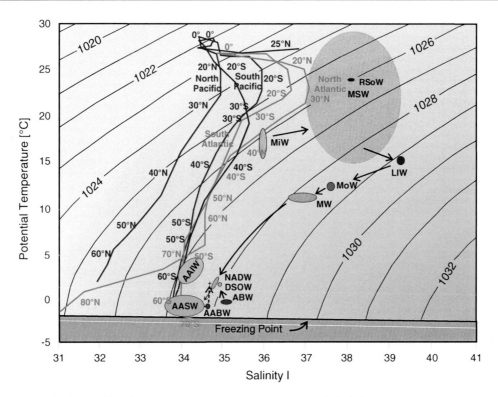

**Fig. 25.9** Temperature and salinity of is major oceanic and marginal sea water masses. 'Potential temperature' is the temperature the water would have if it were at the surface, i.e. eliminating the warming due to pressure. *AABW* Antarctic Bottom Water; *AAIW* Antarctic Intermediate Water; *AASW* Antarctic Surface Water; *ABW* Arctic Bottom Water;

*DSOW* Denmark Strait Overflow Water; *LIW* Levantine Basin (Eastern Mediterranean) Intermediate Water; *MiW* Mediterranean Inflow Water; *MoW* Mediterranean Outflow Water; *MSW* Mediterranean Sea Water (*large pink area*); *NADW* North Atlantic Deep Water; *RSoW* Red Sea Outflow Water

toward the poles, intersecting the surface so that the lightest water is on the surface in the equatorial region and the densest water at high latitudes. The uppermost interior water mass beneath the tropical-subtropical gyres is termed intermediate water. In the South Pacific, South Atlantic and southern Indian Oceans, the intermediate water is cold, low salinity water that sinks at specific sites along and adjacent to the southern Subtropical Front. Most Antarctic Intermediate Water is formed west of southern Chile (for the Pacific) and east of southern Argentina (for the Atlantic). It is spread into all oceans with the Circumpolar Current. In the Northern Hemisphere much of the intermediate water is produced in marginal seas.

The Mediterranean and Red Seas both have strong negative freshwater balances, so that there is surface inflow from the ocean and saline outflow to the ocean at depth. The Mediterranean outflow spreads out at a depth of about 1500 m in the North Atlantic forming a warm saline water mass that extends across the entire North Atlantic and almost to the Equator. The Red Sea outflow forms a warm saline intermediate water mass in the Arabian Sea but can be traced into the Bay of Bengal. In contrast the cold lower salinity outflow from the Sea of Okhotsk forms the basis of the North Pacific intermediate water mass.

However, there is an overall flow of ocean deep waters from the GIN (Greenland-Iceland-Norwegian) Sea to the northern Pacific. This deep-water flow has been termed "The Great Conveyor" by Wally Broecker of Lamont-Doherty Earth Observatory. It is shown in Fig. 25.11.

The northern Atlantic region has several sources of deep water. The GIN Sea basin is isolated from the deepest part of the Arctic Basin, and from the North Atlantic Basin by the Greenland-Scotland Ridge. Cold waters overflow the Ridge into the North Atlantic Basin through channels between Iceland and Scotland and though the Denmark Strait between Iceland and Greenland. Dense warm saline outflow waters from the Mediterranean enter from the east. flow These waters, shown in Fig. 25.12, then mix together and form a dense cold saline water mass, North Atlantic Deep Water which flows south through the western Atlantic basins. This North Atlantic system, with at least four deep water source areas, just happens to work in such a way as to drive the global Conveyor. However, it is very delicately balanced, and may well be prone to disruption.

The North Atlantic Deep Water flows southward through the South Atlantic and into the Southern Ocean, as shown in Fig. 25.13. There it rises, following the isopycnals (density surfaces) to the sea surface in the Weddell Sea. On the sea

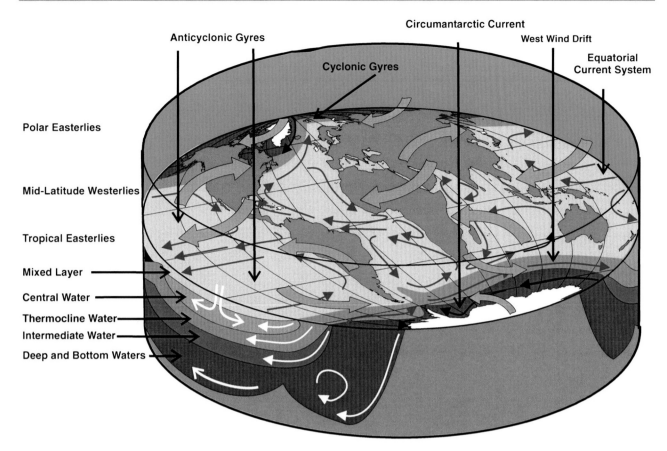

**Fig. 25.10** Schematic representation of Earth's modern atmospheric and ocean circulation. Land areas are shown in *orange*. Major winds are shown by broad *gray arrows*. They drive the ocean currents shown by *narrow blue arrows*. Successively deeper, denser layers of water are shown by increasingly *darker shades of color*

surface it is chilled by the atmosphere and sinks again. Some is so dense it is trapped in the deep basins around Antarctica as Antarctic Bottom Water. However, most becomes slightly less dense Antarctic Deep Water and flows eastward into the depths of the Indian and eventually the Pacific Ocean. It is a curious accident of the modern configuration and circulation of the ocean that much of the modern ocean's deep water forms in the Weddell Sea region through upwelling and chilling of the North Atlantic Deep Water.

Most of the deep water in the ocean forms by sinking in the circum-Antarctic region, spreading into the northern reaches of the Pacific and Indian Oceans.

How fast does the conveyor move? Originally it was thought that the oceans' deep waters must be many hundreds to a thousand years old. New information reveals that they are younger. Radiocarbon ($^{14}$C) dating shows a progression of age along the route of the Great Ocean Conveyor, as shown in Fig. 25.14. The problem of determining when the deep waters of the ocean were last at the surface is not as simple as it sounds. When the water was at the surface, before it sank, it was in equilibrium with atmospheric $CO_2$, and that gas contained 'fresh $^{14}$C.' If nothing else happened

it would be a simple matter to date the water by the amount of $^{14}$C remaining. Unfortunately, living organisms in the surface ocean incorporate $CO_2$ into their tissues, and as they die and their remains sink into the deep sea, their 'fresh $^{14}$C' is carried down, contaminating the signal. Add to this the fact that the $^{14}$C signal in the $CO_2$ in the air is already contaminated by the 'dead carbon' from burning fossils fuels, and that the level of this contamination has changed over the past 200 years, and you realize that getting an exact age for deep water is a daunting task. It now appears that most of the world's deep waters are less than 250 years old.

It is estimated that the long term rate of sinking in the sources around the North Atlantic and Antarctic is about 30 Sverdrups ($30 \times 10^6$ m$^3$/s) and is approximately evenly divided ($\sim 15$ Sverdrups) between the northern and southern hemisphere sources. Although this must be the long term rate, the shorter term rate for the Antarctic over the last 20 years seems to be much slower, about 2–5 Sverdrups, while the North Atlantic continued to supply 15–16 Sverdrups until quite recently. There is now evidence that the rate of production of deep water in the North Atlantic region may be slowing, perhaps by as much as 20 % over the past

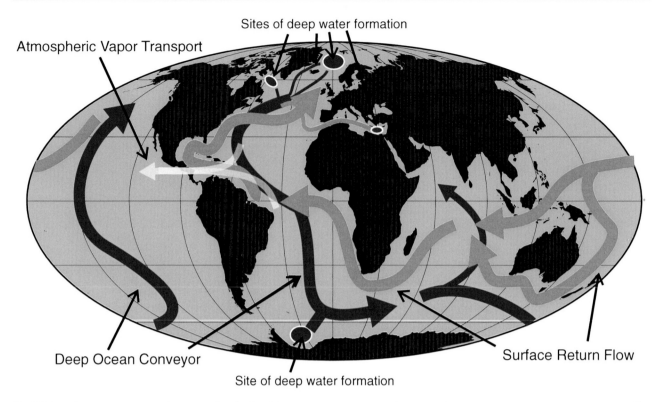

**Fig. 25.11** The Great Conveyor: deep water flow in the global ocean. Two major sites of sinking of deep water from the surface are in the GIN Sea (Greenland-Iceland-Norwegian Sea). Two other lesser sources are shown, one in the Labrador Sea that tends to be periodic, the other producing warm saline dense water in the Eastern Mediterranean

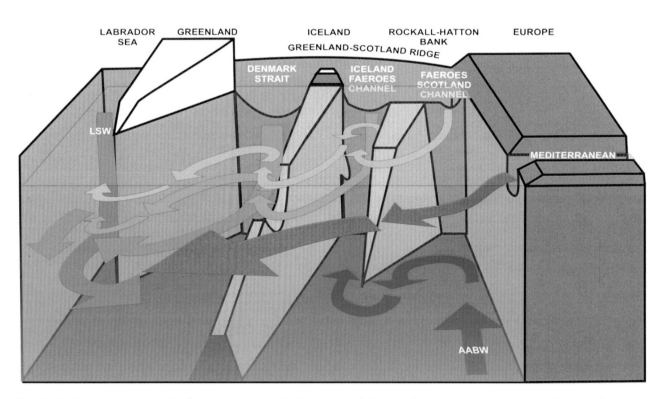

**Fig. 25.12** The complex of ocean interior waters that mix in the North Atlantic basin to form North Atlantic Deep Water (NADW). These are the headwaters of the Great Conveyor. AABW is Antarctic Bottom Water which has made its way northward first through the western basins, then through the Romanche Trench through the Mid-Atlantic Ridge and then through the eastern basins of the North Atlantic. LSW is Labrador Sea Water

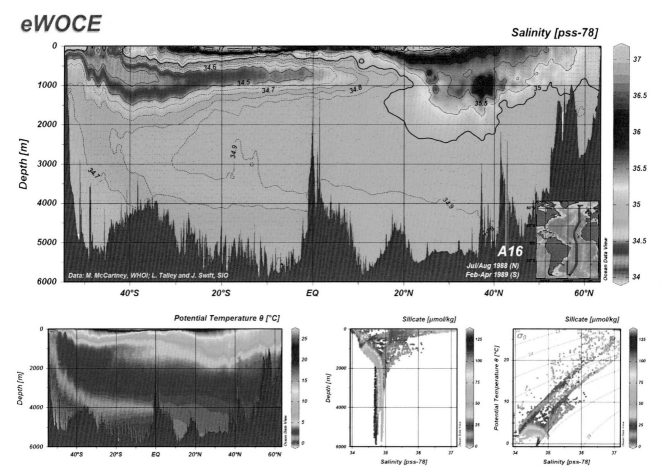

**Fig. 25.13** Salinity, temperature, and dissolve silicate (a phytoplankton nutrient) in a N–S section through the Atlantic. The salinity diagram shows the highly saline Mediterranean water in the North Atlantic (*red* and *yellow*), the North Atlantic Deep Water (Atlantic part of the Great Conveyor) flowing southward (*green*), and the Antarctic Intermediate Water (*blue* and *purple plume*). Antarctic Bottom Water is best seen in the temperature diagram as a *purple plume*. Data compiled by the World Ocean Circulation Experiment (WOCE)

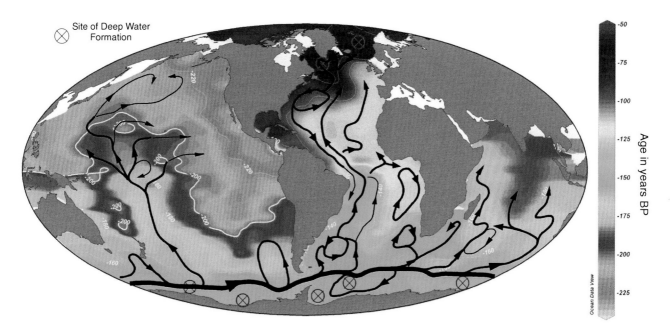

**Fig. 25.14** Ages of ocean deep waters based on [14]C. More detailed paths of ocean deep water flow are shown, along with some additional sites of possible deep water formation not shown in Fig. 25.11. Image courtesy of the CARBOCEAN Project

50 years. In any case, it is clear that the rates of production of deep water from these various sites may vary over time, and from site to site, and some or all may even shut down completely from time to time.

This has important consequences for the climate of the lands surrounding the North Atlantic and GIN Seas (Fig. 25.15).

The tritium produced by American atomic tests in the atmosphere to 1963, along with those from other countries provided us with a tracer to determine the rates of production and downward mixing as deep water formation took place. This tritium source disappeared in 1963 when the Limited Test Ban Treaty was signed by the U.S., Soviet Union, and Great Britain and testing moved mostly underground. France continued atmospheric testing until 1974 and China did so until 1980, but now the tritium, with a half-life of only 12.3 years, has almost completely disappeared. As will be discussed later, the rates of deep water formation and their variability are a matter of great concern and new techniques for their measurement are being explored.

An interesting footnote to the way science works is that the grand scheme of the deep circulation of the oceans was discovered not by physical oceanographers, but by Wally Broecker, a geochemist (Fig. 25.16). Physical oceanographers, those who study and measure the motions of the ocean's waters, had spent much effort trying to measure the motion of the ocean's deep waters. One of the tools used was the 'Swallow Float,' named for the inventor, John Swallow, a very distinguished British oceanographer. John, incidentally, received the Albatross Award at the Joint Oceanographic Assembly in Halifax, Nova Scotia, in 1982. The Swallow Float is a free–drifting instrument that sends back sound signals to a tracking vessel on the surface. When they were first used, it was expected that they would move in a more or less straight and steady line, indicating the direction and speed of the subsurface current. To everyone's surprise, the floats seemed to wander all over the place. The plots of its motions were called 'spaghetti diagrams' because that was what the trace of the float's motion looked like. Once we had observations of the ocean from space, it was evident that the surface currents do not move in straight or even gently curved lines. As mentioned previously, we now know that the water within the major currents actually moves as eddies, usually with a diameter of about 100 km. These so-called 'mesoscale eddies' are now a hot topic of much of physical oceanographic research. Unfortunately observations from space offer no information about the internal circulation of the ocean.

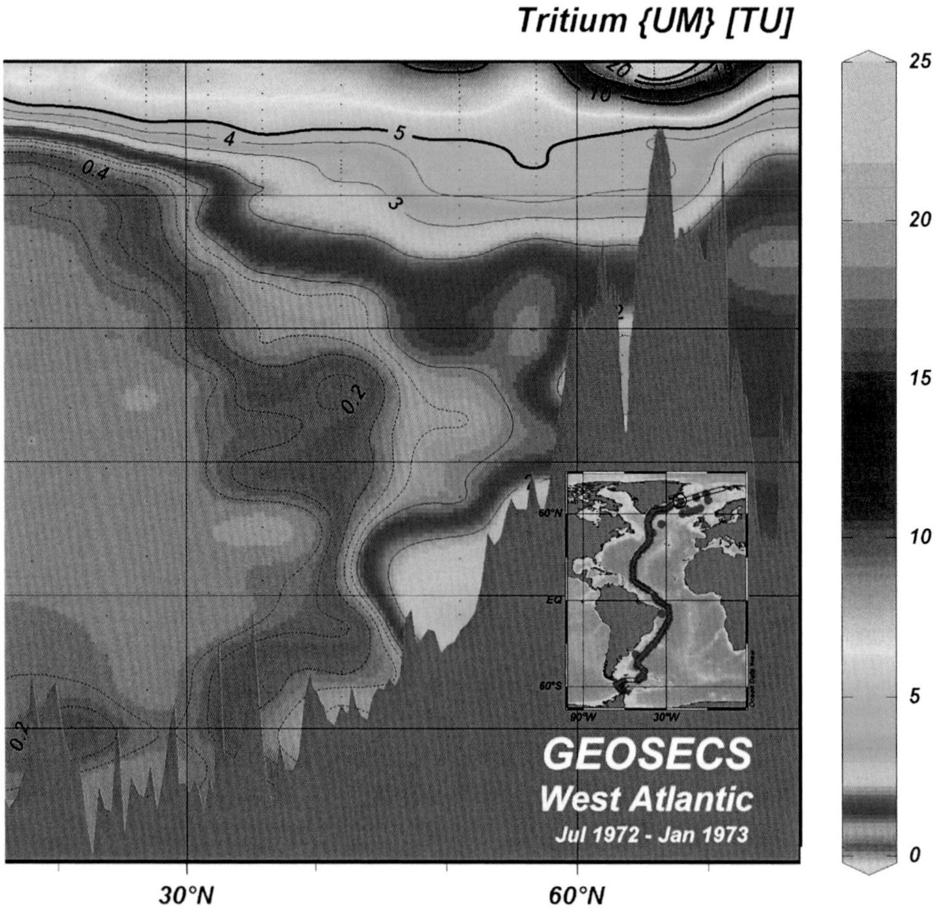

**Fig. 25.15** Tritium from atmospheric bomb tests entered the ocean in the GIN Sea (*right side* of diagram), and is shown cascading down into the North Atlantic basin in 1972–73. The Greenland-Scotland Ridge is the submarine feature above the map. As described in the text, GEOSECS (GEOchemicalSECtionS) was a program to map geochemical properties in the ocean

**Fig. 25.16** Wally Broecker, a marine geochemist who has led the way to our modern understanding of how the Earth-Life system works

Wally Broecker was one of those dedicated scientists who wanted to map the internal chemistry of the ocean to better understand what is going on. A major data gathering project, GEOSECS (for Geochemical Sections) was started in the 1970s. The GEOSECS program consisted of sampling the water column at close intervals at a number of critical locations. The work took over a decade to complete. Geochemical analysis of the samples showed very distinctive trends that had previously been topics of speculation. As Wally put the data together, he realized that a very simple pattern of deep water flow was emerging. Many physical oceanographers doubted the results at first, but now geochemical measurements are routinely used to study the large scale motion of the oceans interior waters. Wally has been a leader in learning about the possible effects of climate change on the ocean. His work has been so outstanding that he is always on the cutting edge of science, and that makes him a good candidate for a modern Galileo.

By this time you may have begun to suspect that the way the ocean carries heat from the equatorial region toward the poles may not be straightforward. If so, you are correct. Figure 25.17 shows the energy transport by the ocean, expressed in petawatts. You will immediately see that only in the Pacific does the energy transport go from the equator toward the Polar Regions. There is no energy transport across the Equator. In the Indian Ocean the energy transport is from north to south, and crosses the Equator. In the

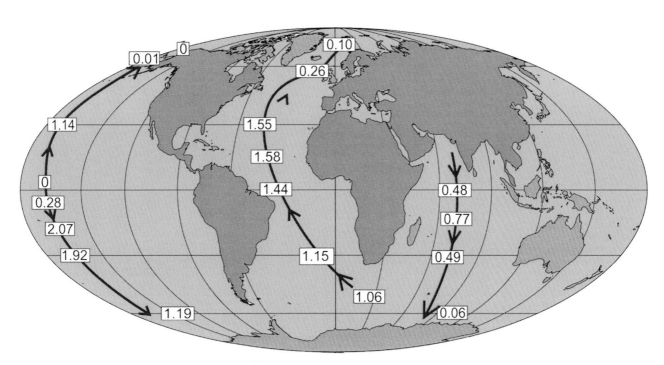

**Fig. 25.17** Energy transport by the surface ocean, expressed in petawatts ($10^{15}$ W)

Atlantic it is from south to north, and a large amount of energy crosses the Equator to warm the North Atlantic, the GIN Sea, and the Arctic. All of this asymmetry is the result of sinking of waters in the northern North Atlantic region and along the Antarctic margin. Again, 'geographic accidents:' particularly the shape and location of the Atlantic Ocean basins and the Greenland-Scotland Ridge determine the way the world works.

## 25.3   Upwelling of Waters from the Ocean's Interior to the Surface

How does the water that sinks into the ocean interior get back to the surface? Some of it simply convects up and down in the high-latitude regions of cyclonic circulation. The ascending water is said to 'upwell.' It ascends in large

amounts where the surface of the ocean is pulled apart by the increasing speed of the westerly winds from their poleward limit around 60° toward their center at 45° latitude, as shown in Fig. 25.5. But what about the deep waters that flow equatorward beneath the tropical-subtropical gyres?

Because of pressure differences in the ocean at depth, the deep water beneath the surface gyres rotates in the opposite direction. Beneath the thin anticyclonic gyres are thick cyclonic gyres, rotating more slowly. They push the deep waters upward, and it gradually mixes in with the overlying waters. However some of it returns along the edges of the gyres, particularly on the eastern sides of the ocean basins where the surface currents are broad and shallow. What happens is shown in Fig. 25.18, a view of sea surface temperatures off southern Africa.

Figure 25.19 is a sketch of the Benguela upwelling system showing how it operates. The Walker circulation in the

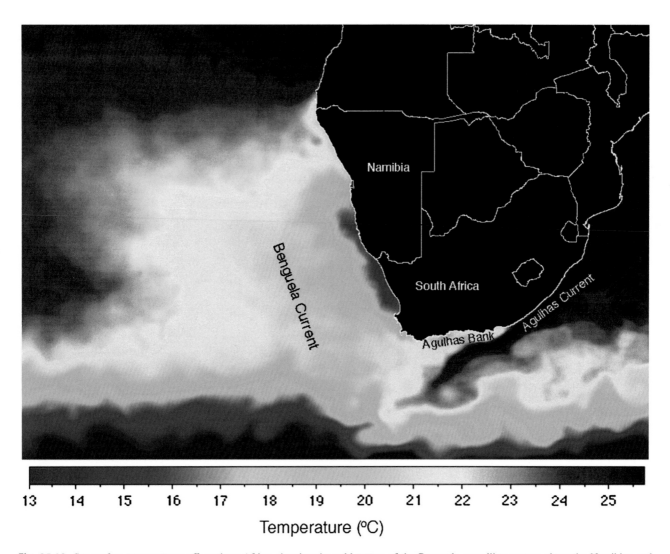

**Fig. 25.18** Sea surface temperatures off southern Africa, showing the cold waters of the Benguela upwelling system along the Namibian and South African margin

**Fig. 25.19** The Benguela upwelling system

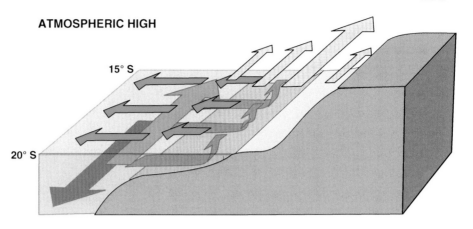

ATMOSPHERIC HIGH

ATMOSPHERIC LOW

15° S

20° S

atmosphere sets up a permanent atmospheric high over the South Atlantic and a low over southern Africa. This results in south to north winds along the coast. These equatorward-blowing winds are intensified by an abrupt slope from the coast to the kilometer-high southern African plateau. Offshore, the Benguela Current, the eastern limb of the South Atlantic gyre, flows from south to north. Beneath it is a countercurrent, the eastern side of the deep water cyclonic gyre, flowing from north to south. The winds induce Ekman transport of the waters along the coast. The Ekman transport is at right angles to the wind, that is, offshore. Deep waters are drawn in to replace the surface waters. These deep waters have high nutrient concentrations, and the upwelling system is rich in both phytoplankton and fish.

The situation off western South America is similar. There the equatorward-flowing Humboldt Current borders a rich upwelling system off the Chilean coast. There are analogous upwelling systems in the Northern Hemisphere, along the coast of California and northwestern Africa, but they tend to be less productive than those in the Southern Hemisphere. The fisheries associated with these upwelling systems are an important source of food.

Another strip of upwelling occurs along the Equator. The northern and southern trade winds blow from northeast to southwest and southeast to northwest respectively. The Ekman transport beneath the winds is 90° to the right of each wind system. The waters in the Southern Hemisphere are forced to the southwest, those in the Northern Hemisphere to the northwest. The result is that the ocean waters along the Equator are pulled apart, allowing the deeper waters to reach the surface. The situation is actually a bit more complicated than this, because the ITCZ, where the trade winds converge is, in fact, north of the Equator. As a result of the interactions between the winds, water, and Coriolis Effect (i.e., rotation of the Earth), there are several zones of upwelling, most of them just north of the Equator.

## 25.4 El Niño—La Niña, the Southern Oscillation and the North Atlantic Oscillation

The El Niño and La Niña phenomena represent breakdowns of the southern hemisphere Walker circulation. The Pacific is so wide, that the usual pattern of an atmospheric low over land and high over the ocean is difficult to maintain. Periodically it breaks down, with major consequences for both the atmospheric and oceanic circulation. One measure of the change is the difference in air pressure between Darwin, in northern Australia, and Tahiti. The changes in pressure are termed the 'Southern Oscillation'. The change is sometimes referred to as 'ENSO' an abbreviation for El Niño/Southern Oscillation. The perturbations of these Southern Hemisphere climate changes extend even into the Northern Hemisphere. Figure 25.13 shows schematically the 'Normal,' 'El Niño' and 'La Niña' conditions.

The warmest ocean waters are in the western Pacific, off Southeast Asia. Normally this is an area of low pressure, generated by the warmth of the waters. The air rises and aloft flows to the east to descend over South America. The trade winds generate equatorial currents that flow from east to west. Along the Equator the winds induce upwelling bringing up cooler, nutrient-rich waters from the depths. The southward flowing western boundary current along the Australian margin is narrow and deep, forcing the thermocline downward. On the eastern margin of the ocean basis, the northward flowing Humboldt Current is shallow and broad, and the thermocline is close to the surface. Winds along South America's western margin blow toward the Equator, forcing upwelling. The Peruvian-Chilean upwelling system has a major climatic effect on the climate of western South America. The cold water does not provide much moisture to the atmosphere, and the onshore Atacama Desert is one of the driest places on Earth. At the same time, the upwelling system with its rich phytoplankton provides food for abundant anchovies and larger fish. It also supports large populations of sea birds (Fig. 25.20).

El Niño means 'the baby boy' and refers to the Christ Child, because the onset of this condition often occurs around Christmas. The low over the western Pacific warm pool weakens, and migrates to the central Pacific. The circulation spits into two latitudinal cells with the air aloft flowing both to the east and the west. The trade winds weaken, and the equatorial upwelling ceases. Warm water covers the equatorial region. The gyral circulation weakens, and off South America the thermocline descends to levels deep enough so that the winds can no longer bring cool nutrient-rich waters to the surface. Without a steady supply of nutrients, the phytoplankton, the base of the food chain, are decimated. The anchovy populations decline sharply, and the larger fish disappear. Without the fish as food, the bird populations are decimated. Onshore the desert regions experience rainfall, sometimes in catastrophic amounts.

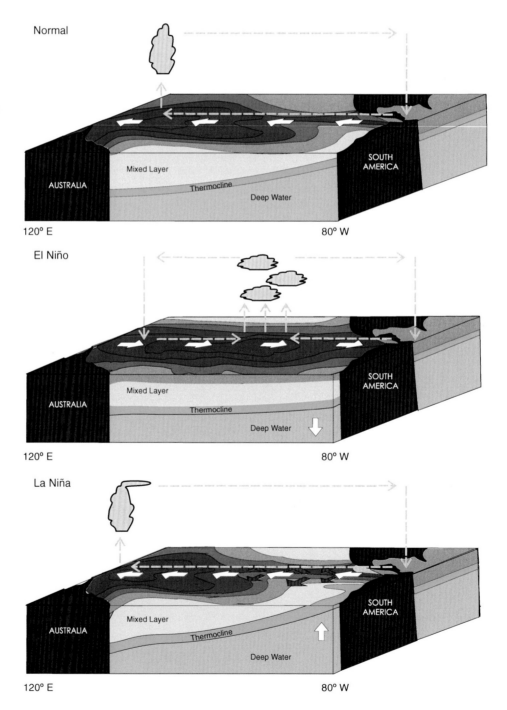

**Fig. 25.20** Atmospheric and oceanic circulation over the South Pacific and Equatorial region showing the 'Normal,' 'El Niño,' and 'La Niña' conditions. Warmer surface waters are *red*, *orange* and *yellow*; cooler waters are *green* and *blue*

The other extreme is La Niña, 'the little girl.' In this condition the western Pacific low moves westward into the Southeast Asian seas, the trade winds intensify, equatorial upwelling increases, and the thermocline may crop out on the ocean surface along the South American margin. The coastal waters off South America receive large amounts of cold nutrient-rich water, the phytoplankton bloom, the anchovies multiply, and larger fish and birds have an abundant food source (Fig. 25.21).

Although their effect is, of course, strongest in western South America, El Niños perturb atmospheric circulation through much of the global climate system. The changes in atmospheric circulation spill over into the Northern Hemisphere and particularly affect the climate in North America. El Niño conditions bring more rain to northern Central America, Mexico and the southern United States. Surprisingly, the largest increases in rainfall occur along the northern margin of the Gulf of Mexico, sometimes resulting in flooding. La Niña conditions result in drier conditions in the same region, producing drought and enhancing the probability of forest fires. The effects extend even into the Atlantic, and to Europe and Africa.

A similar, though less dramatic, change of conditions directly affects the climate of the North Atlantic region. Under normal conditions there is a persistent atmospheric low over Iceland and a high over the Azores. These steer other west to east moving eddies of the mid-latitude atmosphere. When these pressure systems are well developed and strong, with a large pressure difference, the westerly winds are strong. This brings cool summers and wet winters to Europe. However, sometimes the pressure difference weakens and the low moves to the east of Iceland as the high moves west of the Azores. This perturbation, like the Southern Oscillation, was also discovered by Sir Gilbert Walker in the 1920s. It is known as the North Atlantic Oscillation (NAO) and affects the strength of both the westerly winds and storm tracks over Europe, the Mediterranean and North Africa in the winter. Because its effects are mostly on the other side of the Atlantic, the public rarely hears about it in the US.

## 25.5 Kelvin Waves

Kelvin waves are an oceanographic phenomenon resulting from changes in slope of the ocean surface across large basins. The easterly trade winds force water westward in the Equatorial currents, creating a slope on the ocean surface. This can often be of the order of 1 m or even more. Over the Pacific it is steady, constant easterly winds that set up the La Niña condition. When these winds relax (for reasons that are still uncertain) the water that had accumulated as the Pacific's 'western warm pool' can flow back downhill to the east. As they flow to the east they are turned to the south, toward the Equator if in the northern hemisphere, and to the north in the southern hemisphere by the Coriolis Force. The result is a huge low eastward-moving wave is concentrated on the Equator, as shown in Fig. 25.22.

When the wave reaches the eastern side of the ocean basin it splits, with part moving northward and part moving southward, both forced by the Coriolis Effect to hug the coastlines. The influx of warm water causes normally dry coastal areas to receive massive rainfall and experience floods.

This phenomenon occurs periodically on an irregular multi-year basis in the Pacific, but on a smaller scale

**Fig. 25.21** Sea surface temperature changes between El Niño and La Niña conditions, shown as differences (anomalies) from the long-term average

El Niño                                    La Niña

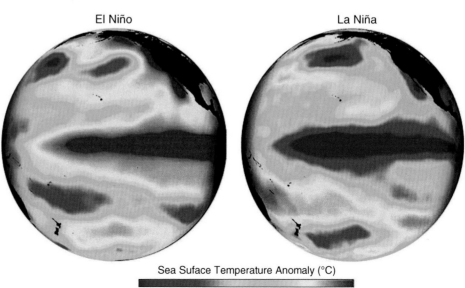

Sea Suface Temperature Anomaly (°C)

-4        -2        0        2        4

**Fig. 25.22** The Kelvin Wave of March 2009 carrying water eastward to make the change from La Niña to El Niño conditions

Sea Level Anomaly (mm)

-180    -140    -100    -60    -20    20    60    100    140    180

annually in the Atlantic where it affects fisheries along the sub-Equatorial coasts.

## 25.6  A New Hypothesis: Very Different Atmospheric and Ocean Circulation on the Warm Earth of the Distant Past

Much of my research has been on the Cretaceous Period, when dinosaurs roamed the lands. The Cretaceous world was different in terms of geography, as shown in Fig. 21.3. The Americas had just started to separate from Europe and Africa. In the mid-Cretaceous, 100 million years ago, the northern Atlantic was a small ocean, about twice the size of the modern Mediterranean and it was centered just south of 30° N latitude. The South Atlantic was a long, narrow ocean, resembling the modern Red Sea, but about 5 times as large.

The first connection between the northern and southern Atlantic was established about this time. The ocean passage around the Antarctic was blocked by Australia, which was just starting to separate from Antarctica. India had separated from Antarctica and started its northward journey to finally collide with Asia about 50 million years ago, resulting in formation of the Himalaya and Tibetan Plateau. A circum-global seaway, the Tethys, was centered at about 20° N latitude and separated the northern continents from the southern continents. Sea level was much higher than today, about 300 m higher, and almost a third of the present land area was covered by ocean water (Fig. 21.3).

Note that in both geographies Earth has circumglobal seaway connecting all parts of the ocean. At present it is around Antarctica. In the Cretaceous it was in the low latitudes north of the Equator. It was named the Tethys (for the Greek Goddess of Rivers and Seas, who was both sister and wife of Oceanus) by Viennese Geologist Eduard Suess in

1893. The geologic evidence is that climatic conditions then were very different from those of today. There was no perennial ice in the Polar Regions except possibly on high mountains, and the yearly average polar temperatures may have been as warm as the modern global average of 15 °C (60 °F). Tropical temperatures may have been much warmer than today, with ocean waters as warm as 42 °C (108 °F). Until very recently modern Tropical Ocean temperatures have never been higher than 29 °C (85 °F).

But more strangely, large areas of the ocean, particularly in the developing Atlantic and Tethys, tended to periodically become anoxic, completely depleted of oxygen. Organic matter settling into the depths from the surface layers was not oxidized, and became incorporated into the sediments. Much organic carbon was deposited in the seas flooding the continental blocks, along the continental margins, and some accumulated even in the deep sea. That organic matter that accumulated then has been transformed by heat and pressure into the petroleum we burn today. More than 2/3 of the petroleum we use comes from deposits of Late Jurassic and Cretaceous age (150–100 Ma). Today, the polar waters sinking into the interior of the ocean contain so much oxygen that anoxic conditions are restricted to a few tiny basins. Even if all the living organisms in the ocean were to be killed at one time there is more than twice as much oxygen in the ocean as would be needed to completely decompose and oxidize their remains.

Today we can recognize the 'deep blue sea,' the open ocean, as being separated from the yellow and brown waters of the coastal regions. The boundary between the open ocean and shelf waters is often marked by a sharp front between the water masses. Off Miami this front separating the water masses is so well defined that a sailboat can have its bow in the clear blue waters of the Gulf Stream while the stern is the less transparent yellow-brown waters of the coast. These 'shelf-break fronts' are a typical feature of the modern ocean, and open-ocean organisms and shelf-sea organisms are often quite different. However, during the Cretaceous such frontal systems did not develop, and the ocean plankton were able to invade the seas on the continental blocks, leaving behind the deposits we know as chalk. Today such deposits, consisting of the remains of coccolithophores and planktonic foraminifera are known only from the deep sea.

Many of us have puzzled over why Cretaceous climate and oceans were so different. I myself spent most of my scientific career (almost 50 years) trying to solve this puzzle. A few years ago I realized that we had been trapped in Charles Lyell's way of thinking—that the present is the key to the past. Almost all geology textbooks assume that the basic patterns of atmospheric and ocean circulation have been very much like they are today. And as long as you believe that, the Cretaceous climate is an insoluble enigma.

An Earth without ice on the poles is a very different kind of planet from the one we know. In fact it is the ice on both poles that stabilizes the modern atmospheric circulation, so that it can vary only within strict limits. The Polar Regions are the coldest places on the planet in their respective winters, and the ice reflects sunlight in the summer, so that they remain cold. This permanent cold means that the air in the Polar Regions is especially dense, and consequently there are permanent high pressure systems at the poles. The heating along the Equator warms the air; it becomes less dense, and forms an equatorial low pressure system. The warm air rises in a low pressure system, and cold air sinks in a high pressure system. If the Earth were not rotating air would rise along the Equator and sink at the poles, and the surface winds would blow from north to south. But the Earth's rotation disturbs this pattern, and we have the 3-cell per hemisphere pattern of atmospheric circulation shown in Figs. 24.6 and 25.2. The boundary between the polar and Ferrell cells moves back and forth with the seasons, but the atmospheric circulation is stabilized by the permanent low at the Equator and permanent highs at the poles.

However, if there were not perennial ice in the Polar Regions, the temperatures there could alternate between cold in winter and warm in summer, and that means that the polar atmospheric pressure systems would change between summer and winter.

The Earth has a geographic peculiarity that has persisted for over 100 million years. As shown in Fig. 25.23, the southern polar region has been occupied by a continent, Antarctica, surrounded by the ocean; the northern polar region has been covered by water almost completely surrounded by land. In fact, the connections with the ocean have sometimes been so tenuous that the 'Arctic Ocean' may at times have been a fresh-water lake! Without ice, the Antarctic continent would lose it warmth in winter and become cold; it would develop an atmospheric high pressure system in winter. But then in summer, with 24 h sunshine, it would become very warm and an atmospheric low pressure system would develop. There was a high in winter, low in summer. In the northern polar region the situation was very different. Ocean waters over the North Pole are surrounded by land. The water has a higher heat capacity than land, so it is slower to warm in response to the energy input from the Sun in summer, and slower to lose its heat by radiation into space in winter. The water being cooler than the land in summer, the North Polar Region would develop a high pressure system in summer, and conversely a low pressure system in winter. The two Polar Regions behave oppositely, with the result that they have high pressure systems and low pressure systems at the same time.

The equatorial low pressure system remains permanent, but the alternation of pressure systems at the poles destabilizes the atmospheric circulation poleward of the Hadley

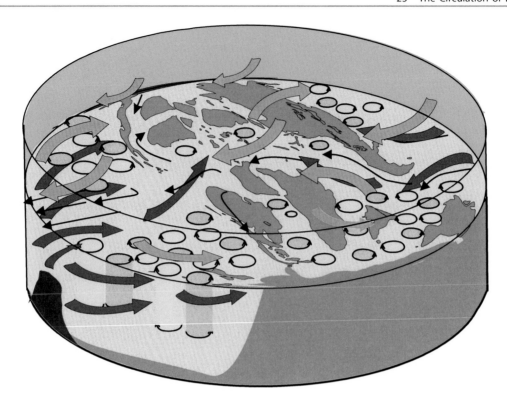

**Fig. 25.23** Highly schematic atmospheric and ocean circulation on the Earth during mid-Cretaceous, 100 million years ago. *Broad gray arrows* between the Equator and 30° N and S represent the atmospheric Hadley Cell circulation at the surface. The higher latitude winds vary with the seasons. The 'Westerlies' would have been well organized only during June-August and weak, variable or even reversing to become Easterlies during the rest of the year. Narrow *light gray arrows* represent the polar easterlies which would have been present only seasonally. Cylinders represent transient mesoscale eddies in the ocean generated by storms. Anticyclonic eddies, which pump water downward, are *pink*. Cyclonic eddies, which pump waters from the ocean interior to the surface are *light blue*. *Ellipses* represent the tops of eddies, their deeper extensions are mostly not shown to simplify the diagram. *Thin black arrows* are ocean surface currents; *broad purple arrows* are possible internal flows

Cells. The Hadley Cells themselves grow larger, while the westerly and easterly winds become seasonal and disorganized. The implications of this are immense. The ocean circulation we see today reflects the effect of the relatively stable winds integrated over time. The stable winds we have today produce the great anticyclonic ocean gyres of the tropics and subtropics, and the cyclonic gyres of the Polar Regions. Without the steady forcing of these wind systems, the highly organized modern ocean circulation would be replaced by a chaotic pattern of giant eddies generated by storms. The trade winds, and the east-to-west flowing equatorial ocean currents would still exist, but at higher latitudes the ocean circulation would consist of giant eddies spinning in both directions depending on how they were generated.

Now comes the big surprise. Without the steady westerly winds, the rigid vertical structure of the ocean we know today would collapse. Today a thin layer of warm water in the tropics and subtropics overlies a huge reservoir of cold deep water. That structure is maintained by the frontal systems that form beneath the westerly winds. Without those steady winds it would break down. It is an ocean as different from today's as one can imagine. No great surface gyres, no subtropical and polar frontal systems, no clear separation between surface and deep waters, no 'Great Conveyor.' Even the way in which deep waters return to the surface would be different. Today this occurs only in the equatorial region, in the Polar Regions, and along the eastern margins of the ocean basins, as described above. In a Cretaceous Ocean upwelling would have continued to be active along the Equator, but elsewhere it would have depended on the development of cyclonic eddies, which pump water upward.

Figure 25.23 is a schematic representation of this hypothetical atmospheric and oceanic circulation in the Cretaceous, showing the 100 Ma paleogeography. Comparison with Fig. 25.10 emphasizes how different the circulation of the atmosphere and ocean may have been.

## 25.7   The Transition from the 'Warm Earth' to the 'Modern' Circulation

There is growing evidence that the transition from the warm Earth ("Cretaceous") to the Modern circulation actually occurred at the end of the Eocene, 35 million years ago. The calcareous nannoplankton disappeared from the Arctic Ocean, indicating the development of perennial sea-ice cover, at the same time ice sheets formed on Antarctica. Both poles became ice-covered at the same time. Unfortunately, there doesn't seem to be a geologic record of an Earth condition when only one pole was ice-covered. It seems likely that this may have been the case for a period of a few hundred thousand years in the transition, but we have not yet recovered a detailed record of that interval in either polar region.

What would happen if we were to somehow suddenly remove the polar ice? The wind systems would start changing with the seasons. Even after the mid-latitude winds became chaotic, the great anti-cyclonic gyres of the ocean would keep flowing under their own momentum, but only for 10–20 years. Then they would gradually decay away into chaotic circulation beneath the shifting winds, being replaced by large eddies formed as storms move over the ocean surface.

We have actually made a good start on an experiment to find out what happens when we remove the polar ice from our planet. The Arctic Ocean is now expected to be seasonally ice-free in a decade or so, and may well be perennially ice free by the end of the 21st century. For science it is a truly grand experiment. We do not know how it will proceed or what the ultimate outcome will be. I only wish that this experiment were being performed on some other planet first.

---

## 25.8   Summary

Earth's two fluids, air and water circulate in response the regional differences in energy received from the Sun, transporting it to regions where it can be more effectively radiated back into space.

The winds drive the surface circulation of the ocean, but because of the rotation of the Earth the general motion of the water is about 45° to the right of the wind in the northern hemisphere, to the left in the southern hemisphere. The result is large scale anticyclonic gyral circulation in the tropics and subtropics, and cyclonic circulation at high latitudes. The anticyclonic gyres force warm less dense water downward onto the cold deep water that fills the ocean basins at depth. The cyclonic circulation in the Polar Regions forces water upward and results in convection from the surface to the bottom. Cold saline waters sinking in the Polar Regions fill the ocean basins. The surficial gyres are separated by frontal systems, the Subtropical and Polar fronts located only about 5° of latitude apart. Waters sinking along and between these fronts form the thermocline that underlies the tropical-subtropical surface gyres. The deep circulation can be regarded as a global conveyor, with dense water sinking in the seas peripheral to the North Atlantic flowing southward to join with water sinking off the Weddell Sea of the Antarctic, though the Indian Ocean and into the Pacific where it slowly returns to the surface. This system is driven by the higher salinity of the Atlantic forced by evaporation from the Atlantic, vapor transport across Central America and precipitation over the eastern Pacific.

Upwelling involves intermediate waters being returned to the surface by offshore drift induced by longshore winds.

The South Pacific experiences periodic changes in circulation in response to changes in the low latitude winds. The phenomenon is called El Niño when the easterly winds relax and warm waters spread eastward as a Kelvin Wave focused on the Equator and shutting off the upwelling systems off South and Central America as it splits and moves north and south along the continental margins. It is called La Niña when the opposite situation occurs and upwelling on the western margin of South America is enhanced.

The loss of one or both polar ice caps will destabilize this system and lead to significantly different atmospheric and ocean circulation.

A Timeline for this chapter:

| | |
|---|---|
| 1735 | George Hadley offers an explanation why the Trade winds, north and south of the Equator, are so constant and why they blow from east to west |
| 1835 | Gaspard Gustave de Coriolis publishes *Sur les équations du mouvement relatif des systèmes de corps*, showing that Newton's laws of motion can be used in a rotating frame of reference if an extra force (now called the 'Coriolis acceleration' or 'Coriolis Force') is added to them |
| 1856 | William Ferrell improves on Hadley's ideas, proposing that there are three atmospheric cells in each hemisphere, a polar cell, an equatorial cell, and an intermediate cell |
| 1902– 1904 | Vagn Walfrid Ekman discovers that the motion of ice on the Arctic waters mover to the right of the wind |
| 1924 | Sir Gilbert Walker describes the peculiar atmospheric circulation of the southern hemisphere, setting the stage for an understanding of the El Niño La Niña phenomenon |
| 1987 | Wallace Broecker proposes that the ocean's deep circulation forms a "Great Conveyor" from the North Atlantic to the Pacific |

If you want to learn more:
Henry Stommel, 1987. *A View of the Sea—A Discussion between a Chief Engineer and an Oceanographer about the Machinery of the Ocean Circulation*, Princeton Science Library, 165 pp.—Absolutely wonderful reading.
Open University, 2007. *Ocean Circulation, 2nd Edition*. Butterworth-Heinemann, 286 pp.—Another great Open University textbook.
Niel Wells, 1998. *The Atmosphere and Ocean: A Physical Introduction*. Wiley, 394 pp.—A more technical account.

Music: Of course: Claude Debussy—La Mer.

Libation: There are some wonderful wines from the Mediterranean coast of France. My favorites are from Bandol, a small fishing village east of Marseille, near Cassis. It is surrounded by vineyards that produce whites, rosés, and wonderful reds. They are all great.

## Intermezzo XXV. Decision Time

After the decision to cancel the Ocean Margin Drilling Program was made, the way was clear to re-open negotiations with the National Science Foundation and our international partners for scientific ocean drilling to succeed the DSDP which would end in 1983. In fact 2 years is a short fuse to develop, organize, and implement such a complex effort.

Meanwhile Ronald Reagan had discovered that the USSR was one of the program's supporters and ordered that they be dropped.

A 'Conference of Scientific Ocean Drilling' to define the program's objectives was held in Austin, Texas, in November 1981. A proposal was prepared by JOIDES and sent to JOI to forward on to the National Science Foundation. As all this was proceeding we hit a snag.

The DSDP had been managed by Scripps Institution of Oceanography in La Jolla, California. The drilling vessel contracted by Scripps was the *GLOMAR Challenger*, owned and operated by Global Marine, Inc., of Houston, Texas. We were told that the White House wanted to have any future continuation of the DSDP be made through a sole-source contract with Scripps and Global Marine. A sole-source contract can only be made when no other organization can provide the necessary service.

But a few years earlier I had visited SEDCO, and knew that they wanted to bid on any future program. Jack Clotworthy warned that trying to go sole-source would surely attract the ire of the Congress and torpedo the proposed program. The matter was brought before the JOI Board of Governors. Bill Nierenberg argued that Scripps was uniquely qualified to operate any future program. He angrily asked how SEDCO had gotten involved, and I recounted my meeting of a couple of years earlier. Finally, the Board agreed that Scripps was probably the most highly qualified potential operator, and would probably win the contract; we had to go out for bids.

In the meantime Mike Halbouty had contacted his old alma mater, Texas A & M, and convinced them that they should think about getting involved. You will remember that I had made a presentation about the OMD program to him when he was Acting Secretary of Energy. Mike understood that scientific ocean drilling was at the cutting edge of the science. He realized that if Texas A & M became the operator of the new program the University would achieve the international recognition it deserved. He must have made a very good presentation to the University administration because when the bids came back in, that from Texas A & M was far and away the best offer. They would build a facility at University expense and provide extensive support. Scripps offered no real competition. Similarly, the offer from SEDCO was significantly better than that from Global Marine.

The new project, which became operational in 1985, was simply the Ocean Drilling Program, ODP. Its operations would continue until 2004. The international partnership no longer included the USSR, but was expanded by ECOD, the European Science Foundation Consortium for Ocean Drilling with Belgium, Denmark, Finland, Iceland, Ireland, Italy, The Netherlands, Norway, Spain, Portugal, Sweden, and Switzerland being members.

I thought I had seen through this difficult transition from DSDP to ODP, and now it was time to do something different. I had accumulated 48 years of service on various JOIDES Committees and Panels. I didn't want to sit back and oversee the new drilling operations from Washington. Ordinarily, if one had served in a position like President of JOI, one could easily be recycled into other D.C. based organizations or into the government, but I wanted to return to science. It was now 1982, the period of expansion of US science was over, and landing a senior faculty position was not going to be easy. I did not want to return to Miami; the Rosenstiel School had a new Dean, and being there could become awkward. I wanted to pursue a new direction, and there was an exciting opportunity for research in Colorado.

In 1982 the National Center for Atmospheric Research in Boulder was being considered for possible expansion to also become the National Center for Oceanographic Computing. I wanted to continue the efforts at climate and ocean modeling we had started a few years earlier, when we had sent Eric Barron to NCAR. The results he had obtained, showing that the warm climate of the Cretaceous had resulted from increased levels of atmospheric $CO_2$, were intriguing. Some in the scientific community were beginning to seriously wonder whether the burning of fossil fuels might eventually change Earth's climate.

As a geologist, I had little chance of getting a research position at NCAR itself, but I had contacts at two nearby universities: the University of Colorado in

Boulder and the Colorado School of Mines in Golden. I wrote to friends there and asked if they knew of any openings. No senior faculty in geology were being sought at either. But Erle Kaufmann at CU Boulder told me the University Museum was looking for a new director.

I had never thought about becoming director of a museum, but I did have a friend who was just that. J. Tuzo Wilson was a famous geophysicist who had become Director of the Ontario Museum in Toronto. I called him up. "Bill, being a museum director is the best job on Earth; you can continue your research and you will meet a host of interesting people." I applied for the job in Colorado.

As I began to think about what I could do as a museum director it occurred to me that if there were going to be changes in Earth's climate, Colorado, with its great elevation differences, would be one of the first places where it would be noticed. On doing a little research I realized that no one at CU Boulder was thinking about climate change. (Susan Solomon's work on the ozone hole would come a few years later.)

As I had my interviews with the faculty of the Museum and University administrators, I talked about climate change and the critical location of Colorado. Only one person took it seriously, Vice Chancellor Milton Lipetz, the person responsible for the hiring. I was 'overqualified' for the position and he was hesitant to offer it to me because he could not match my salary at JOI. I made it clear that salary was not the most important consideration for me, and we made a compromise. I started my new job in the fall of 1982.

I was soon called back to Washington to be the speaker at a dinner for the potential foreign contributors of ODP. It was a joint NSF—State Department—JOI affair, and protocol dictated who sat were. I was impressed to be at the head table. Lev Ropes had made beautiful slides for my presentation and copies were given to each of the participants to use back home. My final statement was simple: "It is not a question whether you can afford to join this program; the question is whether you can afford *not* to join it." That became a slogan. NSF was delighted. Everything looked very promising.

I made another trip to Washington in the fall to explore the possibilities of grants to the Museum. Some colleagues from NSF wanted a private meeting. We met at one of my father's favorite places for private meetings, over lunch at l'Espionage, a small restaurant in the Foggy Bottom area. NSF's message was simple: the foreign partners were interested, but none had yet made a commitment. Somebody needed to kick the ball off dead center. Did I have any ideas? I did.

Over the Christmas-New Year break I made a trip to Europe to visit museums and look into their operations. Museum directors love to talk to other museum directors, as Tuzo had explained, so I was well received everywhere. But on my visit to Germany I had another goal. The Germans were very excited about the drilling program. Their contact person was Eugen Seibold, who I had come to know well over the years. He was the most well-known marine geologist in Germany. Eugen has an ebullient personality and it had always been a joy to work with him. We met in Munich in a little café near the Nationaltheater and I explained the situation to him. He smiled. "The money is already appropriated here, so of course, I will make a big thing of it: Here is our funding for the program, now let's get on with it." A week or two after I got back, I learned that he had done just that, and the other partners were falling into line.

I thought when I left Washington that my involvement in drilling for scientific purposes was over. It wasn't. In 1984 I was appointed to the new JOIDES Advisory Panel on Sediment and Ocean History.

Then in 1985 I was asked to Chair the National Research Council's Committee on Continental Scientific Drilling. It was a case of Project Mohole revisited. The National Academy does not conduct operational scientific programs. It is purely advisory. But we had a number of geoscientists who were interested in drilling holes on land or in lakes for a variety of purposes. They ranged from documenting the history of retreat of the North American ice sheets to exploring the depths of the continental crust. My job was to guide them toward organizing something parallel to the ocean Drilling Program. It took a number of meetings, but finally it came together by founding a non-profit corporation for Drilling, Observation and Sampling of the Earth's Continental Crust (DOSECC), now located in Salt Lake City. That done, our NRC Committee was disbanded in 1988.

In 1983 I was also made a faculty member in the Department of Geological Sciences, and named as a member of CIRES, the cooperative institute between NOAA and CU that had served as a model for CIMAS at the Rosenstiel School. Cooperation with colleagues at NCAR on research flourished. My efforts to encourage museum faculty to seek outside support for their research were not so successful.

I began to attend meetings of Museum Administrators and found them to be a very helpful group. We even received new display cases from a Museum in Calgary which no longer needed them. And I learned about the Museum equivalent to the Rosenstiel Schools yacht donation program. Many donations to Museums are valuable objects given with no strings attached, but there is also the 'conditional gift'. This is usually a work of art or valuable artifact that is given on the condition that the Museum must fulfill some condition, such as its display, within a certain time frame. In most instances the condition is simply that the donor wants the object to be seen, and not just put in storage. However, some of the conditions are used in another way. You may get an offer, as we did at CU, of a valuable object or collection with the stipulation that it must be on permanent display, or even that a new room or wing of the Museum be built to house it. When the new room or wing is not built within a certain time, the objects must be returned to the donor. What happens is this; the donor takes a tax write off when the gift is first made. When he takes it back, he simply forgets to correct that tax write off. By law, the Museum becomes responsible for paying the tax. It seemed that almost everyone at a meeting of Museum administrators had a story to tell about their experiences with this innovative scheme.

Vice Chancellor Milt Lipetz, who had enthusiastically supported me, died suddenly in 1987. His successor had no interest in the CU Museum. In 1988 I would be up for a sabbatical leave, and so I decided to resign as Director, move to the Department of Geological Sciences, and then take my sabbatical.

VLADIMIR VERNASKY PERCEIVES THE EARTH AS A
BALL OF ROCK COVERED BY A THIN VENEER OF LIFE

*The radiations that pour upon the earth cause the biosphere to take on properties unknown to lifeless planetary surfaces, and thus transform the face of the earth. Activated by radiation, the matter of the biosphere collects and redistributes solar energy, and converts it ultimately into free energy capable of doing work on earth.*
Vladimir Vernadsky

It was at one of those convivial evenings in Moscow in 1991 that I was introduced to a whole new world of ideas about how the Earth works. The long dinner, consisting of many small dishes interspersed with a variety of vodkas, whiskey, beer, wine and Pepsi-Cola, led to animated conversation on a seemingly endless series of scientific topics. Tolstoy's descriptions of Russian life are remarkably accurate. My experience was that table conversation does not consist of idle chit-chat, as it so often does in the US, Russians get right into serious topics. And with scientists, it quickly gets into 'what do you think of this or that idea'. I was in Moscow for one of the annual German-Russian Oceanographic-Marine Geological Conferences organized by Jörn Thiede, Director of the Alfred-Wegener Institute in Bremerhaven. These meetings were designed to foster cooperative projects after the collapse of the Soviet Union. It was January, and bitter cold in Moscow. The dinner was at Areg Migdisov's apartment. The guests included his son, Alexander Borisovitch Ronov, Alexander Nikolaievitch Balukhovskiy and several others from the Russian Academy of Sciences' Vernadsky Institute. As I have mentioned before, in the days of the Soviet Union the Vernadsky Institute seemed to be off-limits to foreigners. I had never managed to get there and until now it had not been easy to meet the distinguished geochemists who worked there. I asked the stupidest of questions; "By the way, who was Vernadsky?" There was total shock throughout the room (Fig. 26.1).

"Bill, you don't know Vernadsky? You've never read the book?" I had to admit I knew his name only through the name of the Institute. The dinner went on until just before 1 A.M., so that the guests could still use the Metro to get back to their apartments and me to my hotel. I was filled in on Vernadsky and his ideas in excited conversation liberally lubricated with alcohol. When we decided to break up and get me back to my hotel, Areg and his son insisted on accompanying me. It was

very helpful because when we came up from the Metro station the streets were sheer ice. We made it the few hundred yards to the hotel elbow in elbow slip-sliding along the street. They dropped me off and fortunately there was a taxi at the hotel to get them back home.

As soon as I got back to Kiel, where I was working at the time, I ordered a copy of the book. It had been written in 1926, and translated into French in 1928. But, for reasons not known, it was not translated into English until 1998. It is truly one of the most insightful works on planet Earth. Vernadsky realized that this planet was essentially a symbiosis between a geological entity, the planet, and life. His book was entitled "The Biosphere." The term 'biosphere' had been introduced by the Viennese geologist Eduard Suess in 1885. Vernadsky had met Suess in 1911. He described Earth as a rock ball thousands of kilometers thick ('lithosphere'), partly covered by a layer water about 4 km thick (hydrosphere), and with an incredibly thin film of life (averaging about 1 m thick, the 'biosphere'), beneath its fragile atmosphere (30 km). Vernadsky realized that geological forces had steered and punctuated the evolution of life, but that this tiny film of life had completely altered the planet. Vernadsky explored these interrelations in depth. He described, in the 1920s, the way the world works in terms the rest of us were rediscovering from the 1970s on. The most significant error in Vernadsky's ideas was that he thought that photosynthesis evolved with the earliest life forms. He believed that throughout Earth's history life had captured the energy of the Sun to modify the planet's chemistry. We now know that there were almost 2 billion years of evolution of life before photosynthesis emerged, but since about 2.4 billion years ago, Vernadsky's ideas were correct. He also introduced the term noosphere, the sphere of thoughtful life (humans). He realized that humans were making modifications of the planet far greater and more

**Fig. 26.1**   Vladimir Ivanovich Vernadsky (1863–1945)

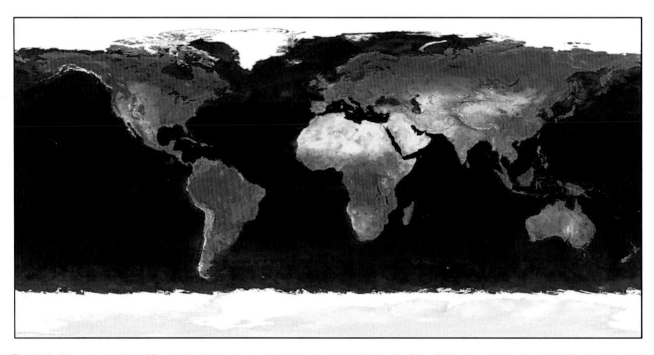

**Fig. 26.2**   Vernadsky's idea of Earth with its water, atmosphere, and the exceedingly thin film of life that has completely modified the nature of the planet's surface. A NASA *Blue* Marble image showing land, vegetation, shallow and deep waters ocean areas, ice sheets and sea ice

rapidly than any of the older life forms that had inhabited the planet (Fig. 26.2).

James Lovelock, in the 1960s had proposed his Gaia hypothesis, that the Earth was the product of the interaction of life and the inorganic world. He had come to this conclusion in working with NASA to determine how life on Mars might be detected. Lovelock concluded that since there was no oxygen in the Martian atmosphere, there was no life (as we know it) on Mars. He went on to discuss the resilience of life on Earth, and its ability to modify the climate and environment to make the planet more suitable for living things. Unfortunately, some of his ideas have been exaggerated to conclude that life can overcome any possible catastrophe. This is probably true of the bacteria. After all there are bacteria that do well in a completely anaerobic environment. Unfortunately the higher life forms, like the plants and animals we see around us, are more sensitive to environmental change.

But for many of us the revolution occurred with a remarkable paper by Wally Broecker published in 1971, entitled *A kinetic model for the chemical composition of seawater*. He argued that because the sediments presently accumulating in the ocean show little evidence of equilibration with the overlying water, the possibility that kinetic factors play an important role in seawater chemical history must be seriously explored. Material balance restrictions are substituted for some of the usual chemical equilibria. The role of organisms is dominant for at least some of the important components of sea salt. If the chemistry of sea water is dependent on rates of supply of individual components, the rate of vertical mixing in the sea, and the type of material formed by organisms, then substantial changes in the chemical composition have almost certainly taken place. Several means by which such changes might be reconstructed from chemical and isotopic measurements on marine sediments are suggested.

Up until that time, those of us in the West had treated life as something wholly separate from the inanimate world. For us, Wally's paper opened the door to an entirely new view of the world, one in which organisms play a critical role in determining the planet's behavior. I find it truly amazing that some of the critics of the modern ideas of the causes of climate change are still living back in the scientific dark ages before this discovery. Even more interesting, this discovery was made in Russia 50 years earlier, and was completely neglected by English-speaking scientists. It was Vernadsky's ideas that had inspired Mikhail Budyko to realize that life, carbon dioxide, and climate were intimately related.

The biological aspects concern four major changes being made by humans. First there are the changes that are occurring on land: (1) replacement of $C_3$ plants by $C_4$ plants; (2) response of land plants to higher levels of $CO_2$, (3) response of land and ocean organisms to warmer temperatures and more chaotic weather, and (4) response of ocean life to the changes in ocean chemistry as it takes up more and more $CO_2$.

## 26.1    Life on Land

The changes in life caused by man are most noticeable on land. Look out your window. Is what you see a pristine natural landscape? I live on the edge of Rocky Mountain National Park. My house does not have a garden or lawn,

just the natural plants. But wait, that is not correct, I have planted a couple of blue spruce that are 'out of place' on our sunny hillside. When I look over at the meadows in the National Park, the willows along the stream banks that were there when I was a boy are gone, eaten by a hungry overpopulation of elk. The elk have multiplied because we have eliminated their predators (wolves and mountain lions) to make life 'safer' for humans and cattle. Even the National Park is no longer like it was before the Europeans came.

The Industrial Revolution and subsequent population growth have had a huge impact on life on land. The world has become very different even from the way it was when Wladimir Köppen first made his climate classification. In response to the ever increasing need for food for the burgeoning human population of the 20th century, vast areas have had their naturally diverse assemblages of plants and animals replaced by a single species, such as *Zea mays* (corn, maize) grown mostly to supply food for animals that humans will eat. Such monocultures make large areas inhospitable to birds and beneficial insects such as bees because the availability of food and pollen is limited to only a week or two. Started by collective farms in the old Soviet Union, North America has become a center for monoculture agribusiness.

## 26.2    C$_3$ and C$_4$ Plants

As you know, $C_3$ and $C_4$ plants use a different metabolism to carry out photosynthesis. The $C_4$ plants, which include the tropical grasses, maize (corn) and some other fast-growing plants, conserve water. Large scale replacement of $C_3$ plants, such as the trees of the rain forests of the Amazon and Central America, with $C_4$ plants to make pastureland for cattle, will tend to dry out areas downwind. One place where this has happened is Costa Rica, where the water from the Atlantic is transported to the Pacific. The transport across Central America takes place in steps, with the moisture from the ITCZ falling as rain, and then being returned to the atmosphere through transpiration by the plants. This may happen several times before the vapor finally makes it over the top of the mountains, and can fall as rain on the Pacific Ocean. Cutting down the rain forest for pastureland has lowered the amount of water vapor transported across the mountains. As a result we can expect that the salinity difference between the Atlantic and Pacific will gradually decrease.

But it is the salinity difference that causes the Great Conveyor to have its origin from the salty waters of the North Atlantic flowing into the Greenland-Iceland-Norwegian ('GIN') Sea, and its end in the lowest salinity area of the world ocean, the North Pacific. If, as expected, the Great Conveyor slows, the strength of the Norway Current will decrease, and the temperate climate of Europe will be replaced by colder, more Arctic, conditions.

## 26.3    Response of Land Plants to Higher Levels of $CO_2$

The simplistic view is that $CO_2$ is a nutrient for plants; they need it for photosynthesis, so the more the better. This has become a slogan for some, but it has little basis in reality.

We already know from studies in the Netherlands, that several species of trees have already reached the limit of their ability to adjust to the higher levels of $CO_2$ in the atmosphere today. What we do not know is what this means for the future. Some plants, like tomatoes produce more fruit with increased $CO_2$. However, if, as many biologists think, the $C_4$ plants evolved to cope with the low levels of $CO_2$ that have prevailed over the last 35 million years, what will be their response to the increasing $CO_2$ concentrations? Some of the $C_4$ plants, such as corn, are major food sources in the United States. While there are data on the response of individual species in laboratory conditions, there are only limited data on the response of entire ecosystems in nature. Some of the experiments suggest that the out-of-place plants we call 'weeds' will do very well with higher levels of $CO_2$. It is just

one of those things we don't know much about. However, one thing is certain. You can't raise the level of $CO_2$ without enhancing the greenhouse effect—global warming.

## 26.4    Response of Land-Based Life to Climate Change

The last time there was rapid climate change on planet Earth was during the deglaciation. The global temperature change was of the order of 4.5 °C ($\sim 8$ °F), from the pre-industrial 15 °C (59 °F) down to 10.5 °C ($\sim 51$ °F). Figure 26.3 shows the temperature differences in the Atlantic Ocean between the 1980s and the Last Glacial Maximum ('LGM') about 21,000 years ago.

When you look at Fig. 26.3, the thing that immediately strikes you is that most of the difference is in the Northern Hemisphere, which is, of course, where the continental ice sheets formed and then melted away. The average rate of temperature change per century is calculated from the temperature difference, assuming that the LGM was 21,000 years ago, and that the ice had finished melting 7,000 years ago. In regions far from the ice sheets, sea level reached its present height about 7,000 years ago. This also happens to be 'when civilization began.' It marks the time

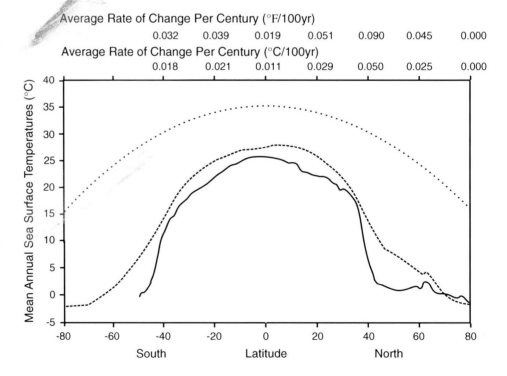

**Fig. 26.3** Annual average sea surface temperature from south to north through the Atlantic Ocean. Modern (1980s) temperatures are shown by the *dashed line*. Temperatures during the Last Glacial Maximum are shown as a *solid line*, based on data from fossils and climate modeling. The average rates of change per century for particular latitudes are shown on the top of the diagram. It assumes that the end of the LGM was 21,000 years ago and the end of the deglaciation was 7,000 years ago. *Dotted line* at top is an estimate of Cretaceous temperatures 90 million years ago

when the climate stabilized. The most rapid rates of temperature change were of the order of 75/1000ths of a degree C (0.14 °F) per century around latitude 40° N, which corresponds to Washington, D.C., the Ohio River Valley, and Boulder, Colorado. It also marks the approximate southern limit of the ice sheets in North America. Temperatures on land at that latitude were somewhat colder than the ocean because of the cold air coming off the ice. It is useful to remember that the tree-line, where forests give way to treeless tundra, approximates the mean annual 0 °C isotherm.

That rate of change is important; if the Earth warms by 2 °C (3.6 °F) during the 21st century, which seems to be a reasonable assumption, the rate of change will be 60 times faster than the average during the deglaciation. From the history of sea level during the deglaciation, which we will discuss later, we know that the deglaciation did not occur smoothly, but that at times it was more rapid, perhaps as much as 4 or 5 times the average. But it is important to realize that what is expected during this century is climate change at a rate unprecedented since an asteroid hit the Earth 65 million years ago. So the question is: how did life on land cope with the climate change during the deglaciation.

Life on land today is dominated by plants. Plants provide the food for animals. But animals are highly mobile; they can go where the food is. Plants are rooted in place. They can migrate by having their seeds transported to new locations, usually either by the wind or by animals. Trees are especially vulnerable to rapid climate change. They require years to grow to maturity and produce seeds. There is another difference between land animals and plants. Plants leave more fossil evidence behind, particularly in the form of pollen grains preserved in bogs and lake sediments. This fossil evidence allows us to know about the migration of plants as the climate changed.

So what happened as the ice sheets grew and decayed? Remember the sawtooth shape of the ice sheet growth and decay curve? The ice sheets grow slowly by accumulation of snow on land that does not completely melt during the summer. It took about 80,000 years for the ice sheets to grow to their greatest extent at the LGM. But deglaciations occur rapidly, the last one in only about 14,000 years. All that has to happen is for the ice to melt, and that can occur much more rapidly than an ice sheet can grow. Trees and other plants were able to cope well with the growth of the ice sheets, although some species became extinct. Others, like the Ponderosa Pines of the lower slopes of the Rocky Mountains, survived in small restricted areas called refugia. During the deglaciation, things were different.

Early in the last century it was thought that the ice sheets grew and decayed slowly, at about the same rate. The sawtooth curve was discovered by Cesare Emiliani in the 1950s. If we look at the distribution of plants in North America before settlement by Europeans and clearing and cultivation of the land, the plants grew in typical communities, now often called 'biomes.' Tropical plants have a foothold in southern Florida, but along the northern margin of the Gulf of Mexico there was an almost continuous forest of pines, the Southern Yellow Pine Forest. Although this forest is actively cut for lumber and to make paper, the yellow pines grow quickly, and much of the forest remains intact today in the form of 'second growth' trees. North of the pine forest, throughout the Midwest, were diverse forests of many species of deciduous trees and undergrowth.

'Deciduous' refers to trees that drop their leaves in the fall. In fact, that is the origin of the term 'fall' for autumn. Autumn is from the French 'automne' and was probably introduced into England by the Norman Conquest; it remains the standard term in England. 'Fall' is a North American term, to remind us of the falling leaves, and probably adopted to distinguish us from the English. Almost all of these deciduous forests were cleared to make way for agriculture. Remember that Thomas Jefferson thought that clearing these forests would improve the climate of North America. Almost none of the original growth remains, but there are some small areas in the Midwest with beautiful second growth trees. In the latter half of the 20th century these forests have returned to large areas of New England as the land is no longer used for agriculture. North of the deciduous forests were (and still are) the 'north woods,' forests of spruce and fir. Still further north these give way to the boreal forest, or taiga, where the evergreen trees are dwarf. The tree-line lies at the northern end of James Bay, the southern embayment of Hudson Bay. North of that there are no trees, only tundra plants.

Early in the last century it was thought that these plant communities had simply migrated south in front of the advancing ice, compressed into much narrower bands north of the Gulf of Mexico, with the tundra in New Jersey, southern Ohio, Indiana and Illinois, and extending west to the Rocky Mountains across Iowa, Kansas and Nebraska. During the glacial retreat, these bands of forests simply expanded back to the positions they occupy today.

We now know much more about the glaciation, thanks to an abundant record of fossil plants in the deposits laid down in lakes and bogs as the ice retreated. We now know that many of the trees were unable to keep up with the rate of climate change. In many areas, the plant communities consisted of groupings that have no modern counterparts. In other words, chaos reigned. By the time some trees reached maturity conditions were no longer optimal for their growth. Plants that could migrate rapidly, annuals and perennials with short lifetimes, were able to survive; others were not able to keep up. After the deglaciation ended 7,000 years ago, the climate stabilized and the plant communities were able to sort themselves out again.

In Europe a special situation existed. The ice sheet over Scandinavia advanced south to the edge of the Central

European massifs, and another smaller ice sheet over the Alps advanced northward out of the Alpine front. The region in between was reduced to tundra. It was in that hostile environment that humans found time to produce the oldest known works of art, recording the animal life of their surroundings in the caves at Altamira and Lascaux.

During the latter part of the deglaciation, many large animals became extinct. The mammoths, giant sloths, large bears, saber-toothed tigers and many other species disappeared. The cause of the extinction remains uncertain. My own view is that humans played the major role. Humans had two hunting methods that were disastrous for many other species: fire and more recently, spears. We know that when the aborigines reached Australia about 40,000 years ago they literally set fire to continent, causing the extinction of many animals and even plants. The use of fire as a hunting tool was widespread, and was a major technique used by native North Americans before the arrival of the European colonists. The large carnivores were probably not a target of humans, but their food supply was. The invention of the spear point had made humans able to attack animals much larger than themselves. It all happened very quickly. Animals that had evolved over millions of years, and that had survived many glaciations and deglaciations suddenly became extinct. In the late 19th century, the westward push of the European settlers resulted in slaughter of the American Bison that reduced its populations to a few herds. The purpose was to eliminate the food supply of the American Indians.

There is considerable objection to the idea that humans were the cause of the extinction of so many of the world's large animals, but I fear that it rests on the idea that humans are basically 'good' and would never do such a thing. Unfortunately, our current experiment on our planet, adding greenhouse gases to the atmosphere at an unprecedented rate, suggests that, although knowledgeable, most of us care no more for the future of the planet than did the unwitting early humans seeking food for their growing populations.

The desire to find a cause for the great Late Pleistocene extinctions other than man is intense. The latest idea, not yet widely discussed in the scientific literature, but reported in newscasts, is that an impacting comet exploded over Canada 11,500 years ago, wiping out the large mammals in a catastrophic extinction event. This would have been a gigantic version of the Tunguska event of June 30, 1908 in Siberia, when an explosion of a meteor or comet fragment in the upper atmosphere leveled about 80 million trees in forests over an area of 2,150 km$^2$ (860 miles$^2$). The Tunguska event, which occurred in an unpopulated area, is now estimated to have occurred at an elevation of 10–15 km above the Earth's surface and to have had an energy about 1,000 times greater than the atomic bomb dropped on Hiroshima. The only surface traces of the proposed 11,500 year event are the circular landforms of the Carolinas, called the Carolina

Bays. Other remnants would have (conveniently) fallen on the Laurentide Ice Sheet and left no trace. If true, this hypothesis would free native North Americans of blame for the extinction of the large mammals, which seems to be its main attraction. However, it fails to explain the extinction of large animals on the other continents during the late deglaciation.

## 26.5    Response of Land Life to Future Climate Change

The largest problem in understanding what may happen in the future is that the global climate is entering uncharted territory. If there is a global temperature increase of 1–2 °C during the 21st century, most of the temperature increases will occur in the mid- and high-latitudes. Such a rate of change is unprecedented in the past. Natural forests will not simply migrate northward, the temperature change is occurring faster than trees can grow to maturity and produce seeds. More likely will be interventions by other organisms that will produce unexpected results. An example is the spread of the Mountain Pine Bark Beetle in Colorado, which has killed millions of trees since 2002. The infestation is attributed to the warmer winters of the first decade of the 21st century. The beetle eggs and larvae will die if there is a period of extremely cold weather, with temperatures of −20 °C (−4 °F) in the fall or −40 °C (−40 °F) in late winter, throughout the affected area. The epidemic of infestation and forest destruction now reaches further northward, even into Canada. The effect is a positive feedback, producing more $CO_2$ as the wood decays or is consumed in fires sweeping through the dead forests.

This raises the question as to whether the regrowth of dead forests and development of new forests can consume an appreciable portion of the $CO_2$ produced by burning fossil fuels. Trees store carbon as wood in their trunks and limbs. Herbaceous plants take up carbon as they grow in the spring and summer, but it is returned as they lose their leaves and decay in the fall and winter. This effect is seen clearly in the ups and downs of the Mauna Loa $CO_2$ concentration data (Fig. 21.15). On a recent trip to China I visited Harbin, the city in the north where many White Russian emigrés had settled after the October Revolution, and which is now famous for its winter ice and snow sculpture festival. I noticed that the roads north of Harbin were lined by four to six rows of young, rapidly growing trees on each side of the roadway—a start on trying to consume the $CO_2$ produced by the vehicles. A hectare of mature forest has about 370 large trees, and they consume about 120 metric tons of $CO_2$ each year. That translates to 53 US tons consumed by 150 trees per acre. The average US household produces about 19 metric tons (21 US tons) per year. Young trees grow faster, but they are smaller, so I would guess that about 10 saplings might consume the same amount of $CO_2$ as a large tree. So, in order to offset its $CO_2$ production, each US household would need to plant about 600 trees each year. On a global scale, the areas of new forest planted each year required to offset the global $CO_2$ increase from burning fossil fuels would be about 500,000 $km^2$ (193,051 $miles^2$) per year. That is an area about twice the size of Colorado, and 20 % larger than California, planted *each year*. And they would have to remain permanently as forest.

## 26.6    Life in the Ocean and Seas

Life beneath the surface of water, whether it is in the open oceans, shallow seas, or even lakes, is fundamentally different from that on land. As on land, it consists of plants that provide the food for animals, and the plant life must always be much more abundant than the animals. If you look around on land, you see plants. Occasionally you may see an animal, but they seem quite rare in comparison with the plants. On the other hand, if you look in the open ocean, you see animals.

## 26.7    Life in the Open Ocean

I will never forget diving in the Gulf Stream off Bimini. When you look down, the waters fade to black; after all, the bottom is about a mile down. Looking horizontally it is as though there is a blue curtain of fog far away. On this dive we were looking for plankton to collect for study. With nothing for reference it is hard to judge the size or distance of anything. I remember seeing some dark shapes appear though the fog, then turn and disappear again. Not being fond of sharks, I was disturbed and just hung motionless in the water. Then the shapes reappeared and came toward me at an incredible speed. They were flattened side-to-side, not top-to-bottom, so I knew they weren't sharks. It was a small school of tuna, each fish almost as big as I was. They came right over to me, looked me over, decided I wasn't edible, and left as quickly as they had appeared. We were actually collecting planktonic foraminifers in jars, but another creature I saw that day was a Ctenophore. It looked like some sort of elongate jellyfish with covered by flickering Christmas-tree lights and until I had a good look at it I had no idea what it was. As alien a creature as you can imagine. On that day of diving I did not see a single plant—all too small to be visible to the naked eye. Diving in the open ocean is not everybody's idea of fun because unless you know what you are looking for there is not much to see, Most of the visible plankton are so small you will not see them if you move, so you just hang there in the water and wait for them to drift past. Diving in shallow water with a visible bottom is a much more entertaining experience.

Although tiny, many of the planktonic organisms create siliceous or carbonate shells of indescribable beauty. Ernst Heinrich Philipp August Haeckel (1834–1919) was a German biologist, naturalist, philosopher, physician, professor, and artist, who discovered, described, illustrated and named thousands of new species, many from the collections of the H.M.S. Challenger expedition. In 1904 he published a book, *Kunstformen der Natur* (Art Forms of Nature) with 100 plates of beautifully arranged drawings. Among them are some of the best pictures of ocean plankton ever made. Three of the plates are reproduced here in Fig. 26.4.

Much of the life in the open ocean is planktonic, that is, it consists of tiny plants and animals that drift passively in the water, too small to be seen. The plankton forms shown in Fig. 26.4a, b are some of the 'primary producers' i.e. plants. Those shown in Fig. 26.4c, d are tiny single-celled animals, protozoans. They catch plants with fine threadlike filaments of protoplasm called pseudopods. The lager animals that can swim are called nekton. The protozoans are food for small marine arthropods, copepods and their somewhat lager relatives, krill. These are in turn food for larger animals such as the rapidly swimming 'arrow worms' *Sagitta,* best seen at night by hanging a light over the side of the boat. Above these in the food chain are the jellyfish, fish, sharks, and whales. The largest animals on Earth, Blue Whales, short circuit the food chain by feeding on krill and other plankton.

Because of the high heat capacity of water, temperature changes in the ocean are not as sudden as on land. Currents changed their position as the Earth cooled and warmed. The plankton, being mostly single-celled plants and animals, evolved in response to the changes. The calcareous nanno-plankton evolved continuously, through a series of forms, over the past few million years. One of the most abundant organisms in the ocean, *Emiliania huxleyi* (Fig. 26.4b) evolved from an ancestral form, *Gephyrocapsa protohuxleyi*, only 268,000 years ago, during one of the colder episodes of the penultimate glaciation. It soon replaced its ancestor. It is a species adapted to change, and has expanded its range to live in all but the coldest waters. It is the only organism other than humans that can change the albedo of the planet. Through its ability to continuously secrete coccoliths made of calcium carbonate, it is also the major agent of removal of $CO_2$ from the ocean. It has recently been discovered that *E. huxleyi* is a very special organism, with great genetic diversity. It has a particularly large 'pan-genome.' This means that there is a set of common genetic information present in all strains. The gene pool varies and depends on the geographic location and the respective living conditions, allowing it to adapt to changing conditions. It reproduces about once a day, greatly enhancing its ability to cope with changing conditions. That is all good news; the bad news is that laboratory experiments indicate that when the ocean becomes more acidic *E. huxleyi* seems to stop making the

**Fig. 26.4 a** Haeckel's Plate 4: Diatoms (opaline silica shells, mostly about 1 mm across), **b** scanning electron micrograph of the calcareous nannoplankton *Emiliania huxleyi* (calcite coccospheres made of coccoliths, each only a few microns across, too small for Haeckel to have seen). **c** Haeckel's Plate 2: Foraminifera, with *Globigerina* in the center (calcite shells called 'tests,' mostly about 1 mm across). **d** Haeckel's Plate 91: Radiolarians (Spumellaria, opaline silica tests, mostly about 1 mm across)

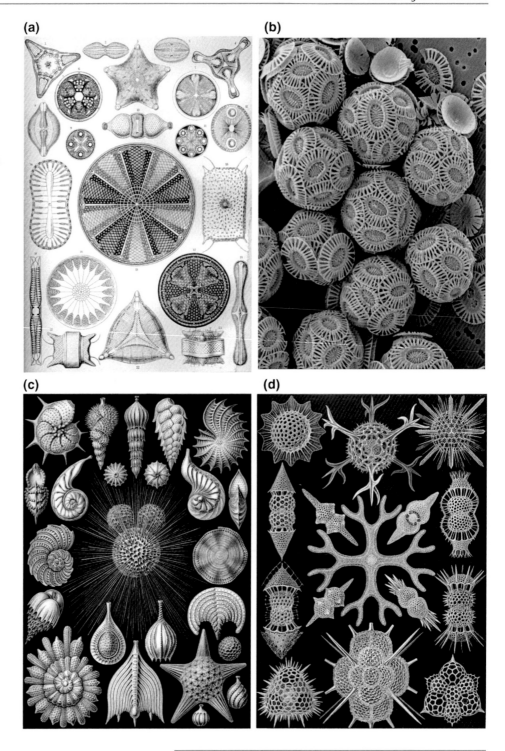

carbonate coccoliths that are such an important part of the carbon cycle. A recent 'bloom' of *Emiliania huxleyi* off southern England is shown in Fig. 26.5.

For organisms living in the open ocean, the glaciations and deglaciations involved movement of their environment with them trapped inside. Independent migration to keep up with the changing environment was not a problem.

## 26.8   Life Along Coasts and in Shallow Seas

Organisms that live on the sea floor are called 'benthos.' In tropical shallow waters you may see marine grasses, land plants that have adapted themselves for life in the sea, and larger algae. But they are a small part of the plant life in the ocean, most of which is microscopic. The animals seem to

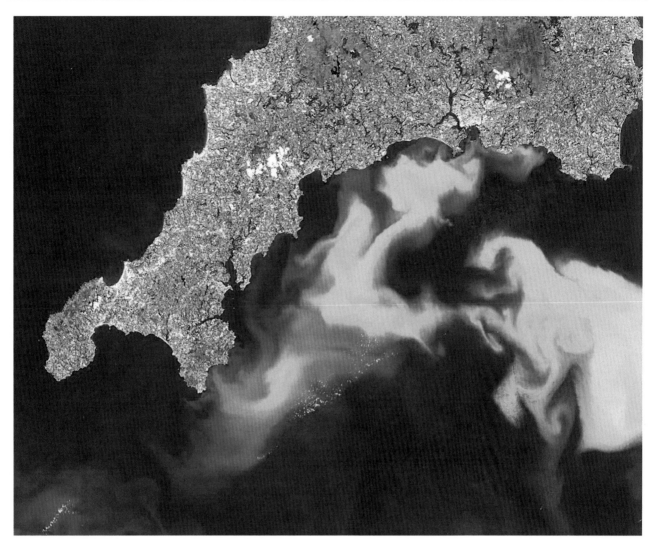

**Fig. 26.5**  NASA image of a bloom of *Emiliani huxleyi* off the south coast of England in 1999. The tiny dots in the picture are mostly fishing boats

dominate and appear to be much more abundant than the plants, but there is a big difference between land and ocean plants. In the oceans most of the plant life is microscopic. The microscopic plants reproduce once a day or sometimes even more often. Because they reproduce so often they provide a rapidly changing food source for the much more obvious longer-lived animals. Cold waters along rocky coasts are characterized by forests of kelp, giant marine algae that grow attached to the rocks. Strands of kelp can be 10 m or more long. Along the California coast you can see the kelp beds reflected as smoothed water on the surface, a short distance offshore. But the giant kelp are restricted to cold waters; equatorward they are replaced by small algae covering the rocks, and marine grasses in the sand and muddy areas. Between the latitudes of about 30° N and S the waters are warm and many of the shallow water organisms secrete calcium carbonate. These build up carbonate banks, like the Bahamas. South Florida is also a carbonate bank,

exposed today just above sea-level. The coasts themselves are often forests of mangroves with many tidal channels that provide space for fish nurseries.

As the glaciation came on, the waters cooled, but more importantly for shallow water organisms, sea level went down. On steep rocky coasts, this was not much of a problem, the organisms, such as the kelp forests, simply migrated downward with the descending sea level. However, where the coast is bordered by a continental shelf, such as off the East Coast of the United States, there was a real problem. As the sea-level declined, less and less space on the gently-sloping shelf was available. At the sea level low-stand about 21,000 years ago, the shore was at the edge of the continental shelf, and the organisms that had inhabited the shelf environments were forced into small areas of refuge. As the deglaciation occurred, sea-level rose rapidly, and quickly new areas were available for colonization. By the time sea level stabilized 7,000 years ago the advancing waters had flooded not only the shelves, but the river valleys that had been cut during the sea level low-stands. Today, these flooded river valleys are called 'estuaries.' They provide a special environment for organisms able to tolerate the salinity changes associated with the fresh water coming from the rivers and the inflows of seawater with the tides. They are among the most productive coastal areas for what we humans call 'seafood.'

The carbonate platforms, like the Bahama Banks (Fig. 26.6), are very different. They are large flat-topped plateaus rising from the deep sea floor or, as is the case for southern Florida, growing outward from the continental margin. Their margins do not slope gently seaward, like the continental shelves; they drop off steeply into the ocean depths. As sea level goes down, these carbonate banks become exposed as a flat land area underlain by limestone; like South Florida and Yucatan today. The organisms that inhabited to tops of the submerged banks had to migrate down the sides into tiny refugia. Carbonate banks, atolls, many tropical islands, and some continental margins are bordered by coral reefs, like the Florida Reef Tract and the Great Barrier Reef of Australia.

There are two major kinds of corals; those that inhabit warm shallow waters, and others that inhabit cold deep waters. The coral animals that inhabit warm shallow waters allow plant cells to inhabit their tissues. The plant produces food for the coral and the coral provides nutrients for the algae. When two organisms live together for mutual benefit, the relationship is termed 'symbiotic.' The shallow water corals which have symbiotic algae are termed 'hermatypic' corals. The other class of corals lives in cool or cold waters below the limit of penetration of light; they do not have symbiotic algae and are termed 'ahermatypic.' There are deep water reefs of ahermatypic corals at the mouths of the Norwegian fjords, but they do not attract many tourists.

As sea level went down, the hermatypic corals migrated downward, forming fringing reefs clinging to the coast. They returned to their former habitats as sea-level rose through the deglaciation. Like land plants, corals are attached to the sea floor. They require rocks or a hard substrate on which to start their growth, and these were abundant as they migrated downward and back up. The only special requirement was that they be able to migrate fast enough to be able to keep up with changes of sea level. Those unable to keep up became extinct.

One Caribbean coral species, *Acropora palmata*, the 'Elkhorn or Moosehorn coral' has been particularly useful in reconstructing the sea-level changes during the deglaciation. It is restricted to waters no more than 5 m (about 16 ft) deep. It grows rapidly, 5–10 cm/year, so it is able to keep up with even the most rapid sea-level change.

## 26.9 Marine Life and the $CO_2$ Connection

You will recall from our discussion of the carbon cycle (Sect. 21.2) that atmospheric $CO_2$ can be removed in two ways: (1) through incorporation in plant and animal tissue, death, and burial in sediments; or (2) over the long term, millions of years, atmospheric $CO_2$, combined with water to make carbonic acid is consumed by weathering of silicate rocks to make limestone ($CaCO_3$). What may come as a surprise is that almost all limestone is made of $CaCO_3$ produced by marine plants or animals. Very little $CaCO_3$ is precipitated directly from seawater.

Much of the ocean is floored by $CaCO_3$. In fact, over half of the solid surface of the planet is covered by $CaCO_3$. In the deep sea it is in the form of calcareous 'oozes' made of coccoliths and the shells of planktonic foraminifera. Could it be that excess $CO_2$ entering the ocean from the atmosphere will dissolve that $CaCO_3$ and allow it to stay there in solution? Or will it combine with the water molecules and accumulate there as carbonic acid, eventually acidifying the ocean and killing off life in the sea? There are two possible outcomes, one benign, one deadly. Which, if either, is likely to happen? The way the system works, although rather complicated, is well worth understanding because it determines how long the human perturbation of $CO_2$ will last and what its effects will be. The ocean system has the capability of regulating $CO_2$ and has the potential to take care of temporary perturbations in a fraction of the time required by the silicate weathering system. It is just a question of how fast the regulation can occur.

Just as land plants and animals exert control over much of the carbon cycle, organisms in the sea do too. If you recall Table 21.1 you will remember that the mass of carbon in living organisms in the ocean ($0.005 \times 10^{15}$ kg) is only a small fraction, about 1 %, of the mass of C in land plants ($0.60 \times 10^{15}$ kg). However, the amount of C dissolved in the ocean ($0.74 \times 10^{15}$ kg) is larger than that in land plants. You will also recall that the mass of inorganic carbon in $CO_2$ dissolved in the ocean is very large ($\sim 39 \times 10^{15}$ kg). It is about 100 times larger, than the mass of carbon in $CO_2$ in the atmosphere ($0.42 \times 10^{15}$ kg). Thus, except for that stored in rocks, the ocean is the largest reservoir of both organic and inorganic carbon.

The biggest difference between land and sea is that the turnover of life in the ocean proceeds at rates far faster than that of life on land. The living, tiny single-celled marine plants may not have as large a mass as the land plants, but they can reproduce once a day or even more often. So oceanic organisms have the potential to play a very important role in the carbon cycle and to regulate atmospheric $CO_2$. In fact the amounts of carbon fixed as living tissue

**Fig. 26.6** The Bahama Banks. Florida is on the *left*, Cuba at the *bottom* of this NASA image. Everything in lighter shades of *blue* was land during the low sea-level stands of the Last Glacial Maximum

each year on land and in the ocean are about the same. Some of the marine plants, the planktonic coccolithophores and benthic green algae, also do something land plants do not do: they precipitate $CaCO_3$ from seawater. At present, for every four atoms of C fixed in living tissue in the sea, one atom of C is fixed in $CaCO_3$. However, over the long-run, it is $CaCO_3$ rock that ultimately becomes the largest storage reservoir of carbon.

To understand how this system works you need to know a little about ocean chemistry. The ocean contains dissolved salts, about 34.7 g/kg of seawater. These dissolved salts are in the form of ions. You will recall that ions are atoms or molecules dissolved in the water, and that each carries an electrical charge, either positive or negative. Every one of the 92 naturally occurring elements occurs dissolved in seawater, but most are in very low concentrations.

A major breakthrough in understanding the chemistry of the oceans came after the British Challenger Expedition (1872–1876) returned from its voyage. It was the first thorough scientific exploration of the ocean, involving circumnavigating the globe and even visiting the Antarctic (Fig. 26.7).

The Challenger brought back 77 pristine seawater samples from around the world. These were analyzed by a German Chemist, Wilhelm (William) Dittmar, who then resided in Glasgow, Scotland. His results were reported in the expedition reports in 1884. He discovered that with one exception, each of the seven major ions in seawater occurred in the same proportions regardless of the salinity of the water. This has become known as the Law of Constant Composition of Seawater. The exception was the calcium

ion, which was slightly more abundant relative to the other ions in samples taken at great depths.

Table 26.1 shows these seven most abundant ions, their abundance in seawater, and their electrical charge: The chloride ($Cl^-$) ion is the most abundant, and is about 16 % more abundant than the sodium ion ($Na^+$).

Dittmar discovered that the calcium ion ($Ca^{++}$) was about 0.5 % more abundant in deep waters than in surface waters. Because it carries a double positive charge, and because the ocean is electrically neutral, he reasoned that the number of

**Table 26.1** The most abundant ions in seawater. The electrical charge equivalent is the ionic charge (+ or ++; − or =) times the number of moles of the ions in a kilogram of seawater

| Ion | Moles/kg seawater | Electrical charge equivalent/kg |
| --- | --- | --- |
| $Na^+$ (Sodium) | 0.470 | 0.470 |
| $K^+$ (Potassium) | 0.010 | 0.010 |
| $Mg^{++}$ (Magnesium) | 0.053 | 0.106 |
| $Ca^{++}$ (Calcium) | 0.010 | 0.020 |
| Total positive ions (Cations) | | 0.606 |
| $Cl^-$ (Chloride) | 0.547 | 0.547 |
| $SO_4^=$ (Sulfate) | 0.028 | 0.056 |
| $Br^-$ (Bromide) | 0.001 | 0.001 |
| Total negative ions (Anions) | | 0.604 |
| Electrical charge discrepancy | Cations − Anions | 0.002 (+) |

**Fig. 26.7** H.M.S. Challenger in 1872. The first vessel designed specifically for oceanographic exploration

hydrogen ions in the deep sea must be less than in the surface waters.

If you recall, it is the concentration of hydrogen ions that defines how alkaline or acid a solution is. The numerical expression for the acid or alkaline nature of a solution is called its pH. You will also recall that the pH of a solution is defined in what seems to many of us to be a strange way. It is the negative logarithm of the hydrogen ion concentration. A neutral solution has a pH of 7. That is, 7 is the negative logarithm of the hydrogen ion concentration. The concentration of hydrogen ions is $10^{-7}$, that is, there is 1 hydrogen ion for every $10^7$ molecules of water, or one hydrogen ion for every 10 million water molecules. An extremely acid solution has a pH of 1 (1 hydrogen ion for every 10 molecules of water; concentrated hydrochloric acid is an example). A very alkaline solution has a pH of 14 (1 hydrogen ion for every $10^{14}$ molecules of water). A pH lower than 7 is acidic, a pH higher than 7 is alkaline. Seawater has a pH ranging between 7.4 and 8.35; it is therefore an alkaline solution.

As noted above, Dittmar reckoned that since the concentration of Ca$^{++}$ ions in the deep-sea is higher than in the surface waters, the concentration of hydrogen ions must be less in order to maintain a neutral electrical charge. Therefore, the water is by definition, more alkaline. He remarked that the 'alkalinity' of the deep sea was higher than that of the surface water.

Since then the term 'alkalinity,' used with reference to seawater, has acquired a very specific meaning. It is the difference in electrical charge between the four major positive ions (Na$^+$, K$^+$, Ca$^{++}$ and Mg$^{++}$) and the three major negative ions (Cl$^-$, SO$_4^=$, and Br$^-$). Overall, seawater has no electrical charge (If it did your toe would get a shock when you dipped it in the ocean at the seashore.). The surplus electrical charge of the major positive ions is balanced by some combination of the negatively charged ions. Here is where CO$_2$ and its aqueous forms come in.

When CO$_2$ enters the ocean, it reacts with the water and is converted to either the bicarbonate (HCO$_3^-$) or carbonate (CO$_3^=$) ion. You will remember this chemical equation, showing that CO$_2$ can exist in three forms in the ocean:

$$CO_2 + H_2O \rightleftarrows H^+ + HCO_3^- \rightleftarrows 2H^+ + CO_3^=$$

Remember that each individual CO$_2$ molecule can readily switch back and forth among these forms, so to keep track of the total of them, marine chemists refer to the 'total dissolved CO$_2$' in the water, denoted as $\Sigma$CO$_2$. It is also important to remember that these are all forms of CO$_2$ that are *not* incorporated into live or dead organic matter. These are 'species' of inorganic CO$_2$. The concentrations of these different forms of CO$_2$ combined with water are shown in Table 26.2. You will see that although the $\Sigma$CO$_2$ can be a large number, there will still be almost no dissolved CO$_2$ as such available for photosynthesis by the plants.

For seawater, the 'alkalinity' is now narrowly defined as the electrical charge on the bicarbonate and carbonate ions required to balance the surplus charge of the positive ions. In Table 26.1, what I have labeled as the 'electrical charge discrepancy' (0.002), is, to a marine chemist, the 'alkalinity.' In modern oceanographic practice, the alkalinity is expressed in 'μ equivalents/ kg'. The 'μ' (micro) means that the 0.002 shown above gets multiplied by 1 million, so you don't need the decimal point. Expressed in these terms, alkalinity in the ocean varies from about 2300 in tropical surface waters, through 2400 in deep waters off the Antarctic, to 2500 in deep waters in the North Pacific. Now the plot thickens, because all of this is related to the way the ocean manages CO$_2$ and how it has begun to respond to the anthropogenic increase in this greenhouse gas.

Dittmar figured that the extra Ca$^{++}$ ions in the deep sea came from dissolution of CaCO$_3$ on the sea floor; another remarkable insight into the way the world works! Then the question arises, what controls whether the CaCO$_3$ will be stable or whether it will dissolve. In fact, there are two minerals that are made of CaCO$_3$, calcite and aragonite (Fig. 26.8).

If you have a mineral collection, you probably have specimens of both calcite and aragonite. They differ in the way the atoms are arranged in the crystal structure. In aragonite they are arranged in a strict rhomboidal pattern. Its crystal structure is technically called orthorhombic. Many cave deposits (stalagmites and stalactites) are made of aragonite. The crystals are often needle-like and make beautiful but very delicate mineral specimens. However, most aragonite crystals are very small, often too small to be seen without a magnifying glass. In shallow tropical seas,

**Table 26.2** The different forms of inorganic CO$_2$ in the ocean, and their relative abundances in shallow and deep waters

| Form (species) | 1 carbon dioxide, 1 water | 1 bicarbonate ion, 1 hydrogen ion | 1 carbonate ion, 2 hydrogen ions | Total dissolved inorganic CO$_2$ |
|---|---|---|---|---|
| Formula | CO$_2$ + H$_2$O | H$^+$ + HCO$_3^-$ | 2H$^+$ + CO$_3^=$ | |
| Abundance in surface waters ($10^{-3}$ mol/kg) | CO$_2$ = 0.01 | HCO$_3^-$ = 1.80 <br> H$^+$ = 1.80 | CO$_3^=$ = 0.20 <br> H$^+$ = 0.40 | 2.01 |
| Abundance in deep waters ($10^{-3}$ mol/kg) | CO$_2$ = 0.03 <br> H$_2$O = 54,000 | HCO$_3^-$ = 2.14 <br> H$^+$ = 2.14 | CO$_3^=$ = 0.06 <br> H$^+$ = 0.12 | 2.23 |

**Fig. 26.8** Crystals of aragonite and calcite. **a** An aragonite needle. These are often exceedingly small, only a few microns in length. **b** An orthorhombic aragonite crystal such as might be found in a cave. **c** A rhomb of calcite. Rhombs may be crystals or formed by breaking along rhomboidally-arranged planes of weakness in the crystal. Such cleavage rhombs are very common mineral specimens. **d** A 'scalenohedral' crystal of calcite

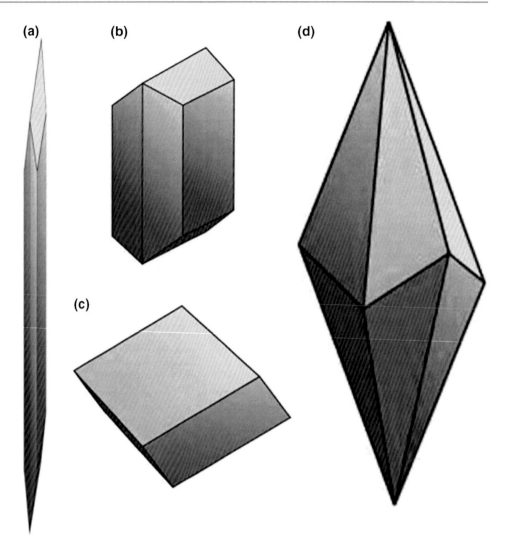

much of the sediment is made of extremely small, long aragonite needles, having the shape shown in Fig. 26.8. In calcite the atoms are arranged to form triangular patterns, but again good crystals are rather rare. If you have a specimen of calcite it is probably labeled 'Iceland spar' and is more or less clear, having a rhomboidal shape. The faces of most specimens are not true crystal faces, but 'cleavage surfaces.' When struck, the mineral breaks along the 'cleavage' planes. These are planes of weakness in the crystal structure, arranged in such a way as to produce the rhombs. Although both of these minerals are made of $CaCO_3$, they have very different solubilities in seawater. Calcite is much more resistant to dissolution than aragonite.

The precipitation of $CaCO_3$ as either of these minerals removes $CO_2$ from the system. If the $CaCO_3$ gets buried in the sediment, it is 'out of the way' for millions or tens or even hundreds of millions of years. This is the end of the road on the carbon cycle, and over Earth history most of the $CO_2$ that has been introduced into the atmosphere from the weathering

of sediments, through volcanic eruptions, and outgassed from the interior of the Earth, has ended up as limestone.

As mentioned above, the solubilities of calcite and aragonite in seawater are different. But both depend on the concentration of the carbonate ion in the water, as shown in Fig. 26.9. To the right of the saturation lines in Fig. 26.9 the minerals are stable and will not dissolve. To the left the minerals go into solution. Note that both saturation lines slope to the right with depth; this means that the solubility of both minerals increases with depth.

You will notice that although the concentration of $CO_3^=$ in the surface waters in the Atlantic and Pacific is about the same, it is much lower in the deeper waters of the Pacific than it is in the Atlantic. This leads to an interesting situation. Aragonite that sinks in the Pacific dissolves and goes back into solution at a depth of about 800 m. The level where it goes from waters being saturated to undersaturated, where it dissolves, is called the Aragonite Compensation Depth. In the Atlantic, aragonite may not dissolve until it

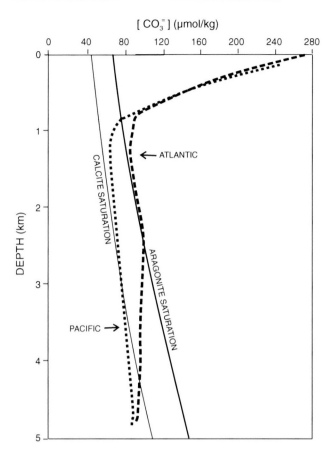

**Fig. 26.9** Saturation levels for calcite and aragonite plotted against the concentration of the carbonate ion [CO₃⁻] and depth. The *dashed line* is a typical depth profile for CO₃⁻ concentration in the tropical Atlantic, the *dotted line* for a profile in the tropical Pacific

reaches a depth of 2,500 m. Calcite particles that sink in the Pacific dissolve below about 2, 700 m, but in the Atlantic they are stable to a depth of over 4,000 m. The level where

calcite dissolves is called the Calcite Compensation Depth (CCD), or because this is where all of the calcium carbonate particles finally disappear, the Calcium Carbonate Compensation Depth (CCCD). Because the concentration of CO₃⁻ in the Atlantic is higher than in the Pacific, much more of the Atlantic Ocean floor is covered by carbonate sediments; it is the 'carbonate ocean.' Fig. 26.10 is a schematic cross section of the South Atlantic at 25° S, showing how much of the sea floor is above the ACD and CCD.

## 26.10 The Role of Plants and Animals in Fixing CaCO₃

Since the upper ocean is supersaturated with respect to both aragonite and calcite, you might think that they would precipitate directly from the seawater. In fact, we had arguments about this through much of the late '60s and '70s. Some geologists working in the Bahamas reported seeing large areas of milky-white water, and found it to be filled with minute aragonite needles. These were called 'whitings.' Usually these observations were made from boats or low-flying planes. Over years of working at Bimini I had never seen the phenomenon and seriously doubted the observations. However, did I see this phenomenon myself on a cruise using the Lerner Laboratory's small research vessel into the interior of the Bank. In the region west of Andros Island, where the water was so shallow the vessel's keel was touching the soft sediment most of the time, we saw a whiting. The milky water patch had a very sharp edge, moving forward with puffs of white cloud appearing one after another. Jumping into the very shallow water to see what was happening I found that there was a huge school of white fish, which turned out to be Bahamian mullet, almost

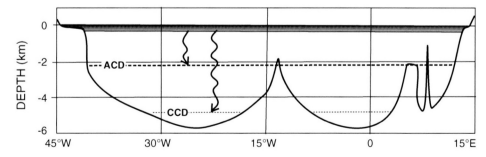

**Fig. 26.10** Schematic cross section of the south Atlantic at 25° S showing the relation of the ACD and CCD to the ocean floor. The *gray bar* at the top of the ocean is the photic zone, the layer through which light capable of supporting photosynthesis penetrates. The wiggly arrows represent particles of aragonite (*left*) and calcite (*right*) settling from their site of production in the photic zone into the ocean depths. The average depth of the Mid-Ocean Ridge is 2.5 km, so only its highest crests, such as that shown here at about 10° W, might be above

the ACD. Some deeper oceanic platforms, such as shown here in the eastern South Atlantic may be sites where aragonite can accumulate. Aragonite is the typical form of CaCO₃ on tropical shelves and shallow carbonate platforms. The CCD is much deeper allowing accumulation of calcitic particles over much of the slopes of the Mid-Ocean Ridge. The Pacific Ocean is deeper than the Atlantic, and its ACD and CCD are shallower, so that carbonate sediments cover much smaller areas of the Pacific Ocean floor invisible against the white bottom. They were stirring up the bottom as they foraged for food. Since the sediment was almost exclusively aragonite needles, the water was indeed filled with aragonite needles. On that cruise we became aware that almost all of the fish we saw were white. Only the sharks had dark topsides and were easy to spot. I became one of those who thought that whitings were caused by the winds or fish stirring up the bottom sediment.

Then, a few years later, in a jet flying from San Juan, Puerto Rico, to Miami, I saw a whiting from 30,000 ft. A huge area of the Great Bahama Bank was covered by milky white water, with many streaks of white extending into the pale blue waters of the Bank, a truly beautiful sight. I became a believer that under the right conditions, aragonite can precipitate directly from seawater. But such inorganic precipitation seems to be a relatively rare phenomenon.

Both calcite and aragonite are secreted by organisms in the sea. The open ocean calcareous plankton, coccolithophores and planktonic foraminifera, both secrete calcite. The coccoliths and shells they produce accumulate over wide areas of the ocean floor as 'Globigerina (a common planktonic foraminifer) ooze, coccolith ooze, or simply 'calcareous ooze.' At present, about 90 % of the fixation of the $CaCO_3$ that accumulates as sediment is carried out by these minute plankton organisms, most of this by the coccolithophores. The most important $CaCO_3$ fixing organism today is our old friend, *Emiliania huxleyi.*

It is thought that deposition of $CaCO_3$ may be related to how these minute marine plants carry out photosynthesis. Remember that free $CO_2$ is very rare in the ocean. To compensate for this, one relatively easy way to get $CO_2$ for photosynthesis is to use the most abundant of the ions, the bicarbonate ion $HCO_3^-$. And to keep the electrical charge neutral, the plant can also utilize one of the abundant positive ions in seawater, $Ca^{++}$ which has a natural affinity for $CO_2$:

$$2 \ HCO_3^- + Ca^{++} \rightarrow CO_2 + H_2O + CaCO_3$$

Having extracted the $CO_2$, it can be used with a water molecule for the photosynthetic production of sugar:

$$CO_2 + H_2O \rightarrow CH_2O + O_2$$

One of the advantages of using $Ca^{++}$ and $HCO_3^-$ together is that it does not change the pH of the surrounding waters. Many marine plants, not just the coccolithophores, produce $CaCO_3$ as a byproduct of photosynthesis. In the coccolithophores the $CaCO_3$ is secreted as calcite coccoliths. The coccoliths themselves are elaborate structures, and we still do not know their real function.

The planktonic foraminifera that inhabit the upper waters of the ocean also secrete calcite in the form of shells that look like tiny globes arranged in a spiral. Hence the name *Globigerina*, of a common form. It turns out that many of the shallow-dwelling planktonic foraminifera harbor symbiotic algae in their protoplasm, and these assist the foraminifer in producing its shell.

The other common carbonate-shelled organisms in the upper ocean are pelagic snails, pteropods. They are especially common in the Southern Ocean, around Antarctica. The shells of pteropods are aragonite. Although pteropods may be responsible for about 1/3 of the $CaCO_3$ that is fixed in the upper ocean, almost none of it becomes incorporated into the sediments on the sea floor. Although the shells of pteropods are stable in the surface waters, they go into solution as they sink to depths of 0.5–2.5 km in the ocean.

In shallow seas, the dominant form of $CaCO_3$ is aragonite. Some of the organisms that secrete aragonite are corals, green benthic algae, and many clams although some clams produce shells containing both aragonite and calcite. Although we usually think of corals as very important because they produce reefs, the much less conspicuous green algae produce most of the carbonate sediment that covers large carbonate platforms like the Bahama Banks. The corals form a barrier wall around the fringe of the bank, and the green algae fill in the center with masses of aragonite needles. Many of the green algae that grow on the shallow sea floor, such as *Penicillus,* which looks very much like a shaving brush, secrete or precipitate aragonite needles on the surface of the cells. It is not known whether they are actively secreted by the plant, or whether they form in the microenvironment on the surface of the cell where it extracts $CO_2$ for photosynthesis from the water, probably by using $HCO_3^-$ and $Ca^{++}$ as shown above. In either case, the aragonite needles accumulate in vast numbers on the sea floor, and the interior of the Bahama Banks are covered by 'muds' made of these needles.

The secretion of aragonite skeletons by hermatypic corals is undoubtedly aided by their symbiotic algae. And the tissues of the largest clam, *Tridacna* (the giant man-eating clams of the old Tarzan films), which produces heavy shells up to a meter across, are packed with symbiotic algae which aid $CaCO_3$ secretion.

But there is even more of a problem with all of this. In order to secrete aragonite or calcite, the animal or plant must create an organic membrane that duplicates the surface of the crystal. It is on this membrane template that the crystal is grown. The process is called 'epitaxial growth.' The problem is that the organisms cannot exactly duplicate the crystal structure with organic molecules. The crystals start to grow, but then they begin to introduce defects into the structure. The process may even need to stop and start over again on a fresh membrane template. Imperfect crystals are more soluble than perfect crystals, but sometimes the organic membranes are still present and they can actually inhibit dissolution. To make things more complicated, the magnesium ion ($Mg^{++}$) is more abundant than $Ca^{++}$ in the ocean, and is almost the same size as the calcium ion. It can substitute for calcium in the crystal structure. If it does, the resulting 'magnesium calcite' is more soluble than pure calcium calcite.

When working on sediments in the Bahamas I became very much impressed that the sediment was made up almost entirely of the shells of organisms that lived on it. But some of the shell materials that should be very abundant were simply not present. One of the most obvious absences was the beautiful calcite spicules produced by the purple sea fans and their relatives, known as 'soft corals'. Although they are made of calcite, they seem to dissolve in the supposedly supersaturated waters as soon as they are released. Study of them revealed they have regularly-spaced crystal defects, and are simply super-soluble. As far as I know, the most perfect calcite is secreted by echinoderms, such as sea urchins and starfish. They have a skeleton made of a myriad of calcareous platelets. Each platelet is a single crystal, but a very strange crystal. On close examination, the platelets look more like Swiss cheese, perforated by irregular holes. These holes prevent the crystals from splitting along cleavage planes like those that are the sides of the calcite rhomb shown in Fig. 26.8c. When slip along one of the cleavage surfaces begins, it soon propagates into a hole and is stopped. The echinoderms have devised a way of making a strong skeletal material with only half the weight it would have if it were solid.

Although coccoliths and the shells of the planktonic foraminifera are both technically calcite, each species produces calcite having a different solubility. As we examine sea-floor samples from greater and greater depths fewer and fewer species are represented by fossils. The pteropods, the pelagic snails that live in the upper layers of the ocean, make their shells of aragonite and occasionally their shells are found on the deep sea sediment surface. But they never survive the solution process long enough to be buried and incorporated into the sediment.

This complexity of solubilities makes it very difficult to predict how fast the ocean can respond to increasing CO$_2$. It is clearly a matter of thousands of years, but just what will happen and how long it will take is a question that is very difficult to answer, because we know very little about how plant and animal life in the ocean will respond to increasing levels of CO$_2$.

Incidentally, aragonite is much less stable under present conditions than calcite. It is generally more soluble in seawater, but it has another peculiarity. When it is exposed to fresh water it recrystallizes to calcite. So when sea level was lower, and the Bahama Banks and many other shallow areas were land, they were exposed to fresh water rain, and became calcite. This process often involves some dissolution and cementation, so that the resulting deposit is what we call limestone. Perhaps the oddest thing is that rain falling on the

islands in the Bahamas percolates through the carbonate sediments converting the aragonite to calcite and cementing the grains together. When it reaches the shore it carries out this same process making a deposit we call 'beach rock.' You can even find broken glass fragments from old bottles cemented into the rock, demonstrating that this process can proceed quite rapidly. Some beach rock deposits from older, lower stands of sea level have been identified by enthusiasts as foundations of buildings or roadways of Atlantis.

## 26.11 How Does the Organic Matter and CaCO$_3$ get to the Sea Floor?

In the 1960s, when we were just beginning to understand how the carbon cycle works in the ocean, many of us thought that when the coccolithophores or planktonic foraminifers died, their protoplasm decayed freeing the carbonate coccoliths or shells to sink through the water column to the sea floor. We actually measured the sinking rates, and found that, for coccoliths, for example, would take a year or more to reach the deep sea floor. They should have been carried all around the ocean by currents. But their fossils were found only on the sea floor beneath the areas they inhabited. Something seemed wrong.

A Japanese colleague, Susumu Honjo, started putting out sediment traps suspended on cables from the ocean floor, to see what the settling material looked like. It did not consist of single coccoliths or shells. It was made up of millimeter-sized particles which were soon identified as fecal pellets.

Come to think of it, I doubt that any plant or animal dies of old age in the sea. Everything gets eaten by some animal. Copepods are tiny shrimp-like animals that are abundant in the shallow waters of the ocean. They graze on the marine plants and also eat animals smaller than themselves. They produce copious amounts of fecal pellets that settle through the water to the sea floor in a matter of days. But here is an important thing to know: organic matter in the fecal pellets has a density almost the same as that of seawater, so those with pure organic matter will sink very slowly. There is plenty of time for them to be decomposed by bacteria that live in the upper waters of the deep ocean below the photic zone. So much oxygen is consumed in this decomposition process that an 'oxygen minimum zone' develops beneath the photic zone.

The deep ocean is rich in oxygen carried down by waters sinking in the Polar Regions. As a result almost all of the organic matter that sinks to the ocean floor is decomposed back into its components, CO$_2$ and H$_2$O, or rather into the bicarbonate, carbonate ions and hydrogen ions. As the deep waters move with the 'giant conveyor' from the North Atlantic through the South Atlantic and Southern Ocean to end up in the deep North Pacific, more and more oxygen is consumed and more and more $CO_2$ produced. The result is that the concentration of the carbonate ion decreases along the way and the alkalinity increases. It is estimated that, of the organic matter that sinks to the ocean floor, only about 1 % gets incorporated into the sediment.

However, the fecal pellets that include $CaCO_3$ sink more rapidly, because calcite and aragonite have a density more than twice that of seawater. The $CaCO_3$ acts as ballast, causing the pellets to sink faster, and many of them reach the sea floor. The dissolution, if it takes place, happens as the pellets decompose on the deep sea-floor. While we know the accumulation rates for calcite on the deep ocean floor, we do not know how much of the calcite originally secreted by coccolithophores and planktonic foraminifera in the near-surface waters this represents. There are many kinds of coccolith that are simply not found in deep sea sediments, so a good guess is that about half of the calcite is more soluble than it should be, and goes back into solution.

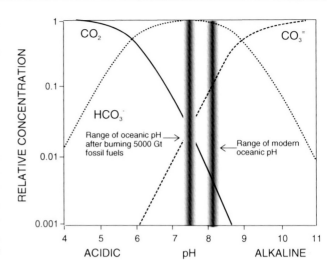

**Fig. 26.11** Relation of the relative abundances of $CO_2$, $HCO_3^-$, and $CO_3^=$ to pH

## 26.12   The Acidification of the Ocean

In response to the buildup of $CO_2$ in the atmosphere, the ocean is currently absorbing about 2 gigatons of C as $CO_2$ more than it releases to the atmosphere each year. Remember that the ocean has already absorbed about half of the $CO_2$ human activity has produced since the industrial revolution began. As the $CO_2$ enters the ocean it is converted to the bicarbonate and carbonate ions.

$$CO_2 + H_2O \rightleftarrows H^+ + HCO_3^- \rightleftarrows 2H^+ + CO3^=$$

These reactions produce positively charged hydrogen ions ($H^+$) that balance the electrical charge of the bicarbonate and carbonate ions. This results in an increase in the hydrogen ion concentration. In other words, the ocean slowly becomes more acidic.

But, as the ocean absorbs more and more $CO_2$ something very interesting happens. As the ocean becomes more acid the proportions of, $HCO_3^-$, $CO_3^=$, and dissolved $CO_2$ change. Today each of their relative abundances differs by about an order of magnitude. This will change as the ocean becomes more acidic. Figure 26.11 shows the relationship between these different species of $CO_2$ and pH.

Figure 26.11 shows that as $CO_2$ is added to the ocean and it becomes more acidic, the abundance of the carbonate ion will decrease while the abundance of the other two species will increase. The reduction in concentration of the $CO_3^=$ ion means that more and more of the ocean will become undersaturated with respect to calcite and aragonite. But something even more interesting happens: the carbonate ion ($CO_3^=$) requires two hydrogen ions ($H^+$) to balance it. As it is converted to bicarbonate ($HCO_3^-$), only one hydrogen ion is

needed to balance it. The drive toward acidity is slowed by this reaction. This effect is called 'buffering' of the system. The addition of $CO_2$ does not drive the ocean toward acidity as fast as one would at first imagine because of this effect. You will also see that as the $CO_3^=$ concentration deceases, less and less is available to be used in this buffering effect. The buffering capacity of the ocean ceases when the pH reaches 6.

There is good evidence that over the past 200 years the pH of the surface ocean has decrease by 0.1. That may not sound like much, but remember that pH is described by a logarithmic scale, and a difference of 0.1 is in fact a 30 % increase in the abundance of $H^+$ ions. Of course, most of this change has occurred during the past 30 years. Future projections are that if the 5000 Gt of readily available fossil fuel are burned, the oceanic pH will be about 7.4, becoming almost neutral.

The good news is that the acid will dissolve the $CaCO_3$ on the ocean floor:

$$CaCO_3 + H^+ + HCO_3^- \rightarrow Ca^{++} + 2HCO_3^-$$

You will notice that this reaction removed a hydrogen ion from the seawater, thereby making it more alkaline, and countering the acidification effect. Thus as $CO_2$ is added to the ocean, we can expect it to dissolve more and more of the vast amounts of $CaCO_3$ on the sea floor. This is a powerful negative feedback, and ultimately it will mitigate the problem of the anthropogenic increase in $CO_2$.

Unfortunately, this will not happen as rapidly as the $CO_2$ enters the ocean. First, the largest area of the ocean in contact with the atmosphere is the warm surface ocean between the subtropical fronts. This is an area of about 275 million $km^2$, about 73 % of the entire ocean surface. It is only about 100 meters (330 ft) deep, and contains only about 5 % of the water in the oceans. Because it is warm, and because $CO_2$ is more soluble in cold water than it is in warm water, it will take up less of the atmospheric $CO_2$ than might be imagined. But there is only a very small surface of $CaCO_3$ for those warm waters to react with. The total area of carbonate banks and reefs is only about 550,000 km$^2$, less than 0.2 % of the area of the warm surface ocean.

In the deep sea, the situation is very different. The area of carbonate sediments on the deep sea floor is about 170 million km$^2$. That is a huge area to support the chemical reactions to neutralize the anthropogenic $CO_2$ pulse. Unfortunately, it will take several hundred years for the $CO_2$ entering the ocean from the atmosphere to reach these sediments and several thousand years for the reactions to occur. This assumes that the deep sea conveyor and internal ocean mixing will continue to function as they have in the past.

The most rapid change in atmospheric and oceanic $CO_2$ in the past was during the deglaciation. The current change is happening 100 times faster. Because the change during the deglaciation was much slower, there was time for the $CO_2$ and $CaCO_3$ deposits to react and for the system to stay in balance.

Perhaps the best way to think of what a change of rate of 100 times means is to consider what would happen to a human child if it aged 100 times as fast as it really does. By the end of the first year, the child would be 100 years old. If it only lived to 75 years, it would have experienced only three of the four seasons of the year, and would not be aware of the seasonal cycle except through stories passed on from generation to generation. Changes in rate have enormous consequences.

As my colleague John Southam told me when we first worked on this many years ago: "Over 10,000 years or so, the $CO_2$ we are adding to the atmosphere and ocean will be neutralized by chemical reaction with the deep-sea sediments. It is only the next few thousand years that are the problem. And it's too bad that the $CO_2$ going into the tropical surface ocean has so little calcium carbonate to react with." If nothing else were to happen, the Earth could return to the climate it had before the industrial revolution. But if the polar ice melts and the ocean warms, the ocean circulation will change, and return to pre-industrial climate may not be possible, even over thousands and thousands of years. If I had to make a guess, I would suppose that the preindustrial climate is something humans will never see again.

## 26.13  What Are the Effects of Acidification on Marine Animals and Plants?

Experiments to determine the effects of increased $CO_2$ on marine plants and animals have just begun. In one test of the effects of increased $CO_2$ on marine phytoplankton, growth of two of the three species tested was inhibited by higher $CO_2$ levels. The third species, *Emiliania huxleyi,* became more prolific, but stopped secreting coccoliths. What this indicates is that overall, the oceans' disposal of $CO_2$ slowed down. Further, without the calcite coccoliths to act as ballast, the settling organic material is more likely to be decomposed high in the water column, adding to the $CO_2$ load there. This would act as a positive feedback, ultimately forcing atmospheric $CO_2$ levels higher.

The aragonite compensation level has already risen to near the ocean surface over part of the Southern Ocean, where pteropods are actively fixing $CO_2$ as aragonite. The higher rates of dissolution there also promote $CO_2$ return to the waters, and promote a positive feedback to the $CO_2$ content of the atmosphere.

It seems likely that the acidification of the ocean will proceed steadily as the ocean absorbs more and more $CO_2$. Nevertheless, it is expected that changes that occur until 2100 are not likely to greatly perturb the ocean system. However, in the centuries after that the ocean will enter a condition it has not had in at least half a million years, and probably not for many million years. We have no idea whether the plankton can evolve rapidly enough to adapt to these new conditions, or whether they will be replaced by other species, or whether the ocean will experience a massive die-off of planktonic life. The latter scenario would create a catastrophic positive feedback raising atmospheric $CO_2$ levels even faster and higher than any current prediction.

Reef-building corals are already responding to the changing climate. It was originally thought that warmer temperatures would enhance carbonate precipitation and benefit coral growth. However, this is clearly not the case. Every time the world was significantly warmer than today, as, for example, during the Eocene and Cretaceous when the poles were ice-free, there were no coral reefs. The corals we see today have evolved since ice appeared on the poles.

About 30 % of the world's coral population has died since 1980. The first sign of trouble is 'bleaching.' Most corals are actually colonies of coral animals. Healthy coral animals have a golden-brown color that comes from symbiotic algae, specialized forms of 'dinoflagellates' that inhabit the coral tissue. In some corals, more than half of the tissue by weight may be symbiotic algae. Bleaching occurs when the animals are stressed and expel the algae. The animal tissue then appears white. Without the symbiotic algae, the coral animals are unable to build skeletons as rapidly; the colony deteriorates, and in most cases, eventually dies. How all this occurs is not yet understood, but two factors are at work, rising temperatures and acidification. Both probably play a role, although just now the temperature effect probably dominates. Acidification further weakens the coral structure and contributes to the destruction of the reefs.

## 26.14  Atmospheric and Oceanic Circulation, Nutrients and the Carbon Cycle in Other Climate States

Chapter 6 introduced the idea that the Earth has four major climate states—Snowball, Icehouse, Greenhouse, and Hothouse.

The snowball state occurred several times during the Precambrian but has not happened since. It involved a rapid loss of the greenhouse effect followed by a long buildup of atmospheric $CO_2$, rapid melting of the snowball and an overshoot to a hot Earth before the system stabilized itself again. The snowball Earth episodes seem to be intimately related to the development of multicellular life on Earth. This is a very active area of ongoing research and new ideas are continuing to be developed.

The greenhouse and icehouse states were Earth's 'normal' conditions during the Phanerozoic and were inherently stable. The greenhouse is Earth's 'default state' and has persisted through about 70 % of Phanerozoic time, the last half billion years. The icehouse state, with ice sheets on one or both poles, has two internal states: glacial and interglacial, reflecting waxing and waning of continental ice sheets. About 25 % of Phanerozoic time has been spent in the icehouse state, and only 10 % of that in the current interglacial state.

The hothouse state, proposed by David Kidder and Tom Worsley of Ohio University, is inherently unstable and lasts for only a few hundred thousand to a million years. Only about 4 % of Phanerozoic time has been spent in the hothouse state. Note that the hothouse condition was included in the greenhouse state of Al Fischer that was described in Chap. 6.

Each of these states has different patterns of atmospheric and oceanic circulation. In the following discussion I have combined my own views on the greenhouse state with those of Kidder and Worsley, and have made some overall generalizations of conditions.

Here I should mention that Tom Worsley was one of my Ph.D. students at the University of Illinois in the late 1960s. Tom was, and is, the ultimate skeptic. He always found a simpler, better way of looking at things, and his solutions to problems have always been highly original. Tom and David Kidder are not part of the mainstream paleoclimate community. Their thinking is quite original and has not been burdened by many of the constraints imposed by assuming that past climates must be analogs of the present. They also include some aspects of climate, such as delivery of nutrients to the ocean, that are not in climate models. I think they are on the right track in their analyses of the differences of the climate states of the Earth's past.

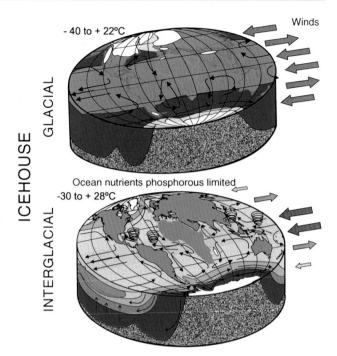

**Fig. 26.12**  The two Icehouse states, glacial and interglacial, of the Phanerozoic. For discussion see text

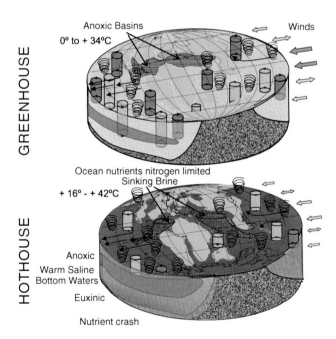

**Fig. 26.13**  The Greenhouse and Hothouse states of the Phanerozoic. For discussion see text

Figures 26.12 and 26.13 show the conditions of the climate states of the Phanerozoic.

The 'icehouse state' is best known because we live in it, albeit in the interglacial sub-state which characterizes only 10 % of the time spent in the icehouse. It is important to realize that this is only about 2.5 % of Phanerozoic time. It is characterized by polar ice sheets, and a very strong polar to

Equator temperature gradient, reflecting a temperature difference on the order of 50–60 °C (=90–108 °F). The winds are driven by the pressure differences at different latitudes, and those pressure differences are a function of the different temperatures. Remember that warm air is less dense than cold air, so cold air forms a 'high' and warm air forms a 'low.' With polar ice ensuring polar high pressure at all seasons, the wind systems are stable throughout the year. The winds drive the ocean currents and create the frontal systems described above. The ocean is filled with cold, oxygen-rich deep water that sinks in the Polar Regions. The result is a highly structured ocean with a strong thermocline separating a thin surficial layer of the warm water in the tropics and subtropics from the cold waters of the ocean interior. The waters of the ocean are almost everywhere well supplied with oxygen; they are highly 'oxic.' However, beneath regions with high organic productivity in the surface waters, some of the oxygen is consumed in decomposing the sinking organic waste. This results in local 'oxygen minima;' the waters are 'hypoxic.' In a very few tiny ocean basins such as the Cariaco Trench on the shelf off Venezuela or the Orca Basin on the Louisiana continental slope, the oxygen is wholly consumed; they are 'anoxic.'

The icehouse state requires low concentrations of atmospheric $CO_2$, probably less than 300 ppmv to maintain cool surface temperatures. The greenhouse effect of $H_2O$ is limited to the lower latitudes. The low atmospheric $CO_2$ is a result of a high consumption rate of atmospheric $CO_2$ through the enhanced weathering of silicates. At the same time the supply rate of $CO_2$ through volcanism is relatively low because of a slower rate of sea-floor spreading and subduction. The availability of silicates for weathering is a function of mountain building in response to continental collisions, like that of India with Asia that began about 50 million years ago. But wait, mountains are high, and the temperatures at high elevations are cold; so how can that draw down atmospheric $CO_2$? The silicate rocks exposed in high mountains are first weathered mechanically, by freezing and thawing, and broken into grains that are then transported by glaciers and finally rivers to lower elevations where the temperatures are warmer. As these grains are spread out over the river flood plains and deltas the reaction with atmospheric $CO_2$ can proceed more rapidly.

As described in Chap. 13, the oscillations between the two icehouse sub-states, glacial and interglacial, are forced by the insolation changes resulting from Milankovitch orbital variations. The insolation changes are reinforced by a variety of feedback mechanisms. It is important to remember that we are currently in a very unusual interglacial, one that, without human intervention, could be expected to last much longer than the usual 10,000 years because the ellipticity of Earth's orbit is approaching a minimum. And it is because of the very unusual climatic stability of this interglacial that 'civilization' has been able to develop.

The major nutrients required for life in the ocean (and on land) are commonly thought to be phosphorous and nitrogen. If you are a gardener, you know all about this. Phosphorous ultimately comes from the weathering of rocks on land. It must be in special molecular compounds, usually combined with oxygen ('phosphate'), to able to be used by plants and animals. Although rivers bring it to the sea, most of that utilized by life in the sea, at least 90 %, is recycled. Most organisms in the ocean die by being eaten, and the waste fecal material is formed into pellets that sink into the depths. As the pellets sink they are decomposed and oxidized, releasing the phosphorous to be returned to the ocean surface waters by the upwelling processes described above. Nitrogen is different. Although the atmosphere is mostly $N_2$, it cannot be utilized in its diatomic form. It must be fixed by combining with hydrogen and/or oxygen to make a compound such as ammonium nitrate ($NH_4NO_3$), used as a fertilizer on land. Some bacteria involved with land plants can do the job, but most of the nitrogen fertilizer used for crops now comes from factories. The industrial production of fixed nitrogen for fertilizer is, after the burning of fossil fuels, probably the second largest anthropogenic perturbation of the Earth's natural systems. Its impacts are just beginning to be known.

In the sea, dissolved nitrate is the form most often used by phytoplankton. If the nitrate supply runs short, special bacteria can synthesize nitrate from dissolved nitrogen, but they do not become active until the supply runs low. But in order to fix nitrogen, another special nutrient is required for the enzyme that makes the process work: iron (Fe). Now there is a real problem, because iron is soluble in acid water but not in alkaline water, and ocean water is alkaline. In those few rivers that are acidic and carry iron in solution to the sea, it precipitates out on contact with the seawater. So how do the bacteria that need it obtain it? From dust. The iron in dust and on the surface of sand grains blown into the ocean serves the purpose. The dust supply depends on the winds.

In the icehouse world glaciers grind up rock and make lots of dust. The high winds erode the dust very effectively and carry it out over the sea, supplying the ocean with an overabundance of iron. With an oversupply of iron, the limiting nutrient in the ocean is phosphorous. This is the situation we have today.

In the greenhouse state there is little or no polar ice. My ideas about atmospheric and oceanic circulation in this state were described in the preceding chapter. Pole to Equator temperature gradients were less, about 34 °C (=61 °F), or less. The winds were weaker. Because of the lesser pressure contrasts due to lesser temperature differences, the pressure contrast due to water vapor content may have become a significant forcing factor. Remember that warm air with a high water vapor content is much lighter and less dense than dry air. There should have been seasonal reversals of high latitude atmospheric pressure systems. This means that except for the tropics, the wind systems would be variable and may have completely reversed with the seasons. The organized ocean currents we see today were largely replaced by transient eddies. In the greenhouse state upwelling of deeper waters could and did occur in the middle of the ocean basins rather than just along the margins. The overall structure of the ocean was much less organized.

The warmer waters of the oceans in the greenhouse state could hold only about half as much dissolved oxygen as those of the icehouse state. Under regions with high organic productivity, the consumption of oxygen in the decomposition of organic matter could effectively deplete the waters of all dissolved oxygen. The waters became 'anoxic.' The excess organic matter sank to lower levels, and settled out on the ocean floor to form organic-rich sediments. This happened periodically in the Cretaceous greenhouse world, especially in the restricted basins of the young, opening North Atlantic Ocean and the Tethys. On at least one occasion the anoxic waters spread into the ancestral Pacific Ocean.

Because of the lower wind speeds and lesser availability of iron-rich dust, nitrate became the limiting nutrient in the ocean. The activity of the nitrogen-fixing bacteria that could produce nitrate was reduced because of the limited iron supply.

The greenhouse state depends on levels of atmospheric $CO_2$ that are at least twice those of the icehouse state, and usually much higher. The higher atmospheric $CO_2$ is the result of more rapid rates of sea-floor spreading and subduction, along with lower rates of silicate weathering as less material is available and exposed to the elements. One of the implications of higher atmospheric $CO_2$ is that the ocean will tend to become less alkaline. This tendency is reinforced by the lower rates of silicate weathering on land. Rivers carrying the products of silicate weathering (e.g. $Ca^{++}$ and $HCO_3^-$) add 'alkalinity' to the ocean. The $CO_2$ absorbed from the atmosphere into the ocean almost immediately combines with $H_2O$ to make $HCO_3^-$ and $H^+$. The $H^+$ makes the ocean more acidic. Under greenhouse conditions, acidification wins out over alkalinization.

The 'hothouse state' requires special conditions. The idea of this special state arose from studies of the extinction at the end of the Permian. This was the greatest extinction in Earth's history, when about 96 % of marine and 70 % of terrestrial organisms died out. Kidder and Worsley became convinced that the end-Permian event was intimately related to the rapid, massive eruption of basalts in Siberia, resulting in the formation of a Large Igneous Province (LIP) known as the Siberian Traps. They believe that very large amounts of $CO_2$ were released into the atmosphere as the basalts erupted. This temporary pulse of $CO_2$, raising atmospheric concentrations to a level of 30 or more times that of today, caused a sharp rise in global temperatures that forced the

global climate system to a new condition. The ocean returned to a system of deep circulation, as in the icehouse, but reversed. The ocean depths were filled with dense, warm, highly saline waters that had formed in the shallow seas of the tropics. Much of this water returned to the surface in the Polar Regions, resulting in extreme warm temperatures in the high latitudes. The warming of the ocean caused thermal expansion of the water (see Table 17.4). A rise of 10 °C (=18 °F) causes water surface in a container originally 1 meter deep to rise by 6 mm. With an average ocean depth of 4 km (13,123 ft), that is a rise of 24 m (78.7 ft). As the hothouse world developed the rise in ocean temperatures could have been even greater. The ensuing transgression of the expanding warmer waters increased the area of shallow seas where more warming and evaporation could take place, an important positive feedback driving the climate toward ever more extreme hothouse conditions. It is the reversal of the deep ocean circulation so that warm water is carried from the tropics through the depths to upwell in the Polar Regions that is critical for producing the hothouse climate state.

However, the reversal of deep ocean heat transport is not all that happened. The warm waters carried less and less oxygen, and waters at depth became increasingly anoxic as sinking organic matter consumed all the available dissolved oxygen in its decomposition. But there was organic matter left over. But there is a second source of oxygen available in seawater, in the sulfate ion ($SO_4^=$), after the chloride ion ($Cl^-$) the second most abundant negative ion in seawater. Using the oxygen of the sulfate ion to oxidize organic matter leaves the $S^=$ ion free to combine with hydrogen ($H^+$) ions to make $H_2S$, a really nasty gas. In very small concentrations, $H_2S$ has the odor or rotten eggs. In greater concentrations it has no odor at all, but is quite deadly. Ocean waters in which this process has taken place and contained dissolved $H_2S$ are said to be 'euxinic.' The term comes from the Latin name for the Black Sea, '*Pontus Euxinis*,' whose deep waters contain large amounts of $H_2S$. One explanation for the extreme nature of the end-Permian extinction is that the large areas of the world ocean became euxinic, and the euxinic waters that upwelled along the coasts fed $H_2S$ into the atmosphere, killing off land life.

To make matters worse, the sulfur left behind as the oxygen of the $SO_4$ is removed can also scavenge iron (Fe) from the water, making pyrite ($FeS_2$). The removal of iron shuts down the activity of the nitrogen-fixing bacteria and the result is a 'nutrient crash.' Life in the ocean becomes almost impossible. Kidder and Worsley gave this complicated scenario the acronym name HEATT, for Haline, Euxinic, Acidic, Thermal, Transgression—certainly a mouthful.

However, because of the high temperatures, weathering rates increase, and this eventually draws down the levels of atmospheric $CO_2$. Because the trigger for the hothouse was a time-limited pulse of $CO_2$ into the atmosphere, the enhanced weathering rate provides a limit on the length hothouse conditions can persist. It is usually less than a million years.

As shown in Fig. 6.2, Kidder and Worsley identified more than a dozen hothouse events in the late Precambrian and Phanerozoic. Most were associated with the volcanism involved in the formation of LIPs, but a few are uncertain. The two youngest LIPs occurred during icehouse conditions and were inadequate to perturb the climate system to the hothouse state.

Our planet has had a fascinating history.

## 26.15  Summary

Life on land and in the sea is experiencing climate change at a rate unprecedented since the catastrophic impact of an asteroid 65 million years ago. The rates of change are of the order of 100 times faster than those accompanying the last deglaciation. Animals, such as insects, will be able to move as rapidly as the climate changes, but many land plants will not be able to keep pace with the change. Destructive infestations and overpopulations of rapidly reproducing, mobile land animals are to be expected. We can also expect to see temporary ecological groupings of plants and animals that have not existed in the past.

Biotechnical engineering may enable genetic alteration of crops to keep up with the changes, but will be complicated by the more chaotic climatic conditions we should expect.

The ocean takes up about half of the extra $CO_2$ being produced by human activity. It will eventually absorb some of the transient pulse of anthropogenic $CO_2$, but only after many thousands or tens of thousands of years. In the meantime, the ocean will become more acidic, and conditions not experienced by ocean animals and plants for millions of years will be reached after 2100. The response of the natural systems to this change is unknown, but the tendency would be for positive feedbacks to develop, reinforcing the effects of the rise of $CO_2$ in the atmosphere.

The circulation of the atmosphere and oceans and with them the global nutrient and carbon cycles have very different characteristics in the Earth's major climate states of the Phanerozoic: Icehouse, Greenhouse, and Hothouse.

A Timeline for this chapter:

| 1872–76 | H.M.S. Challenger collects information and samples from the world's oceans |
|---|---|
| 1904 | Ernst Haeckel completes publication of *Kunstformen der Natur* |
| 1908 | The Tunguska Event devastates forests in Siberia |
| 1926 | Vladimir Vernadsky publishes his book биосфера (The Biosphere) |

(continued)

| 1928 | Vernadsky's book is translated into French |
|---|---|
| 1955 | Cesare Emiliani discovers the sawtooth curve of Pleistocene temperatures |
| 1965 | James Lovelock formulates the GAIA hypothesis while working at the Jet Propulsion Laboratory in Pasadena, CA |
| 1971 | Wally Broecker publishes *A kinetic model for the chemical composition of seawater* |
| 1998 | Vernadsky's book is translated into English—*The Biosphere* |

If you want to know more:

Vladimir I. Vernadsky, 1998. *The Biosphere: Complete Annotated Edition*. Springer, 192 pp.—The unknown classic. Here are so many books about life on land I don't know where to start. It depends on your interests. But for life in the ocean, I recommend:

Carol M. Lalli and Timothy R. Parsons, 1997. *Biological Oceanography: An Introduction*, 2nd Edition. Butterworth-Heinemann, 320 pp.

Music: Stravinsky's The Rite of Spring was used in Disney's original Fantasia sequence portraying the evolution of life on Earth. Its real topic is a bit more serious—a pagan ceremony that ends with a sacrifice. But if you have seen Fantasia, it seems an appropriate piece to listen to again.

Libation: A healthy red wine in celebration of life in its many forms.

**Intermezzo XXVI. Charles Eagle Plume** Only a couple of weeks after I assumed by position as Director of the Museum we had a telephone call from a motel just down the street. In cleaning one of the rooms after the guests had checked out they had found some Indian beadwork and buckskin clothing under one of the beds. The items had cryptic labels attached. They had figured out that these looked like they might be museum specimens and wondered if we might be missing some artifacts.

I immediately got in my car and drove down to take a look. When I saw the cryptic labels, I recognized them. The seemingly meaningless groups of letters were a code that an old friend, Charles Eagle Plume, used to keep track of objects in his American Indian store, Perkins Trading Post. We notified the police in the nearest town, and I called the Trading Post. There was no answer. We waited anxiously while they sent an officer to the store, located on a lonely stretch of Highway 7 just below Long's Peak. Everything looked alright from the front, but what they found in the back was amazing. Charles sold not only Navaho Rugs,

Pueblo Pottery, and silver and turquoise jewelry, but had a large collection of museum quality older artifacts which hung from the ceilings of the showrooms. These were not for sale, they were to be donated to museums after his death. All of the windows of the trading post had iron grilles to prevent burglaries. But what the thieves had apparently done was to put a chain through one of the grilles and attach it to their truck, intending to pull the grille off. However, the grille was more firmly attached than they had thought, and they pulled off an entire section of the back wall of the building. We soon discovered that Charles was away on a buying trip.

The motel owners had a good description of the guests who had rented the room, and a few days later a shop in Denver was offered some of the stolen goods for sale. The labels were still on them, and the shopkeeper recognized them. The thieves were tracked down. They were Native Americans, and claimed that they were simply 'repatriating' the items. Charles was outraged at their actions. "They claimed they were repatriating my grandmother's ceremonial dress from me, her grandson." The 'repatriation' didn't jibe with the fact they had been trying to sell the items to dealers. Most of the items were recovered, and the thieves convicted.

But who was Charles Eagle Plume? I first met Charles in 1939, when I was 5 years old. Charles would give lectures on American Indian customs at the lodges around Estes Park. He wore beautiful costumes for these lectures, changing parts of them as he demonstrated different dances. He also sang Indian songs, although first warning the audience that the Indian scale of notes fit better into the cracks between the keys on a piano. At that first meeting, he allowed me to carry some of the costume materials to the stage. It was a huge honor for a 5 year old.

Charles was an icon in the Estes Park area. He belonged to the Blackfeet tribe (not Blackfoot, they are a different group). He didn't know how old he was; he thought he had been born sometime during the first decade of the 20th century, somewhere in Montana as Charles Burckhardt. His grandmother gave him the Indian name Eagle Plume. He had the sharp facial features characteristic of his tribe, but with one exception. His mother was half Indian, half French, and she had married a German settler who preferred to live with the Indians. Somehow Charles got both of the recessive genes required to have blue eyes. He always seemed to be squinting during his lecture, actually trying to prevent the audience from seeing his blue eyes.

This is a good place to explain the difference between an American Indian and a Native American. American Indians are the people who inhabited North America (excluding Mexico) before the arrival of the Europeans. The term Native Americans includes the original Polynesian inhabitants of Hawaii and American Samoa. Persons of mixed racial heritage, like Charles, were 'breeds': half-breeds, quarter-breeds or whatever, and often not fully accepted into either the European settler or Indian societies.

In his early teens Charles left the reservation. He discovered that Indians were unwelcome in most of Montana and Wyoming, so he kept moving south. He ended up in Estes Park, Colorado, where he heard that a couple who lived about ten miles south of town had a 'Trading Post' and were friendly to Indians. The Perkin's had no children of their own, and took in Charles and another Indian boy, Ray Silver Tongue. During the fall, winter and spring of the 1920s Charles found work in Colorado Springs, and graduated from the High School there in 1928. In the summers he would return to Estes Park. It was then that he began to give the lectures on Indian culture that made him famous. A lady from Illinois, Gertrude Hamilton, was vacationing in Estes Park when she heard him talk. She was so impressed that she encouraged him to go to the University and helped to pay his expenses. He obtained his Bachelor's degree from the University of Colorado in Boulder in 1932. He would then spend his summers working at the Perkins' Trading Post and then began to tour the country giving lectures during the rest of the year. He became one of the family, and ultimately inherited the little store.

Charles' lecture was always very much the same; over the years I had seen it so many times I could have given it myself. But it always seemed completely spontaneous; he was a great orator. But in addition to fascinating the audience with dances and stories he included some comments about the state of Native Americans that were almost always new to the audience. He also liked to point out that the Constitution of the United States drew on that of the Iroquois Confederacy which had been translated into English by Thomas Jefferson.

Few Americans realize that Native Americans did not acquire citizenship in the United States until 1924, and the rights granted them then were limited. They could vote in national, but not in most state elections. New Mexico did not grant its Indians state voting rights until 1962. In fact, during much of the 20th century the reservations were essentially internment camps, Indians were allowed to leave only under certain circumstances. They did not gain full equality until the passing of the Indian Civil Rights Act of 1986.

I spent my high-school summers in Estes Park. In the late 1940s I would drive out to the trading post and we would talk. Charles had his own version of the general American Indian religion: Mother Earth is all that constitutes the planet, including the lands, mountains, waters and all living things. The sky interacts with the Earth, accounting for the Sun, stars, and Moon, the winds and the weather. This is not very different from Vernadsky's idea of Earth as a ball of rock partially covered by water and blanketed by a thin film of life. It also resembles James Lovelock's concept of Gaia—the planet as a living, breathing, thing. You will also recognize this in my references to 'Mother Nature' who may make whatever corrections are necessary to ensure the safety of life on planet Earth. The concept is in many ways the opposite of the anthropocentric Jewish, Christian, and Muslim religions. Learning about it from Charles I have always reveled in the glories of nature and have come to respect our planet as something very special.

When I was living in Washington I started taking my vacations in Estes Park again, and I would go out to the Trading Post and visit with Charles. He had boundless optimism that someday the racial and cultural divides in America would disappear, but it was becoming evident that 'someday' would not occur in his lifetime.

Charles had Indian friends who lived near Boulder, and they were able to track him down and tell him of the robbery. A few months later I had moved to Estes Park. It was an hour long commute to my office in Boulder, but I had wanted to live in the mountains and had a lot of luck in finding just the right home. This meant that Charles and I could meet for lunch in the town every Saturday. He was delighted that I had become Director of the Museum. It would be the recipient of many of the historically important artifacts he had collected over the years. But Charles was growing old, and becoming increasingly frail. He was probably in his 90s although he liked to report with amusement that one newspaper account had given his age as 107. When I visited him just before leaving on my sabbatical he was in a nursing home attached to the local hospital.

WASHINGTON, GONDOLIERS ON THE MALL IN 2150

*I have seen the sea lashed into fury and tossed into spray, and its grandeur moves the soul of the dullest man; but I remember that it is not the billows, but the calm level of the sea, from which all heights and depths are measured.*

President James A. Garfield

'Sea level' is such a commonly used term you may not be aware of the fact that it is only precisely known in a few places. The up and down motions of the water surface with waves and tides and longer-term motions of the land surface make its determination a complicated matter.

## 27.1    What Is Sea Level?

It might seem that sea level at any given place should be easy to define, because all you have to do is go and see where the water is. But do you do this at high tide, low tide, or mid-tide? Most books will tell you that it is the level midway between low and high tides; at many places 'mean low tide' is considered the reference level. But the tides are different each day. This is because the Moon, which, with the Sun, exerts the gravitational pull responsible for the tides, is orbiting the Earth. However, the Moon's elliptical orbit is tilted 5° to the ecliptic (the plane of Earth's orbit around the Sun) and its long axis precesses with a cycle of 18.6 years. During this time the tilt of the Moon's orbit relative to the earth's axis varies from 18° to 28°. This variation has an important effect on the tides, so that their pattern repeats only after 18.6 years!

By convention, mean sea level is now defined as 'the average height of the surface of the sea at a particular location for all stages of the tide over a 19-year period.' This is usually determined from the hourly height readings of a tide gauge at that location. The tide gauge is protected from the effects of ordinary waves, but may be affected by changes in the level of the water surface due to changes in atmospheric pressure. Hopefully those get averaged out over a 19-year period.

As you can guess, the requirement that we have an hourly record at least 19 years long means that the number of places

where we actually know mean sea level is not all that large. And when you examine the records, they are all different. So when someone says sea level is rising, someone else may say that it is not changing, and someone else may say it is really falling. Why?

## 27.2    Why Does Sea-Level Change Differently in Different Places?

There are seven factors determining the position of sea level at any one time: (1) the volume of water in the ocean, (2) the motions of the solid Earth's surface, (3) the gravitational attraction of large masses, such as ice sheets, (4) the speed of rotation of the Earth, and (5) the location of atmospheric high and low pressure systems, (6) the evaporation/precipitation balance, and (7) changes in the volume of the ocean basins.

The first six can be related to climate change: The last is related to plate tectonics. It operates on the very long time scales associated with continental breakups, changes in the length and shape of the mid-ocean ridge system, and the length and shapes of the ocean trenches where subduction takes place.

## 27.3    Changing the Volume of Water in the Ocean

This is the only factor which changes global sea level on a time scale of centuries or longer. There are three ways in which the volume of water in the ocean can change: (1) by changing the mass of water (i.e., the number of molecules of water), (2) by changing the amount of salt dissolved in the water, i.e., the salinity of the ocean, and (3) by changing the temperature of the ocean.

© Springer International Publishing Switzerland 2016
W.W. Hay, *Experimenting on a Small Planet*, DOI 10.1007/978-3-319-27404-1_27

**Table 27.1** Masses, moles, and sea-level equivalents of the major reservoirs of free water on or near the surface of the Earth

| Water reservoir | Mass ($10^{15}$ kg) | Moles ($10^{15}$) | Sea level equivalent (m) |
|---|---|---|---|
| Oceans | 1,371,345.7 | 76,122,436.7 | 3797.69 |
| Groundwater (est.) | 58,613.0 | 3,253,566.5 | 158.14 |
| Ice | 24,110.0 | 1,339,444.4 | 65.05 |
| Rivers & Lakes | 225.0 | 12,489.6 | 0.61 |
| Atmosphere | 13.0 | 721.6 | 0.04 |
| Biosphere | 0.6 | 33.3 | 0.001 |

Sea-level equivalent is the amount sea-level would change if this amount of water were added to the oceans

## 27.4    Changing the Mass of H₂O in the Ocean

Table 27.1 shows the major reservoirs of 'free water' on or near our planet's surface. In contrast to 'free water' there is 'bound water,' that is, water incorporated into the crystals of minerals. There is, for example, bound water in the mineral gypsum, which we also know as Plaster of Paris, or simply the plaster used to cover the walls of your home. Bound water is not readily available for conversion to free water, and in any case its total amount is relatively small.

The amount of water in the ocean can be altered by exchanges among these different reservoirs. However, when you look at the right-hand column, the sea-level equivalent, it is evident that only two of these are really significant, ice, and groundwater.

One of the uncertainties geologists must face is that we do not know whether the total amount of $H_2O$ on planet Earth, both free and bound, remains constant over long periods of time, or whether it changes. There is some evidence that the total volume of water in the oceans has declined with time, at a rate of perhaps $250 \times 10^9$ kg/year. It is assumed that this is due to the subduction of seawater at oceanic trenches, with the water being forced down into the Earth's mantle. That is equivalent to a sea-level fall of 0.0007 mm/year (0.000028 in./year). Clearly, this is a trend we do not need to take into account in considering what may happen to sea level over the next decades to hundreds or even thousands of years. There are, however, two reservoirs of water large enough to change the volume of water in the ocean: ice and groundwater.

## 27.5    Changes in Groundwater Reservoirs

The amount of water in groundwater reservoirs on land is almost twice the amount in ice, but much of it is deeply buried. However, the shallow groundwater reservoirs may contain water equivalent to 10 m of sea level change. These shallow groundwater reservoirs may slowly fill and empty with changing climates. However, changes in the amount of water in these shallow reservoirs is very small in comparison with those resulting from the buildup and decay of ice sheets.

When examining the modern sea level record in detail, the question arises whether the extensive mining of groundwater, has any appreciable effect. 'Mining' of groundwater occurs when the groundwater is extracted, mostly for irrigation purposes, faster than it can be replaced. In the western and southwestern US, groundwater that accumulated during the last glacial is being extensively mined, at a rate of about 35 km³/year. In the former USSR water was taken from the Aral Sea region to the extent that the Sea almost disappeared. The water was taken from the lake itself, from the rivers flowing into it and from the groundwaters in the region surrounding it. During the period of most rapid extraction, around 1960, it was being mined at a rate of about 60 km³/year. These large extractions could account for a sea-level rise of about 0.3 mm/year during the latter half of the 20th century.

## 27.6    Conversion of Ice to Ocean Water

Table 27.2 shows the sea level changes associated with past and present ice masses. The sea-level change between the last glacial and the present is shown in this table. The 145.03 m given as the total in the table differs from the sea level low-stand during the glacial as known from shallow-water reef corals, 135 m. This discrepancy may be because all of the ice sheets did not reach their maximum size at the same time. Some may have already begun to melt while others were still growing. There are two large bodies of ice still in existence, on Antarctica and Greenland. As the table shows, melting all the ice on Greenland would raise global average sea level by about 6 m (19.7 ft). Melting of the entire Antarctic ice sheet would raise global sea level by 59 m (193.6 ft).

Perhaps the most spectacular melting of ice that has been documented is that of the sea-ice cover on the Arctic Ocean. The extent of the ice and its thickness can be sensed by satellites, so we are acquiring a good record of the changes in the Arctic. Both the area and thickness have diminished greatly over that past 30 years. However, it is important to

**Table 27.2** Ice sheets, their areas, thicknesses, volumes, and the sea level change that would result if the ice were melted and the water returned to the ocean

| Region | Area | Thickness | Volume (ice) | Mass of water | =sea level |
|---|---|---|---|---|---|
| | $10^6$ km$^2$ | km | $10^6$ km$^3$ | $10^{15}$ kg | M |
| Antarctic ice sheet | | | | | |
| Present | 13.59 | 2.22 | 30.11 | 27.61 | 73.5 |
| Glacial | 20.40 | 1.75 | 35.70 | 32.74 | 85.7 |
| Greenland ice sheet | | | | | |
| Present | 1.73 | 1.65 | 2.85 | 2.62 | 7.2 |
| Glacial | 2.30 | 1.70 | 3.90 | 3.58 | 9.8 |
| Laurentide ice sheet | | | | | |
| Present | 0 | 0 | 0 | 0 | 0 |
| Glacial | 13.39 | 2.20 | 29.46 | 27.01 | 73.0 |
| Cordilleran ice sheets | | | | | |
| Present | Negligible | Negligible | Negligible | Negligible | Negligible |
| Glacial | 2.37 | 1.5 | 3.56 | 3.26 | 8.9 |
| North European (British-Scandinavian-Barents-Kara) ice sheets | | | | | |
| Present | Negligible | Negligible | Negligible | Negligible | Negligible |
| Glacial | 8.37 | 2.0 | 16.74 | 15.35 | 42.0 |
| Other (Alps, Himalaya, Andes, Iceland, mountain glaciers) | | | | | |
| Present | 0.54 | 0.24 | 0.13 | 0.12 | 0.3 |
| Glacial | 5.2 | 0.30 | 1.56 | 1.43 | 3.9 |
| *Totals* | | | | | |
| *Present* | *30.86* | | *33.14* | *30.38* | *83.10* |
| *Glacial* | *67.03* | | *91.00* | *83.45* | *228.09* |
| *Difference (G–P)* | *36.17* | | *57.86* | *53.06* | *145.03* |

**Table 27.3** Floating ice (sea ice and ice shelves), areas, thicknesses, volumes, and the sea level change that would result if the ice were melted and the water returned to the ocean

| Region | Area ($10^6$ km$^2$) | Thickness (m) | Volume ($10^6$ km$^3$) | Mass ($10^{15}$kg) | =m Sea level |
|---|---|---|---|---|---|
| Arctic sea ice—maximum | 15.00 | 2 | 0.030 | 0.028 | Negligible |
| Arctic sea ice—minimum | 7.00 | 2 | 0.014 | 0.013 | Negligible |
| Arctic sea ice—difference | 8.00 | 2 | 0.016 | 0.015 | Negligible |
| Antarctic sea ice—maximum | 18.00 | 1 | 0.018 | 0.017 | Negligible |
| Antarctic sea ice—minimum | 3.00 | 1 | 0.003 | 0.003 | Negligible |
| Antarctic sea ice—difference | 15.00 | 1 | 0.015 | 0.014 | Negligible |
| Antarctic ice shelves | 1.62 | 475 | 0.732 | 0.671 | 0.035–0.052 |

recall that this is sea-ice, floating on the ocean surface. It has no effect on sea-level as it melts.

It is difficult to imagine the sheer size of the ice sheets that existed during the Last Glacial Maximum. For comparison, the Tibetan Plateau has an area of about $2.5 \times 10^6$ km$^2$ and an average elevation of 5 km. The Laurentide ice sheet of eastern North America, although not as high (average elevation about 1.5 km after isostatic adjustment of the land surface to its weight, but with central elevations probably of the order of 3 km), had an area more than five times that of the Tibetan Plateau. It is even more difficult to imagine that such gigantic geographic features could simply disappear in less than 15,000 years.

Table 27.3 accounts for the ice floating on the sea surface. It is generally assumed that melting of floating ice has no effect on sea level. However, that is not strictly true. The melted ice water may cool the surrounding waters and dilute its salinity.

## 27.7    Storage in Lakes

Lakes are very peculiar features of the Earth's surface. A very few, such as Lake Baikal in Siberia and Lakes Tanganyika, Malawi and others in the East African Rift Valley, and the Dead Sea, are the result of plate tectonics. They occupy sites where the continent is being torn apart, and are often millions of years old. Others, such as the Great Lakes of North America and countless smaller lakes in mountainous and high latitude terrains are the result of glacial erosion. The Great Lakes formed along the margin of the Canadian Shield, a region where the surface is made of hard crystalline rocks like granite and gneiss or dense metamorphic rocks. As the Laurentide ice sheet grew, the ice was unable to erode this hard surface, but when it reached the softer sedimentary rocks that underlie the northeastern and Midwestern United States it was able to dig in and scoop out deep basins. As the ice retreated, these basins were filled with water, forming the lakes we see today. Although these lakes are grand features, the water in them, if added to the ocean would only raise sea level by 0.6 m.

Offsetting the potential sea-level rise due to mining of groundwater is the storage of water in man-made reservoirs and lakes. Most of these reservoirs were built in the latter half of the 20th century. Before the levels of many of these lakes began to fall in the 21st century, they stored 8,400 km$^3$ of water. Filling these reservoirs may have reduced sea level by 0.48 mm/year. Although all of these figures have large uncertainties, it is thought that the net effect of mining groundwater and building reservoirs on global sea level was close to zero.

## 27.8    Effect of Changing the Salinity of the Ocean

Adding salt to water increases its volume, but by an amount less than the original volume of the water and salt added together. This is because of the odd structure of water. As the salt goes into solution, the ions slip in between the water molecules and do not occupy as much space as one might expect. It is very difficult to change the salinity of the ocean as whole, but the extraction of the huge volumes of fresh water as ice during the last glacial did result in an increase in the salinity of seawater. Salinity rose from its present average of about 34.5 parts per thousand to about 36 parts per thousand. Estimates are that removing the fresh water that made the ice sheets that existed at the Last Glacial Maximum should have lowered sea level by almost 139 m. However the increased salinity of the water left behind made it occupy a slightly larger volume that it would have had if the salinity had remained constant, equivalent to about 4 m of sea-level. I am tempted to believe that this is the reason that the glacial

age sea-level lowstand indicated by shallow-water corals is about 135 m rather than 139 m.

The role of salinity changes does make a difference in embayments which are fresher that the open ocean, such as the Baltic Sea. There, increased river flows can raise the local sea-level by several centimeters.

## 27.9    Effect of Warming or Cooling the Ocean

The effect of warming the ocean has a significant effect on sea level. If the warm surface layer of the tropical ocean, 100 m thick, were to warm by 1 °C, sea level would rise by about 27.8 mm. A temperature increase of 1 °C or more may well have taken place in the equatorial region. So it is possible that some of the sea-level rise observed on tropical Pacific and Indian Ocean islands is due simply to the warming of the water.

Things would become very different if the whole ocean were to warm. An increase of 1 °C in the 3700 m of deep water would cause it the water column to expand by about 377 mm, over 1/3 of a meter. If the deep ocean were to take on the temperature of 15 °C, it has had during older times when the poles were ice free, the expansion would raise sea level by about 10 m.

There is a detailed record of temperatures in the upper part of the ocean that goes back to 1961, representing millions of measurements made by oceanographic ships, buoys, and probes. It indicates that between 1961 and 2003 the upper 700 m of the ocean warmed by 0.1 °C. This corresponds to an average expansion raising the sea surface of 0.4 mm/year.

## 27.10    Motion of the Earth's Solid Surface

First of all, you might think that the surface of the ocean is the same distance from the center of the Earth everywhere, so that 'sea level' would be easy to define as that surface. But, as shown in Fig. 27.1, the 'solid Earth' is a layered active planet.

The outer layer of our planet, called the 'crust,' consists mostly of relatively dense black basaltic rocks that form the ocean floor and pink granite and other less dense rocks that form the continents. The lighter continental blocks float like rafts on a sea of basalt. The crust and uppermost mantle form solid but flexible layers, called the 'lithosphere.' Beneath the lithosphere is a layer of much more fluid material, probably made of more dense rock and a small amount of molten material, termed the 'asthenosphere.' When the Earth's surface was loaded by ice, the lithosphere bent downward, and asthenospheric material flowed to peripheral region to form a bulges. When the ice sheets melted, the 'solid Earth' returned to its original configuration. As shown in Fig. 27.2 the ice

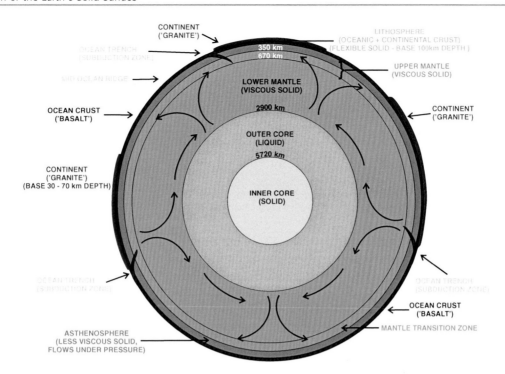

**Fig. 27.1** Cross section of planet Earth, drawn to scale. Distances from the surface to internal boundaries are shown. The upper boundary of Earth's core is almost half of the way down to the center, but the core is only about 16 % of the planet by volume. The *deeper core is solid*; the *outer core* is *liquid* and it is circulation within it that produces Earth's magnetic field. The mantle makes up about 80 % of the planet's volume, and it convects, causing the phenomenon we call plate tectonics. The uppermost part of the mantle, called the 'asthenosphere,' is the most fluid part of the Earth except for the liquid upper core. The outer layer, the 100 km *thick* lithosphere, about 4 % of the planet by volume, is mostly stiff but flexible rock. A tiny fraction of it contains fluid magmas causing volcanism

sheets can build up and decay more rapidly than the solid Earth can respond. Our planet is still responding to the melt-off of the last deglaciation.

The distribution of mass inside our planet is also uneven. As you might guess the density of material inside the earth generally increases downward, but is less where the rocks are warmer, greater where they are cooler. It is the nonuniform distribution of mass within the Earth that causes the slow adjustments we recognize as 'plate tectonics.' The convection of mantle material is ultimately responsible for sea-floor spreading, the drift of the continents, and for the return of crustal material into the Earth's interior at subduction zones.

The uneven distribution of mass means that a surface of equal gravitational attraction, called an 'equipotential surface' is not everywhere the same distance from the center of the Earth. The global shape of this equipotential surface is called the 'geoid.' What may surprise you is that in terms of distance from the center of the Earth, the uneven distribution of mass inside our planet causes large differences in that distance. Sea level over an oceanic trench, like that bordering Indonesia on the south, is up to 100 m closer to the center of the Earth than that in the Central Indian Ocean. Why doesn't the water flow downhill into this depression? Because the trench is underlain by denser rock, the force of

gravity over it is the same as in the Central Indian Ocean. There is no 'downhill;' It's just the shape of the geoid.

It turns out that the uneven distribution of solid material in the Earth actually results in it having a geoid that is roughly pear-shaped. However, the ocean surface closely approximates an equipotential surface in all its detail. It has ups and downs related to gravitational attraction of the mid-ocean ridge, ocean trenches, platforms and even seamounts. The ocean surface also reflects the variations in pressure of the atmosphere in different places and the motion of ocean currents, and will be discussed later. These elevation differences are usually of the order of 20 cm but may be as much as 1 m.

Then, you must surely think—what about the rotation of the Earth; wouldn't centrifugal force make the ocean surface higher on the Equator? The answer is no, because the whole planet has deformed in response to the rotation in such a way that its general shape is not that of a sphere, but an 'oblate spheroid,' slightly flattened at the poles, and in detail shape more like that of a pear. But everything is neatly balanced out. Earth is a smart planet.

All right, now that we understand that sea level is not a constant distance from the center of the Earth, it becomes evident that its determination must be considered a local matter.

**Fig. 27.2** Deformation of the
Lithosphere and Asthenosphere
with loading by an ice sheet (not
to scale). **a** Earth's outer layers
prior to accumulation of ice. **b** As
ice sheet grows the lithosphere
flexes and the asthenosphere
flows to accommodate the load.
**c** At the glacial maximum
adjustment approaches
completion. **d** After rapid melting
of ice sheet, adjustment occurs,
but at a slower rate

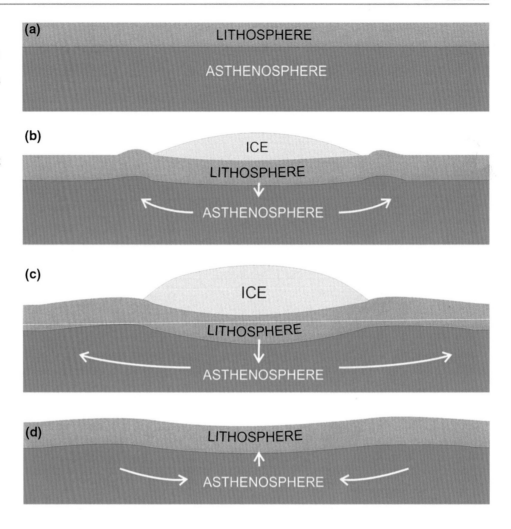

The most rapid large-scale changes in the geoid are the
result of the buildup and decay of the northern hemisphere
ice sheets. As discussed earlier, the weight of the ice sheets
depressed the surface of the continents. The ice weighs about
920 kg/m$^3$. It rests on the relatively stiff crust of the Earth,
the 'lithosphere.' As noted above, the lithosphere is stiff but
flexible. Below about 100 km beneath the Earth's surface is
a layer of mobile rock which is roughly 300 km thick, the
asthenosphere. The asthenospheric material has a density of
about 3,300 kg/m$^3$. It moves to accommodate the additional
weight of the ice, and restore 'isostatic equilibrium.' For
everything to come back into balance, a column of ice 1 km
thick must displace a column of asthenospheric material of
equal weight, which would be 279 m thick. So after this
adjustment has taken place, the top of the ice would be only
721 rather than 1000 m high. As asthenospheric material
flows out from beneath the weight of the ice sheet, the
lithosphere bends as well, like a piece of stiff plastic, so that
a depression surrounds the ice.

Where does the asthenospheric material go? It flows out
from underneath the ice and the depressed lithosphere to
forms a peripheral bulge surrounding the ice sheet. Over

eastern North America, the Laurentide ice sheet extended
south to New York City (Long Island is a terminal moraine,
marking the edge of the ice 20,000 years ago) and across the
middle of the Midwestern states of Ohio, Indiana and Illinois. The depression in front of the ice was several hundred
miles wide. The peripheral bulge extended down to the Gulf
Coast and perhaps even to southern Florida. These were
indeed huge deformations of the continent.

Figure 27.2 shows schematically the development of a
continental ice sheet and deformation of the underlying
lithosphere and asthenosphere. The 100 km thick lithosphere is composed of a thin sedimentary cover overlying
30 km of granite forming the continental block and 70 km
of more dense basaltic rock. It is solid but flexible and
bends in response to the loading of ice. It overlies the 300
+ km thick asthenosphere which, although denser, flows
from beneath the ice sheet to accommodate the load. Flow
from beneath the developing ice sheet is slower than the
rate of growth of the ice, and the result is a bulge of
lithosphere peripheral to the ice as shown in Fig. 27.2b. As
the ice sheet grows the asthenospheric material spreads out
like a wave to greater distances, as shown in Fig. 27.2c. At

the time the Laurentide ice sheet over Northeastern America had reached its greatest extent, its average thickness was 2.2 km. The center, over Hudson Bay, was about 3.5 km thick, depressing the land surface there about 975 m. Because the lithosphere is stiff, the ice sheet would have been surrounded by a narrow trough. Beyond that was a broad peripheral bulge that probably extended throughout the rest of the North American continent. But the growth of the ice sheet was faster than the asthenosphere could respond, so the deformation never came into equilibrium. For that to happen, the ice sheet would have to remain a constant size for at least 20,000 years. The melting of the Laurentide ice sheet occurred much more rapidly than the asthenosphere and lithosphere could respond, as shown in Fig. 27.2d. The remaining depression, where the ice was, is still occupied by water, Hudson Bay.

What controls this complex interaction between ice and Earth? It depends on the viscosity of the asthenospheric material, which moves slower than molasses. The response of the asthenosphere to the addition of weight on the Earth's surface starts immediately, but it takes many thousands of years for everything to come back into equilibrium. The isostatic response to the weight of Lake Mead behind the Hoover Dam was noticed as soon as the lake began to fill but will last for thousands of years. But the complete isostatic adjustment process has been slower than the rate of growth and melting of the ice sheets. The Earth never quite seems to catch up. It is still adjusting to the unloading of the Laurentide and Scandinavian ice sheets. That is, where those ice sheets had been, the land surface is still rising.

Now think about the record of sea level change in the regions occupied by the ice sheet, the marginal trough, and the broad peripheral area of uplift. During the last glaciation sea level fell by about 135 m. But the land surface beneath the center of the ice sheet was depressed almost 1 km. As the ice sheet melted, sea level rose, but the land surface also rose, much further. Relative to the land surface it would appear that there has been a huge fall of sea level. Along the edge of the ice sheet the sea level record would seem to show that there was a fall followed by a rise. In the middle of the peripheral bulge it would appear that sea level rose, but much more than 135 m. Throughout the area occupied by the ice and the surrounding region affected by flow of asthenospheric material, the record of sea level will vary from place to place and may often appear to be quite complex.

Over North America the largest motions are taking place in regions that are largely uninhabited, as around Hudson Bay, but in Europe, it is Scandinavia that is rising. At Stockholm, the isostatic adjustment to the unloading of the ice is still of the order of 10–11 mm/year. The changes in the depth of the passages into the harbor cause ongoing problems.

Most of the sinking of the peripheral bulge along the US Gulf Coast seems to have stopped, but it continues along the southern shore of the Baltic and the North Sea in northern Germany and The Netherlands.

After the reunification of Germany in 1990, I was a periodic visitor to the Institute for Baltic Sea Research in the former Deutsche Demokratische Republik (DDR = East Germany to most Americans). A good friend, Jan Harff, had been appointed Head of the Geological Section. The Institute is in Warnemünde, on the Baltic coast north of Rostock. (If you wonder why so many towns along the German Baltic coast, end with 'münde,' it means 'mouth of a river,' so Warnemünde is the town at the mouth of the Warne river.). In 1993 I spent the summer in Warnemünde working on the sea-level history of the Baltic. It was a topic I had never really studied before, but I wanted to understand more about sea-level change in the more ancient past. I was particularly interested in the effects of loading and unloading by ice sheets. The Baltic Sea is an especially interesting region in this respect because during the deglaciation it was flooded by the sea, then rose above sea level to become a fresh water lake, and then was again flooded by the sea. It records dramatically the complex interplay of global sea-level rise and movement of the land surface.

We had some interesting new data to work with, detailed gravity maps of East Germany and adjacent parts of Poland, the eastern Baltic countries, and Russia. The gravity data were secret during the cold war. Why? Because the tracks of ballistic missiles depend on gravity, and follow the geoid. To know where they will come down requires a detailed knowledge of Earth's gravity field. Incidentally, this sort of information was also considered secret by the US but much of the data for the continental US had been published before there was any thought of missiles. However our military kept what it knew about the gravity field elsewhere closely under wraps.

My investigations ran into an unanticipated obstacle. There was another suite of data that was very curious. The DDR had made a detailed study of the geology of the entire country, drilling a large number of holes and making closely spaced seismic transects across the entire area. They had documented in detail some very strange features, known as 'Rinnen.' These are river valleys cut into the older glacial materials and subsequently filled with sediment. But what is so very odd is their shape. They are 100–200 km long, oriented running away from the old ice edge. They start at the surface at one end, descend to depths as great as 600 m and come back to the surface at the other end. I learned about these from Jan Harff, and went to the Geological Survey of the new German State of Mecklenburg-Vorpommern in Schwerin to meet Werner von Bülow who knew more about them than anyone else. He had a simple solution for the origin of the Rinnen. They were river valleys cut through the peripheral bulge of an earlier glaciation. This meant that the bulge would have arched up the Earth's surface 600 m high. This was a truly revolutionary idea. Most glaciologists assume that such a peripheral bulge was much broader and far lower. Werner von Bülow's idea was that if the ice sheet grew very rapidly, the displaced asthenospheric material would not be able to move away as fast and instead pile up around the edge of the ice forming a high bulge. This may have only happened once in the last several glacial episodes. It became evident that we must be very cautious about interpreting the history of sea-level from observations in areas near the ice sheets.

Clearly, an estimate of global sea-level change must exclude the effects of the moving surface of the solid Earth; the record must be from sites far away from the ice sheets. Two of the favored 'far-field' sites, Barbados and the Sunda Shelf between Indonesia and the Asian mainland, are more stable, but even there local motions need to be taken into account.

The question may also come to mind: what happens if sea level rises and further floods onto the continental blocks? Does isostatic adjustment occur? The answer is yes, and it is even more odd than you might think. If you think about what happens to an oceanic island you come to the conclusion that it records sea level change accurately. The island is attached to the sea floor, and as sea level rises, it acts as a dipstick to record the sea level change. Actually, as more water is added to the ocean basins, the asthenospheric material beneath them will flow out and underneath the continents, raising the areas not covered by water slightly. But at the same time water rises around the edges of the continental block, and floods onto it. A good guess is that if there is 1 m of sea level rise, the Earth will respond isostatically. That is, if you had a measuring stick placed firmly in solid ground and if sea level went up 1 m very quickly, it would immediately show the 1 m rise. But then it would continue to sink another 30 cm over the next few thousand years. For arguments sake, let's say the adjustment occurred in 1,000 years. During that period your post would record an average rate of sea-level rise of 0.3 mm/year, even though the global sea level remained constant.

In addition to isostatic motions of the Earth's 'solid surface' there are motions of the not-so-solid surface where the coast is underlain by unconsolidated sediment. Southern Louisiana is a classic area. Since the postglacial rise in sea-level, the Mississippi River has produced a number of deltaic complexes from near the Texas border to Mississippi. The present delta is just one of many. When the river relocates to a new site, the older deltas begin to compact and sink under their own weight. During the last century, a channel was dug across the continental shelf so that the Mississippi now dumps its sediment load directly into the deep Gulf of Mexico. Now that the supply of sediment to the delta region has been eliminated, the whole landscape of southern Louisiana is slowly subsiding. This gives the appearance of a local rise in sea level. In Louisiana this is not a trivial problem. The state has lost almost 35 miles$^2$ (about 90 km$^2$) of land to the sea each year for the past 50 years. New Orleans has been sinking at a rate of about 3 ft (0.9 m) per century.

These problems associated with delta subsidence are exacerbated by drilling wells for extracting groundwater. The removal of groundwater accelerates the rate of subsidence. This is a very major problem in Bangladesh where the World Bank sponsored a large program intended to provide a safer drinking water supply by drilling water wells in the Ganges Delta. The result has been an acceleration of subsidence rates up to 1 m/year, placing a large part of the population in jeopardy in the event of a major storm.

Finally, there are areas where the Earth's surface is actively moving because of tectonic forces, such as at Barbados where the island is rising at a rate of 0.34 mm/year. More spectacular is the situation in Pozzuoli, a suburb of Naples, where the 'Temple of Serapis' is located. A must-see if you visit Naples. The round Temple was in the middle of a marketplace in Roman times, obviously above sea-level. Today it is flooded to a depth of about half a meter. But more spectacularly, the columns of the Temple are bored by marine mollusks to a height of several meters above present sea level. The Temple has moved down and then back up a distance of several meters since Roman times. This is attributed to the movement of magma beneath the city. Vesuvius is not the only volcanic danger in the Naples area.

You will occasionally hear it argued that sea-level is stable or falling. The global trend of sea-level is upward, but because of independent movements of the Earth's solid surface, there are indeed areas, such as the coast of Sweden, where the local trend is downward.

## 27.11 The Gravitational Attraction of Ice Sheets

You may have heard the story that when the British were mapping India in the middle of the 19th century, the surveyor, George Everest, noticed that their careful survey and astronomical determinations of location did not agree. He discovered that the plumb line, the weight hanging from the surveying instruments and intended to point to the center of the Earth, was not vertical when they were approaching the Himalaya. It became evident that the deflection was due not just to the mountains themselves, but to their roots deep in the Earth's crust.

The glacial ice sheets were huge, massive objects on the Earth's surface. Like the Himalaya, their sheer mass provided gravitational attraction. Since they extended to the ocean, they attracted the surrounding ocean water. You might think, well that effect can't be very big. But it was, being in the order of tens of meters. If a wedge of water had somehow been able to penetrate to the center of Hudson Bay while the Laurentide Ice Sheet was still largely intact it would have been about 85 m higher than the average global sea level due to the gravitational attraction of the ice. More realistically, it was originally thought that for some reason Greenland had risen much more rapidly than North America as the deglaciation proceeded. This was before the gravitational attraction of the ice was taken into account. On the margin of Greenland, the sea-level fall due to the declining gravitational attraction of the ice was about 27 m. The bulge of water around the ice extended out to a distance of about 1000 km, and the water came from the rest of the ocean, lowering sea level there by about 0.26 m.

Today there are two bodies of ice large enough to attract the ocean's waters, Antarctica and Greenland. One of the ways in which the total amount of ice on each of the areas is being tracked is by remote sensing of the sea surface elevation around them. As the ice melts, the amount of water held close around the ice sheets by gravitational attraction decreases; the sea surface relaxes and goes down. Because of this water attracted to the ice sheet, the total global sea-level rise due to both melting of the ice and relaxation of the sea surface would be equal to about 109 % of that due to melting of the ice alone.

## 27.12 Changing the Speed of Earth's Rotation

During the episodes of northern hemisphere glaciation the displacement of water from the world ocean into ice sheets over North America and Europe changed the speed of rotation of the Earth. Just as when an ice-skater moves her arms in during a spin to increase the speed, the displacement of mass toward the Earth's axis or rotation causes it to speed up. This seemingly small correction was needed to be taken into account to get the astronomical cycles to match the geological records in sediments older than half a million years. As the ice sheets melted, the Earth's speed of rotation decreased very slightly.

The first thing to respond to a change in the speed of Earth's rotation is the ocean. If the Earth rotates faster, the centrifugal force will drive water toward the Equator faster than the solid Earth can deform, and sea level there will rise. The body of the planet itself will also respond, but this will take much longer. Thus, as ice sheets grow and decay, there will be changes in Earth's rotation, accompanied by an instantaneous response from the ocean and a delayed response from the Earth as a whole. Since all of these things happen on different time scales, the overall effect gets quite complex. It is only in the last few years that Dick Peltier at the University of Toronto has worked out a solution for this complex interaction of forces. It explains one of the troubling aspects of Holocene sea-level observations: that in some far-field regions, distant from the ice caps, there seems to have been a continual rise in sea-level through the Holocene, or even a maximum of sea-level several meters above that of today some 5,000 years ago. The places where these anomalous sea-level histories are found are on the Atlantic margin of South America, particularly in northern Argentina and southern Brazil, and along the southern margin of Australia. They can now be explained by changes in the rotation speed of the Earth along with all the other isostatic adjustments that accompanied the melting of the ice sheets.

## 27.13 Effect of Winds and Atmospheric Pressure Systems

Winds are the result of movement of air from high to low pressure systems. They exert drag on the ocean surface, and ultimately they drive ocean currents, as described in Chap. 6. The winds and currents result in elevation differences of the sea surface from place to place. This sea surface topography is shown in Fig. 27.3. The total elevation differences are of the order of 20 cm. The topography changes with strength of the winds and with changes in the magnitude and location of the atmospheric highs and lows. These differences in elevation of the sea surface would disappear only if there were no winds or ocean currents.

As shown in Fig. 27.3, the highest areas are in the western Equatorial Pacific. There the elevation above global mean sea level can approach +1 m. This is also the 'Western Pacific Warm Pool' often the warmest body of water in the entire ocean. During an El Niño event the winds which support this elevation difference die down, leaving this mound of warm water free to flow downhill. Because this area is on the Equator, where there is no 'Coriolis Force,' the water can flow back to the east across the Pacific, suppressing the Equatorial upwelling system.

The areas of different colors seen at mid- and higher latitudes are currents and mesoscale eddies. The surface of eddies with cyclonic rotation is depressed; that of the anticyclonic eddies is elevated.

The water would, of course, like to flow downhill, from the elevated areas to the depressed areas. However as soon as it starts to move, except on the Equator, it is turned aside by the 'Coriolis Force,' which, you will recall, is really the effect of the rotation of the Earth. If we were to make a contour map of the ocean surface (a contour is a line of equal elevation), we would find that because of the Coriolis Effect, the water actually flows horizontally along the contours rather than downhill from higher contour to lower contours. Another of the properties of our planet that seem strange to us only because we are not really aware of Earth's rotation.

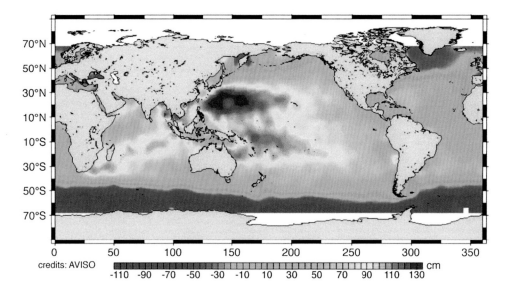

**Fig. 27.3** Topography of the ocean surface during normal conditions, December 2003, based on NASA's TOPEX-Poseidon data. The topographic effects of the geoid and differing gravitational attraction of the mid-ocean ridge, ocean trenches, oceanic plateaus and seamounts has been removed so we see only the effect of winds and ocean circulation on the sea surface. The topographic differences from place to place are generally in the order of 20 cm but may be significantly greater in some areas

It is obvious that the weight of the air above the ocean must affect the position of the water surface. The average pressure of the atmosphere at sea-level was originally thought to be 1 bar, equal to 1,000 millibars or, in metric terms, 1,000 hectopascals. It turned out that the data were based on measurements from low pressure regions of North America and Europe, and the real average sea-level pressure of the atmosphere is 1013.25. It turns out that 1 bar is approximately equal to 10 m of seawater, so it is easy to figure out how the sea surface will respond to changes in atmospheric pressure.

An extreme low atmospheric pressure, as the center of a hurricane, is about 900 millibars (hectopascals). That alone will make the sea surface rise 1 m, and this is what might be expected as the 'storm surge' on a straight coastline. Much higher storm surges occur when the hurricane enters an embayment, where there is no place else for the water to go. Then the combination of the raised sea-surface and wind-forced water can reach as much as 10 m above normal sea level. Although spectacular, Atlantic hurricanes and their Pacific counterparts, typhoons, are ephemeral features and do not affect global sea level per se.

There are some longer term shifts in otherwise stable atmospheric pressure systems that can cause sea level to change over a period of years. One is, of course, the El Niño, La Niña phenomenon of the tropical Pacific. However, the Pacific coastlines are mostly steep, and the sea-level fluctuations associated with these phenomena go largely unnoticed. However, in the North Atlantic there can be multi-year shifts in position of the Azores High and Icelandic Low. The phenomenon is known as the North Atlantic Oscillation, and

was described in more detail in Sect. 25.4. The North Atlantic Oscillation changes the sea-levels observed on shores around the North Atlantic, and since much of the coastline is very gently sloping, it is noticed. More significantly, the level of the Baltic Sea fluctuates as these pressure systems move about. And because the entrance passages are narrow and shallow, it takes a long time for the water to move into or out of the Baltic; it can experience multi-year sea-level variation of 10 cm or more.

## 27.14   Effect of the Evaporation-Precipitation Balance

At present the Earth's atmosphere contains about $13 \times 10^{15}$ kg of water vapor. If for some reason all of it were to be precipitated, sea level would rise by 0.04 m (4 cm, not quite 2 in.). Because of the colder global temperature during the last glacial, the amount of water vapor in the atmosphere was about 2/3 as much, and during the warm Cretaceous it was more, about 1.5 times as much. In terms of global sea level, changing the amount of water in the atmosphere has only a minute effect.

Although it is unimportant in terms of global sea level, evaporation and precipitation are important regionally. Beneath the subtropical high pressure systems of the atmosphere, the descending air is dry and hot. It can evaporate water at rates up to 2 m or more per year! It does just this over the Red Sea which receives almost no rainfall and no runoff from the surrounding land areas. However, the Red Sea is connected to the Indian Ocean through the Strait of

Bab-el-Mendab. Water flows in through the strait to compensate of the water lost to evaporation. As a result, the surface of the Red Sea is only slightly (a couple of centimeters) lower that the Indian Ocean. However, the evaporation of $H_2O$ leaves the salt behind, so the waters of the Red Sea become saltier and denser. As the water from the Indian Ocean flows downhill into the Red Sea, the denser saline waters of the Red Sea flow along the bottom back into the Indian Ocean. As described in Chap. 6, the Red Sea has a 'negative fresh water balance' and is a 'lagoonal sea.'

The same thing happens with the Mediterranean. The evaporation over the Mediterranean averages about 1.5 m/year. Its waters are therefore more saline and its surface slightly lower than the adjacent North Atlantic. As a result Atlantic waters flow into the Mediterranean through the Strait of Gibraltar, at a rate of about 1 Sverdrup ($10^6$ m$^3$/s), roughly equivalent to the flow of all of Earth's rivers combined. There is return flow through the depths of the Strait. The rates are determined by the sea level and density differences between the Mediterranean and the Atlantic. About 5 million years ago the connection between the Mediterranean and Atlantic was interrupted. The Mediterranean dried out. What is amazing to realize is that the evaporation rate over the Mediterranean is so high that without the Gibraltar connection, it could dry out in a couple of thousand years!

In the early part of the 20th century a German architect and engineer, Herman Sorgel, proposed building a dam across the Strait of Gibraltar and using flow to generate electricity. It would have been (and still would be) the greatest building project undertaken by mankind. Such a dam could also be used to regulate sea-level in the Mediterranean as the global sea level rises. The meter rise of global sea-level we may see by 2100 will make the electrical generation potential of such a dam much greater than originally thought. There have been a few updates of the Gibraltar dam project in recent years, and I suspect it might eventually become a reality.

## 27.15   Sea Level Change During the Deglaciation

Records of deglacial sea-level change have been based mostly on reef corals which have a restricted, near-surface depth habitat. With a skeleton of aragonite, fossil corals contain $^{14}C$ that can be dated; the fossil coral skeletons also contain Uranium and Thorium, which provide even more accurate dates.

Much of our knowledge of the deglacial sea-level change comes from fossil *Acropora palmata* reefs off the island of Barbados, in the West Indies. *Acropora palmata* is essentially restricted to depths no greater than 10 m. It can grow up to 14 mm/year and can keep up with most sea-level changes.

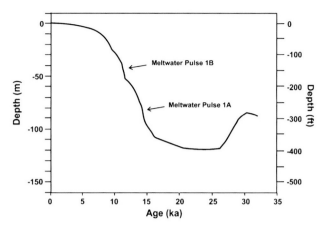

**Fig. 27.4**   The record of deglacial sea-level change at Barbados, based on shallow water dwelling reef corals. The two meltwater pulses represent times when sea-level rose more rapidly

Barbados has the advantage that it was far from the ice sheets and the land surface effects associated with them. It has the disadvantage that it lies over a subduction zone, and is gradually being forced upward at an average rate estimated to be 0.34 mm/year. As shown in Fig. 27.4, the Barbados corals indicate that sea-level change during the last deglaciation began about 21,000 years ago, when sea-level at Barbados was about 120 m lower than it is today. It reached modern levels about 7,000 years ago. The average rate of change was almost 10 m per thousand years; that is 10 mm/year (0.39 in./year). However, the deglacial sea-level rise did not occur smoothly; there were some times of relatively slow rise, and a few episodes of rapid rise. Richard Fairbanks, who first described the Barbados record named the two most obvious sea level rises Meltwater Pulses. Meltwater Pulse 1A occurred between about 14,000 and 13,750 years ago and Meltwater Pulse 1B between 11,000 and 11,500 years ago. These were times when sea level rose more rapidly.

A more detailed estimate of the rates of sea-level rise during times of large-scale meltwater input into the ocean has come from detailed analysis of the nature of the fossil reefs found off a number of the Caribbean islands, shown in Fig. 27.5. Paul Blanchon and John Shaw of the University of Alberta, Canada, found that *Acropora palmata* forms monospecific reefs (that is, it is the only species present) in waters shallower than 5 m depth. However, between 5 and 10 m depth other species of coral are also present. If sea level rises faster than *Acropora palmata* can grow (14 mm/year), the monospecific reefs give way to mixed coral reefs; the *Acropora palmata* reef is effectively drowned. The reefs are completely drowned when the rate of sea-level rise exceeds 45 mm/year (1.8 in./year). They found a record of three major drowning events. These are shown in Fig. 27.5. They termed these 'catastrophic sea-level rise events' (CRE's). Two of them coincide with the 'meltwater

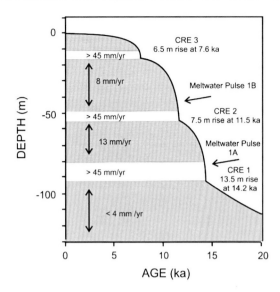

**Fig. 27.5** Composite sea level record from the Caribbean showing different rates of sea level rise and three 'catastrophic rise events' (CRE's) during the last deglaciation. (After Blanchon and Shaw, 1995). 45 mm/year = 4.5 m per century!

pulses' known from Barbados. The third is in the younger part of the record which is not well known at Barbados. The length of these sea-level rise events is not known with certainty, but CRE1, which began about 14,200 years ago, and during which sea-level rose 13.5 m, is estimated to have lasted less than 300 years. CREs 2 and 3 are thought to have lasted about 150 years each.

These rapid episodes of a sea-level rise cannot be due simply to rapid melting of ice. There is no evidence in the fossil record of 'pulses' in the energy coming from the Sun, nor were there any sudden rises in greenhouse gas concentrations. In fact, 50 years ago these pulses would have seemed quite mysterious. The first ideas were that these reflected the rapid emptying of meltwater lakes that had been trapped inland behind the ice-sheets. More recently we have observed another phenomenon which is more likely to be responsible: surges of glaciers at sea level.

One of the factors determining how fast a glacier moves is how well it is attached to the underlying bedrock. If its base is frozen to the bedrock, there may be no motion there at all, and flow will occur at a mid-level within the ice. If this is the case, the ice moves very slowly; typical speeds are from one to a few meters per year. The older literature on glaciers always emphasizes how slowly they move. The term 'glacial progress' has entered the general vocabulary as a synonym for something that moves very slowly. This is because until the late 20th century episodes of fast motion in glaciers were very rare.

Then in the 1970s and early 80s, there was a dramatic increase in the motion of the Columbia Glacier in Alaska. It speeded up to a rate of hundreds of meters per year, a speed

previously known only from dubious anecdotal reports. The Columbia is a tidewater glacier, that is, its mouth is at sea-level; it empties into a bay on the Alaskan Coast. Near its mouth its base is grounded on rock below sea level. When a glacier begins to melt, water trickles down to the base of the glacier, and it may become detached from the bedrock, floating on a layer of frigid water. This is especially true if the glacier is grounded below sea-level, so that it can become detached and start to float. Calving of icebergs off the glacier front lightens it, and the marine waters can penetrate further inland beneath the glacier. As the detachment from the base proceeds, it can surge forward, moving at speeds up to 15 km/year!

Another way in which a glacier can become detached from the bedrock base is through sea-level rise. If the base of the glacier is below sea-level, as many of those in Greenland and West Antarctica are today, a rise of sea-level can simply lift them off the bottom and allow them to surge forward, breaking up as they calve icebergs into the sea. The effect of calving icebergs on sea-level is instantaneous because, although they do not melt immediately, they are floating in the ocean and displace water.

There is indeed evidence for massive numbers of icebergs in the North Atlantic corresponding in time to meltwater pulses 1A and 1B. Pebbles dropped by the icebergs cover wide areas of the ocean floor. As a general scenario for what happened to cause the episodes of rapid sea-level rise, imagine the following: Somewhere, but probably in the margin of the Laurentide ice sheet over Eastern North America, enough water has accumulated beneath the glacier to allow it to surge. It releases flotillas of icebergs as it floats out onto the ocean. This causes sea level to rise slightly. The ice over the Barents Shelf, north of Scandinavia, is grounded below sea-level. The slight sea-level rise buoys it up off the ocean floor and it surges into the ocean, adding to the sea-level rise. In the Antarctic, the rising sea-level causes similar surges. It is a 'chain-reaction' or 'positive feedback' on a global scale. All of the ice that would have otherwise entered the ocean gradually and melted over a period of thousands of years suddenly charges in and melts in the space of a hundred years or so. After this pulse of activity, the sea-level rise slows, and the ice sheets return to a more stable condition. But the ice keeps melting setting the stage for another episode of surges and sea level rise.

The most detailed analyses of sea level change since the LGM have been carried out by William R. (Dick) Peltier and colleagues of the University of Toronto. Figure 27.6 shows a small piece of their work showing sea level curves from six sites displaying very different histories. What is seen is the complex interaction of unloading of the continental blocks due to ice melt, the corresponding increase in volume of water in the ocean, isostatic adjustments in response to both, and changes in the rotation rate of the Earth as mass is displaced from the Polar Regions toward the Equator.

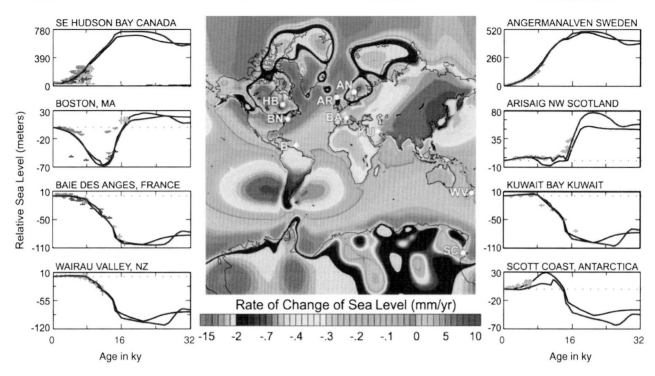

**Fig. 27.6** Sea level curves from six sites and sea level changes since 32 ky ago. Modified after a figure in Peltier, W.R., 2007. *Postglacial coastal evolution: Ice–ocean–solid Earth interactions in a period of rapid climate change.* Geological Society of America Special Publication 426, pp. 5–28

## 27.16    Sea Level During the Holocene

If we look at sea level records for the transition from the deglaciation to the Holocene on apparently stable coastlines far from the glaciated region and its peripheral bulge, we often see an apparent highstand of up to 3 m followed by a declining sea-level, as shown in Fig. 27.7. This is another effect of isostatic adjustment, this time in response to loading the ocean floor with the water from the melted ice. With the additional weight of water, the ocean floor subsides, squeezing asthenospheric material underneath the continents. As a result, the continent rises slightly, and the sea-level, which in reality is stable, appears to go down.

Sea level reached its modern level about 6,000 years ago. That coincides closely with the beginning of 'civilization' as we know it. Much of the human population has always lived close to the ocean. I suspect that it was the stabilization of both sea-level and climate that created the conditions required for the development of urban centers, permitted a division of labor, and led to the rise of what we call civilization.

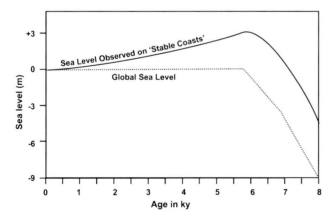

**Fig. 27.7** Effect of global sea-level rise during and after deglaciation on observed sea-level observed on apparently stable coasts far from the glaciated region. The rapid deglacial sea-level rise ended about 6,000 years ago, and has been stable since then. However, on coasts in the tropics sea level apparently reached a highstand at the end of the deglaciation and has declined since then

## 27.17    What Has Happened to Sea Level in the Past Few Centuries?

The oldest tide gauges were installed in the 1700s, but the detailed kinds of tide records used to analyze sea-level change have not been kept until the latter part of the 19th century.

One of the oldest detailed records of sea level stability is provided by the city of Venice. Venice was built on islands, and most of the construction took place during the 12th to

15th centuries. From paintings of the 18th century it is evident that little had changed since the palaces and homes were built along the canals. At Venice, sea level seems to have been quite stable until the 19th and 20th centuries. Today one can see that some of the steps from the canals into the homes are always under water, some walkways along the canals have been elevated, and the grand square, Piazza San Marco, is regularly flooded at high tide. The sea-level rise since the 18th century can be estimated at about 20 cm.

Figure 27.8 shows one of the oldest tide gauge records, for Amsterdam, the 'Venice of the North.' The small black rectangles are yearly averages of the mean tide. The black line is a 19-year running average through the annual data. Amsterdam is built on wooden pilings driven into soft sediment. As long as the pilings are submerged, they do not decay. Control of the water level in the city's canals is critical to the stability of the building's foundations, so there is a complex system of locks to assure stable water levels.

On its face, this diagram would seem to indicate that sea level was stable until early in the 19th century, and that since then it has been rising at an average rate of about 10 mm/year. However, it must be borne in mind that Amsterdam is in an area that was peripheral to the Scandinavian ice sheet and is situated on the ancient Rhine River delta which is still slowly compacting under its own weight. As a result this part of The Netherlands is subsiding at a rate of about 10 mm/year. Taking this into account, the record may indicate that sea-level was lower during the Little Ice Age, which lasted into the 19th century, and was relatively stable until late in the 19th century when sea-level rise began to outstrip the subsidence rate.

Another very long record comes from Kronstadt, Russia. Kronstadt is the naval base in the Gulf of Finland guarding St. Petersburg. Here the record, shown in Fig. 27.9, extends from 1777 to 1993. As in Fig. 27.8, the rectangles are yearly averages of the mean tide and the line is a 19-year running average. The dotted line is the long-term trend. Note the different scale on the left of the two figures. Here too there is

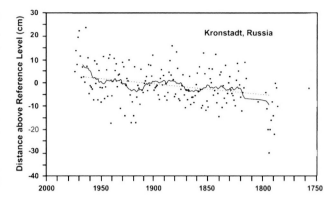

**Fig. 27.9** Tide gauge record for Kronstadt, Russia, from 1777 to 1993

a problem of stability of the solid Earth. Kronstadt lies on the edge of the rising Scandinavian area, with an uplift rate estimated to be close to 0. If sea level were stable, the long-term trend should be downward, because the land is rising. Instead, even here sea level seems to be rising. You will notice that the 19-year average record shows some multi-decadal highs and lows, probably associated with the North Atlantic Oscillation.

A shorter but also very interesting record is from Key West, Florida, shown in Fig. 27.10. Key West is on the southern margin of the solid limestone platform that we know as southern Florida. It was as far as you can get from the ice sheets over northern North America, presumably at the very edge of the peripheral bulge. It is thought that here the isostatic adjustments have gone to completion, so the sea-level record is not influenced by local movements or compaction of the underlying sediment. Again, the rectangles are yearly averages of the mean tide and the line is the 19-year running average through the annual data. Here the record indicates a steady rise of sea-level, at a rate of about 2 mm/year since the early part of the 20th century.

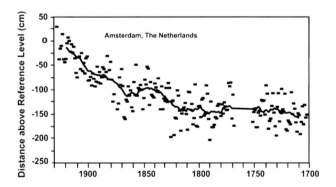

**Fig. 27.8** Tide gauge record for Amsterdam, The Netherlands, from 1700 to 1924

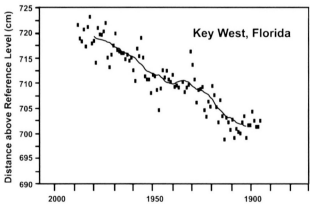

**Fig. 27.10** Tide a gauge record from Key West, Florida. The reference is relative to a tidal benchmark survey marker

## 27.18   The Global Sea Level Record

Geologists often refer to Earth's sea level history as a record of 'global eustasy.' Eustatic sea level is a theoretical concept based on the assumption that at any moment in time sea levels everywhere would be the same; eustatic change must then reflect a change in the quantity of water in the ocean or a change in the capacity of the ocean basins.

In 1977 a new methodology was introduced into geological practice by Peter Vail and co-workers at Exxon Production Research Company (the research division of Exxon) in Houston known as 'Seismic Stratigraphy.' The idea was to use the geologic record of sea level rises and falls, seen on seismic profiles of the continental margins as 'transgressions' and 'regressions' recorded by the sediments, as time lines for geological correlation. Knowing where you are in terms of rock strata is all-important in petroleum exploration. Since the 1920s the microscopic fossils too small to be destroyed by the drill bit had been the tool for establishing the time equivalence of geologic strata. You may recall from the Intermezzi that early in my career I was a micropaleontologist. Until the 1960s the most useful microfossils were the ± 1 mm sized calcium carbonate shells of foraminifera. I was involved in the introduction of nannofossils, also calcareous but a hundred times smaller, into the field of biostratigraphy. In the 1960s I trained a number of graduate students who became micropaleontologists working for industry. One of them, who had become Chief Geologist for a major oil company later told me that seismic stratigraphy worked perfectly—until you drilled a well—then you needed the fossils again.

It was in the late 1960s at a meeting in Vienna that I met another fresh Ph.D. from the University of Stockholm working on the same very minute fossils, 'nannofossils,' that I did, Bilal Haq. Bilal went on to become a research scientist in paleoceanography at the Woods Hole Oceanographic Institution, and then in the 1980s moved to Exxon's Research Center in Houston where he became the guru of seismic sequence stratigraphy and global sea-level change through time. His publications with his Exxon colleagues on the topic are among the most often cited in geology and the sea level curves are often referred to as the 'Haq Curves.' In the late 1980s he became a program director for marine geosciences with the U.S. National Science Foundation. He recently retired and has since moved to the Sorbonne, the University in Paris, France. We have remained great friends over the years. Recently Bilal has become increasingly concerned over the question as to whether the sea level changes recorded in sediment can ever be measured accurately. He has come to the conclusion that the geologic measure of 'eustatic sea level' is a chimera—a concept that we may never be able to quantify accurately. The reason is the many different possible relatively short-term causes of sea-level change introduced earlier in this chapter: changes in the volumes of water in the ocean; in ice on the continents, in groundwater and lakes; changes in the temperature and salinity of ocean waters; motions of the Earth's surface, gravitational attraction of land and ice; changes in the speed of Earth's rotation as ice sheets come and go; changes in atmospheric pressure systems and the winds; and the regional evaporation-precipitation balance. All of these factors act on the sea's base level at different time scales. Bilal is now convinced that all stratigraphic measures of sea-level change are eurybathic (a term he coined for local or relative measures of sea-level changes) and never eustatic.

A glance back at the tide gauge records shown in Figs. 27.8, 27.9 and 27.10 will make you aware that reconstructing a history of global sea level rise over the past 100 years or so is not an easy task. The nature of the records themselves is a complex topic with different technologies being used at different sites. Figure 27.11 shows three recent attempts to solve this problem.

The differences between the three sea-level curves shown in Fig. 27.11 are the result of selection of different tide gauge stations and different analytical techniques. Jevrejeva et al. used a total of 1023 records corrected for isostatic adjustment. The number of useful records varies with time and deceases with age. Their analysis goes back to 1700 when only 3 useful stations were available. They argue that sea level acceleration appears to have started at the end of the 18th century and up to the present has averaged about 0.01 mm/year. They state that sea level rose by 6 cm during the 19th century and 19 cm in the 20th century.

The Church and White analysis is based on records from as many as 399 isostatically adjusted tide gauges, the number varying greatly with time, and on satellite observations since 1993. They estimate the recent rate of rise, from 1993 to 2009 (after correcting for glacial isostatic adjustment) to be 3.2

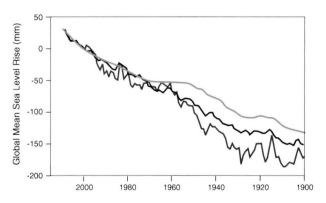

**Fig. 27.11** Three attempts to reconstruct 'global' sea-level curves for the ocean since 1900. The red line is that of Jevrejeva et al. 2008 (Recent global sea level acceleration started over 200 years ago? Geophysical Research Letters, v. 35, L08715, 4 pp); the black line is that of Church & White 2011 (Sea-level rise from the late 19th to the 21st century (Surveys in Geophysics v. 32, 585–602); the green line is that of Hay et al. 2015 (Probabilistic reanalysis of twentieth-century sea-level rise (Nature 517, 481–484)

± 0.4 mm/year based on the satellite data and 2.8 ± 0.8 mm/year from the tide gauge data. They estimate the global average sea-level rise from 1880 to 2009 to have been about 210 mm. The linear trend from 1900 to 2009 is 1.7 ± 0.2 mm/year and since 1961 it is 1.9 ± 0.4 mm/year. The Church and White analysis is essentially that adopted by the latest Intergovernmental Panel on Climate Change (IPCC) Report.

The green line is a reanalysis of the 20th century and younger data by Carling Hay (no relation) and colleagues using a more sophisticated analytical technique ('Kalman smoothing') for 622 tide gauge records. It indicates a mean rate of sea-level rise of 1.2 ± 0.2 mm/year from 1900 to 1990, and a sharply accelerated rate of rise of 3.0 ± 0.7 mm/year between 1993 and 2010. Because of the analytical technique used, I think the Hay et al. analysis is the best estimate of sea level change since 1900. It shows a steady rate of rise from 1900 to about 1940, then a plateau with almost no increase until about 1980, followed by a continuation of the earlier 20th century trend until 1993 when the sharp acceleration begins. As you will see later, in Chap. 30 this is the same pattern shown by global Earth surface temperatures. Although both show the same pattern, there is no relationship whatsoever between the tide gauge and temperature data sets.

Further discussion of sea-level trends of the immediate past with projections into the future is included in Chap. 32.

## 27.19 Summary

The major factors contributing to global sea-level rise since the middle of the 20th century are thought to be thermal expansion of the water and melting of ice. For the latter half of the 20th century, data from Greenland and the Antarctic had large uncertainties. Since 1993 a variety of satellite observations have greatly reduced these uncertainties. Most of the melting has occurred in glaciers and ice caps other than the Antarctic and Greenland. The IPCC Reports indicates that sea-level rise has accelerated during the decade 1993–2003 and since. The estimates and uncertainties associated with the various contributions to sea-level rise are shown in Table 27.4. The observed sea level rise exceeds the sum of the estimates from the major sources by about 0.5 mm/year.

There is no timeline for this chapter.

If you want to know more:
Peltier, W.R., 2007. *Postglacial coastal evolution: Ice–ocean–solid Earth interactions in a period of rapid climate change.* In Harff, J., Hay, W.W., & Tetzlaff, D.M., Coastline Changes: Interrelation of Climate and Geological Processes: Geological Society of America Special Publication 426, pp. 5–28.

**Table 27.4** Contribution of different sources to sea level change during 1961–2003 and 1993–2003

| Source | 1961–2003 | 1993–2003 |
|---|---|---|
| Thermal expansion | 0.42 ± 0.12 | 1.6 ± 0.5 |
| Greenland ice sheet | 0.05 ± 0.12 | 0.21 ± 0.07 |
| Antarctic ice sheet | 0.14 ± 0.41 | 0.21 ± 0.35 |
| Other glaciers and ice caps | 0.50 ± 0.18 | 0.77 ± 0.22 |
| Sum | 1.1 ± 0.5 | 2.8 ± 0.7 |
| Observed | 1.8 ± 0.5 | 3.1 ± 0.7 |

Values are in mm per year

Music: Richard Rogers music for Victory at Sea. Lots of evocations of the ocean's moods.

Libation: Time to open a bottle of a Bordeaux red from the coastal region, from grapes grown a few meters above sea level. Look for something from near Blaye, a town on the west side of the Gironde estuary.

### Intermezzo XXVII. The Zindani Affair

Early in my tenure as Director of the University of Colorado Museum I became involved in a 'religious affair.' First, you need to know that the offices of the Geological Society of America are located in Boulder, Colorado, not far from the University. One of my former colleagues at the University of Illinois in Urbana, Mike Wahl, was Executive Director of the Society, and another old friend Allison R. 'Pete' Palmer was in charge of public relations.

In 1983 two Saudis approached the Geological Society of America to ask it to identify experts on aspects of geology which might relate to the Koran. They were in Boulder, and I was just down the street, so both my and Pete Palmer's names were suggested

By way of background I should mention that at that time there had been news stories that the Saudis wanted to make it possible for science to flourish in the Arab countries as it once had, but the religious authorities stood in the way. The Saudi Royal Family was sponsoring a 'holy man,' Sheikh Abdul Majeed Zindani, for this project.

Mustafa Ahmed and Mohammed Monsour visited me in my office and explained their project. It seemed that there were certain verses in the Koran that were very difficult to understand, and Sheikh Zindani wanted to discuss the matter with scientists. When they found out that I had been dean of an oceanographic institution they were even more insistent. They wanted me to go to Jeddah, Saudi Arabia, to confer with the Sheikh.

I was not enthusiastic, and noncommittal. I had a very busy schedule. They argued that it was a very important project to Muslims. I said I would consider it.

After the meeting I called Cesare Emiliani, and Cesare seemed to be well informed on the Saudis proposed project to reinvigorate science in the Islamic world. He strongly encouraged me to make the trip.

Accordingly, I was flown to Jeddah, and met with the Sheikh for the better part of a week. It was indeed an interesting experience. The Sheikh is a charming gentle man, apparently sincerely seeking advice from knowledgeable scientists. He lived in a small, very neat house, had a neatly trimmed beard and was dressed in a white robe—the very picture of a 'holy man.'

Sheikh Zindani's thesis was that some passages contain information that could not have been known at the time, and must be divine revelation. I was not an expert on classical Greek and Roman and medieval science, so I was at a disadvantage, being ignorant of what was and what was not known at the time of Mohammed in the 7th century.

For me one of the questions concerned a passage in the Koran that seems to refer to internal waves in the ocean, and the idea was that these had been discovered only recently. The Sheik was thinking of the internal waves that occur on the density interface between water masses having two different densities. A few years later a wholly different interpretation of the meaning of the passage would occur to me.

I had seen internal waves while diving in strip mines in Illinois, so I knew that if conditions were right a person in the water might see them. I suggested that perhaps Mohammed was highly intelligent, and a good observer, and had been on a sea voyage. In case you are not aware of it proper Islamists believe that Mohammed was uneducated, illiterate, and that to attribute a high level of intelligence to him is heresy. Of course that makes little sense in view of his accomplishments.

Also, the Sheik and his colleagues insisted that Mohammed never saw the sea; this in spite of the fact that Mecca and Medina are almost in sight of the Red Sea. I suggested that perhaps Mohammed had friends who were observant sailors. Again this was considered heresy. So after one long afternoon on a boat in the hot Sun, with all of my caveats about observations having been rejected, you come to divine inspiration! None of my skepticism was reflected.

A more geological topic was a passage the Sheik interpreted as stating that 'mountains have roots.' He interpreted this as a revelation of what was discovered in the 19th century that mountain ranges are thick isostatically adjusted blocks that have their base well below that of the adjacent continental block. But there is another obvious interpretation if one has been in the field in the deserts of the Middle East. The mountains look as though they rest on the surrounding alluvium, but in a wadi one sees that the solid rocks of the mountain extend below the loose alluvial sands. We dropped that topic.

On the same trip I gave a geological lecture at the university in Jeddah and talking with faculty there I got the impression that the assumption that the goal was to make the pursuit of science safe was essentially correct. However, even today there is a problem about studying geology in Saudi Arabia, since nothing of importance is thought to have happened before the birth of Mohammed. This is why there are no archeological investigations there, and why ARAMCO (the Arabian-American Oil Company) still has few Arab geologists.

When I got back to the US I started to look into how old information about internal waves was, and discovered that the Vikings certainly knew about the effects of the phenomenon, and I wouldn't be surprised if the Greeks/Romans knew about it, and almost certainly the Arabs, who were the best sailors of the 7th century, would have had some experience with this phenomenon. In practical terms where there is a sharp interface within the water, the waves on this interface control the motion of the boat so that it does not correspond to the surface waves or currents.

A couple of years later a conference was organized by Sheikh Zindani in Islamabad bringing together almost all of the non-Islamic scientists he had conferred with. We were asked to prepare papers to be published in the Conference Proceedings, and mine included what I had been able to learn about possible ancient knowledge of internal waves in the ocean. Needless to say it did not get included in the published proceedings. Since then a much simpler explanation of the "internal waves" occurred to me. When you look down in clear, shallow water on a sunny day, you see the effect of diffraction of light passing through the water on the bottom as narrow bright bands separated by broad dark bands. The bright, irregular, rapidly moving stripes are called 'glitter.' They often seem to have little if anything to do with the surface waves that actually produce them. Those might be the "internal waves."

I do not read Arabic, and hence had only the translation provided by Sheik Zindani to go on. I know that Islamic scholars believe the words of the Koran have been preserved exactly, but I would question whether the meaning of the words has remained unchanged. The meanings of words drift with time so that for example, the BBC Productions of Shakespeare have modern English subtitles so one can understand the nuances of meaning. In some cases the meaning of words has changed completely. I have a facsimile copy of Shakespeare's First Folio and also one of the original King James Version of the Bible. Both have archaic language and constructions.

An example of the changing meaning of words is the well-known but cryptic statement "The meek shall inherit the Earth" from the Bible. It has been the topic of innumerable sermons, but there is a small problem. When the King James translation of the original biblical text was made, the word 'meek' had nothing to do with its modern meaning; at the time it meant 'charitable.' I suspect that this may have happened with Arabic texts as well. If this is the case, it makes the modern interpretation of obscure passages of the Koran an exercise in futility. The meeting in Islamabad was an extraordinary experience. I later realized that many of the clerics must have been Taliban. One night at dinner we asked those who were at our table why there were no women present, and were told that it would be inappropriate, since women were like animals, they have no souls. I know from other experience that it is a view held only by the most radical Islamists. Sheik Zindani, as I understand it, is now in Yemen, and was a major supporter of Osama Bin Laden.

As use of the Internet became more widespread over the past couple of decades, some highly edited TV clips of me (and others) have appeared on Islamic websites, Arab TV and even on YouTube. Usually they are confined to my non-committal statement that "I find it interesting that these statements would be in the Koran" and in one instance I am credited with suggesting that they are inspired by a 'divine being.' The latter is taken out of context. We were discussing what would be required to show that a passage implied divine inspiration.

Dan Golden, staff reporter of the Wall Street Journal, did some research and found that Zindani's goal was simply to be able to claim that western scientists believed there were miraculous statements in the Koran. The whole affair was a scam. We were tricked by the charming Sheikh and fooled by the Saudi cover story. Dan's article is available on the Internet at many sites, but here is one that is free: http://www.weatheranswer.com/public/wstjournal%20012302.txt.

In 2011 I was interviewed on this topic by a British computer specialist who has taken it upon himself to contact and interview scientists that had participated in Sheikh Zindani's project. His internet name is The Rationalizer. In the interview I give a more thorough discussion of my experience. It can be seen on YouTube at: http://tinyurl.com/BillHayInterview or http://www.youtube.com/watch?v=Bygb30gAqOs.

Links to the other interviews can be found here: http://www.YouTube.com/user/ThisIsTheTruthUncut.

It is interesting that Islamic clerics try to use science as a recruiting tool. In contrast Christian Fundamentalists regard science as hostile to their faith. In Chap. 4 I recounted how in 1997 I found myself standing on Charles Darwin's grave in Westminster Abbey. The Church of England holds him in the highest regard. In fact, many famous scientists have their final resting place in the Abbey. But a few years later I attended 'Creationist School' at one of our churches in the town where I live. Darwin was considered a tool of the devil, along with Carl Sagan. I must say that the 'teacher' had the most bizarre ideas about evolution I have ever heard. He somehow thought that Darwin had said that an ape had suddenly had a human child. This is exactly the opposite of the slow gradual evolution Darwin proposed. After several sessions someone finally asked if anyone had ever read *The Origin of Species*. No one had; no great surprise. I checked and found we do not even have a copy in our local library.

It distresses me that almost half of Americans believe that the Earth is only 6,000 years old, and about one quarter believe that the Sun revolves around the Earth. Why Babylonian mythology, the origin of those beliefs, is more acceptable than modern science to Americans is a mystery. However, the myths provide a convenient excuse for denying the fact that the outcome of our great uncontrolled experiments on planet Earth may not be beneficial to mankind.

THE '**HOCKEY STICK**' HEARING

# Global Climate Change—The Geologically Immediate Past

> The glacier was God's great plough set at work ages ago to grind, furrow, and knead over, as it were, the surface of the earth.
>
> *Louis Agassiz*

In considering what may happen in the future as a result of the human perturbations of natural processes, it is useful to examine the changes of climate that have occurred in the (geologically) immediate past—during the last glacial, the deglaciation, and the Holocene. Although this is the youngest part of the geologic record our detailed knowledge of it comes not from the deposits preserved on land but from deep-sea and ice cores taken in the past few decades. Although it had been known since the first half of the 20th century that the glacial-Holocene transition was not a smooth change, the discovery that climate change can occur in very short periods of time is quite recent, dating from the 1980s. The following discussion takes up these climate changes in the order in which they became known.

## 28.1 The Climate Changes During the Last Deglaciation

Already by the end of the 19th century it had become evident that the melting of the great ice sheets of the Last Glacial Maximum had not occurred smoothly. In North America a series of moraines marked interruptions in the general retreat of the ice. These interruptions, when the ice advanced again as the climate cooled, were called 'stadials' and the periods of ice retreat, when the climate warmed, were called interstadials. In Europe the situation was different. Central Europe was in a peculiar position, between the great Scandinavian ice sheet in the north and the smaller ice sheet covering the Alps in the south. The land between the ice sheets was covered by tundra vegetation. In the first half of the 20th century, the tundra and the bogs associated with it began to yield an interesting detailed history for the deglaciation. Bogs are widespread in Europe; in Britain they are called 'fens' and in The Netherlands they are 'veen.' The bogs contain pollen from the plant life of the surrounding region, and some of the bogs preserved thousands of years of climate history. One in

France, the 'Grand Pile' has a nearly complete record of the deglaciation. Although the sequence of climatic events is well known, the age dating is somewhat uncertain because of the different local conditions and because of the variety of dating techniques and technologies used, ranging from varves (seasonal layers) in lake sediments, tree rings, including those from logs buried in the bogs, and radioisotopes. In the following paragraph the dates are interpreted from the climate record of the GRIP (GReenland Ice Project) core from central Greenland. In spite of its location, it records each of the northern European events, as shown in Fig. 28.1.

In Northern Europe the last glaciation was known as the Weichselian. Although we now know from sea-level curves that the melting of the ice began about 18,000 years ago, the European bog deposits show that the climate there was quite cold until about 14,500 years ago. One of the common indices for this cold climate is the pollen of a tundra flower, *Dryas octopetala*. The cold period from the height of the Weichselian glaciation until 14,500 years ago has been termed the 'Oldest Dryas Stadial.' Rapid warming between 14,500 and 13,900 years ago marked the Windermere Interstadial, named for the geologic record in the northern English Lake District. This warming episode was too short and cool for trees to become reestablished in northern Europe. However the Windermere deposits contain fossil beetles adapted to warmer conditions. Insect populations, being highly mobile, are able to migrate and keep pace with climate change much more effectively than plants. After a brief cooling, the next warming episode, the Bølling Interstadial, which lasted from 13,750 until about 13,150 years ago, allowed birch trees to recolonize western Central Europe. The Bølling Interstadial is named for the peat deposits found when the Bølling Sø, a shallow lake was partially drained in the 1870s. The lake is in central Jutland, the Danish peninsula on the European mainland. The Bølling warming was interrupted by a short, 250–300 year, cooling episode, known as the Older Dryas Stadial, which ended 12,900 years ago. During the Older Dryas the

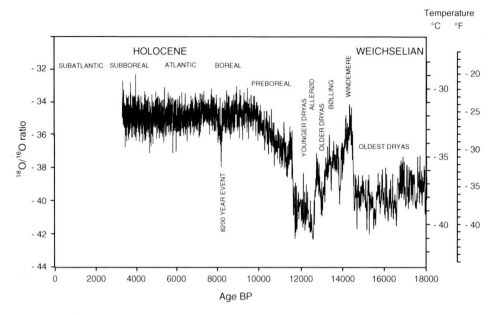

**Fig. 28.1**  Oxygen isotope record of the GRIP (Greenland Ice Project) ice core from central Greenland for the last 18,000 years. Location where the core was taken is shown in Fig. 21.7. The 0 year is 1950. The oxygen isotope ratios approximate relative temperature changes, the more negative the value, the colder the temperature. Present mean annual temperature at the GRIP Summit site is −32 °C (=−25.6 °F). Temperature scale is based on correlation with the $\delta^{18}O$ in modern Antarctic snow data. The deglaciation here was not the relatively smooth transition from the Last Glacial Maximum (18,000 + ky BP) to the Holocene seen in deep sea cores, but shows a series of marked temperature rises and falls. This is because it records air rather than ocean temperatures. The names of northern European climatic events are shown and discussed in the text

Scandinavian glaciers advanced and left a new set of terminal moraines. The warming trend returned with the Allerød Interstadial, from 12,900 to 12,700 years ago, named for the geologic record in a clay pit near the town of Allerød, north of Copenhagen, on the large island of Zealand, Denmark. The geologic record of the clay pit shows that the Older Dryas tundra was replaced by birch and aspen, only to be replaced in turn by yet another return tundra flora, marking a stadial known as the Younger Dryas. The Younger Dryas lasted from about 12,700–11,500 years ago. It marked a 1,200 year long return to the extreme cold, dry conditions that had characterized the last glacial maximum. Temperatures over the North Atlantic declined by about 5 °C (=9 °F); away from the coast the temperature fall may have been as much as 10 °C (=18 °F). Then, 11,500 years ago the temperature rose abruptly and by 9,500 years ago had reached modern levels. The Scandinavian ice sheet had disappeared by about 8,000 years ago; vestiges of the Laurentian ice sheet lasted until about 6,000 years ago. The northern European Holocene is divided into dryer (Boreal, Subboreal) and wetter (Atlantic, Subatlantic) intervals.

This detailed climate history of Central and Northern Europe has its counterpart in North America, although the timing of the individual events differ somewhat, and the names are different. Some of the European climate episodes may have been strictly local. Others seem to have affected the whole North Atlantic region. However, the Younger Dryas, the sharp return to a climate like that of the Last

Glacial Maximum, had the most widespread effect, albeit like the other events it was centered around the North Atlantic. It is clearly evident in deep-sea and ice cores from both the Northern and Southern Hemispheres.

Although oxygen isotope measurements and assemblages of planktonic foraminiferal fossils of many deep-sea cores recovered from the North Atlantic showed the Younger Dryas, the age resolution is rarely fine enough to show the other events. The reason for this is twofold. First, the cores must have a high sedimentation rate to record the event, and second, burrowing organisms, such as worms and shrimp, mix the sediment smearing out the record. The sediments and fossils in North Atlantic deep-sea cores indicate that the Younger Dryas was caused by a collapse of the ocean's thermohaline convection system, the 'Great Conveyor.'

To understand what happened it is important to understand how this very important component of the climate system works today. Figure 28.2 shows the modern ocean circulation in the North Atlantic region and the critical areas were water sinks as the sources of the Great Conveyor. A general discussion of the Great Conveyor was presented in Sect. 25.6; here we will go into more detail to learn how variations in the strength of the sources of deep water affect climate both over the ocean and the adjacent lands.

Three areas that supply water for the Great Conveyor are the GIN (Greenland-Iceland-Norwegian) Sea, the Labrador Sea, and the Mediterranean Sea. Each of these marginal seas

**Fig. 28.2** Modern surface and deep ocean currents in the North Atlantic region. The surface currents are shown in *light blue*, the deep currents in *dark blue*. Regions of sinking are shown as *darker circles* or *ellipses*. The Mediterranean Water that spreads through the North Atlantic at mid-depths is shown in *purple*. The *white area* is the Arctic ice pack, as it has been until very recently. *gray arrows* in the Arctic show the general circulation of the Arctic surface waters and ice in the latter half of the 20th century. *L* is the Laptev Sea, the Arctic's 'ice factory' as discussed in the text

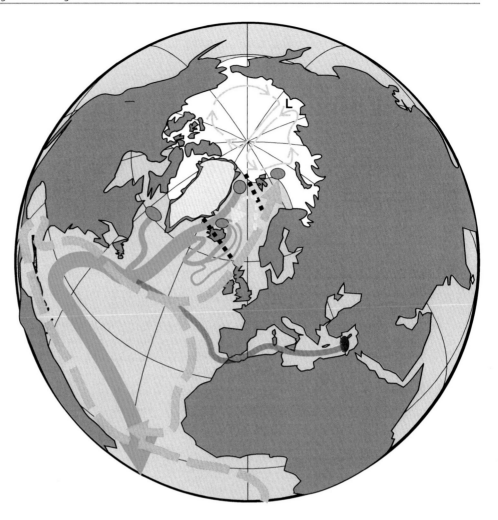

sends water into the North Atlantic where they mix and flow to the south as a water mass known as N̲orth A̲tlantic D̲eep W̲ater (NADW).

The major northern source for deep water to feed the conveyor is in the GIN Sea. The GIN Sea is a deep basin isolated from the North Atlantic by the Greenland-Scotland Ridge and from the Arctic Basin by the Barents Shelf north of Scandinavia. In Fig. 28.2 the Greenland-Scotland Ridge is shown as a black dashed line extending from Greenland to Iceland then to the Faeroe Islands, the Shetland Islands, and finally to Scotland. The southwestern edge of the Barents shelf extends from west of Spitsbergen to Norway. The northern margin of the Barents Shelf runs from north of Spitsbergen to north of Novaya Zemlya. There is a narrow mid-depth connection between the GIN Sea and the Arctic Ocean Basin between northeastern Greenland and the tip of the Barents Shelf just northwest of Spitsbergen, known as Fram Strait. It is named for the *Fram*, one of the vessels to explore the Arctic by Fridtjof Nansen and Otto Sverdrup at the turn of the last century. It was the *Fram* that took Roald Amundsen to Antarctica, landing in 1911 to

begin his trek to become the first explorer to reach the South Pole. The *Fram* is now a Museum, docked in Oslo, and well worth a visit.

While we know that the GIN Sea is the major source of the water that eventually becomes North Atlantic Deep Water (NADW), we are still not entirely sure where it forms. That is because the water probably sinks when the weather gets extremely cold and windy, and no one in their right mind would venture out into it. To understand what is going on, it is also important to know how water gets dense enough to sink to the ocean floor. First, you need to know that the deep water of the world's oceans is remarkably similar everywhere, with a salinity of 34.7 ‰ (‰ = parts per thousand by weight), and a temperature close to 2 °C (=35.6 °F). It must come from surface water source areas where the waters have this salinity and temperature.

Remember from Chap. 17 that one of the oddities of sea water is that, because of its salt content, it simply gets more and more dense as it approaches the freezing point. This is in sharp contrast to fresh water, which reaches its maximum density at 4 °C (=39.2 °F). The freezing point of water with a salinity of 34.7 ‰ is about −2.37 °C (=27.73 °F). You might think that because seawater just gets denser as it gets colder it should sink below the surface, and the whole ocean would have to cool to the freezing point for it to freeze. But another oddity of seawater is that as it cools toward the freezing point, even though its density is increasing, the increase becomes very small. Something else happens that makes it possible for seawater in frigid regions to become denser: sea-ice forms. But how can this happen when the seawater simply becomes denser as its temperature decreases?

When the seawater is so cold, a snowflake falling on it will dissolve in it and lower the salinity of a tiny volume so it can freeze. This tiny bit of ice can grow by adding water molecules onto it. This means that salt is left behind in the liquid seawater, increasing its salinity. As the ice particles grow a small amount of salt, about 7 ‰, becomes incorporated into it. The salt is gradually expelled into internal droplets, and concentrated into brine. The brine droplets are heavier than the ice, and gradually melt their way down and out, delivering the salt to the seawater. In short, sea-ice formation results in making the surrounding water saltier, and that means denser. The sea ice has a much lower salinity than the seawater from which it forms. As it freezes, the salt it expels increases the density of the surrounding seawater so that it can sink into the ocean depths. The process probably occurs in pulses during episodes of what we would call very bad winter weather. Two areas in the GIN Sea are especially suspect: the shelf area off northwestern Iceland, and a region off Greenland in the northwestern GIN Sea.

There is another Arctic area where deep water is formed. This is along the northern margin of the Barents Shelf. What happens is this: as water sinks from the surface in the GIN Sea, it must be replaced. The North Atlantic Drift, which is the name given to the Gulf Steam as it crosses the North Atlantic, splits into two currents, one going to the north, the Norway Current, and the other going to the south, the Canary Current. The Norway Current is drawn northward by the need to replace the waters that sink into the deep ocean interior. It is warm and salty, flowing northward along the coast of Norway, spinning off eddies into the western GIN Sea while the main current flows across the Barents Shelf into the high Arctic. There it becomes rapidly chilled, and as sea-ice formation takes place in the fall, these waters become very cold and dense, sinking into the deep Arctic Basin. In fact, the deep Arctic waters are the coldest (−0.8 °C = 30.6 °F), saltiest (39.95 ‰), densest waters of the world ocean. But they are trapped in the bottom of the 5 km (3.1 mile) deep Arctic Basin because the Fram Strait passage into the GIN Sea is only 2.6 km (1.6 miles) deep. The other connection to the world ocean, the Bering Strait which connects the Arctic to the Pacific, is only 45 m (=148 ft) deep, and was a land bridge between North America and Asia during the last glaciation.

The deep water in the GIN Sea is blocked from flowing into the North Atlantic Basin by the Greenland-Iceland-Scotland Ridge. The deepest passage is the Denmark Strait, between Greenland and Iceland, presently 600 m (=1970 ft) below sea level. The most dense water flows through this passage and down into the deep North Atlantic Basin as what has been termed 'the greatest waterfall on Earth,' unseen because it lies far below the ocean surface. There are other, shallower passages across the ridge between Iceland and the Faeroe Islands off northwestern Scotland. Slightly less dense waters from the GIN Sea flow through these passages and down into the depths of the North Atlantic Basin (Fig. 25.12).

From this discussion you might conclude that the sea-ice cover of the Arctic Ocean is produced by freezing of the seawater throughout the Arctic. This turns out not to be the case. In reality most of the sea ice is formed in a 'sea-ice factory' in the Laptev Sea and has nothing to do with deep water formation. Many Russian rivers flow northward into the Arctic Ocean. It is a strange accident of geography that fully 1/10 of all the river water on Earth flows into the Arctic Ocean, and almost all of it in the short period of only three to four summer months! The Laptev Sea, off Siberia, receives the water from the Lena and several other large rivers. Under the competent leadership of my colleague at GEOMAR, Heidi Kassens, German-Russian expeditions used the Russian fleet of nuclear icebreakers to explore this remote area extensively during the last decade. They found that the river waters dilute the sea water over the thousand kilometer broad Siberian shelf so that it can freeze easily as cold weather comes on in the fall. Massive ice formation takes place, and this drifts out into the Central Arctic. Little was known about the circulation in the Arctic Ocean until it became an area of critical interest to both the US and Soviets during the cold war. We now know that during the last half of the 20th century the general circulation of the Arctic Ocean was in two gyres, shown in Fig. 28.2. These are arranged in such a way as to carry the ice from the Laptev Sea 'factory' throughout the Arctic. The drift of derelict ships lost in the Arctic in the 19th and early 20th century suggests that this circulation pattern has characterized the Arctic for at least 200 years. Incidentally, there is now evidence that the Arctic circulation is changing, with unknown consequences.

Coming back to the origins of the North Atlantic Deep Water (NADW), there is another important site of sinking. It is in the Labrador Sea, between Canada and Greenland. In the winter, the northern extension of the Labrador Sea, Baffin Bay, freezes over and winds coming south between Greenland and Baffin Island become very cold. Sea-ice formation and chilling of the surface waters in the Labrador Sea cause them to become dense, and to sink and flow into the North Atlantic Basin forming a layer between the Denmark Strait and Iceland-Faeroes overflow waters.

To add to the complexity of this situation, there is yet another, very different source of dense water flowing into the

North Atlantic Basin, the Mediterranean. In the Eastern Mediterranean the dry winds coming from the Middle East result in massive evaporation, up to 2 m of water per year. The evaporation causes the salinity of the water to rise to 40 ‰. In the northern part of the Mediterranean, cold winter winds chill the saline water, and cause it to sink into the depths. This cool saline water flows out through the bottom of the Strait of Gibraltar. It flows down the continental slope into the North Atlantic Basin, spreading out on top of the waters from the GIN and Labrador Seas. The Coriolis Force causes some of it to flow northward along the Iberian margin. When it reaches the Iceland-Scotland Ridge it is forced upward, above the denser cold outflow from the GIN Sea. As it rises, it mixes into the warm North Atlantic surface waters flowing northward as the Norway Current. Because the Mediterranean outflow waters are more saline than any other waters in the North Atlantic, the Norway Current becomes saltier than the North Atlantic Drift; the extra salt makes it easier to form dense waters in the GIN Sea and Arctic Basin.

All of these water masses converge and mix in the North Atlantic basin, and flow southward as NADW, the upstream part of the Global Conveyor. The layers become less and less distinct as the waters cross the Equator and flow southward through the South Atlantic.

The situation at the time of the Younger Dryas was very different, as shown in Fig. 28.3 The major ice sheets, although much smaller than at the time of the LGM, were still present. And sea ice was more extensive than in the Holocene. Most importantly, although some dense, salty water was still being supplied from the Mediterranean, there were no northern sites of sinking water to feed the Great Conveyor. What caused the high latitude sites to shut down?

It is generally agreed that the mechanism for cutting off the supply of dense waters from the GIN and Labrador Seas was a lens of fresh glacial meltwater covering the areas where sinking would ordinarily occur. It is thought that there was a sudden massive influx of fresh water, probably resulting from failure of an ice dam holding back one or more large glacial lakes to the west of the ice sheet in North America. The initial proposal was that the outflow was from glacial Lake Agassiz on the southwestern margin of the Laurentide ice sheet, flowing via the St. Lawrence into the North Atlantic. It was then carried across the ocean and into

**Fig. 28.3** The North Atlantic region during the Younger Dryas. Deep water formation turned off. *Dashed pink line* and *arrow* is the Gulf Stream-North Atlantic Drift. *Blue area* west and south of the Laurentide ice Sheet is Lake Agassiz. *Dashed blue line* is hypothetical outflow from Lake Agassiz south of the Laurentide Ice Sheet into the North Atlantic. *Dashed gray line* is the more probable outflow from Lake Agassiz into the Arctic Ocean.

the GIN Sea. More recently evidence has been found to suggest that the fresh water inflow was actually from the western side of the Laurentian ice sheet, into the Arctic Ocean and thence into the GIN Sea. In any case, as the light, low salinity waters covered the northern oceans it became impossible for water to become dense enough to sink, even as sea-ice formed. With no need to replace sinking waters, the Norway Current ceased to flow. It no longer warmed northern Europe, and the climate shifted back to the intense cold of the LGM.

The Younger Dryas lasted about 1,200 years, and ended suddenly as the temperatures in northern Europe rose 3 °C (=5.4 °F) in only 100 years, and continued to rise to reach modern levels in a few hundred years. Even though the Younger Dryas cooled northern temperatures, and the ice sheets grew, it was a transient event. It was already doomed to end as the fresh water lens dissipated and the sinking of waters into the deep ocean began again.

The Younger Dryas was a global event. It is recorded in the Antarctic ice cores, but its effect on southern hemisphere climate was different from that in the north. The Byrd core has a detailed record of the events. The transition from the LGM to the Holocene is much smoother than that recorded in the Greenland cores. However, detailed examination of the record shows that the Bølling/Allerød warm episode of Europe was a cold excursion in the Antarctic. The Younger Dryas, a return to frigid glacial conditions in the North Atlantic, was a warm episode in the Antarctic. The two hemispheres are reacting in an opposite way to the shutdown of deep water formation in the GIN Sea. At first, as you might suspect, it was thought that there must be something wrong with the age correlations of the Greenland and Antarctic cores. That prompted a detailed analysis of the way in which the cores were dated. The Greenland cores were dated by counting the actual annual layers in the ice. The Antarctic cores do not have such distinct layering, and have been dated using radioisotopes. The reexamination resulted in a new method of correlation between Greenland and the Antarctic, using fluctuations in the methane content of the gas bubbles. The new analysis showed that the climatic variations were in fact in the opposite sense. This has become known as the 'seesaw effect.'

When the northern sites of sinking water shut down, the Global Conveyor slowed and perhaps came to a stop. The sinking in the Weddell Sea depends on a supply of cool saline North Atlantic Deep Water from the north. With that supply cut off, the sinking around the Antarctic would shut down too. However, the effects were not instantaneous.

A similar but smaller fresh-water injection event occurred in the 1960s through the 1980s, known as the 'Great Salinity Anomaly.' Temperature and salinity measurements made in the Denmark Strait in 1962 show that everything was normal that year. However, from 1965 to 1971 unusually intense

sea-ice formation took place off northwestern Iceland. This was because the waters had a salinity lower than normal. The source of the fresh water that lowered the salinity is unknown, but it seems to have come out of the Arctic with the southward-flowing East Greenland Current. Because of the lower salinity surface layer, dense water production in the GIN Sea shut down. Low salinity water passed around the southern tip of Greenland into the Labrador Sea in 1968, shutting down the production of Labrador Sea deep water. In 1972 the salinity of Labrador Sea surface waters reached a minimum and the number of icebergs drifting south of 48° N was the largest recorded to that time. There was excitement among the oceanographic community about whether we might observe more than a local climatic effect, but by 1976 deep water production in the Labrador Sea had resumed. Flowing southward out of the Labrador Sea, the low salinity water became entrained in the northern side of the North Atlantic Drift and was off Ireland in 1975. There it split into two bodies, one which was observed flowing south off Iberia in 1977, and the other going north, crossing the Iceland-Scotland Ridge in 1978, and entering the Barents Sea in 1979. Oceanographers waited with great interest to see if the pattern would repeat itself.

Sure enough, a second, less severe 'Great Salinity Anomaly' occurred during the 1980s and another during the 1990s. It is not known whether such events are a normal occurrence, or whether they have been triggered by changes in flow of the Arctic Rivers during the latter half of the 20th century. There is some evidence that the first and second Great Salinity Anomalies may have perturbed northern hemisphere climate, and are reflected in cooling episodes seen in the northern hemisphere temperature record, shown in Fig. 28.4.

Although during the Great Salinity Anomaly parts of the source region of the Great Conveyor were successively shut down, there was almost no effect on the Conveyor itself. Perhaps the reason the Conveyor has been so stable over the past 7,000 years is that it has multiple northern sources, and during this period flows of fresh water into the Arctic region have not been large enough to shut all of them down at the same time. However, one of the most popular catastrophe scenarios for the future is that as the sea-ice, permafrost, and glaciers around the Arctic melt, they might cause a return to conditions like those that existed in the Younger Dryas and put Europe in a deep freeze. This was the major thesis in the film "The Day After Tomorrow," but of course in the film it happened many times faster that it might in the real world.

We know a lot about the Great Salinity Anomaly and its effects on deep-sea circulation because some man-made tracers had been introduced into the surface ocean in the middle of the 20th century. They provided a unique opportunity to investigate the rates of deep water formation and mixing of waters in the deep sea. These tracers were

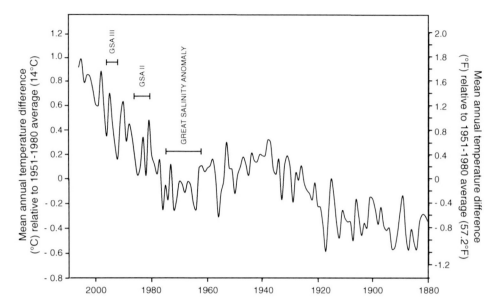

**Fig. 28.4** An annual average northern hemisphere temperature record, showing the times when the Great Salinity Anomalies of the 1960s–1970s, 1980s, and 1990s occurred. The changes in North Atlantic circulation may be partially responsible for the lower temperatures during these times

ozone-destroying hydrofluorocarbons, which began to be introduced during the 1950s, and tritium. Tritium is a third, radioactive isotope of hydrogen, with one proton and two neutrons in the nucleus. It has a short half-life, only 12.32 years. It is produced naturally by interaction of cosmic rays with atoms, mostly nitrogen, in the upper atmosphere. However a large amount of tritium was produced by the (mostly) US and Soviet atmospheric atomic tests from 1951 to 1958. In the last year of the tests, the US and USSR conducted a total of 96 atomic blasts in the atmosphere. This produced an enormous spike of atmospheric tritium which over the next decades was tracked as it entered the ocean.

## 28.2   The Sudden Changes Recorded in Ice Cores: Dansgaard-Oeschger Events

Coring the Greenland ice cap began in the 1960s, but it was not until the late 1970s and 1980s that the full implications of the Greenland climate record were realized. The cores showed that brief episodes of rapid climate change had occurred throughout the last glaciation. Comparing the record in the ice cores with that in deep sea cores, it has become evident that there are two kinds of rapid climate change events that occurred during the last glaciation and the deglaciation: Dansgaard-Oeschger events, when temperatures rose very sharply, then gradually declined, and Heinrich events—when flotillas of icebergs were released into the North Atlantic and resulted in regional cooling. Finally, there was the major perturbation of the climate system near the end of the deglaciation, known as the Younger Dryas episode, described above.

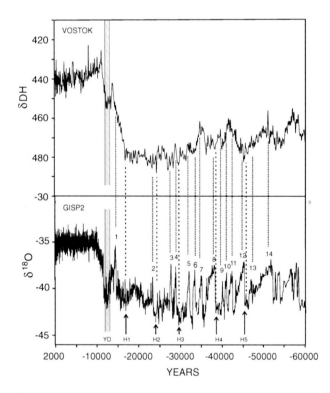

**Fig. 28.5** Comparison of the data for the deuterium-hydrogen ratio (δDH) in the Vostok core (Antarctica) with the oxygen isotope ratio (δ18O) in the GISP2 core Greenland) showing that there is a record of the Dansgaard-Oeschger events in both Polar Regions

Interestingly all of these climate perturbations were centered on the far North Atlantic region, with the effects diminishing with distance from that region. The best record of these events is in the GISP2 ice core, from central Greenland, but most of the events can also be seen in the Vostok ice core from the Antarctic. Figure 28.5 shows climate proxies from both cores, the $\delta^{18}O$ record of the GISP2 core and the Deuterium/Hydrogen (DH) record from the Vostok core. The Antarctic DH record is shown because it is more sensitive to climatic changes than the local $\delta^{18}O$ record.

The Dansgaard-Oeschger events are named for their discoverers, Willi Dansgaard of the University of Copenhagen, Denmark, and Hans Oeschger of the University of Bern, Switzerland (Fig. 28.6). Dansgaard and Oeschger were pioneers in proposing drilling ice cores in Greenland to investigate the paleoclimate, and each has played a very important role in making analyses and interpreting the history recorded by the ice. Starting with the first long ice core, taken at the American Station Camp Century in northwestern Greenland.

Willi Dansgaard developed the techniques for analyzing the gas bubbles in the ice for $^{18}O$ and deuterium, and use of other methods to date the cores precisely. Hans Oeschger pioneered the measurement of gas composition, including $CO_2$ and $CH_4$ in the air bubbles, and developed techniques for measuring the tiny amounts of $^{14}C$ in the gas to date the bubbles.

From the beginning it was evident that there were strange things in the ice cores. The glacial climate recorded in the Greenland ice was more complex than the record in deep sea sediments had indicated. There were instances when the temperature in Greenland suddenly rose, by 5–10 °C (9–18 °F) over a matter of decades, then gradually declined back to typical glacial levels. Two ice cores have played a special role in deciphering the climatic history of Greenland, the GRIP (GReenland Ice Project) Summit core and GISP

(Greenland Ice Sheet Project) 2 core. The locations where these cores were recovered are shown in Fig. 21.7.

The work of Dansgaard and Oeschger on these cores revealed that the periodic instabilities in climate reached back into the last interglacial, and probably earlier. The GRIP Summit core extends back 250,000 years, although the older part of the core is compressed and has only low resolution. It shows 24 instances of sudden temperature rise between 16,000 and 110,000 years ago (Fig. 28.7a). The temperature rises by about 5 °C (=9 °F) in only a few decades, and are followed by gradual cooling, returning to full glacial cold temperatures. The cycles of the last 60,000 years are shown in (Fig. 28.7b). It was quickly realized that these cycles must reflect changes in the ocean circulation in the North Atlantic and GIN Sea. The warm pulses must reflect periodic intrusions of warm North Atlantic water across the Greenland-Scotland Ridge into the GIN Sea.

One of the striking peculiarities of these climate instabilities, which have come to be known as Dansgaard-Oeschger (D/O) events, is that the last one occurred 14,000 years ago, and they have not occurred since. The rapid rise in temperature is termed the 'D/O event' and taken with the slow return to glacial conditions forms the 'D/O cycle.' The GRIP core shows that D/O cycles occurred during the last interglacial, known as the Eemian. Somehow the modern interglacial in which we live, the Holocene, is different. It is the lack of these sudden climate changes that has allowed modern civilization to develop.

The Dansgaard-Oeschger events are characterized by a sudden rise in temperatures in Greenland, and presumably

**Fig. 28.6** Willi Dansgaard (*left*) and Hans Oeschger (*right*)

**Fig. 28.7** The oxygen isotope record of the GRIP Summit ice core. **a** The last 120,000 years; **b** enlargement of the record of the last 60,000 year. The Dansgaard-Oeschger events recognized in 1993 are numbered 1–24. Two other similar climate warming events can be seen in the Eemian Interglacial part of the record. The triangles at the base of (**b**) are the 1,470 frequency thought to characterize the D/O events. Present mean annual temperature at the GRIP Summit site is −32 °C (=−25.6 °F). Temperature scale is based correlation with δ18O in modern Antarctic snow data. For discussion, see text

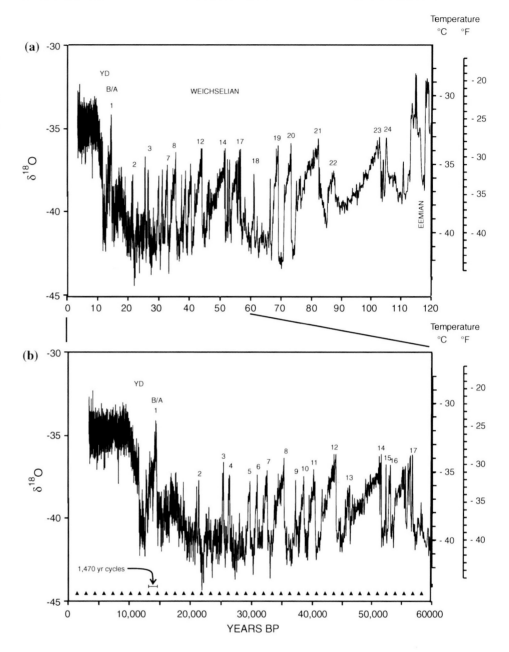

around the entire northern North Atlantic region, followed by a gradual cooling back to glacial conditions. The temperature rise occurs in only 10–30 years, and the ensuing warm excursions are too brief to affect the overall pattern of glacial growth from the Eemian Interglacial to the Last Glacial Maximum. A variety of causes have been proposed, including sudden increases in greenhouse gas concentrations, particularly methane releases from decomposition of gas hydrates. However, there is no evidence of changes in greenhouse gas concentrations large enough to explain the warming recorded in Greenland.

Andrey Ganopolski and Stefan Rahmstorf at the Potsdam Institute for Climate Impact Research just outside Berlin have proposed a simple explanation for the D/O events. The concentration of the climatological effects around the northern North Atlantic point to changes in the source of the 'Great Conveyor,' the ocean's thermohaline circulation system, but in a way very different from what happened during the Younger Dryas. It is thought that during the most of last glacial, the source region for the Great Conveyor had shifted to a position south of the Greenland Scotland Ridge, as shown in Fig. 28.8. The entire region of northern North America, the Labrador Sea, Greenland, the GIN Sea, and northwestern Europe was in a deep freeze. There was almost no fresh water input into the Arctic Ocean or GIN Sea. As sublimation from the ice and snow surfaces removed water

**Fig. 28.8** Conditions in the circum-North Atlantic region during the Last Glacial. The major ice sheets, Cordilleran (*C*), Laurentide (*L*), Greenland (*G*), Scandinavian (*S*), British, and Barents, are shown. The GIN Sea was covered by ice to the Greenland-Scotland Ridge. The North Atlantic Drift crossed the ocean at 45° N. Deep water formation is thought to have taken place south of the Greenland-Scotland Ridge, feeding the glacial-age Great Conveyor

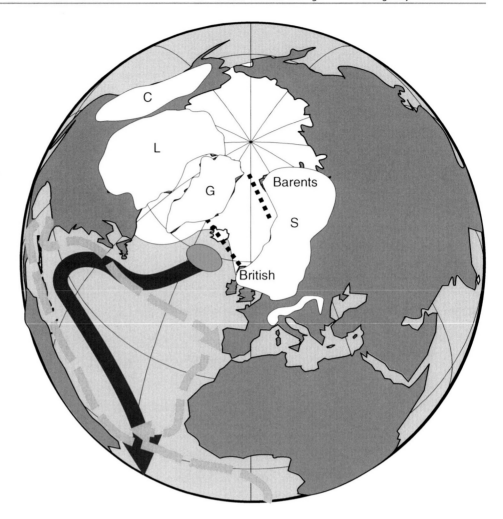

and seasonal sea-ice freezing expelled salt into the surrounding seawater, the salinity in the GIN Sea increased until, eventually, sinking began. This led to the situation shown in Fig. 28.9.

The flux of warm water into the GIN Sea caused the temperature of the surrounding region to rise. This resulted in both greater regional snowfall, and a fresh-water inflow into the Arctic Ocean and the GIN and Labrador Seas. As the fresh-water inflow lowered the salinity, the amount of water sinking into the depths of the GIN Sea decreased. This caused the surface flow across the Iceland-Scotland Ridge to slow, and the regional temperatures gradually fell. Finally the fresh water input ceased, the sinking system shut down, the sea-ice front advanced to the Greenland-Scotland Ridge again. The deep water formation site moved back across the ridge to the northern North Atlantic. The D/O cycle was complete.

A peculiarity of the D/O cycles, is that, when there are very precise time correlations between the Greenland and Antarctic cores, the warming in the northern hemisphere seems to correspond to cooling in the southern hemisphere and vice versa. This phenomenon, already noted in the

discussion of the Younger Dryas, has become known as the interhemispheric seesaw effect.

Although the events seen in Fig. 28.7 were at first thought to occur at irregular intervals, it was realized in the 1990s that the D/O cycles seemed to have a frequency of about 1,500 years. While some seemed to be 1,500 years apart, others were about 3,000 or 4,500 years apart. It was as though there were times when one, two, or even three cycles would be skipped. More recent studies have refined that estimate of the frequency to 1,470 years. Stefan Rahmstorf in Potsdam argues that there is a remarkably precise 'clock' setting the frequency, although a few of the events are 'off' by a couple of hundred years. Figure 28.7b shows only the last 60,000 years of the records but, as noted above, the D/O events can be traced back into the last interglacial in the central Greenland cores. Although there are no D/O events evident in the Greenland cores since the Bølling interstadial event just after 14,000 years ago, North Atlantic deep sea cores have been examined in detail to see if the present interglacial, the Holocene, is indeed free of such climatic variations.

**Fig. 28.9** Conditions in the circum-North Atlantic region during the initial phase of a Dansgaard/Oeschger event. The major ice sheets, Cordilleran (*C*), Laurentide (*L*), Greenland (*G*), Scandinavian (*S*), British, and Barents, are shown. The GIN Sea is partially ice-free in response to an influx of warm saline North Atlantic water to replace that sinking into the depths, feeding the glacial-age Great Conveyor from a site north of the Greenland-Scotland Ridge

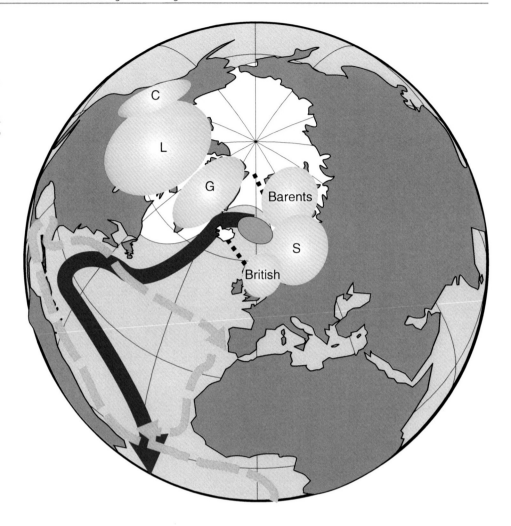

While it is generally accepted that the D/O events during the last glacial reflect relocations of the site of sinking and deep-water formation from south to north across the Greenland-Scotland Ridge, the ultimate cause of the 1,470 periodicity has been the topic of much speculation. There are two possibilities: (1) that the timing is due to some extraterrestrial phenomenon, or (2) that it is inherent to the Earth's climate system.

One of the first suggestions was that there might be a previously unknown ∼1,500 year cycle of solar radiation. Although pure speculation, this has been taken by some as fact. During the 2008 US Presidential Campaign, Republicans claimed it is responsible for the current rise in global temperatures. As soon as the new solar cycle was proposed, the question arose as to why a change in the Sun's output should have its effect concentrated in the northern North Atlantic and not be readily detected elsewhere. Recently a test of the solar irradiance hypothesis has been carried out. You may recall from Sect. 3.7 that we discussed how solar activity affects the production of the radioactive isotopes of carbon and beryllium, $^{14}$C and $^{10}$Be, in the upper atmosphere. Higher solar activity deflects the cosmic rays that

form these isotopes, so that their rate of production drops as solar activity increases. If there were a 1,470 year cycle in solar activity it should be recorded in the abundances of those isotopes over time. A thorough search has been carried out for just such a signal in the isotopic data, and none has been found. There is no evidence for such a cycle in solar activity.

Stefan Rahmstorf believes that the timing is so regular that it might be due to an orbital cycle. The closest one he could find is a lunar cycle of 1,800 years, not close enough.

In the early 1990s, before the regularity of the cycles was fully recognized, Wally Broecker had suggested a cause for periodic climate change within the Earth's climate system itself. He proposed that the glacial world might be interrupted by a 'salt oscillator.' Basically, the idea is this: the sinking of salty waters in the North Atlantic gradually freshens the surface ocean by removing salt. At some point the surface waters become fresh enough that they can no longer become dense enough to sink, and the Conveyor System shuts down. Continuing evaporation over the Atlantic with transport of water through the atmosphere to the Pacific gradually raises the salinity of the Atlantic until finally deep water formation can begin again. This is basically the same argument as that presented by Ganopolski and Rahmstorf, only they emphasized the input of fresh water into the Arctic and GIN Seas while Broecker emphasized the export of salt by the thermohaline convection system.

The argument against a simple periodic salinity cycle is that some modern climate oscillations, notably ENSO (El Niño—Southern Oscillation) and the North Atlantic Oscillation are rather irregular in their timing. I would argue that the D/O events and these modern climate oscillations occur on very different time scales—a few years to a decade versus ~1,500 years. Also, Broecker's salt oscillator is driven by transport of fresh water across Central America, from the Atlantic to the Pacific. This occurs via the most stable part of the climate system, the Hadley Cells, with evaporation beneath the descending limb of the cells and Trade Winds, and return of the fresh water as precipitation in the Intertropical Convergence. The northern hemisphere ice sheets would have acted to make the low-latitude atmospheric circulation system even more stable. I do not find it surprising that the 'salinity oscillator' could force changes in the ocean thermohaline circulation system on a regular basis.

One is reminded of a Shishi-Odoshi, often seen in Japanese gardens. The Shishi-Odoshi is a length of hollow bamboo forming a tube which is blocked at one end and fitted with a pivot near the middle of its length. Water from a spout goes into the open end of the bamboo tube, and when it is almost full it tips on the pivot, emptying the water and falling back, making a 'bonk' sound as it hits a rock in its original position. The name means literally 'deer frightener' and the device was intended to keep deer out of the garden. Where I live in Colorado, the bamboo would probably be eaten by elk in the winter.

## 28.3  The Record of Massive Iceberg Discharges in North Atlantic Deep-Sea and Greenland Ice Cores: The Heinrich Events

There is evidence for another kind of ice-age climatic cyclicity in the North Atlantic. It is in the form of sedimentary layers with large numbers of small rock fragments, whereas the surrounding sediment consists of muds or oozes. These unusual sedimentary layers were originally described by Hartmut Heinrich of the German Hydrographic Office in Hamburg in 1988. They occur not just in a few cores, but in many cores from broad regions of the North Atlantic. It was recognized that these small rocks must have been dropped from icebergs. But they are in such abundance that there must have been huge numbers of icebergs. Now they are described as 'flotillas' or 'armadas' of icebergs. The climatic episodes that produced them are now known as Heinrich events. They occur at a frequency approximately 7,000 years apart (Fig. 28.10).

Where did the icebergs come from and what was their effect on the climate? Many of the rock fragments can be traced back to the Hudson Strait that today connects Hudson Bay and Baffin Bay. During the last glacial the center of the Laurentide Ice Sheet was over southern Hudson Bay, and Hudson Strait served as a passage through which ice could flow, eventually calving into the Labrador Sea. It is thought that the periodic discharge is probably related to warming of the base of the ice sheet by the geothermal heat flux, i.e. the heat coming from the interior of the Earth. It works like this:

As the ice builds up, its base is frozen to the underlying ground. As the ice sheet becomes thick, there is flow from the center to the edges, but the movement occurs at

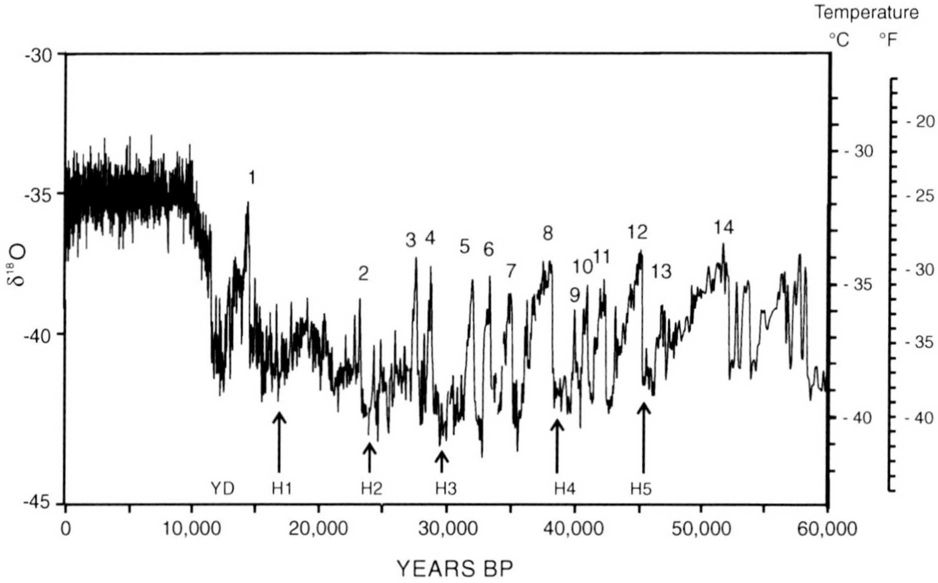

**Fig. 28.10** The oxygen isotope record from the upper part of the GISP2 ice core, central Greenland. Positions of Heinrich events *1–5* are indicated as *H1*, etc. *Numbers* above the record are Dansgaard-Oeschger events. Present mean annual temperature at the GISP site is −32 °C (=−25.6 °F). Temperature scale is based correlation with δ18O in modern Antarctic snow data

mid-levels in the ice. The base remains stuck to the ground. However, ice is a good insulator, and so as it builds up the small heat flux from the interior of the Earth gradually warms its base. When it reaches the freezing point, melting occurs. Lubricated by water, the ice can move, and it surges outward. During Heinrich Events the ice that flowed through Hudson Strait calved as masses of icebergs that drifted south and become entrained in the North Atlantic Drift, and eventually filling the northern North Atlantic. The surge lowers the height of the ice, and reduces the forces that make the ice move. The warm ice at the base of the ice sheet is replaced by cold ice, which freezes solidly to the ground again, and the process starts over again. The cyclicity of the 'Heinrich events' is thus wholly independent of that associated with the D/O events.

As the icebergs melt, they not only drop pebbles and rocks to the sea floor, they cover the surface with fresh water. As this fresh water layer spreads into the eastern North Atlantic and GIN Sea it shuts down the deep water formation, cutting off the flow of water into the Great Conveyor at its source. It is thought that a shut-down of this sort will last for several centuries before the salinity of the surface layers can increase to levels to allow sinking and deep-water formation to resume.

It has recently been discovered that there is an important ocean feedback enhancing the effect of the fresh water discharge onto the North Atlantic. As the surface of the North Atlantic cools, the Intertropical Convergence is forced to the south, to a position near the Equator. The rain falls in the Amazon Basin, and returns to the Atlantic via the Amazon River. This freshens the Atlantic and delays a restart of the deep-water formation. As sinking in the North Atlantic or GIN Sea resumes, the ITCZ moves northward until it is off Central America, and water is again transported as vapor from the Atlantic to the Pacific. This increases the salinity of the Atlantic and reinforces the sinking tendency in the North Atlantic region.

## 28.4   What Did the Deglaciation Look like Outside the North Atlantic Region?

Figure 28.11 shows the correlation between the oxygen isotope records of ice cores taken at GRIP Greenland Summit, Huascaran, at 9° S in the Peruvian Andes, and at EPICA Dome C, in the Antarctic south of Australia. The Antarctic record shows a general rise of temperatures associated with the deglaciation, starting about 17,000 years ago. It also shows a return to cooler temperatures about 12,000 years ago, at the time of the Older Dryas, followed by a brief rise which seems to correspond to the Allerød, followed by a return to cooler conditions at the time of the

Younger Dryas. In the Huascaran ice cap in the Peruvian Andes the Older and Younger Dryas seem to be represented, merged into one cooling event. The Greenland record is characterized by the suddenness of the climatic changes and their magnitude. It is as though the tropical and Antarctic records are damped echoes of the events in the North Atlantic.

The brief 8,200 year cool event, seen clearly in the Greenland ice-core record and found in pollen records throughout Europe (see Fig. 28.12) may also be present in the Huascaran record, but is not obvious in the Antarctic.

Note that the 'Early Holocene Climatic Optimum,' a warm episode centered on 6,000 years ago and until recently thought to be global phenomenon, does not show up in any of these records.

## 28.5   Are There Holocene Climate Cycles?

When looking at climate change during the Holocene, one is immediately struck by the fact that it has been an extraordinarily stable period. The last 8,000 years have, in fact, certainly been the longest period of climatic stability in the last 120,000 years. It may well be the longest period of climatic stability is the past 3.5 million years. Discussions of Holocene climate change have recognized several major trends, originally assumed to be global. However, as we learn more and more about the Holocene, it becomes evident that most if not all of these trends are regional or local, not global. The most 'famous' of these climate variations are the '8200 year event,' the 6,000 year 'Holocene Climatic Optimum,' the 'Medieval Climate Optimum' and the 'Little Ice Age.' The global temperature difference involved in the climate change from the Last Glacial Maximum and the Holocene was about 5 °C (=9 °F). The temperature differences between warm and cold episodes of the Holocene are less than 0.5 °C (less than 1 °F).

Early in the Holocene, about 8,200 years ago there was a brief decline in temperatures around the North Atlantic. This is thought to have been the result of a glacial meltwater pulse, probably as an ice dam failed. The effect was much smaller and shorter-lived than the Younger Dryas. It was recorded by changes in the flora throughout Europe, including the Mediterranean region, and appears prominently on the overall curve for European temperatures, shown in Fig. 28.12, as well as on each of the regional curves.

In any account of Holocene climate you will reference to the 'Climatic Optimum' around 6,000 years ago. Recent studies have shown that the 'climatic optimum' was a regional northern European affair. It was recorded in bog deposits in northeastern Europe; those were among the first to be studied, and it was assumed that they represented a

**Fig. 28.11** Oxygen isotope records of ice cores taken at GRIP (Greenland Summit), Huascaran (Peruvian Andes, 9° S), and EPICA Dome C (Antarctic south of Australia). The positions of the Younger Dryas and the 8,200 year anomaly are marked

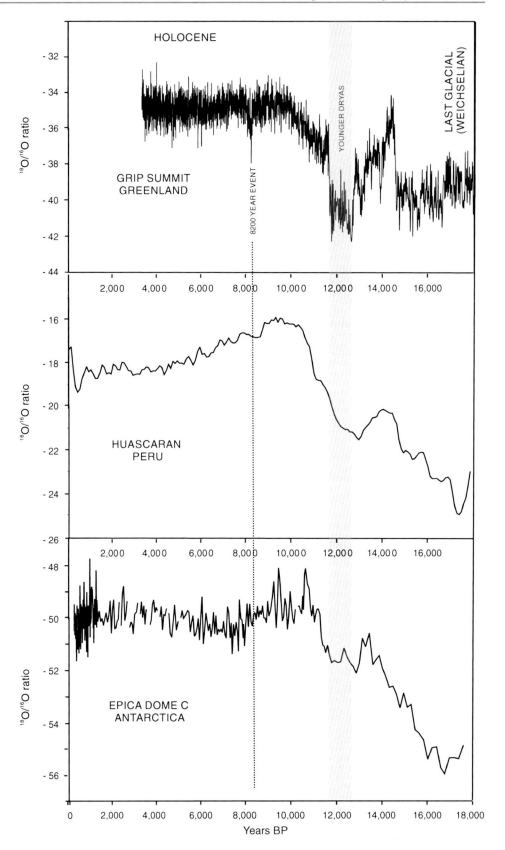

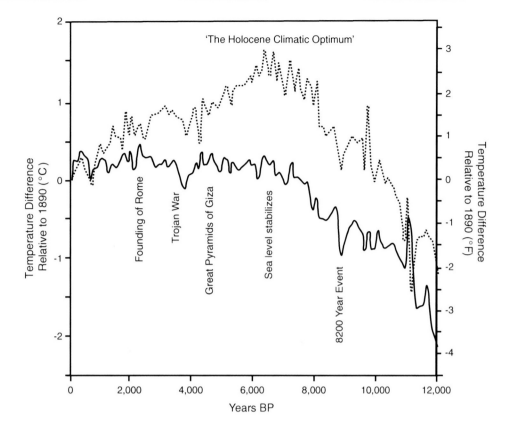

**Fig. 28.12** The Holocene temperature record in Europe, reconstructed from pollen data. *Solid line* is the European average. *Dotted line* is for northeastern Europe (northwestern Russia and Finland)

global signal. It is most spectacularly evident in the records from northwestern Russia and Finland, shown as a dotted line in Fig. 28.12. However, it does not show up elsewhere and was neither a global nor even a general European climatic episode as you can see by comparing the dotted and solid lines in Fig. 28.12.

Much more is known about European climate history than even that of North America. This is both because of the excellent and extensive fossil record there, and because of the intense interest in the ancient history of the European region. One area that has been extensively investigated for climatic variations is the Alps, where advances and retreats of the glaciers can be documented and often more or less accurately dated.

Incidentally, the use of the term 'optimum' in referring to the times of warmest conditions has nothing to do with whether it was actually optimal for life. Clearly, the expression was coined by someone on a cold winter day when they were thinking that they would rather be in Florida.

Gerard Bond of the Lamont-Doherty Earth Observatory in Palisades, New York, and an international group of co-workers have made detailed studies and comparison of cores off Ireland and Greenland looking for clues to changes in the ocean currents since the Younger Dryas. They did indeed find 8 cycles, several of which occur at a frequency approximating the 1,470 year cycle, but others are far off.

Those that do occur 'on schedule' are not sudden warming events, but cold episodes, with the sediment containing evidence for unusually large numbers of icebergs. Although these changes are unlike the D/O events, in that there is no sudden warming followed by gradual cooling, this may be due to the relatively poor time resolution in deep-sea cores. These changes are not apparent in the pollen data from bogs on land.

The record of the last 1800 years has been a topic of intense interest, and the scientific literature is replete with accounts of records for different areas, global compilations, and statistical analyses. A summary by Michael E. Mann of the Pennsylvania State University in 2007 (*Climate over the past two millennia*, Annual Review of Earth and Planetary Science, vol. 35, pp. 111–136) includes the latest graphs and references to the most important literature. It was a paper published in 1998 by Mann, Ray Bradley, both then at the University of Massachusetts, and Malcolm Hughes of the Laboratory for Tree Ring Research at the University of Arizona that included the diagram that became known as 'the hockey stick.' Mann and his colleagues called attention to the 20th century rise in temperature, arguing that this was unprecedented in the last 600 years and appeared to be the result of greenhouse gas additions to the atmosphere. Versions of the diagram were included in the Intergovernmental Panel on Climate Change Report for 2002, and in Al Gore's film *An Inconvenient Truth*. Figure 28.13 shows the main

**Fig. 28.13** The Mann et al. (1998) Northern Hemisphere mean temperature record over the past 600 years, based on tree ring, ice core, ice melt, historical, and other data

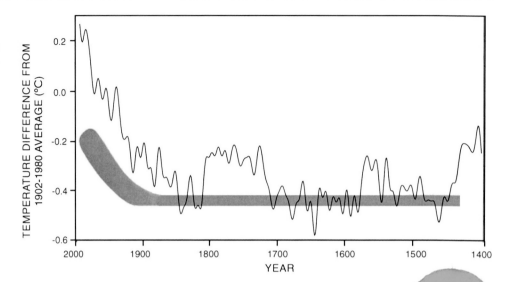

trend line of their figure, with an actual hockey stick for comparison.

In subsequent publications the record was extended back to the year 200, as shown in Fig. 28.14. The Northern Hemisphere and Southern Hemisphere obviously behave differently. This is because the northern hemisphere is largely land and the southern hemisphere largely water. The northern hemisphere data show the classic 'Medieval Climate Optimum' and 'Little Ice Age' while the southern hemisphere shows a series of large oscillations related to alternations between El Niño and La Niña conditions (Fig. 28.14).

The Medieval Climate Optimum (MCO) of 800–1200 AD and the Little Ice Age (LIA), which started at the end of the 15th century, had its maximum in the 17th and 18th centuries, and ended in the 19th century have been suggested as a recent example of a climate cycle. However, unlike the D/O cycles, the warming associated with the beginning of the MCO and the end of the LIA was gradual, not abrupt. In the 1960s it was thought that these warm and cool periods were probably global. It turns out that they were European phenomena. Wally Broecker suggested that the 'Little Ice Age' might have been caused by a partial shut-down of deep water formation in the GIN Sea. Subsequent studies indicate that this is indeed the case.

The variations in temperature in the Southern Hemisphere are much greater and more frequent than in the northern hemisphere. This is a reflection of the different distributions of land and water in the two hemispheres, as shown in Fig. 28.15.

One might expect that, because of the much lower heat capacity of land, the greatest variations would occur in the land (Northern) hemisphere while the water (Southern) hemisphere would be much more stable. On an annual basis this is, of course, true. But what is seen in Fig. 28.14 is not

seasonal variations, but changes over a deca[...] Pacific Ocean does change markedly on this t[...] the occurrence of El Niño and La Niña conditi[...] El Niño the Trade Winds relax as a low dev[...] western Pacific. Warm water from the wester[...] spreads eastward toward South America. It co[...] down the upwelling system along the west [...] America and even affects that in the equatori[...] surface waters of the entire South Pacific bec[...] the end of the El Niño the Trade Winds incre[...] the warm surface waters back westward, the u[...] deeper waters along the South American margi[...] the surface waters of the eastern Pacific beco[...] the opposite extreme, La Niña condition, the[...] intense and the cool waters spread westward o[...] the eastern Pacific. Because the South Pacific occupies fully half of the Southern Hemisphere these changes have a dramatic effect on the hemispheric temperature. What is seen as a major change in Southern Hemisphere temperatures is an oceanographic change in response to the intensity of the southern Trade Winds.

Detailed analysis of the Southern Hemisphere temperature signal suggests two causes for the variations, variations in insolation associated with the sunspot cycle, and variations in the transparency and reflectivity of the atmosphere resulting from volcanic eruptions. Both act in a manner that is counterintuitive. When the intensity of solar radiation is maximal, during sunspot maxima, the South Pacific Ocean tends to go into its La Niña condition, becoming cooler rather than being warmed by the increase solar insolation. Similarly, after a major volcanic eruption which introduces sulfur dioxide and ash into the stratosphere, reflecting sunlight, the Pacific Ocean develops El Niño conditions. The sunspot cycle averages 11 years, and although there is no periodicity in volcanic eruptions, those that affect the South

**Fig. 28.14** Decadal
temperatures over the past
1800 years, based on a variety of
fossil, tree ring, and isotopic data.
The *heavy lines* are running
40 year averages

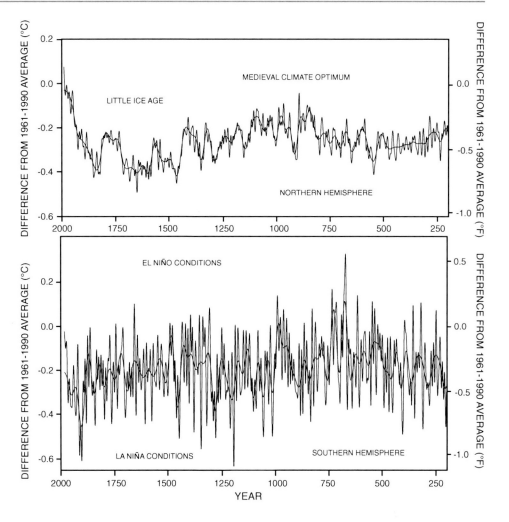

Pacific mostly from volcanoes in Indonesia and the Philippines tend to occur on about the same timescale. The suspicion is that what we see in southern hemisphere temperature variations over the past 1800 years is the change in South Pacific Ocean surface temperatures forced by changes in insolation and major volcanic eruptions.

It has also been suggested that the current warming trend is a D/O event, and not related to any human activity. However, past D/O events never raised the temperature much above the Holocene average. Assuming that the Little Ice Age represents the cool phase of a D/O cycle, the 1,470 year cycle period predicts that the warming trend should not start for another hundred years, after 2100. Instead, the warming trend started in the latter part of the 19th century, so that in this case the periodicity would be off by about 250 years. Moreover, if the warming were due to initiation of a new D/O event, it implies increased sinking in the GIN Sea, drawing warm water northward from the North Atlantic. Instead there is evidence that the rate of sinking in the GIN Sea has declined by about 20 % over the last 50 years.

## 28.6   The Hockey Stick Controversy

The Mann, Bradley, Hughes 1998 paper, published in *Nature*, led to one of the most surreal episodes in modern science. The last year of data on the 'hockey stick' diagram was 1995. The paper was submitted to *Nature* on May 9, 1997. The abstract included the statement that the Northern Hemisphere mean annual temperatures of three of the past 8 years (1995, 1991, and 1990) had been warmer than any year since AD 1400. In 2005 two Canadians, Stephen McIntyre, a retired Mining Executive, who apparently had never published a scientific paper before in his life, and conservative Economist Ross McKitrick published a critique of the 1998 paper claiming that the shape of the 'hockey stick' was "primarily an artifact of poor data handling", obsolete data and incorrect calculation of principal components. They argued that the 15th century had been warmer than today. Their message was that the late 20th century warming was nothing unusual, and certainly not caused by greenhouse gases. They ignored all scientific papers that had been published on the topic since 1999, and the temperature

**Fig. 28.15** Distribution of land and sea on Earth by latitude. Most of the land is in the Northern Hemisphere, greatest expanses of ocean are in the Southern Hemisphere

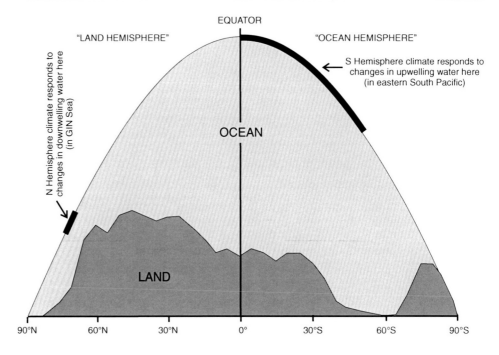

data that had been collected since. They also never mentioned Ray Bradley's excellent 1999 book *Paleoclimatology, Reconstructing Climates of the Quaternary*, which has an extensive discussion of the use of geologic and paleontologic evidence to estimate past temperatures. Nevertheless, their arguments were much appreciated by these in the petroleum and coal industries who feel threatened by the development of alternative energy technologies.

The 15th century marked the transition from the 'Dark Ages' to the Renaissance. Painters of the Renaissance produced portraits of individuals. They all seem to be wearing heavy clothing. Perhaps portrait painting was done only in winter, but then, one must ask, why do summer landscapes appear outside the windows.

The 'controversy' reached a climax on July 19, 2006, when the Subcommittee on Oversight and Investigations of the Committee on Energy and Commerce of the United States House of Representatives held hearings on 'Questions surrounding the 'Hockey Stick' Temperature Studies: Implications for climate change assessments.' The Chairman of the Committee on Energy and Commerce was Joe Barton of Texas' 6th District. The Subcommittee Chair was Ed Whitfield, of Kentucky's 1st District. The parallels with the trial of Galileo Galilei are uncanny.

Galileo's problem was that he had evidence that Nicolas Copernicus had been right is arguing that the Sun, not the Earth was the center of the Universe, and that the Earth orbited the Sun, not the other way round. Copernicus had published his ideas in a book *De revolutionibus orbium coelestium* in 1543, the year of his death. The book was dedicated to Pope Paul III. Copernicus was a Catholic cleric,

and his 'heliocentric theory' aroused much interest among theologians. It was not considered heretical at the time. However, by the late 16th Century the idea of heliocentrism had spread, and a Dominican Monk, Giordano Bruno became one of its champions. Bruno challenged the concept of an Earth-centered Universe as well as several other aspects of Catholic theology. He was charged with heresy by the Roman Inquisition. Cardinal Roberto Francesco Romolo Bellarmino (now Saint Roberto Bellarmine) oversaw the trial. In 1600 Giordano Bruno was burned at the stake in the Campo de' Fiori (the Flower Market) in Rome. Bruno is often cited at the first martyr for science.

About 1609, Galileo, used the newly invented telescope to observe the moons of Jupiter. Seeing these, he came to the conclusion that Nicolas Copernicus had likely been correct in his 1543 publication; the Earth orbits around the Sun. In 1610 Galileo observed the phases of Venus, and it was clear to him that it orbited the Sun. Proof that Copernicus was correct.

Analogously, in 1998, Michael Mann and his colleagues had essentially confirmed the 1896 hypothesis of Svante Arrhenius that the carbon dioxide produced by burning fossil fuels was going to cause warming of the planet. One might also recall that Alfred Wegener's idea of continental drift, proposed in 1912, originally considered a heresy by American geologists, finally gained acceptance as part of the theory of Plate Tectonics in 1967. It seems as though a major new scientific idea requires decades to a century to be accepted.

Now here is where it gets really interesting. In 1611, Cardinal Bellarmine had the mathematicians of the Jesuit

Collegio Romano investigate and certify Galileo's discoveries to be correct. Galileo was elected member of the Accademia dei Lincei, which had been founded in 1603, and is the oldest academy of sciences. He was honored for his discoveries at a banquet in the Collegio Romano in Rome.

Galileo was denounced by Tomasso Caccini in 1614 as a possible heretic. Mann et al. were denounced by McIntyre and McKitrick in 2003. The time constant for identifying a heretic seems to be about 5–7 years.

In 1615 Carmelite Friar Paolo Antonio Foscarini published a letter arguing that the Copernican theory was compatible with scripture. Cardinal Bellarmine wrote to him to be cautious about promoting the Copernican theory, mentioning Galileo in his letter. In the same year, Dominican Friars Niccolo Dorini and Tomasso Caccini filed depositions with the Roman Inquisition, essentially accusing Galileo of heresy. Apparently it takes two to make a good case for heresy. Galileo was put on trial and interrogated by the Roman Inquisition in 1633. Galileo was threatened with torture if he did not recant.

The US Congress moved more quickly. Unfortunately, Mann, like Galileo was unable to be at all of the hearings. As far as we know he was not threatened with waterboarding.

The Roman Inquisition called in a group of outside experts, one of whom, Melchior Inchofer (a name eerily similar to that another of the petroleum industry's attack dogs, Senator James Inhofe), led the group in writing a scathing seven-page denunciation. The Congressional Subcommittee had Statistician Edward Wegman form a group of consultants. They prepared a denunciation of one aspect of the statistical analysis in the Mann et al. 1998 paper. Although using a wholly different technology (computer generated print rather than handwriting), their report was of almost exactly the same length as Inchofer's.

In 1616 Cardinal Bellarmine wrote to Galileo that he had not been on trial or condemned by the Inquisition. However, he also warned Galileo that he was not to discuss the Copernican theory with anyone in any manner. He escaped burning at the stake, but was placed under house arrest for the rest of his life.

We don't know, but this may have something to do with a new, possibly money-making idea. One of the big problems of that age was determining longitude, that is, distance in the east-west direction. Latitude was easy to figure out; all you had to do was to measure the angle of the pole star above the horizon, and that was the latitude. However, to determine longitude you needed to know how far distant from home you were, and to do that you needed to know the difference in time between your location and home. Galileo had figured out that the moons of Jupiter were like a clock in the sky that could be observed from anywhere on Earth. They were eclipsed by the planet at regular intervals. If you could prepare a table of the times of future eclipses of the moons you

had a universal clock and could determine longitude. This would greatly assist in determining sailing times for ships going to the Americas or the Far East. If Galileo could solve this problem it would be worth a lot of money, and the lion's share of the profits would have gone to the Church. Galileo worked on this problem for the rest of his life, but couldn't solve it because there were no clocks on Earth that kept time well enough. John Harrison finally solved the problem by building sea-going clocks early in the 18th century.

To satisfy the political needs of the time, Galileo was forced to make a public recantation in the Convent of Minerva in Rome in 1633. In the intervening 17 years between the start of the Trial and his recantation, a lot had happened. For one, in 1623, his good friend and benefactor from Florence, Cardinal Maffeo Barberini had become Pope Urban VIII. It seems very likely that he personally believed that the Earth orbited the Sun, but the need for stability in the Church at that critical time was far more important.

There is an interesting parallel between acceptance of heliocentrism and global warming. Both were highly inconvenient truths. The 17th century scientific community was convinced by Galileo's argument by his and other observations made shortly after the publication of his book in 1632. Virtually the entire international scientific community had been convinced of global warming by the data available in 2006. Both scientific realities were however, truly inconvenient. In the first half of the 1600s the Catholic Church was embroiled in battle with the Reformation, started by Martin Luther in 1517. Incidentally, Luther and his followers of the time did not believe that the Earth orbited the Sun. For the Catholic Church to have officially recognized that the Earth was not the center of the Universe would have been a disastrous acknowledgment of error. The bloody phase of the Reformation ended with the Treaty of Westphalia in 1648, 6 years after Galileo's death.

Perhaps the most surreal aspect of the US Congressional Hearings in July 2006 was the statement by Chairman Whitfield that the Mann et al. 1998 paper had been instrumental in guiding the Kyoto Treaty of 1997. It was as though he had gotten lost in time. One is reminded of Salvador Dali's painting of melting limp clocks, *The Persistence of Memory*. However the hearings addressed another inconvenient truth, the inconvenience being mainly to short-term profits in the petroleum industry.

The paleoclimate data for the last two millennia have been subjected to more scrutiny and analysis than any other data set I know. The correction proposed by Wegman was found to have no effect on the actual shape of the hockey stick. The contention of McIntyre and McKitrick, that the 15th century was warmer than the late 20th century could not be substantiated. And now almost all of the years since 1998 have turned out to be much warmer than any in the Mann et al. 1998 paper.

On October 27, 2006 Senators Olympia Snowe and Jay Rockefeller wrote to the CEO of ExxonMobil urging him to desist from funding deliberate disinformation about global warming. The Wall Street Journal countered with the argument that the Senators were attacking freedom of speech. One can only assume that misrepresentation of facts or financial situation in a company's Annual Report, for example, should be considered another form of freedom of speech. Since learning that the WSJ regards deliberate disinformation as a form of freedom of speech I have viewed its reporting with some skepticism.

## 28.7  Summary

The deglaciation from the Last Glacial Maximum started about 18,000 years ago. It was essentially completed about 7,000 years ago, marked by the end of 11,000 years of sea level rise. The last vestige of the Laurentide ice sheet was gone 6,000 years ago. Since that time, the Earth's climate has been remarkably stable—the longest period of stability we know of. It has been during this period of climatic stability that civilization developed. I suspect that we will eventually find that there were primitive towns and villages on the older shorelines, but the great development of division of labor and aggregation of humans into cities began only 6000 years ago.

There have been minor perturbations of the climate since. Much of the Sahara was grassland or forest 6,000 years ago, and it rapidly became desert. One hypothesis concerning the origin of civilization is that the Egyptians took advantage of people emigrating from the Saharan region and used them as laborers to build their monumental structures. Roman times were warm and food was plentiful and the great Empire ruled the western world. But then cooling brought hoards of the surrounding peoples, the 'barbarians,' to invade and bring down the empire. The Medieval Climate Optimum was a return to 'good times' in Europe. Unfortunately, in such good times, it was possible to get along without being too clever, hence the lack of scholarly research led to the climatic optimum also being known as the Dark Ages. Cooling brought with it the Plague, but then also the Renaissance. The Little Ice Age also appears to have been a largely European affair. These variations, which drove many of the historical events so familiar to westerners, may reflect changes in the great ocean conveyor which affected climate around the North Atlantic.

However, it is now evident that this long period of climate stability is coming to an end, and not from natural causes. Human activity has so altered the planet that Earth is seeking a new climatic state. The most dramatic changes are occurring in the northern polar region. Some of the short-term possibilities for change are explored in the next chapters.

A Timeline for this chapter:

| ca 750–1250 CE | Europe experiences the Medieval Climate Optimum |
| --- | --- |
| 1500–1900 CE | Europe and North America experience the Little Ice Age |
| 1988 | Hartmut Heinrich notes layers of iceberg debris in North Atlantic sediment cores |
| 1993 | Willy Dansgaard and Hans Oeschger describe the sudden changes in oxygen isotopes in Greenland ice cores |
| 1998 | Michael Mann, Ray Bradley and Malcolm Hughes publish *Global-Scale Temperature Patterns and Climate Forcing Over the Past Six Centuries*, showing the 'hockey stick' shape |
| 2005 | Stephen McIntyre and Ross McKitrick argue that the hockey stick was "primarily an artifact of poor data handling" |
| 2006 | The Subcommittee on Oversight and Investigations of the Committee on Energy and Commerce of the United States House of Representatives holds hearings on 'Questions surrounding the 'Hockey Stick' Temperature Studies: Implications for climate change assessments.' Chairman Whitfield concluded that the Mann et al. 1998 paper had been instrumental in guiding the Kyoto Treaty of 1997 |
| 2006 | Senators Olympia Snowe and Jay Rockefeller wrote to the CEO of ExxonMobil urging him to desist from funding deliberate disinformation about global warming. The Wall Street Journal countered with the argument that the Senators were attacking freedom of speech |

If you want to know more:
William F. Ruddiman, 2007. *Earth's Climate: Past and Future, 2nd Edition*. W. H. Freeman, 388 pp.—An excellent textbook focused on the Quaternary and Holocene.

Music: Ralph Vaughan Williams' Sinfonia Antarctica (Symphony no. 7) is a remarkable piece of music, with reminiscences of the expeditions to that icy continent.

Libation: A healthy dose of a Scotch single malt, and remembering the Scottish Enlightenment, would be a good way to try to forget how stupid our politicians can be.

### Intermezzo XXVII. The 27th International Geological Congress—Moscow
Another interesting extracurricular activity while I was Director of the University of Colorado Museum was involvement in the International Geological Congress held in Moscow in 1984.

You may remember that just after he took office in 1981 Ronald Reagan called on JOI to have JOIDES drop the USSR from membership in the Deep Sea Drilling Project. He was obsessed with the idea that international cooperation in science must be a one-way street for information. On the other hand, he had promoted the meeting with Soviet atmospheric scientists in Leningrad that I recounted in the first Chapter of this book. There, we were supposed to ferret out secrets about climate change that the Soviets knew more about than we did.

Now I have to tell you a wonderful story. You will remember from Intermezzo XXIII that the Ocean Margin Drilling program was offered the *GLOMAR Explorer* as a drilling platform. I hope you recall the whole Howard Hughes-CIA-Submarine Recovery story. In developing the OMD, I would make regular progress reports to the JOIDES Execute and Planning Committees. We not only told them about the offer of the *GLOMAR Explorer*, but with the concurrence of the US Government, we gave them sets of the construction plans for the *Explorer*. Our friends in the Government were hedging their bets, and just in case the OMD project failed to materialize, the *Explorer* might be the ship used for a continuation of the international ocean drilling. Of course, our Soviet colleagues got a set of the plans too.

After I moved to the CU Museum, I gave all of my records of the meetings for the OMD program, along with the construction drawings for the Glomar Explorer to the University Library, thinking they would be archived. About a year later I received a call from a US company interested in possible use of the vessel. They were trying to locate a set of the plans. Those at the NSF had been discarded, as had those at the participating US Institutions. When I called our library to ask about the set I had deposited with them I found that they too had discarded them along with all of the other records I had given them. They had been thrown out without checking back with me; a real loss to science history. But I had a suggestion for locating a set. "I'll bet there is a set in Moscow. Call Nikita Bogdanov at the Academy of Sciences' Institute of the Lithosphere in Moscow and ask if he can send you a copy." Nikita had been the Soviet Union's representative on the JOIDES Planning Committee. He did indeed have his set of the plans and made copies for the US firm.

Just before JOIDES made its break with the USSR, Nikita had asked me to be on the Organizing Committee for the 27th International Geological Congress, to be held in Moscow in 1984. He was in charge of planning the Congress. It was a great honor, but I hesitated to accept. It would mean traveling regularly to Moscow without being on official business for the US. I wondered if that was really a good idea in view of the fact that my co-author on several scientific papers,

Pavel Čepek, had been convicted of crimes against the state in Czechoslovakia and was now living in Germany. Also there was the earlier defection of Jiri Jonas in which I had been involved. That was supposed to be secret, but one never knew. I asked Nikita: "Nikita, just how much do you know about me?" He laughed "Bill, we know more about you than you do yourself." I guess I had been thoroughly vetted by the KGB.

The trips to Moscow began after I had moved to the University of Colorado. We not only had meetings concerning all aspects of the Congress, but I would be invited to different institutions to give science lectures, and we had lots of informal sightseeing trips and meals with fellow scientists in the Academy and the Lomonosov University. And at some point I would be informed that Andrei Monin, who had been on the JOIDES Executive Committee, wanted me to visit him at his Institute, the USSR Academy of Sciences' Shirshov Institute of Oceanology. He would send a car to pick me up after our meetings ended.

In Intermezzo XVII I described how I first met Andrei, but perhaps I should explain why we got along so well. The communist world and the western world were very different places. But from the time I had spent working with Pavel Čepek at the Czechoslovak Academy of Sciences in Prague I had learned how the communist world worked. Just simple things about how things worked in the west, like how dinner was ordered and paid for in the US was very helpful to Andrei and his assistant Igor Mikhaltsev. Knowing how both systems worked made life much easier.

At those meetings in the Shirshov we had wide-ranging discussions not only of advances in our science, but of politics, economics, and the state of the world in general. Then, toward the end of the meeting, when we were well lubricated with the Scotch whiskey Andrei supplied, he would ask me to take a message back to Washington. One time it was "Tell someone who would be interested that we know how to make submarines run silent." I explained that I didn't have any contacts in the Reagan Administration. He just said "But you know your way around."

I called a contact in the National Science Foundation and gave him the message. "Interesting, I'll pass it on." A couple of months later I heard back "That was very interesting. They do indeed know how to make submarines run silent." In case you were wondering, it was one of those silent submarines that became the star of the movie *Red October*.

After Ronald Reagan launched his massive buildup of military hardware, Andrei told me that the USSR would not follow suit. They knew it was a ploy to try to break the Soviet economy, and they had other priorities. There were already more than enough weapons in the world. Soviet scientists were becoming successful in convincing the leadership that nuclear war was unthinkable; it would mean the end of civilization. On the other hand, Pat Robertson was trying to convince Reagan that it was time to fulfill the biblical prophecy of Armageddon by launching an all-out nuclear war. Fortunately calmer heads prevailed in Washington.

Perhaps the most interesting of the messages was "Please tell someone that Mr. Brezhnev would like to meet with Mr. Reagan." I used my NSF contact to pass the message on. A few months later: "When you go back to Moscow, please tell Andrei that Mr. Reagan does not wish to meet with Mr. Brezhnev." It was not until many years later, after the collapse of the Soviet Union, that I learned that Andrei had always been a member of the Central Committee of the Communist Party and that these messages were indeed coming from the top. Why they were not sent through the usual diplomatic channels I don't know.

At the Geological Congress I was given a special role in making a major address to the opening session. It was on the role of Scientific Ocean Drilling in advancing our science. It was broadcast nationwide. I was also interviewed on Soviet TV by one of their best known journalists, Sergei Kapitsa. Nikita Bogdanov wanted to be sure the funding for participation in the Ocean Drilling program would be available should they be allowed to rejoin the program.

I was now familiar enough with Moscow that I didn't use the congress busses to get from one venue to another, but walked or used the Metro or city busses. Over the past couple of years the attitudes of the people on the street had changed. Even with my meager Russian I was aware that there was a lot more discussion about how things were going and a lot more freedom. People often smiled; that was new.

On one occasion I was invited to lunch by two drilling engineers. They had designed the operations for the Kola Superdeep Borehole. The drill site is located on the northern part of the Kola Peninsula on Europe's Arctic margin. The hole was drilled for purely scientific reasons, to explore conditions deep within the continental crust. The granite rock of the crust is exposed at the surface. The Kola Borehole was originally intended to extend to a depth of 15 km (49,000 ft). By late 1983 it had reached a depth of 12 km and drilling was paused to evaluate the conditions in the hole. It had been expected that it would drill through dry rock but instead had encountered

large amounts of water. This was not groundwater, originating at the surface, but water from Earth's interior unable to move upward because of impermeable rock layers.

Most of the Kola well was cored, using the same wireline technique as had been developed for the Glomar Challenger. But there was a big difference in the drilling technique. The Kola well was drilled with retractable drill bits. Instead of needing to pull up the entire drill string to change the bit, my luncheon hosts had invented a folding bit that could be lowered down through the pipe and would then unfold to be like a normal bit at the bottom of the drill string. Similarly, it could be refolded and pulled back up through the pipe when it had worn out. This was a big advance in drilling technology, and my hosts wanted to know about how to market it in the west. Free enterprise on the March.

I was able to participate in field trips to the Crimea, Caucasus, Georgia, Uzbekistan and Azerbaijan. We were in a hotel in the Caucasus the night of August 11, 1984, when Ronald Reagan, thinking the microphone was just being tested made a joke: "My fellow Americans, I'm pleased to tell you today that I've signed legislation that will outlaw Russia forever. We begin bombing in 5 min." The statement was broadcast but the idea that it was just a joke was not made clear. It had enormous impact on the local people who thought we were surely in for a nuclear war. But there was another aspect of the 'joke' that was even more ominous. He said 'Russia' rather than 'the USSR.' There is a huge difference, and it was interpreted as meaning that the US argument was with Russia and the Russian people, not with communism.

From traveling through different parts of the USSR it was clear to me that the genie had been let out of the bottle. People moved about much more freely than before, and topics taboo a decade earlier were openly discussed. And the changes were not just in Moscow, they were everywhere. In short, the same things that I had seen happen in Czechoslovakia in the 1960s were happening throughout the Soviet Union. Change was on its way, and it was unstoppable.

In the fall of 1984, after I returned to the US, the CIA contacted me wanting an interview. I will never forget that lunch in the terrace dining room of the Boulderado Hotel. I told my interviewer about the earlier visits preparatory to the Congress and about how changes that were steadily occurring I had seen so much on the Congress field trips, and had been very impressed with the freedom with which the people were now moving about. I told him about my experiences in Czechoslovakia 20 years earlier. "The same things that happened in Czechoslovakia then are happening in the USSR now. I give the system about five more years, and then it will collapse—in 1989." He looked at me and smiled. I realized that although his notebook was open he had not written anything down for some time. "That's very interesting. Unfortunately I can't report it. It is not what we want to hear." To this day I can still hear him say those words. Our chief intelligence gathering agency was passing on only what someone higher up, presumably the Administration, wanted to hear.

He wasn't kidding. In 1989 the collapse of the Soviet Union caught us completely by surprise.

NO GOOD ANALOG FOR THE FUTURE
TO BE FOUND IN THE PAST

*If you want the present to be different from the past, study the past.*
Baruch Spinoza (1632–1677)

Before trying to make any predictions about the future, let's look into the Earth's past episodes of global warmth to see if they offer any guidance.

We live in one of those brief warm intervals, called interglacials, that have interrupted the 'normal' bipolar glacial state that has characterized the climate of our planet during the past three to four million years. We also know that 'civilization' developed over the past 7,000 years, the longest period of stable climate in the last half million to million years. The climatic changes over the past 7,000 years have been small and regional, not global.

Until the Industrial Revolution, humans were part of the global ecosystem. Special human activities such as using fire as a hunting tool, cultivating plants and animals, tilling the soil for agriculture, building towns and cities, and mining metals were on such a small scale that they had little impact on the global environment. However, I must mention that Cesare Emiliani thought that the reason the present interglacial had not started to end was because of the early human modification of the environment by the use of fire, and Bill Ruddiman believes that humans managed to put just enough $CO_2$ into the atmosphere to counteract the cooling trend of declining insolation due to the Milankovitch orbital cycles (recall Fig. 21.13). But starting with the Industrial Revolution, and particularly during the last century, human activity has been modifying the planet on a large scale. The sprawl of cities has created areas with special microclimates, while the replacement of $C_3$ plants by $C_4$ plants for agricultural purposes, the destruction of forests, the massive movement of soil and rock involved with construction and landscape modification, and large-scale mining—all as a result of rapid growth of the human population, have altered regional climates of the land and coastal regions. But the primary driver of the present perturbation of global climate is the very rapid increase in atmospheric $CO_2$, primarily from burning of fossil fuels.

The question arises, are there analogs for this condition in the geologic past? The human perturbations of the land are obviously unique. Atmospheric $CO_2$ concentrations have been even higher in the past, but only during those times, like the Eocene and Cretaceous, when the Earth lacked polar ice caps. So the answer is—a resounding—NO!

In that case perhaps we should look at past interglacials to see if there any clues about climate change during them that might be useful.

## 29.1 The Eemian, the Last Interglacial

You may recall that in the terminology for marine oxygen isotope stages (MIS) introduced by Emiliani (Fig. 13.14), the warmer times are indicated by odd numbers, and the colder times by even numbers. Thus the modern interglacial is Marine Isotope Stage 1 (MIS 1), and the last glacial is MIS 2. But it turns out that MIS 3 was not an interglacial, but merely a warmer episode in an otherwise glacial climate. It was what is called an 'interstadial.' MIS 4 was a colder episode, but MIS 5 was a real interglacial, at least part of it was. You may also remember that the oxygen isotope record of Emiliani's MIS 5 had a series of bumps, indicating that although overall it was a warmer period, it was punctuated by a series of 'false-starts' back toward a glacial state. Each of the episodes during MIS 5 now has its own Substages name (5a, 5b, 5c, 5d, 5e) shown in the composite record of oxygen isotopes in deep-sea benthic foraminifera known as the 'SPECMAP Stack' (Fig. 13.17). It turns out that only MIS 5e, the oldest and warmest episode of MIS 5, was a real interglacial. It was comparable with the present in terms of temperature, the remaining volume of ice, and sea level. MIS 5e has a special name; it is known internationally as the Eemian or, in North America, the Sangamonian Interglacial.

The Eemian was named by Professor Pieter Harting of the University of Utrecht in 1875 for fossiliferous marine strata recovered from boreholes near the city of Amersfoort, which lies on the Eem River in the Netherlands. Harting did not know these deposits represented an interglacial, but he recognized that the fossil sea shells found in the Eemian beds

© Springer International Publishing Switzerland 2016
W.W. Hay, *Experimenting on a Small Planet*, DOI 10.1007/978-3-319-27404-1_29

were characteristic of waters much warmer than those of the North Sea today. Since his day we have learned much more about the last interglacial.

The land record of the Eemian interglacial is very incomplete, but the record in marine cores is continuous. That from the ice cores from Greenland is complicated by deformation of the ice, but still contains intriguing information.

The record of MIS 5 and the preceding glacial, MIS 6 in the Greenland Ice Core Project (GRIP) ice core, taken at Summit, the highest point on Greenland in 1992, has required a lot of work to straighten out and date. The oxygen isotope record from this core is shown in Fig. 29.1, where it is overlaid on the record of MIS 1 and 2, the Anthropocene-Holocene and the Weichselian, the last glacial. There is striking similarity between the Terminations, the transitions from glacials to interglacials. However, the Holocene and last interglacial are very different. As mentioned before, the present interglacial is the longest period of stable climate in Earth's history over the past million years. The Eemian apparently experienced a series of sudden climate changes that occurred every thousand years or so.

Figure 29.1 shows the δ18O (in this case the $^{18}O{:}^{16}O$ ratio relative to that of average modern marine ocean water) of the ice core. You will immediately notice that the amount of change in the isotope record between glacials and interglacials in much greater than in the cores of marine sediment. If you are really observant, you will notice that the δ18O scale is 'upside down' relative to that used for marine

cores. Yet the shape of the curves looks very much the same. What does all this mean?

In the marine δ18O record based on deep-sea benthic foraminifera, the total change is about 1.8 ‰, whereas in the ice core the total change is about 10 ‰. As you will recall, the change in δ18O in marine waters reflects changes in temperature, ice volume, and the 'vital effect' introduced by different organisms. The effects of transferring ocean water into ice sheets and of forming shells in cooler ocean waters both concentrate the heavier isotope, $^{18}O$, in the shells. This is because as seawater evaporates, the water molecules with the heavier $^{18}O$ isotope are left behind in the ocean while the water molecules with the lighter $^{16}O$ isotope go into the atmosphere and eventually fall out as snow and become incorporated into the ice. As you will recall Chap. 13, when shell formation takes place in the sea, the incorporation of $^{18}O$ in the shell is enhanced in cooler waters. Because we are used to seeing higher temperatures in the upper part of a diagram and colder ones at the bottom, the convention was to plot the δ18O with smaller or negative numbers at the top. Remember that the reason Nick Shackleton decided to use benthic deep-sea foraminifera for his study was because the change in deep water temperature between glacials and interglacials is very small, so that the observed change in δ18O shown in the 'SPECMAP Stack' (Fig. 13.17) reflects changes in global ice volume and sea-level. The change in δ18O in deep-sea benthic foraminifera ranges from 0.8 to −1.2 ‰, a difference of only 2 ‰.

**Fig. 29.1** The oxygen isotope record on the GRIP (Greenland Ice Core Project) ice core with the records for Marine Isotope Stages (MIS) 1 and 2, and 5a–5e and 6 superposed. Data for MIS 1 and 2 are shown as a *thin dotted line*, for MIS 5 and 6 as a *heavy solid line*. Some of the terms used for different parts of these cycles are shown

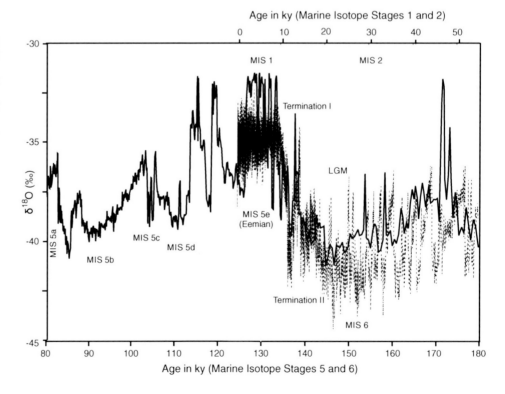

The changes in δ18O in the ice cores represent something very different: the change of δ18O in precipitation (snow) as the heavier $^{18}$O isotope is progressively removed (see Fig. 13.15), and the temperature of snow formation declines over Greenland. This is why the scale for δ18O in the ice cores ranges from −30 to −45 ‰ and is shown with negative values increasing downward. In fact, the glacial-interglacial change in oxygen isotopic composition of seawater damps the curve slightly, reducing the difference between glacials and interglacials by less than 2 ‰. Nevertheless, as can be seen in Fig. 29.1, the overall similarity of the GISP ice core data for the youngest interglacial and last glacial, and for the Eemian interglacial and preceding glacial are striking.

Figure 29.2 is a more detailed view of the last 20,000 years and the interval from 116,000 to 136,000 years, the corresponding parts of the glacial interglacial cycle. The similarities are striking. The transitions from glacial to interglacial conditions, Terminations I and II, both lasted about 10,000 years, and both were characterized by very sudden sharp warming interrupted by a return to glacial climatic conditions that lasted a thousand years of so; the Younger Dryas event had its counterpart during Termination II. At the end of the Terminations, the similarity ends. The GRIP ice core record of δ18O shows limited variability over the past 10,000 years. Other records show the same limited variability, so that we know that the last 8,000 years of Earth's history has been the longest period of stable climate in the past million years. That stable period included the warm Holocene 'Climatic Optimum' (7,000–3,000 B.C.),

when Egypt developed its sophisticated of civilization, probably using immigrant labor from the drying Saharan region. The Roman Period (200 B.C.–400 A.D.) was also warm, and abundant crops facilitated the development of the empire. As the climate cooled and agricultural conditions worsened, the Roman Empire fell under the stress of massive migrations of people seeking better conditions; Europe entered the Dark Ages (400–900 A.D.). Amelioration of the climate brought the Medieval Warm Period (900–1300 A. D.), when agriculture again flourished, followed by the Little Ice Age (1300–1850), and the return to warmer conditions of the past century. However, most of these climate changes were regional, affecting Europe and the Mediterranean in particular. They were probably due largely to variations in the strength of the GIN Sea source of deep water feeding the deep-sea Global Conveyor (see Fig. 25.11). The Greenland ice sheet, drawing the moisture for its snow from the GIN Sea and the North Atlantic, is ideally situated to record these regional variations. They all fall within the band of variations shown in gray Fig. 29.2. The solid line within the gray band is a 50-year average which shows the major trends. If you use your imagination, you may even be able to recognize the larger variations in Holocene-Anthropocene climate listed above.

In contrast, the Eemian oxygen isotope record from Greenland shows much greater variations, up to four times as large as those of the Holocene-Anthropocene. Interestingly, the climatic maxima of the Eemian all seem to go to the same level, with a δ18O value of about 32 ‰, while the

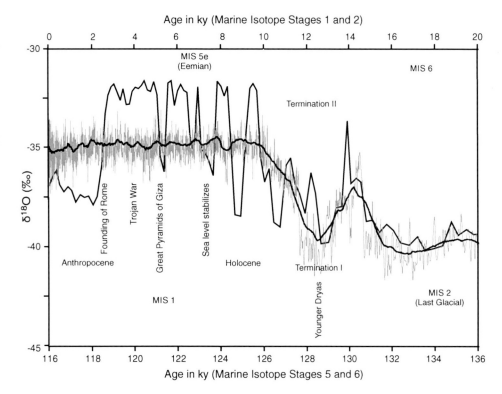

**Fig. 29.2** Oxygen isotope record on the GRIP (Greenland Ice Core Project) ice core with the records for the intervals 0–20,000 years and 116,000–136,000 years superposed. Data for the younger interval are in *gray*, with a 50-year rolling average to eliminate short-term noise shown in *black*; data for the older interval by a *heavy solid line*. Some of the important events during Termination 1 and history of civilization are shown

minima are variable. These Eemian variations appear to represent Dansgaard-Oeschger events, but there must be a word of caution. The Eemian section of the GRIP core lies at a depth of more than 2 km beneath the ice surface, and not far from the bedrock base. It has been disturbed by flowage, and the age scale has been established only with great difficulty. Yet the large variations in oxygen isotope values seem to be real. It is also known from studies of pollen that the climatic variability in Europe during the Eemian was large, much larger than during the Holocene, but not as extreme as the oxygen isotopes on Greenland might indicate. It is thought that what we see in the Greenland record is much greater instability in the widths of the northward flowing warm current along the Norwegian Margin and the southward flowing cold East Greenland Current. The extreme variations in the Greenland record of the Eemian would reflect radical changes in the sources of the water that eventually fell as snow on Greenland.

Further, since there is indeed ice of Eemian age on Greenland, it means that its ice sheet never melted completely away during the last interglacial. This is important to know when we consider what may happen in the future.

The Deuterium/Hydrogen (D/H) ratio in the Vostok ice core provides a record of the changing temperature in the atmosphere over Antarctica. Figure 29.3 shows the Anthropocene-Holocene-Termination I record of D/H variation superposed on that for MIS 5e-Termination II. The left-hand scale is the 'temperature anomaly,' the difference between the temperature recorded in the ice and the temperature averaged over the middle of the 20th century. First, note that the overall temperature difference between the last

glacial and the Holocene is about 8 °C (=14.4 °F). This corresponds to estimates for the difference of global temperature between the last glacial and Holocene. That suggests that we can use the Vostok record to estimate global temperature differences in the past.

You can see that the Holocene-Anthropocene temperatures have been relatively stable since 11,000 years ago. The fluctuations include all of the regional climate changes recorded in Europe. However, the Eemian record is very different. Termination II continued longer than Termination I to peak at about 128,000 years ago with temperatures about 3 °C (=5.4 °F) warmer than the Holocene. That 3 °C may not sound like much until you compare it with the 8 °C (14.4 °F) difference between the glaciated Earth of the Last Glacial Maximum and today. After the peak at 128,000 years ago, the temperature began to decline at an average rate of 0.5 °C/ky (=0.9 °F), with some shorter-term variability generally less than 0.5 °C. If you go back and look at Fig. 21.13 you will see that the pattern of atmospheric $CO_2$ concentrations for terminations III and IV and the initial interglacial conditions associated with them all bear a striking resemblance to the Eemian temperature record. None of the previous interglacials had the stable climate record of the Holocene-Anthropocene. As discussed in Chap. 6, Bill Ruddiman believes that humans, using fire for hunting and developing agriculture, introduced just enough $CO_2$ into the atmosphere to offset the expected Milankovitch orbital cycle-induced cooling trend that should have begun about 8,000 years ago. The massive introduction of fossil fuel $CO_2$ of the recent past has upset that delicate balance.

Looking at the older records of proxy data for temperature from the Vostok ice core (Fig. 29.3), and for ice volume

**Fig. 29.3** Comparison of D/H ratios in the Vostok core, representing atmospheric temperature above Antarctica, between the last 20,000 years (Marine Isotope Stages 1 and 2; Holocene to Last Glacial Maximum) shown as the thin line, and the interval between 120,000 and 140,000 years ago (Marine Isotope Stages 5e and 6; Eemian interglacial and preceding glacial maximum), shown as the *thick line*

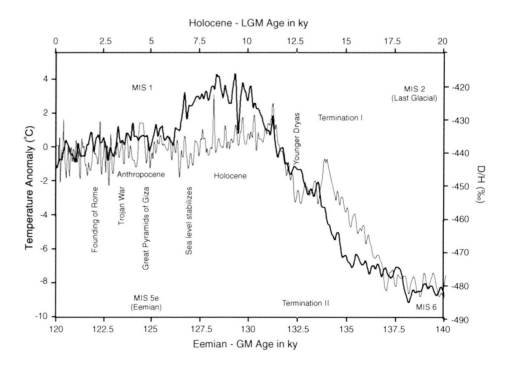

(δ18O) from the composite deep sea record, the 'Specmap Stack' (Fig. 29.4), we can see that not only was the Eemian warmer than the present interglacial, but there was less ice. The penultimate interglacial, MIS 7 was much cooler than the present. The antepenultimate interglacial (MIS 9) was slightly warmer than the present, but the next older interglacial (MIS 11) was similar to MIS 5e in being even warmer and having a significantly smaller glacial ice volume on land.

The maximum atmospheric $CO_2$ concentrations during these older interglacials were: MIS 5e—287 ppm; MIS 7—279 ppm; MIS 9—299 ppm; MIS 11—285 ppm. Prior to the Industrial Revolution, the highest $CO_2$ level during the present interglacial was probably about 280 ppm, toward the lower end of these values; the level in 2014 passed 400 ppm, higher than past interglacial levels.

It is also useful to compare insolation maxima for June at 65° N associated with these times: the present value is a Milankovitch insolation minimum, 484 $W/m^2$; 11,000 years ago, near the end of the deglaciation, insolation was at a maximum of 534 $W/m^2$. Earlier Milankovitch insolation

maxima occurred during MIS 5e, 128,000 years ago, when it reached 554 $W/m^2$: at the beginning of MIS 7, 244,000 years ago, when it was 534 $W/m^2$; during MIS 9, 314,000 years ago, when it was 526 $W/m^2$; and during MIS 11, 411,000 years ago, when it was 528 $W/m^2$.

It seems as though the difference between these interglacials can be explained by the combined effect of insolation and atmospheric $CO_2$ concentrations. Incidentally, each of these unusually warm episodes was shorter than the modern interglacial, lasting only about 5,000–7,000 years (see Fig. 20.13).

## 29.2 Sea Level During Older Interglacials

It was not only warmer during the Eemian, but sea-level was higher. It had been even higher during at least one of the earlier interglacials, as shown in Fig. 29.4. We know this because coral terraces on islands in the Pacific, on Barbados and Bermuda, and in the Bahamas and Florida Keys have

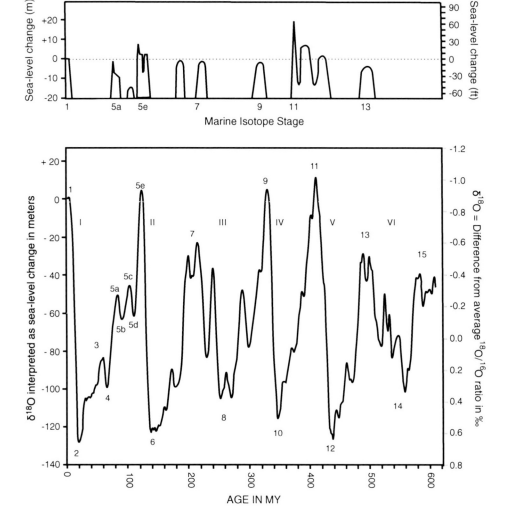

**Fig. 29.4** The oxygen isotope record for deep-sea benthic foraminifera (SPECMAP STACK) interpreted in terms of sea-level change, and the sea-level record from the Bahama Islands based on coral reefs, intertidal notches cut in the rock, caves, and ancient soil horizons (modified after Hearty 1998)

Eemian ages. As discussed in Chap. 13, it was the dating of those terraces in the 1960s that was critical to establishing the age of Emiliani's Marine Isotope Stages and deciphering the role of Milankovitch orbital cycles in climate. Now come the intriguing questions—how much higher were the Eemian and earlier sea levels, and where did the water come from?

The estimates for these past sea-levels come from several different kinds of evidence: (1) coral reefs; (2) wave cut terraces and 'notches' that form directly at sea-level, (3) caves floored by marine sediments, (4) soil horizons, and (5) very detailed records of δ18O where the ice volume signal can be separated from the temperature signal.

Evidence from the Bahamas, Bermuda and Oahu (Hawaii) suggests that the highest sea-level of the past half million years occurred during MIS Stage 11, when it reached 20 m above the modern level. Since complete melting of the Greenland Ice Sheet would raise sea level by only about 7 m, there must have been a major contribution from melting of Antarctic ice.

Eemian sea-level high stands were more modest, with estimates ranging between 2 and 9 m above present. You only have to think about the problem for a moment before you realize that the problem of determining the actual difference between the Eemian and earlier sea levels and that of today is not a simple problem. At Amersfoort in the Netherlands, where the sediments representing the interglacial give the Eemian its name, the deposits are 10–80 m below modern sea-level. This region is, after all, part of the Rhine delta complex, and the sediments are continuously compacting and sinking. Eemian shorelines further inland might sound like a good source for information about sea-level; after all, traces of these shorelines have been found throughout northern Europe. They show that the 'Baltic Sea' then was much larger than it is today. It extended from the North Sea to the east across Denmark and northern Germany, Poland and the Baltic States, and finally across Finland and northern Russia to the White Sea. Scandinavia was an island.

The Eemian sea-level high-stands occurred just at the end of the Termination II deglaciation when isostatic adjustments of the land surface were still going on. While we might be able to figure out where the local sea level was, the evidence is too fragmentary for us to be able to use European shorelines to get an accurate estimate of global sea level. If you think about it, that means we can eliminate all the northern hemisphere high-latitude sea-level records because they would have been affected by isostatic adjustments. Further, we can also eliminate the mid-latitude records as suspect because they would have been affected by the changes in rotation speed of the Earth as polar ice was added to the world ocean. The displacement of this large mass from near the axis of rotation to a greater distance caused the Earth's rotation to slow as it happened. But the isostatic adjustment to the removal of the ice and loading by water, involving flow of viscous mantle material, requires more

than ten thousand years. That leaves the tropics as a good place to look for good records. But even in the tropics the slow isostatic compensation must be taken into account. In fact, as a result of asthenospheric flow resulting from loading the ocean basins with more water, the apparent sea-level record on continental margins and large carbonate platforms, like the Bahamas, will be different from those observed on small oceanic islands which are stuck to the ocean floor and behave as though they were dipsticks. Unfortunately the ocean floor that the oceanic island dipsticks are attached to is also unstable, usually subsiding slowly as it moves away from the mid-ocean ridge, but sometimes rising due to volcanic activity.

One of the first markers of past sea-levels to be used, and still one of the best, were reef corals that live close to the water surface. Coral reefs grow near sea level, but there is only one coral species, *Acropora palmata*, the 'Elkhorn coral' of the Caribbean, that grows up to just below the sea surface. *Acropora palmata* thrives in water less than 3 m (∼10 ft) deep; it typically grows to the low tide level and forms the reef crest. It grows very fast, up to 10 cm/year (∼4 in./year). Unfortunately, it's very open framework means that it usually breaks up and falls apart into rubble as sea level falls. There are only a few places where it seems to be preserved wholly intact. Never the less it is considered a good marker for sea level to within about 2 m (=6.6 ft) and our knowledge of the details of sea-level rise during the last deglaciation depends largely on this species. As discussed in Chap. 13, much of that record comes from Barbados, where the corals have been raised by tectonic uplift. Barbados is on the edge of the Antillean subduction zone, where the Atlantic Plate is sliding beneath the Caribbean plate, and it is gradually rising out of the sea. Unfortunately we do not know just how steady this rise is, but the long-term rate is 0.34 mm/year (=0.013 in./year), and this value is used to make the corrections for determining the position of the fossil reefs in the past. Another place where there is a spectacular record of sea-level stands preserved as raised coral terraces is the Huon Peninsula of Papua New Guinea. The corals are different, but again there are species that grow close to the sea surface. The Huon Peninsula is also on the edge of a subduction zone, and is rising even more rapidly than Barbados. Furthermore, the Huon uplift is known to be discontinuous, associated with earthquakes. Depending on location, the uplift rate averages 1.9–3.3 mm/year (=0.075-0.13 in./year) over intervals of a few thousand years. The uncertainties of these uplift rates make it difficult to give precise estimates of Eemian sea-level, which was only a few meters above that of today.

Just when everyone was getting utterly frustrated by the uncertainties in determining Eemian sea levels, initially unrelated studies in a new area have provided an answer. Detailed and precise records of sea-level changes during the

Eemian comes from integrated studies of both fossil coral reefs and oxygen isotopes in deep sea cores taken in the Red Sea. This study, published in January 2008, was carried out by a European group led by Eelco Rohling of the British National Oceanography Centre in Southampton. Incidentally, Eelco had been one of the very bright Dutch geoscience students I met when I was Guest Professor in Utrecht. Their recent work (Fig. 29.5) shows that there was an initial sea-level high stand of about +7 m (=+23 ft) early in the Eemian, followed by a fall to about +2 m (=+6.6 ft) about a thousand years later, then culminating in a +12 m (= +39.4 ft) highstand after roughly another thousand years. This highstand was followed by a fall, another lesser rise, and then a further decline as the ice began to build up leading to the last glacial. The entire episode, when sea levels were higher than today lasted only about 4,000 years. Unlike the modern interglacial with its sea-level stability over 6,000 years, Eemian sea-levels were quite unstable and the rates of sea-level rise and fall were of the order of 1–2.5 m (=3.3–8.2 ft) per century.

The important thing to realize is that the stable sea level of the Holocene reflects stability of the Greenland and Antarctic Ice Sheets. The variable and higher sea-levels of the Eemian and older interglacials indicate that at least parts of these ice sheets melted. Until recently it was thought that the major contribution came from catastrophic collapse of the West Antarctic Ice Sheet. 'Collapse' here is in the geologic sense, meaning that it occurred in a few hundred to a couple of thousand years. But 'collapse' is a very dramatic word that can be seized upon by the director of a catastrophe film, and in at least one movie about the future, the entire Greenland ice sheet slides off into the sea overnight. If an ice sheet collapses, you will still be able to walk to higher ground.

Before 1970 and the recovery of long ice cores, our knowledge of the history of the Antarctic and Greenland Ice Sheets was almost nil. When it became evident that some

interglacial sea levels had been significantly higher than today, there seemed a simple explanation: sea-level rise had destabilized the West Antarctic Ice Sheet and large parts of it had glided off into the sea.

If you go back and look at Fig. 17.16 you will see that much of western Antarctica lies below sea level. Most of the ice sheet covering it rests on bedrock, but around the edge of the continent there is a fringe of the ice sheet that is floating on seawater. The original idea of catastrophic collapse of the West Antarctic Ice Sheet was very simple: the larger ice sheet that would have existed during the glacial would have been very sensitive to sea-level rise. As sea level rose, it lifted more and more of the fringing edge of the sheet off the bedrock, forming floating ice shelves. The floating shelves then broke up, and the reduced friction beneath the ice streams made it much easier for the ice further inland to move seaward. As the inland ice flowed into the sea and melted, sea-level rose further, lifting more of the ice sheet off its bedrock base, and allowing the inland ice to flow even faster seaward. It was a classic example of a positive feedback mechanism that resulted in ever faster sliding of the ice into the sea—the idea became known as 'ice-sheet collapse.' It was thought that this was a special process that could only apply to the West Antarctic Ice Sheet, because the East Antarctic and Greenland ice Sheets have their bases above sea-level even today, and should not be susceptible to this phenomenon.

In the 1980s it was recognized that this positive feedback could explain sudden rises in sea-level that occurred during the Termination I deglaciation. Any suggestion that these pulses of rapid sea-level rise could be due to enhanced melting of the in-place ice sheets could be easily ruled out because of the enormous amounts of energy that would have been required. You will recall from Chap. 27 (Fig. 27.5) that there were three of these 'Catastrophic Sea-Level Rise Events' ('CRE's') or 'Meltwater Pulses' with the largest

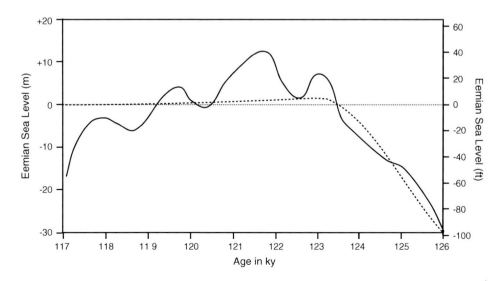

**Fig. 29.5** Eemian sea-level curve (*solid line*) based on oxygen isotopes from deep-sea cores and data from fossil reefs in the Red Sea region compared to the Holocene sea-level curve (*dotted line*)

causing sea-level to rise by 13.5 m (44.3 ft) in less than 300 years. That is a rise rate of 4.5 cm/year (=1.8 in./year). There are three candidates for causing such rapid sea-level rise through catastrophic collapse: the West Antarctic Ice Sheet, the eastern part of the Laurentide Ice Sheet, and the Barents-Kara Ice Sheet north of Scandinavia and Russia.

Today the West Antarctic Ice Sheet has an area of $2.11 \times 10^6$ km$^2$ (=815,000 miles$^2$) and a volume of $3.40 \times 10^6$ km$^3$ (=816,000 miles$^3$); if it were to disappear completely, it would raise sea level by about 8.5 m (=27.9 ft). It is held in place by ice shelves that have an area of $1.32 \times 10^6$ km$^2$ (=510,000 miles$^2$) and a volume of $0.66 \times 10^6$ km$^3$ (=158,000 miles$^3$); since they are floating, their effect on sea level if they were to melt would be negligible. However, if the ice shelves were to break up and drift away, allowing the inland ice to flow into the sea, it has been estimated this could result in a 6-m (=19.7 ft) sea-level rise. The length of time it would take for this to occur is uncertain, but has been thought to be on the order of hundreds to possibly a thousand years.

However, in the past the West Antarctic ice sheet was larger, and having advanced over a narrow continental shelf, was even more unstable. At the Last Glacial Maximum it is thought to have covered an area of $9.90 \times 10^6$ km$^2$ (=3,822,000 miles$^2$) and had a volume of $7.83 \times 10^6$ km$^3$ (=1,879,000 miles$^3$). The change in the West Antarctic from the LGM to present accounts for 7.9 m (=25.9 ft) of the deglacial sea-level rise. Because of its inherent instability it is the most likely contributor to the meltwater pulses, but its total volume change is too small to account for all of any of them. There must have been other unstable areas.

We already know from the Heinrich Events, the periodic flotillas of icebergs that came out of the Hudson Strait and the Labrador Sea, that the Laurentide Ice Sheet was inherently unstable. Most of the Heinrich Events occurred during the last glacial, and although spectacular for their effects on North Atlantic circulation, the amount of water in the icebergs then was not great enough to leave a record on the sea-level curve. These events may well have been triggered by sea-level rise lifting the ice off its bedrock base, and penultimate event H1, which has left a very widespread trace, seems to have coincided with Meltwater Pulse 1A. Again, it is unlikely that all of the water required for Meltwater Pulse 1A could have come only from the Laurentide Ice Sheet. Instead, it is more likely that the sudden sea-level rise triggered by ice-sheet collapse in one area acted as a trigger for ice-sheet collapse in another area—another case of positive feedback resulting in rapid dramatic changes during the deglaciation.

The third possible source for meltwater pulses is the Barents Ice Sheet. It has only recently been recognized that the broad Barents-Kara shelf north of Norway, Finland and Russia was covered by ice during the last glacial, and late in the deglacial sea-level rise it would have become unstable. Its collapse may have been responsible for 'CRE 3' or the 8,000 year cold spell that affected the European area.

It seems certain that ice-sheet collapse has occurred in the past, but the details still need to be worked out, and that requires new information, new techniques, or new ideas.

In summary, our search for an analog to the modern interglacial has shown that, even discounting the perturbations by human activity, it appears to be unique. It is useful to recall the work of Andre Berger and Marie-France Loutre of the Université Catholique de Louvain (Fig. 13.21) that the present interglacial is very odd in that without human intervention it would last for an unprecedented 80,000 years. It is the first interglacial in more than a million years that happens to coincide with an insolation minimum in the orbital eccentricity cycle.

## 29.3   Arctic Sea-Ice During Interglacials

Finally, what do we know about the sea-ice cover of the Arctic Ocean during the glacials and interglacials. There is general agreement that during the glacials the sea-ice cover was much thicker than during the interglacials. 'Much thicker' means thick enough so that insufficient light penetrated to enable active photosynthesis beneath the ice. That probably means a thickness in the order of 8–12 m ($\sim$26–40 ft). That is probably about as thick as sea-ice can get beneath reasonable polar temperatures without some extraordinary circumstance. The thickest known ice cover is that of Lake Vida in the Dry Valleys near McMurdo Sound in Antarctica. There the ice is 19 m (=62.3 ft) thick and overlies a brine with a salinity seven times that of seawater. It is thought that the Lake Vida ice took a very, very, long time to form.

During the glaciations, the ice cover of the Arctic remained intact; there were no leads between ice floes or other open waters. During the deglaciation, the ice-cover began to break up, and icebergs from the adjacent glaciated areas were able to move through the Arctic Ocean. Some of these icebergs must have been enormous, because they left tracks on the sea floor a kilometer (0.6 mile = 547 fathoms) below the sea surface as they drifted across Lomonosov Ridge in the central Arctic. If kilometer-deep icebergs seem outrageous, bear in mind that one iceberg spotted in the North Atlantic rose 170 m (=558 ft) out of the water. Remember that 9/10ths of the iceberg is below water, so its base was about 1.5 km (almost a mile) deep.

Information about sea-ice conditions in the Arctic during the interglacials is less certain. Many hundreds of deep-sea cores have been taken in the Arctic. Most are barren of fossils, but a few do contain tiny calcareous fossils that indicate there were some patches of open water. There are two kinds of fossil that provide important clues about the sea-ice cover: one is distribution of variety of a planktonic foraminifer with a name much larger that it is, *Neogloboquadrina dutertrei*, and the other the distribution of coccolithophorids like our old friend *Emiliania huxleyi*. Specimens of the foraminifer have been found in the sediments deposited during the present interglacial and during MIS 5, possibly during both Substages 5a and 5e. *N. dutertrei* lives in the GIN Sea, and would be carried into the Arctic by the currents flowing in through Fram Strait and across the Barents shelf. The problem is that it reproduces only once every few months or so. We don't know whether the fossil shells represent animals that were alive and thriving or whether they simply drifted in and their shell wound up being deposited in the sediment. The shells of these foraminifera have been found in sediments deposited during the Eemian northwest of Greenland. Their occurrence is interpreted as indicating that there were periodically areas of open water in a region that until today has remained ice-covered even in late summer. I tend to put more trust in the evidence from fossil coccolithophores because their life cycle plays out in a day or so, and they require light for photosynthesis. Their coccoliths are found in sediments deposited earlier in the present interglacial in the central Arctic, perhaps during the 'climatic optimum' around 6,000 years ago. Coccoliths have also been found in sediments thought to represent the Eemian. The interglacial sediments in the central Arctic accumulated at a rate of 1–2 cm (=0.4–0.8 in.) per thousand years, and if there were a long period when the Arctic was ice-free in summer, we should see it recorded as a coccolith-rich sediment layer at least several millimeters thick. But such a layer has never been found, and what we see instead are isolated occurrences of coccoliths. In the words of Gunilla Gard of the University of Stockholm, the leading expert on these matters: "The presence of nannofossils does not imply that sea ice was absent in the central Arctic Ocean, but that large leads of open water were present between scattered ice floes…. The extent of the ice-free waters is not known."

The best guess is that the Arctic Ocean has not been seasonally ice-free for at least the last several million years. A few years ago, with my colleagues in Kiel, I had the opportunity of reviewing all the available evidence for coccolithophores in the Arctic. We found that they seem to have been abundant in the Arctic through all of the Cretaceous and early Tertiary, indicating ice-free conditions. This fits in well with other evidence, such as the presence of crocodiles, that even wintertime temperatures on the shores of the Arctic Ocean never reached freezing. But the record of calcareous nannoplankton ceases abruptly at the end of the Eocene, 34 million years ago, suggesting that a very dramatic change in climate took place then. This corresponds to the sudden replacement of forests typical of warm climates by evergreen forests indicating colder conditions in North America, Europe, and northern Asia. My own opinion is that perennial sea-ice cover appeared on the Arctic Ocean at the end of the Eocene, and that the northern polar region has not been ice-free since, even during the Eemian.

## 29.4   The Paleocene-Eocene Thermal Maximum

The only event that may be analogous to what is happening today occurred in the more distant geologic past, about 55.8 million years ago. However, it happened much more slowly.

In the Southern Hemisphere summer of 1986–87, the Ocean Drilling Program's *JOIDES Resolution* drilled and cored a number of sites in the Weddell Sea region off Antarctica south of Africa. The goal had been to document the onset and development of the glaciation of Antarctica, and Jim Kennett of the University of California at Santa Barbara, an expert in this field, led the expedition. But, as is so often the case in exploring new territory, the cores contained something quite unexpected: older sediments, from before the time Antarctica became covered with ice. One of my former students, Woody Wise, was on that leg and remembers the great surprise when these older sediments were recovered. Among other things they revealed that during the earlier Tertiary, the Eocene and Paleocene, and Late Cretaceous the ocean temperatures around Antarctica were much warmer than today, almost tropical. One of the drill sites, ODP 690, was located on Maud Rise, a submarine plateau just north of the Queen Maud Land sector of Antarctica. It had an extraordinarily complete stratigraphic section of the older Tertiary and Late Cretaceous sediments, with very well preserved fossils.

The detailed analyses of these deep -sea cores always takes a few years to complete, and in 1991 Jim Kennett and Lowell Stott reported that the oxygen isotopes showed that there had been a sudden increase on ocean temperatures of more than 4 °C (≥F) over a period of a few tens of thousands of years, and that the temperature pulse decayed away following a classic exponential curve over the next 200,000 years. This event straddled the Paleocene-Eocene boundary, 55.8 million years ago. Accordingly, it has become known as the Paleocene-Eocene Thermal Maximum ('PETM').

The discovery of the temperature pulse led to many other investigations, looking for evidence for it in other regions, and for a possible cause. We now know that the global temperature increase was about 6 °C (11 °F) with polar temperatures increasing by 8 °C (14 °F). The temperature rise may have occurred in a time as short as 2,000 years.

Remember that the last glaciation reduced global temperatures by only about 6 °C, so a +8 °C increase in high latitude ocean temperatures during a time when the Earth was free of ice means that the planet went from being warm to hot. A 6 °C increase in global temperature is what we can expect if we burn all the available fossil fuels.

In 1997, Gerald Dickens and colleagues at the University of Michigan published a paper with the probable cause: *A blast of gas in the latest Paleocene: Simulating first-order effects of massive dissociation of oceanic methane hydrate*. They had analyzed ODP 690 and another ODP core for carbon isotopes and found that there was a decrease in the abundance of the $^{13}C$ isotope that paralleled the temperature curve.

All living organisms discriminate between the two common isotopes of carbon, $^{12}C$ and $^{13}C$. The ratio in organic matter is an important clue as to what produced it. In the same way that the ratio of the oxygen isotopes $^{18}O$ to $^{16}O$ are given as δ18O in per mille (‰) relative to a standard, the difference in $^{13}C$ to $^{12}C$ relative to a standard. The standard for both oxygen and carbon isotopes is (or was, until the samples ran out) a fossil, the Pee Dee Belemnite from Cretaceous rocks in North Carolina. It was what the Chicago isotope group, Harold Urey, Hans Suess, Sam Epstein, and Cesare Emiliani used. Now the standard is carbonate with that same isotopic ratio made available to researchers through the International Atomic Energy Agency in Vienna, Austria.

The δ13C of atmospheric oxygen is about −6 ‰. The change Dickens and colleagues found recorded in deep sea sediment was −2.5 ‰. This implies that a massive amount of isotopically light, i.e. $^{12}C$ enriched, gas must have been released to the atmosphere. The most likely culprit; methane from methane clathrates in sediment of the sea floor. Why methane?

Methane is produced by bacteria and in doing so they selectively discriminate between the $^{12}C$ and $^{13}C$ isotopes in such a way as to strongly enrich the gas in $^{12}C$ and deplete it in $^{13}C$. The methane in clathrates that can be sequestered in sediments of the sea floor is so strongly depleted in $^{13}C$ that their δ13C is about—60 ‰. By comparison, $C_3$ plants have a δ13C ranging between −22 and −30 ‰; $C_4$ plants range between −10 and −14 ‰. Even so, the amount of methane clathrate released had to be enormous to have such a large effect on the δ13C record, of the order of $1.4 \times 10^{18}$ g of $CH_4$. That is about a tenth of the estimated $11 \times 10^{18}$ g of methane clathrates in the sea floor today. Once in the atmosphere the methane would be oxidized to $CO_2$ in a relatively short time. That is important because, remember that the greenhouse potential of methane is about 25 times that of $CO_2$. It is estimated that atmospheric $CO_2$ levels may have risen to as high as 2,000 ppmv. All that happened in perhaps 2,000 years. Today we are headed toward a similar rise in $CO_2$ levels, but in about 200 years. That is 10 times faster than during the PETM. Remember that in change, rate is important. The result is a difference like that between your car hitting a concrete wall at 10 mph, which you will easily survive, versus 100 mph.

We know that the PETM resulted in the extinction of many animals, but also created the opportunity for many new kinds of mammal to evolve over the next few hundred thousand years. By wiping out much of the fauna, it opened 'ecospace' that new organisms could fill. It was a time of drought on land, and coastal oceans too warm for many marine animals. The documentation of the event and its consequences is an exciting ongoing area of research. It is the closest parallel to our human perturbation of the climate, but it happened at a time when there was no polar ice, and much more slowly than our grand uncontrolled experiment.

There may have been older, similar events. Chinese colleagues will soon be publishing results of research that suggest sudden increases in atmospheric $CO_2$ followed by gradual falls in the Cretaceous.

## 29.5   Summary

This summary can be quite short. The older interglacials do not tell us much about what the future may be like.

The Paleocene-Eocene Thermal Maximum (PETM) as the largest and best known of the pre-Late Neogene climatic events. The PETM was marked by a rise in polar temperatures of 8 °C with a global increase of about 6 °C; it was initiated over a period perhaps as short as 2 ky and lasted about 200 ky. It was clearly a greenhouse event, widely thought to be due to a massive release of submarine methane from decomposition of clathrates resulting from either ocean warming or submarine landslides initiated by earthquakes. But it occurred when Earth's climate was very different from that of today.

To paraphrase Yogi Berra—this is not like déjà vu all over again. We are headed into uncharted territory.

There is no timeline for this chapter.

If you want to know more:

Bill Ruddiman's 2007 book *Earth's Climate: Past and Future*, cited at the end of the last chapter is highly relevant, but he has another that is more specific:

William F. Ruddiman, 2005. *Plows, Plagues and Petroleum —How Humans Took Control of Climate*. Princeton University Press, 226 pp.

Music: Vincent d'Indy: Symphony on a French Mountain Air (*Symphonie sur un chant montagnard français*, aka *Symphonie cévenole*)—a piece of music I have always found fascinating.

Libation: A nice Californian Zinfandel—made from a black grape variety once native to Croatia but wiped out there by the *phylloxera* epidemic of the late 19th century (a few live survivor vines were discovered there in 2001). It flourishes in California and has a rich deep flavor I especially like.

**Intermezzo XXIX. My Second Sabbatical**

In 1988–89 I took my second Sabbatical Leave. Faculty are supposed to take a Sabbatical every seventh year. You are relieved of teaching duties and supposed leave the University and spend a year (at half salary) or a semester (at full salary) working on your research or delving into new areas of knowledge. My last Sabbatical had been when I was at the University of Illinois in 1967–68, 20 years earlier. With my dual arrangement between Illinois and Miami, spending a month each year at University College in London, and then being full time in the graduate school atmosphere of RSMAS before I became Dean, it had been rather like I was on Sabbatical every year.

I wanted to spend another year in my favorite city, Munich, and I was welcomed into the Institute of Paleontology in the Richard-Wagner-Strasse by my old friend from my 1959–60 Postdoctoral year in Basel, Switzerland, Dieter Herm. He was now Professor and Head of the Institute of Paleontology. I was given an office, and keys to the building. I had several years of research done with colleagues at NCAR to write up, and I wanted to develop a better set of maps of the Earth's geography in ancient times.

In 1987 I had made arrangements from one of my Master's Degree candidates in Geological Sciences at CU Boulder to spend a couple of years of his graduate school experience abroad, in Munich. Chris Wold didn't know German before he went, and he spent the summer of 1987 in one of the intensive language courses 'Deutschkurse für Ausländer,' taught by the University for foreign students. It was very much like what I had experienced in 1957. In 3 months Chris had mastered the language well enough to be able to register in Ludwig-Maximillians-Universität zu München, the University of Munich as a regular student. Because he was working on plate tectonic maps he soon found a home in the Institute of Geophysics. He would stay on in Munich during my sabbatical year.

Chris lived with a family in Schwabing, the northern suburb where many of the University students lived. Before I arrived he found a small apartment for me not far away. Working together, we made a lot of progress on the maps. If you are interested you can see where it all led a decade or so later with involvement of many other students and colleagues: http://www.odsn.de/odsn/services/paleomap/paleomap.html.

1989 started out as a rather normal year. But during the spring a number of things happened. Mikhail Gorbachev had become General Secretary of the Communist Party of the Soviet Union in 1985. In 1988

he had become Head of State. He was promoting *perestroika*—restructuring of the Soviet political and economic system, and *glasnost*—openness and free discussion.

On March 26, 1989 the first free elections took place in the Soviet Union. Boris Yeltsin was elected president of the Russian Federation. New faces filled 1,500 of more than 2,000 seats in the new Congress of People's Deputies.

On April 7 I traveled from Munich to Halle an der Saale in the German Democratic Republic (DDR = East Germany) to attend a meeting of the Deutsche Akademie der Naturforscher Leopoldina (literally translated that is 'The German Academy of Researchers of Nature.' I had been elected to the Leopoldina in 1986. It is the oldest Academy of Natural Sciences, founded in 1652 in the Free Imperial City of Schweinfurt. Since 1878, it has been seated in Halle on the Saale River.

Although I had been to Berlin and had seen East Berlin, this was my first time in the DDR proper. There was the usual inspection of the train at the border, and my invitation from the Leopoldina earned me a salute form the border guards. Arriving in Halle I was able to wander around a bit before the meeting. The city had that same gray pall as Prague. Almost everything seemed in bad need of repair. The meeting was a wonderful experience. The word 'German' in the Academy's name was used loosely. It was in fact the Academy of Sciences for all of German-speaking central Europe, including Austria and Switzerland. It had many members from more distant regions as well. It was quite free from interference by the DDR government, which treated it as a hands-off national treasure they had inherited. I had never met so many well-known scientists in one place in my life.

We had excursions so that we could see more of the country. One was to Leipzig, to the Thomasanerkiche, the church where Johann Sebastian Bach had been organist. Its famous choir gave a concert for us was a memorable experience, It was made especially memorable for me because my chair was on Bach's gravestone

Then, later in April Chinese students began demonstrating for democracy. The protest movement centered in Tiananmen Square, Beijing. The demonstrations continued until the government cracked down with military force on June 4.

But things were becoming loose elsewhere. On June 16, Hungarians paid homage to former premier Imre Nagy who had led the revolt against the Soviets in 1956. At least 250,000 people attended the ceremonial reburial of PM Imre Nagy and four others who had been hanged 31 years earlier and buried face down in unmarked graves. Hungary's communist leadership and the democratic opposition began negotiations on a transition to democracy. Communist Eastern Europe was starting to become unraveled.

In Poland Lech Walesa's Solidarity movement and gaining ground, and on June 30, General Wojciech Jaruzelski announced he would not run for Poland's new presidency.

I returned to the US in July to attend the 28th International geological Congress being held in Washington, D.C. With European colleagues I watched the news as events developed.

In July President George H.W. Bush began a visit to Poland, continuing on to Hungary. He gave a speech that stunned Europeans but was not widely reported in the US. He said that Lech Walesa, a trade unionist, was not much better than the old Communists. There was no European response for several days, and then the German Foreign Minister, Hans-Dietrich Genscher announced that Germany would begin supplying aid to Poland in its transition to Democracy. That made it clear that the liberalization movements in Eastern Europe were not going to be stopped.

I was back at my home in Estes Park when, on August 11, Poland's Solidarity-dominated Senate adopted a resolution expressing sorrow for the nation's participation in the 1968 Soviet-led invasion of Czechoslovakia.

A week later, members of Hungary's opposition organized the "Pan-European Picnic" on the border with Austria to press for greater political freedom and promote friendship with their Western neighbors. Among the 10,000 participants were 600 East Germans vacationing in Hungary. They took advantage of the occasion to escape to Austria.

A few days later, on August 23, two million people joined hands to form a human chain over 600 km (373 miles) long across the three Baltic states: Estonia, Latvia, and Lithuania. It marked the 50th anniversary of the Molotov-Ribbentrop Pact of 1939, when the Soviet Union and Germany agreed on spheres of interest in Eastern Europe, leading to the occupation of these three states. That same day Hungary removed its physical border defenses with Austria.

I was back in Europe, attending a Conference on Paleoceanography in Cambridge, England, in September when things really began to come apart. Hungary gave permission for thousands of East

German refugees and visitors to immigrate to West Germany. More than 13,000 East German tourists in Hungary escaped to Austria and proceeded on to West Germany. Thousands of East Germans who had sought refuge in West German embassies in Czechoslovakia and Poland began emigrating under an accord between Soviet bloc and NATO nations. I flew back to the US.

At the beginning of October I was back in Germany attending geological meetings when the Czech Government agreed to let the Germans in the embassies leave for the West. Then the demonstrations in Leipzig began, with 10,000 people in the streets demanding reforms. Things reached a climax on December 3, when special convoys of trains carrying East Germans from Prague to West Germany passed through Dresden and Leipzig. That evening I had been at dinner with friends in a village north of Munich. On the drive back into the city we listened to the radio account of what was happening as the trains passed through those cities. It was evident that massive change was on its way. The next week I was back in Colorado.

Events continued to unfold. On October 23, Hungary proclaimed itself a Republic and declared an end to communist rule. The next day Egon Krenz assumed the chairmanship of the Council of State in East Germany replacing Erich Honecker, who had always been severely out of touch with reality. A few days later Krenz delivered a nationally broadcast speech in which he promised sweeping economic and political reforms and called on East Germans to stay put and not to emigrate. He then ousted the old guard of the ruling Politburo, replacing them with reformers.

Then came that momentous day, November 9, when Communist East Germany threw open its borders, allowing citizens to travel freely to the West. Joyous Germans danced atop the Berlin Wall. I had been there in 1961 when the wall went up. I so wished I could have been there to see its fall.

The story of the fall of the wall has only become fully known in recent years. In fact it seems to have been a misunderstanding. The border crossings were to be open only to those East Germans who had been in the embassies in Prague, or in Hungary and who now had permission to emigrate. The new rules were to go into effect a few days later. Somehow the details got left out of the internal communications. When asked at a press conference that day when the crossing points would be opened, Günter Schabowski, the Communist Party boss in East Berlin said, "As far as I know effective immediately, without delay." He had forgotten to mention it was only for those East Germans who could prove they had been in Czechoslovakia or Hungary and had permission to emigrate.

The Border Guards simply heard that the border was open to anyone, immediately.

There are films about all this. I can highly recommend: *Good Bye Lenin*—a hilarious account of the changes 1989.

On November 23 at least 300,000 people jammed my old haunt Václavské námestí (Wenceslas Square) in Prague. They demanded democratic reforms in Czechoslovakia. The next day the hardline Communist party leadership resigned. The 'velvet revolution' was under way.

On December 28, Alexander Dubček, the reform-minded Czechoslovak Communist leader deposed in the 1968 in the Soviet invasion, was named chairman of the country's parliament. The next day Playwright Vaclav Havel was elected president of Czechoslovakia, George H.W. Bush was astonished. He said he simply couldn't understand how any country could elect someone other than a professional politician to be its leader.

DANIEL FARENHEIT CALIBRATING A THERMOMETER

*For humans, the Arctic is a harshly inhospitable place, but the conditions there are precisely what polar bears require to survive—and thrive. 'Harsh' to us is 'home' for them. Take away the ice and snow, increase the temperature by even a little, and the realm that makes their lives possible literally melts away.*
Sylvia Earle

Before discussing the future, it is useful to learn about what we know about the Earth's temperature and climate history of the past century and a half, how we know it, and how long we have known it. The main factors used to describe climate are temperature and precipitation. Temperature is measured with thermometers and precipitation by rain and snow gauges, both of which must be located so they are not influenced by local features. However, precipitation is especially difficult to measure when it comes in the form of snow. Wind may empty or overfill the collecting box when the region is blanketed in snow. Hence, the variations in 'climate' are often reduced simply to a temperature record.

## 30.1 The Development of the Thermometer

Although devices intended to measure temperature, called thermoscopes, had been developed in the late 16th century, they were more like toys than scientific instruments, and each had its own scale. The reproducible numerical measurement of temperature began in 1654 when Ferdinando II de' Medici invented the sealed glass tube partially filled with a liquid which would expand or contract with temperature change independent of atmospheric pressure. In his case, the tubes contained alcohol. In 1665 Christiaan Huygens, better known for his work in optics and astronomy, suggested using the freezing and boiling points of water as references for calibrating the thermometer. He understood that the divisions of scale must be defined by two points, but at the time there was no agreement on what numbers would be given the two points. These primitive alcohol devices spread through Western Europe, and a number of them were used to compile the early data for what was to become the longest continuous instrumental temperature record, that of Central

England which goes back to 1659. It is shown in Figs. 14.1 and 30.3.

The substitution of mercury for alcohol by Daniel Gabriel Fahrenheit (Fig. 30.1) in 1714 made it possible to make temperature measurements with greater precision because it was easier to obtain pure mercury (alcohol has an affinity for water), and because the expansion of mercury with increasing temperature is more constant than is that of alcohol. Fahrenheit was born in Danzig, on the Baltic coast, but spent most of his life in Amsterdam, in The Netherlands. His original temperature scale (F, without the degree sign), published in 1724, had three, rather than two reference points: 0 F being the coldest temperature measured for a mixture of equal parts water and ammonium chloride, 30 F being the freezing point of fresh water, and 90 F supposedly being the temperature of his wife's armpit (but note that $3 \times 30 = 90$). At the time it was thought that the body temperatures of men and women were different and that of women was more constant. On his original scale, boiling point of water would have been about 198.75 F. Clearly, two of the temperatures cited by Fahrenheit must have been used as the reference points and the third was determined from them. Unfortunately, no one knows for sure which of the three the two defining points were. And remember, you probably learned in school that 0 °F is the freezing point of water saturated with salt, the NaCl kind that is our common table salt. That is not what Fahrenheit used to define his 0.

Problems were soon recognized with his temperature scale, and in cooperation with other researchers it was revised so that the freezing point of water became 32°. There are some interesting stories about why 32 was chosen as the freezing point of water. One of the more interesting speculations is that someone involved in redefining the scale was a Freemason and the 32 refers to the degrees of enlightenment

**Fig. 30.1**  Daniel Gabriel Fahrenheit (1686–1736)

of the Masons. This would also explain why they chose the term 'degree' for the units of the scale. The human body temperature then became defined as 96° (3 × 32 = 96), which suggests to me that the scale was originally based on the freezing point of brine and the freezing point of fresh water. Subsequent revision of the scale was based on the idea that the difference between the freezing and boiling points of fresh water on the Fahrenheit scale of the time was about (but not exactly) 180°. It was decided to use these two points to define the scale and make them exactly 180° apart. Why 180°? Because it is evenly divisible by 12, just as there are 12 in. to the foot. Apparently it made sense at the time. The 'normal' temperature of the human body then became 98.6 °F (plus or minus a few tenths of a degree depending on what time of the day it is and what you have been doing). The confusion didn't end for a while because, although the freezing point of water is only slightly dependent on pressure, the boiling point varies significantly with declining atmospheric pressure at higher elevations, so the Swiss, for example, got confused. Fahrenheit's scale was adopted as standard in the Netherlands, Britain, and the United States. The French and many other countries considered it a curiosity and never used it. The United States is one of only two countries that still use the Fahrenheit scale, the other being Belize, which presumably keeps it to please American tourists.

In 1742 Anders Celsius (Fig. 30.2), a Swedish astronomer, devised a simpler temperature scale, with 100° being the freezing point and 0° the boiling point of water at sea-level atmospheric pressure. It was often referred to as the "centigrade scale" because it is defined by two points 100 units apart. It was not long before the calibration was

reversed, so that 0° became the freezing point of water. Like the metric system, which uses the base 10, the use of Celsius' scale, abbreviated 'C' which could stand for both 100 or for 'Celsius,' was spread by Napoleon. If Waterloo had turned out differently, we too would consider the Fahrenheit scale a curiosity.

In 1848, Lord Kelvin proposed an 'absolute temperature scale' based on the physical science of thermodynamics. In his scale 0 is the temperature where molecular motion ceases, but the spacing of the units is the same as in Celsius' scale. The 'absolute temperature scale' has since been named in his honor: the Kelvin scale (K), and does not use the degree signs. Originally, the second defining point of the Kelvin scale was the freezing point of water, then thought to be 273 K, but later found to be 273.15 K. The second point was later redefined as the triple point of water, where the three phases of $H_2O$, water, ice, and vapor, can all coexist (refer to Chap. 17 if you have forgotten all this). This is now known to be 273.16 K. So water freezes at 273.15 K and boils at 373.15 K at 'standard pressure.' For thermodynamic purposes, such as calibrating thermometers, there is general agreement that 'standard pressure' 100 kPa (=14.504 psi, 0.986 atm), but for other purposes, the (US) National Institute of Standards and Technology (NIST), formerly the National Bureau of Standards, shows its independence from the rest of the world by using 101.325 kPa (=14.696 psi, 1 atm). Incidentally,

**Fig. 30.2**  Anders Celsius (1701–1744)

climatologists don't bother with the decimal fractions and still set 0 °C equal to 273 K. Close enough for now.

Another method of measuring temperature was developed incidentally by the Englishman John Harrison in his pursuit of building a chronometer accurate enough to determine longitude, mentioned briefly in Chap. 3. It is an interesting story. In the Northern Hemisphere latitude can be determined easily by measuring the angle of elevation of the pole star (Polaris) above the horizon. It is not quite so easy in the Southern Hemisphere where there is no pole star, and in reality Polaris is not exactly above the North Pole, but it's close. Determining longitude is a different matter altogether. You may recall that one of the mitigating circumstances in the sentencing of Galileo, after his trial for heresy, to house arrest rather than burning at the stake was that he was working on a way of determining longitude using the moons of Jupiter as a clock in the sky. He failed because he did not have a time-keeping device accurate enough to record the eclipses of Jupiter's moons with the necessary precision.

The matter was of such importance to Britain as a sea-faring nation that in 1714 the Parliament passed the "Longitude Act" which would award a prize of £ 20,000 to anyone who could find a method of determining longitude to an accuracy of half a degree of a great circle. Determining longitude requires that one know the exact time back in Greenwich, England, where the Royal Astronomical Observatory is located, and which is at longitude 0°. If the time at your location at sea is noon (measured by the Sun being overhead) and you have a clock on board that tells you it is midnight in Greenwich, you are halfway around the Earth at 180°. The problem was that the ship's motion made it impossible to use an ordinary 18th century clock, which kept time with a pendulum, on board. John Harrison, son of a carpenter, had a fascination with clocks and set out in 1727 to make one that could claim the prize. He devised clocks that used springs rather than a pendulum. Many clocks later, in 1773, he was finally awarded the prize. He died 3 years later at the age of 83. One of the many problems Harrison had to solve in making his sea-going chronometers was to correct for the effect of changes in temperature in the mechanism. He did this by bonding together two strips of metal with different coefficients of expansion. The 'bimetallic strip' would bend if the temperature changed and keep the clock accurate.

The bimetallic strip was then used by others, notably the American inventor David Rittenhouse and the Swiss physicist Johann Lambert to make a new kind of thermometer, based on how far the strip bent. If you used a long straight bimetallic strip and put a pen in the end you have a maximum-minimum thermometer, and if you put a slowly rotating recording drum under the pen you have a continuous record of temperature changes. These were around in the 19th century, but became commonplace after World War II.

If you have a 'dial' thermometer (which looks like a clock) in your house or on your patio you have one of these thermometers. They use a long bimetallic strip wound into a coil. I have one in my sauna; being a mechanical device, it needs to be tapped once in a while to get the correct temperature. The bimetallic strip is also what is in a thermostat; the bending can make or break an electrical connection.

A great leap forward in temperature measurement came with the development of thermometers using electrical resistors that are sensitive to temperature. These can offer much greater precision than either the bulb or bimetallic strip thermometers. If you have a digital readout thermometer, you have one of these.

In meteorological stations the good old bulb thermometer filled with mercury has been the standard, but these are being replaced by the newer electrical resistance types.

## 30.2 Keeping Records

Although temperature record keeping began in northern Italy, England holds the prize for the longest record. Termed the Central England Temperature Record (Fig. 30.3), it is a compilation of measurements from many sites in rural England. Unfortunately, Central England is not a favorable place to look for 'global trends' because it is strongly influenced by the North Atlantic Drift, the warm continuation of the Gulf Stream. The unusually moderate English climate is the source of the term 'temperate.'

To understand what might be happening to the planet as a whole, data coverage needed to be expanded. The International Meteorological Organization (IMO), later to become the World Meteorological Organization (WMO) was founded in 1873 to collect global data on temperature and precipitation. One of the first persons to make use of global data was a young German-Russian meteorologist, Wladimir (Vladimir) Köppen, who you may remember as the father-in-law of Alfred Wegener. From examining the records of over 100 stations world-wide he found useful data from about 1840 on. By 1880 he was able to make a cogent argument that sunspot cycles had an effect on global climate. By then he had moved to Austria, where he pursued his interests at the University of Graz. In fact, the records of the middle of the 19th century do seem to reflect the 11 year sunspot cycle. However, this correlation is not obvious in younger data. It would be almost 60 years later before a second attempt to construct a global temperature history would be made, by Guy Stewart Callendar. Köppen's most lasting contribution was in using the meteorological data to devise a scheme for classifying Earth's climates. With some modifications his climate classification is still widely used today.

When, at the end of the 19th century, Svante Arrhenius called attention to the potential for increasing atmospheric

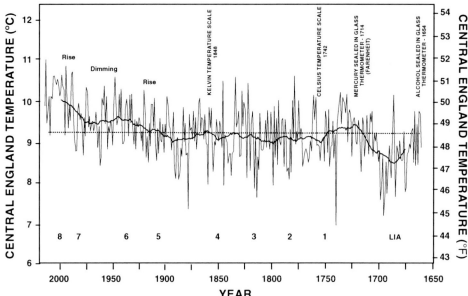

**Fig. 30.3** The Central England temperature record, from 1659 through 2015 based on the UK meteorological office Hadley Centre Central England Temperature (HadCET) dataset. Note that in this, as in the other diagrams in this book, I have followed the geologists' convention of showing young on the *left* and old on the *right*; backwards from the usual meteorologists' representation of the data. The *thin line* is the average temperature in Central England over this period (9.21 °C = 48.59 °F). The *heavy line* is a 30 year moving average through all of the values; a 30 year average is a widely accepted definition of 'climate.' Several important events related to the regional climate of western Europe are noted: (*LIA*) an especially cold period in the middle of the Little Ice Age which is considered to have lasted until the end of the 19th century; (*1*) the Irish famine of 1740–41; (*2*) the 'worst of times' according to Charles Dickens in *A Tale of Two Cities*—the years of poor harvests leading to the French Revolution of 1789; (*3*) 1816, the 'Year Without a Summer' after the eruption of Tambora in Indonesia in 1815; (*4*) the Great Famine of Ireland (1845) which had major emigration to the US as a result; (*5*) end of the 'Little Ice Age;' (*6*) the 1900–1940 temperature rise; (*7*) the episode of global dimming cooling of the 1950s through 70s that suggested to some we were going into another ice age; (*8*) the modern temperature rise

carbon dioxide levels to raise the temperature of the planet through the greenhouse effect, it was thought that it would not be come noticeable until some centuries in the future. In fact, at the time, Europe was gradually recovering from the "Little Ice Age," the period of colder temperatures that lasted from the fifteenth through the middle of the nineteenth century, with exceptionally cool episodes around 1650, 1770, and 1850. In the early part of the 20th century the idea that $CO_2$ might change climate fell into disrepute, mostly because of the argument that both water vapor and $CO_2$ absorb the same wavelengths of radiation from the Earth. Now we realize that at that time, there was a general ignorance of the complexity of the passage of both short- and long-wavelength radiation through the atmosphere, and the argument that $CO_2$ was unimportant was based on false premises.

A major advance in understanding the global situation occurred in 1923 when the International Meteorological Organization (IMO), which had been founded in 1873, initiated a program of data compilation known as the World Weather Records (WWR). This resulted in 1,196 pages of monthly temperature, precipitation and air pressure records that had been kept by hundreds of meteorological stations around the world. Unfortunately, the records differ greatly in length and quality.

The temperature of greatest interest to us is the 'surface air temperature' or SAT. Discussions of 'global warming' are centered on what is happening to surface air temperatures. Unfortunately, there is no convention on how one measures the SAT, or even what is meant by the term. During daylight hours, the land surface itself is usually warmer than the air immediately above it, and if there is no wind there is often a steep temperature gradient through the air just above the ground. As the NASA/GISS website points out, the ideal would be to have an average of the air temperature over the first 50 ft above the ground (15.24 m), using 50 sensors spaced 1 ft (30.5 cm) apart. They also note that this is never going to happen. So we have to settle for what we've got: most of the meteorological thermometers are 1–2.5 m (~3–8 ft) above the ground. Why that elevation? It has no basis in science, but it means that the observer doesn't have to stoop over or climb a ladder to make the reading. Most meteorological stations are set up like this to make everything easy for the observer. It would be nice to have hourly readings of the temperature, but at many stations they are taken every 6 h, and some simply note the maximum and minimum temperatures recorded over a 24-hour period. The averages of the observations made with different frequency are, strictly speaking, not directly comparable. In spite of all these apparent problems, different compilations show a remarkably consistent pattern.

Following the agreements made early in the last century, the data forwarded for inclusion in the WWR are monthly averages. Today with computer networking and large data storage devices it would be possible to keep all the daily or even hourly records, and some of this is being done by a NOAA scientist as an informal educational experiment using grade-school and high-school students from around the world.

The first modern attempt to compile a long-term temperature record was made in the middle of the last century. A British steam engineer, Guy Stewart Callendar, was firmly convinced that the addition of anthropogenic $CO_2$ to the atmosphere could affect the planet's climate. In 1939 he made use of the WWR compilation and other data. He published a diagram showing a gradual but steady increase in temperature starting in 1900 and continuing through the 1930s. He used information from about 200 stations world-wide, selected using a $5° \times 5°$ longitude/latitude grid. Callendar calculated that a doubling of atmospheric $CO_2$ levels from that of the pre-industrial era would result in a $2°$ C ($3.6°$F) increase in global temperature. Seventy years later on, we haven't improved much on that estimate. The older WWR data were digitized in the early 1960s by the US National Climatic Data Center, now part of the National Oceanic and Atmospheric Agency (NOAA). The WWR project has continued under the auspices of the WMO with regular publication of digital updates for thousands of stations worldwide.

Callendar was an engineer, not a meteorologist, and his ideas were received with considerable skepticism by the climate community. Like Alfred Wegener, the meteorologist who argued for the idea that the continents had moved, an idea vehemently rejected by American geologists for 60 years, he came from the wrong field to be taken seriously.

A major problem in understanding how the greenhouse effect works was that no one understood the basic physics involved. The physicists' view of the world had been originally set by Isaac Newton in the 18th century. It received radical revision by Albert Einstein's Theory of Relativity in the early 20th century. The second radical revision occurred a few years later with Werner Heisenberg's discovery that really small objects, like molecules, don't seem to know about relativity and play the game under their own special set of rules known as quantum mechanics. Heisenberg introduced the "Uncertainty Principle" which states that for an atom or molecule, it is not possible to know both its position and speed at the same time. Even today, few meteorologists are well versed in the quantum mechanical theoretics needed to comprehend just how greenhouse gases do their job.

As a consequence, it remained for a mathematically inclined academic meteorologist who had also worked in industry, Gilbert Plass, to develop a more realistic account of 'radiative transfer' through the atmosphere which made sense to the more astute in the meteorological community. He published two classic papers on $CO_2$ and climate in 1956. He had been able to take advantage of the newly developed electronic computers of that age to assist his calculations. He concluded that humans were indeed conducting an uncontrolled experiment on Earth's atmosphere, but that it might be several generations before the outcome would become apparent.

In 1950, Hurd C. Willett, Professor of Meteorology at M. I.T., used the WWR data from 129 of the stations to construct a global temperature record from 1845 to 1940. He realized that the data cannot be simply averaged. The meteorological stations are not evenly spaced over the surface of the globe, and he selected one station from each $10° \times 10°$ latitude-longitude area for which meteorological data were available. It should be noted that the total of $10° \times 10°$ areas for planet Earth is 1296, so his data set represented 10 % of the planet, mostly the land areas of the middle and lower latitudes of the northern hemisphere.

In 1957, Roger Revelle and Hans Suess, both at the Scripps Institution of Oceanography, published what has become the classic paper on the possible future effects of increasing atmospheric $CO_2$. Roger had been trained as a geologist, but moved into oceanography. Hans Suess, trained as a physicist in Austria, had been involved with the German attempt to produce heavy water in Norway during WWII. After the war he emigrated to the US and joined Harold Urey's group (which as you will recall included Samuel Epstein, who devised the method of using oxygen isotopes in shells to determine the temperature of the water in which they formed, and Cesare Emiliani, who extended use of this method to determine paleotemperatures of the ocean during the geologic past) at the University of Chicago. Suess had worked on $^{14}C$ dating, and had noted that in young samples the radioactive isotope was less common than expected. He attributed the decline to dilution of the $^{14}C$ through the introduction of 'dead' (=stable, non-radioactive) $^{12}C$ from burning of fossil fuels into the atmosphere and then into plants and animals; this has become known as the "Suess effect." Revelle and Suess were particularly concerned with how the ocean might take up some fraction of the anthropogenic $CO_2$ being introduced into the atmosphere. They referred to the increasing level of $CO_2$ in the atmosphere due largely to burning of fossil fuels as the "Callendar Effect." Their 1957 paper included the statement

> Within a few centuries we are returning to the atmosphere and oceans the concentrated organic carbon stored in sedimentary rocks over hundreds of millions of years. This experiment, if adequately documented, may yield a far-reaching insight into the processes determining weather and climate.

In one of the odd strokes of fate, their paper appeared early in what turned out to be a 35-year decline of northern hemisphere temperatures. A few years after their paper was published, updated versions of hemispheric and global temperature trends appeared in the scientific literature. Callendar revised his temperature curve in 1961, this time using data from over 400 stations and following Willett's technique of using stations characteristic of a large area to bring it up to 1950. That same year a student of Willett's, J. Murray Mitchell, Jr., brought the curve up to date, that is to 1957. He used between 179 and 118 stations world-wide for the younger part of his compilation and discovered that the warming trend that had been so obvious in the 1900–1940 data had apparently reversed. It is useful to remember that at that time the station data had to be entered into the database using punch-cards. Younger readers will need to read up on the history of computing to find out what a laborious and time-consuming process that was.

In the meantime, in Leningrad, Mikhail Budyko used a different technique, using the data to make global maps of seasonal temperature distributions and then obtaining hemispheric and global averages by calculating the areas at each temperature. In those days scientists in the USSR determined the areas by using equal-area map projections, and then cutting up the maps using scissors, and weighing the pieces; again, this was a difficult, time-consuming procedure, especially because it was before the advent of copying machines.

## 30.3 The 20th Century Pattern of Temperature Change

For many local and regional temperature records the 20th century shows a distinctive pattern, shown in Fig. 30.4. From the beginning of the century until about 1940 there is a gradual rise. This is what Guy Callendar documented in his 1938 paper citing it as evidence for warming due to increasing atmospheric $CO_2$. Just after he published his paper there began a gradual decline in Britain and much of the US that lasted until the mid-1970s. Since then there has been a return to rising temperatures. In the US this resumption of the temperature rise was initially at about the

same rate as that of 1900–1940, but since 1990 it has begun to accelerate. Incidentally, the change from the mid-century plateau back to a rise occurred about 1976, the year Al Gore, who had attended Roger Revelle's lectures about $CO_2$ and climate at Harvard, was elected to the US Congress.

To some, particularly geologists familiar with the Milankovitch insolation curves, this indicated that we might be starting to see the return into an ice age. We are, after all, at a northern hemisphere Milankovitch insolation minimum. However, investigation of records of the amount of insolation reaching Earth's surface shows that there was a decline starting in the 1940s. The decline can be attributed to lowered transparency of the atmosphere due to increasing atmospheric pollution, mostly from industrial development and automobiles. The period 1940–1980 is now referred to as an episode of 'global dimming.' During that time the reduction of insolation reaching Earth's surface masked the warming effect of increasing greenhouse gas concentrations. In 1975, the average atmospheric $CO_2$ concentration over the year was 331.1 ppm, an increase of 18 % over its pre-industrial level. In April 2014 the atmospheric $CO_2$ concentration passed 400.0 ppm almost 47 % over the pre-industrial level.

However, then, for the next 40 years temperatures seem to have stabilized or in some cases even declined. Since about 1980 temperatures have again risen, with some records suggesting an accelerating pace.

The rises have been attributed to increasing greenhouse gas concentrations, mostly $CO_2$. The cause of the mid-century episode of stabilization-decline was a matter of controversy in the 1970s.

## 30.4 Some Examples of Local Temperature Records

To get an idea of what the original records are like, Figs. 30.5 through 30.13 show plots of average annual temperatures from ten meteorological stations with long records located around the world. These records are longer than most, but give an idea about how different the weather and temperature trends can be in different parts of the world. The common feature to all the records is the 'variability' of the temperature from year to year. Variability is a characteristic of weather. Remember Ed Lorenz's discovery when he was using one of the early computers to try to predict the weather—it is a chaotic system. This variability occurs on a daily scale—nighttime versus daytime temperature, on a scale of a few days, as weather fronts move through, on a seasonal scale—spring, summer, fall, winter, and on an annual scale. These figures show annual average temperatures, and if you look carefully you will see that they all have one thing in common

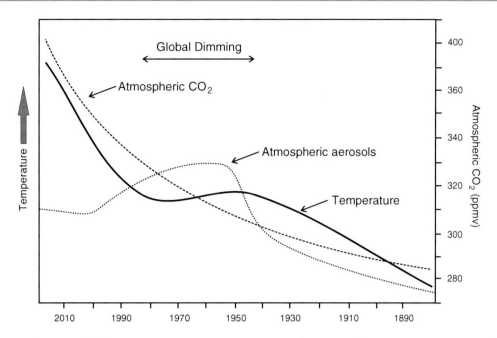

**Fig. 30.4** The general pattern of 20th century temperatures seen in many parts of the world, with a rise, pause, and rise. Also shown are the increase in atmospheric $CO_2$ concentration and an estimate of changes in the level of atmospheric aerosols. The 1940–1980 temperature plateau is now referred to as an episode of 'global dimming,' when the increase in amount of atmospheric aerosols, mostly anthropogenic pollutants introduced by industrial expansion, reduced the amount of insolation reaching Earth's surface, masking the greenhouse effect of $CO_2$

—it is very common for a warm year to be followed by a cold year and vice versa. There are only a few instances where there is a short-term trend, that is for a cold year to be followed by a warmer year followed by an even warmer year. Such a trend over more than three years is very rare. The variability is like the trees in a forest. It can be difficult to see the forest because of all the trees. So the problem becomes one of devising a method of seeing the general trends through all the details of the variability. Each of the figures uses a different approach. If you go back and look at Fig. 30.3, the long-term temperature record for Central England, you will see one time-honored way of looking for the trends, the moving average. In that figure the solid black line is the average of 30 years of data starting from the oldest and moving year by year to the youngest. Remember that a 30-year average is thought to represent the 'climate.' When you compare the 30-year average with the annually varying temperatures you get a clear understanding of the expression "the climate is what you expect; the weather is what you get."

Start our world tour by taking a look at Fig. 30.5, the record from Boulder, Colorado, about an hour's drive from where I live. Boulder is the home of the largest campus of the University of Colorado, the National Center for Atmospheric Research (NCAR), a major facility of the National Bureau of Standards, and many laboratories of the National Oceanic and Atmospheric Agency (NOAA). The town has more climate scientists than anywhere else on Earth, but one of the strangest temperature records. It shows some remarkable variations: a very cold year, 1929, and the very warm year 1934 that occurred during the middle of the 'dust bowl' stands out conspicuously. In fact, for the 48 contiguous United States, 1934 was essentially tied with 1998 as the warmest year on record, but both records have been surpassed by 2006 and 2012. The early 1930s were particularly warm in the US, not reflecting only the hot summer temperatures, but the unusually mild winters as well.

Superimposed on the annual temperature record is a dashed straight 'linear regression' trendline representing a mathematical fit to all the data. The slope of the line is the average rate of change with time. The trendline shows a very slight upward (warming) tendency. The oldest part of the record suggests a cooling trend, the middle a warming then cooling trend, and the youngest part of the record shows a marked warming. If we didn't know anything else about what is going on, this record would seem to indicate that not much is happening; just the ups and down one might expect as 'natural variability.'

Although linear regression trendlines may seem like a good way to look at the records, they don't make much sense for many of the records outside eastern North America and Britain. I tried making different kinds of regressions and fits of curves to the data to see if I could find a better way to look at it. After some fiddling around, I found one that more closely resembles the way a geologist thinks of time—a

**Fig. 30.5** Mean annual temperature record for Boulder, Colorado, from 1898 to 2015, with several gaps Note the great variation from year to year and the lack of cold years after 1993. Also shown is a straight regression trendline through all the data and a fourth order polynomial curve fit to the data; all explained in the text below

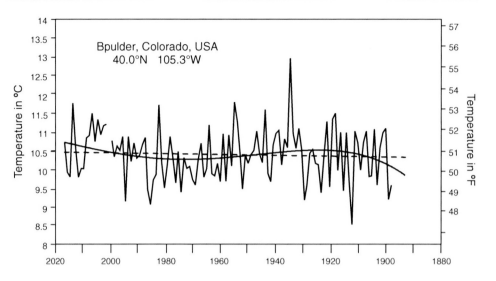

'polynomial curve' fit to the data. A fourth order polynomial is shown in this figure. For the mathematically inclined, a fourth order polynomial has the form

$$y = mx^4 - nx^3 + ox^2 - px - q$$

The record conforms to what has become the iconic American view of climate history of the late 20th century. Based largely on the temperature records from the 48 contiguous states, the view was that there had been a rise over the earlier part of the 20th century. In the Boulder record there is one highly relevant oddity—there have been no 'very cold years' after 1993. The lack of very cold winter weather since then has allowed the Mountain Pine Beetle, *Dendroctonus ponderosae*, to flourish and devastate the lodgepole pine forests of the Rockies. Along with the warmer temperatures in Colorado has come a drought which weakened the pinion pine forests and aspen groves of the American southwest and left them vulnerable to assault by insects. The overall result has been a massive die-off of the forests which extends from the southwestern US into British Columbia.

Figure 30.6 shows the continuous record from the meteorological station at Hohenpeissenberg in Bavaria. Hohenpeissenberg is the name given to both a small town and a hill just southwest of Munich, in the foreland in front of the Alps. The Meteorological Observatory is located on top of the hill. It has one of the longest, best-kept temperature records in Germany. It is close to the center of Western Europe. Again, the temperature record is overlain by the two trendlines discussed above. The impression is that the record is similar to that of Boulder, but the 1940–1975 decline is not as pronounced.

When I was a student in Munich in 1953–54, I sat in on a course in forestry, a specialty in Germany. The course

was taught in the University's main building on the Ludwigstrasse. Most of the building had been burned out, and was being reconstructed. There was a large hole in the roof and ceiling over the lecture hall where a bomb had come through but had not detonated. Because of the hole, the room, like a number of others in the University in those days, was not heated. That winter was one of the coldest of the decade. I remember two things about that course very clearly: (1) I learned that it is difficult to take notes wearing a heavy coat, scarf, wool cap and woolen gloves, and (2) the lecturer emphasized the difference between the climate in the city and the surrounding area. The city is warmer, and the snowfall in the periphery is sometimes cold rain in the city. Even in the 19th century it was recognized that cities modify climate, and so the regional Meteorological Station had been established well outside the city. The record from Hohenpeissenberg (Fig. 30.6) is a classic 'rural' station site.

The Hohenpeissenberg record is similar to that at Boulder, but it looks as though the entire record has been tilted so that it slopes upward to the left. The overall trend is warming, with the earlier part being gradual and the later part being more rapid. You can see that cold year when I was a student in Munich in the early 50s in the Hohenpeissenberg record. In recent years I have been able to be in Munich for the Christkindl Markt, the Christmas Fair that takes place during Advent. It used to always be a romantically snowy affair, with the falling snowflakes tempered by the warm Glühwein. But in recent years it often takes place in cold drizzling rain. The Glühwein still keeps the body warm.

Going further to the east, Fig. 30.7 shows the record for St. Petersburg, Russia, formerly Leningrad, USSR, as it was called when we had our meeting on climate change there in

**Fig. 30.6** Mean annual temperature record for Hohenpeissenberg, Bavaria, Germany from 1881 to 2015. Trendline and curve as in Fig. 30.5

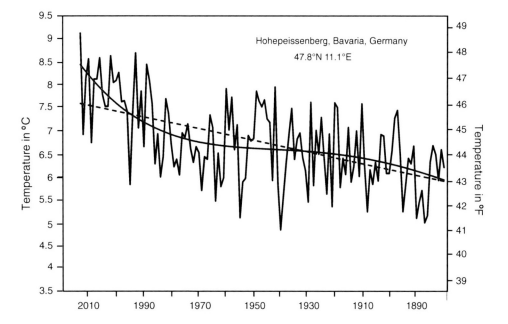

**Fig. 30.7** Mean annual temperature record for St. Petersburg, Russia, from 1881 to 2015. Trendline and curve as in Fig. 30.5

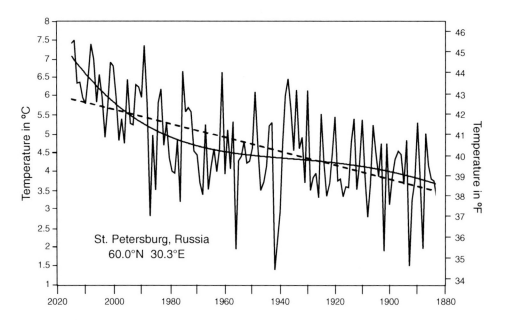

1982. The overall trendline shows a distinct rise of about 3 ° F (1.7 °C); the shorter-term trendlines are steeper, but is as though their starting points are reset in 1940 and 1975. It does not show the 'American' pattern represented by Boulder. One might begin to suspect that the pattern of rise-fall-rise noted in the US might be a regional phenomenon. Note the much greater variability of the temperature from year to year, characteristic of the more continental climate of Russia. Being Mikhail Budyko's home town, he was thoroughly familiar with its long-term temperature history of gradual rise, but when we met, the sharp rise after 1975 had just begun. The record from Moscow, not shown here, is similar.

The record for Beijing, Fig. 30.8, on the other end of the Eurasian landmass, is strikingly similar to that from Germany and Russia, but shows an even sharper rise after 1975. However this is undoubtedly partly due to the growth of the city, industrial development, and desertification as the Gobi Desert to the west encroaches on the city. We can be sure that this record shows the anthropogenic effects of urbanization.

What do records from tropics look like? The city of Cuiaba is capital of the Brazilian state of Mato Grosso and is in the very center of South America. Its temperature record, shown in Fig. 30.9, would seem to indicate some sort of cyclic variation in the climate. The overall trend for 1901–1940 is a decline, but on closer inspection there was a very

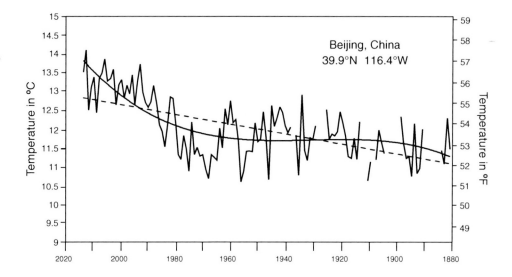

**Fig. 30.8** Mean annual temperature record for Beijing, China, from 1881 to 2015, with some gaps during troubled times. Trendline as in Fig. 30.5

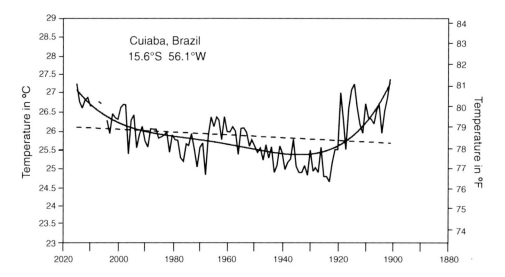

**Fig. 30.9** Mean annual temperature record for Cuiaba, Mato Grosso, Brazil, from 1901 to 2015, with a gap for 2005. Trendline and curve as in Fig. 30.5

sharp decline between 1915 and 1920, followed by a rather steady rise from 1940 until 1965, when there was another sharp decline. Since then there has been a rise at a rate comparable to that observed in many other parts of the world. It is as though the temperature rise over the first half of the 20th century that had so impressed Guy Callendar was simply delayed by a few decades in the Mato Grosso.

Note that there is no long-term upward trend. Does this mean that 'global warming' is a fiction? No, it means that the tropics already have their own greenhouse effect from water vapor. Wherever the humidity is high the vapor greenhouse effect dominates over the effect of increasing the levels of other greenhouse gases in the atmosphere.

The city of Quixeramobim in the State of Ceará, northeastern Brazil is only 5° south of the Equator. Its temperature record, shown in Fig. 30.10, is almost flat until 1950 with a rather steady rise since. Again, it is as though the temperature rise over the first half of the 20th century that had so

impressed Guy Callendar was simply delayed by a few decades in this part of Brazil.

Figure 30.11 shows the record from Alice Springs, in the middle of Australia. At 29° 49′ S, Alice Springs is in the middle of the Australian Outback desert close to the Tropic of Capricorn. The older part of the record shows a slight decline, and there has been rapid warming since 1975. There were a number of warm years in the first half of the last century that were almost as warm as those of the last decade.

What is striking about these tropical and southern hemisphere records is that they are completely different from that for the US. Instead of up-down-up, the temperatures over our selected intervals go down-up-up.

What do records from ocean islands look like? First of all, the variability from one year to the next is less than it is for the continental sites. This is because of the stabilizing influence of the surrounding ocean. Secondly, they tell us about the changing temperature of the ocean itself.

**Fig. 30.10** Mean annual temperature record for Quixeramobim, Brazil (5.2° S, 39.3° W) from 1896 to 2015. Trendline and curve as in Fig. 30.5

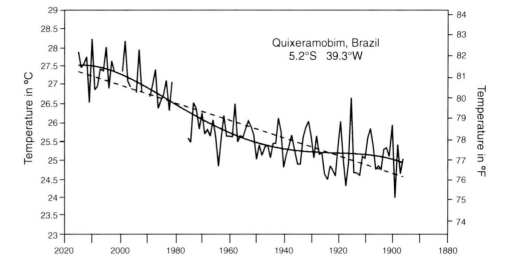

**Fig. 30.11** Mean annual temperature record for Alice Springs, Australia, from 1881 to 2015. Trendline and curve as in Fig. 30.5

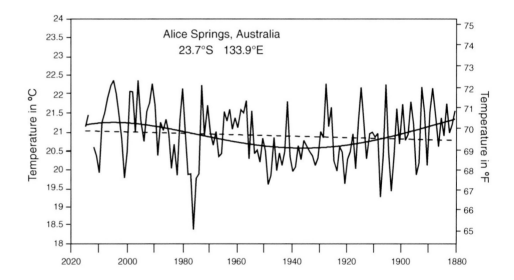

Figure 30.12 is the record from Key West. At the end of the Florida Keys, Key West is the southernmost point in the 48 continuous states. It is bathed by waters of the Gulf Stream as it leaves the Gulf of Mexico, and so it is actually giving us information about the ocean temperature in that region. There is an overall long-term temperature increase, but since 1975 the rate of increase has declined. Not evident on this record are the hurricane season temperatures which influence storm intensity. Hurricane Katrina occurred in 2005, the next to last point on the line; it was a relatively cool year overall, but with extraordinarily high temperatures during the month of August.

Similarly, the record in Fig. 30.13 is Hawaii, and represents both what is happening in the tropical North Pacific and the climate of the large island. It is very different from most of the other records we have examined. It shows a very sharp temperature rise prior to 1940, a sharp decline until

1955, another sharp rise to 1975, and then a slow rise to present.

Both the Key West and the Honolulu records show much less variability than those from the continent. That is, of course, due to the stabilizing influence of the water with its high heat capacity and corresponding resistance to temperature change.

One thing that is obvious in all of these records is that it is very common for a warm year to be followed by a cold year. Temperature 'trends' with 3 or 4 years of successive temperature increases or declines are relatively rare. These ups and downs are the annual 'variability' or 'noise' in the climate system and reflect the chaotic behavior inherent in weather even over the course of a year. Remember the old adage "the climate is what you expect; the weather is what you get." Also recall that the 'climate' is usually considered to be a 30 year average of the weather.

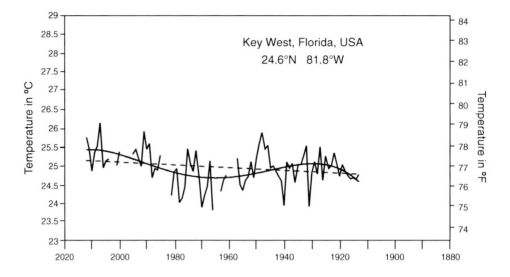

**Fig. 30.12** Mean annual temperature record for Key West, Florida, USA, from 1913 to 2012. Trendlines and curve as in Fig. 30.5

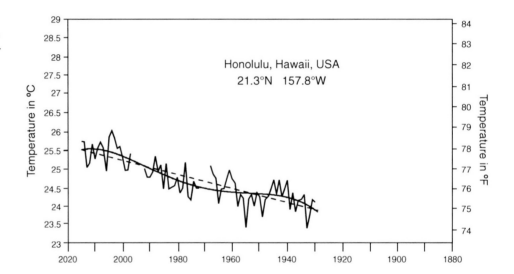

**Fig. 30.13** Mean annual temperature record for Honolulu, Hawaii, USA, from 1929 to 2015. Trendline and curve as in Fig. 30.5

It is also obvious that the 'American sequence' of up-down-up is a regional phenomenon. The one common property of these records is that since 1975, the temperature trend is upward. In looking over many individual records, I did find one that does not obey this general rule. It is that for Washington, District of Columbia, USA, shown in Fig. 30.14.

It is somehow fitting that the place of greatest political skepticism about the global temperature trend should be one of the rare places where the warming trend of the last 30 years is not evident. Washington is an unusual city, completely different from others in the US. It was originally laid out in 1791 by Pierre L'Enfant, to have broad streets, a height limit on its buildings, and many open spaces. The District of Columbia was redesigned at the beginning of the 20th century to be the unique, monumental city it has

become. Its climatology is different from most urban areas. It is famous for being hot and humid, and until recently it was considered a 'hardship post' for British Embassy personnel.

For Washington, DC, there was a general long-term rise until about 1950, then a slight decline until 1970, followed by a sharp rise until 1975. Then something very unusual happens: the younger mean annual temperature trend is highly variable. This is often cited as the source of the problem many in the US Government have had with discussions of 'global warming.' For politicians living in Washington the local long term record is indecisive. The even temperature over the past 30 years may reflect both the development of the underground Metro which has reduced the automobile traffic in the city and the creation of more open space.

**Fig. 30.14** Mean annual temperature record for Washington, DC, USA, from 1881 to 2013. Trendline and curve as in Fig. 30.5

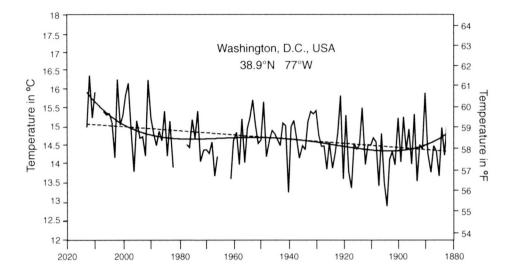

## 30.5    Regional Temperature Compilations for the United States

The US National Climate Data Center (NCDC) makes compilations of US temperature and precipitation data for cities, states, 9 regions, and the 48 contiguous states, on a monthly, seasonal, and annual basis. The compilations are averages of meteorological observation stations with records going back to 1895. The averages are not simply the average of all temperature records in a given area, but are 'area weighted'—the area represented by each station is taken into account. The data set from which these compilations are made is called the U.S. Historical Climate Network (USHCN) and consists of about 1200 stations located mostly in rural areas and small towns. Figure 30.15 shows the regions used by the USHCN.

The regional compilations are shown in Figs. 30.16, 30.17, 30.18 and 30.19, from the Atlantic margin, through the eastern and western central US to the Pacific states.

Now, let's look at some of these area averages. What is particularly interesting about regional compilations is that they can be very different, and if this were all the information one had, one could come to different conclusions about how the climate is changing. Furthermore, different time periods can show very different trends.

Some of them, particularly those in the central and western US, show the famous 1934 'hottest year' anomaly that was often cited as evidence that there is no such thing as 'global warming.' It has been surpassed twice in the US contiguous 48 states records, in 1998 and 2006. In any discussion about global warming it is important to remember that the 48 contiguous United States cover only about 1.6 %

**Fig. 30.15** Regions used for NOAA's USHNC historical temperature compilations

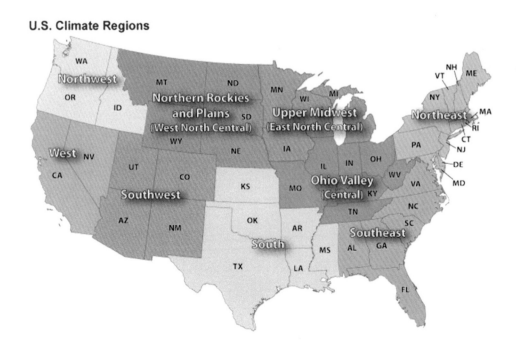

**Fig. 30.16** Two regional mean annual temperature compilations for the Atlantic states. Again, as for Fig. 30.5, the *straight dashed line* is linear trendline fit to the data. The *broad curve* is a fourth order polynomial fit to the data. The *upper compilation* is for the southeastern states from Virginia to Florida, and the Gulf states from Florida to Mississippi. The *lower one* for the northeastern states from Maryland through New England

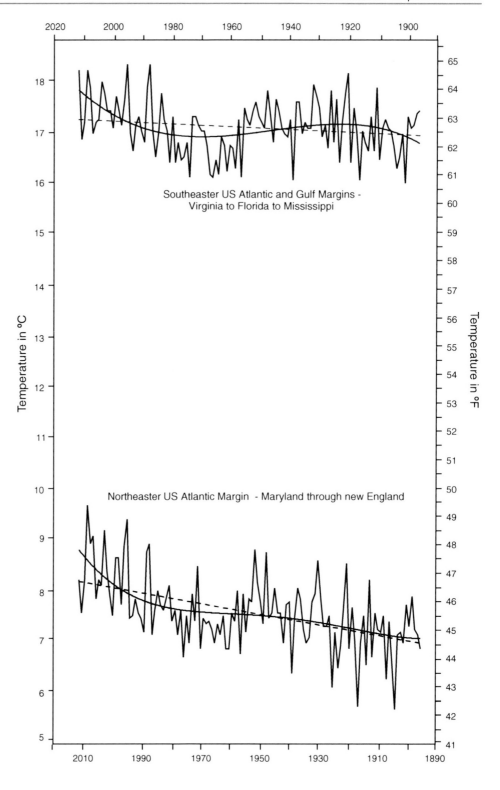

of Earth's surface, so the assumption that anything in the US record is representative of Earth as a whole is fallacious. The climate change denial community has yet to understand that there is a difference between the US and planet Earth.

The two regions on the US Atlantic seaboard have very different temperatures in large part due to the Atlantic. The southeastern states are bordered by the warm northward flowing Gulf Stream. They show the lowest temperature

**Fig. 30.17** Three regional temperature compilations for the eastern central US. The *uppermost* is for the 'South.' The *middle record* is for the 'Ohio Valley.' The *lower curve* is the 'Upper Midwest'. Trendline and curve as in Fig. 30.5

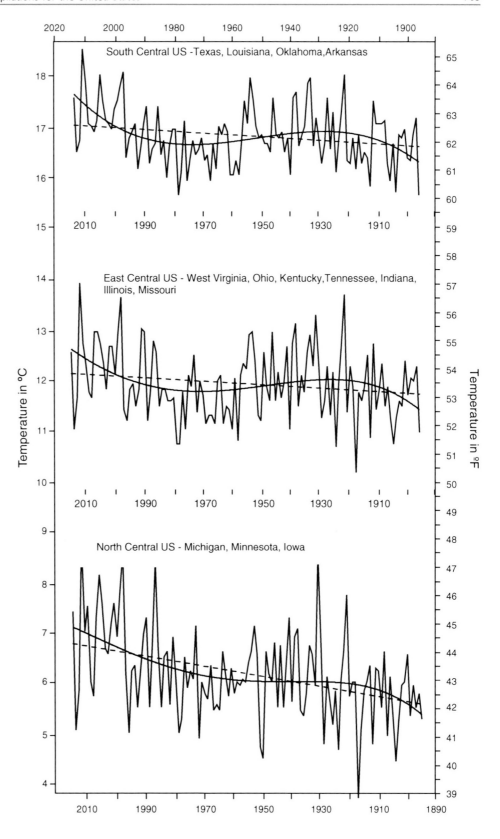

increase during the 20th century. However, the Gulf Stream turns abruptly eastward at the latitude of Virginia. The northeastern states are bordered by cold Labrador Current flowing to the south. Washington, D.C., is located at the boundary between these two regional compilations.

**Fig. 30.18** Two regional temperature compilations for the western US, the Southwestern US states of Arizona, New Mexico, Utah and Colorado, and the Northern Rockies and Plains states of Wyoming, Montana north and South Dakota and Nebraska. Trendline and curve as in Fig. 30.5

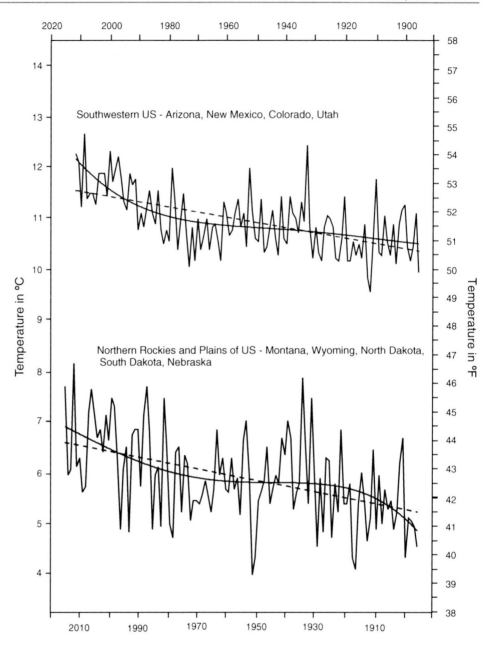

These regional data show the rise-fall-rise pattern noted at many sites in the US. However, of the nine regional records for the US, it is only that for the southeastern states which suggests a long term temperature decline. This is because this region is humid, and the greenhouse effect of the water vapor has thus far overwhelmed that of increasing $CO_2$. All of the other compilations show a long term trend upward, but to varying degrees.

Figure 30.17 shows compilations for the eastern central United States, from Texas to the Canadian border. As might

be expected in the interior of the continent, the temperature contrasts are extreme.

Figure 30.18 shows compilations for the western central United States, from the Mexican to the Canadian border. Again as might be expected in the interior of the continent, the temperature contrasts are extreme.

Figure 30.19 shows the regional temperature records for the western US, the states bordering the Pacific and those immediately inland from them. This includes all of the states on or west of the continental divide. The north-south

**Fig. 30.19** Three regional temperature compilations for the western US. That at the *top* is for California and Nevada. The *bottom compilation* is for the northwestern states, Washington, Oregon, and Idaho. Trendline and curve as in Fig. 30.5

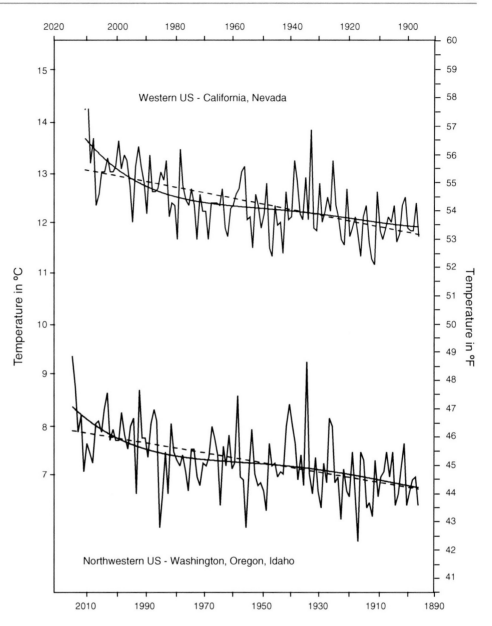

temperature spread is much less than for the Atlantic and Central compilations. The cool Pacific Ocean and the high altitudes of these states are important climatic factors.

The United States is probably still the largest area on Earth with consistent standards for making meteorological observations. The NCDC and NASA-GISS compilations for the 48 contiguous states since 1895 are compared in Fig. 30.20. The NASA-GISS version of the mean annual temperatures of the 48 contiguous United States from 1880 to 2015 is shown in Fig. 30.21. To make averages for larger areas requires international cooperation, coordination and standardization of the data.

## 30.6   Larger Regional and Global Compilations

Philip Jones at the University of East Anglia in Britain published the first report of their group's efforts to determine a century-long term history of temperature of the Northern Hemisphere in 1982. They made a major innovation, termed the anomaly method. The problem with simple averaging of temperatures from different stations is that very large or very small numbers will strongly affect the result. They realized that what is really important is to determine the long-term changes. To do this they selected a 25 year reference period, and subtracted the average temperature of the reference period for each station from the temperature for each month or year. The reference period average temperature then becomes 0, and the difference is called the climate anomaly. Over the next decade their system became ever more sophisticated. They started with a crude grid based on about 300–1300 northern hemisphere meteorological stations, depending on the time interval. This was followed in 1986 by an analogous compilation for the Southern Hemisphere. More recently, the original $5° \times 10°$ latitude-longitude grid used as a base has been reduced to $1° \times 1°$, and the number of stations increased to more than 2000 and the reference period set at 1961–1990. The areal average temperatures are calculated as weighted averages dependent on the distance of each station from nearest grid point. Because the meteorological stations are on land, only a small part of the Earth is represented by these compilations. To get a better estimate for the planet as a whole it is necessary to add temperature data for the oceanic regions.

These records illustrate the problems faced by those interested in making regional and global compilations to search for long-term trends. At the participating meteorological stations measurements are taken at different times of the day, using different instruments reflecting the technological development by differently trained personnel. The records are of different lengths, and there are some gaps, even in records from places one would not expect them, such as Boulder, Colorado, where the US National Center for Atmospheric Research is located. Many of the stations have been moved from one place to another to try to avoid the effects of urban or agricultural development. Not only the geographical location, but the elevation of the station may have changed. Observation times for individual stations may have changed over the years. The methods of calculating daily and monthly temperature averages may differ and may have changed with time. Remember that the long-term trends climatologists are looking for are of the order of a few tenths of a degree per century, and the daily and seasonal variations are tens of degrees. It's a bit like looking for a needle in a haystack.

From the individual station data, the goal is to decide what area of the Earth's surface each station represents. Fortunately, the development of computers has progressed to the point where the enormous databases can be manipulated digitally. Four major centers have undertaken to work through the WMO data now which includes 11,494 observing stations. Each of the centers, the University of East Anglia and Hadley Center in the United Kingdom, led by Phil Jones; the National Climatic Data Center (NCDC) of the US National Atmospheric and Oceanic Agency (NOAA), led by Tom Peterson; the Goddard Institute for Space Science (GISS) of the US National Aeronautics and Space Administration (NASA), led by James Hansen; and

the State Hydrologic Institute in St. Petersburg, Russia (originally led by Konstantin Vinnikov), produces its own versions of regional, hemispheric, and global averages for the months, seasons, and entire years. Currently each of these institutions uses 4,000–5,000 of the WMO meteorological stations to develop regional and global temperature records. The Hadley Center record is a synthesis of about 3,700,000 measurements, and the other compilations use similarly large datasets.

The WMO meteorological stations are on land; excluding Antarctica the land area is about 26.4 % of the Earth's surface. However, most of the stations are in the Northern Hemisphere, and most of those are in eastern North America and Europe. It is evident that because of this uneven distribution, determining a land-only temperature history is a major undertaking, and each of the four groups has chosen a different way of calculating large-scale monthly and annual averages.

Unfortunately, determining the area represented by each station is just the beginning of the analysis. The data for each station must be examined to see if there are obvious errors. Records older than 1880 are particularly difficult to interpret, and except for special regions, the syntheses usually start with that date. Sometimes there are obvious errors that can be traced back to the original record, when the writing was illegible or simply mistaken. The number of such spurious records is very small, and eliminating them often has almost no effect on the larger compilations, but critics find the ongoing revisions annoying. Each of the groups has its own method of figuring out which records are suspect, and the criteria for deciding whether to delete the record entirely or how to make a reasonable correction.

Some of the meteorological observation sites are in or near cities, and cities themselves produce a local microclimate, usually warmer than the countryside. Some critics have argued that the global temperature record is biased and that what it really shows is population growth and the growth of urban areas. It is certainly true that as urban areas, particularly in the United States, spread and join together as metropolitan areas, they modify the climate of increasingly large regions. However, the compilations made by the groups mentioned above have already taken this effect into account, and specific bias is removed for the compiled records. Changes in agricultural practices can also affect the temperature. Deforestation results in warming and irrigation results in cooling.

Most of the analyses use some sort of averaging of the data within a 'square' defined by latitude and longitude. Callendar had used $10° \times 10°$ 'squares' in his analyses. I enclose the word 'squares' in quotes because, except at the Equator, these are not really squares but rectangles. Lines of latitude, or 'parallels' are equidistantly spaced, but lines of longitude, or meridians, converge and meet at the poles. More recent studies have continued to use a latitude-longitude grid as the basis for averaging.

Now you will ask, how big is a latitude/longitude square? Here we need to go back to Chap. 2, and recall how the meter/kilometer and foot/yard/mile are related to the size of the Earth. As you will recall, there are 60' (minutes) in 1° (degree) and 60" (seconds) in 1' of arc on a circle. Remember that the concept of the nautical mile (NM) goes back to ancient times, perhaps to Pythagoras, who figured out that the Earth was spherical, or to Eratosthenes who is credited with having made the first estimate of its circumference. It was to be directly related to the size of the Earth; it was intended to be 1' of longitude at the Equator or 1' of latitude, so that 60 NM would equal one degree of latitude. Now that we know more about the size and shape of the Earth, we know that it is not a perfect sphere and that at the Equator 1' of arc is actually 1.0018 NM. However, a NM is almost exactly 1 min of latitude at latitude 45°. If you are clever you will figure out that if a minute of longitude is 1 NM at the Equator (latitude 0°) and 0 NM at the pole (latitude 90°) this sounds like a trigonometric function. What is 1 at 0° and 0 at 90°? A cosine function! What would be the distance between two meridians 1 min apart at 40°, the latitude of Boulder, Colorado? The cosine of 40° is 0.7660, so it would be 0.7660 nautical miles, so a 1' × 1' rectangle at this latitude would have an area of 0.7660 square NM.

Several of the global land temperature calculations average over 5° × 5° rectangles. At 40° latitude, such a rectangle would have an area of 6972 square NM. But we don't use nautical miles on land—so what is this in more familiar terms?

On land, we in the United States use 'statute miles,' the 5,280-ft miles originally defined by Elizabeth I of England and used in the United States and still in the United Kingdom although officially the UK has gone to the metric system. This 'Imperial' measurement system doesn't relate directly to the size of the Earth, and the standard of length is not the foot, as you may have been taught, but the yard. Unfortunately, the standard 'yard' is slightly different in Britain and the United States, so a British mile is not exactly the same length as a US mile although both are 5,280 ft. In terms of 'US statute miles' (or simply miles) the area of a 5° × 5° rectangle at 40° latitude is about 89,769 miles$^2$ (a little larger than Minnesota, which is 86,943 miles$^2$).

When the metric system was devised in 1791, shortly after the French Revolution, measures of length were to be based on the size of the Earth. The meter was intended to be one 10 millionth of the distance from the Equator to the pole along the meridian that passes through Paris, France. Two marks were inscribed on a platinum bar, kept at the Bureau International des Poids et Mesures (International Bureau of Weights and Measures) in Sèvres, just outside Paris. Later it was discovered that the original calculation of the Equator pole distance was wrong because they had not known about the flattening of the Earth at the poles, so the platinum bar became the standard for the meter rather than the Earth itself. If you divide 10,000,000 by 90° you get 111,111 m per degree of latitude. Taking the flattening of the Earth into account, the exact distance between two degrees of latitude centered on the latitude of Paris, 48° 52' N, is 111.207 km. It's off by less than 1 part in a thousand, which is not bad for the 18th century when nobody had been near either pole. A 5° × 5° rectangle at 40° latitude is about 232,932 km$^2$, or about half the size of France.

By now you get the idea: averaging is going to be over areas large enough that local weather conditions will not play a role, and that there will be enough meteorological stations to provide a good sampling of the climatic conditions.

Another apparent oddity in the search for changes in regional and global temperature with time: if you look at any of the analyses of long-term temperature change, as we will later in this chapter, you will find that none of them show the absolute global temperature change over time, but relative change with respect to some base period. The base period is usually taken to be about 30 years long. Why 30 years? You may recall that 'the climate is what you expect; the weather is what you get.' In the middle of the last century it was decided that a 20–30 years average would be a good way to define the climate. That was when no one except Guy Callendar believed that the climate was changing over longer timescales. The base periods used by the climate research centers are different. The University of East Anglia and Hadley Center originally used 1951–1970, but has subsequently updated it to 1961–1990. 1961–1990 is also used by NOAA's NCDC. NASA/GISS uses 1951–1980. The Russian State Hydrological Institute uses 1951–1975. Not to worry, as they say in England, a base period is a base period.

The next step in the averaging process is not so obvious until you think about it. If you knew nothing else you could already guess that the temperature at a meteorological station is going to depend on the latitude (the Equator is warm and the pole is cold), and elevation (the temperature goes down as you go up). Superimposed on these general trends is the variability with time of the day, local weather conditions, the seasons, and the variations from year to year. Furthermore, the meteorological stations are not spaced on a grid, but are located for the convenience of weather observers. What we are looking for is long-term trends, which are much smaller than the short-term variability, over large areas. Here is the nature of the problem: imagine a mountainous area adjacent to the coast. All the towns and villages are along the coast, and that is where the meteorological stations are too. There are just a few in the much larger mountainous region. If I just average the temperatures recorded at each station I will get an average temperature that is typical of the coast, but not of the region as a whole. In fact, if I am a climatologist looking for long-term change, what I want to know is not what the average temperature is over time, but whether the temperature trends up or down. The British group decided that what they really wanted to know is the difference in temperature from the average of the base period, say for a given month of the year, or over the average of a whole year. They set the long-term base period average to be 0, and call the difference from this long term average the 'temperature anomaly.' The 0 has nothing to do with 0 °C or 0 °F; it is simply the average temperature of the base period. They call this the Climate Anomaly Method (CAM).

With the average temperature for each station for the base period set to 0, the deviations from that average are the 'temperature anomalies' for each station. This eliminates the effects of peculiarities of the locations, such as different elevations or surroundings of the meteorological stations. Remember, it is not the average regional or global temperature we are after, it is the long-term change in temperature, and that is what the 'anomaly' represents. The station temperature anomaly values are averaged together for all stations within a 5° × 5° latitude/longitude grid box to get a hemispheric or global average of the land temperature anomaly. Of course, the number of stations varies between grid boxes and by year within the grid boxes, but there is a thing in statistics called the Law of Large Numbers, which says that when you average an increasingly large sample of data, the average becomes ever closer to the true number. How close the sample average is to the real value depends, of course, on how large the sample is and how big the total spread of numbers is. I learned about this in a wholly different geological context, and was amazed to find that after we had calculated the average of seven values we had a very close approximation to the real value. The 5° × 5° latitude/longitude grid boxes usually have a hundred or more station values, so even some spurious data have almost no effect on the overall average.

Incidentally, if you want a global reference for '0,' the East Anglia group has estimated the global average temperature over the period 1961–1990 to be almost exactly 14 °C (=57.2 °F). You will note that this is 1 °C less than the 15 °C that is in every textbook and that we used in our discussion of energy balance. The 15 °C was a much earlier educated guess; not bad for the age before computers were around to handle large data sets.

The NCDC group has used a technique called 'first difference time series' which, when I first heard about it, carried no meaning to me at all. What they mean is that they take the calendar-month differences in temperature between successive years of station data. For example, when creating a station's first difference series for mean January temperature, the station's January 1880 temperature is subtracted from the station's January 1881 temperature to create a January 1881 first difference value. First difference values for subsequent years are calculated in the same fashion by subtracting the station's preceding year temperature from the current year temperature for all available years of station data.

The difference is the change over one year; if you are a fan of calculus, this is an approximation of the first derivative of the temperature data. All of the 'first difference' data within a 5° × 5° latitude/longitude square are then averaged, and these areal averages are added and again averaged to make regional, hemispheric, or global temperature time series.

The State Hydrographic Institute in St. Petersburg, Russia, uses a statistical technique called 'optimal averaging.' Most of my Russia friends have a knowledge of math which is way, way beyond me, and 'optimal averaging' is no exception. Suffice it to say, it is a very sophisticated way to analyze irregularly spaced, sometimes incomplete data.

The NASA group at GISS started out using an averaging method similar to that of the British group, but it has evolved into a very different method. Now they overlay the Earth with a 2° × 2° grid, and incorporate the data from all stations within 1200 km (750 miles) of the grid-point. The value assigned to the grid-point is a weighted average depending on its distance from the point. That is to say, in calculating the average, a station within 1 km of the point is 1200 times more important than one on the periphery of the circle, 1200 km away. Now here is what gets very interesting. At the Equator, intersections of the 2° × 2° grid are 222.2 km (133.3 miles) apart, and at 40° latitude they are 222.2 km apart in the NS direction, but only 167.6 km (=100.5 miles) apart in the EW direction. But because all the stations within a circle of 1200 km of the grid point will be included in the averaging process, the area over which the averaging takes place remains a constant 7,540 km$^2$ (=2911 miles$^2$). Many stations will be included in more than one of these areas but their weight in the averaging processes will differ in each case because it depends on the distance from the grid point. I call that clever.

To understand the differences between the two US methodologies, the NASA-GISS compilation for the contiguous 48 United States since 1895 is shown in Fig. 30.20. The NCDC data (1895–2015) are superimposed on it as a dashed line, but it is not visible because the differences are smaller than the width of the printed line. Note that the NASA-GISS record actually goes back to 1880, including older data than that used in the NCDC compilation; that more complete record is shown in Fig. 30.21. There is one difference that attracted considerable attention: the anomaly values for 1998 and 2006 are reversed. In the NCDC compilation the value for 1998 is 2.21 °F and that for 2006 is 2.23 °F whereas in the NASA compilation the value for 1998 is 1.3241 °C and that for 2006 is 1.3026 °C. Thus

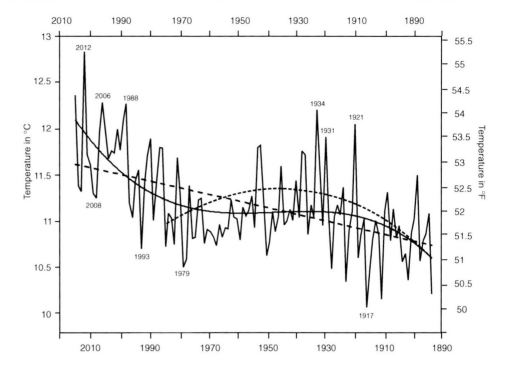

**Fig. 30.20** The NASA-GISS and NCDC temperature records for the contiguous 48 US states since 1895 superposed. Trendlines and curve as in Fig. 30.5

(until 2012) for the NCDC the warmest year on record was 2006, for NASA it was 1998. Climate change denier conspiracy theorists went wild. Ignoring the fact that two different sets of stations were used and the differences are in the second decimal place, they argued that someone was fudging the data They had asserted that the decline in temperature from 1998 to 2005 was poof that 'global warming' had ended and an episode of cooling had begun. The 1998 versus 2006 controversy was a convenient diversion, until 2012 when a new much higher record was set.

Figure 30.20 shows two fourth order polynomial fit to the data. The solid line is fit through all of the data. The dotted curve is fit through the data that would have been available in 1980. The difference between these two curves is striking and explains a controversy of the 1970s. At that time many geologists (including Cesare Emiliani) thought that the declining temperatures indicated that we might be seeing the beginning of an overdue ice age. After all, interglacials typically last for 10,000 years, and we were already 11,000 years into this one. Based on what was known of the temperature records at the time, it seemed quite reasonable. Remember that the major global compilations were yet to come.

While the geologists were convinced that Milankovitch forcing would drive the climate, climatologists themselves recognized that the increasing concentration of $CO_2$ in the atmosphere made it highly unlikely that the decline would continue. The cause for the decline was already known to many of those studying changes in the atmosphere, but that knowledge was not widely disseminated. It was the introduction of water vapor into the stratosphere by jet aircraft

and pollution of the troposphere with aerosols as heavy industries and agricultural development expanded during the economic boom following World War II. Jet aircraft had two effects; first they introduced the water vapor that was a by-product of fuel combustion into the stratosphere. Some of this was in the form of reflective contrails, increasing Earth's albedo. But then these contrails also evaporated into water vapor. If you look at Fig. 20.1 you will see that water vapor has three broad absorption/emission bands in the infrared part of the solar spectrum. The vapor intercepts some of the infrared in the stratosphere, so that it does not reach Earth's surface. That caused the temperature decline. During the 1950s through 1970s aerosol pollution from automobiles, factories and mechanized agriculture increased. Smoke was good; it was a sign of an improving economy. Paintings from the USSR from that period often show chimneys belching black smoke as a symbol of industrial growth. It might be called the 'Era of Smog.' We now call this the episode of 'global dimming,' referring to the fact that less solar energy was reaching Earth's surface.

NASA scientist James Hansen has led the way in the US in making global compilations of the temperature records. Two recent NASA-GISS compilations for the global temperature history are shown in Figs. 30.22 and 30.23. The first, Fig. 30.22, uses only data from meteorological stations, mostly on land, but a few from buoyed ships at sea. The averaging is done using the very clever method described above. However, it is important to realize that this record is based on data from only about 29 % of Earth's surface. You can see the episode of global dimming from 1940 to 1980.

**Fig. 30.21** The NASA-GISS version of the mean annual temperatures of the 48 contiguous United States from 1880 to 2015. Trendline and curve as in Fig. 30.5

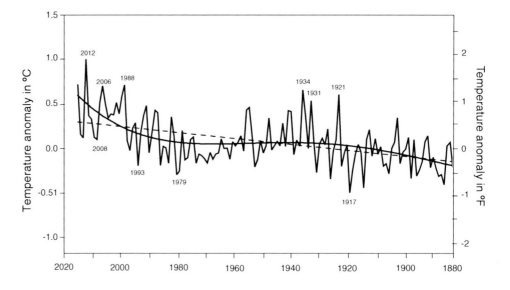

**Fig. 30.22** The NASA-GISS global temperature record from 1880 to 2015 based solely on meteorological stations. Some of the times of extreme temperatures in the US are noted. Linear and fourth order polynomial fits to the data are shown

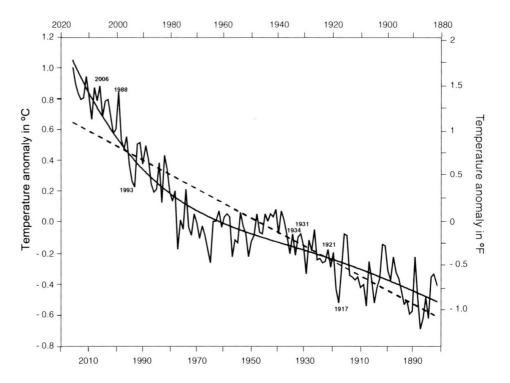

In Fig. 30.22 the years noted for their extreme temperatures in the 48 contiguous United States are indicated. With the exception of the cold year 1917, and the warm years 1998 and 2006 the US extremes are not part of the global pattern.

To make a better global estimate, data taken by from ships at sea and more recently from satellites must be included. When this is done, the global temperature history looks different (Fig. 30.23).

In this record, both the overall trend and the year to year variations are damped, which is what one expects because of the resistance of the ocean to temperature change.

A more revealing analysis of what is happening is evident from the NASA-GISS compilations for specific latitude bands. The Earth can be divided into three major areas:

(1) The region between the Tropic of Cancer and the Tropic of Capricorn at 26° 16′ N and S respectively. For

convenience, NASA-GISS uses 24° N and S as an approximation. This is the warmest part of the Earth which never experiences freezing temperatures except at high altitudes. The total area is 214,732,000 km$^2$, or 42.1 % of Earth's surface. It is 25 % land, 75 % ocean.

(2) The rest of the Northern Hemisphere, from 24° to 90° N. This includes the mid and high latitudes. The total area is 147,628,000 km$^2$ or 28.9 % of the planet's surface. It is 49 % land, 51 % ocean.

(3) The rest of the Southern Hemisphere, from 24° to 90° S. This includes the mid and high latitudes with a total area is 147,654,000 km$^2$ or again 28.9 % of the planet's surface. It is 16 % land and 84 % ocean.

The slight areal differences between these two regions are because the Earth is neither spherical nor spheroidal, but slightly pear shaped. Don't worry about it; you will never notice the difference on an ocean voyage.

The Equatorial-Tropical temperature record of Fig. 30.24 shows some tendency towards a plateau in the mid-century, but it is poorly defined amidst the general variability. The overall and year to year variability is limited because 75 % of this area is ocean.

Because of its overall warm temperatures and great water surface area, the water vapor content of the atmosphere is generally high, and dominates over any effect of other greenhouse gases. Nevertheless, a long-term upward temperature trend is evident. The overall rate of temperature increase is about 0.41 °C (=0.74 °F) per decade.

The area shown in Fig. 30.25 is almost half land, half ocean. It includes most of the industrialized world. It shows the greatest temperature variability, both on a year to year basis and over the long term. Because of its great land area and generally cooler temperatures, there is less water vapor in the air and the greenhouse effect of the increase in atmospheric $CO_2$ is more pronounced than elsewhere on the planet. The decline of the 1950s–1970s is quite evident, though not as extreme as that for the contiguous 48 United States.

A comparison with the same area of the Southern Hemisphere, 24° S to the pole, is shown in Fig. 30.26. This region is 84 % ocean and has virtually no industry. It shows a steady upward temperature trend since 1900. The overall rate of temperature increase is about 0.054 °C (=0.10 °F) per decade.

All of this makes good sense. It is exactly what one would expect if a long-lived greenhouse gas were being added and mixed throughout the atmosphere. The rate of temperature increase is greatest where the water vapor greenhouse effect is minimal. That is where the effect of a well-mixed greenhouse gas would be maximal. The temperatures of the higher latitudes are increasing faster than those of the lower latitudes. The polar to Equator temperature gradients are lessening. The Earth is becoming more 'equable'—that is, the temperature differences are gradually evening out.

A more equable climate, with lesser temperature differences from place to place may sound like a good thing, but

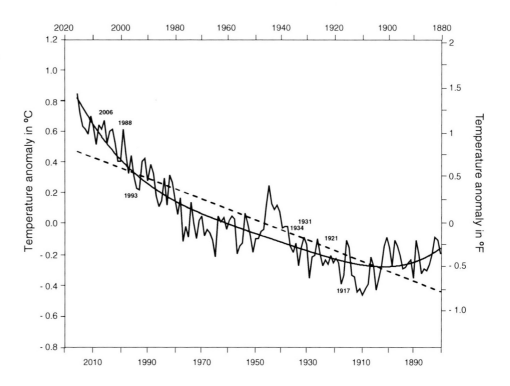

**Fig. 30.23** NASA-GISS global temperature compilation for the tropics using data from both land and oceans from 1880 through 2015. Some of the times of extreme temperatures in the US are noted. Linear and polynomial fits to the data as in Fig. 30.5

**Fig. 30.24** The NASA-GISS
temperature record for the
Equatorial-Tropical region
between 24° N and S from 1880
through 2015. Linear and
polynomial fits to the data as in
Fig. 30.5

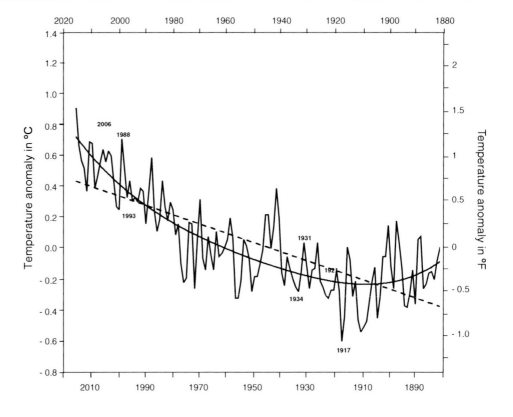

**Fig. 30.25** The NASA-GISS
temperature record for the extra
tropical Northern Hemisphere,
from 24° N to the Pole from 1880
through 2015. Linear and
polynomial fits to the data as in
Fig. 30.5

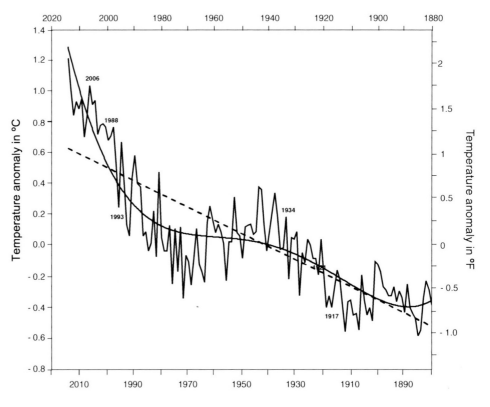

what it means in reality is that the atmosphere will become more unstable as the planet shifts toward a new climatic state. An unstable atmosphere means more extreme weather and unexpected devastating storms.

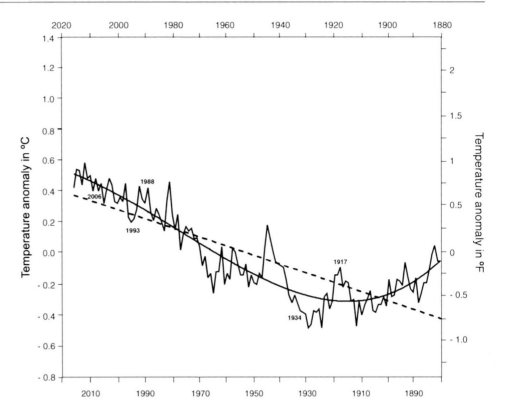

**Fig. 30.26** The NASA-GISS temperature record for the extra tropical Southern Hemisphere, from 24° N to the Pole from 1880 through 2015. Linear and polynomial fits to the data as in Fig. 30.5

Although over the next centuries the Earth will continue to warm, Tom Friedman's 'global weirding' is a better description of how the climate change will be perceived by humans.

## 30.7 Summary

Crude devices to measure temperature were developed in the latter part of the 16th century. In the middle of the 17th century Ferdinando II de Medici invented a sealed glass tube partially filled with a liquid which would expand or contract with temperature changes. Keeping regular records of temperature began in England in 1959. In 1714 Gabriel Fahrenheit improved the thermometer by substituting mercury for alcohol, and established a scale, albeit a rather complicated one. In 1742 Anders Celsius proposed a scale with 100 units between the freezing and boiling points of water. Thermometers made of bimetallic strips were introduced in the 18th century. Although local temperature records were kept at many places, there was little effort to bring them together until 1923, when the International Meteorological Organization initiated the program of data compilation known as the World Weather Records (WWR).

British engineer Guy Callendar was the first to compile a regional temperature record from the WWR reports. It was for Britain, and he noticed that the temperature had been rising rather steadily since 1900. The first attempt at a global temperature record was made by Hurd Willet in 1950, just in time to begin to capture the period of stable or declining temperatures that was to last for 20 years. During that time geologists thought we might be entering another ice age, but some climatologists suspected that any decline in temperature would soon be overwhelmed by the greenhouse effect of burning fossil fuels and releasing their $CO_2$ into the atmosphere. Gilbert Plass, Roger Revelle, and others began to be concerned about the 'great uncontrolled experiment' being performed on our planet.

In the 1970s the use of computers made it feasible to make much more extensive compilations of the temperature records, and institutions in the United States, Britain, and the USSR/Russia have developed their own methods of selecting and averaging the data.

Temperature records for individual stations show considerable variability from year to year. Often a warm year will be followed by a cold year and vice versa. Multi-year trends are relatively uncommon.

In the United States, The National Oceanic and Atmospheric Administration's National Climate Data Center (NCDC) keeps local records and compiles them into state, regional, and national databases. They routinely make compilations for ten regions in the 48 contiguous states. The data for the Atlantic states show the strong influence of the ocean currents on coastal state temperatures. Those for the middle of the country show the greater south-north change and the extreme variability from year to year. Those for the

Pacific states reflect the moderation of the Pacific Ocean and the higher regional elevation. The record for the 48 states as a whole shows a rise from the end of the 19th century, a plateau from about 1945–1975, and then a steep rise toward the present. The plateau, now known as the mid-century episode of global dimming, is thought to have been caused by increased introduction of aerosols into the atmosphere as manufacturing expanded, agriculture became more mechanized, and automobiles became more common.

Global compilations of the data from meteorological stations show the classic upward trend, plateau, and then a steep upward trend through the 20th century. These stations are almost all on land, and the vast majority of them are in the northern hemisphere.

Global compilations that include data from the oceans show a steadier upward trend, but with a more damped signal because of the oceans' resistance to temperature change.

Compilations for the Equatorial-Tropical region and the mid to high latitude regions poleward of the tropics show that the rise-plateau-rise pattern is largely a northern hemisphere phenomenon, due to the introduction of jet aircraft and the economic boom after World War II. The overall rise in temperatures is most prominent in the mid and high latitudes. This is what is to be expected from adding a well-mixed greenhouse gas to the planetary atmosphere dominated in the lower latitudes by a water vapor greenhouse. The Earth's climate is becoming more equable, but this transition introduces instability and implies more local weather extremes and more large storm events.

All is going as is to be expected from the ongoing injection of fossil fuel $CO_2$ into the atmosphere.

A Timeline for this chapter:

| ca. 1592 | Galileo invents the thermoscope |
|---|---|
| 1654 | Ferdinando II de' Medici invented the sealed glass tube thermometer |
| 1659 | Regular record keeping of temperature begins in England |
| 1665 | Christian Huygens proposes the freezing and boiling points of water as references for a temperature scale |
| 1714 | Gabriel Fahrenheit substitutes mercury for alcohol in the sealed glass thermometer |
| 1714 | British Parliament passes the "Longitude Act" offering a prize for discovering a way of accurately determining longitude |
| 1724 | Fahrenheit publishes his temperature scale with three reference points |

(continued)

(continued)

| 1727 | John Harrison takes up the challenge to find a way of determining longitude using new kinds of clocks— sea-going chronometers |
|---|---|
| 1742 | Anders Celsius proposes a temperature scale with 100 units between the freezing and boiling points of water |
| ca. 1750? | John Harrison invents the bimetallic strip to correct his clocks for temperature changes |
| 1767 | David Rittenhouse in the US and Johann Lambert in Switzerland independently invent the bimetallic strip thermometer |
| 1773 | John Harrison is awarded the Longitude Prize |
| 1776 | John Harrison dies at age 83 |
| 1800–1815 | Napoleon spreads the use of the 100 unit thermal scale throughout continental Europe |
| 1848 | Lord Kelvin proposes an 'absolute' temperature scale |
| 1873 | The International Meteorological Organization (IMO) is founded |
| 1880 | Vladimir Köppen makes a compilation of temperature data collected since 1840 |
| ca. 1888 | Standardized temperature measurements become collected at many stations |
| 1896 | Svante Arrhenius publishes *On the Influence of Carbonic Acid in the Air upon the Temperature of the Ground* |
| 1923 | The IMO initiates the program of data compilation known as the World Weather Records (WWR) |
| 1939 | Guy Callendar publishes the first analysis of WWR data, notes the temperature rise and suggests $CO_2$ might be the cause |
| 1947 | The IMO is adopted by the United Nations as a specialized agency, the World Meteorological Organization (WMO) |
| 1950 | Hurd Willett uses WWR data from 129 stations to construct a global temperature record from 1845 to 1940, noting that the data cannot be simply averaged. He devises an areal averaging technique |
| 1951 | United States National Climatic Data Center (NCDC) is founded. Located in Asheville, North Carolina, it is the world's largest active archive of weather data |
| 1956 | Gilbert Plass proposes that mankind is performing an uncontrolled experiment on planet Earth by increasing atmospheric $CO_2$ |
| 1957 | Roger Revelle and Hans Suess suggest that we may learn a lot about climate change from the great uncontrolled experiment. They refer to the increasing level of $CO_2$ in the atmosphere due largely to burning of fossil fuels as the "Callendar effect" |

(continued)

(continued)

| 1961 | Callendar revises his temperature curve using Willett's averaging technique |
| ca. 1945–1975 | Northern Hemisphere temperatures stop increasing, and some areas show a decline |
| 1968 | Al Gore attends Roger Revelle's lectures on climate at Harvard |
| 1970 | The US National Oceanic and Atmospheric Administration (NOAA) is founded |
| 1972 | The Climatic Research Unit (CRU) is established in the School of Environmental Sciences (ENV) at the University of East Anglia in Norwich, England in 1972 |
| 1981 | James Hansen becomes Director of the NASA Goddard Institute for Space Studies in New York City |
| 1990 | The Hadley Centre for Climate Prediction and Research is founded, located with the British Meteorological Office (the 'Met Office') in Exeter, England |
| 2015 | The globally warmest year since instrumental measurements began |

If you want to know more:
W.E. Knowles Middleton, 2002. *A History of the Thermometer and Its Use in Meteorology*. Johns Hopkins University Press, 262 pp.—A fascinating account.

Music: Perhaps after all this, a soothing piece with lots of ups and downs: Brahms' Symphony NO. 4.

Libation: Salute warmer weather with a gin and tonic. I prefer Gordon's gin and Schweppes Tonic Water. Alternatively you might want to try a Pimm's Cup

## Intermezzo XXX. A European Odyssey

Movement toward a European Union Began with the Treaty of Rome in 1957. The European Economic Union of 1958, included Belgium, France, Italy, Luxembourg, the Netherlands, and West Germany. In 1973 Denmark, Ireland, and the United Kingdom joined; Greece joined in 1981, Portugal and Spain in 1986. In 1985, the Schengen (Luxembourg) Agreement paved the way for open borders without passport controls. Making frequent trips to Europe I had followed all this with great interest.

The events of 1989 changed everything. Recognizing that it was a once-in-a-lifetime opportunity to see an epic event in history, I resolved to seek a faculty position there. I wrote letters to many contacts, and applied for more than one faculty position. An old friend, Jörn Thiede, made an immediate offer.

Jörn, who had been born in Berlin, had an illustrious academic career as a marine geologist. He had

gotten his Ph.D. in Kiel under Eugen Seibold, the person I had contacted to jump-start the Ocean Drilling Program when its future was uncertain. Jörn and I had one thing in common, both of us had moved around a lot. Jörn had held academic appointments in Universities in Aarhus (Denmark), Bergen (Norway), Oregon (USA), and Oslo. He had returned to the University of Kiel as Professor in 1982, and in 1987 had founded its new marine Geological Institute, GEOMAR. In 1988 Jörn had received a Leibnitz Prize from the German Science Foundation, an award of several million marks to foster research as he saw fit. He invited me to come to GEOMAR whenever I wanted on a flexible schedule. He also offered to fund Chris Wold's Ph.D. studies in Kiel.

We arrived in Kiel in the summer of 1990 by a circuitous route, first attending a Symposium of the International Association of Mathematical Geology in Güstrow in the DDR. It was organized by Jan Harff. Jan had been lucky enough to be able to spend a year with the Mathematical Geology group at the University of Kansas a couple of years earlier. I had met him there and we had become good friends.

Chris and I flew into Munich June 10, and after visiting friends there, rented a car and drove to Berlin. After the Soviet's 1948–49 Blockade of Berlin, a few Autobahns had been opened for western travel to and from Berlin, but you could not leave the highway. There were very high guard towers along the entire route to ensure that no one did. When we left Berlin for Güstrow, which is about 125 miles to the north, we entered the territory of the DDR. The old border crossing facilities were formidable but the formalities were gone. At the border the roadway spread into many lanes separated by booths for the border guards, and overhead was a forest of lights on tall poles. But the wall had fallen and now everyone was simply waved through. A year earlier we would have spent an hour or so getting through. After the meeting in Güstrow we went on to Kiel where Jörn welcomed us.

October 3, 1990 was 'Reunification Day,' when the border between East and West Germany disappeared. Chris and I wanted to see what it would be like, so we spent the night of October 2 in a hotel in Ratzeburg, just on the west side of the border. The next morning we drove east. A few miles out of town we came to the border crossing site at Mustin. The cleared swath of land on the east side of the old border was all that was left of the complex of fences and guard posts. But there were warning signs that the minefield along the border had not yet been cleared. We drove into Schwerin, and mingled with the crowds of former East

Germans, and curious visitors who had driven in from the west. With reunification, the former East Germany was now part of the European Union.

It didn't take long for the West Germans to discover that East Germany was a massive expanse of deferred maintenance. It was a giant Potemkin Village. Off the main roads many houses were dilapidated and much in need of repair. To bring the east up to something resembling western living standards required an enormous investment from the West Germans. They did it largely by raising the tax on gasoline, and almost no one objected. They realized it would also reduce dependence on Middle Eastern oil. It took more than a decade.

I returned to the University of Colorado for the spring semester of 1991 and returned to GEOMAR the next fall as an Alexander von Humboldt Senior Research Scientist for the academic year 1991–1992. From then until 1998 I would spend most of the year in Europe but the spring semester in Colorado. It was rather like my Illinois-Miami arrangement of the late 1960s early 1970s, but came to involve several other Universities.

It turned out that my part-time move to Europe was very important in enabling me to continue my work toward understanding past 'warm Earth' conditions. The National Science Foundation had decided to emphasize research on the youngest part of geologic history, when the Earth was oscillating between glacial and interglacial states. There were no new funds for this, so it meant that there was much less funding available for research on older 'warm Earth' times. Only about 15 % of the proposals submitted to the NSF receive funding. In Germany, NSF's equivalent, the Deutsche Forschungs Gemeinschaft remained open to all topics, and new initiatives received extra funding. I am sure that Jörn expected me to turn my attention to the younger geologic past, and I always thought I would solve the Cretaceous climate problem with just a few more months work. I am still trying to understand it and all of its ramifications.

Jörn and I shared an interest in building bridges to the East. Jörn set up cooperative programs with scientific institutions in the USSR. I accepted a position as Visiting Professor in the Department of Marine Geology of the Institut für Ostseeforschung (Institute for Baltic Sea Research) for the summer of 1993. It is located in Warnemünde, a resort town on the Baltic coast just north of Rostock. On weekends I got to see a lot of the northern coastal area of the former East Germany. It had suffered very little damage during the

war. The towns were small, but each had a huge church, made of a very solid stone base made of the ice-transported glacial erratics that once littered the landscape. This fortress-like Romanesque base is surmounted by soaring gothic architecture made of bricks All of these were built from the 11th to 13th centuries; large enough to house the entire local population in case of emergencies. The Vikings of northern Germany and Denmark had been (forcibly) converted to Christianity at the end the 10th century, but there were more heathens across the Baltic.

Spending the spring semesters in Boulder, and with GEOMAR as my European base, I spent some fall semesters at other Universities: 1983 at the University of Utrecht's Institute of Earth Sciences, on its new campus at the Uithof, just outside city. It was a wonderful time, but I think I learned more from them than they did from me. It was the year when the geologists of the Dutch Universities were discussing banding together to pool resources. This innovative cooperative venture has become known as The Netherlands School of Sedimentary Geology (NSG) and I was put on its Advisory Board. The NSG revolutionized graduate study in this field. Instead of the usual semester-long courses, the NSG has intensive all-day-long short-courses lasting a few days to a week. In between the short courses the graduate students can pursue their research without interruption. It works very well, and over the years I have been privileged to teach a number of the short courses. The fall of 1995 I was Visiting Professor in the magnificent new building of the Institut für Paläontologie, at the University of Vienna in Austria,

In the fall of 1996 I was Visiting Professor at the Ernst-Moritz-Arndt University, in Greifswald in the northwest corner of the new Germany, just west of the border with Poland. It was founded by Papal Decree in 1456. It is where my 'geologic rates' hero Sergei von Bubnoff taught. It was one of the most exciting groups of students I ever had. In geology much depends on your experience in the field. Seeing and sampling the rocks and looking at the geology of large areas is far more instructive than reading descriptions in the literature. Many of the students had started their studies before the fall of the Berlin Wall. Their field excursions in the days of the DDR had been to places in the east as far away as Central Asia,

In addition to living in Kiel, Utrecht, Warnemünde, Vienna, and Greifswald, I was able to develop close cooperation with colleagues in Slovakia, Hungary, Russia, Switzerland, France, Italy, Great Britain, and finally China. There were annual meetings of

European societies in Strasbourg and Nice, France, and I took my vacations in the surrounding countryside.

As I mentioned earlier, I thought that when I left Washington my role in Scientific Ocean Drilling was over. But in 1984 I was appointed to the JOIDES Advisory Panel on Sediment and Ocean History, serving until 1987. Then in 1982 I was appointed to JOIDES' Advisory Panel on Sedimentary Geochemistry and Physical Processes, serving as Chairman from 1994–97. In 1996 I was asked to Chair a Planning Group on Antarctic Ocean Drilling. My response was "I know nothing about the Antarctic." The reply "That is why we want you to Chair this Committee. You don't have any preconceived notions." Over the next few years I learned a lot about the Antarctic, recognized it had a far more interesting history than the general science community thought, and we were able to work out a plan for adding to knowledge of its history.

In 1998 I was being considered for Department Chair at the University of Colorado. But Jörn Thiede and Helmut Beiersdorf of the German Geological Survey had other plans for me. Helmut was the German representative on the JOIDES Executive Committee, and they wanted me to accept a full-time appointment in GEOMAR and to serve as the German Representative on the JOIDES Scientific Committee (the ODP equivalent of the older JOIDES Planning Committee). The next year, 1999, the Chairmanships of the Executive and Science Committees would rotate to Germany, and it would be a critical time because the ODP was scheduled to end in 2004. Because lead times of 5 years are needed for such complex international projects, planning for a future program of scientific drilling in the ocean would need to start in 1999.

So, in 1998 I retired from the University of Colorado, and moved full time to GEOMAR. Jörn Thiede had just accepted the role of Director of the Alfred-Wegener-Institute in Bremerhaven. The Wegener Institute is devoted to polar research, one of Jörn's first loves. I took over his Professorship in Kiel and finally retired in 2002.

I was very surprised to learn that no planning for the long term future had been started at JOI in Washington. Then in 2000 we learned that someone in the US Government had been making plans. Japan had approached the US with a proposal for a joint Japanese-US Ocean Drilling Program using a huge new vessel the Japanese would construct. The other foreign partners would not be involved. Helmut Beiersdorf and I had heard the news just before

attending a meeting of the European Consortium for Ocean Drilling to be held in Strasbourg, France. We were sitting in the railway car on the way to Strasbourg when we both had the same idea. Europe must go it alone on its own on a new program. When we arrived in Strasbourg we met the French representatives. Before dinner was over we had two European countries on board.

The next few of years involved many complex negotiations. Japan built its drilling vessel, equipped with a riser for return circulation of the drilling fluid, the *Chikyū*. The United States refurbished the *JOIDES Resolution*, but budget cuts have restricted its operations to only part of the year. The European Consortium for Ocean Drilling (ECOD) became the European Consortium for Research Ocean Drilling (ECORD), and uses ships of opportunity to carry out its program. By doing this was it was able to achieve one of our long term goals, drilling in the Arctic Ocean. We now have an idea of the long term (50+ million year) history of the Arctic which is invaluable in understanding what is happening today. These three scientific drilling programs are loosely affiliated as the Integrated Ocean Drilling Program (IODP).

For over 30 years I have been trying to understand how the climate system works on a warm Earth. I thought it was just a matter of a few small differences. But over the years it has become evident that it is not just a minor modification of the present system. Many things are wholly different. And our present perturbation has no analog in the past.

I can only express my gratitude to the many colleagues in these different institutions who have contributed to putting the pieces of the puzzle together.

Since 2003 I have made my home in Colorado, but return to Europe several times a year to teach short courses and attend scientific meetings.

Finally, I am grateful for the recent years I was able to spend in Europe. The German towns and cities in the east had been restored to much of their former beauty. Prague is no longer gray, but a vibrant colorful city. I had the opportunity to work with colleagues in Slovakia and Hungary, and had seen the great transformations there. The Euro had been introduced as a common currency. The borders within the European Union are now only on the maps; no border crossing formalities anymore. High-speed trains on all-new rail lines cut travel times sharply. Great new bridges now connect the Danish Islands and Denmark with Sweden.

The great cultural amalgamation of Europe that many anticipated has not occurred. For a while it was thought that Europe might become homogenized, with a single language and a uniform culture. The opposite has happened. Formerly suppressed languages and dialects are blossoming. One hears Catalan in northern Spain, Provençal in southeastern France, Slovak instead of Czech in Slovakia after the break with the Czech Republic, Ladin in South Tyrol, Romanic in Switzerland, and a wealth of other local dialects. People are proud of their heritage.

COMPUTER MODELS DEVELOP A PLETHORA OF PREDICTIONS
OF THE FUTURE

# The Future

*Making predictions is very difficult, especially about the future.*
Variously attributed to Niels Bohr, Yogi Berra, Mark Twain, Enrico Fermi, Groucho Marx, Albert Einstein, Freeman Dyson, Woody Allen, Confucius, and others

The first person to realize that human activity has the potential to substantially change the world may well have been Thomas Jefferson. The third U.S. president was very much interested in measuring and tracking temperature. He believed that the climate of North America might be modified to resemble that of more temperate Britain by cutting down forests to make way for agriculture. In particular he thought that the unpleasant summer temperatures and humidity of Virginia could be reduced to make for a more livable environment.

Scientific awareness that human activities might affect the whole of planet Earth is only just over a century old. Swedish scientist Svante Arrhenius 1896 article *On the influence of carbonic acid in the air upon the temperature of the ground*, and his 1908 book *Worlds in the Making* suggested that we might expect a gradual warming of our planet with a potential for beneficial results for agriculture at higher latitudes. Arrhenius calculated that a doubling of the $CO_2$ in the atmosphere would raise Earth's average temperature by 5–6 °C. That estimate is still used today.

The only researcher to seriously follow up on Arrhenius' idea was British steam engineer Guy Callendar. He tried to track the atmospheric increase in $CO_2$ and relate it to regional and global temperatures is a series of papers published between 1938 and 1957. His work was largely discounted by the meteorological community because he was, after all, an engineer and not a scientist.

It was almost by accident that Callendar's work eventually received wider attention. In 1955 Canadian physicist Gilbert Plass published an article titled *The carbon dioxide theory of climate change*. He reckoned that a doubling of atmospheric $CO_2$ would result in a 3–4 °C temperature rise, most in the next century.

In 1957 oceanographers Roger Revelle and Hans Suess published their now famous paper with the confusing title *Carbon dioxide exchange between the atmosphere and ocean and the question of an increase of atmospheric $CO_2$ in the past decades*. They named the global warming phenomenon the 'Callendar Effect,' and noted that if we keep careful records we will learn a lot about climate change.

However, as discussed in the last chapter, scientists detected almost no rise in regional or global temperatures from about 1940 onward until 1980. We now refer to that period as an episode of 'global dimming' when atmospheric aerosol concentrations, mostly pollutants, inhibited transmission of solar radiation through the atmosphere thereby lowering insolation at Earth's surface. At the same time some scientists thought that because the insolation for the Northern Hemisphere was at a Milankovitch minimum, Earth might be headed toward another ice age. In the midst of this period of uncertainty, in 1975, Wally Broecker published *Climatic change: Are we on the brink of a pronounced global warming* which included this statement:

> If man-made dust is unimportant as a major cause of climatic change, then a strong case can be made that the present cooling trend will, within a decade or so, give way to a pronounced warming induced by carbon dioxide. By analogy with similar events in the past, the natural climatic cooling which, since 1940, has more than compensated for the carbon dioxide effect, will soon bottom out. Once this happens, the exponential rise in the atmospheric carbon dioxide content will tend to become a significant factor and by early in the next century will have driven the mean planetary temperature beyond the limits experienced during the last 1000 years.

In Chap. 1 I noted that it was the Arctic sea ice meltback of 2007 that grabbed my attention and made me realize that the projected climate change expected from rising $CO_2$ levels was happening far ahead of schedule. I had devoted 40 years to working on understanding the warm climates that dominated 80 % of Earth's history, and it took me 4 years to write the first edition of this book. But as soon as it was published it was out of date. A plethora of positive feedbacks have appeared to accelerate the effects of increasing atmospheric $CO_2$ concentrations as have the refusals of the governments of the United States, India and (until recently) China to participate in the international efforts to reduce the rates of $CO_2$ and other greenhouse gas

© Springer International Publishing Switzerland 2016
W.W. Hay, *Experimenting on a Small Planet*, DOI 10.1007/978-3-319-27404-1_31

emissions. U.S. emissions did decline somewhat after 2008, but not because of any plan. Rather, the emissions decline was simply one of the effects of the general economic decline of the Great Recession.

Although the increase in atmospheric $CO_2$ levels has been the focus of much media attention, it is only one of a number of anthropogenic activities that are changing the natural environments of planet Earth. I now identify eight human-driven factors that are forcing climatic and environmental change: (1) increased concentrations of atmospheric greenhouse gases; (2) introduction of ozone-destroying compounds into the atmosphere; (3) introduction of aerosols into the atmosphere; (4) replacement of $C_3$ plants by $C_4$ plants as tropical forests are cut and burned to make way for pastureland; (5) development of large areas of plant monocultures as a way to increase agricultural production (6) widespread construction and urban sprawl; (7) overuse of fertilizers and pesticides in agriculture, and (8) mining of minerals for fertilizers and metals.

All of these activities are, of course, closely related to the root cause of the planet's major environmental problems: the growth of the human population of the planet. Of these eight factors, only the first has been seriously addressed in attempts to model future climate change. Greenhouse gas concentrations have a global effect. The other factors have regional effects that may have global implications.

## 31.1  Predictions of the Future—Models

Predictions are made with 'models.' They can be conceptual models that simply involve ideas, like those of Malthus concerning the growth of population to a level where starvation becomes the limiting factor. They can also be simple mathematical models with a few rules, like Verhulst's logistic equation for the growth of population to a stable level where birth and death rates are equal (Sect. 3.13; Fig. 3.16). Or they can be more complicated mathematical models designed to test the differences in outcomes caused by differences in inputs or assumptions, such as those shown in Figs. 11.7 and 11.8.

The invention of the computer opened new vistas in modeling. When I joined the faculty of the University of Illinois in 1960, the school was a world leader in computer development with its ILLIAC (Illinois Automatic Computer). ILLIAC could do some of the things you can do with your iPhone today, but much more slowly. Information was fed to the ILLIAC by punch cards (cards with perforations representing instructional codes)—a bit of technology so out-of-date that even many older readers may not remember them.

It was in 1963 that I tried to use ILLAC for a really original type of task: to see if the machine could classify some of the tiny fossils we used in micropaleontology, planktonic foraminifera, into species. Hanspeter Luterbacher, one of my colleagues during my post-doctoral year at the University of Basel in Switzerland, offered to collect and supply the data. He measured a number of different parameters, including width, height, number of chambers, presence or absence of a keel, etc. of 1,000 specimens of the foraminifera and sent the data to me. Figure 31.1 shows the shell of one of the types of foraminifera he measured for me. I hired an undergraduate student to transfer the data onto punch-cards and then fed the information into ILLIAC. I got the enormous paper output of the computer's attempt at species classification the day before I was to fly to Europe for field work that summer. I stuffed the output into my hand luggage. After we were on our way from Chicago to Paris and

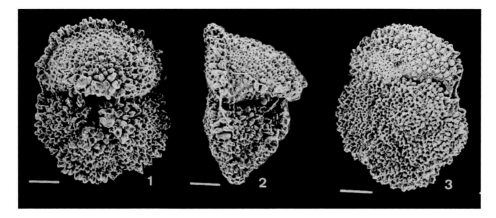

**Fig. 31.1** Different perspectives of a shell ('test') of a type of Paleocene planktonic foraminifer measured by Hanspeter Luterbacher in the early 1960s. This species was known then as *Globorotalia aequa* but is now called *Morozovella aequa*. Figure from the Smithsonian Institution's Atlas of Planktonic Foraminifera. *Lines* represent 1 μm

I had ordered a gin and tonic I pulled out the output and started to look it over. It took only a few minutes to realize that the computer had chosen to divide the data into three categories ('species')—merely according to size. ILLIAC's classification had no relation to the species recognized by micropaleontologists. When I showed the results to Hanspeter he was not surprised and he noted that a potential technological threat to human micropaleontologists was gone. The idea to let computers do such species classification even with automated observations has been pursued by a number of micropaleontologists but so far without much success (although we now have 'face recognition' technology for police work). The human brain is able to handle some very complex problems with great efficiency and useful insights unmatched by computers. That first experience has always made me skeptical of the results of computer modeling.

Model projections of the size of the human population are relatively simple. They depend on birth rates and death rates, which can be divided into mortality rates for infants and older people. Changing these rates over time results in different projections for future population. Such projections do not include the possibilities of major reductions in population size, such as from nuclear war or epidemics.

My first experience with climate models came in 1978 when we sent one of my graduate students at the Rosenstiel School in Miami, Eric Barron, to the National Center for Atmospheric Research (NCAR) in Boulder, Colorado to work with atmospheric scientist Warren Washington, one of the pioneers in developing computer models for climate studies. NCAR is a central computing facility for a number of US academic institutions with programs in Atmospheric Science. Eric and Warren used NCAR's climate model to simulate the warm Earth of the mid-Cretaceous, 90 million years ago. There were many features of the Cretaceous Earth that were very different from today; different positions of the continent, different distribution of mountains, a sea level so high that about one third of the continental blocks were flooded, and a higher concentration of greenhouse gases. The models indicated that the greenhouse effect was decisive for explaining the warm climate. Eric's doctoral thesis work became a classic series of papers.

Computer models of climate are a special category of models that involve much complexity. Some geologists have used conceptual models to try to understand Earth's past climates, but these are at best educated guesses, Mathematical models of climate started out with the basic equations governing Earth's energy balance, involving insolation and albedo with the planet considered as a single point. However both of these factors vary from region to region, requiring two dimensional modeling. Incorporating atmospheric complexity adds a third dimension. The greenhouse effect adds further complexity and, except for $CO_2$, the different greenhouse gases are not evenly distributed on a global scale. Still

other factors are land use changes by humans, which have both global and regional effects. The amount of computer time required for a model increases with the number of factors considered, with the three dimensional spatial resolution, and with the number of iterations per model day.

In 1981 I moved to the University of Colorado in Boulder, to serve as Director of the University Museum. There I was able to get much more directly involved in the advanced modeling work being conducted at NCAR. Eric Barron was then a postdoctoral researcher, and Stephen Schneider and Starley Thompson were scientific investigators interested in Earth's past. Stephen was especially interested in ancient climates, and as mentioned in Chap. 14 I used his book *The Coevolution of Climate and Life* in a course I taught. Starley was interested in, among other things, exploring factors influencing Earth's radiation balance. One day we were comparing the Cretaceous climate with the modern climate in the simulations made by NCAR's sophisticated climate model, CCMI (Community Climate Model I). CCMI was, as you might guess from its name, a modeling program to which many scientists had contributed parts. One of my colleagues at Miami had told me that some parts were as refined as a Mercedes-Benz, others more like a pogo-stick. In looking at the simulation of today's climate I noticed that the warmest temperatures were along the Equator when they should have been in the next set of grid cells, centered on 5° N, because the uneven distribution of land and sea forces the modern heat equator to be north of the geographic Equator. Something was wrong but it was not easy to figure out where the problem lay. CCM1 was many thousands of line of computer programming long and no one knew the entire thing.

After some days of discussion we decided to try something entirely new. I wrote a proposal to the National Science Foundation to request funds for a very bright young climate model developer, David Pollard, to write a climate modeling program specifically for simulating ancient climates. He would be the sole programmer, writing it from start to finish. The funds were granted, and the result was a new model, called GENESIS. The project took a number of years to complete and the formal release of GENESIS came in 1991. This model has undergone many subsequent modifications and improvements, and it is still widely used today. I myself have never dared to write a line of program for climate modeling; I have simply been an enabler and a user.

Atmospheric models often use a 'slab ocean' which—in its simplest form is only a motionless surface layer of water. However, there are also some ocean models driven by the atmospheric circulation. Atmosphere pressure systems are hundreds of kilometers across, but the mesoscale eddies of the ocean are much smaller and modeling them requires a higher spatial resolution.

Accounting for changes of vegetation on land in response to climate change further adds to the complexity, and a number of experts have devised modeling schemes ranging from simple to highly complex. The usual method is to first let the climate model come to equilibrium, and then run the vegetation model under that climate. The vegetation will change, and the new vegetation becomes a new 'initial condition' for the climate model. The climate and vegetation models work back and forth with each other in this way until stability is achieved.

To set up a climate model, the distribution of land, with its topography and vegetation cover, and the sea, with either simplified or more realistic bathymetry, provide the 'initial conditions' at whatever spatial resolutions are used. Insolation, albedo, and greenhouse gas concentrations are the 'forcing factors.' Each addition of a factor or increase in spatial resolution causes a major increase in the computing time required. It is no wonder that many of the world's largest computers are devoted to modeling climates. Every increase in complexity or resolution adds to the computing time in an exponential manner. Thus, as we strive to become more realistic and accurate in our modeling and predictions, we need larger supercomputers.

After each addition of a complex variable the models are given test runs to determine how well they simulate the modern weather and climate. Model runs are divided into two parts. First the 'initial conditions' for a run are specified for every parameter at every location. Then, the model run is started with a 'spin-up,' typically 30 years of model simulations. At the end of the spin-up the model is expected to settle into the pattern of random variability we would call the weather, rather than drifting toward another set of conditions. Once the model seems stable it is run for another 30 years of simulations of the weather; these results are then averaged as the climate. The most sophisticated models allow for the forcing factors to change over time, resulting in models of climate evolution over time.

There are many other factors that affect regional climates, such as fires, industrial aerosol pollution, and changes in an area's wildlife composition. Climate models are becoming increasingly complex, requiring ever larger computing capabilities as we strive for better predictions. There is, of course, a trade-off in this process. Given that the computing time is limited we have a choice. We can run either a few high resolution, highly sophisticated models, or a larger number of lower resolution, less complex models. The number of climate models being used throughout the world today probably exceeds 100—all programmed differently, with different initial conditions and forcing factors, and all producing somewhat different results for a given set of input values. Here it is useful to recall the discussion in Chap. 14

about the mathematical chaos inherent in many of the basic equations describing atmospheric and ocean dynamics. Many studies of future climate now incorporate averages of different models, as 'ensemble results.'

## 31.2   What Is Expected in the 21st Century—Human Population

The stress on Earth's environments is being caused by the very rapid expansion of a single species, *Homo sapiens,* whose requirements are altering the planet's land surface, atmosphere and oceans. We are well along in the process of 'anthropoforming' Earth. Prior to the Industrial Revolution the total human population was probably slightly more than 1 billion people (see Figs. 3.17 and 3.18), and this number is believed to have increased very slowly and gradually during the 19th century and first part of the 20th century. After the Second World War and the formation of the United Nations more precise population numbers became available as census-taking became a truly global activity. The post-war statistics revealed that the world population began a sharp increase, as shown in Fig. 31.2. This increase was probably due to the introduction of antibiotics such as penicillin, resulting in fewer deaths from disease, and of nitrate fertilizers resulting in an increased food supply.

As noted previously, the population growth rate depends on two factors—the birth rate and the death rate. Before instituting a one-child per family law in 1980 China's birth rate was about 4.2 children per woman. The birth control law was instituted partly in response to the Great Famine of 1958–1961, which cost the lives of 15–45 million people that the nation was unable to feed. By 2012 China's birth rate had dropped to 1.66 children per woman. In 2014 it was so low that the Chinese Government was relaxing the law. In the United States the birth rate was about 1.9 children per woman in 2014. European birth rates are generally about 1.2. People have the longest lifespans in Europe where average longevity reaches the 80s. Average longevity in the United States is 79 and in China 74. By contrast, in much of sub-Saharan Africa the average life span is less than 50 years.

Why do people have any given number of children? One widely accepted idea is that in earlier times people would have a large number of children because many of them would die young, and it was essential to have enough survive to support their parents in old age. After WWII a substantial decline in infant mortality occurred. As people realized that most of their children would now live to maturity, large family sizes became a liability, so birth rates declined. In mainland Europe and Russia, almost all counties

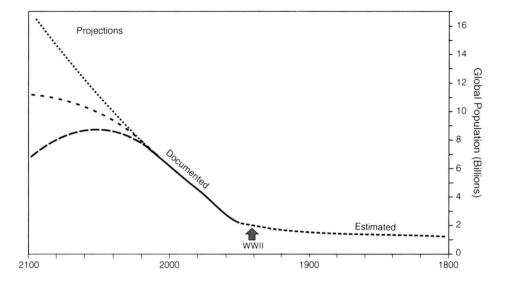

**Fig. 31.2** The United Nations 2015 account of past and projected growth of Earth's human population. Projections are determined solely by changes in birth and death rates. The three shown projections are termed high, medium (most likely), and low

now have negative population growth rates of about—0.5 % per year, even though people are living longer lives. The high population growth trends in Latin America began to change in the late 20th century and growth rates there are now on the order of 1.2 % per year (representing a doubling time of about 60 years). Many other counties, particularly in the Middle East and Africa continue to have high population growth rates of 2–4 % per year (doubling times of 36–18 years). Those high rates will likely continue to destabilize the social and economic structures of those regions for years to come.

Figure 31.2 shows the growth of the human population since 1800, along with three UN projections for the future to 2100. The high projection, reaching a global population of 16 billion in 2100 is simply a projection of the current growth rate. The middle projection, reaching 11 billion by the end of the century, assumes a gradual slowing of the growth rate. The low projection, reaching a high of 9 billion in the middle of the century and a decline back to 6 billion by the end of the century assumes that the need for smaller family sizes will become apparent everywhere. The middle projection is generally considered most likely.

These projections raise the question: what is the human carrying capacity of planet Earth? 'Carrying capacity' is a term used in ecological studies to define the maximum number of individuals of a species that can be supported indefinitely in a given environment. Although it may be relatively easy to evaluate the factors controlling most animal populations, such is not the case for humans. The human population depends not only on natural environmental constraints but also on economics, cultural values, and politics. Various authors have estimated the carrying capacity to be from 1 billion to 1 trillion people.

What exactly limits populations? In 1925 geologist/polymath Albrecht Penck formulated the 'Law of the Minimum' to describe how populations are limited by whatever needed resource is in least supply. The primary factors limiting the human population are now thought to be food and water. Food was once thought to be the most important limiting factor. However that idea did not take into account the widespread mechanization of farming, use of chemical fertilizers, and conventional or genetic improvement of crops to produce greater yields. In recent years we have come to realize that water supplies are finite and may be more important than food as a limiting factor. Excluded from these calculations has been the quality of life, which can be thought of as the share of resources available to each person. Is it 'better' to have more people living at a subsistence level or fewer people enjoying a more advanced lifestyle? Also excluded have been the effects of anthropogenic climate and environmental change.

There is a trade-off between population and life-style. My own guess is that a global population between 1 billion and 2 billion would allow for very pleasant lifestyles for most people. That guess is based on nothing more than my personal experiences and observations when I was young. But I was backed up in my 'guesstimate' by a taxi driver in Munich in 2014. He was driving me to the Nationaltheater where the Bavarian State Opera was going to perform one of my favorites, Puccini's *La Bohême*. We got stuck in traffic and the driver, who was from Afghanistan and studying at the Technical University, volunteered "I think when the oil runs out the world's population will go back to 1–2 billion." Despite the traffic we made it to the opera with time to spare. Later, in reviewing the literature on this subject I found that ecologist and agricultural scientist David Pimentel and colleagues published an excellent paper in 1994 titled *Natural resources and an optimum human population* (Population and Environment, v. 15, pp. 347–369), Although published only 20 years ago, their assessment of availability of natural resources is already out of date, but their conclusion is not: an optimal human population would be 1 billion–2 billion. So there you have it: three wholly independent sources using completely different criteria arriving at the same answer.

The problem is quite simply that we have already exceeded Earth's carrying capacity for humans. As Cesare Emiliani pointed out toward the end of his 1992 book *Planet Earth*, there are just too many people.

The United Nations' low population projection assumes a gradual reduction of birthrates leading to a slow decline of the global population after 2040. Although such reductions have happened regionally, they has never before happened on a global scale. History tells us that there are only three ways the human population can be drastically reduced: war, famine, and disease.

Ordinary wars do not affect the global population, but thermonuclear war might. There are some religious zealots who think we need a thermonuclear war to 'fulfill Biblical prophecy.' In the 1980s television evangelist Pat Robertson tried unsuccessfully to convince President Ronald Reagan that it was time for Armageddon.

Famine is one of Mother Nature's ways of making population corrections. However, it is mostly regional in effect. Although there are isolated areas of famine today, the local population loss scarcely slows global population growth. However, shortages of food brought on by catastrophic weather events could have larger impacts on population in the future.

Disease is a tried and tested way to reduce global population. The Black Plague of the 13th century probably reduced the Eurasian population by half. The introduction of smallpox and other European diseases into the Americas was even more effective, with estimates as high as 80–90 % of the native population lost in some regions. I expect that the development of a highly contagious deadly new virus will eventually make a major population correction. And it is not just the direct effect of the disease itself that will be important. Rather, the general breakdown of society and the loss of transportation and food supplies will likely magnify the effect of the epidemic.

When might we see such a devastating epidemic? I have no idea, but clearly we are set up for it to have maximum effect. With modern air travel, it could make its way around the world in days. I don't think the world will reach a population of 9 billion in 2050 as is forecast in some projections. I think Mother Nature will make her move before then.

In 1826 Mary Shelley, who had written *Frankenstein* during the cold summer of 1816 when the ash and gases from the volcanic eruption of Mount Tambora in Indonesia blanketed the Earth, wrote *The Last Man*. It is one of the first science fiction novels about the future, describing the human population of the Earth being decimated to extinction by a plague that starts in an unknown place in the latter part of the

21st century. By 2092 the plague has spread across Europe and the Americas. The book ends with the last living man setting to sea in a boat in 2100. *The Last Man* received bad reviews when published. After all, such events were unthinkable and impossible! The book was not republished until the 1960s when it began to seem more relevant and plausible. Today's global population is far larger and the means of spreading disease are far more efficient than what Mary Shelley had imagined. It's a great novel that you really should put on you reading list.

What might happen after a future population correction? Al Bartlett, a Professor of Physics at the University of Colorado at Boulder, was very much concerned about human population and the future. He often said that a lot depends on whether humans ever come to understand the implications of exponential growth. Al passed away in 2013 at age 90, a great loss.

## 31.3  Environmental Variability and Tipping Points

The range of variability is an important descriptor of any system. It is especially important in any discussion of environmental change. The classic expression 'The climate is what you expect—the weather is what you get' is an excellent example. The 'climate' for any given place can be described by a 30-year average of temperature and rainfall amounts, along with an indication of their variability, such as a statistical measure of standard deviation from the average. Such a description makes the assumption that there is no discernible long-term trend to the climate. Most TV meteorologists learned in college that the weather is variation about a stable mean value, and as a result many of them are skeptics of climate change. Accepting the idea of a gradual shift in the average over time or of an increase in the variability makes it more difficult to predict the future.

Figure 31.3 is a simplification of the system governing the Earth's four major climate states. The two variable forcing factors shown are those that have governed Earth's climate for roughly the past 100 million years: changing insolation at high northern latitudes and atmospheric greenhouse gas ($CO_2$) concentrations.

You can think of the climate system as being like a board with four depressions that can be tilted back and forth and to the right or left and a ball that can roll about, as shown in Fig. 31.3. If the board is tilted to the right, representing Milankovitch orbital fluctuations affecting insolation, the ball can land in one of the right-hand depressions, such as the interglacial where it is today. If the board is tilted to the left the ball can land in one of the left-hand depressions such as the glacial where it was 25,000 years ago. The board can also tilt front to back. But if greenhouse gas concentrations are

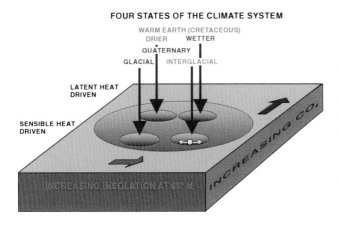

**Fig. 31.3** Earth's four most common climate states and the two factors controlling them. For discussion see text

low the ball can only oscillate between the glacial-interglacial depressions. If greenhouse gas concentrations increase beyond where they have been for the past few million years (below 300 ppmv) the ball can roll backward and drop into one of the greenhouse Earth depressions.

The movements of the ball represent the variability inherent in the weather and climate system. If the variability becomes extreme the ball might roll out of the depression it is in and drop into another. This change represents the climate shifting from one state to another. Increasing variability in the weather/climate system is reflected by extremes of temperature, precipitation or other factors.

In the Annual Meeting of the American Association for the Advancement of Science in Washington DC in 1982 German climatologist Herman Flohn made the prophetic comment:

> Long before global warming due to $CO_2$ can be proven, its effect will be noticed by extremes of weather, setting local records such as the highest and lowest temperatures, greatest and least rain and snowfall amounts, most powerful storms, unusually large numbers of tornados. These indicate that the climate system is becoming unstable in response to the greenhouse effect.

At a recent meeting called to discuss ways of communicating ideas on climate science to the general public I found that many climate scientists were opposed to telling the public that increasing numbers of extreme climate events are evidence of changing climate. That is because there is inherent instability and variability in the climate system and thus we might just seeing a little more variability than in the past. But I also discovered that the idea of Earth having a climate state different from that of today—like the warm Cretaceous that has been Earth's preferred state for 80 % of the last half-billion years of its history—was unknown to my colleagues. Hermann Flohn's idea that unusual extreme weather might be our planets way of signaling a shift from one climate state to another was a new concept to them.

One of the most popular ways of identifying the dangers of climate change is in terms of 'tipping points.' Everyone has an intuitive feeling of what a tipping point is—when you go past a tipping point things change. It is like going over the rim of one of the four small depressions in Fig. 31.3. However, it is difficult to define exactly what a tipping point is in the climate system or to predict exactly what change would occur after passing it. An article published in 2008 in the Proceedings of the National Academy of Sciences (PNAS) reported on a conference held in Berlin in 2007. There an eminent group of climate scientists agreed on a mathematical definition of the term 'tipping point.' They listed a number of elements of climate change that are sensitive enough that they might 'tip' to a qualitatively different state within the next century. The mathematical formulation is a bit technical and I will not repeat it here; it is openly available at the PNAS website by searching for volume 105, no. 6, p. 1786 (Lenton et al.). Supplementary material is available at the same website. It is important to realize that some terms in the equations used to define 'tipping points' are not fully understood.

## 31.4   The Anthropogenic Factors Causing the Earth's Climate and Environment to Change

The eight previously mentioned human-driven factors that are forcing climatic and environmental change listed previously can be divided into those with global or regional effects:

Global Effects:

(1) Increased concentrations of atmospheric greenhouse gases: including $H_2O$, $CO_2$, $NO_3$, $CH_4$, and some synthetic gases. Of these $H_2O$ vapor is the most powerful, but its concentration depends very largely on temperature although water surface areas available for evaporation and the nature of vegetation can be regionally important.

(2) Introduction of ozone-destroying compounds into the atmosphere. The phase-out of chlorofluorocarbons is well under way, but fertilizer-use related levels of $NO_3$ are increasing.

Regional Effects:

(3) Introduction of pollutant aerosols into the atmosphere;

(4) Replacement of $C_3$ by $C_4$ plants as tropical forests are cut and burned to make way for pastureland. This vegetation change alters the evaporation/precipitation balance as well as albedo.

(5) Development of large areas of plant monocultures to increase agricultural production, often with irrigation by groundwater. These disturb the overall water balance as well as the seasonal albedo variations.

(6) Construction and urban sprawl affecting both albedo and water balance.

(7) Overuse of fertilizers and pesticides in agriculture upsetting the ecological balances in lakes and coastal ocean area.

(8) Mining of minerals for fertilizers and metals introducing life-threatening materials into the environment.

All of these topics are important. A number of them have been discussed in the reports of the Club of Rome. However, because of their global effects greenhouse gas emissions have received the most attention and are the topic given the most coverage in the reports of the Intergovernmental Panel on Climate Change (IPCC).

## 31.5   The Reports of the Club of Rome

In April 1968, a small international group of intellectuals representing the fields of diplomacy, industry, academia and civil society met at a villa in Rome. They had been invited by Italian industrialist Aurelio Peccei and Scottish scientist Alexander King to discuss the dilemma of prevailing short-term thinking in international affairs amid concerns regarding unlimited resource consumption in an increasingly interdependent world. Out of this meeting The Club of Rome was born. Additional like-minded individuals soon joined the group.

In 1972 the group issued the first Report of the Club of Rome, titled *The Limits to Growth*, prepared by a group of systems scientists from the Massachusetts Institute of Technology. That report described computer modeling of unchecked economic and population growth with finite resource supplies and explored a number of scenarios concerning the choices open to society to reconcile sustainable progress within environmental constraints. The report's findings are diagrammed in Fig. 31.4.

*The Limits to Growth* was the first documented warning that there was trouble ahead and that the 21st century would be a crucial time in human history. When the report was released it was thought that resource depletion would be the main critical factor affecting mankind. You can see from Fig. 31.4 that the contributors to *Limits to Growth* expected the first half of the 21st century to be a time of major change.

In 2002 the Club of Rome issued a follow-up second report titled *Limits to Growth—The 30-Year Update*. It recognized the complexity of the problems facing humankind, although it did not treat the threats posed by climate change in depth. It describes 11 different scenarios for the future, two of which are depicted in Figs. 31.5 and 31.6

*Limits to Growth—The 30-Year Update* discusses two important aspects of human life: 'Human welfare' and the 'Human ecological footprint.' Human welfare is defined as "the quality of life of the average global citizen in its broadest sense, including both material and immaterial components." The human ecological footprint is defined as "the total environmental impact placed on the global resource base end ecosystem by humanity." A comparison of these factors for scenarios 1 and 9 is shown n Fig. 31.7.

The Club of Rome's studies do not discuss or take into account climate change in detail. I had the opportunity to meet some of the Club of Rome's contributors in November, 2012 at a conference organized by the Accademia Lincei in Rome titled *Anthropogene—Natural and Anthropogenic Modifications of the Fragile Equilibrium of the Earth*. After my talk on the acceleration of the rate of climate change one contributor remarked—"It's worse than we thought."

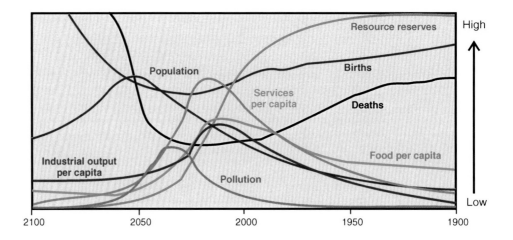

**Fig. 31.4** The Club of Rome's 1972 predictions of changes in different factors affecting the human population of Earth

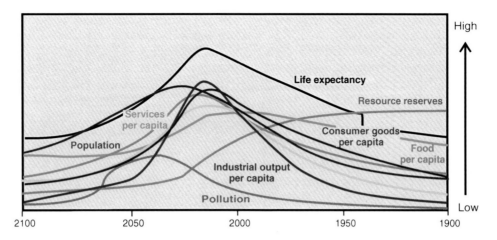

**Fig. 31.5**  Scenario 1 from *Limits to Growth—The 30-Year Update*. This scenario assumes a continuation of 'business as usual' as it was during the 20th century. A crash in population and living standards occurs towards the middle of the 21st century

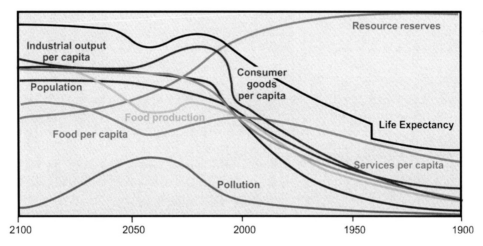

**Fig. 31.6**  Scenario 9 from *Limits to Growth—The 30-Year Update*. In this optimistic scenario the world seeks a stable population and stable industrial output per person, taking into account the further development of pollution, resource, and agricultural technologies from 2002. In a sustainable society, according to this scenario nearly 8 billion people can have a high quality of life

**Fig. 31.7**  Comparison of the human welfare (*solid lines*) and ecological footprints (*dashed lines*) for *Limits to Growth—The 30-Year Update*. The Club of Rome's scenario 1 is *red* and scenario 9 is *green*

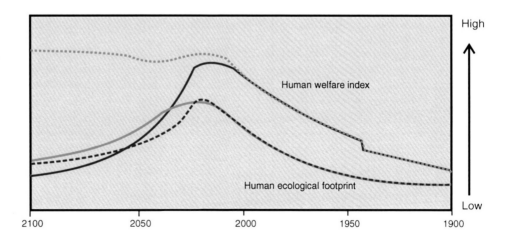

## 31.6    The Intergovernmental Panel on Climate Change—The 'Official' Scenarios for Future Increases in Greenhouse Gas Concentrations

The Intergovernmental Panel on Climate Change was created in 1988. It was set up by the World Meteorological Organization (WMO) and the United Nations Environment Program (UNEP) to prepare assessments on all aspects of climate change and its impacts, based on available scientific information and with a view of formulating realistic response strategies.

Today the IPCC's role is "…to assess on a comprehensive, objective, open and transparent basis the scientific, technical and socio-economic information relevant to understanding the scientific basis of risk of human-induced climate change, its potential impacts and options for adaptation and mitigation. IPCC reports should be neutral with respect to policy, although they may need to deal objectively with scientific, technical and socio-economic factors relevant to the application of particular policies."

In the late 20th century a special subcommittee of the IPCC, Working Group III, defined a series of scenarios of increasing greenhouse gas concentrations to facilitate discussions of the effects. These were described in the IPCC's *Special Report on Emissions Scenarios* (SRES) published in 2000. The scenarios, with labels stating with A or B, were based on estimates of carbon emissions that would be produced by populations varying in size and economic condition as well as in political decisions and policies.

In the IPCC's Fifth Assessment Report (AR5) published in 2014 the SRES scenarios from 2000 have been replaced by four greenhouse gas concentration (not emission) trajectories termed Representative Concentration Pathways (RCPs). However the 2000 scenarios are still in common use in discussions of the future so a brief account of them is in order here.

The 2000 report defined two major groups of scenarios, A and B, each with several variants described in Table 31.1 and Figs. 31.8 and 31.9. The A scenarios assume a world with rapid traditional economic growth and less concern for environmental consequences. The B scenarios are of a more ecologically sensitive world. The A1 scenarios are of a more integrated world characterized by: rapid economic growth; a global population reaching 9 billion in 2050 and then gradually declining; quick spread of new and efficient technologies; and world-wide convergence of income and ways of life with extensive social and cultural interactions. The A1 scenarios are divided into sub-scenarios based on technological emphasis: A1FI has an emphasis on fossil-fuels; A1B has a balanced emphasis on all energy sources; and A1T has an emphasis on non-fossil energy sources. A1B is commonly used for future projections although the largest player in the game, the United States, is still following the A1F1 scenario.

The A2 scenarios assume a more divided world characterized by independently operating, self-reliant nations, continuously increasing populations, regionally oriented economic development, and slower, more fragmented technological changes and improvements to per capita income. These scenarios are also commonly used for future projections to 2100, beyond which a human population (over 30 billion in 2200) and energy consumption levels simply become unreal.

The B1 scenarios are of a world more integrated, and more ecologically friendly characterized by rapid economic growth (as in A1), but with rapid changes towards a service and information economy; a global population rising to 9 billion in 2050 and then declining as in A1; reductions in the intensity of use of materials and the introduction of clean, resource efficient technologies: and emphasis on global solutions to economic, social and environmental stability. B1 is also taken as a frequently used basis for modeling future climate change.

The B2 scenarios are of a more divided but more ecologically friendly world, characterized by: continuously increasing population, but at a slower rate than in A2; emphasis on local rather than global solutions to economic, social and environmental stability; intermediate levels of economic development; and slower, more fragmented technological change than in B1 or A1.

Projections for some of the 2000 scenarios are shown in Table 31.1

Figure 31.8 shows $CO_2$ emissions scenarios; Emissions are from all sources, although they are strongly dominated by burning fossil fuels. There is however, a problem is determining current and past emissions because data from some countries are not necessarily systematic and comparable and data from other countries are simply not reported.

Another problem with $CO_2$ emissions scenarios is that there is no direct relationship between emissions and atmospheric $CO_2$ concentrations. Some of the emissions are consumed in the regrowth of forests, some is absorbed by the ocean and other sinks. Furthermore, there is no direct link between emissions and absorption rates, rather a complex suite of interactions occurs as discussed in Chap. 17. There were several additional problems with the scenario development taken by the IPCC in 2000. The projections of both population growth and economic conditions had become out of date by 2010. Trying to figure out what political decisions and policies might be made and when they might be enacted turned out to be a hopeless task. Even the emissions scenarios themselves soon became out of date for reasons not even considered in 2000, such as Europe's success

**Table 31.1** Some scenarios of the IPCC's Working Group III for increasing greenhouse gas concentrations developed in the late 1990s and published in 2000 as the *Special Report on Emissions Scenarios*

| Year | If CO$_2$ emissions had stabilized at 2000 levels' | | | Scenario B1 | | Scenario A1B | | Scenario A2 | |
|---|---|---|---|---|---|---|---|---|---|
| | 2000 | 2050 | 2100 | 2050 | 2100 | 2050 | 2100 | 2050 | 2100 |
| Population (billions) | 6.3 | ? | ? | 8.7 | 7.0 | 8.7 | 7.1 | 11.3 | 15.1 |
| World Gross World Product (GWP) Trillions of $ | 41 | ? | ? | 136 | 328 | 181 | 529 | 82 | 243 |
| CO$_2$ concentration (ppmv) | 370 | 370 | 370 | 480 | 540 | 640 | 690 | 650 | 840 |
| Temperature increase (°C) | *0.6–0.7* | 0.3–0.6 | 0.4–0.6 | 0.7–1.2 | 1.1–1.5 | 1.2–1.9 | 1.9–2.6 | 1.1–1.8 | 2.2–3.5 |
| Temperature increase (°F) | *1.1–1.2* | 0.5–1.1 | 0.7–1.1 | 1.3–2.2 | 2.0–2.7 | 2.2–3.4 | 3.4–4.7 | 2.0–3.2 | 4.0–6.3 |
| Thermal expansion of seawater (cm) | 0 | 5–7 | 8–10 | 9–12 | 14–18 | 9–18 | 18–26 | 10–18 | 14–30 |

*Note* '?' symbol signifies that no data/estimates are available

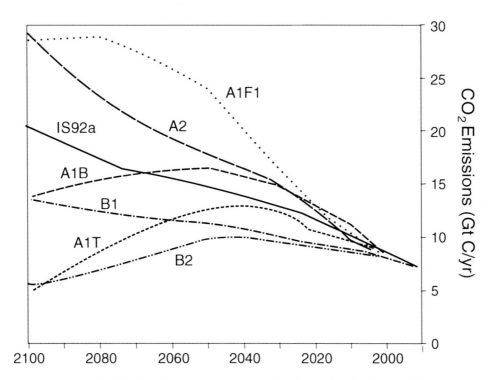

**Fig. 31.8** IPCC CO$_2$ Emissions scenarios. The *A* and *B* scenarios are discussed in the text. The IS92a scenario is from an IPCC 1990 report and assumes a global population of 11.3 billion in 2100. It has been used as a base in many climate models. GtC/year = gigatons of carbon per year

in promoting solar and wind energy and the United States' unintentional reduction of emissions because of the Great Recession of 2008–2009.

These problems led the IPCC to change its basis for developing future scenarios. The panel decided to work backward from radiative forcing of Earth's surface temperatures through atmospheric greenhouse gas concentrations, then back to emissions and the causative population and economic dynamics. This approach allows investigators to examine the effect of different degrees of global warming

and to determine how the relevant population and economic factors might change or be changed. Political considerations and policies are not included in the revised scenarios. Rather, the scenarios suggest courses of action and timing. In effect the revisions provide a better guide to the consequences of inaction as well as a basis for the planning of remedial actions. My personal opinion is that this is a more logical 'scientific' approach.

Excellent data for atmospheric CO$_2$ concentrations exist as a result of the work of chemist Charles David Keeling,

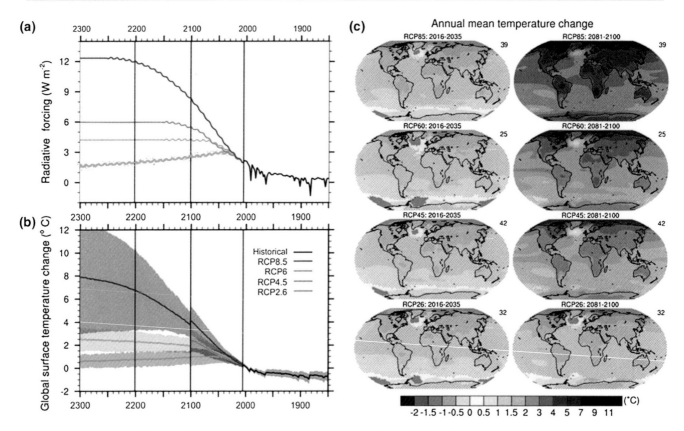

**Fig. 31.9 a** Historical and projected changes in radiative forcing for the historical record and the four RCP scenarios. **b** Simulated global average annual temperatures for the historical record and the four RCP scenarios. **c** Maps of model-simulated annual mean temperature anomalies for averages of the years 2016–2035 and 2081–2100 over that of the base period 1986–2005. Adapted from the IPCC Reports

who initiated the program of regular routine measurements at Mauna Loa in 1958. Systematic data for methane ($CH_4$) and nitrous oxide ($N_2O$) have been collected since the mid-1970s. In the IPCC's Fifth Assessment Report (AR5) published in 2014 the projections for the future are 'Representative Concentration Pathways' (RCPs) defined by four greenhouse gas concentration (not emission) trajectories and named according to their radiative forcings. These RCPs were developed and analyzed through studies of hundreds of simulations of future climate published in the peer-reviewed literature, not only those published by scientists working with the IPCC.

Radiative forcing is defined as the difference between the insolation absorbed by the Earth and the energy radiated back to space. Technically, radiative forcing is usually quantified at the tropopause in units of watts per square meter of Earth's surface. Positive forcing (more incoming energy) warms the system, while negative forcing (more outgoing energy) cools it. In the context of the IPCC reports the causes of radiative forcing are changes in the concentrations of greenhouse gases and aerosols, as well as changes in land use which affect albedo.

The four RCPs have official names according to their radiative forcing, and nicknames related to their origin:

RCP8.5 has a radiative forcing of 8.5 watts per square meter ($W/m^2$) in 2100; its nickname is MESSAGE. It was developed by the International Institute for Applied Systems Analysis in Austria. It is essentially 'business as usual' with no significant attempts at mitigation.

RCP6 has a radiative forcing is 6 $W/m^2$ in 2100, nicknamed AIM; developed by the National Institute for Environmental Studies in Japan. It represents stabilization of radiative forcing at 6 $W/m^2$ in 2150 without an intermediate overshoot.

RCP4.5's radiative forcing is 4.5 $W/m^2$ in 2100, nicknamed GCAM; developed by The US Pacific Northwest National Laboratory's Joint Global Change Research Institute. It represents stabilization of radiative forcing at 4.5 $W/m^2$ in 2150 without an intermediate overshoot.

RCP2.6's radiative forcing is 2.6 $W/m^2$ in 2100, nicknamed IMAGE; developed by the PBL Netherlands Environmental Research Agency.

Each of the RCPs has an extensive database to be used in developing initial conditions and forcing factors for climate model runs (Table 31.2).

Figure 31.9a is a graphical representation of the radiative forcing for the four RCPs relative to pre-industrial values using records back to 1900. Radiative forcing during the

**Table 31.2** RCPs, timing of their radiative forcing, atmospheric greenhouse gas concentrations in terms of $CO_2$ equivalents, global temperature anomalies, pathways, and related emissions scenarios from the 2000 IPCC Working Group Report

| Name | Radiative forcing | $CO_2$ equivalent (ppmv) | Temperature anomaly (°C) | Temperature anomaly (°F) | Pathway | SRES temperature anomaly equivalent |
| --- | --- | --- | --- | --- | --- | --- |
| RCP8.5 | 8.5 W/m² in 2100 | 1370 | 4.9 | 8.8 | Rising | SRES A1F1 |
| RCP6.0 | 6.0 W/m² post 2100 | 850 | 3.0 | 5.4 | Stabilization without overshoot | SRES B2 |
| RCP4.5 | 4.5 W/m² post 2100 | 650 | 2.4 | 4.3 | Stabilization without overshoot | SRES B1 |
| RCP2.6 | 3 W/m² before 2100 declining to 2.5 W/m² by 2100 | 490 | 1.5 | 2.7 | Reaches Peak about 2040, then declines | None |

Last Glacial Maximum, 24,000 years ago, is estimated to have been −4 to −6 W/m². Figure 31.9b shows the global temperature change that these radiative forcing would produce, with potential ranges of error for each scenario shown by pale shading. Global temperatures of the Last Glacial Maximum were about 6 °C below those of the last millennium. Figure 31.9c shows maps of simulated temperature increases over the averages of a base period 1986–2005 for combined averages of the years 2016–2035 and 2081–2100.

The RCP scenarios are defined in terms of greenhouse gas concentrations expressed as $CO_2$ equivalents. The major greenhouse gases are $CO_2$, $CH_4$ (methane) and $N_2O$ (nitrous oxide), discussed in Chaps. 21 and 22. Figure 31.10 shows the trajectories of their concentrations for the four RCP scenarios.

These atmospheric concentrations are derived from emissions but taking into account the possible sinks for these gases other than the atmosphere. Emission scenarios for the four RCPs are shown in Fig. 31.11.

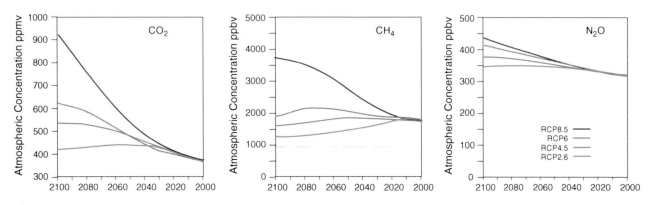

**Fig. 31.10** RCP trajectories for atmospheric concentrations of the major greenhouse gases to 2100

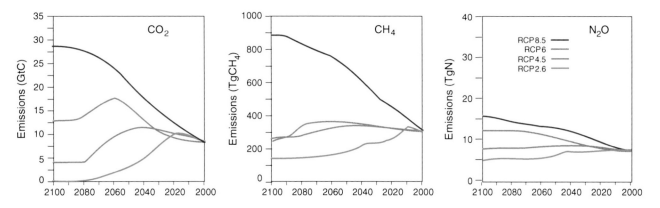

**Fig. 31.11** RCP trajectories for emissions of the major greenhouse gases to 2100

**Fig. 31.12** RCP trajectories for emissions of the major atmospheric pollutants affecting radiative balance of sulfur dioxide and nitrogen dioxide to 2100

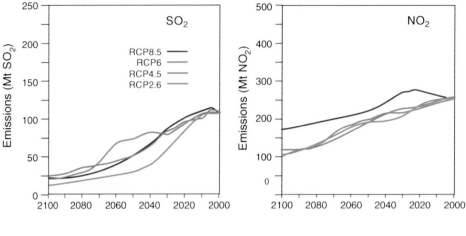

**Fig. 31.13** **a** Global primary energy consumption in exajoules (EJ = $10^{18}$ J). **b** Global consumption of petroleum in terms of exajoules of energy produced. The RCP 8.5 and RCP 6 scenarios assume peak oil production will be reached about 2070

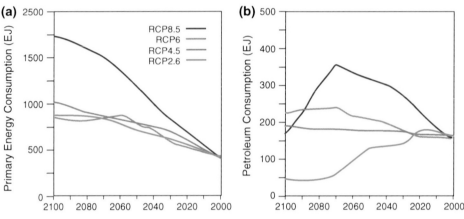

Radiative forcing is also influenced by atmospheric pollutants that reduce the transmission of insolation. The two main pollutants that have become widely distributed on a global scale are sulfur dioxide ($SO_2$) and the reddish brown gas nitrogen dioxide ($NO_2$), both produced by industrial processes. Both are hazardous to life. Figure 31.12 shows projections for future emissions of these gases.

The emissions are a function of energy consumption, depicted in Fig. 31.13a. The supplies of coal, petroleum and natural gas are finite, although technological developments can make exploitation of previously unrecoverable deposits possible. The continuing increase in use of these deposits for energy through the rest of this century will likely result in the catastrophic climate change of the RCP8.5 scenario. Figure 31.13b shows the consumption of petroleum suggested by the four RCP scenarios.

Figure 31.14 shows the changes in energy sources expected by 2100 as implied by the four RCP scenarios. For RCP8.5 almost two thirds of the energy is produced by burning fossil fuels, and most of that is from coal. Coal reserves are fully adequate to enable this scenario. Nuclear energy is also a major energy source in this scenario. The term 'bio-energy' refers to non-fossil biological materials. In the late 18th and early 19th centuries the Industrial

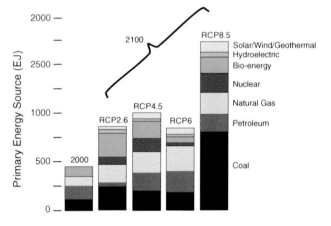

**Fig. 31.14** The various sources of energy used by humans in 2000 and according to projections of the four RCP scenarios

Revolution was powered by wood, leading to the decimation of forests. Wood is still a primary energy source in many underdeveloped countries. Ethanol produced from corn or grain is a technologically advanced bio-energy source. The term bio-energy generally implies no net contribution of $CO_2$ to the atmosphere, although this may not be the case if forests are not regrown. Hydropower was developed extensively in the 20th century, and little remains to be exploited.

**Fig. 31.15** Past and projected areas devoted to growing crops according to the four RCP scenarios. 100 ha = 1 km$^2$

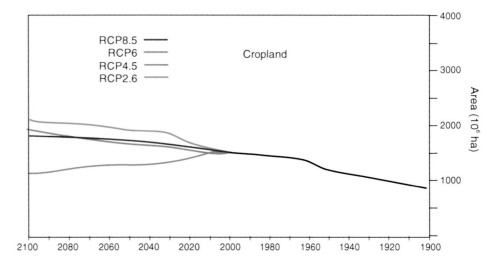

**Fig. 31.16** Past and projected areas devoted to pasture for dairy and food according to the four RCP scenarios. 100 ha = 1 km$^2$

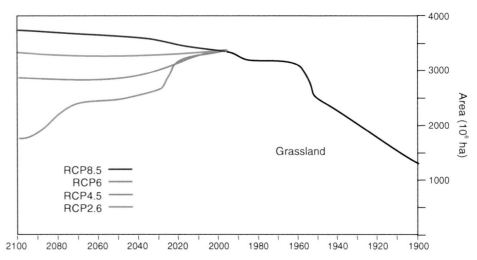

The contributions from solar, wind and geothermal energy are shown in Fig. 31.14 as relatively minor components of the mix, but I suspect they are grossly underestimated. As long as fossil fuels remain cheap and there is no penalty for dumping greenhouse gases into the atmosphere alternative non-polluting energy sources are at a disadvantage.

Changes in land use are an important factor in the RCP scenarios because they directly affect albedo. Figures 31.15, 31.16 and 31.17 show historical and projected changes in croplands, grasslands used for cattle, sheep and other dairy and food animals, and vegetation not used for food production. All of these projections imply a gradually increasing efficiency of food utilization and less waste. A decrease in grassland concomitant with an increase in cropland implies a shift to a more vegetarian diet. The area units in the figures are given as hectares, the common measure of agricultural land. There are 100 ha in a square kilometer.

Inasmuch as we are working backwards from atmospheric greenhouse gas concentrations we can now critically examine the projections of human populations implied by the four RCP scenarios. Past history and future projections

of the global population by the United Nations and for the four RCPs are shown in Fig. 31.18. The UN's high population estimate would correspond to a radiative forcing far higher than 8.5 W/m$^2$ and is probably unrealistic. Similarly, the UN's low population estimate is unrealistic. A major nuclear war or epidemic could drastically change the shape of the future population projection curves.

The RCP scenarios are also related to overall economic heath as reflected by national 'Gross Domestic Product" (GDP) and the global equivalent Gross World Product (GWP), shown in Fig. 31.19. The implication is that a smaller global population will be more economically productive and wealthier than a larger population.

## 31.7 Comparison of Climate Models Based on SRES and RCP Scenarios

How different are the results of climate model based on the SRES and RCP scenarios? Figures 31.20 and 31.21 show comparisons of the SRES and RCP temperature and

**Fig. 31.17** Past and projected areas covered by vegetation not directly involved in producing food for human uses, including forests, shrublands and steppes according to the four RCP scenarios. 100 ha = 1 km²

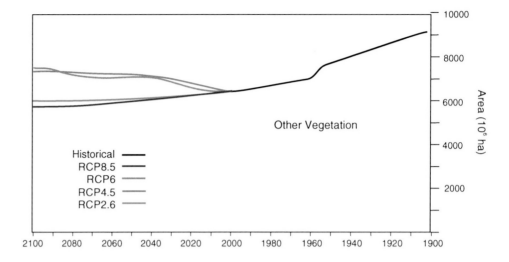

**Fig. 31.18** United Nations' and RCP projections of global population growth along with documented and estimated historical populations

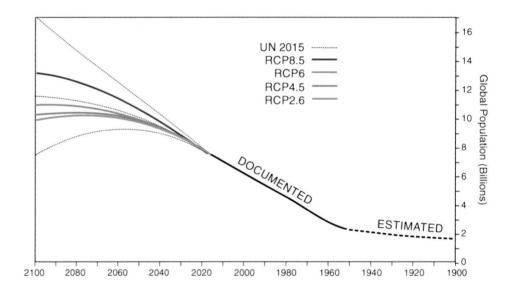

**Fig. 31.19** Historical data and RCP projections for Gross World Product valued in term of trillions of US $ in the year 2000

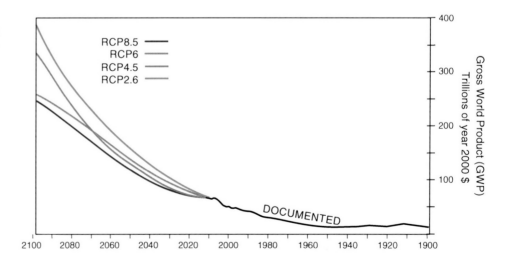

precipitation projections developed by the Coupled Model Intercomparison Project (CMIP), a project of the World Climate Research Program. The results are ensembles, averages of a large number of models run by different institutions. CMIP3 was developed in 2006 using the SRES scenarios, and CMIP5 was completed in 2014.

## 31.8 The Kyoto Protocol

The IPCC provided guidance for the Kyoto Protocol, an international agreement reached in 1997 linked to the United Nations Framework Convention on Climate Change. The Protocol commits its signatories, upon approval by their governments, to set internationally binding emission reduction targets for slowing the rate of rise in atmospheric greenhouse gas concentrations. The targets depend in part on each country's emission rates and infrastructure. Subsequent meetings have refined the Protocol's targets, with some being significantly raised as technological solutions are developed. Although the United States signed the Protocol under the Clinton administration in 1998, the Bush administration withdrew from it in 2001. Canada withdrew in 2012. As of 2015 there are 191 individual counties and the European Union (consisting of 28 additional counties) participating in the program. The only exceptions among the community of nations are the United States, Canada, and South Sudan.

## 31.9 A Critique of the IPCC Reports

The IPCC Reports are the direct and indirect work of many hundreds of climate scientists. My concern with them is that they represent an extremely conservative view of the future. They rely on undisputed numerical data already in hand. Unfortunately there are a number of aspects of Earth science in which we are still learning and ideas are changing at a rapid pace in those areas. Geology was traditionally thought to be an area of science devoted to processes that operated on such long timescales that, except in special cases such as earthquakes and major floods, change could not be observed in a human lifetime. Today, however, change is happening on an accelerating timescale. The next chapter explores the topic of future climatic change in its geologic context.

## 31.10 Summary

Since the early 1970s computer models of varying complexity have been used to make projections of different aspects of the future of both societies and climate. Computer simulations of climate change commonly assume that change will be gradual, not with the relatively rapid transitions from one state to another that have happened in the geologic past. This may explain why they generally predict slower change than is actually occurring.

Eight human-driven factors that are forcing climatic and environmental change: (1) increased concentrations of atmospheric greenhouse gases; (2) introduction of ozone-destroying compounds into the atmosphere; (3) introduction of pollutant aerosols into the atmosphere; (4) replacement of $C_3$ plants by $C_4$ plants; (5) development of large areas of plant monocultures to increase agricultural production; (6) construction and urban sprawl affecting albedo and water balance; (7) overuse of fertilizers and pesticides in agriculture; and (8) mining of minerals for fertilizers and metals.

Two major efforts have been directed toward studying these factors and their effects on the future of our planet—those of the Club of Rome and the United Nation's Intergovernmental Panel on Climate Change (IPCC).

The Club of Rome was founded on 1968, before the possibility of future climate change was recognized. It published *The Limits to Growth* in 1972, and a 30 year update in 2002. The later report has describes eleven scenarios based on availability of natural resources, population growth, food supplies and other factors, but gives little attention to the effects of climate change.

The IPCC, founded in 1988 has issued a series of reports, with the most recent published in 2014. The IPCC includes a large number of climate scientists collaborating in working groups which have developed a number of scenarios for the future, based largely on changes in the carbon dioxide content of the atmosphere. The IPCC's most pessimistic scenario, based on continuing 'business as usual' as it has been in the late 20th and early 21st centuries projects increases of global temperatures to levels barely tolerable for humans. The IPCC's most optimistic scenario indicates that we might be able to limit the global temperature rise to 2.7 ° C at the middle of the 21st century and then to get the temperature to decline slightly thereafter. That scenario

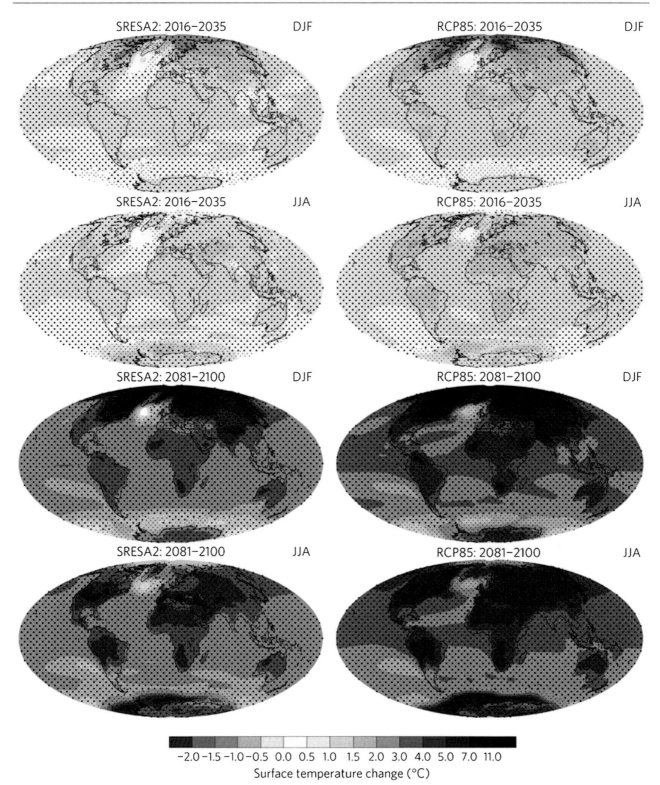

**Fig. 31.20** Comparison of future temperatures based on ensembles of climate models run using databases of the RCP and SRES scenarios. After Huber, M., & Knutti, R., 2014. Natural variability, radiative forcing and climate response in the recent hiatus reconciled. *Nature Geoscience*, v. 7, 651–656

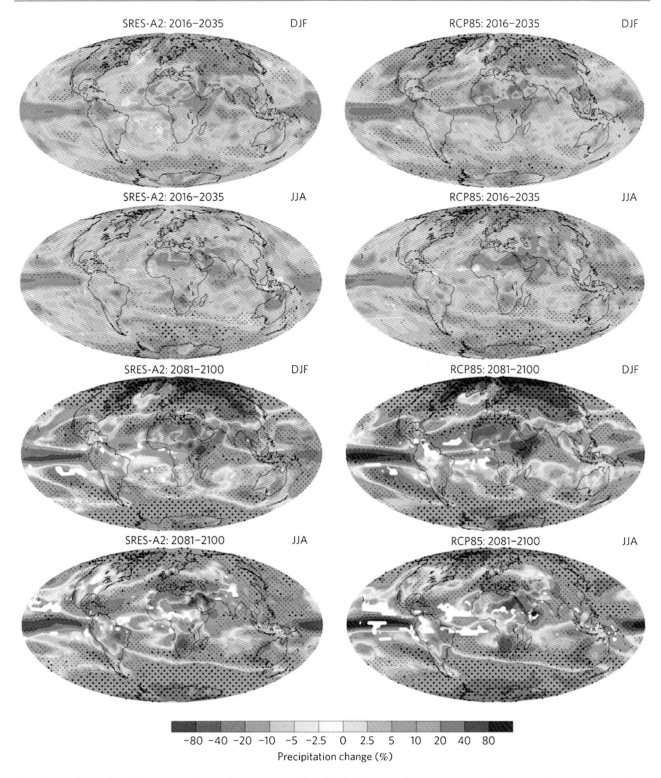

**Fig. 31.21** Comparison of future precipitation based on ensembles of climate models run using databases of the RCP and SRES scenarios. After Huber, M., & Knutti, R., 2014. Natural variability, radiative forcing and climate response in the recent hiatus reconciled. Nature Geoscience, v. 7, 651–656

would require that major actions to limit greenhouse emissions must be undertaken now and be rapidly expanded in the near future.

The IPCC provided guidance for the Kyoto Protocol of 1997, an international agreement linked to the United Nations Framework Convention on Climate Change. The

Protocol commits its parties to set internationally binding emission reduction targets to slow the rate of rise in atmospheric greenhouse gas concentrations. 191 counties and the European Union participate in the program. The United States, Canada, and South Sudan do not.

A Timeline for this chapter:

| 1951 | The United Nations begins to make projections for the future global population |
| 1963 | The author uses ILLIAC to try to classify planktonic foraminifera—unsuccessfully |
| 1968 | The group that will form the Club of Rome has their first meeting |
| 1972 | The Club of Rome issues *Limits to Growth* |
| 1988 | The UN creates the Intergovernmental Panel on Climate Change |
| 1990 | The IPCC issues its First Assessment Report |
| 1995 | The IPCC issues its Second Assessment Report |
| 1997 | The Kyoto Protocols, an international agreement to address future climate change, are adopted |
| 2000 | A special committee of the IPCC issues the *Special Report on Emissions Scenarios* |
| 2001 | The Bush administration rejects the Kyoto Protocols |
| 2001 | The IPCC issues its Third Assessment Report |
| 2007 | The IPCC issues its Fourth Assessment Report |
| 2007 | The IPCC and former Vice President Al Gore share the Nobel Peace Prize |
| 2013-2014 | The IPCC issues its Fifth Assessment Report |

If you want to know more: All of the IPCC Reports are available online from their website at www.ipcc.ch.

Music: Tchaikovsky's Symphony No. 6 is good for contemplation.

Libation: A Cabernet Sauvignon would go well now.

## Intermezzo XXXI. Return to the United States

I returned to the US in 2002. I found it was very different from when I had left and began spending most of my time in Europe in the early 1990s. It was as though in the United States time had stopped in the 1980s. Deferred maintenance had become the rule. But more significantly, many of the small locally-owned stores of the town centers have been replaced by suburban shopping malls with chain stores and franchises. What Texas Governor Rick Perry calls 'Vulture Capitalism' has ruled since the turn of the century. The idea of long term planning seems to have been forgotten.

Part of the problem was, of course, the huge national debt run up under the Reagan and first Bush administrations through increasing spending while reducing income through tax cuts for the wealthy.

'Trickle down' economics, termed 'Voodoo economics' by George H.W. Bush before he became Vice-President, didn't work to grow the economy. Bill Clinton had managed to get things under control and start showing surpluses to pay down the debt, but his efforts were demolished by George W. Bush with huge tax cuts for the wealthy and war in the Middle East 'put on our Chinese credit card.' In the US the devaluation of the dollar by 30 % in 2002–2003 went unnoticed except for those of us who traveled back and forth to Europe.

When my father's Life Insurance Company was sold to buyers from New York in 1958, he had noticed what he thought was an alarming trend. They were more interested in high salaries for upper management than for return on investment to the stockholders or services to the insured. He told me of his concerns many years later, when the trend had become widely apparent. The growing income disparity between executive and faculty salaries in the Universities had become apparent to me in the 1980s and by 2000 it was becoming extreme. The trend has continued, and now the United States leads the developed world in income disparity. In Kiel, the Rector, head of the University, was simply a faculty member with a modest supplement to his salary for the many extra duties he performed.

I had already learned when I went to the University of Colorado that a verbal agreement was worthless. But I soon learned that in the twentieth century financial arrangements, such as mortgages, had fine print clauses that even lawyers might not understand. The result was the US financial meltdown of 2008.

What surprised me most was the spread of fundamentalist religion. The benevolent Christianity I had known in college and as a young man had been largely replaced by the Televangelist messaging that God wants you to be rich no matter how you do it. Fiscal morality was becoming a thing of the past.

In order to try to understand the new religion, I attended Creationist School in one of our local churches. There I learned that Carl Sagan was an agent of the Devil. Most surprising was that I learned that Charles Darwin had argued that humans had evolved from apes in only one (or at most two) generations. How could it be otherwise, since the Earth was only 6,000 years old? Someone in the audience asked if anyone had read Darwin's 'The Origin of Species.' No one had (except me, and I kept my mouth shut because I was already being eyed with suspicion). I checked with our local library and found they did not have a copy, so I gave them one. Apparently some members of the library advisory committee found it to be too radical and the copy I gave was sold at the next book sale.

Colorado Springs is a center for Creationist activities, and the leader of the school I attended was from one of the organizations there. He relied heavily on presentations prepared by an organization on the west coast, and much had to do with 'Intelligent Design.' The idea is that life is so complex that only an intelligent designer could have created it, i.e. God. If you have no concept of the abyss geologic time it would seem to make sense. However, recently an article by R. John Ellis with the interesting title 'Tackling unintelligent design' appeared in Nature (2010, v. 463, pp. 164–165) pointing out that the key enzyme in photosynthesis, Rubisco, is a highly inefficient relic of a bygone age, the Precambrian, when conditions on Earth were very different. Although it has not evolved to keep up with the changes that have occurred since then, it might be possible to genetically manipulate it to make it fit for the modern world.

The election of Barak Obama seemed to be a major milestone in our country's history. It appeared that the racism of the 20th century had at last disappeared. That misconception was soon dispelled, and blatant racism has become rampant. It was never dead, only sleeping.

The Supreme Court, once a seat of great wisdom, has become politicized and passes down decisions that surpass all understanding. The idea that corporations are people is bizarre, since corporations were set up specifically to avoid the liabilities of individuals. The idea that corporations (other than churches) can have religious beliefs and impose these on the corporation's employees boggles the mind. The 'Citizens United' decision made what in every other country is called 'bribery of public officials' perfectly acceptable in the United States. The only restriction is that you should wait 24 h after you make your gift to a politician (usually through his/her re-election committee) before you make your request for special favors.

Another Supreme Court malfunction was when they stated that they simply didn't understand those words about a 'militia' in the first part of the Second Amendment of the Constitution: "A well regulated Militia, being necessary to the security of a free State, the right of the people to keep and bear Arms, shall not be infringed." That sounds like how the only other democracy that existed when the colonies declared their independence from Britain, Switzerland, handled the problem of its defense. Switzerland has an army consisting of all male citizens from the time they are in their late teen years until they are into their fifties. They are issued rifles or other weapons, which they keep at home. Certainly the founding fathers did not envision semi-automatic weapons being in the hands of anyone who might want to buy one. As a result, the United States has a very special 'gun problem' with large numbers of citizens killed each year, and mass shootings almost a weekly affair. The police now assume that anyone they stop might be armed and dangerous, and much of the public no longer trusts the police.

The biggest problem, however, is that religious fundamentalists have made science in particular, and education in general, villains. They have succeeded in electing politicians who cut the budgets for funding of scientific research and education. In the latter half of the 20th century, the United Sates was the world leader in science; now that honor is passing onto other countries.

I return to Europe usually two or more times each year to attend scientific conferences and teach intensive 'short courses' on Earth's past and possible future climate history. The short courses consist of about 6 h of lectures per day for a week. I have taught these many times at the Universities of Utrecht and Amsterdam, where students have come from all over Europe, and also in Frankfurt-am-Main, Bremen, and Vienna. I have also taught short courses in China, in Shanghai and Beijing.

Science is alive and well abroad.

PASSING AN UNCERTAIN FUTURE
TO COMING GENERATIONS

# Global and Regional Aspects of Climate Change

*One of the big questions in the climate change debate: Are humans any smarter than frogs in a pot? If you put a frog in a pot and slowly turn up the heat, it won't jump out. Instead, it will enjoy the nice warm bath until it is cooked to death. We humans seem to be doing pretty much the same thing.*
Jeff Goodell

Most climatologists think of change in terms of a continuum of climates. The geological evidence is that Earth has two very different preferred states—Icehouse and Greenhouse, with each of those having two sub-states dependent on Milankovitch insolation changes. The Icehouse state alternates between 'Glacial' and 'Interglacial' sub-states, with about 90 % of the time in the glacial sub-state. The alternations of the Greenhouse state are far more subtle, between slightly warmer-wetter and slightly cooler-drier sub-states, which are probably roughly equally divided in time. Warming the Earth's surface and troposphere with greenhouse gases when it is in an interglacial sub-state leaves the planet nowhere to go except to transition to the greenhouse state. In the past, transitions between states required a few million years, but the Earth has never experienced such a rapid change in atmospheric greenhouse gas concentrations as is occurring today. The current rate of increase of $CO_2$ is about 300 times faster than the most rapid previous change, during a Milankovitch-forced Termination from glacial to interglacial, in the late Quaternary.

## 32.1    The Long Term View

The last 8,000 years since the end of the deglaciation are the longest period of climatic stability in at least the last half million years or more likely millions of years. As discussed in Chap. 21, this Interglacial is unlike any of the earlier three in that its temperatures have remained almost constant rather than reaching a peak and then declining (see Fig. 21.13). It appears that after the end of the deglacial sea level rise and prior to the Industrial Revolution, humans quite accidentally introduced just enough $CO_2$ into the atmosphere to counteract the Milankovitch decline in northern hemisphere insolation.

Figure 32.1 is a version of the temperature reconstruction of Michael Mann and colleagues for the last 1300 years. Until 1900 the variations in temperature seem to have been no more than 0.2 °C. Although there were some local climate variations, such as periods of drought, it was during this long period of climate stability that human civilizations developed. All that began to change a century after the beginning the Industrial Revolution and the rate of change has been accelerating since.

## 32.2    Milankovitch Insolation

Where are we in terms of the Milankovitch insolation cycles that have driven the alternations of glacial and interglacial states over the past few million years? Recall from Chap. 7 that it is the insolation at the latitude of the Arctic Circle that sets the pace of glacial-interglacial climate change. In that chapter I noted that we are in a special place in that cycle. As shown in Fig. 32.2 we are still close to a minimum of solar insolation for the Arctic. The summer and winter solstices occur only about 10 days before Earth's aphelion (caloric winter) and perihelion (caloric summer) respectively. Assuming today's annual average insolation to be 340 W/m$^2$, caloric summer insolation is 351.7 W/m$^2$ and caloric winter insolation is 328.9 W/m$^2$— a difference of about 6.5 % (see also Fig. 32.18).

In Chap. 1 I noted that Mikhail Budyko referred to the Arctic region as 'Earth's Achilles Heel.' Since we are at a point in the Milankovitch cycles when the Arctic is receiving minimal insolation, our grand uncontrolled experiment with changing Earth's atmospheric greenhouse gas concentrations should have a minimal effect on the Arctic. Unfortunately, since the latter part of the 20th century the Arctic is changing to a condition it has not had for at least the past half million years.

© Springer International Publishing Switzerland 2016
W.W. Hay, *Experimenting on a Small Planet*, DOI 10.1007/978-3-319-27404-1_32

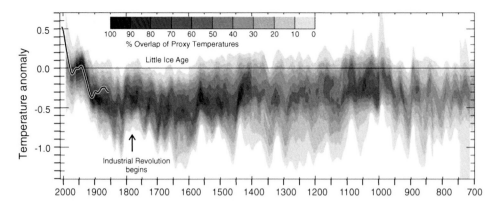

**Fig. 32.1** The instrumental temperature record (*black line* at *left*) and temperatures reconstructed from a wide variety of proxy indicators such as tree rings, pollen, etc. for the Northern Hemisphere. After Fig. 6.10 in the IPCC Fourth Assessment Report: Climate Change 2007—The Physical Science Basis

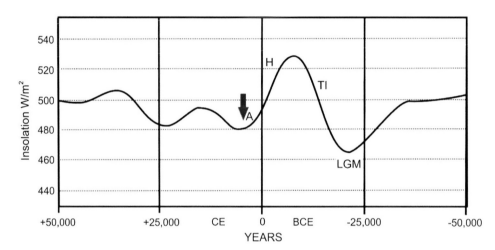

**Fig. 32.2** Insolation at 65° N from 50,000 years in the past to 50,000 years in the future. Note that we are now (*red arrow*) approaching an insolation minimum. *LGM* Last Glacial Maximum; *TI* Termination I (Last Deglaciation); *H* Holocene; *A* Anthropocene

## 32.3    The Decline in Arctic Summer Sea Ice

Among all of the possible tipping points in terms of their importance to climate change I would rank the loss of the sea-ice cover of the Arctic Ocean first. It is the one element of the climate system that has been most affected by greenhouse warming to date, and it is the one that has the greatest potential to make the effects of global warming irreversible. When you think about it, it seems incredible that in our grand uncontrolled experiment on planet Earth we unknowingly hit upon the easiest way to change the planet on an enormous scale—by removing ice from one of its poles. Humans have inadvertently discovered how to do this in the simplest possible way—by injecting into the planet's atmosphere greenhouse gases that differentially affect the Polar Regions.

The Vikings found and settled Iceland around 870, during Europe's 8th and 9th century Medieval Climate Optimum. In 982 CE Eric the Red (Eirík Rauði) was 'invited' to leave Iceland for 3 years after he had murdered some of his neighbors. Eric the Rowdy might have been a more accurate description of him, but *rauði* is Icelandic for red, referring to his beard. He sailed westward and explored Greenland in 983 and 984, perhaps reaching as far as 70° North on the island's western side. At the time this area was uninhabited; the Inuits (Eskimos) who live there today came later. The name 'Greenland' is deceptive, just as Eric intended it to be when he gave the land that name. He wanted to attract settlers to the colony he was founding there.

The Icelandic sagas say that 25 ships with colonists left Iceland with Erik the Red in 985, but that only 14 of them made it safely in Greenland. Other colonists followed. The sailing directions in the Icelandic Sagas say to sail west and look for mountains covered by ice, then go south and round the tip of the island to find the colony. If instead of mountains with ice you come upon land with a lot of small trees you missed Greenland and went too far west. The Norse settled in three separate ice-free locations on the western side of Greenland, from near the southern tip to just south of the Arctic Circle. At their height, the settlements probably had a population of about 5,000. In the year 1000, Erik's son, Leif Eirikson, left the settlement to explore Vinland, assumed to be Newfoundland. The Norse Greenland colony died out in the early part of the 15th century.

The Icelandic 'Vinland Sagas' contain descriptions of the coast of northeastern North America that are so tantalizing that I (following many others) have tried, in vain, to locate the site of a small bay where the explorers set up a small base on shore. The saga states that their ship could get into the bay only at high tide. They did not try to set up a permanent colony; rather they cut down maples to take the beautiful wood back to Greenland, Iceland and even Norway. Unfortunately 1,000 years of isostatic rebound after melting off of the Laurentian ice sheet has changed the North American coastline so much that the base's location remains elusive. A number of possible Viking sites in North America have been described, but the only undisputed one is remains of a Norse village discovered in 1960 on the north tip of Newfoundland, called L'Anse aux Meadows.

Exploration of the Arctic began in earnest in the 16th century—but with very little success until the 20th century. The main reason for exploring the Arctic was commercial rather than scientific—to find a shorter route to China. The obstacle was the Arctic wasteland of ice. The early expeditions sought a 'Northeast Passage' from the GIN (Greenland-Iceland-Norwegian) Sea to the Pacific by sailing north of Russia. Henry Hudson tried this in 1607 only to find he could not get north of Spitsbergen. The 'Northeast Passage' expeditions were so disastrous that interest switched to the "Northwest Passage" north of Canada. Hudson turned his attention westward, seeking a 'Southwest Passage,' leading to the exploration of the American coast north of Jamestown. Hudson discovered the river that now bears his name in what is now eastern New York State, but no route to China. In 1610 he sailed further northward and discovered the bay that bears his name, too late to avoid the winter freeze-up.

Hudson's crewmen lacked his enthusiasm for adventure, and after overwintering in the bay they left Henry and a few others in a skiff and sailed back to England. Hudson was never heard from again. He is presumed to have died in 1611.

A number of later expeditions sought a route though the Arctic, but all ended in failure until near the end of the 19th century. In 1878, Finnish-Swedish explorer Adolf Erik Nordenskiöld made the first trip through the Northeast Passage in the steamship *Vega*, a feat that was not to be repeated for many years. Roald Amundsen made the first trip through the Northwest Passage in a small, steel, shallow draft (0.9 m = 3 ft) vessel designed for seal hunting, the *Gjøa*. Amundsen's passage took 3 years (1903–1906) to complete. The important thing to realize is that no explorers succeeded in penetrating the Arctic wasteland until the 20th century. The first ship to reach the North Pole was the US nuclear submarine *Nautilus* in 1958. You can now visit the North Pole each summer as a tourist on board a Russian nuclear powered ice-breaker.

The Northeast and Northwest Passages were explored by ice-breakers toward the end of the 20th century. Both became passable through ice-fee waters for the first time in the summer of 2008.

How long has the ice been on the Arctic Ocean? Serendipity strikes again. At GEOMAR in Kiel in 2003 we were working on developing a searchable database containing all the data on fossils recovered by the Deep Sea Drilling Project and its successor, the International Program of Ocean Drilling. These data are almost entirely records of calcareous nannoplankton, planktonic foraminifera, diatoms, and radiolarians. I stumbled across a paper by Gunilla Gaard at the University of Stockholm on fossils of calcareous nannofossils in sediment cores taken from the floor of the Arctic Ocean. The fossils she described were 'reworked,' that is, eroded from older sediments, transported, and redeposited into younger sediment. No matter, these reworked fossils show that the organisms producing them were present somewhere in the high Arctic. There are virtually no calcareous nannofossils discovered that are younger than the end of the Eocene, 34 million years ago.

I learned on 'The Great Wild Goose Chase' in 1964 (see Reminiscence by Professor Emeritus George W. Swenson at: www.ece.illinois.edu/about/history/reminiscence/goose.asp) that the coccolithophores that produce these nannofossils do not live beneath ice. I came to the conclusion that the ice cover of the Arctic Ocean has not been significantly breached since the end of the Eocene. A similar conclusion was reached by the Arctic Coring Expedition (ACEX) which recovered cores from Lomonosov Ridge in the summer of 2004. The Arctic sea-ice has been there for a long time.

My own interest in the Arctic was piqued on that trip to Leningrad in 1982 to meet with Mikhail Budyko. It was not only his discussion that made me think I should learn more. On an afternoon off, I was enjoying myself, walking down the Nevsky Prospect, when I noticed a big bookstore across from the Kazan Cathedral. I went in, and there on one of the front tables were copies of a huge Atlas of the Arctic Ocean (Алас Океанов: Северни Ледовити Океан). It was the third volume of the series of Ocean Atlases published by the Soviet Admiralty. I had been given copies of the earlier volumes (for the Atlantic-Indian and Pacific Oceans) by colleagues from the Soviet Academy of Science's Institute of Oceanology in Moscow when they had made visits to the United States. These are spectacular atlases with all sorts of information, including such esoteric items as maps of the speed of sound in the water at different levels throughout the Atlantic, Indian, and Pacific Oceans. I had that information checked by one of my colleagues in Miami who worked on a classified project on that topic. He studied the maps, and came back laughing, "Looks fine, I didn't see any obvious errors. Of course this is all top secret in the US." Knowledge

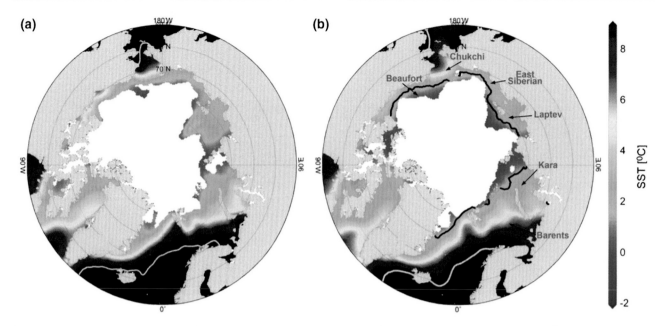

**Fig. 32.3** Water temperatures in the Arctic Ocean. **a** Shows the average sea ice and water temperatures in September 1982–2010. **b** Shows the sea ice and water temperatures in September 2014 along with the commonly used names of important areas of the Arctic Ocean some of which are mentioned in the text. The *black line* indicates the average September 1982–2010 ice edge. Modified after a figure in Timmermans and Proschutinsky (2015) Sea Surface Temperature, Arctic Report Card 2015, NOAA. http://www.arctic.noaa.gov/reportcard/sea_surface_temperature.html

of the speed of sound in the water is critical to sonar operations for locating submarines.

Other American colleagues were suspicious of these Soviet atlases, expressing concern that they did not show the individual stations where measurements had been made. Later, after the political changes in 1989, we learned why. There were so many measurement stations that the maps would have been black with station locations.

In the Leningrad bookstore I picked up a copy of the Arctic Atlas (it weighed about 10 pounds) and walked over to the cashier. I had no idea whether I, as a foreigner, would be allowed to buy it. But I was. It cost the grand sum of 4 rubles (about $5 in US currency at the time). I caught a taxi back to our hotel and inspected the atlas in detail. I found it to be incredibly comprehensive, and it is still probably the most complete compendium of knowledge on the Arctic. It contains an enormous amount of information on the climatology and oceanography of the Arctic—current up to the time of its publication. One important thing to realize is that the Arctic seems to have changed little in the 30 years after World War II—but a lot has changed since.

The Arctic is becoming warmer. Figure 32.3 is from NOAA's Arctic Report Card: Update for 2014. August is the Arctic's warmest month although the sea ice minimum occurs in September. The white area is the sea ice cover in August, and you will notice that in 2014 much of the frigid water that surrounded the ice in earlier years is gone, replaced by warmer 2–3 °C waters.

It was the extraordinary meltback of Arctic sea ice in 2007 that moved me to write the first edition of this book. We had discussed the possibility of the sea ice melting at the meeting in Leningrad in 1982, but a large meltback was not expected until the end of the 21st century. The 2007 meltback was 90 years ahead of schedule, indicating that the rate of climate change was being accelerated. Figure 32.4 shows that dramatic event.

There is a growing consensus that we have passed the Arctic ice's tipping point. One of the nasty properties of 'tipping points' is that you don't know you went past it until you are so far past that there is no possibility of return. If you perturb a system and then try to reverse the process, that reversal is easy at first. But it becomes more difficult as you approach the tipping point. It is problematic at the tipping point, and once you are clearly past the tipping point it becomes impossible. The tipping point becomes a point of no-return.

Figure 32.5 shows the modeled and observed minimum areas of Arctic sea ice extent, occurring around mid-September each year. The results of 13 climate models are shown as dotted lines. The heavy black line is the ensemble mean average of the models. Actual measured data for the entire Arctic were not possible until the advent of extensive aircraft observations in 1953 and extensive satellite observations in 1979. The data track the model ensemble mean until 1982 when the observations revealed a sharp decline in area.

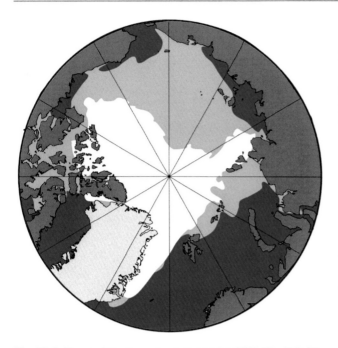

**Fig. 32.4** Extent of Arctic sea ice in September 2007. The *light blue area* is the average extent of sea ice in 1979–2002; the *white area* is the sea ice extent in 2007. In the years after that slightly more sea-ice cover appeared until 2012 when an even larger melt-back occurred. Despite the temporary increases in area, the sea ice has steadily become thinner. Both the Northwest and Northeast passages from the Atlantic to the Pacific have opened at least briefly each year, since 2008. The edge of this map is the northern polar circle

It is an extraordinary coincidence that when our group of American climate scientists was in Leningrad meeting with Mikhail Budyko and his colleagues, the Arctic went over its ice-cover tipping point, yet it was a quarter of a century later that the effects became apparent in the form of the extraordinary 2007 melt-back. Compared with the model projections, this melt-back occurred 45 years ahead of schedule. The sea ice melt-backs from 2008 to 2011 were not as severe in terms of area although the ice continued to thin. In 2012, however, there was another major meltback—and it dwarfed that of 2007. It was 60 years ahead of the ensemble schedule. Since then there has been another recovery, but the steady downward trend is obvious. The dotted red line is a curve drawn though the data suggesting that we may see a seasonally ice-free Arctic by 2040. According to data analyzed by the US Military, the Arctic may, by then, be ice-free for as long as 4 months each year.

Figure 32.6 is a satellite image of the Arctic in September 2012, showing the largest melt-back to date. In this figure the shades of color, from white to sky blue, indicate the proportion of the water covered by sea-ice, ranging from 100 to 10 %.

More significant than the areal extent of the ice is the fact that it is becoming thinner and thinner, as indicated by the declining ice volumes shown in Fig. 32.7. Gradual ice thinning had been detected by ice-breakers since at least the 1980s. As satellite technology developed further and it

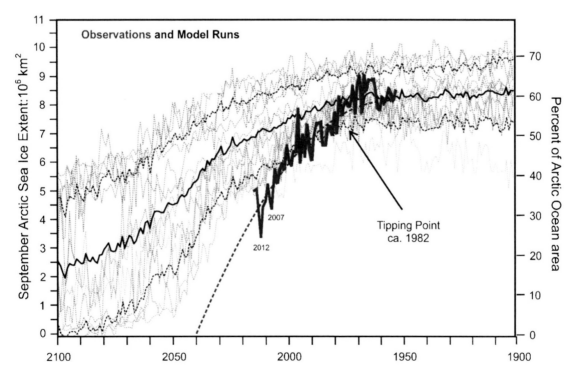

**Fig. 32.5** The areal extent of sea ice in the Arctic. The 1900–2100 ensemble average of 13 climate models is shown as a *solid black line* with the *dashed lines* above and below representing 2 standard deviations (95 % of all the calculated data). The total area of the Arctic Ocean is $14 \times 10^6$ km². Data from the National Snow and Ice Data Center (NSIDC) at the University of Colorado at Boulder

became feasible to measure ice thickness from space, we were able to obtain more detailed records of this variable. The decline in ice volume began to accelerate in about 1990, and this decline has continued to become more and more rapid. This is additional evidence that the Arctic is past its climatic tipping point.

All of this is especially alarming because a loss of Arctic sea-ice of the magnitude occurring today does not seem to have happened during previous interglacials—not even during the Eemian, the last interglacial, when polar temperatures may have been 2 °C warmer than in the pre-industrial era and when sea-levels were a few meters higher (indicating that more of the ice on land had melted). As discussed in Chap. 27, the best guess is that during the Eemian there may have been extensive leads of open water between the ice floes, but most of the sea-ice cover was still there.

Why didn't more of the sea ice melt in older interglacials? The forcing of climate in those past episodes was mostly by changes in insolation related to the Milankovitch orbital variations you read about in Chap. 13. $CO_2$ levels were controlled largely by outgassing from the ocean as it warmed, and never exceeded 300 ppmv. The forcing of climate today, by contrast, is due to a wholly different phenomenon, the rapid increase in atmospheric $CO_2$ to levels not seen for the past 35 million years. Different means

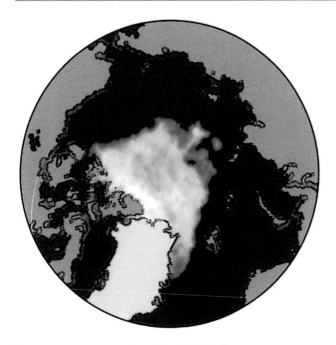

**Fig. 32.6**  The Arctic in September 2012. Ice-free open ocean waters are *dark blue*. Land masses are outlined with *solid black lines*. *White* indicates 100 % ice cover, essentially restricted to the area northwest of Greenland. *Shades* from very *pale* to *sky blue* indicate decreasing ice cover, from 90 to 10 %. Both the Northwest and Northeast passages are open

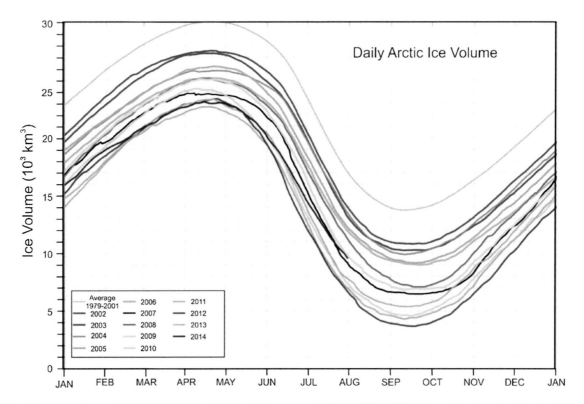

**Fig. 32.7**  Changing volumes of Arctic sea ice from 2002 to 2014 compared with the 1979–2001 average. Data from the Pan-Arctic Ice Ocean Modeling and Assimilation System (PIOMAS) at the Polar Science center at the University of Washington

of forcing climate change result in a different kind of climate change. It's as simple as that.

A very large part of the heat produced by the greenhouse effect in the Polar Regions has been consumed in melting ice. Remember from Chap. 17 that the energy required to melt a cubic meter of ice at 0 °C is $333 \times 10^3$ J/kg while that required to warm water 1 °C is only $4.18 \times 10^3$ J/kg. In other words, the same amount of energy involved in melting 1 kg of ice could also warm the resulting water from 0 to 79.7 °C. The consumption of energy in the phase transition from ice to water is a very powerful climate stabilizing mechanism. The Arctic is particularly sensitive because the sea-ice on the Arctic Ocean was only 3 m thick when the current warming began, We can expect the warming of the Arctic to accelerate as less and less of the incoming energy is used in melting sea ice.

Incidentally, it is important to remember that the Arctic sea-ice is floating on the ocean. Thus, as it melts it has no effect on global sea-level.

## 32.4   Feedbacks in the Arctic

The sudden, unexpected, acceleration of Arctic sea ice melt since the latter part of the last century indicates that positive feedback mechanisms have kicked in. Some of these mechanisms are (1) increased water vapor content of the air; (2) changing albedo as the sea-ice melts and darkens and more of the water surface is exposed; (3) waves on the water surface enhancing the absorption of insolation energy; this melts more of the sea ice overall, including the older sea ice, and results in the production of more young sea ice; (4) a change in the winds over the Arctic, which can break up the sea ice (5) an increased inflow of fresh water into the Arctic Basin, and (6) release of methane from melting permafrost; there may be others yet to be documented.

For the first feedback mechanism, consider what is happening to the water vapor content of the atmosphere of the Northern Polar Region. When ice is present, it acts as an insulator, so that the water overlain by the ice is at a temperature slightly below 0 °C the air above the ice is about −20 °C. A general characteristic of the air immediately over the ocean is that the air takes on the temperature of the water. To overcome this tendency, a flow of air from another area (advection) is required. As discussed in Chap. 16, the vapor content of air can double with every 10 °C rise in temperature. As the insulating ice melts off the sea surface in the Arctic, the vapor content of the air will jump by a factor of four. That means that the $H_2O$ component of the greenhouse effect essentially quadruples.

The second feedback refers to the changing albedo of the Arctic. As you will recall from Sect. 28.1, much of the sea-ice in the Arctic is formed in the Laptev Sea 'ice factory.' Mud from the Lena River gets incorporated into the ice. Then as the ice melts, the water runs off, leaving behind the muddy sediment that had been in it to accumulate on the sea-ice surface. This sediment darkens the surface of the ice, promoting more absorption of heat. Another source of darkening matter in the sea ice was previously thought to be atmospheric aerosol pollutants coming from factories in Europe and North America settling onto the surface of the ice. However, it is now recognized that wildfires in the tundra and boreal forest surrounding the Arctic Ocean are at their highest level in the last 10,000 years, and much of the charcoal from these fires is settling onto the sea ice. The rapid darkening of the Arctic sea-ice has been noted for over a decade.

Regarding the third feedback, the Arctic water, in contrast to the white ice, appears black from space. This sets the stage for an extreme change in the albedo of the Northern Polar Region. The effect is further enhanced by the increase in day length to 24 h of sunshine for most of 3 months of the year. There is however a potential negative feedback because of the reflection of light off the water. Total reflection of sunlight off a smooth water surface occurs at an angle of incidence of 53.06°. Because of the low angle of the Sun's rays, about 47° at the Arctic Circle and 23.5° at the North Pole during the Northern Hemisphere's summer solstice, the incoming energy would simply bounce off smooth water.

The fourth feedback is that the surface of the water is roughened by waves under the winds. Less sunlight will be reflected and more and more absorption of solar energy will occur, as shown in Fig. 32.8.

When the sea-ice cover was extensive, as it was in the past, the water was exposed only in the leads between ice floes and large waves could not develop even if there were strong winds. However, now that much of the Arctic is ice-free in the late summer, the winds can generate large waves, some up to 5 m (16.5 ft) high have been reported, and absorption of energy from sunlight is much more efficient. Figure 32.9 shows the wave heights determined by satellite measurements during the late summer of 2012. The areas of greatest disturbance were in the Beaufort Sea, north of Alaska, and off southeastern Greenland.

One problem in incorporating these effects into climate modeling is that we have no extensive database to work from; the phenomena are too new.

A fifth positive feedback is that the waves act to break up the existing sea-ice into smaller pieces, making it easier for the ice to be carried away and out of the Arctic by ocean currents. Figure 32.10 is an amazing image of the East Greenland Current in the summer of 2014, choked with small fragments of sea ice being flushed from the Arctic Ocean through the Fram Strait between Greenland and Spitsbergen.

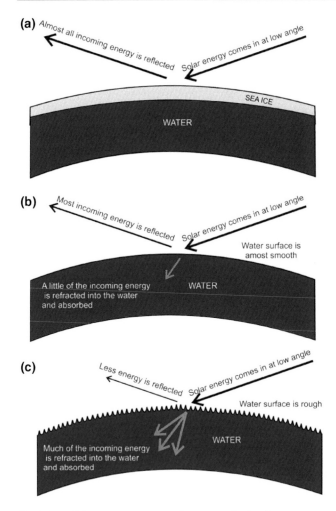

**Fig. 32.8** Differences in a sea-ice covered Arctic versus a water-covered Arctic. Note that at summer solstice incoming solar rays are at angle of 23.5° above the horizon at the North Pole and 47° at the Arctic Circle, and the Sun does not set. **a** With sea-ice cover much of the incoming solar energy is reflected back into space. **b** Without sea-ice cover, an almost smooth water surface would reflect most of the incoming solar energy. **c** With a rough water surface with waves much of the insolation is refracted into the water where it is absorbed as heat

A sixth positive feedback is the 'young sea ice effect.' A couple of decades ago, most of the sea ice in the Arctic was 'multi-year ice.' That is, it had frozen, partly melted, re-froze, and then re-froze over and over again so that much of it had an age of approximately 5–10 years. The older the ice is, the more salt has been expelled and the more resistant it is to melting. By 2012 most of the ice was only 2 years old, and since then most of it has become only one season old. The young ice contains more salt and is more susceptible to melting. The dramatic change over the past decade is shown in Fig. 32.11.

The effect of losing older sea ice is somewhat counterbalanced by changes in the reflectivity of the ice as it ages. As shown in Fig. 32.12, snow-covered ice reflects more than 80 % of the incoming radiation. However, bare first-year ice

reflects only about half and blue older ice only about a third of the incoming radiation. Water reflects just a small fraction of the incoming radiation, but this increases sharply as the Sun's angle becomes very low. Of course, the reflectivity of the water depends on how rough the surface is. Narrow leads between ice floes cannot develop waves, but as more and more of the Arctic Ocean surface is exposed, waves become increasingly important in the reflection/absorption budget. Since we have not had large ice-free areas in the high Arctic until the last few years, we do not have much in the way of real data to guide us in making models or projections of the rate of future changes.

A seventh positive feedback factor is a change in the pressure system and winds over the Arctic. The ongoing warming of the Arctic has weakened the perennial atmospheric high pressure system of the Northern Polar Region. That high pressure system has been stable for millions of years, and it has been responsible for the stability of the westerly winds at mid-latitudes. As the northern polar high becomes less intense or even begins to reverse into a low as the sea-ice disappears in the late summer and fall, the westerly winds become weaker and more unstable. In August of 2012, the polar high reversed to became a low pressure system, shown in Fig. 24.12. By September 2012, the Arctic low had forced the perennial high pressure system over the Greenland Ice Sheet to the south, creating a system of easterly winds over the northern North Atlantic. As Hurricane Sandy moved northward it encountered these easterly winds and was diverted from what in the past would have been a path to the east over the North Atlantic Ocean. Instead, Sandy turned west and wreaked havoc in New Jersey and New York.

As discussed in Chap. 24, the boundary between the Ferrell and Polar atmospheric cells is marked by the high-latitude jet stream. The jet stream is responsible for forcing storm tracks, the succession of low and high pressure systems, around the mid-latitudes of the Northern Hemisphere. It undulates as north-south waves ('Rossby waves') that ordinarily circle the Earth in a week or so. With the warming of the Arctic the amplitude of the Rossby wave undulations of the jet stream increases and their west—to east motion may slow or stop. In the winter of 2014–2015 the north-south extent of the undulations was very large, and then they stopped. Polar air streamed southward over the eastern United States. There the denser frigid air met and slid beneath the lighter warm moisture-laden Atlantic air flowing in from the east. At the time, the Atlantic waters off the United States northeast coast were 6 °C (9.6 °F) warmer than in the 20th century. The result was repeated blizzard conditions and record snowfalls along the northern US east coast. As the Arctic warming continues the effects of the changing climate can be expected to be felt further and further south.

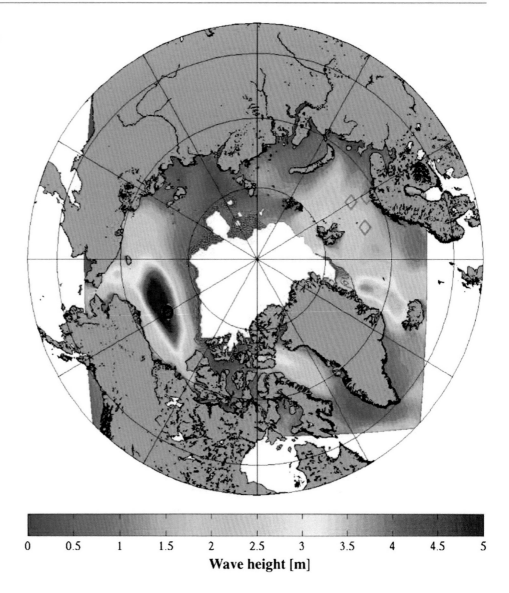

**Fig. 32.9** Calibrated model of wave heights in the Arctic Ocean in September 2012. *Black circle* is monitoring state used to calibrate the model. From Thomson and Rogers (2014). Swell and sea in the emerging Arctic Ocean. Geophysical Research Letters v. 41, 3136–3140

**Wave height [m]**

An eighth positive feedback is the increasing flow of rivers into the Arctic, particularly rivers from Siberia. The increased water comes both from the melting of snowfields and glaciers in the rivers' source areas and from the melting permafrost. You might expect that if more fresh water enters the Arctic from rivers and melting of sea-ice, the out flow of this fresh water into the GIN Sea and North Atlantic would hold back the flow of warm saline waters from the North Atlantic toward the Arctic. However, the opposite is true.

There is a very peculiar phenomenon that takes place when two bodies of seawater, one fresher that the other, are connected. It was discovered by Luigi Marsigli in 1679. Marsigli had been sent by the Venetian Republic as a 'diplomat' (i.e. spy) to Constantinople. He gathered information on the Turkish military forces, but more importantly for us he observed something strange in the Bosporus. The Bosporus strait connects the Black Sea with the Sea of Marmora, which is in turn connected through the Dardanelles with the

Mediterranean's Aegean Sea. The Black Sea is less saline than the Mediterranean, about 27 ‰ versus almost 36 ‰ respectively. Although the Black Sea lies at a latitude where evaporation dominates over precipitation, it is fed with fresh water by several major rivers, the Danube, Dnieper, Don, and Rioni. We now know that the surface of the Black Sea is about 30 cm higher than that of the Sea of Marmora at the other end of the Bosporus. The surface waters of the Bosporus flow downhill from the Black Sea to the Sea of Marmora. Marsigli observed that the fishing boats drifted with the surface current until their fishing nets were lowered to the bottom. Then, the boats began to drift upstream, against the current. Marsigli correctly concluded that there was an undercurrent flowing from the Sea of Marmora into the Black Sea. He built a model to find out how this happened. It consisted of an aquarium-like box with a glass side and a partition in the middle. He poured water from the Black Sea in one side, and water from the Sea of Marmora in the other side. Then the

**Fig. 32.10** NASA image of the East Greenland Current carrying wave-fragmented sea ice from the Arctic Ocean through Fram Strait into the GIN (Greenland-Iceland-Norwegian) Sea on August 18, 2014. The scale is approximate

partition was removed and the Black Sea water flowed over on top of the Sea of Marmora water, and the Sea of Marmora water flowed on the bottom underneath the Black Sea water. As far as we know this was the first experiment in physical oceanography. Why does this two-way flow happen?

The Bosporus is only about 1 km wide at its narrowest point. The depth at which the flow reverses is about 33 m. This type of water flow is characteristic of estuaries, embayments and even ocean basins where density differences exist. Figure 32.13 shows the situation at the two ends of the Bosporus. The waters of the Sea of Marmora and the Black Sea are represented by two columns. The waters of the Sea of Marmora are more saline and hence denser than those of the Black Sea. Because the surface waters of the Black Sea are less

dense its surface is higher than that of the Sea of Marmora. However as we move downward though the water columns, the slopes of the surfaces of equal pressure from the less dense shallow water to the more dense deep water become less and less steep until, at some depth, the surface becomes horizontal. Below that depth, the slope reverses. It is downward from the Sea of Marmara to the Black Sea. The water wants to flow down the sloping lines. The horizontal line, known as the 'level of no motion' separates the less dense surface and more dense subsurface waters. Along the bottom of the Bosporus the flow is from the Sea of Marmora to the Black Sea. About twice as much water flows out from the Black Sea as flows in from the Sea of Marmara reflecting the large inflows from the large watershed draining into the Black Sea.

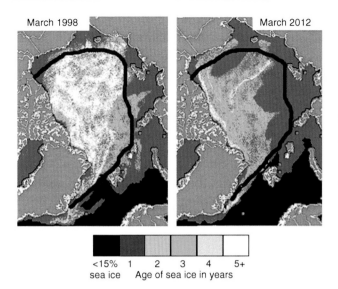

March 1998          March 2012

<15%    1    2    3    4    5+
sea ice    Age of sea ice in years

**Fig. 32.11** The young sea ice effect. Since the end of the 20th century most of the sea ice older than 5 years has disappeared

In the narrow Bosporus, the flows are stacked; surface flow is out of the Black Sea, the bottom flow is into the Black Sea. The same is true of the Strait of Gibraltar where surface waters flow from the Atlantic into the Mediterranean, and though the Strait of Bab-el-Mandeb where lower salinity waters from the Gulf of Aden flow into the hypersaline Red Sea.

However, if the strait connecting bodies of less dense and more dense seawater is wider, the Earth's rotation (the 'Coriolis Effect') acts to separate the flows on the surface. In the northern hemisphere the water is forced to the right of the direction of flow. As an example, Hudson Bay is much fresher than Baffin Bay with which it is connected through James Strait. The outflow from Hudson Bay is along the south side of James Strait, the inflow along the north side.

The same thing is happening in the Arctic. As the summertime flow of rivers into the Arctic Ocean increases, this flow, along with the fresh water from the melting sea-ice, forms a fresh water lid over much of the Arctic. This fresher

water is, of course, less dense than normal seawater. Because of its lower density, the surface of the fresher Arctic waters is higher than that of the North Atlantic, and the water naturally wants to flow downhill. But at depth there is a level-of-no-motion, and below that warm saline water from the North Atlantic flows northward toward the Arctic. However, again, the effect of the rotation of the Earth is to deflect the flow to the right. The water leaving the Arctic flows southward along the eastern margin of Greenland. It is balanced by more dense inflowing waters, the warm saline Norway Current. The melting of ice freshens the Arctic, and as the surface of the Arctic Ocean rises, it forces the Norway Current to strengthen and bring more warm water into the Arctic, where it can melt more ice. This is another powerful positive feedback warming the Arctic (Fig. 32.14).

The odd thing is that, even as the sinking of cold saline water into the depths of the GIN Sea slows, the transport of warm saline water into the Arctic stays the same or even increases in response to the increased outflow of Arctic rivers. The prospect of Europe going into a deep freeze, an idea popular among some scientists 20 years ago, may not happen at all, but regional snowfall patterns could change substantially.

As the ice-free state of the Arctic Ocean extends further and further into the fall, the relatively warm surface waters will supply large quantities of vapor to the atmosphere. As winter begins this moisture will fall out as snow, accumulating where the land is coolest. If the north-south extent of the Rossby waves were minimal this site might well be limited to northeastern Canada, and it has been suggested that this could bring on another glaciation of North America.

This idea echoes a suggestion made in the 1970s to explain the alternation of glacials and interglacials. The hypothesis was that the Arctic Ocean became ice-free during each of the interglacials, and served as the moisture supply to provide the snow that accumulated over northeastern Canada to form the nucleus of the Laurentide ice-sheet and starting the next glacial. As the Laurentide, and then the

**Fig. 32.12** Albedo of sea-ice in different conditions and of open smooth water. Modified after Fig. 3 in Perovich, D.K., 1996, The optical properties of sea ice. U.S. Army Corps of Engineers Cold Regions Research and Engineering Laboratory Monograph, 23 pp

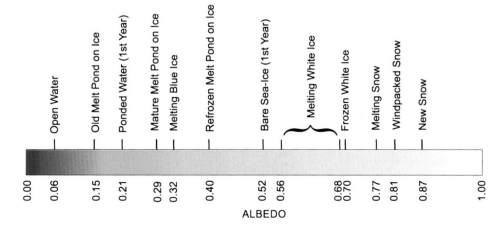

ALBEDO

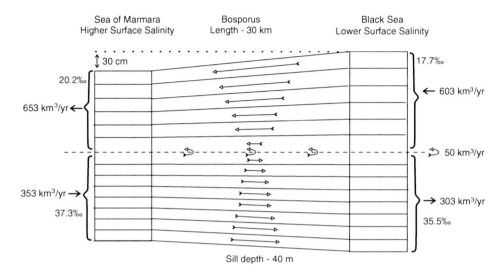

**Fig. 32.13** Flows of water in the Bosporus. This narrow channel connects the Black Sea, with surface waters having a salinity of about 17.7 ‰ and a density of 1,017 kg/m³, and the Sea of Marmora with a surface water salinity of 20.2 ‰, and a density of 1,025 kg/m³. The Black Sea outflow overlies an inflow of much dense subsurface water from the Sea of Marmara. The volumetric excess of outflow over inflow reflects the large inflows of river water into the Black Sea. Note that the surface salinities of both water bodies are much lower than that of the open ocean (34.5 ‰). Relative thicknesses of the two flows are not to scale. Data from Beşiktepe et al. (1994), the circulation and hydrography of the Marmara Sea. Progress in Oceanography, v. 34, 285–334

Scandinavian ice-sheets grew, the Arctic Ocean froze over, cutting off the moisture supply. The declining supply of snow caused these ice sheets to gradually decay, leading to an interglacial. This hypothesis was developed when scientific knowledge of the Milankovitch orbital cycles and the history of the ice sheets was in its infancy. The hypothesis was discarded as more information became available, particularly when it became evident that the Arctic sea-ice had not melted off during the interglacials.

In summary, the meltback of Arctic Ocean sea-ice is due in first order to the enhanced polar greenhouse effect of atmospheric $CO_2$, reinforced by the change to a lower albedo as the melting reflective ice exposes water which absorbs solar radiation. These two warming effects are further reinforced from year to year. Because the sea-ice that now forms during the winter is thin, it melts much more readily than the multi-year ice it replaces. The area of meltback is increasing rapidly. The shifts in atmospheric pressure systems over the Arctic, as the perennial high is replaced by lows over the exposed water, change the circulation in the Arctic; the ocean currents in the region are becoming swifter. This is occurring in such a way as to flush sea-ice floes with much of the older ice out of the Arctic via the East Greenland Current through the GIN Sea into the North Atlantic where the last of the ice will finally melt. The increased flow of rivers into the Arctic in summertime has the effect of drawing more warm water from the North Atlantic into the Arctic via the Norway Current.

Each of these effects acts as a positive feedback to speed the warming of the Arctic much more rapidly than anyone had anticipated. The melting of sea ice due to the greenhouse effect and the positive feedback of reduced albedo as the ice melts was predicted by Mikhail Budyko, but the other effects have been a surprise. Climate modelers are having a hard time keeping up with Mother Nature.

Finally, remember that most of the 'global warming' energy being taken up in the Arctic is being absorbed by the phase transformation of ice into water, which does not involve a change of temperature. When the ice is gone, the temperatures in the Arctic should increase much more rapidly than they have up to now. We can only marvel at what a sensitive system Earth's climate is. Too bad we are learning about all this as it happens.

## 32.5    Arctic Permafrost

Not all of Earth's ice is visible from space. A very large area of the Arctic, beneath both the land and shelf seas, is underlain by permafrost—permanently frozen ground. The amount of water tied up in permafrost is estimated to be 300,000 km³ (≈72,000 miles³). The maximum thicknesses of permafrost are in Siberia, about 1,500 m (≈5,000 ft); in Alaska the maximum thickness is about 600 m (≈2000 ft). The entire permafrost region also contains large amounts of methane. In Siberia the permafrost also provides a seal over large reserves of natural gas (mostly methane). The Arctic permafrost is melting rapidly but it will not cause an appreciable rise in global sea level, because as it melts, the land surface subsides, accommodating the water (Fig. 32.15).

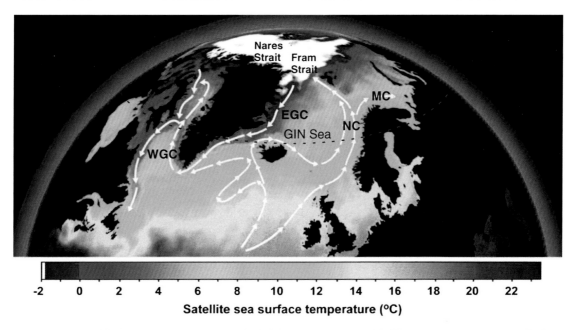

**Fig. 32.14** Ocean currents of the North Atlantic and Sub-Arctic. *WGC* West Greenland Current. *EGC* East Greenland Current. *NC* Norway Current. *MC* Murman Current. Modified from a diagram prepared by the European SEOS (Science Education through Earth Observation for High Schools) Project with data from NOCS/Euro-Argo

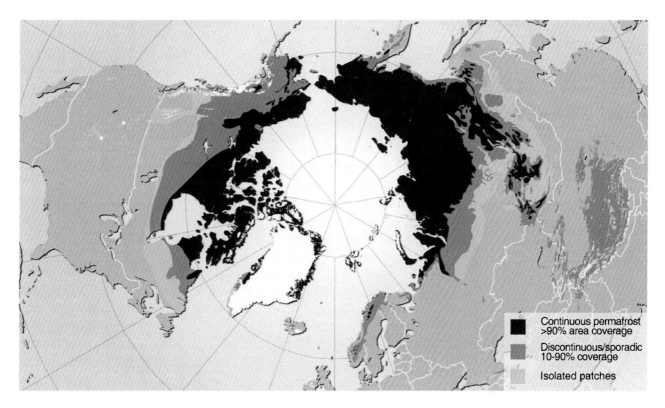

**Fig. 32.15** Permafrost surrounding the Arctic Ocean. Image by Hugo Ahlenius, UNEP/GRID-Arendal via NASA

The major environmental effect of melting permafrost is release of methane, a powerful greenhouse gas. Figure 32.16 shows the dramatic increase in methane release from 2008 to 2011. This acceleration of methane release was so rapid that it caused reanalysis of the satellite data to ensure that there were no errors. There were none, the increase is real. The recent methane release from permafrost concentrates this greenhouse gas in the Earth's most sensitive area, Budyko's

**Fig. 32.16** Increase in methane concentrations in the Arctic troposphere, thought to be due to releases from land and subsea permafrost and possible methane clathrates on the ocean floor. Modified after images prepared by Leonid Yurganov, JCET, UMBC, with IASI/METOR satellite data EUMETSAT

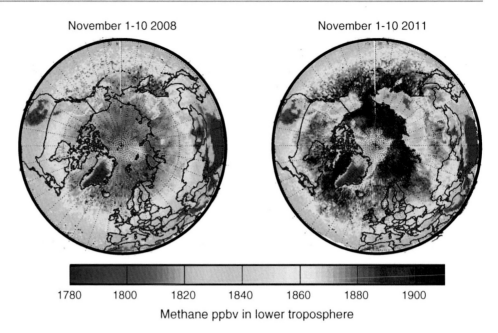

November 1-10 2008                    November 1-10 2011

1780    1800    1820    1840    1860    1880    1900

Methane ppbv in lower troposphere

Achilles Heel of the climate system, the Arctic. Its increase there should greatly accelerate climate change in the northern hemisphere.

## 32.6  The Arctic Tundra and Boreal Forest

The word 'tundra' comes from the Russian тундра; it is an environment where the tree growth is hindered by low temperatures and short growing seasons. The boundary between the tundra and Boreal Forest approximates the mean annual 0 °C isotherm. The vegetation of the Arctic tundra consists of dwarf shrubs, sedges, grasses, mosses, and lichens amongst bogs and peat land. As shown in Fig. 32.17, it extends across Siberia, Alaska, and Canada. Trapped within it are unknown but probably large quantities of the methane clathrates discussed in Chap. 22. The methane was formed from decomposition of organic matter produced by the tundra bogs. All of the tundra is underlain by permafrost which has been stable for millennia, but has now started to melt. Structures built on it are elevated above the ground to prevent melting of the permafrost, but in recent years structural problems resulting from melting as the summers become warmer have become widespread.

The boreal forest forms a green belt surrounding the Arctic tundra. Because of its lower albedo it accepts thermal energy from the Sun and acts to sharpen the boundary between the snow-covered tundra and sea-ice of the Arctic and the forested sub-Arctic. It acts to isolate the Arctic from the regions further south. However, as the Arctic warms, increased water stress and summer heat stress may cause increased mortality and vulnerability to disease. The boreal forest has no protection against the pine bark beetle

devastating the temperate forests further south. There is real concern that the boreal forest could experience a major dieback when the beetle populations reach it. The result would be a transition to open woodlands or grasslands with higher albedo, further destabilizing the Arctic climate. Studies suggest a threshold for boreal forest dieback at 3 °C global warming, but these did not take into account the possibility of beetle infestation.

## 32.7  The Polar Ice Sheets

First, some terminology: 'ice sheet' refers to an ice mass covering more than 50,000 km$^2$ ($\approx$20,000 miles$^2$), the Earth's ice sheets are listed in Table 32.1; 'ice cap' is a mound of ice covering less than 50,000 km$^2$, such as Iceland's Vatnajökull or the Barnes Ice Cap of Baffin Island; 'ice shelf' is a thick floating slab of ice, the seaward extension of a glacier; 'glacier' is an ice stream, usually bounded on both sides by mountains or other topographic features. An overall detailed inventory of Earth's ice is given in Table 17.5.

Before reading further you may wish to go back and review the discussion of ice on land in Sect. 17.12.

It is interesting to note that while the Arctic is currently receiving the minimum insolation of the Milankovitch orbital cycle, the Antarctic is receiving maximum insolation, as shown in Fig. 32.18. The insolation conditions for melting Antarctic ice are quite favorable. The next few thousand years will see maximal insolation at all southern high latitudes during the southern hemisphere summer.

Table 32.3 shows that the areas, ice thicknesses, and volumes of the Greenland and West Antarctic ice sheets are

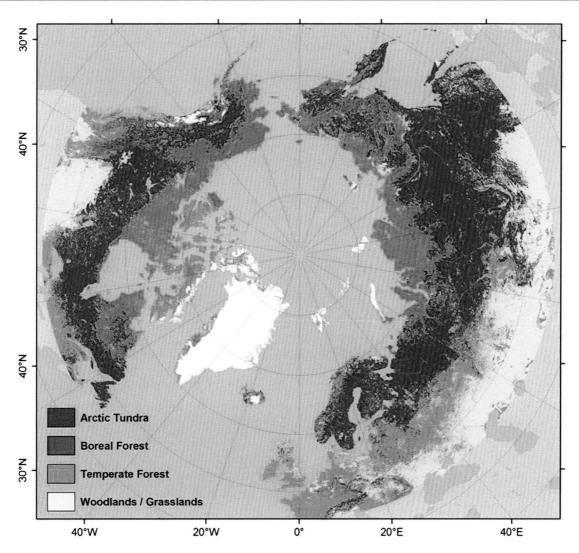

**Fig. 32.17** The Arctic tundra, boreal forest, temperate forests and woodlands-grasslands of the Arctic region. Image courtesy of the Bonanza Creek Long Term Ecological Research Project

**Table 32.1** Antarctic and Greenland ice areas, thicknesses, volumes, sea level equivalents, and energy required to melt

| Region | Area ($10^6$ km$^2$) | Ice thickness (km) | Volume (ice) ($10^6$ km$^3$) | =Sea level (m) | =Sea level (ft) | Energy to melt ($10^{21}$ J) |
|---|---|---|---|---|---|---|
| Greenland | 1.73 | 1.52 | 2.90 | 7.08 | 23.4 | 887.0 |
| West Antarctica | 1.81 | 1.78 | 3.22 | 7.86 | 25.9 | 986.7 |
| East Antarctica | 9.86 | 2.63 | 25.92 | 63.25 | 209 | 7929 |
| Antarctic Peninsula | 0.30 | 0.60 | 0.18 | 0.44 | 1.4 | 56.1 |
| Total | 13.70 | | 32.22 | 78.63 | 259.7 | 9858.8 |

comparable. Were it to melt, the West Antarctic ice sheet would raise sea-level about 8 m. This seems to have happened during the last interglacial, the Eemian, as discussed in Chap. 29 (see Fig. 29.5). Much of the West Antarctic Ice Sheet rests on a rock surface below sea level, so that it is not as solidly grounded as the East Antarctic Ice Sheet. But its ice streams are blocked by floating ice shelves pinned in position by submarine rises and islands.

**Fig. 32.18** Milankovitch Polar insolation curves for the past and future 50,000 years. At present we are at maximum insolation for 60° S to the South Pole

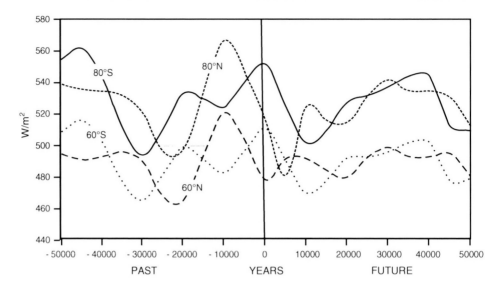

## 32.8    The Greenland Ice Sheet

The only ice sheet in the northern hemisphere is that on Greenland, shown in Fig. 32.19. The first question that come to mind is—why didn't it melt away during the deglaciation along with all the other northern ice sheets? The answer is that it is locked in the Arctic's ice box. Its east cost is bathed by the frigid waters of the East Greenland Current flowing out of the Arctic Ocean through the Fram Strait. The East Greenland Current rounds the tip of southern Greenland and then flows northward as the West Greenland Current. It is the result of the powerful high-latitude Coriolis Effect that

**Fig. 32.19** Greenland Ice Sheet. Most of the surface is over 2 km high, bordered by sharply descending edges feeding into ice stream glaciers. Jakobshavn Isbrae is famous as holding the world speed record for a glacier. The seaward end of the Petermann Glacier has been experiencing extraordinary breakup. Image from NASA

makes these very cold waters hug the Greenland coast. It is a 'geographic accident.'

Thirty years ago it was thought that it would take a very long time for the Greenland ice sheet to melt away. The general understanding was that since most of the surface of the ice sheet is at high elevations only the fringes would be exposed to air temperatures above freezing. In fact, the 'northern dome' in east-central Greenland, which is the area of maximum thickness, is more than 3,000 m (10,000 ft) above sea level. Mean annual temperature there was typically −31 °C (−24 °F). Summer temperatures (July) were always well below 0 °C. It was not until 2012 that there were a few days when the summer reached temperatures above freezing. It was expected that the ice could only be lost by slowly flowing down glaciers to lower elevations or into the sea.

The flow of glaciers and ice sheets was discussed Chap. 17. Glacial ice ordinarily flows slowly downhill as a stream but in an ice sheet it behaves in a somewhat peculiar way. As shown in Fig. 17.14, the flow in an ice sheet depends on whether the base is frozen to the underlying rock, or is near or above the melting point. If the ice is frozen to the rock, the basal ice does not move at all, and most of the movement takes place in the mid-level of the ice sheet, less at the surface. If, however, the base is at or near the freezing point, it can become highly mobile, resulting in a 'surge.'

What has been happening in Greenland over the past few decades was unexpected. The area of melting ice has spread as warming causes summer above-freezing temperatures to move upward, as shown in Fig. 32.20. The meltwater can accumulate on the surface, forming lakes, and can run downhill as rivers on the ice surface, but it soon finds a passage down to the base of the ice sheet, a 'moulin.' At the base of the ice sheet, the water acts as a lubricant. This allows the flows of ice as glaciers from the ice sheet to the sea to move very rapidly. Typical motion of a major glacier is of the order of 1 m per day, but when a surge occurs they can accelerate to 20–30 m per day (≈65–100 ft per day). In summer 2012, Greenland's Jakobshavn Isbrae Glacier set a new speed record advancing 46 m per day (≈151 ft per day). Glaciers that empty into the sea can be affected by offshore islands that may block their progress. The Petermann Glacier of northwest Greenland overlies a water-filled subglacial canyon leading back into Greenland's interior. In 2010 a 260 km$^2$ (100 miles$^2$) iceberg broke off its floating seaward end.

The shape of Greenland beneath the ice is not what you might expect. It is a trough surrounded by marginal mountains as shown in Fig. 32.21. Over thousands of years isostatic adjustment in response to unloading by removing the ice would bring the floor of the trough upward somewhat, but the overall shape as an elongate bowl would not change. Removal of the ice without isostatic adjustment would leave the trough region below sea level and flooded as a bay.

How much would melting off of the Greenland ice sheet raise sea level? The usual answer is about 7 m (21 ft), but this does not include something often overlooked—a large amount of the Earth's water is attracted to Greenland by the gravitational attraction of its ice sheet, enough to alter the global sea level by 1 m. It has been thought that melting of the Greenland ice sheet would require many thousands of years, but in the Eemian there was a sudden rise of sea level of the order of 10 m, which, it is thought, could only have come from both Greenland and West Antarctica.

## 32.9  The Antarctic Ice Sheets

Although a hypothetical continent was shown on many earlier maps, the first actual sighting of Antarctica was in 1820, and significant exploration of the continent did not begin until the 20th century. Regular record keeping began after World War II.

Antarctica has two major ice sheets and one minor one. They are shown in Figs. 17.15 and 32.22. The largest, the East Antarctic Ice Sheet (EAIS) is grounded on the main continental block, and its base is mostly above sea level, as shown in Fig. 17.16. An easy way to remember what is east and west in the Antarctic is to remember that East Antarctica lies south of the Atlantic and Indian Oceans, and Australia, while West Antarctica lies south of the eastern Pacific.

As shown in Table 32.1, its volume is about 26 million cubic kilometers. Because the EAIS has been thought to be solidly grounded on land, it has been considered to be quite stable. The West Antarctic Ice Sheet (WAIS) lies on the mountainous northwestern margin of the continent. It is much smaller; its volume is about 3.2 million cubic kilometers. However, much of its base is below sea level, in places down to depths of 2.5 km. Between these two ice sheets is an ice-filled shallow trough that extends from the Weddell Sea to the Ross Sea. The next few thousand years The third 'ice sheet' is in reality a group of glaciers occupying the Antarctic peninsula which separates the Weddell Sea south of the Atlantic, from the Bellingshausen Sea south of the eastern Pacific.

A unique feature of the Antarctic are the ice shelves. These are tongues of glacial ice pushed out into the sea where they are pinned in place by submarine rises and islands. They are often of the order of 600 m (≈2,000 ft) thick but may be as much as 2,000 m (≈6,600 ft) thick.

As shown earlier in Fig. 32.18, while the Arctic is currently receiving the minimum insolation of the Milankovitch orbital cycle, the Antarctic is receiving maximum insolation. The insolation conditions for melting South Polar ice are quite favorable. The next few thousand years will see maximal insolation at all southern high latitudes during the Southern Hemisphere summer.

**Fig. 32.20** Recent ice melt on Greenland in terms of area and number of days for 2011–2014. During the very warm summer of 2012 the ice melt area covered almost all of Greenland. Figure credit: From NSIDC (National Snow and Ice data Center) prepared by Thomas Mote, University of Georgia

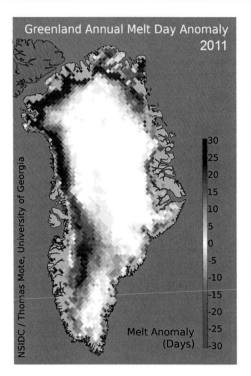

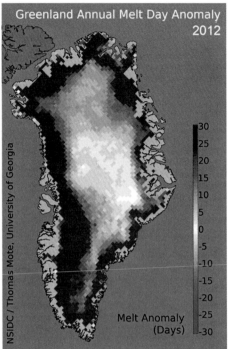

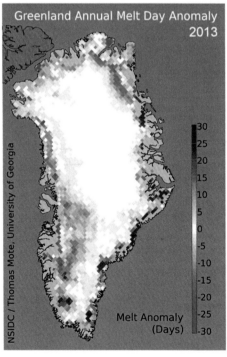

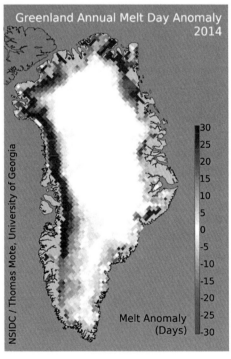

Table 32.1 shows that the areas, ice thicknesses, and volumes of the West Antarctic and Greenland ice sheets are comparable. Were it to melt, the West Antarctic ice sheet would raise sea-level about 8 m. Much of the West Antarctic Ice Sheet rest on a rock surface below sea level, so that it is not as solidly grounded as the East Antarctic Ice Sheet, but its ice streams are blocked by floating ice shelves pinned in position by submarine rises and islands.

Forty years ago John Mercer at the University of Colorado's Institute for Arctic and Alpine Research recognized the vulnerability of West Antarctica, and spoke of a possible 'collapse' of the West Antarctic ice sheet leading to rapid sea level rise over the course of a century. While many glaciologists believed that the West Antarctic ice sheet was indeed highly vulnerable to climate change, they questioned the suggested time scale. The problem in understanding the

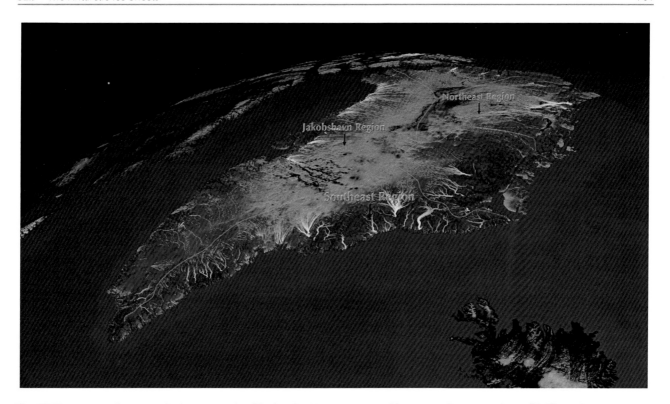

**Fig. 32.21** Greenland's topography beneath the ice. The ice sheet is transparent and ice streams from it are shown. Divides on the ice sheet are shown as *yellow lines*. Image made by NASA's Scientific Visualization Studio

dynamics of a rapid demise of the West Antarctic ice sheet is that it may involve processes about which we know very little. The estimates of time required for the ice sheet to disintegrate depend on guesses about how rapidly the ice shelves might break up and how fast the ice streams might flow.

The ice shelves, long thought to be relatively stable features, have recently shown an unexpected vulnerability to warming ocean waters. We have been surprised to learn that the ice shelves are being melted from below. Huge ice islands break off from them, and as they disintegrate the velocities of the inland ice streams toward the coast increase by factors of 2–4 or more. Conservative estimates for the time required for the 'collapse' of the West Antarctic Ice Sheet are now 500 years or less. As in the case of Greenland, new processes favoring accelerated ice mass loss are being discovered, and Mercer's original estimate may not be that far off the mark. Another of Mother Nature's surprises!

The Antarctic Peninsula lies directly south of South America. With South America they form the margins of the Drake Passage, a choke valve on the Circum-Antarctic Current. It is the most sensitive area of the Antarctic because, lying at lower latitude, it is bordered by ocean waters that are experiencing rapid warming As shown in Fig. 32.23, the waters surrounding the peninsula are warming-up about 2.5 °C (=4.5 °F) since 1950. It is probably the fastest-warming area on Earth. The ice shelves along the peninsula, which have mostly been stable for the last 12,000 years, are breaking up rapidly. The Larsen Ice Shelf is divided into three parts, A, B, and C. The smallest, Larsen A is thought to have formed only about 4,000 years ago and to have been stable until it disintegrated in 1995. Larsen B and C have apparently been stable throughout the last 12,000 years since the deglacial sea level rise ended.

Over a period of a few months in early 2002, a piece measuring approximately 3,250 km$^2$, an area comparable to the state of Rhode Island, broke off the Larsen B ice shelf on the east side of the peninsula and shattered, releasing 720 billion tons of ice into the Weddell Sea. At the time it was the largest single disintegration event in 30 years of monitoring of the ice shelves. In March 2008, the Wilkins Ice Shelf adjacent to the southwest margin of the peninsula Antarctica retreated by more than 400 km$^2$ (160 miles$^2$) in pieces up to 200 m ($\sim$660 ft) thick. Satellite measurements by ESA's CryoSat-2 show that the West Antarctic Ice Sheet (WAIS) is losing more than 150 km$^3$ (36 miles$^3$) of ice each year, mostly from breakup of the floating ice shelves.

In recent years there has been an extraordinary acceleration of the speeds of glaciers in the Amundsen Sea sector. The Pine Island, Thwaites and Smith Glaciers, which discharge into the embayment just above the words Amundsen Sea in Fig. 32.22, are losing about 60 % more ice than is being replaced by snowfall. There is real concern that the entire West Antarctic Ice Sheet may be starting to disintegrate.

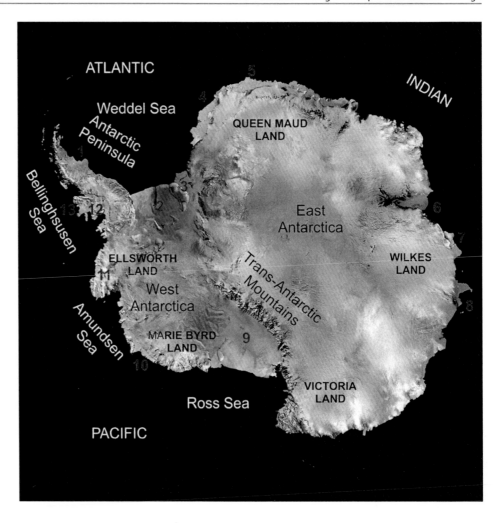

**Fig. 32.22** NASA' Blue Marble view of Antarctica. Names of some areas often discussed in climate change literature are shown. The three large ice shelves are: *1* Larsen Ice Shelf (48,600 km²); *2–3* Ronne-Filchner Ice Shelf (422,420 km²); *4* Riiser-Larsen Ice Shelf; (48,180 km²); *5* Fimbul Ice Shelf (41,060 km²); *6* Amery Ice Shelf (62,620 km²); *7* West Ice Shelf (16,370 km²); *8* Shackleton Ice Shelf (33,820 km²); *9* Ross Ice Shelf (472,960 km²); *10* Getz Ice Shelf (32,810 km²); *11* Abbott Ice Shelf (27,000 km²); *12* George V Ice Shelf (23,880 km²); *13* Wilkins Ice Shelf (13,680 km²)

East Antarctica is very different. With an average elevation of about 2300 m (≈7,500 ft) and a large crest region over 4,000 m (≈13,000 ft) high, summer temperatures are almost everywhere well below 0 °C. The highest temperature ever recorded in the crest region, at the South Pole, was −12.3 °C (9.9 °F) on 25 December 2011. Winter temperatures in central East Antarctica are around −80 °C. Under such frigid conditions it could be expected that the flow of the ice would be minimal. However, it has been discovered that the ice is being warmed from below by the geothermal heat flux and over broad areas the basal ice must be close to the freezing point. The temperature gradient is shown in Fig. 17.4. It is becoming evident that there are extensive subglacial lakes beneath the East Antarctic Ice Sheet, with more being discovered each year. We now realize that large areas of it are not frozen to the bedrock, but are resting on water. The lakes' water serves as a lubricant, and there are already some fast flowing ice streams in East Antarctica. In some areas the surface of the ice is actually rising as it moves over obstacles.

At the coast the drainage of ice from East Antarctica has been inhibited by ice shelves blocking the major outlets.

These are at higher latitudes than those of the Antarctic Peninsula, and the surrounding ocean waters are just beginning to warm. Yet incidents of disintegration have begun to occur and their frequency is increasing. The Amery ice shelf blocks the seaward flow of several major glaciers draining Wilkes Land, including the very large Lambert Glacier. It is known that the ice shelf is melting from below, and NASA imagery from January 2012 show that large cracks are developing and breakoff of a piece with an area of about 1,000 km² (≈400 miles²) is expected soon. It is expected that if the Amery Ice Shelf breaks up the glaciers that have been feeding it will surge.

Although the rest of the planet is gradually warming, East Antarctica is cooling as shown in Fig. 32.23. It is too slow to have any direct response to the increase in atmospheric $CO_2$ for two reasons: The first is the height of the ice sheet itself: with an average elevation of 2300 m (≈7,500 ft) 20 % of the atmosphere (and its $CO_2$) are below it. The second reason is counterintuitive. The ozone hole, still well developed over the Antarctic but gradually closing as the phased-out chlorofluorocarbons decay away, has altered the nature of the southern polar tropopause in such a way as to allow

**Fig. 32.23** Current temperature trends over and around Antarctica. Image attributed to NASA

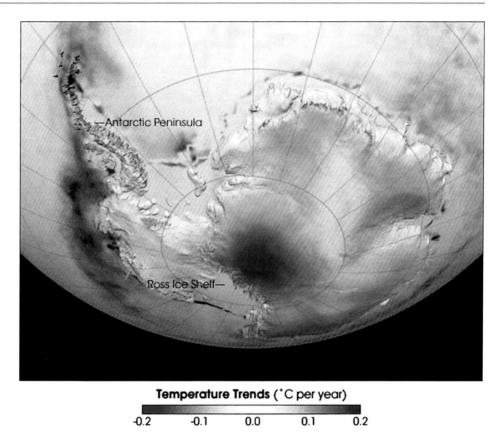

**Temperature Trends (°C per year)**

-0.2          -0.1          0.0          0.1          0.2

exchange of air with the frigid stratosphere, so that the convecting troposphere over Antarctica is much cooler than it would otherwise be. As the ozone hole closes, the air over Antarctica should become warmer.

David Pollard and Robert DeConto developed a model for Antarctic ice sheet growth and decay with changing levels of atmospheric $CO_2$. Neglecting the effect of orbital variations, and assuming gradually declining levels of $CO_2$, they found that incipient ice sheet growth is initiated when then the atmospheric $CO_2$ concentration falls to $3 \times$ its pre-industrial level ($3 \times 280$ ppmv = 840 ppmv) and is complete when the level is $2 \times$ pre-industrial (560 ppmv). Their model assumed that this drop in $CO_2$ occurred over a period of 5 million years. However, in melting the ice sheet they found that there is a hysteresis effect. Hysteresis is defined as the lagging of an effect behind its cause. Melting their ice-covered Antarctica required that the $CO_2$ level rise to $3.2 \times$ pre-industrial (896 ppmv) and was completed when the concentration rises to $3.5 \times$ pre-industrial levels (980 ppmv), over a period of 1.5 million years. If the present rate of increase of atmospheric $CO_2$ is projected into the future, we can expect to see $CO_2$ levels at 900, going to 1000 ppmv about the middle of the next century. Large scale melting of East Antarctica may be postponed until then. If the past is a good guide, this is probably a best case scenario. It assumes that nothing else will go wrong.

However, today the loss of the ice sheet depends not on the concentration of $CO_2$ in the atmosphere, but on the thermal and mechanical inertia of the ice itself. All one can say with certainty is that if we burn the entire fossil fuel reserves of 5000 gigatons, atmospheric $CO_2$ levels will stay high enough long enough to eliminate the East Antarctic ice sheet, although it may take several thousand years. At this time there is not enough known about the behavior of such a large mass of polar ice to make a realistic estimate of the time required for its demise.

## 32.10   Temperature

Two effects of global warming should be apparent from a purely theoretical point of view. There will be a gradual overall increase in global temperatures, and a particularly strong increase in the Northern Polar Region. Both of these phenomena are clearly represented in the changes since the middle of the 20th century, and documented in Chap. 30.

The global effects tend to be subtle. After all, what does a 1 or 2 °C rise in global temperatures really mean? How would we notice it? For one thing, with local exceptions in space and time, winters tend to be less severe, and the summers may have days that set new temperature records. Mountain glaciers almost everywhere are in rapid retreat.

However, the most important effect is in increasingly chaotic weather, with record or near-record snowfalls, such as those of Spring 2008 in affecting large areas of southern China and the winter of 2014–2015 along the U.S. East Coast. Hurricanes and typhoons have become more intense, not by having higher wind speeds but by becoming larger in terms of area and overall kinetic energy. Out-of-season and out-of-place tornados are becoming more common. Heat waves are setting new records. A long-term gradual trend that can be expected is the expansion of the northern hemisphere desert zone to the north. What is currently thought of as an ongoing drought in the southern and western United States may not be temporary; it may reflect a new persistent condition. It is useful to recall that it does not require long-term drought to wreak havoc with food supplies. In 1972 a late freeze in the Soviet Union damaged the wheat crop. The decision was made not to replant, but to count on recovery of the plants to produce a harvest. It may have actually been too late to replant, but in any case the wheat did not mature, and the overall food supply for that year was severely impacted. The US was able to sell its excess grain, but the backup supplies of food that existed before 1972 have never been reached again. A single unique spell of bad weather can cause very serious damage to the food supply.

One of the major effects of warming is already being felt in many areas, reduction in the water supply. With few exceptions, the world's major rivers originate in snow pack and glaciers in mountainous regions. It is the melting snow pack that provides a continuous flow through the summer and into fall. Much of the snow and ice is disappearing, and we can expect to see increasingly long low-water levels in late summer and fall.

Another effect, wholly unanticipated, until a few years ago is the response of insects. As winter temperatures become less extreme, insect eggs and larvae survive in numbers unknown in recent centuries. Forests in the Rocky Mountains are being devastated by the pine bark beetle. This effect was first brought home to the public in Al Gore's film, "An Inconvenient Truth," and since the film was produced vast areas of temperate forest have been killed. The warmer spring and summer nights mean that insects can be more active and require more food.

You may hear that rising atmospheric $CO_2$ concentrations are good for agriculture because plants use it for photosynthesis. Greenhouse experiments had indicated that the growth of some plants is enhanced by increasing the $CO_2$ content of the air. This fact has long been used by the growers in the Netherlands to increase the production of tomatoes. Unlike the outside world, the greenhouses are insect-free. There is now good evidence that in the open, the 'fertilization' of plants by increased atmospheric $CO_2$ is offset and reversed by insect damage.

You might expect that because the amount of water the air can hold increases dramatically with temperature, that warmer weather would bring with it greater rainfall, and hence the climate would generally become more humid. That may indeed be true over the ocean, but conditions on land are a different matter. Away from the coasts, the water that falls from rain comes from transpiration of plants. Over the mid-latitude regions the extensive replacement of $C_3$ with $C_4$ plants has reduced transpiration, so that rainfall becomes less. Further, evaporation goes on continuously, while rainfall is sporadic. The result is that with warmer temperatures and the present plant cover, the climate will be generally drier. Less and less water will be available. This effect is already evident in the water levels in reservoirs in the US west.

The US may experience a greater than average change to drier conditions because we have been mining groundwater in the Great Plains for irrigation. The groundwater being actively mined is left over from the last ice age, and is not being replaced at a rate even approaching the rate of removal. The method of irrigation has been to spray the water into the dry air, where much of it evaporates. That lost water contributed to the amount in the atmosphere available for rainfall elsewhere. As the groundwater supply is exhausted, and the irrigation shut down, the drying effect of warmer weather will be compounded.

The general trends, most pronounced in the middle and high latitudes of the Northern Hemisphere, and which have already become apparent in the past few decades, are: somewhat warmer days and much warmer nights, resulting in a decrease in the diurnal temperature range; more days with summer-like conditions; fewer days with frost and ice, with a correspondingly longer growing season; longer warm spells, sometimes turning into heat waves, fewer and shorter cold spells; and a general increase in total precipitation over the year. Unfortunately, the evaporation rate also increases as the temperatures rise, so that in many areas the net of precipitation and evaporation is dryer conditions.

As the growing season at high latitudes lengthens and aridity spreads northward in North America, we can expect that more and more of the food supply will come from Canada and Siberia. However these regions have been occupied by the boreal forest and tundra and soil conditions are very different from those of todays' food-producing region.

However, the minimal impact RCP2.6 discussed in Chap. 31 assumes a temperature increase to 2.6 °C (4.7 °F) by the end of the century, already above the 2 °C (3.6 °F) often cited as the maximum 'safe' increase. All of the other RCP scenarios being modeled are significantly worse. It is important to recall that the global temperature change from the last glacial maximum to the modern interglacial was no more than 6 °C (10.8 °F) and recent studies suggest it may

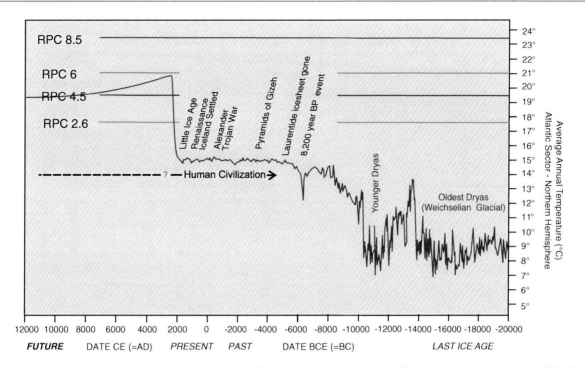

**Fig. 32.24** Global temperatures past, present and future. Glacial temperatures are thought to have been about 9 °C (≈48 °F); Holocene temperature have hovered around 15 °C (59 °F) for 8,000 years, the longest period of stable climate in at least the last half million years. The temperature variations during the last glacial are based on the Greenland ice cores, and best reflect the Atlantic sector of the Northern Hemisphere We are at the beginning of the temperature increase indicated by the *red line*. The magnitudes of the increase used in the IPCC RCPs are indicated

have been as little as 4 °C (7.2 °F). Global temperatures during the warm Cretaceous are thought to have been about 4 °C warmer than the Holocene average of 15 °C (59 °F). The other three RCP's used by the IPCC arrive at an Earth with 4.5 °C (8.1 °F), 6 °C (10.8 °F) and 8.5 °C (15.3 °F) global temperature increases.

Figure 32.24 shows past and possible future temperatures using an estimate for business as usual but assuming that we run out of things to burn in about 2200 (red line) and the limits used for the four RCP scenarios.

One aspect of global warming is not often taken into account. You may hear "dinosaurs liked a warm Earth and so will you." Not necessarily. We are mammals, and our bodies are temperature regulated. The core temperature for humans is 37 °C (98.7 °F). Skin temperature is slightly lower. In warm or hot weather our bodies regulate the core temperature by evaporation (perspiration). If you are unable to perspire and the human body temperature rises to 41 °C (106 °F) you will die in minutes. I know about this from personal experience. A few years ago I was in a hot tub that had overheated to 105 °F for about 30 min and was on the verge of fainting, and was almost too weak to get out. Fortunately, I was able to work my way out enough to collapse on the floor, but it took an hour to work my way over a few feet to a telephone. When the ambulance arrived an hour or

so after I had gotten out of the tub my blood pressure was 60/40, I was almost dead.

What is dangerous is not the temperature alone; you can survive the heat in a desert. It is the combination of heat and high humidity. One common measure of the combination of these factors is 'wet-bulb' temperature, the temperature measured when the bulb is covered with a wet cloth to simulate 100 % humidity. Figure 32.25 shows present day wet bulb temperatures. Humans and most other mammals (we all have similar core temperatures) cannot survive for long if the wet-bulb temperature exceeds 35 °C (95 °F). These conditions are rare on Earth today, but may occur during heat waves, causing loss of lives due to 'heat stroke.'

Figure 32.26 shows what the wet bulb temperatures of the Earth would look like with a global temperature increase of 10 °C. Large areas would be uninhabitable by humans and most other mammals. This may explain why the highest ICCP RCP is for a temperature rise of 8.5 °C. Even at that temperature there would be frequent episodes of deadly wet-bulb temperatures.

While models can suggest how the major changes in climate will occur, they do not predict how the chaos in the system will increase the frequency of short-term catastrophic events, such as extreme winds, tornados, and flash floods.

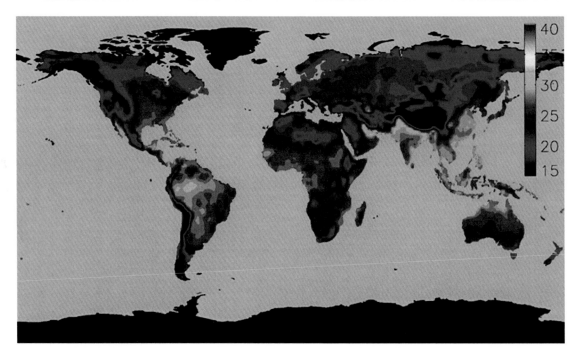

**Fig. 32.25** Global wet-bulb temperatures today. Figure courtesy Steve Sherwood, Climate Change Research Centre, University of New South Wales

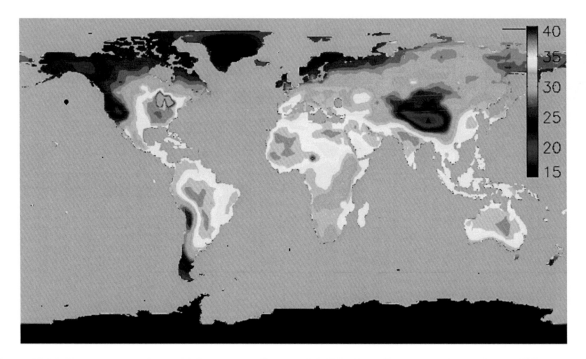

**Fig. 32.26** Wet bulb temperatures with a global temperature increase of 10 °C. Most of eastern North America would be uninhabitable. Figure courtesy Steve Sherwood, Climate Change Research Centre, University of New South Wales

## 32.11   Sea Level

The factors controlling sea level were discussed in Chap. 27, and Fig. 27.11. During the late 19th through most of the 20th century the rate of rise averaged about 1.2 mm year,

but since 1990 the rate has increased to about 3.0 mm/year. Where has the water come from? The melting of Arctic sea ice and Antarctic ice shelves has no effect since they are floating on water. Mining of groundwater has added water to earth's surface, but construction of dams and reservoirs has reduced the amount reaching the ocean, and the two are thought to have approximately balanced out. It is thought that until now most of the current rise is from melting of temperate and low latitude mountain glaciers with some contribution from Greenland. In addition to added water, sea level rise is being caused by the thermal expansion of seawater as it warms.

The sea level problem is exacerbated by the fact that some human activities actually lower the land surface in coastal regions. The World Bank, in an effort to supply the people of Bangladesh with safer drinking water, financed drilling of many water wells in the Ganges-Brahmaputra Delta. An unanticipated result is that the surface of the delta is subsiding very rapidly as the groundwater is withdrawn. Similarly, extraction of petroleum and natural gas in the Mississippi Delta region are accelerating the rate of subsidence due to natural compaction of the delta sediments.

The 2007 IPCC report projected a rise of no more than 20–50 cm during the 21st century assuming that the rate of melting of ice will remain constant at the rates that had been observed through 2003. However, since 2003, the rate of melting of Greenland ice has more than doubled, so that at its 2006 rate, Greenland alone would account for 30 cm of sea level rise by the end of the century. None of the estimates

made then took into account possible surges of Greenland and West Antarctic ice into the sea.

The 2104 IPCC Report re-evaluated that problem and the possible range of sea level rise for the four RCP Scenarios range from 0.26 to 0.82 m by then end of the century, with mean estimates ranging from 0.40 to 0.63 m. As mentioned previously, the IPCC is extremely conservative in its estimates. It used a now outdated estimate of a relatively high rate of sea level rise during most of the 20th century and underestimated the acceleration since 1990. Further its estimates do not take into account the acceleration of melt processes in Greenland and West Antarctica that are now taking place.

Comparison with the events during the last interglacial, the Eemian, now seem appropriate. During that time there were rapid rises of sea level at rates at least 1 m per century, and the highstands were 6–9 m above that of today. That happened when global temperatures were only 1 °C higher than today. With the much more intense climatic greenhouse forcing currently underway, it is likely that similar sea level rise will occur. Sea level rise of 1 m or more by then end of the 21st century now seems likely.

Figure 32.27 shows the southeastern United States, one of the areas that will be strongly impacted by sea level rise. It

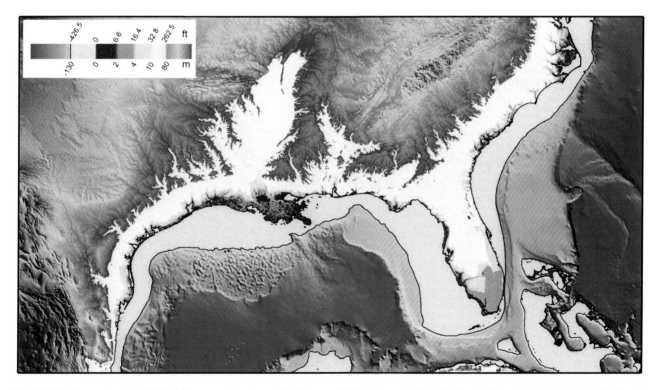

**Fig. 32.27** The southeastern United States; past and possible future changes in sea level. During the Last Glacial maximum, 24,000 years ago, sea level was about 130 m lower than it is today; the *beige* shows the area off the continental shelf exposed as land at that time. Sea level stabilized about 6,000 years ago and remained stable until a century or so ago when it began to slowly rise. Sea level rise has accelerated since 1990 and may well reach +1 m by the end of this century and +2 m by the middle of the next century. A rise of +10 m could occur through melting of much of the Greenland and West Antarctic ice sheets. Complete melting of these and the East Antarctic ice sheet would raise sea level another 80 m, making St. Louis a seaport

may well be that by the end of this century the Louisiana coastline will have retreated to Baton Rouge. Much of North Carolina and Florida are less than 2 m (6.6 ft) above sea level, and we can expect them to be gradually flooded.

It is obvious that southern Florida is particular vulnerable to flooding by sea-level rise. However that region may become uninhabitable sooner because it faces another problem. Almost all of the Florida Peninsula is underlain by very porous carbonate rock. The deeper strata are filled with sea water, but it is overlain by a shallow lens of fresh water. This lens is South Florida's fresh water supply. Here is the problem: a cubic meter of that fresh water in the lens weighs about 997 kg; it overlies salty sea water that weighs about 1027 kg/m$^3$, a difference of 30 kg/m$^3$ (see Table 17.4). If there were no fresh-water lens, sea water would intrude the porous carbonate rock and if the land surface were 2 m above sea level, the salt water in the rock would be 2 m below the surface. But if there is a fresh water lens, filling all the pore space and extending up to 1 m below the land surface, the fresh water, being lighter, will depress the top of the salt water lens. In the case of South Florida, where the density difference is 30 kg/m$^3$, the top of the lens would be at 1 m below the land surface and the fresh to salt water interface would be 30 m below. But if sea level rises, it pushes up that interface lifting the fresh water lens. As it moves upward, the top of the fresh water lens has nowhere to go; it will flow out on the surface as springs. If equilibrium is reestablished with the top of the fresh water lens 0.5 m below the land surface (0.5 m the above the new sea level) the lens will be only half as thick. In south Florida these lenses of fresh water support very large populations; they essentially store the rainfall like a reservoir on solid land would. As Miami expanded, canals were cut and the top of the fresh water lens deliberately lowered to provide more dry surface space for real estate development. Today the situation is becoming critical; almost all of the fresh water supply is already being used, and as the volume of the fresh water lens declines with sea level rise the amount of water available per person decreases. There is no simple engineering fix to South Florida's problem.

Not everywhere has been so slow to recognize the problem. In Kiel, Germany, where I worked for a number of years at GEOMAR's facility on the harbor, new construction since the 1990s has been based on the assumption of a 2 m sea-level rise within the presumed 200 year lifetime of the building. I would think it prudent to assume a sea level rise of at least 1 m by the end of this century. A sea level rise rate of that order of magnitude is reasonable to expect for many centuries into the future.

Figure 32.28 shows that effects of sea level rise in Western Europe. The low-lying regions of Belgium, The Netherlands, northern Germany, Denmark, Poland and northeastern Italy are especially vulnerable and would be most strongly affected.

Figure 32.29 shows the effects of sea level change in part of eastern Asia. During the Last Glacial Maximum much of the regions was exposed as land, and significant area would be flooded by rising sea levels.

Sea-level rise will affect a large part of the human population. Table 32.2 shows the current populations living in coastal areas that would be impacted by sea-level rise. The most strongly affected areas are in southern Asia. Clearly, sea-level rise will displace large numbers of people and force the relocation of many coastal cities. The IPCC has projected a sea-level rise of 0.5–1 m by the year 2100. Rates of melting of ice have increased significantly since those projections were made and I expect that the rise of sea level by the end of the century will be more than 1 m.

A sea level rise of only 1 m will displace over 100 million people globally, but only about 2.6 million in the United States, mostly those in south Florida. Those estimates assume that coastal populations will not increase during the rest of this century. That displacements of this sort can easily occur became evident with Hurricane Katrina, where about 400,000 displaced persons have not returned.

A friend who has worked for the World Bank on this topic suggests that the sea level problem will not become evident gradually, but through a catastrophic event such as a major hurricane, typhoon, or cyclone causing such devastation and loss of life that it will become apparent that the area must be abandoned.

## 32.12  The Thermohaline Circulation

Another critical factor in global change is the possibility of change in the global thermohaline circulation of the ocean. It has long been speculated that one of the early outcomes of global warming might be a shutdown of the global thermohaline conveyor. An analogy was drawn with the Younger Dryas event during the last glacial termination, when glacial conditions returned to Europe for about a thousand years. However, we have learned that the system is much more complex than originally thought, and predictions about the behavior of the thermohaline system in the future are only speculative. Much will depend on the balance between melting of Arctic sea-ice and the changing Arctic albedo, the fresh water input into Arctic Ocean and subsequently into the Greenland-Iceland-Norwegian Sea, the development of atmospheric low pressure systems over the Arctic, and stability of the northern hemisphere westerly winds. It is not clear that there is any geologic analog for an Earth with unipolar ice.

The thermohaline circulation system is very delicately balanced. It depends on sinking of cold saline water in the GIN Sea, but we do not know exactly where or when that takes place. That water must then overflow the Greenland-Scotland Ridge and in the depths of the North

**Fig. 32.28** The effects of sea-level change in Western Europe. Areas exposed at Last Glacial Maximum are not shown because of ice cover and isostatic adjustments at that time

Atlantic meet and mix with more saline warmer waters that originate in the western Mediterranean. These waters then flow southward as North Atlantic Deep Water toward the Antarctic where they resurface, become further chilled and sink again. There are so many things that can go wrong in this delicately balanced system. Furthermore, it is difficult to know whether the system is working or not. In the last century we had H-bomb tritium and chlorofluorocarbons as tracers, but those have now dissipated.

What we do know is that about 93 % of the excess energy the Earth has absorbed as a result of the enhanced greenhouse effect has gone into warming the ocean. During what appeared to be a cessation or slowdown of global warming in the late 1990s and early 2000s has turned out to have been a switch from warming the land and ocean surface to heat being transferred to the deeper layers of the upper ocean and further into the ocean's depths. Figure 32.30 shows the global increase in heat (energy) content of the ocean since

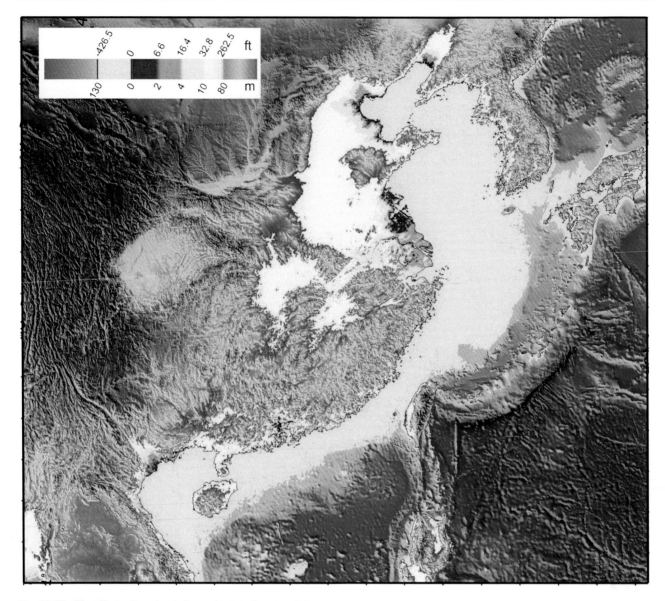

**Fig. 32.29**  The effects of sea-level change in part of eastern Asia

**Table 32.2**  Sea-level rise, areas inundated and their current populations for the entire Earth and for the southeastern United States

| Sea-level rise (m) | Inundated area ($10^3$ km²), global | Current population, millions, global | Inundated area ($10^3$ km²) Southeastern US | Current population, millions, Southeastern US |
|---|---|---|---|---|
| 1 | 1054.99 | 107.94 | 62.28 | 2.640 |
| 2 | 1312.97 | 175.10 | 104.51 | 5.493 |
| 3 | 1538.58 | 233.99 | 137.35 | 8.707 |
| 4 | 1775.24 | 308.8 | 164.76 | 13.217 |
| 5 | 2004.37 | 376.26 | 190.62 | 17.086 |
| 6 | 2193.3 | 431.44 | 210.40 | 19.271 |

Numbers at different sea levels are cumulative

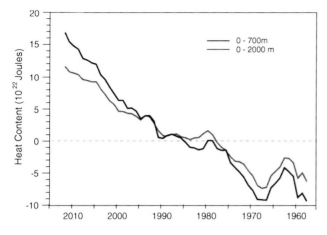

**Fig. 32.30** Increase of heat content of the global ocean. The *lines* are 5-year averages from 1955–1960 through 2009–2013. Modified from a figure prepared by NOAA/NESDIS/MODC Ocean Climate Laboratory

1957. Significant storage of energy in the ocean appears to have begun about 1970.

It now seems inevitable that by 2050 sea-ice will be gone from the Arctic over much of the year. The Antarctic will remain ice covered, although the ice will be melting. The planet will enter a new state, with the southern polar region perennially ice-covered and the northern polar region seasonally ice free.

The prospect of a southern hemisphere with relatively stable atmospheric and ocean circulation and a northern hemisphere with unstable variable conditions is a challenge for both atmospheric and ocean modeling. It is likely to be a system with far more chaotic weather than Earth has experienced for millions of years.

## 32.13  El Niño and the Southern Oscillation

There is growing evidence that a greater input of energy to the surface of the South Pacific Ocean has a counterintuitive effect. Instead of the ocean's surface waters becoming warmer, a La Niña condition develops. As you will recall from Sect. 25.5 that during a La Niña the upwelling of cold deep waters on the western margin of South America intensifies. These spread out, cooling the eastern part of the South Pacific. Remember that the air over the water generally takes on the temperature of the water. The cooler air can hold less moisture. The cooling effect crosses the Equator with the result that less water is available as rainfall in the southwestern part of North America. La Niña conditions mean drought in the American southwest. On the other hand it usually means greater rainfall in the US southeast. It also results in atmospheric pressure systems that change the flow of the northern hemisphere westerly winds. This results in a shift in the position of the Iceland High and Azores Low that

we call the North Atlantic Oscillation. One recalls John Muir's famous saying "When we try to pick out anything by itself, we find it hitched to everything in the universe."

During a La Niña the easterly winds drive Pacific water westward into the 'Pacific Warm Pool' in the South China and Philippine Seas. There the surface rises to a level a meter or so higher than normal. When the winds relax this warm water comes back across the Pacific as a Kelvin Wave, locked on the Equator by the Coriolis Effect. When the wave reaches the Americas is splits, with part going north and south. The Coriolis Effect causes it to hug the coasts. In 2015 this is happening on a grand scale, as shown in Fig. 32.31, but the major rainfall is expected later in the year.

While most specialists think that some change in frequency and/or magnitude of El Niño-La Niña events is likely, there is no consensus on what may happen. The warming of the Pacific could cause a permanent deepening of the thermocline in the eastern Equatorial Pacific, favoring El Niños. Others propose that stronger warming of the West Equatorial Pacific than the East, which seems to be occurring, will result in enhanced easterly winds and thereby reinforcing the upwelling of cold water in the Eastern Equatorial Pacific as stronger La Niñas. As my atmospheric science colleague in Miami, Eric Kraus, used to say: "We are sure there will be a change, we just don't know the sign (= + or − ?) of the change."

## 32.14  Ocean Acidification

The ocean is becoming more acid. Figure 32.32 shows the saturation state with respect to aragonite, the mineral used by shallow water corals to build their skeletons, in surface waters of the world's oceans. For a review of acidification and aragonite saturation, refer to the discussions in Chap. 26. Figure 32.32a, shows that in 2005 the major areas where conditions are becoming marginal for reef corals are the Red Sea, Persian Gulf, the southwest Pacific, and western Caribbean and Gulf of Mexico. Indeed, many of the reefs in these areas are deteriorating but the current question is whether this is due to acidification or to warming of the waters or both. The projections (Fig. 32.32b, c) are for rapid deterioration of most of the world coral reefs by the middle of the 21st century.

Another concern is about the fate of the ocean plankton. At present the ocean serves as a sink for atmospheric $CO_2$. As described in Chap. 26 the $CO_2$ reacts with the water to make carbonate or bicarbonate ions. These are utilized in metabolism of the calcareous phytoplankton which then precipitate the $CO_2$ as $CaCO_3$, in most cases as calcite. The phytoplankton $CaCO_3$ is ingested by animals such as copepods and incorporated into fecal pellets. These are heavier than seawater and sink from the surface waters into

## El Niño Arrives in 2015

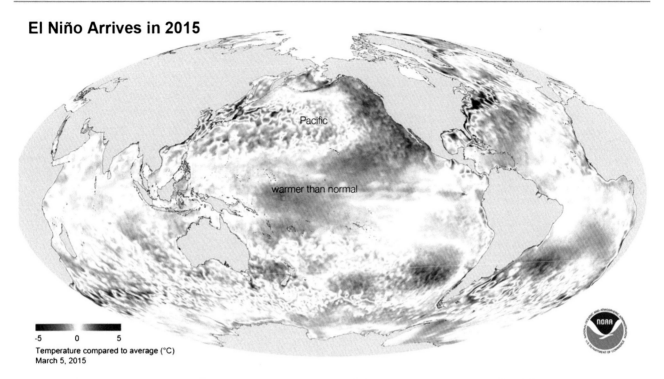

Temperature compared to average (°C)
March 5, 2015

**Fig. 32.31** El Niño warmed the waters off western North America but through the summer caused only sporadic rainfall. A more intense effect is expected in the winter of 2015–2016

the depths of the ocean. This short-circuits the $CO_2$ path through the ocean giving it a direct route from the atmosphere to the deep sea where it can become incorporated into deep sea sediments.

If we try changing the pH of sea water in a tank with living marine organisms we usually find that they do not respond well. Commonly they die. But we don't know how ocean life will respond to the gradual change that is occurring. Nothing like this has happened since the Paleocene-Eocene Thermal Maximum, and then it happened on a much slower timescale. In one laboratory experiment, *Emiliania huxleyi*, the most abundant phytoplankton organism, stopped making carbonate coccoliths in response to acidification of the water. We have no idea whether this might also happen under natural conditions. However, if it does it would stop the short path for $CO_2$ from the atmosphere to the deep sea via rapidly settling fecal pellets discussed in Chap. 26 and be a powerful feedback increasing atmospheric $CO_2$ levels.

### 32.15   Anoxia in the Ocean

The increased use of fertilizers has introduced another problem. Much of the fertilizer intended for food production on land is not utilized by the plants for which it was intended and washes into streams and rivers. On reaching a lake or the ocean they fertilize the plankton causing a bloom which may consume all of the available the oxygen causing hypoxic or anoxic conditions deadly to higher organisms. One of the places this is occurring on a grand scale is the where the Mississippi River enters the Gulf of Mexico. Figure 32.33 shows the areas of farmland and cities whose excess fertilizers and waste products drain into the Mississippi, and the regions of hypoxia and anoxia in the waters off the Gulf coast.

On a larger scale, the possibility of widespread anoxic conditions and a large-scale die-off in the ocean is that it sets in motion a large positive feedback. As anoxic conditions spread, the decomposition of dead organic matter consumes oxygen, and more and more organisms die from the increasing anoxia. From my own research I would argue that it has already happened once, as a result of the asteroid impact at the end of the Cretaceous. So it is a distinct possibility. It would not only be catastrophic for life in the ocean, but for life on land as well.

### 32.16   The Sahara Sahel

The Sahara and the semi-arid Sahel region south of it have had a long history of aridity interrupted by shorter-lived humid conditions. At the end of the last deglaciation, about 8.500 BCE the area became more humid, and the landscape was one of vegetated steppes. They supported significant populations of large animals and a society of hunter-gatherer

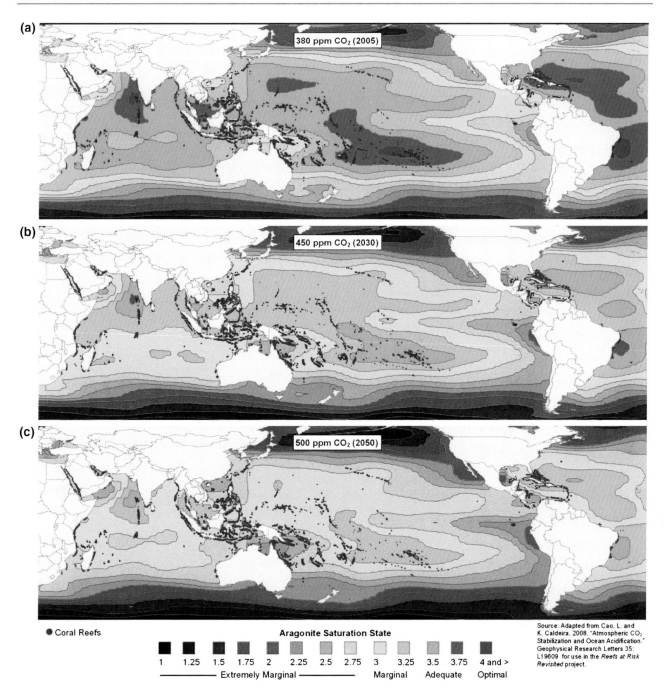

**Fig. 32.32** Present and predicted future acidification of the world's oceans. Figure prepared by the Reefs at Risk Revisited Project

humans. Then, about 5,300 BCE the regions began to dry, resulting in an exodus of the human population to the refuge of the Nile Valley.

The drying reflects final stabilization of the descending limb of the atmosphere's northern hemisphere Hadley circulation cell at about 30° N. The descending air contains no moisture, and becomes very warm as the pressure increases, following the dry adiabats shown in Fig. 20.2 downward. Over the past years, the arid conditions have expanded southward as the atmospheric Hadley cells gradually

expand. This expansion is noted as both the Sahel drought of Africa and the ongoing drought in southwestern North America. It has generally been though that this expansion of arid conditions will simply continue as the planet warms. However, there is a problem. As discussed in Chap. 24, the Hadley circulation of the atmosphere carries energy toward the Equator. The ocean circulation does the opposite, and transports heat poleward. It would seem that there must be a limit as to how far poleward the Hadley circulation can expand.

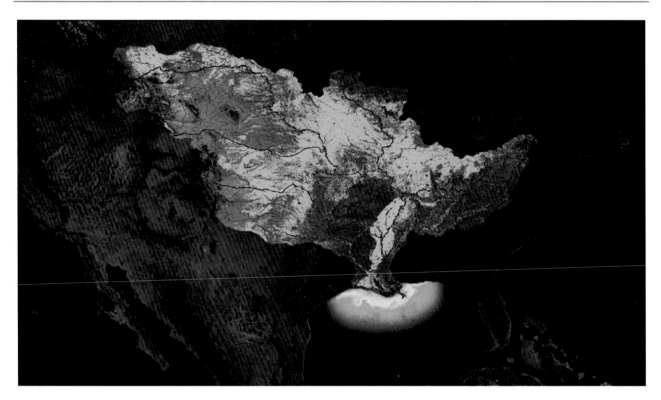

**Fig. 32.33** Areas of farmland (*green*) and cities (*pink*) in the Mississippi River watershed, and the areas of hypoxic (*blue*) and anoxic (*yellow*) waters. Image prepared by NOAA

Recent studies of Cretaceous desert deposits in Asia by Hitoshi Hasegawa of the University of Tokyo and colleagues from Japan, China, Mongolia, and Thailand show that when the Earth warms beyond a certain point, the Hadley Circulation suddenly contracts. The descending limb jumps from about 30° N and S back to about 15° N and S. They believe that this happens when atmospheric $CO_2$ levels reach about 1,000 ppmv. This involves a major reorganization of the atmospheric circulation, and a relocation of the Earth's desert regions. The process of reorganization must involve very chaotic weather patterns. It is likely that we will reach 1,000 ppmv early in the next century if current trends in fossil fuel usage continue.

## 32.17    Amazonia

Much of the precipitation in the Amazon basin is recycled. The natural jungle vegetation is $C_3$ plants. The region is currently undergoing large-scale deforestation with the $C_3$ plants being replaced by $C_4$ grasses to serve as pastureland. This creates two problems. First, the soils of the Amazon basin are very much depleted in the nutrients plants require. In effect, the nutrients are in the plants themselves. Burning the original vegetation releases some of the nutrients into the air, and removes them from the basin. Harvesting crops

removes more. Secondly, the replacement of $C_3$ with $C_4$ plants reduces the transpiration and results in 20–30 % reductions in precipitation, a lengthening of the dry season, and increases in summer temperatures. As a result, a dieback of the remaining Amazon rainforest has been predicted with the region reverting to arid conditions and possibly becoming desert.

## 32.18    Methane Clathrates

One topic that has been of considerable concern to geologists but rarely considered by climatologists is the possibility of decomposition of the solid gas hydrates, clathrates, not only in the permafrost regions but in the sediments of the ocean margins as Earth warms. If you go back and look at Table 21.1, you will see that there is about 20 times as much methane tied up in clathrates as there is $CO_2$ in the atmosphere. Since methane is about 23 times as effective as a greenhouse gas as $CO_2$, the possible release of this trapped gas into the atmosphere is a matter of real concern. As mentioned above, increasing amounts of methane are bubbling to the surface as the permafrost melts. Presumably, the addition of methane from melting permafrost will continue to be a steady but minor increment to the atmospheric greenhouse gas inventory. What concerns geologists most is the possibility of a massive release of methane as clathrates decompose. As discussed at the end of Chap. 29, such an event occurred naturally about 56 million years ago, creating the Paleocene-Eocene Thermal Maximum when global temperatures rose by 6 °C. That event involved release of about one tenth of the clathrates thought to be stored in continental margin sediments today.

Decomposition of methane clathrates might have another environmental effect—destabilization of the continental slope. The clathrates act as cement, holding the sediment in place. But it is a very peculiar kind of cement. It disappears when warmed. If the clathrate cement disappears, the result can be large landslides of slope and shelf sediments into the deep sea producing tsunamis.

This seems to have happened on the Norwegian margin during the last deglaciation. The Storegga Slides occurred around 6100 BCE, during the time of the local thermal optimum (see Fig. 28.12). There are thought to have been three separate events that occurred in rapid sequence and involved failure along 290 km of the Norwegian coastal shelf. They resulted in a total volume of 3,500 km³ of debris sliding down into the depths of the GIN Sea. The resulting tsunami was 21 m (70 ft) high on the eastern coast of Scotland, some 1400 km (850 miles) away.

Clathrates are an important component of Atlantic margins sediments. When and how they might disappear and leave large volumes of slope and shelf sediments unsupported is unknown. But as the ocean warms we may find out.

## 32.19 Critical Tipping Points

One of the most popular ways of identifying the dangers of climate change is in terms of 'tipping points.' Everyone has an intuitive feeling of what a tipping point is—when you go past a tipping point things change. It gets much more difficult if one wants to define exactly what a tipping point is and what change will occur. An article published early in 2008 in the Proceedings of the (US) National Academy of Sciences (PNAS) reported on a conference held in Berlin in 2007 where an eminent group of climate scientists agreed on a mathematical definition of the term 'tipping point' and listed a number of elements of climate change that are sensitive enough that they might 'tip' to a qualitatively different state in the next century. The mathematical formulation becomes a bit technical and I will not repeat it here; it is openly available at the PNAS website, volume 105, no. 6, p. 1786 (Lenton et al.), along with supplementary material available at the same website. It is important to realize that some of the terms in the equations used to define 'tipping points' are still very poorly known.

Table 32.3 shows the tipping elements identified by the group at the Berlin meeting. I discuss each of these in the sections below. You will note that the first four elements are associated with the Polar Regions. Remember that because they are the coldest places on Earth, the greenhouse effect of water vapor is minimal, and atmospheric $CO_2$ dominates.

In a way it is ironic that the top three items on the Tipping Points list should be associated with the Polar Regions. The Polar Regions are defined as the parts of the Earth poleward of the polar circles at 66° 33′ 38″ N and S. They are almost uninhabited and until very recently *terra incognita*. Most of us have a distorted impression of the Polar Regions because of the maps we commonly see and use. The most common sort of wall map in classrooms and children's rooms in many American homes shows the Earth in Mercator projection. Greenland appears to be much larger than the United States, and about the same size as South America or Africa. Only parts of Antarctica and the Arctic Ocean can be shown because on a Mercator map the poles plot infinitely far from the Equator. For this reason, the Polar Regions, appear to be much larger than they really are. Why do we use this map projection to show the Earth to our kids? Tradition. It was originally created by Gerardus Mercator in 1569 for navigation at sea. It is the only map for which a constant ship's course or bearing plots as a straight line. Other than that, it is a highly distorted representation of the Earth.

In reality, each of the Polar Regions represents only about 7 % of the area of the Earth's surface. In spite of their small area they are extremely important in terms of their influence on the planet's climate. As discussed in Chap. 24, the stability of the winds and ocean current systems depends on ice-cover at the Polar Regions. In addition, it is the on-land ice sheets in the Polar Regions that have the greatest potential to change sea level.

Because of their inhospitable nature, the Polar Regions are the least-known parts of Earth's surface. Until the middle part of the 20th century, very few humans had ventured beyond the edges of these areas. Knowledge of the Polar Regions was obtained by expeditions through the sea-ice region, and onto the continents and islands. Although these daring explorations attracted great public interest, they were infrequent, and provided only spotty information. After World War II exploration of both the Arctic and Antarctic began in earnest. From the 1950s on the US established stations on ice islands drifting in the Arctic, and we began to learn about conditions in winter. It became possible to navigate the Arctic Ocean beneath the sea-ice using nuclear submarines, and technology developed to allow them to surface through the ice. Much more information about the sea-floor topography, oceanography and climate were collected, but the data were classified and kept secret, unavailable to the general scientific community. We

**Table 32.3** Tipping elements of the climate system

| Tipping element | What happens | Feedback to global warming |
|---|---|---|
| Arctic summer sea-ice | Decline in area of sea-ice—PASSED IN 1982 | Positive, strong |
| Greenland ice sheet | Decline in ice volume | Positive |
| West Antarctic ice sheet | Decline in ice volume | Positive |
| Atlantic thermohaline circulation | Atlantic part of the Great Conveyor slows | ? |
| El Niño—southern oscillation | El Niños change in amplitude and frequency | ? |
| Indian summer monsoon | Rainfall decreases | ? |
| Sahara/Sahel and West African monsoon | Precipitation increases and vegetation returns | Negative |
| Amazonian Rainforest | Jungle is replaced by grassland and desert | Positive |
| Boreal Forest | Tree-covered area decreases | Positive |
| Tundra | Tundra replaced by boreal forest | Positive |
| Permafrost | Permafrost melts releasing methane and carbon dioxide | Positive, strong |
| Antarctic bottom water formation | Slows, changing Southern Hemisphere ocean circulation | ? |
| Marine methane hydrate release | Greatly enhances greenhouse effect | Positive, catastrophic |
| Ocean anoxia | Die-off of marine life releases vast amounts of carbon dioxide | Positive, catastrophic |

Modified after Lenton et al. (2008)

assumed that the Soviets were doing the same things, but little was known. In the 1970s the Soviet Union built a fleet of nuclear icebreakers which were used not only to ensure that their Arctic ports, Murmansk, Arkhangelsk, and others, were open in winter, but also for scientific exploration of the Arctic.

Real knowledge of the Polar Regions came with the development of Earth Remote Sensing Satellites. The Landsat series in began in 1972. At first the imagery was only of selected areas, but starting in 1979 meteorological satellites began to provide regular overviews of sea-ice conditions of the entire Arctic. The sophistication of these observations increased markedly in the 1990s with increased spatial resolution, sensing of topographic relief, and even the capability of determining sea-ice thickness. It is important to realize that our knowledge of climatic conditions in the high Polar Regions covers less than three decades. If it were not for Earth Satellites we would be largely ignorant of the

significant changes occurring there. It is a case of development of a critical technology 'just in time' to observe major changes in the Polar Regions.

## 32.20  Mother Nature's Remedy

Apparently our greed for short-term profit trumps any concern for long-term consequences. But Mother Nature may have her own answer. The crux of the problem is not the increased concentrations of atmospheric greenhouse gases caused by the use of fossil fuels, or our replacement of $C_3$ by $C_4$ plants to increase food production, or construction and urban sprawl, or the overuse of fertilizers and pesticides in agriculture, or the pollution associated with mining. The problem is that we have exceeded Earth's carrying capacity for humans. As Cesare Emiliani pointed out toward the end of his book *Planet Earth*, there are just too many people.

There are three ways the human population can be drastically reduced. These have been discussed before, but it is worth mentioning them again.

Ordinary war is not an effective means of reducing populations, but thermonuclear war would be. As mentioned before, some religious zealots think we need a thermonuclear war to 'fulfill Biblical prophecy.' Pat Robertson tried unsuccessfully to convince Ronald Reagan that it was time for Armageddon.

Famine is one of Mother Nature's ways of making a correction. However, it is mostly regional. Although we have areas of famine today, the local population loss scarcely slows global population growth. However, shortages of food brought on by catastrophic weather events may have a larger impact in the future.

The third was is tried and tested: disease. The Black Plague of the 13th century probably reduced the Eurasian population by half. The introduction of smallpox and other European diseases into the Americas was even more effective. I expect that there will be the appearance of a highly contagious deadly new virus that will make the correction. It is not just the effect of the disease itself that will be important, but the breakdown of society with a loss of transportation and food supplies that will magnify the effect of the epidemic.

When will we have a devastating epidemic? I have no idea, but clearly we are set up for it to have maximum effect. With modern air travel, it could make its way around the world in days. I don't think the world will reach a population of 9 billion in 2050. I think Mother Nature will make her move before then.

What about after the 'correction'? Al Bartlett would say that a lot depends on whether humans ever come to understand the implications of exponential growth.

## 32.21 Summary

This has been a long list of unpleasant topics. Summarizing them might lead to depression. Predictions are much too vague to make a timeline but the implications of Fig. 32.25 are well worth contemplating. It's time to open a good bottle of wine.

## 32.22 An Afterthought

While writing this book I have been tempted to point out some of the nonsense spread by the global warming—climate change denial community. They thrive on an uninformed public. I have made a few comments here and there, and I hope through reading this book you are now in a position to debunk the arguments yourself.

But what you need to know is that global warming—climate change denial is a very lucrative business. Counting the television ads promoting greater energy use, new fossil fuel sources, and 'clean coal,' there is significantly more money spent on disinformation and denial than on climate research. When you hear someone say "I have received no payment for my testimony" remember what ex Arkansas Senator Dale Bumpers said in wrapping up opening arguments for the President's team during Bill Clinton's impeachment trial in 1998: "When you hear somebody say, 'This is not about sex,' it's about sex." In climate change denial, it's about money. Some of the deniers even set up dummy investment corporations to pass the money through and yet maintain' plausible deniability' that they are receiving no funds for their work.

For an illuminating discussion of how this disinformation campaign works, read James Hoggan's 2009 book *Climate Cover Up*. Hoggan is not a scientist, he is in the public relations industry and became curious when he realized that the same experts testifying that global warming is not real had also been experts testifying that smoking cigarettes does not cause cancer. One thing led to another and he tells a great story of discovery, or rather 'uncovery.'

There are three websites I can recommend: www. realclimate.org; www.skepticalscience.com; are great for their coverage of both the science and the disinformation campaign, and thinkprogress.org. Joe Romm, its manager, keeps track of new developments in science, science policy and politics

A Timeline for this chapter:

| | |
|---|---|
| 985 | Eirík Rauði brings colonists to Greenland |
| 1400+ | Greenland colony dies out |
| 1610 | Henry Hudson discovers Hudson Bay |

(continued)

(continued)

| | |
|---|---|
| 1878 | Adolf Erik Nordenskiöld makes the first trip through the Northeast Passage in the steamship *Vega* |
| 1903–1906 | Roald Amundsen makes the first trip through the Northwest Passage in the *Gjøa* |
| 1958 | The US nuclear submarine *Nautilus* surfaces at the North Pole |
| 1968 | John Mercer warns of possible collapse of the West Antarctic ice sheet |
| 1975 | After the 1940–1975 pause due to global dimming, a global rise of temperatures starts |
| 1982 | American climate scientists meet with Michael Budyko in Leningrad. He expresses his concern that we have already set in motion processes that will result in melting of the Arctic sea ice and noted that the Arctic is the 'Achilles Heel' of the climate system. |
| 1982 | The Arctic sea ice begins its decline |
| 1990 | Sea level rise begins to accelerate |
| 1993 | Michael Mann and colleagues begin to publish papers on the reconstruction of temperatures for the past hundreds and then thousand years |
| 2003 | Hay and colleagues propose that the Arctic Ocean has been frozen over since the end of the Eocene, 34 million years ago |
| 2007 | First great Arctic sea ice meltback |
| 2011 | Large discharges of methane noted in the Arctic |
| 2012 | Second great Arctic sea ice meltback |
| 2012 | Waves up to 5 m high recorded in the Arctic Ocean north of Alaska |
| 2012 | Greenland's Jakobshavn Isbrae Glacier sets a new speed record advancing 46 m per day ($\approx$151 ft per day). |
| 2012 | In August the perennial Arctic atmospheric high pressure system becomes a low |
| 2012 | The atmospheric high over Greenland moves southward |
| 2012 | Hurricane Sandy turns left into New Jersey |
| 2012 | Ice melt over Greenland reaches all but the highest elevations |
| 2040 | According to US military analysts the Arctic may be ice-free for up to four months per year |

If you want to know more:

There are many books about climate change and the impending crisis. One I find especially informative and readable is

David Archer and Stefan Rahmstorf, 2010. The Climate Crisis—An Introductory Guide to Climate Change. Cambridge University Press. 249 pp.

Music: Giuseppe Verdi's *Requiem*. One of my favorites.

Libation: Time to open that bottle of great wine, Scotch whiskey, or French brandy you have been harboring for celebrating finishing this book.

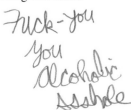

## Intermezzo XXXII. Final Thoughts

What are my personal thoughts about the future? Contrary to most of my colleagues studying climate change, I believe we went over the major tipping point for climate change all the way back in 1982 when the Arctic sea ice began to melt significantly. The future of our planet may have been determined then, before most were even aware that there might be a problem. At the meeting in Leningrad that year, Mikhail Budyko seemed sure we were headed toward an ice-free Arctic Ocean.

The root cause of the environmental problems in all their ramifications, global population growth, is gradually slowing, but not enough to prevent major problems unless Mother Nature steps in and makes a reduction on her own. My expectation is that this will happen in the next few decades either by pandemic or shortages of food supplies.

To my mind, the brightest spot is China. They recognized their population problem, which was a major factor in the Great Chinese Famine' of 1959–1961 and took steps to correct it. As a result, the Chinese population has reached its peak and is now declining. Although officially 'communist' they have an economy much like that of the U.S. after the Great Depression and recovery after World War II. It is now a major industrial country with a burgeoning middle class and bright future.

I have recently had the opportunity to work with Chinese colleagues, having visited a number of times since the late 1990s, attending meetings of international science working groups and teaching short courses. My first trip to China was in September, 2001, to give a lecture at a conference on marine geology held at Tongji University and organized by Pinxian Wang. I also taught a 2 day short course. I had known Pinxian for a number of years as he was a doctoral student in Kiel. Tongji and Christian Albrecht's University in Kiel have a long relationship that goes back to the middle of the 20th century. I was amazed at all the new buildings everywhere, the beautiful restoration of the 'Bund' the European part of the city along the Yangtze, and the spectacular skyscrapers built in the former swamplands across the river from the older city center.

My next visit was in September 2004, starting with a meeting of an international research group in Beijing. This time I took the opportunity to travel to a number of places popular with tourists. I started in Guilin, a small tourist town where you start the boat trip on the Li River through the spectacular karst mountain landscape. It is not a large city, and most inhabitants get around on motor scooters rather than automobiles. The scooters are all electric and make no sound. My first experience with a bustling city with silent streets. The boat trip ended in Yangzhou and I had a car and driver to show me the sights of the local countryside. For me the highlight of my trip was Dali, a picturesque small medieval town that had been badly damaged in an earthquake a few years earlier. All the damage had been lovingly repaired, and it was beautiful to walk around in. I also went to see the stone forest (Shi Lin) and Lijiang, another town to walk around in. Then I went to Xian, with its huge system of city walls built about the same time as those of Rome, but far larger and more impressive, and saw the terra-cotta army warriors in the tomb of Qin Shi Huangdi. I visited Chongqing and took the Yangtze River boat trip through the Three Gorges ending at the new dam of that name. I ended my trip back in Beijing visiting the Forbidden City, temples, the Great Wall and a number of other sights. Few of the buildings in Beijing appeared to be more than 20 years old.

Much of my tourism trip was made with a car and driver, so I saw a lot of the non-tourist countryside. We stopped at local markets, and I began to gain an appreciation of the great variety of vegetables and fruits available in China, many of which are still unknown in the west. My trip cost a fraction of what a similar expedition in Europe would cost.

In 2007 our IGCP group met in late August in Daging, China to discuss research on the Songliao Basin. The basin is the site of a very large lake during the Cretaceous which is China's major petroleum resource. Most of the participants met first in Beijing, traveled together by train to Harbin, and then by bus to Daging. A field trip through northeastern China led us up to the Russian border. I was especially impressed by the plantings of shrubs and trees alongside all of the roads we traveled.

My most recent trip was in September, 2015 to attend another international working group meeting in Nanjing and afterward I taught a three week-long short course at the University of Geosciences in Beijing. We traveled from Nanjing to Beijing by 300 km/h (180 mph) high-speed train. Education is something the Chinese prize very highly; the Universities are not free, but tuition is low and easily affordable, and student dormitories are very inexpensive. As in Europe, there is no need for anyone to take out a loan to get a higher education.

The Chinese have found innovative solutions to problems like urban sprawl. The areas of cities are limited to prevent the loss of farmland, and so the huge populations are accommodated in clusters of high-rise apartment buildings separated by shopping streets and open spaces. Rather than have a central area with very tall buildings and canyon-like streets, Chinese cities are really groups of 'subcities,' rather like Los Angeles but without the sprawl of single or two-story houses. The roadways are lined with flowers, shrubs and trees, all of which help reduce pollution. Lots of solar panels on rooftops, and the government has announced an ambitious plan to dramatically reduce $CO_2$ emissions. The Chinese have built a high-speed rail network across much of the country in only 10 years. By comparison, the United States is becoming a museum of what the world was like in the 1980s.

As a final word, if you have read these Intermezzi you will have discovered that the academic life is not exactly the quiet cloistered life it is reputed to be. Serendipity makes for chance meetings and opens opportunities if one chooses to take advantage of them. And there is nothing more fun than to have your favorite hobby be your profession.

The world is changing very rapidly. But climate change hangs over our heads like the Sword of Damocles. We'll just have to see what the future brings.

# Name Index

© Springer International Publishing Switzerland 2016
W.W. Hay, *Experimenting on a Small Planet*, DOI 10.1007/978-3-319-27404-1

# Subject Index

# Index of Works Mentioned